Ionic Liquids in Synthesis

Edited by
Peter Wasserscheid and
Tom Welton

Further Reading

Endres, F., MacFarlane, D., Abbott, A. (Eds.)

Electrodeposition in Ionic Liquids

2007
ISBN 978-3-527-31565-9

Sheldon, R. A., Arends, I., Hanefeld, U.

Green Chemistry and Catalysis

2007
ISBN 978-3-527-30715-9

Loupy, A. (Ed.)

**Microwaves in Organic Synthesis
Second, Completely Revised and Enlarged Edition**

2006
ISBN 978-3-527-31452-2

Ionic Liquids in Synthesis

Second, Completely Revised and Enlarged Edition

Volume 1

Edited by
Peter Wasserscheid and Tom Welton

WILEY-VCH Verlag GmbH & Co. KGaA

The Editors

Prof. Dr. Peter Wasserscheid
Friedrich-Alexander-Universität
Lehrstuhl für Chemische Reaktionstechnik
Institut für Chemie und Bioingenieurwesen
Egerlandstr. 3
91058 Erlangen
Germany

Prof. Dr. Tom Welton
Imperial College of Science,
Technology and Medicine
Department of Chemistry
South Kensington
London, SW7 2AZ
United Kingdom

1st Edition 2008
1st Reprint 2008

■ All books published by Wiley-VCH are carefully produced. Nevertheless, authors, editors, and publisher do not warrant the information contained in these books, including this book, to be free of errors. Readers are advised to keep in mind that statements, data, illustrations, procedural details or other items may inadvertently be inaccurate.

Library of Congress Card No.: applied for

British Library Cataloguing-in-Publication Data
A catalogue record for this book is available from the British Library

Bibliographic information published by the Deutsche Nationalbibliothek
Die Deutsche Nationalbibliothek lists this publication in the Deutsche Nationalbibliografie; detailed bibliographic data are available on the Internet at <http://dnb.d-nb.de.>

© 2008 WILEY-VCH Verlag GmbH & Co. KGaA, Weinheim

All rights reserved (including those of translation into other languages). No part of this book may be reproduced in any form – by photoprinting, microfilm, or any other means – nor transmitted or translated into a machine language without written permission from the publishers. Registered names, trademarks, etc. used in this book, even when not specifically marked as such, are not to be considered unprotected by law.

Composition   Aptara, New Delhi, India
Printing   Betz-Druck GmbH, Darmstadt
Bookbinding   Litges & Dopf GmbH, Heppenheim
Cover Design   Adam-Design, Weinheim
Wiley Bicentennial Logo   Richard J. Pacifico

Printed in the Federal Republic of Germany
Printed on acid-free paper

ISBN 978-3-527-31239-9

Contents

Preface to the Second Edition *xv*

A Note from the Editors *xix*

Acknowledgements *xix*

List of Contributors *xxi*

Volume 1

1 **Introduction** *1*
John S. Wilkes, Peter Wasserscheid, and Tom Welton

2 **Synthesis and Purification** *7*
2.1 Synthesis of Ionic Liquids *7*
Charles M. Gordon and Mark J. Muldoon
2.1.1 Introduction *7*
2.1.2 Quaternization Reactions *9*
2.1.3 Anion-exchange Reactions *13*
2.1.3.1 Lewis Acid-based Ionic Liquids *13*
2.1.3.2 Anion Metathesis *14*
2.1.4 Purification of Ionic Liquids *18*
2.1.5 Improving the Sustainability of Ionic Liquids *20*
2.1.6 Conclusions *23*
2.2 Quality Aspects and Other Questions Related to Commercial Ionic Liquid Production *26*
Markus Wagner and Claus Hilgers
2.2.1 Introduction *26*
2.2.2 Quality Aspects of Commercial Ionic Liquid Production *27*
2.2.2.1 Color *28*
2.2.2.2 Organic Starting Material and Other Volatiles *29*
2.2.2.3 Halide Impurities *30*
2.2.2.4 Protic Impurities *32*

2.2.2.5	Other Ionic Impurities from Incomplete Metathesis Reactions	33
2.2.2.6	Water	33
2.2.3	Upgrading the Quality of Commercial Ionic Liquids	34
2.2.4	Novel, Halide-Free Ionic Liquids	34
2.2.5	Scale-up of Ionic Liquid Synthesis	36
2.2.6	Health, Safety and Environment	37
2.2.7	Corrosion Behavior of Ionic Liquids	41
2.2.8	Recycling of Ionic Liquids	42
2.2.9	Future Price of Ionic Liquids	43
2.3	Synthesis of Task-specific Ionic Liquids	45
	James H. Davis, Jr., updated by Peter Wasserscheid	
2.3.1	Introduction	45
2.3.2	General Synthetic Strategies	47
2.3.3	Functionalized Cations	48
2.3.4	Functionalized Anions	53
2.3.5	Conclusion	53
3	**Physicochemical Properties**	**57**
3.1	Physicochemical Properties of Ionic Liquids: Melting Points and Phase Diagrams	57
	John D. Holbrey and Robin D. Rogers	
3.1.1	Introduction	57
3.1.2	Measurement of Liquid Range	59
3.1.2.1	Melting Points	60
3.1.2.2	Upper Limit – Decomposition Temperature	60
3.1.3	Effect of Ion Sizes on Salt Melting Points	62
3.1.3.1	Anion Size	63
3.1.3.2	Mixtures of Anions	64
3.1.3.3	Cation Size	65
3.1.3.4	Cation Symmetry	66
3.1.3.5	Imidazolium Salts	67
3.1.3.6	Imidazolium Substituent Alkyl Chain Length	68
3.1.3.7	Branching	69
3.1.4	Summary	70
3.2	Viscosity and Density of Ionic Liquids	72
	Rob A. Mantz and Paul C. Trulove	
3.2.1	Viscosity of Ionic Liquids	72
3.2.1.1	Viscosity Measurement Methods	73
3.2.1.2	Ionic Liquid Viscosities	75
3.2.2	Density of Ionic Liquids	86
3.2.2.1	Density Measurement	86
3.2.2.2	Ionic Liquid Densities	86
3.3	Solubility and Solvation in Ionic Liquids	89
	Violina A. Cocalia, Ann E. Visser, Robin D. Rogers, and John D. Holbrey	

3.3.1	Introduction	89
3.3.2	Metal Salt Solubility	90
3.3.2.1	Halometallate Salts	90
3.3.2.2	Metal Complexes	91
3.3.3	Extraction and Separations	92
3.3.4	Organic Compounds	96
3.3.5	Conclusions	101
3.4	Gas Solubilities in Ionic Liquids	103

Jessica L. Anderson, Jennifer L. Anthony, Joan F. Brennecke, and Edward J. Maginn

3.4.1	Introduction	103
3.4.2	Experimental Techniques	104
3.4.2.1	Gas Solubilities and Related Thermodynamic Properties	104
3.4.2.2	The Stoichiometric Technique	106
3.4.2.3	The Gravimetric Technique	107
3.4.2.4	Spectroscopic Techniques	107
3.4.2.5	Gas Chromatography	108
3.4.3	Gas Solubilities	108
3.4.3.1	CO_2	109
3.4.3.2	Reaction Gases (O_2, H_2, CO)	117
3.4.3.3	Other Gases (N_2, Ar, CH_4, C_2H_6, C_2H_4, H_2O, SO_2, CHF_3, etc.)	121
3.4.3.4	Mixed Gases	122
3.4.3.5	Enthalpies and Entropies	123
3.4.4	Applications	123
3.4.4.1	Reactions Involving Gases	124
3.4.4.2	Gas Storage	125
3.4.4.3	Gas Separations	125
3.4.4.4	Extraction of Solutes from Ionic Liquids with Compressed Gases or Supercritical Fluids	126
3.4.5	Summary	126
3.5	Polarity	130

Tom Welton

3.5.1	Microwave Dielectric Spectroscopy	131
3.5.2	Chromatographic Measurements	131
3.5.3	Absorption Spectra	133
3.5.4	Antagonistic Behavior in Hydrogen Bonding	136
3.5.5	Fluorescence Spectra	137
3.5.6	Refractive Index	137
3.5.7	EPR Spectroscopy	138
3.5.8	Chemical Reactions	138
3.5.9	Comparison of Polarity Scales	138
3.5.10	Conclusions	140
3.6	Electrochemical Properties of Ionic Liquids	141

Robert A. Mantz

3.6.1	Electrochemical Potential Windows	*142*
3.6.2	Ionic Conductivity	*150*
3.6.3	Transport Properties	*165*

4 Molecular Structure and Dynamics *175*

4.1 Order in the Liquid State and Structure *175*
Chris Hardacre
4.1.1 Neutron Diffraction *175*
4.1.2 Formation of Deuterated Samples *176*
4.1.3 Neutron Sources *177*
4.1.3.1 Pulsed (Spallation) Neutron Sources *177*
4.1.3.2 Reactor Sources *178*
4.1.4 Neutron Cells for Liquid Samples *178*
4.1.5 Examples *178*
4.1.5.1 Binary Mixtures *179*
4.1.5.2 Simple Salts *182*
4.1.6 X-ray Diffraction *184*
4.1.6.1 Cells for Liquid Samples *184*
4.1.6.2 Examples *185*
4.1.7 Extended X-ray Absorption Fine Structure Spectroscopy *190*
4.1.7.1 Experimental *191*
4.1.7.2 Examples *193*
4.1.8 X-ray and Neutron Reflectivity *199*
4.1.8.1 Experimental Set-up *199*
4.1.8.2 Examples *200*
4.1.9 Direct Recoil Spectrometry (DRS) *201*
4.1.9.1 Experimental Set-up *202*
4.1.9.2 Examples *202*
4.1.10 Conclusions *203*
4.2 Computational Modeling of Ionic Liquids *206*
Patricia A. Hunt, Edward J. Maginn, Ruth M. Lynden–Bell, and Mario G. Del Pópolo
4.2.1 Introduction *206*
4.2.1.1 Classical MD *209*
4.2.1.2 *Ab initio* Quantum Chemical Methods *210*
4.2.1.3 *Ab initio* MD *211*
4.2.1.4 Using *Ab Initio* Quantum Chemical Methods to Study Ionic Liquids *211*
4.2.2.1 Introduction *211*
4.2.2.2 Acidic Haloaluminate and Related Melts *212*
4.2.2.3 Alkyl Imidazolium-based Ionic Liquids *214*
4.2.2.4 The Electronic Structure of Ionic Liquids *218*
4.2.3 Atomistic Simulations of Liquids *220*
4.2.3.1 Atomistic Potential Models for Ionic Liquid Simulations *221*

4.2.3.1	Atomistic Simulations of Neat Ionic Liquids – Structure and Dynamics *226*	
4.2.4	Simulations of Solutions and Mixtures *236*	
4.2.5	Simulations of Surfaces *239*	
4.2.6	*Ab initio* Simulations of Ionic Liquids *239*	
4.2.7	Chemical Reactions and Chemical Reactivity *244*	
4.3	Translational Diffusion *249*	
	Joachim Richter, Axel Leuchter, and Günter Palmer	
4.3.1	Main Aspects and Terms of Translational Diffusion *249*	
4.3.2	Use of Translational Diffusion Coefficients *251*	
4.3.3	Experimental Methods *252*	
4.3.4	Results for Ionic Liquids *254*	
4.4	Molecular Reorientational Dynamics *255*	
	Andreas Dölle, Phillip G. Wahlbeck, and W. Robert Carper	
4.4.1	Introduction *255*	
4.4.2	Experimental Methods *256*	
4.4.3	Theoretical Background *257*	
4.4.4	Results for Ionic Liquids *258*	
4.4.5	Chemical Shift Anisotropy Analysis *261*	
4.4.6	Stepwise Solution of the Combined Dipolar and NOE Equations *261*	
4.4.7	NMR–Viscosity Relationships *264*	
5	**Organic Synthesis** *265*	
5.1	Ionic Liquids in Organic Synthesis: Effects on Rate and Selectivity *265*	
	Cinzia Chiappe	
5.1.1	Introduction *265*	
5.1.2	Ionic Liquid Effects on Reactions Proceeding through Isopolar and Radical Transition States *268*	
5.1.2.1	Energy Transfer, Hydrogen Transfer and Electron Transfer Reactions *268*	
5.1.2.2	Diels–Alder Reactions *272*	
5.1.2.3	Ionic Liquid Effects on Reactions Proceeding through Dipolar Transition States *274*	
5.1.3.1	Nucleophilic Substitution Reactions *275*	
5.1.3.2	Electrophilic Addition Reactions *284*	
5.1.3.3	Electrophilic Substitution Reactions *287*	
5.1.4	Conclusions *289*	
5.2	Stoichiometric Organic Reactions and Acid-catalyzed Reactions in Ionic Liquids *292*	
	Martyn Earle	
5.2.1	Electrophilic Reactions *294*	
5.2.1.1	Friedel-Crafts Reactions *294*	
5.2.1.2	Scholl and Related Reactions *310*	
5.2.1.3	Cracking and Isomerization Reactions *312*	

5.2.1.4	Electrophilic Nitration Reactions	315
5.2.1.5	Electrophilic Halogenation Reactions	316
5.2.1.6	Electrophilic Phosphylation Reactions	318
5.2.1.7	Electrophilic Sulfonation Reactions	318
5.2.2	Nucleophilic Reactions	319
5.2.2.1	Aliphatic Nucleophilic Substitution Reactions	319
5.2.2.2	Aromatic Nucleophilic Substitution Reactions	326
5.2.3	Electrocyclic Reactions	327
5.2.3.1	Diels-Alder Reactions	327
5.2.3.2	Hetero Diels-Alder Reactions	330
5.2.3.3	The Ene Reaction	332
5.2.4	Addition Reactions (to C=C and C=O Double Bonds)	334
5.2.4.1	Esterification Reactions (Addition to C=O)	334
5.2.4.2	Amide Formation Reactions (Addition to C=O)	335
5.2.4.3	The Michael Reaction (Addition to C=C)	336
5.2.4.4	Methylene Insertion Reactions (Addition to C=O and C=C)	339
5.2.4.5	Addition Reactions Involving Organometallic Reagents	340
5.2.4.6	Miscellaneous Addition Reactions	344
5.2.5	Condensation Reactions	345
5.2.5.1	General Condensation Reactions	345
5.2.5.2	The Mannich Reaction	349
5.2.6	Oxidation Reactions	350
5.2.6.1	Functional Group Oxidation Reactions	350
5.6.6.2	Epoxidation and Related Reactions	353
5.2.6.3	Miscellaneous Oxidation Reactions	355
5.2.7	Reduction Reactions	356
5.2.8	Miscellaneous Reactions in Ionic Liquids	358

Volume 2

5.3	Transition Metal Catalysis in Ionic Liquids	369
	Peter Wasserscheid and Peter Schulz	
5.3.1	Concepts, Successful Strategies, and Limiting Factors	372
5.3.1.1	Why Use Ionic Liquids as Solvents for Transition Metal Catalysis?	372
5.3.1.2	The Role of the Ionic Liquid	377
5.3.1.3	Methods for Analysis of Transition Metal Catalysts in Ionic Liquids	383
5.3.2	Selected Examples of the Application of Ionic Liquids in Transition Metal Catalysis	390
5.3.2.1	Hydrogenation	390
5.3.2.2	Oxidation Reactions	405
5.3.2.3	Hydroformylation	410
5.3.2.4	Heck Reaction and Other Pd-catalyzed C–C-coupling Reactions	419
5.3.2.5	Dimerization and Oligomerization Reactions	430
5.3.2.6	Olefin Metathesis	441

5.3.2.7	Catalysis with Nanoparticulate Transition Metal Catalysts	*444*
5.3.3	Concluding Remarks: "Low-hanging Fruits" and "High-hanging Fruits"— Which Transition Metal Catalyzed Reaction Should Be Carried Out in an Ionic Liquid?	*448*
5.4	Ionic Liquids in Multiphasic Reactions	*464*
	Hélène Olivier-Bourbigou and Frédéric Favre	
5.4.1	Multiphasic Reactions: General Features, Scope and Limitations	*464*
5.4.2	Multiphasic Catalysis: Limitations and Challenges	*465*
5.4.3	Why Ionic Liquids in Mutiphasic Catalysis?	*466*
5.4.4	Different Technical Solutions to Catalyst Separation through the Use of Ionic Liquids	*469*
5.4.5	Immobilization of Catalysts in Ionic Liquids	*473*
5.4.6	The Scale-up of Ionic Liquid Technology from Laboratory to Continuous Pilot Plant Operation	*476*
5.4.6.1	Dimerization of Alkenes Catalyzed by Ni complexes	*477*
5.4.6.2	Alkylation Reactions	*483*
5.4.6.3	Industrial Use of Ionic Liquids	*485*
5.4.7	Concluding Remarks and Outlook	*486*
5.5	Task-specific Ionic Liquids as New Phases for Supported Organic Synthesis	*488*
	Michel Vaultier, Andreas Kirschning, and Vasundhara Singh	
5.5.1	Introduction	*489*
5.5.2	Synthesis of TSILs	*490*
5.5.2.1	Synthesis of TSILs Bearing a Hydroxy Group	*491*
5.5.2.2	Parallel Synthesis of Functionalized ILs from a Michael-type Reaction	*495*
5.5.2.3	Synthesis of TSILs by Further Functional Group Transformations	*496*
5.5.2.4	Loading of TSIL Supports	*500*
5.5.3	TSILs as Supports for Organic Synthesis	*501*
5.5.3.1	First Generation of TSILs as New Phases for Supported Organic Synthesis	*503*
5.5.3.2	Second Generation of TSILs: The BTSILs	*510*
5.5.3.3	Reactions of Functionalized TSOSs in Molecular Solvents	*515*
5.5.3.4	Lab on a Chip System Using a TSIL as a Soluble Support	*523*
5.5.4	Conclusion	*523*
5.6	Supported Ionic Liquid Phase Catalysts	*527*
	Anders Riisager and Rasmus Fehrmann	
5.6.1	Introduction	*527*
5.6.2	Supported Ionic Liquid Phase Catalysts	*527*
5.6.2.1	Supported Catalysts Containing Ionic Media	*527*
5.6.2.1.1	Process and engineering aspects of supported ionic liquid catalysts	*528*
5.6.2.1.2	Characteristics of ionic liquids on solid supports	*529*
5.6.2.2	Early Work on Supported Molten Salt and Ionic Liquid Catalyst Systems	*531*
5.6.2.2.1	High-temperature supported molten salt catalysts	*531*

5.6.2.2.2	Low-temperature supported catalysts *533*	
5.6.2.3	Ionic Liquid Catalysts Supported through Covalent Anchoring *534*	
5.6.2.3.1	Supported Lewis acidic chlorometalate catalysts *534*	
5.6.2.3.2	Neutral, supported ionic liquid catalysts *537*	
5.6.2.4	Ionic Liquid Catalysts Supported through Physisorption or via Electrostatic Interaction *540*	
5.6.2.4.1	Supported ionic liquid catalysts (SILC) *540*	
5.6.2.4.2	Supported ionic liquid phase (SILP) catalysts incorporating metal complexes *543*	
5.6.2.4.3	Supported ionic liquid catalyst systems containing metal nanoparticles *552*	
5.6.2.4.4	Supported ionic liquid catalytic membrane systems containing enzymes *554*	
5.6.3	Concluding Remarks *555*	
5.7	Multiphasic Catalysis Using Ionic Liquids in Combination with Compressed CO_2 *558*	
	Peter Wasserscheid and Sven Kuhlmann	
5.7.1	Introduction *558*	
5.7.2	Catalytic Reaction with Subsequent Product Extraction *560*	
5.7.3	Catalytic Reaction with Simultaneous Product Extraction *561*	
5.7.4	Catalytic Conversion of CO_2 in an Ionic Liquid/$scCO_2$ Biphasic Mixture *562*	
5.7.5	Continuous Reactions in an Ionic Liquid/Compressed CO_2 System *562*	
5.7.6	Concluding Remarks and Outlook *567*	
6	**Inorganic Synthesis** *570*	
6.1	Directed Inorganic and Organometallic Synthesis *569*	
	Tom Welton	
6.1.1	Coordination Compounds *569*	
6.1.2	Organometallic Compounds *570*	
6.1.3	Formation of Oxides *572*	
6.1.4	Other Reactions *574*	
6.1.5	Outlook *574*	
6.2	Inorganic Materials by Electrochemical Methods *575*	
	Frank Endres and Sherif Zein El Abedin	
6.2.1	Electrodeposition of Metals and Semiconductors *576*	
6.2.1.1	General Considerations *576*	
6.2.1.2	Electrochemical Equipment *577*	
6.2.1.3	Electrodeposition of Less Noble Elements *578*	
6.2.1.4	Electrodeposition of Metals That Can Also Be Obtained From Water *582*	
6.2.1.5	Electrodeposition of Semiconductors *585*	
6.2.2	Nanoscale Processes at the Electrode/Ionic Liquid Interface *587*	
6.2.2.1	General Considerations *587*	

6.2.2.2	The Scanning Tunneling Microscope	587
6.2.2.3	Results	589
6.2.3	Summary	604
6.3	Ionic Liquids in Material Synthesis: Functional Nanoparticles and Other Inorganic Nanostructures	609
	Markus Antonietti, Bernd Smarsly, and Yong Zhou	
6.3.1	Introduction	609
6.3.2	Ionic Liquids for the Synthesis of Chemical Nanostructures	609
7	**Polymer Synthesis in Ionic Liquids**	**619**
	David M. Haddleton, Tom Welton, and Adrian J. Carmichael	
7.1	Introduction	619
7.2	Acid-catalyzed Cationic Polymerization and Oligomerization	619
7.3	Free Radical Polymerization	624
7.4	Transition Metal-catalyzed Polymerization	627
7.4.1	Ziegler–Natta Polymerization of Olefins	627
7.4.2	Late Transition Metal-catalyzed Polymerization of Olefins	628
7.4.3	Metathesis Polymerization	630
7.4.4	Living Radical Polymerization	631
7.5	Electrochemical Polymerization	633
7.5.1	Preparation of Conductive Polymers	633
7.6	Polycondensation and Enzymatic Polymerization	634
7.7	Carbene-catalyzed Reactions	635
7.8	Group Transfer Polymerization	636
7.9	Summary	637
8	**Biocatalytic Reactions in Ionic Liquids**	**641**
	Sandra Klembt, Susanne Dreyer, Marrit Eckstein, and Udo Kragl	
8.1	Introduction	641
8.2	Biocatalytic Reactions and Their Special Needs	641
8.3	Examples of Biocatalytic Reactions in Ionic Liquids	644
8.3.1	Whole Cell Systems and Enzymes Other than Lipases in Ionic Liquids	644
8.3.2	Lipases in Ionic Liquids	651
8.4	Stability and Solubility of Enzymes in Ionic Liquids	655
8.5	Special Techniques for Biocatalysis with Ionic Liquids	657
8.6	Conclusions and Outlook	658
9	**Industrial Applications of Ionic Liquids**	**663**
	Matthias Maase	
9.1	Ionic Liquids in Industrial Processes: Re-invention of the Wheel or True Innovation?	663
9.2	Possible Fields of Application	664
9.3	Applications in Chemical Processes	666
9.3.1	Acid Scavenging: The BASIL™ Process	666

9.3.2	Extractive Distillation 669
9.3.3	Chlorination with "Nucleophilic HCl" 670
9.3.4	Cleavage of Ethers 672
9.3.5	Dimerization of Olefins 673
9.3.6	Oligomerization of Olefins 673
9.3.7	Hydrosilylation 674
9.3.8	Fluorination 675
9.4	Applications in Electrochemistry 675
9.4.1	Electroplating of Chromium 675
9.4.2	Electropolishing 676
9.5	Applications as Performance Chemicals and Engineering Fluids 677
9.5.1	Ionic Liquids as Antistatic Additives for Cleaning Fluids 677
9.5.2	Ionic Liquids as Compatibilizers for Pigment Pastes 678
9.5.3	Ionic Liquids for the Storage of Gases 679
9.6	FAQ – Frequently Asked Questions Concerning the Commercial Use of Ionic Liquids 681
9.6.1	How Pure are Ionic Liquids? 681
9.6.2	Is the Color of Ionic Liquids a Problem? 682
9.6.3	How Stable are Ionic Liquids? 682
9.6.4	Are Ionic Liquids Toxic? 683
9.6.5	Are Ionic Liquids Green? 684
9.6.6	How Can Ionic Liquids be Recycled ? 684
9.6.7	How Can Ionic Liquids be Disposed Of? 685
9.6.8	Which is the Right Ionic Liquid? 686

10 **Outlook**   689
Peter Wasserscheid and Tom Welton

Index   705

Preface to the Second Edition

"And with regard to my actual reporting of the events [. . .], I have made it a principle not to write down the first story that came my way, and not even to be guided by my own general impressions; either I was present myself at the events which I have described or else I heard of them from eye-witnesses whose reports I have checked with as much thoroughness as possible. Not that even so the truth was easy to discover: different eye-witnesses give different accounts of the same events, speaking out of partiality for one side or the other or else from imperfect memories. And it may well be that my history will seem less easy to read because of the absence in it of a romantic element. It will be enough for me, however, if these words of mine are judged useful by those who want to understand clearly the events which happened in the past and which (human nature being what it is) will, at some time or other and in much the same ways, be repeated in the future. My work is not a piece of writing designed to meet the taste of an immediate public, but was done to last for ever."

The History of the Peloponnesian War (Book I, Section 22),
Thucydides (431–413 BC), translated by Rex Warner

Almost five years ago to this day, I wrote the preface to the first edition of this book (which is reproduced herein, meaning I don't have to repeat myself). I was honoured to be asked to do it, and it was an enjoyable task. How often do we, as scientists, get the privilege to write freely about a subject close to our hearts, without a censorious editor's pen being wielded? This is a rite of passage we more normally associate with an arts critic. So when Peter and Tom asked me to write the preface for the second edition, I was again flattered, but did wonder if I could add anything to what I originally wrote.

I was literally shocked when I read my original preface—was this really written only five years ago? How memory distorts with time! The figure illustrating the publication rate, for example—was it only five years ago that we were in awe of the fact that there was a "burgeoning growth of papers in this area"—when the total for 1999 was almost as high as 120! Even the most optimistic of us could not have anticipated how this would look in 2007 (see Figure 1). Approximately two thousand papers on ionic liquids appeared in 2006 (nearly 25% originating in China), bringing the total of published papers to over 6000 (and of these, over 2000 are concerned

---

Ionic Liquids in Synthesis, Second Edition. P. Wasserscheid and T. Welton (Eds.)
Copyright © 2008 WILEY-VCH Verlags GmbH & Co. KGaA, Weinheim
ISBN: 978-3-527-31239-9

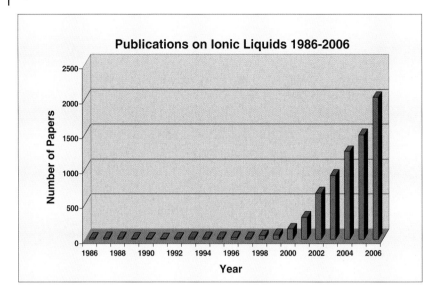

with catalysis)—-and there are also over 700 patents! That is 40 papers appearing per week—more than were being published annually a decade ago. And, on average, a review appears every two to three days. That means there is one review being published for every 20 original papers. If one assumes the garbage factor[1] runs at about 90% (a generous assumption), that means there is a review being published for every two valuable original contributions.

This is a bizarre and surreal situation, which seems more appropriate to a Kurt Vonnegut[2] novel—did buckminsterfullerene and superconductivity have the same problem? And how many papers within this annual flood of reviews say anything critical, useful, or interesting? How many add value to a list of abstracts which can be generated in five minutes using SciFinder or the ISI Web of Knowledge? How many of them can themselves be categorised as garbage? It is the twenty-first century—if a review is just an uncritical list of papers and data, what is its value?

So, am I being cynical and judgemental when I state that 90% of the published literature on ionic liquids adds little or no useful information? The PhD regulations for my University state that a satisfactory thesis must:

(1) Embody the results of research which make a distinct contribution to scholarship and afford evidence of originality as shown by the discovery of new facts, the development of new theory or insight or by the exercise of independent critical powers; and
(2) contain an acceptable amount of original work by the candidate. This work must be of a standard which could be published, either in the form of articles

---

[1] Discussed in the Preface to the First Edition.
[2] Sadly, he died in April 2007.

in appropriate refereed journals or as the basis of a book or research monograph which could meet the standards of an established academic publisher.

Well, clearly (2) is not evidence of (1); examination of the published literature undoubtedly demonstrates that "the results of research which make a distinct contribution to scholarship and afford evidence of originality as shown by the discovery of new facts, the development of new theory" is no longer a criterion for publication in refereed journals. If it was, would we find multiple publication of results from the same authors, or (frighteningly common) publication of work already published elsewhere by another, frequently uncited, group? Would papers on ionic liquids still be appearing where there is no report of the purity or water content of the ionic liquids, where claims of autocatalytic effects from the solvent appear based on reactions carried out in hexafluorophosphate or tetrafluoroborate ionic liquids (which contain HF), where physical properties are reported on impure materials, if the publications were properly refereed? I reject many of the papers which cross my (electronic) desk on these grounds when submitted to the ACS or RSC; months later I will see these papers appear, largely unchanged, in the pages of commercial journals—clearly, you can't keep a bad paper down—publish, and be damned! I have actually heard scientists say "I can't be expected to keep on top of the literature when it is appearing so rapidly." Well, sorry, yes you can—it is your job and duty as a scientist to know the published literature. It has never been easier to keep up-to-date with the literature, but finding and downloading a .pdf file is not the same as reading it!! With 2000 papers appearing in 2006 (and will anyone bet against over 2500 in 2007?), we must exercise our critical faculties to the full; we much teach our students, colleagues and collaborators to look for experimental evidence, not unsubstantiated claims. The field of ionic liquids is vibrant, fascinating, and rewarding, and offers a phenomenal opportunity for new science and technology, but we must guard, as a community, against it getting a reputation (as green chemistry has already gained) for being an area of soft publications by mediocre scientists. And the attacks and carping criticism have started; Murray, in an editorial in the ACS journal *Analytical Chemistry* [*Anal. Chem.*, **2006**, *78*, 2080], rubbished both the areas of ionic liquids and green chemistry; although he later published a mealy-mouthed, insincere apology at the end of a response from Robin Rogers and myself [*Anal. Chem.*, **2006**, *78*, 3480–3481], it is clear that this will not be the last emotive, rather than logical, attack on the field. There are hundreds of outstanding papers being published annually in this area—they must not be tarnished by the hundreds of reports of bad science.

So, having vented my spleen, how do these rhetorical comments relate to this book, which has grown dramatically in size (but, thankfully, not to a size reflecting the growth of the number of publications) since the First Edition? The number of chapters and sections in the Second Edition reflect the broadening of the applications of ionic liquids; wherever a conventional fluid can be used, the option for replacing it with an ionic liquid exists. The present chapters are written from a depth of understanding that did not exist five years ago. Today, there are over a dozen extant industrial processes; in 2002, there were none in the public domain. This has been achieved by ongoing synergistic collaborations between industry and academia, and not by the literally fantastic views expressed recently in an article entitled "Out of the Ivory Tower" [P.L. Short, *Chem. Eng. News*, **2006**, *84* (24th April)] [15–21]. The field has expanded and matured, and so has this Second Edition. The team of expert writers remains impressive—these are authors who are at the top of their field. The chapters radiate the informed writing of specialists; their wisdom is generously shared with us. The editors have performed a Herculean task in bringing this all together in a coherent and smooth account of the whole field as it stands today (although, at the current rate, the total number of papers published will rise above 10000 by 2009). If there is to be a Third Edition, and we will need one, it will have to be in two volumes! So let us hope this book is read by all practitioners of the field—by some for enjoyment, by all for insight and understanding, and by some as a bible. The field continues to expand and intrigue—by the time this book is in print, nearly one thousand more papers will have appeared—this textbook will remain the rock upon which good science will be built. To return to thoughts expressed over two thousand years ago, it will be enough *"if these words [. . .] are judged useful by those who want to understand clearly the events which happened in the past and which (human nature being what it is) will, at some time or other and in much the same ways, be repeated in the future. My work is not a piece of writing designed to meet the taste of an immediate public, but was done to last for ever."*

K.R. Seddon
April, 2007

A Note from the Editors

This book has been arranged in several chapters that have been prepared by different authors, and the reader can expect to find changes in style and emphasis as they go through it. We hope that, in choosing authors who are at the forefront of their particular specialism, this variety is a strength of the book.

In addition to the subjects covered in the first edition we have added five new chapters describing newly emerging areas of interest for ionic liquids in synthesis. The book now ranges from the most fundamental theoretical understanding of ionic liquids through to their industrial applications.

In order to cover the most important advances we allowed the book to double in length. Yet, due to the explosion of interest in the use of ionic liquids in synthesis it has not been possible to be fully comprehensive. Consequently, the book must be didactic with examples from the literature used to illustrate and explain. We hope that no offence is caused to anyone whose work has not been included. None is intended.

Naturally, a multi-authored book has a time gap between the author's submission and the publication which can be different for different contributions. However, this was the same for the first edition of this book and did not harm its acceptance.

Acknowledgements

We would like to sincerely thank everyone who has been involved in the publication of this book. All our authors have done a great job in preparing their chapters and it has been a pleasure to read their contributions as they have come in to us. We are truly grateful for them making our task so painless. We would also like to thank the production team at VCH-Wiley, particularly Dr. Elke Maase, Dr. Rainer Münz and Dr. Tim Kersebohm.

Finally in a project like this, someone must take responsibility for any errors that have crept in. Ultimately we are the editors and this responsibility is ours. So we apologize unreservedly for any mistakes that have found their way into this book.

August 2007

Peter Wasserscheid, Tom Welton

List of Contributors

Jessica L. Anderson
University of Notre Dame
Notre Dame, IN 46556
USA

Jennifer L. Anthony
University of Notre Dame
Notre Dame, IN 46556
USA

Markus Antonietti
Max Planck Institute of Colloids
and Interfaces
Research Campus Golm
14424 Potsdam
Germany

Joan F. Brennecke
University of Notre Dame
Notre Dame, IN 46556
USA

Adrian J. Carmichael
University of Warwick
Dept. of Chemistry
Coventry CV4 7AC
UK

W. Robert Carper
Wichita State University
Dept. of Biophysical and Physical
Chemistry
206 McKinley Hall
1845 Fairmount
Wichita, KS 67260-0051
USA

Cinzia Chiappe
Università di Pisa
Dipartimento di Chimica
Bioorganica e Biofarcia
Via Bonanno Pisano 33
56126 Pisa
Italy

Violina A. Cocalia
Cytec Industries Inc.
Mining Chemicals Department
1937 West Main Street
Stamford, CT 06904
USA

James H. Davis, Jr.
University of South Alabama
Dept. of Chemistry
Mobile, AL 36688-0002
USA

Mario G. Del Pópolo
Queen's University Belfast
Atomistic Simulation Centre
School of Mathematics and Physics
Belfast BT7 1NN
Northern Ireland, UK

Andreas Dölle
RWTH Aachen
Institute of Physical Chemistry
Templergraben 59
52062 Aachen
Germany

Susanne Dreyer
University of Rostock
Department of Chemistry
Albert-Einstein-Str. 3a
18059 Rostock
Germany

Martyn Earle
The Queen's University
School of Chemistry
Stransmills Rd.
Belfast BT9 5AG
Northern Ireland
UK

Marrit Eckstein
RWTH Aachen
Institute for Technical and
Macromolecular Chemistry
Worringerweg 1
52074 Aachen
Germany

Sherif Zein El Abedin
Clausthal University of Technology
Faculty of Natural & Material Sciences
Robert-Koch-Str. 42
38678 Clausthal-Zellerfeld
Germany

Frank Endres
Clausthal University of Technology
Institute of Metallurgy
Robert-Koch-Str. 42
38678 Clausthal-Zellerfeld
Germany

Frédéric Favre
Institut Francais du Pétrole
IFP Lyon
69390 Vernaison
France

Rasmus Fehrmann
Technical University of Denmark
Department of Chemistry
Building 207
2800 Kgs. Lyngby
Denmark

Charles M. Gordon
Pfizer Global Research
and Development
Ramsgate Road
Sandwich
Kent CT13 9NJ
UK

David M. Haddleton
University of Warwick
Dept. of Chemistry
Coventry CV4 7AC
UK

Chris Hardacre
Queen's University Belfast
School of Chemistry and Chemical
Engineering
Stranmillis Road
Belfast BT9 5AG
Northern Ireland
UK

Claus Hilgers
Solvent Innovation GmbH
Nattermannallee 1
50829 Köln
Germany

John D. Holbrey
Queen's University of Belfast
QUILL, School of Chemistry
and Chemical Engineering
David Keir Building
Stranmillis Road
Belfast BT9 5AG
Northern Ireland
UK

Patricia A. Hunt
Imperial College of Science,
Technology and Medicine
Department of Chemistry
South Kensington
London, SW7 2AY
UK

Andreas Kirschning
University of Hannover
Institute of Organic Chemistry
Schneiderberg 1b
30167 Hannover
Germany

Sandra Klembt
University of Rostock
Dept. of Chemistry
Albert-Einstein-Str. 3a
18059 Rostock
Germany

Udo Kragl
University of Rostock
Dept. of Chemistry
Albert-Einstein-Str. 3a
18059 Rostock
Germany

Axel Leuchter
RWTH Aachen
Institut für Physikalische Chemie
Templergraben 59, Raum 304
52056 Aachen
Germany

Ruth M. Lynden-Bell
University of Cambridge
University Chemical Laboratory
Lensfield Road
Cambridge, CB2 1EW
UK

Matthias Maase
BASF AG
Global New Business Development
Chemical Intermediates for Industrial
Applications
CZ/BS – E 100
67056 Ludwigshafen
Germany

Edward J. Maginn
University of Notre Dame
Notre Dame, IN 46556
USA

Robert A. Mantz
U.S. Army Research Laboratory
2800 Powder Mill Rd
Adelphi, MD 20783-1197
USA

Mark J. Muldoon
Queen's University Belfast
School of Chemistry and Chemical
Engineering
Stranmillis Road
Belfast, BT9 5AG
Northern Ireland
UK

Hélène Olivier-Bourbigou
Institut Francais du Pétrole (IFP)
Division Cinétique et Catalyse
1 / 4 Avenue de Bous Préau
92852 Rueil-Malmaison
France

Günter Palmer
RWTH Aachen
Institut für Physikalische Chemie
Landoltweg 2
52056 Aachen
Germany

Joachim Richter
RWTH Aachen
Institut für Physikalische Chemie
Landoltweg 2
52056 Aachen
Germany

Anders Riisager
Technical University of Denmark
Department of Chemistry
Building 207
2800 Kgs. Lyngby
Denmark

Robin D. Rogers
The University of Alabama
Department of Chemistry
Box 870336
Tuscaloosa, AL 35487-0336
USA

Peter Schulz
Friedrich-Alexander-Universität
Lehrstuhl für Chemische
Reaktionstechnik
Institut für Chemie und
Bioingenieurwesen
Egerlandstr. 3
91058 Erlangen
Germany

Vasundhara Singh
University College of Engineering
Punjabi University
Reader in Chemistry
Department of Basic and Applied
Sciences
Patiala, 147002
India

Bernd Smarsly
Max Planck Institute of Colloids and
Interfaces
Research Campus Golm
14424 Potsdam
Germany

Paul C. Trulove
Centre for Green Manufacturing
Department of Chemistry
United States Naval Academy
572 Holloway Road
Annapolis, MD 21402-5026
USA

Michel Vaultier
Univ. Rennes
Groupe Rech. Physicochim. Struct.
CNRS
35042 Rennes
France

Ann E. Visser
Savannah River National Laboratory
Aiken, SC 29808
USA

Markus Wagner
Solvent Innovation GmbH
Nattermannallee 1
50829 Köln
Germany

Phillip G. Wahlbeck
Wichita State University
Department of Chemistry
Wichita, KS 67260–0051
USA

Peter Wasserscheid
Friedrich-Alexander-Universität
Lehrstuhl für Chemische
Reaktionstechnik
Institut für Chemie und
Bioingenieurwesen
Egerlandstr. 3
91058 Erlangen
Germany

Tom Welton
Imperial College of Science,
Technology and Medicine
Department of Chemistry
South Kensington
London, SW7 2AY
UK

John S. Wilkes
Department of Chemistry
U.S. Air Force Academy
2355 Fairchild Drive
Colorado 80840
USA

Yong Zhou
Max Planck Institute of
Colloids and Interfaces
Research Campus Golm
14424 Potsdam
Germany

1
Introduction

John S. Wilkes, Peter Wasserscheid, and Tom Welton

Ionic liquids may be viewed as a new and remarkable class of solvents, or as a type of materials that has a long and useful history. In fact, ionic liquids are both, depending on your point of view. It is absolutely clear that whatever "ionic liquids" are, there has been an explosion of interest in them. Entries in Chemical Abstracts for the term "ionic liquids" were steady at about twenty per year through 1995, but grew to over 140 in the year 2000 and to more than 1500 in 2005. The reason for the increased interest is clearly due to the realization that these materials, formerly used for specialized electrochemical applications, may have greater utility as solvents for reactions and materials processing, as extraction media or as working fluids in mechanical applications, to name just a few of the more recent applications of ionic liquids.

For the purposes of discussion in this volume we will define ionic liquids as salts with a melting temperature below the boiling point of water. That is an arbitrary definition based on temperature, and says little about the composition of the materials themselves, except that they are completely ionic. In reality, most ionic liquids in the literature that meet our present definition are also liquids at room temperature. The melting temperature of many ionic liquids can be problematic, since they are notorious glass-forming materials. It is a common experience to work with a new ionic liquid for weeks or months only to find one day that it has crystallized unexpectedly. The essential feature that ionic liquids possess is one shared with traditional molten salts – a very wide liquidus range. The liquidus range is the span of temperatures between the melting point and boiling point. No molecular solvent can match the liquidus range of ionic liquids or molten salts, except perhaps some liquid polymers. Ionic liquids differ from molten salts just in where the liquidus range is in the scale of temperature.

There are many synonyms used for ionic liquids that complicate a literature search. "Molten salts" is the most common and most broadly applied term for ionic compounds in the liquid state. Unfortunately the term "ionic liquid" was also used to mean "molten salt" long before there was much literature on low melting salts. It may seem that the difference between ionic liquids and molten salts is just a matter of degree (literally); however the practical differences are

Ionic Liquids in Synthesis, Second Edition. P. Wasserscheid and T. Welton (Eds.)
Copyright © 2008 WILEY-VCH Verlags GmbH & Co. KGaA, Weinheim
ISBN: 978-3-527-31239-9

sufficient to justify a separately identified area for the salts that are liquid around room temperature. That is, in practice the ionic liquids may usually be handled like ordinary solvents. There are also some fundamental features of ionic liquids, such as strong ion–ion interactions that are not often seen in higher temperature molten salts. Synonyms in the literature for materials that meet the working definition of ionic liquid are: "room temperature molten salt," "low temperature molten salt," "ambient temperature molten salt," and "liquid organic salt."

Our definition of an ionic liquid does not answer the general question, "What is an ionic liquid?" This question has both a chemical and a historical answer. The details of the chemical answer are the subject of several subsequent chapters in this book. The general chemical composition of ionic liquids is surprisingly consistent, even though the specific composition and the chemical and physical properties vary tremendously. Most ionic liquids have an organic cation and an inorganic polyatomic anion. Since there are many known and potential cations and anions, the potential number of ionic liquids is huge. Discovering a new ionic liquid is relatively easy, but determining its usefulness as a solvent requires a much more substantial investment in determination of physical and chemical properties. The best trick would be a method for predicting an ionic liquid composition with a specified set of properties. That is an important goal that still awaits a better fundamental understanding of structure–property relationships and the development of better computational tools. I believe it can be done.

The historical answer to the nature of the present ionic liquids is somewhat in the eye of the beholder. The very brief history presented here is just one of many possible ones, and is necessarily biased by the point of view of just one participant in the development of ionic liquids. The earliest material that would meet our current definition of an ionic liquid was observed in Friedel–Crafts reactions in the mid-19th century as a separate liquid phase called the "red oil." The fact that the red oil was a salt was determined more recently when NMR spectroscopy became a commonly available tool. Early in the 20th century some alkylammonium nitrate salts were found to be liquids [1], and more recently liquid gun propellants have been developed using binary nitrate ionic liquids [2]. In the 1960s John Yoke at Oregon State University reported that mixtures of copper(I) chloride and alkylammonium chlorides were often liquids [3]. These were not as simple as they might appear, since several chlorocuprous anions formed, depending on the stoichiometry of the components. In the 1970s Jerry Atwood at the University of Alabama discovered an unusual class of liquid salts he termed "liquid clathrates" [4]. These were composed of a salt combined with an aluminum alkyl, which then forms an inclusion compound with one or more aromatic molecules. A formula for the ionic portion is $M[Al_2(CH_3)_6X]$, where M is an inorganic or organic cation and X is a halide.

None of the interesting materials just described are the direct ancestors of the present generation of ionic liquids. Most of the ionic liquids responsible for the burst of papers in the last several years evolved directly from high temperature molten salts, and the quest to gain the advantages of molten salts without the disadvantages. It all started with a battery that was too hot to handle.

Fig. 1.1 Major (Dr.) Lowell A. King at the U.S. Air Force Academy in 1961. He was an early researcher in the development of low temperature molten salts as battery electrolytes. At that time "low temperature" meant close to 100°C, compared to many hundreds of degrees for conventional molten salts. His work led directly to the chloroaluminate ionic liquids.

In 1963 Major (Dr.) Lowell A. King (Fig. 1.1) at the U.S. Air Force Academy initiated a research project aimed at finding a replacement for the LiCl–KCl molten salt electrolyte used in thermal batteries.

Since then there has been a continuous molten salts/ionic liquids research program at the Air Force Academy, with only three principal investigators – King, John Wilkes (Fig. 1.2), and Richard Carlin. Even though the LiCl–KCl eutectic mixture has a low melting temperature (355°C) for an inorganic salt, the temperature causes materials problems inside the battery, and incompatibilities with nearby devices. The class of molten salts known as chloroaluminates, which are mixtures of alkali halides and aluminum chloride, have melting temperatures much lower than nearly all other inorganic eutectic salts. In fact NaCl–AlCl$_3$ has a eutectic composition with a 107°C melting point, very nearly an ionic liquid by our definition [5]. Chloroaluminates are another class of salts that are not simple binary mixtures, because the Lewis acid–base chemistry of the system results in the presence of the series of anions Cl$^-$, [AlCl$_4$]$^-$, [Al$_2$Cl$_7$]$^-$, and [Al$_3$Cl$_{10}$]$^-$ (although fortunately not all of these in the same mixture). Dr. King taught me a lesson that we should take heed of with the newer ionic liquids – if a new material is to be accepted as a technically useful material the chemists must present reliable data on the chemical and physical properties needed by engineers to design processes and devices. Hence, the group at the Air Force Academy in collaboration with several other groups determined the densities, conductivities, viscosities, vapor pressures, phase equilibria and

Fig. 1.2 Captain (Dr.) John S. Wilkes at the U.S. Air Force Academy in 1979. This official photo was taken about the time when he started his research on ionic liquids, then called "room temperature molten salts."

electrochemical behavior of the salts. The research resulted in a patent for a thermal battery using the NaCl–AlCl$_3$ electrolyte, and a small number of the batteries were manufactured.

Early in their work on molten salt electrolytes for thermal batteries, the Air Force Academy researchers surveyed the aluminum electroplating literature for electrolyte baths that might be suitable for a battery with an aluminum metal anode and chlorine cathode. They found a 1948 patent describing ionically conductive mixtures of AlCl$_3$ and 1-ethylpyridinium halides, mainly bromides [6]. Subsequently the salt 1-butylpyridinium chloride – AlCl$_3$ (another complicated pseudo-binary) was found to be better behaved than the earlier mixed halide system, so the chemical and physical properties were measured and published [7]. I would mark this as the start of the modern era for ionic liquids, because for the first time a wider audience of chemists started to take interest in these totally ionic, completely nonaqueous new solvents.

The alkylpyridinium cations suffer from being relatively easy to reduce, both chemically and electrochemically. Charles Hussey (Fig. 1.3) and I set out a program to predict cations more resistant to reduction, synthesize ionic liquids based on those predictions, and electrochemically characterize them for use as battery electrolytes.

We had no good way to predict if they would be liquid, but we were lucky that many were. The class of cations that were the most attractive candidates was the dialkylimidazolium salts, and the 1-ethyl-3-methylimidazolium, [EMIM], was our particular favorite. [EMIM]Cl mixed with AlCl$_3$ made ionic liquids with melting temperatures below room temperature over a wide range of compositions [8]. We determined chemical and physical properties once again, and demonstrated some new battery concepts based on this well-behaved new electrolyte. We and others

Fig. 1.3 Prof. Charles Hussey of the University of Mississippi. The photo was taken in 1990 at the U.S. Air Force Academy while he was serving on an Air Force Research active duty assignment. Hussey and Wilkes collaborated in much of the early work on chloroaluminate ionic liquids.

also tried some organic reactions, such as Friedel–Crafts chemistry, and found the ionic liquids to be excellent as both solvent and catalyst [9]. They appeared to act like acetonitrile, except that they were totally ionic and nonvolatile.

The pyridinium- and the imidazolium-based chloroaluminate ionic liquids share the disadvantage of being reactive with water. In 1990 Mike Zaworotko (Fig. 1.4) took a sabbatical leave at the Air Force Academy, where he introduced a new dimension to the growing field of ionic liquid solvents and electrolytes.

His goal for that year was to prepare and characterize salts with dialkylimidazolium cations, but with water-stable anions. This was such an obviously useful idea that we marveled that neither we nor others had tried to do this already. The preparation chemistry was about as easy as the formation of the chloroaluminate salts, and could be done outside the glove box [10]. The new tetrafluoroborate, hexafluorophosphate, nitrate, sulfate, and acetate salts were stable (at least at room temperature) towards hydrolysis. We thought of these salts as candidates for battery electrolytes, but they (and other similar salts) have proven more useful for other applications. Just as Zaworotko left, Joan Fuller came to the Air Force Academy, and spent several years extending the catalog of water stable ionic liquids, discovering better ways to prepare them, and testing the solids for some optical properties. She made a large number of ionic liquids from the traditional dialkylimidazolium cations, plus a series of mono- and tri-alkylimidazoliums. She combined those cations with the water stable anions mentioned above *plus* the additional series bromide, cyanide, bisulfate, iodate, trifluoromethanesulfonate, tosylate, phenylphosphonate and tartrate. This resulted in a huge array of new ionic liquids with anion sizes ranging from relatively small to very large.

Fig. 1.4 Dr. Michael Zaworotko from Saint Mary's University in Halifax, Nova Scotia. He was a visiting professor at the U.S. Air Force Academy in 1991, where he first prepared many of the water-stable ionic liquids popular today.

It seems obvious to me and most other chemists that the table of cations and anions that form ionic liquids can and will be extended to a nearly limitless number. The applications will be limited only by our imagination.

References

1. Walden, P., *Bull. Acad. Imper. Sci. (St. Petersburg)* **1914**, 1800.
2. CAS Registry Number 78041-07-3.
3. Yoke, J. T.; Weiss, J. F.; Tollin, G., *Inorg. Chem.* **1963**, *2*, 1210–1212.
4. Atwood, J. L.; Atwood, J. D., *Inorganic Compounds with Unusual Properties*, Advances in Chemistry Series No. 150, American Chemical Society: Washington, DC, **1976**, pp. 112–127.
5. For a review of salts that were formerly thought of as low temperature ionic liquids see Mamantov, G., Molten salt electrolytes in secondary batteries, in *Materials for Advanced Batteries*, Murphy, D. W.; Broadhead, J.; Steele, B. C. H. (Eds.), Plenum Press, New York, **1980**, pp. 111–122.
6. Hurley, F. H., U.S. Patent 4 446 331, **1948**; Wier, T. P. Jr.; Hurley, F. H., U.S. Patent 4 446 349, **1948**; Wier, T. P. Jr., U.S. Patent 4 446 350, **1948**; Wier, T. P. Jr., U.S. Patent 4446 350, **1948**.
7. Gale, R. J.; Gilbert, B.; Osteryoung, R. A., *Inorg. Chem.*, **1978**, *17*, 2728–2729; Nardi, J. C.; Hussey, C. L.; King, L. A., U.S. Patent 4 122 245, **1978**.
8. Wilkes, J. S.; Levisky, J. A.; Wilson R. A.; Hussey, C. L., *Inorg. Chem.*, **1982**, *21*, 1263.
9. Boon, J.; Levisky, J. A.; Pflug, J. L.; Wilkes, J. S., *J. Org. Chem.*, **1986**, *51*, 480–483.
10. Wilkes, J. S.; Zaworotko, M. J., *J. Chem. Soc., Chem. Commun.*, **1992**, 965–967.

2
Synthesis and Purification

2.1
Synthesis of Ionic Liquids

Charles M. Gordon and Mark J. Muldoon

2.1.1
Introduction

The increasing interest in ionic liquids, especially in the light of their recent widespread commercial availability, has resulted in further developments in their synthesis and purification. In particular, this has required a shift towards improving the standard synthetic procedures to ensure consistency in the quality of the materials. The majority of research papers still report the use of better understood ionic liquids such as those based on 1,3-alkylimidazolium cations and anions such as $[PF_6]^-$ and $[(CF_3SO_2)_2N]^-$ ($[Tf_2N]^-$). However, in order to improve the chances of large-scale commercial applications, the efficiency of synthetic procedures, ionic liquid toxicity and biodegradation have all become important topics. This chapter will cover the important areas related to the general synthetic methods that are applicable to the most commonly used ionic liquids. The issue of purity and purification of ionic liquids will also be discussed, as this is an area that is of great consequence when the physical properties of ionic liquids are being investigated, and will be essential as further large-scale applications are developed. The chapter will also highlight the environmental concerns related to ionic liquids and the recent developments to improve the sustainability of these materials. The aim is to provide a summary for new researchers in the area that can point to the best preparative methods, and the potential pitfalls, as well as helping established researchers in the area to refine the methods used in their laboratories.

The story of ionic liquids is generally regarded as beginning with the first report of the preparation of ethylammonium nitrate in 1914 [1]. This species was formed by the addition of concentrated nitric acid to ethylamine, after which the water was

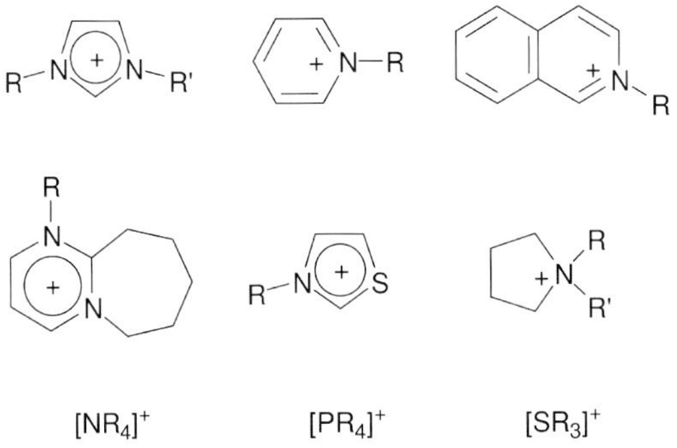

Fig. 2.1-1 Examples of cations commonly used for the formation of ionic liquids.

removed by distillation to give the pure salt that was liquid at room temperature. The protonation of suitable starting materials (generally amines and phosphines) still represents the simplest method for the formation of such materials, but unfortunately it can only be used for a small range of useful salts. The possibility of decomposition via deprotonation has severely limited the use of such salts, and so more complex methods are generally required. Probably the most widely used salt of this type is pyridinium hydrochloride; the applications of this salt may be found in a thorough review by Pagni [2].

Thus, most ionic liquids are formed from cations that have not been obtained by protonation of a nucleophile. A summary of the applications and properties of ionic liquids may be found in a number of review articles [3]. The most common classes of cations are illustrated in Fig. 2.1-1, although low melting point salts based on other cations such as complex polycationic amines [4], and heterocycle-containing drugs [5] have also been prepared. The synthesis of ionic liquids can generally be split into two steps: the formation of the desired cation, and anion exchange where necessary to form the desired product. In some cases only the first step is required, as with the formation of ethylammonium nitrate. In many cases the desired cation is commercially available at reasonable cost, most commonly as a halide salt, thus requiring only the anion exchange reaction. Examples of these are 1-butyl-3-methylimidazolium chloride ([BMIM]Cl), the symmetrical tetraalkylammonium and tetraalkylphosphonium salts as well as trialkylsulfonium iodide.

This chapter will concentrate on the preparation of ionic liquids based on 1,3-dialkylimidazolium cations, as these have dominated the area over the last twenty years. The techniques discussed in this chapter are, however, generally applicable to the other classes of cations indicated in Fig. 2.1-1. The original decision by Wilkes et al. to prepare 1-alkyl-3-methylimidazolium ([RMIM]$^+$) salts was prompted by the requirement for a cation that had more negative reduction potential than Al(III) [6]. The discovery that the imidazolium-based salts also generally displayed lower melting points than the 1-alkylpyridinium salts used prior to this cemented their position as the cations of choice since this time. Indeed, the method reported by

Wilkes et al. for the preparation of the [RMIM]Cl/AlCl$_3$ based salts remains very much that employed by most workers to this day.

2.1.2
Quaternization Reactions

The formation of the cations may be carried out either via protonation with a free acid, as noted above, or by quaternization of an amine, phosphine or sulfide, most commonly using a haloalkane or dialkylsulfates. The protonation reaction, as used in the formation of salts such as ethylammonium nitrate, involves the addition of 3 M nitric acid to a cooled aqueous solution of ethylamine [7]. A slight excess of amine should be left, which is removed along with the water by heating to 60 °C *in vacuo*. The same general process may be employed for the preparation of all salts of this type, but when amines of higher molecular weight are employed, there is clearly a risk of contamination by residual amine. A similar method has been reported for the formation of low melting, liquid crystalline, long alkyl chain substituted 1-alkylimidazolium chloride, nitrate and tetrafluoroborate salts [8]. Here, a slight excess of acid was employed as the products were generally crystalline at room temperature. In all cases it is recommended that the addition of acid is carried out with cooling of the amine solution, as the reaction can be quite exothermic.

The alkylation process to form halide salts possesses the advantages that (i) a wide range of cheap haloalkanes are available, and (ii) the substitution reactions generally occur smoothly at reasonable temperatures. Furthermore, the halide salts formed can be easily converted to salts with other anions. Although this section will concentrate on the reaction of simple haloalkanes with the amine, more complex side chains may be added, as will be discussed later in this chapter. The quaternization of amines and phosphines with haloalkanes has been known for many years, but the development of ionic liquids has resulted in several recent developments in the experimental techniques used for the reaction. In general, the reaction may be carried out using chloroalkanes, bromoalkanes and iodoalkanes, with the reaction conditions required becoming steadily gentler in the order Cl → Br → I, as is expected for nucleophilic substitution reactions. Fluoride salts cannot be formed in this manner.

In principle, the quaternization reactions are extremely simple: the amine (or phosphine) is mixed with the desired alkylating agent, and the mixture is then stirred and heated. The following section refers to the quaternization of 1-alkylimidazoles, as these are the most common starting materials. The general techniques are similar, however, for other amines such as pyridine [9], isoquinoline [10], 1,8-diazabicyclo[5,4,0]-7-undecene [11], 1-methylpyrrolidine [12], and trialkylamines [13], as well as for phosphines [14]. The reaction temperature and time are very dependent on the alkylating agent employed, chloroalkanes being the least reactive and iodoalkanes the most. The reactivity of the haloalkanes also generally decreases with increasing alkyl chain length. As an illustration, in the authors' laboratory it is generally found to be necessary to heat 1-methylimidazole with chloroalkanes to about 80 °C for 2–3 days to ensure complete reaction. The equivalent reaction with bromoalkanes is usually complete within 24 h, and can be achieved using lower temperatures (ca. 50–60 °C). In the case of bromoalkanes we have found that care must be taken with large-scale reactions, as a strong exotherm can occur as the reaction rate increases. Besides the obvious safety implications, the excess heat generated can result in discoloration of the final product. The reaction with iodoalkanes, dimethylsulfate and diethylsulfate can often be carried out at room temperature, but the iodide salts formed are light sensitive, requiring shielding of the reaction vessel from bright light.

A number of different protocols have been reported, but most researchers use a simple round bottomed flask/reflux condenser experimental setup for the quaternization reaction. If possible, the reaction should be carried out under dinitrogen or some other inert gas in order to exclude water and oxygen during the quaternization. Exclusion of oxygen is particularly important if a colorless halide salt is required. Alternatively, the haloalkane and 1-methylimidazole are mixed in Carius tubes, degassed via freeze–pump–thaw cycles, and then sealed under vacuum and heated in an oven for the desired period. The preparation of salts with very short alkyl chain substituents, for example [EMIM]Cl, is more complex, however, as chloroethane has a boiling point of 12 °C. Such reactions are generally carried out in an autoclave, with the chloroethane cooled to below its boiling point before being added to the reaction mixture. In this case, the products should be collected at high temperature, as these halide salts are generally solids at room temperature.

In general, the most important requirement is that the reaction mixture is kept free of moisture, as the products are often extremely hygroscopic. The reaction may be carried out without the use of a solvent, as the reagents are generally liquids and mutually miscible, while the halide salt products are usually immiscible in the starting materials. A solvent is often used, however, examples of which include the alkyl halide itself [6], 1,1,1-trichloroethane [15], and toluene [16], although no particular advantage appears to accrue with any specific one. Ethyl ethanoate has also been widely employed [17], but may undergo base-catalyzed hydrolysis so should be used with caution. The unifying factor for all of these is that they are immiscible with the halide salt product, which will thus form as a separate phase. The effect of solvent was recently examined in more detail when the kinetics of a single-phase reaction was compared to those of a biphasic system for the synthesis of [BMIM]Cl [18]. The rate of a stirred solvent-free reaction that became biphasic when conversion exceeded 8% was almost the same as that in a single-phase system containing 20 vol% ethanol. The same authors examined the solvent-free synthesis in a continuous mode using a tubular reactor, and found the residence time to be equivalent to the reaction time in the batch process. Such studies might be particularly applicable to larger scale commercial synthesis. A practical advantage of using a solvent in which the product is immiscible is that removal of the majority of the excess solvent and starting material can be achieved simply by decantation. In all cases, however, after reaction is complete and the solvent is decanted, it is necessary to remove all

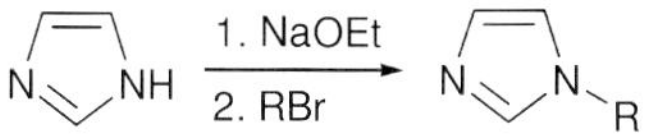

Scheme 2.1-1

excess solvent and starting material by heating the salt under vacuum. Care should be taken at this stage, especially in the case of halide salts, as overheating can result in a reversal of the quaternization reaction. It is not advised to heat the halide salts to temperatures greater than about 80 °C.

The halide salts are generally solids at room temperature, although examples such as the $[C_n\text{MIM}]^+$ salts where $n = 6–8$ remain as viscous oils, even at room temperature. Crystallization can take some time to occur, however, and many salts remain as oils even when formed in good purity. Purification of the solid salts is best achieved by recrystallization from a mixture of dry acetonitrile and ethyl ethanoate. In the case of salts that are not solid, it is advisable to wash the oil as well as possible with an immiscible solvent such as dry ethyl ethanoate or 1,1,1-trichloroethane. If the reactions are carried out on a relatively large scale, even if a recrystallization step is carried out, it is generally possible to isolate product yields of >90%, making this an extremely efficient reaction. A drybox is not essential, but can be extremely useful for handling the salts as they tend to be very hygroscopic, particularly when the alkyl chain substituents are short. In the authors' experience, solid 1-alkyl-3-methylimidazolium halide salts can form as extremely hard solids in round bottomed flasks. Therefore, if a drybox is available the best approach is often to pour the hot salt into shallow trays made of aluminum foil. Once the salt cools and solidifies, it may be broken up into small pieces to aid future use.

The thermal reaction has been used in almost all reports of ionic liquids, being easily adapted to large-scale processes, and providing high yields of products of acceptable purity with relatively simple methods. A number of reports have also examined the use of microwave irradiation, giving high yields with very short reaction times (minutes rather than hours) and using solvent-free conditions [19]. However, the use of microwave irradiation always brings the risk of overheating (see above) and microwave technology in itself has still not been widely demonstrated for large-scale production, therefore commercial preparation of ionic liquids using microwaves would seem unlikely to happen in the near future.

By far the most common starting material is 1-methylimidazole. This is readily available at a reasonable cost, and provides access to a great number of cations likely to be of interest to most researchers. There is only a limited range of other N-substituted imidazoles commercially available and many are relatively expensive. The synthesis of 1-alkylimidazoles may be achieved without great difficulty, however, as indicated in Scheme 2.1-1.

A wider range of C-substituted imidazoles is commercially available, and the combination of these with the reaction in Scheme 2.1-1 allows the formation of many different possible starting materials. In some cases, however, it may still be necessary to carry out synthesis of the heterocycle from first principles.

Scheme 2.1-2

Relatively little has been reported regarding the determination of the purity of the halide salts besides standard spectroscopic measurements and microanalysis. This is largely because the halide salts are rarely used as solvents themselves, but are generally simply a source of the desired cation. Also, the only impurities likely to be present in any significant quantity are unreacted starting materials and residual reaction solvents. Thus, for some applications it is sufficient to ensure, using ^{1}H NMR spectroscopy, that they are free of these. The removal of the haloalkanes and reaction solvents is generally not a problem, especially for the shorter chain haloalkanes that are relatively volatile. On the other hand, the presence of even small quantities of unreacted 1-methylimidazole (a coordinating base) could cause problems in many applications. Furthermore, its high boiling point (198 °C) means that it can prove difficult to remove from ionic liquids. Holbrey et al. have reported a simple colorimetric determination based on the formation of the blue $[Cu(MIM)_4]^{2+}$ ion that is sensitive to 1-methylimidazole in the concentration range 0–3 mol% [20]. Although this does not solve the problem, it does allow samples to be checked before they are used, or for the progress of a reaction to be monitored.

It should again be emphasized that not only halide salts may be prepared in this manner, quaternization reactions can be carried out using methyl or ethyl triflate, methyl trifluoroacetate [15], methyl tosylates [21], and octyl tosylate [22]. All these alkylation agents have been used for the direct preparation of ionic liquids and, in principle, any alkyl compound containing a good leaving group may be used in this manner. Holbrey et al. reported the preparation of ionic liquids with alkyl sulfate anions using this methodology [23]. Dimethyl sulfate or diethyl sulfate were used to prepare a range of alkylimidazolium ionic liquids that were, in many cases, liquid at room temperature, as shown in Scheme 2.1-2. The synthesis can be carried out solvent free or using a solvent in which the product is immiscible. In all such direct alkylation reactions, care should be taken during addition of the alkylating agent. The addition should be slow and under an inert atmosphere to a cooled solution. Such reactions are highly exothermic and the reagents can be sensitive to hydrolysis. Caution must also be exercised when using these types of alkylating agents as many are known to be highly toxic and carcinogenic. Therefore a small excess of nucleophile is advised to avoid traces of the alkylating agent in the product. However, it is important to state that in the case of dialkyl sulfates as alkylating agents, the resultant alkyl sulfate anions are non-toxic.

This approach has the major advantage of generating the desired ionic liquid with no side products and, in particular, no halide ions. At the end of the reaction it is only necessary to ensure that all remaining starting materials are removed either by washing with a suitable solvent or *in vacuo*.

Supercritical carbon dioxide (scCO$_2$) is a recognized green alternative to organic solvents and ionic liquid/scCO$_2$ biphasic systems have become a potential answer to the problem of product extraction from ionic liquids. This is because although scCO$_2$ dissolves in ionic liquids, ionic liquids do not extract into the scCO$_2$ phase, allowing clean product extraction [24]. Wu et al. demonstrated that these features could be exploited for the synthesis of ionic liquids in scCO$_2$ [25]. [BMIM]Br and 1,3-dimethylimidazolium triflate ([DMIM][TfO] were prepared in scCO$_2$ in 100% yield. The starting materials are soluble in the scCO$_2$ phase and as the reaction proceeded an insoluble ionic liquid phase formed. In the synthesis of [BMIM]Br, 1-methylimidazole was reacted with a 20 mol% excess of 1-bromobutane at 70 °C and 15 MPa CO$_2$ pressure. The yield reached 100% within 48 h and the excess 1-bromobutane was extracted cleanly by scCO$_2$ at 50 °C and 15 MPa, collected in a cold trap, and could be recycled for future reactions. For the preparation of [DMIM][TfO], 1-methylimidazole was reacted with around 10 mol% excess methyl trifluoromethanesulfonate. Due to the more reactive nature of methyl trifluoromethanesulfonate the reaction was complete within 2 h at 32 °C and 10 MPa. ScCO$_2$ is a more environmentally friendly alternative to organic solvents and may also be more efficient at producing very pure ionic liquids. This is because the extraction of excess starting material using scCO$_2$ results in no cross-contamination, unlike when washing with organic solvents such as ethyl ethanoate or 1,1,1-trichloroethane.

2.1.3
Anion-exchange Reactions

The anion-exchange reactions of ionic liquids can be divided into two distinct categories: direct reaction of halide salts with Lewis acids, and the formation of ionic liquids via anion metathesis. These two approaches will be dealt with separately as quite different experimental methods are required for each.

2.1.3.1 Lewis Acid-based Ionic Liquids

The formation of ionic liquids by the reaction of halide salts with Lewis acids (most notably AlCl$_3$) dominated the early years of this area of chemistry. The great breakthrough came in 1951 with the report by Hurley and Weir on the formation of a salt that was liquid at room temperature based on the combination of 1-butylpyridinium with AlCl$_3$ in the relative molar proportions 1:2 (X = 0.66) [26].[1]

More recently, the groups of Osteryoung and Wilkes developed the technology of room temperature chloroaluminate melts based on 1-alkylpyridinium [27] and [RMIM]$^+$ cations [6]. In general terms, the reaction of a quaternary halide salt Q$^+$X$^-$ with a Lewis acid MX$_n$ results in the formation of more than one anion species, depending on the relative proportions of Q$^+$X$^-$ and MX$_n$. Such behavior

---

[1] Compositions of Lewis acid-based ionic liquids are generally referred to by the mole fraction of monomeric acid present in the mixture.

can be illustrated for the reaction of [EMIM]Cl with $AlCl_3$ by a series of equilibria (1)–(3).

$$[C_2mim]^+Cl^- + AlCl_3 \rightleftharpoons [C_2mim]^+[AlCl_4]^- \qquad (2.1\text{-}1)$$

$$[C_2mim]^+[AlCl_4]^- + AlCl_3 \rightleftharpoons [C_2mim]^+[Al_2Cl_7]^- \qquad (2.1\text{-}2)$$

$$[C_2mim]^+[Al_2Cl_7]^- + AlCl_3 \rightleftharpoons [C_2mim]^+[Al_3Cl_{10}]^- \qquad (2.1\text{-}3)$$

When [EMIM]Cl is present in a molar excess over $AlCl_3$, only equilibrium (2.1-1) needs to be considered, and the ionic liquid is basic. When, a molar excess of $AlCl_3$ over [EMIM]Cl is present on the other hand, an acidic ionic liquid is formed, and equilibria (2.1-2) and (2.1-3) predominate. Further details of the anion species present may be found elsewhere [28]. The chloroaluminates are not the only ionic liquids prepared in this manner. Other Lewis acids employed include $AlEtCl_2$ [29], BCl_3 [30], CuCl [31], $SnCl_2$ [32], and $FeCl_3$ [33]. In general, the preparative methods employed for all of these salts are similar to those indicated for $AlCl_3$-based ionic liquids, as outlined below.

The most common method for the formation of such liquids is simple mixing of the Lewis acid and the halide salt, with the ionic liquid forming on contact of the two materials. The reaction is generally quite exothermic, which means that care should be taken when adding one reagent to the other. Although the salts are relatively thermally stable, the build-up of excess local heat can result in decomposition and discoloration of the ionic liquid. This may be prevented either by cooling the mixing vessel (often difficult to manage in a drybox), or by adding one component to the other in small portions to allow the heat to dissipate. The water-sensitive nature of most of the starting materials (and ionic liquid products) means that the reaction is best carried out in a drybox. Similarly, the ionic liquids should ideally also be stored in a drybox until use. It is generally recommended, however, that only enough liquid is prepared to carry out the desired task, as decomposition by hydrolysis will inevitably occur over time unless the samples are stored in vacuum-sealed vials.

Finally in this section, it is worth noting that some ionic liquids have been prepared by the reaction of halide salts with metal halides that are not usually thought of as strong Lewis acids. In this case only equilibrium (2.1.1) will apply, and the salts formed are neutral in character. Examples of these include salts of the type $[RMIM]_2[MCl_4]$ (R = alkyl, M = Co, Ni) [34], and $[EMIM]_2[VOCl_4]$ [35]. These are formed by the reaction of two equivalents of [EMIM]Cl with one equivalent of MCl_2 and $VOCl_2$ respectively.

2.1.3.2 Anion Metathesis

The first preparation of air- and water-stable ionic liquids based on 1,3-dialkyl-methylimidazolium cations (sometimes referred to as "second generation" ionic liquids) was reported by Wilkes and Zaworotko in 1992 [36]. This preparation involved a metathesis reaction between [EMIM]I and a range of silver salts ($Ag[NO_3]$, $Ag[NO_2]$, $Ag[BF_4]$, $Ag[CH_3CO_2]$ and $Ag_2[SO_4]$) in methanol or aqueous methanol

Table 2.1-1 Examples of ionic liquids prepared by anion metathesis

Salt	Anion source	Reference
[cation][PF_6]	HPF_6	9, 17, 37, 38
[cation][BF_4]	HBF_4, NH_4BF_4, $NaBF_4$	36–40
[cation][$(CF_3SO_2)_2N$]	$Li[(CF_3SO_2)_2N]$	15, 38
[cation][(CF_3SO_3)]	$CF_3SO_3CH_3$, $NH_4[(CF_3SO_3)]$	15, 41
[cation][CH_3CO_2]	$Ag[CH_3CO_2]$	36
[cation][CF_3CO_2]	$Ag[CF_3CO_2]$	36
[cation][$CF_3(CF_2)_3CO_2$]	$K[CF_3(CF_2)_3CO_2]$	15
[cation][NO_3]	$AgNO_3$, $NaNO_3$	15, 38, 42
[cation][$N(CN)_2$]	$Ag[N(CN)_2]$	43
[cation][$CB_{11}H_{12}$]	$Ag[CB_{11}H_{12}]$	44
[cation][$AuCl_4$]	$HAuCl_4$	45

solution. The very low solubility of silver iodide in these solvents allowed its separation simply by filtration, and removal of the reaction solvent allowed isolation of the ionic liquids in high yield and purity. This method remains the most efficient for the synthesis of water-miscible ionic liquids, but is obviously limited by the relatively high cost of silver salts, not to mention the large quantities of solid by-product produced. The first report of a water-insoluble ionic liquid was two years later, with the preparation of [EMIM][PF_6] from the reaction of [EMIM]Cl and HPF_6 in aqueous solution [37]. The protocols reported in the above two papers have stood the test of time, although subsequent authors have suggested refinements of the methods employed. Most notably, many of the [EMIM]$^+$-based salts are solid at room temperature, facilitating purification which may be achieved via recrystallization. In many applications, however, a product is required that is liquid at room temperature, so most researchers now employ cations with 1-alkyl substituents of chain length 4 or greater, which results in a considerable lowering in melting point. Over the past few years, an enormous variety of anion exchange reactions has been reported for the preparation of ionic liquids. Table 2.1-1 gives a representative selection of both commonly used, and more esoteric examples, along with references that give reasonable preparative details.

The preparative methods employed generally follow similar lines, however, and representative examples are therefore reviewed below. The main goal of all anion-exchange reactions is the formation of the desired ionic liquid uncontaminated with unwanted cations or anions, a task that is easier for water immiscible ionic liquids. It should be noted, however, that low melting salts based on symmetrical onium cations have been prepared using anion-exchange reactions for many years. For example, the preparation of tetrahexylammonium benzoate, a liquid at 25 °C, from tetrahexylammonium iodide, silver oxide and benzoic acid was reported as early as 1967 [46]. The same authors also commented on an alternative approach involving the use of an ion-exchange resin for the conversion of the iodide salt to hydroxide, but concluded that this approach was less desirable. Low melting salts based on cations such as tetrabutylphosphonium [47] and trimethylsulfonium [48] have also been produced using very similar synthetic methods.

To date, surprisingly few reports of the use of ion-exchange resins for the large-scale preparation of ionic liquids have appeared in the open literature, to the best of our knowledge. One exception is a report by Lall et al. regarding the formation of phosphate-based ionic liquids with polyammonium cations [4]. Wasserscheid and Keim have suggested, however, that this might be an ideal method for their preparation in high purity [3f].

As the preparation of water-immiscible ionic liquids is considerably more straightforward than the preparation of their water-soluble analogues, these methods will be considered first. The water solubility of the ionic liquids is very dependent on both the anion and cation present, and in general will decrease with increasing organic character of the cation. The most common approach for the preparation of water-immiscible ionic liquids is first to prepare an aqueous solution of a halide salt of the desired cation. The cation exchange is then carried out using either the free acid of the appropriate anion, or a metal or ammonium salt. Where available, the free acid is probably to be favored, as it leaves only HCl, HBr or HI as the by-product, easily removed from the final product by washing with water. It is recommended that these reactions are carried out with cooling of the halide salt in an ice bath, as the addition of a strong acid to an aqueous solution is often exothermic. In cases where the free acid is unavailable, or inconvenient to use, however, alkali metal or ammonium salts may be substituted without major problems. It may also be preferable to avoid using the free acid in systems where the presence of traces of acid may cause problems. A number of authors have outlined broadly similar methods for the preparation of $[PF_6]^-$ and $[(CF_3SO_2)_2N]^-$ salts that may be adapted for most purposes [15, 17].

When free acids are used, the washing should be continued until the aqueous residues are neutral, as traces of acid can cause decomposition of the ionic liquid over time. This can be a particular problem for salts based on the $[PF_6]^-$ anion which will slowly form HF, particularly on heating, if not completely acid free. When alkali metal or ammonium salts are used, it is advisable to check for the presence of halide anions in the washing solutions, for example by testing with silver nitrate solution. The high viscosity of some ionic liquids makes efficient washing difficult, even though the presence of water results in a considerable reduction the viscosity. As a result, a number of authors have recently recommended dissolving these liquids in either CH_2Cl_2 or $CHCl_3$ prior to carrying out the washing step.

The preparation of water-miscible ionic liquids can be a more demanding process, as the separation of the desired and undesired salts may be complex. The use of silver salts described above allows the preparation of many salts in very high purity, but is clearly too expensive for large-scale use. As a result, a number of alternative protocols have been developed that employ cheaper salts for the metathesis reaction. The most common approach remains to carry out the exchange in aqueous solution using either the free acid of the appropriate anion, the ammonium salt, or an alkali metal salt. When using this approach, it is important that the desired ionic liquid can be isolated without excess contamination from unwanted halide-containing by-products. A reasonable compromise has been suggested by Welton et al. for the preparation of [BMIM][BF$_4$] [40]. In this approach, which could in principle be adapted to many other water-miscible systems, the ionic liquid is formed by metathesis between [BMIM]Cl and HBF$_4$ in aqueous solution. The product is extracted into CH$_2$Cl$_2$, and the organic phase is then washed with successive small portions of deionized water until the washings are pH neutral. The presence of halide ions in the washing solutions can be detected by testing with AgNO$_3$. The CH$_2$Cl$_2$ is then removed on a rotary evaporator, and the ionic liquid then further purified by mixing with activated charcoal for 12 h. Finally, the liquid is filtered through a short column of acidic or neutral alumina and dried by heating *in vacuo*. Yields of around 70% are reported when this approach is carried out on large (~1 molar) scale. Although the water wash can result in a lowering of the yield, the aqueous wash solutions may ultimately be collected together, the water removed, and the crude salt added to the next batch of ionic liquid prepared. In this manner, the amount of product loss is minimized, and the purity of the ionic liquid prepared appears to be reasonable for most applications.

Alternatively, the metathesis reaction may be carried out entirely in an organic solvent such as CH$_2$Cl$_2$, as described by Cammarata et al. [38], or acetone, as described by Fuller et al. [41]. In both of these systems the starting materials are not fully soluble in the reaction solvent, so the reaction is carried out as a suspension. In the case of the CH$_2$Cl$_2$ process, the reaction was carried out by stirring the 1-alkyl-3-methylimidazolium halide salt with the desired metal salt at room temperature for 24 h. The insoluble halide by-products were then removed by filtration. Although the halide by-products have limited solubility in CH$_2$Cl$_2$, they are much more soluble in the ionic liquid/CH$_2$Cl$_2$ mixture. Thus, when this method is employed it is important that the CH$_2$Cl$_2$ extracts are washed with water to minimize the halide content of the final product. This approach clearly results in a lowering of the yield of the final product. Therefore care must be taken that the volume of water used to carry out the washing is low. Lowering the temperature of the water to near 0 °C can also reduce the amount of ionic liquid loss. The final product can be purified by stirring with activated charcoal followed by passing through an alumina column, as described in the previous paragraph. This process was reported to give final yields in the region of 70–80%, and was used to prepare ionic liquids containing a wide variety of anions ([PF$_6$]$^-$, [SbF$_6$]$^-$ [BF$_4$]$^-$, [ClO$_4$]$^-$, [CF$_3$SO$_3$]$^-$, [NO$_3$]$^-$ and [CF$_3$CO$_2$]$^-$). In the case of the acetone route, [EMIM]Cl was stirred with [NH$_4$][BF$_4$] or [NH$_4$][CF$_3$SO$_3$] at room temperature for 72 h. In this case all starting materials were only slightly soluble in the reaction solvent. Once again, the insoluble [NH$_4$]Cl by-product was removed by filtration. No water wash was carried out, but trace organic impurities were removed by stirring the acetone solution with neutral alumina for 2 h after removal of the metal halide salts by filtration. The salts were finally dried by heating at 120 °C for several hours, after which they were analyzed for purity by electrochemical methods, giving quoted purities of at least 99.95%.

2.1.4
Purification of Ionic Liquids

The lack of significant vapor pressure prevents the purification of ionic liquids by distillation. The counterpoint to this is that any volatile impurity can, in principle, be separated from an ionic liquid by distillation. In general, however, it is better to remove as many impurities as possible from the starting materials and, where possible, to use synthetic methods that either generate as few side products as possible, or allow their easy separation from the final ionic liquid product. This section will first describe the methods employed to purify starting materials, and then move onto methods used to remove specific impurities from the different classes of ionic liquids.

The first requirement is that all starting materials used for the preparation of the cation should be distilled prior to use. The authors have found the methods described by Amarego and Perrin to be suitable in most cases [49]. For example, in the preparation of [RMIM]$^+$ salts we routinely distil the 1-methylimidazole under vacuum from sodium hydroxide, and then store any that is not used immediately under nitrogen in the refrigerator. The haloalkanes are first washed with portions of concentrated sulfuric acid until no further color is removed into the acid layer, then neutralized with NaHCO$_3$ solution and deionized water, and finally distilled before use. All solvent used in quaternization or anion-exchange reactions should also be dried and distilled before use. If these precautions are not taken it is often difficult to prepare colorless ionic liquids. In cases where the color of the ionic liquids is less important, the washing of the haloalkane may be unnecessary, as the quantity of colored impurity is thought to be extremely low, and thus will not affect many potential applications. It has also been observed that, in order to prepare AlCl$_3$-based ionic liquids that are colorless, it is usually necessary to sublime the AlCl$_3$ prior to use (often more than once). It is recommended that the AlCl$_3$ should be mixed with sodium chloride and aluminum wire for this process [27b].

AlCl$_3$-based ionic liquids often contain traces of oxide ion impurities, formed by the presence of small amounts of water and oxygen. These are generally referred to as [AlOCl$_2$]$^-$, although ^{17}O NMR measurements have indicated that a complex series of equilibria are in fact occurring [50]. It has been reported that these can be efficiently removed by bubbling phosgene (COCl$_2$) through the ionic liquid [51]. In this case the by-product of the reaction is CO$_2$, and thus easily removed under vacuum. This method should be approached with caution due to the high toxicity of phosgene, and more recently an alternative approach using the less toxic triphosgene has also been reported [52]. In the presence of water or other proton sources, chloroaluminate-based ionic liquids may contain protons, which will behave as a Brønsted superacid in acidic melts [53]. It has been reported that these may be removed simply by the application of high vacuum ($<5 \times 10^{-6}$ Torr) [54].

Purification of ionic liquids formed by anion metathesis can throw up a different set of problems, as already noted in Section 2.1.3.2. In this case the most common impurities are halide anions, or unwanted cations inefficiently separated from the final product. The presence of such impurities can be extremely detrimental to the performance of the ionic liquids, particularly in applications involving transition metal-based catalysts, which are often deactivated by halide ions. In general this is much more of a problem in water-soluble ionic liquids as the water-immiscible salts can generally be purified quite efficiently by washing with water. The methods used to overcome this problem have already been covered in the previous section. The problems inherent in the preparation of water-miscible salts have been highlighted by Seddon et al. [42], who studied the Na^+ and Cl^- concentrations in a range of ionic liquids formed by the reaction of [EMIM]Cl and [BMIM]Cl with $Ag[BF_4]$, $Na[BF_4]$, $Ag[NO_3]$, $Na[NO_3]$ and HNO_3. They found that the physical properties such as density and viscosity of the liquids can be radically altered by the presence of unwanted ions. The results showed that all preparations using Na^+ salts resulted in high residual concentrations of Cl^-, while the use of Ag^+ salts gave rise to much lower levels. The low solubility of NaCl in the ionic liquids indicates, however, that the impurities arise because with the Na^+ salts the reaction does not proceed to completion. Indeed, it was reported that unreacted [BMIM]Cl was isolated by crystallization from $[BMIM][NO_3]$ in one case. A further example of the potential hazards of metal-containing impurities in ionic liquids is shown when $[EMIM][CH_3CO_2]$ is prepared from [EMIM]Cl and $Pb[CH_3CO_2]_4$ [55]. The resulting salt has been shown to contain ca. 0.5 M residual lead [56].

In practical terms, it is suggested that, in any application where the presence of halide ions may cause problems, the concentration of these be monitored to ensure the purity of the liquids. This may be achieved either by the use of an ion-sensitive electrode, or by using a chemical method such as the Vollhard procedure for chloride ions [57]. Seddon et al. report that effectively identical results were obtained using both methods [42].

Most ionic liquids based on the common cations and anions should be colorless, with minimal absorbance at wavelengths greater than 300 nm. In practice the salts often take on a yellow color, particularly during the quaternization step. The amount of impurity causing this is generally extremely small, being undetectable using ^1H NMR or CHN microanalysis, and in many applications the discoloration may not be of any importance. This is clearly not the case when the solvents are required for photochemical or UV–visible spectroscopic investigations, however. To date, the precise origin of these impurities has not been determined, but it seems likely that they arise from unwanted side reactions involving oligomerization or polymerization of small amounts of free amine, or else from impurities in the starting materials. Where it is important that the liquids are colorless, however, the color may be minimized by following a few general steps:

- All starting materials should be purified as discussed above [49].
- The presence of traces of acetone can sometimes result in discoloration during the quaternization step. Thus, all glassware used in this step should be kept free of this solvent.
- The quaternization reaction should be carried out either in a system that has been degassed and sealed under nitrogen, or else under a flow of inert gas such as nitrogen. Furthermore the reaction temperature should be kept as low

as possible (no more than ca. 80 °C for Cl^- salts, and lower for Br^- and I^- salts).

If the liquids remain discolored even after these precautions, it is often possible to further purify them by first stirring with activated charcoal, then passing the liquid down a short column of neutral or acidic alumina as discussed in Section 2.1.3.2 [38].

Clearly, the impurity likely to be present in largest concentrations in most ionic liquids is water. The removal of other reaction solvents is generally easily achieved by heating the ionic liquid under vacuum. Water is generally one of the most problematic solvents to remove, and it is generally recommended that ionic liquids are heated to at least 70 °C for several hours with stirring to achieve an acceptably low degree of water contamination. Even water-immiscible salts such as $[BMIM][PF_6]$ can absorb water when exposed to the air [38, 58]. Thus it is advised that all liquids are dried directly before use. If the amount of water present is of importance, this may be determined either using Karl–Fischer titration, or a less precise determination may be carried out using IR spectroscopy.

2.1.5
Improving the Sustainability of Ionic Liquids

As the interest in ionic liquids increases, their 'green' credentials are increasingly being called into question. To address these issues, improvements in the methodology for the preparation of ionic liquids are increasingly being accompanied by investigations of their toxicity and potential environmental impact. Such information is vital in a situation where these liquid materials find widespread commercial use nowadays.

For ionic liquids to be truly considered as green solvents, their life cycle analysis has to be taken into account. In an important preliminary study combining computational and experimental methods, Kralish et al., examined the energetic, environmental and economic impact of ionic liquids [59]. The cumulative energy demand (CED), the environmental impact and the economics of preparing ionic liquids were all considered. The CED takes into account the supply of raw materials, recycling and disposal of chemicals, the energies required for heating, stirring, pumping and providing cooling water. In the synthesis of ionic liquids the choice of cation, anion and reaction solvents was found to have a large impact on the CED, toxicity and economics. This study highlights the fact that for an imidazolium-based ionic liquid such as $[BMIM][BF_4]$ nearly 50% of the total energy needed for its preparation derives from the required supply of 1-methylimidazole. This suggests that, where technically feasible, cheaper cations such as tetraalkylammonium should be employed, as the manufacture of tertiary amines is many times lower in both energy and cost. The toxicity and environmental persistence of anions such as $[BF_4]^-$ was also highlighted. Clearly many different factors are important if commercial applications are to be implemented. Therefore, researchers should be aware of such information when choosing the type of ionic liquid and synthetic methods, even at the very early stages of development. As mentioned earlier, researchers have examined using microwave irradiation [19] and the use of scCO$_2$ [25] to improve the efficiency of ionic liquid synthesis. Perhaps in future critical studies, the CED and economics of such methods could be compared with more traditional methods.

Another approach to improving the sustainability of ionic liquids is to use biorenewable feedstocks for their synthesis. Handy et al. reported the synthesis of imidazolium ionic liquids derived from fructose, as shown in Scheme 2.1-3 [60].

To show their suitability as replacements for conventional imidazolium-based ionic liquids the authors demonstrated that these protic ionic liquids performed well as solvents for the Heck reaction. The alcohol functionality also has the benefit that it can be used to tether substrates, allowing these ionic liquids to be used as recyclable homogenous supports, an alternative to traditional polymer resins [61].

Detailed reports on the toxicity and potential environmental impact of commonly used ionic liquids have been published only recently [62]. The information from such studies will probably be crucial in selecting ionic liquids for large-scale applications. One method of addressing the potential toxicity issues is to use ions of known toxicity, an area of increasing interest. Biologically acceptable anions such as chloride, lactate and acetate have been used in many cases and, as mentioned earlier, methyl and ethyl sulfate ionic liquids offer another non-toxic, non-halogenated alternative. As these short chain alkyl sulfate anions are sensitive to hydrolysis, Wasserscheid et al. suggested the use of ionic liquids based on the octylsulfate anion that solve this stability problem [63]. This particular anion, like many other long chain alkyl sulfates, is already in wide-scale use for detergent and cosmetic applications, and therefore has known toxicological behavior. [BMIM][n-C$_8$H$_{17}$OSO$_3$] was prepared by anion exchange between [BMIM]Cl and [Na][n-C$_8$H$_{17}$OSO$_3$] in water at 60 °C, giving a liquid of >87% purity, the main impurities being water soluble salts such as Na$_2$SO$_4$. The water was slowly removed under vacuum and a white solid precipitated. Dichloromethane was then used to extract the product from the mixture and the white solid removed by filtration. The filtrate was then washed several times with portions of water until the water washings were chloride free (AgNO$_3$ test). The dichloromethane was then removed under vacuum (and collected for subsequent reuse) giving an ionic liquid with a melting point of 34–35 °C. The hydrolytic stability was compared to methyl and ethyl sulfate anions by heating the ionic liquids in water at 80 °C. While the short chain sulfate ionic liquids gave an acidic pH within

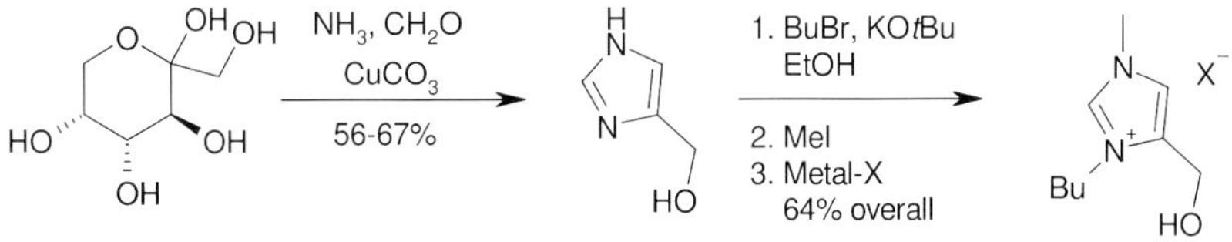

Scheme 2.1-3

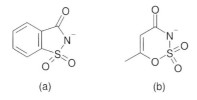

Fig. 2.1-2 (a) saccharinate (b) acesulfamate.

30 min, [BMIM][n-$C_8H_{17}OSO_3$] showed no lowering of the pH after 8 h under the same conditions.

In an innovative approach to the generation of non-toxic ionic liquids, Davis and coworkers prepared ionic liquids with anions based on non-nutritive sweeteners of known toxicity [64]. Saccharinate and acesulfamate anions (Fig. 2.1-2) were paired with a range of cations including a known non-toxic cation, butyryl choline. The ionic liquids based on this cation had melting points of 49 °C (saccharinate) and 89 °C (acesulfamate).

Ionic liquids containing these non-toxic anions were water soluble, and were prepared by anion exchange of the halide salt of the cation with the sodium salt of the sweetener in acetone, or in the case of the butyryl choline salts, using 50/50 (v/v) water–acetone and the silver salt of sweetener (prepared by exchanging the sodium or potassium salts with $AgNO_3$). In all cases the precipitated metal halide by-product was removed by filtration. The ionic liquids that were solids at room temperature were recrystallized from ethanol or methanol, while those that were glasses or liquid at room temperature were purified by column chromatography using dichloromethane and acetonitrile. In a more recent report, the acesulfamate anion was combined with long chain phosphonium cations to produce ionic liquids that were immiscible with water [65]. In this case the anion exchange could be carried out in water and the halide impurities removed by aqueous extraction.

Ohno and coworkers also produced a range of ionic liquids using anions of known toxicity and biodegradability using 20 natural amino acids combined with the [EMIM]$^+$ cation [66]. The salts were prepared using a neutralization method that has been rarely exploited in ionic liquid synthesis so far. First, [EMIM][OH] was prepared from [EMIM]Br using an anion-exchange resin. An aqueous solution of [EMIM][OH] was then added dropwise to a small excess of amino acid in aqueous solution, then stirred for 12 h. After the water was removed under vacuum, 90/10 (v/v) acetonitrile/methanol was added and the mixture stirred vigorously before removal of the excess amino acid by filtration. Amino acid-based ionic liquids not only use anions of known toxicity and biodegradability, but may have solvent properties useful for various applications.

Commercial suppliers of ionic liquids are also preparing salts based on ions of known toxicity. For example, Sachem produce Terrasail™ ionic liquids based on the docusate (dioctylsulfosuccinate) anion (Fig. 2.1-3) [67]. Some ionic liquids containing this anion have been found to be miscible with hydrocarbon solvents,

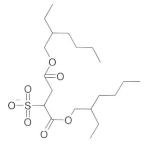

Fig. 2.1-3 Docusate anion.

unlike most other ionic liquids [68]. Another example is Solvent Innovation GmbH, that has introduced their ECOENG™ and AMMOENG™ ionic liquid families, ionic liquids based on alkylsulfate and alkoxysulfate anions and ethylene glycol based ammonium cations, respectively [69]. In these ionic liquids ions that have already been studied with regards to toxicity and are already utilized in other industries are applied to form ionic liquids.

A further very important issue that will determine the potential future applications of ionic liquids is their biodegradation. Ionic liquids are structurally related to many commercial surfactants, and Gathergood and coworkers took the lessons learned from the design of biodegradable surfactants to examine imidazolium ionic liquids with ester and amide functionality on the cation substituents [70]. The ionic liquids were prepared via standard methods using ester or amide derivatives of bromoacetic acid. It was found that materials possessing an ester group on the cation showed the greatest biodegradation, while the octyl sulfate anion was the most easily degraded of the anions studied. Just as this area is now important in the commercial design of surfactants, it is imperative that there is continued effort in the area of designing biodegradable ionic liquids.

2.1.6
Conclusions

It is hoped that this chapter will give the reader a better appreciation of the range of ionic liquids that have already been prepared, as well as a summary of the main techniques involved and the potential pitfalls. While the basic chemistry involved is relatively straightforward, the preparation of ionic liquids of known purity may be less easily achieved, and it is hoped that the ideas given here may be of assistance to the reader. It should also be noted that many of the more widely used ionic liquids are now commercially available from a range of suppliers [67, 69, 71].

References

1. P. Walden, *Bull. Acad. Imper. Sci. (St. Petersburg)*, **1914**, 1800.
2. R. M. Pagni, *Advances in Molten Salt Chemistry* 6, G. Mamantov, C. B. Mamantov, J. Braunstein (Eds.), Elsevier: New York, **1987**, pp. 211–346.

3. (a) C. Chiappe, D. Pieraccini, *J. Phys. Org. Chem.*, **2005**, *18*, 275; (b) T. Welton, *Coord. Chem. Rev.*, **2004**, *248*, 2459; (c) J. Dupont, R. F. de Souza, P. A. Z. Suarez, *Chem. Rev.*, **2002**, *102*, 3667; (d) C. M. Gordon, *Appl. Catal. A, General*, **2001**, *222*, 101; (e) R. A. Sheldon, *Chem. Commun.*, **2001**, 2399; (f) P. Wasserscheid, W. Keim, *Angew. Chem., Int. Ed.*, **2000**, *39*, 3772; (g) J. D. Holbrey, K. R. Seddon, *Clean Prod. Processes*, **1999**, *1*, 223; (h) T. Welton, *Chem. Rev.*, **1999**, *99*, 2071.
4. S. I. Lall, D. Mancheno, S. Castro, V. Behaj, J. L. I. Cohen, R. Engel, *Chem. Commun.*, **2000**, 2413.
5. J. H. Davis, Jr., K. J. Forrester, T. Merrigan, *Tetrahedron Lett.*, **1998**, *39*, 8955.
6. J. S. Wilkes, J. A. Levisky, R. A. Wilson, C. L. Hussey, *Inorg. Chem.*, **1982**, *21*, 1263.
7. D. F. Evans, A. Yamouchi, G. J. Wei, V. A. Bloomfield, *J. Phys. Chem.*, **1983**, *87*, 3537.
8. C. K. Lee, H. W. Huang, I. J. B. Lin, *Chem. Commun.*, **2000**, 1911.
9. C. M. Gordon, J. D. Holbrey, A. R. Kennedy, K. R. Seddon, *J. Mater. Chem.*, **1998**, *8*, 2627.
10. A. E. Visser, J. D. Holbrey, R. D. Rogers, *Chem. Commun.*, **2001**, 2484.
11. T. Kitazume, F. Zulfiqar, G. Tanaka, *Green Chem.*, **2000**, *2*, 133.
12. D. R. MacFarlane, P. Meakin, J. Sun, N. Amini, M. Forsyth, *J. Phys. Chem. B*, **1999**, *103*, 4164.
13. J. Sun, M. Forsyth, D. R. MacFarlane, *J. Phys. Chem. B*, **1998**, *102*, 8858.
14. C. J. Bradaric, A. Downard, C. Kennedy, A. J. Robertson, Y. Zhou, *Green Chem*, **2003**, *5*, 143.
15. P. Bonhôte, A.-P. Dias, N. Papageorgiou, K. Kalyanasundaram, M. Grätzel, *Inorg. Chem.*, **1996**, *35*, 1168.
16. P. Lucas, N. El Mehdi, H. A. Ho, D. Bélanger, L. Breau, *Synthesis*, **2000**, 1253.
17. J. G. Huddleston, H. D. Willauer, R. P. Swatlowski, A. E. Visser, R. D. Rogers, *Chem. Commun.*, **1998**, 1765.
18. A. G. Böwing, A. Jess, *Green Chem.*, **2005**, *7*, 230.
19. (a) R. S. Varma, V. V. Namboodiri, *Chem. Commun.*, **2001**, 643; (b) B. M. Khadilkar, G. L. Rebeirio, *Org. Proc. Res. Develop.*, **2002**, *6*, 826; (c) M. C. Law, K. Wong, T. H. Chan, *Green Chem.*, **2002**, *4*, 328; (d) M. Deetlefs, K. R. Seddon, *Green Chem.*, **2003**, *5*, 181.
20. J. A. Holbrey, K. R. Seddon, R. Wareing, *Green Chem.*, **2001**, *3*, 33.
21. H. Waffenschmidt, Dissertation, RWTH Aachen, Germany, **2000**.
22. N. Karodia, S. Guise, C. Newlands, J.-A. Andersen, *Chem. Commun.*, **1998**, 2341.
23. J. D. Holbrey, W. M. Reichert, R. P. Swatloski, G. A. Broker, W. R. Pitner, K. R. Seddon, R. D. Rogers, *Green Chem.*, **2002**, *4*, 407.
24. L. A. Blanchard, D. Hancu, E. J. Beckman, J. F. Brennecke, *Nature*, **1999**, *399*, 28.
25. W. Wu, W. Li, B. Han, Z. Zhang, T. Jiang, Z. Liu, *Green Chem.*, **2005**, advanced article.
26. F. H. Hurley, T. P. Wier, *J. Electrochem. Soc.*, **1951**, *98*, 203.
27. (a) H. L. Chum, V. R. Koch, L. L. Miller, R. A. Osteryoung, *J. Am. Chem. Soc.*, **1975**, *97*, 3264; (b) J. Robinson, R. A. Osteryoung, *J. Am. Chem. Soc.*, **1979**, *101*, 323.
28. H. A. Øye, M. Jagtoyen, T. Oksefjell, J. S. Wilkes, *Mater. Sci. Forum*, **1991**, *73–75*, 183.
29. (a) Y. Chauvin, S. Einloft, H. Olivier, *Ind. Eng. Chem. Res.*, **1995**, *34*, 1149; (b) B. Gilbert, Y. Chauvin, H. Olivier, F. DiMarco-van Tiggelen, *J. Chem. Soc., Dalton Trans.*, **1995**, 3867.
30. S. D. Williams, J. P. Schoebrechts, J. C. Selkirk, G. Mamantov, *J. Am. Chem. Soc.*, **1987**, *109*, 2218.
31. Y. Chauvin, H. Olivier-Bourbigou, *CHEMTECH*, **1995**, *25*, 26.
32. G. W. Parshall, *J. Am. Chem. Soc.*, **1972**, *94*, 8716.
33. M. S. Sitze, E. R. Shreiter, E. V. Patterson, R. G. Freeman, *Inorg. Chem.*, **2001**, *40*, 2298.
34. P. B. Hitchcock, K. R. Seddon, T. Welton, *J. Chem. Soc., Dalton Trans.*, **1993**, 2639.
35. P. B. Hitchcock, R. J. Lewis, T. Welton, *Polyhedron*, **1993**, *12*, 2039.
36. J. S. Wilkes, M. J. Zaworotko, *Chem. Commun.*, **1992**, 965.
37. J. Fuller, R. T. Carlin, H. C. DeLong, D. Haworth, *Chem. Commun.*, **1994**, 299.

38. L. Cammarata, S. Kazarian, P. Salter, T. Welton, *Phys. Chem. Chem. Phys.*, **2001**, *3*, 5192.
39. J. D. Holbrey, K. R. Seddon, *J. Chem Soc., Dalton Trans.*, **1999**, 2133.
40. N. L. Lancaster, T. Welton, G. B. Young, *J. Chem. Soc., Perkin Trans. 2*, **2001**, 2267.
41. J. Fuller, R. T. Carlin, *Proc. Electrochem. Soc.*, **1999**, *98*, 227.
42. K. R. Seddon, A. Stark, M.-J. Torres, *Pure Appl. Chem.*, **2000**, *72*, 2275.
43. D. R. McFarlane, J. Golding, S. Forsyth, M. Forsyth, G. B. Deacon, *Chem. Commun.*, **2001**, 2133.
44. A. S. Larsen, J. D. Holbrey, F. S. Tham, C. A. Reed, *J. Am. Chem. Soc.*, **2000**, *122*, 7264.
45. M. Hasan, I. V. Kozhevnikov, M. R. H. Siddiqui, A. Steiner, N. Winterton, *Inorg. Chem.*, **1999**, *38*, 5637.
46. C. G. Swain, A. Ohno, D. K. Roe, R. Brown, T. Maugh, II, *J. Am. Chem. Soc.*, **1967**, *89*, 2648.
47. R. M. Pomaville, S. K. Poole, L. J. Davis, C. F. Poole, *J. Chromatogr.*, **1988**, *438*, 1.
48. H. Matsumoto, T. Matsuda, Y. Miyazaki, *Chem. Lett.*, **2000**, 1430.
49. W. L. F. Armarego, D. D. Perrin, *Purification of Laboratory Chemicals*, 4th Edn., Butterworth-Heinemann: London, **1997**.
50. T. A. Zawodzinski, R. A. Osteryoung, *Inorg. Chem.*, **1990**, *29*, 2842.
51. (a) A. K. Abdul-Sada, A. G. Avent, M. J. Parkington, T. A. Ryan, K. R. Seddon, T. Welton, *Chem. Commun.*, **1987**, 1643; (b) A. K. Abdul-Sada, A. G. Avent, M. J. Parkington, T. A. Ryan, K. R. Seddon, T. Welton, *J. Chem. Soc., Dalton Trans.*, **1993**, 3283.
52. A. J. Dent, A. Lees, R. J. Lewis, T. Welton, *J. Chem. Soc., Dalton Trans.*, **1996**, 2787.
53. (a) G. P. Smith, A. S. Dworkin, R. M. Pagni, S. P. Zing, *J. Am. Chem. Soc.*, **1989**, *111*, 525; (b) G. P. Smith, A. S. Dworkin, R. M. Pagni, S. P. Zing, *J. Am. Chem. Soc.*, **1989**, *111*, 5075; (c) S.-G. Park, P. C. Trulove, R. T. Carlin, R. A. Osteryoung, *J. Am. Chem. Soc.*, **1991**, *113*, 3334.
54. M. A. M. Noel, P. C. Trulove, R. A. Osteryoung, *Anal. Chem.*, **1991**, *63*, 2892.
55. B. Ellis, *Int. Pat.*, WO 96/18459, **1996**.
56. J. T. Hamill, C. Hardacre, M. Nieuwenhuyzen, K. R. Seddon, S. A. Thompson, B. Ellis, *Chem. Commun.*, **2000**, 1929.
57. A. I. Vogel, *A Textbook of Quantitative Inorganic Analysis*, 3rd Edn., Longmans, Green and Co., London, **1961**.
58. C. D. Tran, S. H. D. Lacerda, D. Oliveira, *Appl. Spectrosc.*, **2003**, *57*, 152.
59. D. Kralisch, A. Stark, S. Körsten, G. Kreisel, B. Ondruschka, *Green Chem.*, **2005**, *7*, 301.
60. S. T. Handy, M. Okello, G. Dickenson, *Org. Lett.*, **2003**, *5*, 2513.
61. S. T. Handy, M. Okello, *Tetrahedron Lett.*, **2003**, *44*, 8399.
62. (a) D. Jastroff, K. Mölter, P. Behrend, U. Bottin-Weber, J. Filser, A. Heimers, B. Ondruschka, J. Ranke, M. Schaefer, H. Schröder, A. Stark, P. Stepnowski, F. Stock, R. Störmann, S. Stolte, U. Welz-Biermann, S. Ziegert, J. Thöming, *Green Chem.*, **2005**, *7*, 362; (b) K. M. Docherty, C. F. Kulpa, Jr., *Green Chem.*, **2005**, *7*, 185; (c) A. Latała, P. Stepnowski, M. Nędzi, W. Mrozik, *Aquat. Tox.*, **2005**, *73*, 91; (d) R. J. Bernot, M. A. Brueseke, M. A. Evans-White, G. A. Lamberti, *Environ. Toxicol. Chem.*, **2004**, *24*, 87.
63. P. Wasserscheid, R. van Hal, A. Bösmann, *Green Chem.*, **2002**, *4*, 400.
64. E. B. Carter, S. L. Culver, P. A. Fox, R. D. Goode, I. Ntai, M. D. Tickell, R. K. Traylor, N. W. Hoffman, J. H. Davis, Jr., *Chem. Commun.*, **2004**, 630.
65. J. Pernak, F. Stefaniak, J. Węglewski, *Eur. J. Org. Chem.*, **2005**, 650.
66. K. Fukumoto, M. Yoshizawa, H. Ohno, *J. Am. Chem. Soc.*, **2005**, *127*, 2398.
67. www.sacheminc.com
68. J. H. Davis, Jr., P. A. Fox, *Chem. Commun.*, **2003**, 1209.
69. www.solvent-innovation.com.
70. N. Gatherhood, M. T. Garcia, P. J. Scammells, *Green Chem.*, **2004**, *6*, 166; M. T. Garcia, N. Gatherhood, P. J. Scammells, *Green Chem.*, **2004**, *7*, 9.
71. (a) www.covalentassociates.com; (b) www.sigmaaldrich.com; (c) www.merck.de; (d) www.bioniqs.com; (e) www.cytec.com; (f) www.fisher.co.uk/chemicals.

2.2
Quality Aspects and Other Questions Related to Commercial Ionic Liquid Production

Markus Wagner and Claus Hilgers

2.2.1
Introduction

From Section 2.1 it has become very clear that the synthesis of an ionic liquid, in general, is quite simple organic chemistry while the preparation of an ionic liquid of a certain quality requires some know-how and experience. Since both distillation and crystallization cannot be easily used to purify ionic liquids after their synthesis (even though the distillation of some specific ionic liquids has been described recently [1]) maximum care has to be taken before and during the ionic liquid synthesis to obtain the desired quality.

Back in the 1990s, the know-how to synthesize, purify and handle ionic liquids was treated somewhat like a "holy grail". Indeed, only a small number of specialized industrial and academic research groups were able to prepare and handle the highly hygroscopic chloroaluminate ionic liquids, which were the only ionic liquid systems available in larger amounts up to the mid-nineties. For example, acidic chloroaluminate ionic liquids have to be stored in glove boxes to prevent their contamination with traces of water. Water impurities are known to react with the anions of the melt with release of superacidic protons. These cause unwanted side reactions in many applications and possess a considerable potential for corrosion (a detailed description of protic and oxidic impurities in chloroaluminate melts is given in Welton's review article from 1999 [2]). Without a doubt, this need for very special and expensive handling protocols has hampered the commercial production and distribution of chloroaluminate ionic liquids, even up to now.

The introduction of the more hydrolysis stable tetrafluoroborate [3] and hexafluorophosphate systems [4] and especially the development of their synthesis by metathesis reaction from alkali salts [5] can be regarded as a first key step towards commercial ionic liquid production.

However, it still took its time. Prior to the foundation of Solvent Innovation [6] in November 1999 the commercial availability of ionic liquids was indeed very limited. Only a small number of systems could be purchased from Sigma-Aldrich in quantities up to 5 g [7].

In the last six years we have witnessed several small companies entering the field of ionic liquids. The leading players in this field today, such as Degussa, BASF, Merck and Solvent Innovation have further increased their capacities within the last two years and have in-house production capacities for ionic liquids on the multi-ton level. Hence, those companies which are flexible in production campaign scheduling have reduced the lead-time for standard ionic liquids, even on ton scales, to a few weeks.

In addition, the distribution of ionic liquids by Fluka, Acros Organics, Kanto and Tokyo Kasei assures today a much better availability of many ionic liquids on a rapid-delivery basis.

From discussion with many people now working with ionic liquids we know that, at least for the start of their work, the ability to buy an ionic liquid was important. In fact, a synthetic chemist searching for the ideal solvent for his specific application usually takes solvents which are ready for use on the shelf of his laboratory. The additional effort of synthesizing a new special solvent can be rarely justified, especially in industrial research. Of course, this is not only true for ionic liquids. It is very likely that nobody would use acetonitrile as a solvent in the laboratory if one had to synthesize it before use.

Thus, commercial availability of ionic liquids was a key factor for the actual success of the ionic liquid methodology. Apart from the matter of lowering the "activation barrier" for those synthetic chemists interested in entering the field, it allows access to ionic liquids for those communities who traditionally do not focus on synthetic work. Physical chemists, engineers, electrochemists and scientists interested in developing new analytical tools are among those who have already developed many new exciting applications using ionic liquids [8].

2.2.2
Quality Aspects of Commercial Ionic Liquid Production

As mentioned above, ionic liquids are now available from a number of commercial manufacturers and from many suppliers. Still, one should not forget that an ionic liquid is a quite different product compared to traditional organic solvents, simply because it cannot be easily purified by distillation, due to its very low volatility. This, combined with the fact that small amounts of impurities influence the ionic liquid's properties significantly [9] makes the quality of an ionic liquid a quite important consideration for many applications.

For academic researchers it is possible and desirable to obtain perfectly pure ionic liquids with negligible amounts of impurities. However, such efforts are simply not necessary for most applications and would significantly add to the cost of ionic liquids when produced on a commercial scale. Thus, the authors have taken the approach that once a customer has decided in which way to use a certain ionic liquid in a specific application the set of specification parameters (purity, by-products, halide content, water content, color) will have to be defined in detail together.

In other words, the desire for absolute quality of the product and the need for a reasonable price have to be reconciled. This, of course, is not new. If one looks into the very similar business of phase transfer catalysts or other ionic modifiers (e.g. commercially available ammonium salts) one rarely finds absolutely pure materials. Sometimes, the active ionic compound is only present in about 85% quantity. However, and this is a crucial point, the product is well specified, the nature of the impurities is known and the quality of the material is absolutely reproducible from batch to batch.

From our point of view, this is exactly what commercial ionic liquid production is about. Commercial producers try to make ionic liquids in the highest quality that can be realized at reasonable cost. For some ionic liquids they can guarantee purities greater 99%, for others perhaps only 95%. However, by indicating the nature and the amount of the impurities to the customer the latter can decide what kind of purity grade is needed, given that the customer does not have the opportunity to purify the commercial material himself. Customers who really need a purity of greater than 99.99% should discuss their requested specification and the methods to assess this purity with the supplier of choice.

The following sections attempt to comment upon common impurities in commercial ionic liquid products and their significance for known ionic liquid applications. The aim is to help the reader to understand the significance of different impurities for their application. Due to their great hydrolytic instability, chloroaluminate ionic liquids will not be treated here.

2.2.2.1 **Color**
Reading the literature one gets the impression that ionic liquids are all colorless and look almost like water. However, most people who start ionic liquid synthesis will probably get a highly colored product at first and even most suppliers of ionic liquids used to make ionic liquids that were more or less yellowish in color. The exact chemical nature of the colored impurities in ionic liquids is still unclear, but it is probably a mixture of traces of compounds originating from the starting materials, oxidation products and thermal degradation products of the starting materials. Sensitivity to coloration during ionic liquid synthesis can vary significantly with the type of cation and anion of the ionic liquid. For instance, pyridinium salts tend to form colored impurities more easily than imidazolium salts.

In Section 2.1, methods to produce colorless ionic liquids have been described excellently and it has become obvious that freshly distilled starting materials and low temperature processing during the synthesis and drying steps are key aspects to avoid coloration of the ionic liquid.

From a commercial point of view it is possible to obtain colorless ionic liquids. This can be achieved when sophisticated techniques are applied or special care is taken during and after the their synthesis. Nevertheless, these labor intensive procedures (some examples are described in Section 2.1) add to the cost of ionic liquids and this of course has to be weighed against the benefits of colorless ionic liquids.

For a commercial producer three points are important to be mentioned in this context.

1 The colored impurities are usually present only in very low trace amounts. It is impossible to detect them with NMR or analytical techniques other than UV–VIS spectroscopy (hence the difficulty in determining the chemical structure of the colored impurities).
2 For almost all applications involving ionic liquids the color is not the *crucial* parameter. For example in catalytic applications it appears that the concentration of

Fig. 2.2-1 ECOENG™ 41M (1-butyl-3-methylimidazolium 2-(2-methoxyethoxy) ethyl sulfate) produced by two different routes.

colored impurities is significantly lower than the concentration of catalysts commonly used. Exceptions are, of course, any application where UV spectroscopy is used for product or catalyst analysis and for all photochemical applications.
3 Prevention of coloration of the ionic liquid may, for some substances, not be compatible with the aim of a rational economic ionic liquid production. Additional distillative cleaning of the feedstocks consumes time and energy and additional cleaning by e.g. chromatography after synthesis is also a time-consuming step. The most important restriction is, however, the need to perform the synthesis (mainly the alkylation reaction) at the lowest temperature possible and thus at the lowest rate. This requires long reaction times leading to high production costs. However, it is important to state that it is indeed feasible to obtain colourless ionic liquids through process improvements as can be seen in Fig. 2.2-1 that shows the same ionic liquid produced by different routes.

Therefore, a compromise between coloration and economics is necessary in commercial ionic liquid production. Additional de-coloration-steps are usually performed by the IL-suppliers only on special request! We expect that the market for colorless ionic liquids will be comparatively small, if the same substance can be made slightly colored at a much lower price.

2.2.2.2 Organic Starting Material and Other Volatiles

The presence of volatile impurities in an ionic liquid may have different origins. It may result from solvents used in the extraction steps during their synthesis, from unreacted starting materials of the alkylation reaction (to form the ionic liquid's cation) or from any volatile organic compound previously dissolved in the ionic liquid.

In theory, volatile impurities can be easily removed from the non-volatile ionic liquid by simple evaporation. However, sometimes this process can take a considerable time. Factors that influence the time required for the removal of all volatiles from an ionic liquid (under a given temperature and pressure) are the following: (i) the amount of volatiles; (ii) their boiling points; (iii) the viscosity of the ionic liquid, (iv) the surface of the ionic liquid and – probably most important – (v) their interactions with the ionic liquid.

An alkylating agent, whether used in excess or as a limiting reagent, is typically sufficiently volatile to be efficiently removed under vacuum. Other starting materials may lack sufficient volatility to be efficiently removed under vacuum. A typical example of a volatile impurity that can be found as one of the main impurities in ionic liquids with 1-alkyl-3-methylimidazolium cations is the methylimidazole starting material. Due to its high boiling point (198 °C) and its strong interaction with the ionic liquid, this compound is very difficult to remove from an ionic liquid, even at elevated temperatures and high vacuum. Therefore, it is important to make sure, by appropriate alkylation conditions, that no unreacted methylimidazole is left in the final product.

Traces of bases such as methylimidazole in the final ionic liquid product may play an unfavorable role in some common applications of ionic liquids (e.g. in transition metal-mediated catalysis). Many electrophilic catalyst complexes will coordinate the base in an irreversible manner and will be deactivated in this way.

A number of different methods to monitor the amount of methylimidazole left in a final ionic liquid are known. NMR spectroscopy is used by most academic groups but can have a detection limit of about 1–2 mol%. The photometric analysis described by Holbrey, Seddon and Wareing has the advantage of being a relatively quick method that can be performed with standard laboratory equipment [10], making it particularly suitable for monitoring the methylimidazole content during commercial ionic liquid synthesis. The method is based on the formation and colorimetric analysis of the intensely colored complex of 1-methylimidazole with copper (II) chloride but suffers from lack of precise quantification.

In order to obtain a more complete picture of the organic impurities in ionic liquids the leading academic and industrial players have now established a comprehensive set of state-of-the-art analytical methods. Today, there is much agreement that GC-headspace – in combination with HPLC – techniques are the analytical methods of choice to quantify organic impurities, such as e.g. the content of methylimidazole impurities, down to the ppm range.

2.2.2.3 Halide Impurities

Many ionic liquids, including the still widely used tetrafluoroborate and hexafluorophosphate systems, are synthesized in a two-step synthesis. In the first step a nucleophile, such as a tertiary amine or phosphine, is alkylated to form the cation. For this reaction alkyl halides are frequently used as alkylating agents, forming the halide salts of the desired cation. To obtain a non-halide ionic liquid, the halide anion is exchanged in a second reaction step. This can either be realized by addition of the alkali salt of the desired anion (with precipitation of the alkali halide salt) or by reaction with a strong acid (with removal of the corresponding hydrohalic

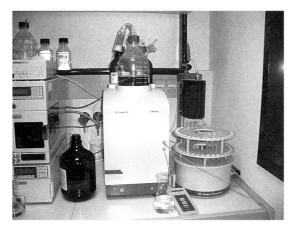

Fig. 2.2-2 Ion chromatography – the method of choice for the quantification of ionic impurities in ionic liquids.

acid) or by using an ion exchange resin (for more detail see Section 2.1). Alternative synthetic procedures involving the use of silver [3] or lead salts [11] are – at least from our point of view – not acceptable for commercial ionic liquid production.

Halide impurities are probably the most studied of the four general categories of impurities common to ionic liquids and, besides electrochemical analysis, two methods are currently being used to determine the level of residual halide impurities in ionic liquids [12]. The titration of the ionic liquid with $AgNO_3$ is still widely used but suffers from a certain solubility of AgCl in the ionic liquid under investigation. This method can be enhanced by the Vollhard method for chlorine determination where the chloride is first precipitated with excess $AgNO_3$ followed by back-titration of the mother liquor with aqueous potassium thiocyanate [13]. This method uses a visual endpoint through the formation of a complex between thiocyanate and an iron (III) nitrate indicator.

The method of choice, which has been established in several companies within the last three years, uses ion chromatography (see Fig. 2.2-2). This comparatively convenient method can achieve detection limits in the low ppm range [6]. For applications where an even lower chloride content is required, ICP-MS (inductively coupled plasma-mass spectrometry), which has a detection limit at the 10 ppb level, can be used [14].

All the halide exchange reactions mentioned above proceed more or less quantitatively, causing a residual amount of halide impurities in the final product. The choice for the best procedure to obtain a complete exchange depends mainly on the nature of the ionic liquid that is being produced. Unfortunately, there is no general method to obtain a "halide-free" ionic liquid that can be used for all types of ionic liquid. For two defined examples, the synthesis of $[BMIM][(CF_3SO_2)_2N]$ and the synthesis of $[EMIM][BF_4]$, this will be explained in a little more detail.

$[BMIM][(CF_3SO_2)_2N]$ has a miscibility gap with water (about 1.4 mass% of water dissolves in the ionic liquid [15]) and shows high stability to hydrolysis. Therefore, it is very easy to synthesize this ionic liquid in a halide-free quality. Following a procedure described first by Bonhôte and Grätzel [15], [BMIM]Cl (obtained by the alkylation of methylimidazole with butylchloride) and Li[(CF$_3$SO$_2$)$_2$N] are dissolved in water. Upon mixing the aqueous solutions, the ionic liquid is formed as a second layer. After separation from the aqueous layer, the ionic liquid can be easily washed with water and subsequently dried so that an ionic liquid can be obtained where no halide can be detected by ion chromatography, titration with AgNO$_3$ or by electrochemical analysis.

The halide-free preparation of [EMIM][BF$_4$] is significantly more difficult. Since this ionic liquid mixes completely with water and cannot be re-extracted from aqueous solutions using organic solvents, the removal of the halide ions by washing with water is not an option. Metathesis reaction in specially dried acetone or CH$_2$Cl$_2$ is possible but suffers from the low solubility of NaBF$_4$ in these solvents, resulting in long reaction times. Therefore, the exchange reaction in this type of suspension tends to be incomplete, even after long reaction times, when carried out on a larger scale. Consequently, to synthesize completely halide-free [EMIM][BF$_4$] special procedures have to be applied. Two examples are synthesis using an ion-exchange resin [16] or the direct alkylation of ethylimidazole with Meerwein's reagent [Me$_3$O][BF$_4$] [17]. In this context it should be noted that there is quite a difference between an ionic liquid made in a metathesis step from a halide ionic liquid, with subsequent removal of residual halide by washing, and one where no alkyl halide was involved during the course of its synthesis.

In contrast to colored impurities, residual halide can seriously affect the usefulness of the material as solvent for a given chemical reaction. Apart from the point that some physicochemical properties are largely dependent on the presence of halide impurities (as demonstrated by Seddon et al. [9]) the latter can act as a catalyst poison [18], stabilizing ligand [19], nucleophile or reactant, depending on the chemical nature of the reaction. Consequently, it is necessary to have a halide-free ionic liquid to investigate its properties for a given reaction, especially in catalysis, where the amount of catalyst used can be in the range of the concentration of the halide impurities in the ionic liquid.

2.2.2.4 Protic Impurities

Protic impurities have to be considered for two groups of ionic liquids: those that have been produced by an exchange reaction using a strong acid (e.g. often the case for [BMIM][PF$_6$]), and those that are sensitive towards hydrolysis. In the latter case the protons may originate from hydrolysis of the anion forming an acid that may be dissolved in the ionic liquid.

For ionic liquids that do not mix completely with water (and which show sufficient hydrolysis stability) there is an easy test for acidic impurities. The ionic liquid is added to water and a pH test of the aqueous phase is carried out. If the aqueous phase is acidic the ionic liquid should be washed with water to the point where the washing water becomes neutral. For ionic liquids that mix completely with water we recommend a standardized, highly proton sensitive test reaction to check for protic impurities.

Obviously, the check for protic impurities becomes crucial if the ionic liquid is to be used for applications where protons are known to be active compounds. For some organic reactions, one has to be sure that an "ionic liquid effect" does not turn out to be a "protic impurity effect" at some later stage of the research!

2.2.2.5 Other Ionic Impurities from Incomplete Metathesis Reactions

Apart from halide and protic impurities, ionic liquids can also be contaminated with other ionic impurities from the metathesis reaction. This is especially likely if the alkali salt that is used in the metathesis reaction shows significant solubility in the ionic liquid formed. In this case the ionic liquid can contain significant amounts of the alkali salt. While this may not be a problem even for some catalytic applications (since the presence of the alkali cation may not effect the catalytic cycle of a transition metal catalyst) it is of great relevance for the physicochemical properties of the ionic liquid.

In this context it is important to note that the detection of this kind of alkali cation impurity in ionic liquids is not easy with the traditional methods for reaction control of ionic liquid synthesis (e.g. conventional NMR spectroscopy). More specialized procedures are required to quantify the amount of alkali ions in the ionic liquid or the quantitative ratio of organic cation to anion. Quantitative ion chromatography is probably the most powerful tool for this kind of quality analysis.

Due to the analytical problems described we expect that some of the disagreements in the literature (concerning mainly the physicochemical data of some tetrafluoroborate ionic liquids) may originate from differing amounts of alkali cation, halide anion and residual water impurities in the ionic liquids analyzed.

2.2.2.6 Water

Without special drying procedures and completely inert handling, water is omnipresent in ionic liquids. Even the apparently hydrophobic ionic liquid [BMIM][$(CF_3SO_2)_2N$] saturates with about 1.4 mass% of water [15], which is a significant molar amount. For more hydrophilic ionic liquids the water uptake from air can be much higher. Imidazolium halide salts in particular are known to be extremely hygroscopic, which is one of the reasons why it is so difficult to make completely proton-free chloroaluminate ionic liquids.

For commercial ionic liquid production this clearly means that all products contain some higher or lower amount of water. Depending on the production conditions and the logistics, the ionic liquids can be expected to come into some contact with traces of water.

The presence of water in an ionic liquid may be a problem for some applications, but not for others. However, in all cases one should know the approximate amount of water in the ionic liquid used. Depending on the application, one should be aware of the fact that water in the ionic liquid may not be inert. Moreover, the presence of water can have significant influence on the physicochemical properties of the ionic liquid, on its stability (e.g. wet tetrafluoroborate and hexafluorophosphate ionic liquids will undergo anion hydrolysis with formation of protic impurities) and on the reactivity of catalysts dissolved in the ionic liquid.

2.2.3
Upgrading the Quality of Commercial Ionic Liquids

Prior to research carried out with ionic liquids, one should be aware of the level of impurities in different ionic liquids. As already mentioned, a good commercial ionic liquid may be colored and may contain some traces of water. However, it should be free of organic volatiles, halides (if not a halide ionic liquid) and all ionic impurities. Some commercial suppliers give their typical specifications on their homepages or in the accompanying product information. In our experience the quality level that has been achieved through continuous process improvement within the last three years is absolutely sufficient for most applications. For those customers having special quality requirements with regard to their projects we recommend they either discuss their special needs with their supplier of choice or perform a quality upgrade themselves. This quality upgrade can be achieved as follows.

To remove water, commercial ionic liquids used for fundamental research purposes should be dried at 60 °C over night *in vacuo*. The water content should be checked prior to their use. This can be done qualitatively with infrared spectroscopy or cyclic voltammetry or quantitatively by Karl–Fischer titration. If the ionic liquids cannot be dried to zero water content for any reason, the water content should always be mentioned in all descriptions and documentation of the experiments to allow proper interpretation of the results obtained.

Regarding the color, we only see a need for colorless ionic liquids in very specific applications which were described earlier. One easy treatment that often reduces the coloration of especially imidazolium ionic liquids quite impressively is purification by column chromatography/filtration over silica 60. For this purification method, the ionic liquid is dissolved in a volatile solvent such as CH_2Cl_2. Usually most of the colored impurities stick to the silica while the ionic liquid is eluted with the solvent. By repeating the process several times a seriously colored ionic liquid can be converted to an almost completely colorless material.

2.2.4
Novel, Halide-Free Ionic Liquids

The "work horses" in the field of ionic liquids research used to be [BMIM][PF_6] and [BMIM][BF_4] which are still widely used because of their relatively easy preparation and purification. However, these two ionic liquids suffer from the instability of their anions to hydrolysis, which will restrict their use to applications under anhydrous conditions since their reaction with water releases toxic and highly corrosive HF. Other ionic liquids that are stable to hydrolysis, such as those carrying the bis(trifluoromethylsulfonyl)imide ion have subsequently been developed but these systems suffer from the intrinsically high price of the anion.

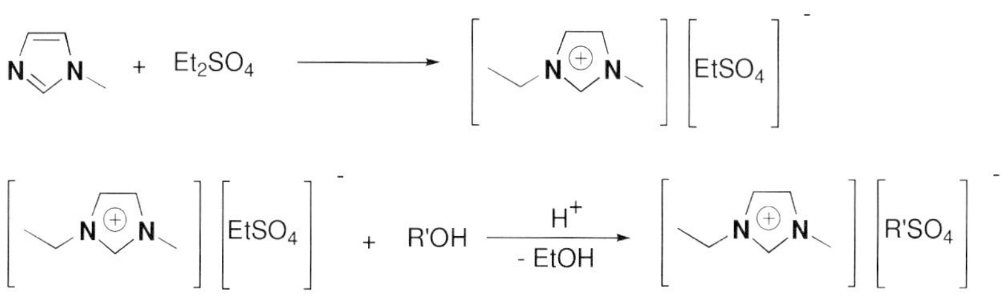

2.2 Quality Aspects and Other Questions Related to Commercial Ionic Liquid Production

Scheme 2.2-1 Two-step synthesis of [EMIM][R'SO$_4$] ionic liquids by a quaternization/transesterification sequence.

Taking these considerations into account, ionic liquids with halogen-free anions have become an important area of research. Several of these systems have been described such as alkyl sulfonates, alkyl sulfates and dialkyl phosphates. One representative, namely ECOENG™ 212 (1-ethyl-3-methylimidazolium ethylsulfate), has become a commercial ionic liquid (produced on a ton scale) with a set of toxicological data generated for the registration process indicating that this ionic liquid is a safe and non-toxic chemical [6]. To extend the variety of these ionic liquids, Wasserscheid et al. have developed a very elegant way to obtain long-chain alkyl sulfates [20] that were previously either not accessible or difficult to synthesize. This two-step synthesis is based on a quaternization step with dialkyl sulfate as alkylating agent, followed by a transesterification reaction of the alkyl sulfate anion (Scheme 2.2-1).

In the first step an amine such as methylimidazole is alkylated, e.g. with diethyl sulfate. This is followed by a transesterification step where a long chain alcohol and an acidic catalyst such as methanesulfonic acid are added to the intermediate obtained after quaternization. In order to shift the equilibrium of this reaction to the products, ethanol liberated in the transformation must be efficiently removed by applying mild vacuum. Scheme 2.2-2 shows some of the alcohols converted in this manner.

It is interesting to note that all the modified alkyl sulfate ionic liquids obtained with these alcohols show attractively low viscosities and their melting points were all below 0 °C. Thus, this synthesis sequence created a new class of halogen-free ionic liquids with attractive physicochemical properties. No organic solvent is needed in

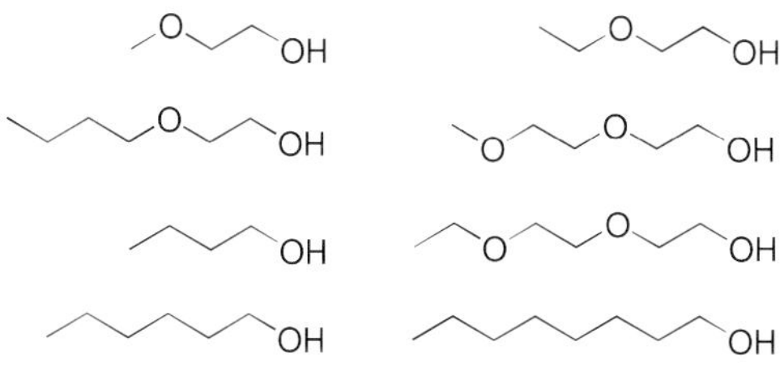

Scheme 2.2-2 Alcohols applied in the transesterification of [EMIM][EtSO$_4$].

the whole reaction sequence. Moreover, the reaction sequence is very atom-efficient with ethanol being the only by-product. Last but not least, these novel ionic liquids do not contain any halide impurities, as the whole synthesis is halogen-free which makes these products highly suitable for a wide variety of technical applications.

2.2.5
Scale-up of Ionic Liquid Synthesis

When research into the development and applications of ionic liquids started in the beginning of the 1990s, only a few people imagined the success this technology would achieve. Thus efficient scale-up of ionic liquid production was not necessarily a major area of concern in those days. This has changed considerably, since many historic ways to produce ionic liquids proved to be impractical on a larger scale. For example the use of expensive anion sources, such as silver salts [3], is prohibitive for the economically efficient production of ionic liquids. Furthermore, reaction sequences where very long reaction times are necessary for the alkylation step contribute significantly to the price of ionic liquids for large-scale production so that 1-methyl-3-octylimidazolium chloride ([OMIM]Cl) will always be more expensive than 1-butyl-3-methylimidazolium chloride [BMIM]Cl.

Ionic liquids are often described as "green" solvents. However, many classical two-step syntheses involve a metathesis step for the formation of $[BF_4]^-$ and $[PF_6]^-$ salts and require a significant amount of organic solvents. Furthermore, the halide salt formed as by-product in the metathesis step is nothing but waste and has to be disposed of. Other concerns about whether the traditional synthesis routes for ionic liquids are "green processes" are imposed by the washing procedures which are often necessary to remove residual halide. Usually the latter are associated with significant product loss into the washing phase. All these drawbacks should be avoided for a good synthesis on a larger scale.

Commercial manufacturers try to focus on economically efficient routes and have developed some alternative approaches over the last years that can indeed solve some of these difficulties.

The best alternative to the two-step synthesis is direct alkylation. Therefore a lot of different alkylating agents such as dialkyl sulfates, trialkyl phosphates or alkyl triflates – to name just a few – have been used to alkylate e.g. N-methylimidazole. The main advantage is that no by-products are formed in such direct alkylation due to the perfect atom efficiency. Moreover, as almost all alkylating agents are liquids, no complicated handling of solids is usually necessary. Ionic liquids obtained by direct alkylation are commercially available from most of the manufacturers and we expect larger potential for price reductions for these compounds compared to $[BF_4]^-$ and $[PF_6]^-$ salts. One of these ionic liquids, 1-ethyl-3-methylimidazolium ethyl sulfate (Solvent Innovation's trade name: ECOENG™ 212) has already been efficiently produced on a ton scale.

For the scale-up of ionic liquid synthesis by direct alkylation the most important aspects to consider are heat management (alkylation reactions are exothermic!) and

Fig. 2.2-3 100 liter reactor for ionic liquid production at Solvent Innovation.

proper mixing. For both of these the proper choice of the reactor set-up is of crucial importance. Fig. 2.2-3 shows one of Solvent Innovation's several production plants.

In this context, it is worth commenting on the different approaches that commercial manufacturers take to the ionic liquids market. Global players, such as Degussa, BASF or Merck, are keen to identify truly large-scale applications that require ionic liquids in large ton-scale quantities. Other players, like us, who profit from flexible production campaign scheduling, focus on the development of mid-range applications where ionic liquids are needed on the multi-100 kg to double-digit ton scale. Small commercial suppliers aim to find niche markets for tailor-made ionic liquids suitable for their small-scale production.

2.2.6
Health, Safety and Environment

The lack of any measurable vapor pressure distinguishes ionic liquids from many conventional organic solvents that emit volatile organic compounds (VOC). Near-zero VOC emissions provide the basis for clean manufacturing, a highly desirable goal for the chemical industry. However, even if ionic liquids do not evaporate and contribute to air pollution most of them are water soluble and can enter the environment via this path. This raises the question of registration of these new materials and generation of HSE data, as detailed studies giving information about toxicity, biodegradability and ecotoxicity are a must for the broad implementation of ionic liquids into large-scale applications.

Until recently information on these issues was relatively sparse, but Jastorff and coworkers have recently published an excellent and comprehensive overview on this topic [21]. They have reported on an environmental risk analysis on $[C_4MIM][BF_4]$

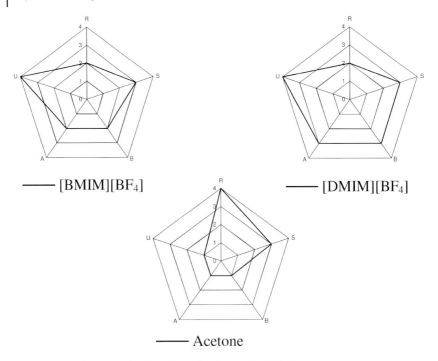

Fig. 2.2-4 Five-dimensional risk profiles (taken from VCH: Multiphase Homogeneous Catalysis, **2005**, p. 597).

and $[C_{10}MIM][BF_4]$ in which five ecotoxicological indicators [(a) release; (b) spatiotemporal range (c) bioaccumulation (d) biological activity and (e) uncertainty] were assigned on a basis from 1 to 4 (1 = very low risk, 4 = very high risk). The risk of release was assigned as very low since ionic liquids do not have any measurable vapor pressure. However, their water solubility, even when it is low, offers the potential for release through waste water. Assessment of the spatiotemporal range is limited by the lack of quantitative information about ionic liquid decomposition and biodegradability. The biological activity and bioaccumulation of the two abovementioned ionic liquids showed "rather high risks" for $[C_{10}MIM][BF_4]$ and "rather low risks" for $[C_4MIM][BF_4]$, demonstrating a clear correlation between the length of the alkyl chain and ionic liquid toxicity, as can be seen in Fig. 2.2-4.

This observation was supported by subsequent evaluations from the same group [22] who conducted a comprehensive study on the biological effects of 1,3-dialkylimidazolium ionic liquids on luminescent bacteria (*Vibrio fischeri*) and two mammalian cell lines (IPC-81 leukemia cells and C_6 glioma cells). They investigated a set of $[C_3$-$C_{10}MIM]^+$ cations combined with Cl^-, Br^-, $[BF_4]^-$, and $[PF_6]^-$ anions and showed that the 1,3-dialkylimidazolium cation is an important determinant of ionic liquid toxicity. Toxicity increased as alkyl chain length increased in all assays whereas no general systematic effect could be contributed to the ionic liquid anion

Fig. 2.2-5 ECOENG™ 212 – an ionic liquid available on ton-scale.

(although in most cases the $[PF_6]^-$ anion proved to be slightly more toxic than the other anions).

In another study, the toxicological properties of different cations were compared [23]. The effect of dialkylimidazolium, pyridinium and phosphonium ionic liquids on acetylcholine esterase activity was measured. This enzyme was chosen because it is an important component of the central nervous system (CNS) of all higher organisms, including humans.

In the last year we have witnessed registration procedures for two ionic liquids by the leading companies in this field (Table 2.2-1).

Solvent Innovation, in cooperation with Degussa, has obtained an ELINCS-Listing for 1-ethyl-3-methylimidazolium ethylsulfate (ECOENG™ 212) (Fig. 2.2-5) [6]. Toxicity studies were conducted and showed an LD_{50} (oral, Wistar rats) > 2000 mg kg^{-1}. Acute toxicities studies in Daphnia magna induced no immobilization after 48 h, at 100 mg l^{-1}.

BASF has obtained an ELINCS-listing for another imidazolium-based ionic liquid, 1-butyl-3-methylimidazolium chloride. (see Chapter 9 by M. Maase). This ionic liquid had to be classified as toxic, but this does not of course prevent its use in applications where no direct personal contact with this ionic liquid is necessary. Air Products for example recently disclosed an exciting application of [BMIM]Cl used in a special formulation as a medium for the storage of hazardous gases [24] where toxicity of this ionic liquid is not prohibitive for its use. The results of both studies

Table 2.2-1 Toxicological properties of two commercial ionic liquids

	ECOENG™ 212	BASIONIC ST 70
Manufacturer	Solvent Innovation	BASF
Chemical name	[EMIM][EtOSO$_3$]	[BMIM][Cl]
Acute oral toxicity	LD$_{50}$ >2000 mg kg^{-1} not harmful	LD$_{50}$ >50< 300 mg kg^{-1} toxic
Skin irritation	non-irritant	irritant
Eye irritation	non-irritant	Irritant
Sensitization	non-sensitizing	non-sensitizing
Mutagenicity	non-mutagenic	non-mutagenic
Biological degradability	not readily degradable	not readily degradable
Toxicity to daphniae	EC$_{50}$ (48 h) >100 mg l^{-1} acutely not harmful	EC$_{50}$ (48 h) >6.3 mg l^{-1} acute toxic
Toxicity to fish		LC$_{50}$ (96 h) >100 mg l^{-1} acutely not harmful

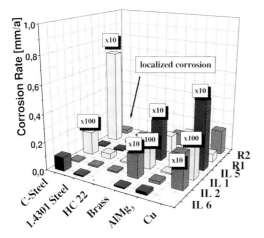

Fig. 2.2-6 Materials corrosion in water-free ionic liquids (ILs) and reference media. (R1: Water with 250 ppm chloride, R2: uninhibited 1:1 water/glycol mixture.)

support again the correlation between alkyl chain length and ionic liquid toxicity. Merck is currently in the registration process for 1-ethyl-3-methylimidazolium chloride and it will be interesting to see the results as the influence on the toxicity caused by an $[EMIM]^+$ vs. $[BMIM]^+$ cation and chloride vs. ethyl sulfate anion can then be assessed.

The authors expect that more ionic liquids will enter the market in large volumes within the next two to three years. HSE considerations will be an important criterion in selection and exclusion of specific ionic liquid candidates for future large scale, technical applications.

2.2.7
Corrosion Behavior of Ionic Liquids

For industrial implementation the corrosion behavior of ionic liquids must of course be investigated as materials corrosion can be a risk to plant integrity, reduce plant efficiency and may require costly maintenance. A recently published report describes the corrosion behavior of a number of metallic materials in seven commercially available ionic liquids in the absence and presence of water under flow conditions at temperatures up to 90 °C (Scheme 2.2-3, Fig. 2.2-6) [25].

The best performance for all IL media tested was observed with stainless steel type 304 which proved to be resistant in all water-free and water-diluted systems. The Ni-based alloy C22 generally also proved to have a high resistance to most IL-containing media. For carbon steel and aluminum the corrosiveness of IL media depends strongly on the chemical structure of the cationic moiety and the nature of the anion in the ionic liquid. In "water-free" quality (water-content below 1%)

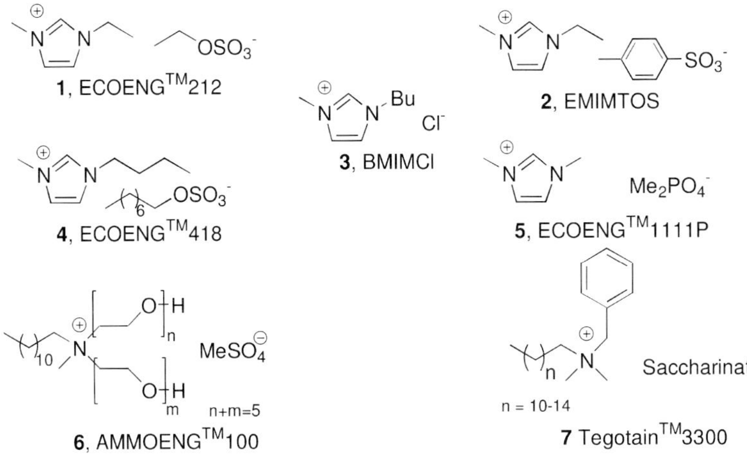

Scheme 2.2-3 Ionic liquids tested.

the tosylate anion for example shows corrosiveness with corrosion rates between 15 and 25 mm a^{-1}. On the other hand, with ethyl sulfate as the anion, carbon steel is not attacked.

Furthermore, some of the corrosion problems described in this publication could be overcome in the meantime by further significantly reducing the water content! Recent investigations in our laboratories with ECOENG™ 212 have shown that corrosion rates are significantly lower when the water content is below 1000 ppm and this seems to apply for other ionic liquids as well [26].

The effect of water addition was also investigated for the ILs 2 and 6. It was observed that the corrosiveness of IL media increased significantly with the addition of 10% water. While sulfuric acid could possibly be formed in both IL media it seems that the imidazolium ion is a better corrosion inhibitor for steel than the PEG-modified quaternary ammonium moiety in IL 6.

These findings led to the development of an inhibition concept and in cases where corrosion was extensive, e.g. for copper, inhibition was possible by addition of small amounts of 1H-benzotriazole.

2.2.8
Recycling of Ionic Liquids

As ionic liquids are still expensive in comparison with conventional molecular solvents, their efficient recycling is an important issue that addresses the economics of their use, especially in large-scale applications. The most studied examples are the 1,3-dialkylimidazolium-based ionic liquids where the ionic solvent is usually recycled through several cycles of the reaction. Numerous examples describe the immobilization of a transition metal catalyst in the ionic liquid phase of a biphasic system. A variety of palladium coupling reactions, such as the Heck [27], Suzuki

[28], Stille [29] and Sonogashira [30] reactions or the palladium-catalyzed oxidation of benzaldehydes and alkylbenzenes [31], have demonstrated excellent recyclability of both the catalyst and the ionic liquid. Recycling protocols usually make use of the low solubility of the ionic liquids in organic solvents such as toluene. This allows products and residual organics to be extracted using an organic solvent while salt by-products present in water-immiscible ionic liquids can be washed out with water (water-miscible ionic liquids on the other hand are more difficult to recycle as inorganic by-products cannot be easily removed). Furthermore, if the reaction products are sufficiently volatile they can be distilled directly from the ionic liquid, since ionic liquids are inherently non-volatile.

It is noteworthy that the efficiency of the recycling process for various ionic liquids varies from quite poor [27] to very good [28–31]. The quality of the recovered ionic liquid can then easily be assessed by the methods described in Section 2.2.2.

Another elegant method to recycle ionic liquids is to extract them with supercritical CO_2, as demonstrated in the pioneering work of Brennecke et al. [32]. They demonstrated that $scCO_2$ exhibits high solubility in ionic liquids whereas ILs show generally no detectable solubility in CO_2 (the solubility was determined to be less than 5×10^{-7} mole fraction [33]). Thus $scCO_2$ can be used to extract organic species from an IL solution without cross-contamination of the gas phase with IL, as was demonstrated with naphthalene as a solute [32]. Recovery of the organic solute was near quantitative and achieved by simple expansion of the gas phase downstream. This efficient exchange between the two phases, coupled with the lack of mutual solubility seems to be ideally suitable for product separation and recycling of ionic liquids.

Additionally, membrane techniques have been applied to perform fine separation of undesirable constituents and these show promising results for the selective removal of volatile solutes from ionic liquids [34].

To summarize, several procedures for the recycling of ionic liquids have been reported. Depending on the ionic liquid used and the reaction performed in it, a variety of methods for recycling are possible. By picking the right purification steps, an individually optimized work-up procedure can be obtained.

However, if the customer does not feel comfortable with the task of recycling the ionic liquid there is a further option at hand: why not rent or lease the ionic liquid rather than buy it? The customers would in this case perform their reaction and send the, probably impure, ionic liquid back to the supplier, who has the expertise to recycle and clean it up. This scenario could be interesting from an economic point of view for truly commercial applications on a large scale.

2.2.9
Future Price of Ionic Liquids

Why are ionic liquids still comparatively expensive? The price of ionic liquids is determined by many parameters such as personnel, overheads and real production costs. As one can imagine the first small quantities of new substances reflect more a prototype status than future bulk chemical prices. An impressive example is

Table 2.2-2 Price indication for ECOENG™212[a].

Quantity	Price per kg (€)
1 kg	153
10 kg	95
100 kg	80
1 t	40
100 t	20–30

[a] Prices taken from Solvent Innovation.

butadiene which was quite expensive only some 40 years ago and is now produced on a million-ton scale each year for less than 1 € per kg. Similar rules for economy-of-scale can also be applied to ionic liquids. The average price of more than 1 000 € per kg back in 1999 has already gone down to a few hundred € per kg, even on comparatively small scales. On a ton scale some ionic liquids are already available for less than 50 € per kg.

In this context it is important to note that future prices for ionic liquids will not only be determined by the price of the starting materials, besides those variable costs, factors such as complexity of manufacturing processes, certain quality requirements, IP-costs together with the notification costs for new substances, contribute significantly to the price of ionic liquids. Furthermore, operational costs such as rent for production facilities, deviation of production equipment, waste disposal and overheads, to name but a few, need to be incorporated into the selling prices for ionic liquids.

However, price reductions for ionic liquids will continue in the future in relation to the quantities needed. An example of what one might expect for the future is shown in Table 2.2-2.

Taking all the previously mentioned factors into account, we expect, on a mid-term time scale, a range of ionic liquids being commercially available for € 25–50 per kg on a multi-ton scale. The first large-scale production on a single-digit ton-scale has already taken place and we expect a significant increase in large-scale production within the next two to three years in order to meet the continuously rising market demands.

References

1. (a) M. J. Earle, J. M. S. S. Esperanca, M. A. Gilea, J. N. Canongia Lopes, L. P. N. Rebelo, J. W. Magee, K. R. Seddon, J. A. Wildegren, *Nature* **2006**, *439*(16), 831–834; (b) P. Wasserscheid, *Nature* **2006**, *439*(16), 797.
2. T. Welton, *Chem. Rev.* **1999**, *99*, 2071–2083.
3. J. S. Wilkes, M. J. Zaworotko, *J. Chem. Soc., Chem. Commun.* **1992**, 965–967.
4. J. Fuller, R. T. Carlin, H. C. de Long, D. Haworth, *J. Chem. Soc., Chem. Commun.* **1994**, 299–300.
5. P. A. Z. Suarez, J. E. L. Dullius, S. Einloft, R. F. de Souza, J. Dupont, *Polyhedron* **1996**, *15*(7), 1217–1219.

6. Solvent Innovation GmbH, Cologne (www.solvent-innovation.com).
7. M. Tinkl, *Chem. Rundschau* **1999**, *2*, 59.
8. For example: (a) D. W. Armstrong, L. He, Y.-S. Liu, *Anal. Chem.* **1999**, *71*, 3873–3876; (b) D. W. Armstrong, *Anal. Chem.* **2001**, *73*, 3679–3686; (c) F. Endres, *Phys. Chem. Chem. Phys.* **2001**, *3*, 3165–3174; (d) C. Ye, W. Liu, Y. Chen, L. Yu, *Chem. Commun.* **2001**, 2244–2245; (e) C. Jork, M. Seiler, Y. A. Beste, W. Arlt, *J. Chem. Eng. Data*, **2004**, *49*, 852–857; (f) Mu Z., Zhou F. et al. *Tribology Int.* **2005**, *38*, 725–731.
9. K. R. Seddon, A. Stark, M. J. Torres, *Pure Appl. Chem.* **2000**, *72*, 2275–2287.
10. J. D. Holbrey, K. R. Seddon, R. Wareing, *Green Chem.* **2001**, *3*, 33–36.
11. B. Ellis, WO 9 618 459 (to BP Chemicals Limited, UK) **1996** *Chem. Abstr.* **1996**, *125*, 114635.
12. C. Villagran, M. Deetlefs, W. R. Pitner, C. Hardacre, *Anal. Chem.* **2004**, *76*, 2118–2123.
13. A. I. Vogel, *A Textbook of Quantitative Inorganic Analysis*, 3rd edn., Longmans, Green and Co.: London, **1961**.
14. K. McCamley, N. A. Warner, M. L. Lamoureux, P. J. Scammels, R. D. Singer, *Green Chem.* **2004**, *6*, 341–344.
15. P. Bonhôte, A.-P. Dias, N. Papageorgiou, K. Kalyanasundaram, M. Grätzel, *Inorg. Chem.* **1996**, *35*, 1168–1178.
16. H. Waffenschmidt, Dissertation, RWTH Aachen, **2000**.
17. C. Hilgers, Dissertation, RWTH Aachen, **2001**.
18. For example: Y. Chauvin, L. Mußmann, H. Olivier, *Angew. Chem. Int. Ed. Engl.* **1995**, *34*, 2698–2700.
19. For example: C. J. Mathews, P. J. Smith, T. Welton, A. J. P. White, D. J. Williams, *Organometallics* **2001**, *20*(18), 3848–3850.
20. P. Wasserscheid, S. Himmler, S. Hörmann, R. van Hal, P. Schulz, *Green Chem.*, submitted for publication.
21. B. Jasstorff et al. in *Multiphase Homogenous Catalysis*. Vol. 2, Wiley-VCH, **2005**, 588–600.
22. J. Ranke, K. Mölter, F. Stock, U. Bottin-Werner, J. Poczubutt, J. Hoffmann, B. Ondruschka, J. Filser, B. Jastorff, *Ecotoxicol. Environ. Saf.*, **2004**, *58*, 396.
23. F. Stock, J. Hoffmann, J. Ranke, R. Störmann, B. Ondruschka, B. Jastorff, *Green Chem.* **2004**, *6*, 286.
24. Air Products, US Patent 20040206241.A1.
25. M. Uerdingen, C. Treber, M. Balser, G. Schmitt, C. Werner, *Green Chem.*, **2005**, *7*, 1–6.
26. M. Uerdingen, unpublished results.
27. K. R. Seddon, A. J. Carmichael, M. J. Earle, J. D. Holbrey, P. B. McCormac, *Org. Lett.* **1999**, *1*, 997.
28. T. Welton, P. J. Smith, C. J. Mathews, *Chem. Commun.* **2000**, 1249.
29. S. T. Handy, X. Zhang, *Org. Lett.* **2001**, *3*, 233.
30. T. Fukuyama, M. Shimnen, S. Nishitani, M. Sato, I. Ryu, *Org. Lett.* **2003**, *4*, 1691.
31. K. R. Seddon, A. Stark, *Green Chem.* **2002**, *4*, 119.
32. L. A. Blanchard, D. Hancu, E. J. Beckman, J. F. Brennecke, *Nature*, **1999**, *399*, 28.
33. L. A. Blanchard, Z. Y. Gu, J. F. Brennecke, *J. Phys. Chem. B*, **2001**, *105*, 2437.
34. Solvent Innovation WO 2003.039.719 (2003).

2.3
Synthesis of Task-specific Ionic Liquids

James H. Davis, Jr., updated by Peter Wasserscheid

2.3.1
Introduction

Early studies probing the feasibility of conducting electrophilic reactions in chloroaluminate ionic liquids (ILs) demonstrated that the ionic liquid could act as both solvent and catalyst for the reaction [1–3]. The success of these efforts hinged upon the capacity of the salt itself to manifest the catalytic activity necessary to promote the reaction. Specifically, it was found that the capacity of the liquid to function as an electrophilic catalyst could be adjusted by varying the $Cl^-/AlCl_3$ ratio of the complex anion. Anions that were even marginally rich in $AlCl_3$ catalyzed the reaction.

Despite the utility of chloroaluminate systems as a combination of solvent and catalyst in electrophilic reactions, subsequent research on the development of newer ionic liquid compositions focused largely on creating liquid salts that were water stable [4]. To this end, new ionic liquids were introduced that incorporated tetrafluoroborate, hexafluorophosphate and bis(trifluoromethylsulfonyl)amide anions. While these new anions generally imparted to the ionic liquid a high degree of water stability, the functional capacity inherent in the IL due to the chloroaluminate anion was lost. Nevertheless, it is these water stable ionic liquids that have become the *de rigueur* choices as solvents for contemporary studies of reactions and processes in these media. [5].

A 1999 report of the formation of ionic liquids by relatively large, structurally complex ions derived from the antifungal drug miconazole re-emphasized the possibilities for formulating salts that remain liquids at low temperatures, even while incorporating functional groups in the ion structure [6]. This prompted the introduction of the concept of "task-specific" ionic liquids [7]. Task-specific ionic liquids (TSILs) may be defined as ionic liquids in which a functional group is covalently tethered to the cation or anion (or both) of the IL. Further, the incorporation of this functionality should imbue the salt with a capacity to behave not only as a reaction medium but also as a reagent or catalyst in some reaction or process. The definition of TSIL also extends to "conventional" ionic liquids to which ionic solutes are added to introduce a functional group into the liquid. Logically, when added to a "conventional" ionic liquid, these solutes become integral elements of the overall "ion soup" and must then be regarded as an element of the ionic liquid as a whole, making the resulting material a TSIL.

Viewed in conjunction with the solid-like non-volatile nature of ionic liquids, it is apparent that TSILs can be thought of as liquid versions of solid-supported reagents. However, unlike solid supported reagents, they possess the added advantages of kinetic mobility of the grafted functionality and an enormous operational surface area (Fig. 2.3-1). It is this combination of features that makes TSILs an aspect of ionic liquids chemistry that is poised for explosive growth.

Conceptually, the functionalized ion of a TSIL can be regarded as possessing two elements. The first is a core that bears the ionic charge and serves as the locus for the second element, the substituent group. Established TSILs are largely species in which the functional group is cation-tethered. Consequently, our discussion of TSIL synthesis from this point will mainly stress the synthesis of salts possessing functionalized cations, though the general principles outlined are pertinent to the synthesis of functionalized anions as well.

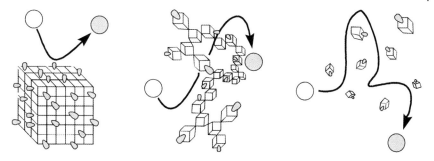

Fig. 2.3-1 Substrate interactions with (left to right) solid-supported reagent; polymer gel-supported reagent; task-specific ionic liquid.

2.3.2
General Synthetic Strategies

The incorporation of functionality into an ion intended for use in formulating an ionic liquid is usually a multi-step process. Consequently, a number of issues must be considered in planning the synthesis of the ion. The first of these is the choice of the cationic core. The core of TSIL cations may be as simple as a single atom such as N, P or S, as found in ammonium, phosphonium or sulfonium ions, respectively, or the core of the ion may be (and frequently is) a heterocycle such as imidazole or pyridine. The choices made in this regard will play a large role in both the chemical and physical properties of the resulting salt. For example, ionic liquids incorporating phosphonium cations generally exhibit the greatest thermal stability, but also commonly possess higher melting points than salts of other cations [8]. Thus, if the desired ionic liquid is to be used in a process that is conducted at 0 °C, building the cation core around a phosphonium ion may prove especially challenging. If the ionic liquid is to be used in a metal catalyzed reaction, the use of an imidazolium-based ionic liquid might be critical, especially in the light of the possible involvement in some reactions of imidazolylidene carbenes originating with the IL solvent [9].

The second element of general importance in the synthesis of a task-specific ionic liquid is the source of the functional group that is to be incorporated. Key to success here is the identification of a substrate that contains two functional groups with different reactivity, one which allows the attachment of the substrate to the core, and the other which either is the functional group of interest or is modifiable to the group of interest. Functionalized alkyl halides are commonly used in this capacity, although the triflate esters of functionalized alcohols work as well.

The choice of reaction solvent is also of concern in the synthesis of new TSIL. Toluene and acetonitrile are the most widely used solvents, the choice in any given synthesis being dictated by the relative solubility of the starting materials and products. The use of volatile organic solvents in the synthesis of ionic liquids is decidedly the least "green" aspect of their chemistry. Notably, developments in the area of the solventless synthesis of ionic liquids promise to improve this situation [10].

In the case of the TSIL carrying a functionalized cation the choice of the anion that is to ultimately be an element of the ionic liquid is of particular importance. Perhaps more so than any other single factor, it appears that the anion of the ionic liquid exercises a significant degree of control over the molecular solvents (water, ether, etc.) with which the IL will form two-phase systems. For example, nitrate salts are typically water miscible and those of hexafluorophosphate are not; those of tetrafluoroborate may or may not be, depending on the nature of the cation. Certain anions such as hexafluorophosphate are subject to hydrolysis at higher temperatures, while those such as bis(trifluoromethylsulfonyl)amide are not, but are significantly more expensive. Additionally, the cation of the salt used to perform any anion metathesis is important. While salts of potassium, sodium and silver are routinely used for this purpose, the use of ammonium salts in acetone is frequently the most convenient and least expensive approach.

2.3.3
Functionalized Cations

Although the first ionic liquid expressly categorized as being "task-specific" featured the incorporation of function within the cation core, subsequent research has focused on the incorporation of functionality into a branch appended to the cation [11]. In this fashion, a great number of the task-specific ionic liquids have been prepared that have been built-up from 1-methyl and 1-butylimidazole. By far the most of these ionic liquids have been prepared by quaternization of the aforementioned imidazoles with a functionalized alkyl halide to afford the corresponding functionalized imidazolium halides in usually good yield (Scheme 2.3-1).

For example, Bazureau has used this approach to prepare imidazolium ions with appended carboxylic acid groups, which have been used as replacements for solid polymer supports in the heterogeneous phase synthesis of small organic molecules via Knoevenagel and 1, 3-dipolar cycloaddition reactions [12].

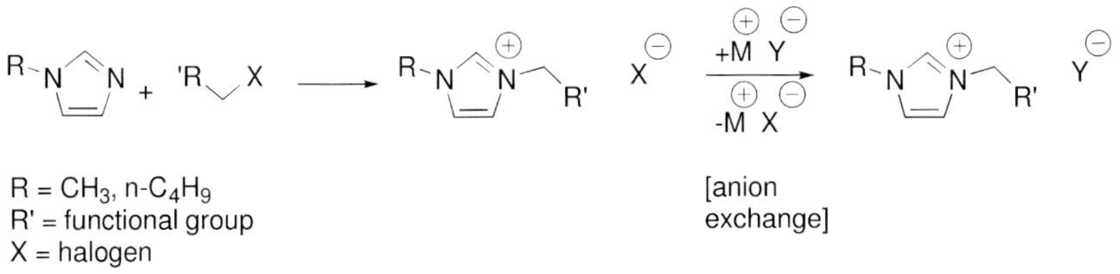

R = CH_3, n-C_4H_9
R' = functional group
X = halogen

[anion exchange]

Scheme 2.3-1 General synthesis of task specific ionic liquids with functionalized cations from 1-alkylimidazoles. The preparation of functionalized pyridinium, phosphonium, etc. cations may be accomplished in like fashion.

2.3 Synthesis of Task-specific Ionic Liquids

Another commercially available imidazole "scaffold" upon which a number of other functionalized cations have been constructed is 1-(3-aminopropyl)imidazole. The appended amino group in this material is a versatile reactive site that lends itself to conversion into a variety of derivative functionalities (Scheme 2.3-2).

Treatment of 1-(3-aminopropyl)imidazole with isocyanates and isothiocyanates gives urea and thiourea derivatives [13]. These elaborated imidazoles can then be quaternized at the ring nitrogen by reaction with alkyl iodides to produce the corresponding N(3)-alkylimidazolium salts. Because of a competing side reaction arising from the interaction of the alkylating species with the urea or thiourea groups, the reactions must be conducted within relatively narrow temperature and solvent parameters (below reflux in acetonitrile). Similar care must be exercised in the synthesis of IL cations with appended acetamide and formamide groups.

A variation on this overall synthetic approach allows the formation of related TSIL ureas by first converting 1-(3-aminopropyl)imidazole into an isocyanate, then reacting it with an amine and alkylating agent. The latter approach has been used to append both amino acids and nucleic acids onto the imidazolium cation skeleton [14].

The incorporation of more "inorganic" appendages into TSIL cations has also been achieved using 1-(3 aminopropyl)imidazole. Phosphoramide groups are readily synthesized via reaction of phosphorous(V) oxyhalides and primary or secondary amines. Using just such an approach, 1-(3-aminopropyl)imidazole was allowed to react with commercially available $(C_6H_5)_2POCl_2$ in dichloromethane. After isolation, the resulting phosphoramide was then quaternized at the imidazole N(3) position by reaction with ethyl iodide (Scheme 2.3-2). The viscous, oily product was found to mix readily with more conventional ionic liquids such as [HMIM][PF$_6$], yielding a more tractable material. This particular TSIL has been used to extract a number of actinide elements from water. Similarly, thiourea-appended TSIL, discussed earlier,

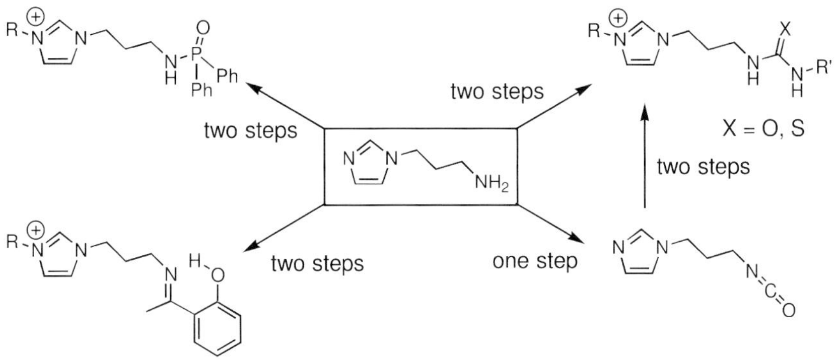

Scheme 2.3-2 Representative syntheses of functionalized cations for task-specific ionic liquids, beginning with 1-(3-aminopropyl)imidazole. Step one of the synthetic transformations is the conversion of the primary amine moiety to the functional group of interest. Step two of the process is the quaternization of the imidazole ring by alkylation at N(3).

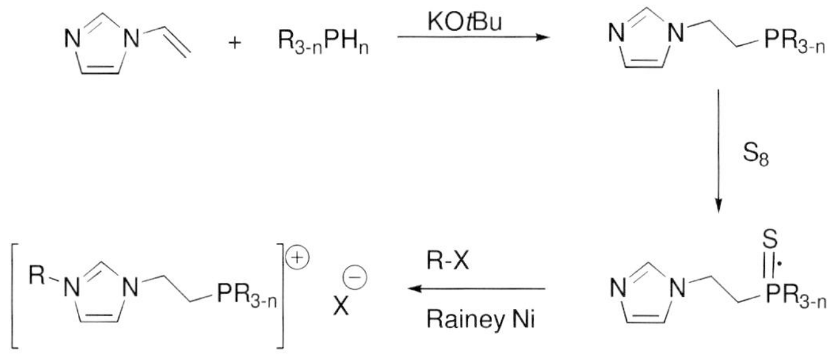

Scheme 2.3-3 Synthesis of phosphine-appended imidazolium salts. Combination of these species with the conventional ionic liquid [BMIM][PF$_6$] and Rh(I) gives rise to a task-specific ionic liquid active for the hydroformylation of 1-octene.

have been used for the extraction of Hg^{2+} and Cd^{2+} from IL-immiscible aqueous phases.

While certain TSIL have been developed to pull metals into the IL phase, others have been developed to keep metals in an IL phase. The use of metal complexes dissolved in IL for catalytic reactions has been one of the most fruitful areas of ionic liquid research to date. Still, these systems have a tendency to leach dissolved catalyst into the co-solvents that are used to extract the product of the reaction from the ionic liquid. Consequently, Wasserscheid et al. have pioneered the use of TSILs based on the dissolution into a "conventional" IL of metal complexes that incorporate charged phosphine ligands in their structure [16–18]. These metal complex ions become an integral part of the ionic medium, and remain there when the reaction products arising from their use are extracted into a co-solvent. Certain of the charged phosphine ions that are the basis of this chemistry (e.g., $P(m\text{-}C_6H_4SO_3^-Na^+)_3$) are commercially available, while others are prepared using established phosphine synthetic protocols.

An example of this approach to TSIL formulation is the synthesis from 1-vinyl imidazole of a series of imidazolium cations with appended tertiary phosphine groups (Scheme 2.3-3). The resulting phosphines are then coordinated to a Rh (I) organometallic and dissolved in the conventional ionic liquid [BMIM][PF$_6$], the mixture constituting a TSIL. The resulting system is active for the hydroformylation of 1-octene, with no observable catalyst leaching [17]. More examples of phosphine functionalized ionic liquids that have been reported later to improve the immobilization of transition metals in ionic liquids are given in Chapter 5, Section 5.4.

Task-specific ionic liquids designed for the binding of metal ions need not be only monodentate in nature. Taking a hint from classical coordination chemistry, a bidentate TSIL has been prepared and used in the extraction of Ni^{2+} from an aqueous solution. This salt is readily prepared in a two-step process. First, 1-(3-aminopropyl)imidazole is condensed under Dean–Stark conditions with 2-salicylaldehyde, giving the corresponding Schiff base. This species is readily

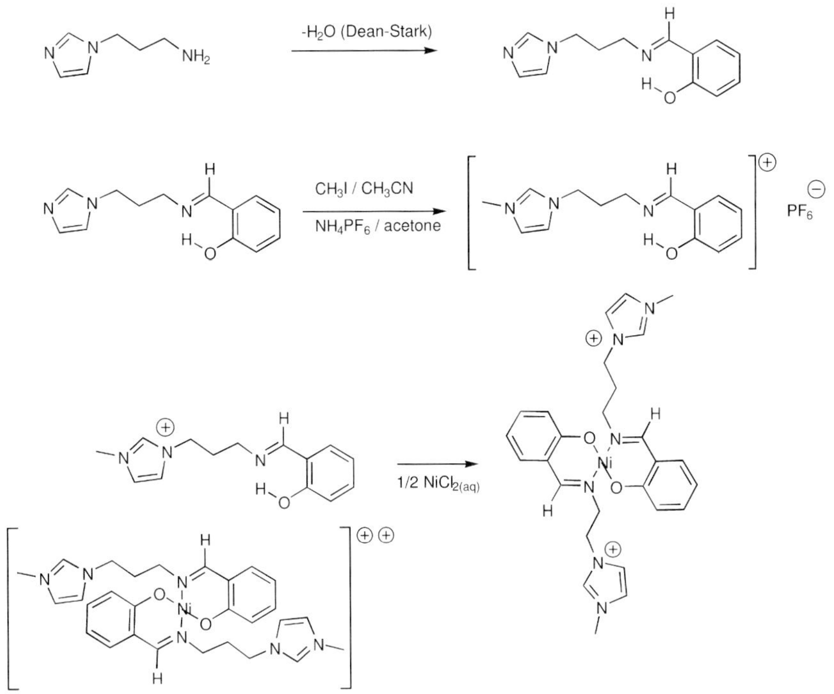

Scheme 2.3-4 Synthesis and utilization for metal extraction of a chelating task-specific ionic liquid.

alkylated in acetonitrile to form the imidazolium salt. Mixed as the $[PF_6]^-$ salt in a 1:1 (v/v) fashion with [HMIM][PF_6], this new TSIL quickly decolorizes green, aqueous solutions containing Ni^{2+} with which it is placed into contact, the color moving completely into the ionic liquid phase (Scheme 2.3-4).

The types of functional groups incorporated into TSIL need not be limited to those based upon nitrogen, oxygen or phosphorus. For example, ionic liquids have been reported that contain imidazolium cations with long, appended fluorous tails (Fig. 2.3-2). Though these species are not liquids at room temperature (melting in the 60–150 °C range), they nevertheless exhibit interesting chemistry when "alloyed" with conventional IL. While their solubility in conventional ionic liquids is rather limited (saturation concentrations of about 5 mM), the TSILs apparently form fluorous micelles in the latter. Thus, when a conventional ionic liquid doped with the fluorous TSIL is mixed with perfluorocarbons, extremely stable emulsions can be formed. These may be of use in developing two-phase fluorous–ionic liquid reaction systems [19]. As with many other TSIL reported thus far, these compounds are prepared by the direct reaction of 1-alkylimidazoles with a (polyfluoro)alkyl iodide, followed by anion metathesis.

Based on the same methodologies as the pioneering examples a plethora of other TSILs with functionalized imidazolium cations have been published in the last

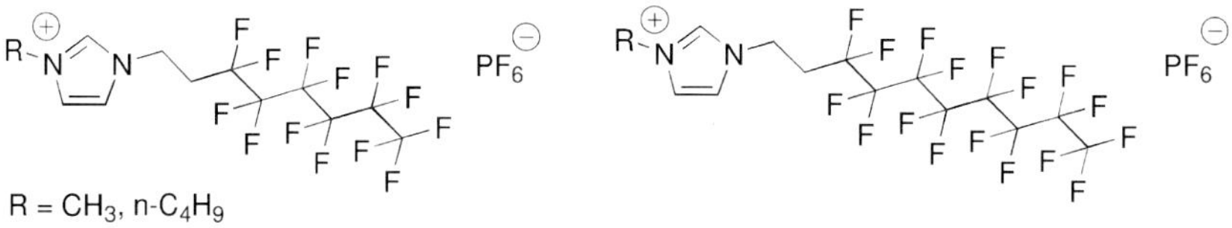

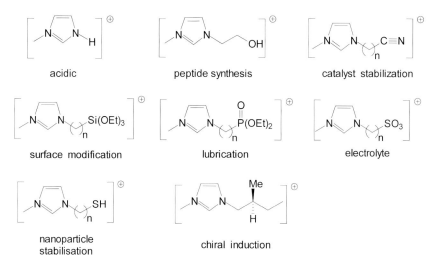

Fig. 2.3-2 Imidazolium-based task-specific ionic liquids with cation-appended fluorous tails.

Fig. 2.3-3 Further examples of mono-functionalized imidazolium cations and their potential applications.

four years. Fig. 2.3-3 presents a selection of TSILs carrying mono-functionalized 1,3-dialkylimidazolium cations and indicates the application area that these TSILs have been developed for. The figure has been adapted from an excellent, recent review by Dyson and coworkers on the design and application of functionalized ionic liquids [20].

It is noteworthy that deprotonation of imidazole with NaH or KH, followed by addition of two equivalents of a functionalized alkyl halide, as well as reaction of 1-trimethylsilylimidazole with two equivalents of a functionalized alkyl halide, also gives access to 1,3-bis-functionalized imidazolium halides [21]. The latter can be further converted by anion exchange.

While the overwhelming bulk of research on and with TSILs has been carried out on imidazolium-based systems, there is little obvious reason for this to remain the case. Rather, due to the relatively higher cost of commercial imidazole starting materials, economic considerations would suggest that future research places more emphasis on less-costly ammonium and phosphonium-based systems. Indeed, it is

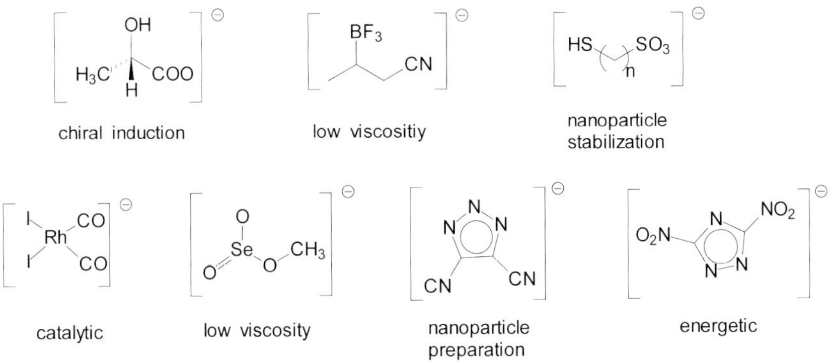

Fig. 2.3-4 Selection of reported functionalized anions for the synthesis of TSILs.

notable that a huge number of functionalized phosphonium salts (mostly halides) are in the literature, having been synthesized over the past forty-odd years as Wittig reagent precursors [22]. Many of these compounds will likely be found to give rise to ionic liquids when the cations are paired with an appropriate anion. In similar fashion, large numbers of natural products exist which are quaternized (or quarternizable) ammonium species that incorporate elsewhere in their structure other, useable functional groups. Many of these molecules are optically active, and could form the basis of entirely new TSIL systems for use in catalysis and chiral separations.

2.3.4
Functionalized Anions

Compared to TSILs with functionalized cations, much less effort has been devoted to the synthesis of ionic liquid systems with functionalized anions. Some reported examples with their potential applications are presented in Fig. 2.3-4, more details can be found in Ref. [20] and the literature cited therein.

2.3.5
Conclusion

Clearly, the potential for the development of new TSILs is limited only by the imagination of the chemists working in the area. In addition to TSILs with either functionalized cations or functionalized anions, the first examples of "dual-functionalized ionic liquids" in which both cation and anion contain functionalities have also been reported [23]. So far, the combination of several functionalized cations and anions in an "ion soup" that would allow a tuning and further increase of task-specificity by playing with different concentrations of the individual functionalized ions has remained completely unexplored.

However, to fully explore the potential of task-specific ionic liquids in the future the influence of the different functional groups on the TSIL's physicochemical properties needs to be much better understood. As a prerequisite for the development of suitable models a much better data set is required. This is not a trivial aspect as physicochemical properties of newly synthesized TSILs have, so far, not been routinely or systematically reported.

Some potential applications for TSILs have been briefly highlighted in Figs. 2.3-3 and 2.3-4. Many more examples can be found throughout this book. The reader interested in catalytic applications of TSILs is referred to Chapter 5, Section 5.3 for more details. Section 5.5 describes explicitly the role of task-specific ionic liquids as new liquid supports in combinatorial syntheses. This section also provides more details on the synthetic procedures leading to the specific functionalized ionic liquids that have turned out to be particularly suitable for this purpose. While Section 5.6 expands on the role of alkoxysilyl functionalized ionic liquids for surface modification in the preparation of supported ionic liquid phase (SILP) catalysts, Section 6.3 is devoted to the synthesis of nanoparticles and nanostructures in which TSILs often play a decisive role as templates or particle stabilizing agents.

References

1. Boon, J. A., Levisky, J. A., Pflug, J. L., Wilkes, J. S., *J. Org. Chem.* **1986**, *51*, 480.
2. Chauvin, Y., Hirschauer, A., Olivier, H., *J. Mol. Catal.* **1994**, *92*, 155.
3. Boon, J. A., Lander, S. W. Jr., Leviski, J. A., Pflug, J. L., Skrzynecki-Cooke, L. M., Wilkes, J. S., *Proc. Electrochem. Soc. 87-7 (Proc. Joint Int. Symp. Molten Salts)* **1987**, 979.
4. Wilkes, J. S., Zaworotko, M. J., *J. Chem. Soc., Chem. Commun.* **1990**, 965.
5. Welton, T., *Chem. Rev.* **1999**, *99*, 2071.
6. Forrester, K. J., Merrigan, T. L., Davis, J. H. Jr., *Tetrahedron Lett.* **1998**, *39*, 8955.
7. Wierzbicki, A., Davis, J. H. Jr., *Proceedings of the Symposium on Advances in Solvent Selection and Substitution for Extraction*, March 5–9, **2000**, Atlanta, Georgia. AIChE, New York: **2000**.
8. Karodia, N., Guise, S., Newlands, C., Andersen, J. -A., *Chem. Commun.* **1998**, 2341.
9. Mathews, C. J., Smith, P. J., Welton, T., White, A. J. P., Williams D. J., *Organometallics* **2001**, *20*, 3848.
10. Varma, R. S., Namboodiri, V. V., *Chem. Commun.* **2001**, 643.
11. Forrester, K. J., Davis, J. H. Jr., *Tetrahedron Lett.* **1999**, *40*, 1621.
12. Fraga-Dubreuil, J., Bazureau J. P., *Tetrahedron Lett.* **2001**, *42* 6097.
13. Visser, A. E., Swatloski, R. P., Reichert, W. M., Mayton, R., Sheff, S., Wierzbicki, A., Davis, J. H. Jr., Rogers, R. D., *Chem. Commun.* **2001**, 135.
14. Davis, J. H. Jr., Bates, E. D., unpublished results.
15. Davis, J. H. Jr., Working Salts: Syntheses and Uses of Ionic Liquids Containing Functionalised Ions, *ACS Symp. Ser.* **2002**, *818*, 247–258.
16. Brasse, C. C., Englert, U., Salzer, A., Waffenschmidt, H., Wasserscheid, P., *Organometallics*, **2000**, *19*, 3818.
17. Kottsieper, K. W., Stelzer, O., Wasserscheid, P., *J. Mol. Catal. A.*, **2001**, *175*, 285.
18. Brauer, D. J., Kottsieper, K. W., Liek, C., Stelzer, O., Waffenschmidt, H., Wasserscheid, P., *J. Organomet. Chem.*, **2001**, *630*, 177.

19. Merrigan, T. L., Bates, E. D., Dorman, S. C., Davis, J. H. Jr. *Chem. Commun.* **2000**, 2051–2052.
20. Fei, Z., Geldbach, T. J., Zhao, D., Dyson, P. J., *Chem. Eur. J.* **2006**, *12*, 2122.
21. (a) Harlow, K. J., Hill, A. F., Welton, T., *Synthesis* **1996**, *6*, 697; (b) Dzyuba, S. V., Batsch, R. A., *Chem. Commun.* **2001**, 1466.
22. Johnson, A. W., *Ylides and Imines of Phosphorus*, Wiley-Interscience, New York: **1993**.
23. Zhao, D., Fei, Z., Ohlin, C. A., Laurenczy, G., Dyson, P. J., *Chem. Commun.* **2004**, 2500.

3
Physicochemical Properties

3.1
Physicochemical Properties of Ionic Liquids: Melting Points and Phase Diagrams

John D. Holbrey and Robin D. Rogers

3.1.1
Introduction

What constitutes an *ionic liquid*, as distinct from a *molten salt*? It is generally accepted that ionic liquids have relatively low melting points, ideally below ambient temperature [1,2]. The distinction is arbitrarily based on the salt exhibiting liquidity at, or below, a given temperature, which is often conveniently taken to be 100 °C. However, it is clear from observation, that a major distinction between the materials of interest today, as ionic liquids, and more specifically as room-temperature ionic liquids, and conventional molten salts, is that ionic liquids generally contain organic, rather than inorganic cations. This allows a convenient differentiation without concern that some 'molten salts' may have lower melting points than some 'ionic liquids'.

Although the distinctions of 'high temperature' and 'low temperature' are entirely subjective, depending to a great extent on the experimental context. If we exclusively consider ionic liquids to incorporate an organic cation, and limit further the selection of salts to those that are liquid below 100 °C, a large range of materials is still available for consideration.

The utility of ionic liquids can primarily be traced to the work of Osteryoung et al. [3] and Wilkes and Hussey [4–6] who pioneered the use of *N*-butylpyridinium and 1-ethyl-3-methylimidazolium-containing ionic liquids as liquid electrolytes for electrochemical studies. These studies have strongly influenced the choice of ionic liquids for subsequent research [7] and the vast majority of work published on room-

Ionic Liquids in Synthesis, Second Edition. P. Wasserscheid and T. Welton (Eds.)
Copyright © 2008 WILEY-VCH Verlags GmbH & Co. KGaA, Weinheim
ISBN: 978-3-527-31239-9

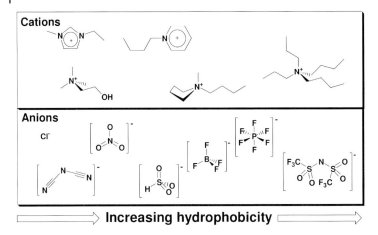

Fig. 3.1-1 Examples of common cation and anion pairs used in the formation of ionic liquids, and general progression of changes in IL properties with anion type.

temperature ionic liquids to date, relates to systems containing *N*-butylpyridinium and 1-ethyl-3-methylimidazolium (EMIM) cations.

Ionic liquids containing other classes of organic cations are known, and in addition to pyridinium and imidazolium systems, quaternary ammonium, phosphonium, pyrrolidinium, and sulfonium cations combined with a variety of anions (Fig. 3.1-1 provides some common examples) have been reported and have been studied for applications in electrochemistry [7, 8] and synthesis [9–11].

Since ionic liquids are simply salts, defined by the sole characteristic of having a low melting point, it should not be surprising to discover that many examples of low melting salts, ionic liquids, have utility other than as liquid materials, for example as electrolytes, phase transfer reagents [12], surfactants [13], and fungicides and biocides [14, 15] and many emerging examples of new ionic liquid materials are coming from the reinvestigation of chemical inventories in these fields.

The wide liquid ranges exhibited by ionic liquids, combined with low melting points, and the potential for tailoring size, shape and functionality offer opportunities for control in reactivity unobtainable with molecular solvents. It is worth noting that quaternary ammonium, phosphonium, and related salts are being widely reinvestigated [16–18] as the best ionic liquid choice for particular applications, particularly in synthetic chemistry. Protic ionic liquids, formed from acid–base reactions, are also finding renewed interest, especially in the emerging areas of hydrogen transport and fuel cell technologies. Changing ion type, substitution and composition produces new ionic liquid systems, each with a unique set of properties that can be explored and applied to the issues. With the potential large matrix of both anions and cations, it becomes clear that it will be impossible to screen any particular reaction in all the ionic liquids, or even all within a subset containing only a single anion or cation. Work is clearly needed to determine how the properties of ionic liquids vary as a function of anion/cation/substitution patterns, etc., and to establish which, if any, properties change in systematic (that is, predictable) ways.

The simplest ionic liquids are comprised of a single cation and single anion. More complex examples can also be considered by combining greater numbers of cations and/or anions, or when complex anions are formed as the result of equilibrium processes, as can be seen in Eq. (3.1-1):

$$\text{Cl}^- + \text{AlCl}_3 \rightarrow [\text{AlCl}_4]^- + \text{AlCl}_3 \rightarrow [\text{Al}_2\text{Cl}_7]^- \qquad (3.1\text{-}1)$$

Chloroaluminate(III) ionic liquid systems are perhaps the most well established and have been most extensively studied in the development of low melting organic ionic liquids with particular emphasis on electrochemical and electrodeposition applications, transition metal coordination chemistry, and in applications as liquid Lewis acid catalysts in organic synthesis. Variable and tunable acidity, from basic through neutral to acidic enables some very subtle changes in transition metal coordination chemistry. The melting point for [EMIM]Cl/AlCl$_3$ mixtures can be as low as $-90\,°C$, and the upper liquid limit almost $300\,°C$ [4].

The following discussion will concern the liquid ranges made available in different ionic liquids, as a factor of cation and anion structure and composition. In particular, the structural features of cation and anion that promote these properties (while providing other desirable, and sometimes conflicting, characteristics of the liquid, for example low viscosity, chemical stability, etc.) and variation in liquid range and stability will be the focus of this chapter.

The general observations that will be made, regarding structural influences on melting points are transferable across cation type, and apply in each case. The primary focus will be on 1-alkyl-3-methylimidazolium cations, coupled with simple organic and inorganic anions. Complex anions, such as mixed X^-/MX_n systems, will be mentioned as will other series of cations (including some examples from tetraalkylammonium salts).

3.1.2
Measurement of Liquid Range

The liquid range exhibited by ILs can be much greater than that found in common molecular solvents; for example water has a liquid range of $100\,°C$ (0 to $100\,°C$), and dichloromethane has a liquid range of $145\,°C$ (-95 to $40\,°C$) at ambient pressure. The lower temperature limit, solidification (either as crystallization or glassification) is governed by the structure and interactions between the ions. Ionic liquids, comprised of totally ionized components and having relatively weak ion–ion pairing (in comparison to molten salts), have little measurable vapor pressure and thus, in contrast to molecular solvents, the upper limit of the liquid phase for fully ionic liquids is usually that of thermal decomposition rather than vaporization.

3.1.2.1 Melting Points

The solid–liquid transition temperatures to ionic liquids can (ideally) be below ambient and as low as −100 °C. The most efficient method for measuring the transition temperatures is by differential scanning calorimetry (DSC). Other methods have been used including cold-stage polarizing microscopy, NMR, and X-ray scattering.

The thermal behavior of many ionic liquids is relatively complex. For a typical IL, cooling from the liquid state causes glass formation at low temperatures; solidification kinetics are slow. On cooling from the liquid, the low-temperature region is not usually bounded by the phase diagram liquidus line, but rather is extended down to a lower temperature limit imposed by the glass transition temperature [19]. This tendency is enhanced by the addition of lattice-destabilizing additives, including organic solutes, and by mixing salts. Solidification (glass) temperatures recorded on cooling are not a true measure of either heating T_g, or melting points and represent a kinetic transition. Thermodynamic data must be collected in heating mode to obtain reproducible results. So, in order to obtain reliable transition data, long equilibration times, using small samples that allow rapid cooling, are needed to quench non-equilibrium states in mixtures. Formation of metastable glasses is common in molten salts. In many cases, the glass transition temperatures are low; for 1-alkyl-3-methylimidazolium salts, glass transition temperatures recorded are typically in the region −70 to −90 °C. Heating from the glassy state yields, in many cases, an exothermic transition, associated with sample crystallization, followed by subsequent melting.

In some cases, there is evidence of multiple solid–solid transitions, either crystal–crystal polymorphism (seen for Cl^- salts [20]) or more often probably formation of plastic crystal phases – indicated by solid–solid transitions that consume a large fraction of the enthalpy of melting [21], which also leads to low energy melting transitions. The overall enthalpy of the salt can be dispersed into a large number of fluxional modes (vibration and rotation) of the organic cation, rather than into enthalpy of fusion. Thus, energetically, crystallization is often not overly favored.

3.1.2.2 Upper Limit – Decomposition Temperature

The upper limit of the liquid range is usually bounded by the thermal decomposition temperature of the ionic liquid, since most ionic liquids exhibit extremely low vapor pressure to their respective decomposition temperatures. In contrast to molten salts, which form tight ion pairs in the vapor phase, the reduced Coulombic interactions between ions energetically restricts the ion pair formation required for volatilization of ILs, leading to low vapor pressures. This leads to high upper temperature limits, defined in many cases by decomposition of the IL rather than vaporization. However, it has also been demonstrated recently that certain, thermally very stable, ionic liquids do evaporate under harsh temperature and vacuum conditions [22]. It has even been demonstrated that mixtures of such ionic liquids can be separated by distillation.

The nature of the ionic liquids, containing organic cations generally restricts upper stability temperatures, up to 350 °C, where pyrolysis occurs if no other lower temperature decomposition pathways are accessible [23]. In most cases,

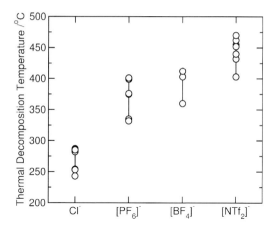

Fig. 3.1-2 Thermal decomposition temperature ranges (in °C, obtained from TGA on-set measurements) for ionic liquids containing 1-alkyl-3-methylimidazolium cations [23–27]. The thermal stability of the ionic liquids depends on the coordinating ability of the anion.

decomposition occurs with complete mass loss and volatilization of the component fragments. Grimmett et al. have studied the decomposition of imidazolium halides [23], and identified the degradation pathway as E2 elimination of the N-substituent. For protic ionic liquids, decomposition at lower temperatures occurs through shifts in the proton transfer equilibrium between salt form and parent acid/base pairs. Since, in many cases, the thermal stability of the 1:1 salt tends to be greater than that of the respective acids and bases, loss of the volatile components and their ability to recombine on condensation leads to the apparent ability to distill some ionic liquids, a process which should be familiar to anyone who has tried to determine the melting point of ammonium chloride!

If decomposition temperatures for a range of ILs with differing anions are compared, the stability of the ionic liquid is inversely proportional to the tendency to form a stable alkyl-X species. As can be seen from TGA decomposition data for a range of $RMIM^+$ salts (Fig. 3.1-2) collected by anion type, the decomposition temperatures vary with anion type and follow the general stability order, $Cl^- < [BF_4]^- \sim [PF_6]^- < [NTf_2]^-$, so that ionic liquids containing weakly nucleophilic anions are most stable to high temperature decomposition [24–28]. Note that all data given in Fig. 3.1-2 have been obtained by TGA-onset measurements.

The determination of the "right" thermal decomposition temperature of an ionic liquid is not trivial. It is quite obvious from the above mentioned aspects that the thermal stability determined in a TGA experiment will be a strong function of the ionic liquid's quality, with many impurities significantly reducing the stability. Moreover, Ngo et al. [25] have shown that the thermal decomposition of ionic liquids, measured by TGA, varies depending on the sample pans used; in some cases increased stabilization of up to 50 °C was obtained on changing from aluminum to alumina sample pans. Finally, and most importantly, the thermal decomposition of an ionic liquid was found to be a time-dependent process. Thus, TGA experiments with different heating rates have led to largely varying decomposition temperatures. It has been found that the long-term thermal stability of ionic liquids can be more than 50 °C lower than the decomposition temperatures obtained by step-tangent analysis of rising temperature scanning differential thermal analysis (TGA-onset) experiments. MacFarlane and coworkers have shown, for example, that the temperature at which 1% thermal degradation of [EMIM][(CF$_3$SO$_2$)$_2$N] occurs is 307 °C if the ionic liquid is exposed to thermal stress for 1 h while it is only 251 °C if the ionic liquid is exposed for 10 h [29]. Extrapolation to even longer times confirmed that this specific ionic liquid can be indeed regarded as long-term stable at a temperature of 250 °C. These findings have to be considered for a proper selection of ionic liquids for practical high temperature applications.

3.1.3
Effect of Ion Sizes on Salt Melting Points

It is well known that the characteristic properties of ionic liquids vary with the choice of anion and cation. The structure of an ionic liquid directly impacts upon its properties, in particular, the melting point and liquid ranges. The underlying principles behind the drive to reduce the melting points (and thus operational range limits) for battery electrolytes have been described elsewhere [4]. Exploiting the changes in these characteristics enables the design of ILs with a wide range of properties.

Changes in size, shape, and character of the component ions, most particularly with the introduction of charge dispersed or non-charged components, influence the melting points of salts with the reduction in melting point resulting from disruption of crystal packing and reduction of the crystal lattice energy.

The dominant force in ionic liquids is the Coulombic attraction between ions. The Coulombic attraction terms are given by Eq. (3.1-2)

$$E_c = MZ^+Z^-/4\pi\varepsilon_o r \tag{3.1-2}$$

where Z^+ and Z^- are the ion charges, and r is the inter-ion separation.

The overall lattice energies of ionic solids, as treated by the Born–Landé or Kaputinskii equations, then depend on (i) the product of the net ion charges, (ii) the ion–ion separation, and (iii) the packing efficiency of the ions (reflected in the Madelung constant, M, in the Coulombic energy term). Thus, low melting salts should be most preferred when the charges on the ions are respectively ±1 and when the size of the ions is large, thus ensuring that the inter-ion separation (r) is also large. In addition, large ions enable charge delocalization, further reducing overall charge density.

This can be illustrated for a series of sodium salts, shown in Table 3.1-1, in which the size of the anion is varied. As the size of the anion is increased, the melting point of the salt decreases, reflecting the weaker Coulombic interactions in the crystal lattice. On increasing the thermochemical radius of the anion, from Cl$^-$ to

Table 3.1-1 Melting points (°C) and thermochemical radii of the anions Å for Na$^+$ and [EMIM]$^+$ salts. The ionic radii of the cations are 1.2 Å (Na$^+$) and 2 × 2.7 Å ([EMIM]$^+$, non-spherical)

		Melting point	
X$^-$	r	NaX	[EMIM]X
Cl$^-$	1.7	801	87
[BF$_4$]$^-$	2.2	384	6
[PF$_6$]$^-$	2.4	>200	60
[AlCl$_4$]$^-$	2.8	185	7

[BF$_4$]$^-$ to [PF$_6$]$^-$ to [AlCl$_4$]$^-$, the melting points of the sodium salts are reduced from 801 to 185 °C. The results from the sodium salts can be roughly extrapolated to room temperature and indicate that, in order to obtain a salt that would melt at room temperature, the anion would be required to have a radius in excess of about 3.4–4 Å [30]. Large anions are, in general, non-spherical and have significant associated covalency. A similar increase is observed on increasing the cation size, for example by moving down a group in the periodic table. Lithium salts tend to be higher melting, that the sodium or cesium analogs. If the charge on the ion can also be delocalized or if the charge-bearing regions can be effectively isolated in the interior of the ionic moiety, then Coulombic terms are further reduced.

Reduction in melting point can, simplistically, be achieved by increasing the size of the anion or the cation. Ionic liquids contain organic cations that are large in comparison to the thermodynamic radii of inorganic cations. This results in significant reductions in the melting points for the organic salts, as illustrated by the [EMIM]$^+$ examples in Table 3.1-1. The Coulombic attraction terms for ionic liquids are of comparable magnitude to the intermolecular interactions in molecular liquids.

The Coulombic interaction terms for ionic liquids are relatively small compared to those in analogous inorganic salts as a result of size and charge differences, and the presence of flexible substituents which contribute charge separation, disruption of lattice isotropy and introduce many rotational and vibrational modes of freedom in these ions. This introduction of ion flexibility, inefficiencies of packing, and non-charge bearing hydrocarbon groups, provides significant reductions in the salt lattice energies which can lead to low melting points and, in many cases, glass formation with marked inhibition of crystallization.

3.1.3.1 Anion Size

As shown above, increases in anion size lead to reductions in the melting points of salts by reducing the Coulombic attraction contributions to the lattice energy of the crystal and increasing the covalency of the ions. In general, for ionic liquids, increasing anion size leads to lower melting points, as can be seen for a selection of [EMIM]X salts in Table 3.1-2.

Table 3.1-2 [EMIM]X salts and melting points, illustrating anion effects

Anion, [X]	Melting point (°C)	Reference
Cl^-	87	4
Br^-	81	23
I^-	79–81	23
$[BF_4]^-$	15	6, 27, 31
$[AlCl_4]^-$	7	4, 6
$[GaCl_4]^-$	47	32
$[AuCl_3]^-$	58	33
$[PF_6]^-$	62	34
$[AsF_6]^-$	53	34
$[NO_3]^-$	38	35
$[NO_2]^-$	55	35
$[CH_3CO_2]^-$	ca. −45	35
$[SO_4]\cdot 2H_2O^{2-}$	70	35
$[CF_3SO_3]^-$	−9	26
$[CF_3CO_2]^-$	−14	26
$[N(SO_2CF_3)_2]^-$	−3	26
$[N(CN)_2]^-$	−21	36
$[CB_{11}H_{12}]^-$	122	37
$[CB_{11}H_6Cl_6]^-$	114	37
$[CH_3CB_{11}H_{11}]^-$	59	37
$[C_2H_5CB_{11}H_{11}]^-$	64	37

Ionic liquids containing carborane anions, described by Reed et al. [37], contain large, near spherical anions with highly delocalized charge distribution. These ionic liquids have low melting points, compared to the corresponding lithium and ammonium salts, but the melting points are higher then might be anticipated from a simplistic model based solely on comparison of the size of the anions. Similarly, it should be noted that [EMIM][PF$_6$] [35] appears to have a higher melting point than would be anticipated. In other large anions, for example tetraphenylborate [41], additional attractive interactions such as aromatic $\pi-\pi$ stacking can lead to increased melting points.

Anion and cation contributions cannot be taken in isolation; induced dipoles can increase melting points through hydrogen-bonding interactions, seen in the crystal structures of [EMIM]X (X = Cl, Br, I) salts [38] and absent from the structure of [EMIM][PF$_6$] [34]. In addition to increasing ion–ion separations, larger (and in general, more complex) anions can allow greater charge delocalization. For bis(trifluoromethanesulfonyl)amide salts [26, 39, 40], this is affected by the $-SO_2CF_3$ groups which effectively provide a steric block, isolating the delocalized [S–N–S]$^-$ charged region in the center of the anion.

3.1.3.2 Mixtures of Anions

Complex anions, formed when halide salts are combined with Lewis acids (e.g., Cl$^-$/AlCl$_3$) lead to ionic liquids with reduced melting points through the formation

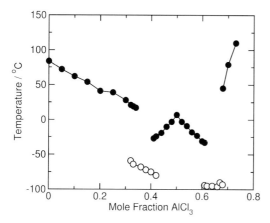

Fig. 3.1-3 Phase diagram for [EMIM]Cl/AlCl$_3$: (●) melting and freezing points; (○) glass transition points [4, 6].

of eutectic compositions [8]. The molar ratio of the two reactants can influence the melting point of the resultant mixed salt system through speciation equilibria. For [EMIM]Cl/AlCl$_3$, an apparently simple phase diagram for a binary mixture forming a 1:1 compound and exhibiting two eutectic minima is formed with a characteristic W-shape to the melting point transition [4, 6] (Fig. 3.1-3). Polyanionic species including [Al$_2$Cl$_7$]$^-$ and [Al$_3$Cl$_{10}$]$^-$ have been identified. Only at 50% composition is the pure compound [EMIM][AlCl$_4$] formed.

The two eutectic minima, corresponding to 1:2 and 2:1 compositions, result in liquids with very low solidification temperatures; the glass transition temperature for [EMIM]Cl/AlCl$_3$ (33:66%) is −96 °C. Similar compositional variation should be anticipated in the phase diagrams of other metal halide ionic liquids, for example the phase behavior of [EMIM]Cl/CuCl [40, 41] and [EMIM]Cl/FeCl$_3$ [42] show similar behavior. For the [EMIM]Cl/CuCl system, the lower freezing temperatures in the basic and acidic regions are −40 and −65 °C, respectively. More complex phase diagrams have been reported, for example in [HPy]Cl/ZnCl$_2$ [19], where a range of multinuclear zinc halide anions can exist.

The presence of several different anions in these ionic liquids has the effect of significantly decreasing the melting point. Considering that the formation of eutectic mixtures of molten salts is widely used to obtain lower melting points, it is surprising that little effort has been put into identifying the effects of mixtures of cations or anions on the physical properties of other ionic liquids [17].

3.1.3.3 Cation Size

The size and shape of cations in ionic liquids are important in controlling the melting points of the salts. The formation of low melting ionic liquids results from reduction in the magnitude of the Coulombic attraction term, and disruption of packing efficiency. On a simple basis, large ions tend to lead to reductions in the melting

Table 3.1-3 Changes in melting points for symmetric tetraalkylammonium bromide salts with increasing size of alkyl substituents

Cation	Melting point (°C)
[NMe₄]	>300
[NEt₄]	284
[NBu₄]	124–128
[NHex₄]	99–100
[NOct₄]	95–98

points. Tetraalkyllammonium and phosphonium salts are the best examples of salts containing large symmetric cations with alkyl-shielded charge. The cation radius, r, is large, and the salts display correspondingly lower melting points than the Group 1 analogs. For example, the melting point reduction for a series of tetraalkylammonium bromide salts with increasing cation size is shown in Table 3.1-3.

In addition to the increase in apparent cation size arising from increasing the length of the substituents, the conformational flexibility of linear alkyl-substituents, which have many rotational degrees of freedom, allows the alkyl chains to 'melt' at temperatures below the melting point, leading to solid–solid polymorphic transitions. For example, the tetrabutylammonium thiocyanate ionic liquid (mp 49.5 °C) has a number of solid–solid transitions associated with changes in alkyl chain conformation [45], which change the density of the solid below the melting point.

3.1.3.4 Cation Symmetry

The melting points of organic salts have an important relationship to the symmetry of the ions: increasing symmetry in the ions increases melting points by enabling more efficient ion–ion packing in the crystal cell. Conversely, reducing the symmetry of the cations causes a distortion from ideal close-packing of the ionic charges in the solid state lattice, a reduction in the lattice energy, and depression of melting points. Changing from spherical, or high symmetry ions such as Na^+ or $[NMe_4]^+$ to lower symmetry ions such as the imidazolium cations distorts the Coulombic charge distribution. In addition, cations, such as the imidazolium cations contain alkyl groups that do not participate in charge delocalization.

Reduction in cation symmetry (ideally to C_1) lowers the freezing point and markedly expands the range of room temperature liquid salts. Table 3.1-4 shows the effect of symmetry for a series of NR_4X salts, in which all the cations contain 20 carbon atoms in the alkyl substituents. Room-temperature liquids are obtained for the salts; $N_{6554}Br$, $N_{10,811}Br$, $N_{6644}Br$, $N_{8543}ClO_4$, $N_{10,811}ClO_4$, $N_{9551}ClO_4$, and $N_{8651}ClO_4$, whereas the salts containing cations with high symmetry have much higher melting points. It can be seen that the melting points of these isomeric salts vary by over 200 °C depending on the symmetry of the cation.

Table 3.1-4 Effect of cation symmetry on melting points of isomeric tetraalkylammonium salts. In each case, the cation (designated $[N_{nmop}]^+$) has four linear-alkyl substituents containing a total of 20 carbons [44]. Salts that are liquid at room temperature are indicated by *l*

Cation ($[N_{nmop}]^+$)	Br^-	$[ClO_4]^-$	$[BPh_4]^-$
5555	101.3	117.7	203.3
6554	83.4		
6644	83.0		
8444	67.3		
8543	*l*	109.5	
6662	46.5		
7733	*l*	45–58	138.8
8663	*l*	*l*	110.2
7751	*l*	104	
8651	*l*		
9551	*l*		
9641	*l*	*l*	
11333	67–68	65.5	
11432	*l*		
8822	62		
9821	*l*		
13331	71–72	52–53	
9911	*l*		
10811	*l*	*l*	
14222	170	152	
16211	180	155	
17111	210	205	

3.1.3.5 Imidazolium Salts

Changes in the ring substitution patterns can have significant effects on the melting points of imidazolium salts, beyond those anticipated by simple changes in symmetry or H-bonding interactions (i.e., substitution at the C-(2,4,5) positions on an imidazolium ring, affects packing and space-filling of the imidazolium cations). For example, substitution at the C(2)-position of the imidazolium ring increases the melting points of the salts. This is not necessarily an obvious or straightforward result, but may be caused by changes in the cation structure that can induce aromatic stacking or methyl-π interactions between cations. The introduction of other functionalities around the periphery of the ions can also change the interactions between ions. In most cases, additional functions, such as ether groups, increase the number of interactions, and thus increase melting points.

3.1.3.6 Imidazolium Substituent Alkyl Chain Length

The data in Table 3.1-4 illustrate the changes in melting points that can be achieved by changing the symmetry of the cation. $[RMIM]^+$ salts, with asymmetric N-substitution have no rotation or reflection symmetry operations. Changing the alkyl chain substitution on one of the ring hetero-atoms does not change the symmetry of the cation. However, manipulation of the alkyl chain can produce major changes in the melting points, and on the tendency of the ionic liquids to form glasses rather than crystalline solids on cooling, by changing the efficiency of ion packing.

Characteristic changes in the melting points of organic salts with simple changes in a single alkyl-chain substituent are shown in Fig. 3.1-4 for 1-alkyl-3-methylimidazolium tetrafluoroborate ionic liquids ([RMIM][BF$_4$]). This highlights the two competing effects on the melting points of changing alkyl-group substituents. It is immediately noticeable, that increasing the substituent length initially reduces the melting point of the IL, with the major trend towards glass formation on cooling for $n = 4$–10. On extending the alkyl chain lengths beyond a certain point (which for alkyl-methylimidazolium salts, is around 8–10 carbons), the melting points of the salts start to increase again with increasing chain length, as van der Waals interactions between the long hydrocarbon chains contribute to local structure by inducing microphase separation between the covalent, hydrophobic alkyl chains and charged ionic regions of the molecules. Initial lengthening of the substituent leads to a reduction in the melting points through destabilization of Coulombic packing, and a trend towards glass formation. However, further increases in substitution lead to increased attractive van der Waals forces between the hydrocarbon chains and increased structural ordering which can be seen with the re-emergence of higher melting points and the formation of structured liquid crystalline materials. The intermediate region (the well in Fig. 3.1-4) represents glass-forming materials in which crystallization is inhibited through the attractive and dispersive terms and the presence of many rotational modes that provide a wide range of ion conformers with similar local energy minima.

A consideration of the changes in molecular structure, and of the underlying effects that this will have in both the liquid and crystal phases helps to rationalize changes in melting points with substitution. The crystalline phases of the IL are dominated by Coulombic ion–ion interactions, comparable to those in typical salt crystals although, since the ions are larger, the Coulombic interactions are weaker (decreasing with r^2). An effect of this is that many organic salts (including 'ionic liquids') crystallize with simple salt-like packing of the anions and cations.

Increased asymmetric substitution on 1-alkyl-3-methylimidazolium salts increases the asymmetric disruption and distortion of the Coulombic packing of ions, leading to substantial decreases in the melting point as the efficiency of packing and crystallization is reduced. This results in (i) melting point reduction and (ii) a pronounced tendency for glass formation on cooling, rather than crystallization, on extending the alkyl substituents. This is indicative of inefficient packing within the crystal structures which is a function of the low-symmetry cations employed. Increasing alkyl chain substitution can also introduce other rheological changes in the ionic liquids including increased viscosity, reduced density, and increased lipophilicity, which must also be considered.

The incorporation of alkyl substituents of increasing chain length in a non-symmetrical arrangement on the ions leads to the introduction of 'bulk' into the crystalline lattice that disrupts the attractive charge–charge lattice. Relatively short

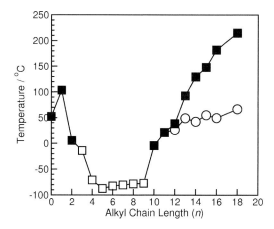

Fig. 3.1-4 Changes in melting points for 1-alkyl-3-methylimidazolium tetrafluoroborate [26] ionic liquids as a function of chain length, showing true melting points (■) and glass transitions (□) with the general trends for decreasing melting points with $n = 1$ to $n = 8$, followed by an increase in melting point and formation of a smectic liquid crystal phase (o) with increasing n clearly visible.

alkyl chains act in this manner, as a buffer, and do not pack well into the available space in the crystalline lattice; high rotational freedom leads to low occupation densities over a relatively large volume of space. This free rotation volume probably gives rise to the 'void-space' considered by Brennecke [48] to explain the extraordinary propensity for sc-CO_2 to dissolve in ILs without substantially changing the volume of the liquid phase.

3.1.3.7 Branching

Table 3.1-5 provides data for a series of ionic liquids, where the only difference is the degree of branching within the alkyl chain at the imidazolium ring 3-position. The melting points and enthalpies for three isomeric 1-butyl-3-methylimidazolium hexafluorophosphate salts, [BMIM][PF_6], increase with the degree of chain branching, reflecting the changes in efficiency of the crystal packing as free-rotation volume decreases and atom density is increased. The same effects are also observed for the two isomers of 1-propyl-3-methylimidazolium hexafluorophosphate, [PMIM][PF_6] [25].

3.1.4
Summary

Liquid structure is defined by short-range ordering, with long-range disorder. The short-range (near-neighbor) structuring of the liquids is a combination of dominant Coulombic charge–charge attractions balanced against the rotational and vibrational freedom of the ions. Changes in the degrees of freedom and increases in

Table 3.1-5 Melting points and heats of fusion for isomeric [BMIM][PF$_6$] [45] and [PMIM][PF$_6$] [24] ionic liquids, showing increased melting point and crystal stability with the degree of branching in the alkyl substituent

N(1)-substitution	Melting point (°C)	ΔH_{fusion} (kJ mol^{-1})
n-butyl	6.4	31
sec-butyl	83.3	72
tert-butyl	159.7	83
n-propyl	40	
isopropyl	102	

non-participating portions of the cation that do not contribute to Coulombic stabilization of the crystal in a salt-like lattice, lead to a decrease in melting points and heats of formation. At longer chain lengths amphiphilic nature is generated, leading to hydrophobic van der Waals contributions and formation of typically bilayer lattices.

The effects of cation symmetry are relatively clear; the melting points of symmetrically substituted 1,3-dialkyl cations are higher than those of the unsymmetrical cations, and continue to decrease with increasing alkyl substitution, up to the critical point around 8–10 carbons, then increase with increasing additional substitution. Both alkyl-substitution and ion asymmetry interfere with efficient packing of ions into a crystalline lattice based on Coulombic attractions. However, there appears to be no simple correlation with hydrogen-bonding ability. The absence of strong H-bonding is certainly a major contributor to low melting points, but ionic liquids containing strongly H-bonding anions (e.g. [CH$_3$COO]$^-$) have similar melting points to those incorporating anions that are highly delocalized and unable to H-bond (e.g., [NTf$_2$]$^-$). Similarly, C(2)-substitution of imidazolium rings might be anticipated to suppress melting points, by suppressing hydrogen-bonding. This does not appear to be the case, with significant increases in melting points with C(2)-substitution. This implies that the effects of van der Waals interactions via the methyl group, or methyl-π interactions, etc. are more important than the electrostatic interactions via the C(2)-hydrogen.

Hagiwara and Ito [49] and Bonhôte et al. [26] have indicated that there appears to be no overall correlation between the composition of an IL and its melting point, based on non-systematic changes in cation substitution and anion types. Ngo et al. [24] indicate that the melting points decrease with incorporation of larger, more asymmetrical cations. Highly fluorinated anions, [BF$_4$]$^-$, [PF$_6$]$^-$, [NTf$_2$]$^-$, [CF$_3$COO]$^-$, etc., are generally liquid to low temperatures, forming glasses on solidification (slow crystallization on heating prior to melting is often observed). However, Katritzky et al. [50,51] have started to show that the physical properties of imidazolium and pyridinium salts (including ionic liquids) can be modeled using QSPR and CODESSA computational methods, allowing the prediction of melting points with reasonable confidence.

It is important that the forces and interactions that govern the melting points of ionic liquids are not considered in isolation; these interactions also control the dissolution and solubility of other components in the ionic liquids. For example, if there is a requirement for an ionic liquid to have strong H-bond accepting character (in the anion), then it should be anticipated that this will also lead to hydrogen-bonding interactions between ions, resulting in greater attractive forces and elevated melting points.

References

1. Seddon, K. R., *J. Chem. Technol. Biotechnol.* **1997**, *68*, 351.
2. Rogers, R. D., *Green Chem.* **2000**, *5*, G94.
3. Osteryoung, R. A., Gale, R. J., Robinson, J., Linga, H., Cheek, G., *J. Electrochem. Soc.* **1981**, *128*, 79.
4. Wilkes, J. S., Levisky, J. A., Wilson, R. A., Hussey, C. L., *Inorg. Chem.* **1982**, *21*, 1263.
5. Fannin, A. A., King, L. A., Levisky, J. A., Wilkes, J. S., *J. Phys. Chem.* **1984**, *88*, 2609.
6. Fannin, A. A., Floreani, D. A., King, L. A., Landers, J. S., Piersma, B. J., Stech, D. J., Vaughn, R. L., Wilkes, J. S., Williams, J. L., *J. Phys. Chem.* **1984**, *88*, 2614.
7. Hussey, C. L., *Adv. Molten Salt Chem.* **1983**, *5*, 185.
8. Cooper, E. I., Sullivan, E. S. M., *Eighth International Symposium on Molten Salts*, Vol. 92-16; The Electrochemical Society, Pennington, NJ, **1992**, p. 386.
9. Welton, T., *Chem. Rev.* **1999**, *99*, 2071.
10. Holbrey, J. D., Seddon, K. R., *Clean Prod. Proc.* **1999**, *1*, 233.
11. Keim, W., Wasserscheid, P., *Angew. Chem. Int. Ed.* **2000**, *39*, 3772.
12. Albanese, D., Landini, D., Maia, A., Penso, M., *J. Mol. Catal. A* **1999**, *150*, 113.
13. Blackmore, E. S., Tiddy, G. J. T., *J. Chem. Soc., Faraday Trans. 2* **1988**, *84*, 1115.
14. Pernak, J., Krysinski, J., Skrzypczak, A., *Pharmazie* **1985**, *40*, 570.
15. Pernak, J., Czepukowicz, A., Pozniak, R., *Ind. Eng. Chem. Res.* **2001**, *40*, 2379.
16. Sun, J., Forsyth, M., MacFarlane, D. R., *J. Phys. Chem. B* **1998**, *102*, 8858.
17. MacFarlane, D. R., Meakin, P., Sun, J., Amini, N., Forsyth, M., *J. Phys. Chem. B* **1999**, *103*, 4164.
18. Abdallah, D. J., Robertson, A., Hsu, H.-F., Weiss, R. G., *J. Am. Chem. Soc.* **2000**, *122*, 3053.
19. Easteal, E. J., Angell, C. A., *J. Phys. Chem.* **1970**, *74*, 3987.
20. Holbrey, J. D., Reichert, W. M., Nieuwenhuyzen, M., Johnson, S., Seddon, K. R., Rogers, R. D., *Chem. Commun.* **2003**, 1636.
21. Hardacre, C., Holbrey, J. D., McCormac, P. B., McMath, S. E. J., Nieuwenhuyzen, M., Seddon, K. R., *J. Mater. Chem.* **2001**, *11*, 346.
22. Earle, M. J., Esperanca, J. M. S. S., Gilea, M. A., Canongia Lopes, J. N., Rebelo, L. P. N., Magee, J. W., Seddon, K. R., Widegren, J. A., *Nature* **2006**, *439*, 831.
23. Chan, B. K. M., Chang, N.-H., Grimmet, M. R., *Aust. J. Chem.* **1977**, *30*, 2005.
24. Huddleston, J. G., Visser, A. E., Reichert, W. M., Willauer, H. D., Broker, G. A., Rogers, R. D., *Green Chem.* **2001**, 156.
25. Ngo, H. L., LeCompte, K., Hargens, L., McEwan, A. B., *Thermochim. Acta* **2000**, 357–358, 97.
26. Bonhôte, P., Dias, A. P., Papageorgiou, N., Kalyanasundaram, K., Grätzel, M., *Inorg. Chem.* **1996**, *35*, 1168.
27. Holbrey, J. D., Seddon, K. R., *J. Chem. Soc., Dalton Trans.* **1999**, 2133.
28. Takahashi, S., Koura, N., Kohara, S., Saboungi, M.-L., Curtiss, L. A., *Plasmas Ions* **1999**, *2*, 91.
29. Baranyai, K. J., Deacon, G. B., MacFarlane, D. R., Pringle, J. M., Scott, J. L., *Aust. J. Chem.* **2004**, *57*, 145–147.
30. Rooney, D. W., Seddon, K. R., Ionic Liquids, in *Handbook of Solvents*; Wypych, G. (Ed.), ChemTec, Toronto, **2001**, p. 1459.

31. Fuller, J., Carlin, R. T., Osteryoung, R. A., *J. Electrochem. Soc.* **1997**, *144*, 3881.
32. Wicelinski, S. P., Gale, R. J., Wilkes, J. S., *J. Electrochem. Soc.* **1987**, *134*, 262.
33. Hasan, M., Kozhevnikov, I. V., Siddiqui, M. R. H., Steiner, A., Winterton, N., *Inorg. Chem.* **1999**, *38*, 5637.
34. Fuller, J., Carlin, R. T., De Long, H. C., *Chem. Commun.* **1994**, 229.
35. Wilkes, J. S., Zaworotko, M. J., *J. Chem. Soc., Chem. Commun.* **1992**, 965.
36. MacFarlane, D. R., Golding, J., Forsyth, S., Forsyth, M., Deacon, G. B., *Chem. Commun.* **2001**, 1430.
37. Larsen, A. S., Holbrey, J. D., Tham, F. S., Reed, C. A., *J. Am. Chem. Soc.* **2000**, *122*, 7264.
38. Elaiwi, A., Hitchcock, P. B., Seddon, K. R., Srinivasan, N., Tan, Y. M., Welton, T., Zora, J. A., *J. Chem. Soc., Dalton Trans.* **1995**, 3467.
39. Golding, J. J., MacFarlane, D. R., Spicca, L., Forsyth, M., Skelton, B. W., White, A. H., *Chem. Commun.* **1998**, 1593.
40. Noda, A., Hayamizu, K., Watanabe, M., *J. Phys. Chem. B* **2001**, *105*, 4603.
41. Suarez, P. A. Z., Dupont, J., de Souza, R. F., Burrow, R. A., Kintzinger, J.-P., *Chem. Eur. J.* **2000**, 2377.
42. Bolkan, S. A., Yoke, J. T., *J. Chem. Eng. Data.* **1986**, *31*, 194.
43. Bolkan, S. A., Yoke, J. T., *Inorg. Chem.* **1986**, *25*, 3587.
44. Sitze, M. S., Schreiter, E. R., Patterson, E. V., Freeman, R. G., *Inorg. Chem.* **2001**, *40*, 2298.
45. Coker, T. G., Wunderlich, B., Janz, G. J., *Trans. Faraday Soc.* **1969**, *65*, 3361.
46. Gordon, J. E., SubbaRao, G. N., *J. Am. Chem. Soc.* **1978**, *100*, 7445.
47. Carmichael, A. J., Hardacre, C., Holbrey, J. D., Nieuwenhuyzen, M., Seddon, K. R., *Eleventh International Symposium on Molten Salts*, Vol. 99-41, Truelove, P. C., De Long, H. C., Stafford, G. R., Deki, S. (Eds.), The Electrochemical Society, Pennington, NJ, **1999**, 209.
48. Blanchard, L. A., Gu, Z., Brennecke, J. F., *J. Phys. Chem. B* **2001**, 2437.
49. Hagiwara, R., Ito, Y., *J. Fluor. Chem.* **2000**, *105*, 221.
50. Katritzky, A. R., Lomaka, A., Petrukhin, R., Jain, R., Karelson, M., Visser, A. E., Rogers, R. D., *J. Chem. Inf. Comput. Sci.* **2002**, *42* (1), 71.
51. Katritzky, A. R., Jain, R., Lomaka, A., Petrukhin, R., Karelson, M., Visser, A. E., Rogers, R. D., *J. Chem. Inf. Comput. Sci.* **2002**, *42* (2), 225.

3.2
Viscosity and Density of Ionic Liquids

Rob A. Mantz and Paul C. Trulove

3.2.1
Viscosity of Ionic Liquids

The viscosity of a fluid arises from the internal friction of the fluid, and it manifests itself externally as the resistance of the fluid to flow. With respect to viscosity there are two broad classes of fluids, Newtonian and non-Newtonian. Newtonian fluids have a constant viscosity regardless of strain rate. Low molecular weight pure liquids are examples of Newtonian fluids. Non-Newtonian fluids do not have a constant viscosity and will either thicken or thin when strain is applied. Polymers, colloidal suspensions, and emulsions are examples of non-Newtonian fluids [1]. To date, researchers have treated ionic liquids as Newtonian fluids, and no data have so far been published to indicate that there are ionic liquids that are non-Newtonian.

Experimentally determined viscosities are generally reported as either absolute viscosity (η) or kinematic viscosity (υ). Kinematic viscosity is simply the absolute viscosity normalized by the density of the fluid. The relationship between absolute viscosity (η), density (ρ), and kinematic viscosity (υ) is given by:

$$\frac{\eta}{\rho} = \upsilon \qquad (3.2\text{-}1)$$

The unit of absolute viscosity is the Poise (P, g cm^{-1} s^{-1}) while the unit for kinematic viscosity is the Stoke (St, cm^2 s^{-1}). Because of the large size of these viscosity units, absolute viscosities for ionic liquids are usually reported in centipoises (cP) and kinematic viscosities in centistokes (cSt).

3.2.1.1 Viscosity Measurement Methods

The viscosities of ionic liquids have normally been measured using one of three methods: falling or rolling ball, capillary, or rotational. Falling ball viscometers can easily be constructed from a graduated cylinder and appropriately sized ball bearings. The ball bearing material and the diameter can be varied. The experiment is conducted by filling the graduated cylinder with the fluid to be investigated and carefully dropping the ball through the fluid. After allowing the ball to reach steady state the velocity is measured. The absolute viscosity can then be calculated using Stokes' law [1]:

$$\eta = \left(\frac{2}{9}\right)\frac{(\rho_s - \rho)\, g\, R^2}{\upsilon} \qquad (3.2\text{-}2)$$

where η is the absolute viscosity, ρ_s is the density of the ball, ρ is the density of the fluid, g is the gravity constant (980 cm s^{-2}), R is the radius of the ball, and υ is the steady-state velocity of the ball. Commonly a falling or rolling ball viscometer is calibrated with a standard fluid that is similar in viscosity to the fluid of interest, and an instrument constant is then determined (k). Comparisons between the standard fluid and the unknown fluid can then be made using:

$$\upsilon = k\,(\rho_s - \rho)\,\theta \qquad (3.2\text{-}3)$$

where θ is the time of fall between two fiducial marks on the viscometer tube. This technique does have several limitations; the fluid must be Newtonian; the density of the fluid must be known, and the downward velocity of the ball should not exceed $\sim$1 cm s^{-1} to aid in time measurement. The falling ball method is generally used to measure absolute viscosities from 10^{-3} to 10^7 P [2].

Capillary viscometers are simple and inexpensive. They are normally constructed from glass and resemble a U-tube with a capillary section between two bulbs. The initial design originated with Ostwald, however the Cannon–Fenske moves the bulbs into the same vertical axis and is more common today. Capillary viscometers are normally immersed in a constant-temperature bath to regulate the sample temperature precisely during the experiment. To determine the viscosity, the fluid in the viscometer is drawn into the upper bulb using a vacuum. The vacuum is released and the time for the fluid to fall past the marks, above and below the bulb, is measured. The main driving force for flow in this type of viscometer is gravity, although pressure can be applied to one side of the viscometer to provide an additional driving force (i.e., increased head pressure) [1]. Since the driving pressure is governed by the difference in heights of the liquid in the viscometer, it is important to always use the same volume of liquid in each experiment. The kinematic viscosity can be calculated using [2]:

$$\nu = \frac{[\pi g(z_1 - z_2) D^4]}{128 L V_o} \Delta t \qquad (3.2\text{-}4)$$

where $(z_1 - z_2)$ is the change in height, D is the capillary inner diameter, L is the length of the capillary, and V_o is the volume between the fiducial marks. This equation only holds as long as the liquid behaves as a Newtonian fluid and the length to diameter ratio of the capillary tube is large.

Capillary viscometers measure the kinematic viscosity directly because the head pressure is generated by the weight of the fluid. In order to convert to absolute viscosity, the kinematic viscosity must be multiplied by the fluid density (Eq. (3.2-1)). Obviously, this requires additional experiments to determine fluid density so that the absolute viscosity can be calculated. The capillary type viscometer is normally used to measure kinematic viscosities spanning 4×10^{-3} to 1.6×10^2 St, with experimental times ranging from 200 to 800 s [1]. This range of kinematic viscosities corresponds to absolute viscosities of 6×10^{-3} to 2.4×10^2 P, assuming an average ionic liquid density of 1.5 g cm^{-3}.

The last type of widely used viscometer is the rotational viscometer. These can adopt a variety of geometries including concentric cylinders, cone and plate, and parallel disks. Of the three geometries, concentric cylinders is the most common, because it is well suited for low viscosity fluids [2]. Rotational viscometers consist of two main elements: a rotating element and a fixed element. The liquid to be measured is placed in the space between the two elements. The viscosity is determined by measuring the torque transferred between the two elements by the liquid. For the concentric cylinder geometry, the outer cylinder is often rotated at a fixed speed and the torque is measured on the fixed center cylinder immersed in the liquid. By measuring the angular speed of the rotating cylinder and the torque on the fixed cylinder, the fluid viscosity can be calculated using [2]:

$$\eta = \left\{ \frac{[\beta^2 - 1]}{[4\pi R_2^2 L_e]} \right\} \frac{T}{\omega_2} \qquad (3.2\text{-}5)$$

where β is the ratio of the cylinder radii, R_2 is the radius of the outer cylinder, L_e is the effective length of the cylinder, T is the torque applied to the rotating cylinder, and ω_2 is the rotational speed of the outer cylinder (2). The effective length of the cylinder (L_e) consists of the immersion depth of the center cylinder plus an end effect correction. This equation requires β be less than 1.2.

All three methods discussed above appear to provide equally high quality ionic liquid viscosity data. However, the rotational viscometer has the potential to provide additional information concerning the Newtonian behavior of the ionic liquids. The capillary method has been by far the most commonly used to generate the ionic liquid viscosity data found in the literature. This is probably due to its low cost and relative ease of use.

3.2.1.2 Ionic Liquid Viscosities

As a group, ionic liquids are more viscous than most common molecular solvents. Ionic liquid viscosities at room temperature range from a low of around 10 cP to values in excess of 500 cP. For comparison purposes the viscosities of water, ethylene glycol, and glycerol at room temperature are 0.890, 16.1, and 934 cP, respectively [3]. The room-temperature viscosity and density data for a wide variety of ionic liquids are listed in Tables 3.2-1–3.2-3. These tables are organized by the general type of ionic liquid. Table 3.2-2 contains data for non-haloaluminate alkylimidazolium ionic liquids; Table 3.2-2 contains data for the haloaluminate ionic liquids; and Table 3.2-3 contains data for other types of ionic liquids. There are multiple listings for several of the ionic liquids in these tables. These represent measurements by different researchers and were included to help emphasize the significant variability in the viscosity data found in the literature.

The viscosity of many ionic liquids is strongly dependent on temperature [4, 5]. For example, the viscosity of a 1-butyl-3-methylimidazolium hexafluorophosphate increases 27% with a 5 K change between 298 and 293 K [4]. Clearly some of the apparent variability in the literature data seen in Tables 3.2-1–3.2-3 may have resulted from errors associated with control of experimental temperature. However, much of this variability is likely the result of impurities in the ionic liquids. Recent work in the non-haloaluminate alkylimidazolium ionic liquids has established the ubiquitous nature of impurities in these ionic liquids, and demonstrated the dramatic impact relatively small amounts of impurities can have on ionic liquid viscosity [6, 7]. In one study a series of ionic liquids were prepared and purified using a variety of techniques. They were then analyzed for impurities and their physical properties evaluated. Chloride concentrations up to 6 wt.% were found for some of the preparation methods. Residual chloride concentrations between 1.5 and 6 wt.% raised the observed viscosity between 30 and 600% [6]. This work also showed the strong propensity of the non-haloaluminate alkylimidazolium ionic liquids to absorb water from laboratory air, and the significant solubility of water in these same ionic liquids (up to 14 wt.% for one of the ionic liquids under investigation). Surprisingly, as little as 2 wt.% (20 mol%) water (as well as other cosolvents, vide infra) reduced the observed viscosity of [BMIM][BF$_4$] by more than 50%. Given this information, it is highly likely that many of the ionic liquids listed in the tables contain significant concentrations of impurities (especially water). This, in turn, complicates the evaluation of the literature data, and, consequently, any conclusions drawn below must be used with care.

Within a series of non-haloaluminate ionic liquids containing the same cation, changing the anion clearly impacts the viscosity (Tables 3.2-1 and 3.2-3). The general order of increasing viscosity with respect to the anion is: bis(trifluoromethylsulfonyl)amide, tetrafluoroborate, trifluoromethyl acetate, triflate, heptafluoroethylsulfonate, heptafluoropropyl acetate, methyl acetate, mesylate, nonafluorobutylsulfonate. Obviously, this trend does not exactly correlate with anion size. This may be

Table 3.2-1 Viscosity and density data for non-haloaluminate alkylimidazolium-based ionic liquids

Cation	Anion	Temperature, K	Viscosity (η), cP	Density, g cm^{-3}	Reference
MIM$^+$	$(CF_3SO_2)_2N^-$	298	81		27
MIM$^+$	$(C_2F_5SO_2)_2N^-$	298	218		27
MMIM$^+$	$(CF_3SO_2)_2N^-$	293	44	1.559	28
2-M-MIM$^+$	BF_4^-	298	100		27
2-M-MIM$^+$	$(CF_3SO_2)_2N^-$	298	100		27
EIM$^+$	BF_4^-	298	41		27
EIM$^+$	ClO_4^-	298	112		27
EIM$^+$	PF_6^-	298	550		27
EIM$^+$	$CF_3SO_3^-$	298	58		27
EIM$^+$	$(CF_3SO_2)_2N^-$	298	54		27
EIM$^+$	$(C_2F_5SO_2)_2N^-$	298	133		27
2-M-EIM$^+$	BF_4^-	298	67		27
2-M-EIM$^+$	$(CF_3SO_2)_2N^-$	298	69		27
2-M-EIM$^+$	$(C_2F_5SO_2)_2N^-$	298	186		27
2-E-C$_6$H$_5$CH$_2$IM$^+$	$(CF_3SO_2)_2N^-$	298	252		27
2-E-C$_6$H$_5$CH$_2$IM$^+$	$(C_2F_5SO_2)_2N^-$	298	552		27
EMIM$^+$	BF_4^-	298	34	1.240	29
EMIM$^+$	BF_4^-	298	32	1.279	30
EMIM$^+$	BF_4^-	299	43		31
EMIM$^+$	BF_4^-	298	38	1.285	32
EMIM$^+$	BF_4^-	293	67		33
EMIM$^+$	$CH_3CO_2^-$	293	162		28
EMIM$^+$	$CF_3CO_2^-$	293	35	1.285	28
EMIM$^+$	$C_3F_7CO_2^-$	293	105	1.450	28
EMIM$^+$	$CH_3SO_3^-$	298	160	1.240	34
EMIM$^+$	$CF_3SO_3^-$	293	45	1.390	28
EMIM$^+$	$CF_3SO_3^-$	298	43	1.380	34
EMIM$^+$	$CF_3SO_3^-$	293	50		33
EMIM$^+$	$(CF_3SO_2)_2N^-$	293	34	1.520	28
EMIM$^+$	$(CF_3SO_2)_2N^-$	298	31	1.518	30
EMIM$^+$	$(CF_3SO_2)_2N^-$	298	34	1.510	35
EMIM$^+$	$(CF_3SO_2)_2N^-$	299	28		31
EMIM$^+$	$(CF_3SO_2)_2N^-$	298		1.519	36
EMIM$^+$	$(CF_3SO_2)_2N^-$	298	34	1.520	37
EMIM$^+$	$(CF_3SO_2)_2N^-$	298	32		38
EMIM$^+$	$(C_2F_5SO_2)_2N^-$	299	61		31
EMIM$^+$	$(C_2F_5SO_2)_2N^-$	298		1.519	39
EMIM$^+$	$(CF_3SO_2)(CF_3CO)N^-$	298	25	1.460	37
EMIM$^+$	$(CN)_2N^-$	298	21	1.060	40
EMIM$^+$	$CH_3BF_3^-$	298	47	1.154	41
EMIM$^+$	$C_2H_5BF_3^-$	298	72	1.133	41
EMIM$^+$	n-$C_3H_7BF_3^-$	298	54	1.107	41
EMIM$^+$	n-$C_4H_9BF_3^-$	298	83	1.082	41
EMIM$^+$	n-$C_5H_{11}BF_3^-$	298	88	1.065	41
EMIM$^+$	n-$CH_2CHBF_3^-$	298	41	1.161	41
EMIM$^+$	1.3 HF/1HF$_2^-$	298	5		42

(Continued)

Table 3.2-1 (Continued)

Cation	Anion	Temperature, K	Viscosity (η), cP	Density, g cm^{-3}	Reference
HO(CH$_2$)$_2$MIM$^+$	BF$_4^-$	293	91	1.330	43
HO(CH$_2$)$_2$MIM$^+$	PF$_6^-$	293	149	1.480	43
2-M-EMIM$^+$	(CF$_3$SO$_2$)$_2$N$^-$	293	88	1.495	28
5-M-EMIM$^+$	CF$_3$SO$_3^-$	293	51	1.334	28
5-M-EMIM$^+$	(CF$_3$SO$_2$)$_2$N$^-$	293	37	1.470	28
EEIM$^+$	CF$_3$CO$_2^-$	293	43	1.250	28
EEIM$^+$	CF$_3$SO$_3^-$	293	53	1.330	28
EEIM$^+$	(CF$_3$SO$_2$)$_2$N$^-$	293	35	1.452	28
5-M-EEIM$^+$	(CF$_3$SO$_2$)$_2$N$^-$	293	36	1.432	28
PMIM$^+$	(CF$_3$SO$_2$)$_2$N$^-$	299	60		31
PMIM$^+$	(CF$_3$SO$_2$)$_2$N$^-$	298		1.475	36
MeOEtMIM$^+$	BF$_4^-$	293	263	1.260	43
MeOEtMIM$^+$	PF$_6^-$	293	284	1.400	43
MeOEtMIM$^+$	CF$_3$SO$_3^-$	293	74	1.364	28
MeOEtMIM$^+$	(CF$_3$SO$_2$)$_2$N$^-$	293	54	1.496	28
BMIM$^+$	BF$_4^-$	298	115	1.140	44
BMIM$^+$	BF$_4^-$	293	154		33
BMIM$^+$	BF$_4^-$	298	219	1.120	45
BMIM$^+$	BF$_4^-$	298	92		46
BMIM$^+$	PF$_6^-$	293		1.363	47
BMIM$^+$	PF$_6^-$	298	207		4
BMIM$^+$	PF$_6^-$	298	320	1.370	44
BMIM$^+$	PF$_6^-$	298	270	1.368	48
BMIM$^+$	PF$_6^-$	298		1.368	36
BMIM$^+$	PF$_6^-$	293	371		33
BMIM$^+$	PF$_6^-$	298	450	1.360	45
BMIM$^+$	PF$_6^-$	298	173		46
BMIM$^+$	CH$_3$CO$_2^-$	298	440		38
BMIM$^+$	CF$_3$CO$_2^-$	293	73	1.209	28
BMIM$^+$	CF$_3$CO$_2^-$	298		1.296	39
BMIM$^+$	CF$_3$CO$_2^-$	298	70		38
BMIM$^+$	C$_3$F$_7$CO$_2^-$	293	182	1.333	28
BMIM$^+$	CF$_3$SO$_3^-$	293	90	1.290	28
BMIM$^+$	CF$_3$SO$_3^-$	298	90		46
BMIM$^+$	C$_4$F$_9$SO$_3^-$	293	373	1.427	28
BMIM$^+$	C$_4$F$_9$SO$_3^-$	293	99		39
BMIM$^+$	(CF$_3$SO$_2$)$_2$N$^-$	293	52	1.429	28
BMIM$^+$	(CF$_3$SO$_2$)$_2$N$^-$	293		1.436	39
BMIM$^+$	(CF$_3$SO$_2$)$_2$N$^-$	298		1.436	36
BMIM$^+$	(CF$_3$SO$_2$)$_2$N$^-$	298	69	1.430	45
BMIM$^+$	(CF$_3$SO$_2$)$_2$N$^-$	298	47		46
BMIM$^+$	NO$_3^-$	293	266		33
BMIM$^+$	ICl$_2^-$	298	50	1.780	49
BMIM$^+$	Br$_3^-$	298	93	1.702	49
BMIM$^+$	IBr$_2^-$	298	57	1.547	49
BMIM$^+$	Cl$^-$	293	40890		33
BMIM$^+$	Cl$^-$	298		1.080	45

(Continued)

Table 3.2-1 (Continued)

Cation	Anion	Temperature, K	Viscosity (η), cP	Density, g cm^{-3}	Reference
BMIM$^+$	I$^-$	298	1110	1.440	45
BMIM$^+$	SbF$_6^-$	298	108		46
iBMIM$^+$	(CF$_3$SO$_2$)$_2$N$^-$	293	83	1.428	28
2-M-BMIM$^+$	BF$_4^-$	298	243		46
2-M-BMIM$^+$	(CF$_3$SO$_2$)$_2$N$^-$	298	88		46
BEIM$^+$	CF$_3$CO$_2^-$	293	89	1.183	28
BEIM$^+$	CH$_3$SO$_3^-$	298		1.140	34
BEIM$^+$	CF$_3$SO$_3^-$	298		1.270	34
BEIM$^+$	C$_4$F$_9$SO$_3^-$	293	323	1.427	28
BEIM$^+$	(CF$_3$SO$_2$)$_2$N$^-$	293	48	1.404	28
PMIM$^+$	PF$_6^-$	298		1.326	36
PMIM$^+$	(CF$_3$SO$_2$)$_2$N$^-$	298		1.403	36
2-M-PMIM$^+$	(CF$_3$SO$_2$)$_2$N$^-$	299	60		31
MNPIM$^+$	BF$_4^-$	295	4638	1.230	32
MNPIM$^+$	(CF$_3$SO$_2$)$_2$N$^-$	295	161	1.500	32
HMIM$^+$	BF$_4^-$	293	314		33
HMIM$^+$	PF$_6^-$	298		1.307	47
HMIM$^+$	PF$_6^-$	298		1.302	39
HMIM$^+$	PF$_6^-$	298		1.292	36
HMIM$^+$	PF$_6^-$	293	690		33
HMIM$^+$	PF$_6^-$	298	585	1.290	45
HMIM$^+$	(CF$_3$SO$_2$)$_2$N$^-$	298		1.373	39
HMIM$^+$	(CF$_3$SO$_2$)$_2$N$^-$	298		1.372	36
HMIM$^+$	(CF$_3$SO$_2$)$_2$N$^-$	298	68		38
HMIM$^+$	NO$_3^-$	293	804		33
HMIM$^+$	Cl$^-$	293	18000		33
HMIM$^+$	Cl$^-$	298	716	1.030	45
2-M-HMIM$^+$	(CF$_3$SO$_2$)$_2$N$^-$	298	131		38
MeO(CH$_2$)$_2$O(CH$_2$)$_2$MIM$^+$	BF$_4^-$	293	263	1.220	43
MeO(CH$_2$)$_2$O(CH$_2$)$_2$MIM$^+$	PF$_6^-$	293	426	1.320	43
MeO(CH$_2$)$_2$O(CH$_2$)$_2$MIM$^+$	Cl$^-$	293	613	1.140	43
HpMIM$^+$	PF$_6^-$	298		1.262	36
HpMIM$^+$	(CF$_3$SO$_2$)$_2$N$^-$	298		1.344	36
OMIM$^+$	BF$_4^-$	293	439		33
OMIM$^+$	PF$_6^-$	298		1.237	47
OMIM$^+$	PF$_6^-$	298		1.233	39
OMIM$^+$	PF$_6^-$	298		1.237	36
OMIM$^+$	PF$_6^-$	293	866		33
OMIM$^+$	PF$_6^-$	298	682	1.220	45
OMIM$^+$	CF$_3$SO$_3^-$	293	492		33
OMIM$^+$	(CF$_3$SO$_2$)$_2$N$^-$	298	93	1.317	39
OMIM$^+$	(CF$_3$SO$_2$)$_2$N$^-$	298		1.320	36
OMIM$^+$	NO$_3^-$	293	1238		33
OMIM$^+$	Cl$^-$	293	33070		33
OMIM$^+$	Cl$^-$	298	337	1.000	45
NMIM$^+$	PF$_6^-$	298		1.212	36
NMIM$^+$	(CF$_3$SO$_2$)$_2$N$^-$	298		1.299	36

(Continued)

Table 3.2-1 (Continued)

Cation	Anion	Temperature, K	Viscosity (η), cP	Density, g cm^{-3}	Reference
DMIM$^+$	BF$_4^-$	293	928		33
DMIM$^+$	CF$_3$SO$_3^-$	293	981		33
DMIM$^+$	(CF$_3$SO$_2$)$_2$N$^-$	298		1.271	36
DEIM$^+$	CF$_3$SO$_3^-$	293		1.100	34
CF$_3$CH$_2$MIM$^+$	(CF$_3$SO$_2$)$_2$N$^-$	293	248	1.656	28
ECNMIM$^+$	BF$_4^-$	293	66	2.150	46
PCNMIM$^+$	BF$_4^-$	293	230	1.870	46
BCNMIM+	Cl$^-$	293	5222	1.610	46
BCNMIM$^+$	PF$_6^-$	293	2181	1.990	46
BCNMIM+	BF$_4^-$	293	553	1.710	46
MTMSiMIM$^+$	BF$_4^-$	295	631	1.220	32
MTMSiMIM$^+$	(CF$_3$SO$_2$)$_2$N$^-$	295	98	1.460	32
Ph(CH$_2$)MIM$^+$	(CF$_3$SO$_2$)$_2$N$^-$	298		1.491	36
Ph(CH$_2$)$_2$MIM$^+$	(CF$_3$SO$_2$)$_2$N$^-$	298		1.470	36
Ph(CH$_2$)$_3$MIM$^+$	PF$_6^-$	298		1.407	36
Ph(CH$_2$)$_3$MIM$^+$	(CF$_3$SO$_2$)$_2$N$^-$	298		1.455	36

due to the importance of the effect of other anion properties on the viscosity, such as their ability to form weak hydrogen bonds with the cation.

The viscosity of the non-haloaluminate ionic liquids is also affected by the identity of the organic cation. For ionic liquids with the same anion the trend is: larger alkyl substituents on the imidazolium cation lead to more viscous fluids. For instance, the non-haloaluminate ionic liquids composed of substituted imidazolium cations and the bis(trifluoromethylsulfonyl)amide anion exhibit an increase in viscosity from [EMIM]$^+$, [EEIM]$^+$, [EMM(5)IM]$^+$, [BEIM]$^+$, [BMIM]$^+$, [PMMIM]$^+$, to [2-M-EMIM]$^+$ (Table 3.2-1). Using the size of the cation as the sole criterion, the [BEIM]$^+$ and [BMIM]$^+$ cations from this series would appear to be transposed and the [2-M-EMIM]$^+$ would have been expected much earlier in the series. Given the limited data set, potential problems with impurities, and experimental differences between laboratories, we are unable to propose an explanation for the observed disparities.

The haloaluminate ionic liquids are prepared by mixing two solids, an organic chloride and an aluminum halide (e.g., [EMIM]Cl and AlCl$_3$). These two solids react to form ionic liquids with a single cation and a mix of anions. The anion composition depends strongly on the relative molar amounts of the two ingredients used in the preparation. The effect of anionic composition on the viscosity of haloaluminate ionic liquids has long been recognized. Figure 3.2-1 shows the absolute viscosities of the [EMIM]Cl-AlCl$_3$ ionic liquids at 303 K over a range of compositions. When the [EMIM]Cl is below 50 mol% the viscosity is relatively constant, only varying from 14 to 18 cP. However, when the [EMIM]Cl exceeds 50 mol%, the absolute viscosity begins to increase, eventually rising to over 190 cP at 67 mol% [EMIM]Cl [8]. This dramatic increase in viscosity is strongly correlated to the corresponding growth in

Table 3.2-2 Viscosity and density data for binary haloaluminate ionic liquids

Ionic liquid system	Cation	Anion(s)	Temp, K	Viscosity (η), cP	Density, g cm^{-3}	Reference
34.0–66.0 mol% MMIMCl–AlCl$_3$	MMIM$^+$	Al$_2$Cl$_7^-$	298	17	1.404	8a
34.0–66.0 mol% EMIMCl–AlCl$_3$	EMIM$^+$	Al$_2$Cl$_7^-$	298	14	1.389	8a
50.0–50.0 mol% EMIMCl–AlCl$_3$	EMIM$^+$	AlCl$_4^-$	298	18	1.294	8a
60.0–40.0 mol% EMIMCl–AlCl$_3$	EMIM$^+$	Cl$^-$, AlCl$_4^-$	298	47	1.256	8a
34.0–66.0 mol% EMIMBr–AlBr$_3$	EMIM$^+$	Al$_2$Br$_7^-$	298	32	2.219	50a,b
60.0–40.0 mol% EMIMBr–AlBr$_3$	EMIM$^+$	Br$^-$, AlBr$_4^-$	298	67	1.828	50a,b
40.0–60.0 mol% PMIMCl–AlCl$_3$	PMIM$^+$	AlCl$_4^-$, Al$_2$Cl$_7^-$	298	18	1.351	8a
50.0–50.0 mol% PMIMCl–AlCl$_3$	PMIM$^+$	AlCl$_4^-$	298	27	1.262	8a
34.0–66.0 mol% BMIMCl–AlCl$_3$	BMIM$^+$	Al$_2$Cl$_7^-$	298	19	1.334	8a
50.0–50.0 mol% BMIMCl–AlCl$_3$	BMIM$^+$	AlCl$_4^-$	298	27	1.238	8a
34.0–66.0 mol% BBIMCl–AlCl$_3$	BBIM$^+$	Al$_2$Cl$_7^-$	298	24	1.252	8a
50.0–50.0 mol% BBIMCl–AlCl$_3$	BBIM$^+$	AlCl$_4^-$	298	38	1.164	8a
33.3–66.7 mol% MPCl–AlCl$_3$	MP$^+$	Al$_2$Cl$_7^-$	298	21	1.441	51
33.3–66.7 mol% EPCl–AlCl$_3$	EP$^+$	Al$_2$Cl$_7^-$	298	18	1.408	51
33.3–66.7 mol% EPBr–AlCl$_3$	EP$^+$	Al$_2$Cl$_x$Br$_{7-x}^-$	298	22	1.524	51
33.3–66.7 mol% EPBr–AlCl$_3$	EP$^+$	Al$_2$Cl$_x$Br$_{7-x}^-$	298	25		52
33.3–66.7 mol% EPBr–AlBr$_3$	EP$^+$	Al$_2$Br$_7^-$	298	50	2.200	26
33.3–66.7 mol% PPCl–AlCl$_3$	PP$^+$	Al$_2$Cl$_7^-$	298	18	1.375	51
33.3–66.7 mol% BPCl–AlCl$_3$	BP$^+$	Al$_2$Cl$_7^-$	298	21	1.346	51
33.3–66.7 mol% PMPCl–AlCl$_3$	N-propyl-4-methyl P$^+$	Al$_2$Cl$_7^-$	298	18		46
40–60 mol% O-methylisourea HCl–AlCl$_3$	protonated O-methylisourea	Al$_2$Cl$_7^-$	298	27		46
40–60 mol% acetamidine HCl–AlCl$_3$	protonated acetamidine	Al$_2$Cl$_7^-$	298	36		46
33.3–66.7 mol% phenyltrimethylammonium Cl–AlCl$_3$	phenyltrimethylammonium	Al$_2$Cl$_7^-$	298	33		46
33.3–66.7 mol% trimethylsulfonium Br–AlCl$_3$	trimethylsulfonium	Al$_2$Cl$_x$Br$_{7-x}^-$	298	48		46
33.3–66.7 mol% benzyltrimethylammonium Cl–AlCl$_3$	benzyltrimethylammonium	Al$_2$Cl$_7^-$	298	64		46
33.3–66.7 mol% tetrabutylphosphonium Cl–AlCl$_3$	tetrabutylphosphonium	Al$_2$Cl$_7^-$	298	90		46
33.3–66.7 mol% benzyltriethylammonium Cl–AlCl$_3$	benzyltriethylammonium	Al$_2$Cl$_7^-$	298	152		46
33.3–66.7 mol% benzyltri-n-propylammonium Cl–AlCl$_3$	benzyltri-n-propylammonium	Al$_2$Cl$_7^-$	298	226		46
40–60 mol% ethylbenzimidate HCl–AlCl$_3$	protonated ethylbenzimidate	Al$_2$Cl$_7^-$	298	223		46

Table 3.2-3 Viscosity and density data for other room-temperature ionic liquids

Cation	Anion	Temperature, K	Viscosity (η), cP	Density, g cm^{-3}	Reference
Ammonium					
$(CH_3)_3(C_2H_5)N^+$	$(CF_3SO_2)(CF_3CO)N^-$	298	51	1.400	37
$(CH_3)_3(allyl)N^+$	$(CF_3SO_2)(CF_3CO)N^-$	298	42	1.380	37
$(CH_3)_3(isopropyl)N^+$	$(CF_3SO_2)(CF_3CO)N^-$	298	90	1.410	37
$(CH_3)_3(propyl)N^+$	$(CF_3SO_2)_2N^-$	298	72	1.440	37
$(CH_3)_3(propyl)N^+$	$(CF_3SO_2)(CF_3CO)N^-$	298	45	1.380	37
$(C_2H_5)_4N^+$	$(CF_3SO_2)(CF_3CO)N^-$	298	60	1.370	37
$(n-C_3H_7)(CH_3)_3N^+$	$(CF_3SO_2)_2N^-$	298	72	1.440	35
$(butyl)_4N^+$	docusate	298	12100		38
$(n-C_6H_{13})(C_2H_5)_3N^+$	$(CF_3SO_2)_2N^-$	298	167	1.270	25
$(n-C_8H_{17})(C_2H_5)_3N^+$	$(CF_3SO_2)_2N^-$	298	202	1.250	25
$(n-C_8H_{17})(C_4H_9)_3N^+$	$(CF_3SO_2)_2N^-$	298	574	1.120	25
$(CH_3)_3(CH_3OCH_2)N^+$	$(CF_3SO_2)_2N^-$	298	50	1.510	35
$(CH_3)(C_2H_5)_2(C_4H_9)N^+$	$CF_3BF_3^-$	298	210	1.180	53
$(CH_3)_2(C_2H_5)(CH_3OC_2H_4)N^+$	$CF_3BF_3^-$	298	97	1.270	53
$(CH_3)(C_2H_5)_2(CH_3OC_2H_4)N^+$	$CF_3BF_3^-$	298	108	1.250	53
$(C_2H_5)_3(CH_3OC_2H_4)N^+$	$CF_3BF_3^-$	298	151	1.220	53
$(CH_3)(C_2H_5)_2(C_4H_9)N^+$	$C_2F_5BF_3^-$	298	104	1.250	53
$(CH_3)_2(C_2H_5)(CH_3OC_2H_4)N^+$	$C_2F_5BF_3^-$	298	58	1.330	53
$(CH_3)(C_2H_5)_2(CH_3OC_2H_4)N^+$	$C_2F_5BF_3^-$	298	68	1.310	53
$(C_2H_5)_3(CH_3OC_2H_4)N^+$	$C_2F_5BF_3^-$	298	87	1.280	53
$(CH_3)_2(C_2H_5)(CH_3OC_2H_4)N^+$	$n-C_3F_7BF_3^-$	298	76	1.410	53
$(CH_3)_2(C_2H_5)(CH_3OC_2H_4)N^+$	$n-C_3F_7BF_3^-$	298	70	1.390	53
$(CH_3)(C_2H_5)_2(CH_3OC_2H_4)N^+$	$n-C_3F_7BF_3^-$	298	88	1.370	53

(Continued)

Table 3.2-3 (Continued)

Cation	Anion	Temperature, K	Viscosity (η), cP	Density, g cm^{-3}	Reference
$(C_2H_5)_3(CH_3OC_2H_4)N^+$	n-$C_3F_7BF_3^-$	298	91	1.340	53
$(CH_3)_2(C_2H_5)(CH_3OC_2H_4)N^+$	n-$C_4F_9BF_3^-$	298	102	1.450	53
$(CH_3)(C_2H_5)_2(CH_3OC_2H_4)N^+$	n-$C_4F_9BF_3^-$	298	118	1.420	53
$(C_2H_5)_3(CH_3OC_2H_4)N^+$	n-$C_4F_9BF_3^-$	298	135	1.400	53
$(CH_3)_2(C_2H_5)(CH_3OC_2H_4)N^+$	BF_4^-	298	335	1.210	53
$(CH_3)(C_2H_5)_2(CH_3OC_2H_4)N^+$	BF_4^-	298	426	1.200	53
$(CH_3)(C_2H_5)_2(C_3H_7)N^+$	$(CF_3SO_2)_2N^-$	298	94	1.420	53
$(CH_3)(C_2H_5)_2(C_4H_9)N^+$	$(CF_3SO_2)_2N^-$	298	120	1.380	53
$(CH_3)_2(C_2H_5)(CH_3OC_2H_4)N^+$	$(CF_3SO_2)_2N^-$	298	60	1.450	53
$(CH_3)(C_2H_5)_2(CH_3OC_2H_4)N^+$	$(CF_3SO_2)_2N^-$	298	69	1.420	53
$(C_2H_5)_3(CH_3OC_2H_4)N^+$	$(CF_3SO_2)_2N^-$	298	85	1.400	53
$(C_2H_5)_3(menthoxymethyl)N^+$	$(CF_3SO_2)_2N^-$	303	876	1.250	54
$(C_2H_5)_2(CH_3)(menthoxymethyl)N^+$	$(CF_3SO_2)_2N^-$	303	754	1.260	54
$(C_2H_5)(CH_3)_2(menthoxymethyl)N^+$	$(CF_3SO_2)_2N^-$	303	714	1.270	54
$(C_4H_9)(CH_3)_2(menthoxymethyl)N^+$	BF_4^-	303	745	1.240	54
$(C_6H_{13})(CH_3)_2(menthoxymethyl)N^+$	BF_4^-	303	774	1.210	54
$(C_7H_{15})(CH_3)_2(menthoxymethyl)N^+$	BF_4^-	303	787	1.190	54
$(C_8H_{17})(CH_3)_2(menthoxymethyl)N^+$	BF_4^-	303	806	1.180	54
$(C_9H_{19})(CH_3)_2(menthoxymethyl)N^+$	BF_4^-	303	829	1.170	54
$(C_{10}H_{21})(CH_3)_2(menthoxymethyl)N^+$	BF_4^-	303	840	1.150	54
$(C_{11}H_{23})(CH_3)_2(menthoxymethyl)N^+$	BF_4^-	303	844	1.140	54
(3-chloro-2-hydroxypropyl)$(CH_3)_3N^+$	$(CF_3SO_2)_2N^-$	298	237		46
Pyridinium					
EP^+	$EtSO_4^-$	298	137		38
1-ethyl-3-methyl P^+	$EtSO_4^-$	298	150		38

(Continued)

BP$^+$	BF$_4^-$	298	103	1.220	30
BP$^+$	(CF$_3$SO$_2$)$_2$N$^-$	298	56.8	1.449	30
1-butyl-3-methyl P$^+$	BF$_4^-$	298	177		38
1-butyl-3-methyl P$^+$	(CF$_3$SO$_2$)$_2$N$^-$	298	63		38
1-hexyl P$^+$	(CF$_3$SO$_2$)$_2$N$^-$	298	80		38
1-hexyl-3-methyl P$^+$	(CF$_3$SO$_2$)$_2$N$^-$	298	85		38
1-hexyl-3,5-dimethyl P$^+$	(CF$_3$SO$_2$)$_2$N$^-$	298	104		38
1-hexyl-2-ethyl-3,5-dimethyl P$^+$	(CF$_3$SO$_2$)$_2$N$^-$	298	245		38
1-hexyl-2-propyl-3,5-dimethyl P$^+$	(CF$_3$SO$_2$)$_2$N$^-$	298	206		38
1-hexyl-4-(dimethylamino) P$^+$	(CF$_3$SO$_2$)$_2$N$^-$	298	111		38
1-hexyl-3-methyl-4-(dimethylamino) P$^+$	(CF$_3$SO$_2$)$_2$N$^-$	298	112		38
1-octyl-3-methyl P$^+$	(CF$_3$SO$_2$)$_2$N$^-$	298	112		38
Pyrrolidinium					
BM pyrrolidinium	(CF$_3$SO$_2$)$_2$N$^-$	298	76		46
1,1-dimethyl-pyrrolidinium	(CF$_3$SO$_2$)(CF$_3$CO)N$^-$	298	80	1.430	37
1-propyl-1-methyl-pyrrolidinium	(CF$_3$SO$_2$)$_2$N$^-$	298	63	1.450	55
1-propyl-1-methyl-pyrrolidinium	(CN)$_2$N$^-$	298	45	0.920	40
1-butyl-1-methyl-pyrrolidinium	(CF$_3$SO$_2$)$_2$N$^-$	298	85	1.410	55
1-butyl-1-methyl-pyrrolidinium	(CN)$_2$N$^-$	298	50	0.950	40
1-hexyl-1-methyl-pyrrolidinium	(CN)$_2$N$^-$	298	45	0.920	40
Sulfonium					
(CH$_3$)$_3$S$^+$	HBr$^-$	298	20.5	1.740	26
(CH$_3$)$_3$S$^+$	HBr$^-$/H$_2$Br$_3^-$	298	8.3	1.790	26
(CH$_3$)$_3$S$^+$	Al$_2$Cl$_7^-$	298	39.3	1.400	26
(CH$_3$)$_3$S$^+$	Al$_2$Cl$_6$Br$^-$	298	54.9	1.590	26
(CH$_3$)$_3$S$^+$	Al$_2$Br$_7^-$	298	138	2.400	26

(Continued)

Table 3.2-3 (Continued)

Cation	Anion	Temperature, K	Viscosity (η), cP	Density, g cm^{-3}	Reference
(CH$_3$)$_3$S$^+$	(CF$_3$SO$_2$)$_2$N$^-$	318	44	1.580	35
(CH$_3$)$_3$S$^+$	N(CN)$_2^-$	293	27.2		56
(CH$_3$)(C$_2$H$_5$)$_2$S$^+$	N(CN)$_2^-$	293	22.9		56
(CH$_3$)(C$_3$H$_7$)$_2$S$^+$	N(CN)$_2^-$	293	29.5		56
(CH$_3$)(C$_4$H$_9$)$_2$S$^+$	N(CN)$_2^-$	293	60		56
(CH$_3$)$_2$(C$_2$H$_5$)S$^+$	N(CN)$_2^-$	293	25.3		56
(C$_2$H$_5$)$_3$S$^+$	(CF$_3$SO$_2$)$_2$N$^-$	298	30		35
(C$_2$H$_5$)$_3$S$^+$	(CF$_3$SO$_2$)$_2$N$^-$	298	33	1.460	46
(C$_2$H$_5$)$_3$S$^+$	N(CN)$_2^-$	293	20.9		56
(C2H$_5$)(C$_3$H$_7$)$_2$S$^+$	N(CN)$_2^-$	293	29.4		56
(CH$_3$)$_2$(C$_4$H$_9$)S$^+$	N(CN)$_2^-$	293	51.7		56
(n-C$_4$H$_9$)$_3$S$^+$	(CF$_3$SO$_2$)$_2$N$^-$	298	75	1.290	35
dimethylpropargyl S$^+$	(CF$_3$SO$_2$)$_2$N$^-$	298	103		46
(ethoxycarbonylmethyl) Dimethyl S$^+$	(CF$_3$SO$_2$)$_2$N$^-$	298	162		46
Phenyltrimethylammonium					
PTA$^+$	Cl$^-$/AlCl$_4^-$	298	30		46
Diacetone acrylamide					
DA$^+$	CH$_3$COO$^-$	293	27.2		57
DA$^+$	CF$_3$CO$_2^-$	293	6.1		57
DA$^+$	BF$_4^-$	373	25.5		57
DA$^+$	Cl$^-$	358	18.5		57
Nicotinic acid esters					
1-ethyl-nicotinic acid ethyl ester	EtSO$_4^-$	298	3173		38
1-butyl-nicotinic acid butyl ester	(CF$_3$SO$_2$)$_2$N$^-$	298	531		38
Miscellaneous					
acetylcholine	(CF$_3$SO$_2$)$_2$N$^-$	298	154		46
1-ethylthiazolium	CF$_3$SO$_3^-$	298		1.500	34

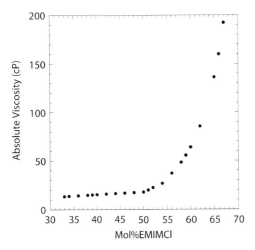

Fig. 3.2-1 Change in the absolute viscosity (cP) as a function of the [EMIM]Cl mol% in an [EMIM]Cl-AlCl$_3$ ionic liquid at 303 K.

chloride ion concentration as the [EMIM]Cl mol% increases, and appears to be the result of hydrogen bonding between the Cl$^-$ anions and the hydrogen atoms on the imidazolium cation ring [9–12].

The size of the cation in the chloroaluminate ionic liquids also appears to have an impact on the viscosity. For ionic liquids with the same anion(s) and compositions, the trend is for greater viscosity with larger cation size (Table 3.2-2). An additional contributing factor to the effect of the cation on viscosity is the asymmetry of the alkyl substitution. Highly asymmetric substitution has been identified as important to obtain low viscosities [8].

The addition of cosolvents to ionic liquids can result in dramatic reductions in the viscosity without changing the cations or anions in the system. The haloaluminate ionic liquids present a challenge due to the reactivity of the ionic liquid. Nonetheless, several compatible co-solvents have been investigated, including benzene, dichloromethane, and acetonitrile [13–17]. The addition of as little as 5 wt.% acetonitrile or 15 wt.% of benzene or methylene chloride was able to reduce the absolute viscosity by 50% for [EMIM]Cl-AlCl$_3$ ionic liquids with less than 50 mol% AlCl$_3$ [13]. Non-haloaluminate ionic liquids have also been studied with a range of co-solvents including water, acetone, ethanol, methanol, butanone, ethyl acetate, toluene, and acetonitrile [6, 18–22]. The ionic liquid response is similar to that observed in the haloaluminate ionic liquids. The addition of as little as 20 mol% co-solvent reduced the viscosity of a [BMIM][BF$_4$] melt by 50% [6].

3.2.2
Density of Ionic Liquids

Densities are perhaps the most straightforward and unambiguous physical property of ionic liquids to determine. Given a quality analytical balance and good volumetric glassware the density of an ionic liquid can be measured gravimetrically (i.e., the sample can be weighed).

3.2.2.1 Density Measurement

To measure the density properly with a minimal amount of sample, a pycnometer should be employed. A pycnometer removes the ambiguity of measuring the bottom of the meniscus in a piece of glassware calibrated with aqueous solutions that have potentially very different surface tensions. The most common types of pycnometers are the Ostwald–Sprengel and the Weld or stopper pycnometer. These devices are generally constructed of glass and consist of a reservoir connected to a capillary or capillaries with fiducial marks. The pycnometer is weighed while empty, filled with the fluid of interest, and allowed to thermally equilibrate. The fluid above the fiducial marks is removed and the pycnometer is weighed [23, 24]. Pycnometers must be calibrated prior to use to determine the exact volume. The density is then calculated by dividing the mass of the fluid by the pycnometer volume.

3.2.2.2 Ionic Liquid Densities

The reported densities of ionic liquids vary between 1.12 g cm^{-3} for [(n-C_8H_{17})(C_4H_9)$_3$N][(CF_3SO_2)$_2$N] and 2.4 g cm^{-3} for a 34–66 mol% [(CH_3)$_3$S]Br-AlBr$_3$ ionic liquid [25, 26]. The ionic liquid densities appear to be the least sensitive physical property to variations in temperature. For example, a 5 K change in temperature from 298 to 303 K results in only a 0.3% decrease in the density for a 50.0–50.0 mol% [EMIM]Cl–AlCl$_3$ [8]. In addition, the impact of impurities appears to be far less dramatic than in the case of viscosity. The density of ionic liquids seems to vary linearly with wt.% of impurities. For example, 20 wt.% water (75 mol%) in [BMIM][BF$_4$] results in only a 4% decrease in density [13].

Within the binary haloaluminate ionic liquids, increasing the mole percent of the imidazolium salt decreases the density of the liquid (Table 3.2-2). The bromoaluminate ionic liquids are substantially denser than their chloroaluminate counter parts, being between 0.57 and 0.83 g cm^{-3} denser than the analogous chloroaluminate ionic liquids (Table 3.2-2). Variation of the substituents on the imidazolium cation in the chloroaluminate ionic liquids has been shown to affect the density based upon the cation size [8].

Within a series of non-haloaluminate ionic liquids containing the same cation species, increasing anion mass corresponds to increasing ionic liquid density (Tables 3.2-1 and 3.2-3). Generally, the order of increasing density for ionic liquids composed of a single cation is mesylate ≈ tetrafluorborate < trifluoromethyl acetate < triflate < heptafluoropropyl acetate < bis(trifluoromethylsulfonyl)amide.

The density of the non-haloaluminate ionic liquids is also affected by the identity of the organic cation. Like the haloaluminate ionic liquids, the density decreases as the size of the cation increases. For instance, in the non-haloaluminate ionic liquids composed of substituted imidazolium cations and the triflate anion the density decreases from 1.390 g cm^{-3} for [EMIM]$^+$ to 1.334 g cm^{-3} for the [EMM(5)IM]$^+$, to 1.330 g cm^{-3} for [EEIM]$^+$, to 1.290 g cm^{-3} for [BMIM]$^+$, and 1.270 g cm^{-3} for [BEIM]$^+$ (Table 3.2-1).

References

1. J. R. Van Wazer, J. W. Lyons, K. Y. Kim, R. E. Colwell, *Viscosity and Flow Measurement: A Laboratory Handbook of Rheology*, Interscience Publishers-John Wiley, New York, **1963**, Ch. 2, 4, 5.
2. G. E. Leblanc, R. A. Secco, M. Kostic, in *Mechanical Variables Measurement: Solid, Fluid, and Thermal*, J. G. Webster (Ed.), CRC Press: Boca Raton, **2000**, Ch. 11.
3. *Handbook of Chemistry and Physics*, 82nd Edn., D. R. Linde (Ed.), CRC Press, New York, **2001**, pp. 6-182–6-186.
4. S. N. Baker, G. A. Baker, M. A. Kane, F. V. Bright, *J. Phys. Chem. B*, **2001**, *105*, 9663.
5. K. R. Harris, L. A. Woolf, M. Kanakubo, *J. Chem. Eng. Data*, **2005**, *50*, 1777.
6. K. R. Seddon, A. Stark, M. J. Torres, *Pure Appl. Chem.*, **2000**, *72*, 2275.
7. J. A. Widegren, A. Laesecke, J. W. Magee, *ChemComm*, **2005**, 1610.
8. (a) J. S. Wilkes, J. A. Levisky, R. A. Wilson, C. L. Hussey, *Inorg. Chem.*, **1982**, *21*, 1263; (b) A. A. Fannin Jr., D. A. Floreani, L. A. King, J. S. Landers, B. J. Piersma, D. J. Stech, R. J. Vaughn, J. S. Wilkes, J. L. Williams, *J. Phys. Chem.*, **1984**, *88*, 2614.
9. C. J. Dymek, D. A. Grossie, A. V. Fratini, W. W. Adams, *J. Mol. Struct.*, **1989**, *213*, 25.
10. C. J. Dymek, J. J. Stewart, *Inorg. Chem.*, **1989**, *28*, 1472.
11. A. G. Avent, P. A. Chaloner, M. P. Day, K. R. Seddon, T. Welton, in *Proceedings of the Seventh International Symposium on Molten Salts*, C. L. Hussey, J. S. Wilkes, S. N. Flengas, Y. Ito (Eds.), The Electrochemical Society: Pennington NJ, **1990**, Vol. 90-17, pp. 98–133.
12. A. G. Avent, P. A. Chaloner, M. P. Day, K. R. Seddon, T. Welton, *J. Chem. Soc., Dalton Trans.*, **1994**, 3405.
13. R. L. Perry, K. M. Jones, W. D. Scott, Q. Liao, C. L. Hussey, *J. Chem. Eng. Data*, **1995**, *40*, 615.
14. Q. Liao, C. L. Hussey, *J. Chem. Eng. Data*, **1996**, *41*, 1126.
15. N. Papageorgiou, Y. Athanassov, M. Armand, P. Bonhôte, H. Pattersson, A. Azam, M. Grätzel, *J. Electrochem. Soc.*, **1996**, *143*, 3099.
16. R. Moy, R. P. Emmenegger, *Electrochim. Acta*, **1992**, *37*, 1061.
17. J. Robinson, R. C. Bugle, H. L. Chum, D. Koran, R. A. Osteryoung, *J. Am. Chem. Soc.*, **1979**, *101*, 3776.
18. J. Wang, Y. Tian, Y. Zhao, K. Zhuo, *Green Chem.*, **2003**, *5*, 618.
19. J. Zhang, W. Wu, T. Jiang, H. Gao, Z. Liu, J. He, B. Han, *J. Chem. Eng. Data*, **2003**, *48*, 1315.
20. A. Chagnes, A. Tougui, B. Carŕe, N. Ranganathan, D. Lemordant, *J. Solution Chem.*, **2004**, *33*, 247.
21. J. Wang, A. Zhu, Y. Zhao, K. Zhuo, *J. Solution Chem.*, **2005**, *34*, 585.
22. H. Xu, D. Zhao, P. Xu, F. Liu, G. Gao, *J. Chem. Eng. Data*, **2005**, *50*, 133.
23. H. Eren, in *Mechanical Variables Measurement: Solid, Fluid, and Thermal*, J. G. Webster (Ed.), CRC Press: Boca Raton, **2000**, Ch. 2.
24. D. P. Shoemaker, C. W. Garland, J. I. Steinfeld, J. W. Nibler, *Experiments in Physical Chemistry*, 4th edn., McGraw-Hill: New York, **1981**, exp. 11.
25. (a) J. Sun, M. Forsyth, D. R. MacFarlane, *Molten Salt Forum*, **1998**, *5–6*, 585; (b) J. Sun, M. Forsyth, D. R. MacFarlane, *J. Phys. Chem. B.*, **1998**, *102*, 8858.
26. M. Ma, K. E. Johnson, in *Proceedings of the Ninth International Symposium on Molten Salts*, C. L. Hussey, D. S. Newman, G. Mamantov, Y. Ito (Eds.), The Electrochemical Society: Pennington NJ, **1994**, Vol. 94-13, pp. 179–186.
27. H. Ohno, M. Yoshizawa, *Solid State Ionics*, **2002**, *154–155*, 303.
28. P. Bonhôte, A.-P. Dias, N. Papageorgiou, K. Kalyanasundaram, M. Grätzel, *Inorg. Chem.*, **1996**, *35*, 1168.
29. (a) J. Fuller, R. T. Carlin, R. A.

Osteryoung, *J. Electrochem. Soc.*, **1997**, *144*, 3881; (b) J. Fuller, R. A. Osteryoung, R. T. Carlin, *Abstracts of Papers*, 187th Meeting of The Electrochemical Society, Reno, NV, **1995**, Vol. *95-1*, p. 27.
30. A. Noda, M. Watanabe, in *Proceedings of the Twelfth International Symposium on Molten Salts*, P. C. Trulove, H. C. De Long, G. R. Stafford, S. Deki (Eds.), The Electrochemical Society: Pennington NJ, **2000**, Vol. 99-41, pp. 202–208.
31. A. B. McEwen, H. L. Ngo, K. LeCompte, J. L. Goldman, *J. Electrochem. Soc.*, **1999**, *146*, 1687.
32. H. Shirota, E. Castner Jr., *J. Phys. Chem. B*, **2005**, *109*, 21576.
33. K. Seddon, A. Stark, M.-J. Torres, in *ACS Symposium Series 819 "Clean Solvents: Alternative Media for Chemical Reactions and Processing*, American Chemical Society, Washington DC, **2002**, pp. 34–49.
34. E. I. Cooper, E. J. M. O'Sullivan, in *Proceedings of the Eighth International Symposium on Molten Salts*, R. J. Gale, G. Blomgren (Eds.), The Electrochemical Society: Pennington NJ, **2000**, Vol. 92-16, pp. 386–396.
35. H. Matsumoto, M. Yanagida, K. Tanimoto, M. Nomura, Y. Kitagawa, Y. Miyazaki, *Chem. Lett.*, **2000**, 922.
36. S. Dzyuba, R. Bartsch, *ChemPhysChem*, **2002**, *3*, 161.
37. H. Matsumoto, H. Kageyama, Y. Miyazaki, *ChemComm*, **2002**, *16*, 1726.
38. J. M. Crosthwaite, M. J. Muldoon, J. K. Dixon, J. L. Anderson, J. F. Brennecke, *J. Chem. Thermodynam.*, **2005**, *27*, 559.
39. S. H. Lee, S. B. Lee, *Chem. Commun.*, **2005**, *27*, 3469.
40. D. R. MacFarlane, J. Golding, S. Forsyth, M. Forsyth, G. B. Deacon, *Chem. Commun.*, **2001**, *16*, 1430.
41. Z-B. Zhou, H. Matsumoto, K. Tatsumi, *ChemPhysChem*, **2005**, *6*, 1324.
42. M. Ue, M. Takeda, A. Toriumi, R. Kominato, Y. Hagiwara, Y. Ito, *J. Electrochem. Soc.*, **2003**, *150*, A499.
43. L. C. Branco, J. N. Rosa, J. J. Moura Ramos, C. A. M. Afonso, *Chem. Eur. J.*, **2002**, *8*(16), 3671.
44. D. Zhao, Z. Fei, R. Scopelliti, P. J. Dyson, *Inorg. Chem.*, **2004**, *43*, 2197.
45. J. G. Huddleston, A. E. Visser, W. M. Reichert, H. D. Willauer, G. A. Broker, R. D. Rogers, *Green Chem.*, **2001**, *3*, 156.
46. O. O. Okoturo, T. J. VaderNoot, *J. Electroanal. Chem.*, **2004**, *568*, 167.
47. S. Chun, S. V. Dzyuba, R. A. Bartsch, *Anal. Chem.*, **2001**, *73*, 3737.
48. J. Wang, A. Zhu, Y. Zhao, K. Zhuo, *J. Soln. Chem.*, **2005**, *34* (5), 585.
49. A. Bagno, C. Butts, C. Chiappe, F. D'Amico, J. C. D. Lord, D. Pieraccini, F. Rastrelli, *Org. Biomol. Chem.*, **2005**, *3* (9), 1624.
50. (a) J. R. Sanders, E. H. Ward, C. L. Hussey, in *Proceedings of the Fifth International Symposium on Molten Salts*, M.-L. Saboungi, K. Johnson, D. S. Newman, D. Inman (Eds.), The Electrochemical Society: Pennington NJ, **1986**, Vol. 86-1, pp. 307–316; (b) J. R. Sanders, E. H. Ward, C. L. Hussey, *J. Electrochem. Soc.*, **1986**, *133*, 325.
51. R. A. Carpio, L. A. King, R. E. Lindstrom, J. C. Nardi, C. L. Hussey, *J. Electrochem. Soc.*, **1979**, *126*, 1644.
52. V. R. Koch, L. L. Miller, R. A. Osteryoung, *J. Am. Chem. Soc.*, **1976**, *98*, 5277.
53. Z-B. Zhou, H. Matsumoto, K. Tatsumi, *Chem. Eur. J.*, **2005**, *11*, 752.
54. J. Pernak, J. Feder-Kubis, *Chem. Eur. J.*, **2005**, *11*, 4441.
55. D. R. MacFarlane, P. Meakin, J. Sun, N. Amini, M. Forsyth, *J. Phys. Chem. B.*, **1999**, *103*, 4164.
56. D. Gerhard, S. C. Alpaslan, H. J. Gores, M. Uerdingen, P. Wasserscheid, *Chem. Comm.*, **2005**, *40*, 5080.
57. Q. H. Cai, Y. K. Shan, L. Y. Dai, B. X. Su, M. Y. He, *Chinese Chem. Lett.*, **2003**, *14*, 523.

3.3
Solubility and Solvation in Ionic Liquids

Violina A. Cocalia, Ann E. Visser, Robin D. Rogers, and John D. Holbrey

3.3.1
Introduction

The interest in using ionic liquid (IL) media as alternatives to traditional organic solvents in synthesis [1–4], in liquid/liquid separations from aqueous solutions [5–9], and as liquid electrolytes for electrochemical processes including electrosynthesis, focuses on the unique combination of properties exhibited by ILs that differentiate them from molecular solvents, despite the apparent paradoxical desire to utilize ILs as replacements for molecular solvents in these systems.

ILs are generally considered to be highly polar, yet often weakly coordinating, solvents. Solvatochromatic studies indicate that ILs have polarities similar to those of short-chain alcohols and other polar, aprotic solvents (DMSO, DMF etc.) [10–14]. That is, their polarity is intermediate between water and chlorinated organic solvents and varies within the region depending on the nature of the IL components. By changing the nature of the ions present in an IL, the resulting properties of the IL can be changed. For example, the solubility of water in ionic liquids can be varied from complete miscibility to almost total immiscibility, by changing the anion from e.g. Cl^- to $[PF_6]^-$. Similarly, lipophilicity of ILs is modified by the degree of cation substitution. Primary solvent features of ILs are the ability for H-bond donation from the cation to polar or dipolar solutes, H-bond accepting functionality in the anion (this is variable, e.g. Cl^- is a good H-bond acceptor whereas $[PF_6]^-$ is poor), and $\pi-\pi$ or $C-H\cdots\pi$ interactions (which enhance aromatic solubility). ILs tend to be immiscible with alkanes and other non-polar organic solvents and hence can be used in two-phase systems. Similarly, ILs can be designed that are hydrophobic and can be used in aqueous/IL biphasic systems.

The solubility of both organic compounds and metal salts in ILs is important with regard to chemical synthesis, especially in Green Chemistry, with an emphasis on catalytic processes. Not only must reagents and catalysts be sufficiently soluble in the solvent, but differential solubility of reagents, products, and catalysts is required in order to enable effective separation and isolation of products. As well as requiring a knowledge of solute solubility in ILs, to assess the relative merits of a particular IL for chemical or separation processes, relative solubility and partitioning information about the preference of the solutes for IL phases relative to extractants is needed in order to design systems in which both reactions and extractions can be performed efficiently. However, there is only limited systematic data on these properties in the literature. In many cases, solutes and solvents are described as immiscible in a particular IL on the basis of observation that two phases are formed, rather than compositional analysis to determine the limits of solubility or co-miscibility.

ILs have been investigated as alternatives to traditional organic solvents in liquid/liquid separations. Reports highlighting separations based on ILs [5, 15–18] for implementation into industrial separation systems demonstrate the design principles, and have identified hydrophobic ILs as possible replacements for volatile organic compounds (VOCs) in aqueous/IL biphase separations schemes. Other work on novel solvent media has shown how supercritical water [19, 20], $scCO_2$ [21], and fluorous phases [22] can be used in an effort to broaden the scope of possibilities available for more environmentally responsible processes. From the synthetic perspective, desirable features of an IL are: (i) catalyst solubility, enabling high catalyst capacity, immobilization to extraction processes; (ii) reagent solubility, ideally in high capacity, and (iii) product extractability.

Extractions and separations in two-phase systems require knowledge of the miscibility and immiscibility of ILs with other solvents compatible with the process. These are most usually IL/aqueous biphase systems in which the IL is the less polar phase and organic/IL systems in which the IL is used as the polar phase, respectively. Under these two-phase systems, both extraction to, and from, the IL phase is important.

3.3.2
Metal Salt Solubility

Metal ion solubility for synthesis and catalysis principally requires sufficient solubility of the catalytic transition metal complex for immobilization in the IL phase relative to an extraction process or phase. Solubilization of metal ions in ILs can be separated into processes involving the dissolution of simple metal salts (often through coordination with anions from the ionic liquid) and the dissolution of metal coordination complexes in which the metal coordination sphere remains intact.

3.3.2.1 Halometallate Salts

The formation of halometallate ionic liquids by the equilibrium reactions of organic halide salts with metal halide compounds is well established and has been reviewed by Hussey [23]. A wide range of metallates have been prepared and investigated, primarily as liquid electrolytes for electrochemistry and battery applications, and for electroplating, electrowinning, and as Lewis-acid catalysts for chemical synthesis. In particular, acidic tetrachloroaluminate(III) IL have been used in place of solid $AlCl_3$ heterogeneous catalysts, with the IL acting as a liquid catalyst and with product separation from the IL encouraged by differential solubility of the reagents and products.

Many simple metal compounds (salts) are dissolved in 'basic' ionic liquids, containing coordinating Lewis base ligands, by complexation mechanisms; most metal halides can be dissolved in chloride-rich ILs as chloro-containing metallate species. This is the basis for the formation of chlorometallate ionic liquids, containing metal-complex anions. Among the more recent examples of metal-containing ILs is the gold-containing [EMIM][$AuCl_4$] [24]. Seddon and Hussey have investigated the dissolution of many transition metals in halometallate ILs, for example see Refs. [25–27].

Metal halide compounds can be dissolved in basic tetrachloroaluminate ionic liquids, but in many cases, can be precipitated from acidic ILs. For example, crystals of [EMIM]$_2$[PdCl$_4$] were obtained by dissolving PdCl$_2$ in acidic [EMIM]Cl/AlCl$_3$ IL [28]. This reflects changes in ligating ability of the predominant anions present in tetrachloroaluminate ILs on changing from the basic regime (Cl$^-$) through to acidic ([AlCl$_4$]$^-$). Simple metal salts can also be dissolved in other ionic liquids, containing coordinating anions such as nitrates.

3.3.2.2 Metal Complexes

The principle examination of metal complex solubility in ILs has stemmed from the specific requirements for transition metal catalysis. The most effective method is through selective solubility and immobilization of the catalyst in the reacting phase, allowing product separation (with no catalyst leaching) into a second, extracting phase. In the context of emerging separations and extraction investigations and homogeneous catalysis, efficient recycling of metal catalysts is an absolute necessity. Systematic studies of metal complex solubility in ILs have not been reported and warrant investigation.

As a set of general observations; (i) ionic *compounds* are generally poorly soluble in ILs; (ii) ionic *complexes* are more soluble; (iii) compounds are solubilized by complexation; (iv) the peripheral environment of the ligands is important in affecting solubility, and can be modified to enable better solubility. Solubility depends on the nature of the IL and solvation or complex formation, most metal ions display preferential partitioning to water and are, hence, less soluble in the IL than in water.

Simple metal compounds are poorly soluble in non-coordinating ILs. By adding lipophilic ligands, the solubility of metal ions in ILs can be increased. However, enhancing lipophilicity also increases the tendency for the metal complex to leach to less-polar organic phases.

Ionic complexes tend to be more soluble in ILs than neutral complexes. Representative examples of transition metal salts and complexes that have been used as homogeneous catalysts in IL systems include [LNiCH$_2$CH$_3$][AlCl$_4$] (where L is PR$_3$) used in the Difasol olefin oligomerization process, [Rh(nbd)(PPh$_3$)$_2$][PF$_6$] [29], [Rh(cod$_2$)][BF$_4$]$_2$ [30], and [H$_4$Ru(η^6-C$_6$H$_6$)$_4$][BF$_4$]$_2$ [31] complexes, which have been described as catalysts for hydrogenation reactions. Chauvin et al. noted that in catalytic hydrogenation studies, neutral catalysts, e.g. Rh(CO)$_2$(acac), are leached into the organic phase, whereas charged species are maintained in the IL phase [32].

The precipitation of neutral complexes from solution, or extraction into a secondary phase has enormous implications in the design of two-phase catalytic systems (to eliminate catalyst leaching) and in extractions (where selective extraction from either aqueous or organic phases is required, followed by controlled stripping of metals from the IL phase for recovery). Metal ion solubility in ILs can be increased by changing the complexing ligands present, for example using soluble organic complexants such as crown ethers, or by modifying the ligands to increase the solubility in ILs.

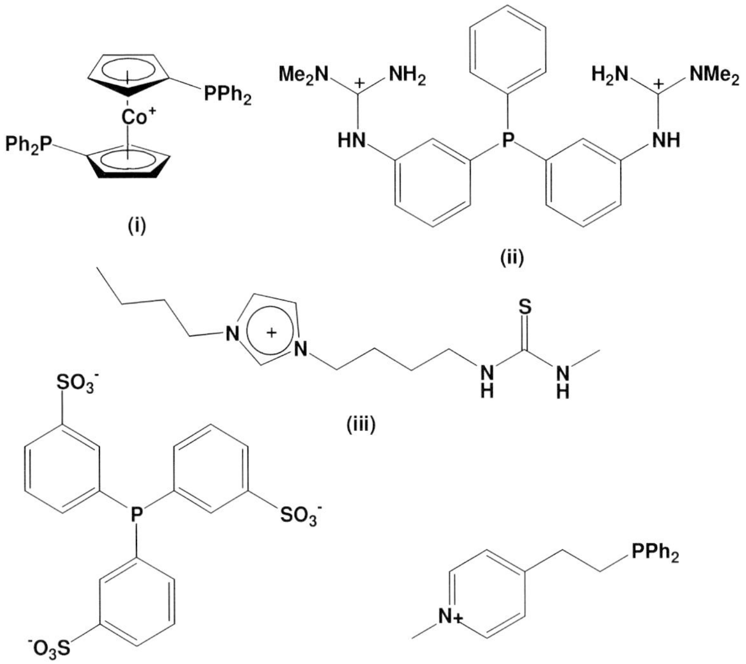

Fig. 3.3-1 Incorporating groups with high affinity for ILs including cobaltacenium (i), guanadinium (ii), sulfonate (iv), and pyridinium (v), or even that are ionic liquid moieties such as imidazolium (iii), as peripheral functionalities on coordinating ligands increases the solubility of transition metal complexes in ILs.

Chauvin showed that sulfonated triphenylphosphine ligands e.g. tppts and tppms [(iv) in Fig. 3.3-1], prevented leaching of neutral Rh hydrogenation catalysts from ILs [29]. However, Cole-Hamilton and coworkers [33] have noted that the solubility of Rh-tppts complexes in ILs is low. Wasserscheid and coworkers [34] and Olivier-Bourbigou and coworkers [35] have demonstrated that adding cationic functionality to the periphery of otherwise neutral ligands can be used to increase the solubility and stability of metal complexes in the IL phase relative to leaching into an organic extractant phase [(i), (ii), (iv) and (v) in Fig. 3.3-1]. This approach mimics that taken to confer greater water solubility onto metal complexes for aqueous-biphasic catalysis, and is equivalent to the task-specific ionic liquid (TSIL) approach of Davis, Rogers and coworkers [36] for enhanced metal transfer and binding in IL phases for extractions, [(iii) in Fig. 3.3-1]. More details about the immobilization of transition metal complexes in ionic liquids are found in Chapter 5, Section 5.4.

ILs have also been used as inert additives to stabilize transition metal catalysts during evaporative work-up of reactions in an organic solvent system [37, 38]. The extremely low volatile IL component solubilizes the catalyst upon concentration and removal of organic solvent and products, thereby preventing catalyst decomposition and enabling recycling and reuse of catalysts in batch processes.

3.3.3
Extraction and Separations

Studies of extraction and separations provide information on the relative solubility of solutes between two phases, e.g. partitioning data which is required in order to design systems in which a solute is either selectively extracted from, or immobilized in, one phase. Liquid/liquid separation studies of metal ions are principally concerned with aqueous/organic two-phase systems, with relevance for extraction and concentration of metal ions in the organic phase. In terms of IL/aqueous partitioning, there is considerable interest in replacing organic extracting phases with ILs for recovery of metals from waste water, in mining, nuclear fuel and waste reprocessing, and immobilizing transition metal catalysts. In IL/aqueous extraction systems, the hydrated nature of the metal ions makes the use of complexing agents with preferential (and ideally complete selectivity) for the IL phase necessary to achieve metal ion complexation and extraction to the IL phase. In contrast to systems comprised of purely molecular solvents, ILs have two components, a cation and an anion, which do not necessarily play a passive role in extraction, a fact highlighted in the recent literature discussed below.

Metal ion separations in ILs have been of interest ever since the first reports of using extractants to partition metal ions from aqueous media to ILs appeared in 1999 [8, 9] and extremely high distribution ratios were obtained, e.g. for Sr^{2+} extraction with dicyclohexano-18-crown-6 from an aqueous solution into ILs [9]. Dietz et al. showed that the origin of this enhancement is the change of extraction mechanism from ion pairing in traditional solvents to cation exchange in ILs, with the transfer of the IL cation into the aqueous phase [39]. A wide range of conventional and less conventional extractants and ligand modifications have been used to control the distribution and selectivity of the metal ions for one phase over the second [9, 15–17, 40]. It is important to understand that speciation, complexation and distributions of metals in two-phase systems containing one IL phase can be substantially different to systems with conventional solvents, although this is not always the case.

Visser et al. [41] studied uranyl extraction with CMPO (carbamoylmethylphosphine oxide) and TBP (tri-n-butylphosphate) into [BMIM][PF_6] and [OMIM][Tf_2N] (Tf_2N = bis(trifluoromethanesulfonyl)imide) from aqueous acidic media and also showed enhancement for uranyl ion extraction into ILs when compared to dodecane. Extended X-ray absorption fine structure (EXAFS) measurements (Fig. 3.3-2) and UV/vis spectroscopy (Fig. 3.3-3) showed that uranyl-CMPO complexes in [BMIM][PF_6] and [OMIM][Tf_2N] are similar, but different from those in dodecane.

These results for extracted uranyl species demonstrated that the mechanism of extraction in the ILs involves the loss of the IL cation into the aqueous phase, as previously reported for Sr^{2+} extraction into ILs [39]. Further investigations have been carried out to try to understand if this phenomenon is a general one for

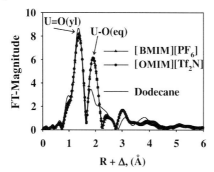

Fig. 3.3-2 Fourier transform magnitude of the k^3 weighted L_3 EXAFS of the UO_2^{2+} complexes in dodecane, [BMIM][PF$_6$], and [OMIM][Tf$_2$N]. From Ref. [41].

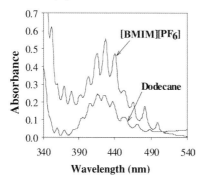

Fig. 3.3-3 Absorption spectra of dodecane and [BMIM][PF$_6$] solutions containing 0.1 M CMPO and 1 M TBP, after contact with 20 mM $UO_2(NO_3)_2$ in 1 M HNO_3. From Ref. [41].

metal ion separation into ILs. In contrast, recent studies show, for extraction of trivalent lanthanides using Htta {1-(2-thienyl)-4.4,4-trifluoro-1,3-butanedione} in [BMIM][Tf$_2$N], an anionic exchange mechanism where [Ln(tta)$_4$]$^-$ anions are exchanged with the [Tf$_2$N]$^-$ anions from the IL, this mechanism being not very common in traditional solvents [42].

The cation and anion exchange mechanisms are not advantageous due to the loss of IL components into the aqueous phase. Therefore efforts have been oriented toward rationally controlling the extraction mechanisms for metal ions. In this context, Dietz et al. showed a change in the mechanism from cation exchange to solvent extraction with increasing the alkyl chain in the IL, but found that the coordination mode for the extracted metal is different for the short and long alkyl imidazolium ILs [43]. Also, by addition of Na[BPh$_4$], as a sacrificial cation source, to the IL phase the imidazolium cation loss has been reduced by 24% due to the competition between the highly hydrated sodium cation and imidazolium cation to migrate in the aqueous phase [44]. In an effort to find metal ion coordination environments that would facilitate practical extractions, Cocalia et al. [45] studied the

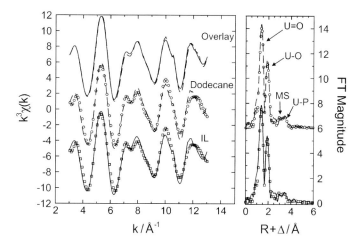

Fig. 3.3-4 k^3-Weighted uranium L_3-edge EXAFS of the UO_2^{2+}-HDEHP complex extracted into [DMIM][Tf_2N] (—) and dodecane (– –). From Ref. [45].

coordination and extraction behavior of Am^{3+} and UO_2^{2+} with Cyanex-272 and di-(2-ethylhexyl) phosphoric acid (HDEHP) into [DMIM][Tf_2N], from acidic aqueous solutions. A comparison of the metal ions distribution ratios patterns in the IL and dodecane shows similar trends at low acidities, indicating that the extraction mechanism is the same in the IL and dodecane. Coordination studies of the metal ions using EXAFS (Fig. 3.3-4) and UV/vis (Fig. 3.3-5) in the IL and traditional solvent, showed that the coordination mode does not change upon changing the identity of the solvent.

Another developing approach for using ILs in metal ions separation was first initiated by Davis, who introduced the term task-specific ionic liquid (TSIL) to describe ILs prepared using the concept of increasing the IL affinity of the extractants by incorporating the complexing functionality as an integral part of the IL [36].

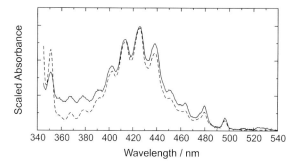

Fig. 3.3-5 Absorption spectra of (—) IL and (- - -) dodecane solutions containing 1 M HDEHP, after contact with 0.06 M UO_2^{2+}/0.25 M HNO_3. From Ref. [45].

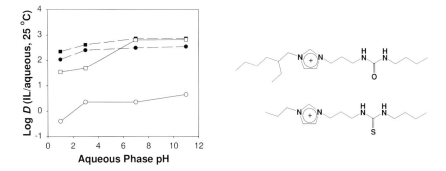

Fig. 3.3-6 Hg^{2+} (■) and Cd^{2+} (○) distribution ratios between IL and aqueous phases with 1:1 [BMIM][PF$_6$] and urea-(dashed lines) or thiourea- (solid lines) appended TSIL as the extracting phase [36].

Thus ILs with built-in extracting capability, and differentiation from both aqueous and organic phases, through modification of the IL co-miscibility can be prepared. These TSIL can be used as an IL extracting phase, or may be mixed with a second, more conventional IL to modify the rheological properties. Metal ion ligating groups are incorporated into the cationic moiety of the IL by tethering to the imidazolium cation. Thioether, urea, and thiourea derivatized imidazolium ILs have been investigated as Hg^{2+} and Cd^{2+} extractants, see Fig. 3.3-6 [36]. Comparison of the distribution ratios reveals that extractability of Hg^{2+} is typically higher and a change in the aqueous phase pH has only a slight effect on the partitioning.

In a liquid/liquid system using ILs as alternatives to organic solvents, the tendency for metal ions to remain in the aqueous phase can be offset through the presence of either organic or inorganic extractants or TSIL in the system. These extractants serve to modify the hydration environment of metal ions through complexation with ligating functional groups and increase metal ion partitioning to the IL phase. When designing an IL, fine-tuning the properties can be achieved by changing the cation substituent groups, anion identity, or by mixing two types of IL with differing, but defined characteristics.

3.3.4
Organic Compounds

Interest and the approaches to organic compound separations from ionic liquids emerged initially from the observation that many hydrocarbons were only poorly soluble in ionic liquids, enabling reaction and facile extraction processes to be achieved [32, 46]. The advent of air-stable hydrophobic ionic liquids [10, 47] opened up the potential to use ionic liquids in place of volatile organic solvents for aqueous/organic liquid–liquid extraction systems.

The basis for the utility of ionic liquids as VOC replacements in liquid–liquid separations came from the investigation of organic solute distribution coefficients between [BMIM][PF$_6$] and water [5] in which it was shown that the distribution coefficients could be correlated with the corresponding octanol/water partition coefficient k_{ow}, and that the linear relationship allowed partitioning of organic solutes in ionic liquid biphasic systems to be modeled and predicted using Abraham's LSER approach. That is, that solubility and partitioning can be explained in terms of solute and solvent contributions from molecular volume, hydrogen-bond donor and acceptor and polarizability characteristics.

Studies have been extended to ionic liquid/organic solvent biphases and to liquid–gas interfaces [48] using similar approaches with considerable success. Partitioning and solubility of specific organic compounds in ILs depend on the solubilizing interactions between the solute and IL components. In general, ILs behave as moderately polar organic solvents with respect to organic solutes, however, in contrast to the organic solvents, they are commonly compared to, ILs are poorly solvating.

Armstrong and coworkers [48] have investigated the interactions of solutes with ILs, using the ILs as stationary phases for gas–liquid chromatography (GLC) and have shown that ILs appear to act as low-polarity phases in their interactions with non-polar compounds, and the solubility increases with increasing lipophilicity (alkyl-chain length, etc.). Hydrocarbons are poorly soluble in most ILs, but they are not insoluble. For example, alkanes are essentially insoluble in all ILs, while the solubility of alkenes is low [49], but increases with increased alkyl chain substitution in the IL and with delocalization of charge in the anion. The solubility of short-chain hydrocarbons over oligomers and polymers is the basis for the efficient separation of products from alkylation, oligomerization, polymerization [50] and hydrogenation reactions described using IL catalyst systems. The lipophilicity of the IL, and the solubility of non-polar solutes can be increased by adding additional non-polar alkyl functionality to the IL, thus reducing further coulombic ion–ion interactions. Polar molecules, or those containing strong proton donor functionality (for example phenols, carboxylic acids, diols), also interact strongly with ILs. Compounds with weak proton donor/acceptor functions (for example aromatic and aliphatic ketones, aldehydes and esters) appear to interact with the ionic liquids through induced ion–dipole, or weak van der Waals interactions.

Bonhôte and coworkers [10] reported that ILs containing triflate, perfluorocarboxylate and bistriflylimide anions were miscible with liquids of medium-high dielectric constant (ε), including short-chain alcohols, ketones, dichloromethane, and THF, and were immiscible with low dielectric constant materials such as alkanes, dioxane, toluene, and diethylether. It was noted that ethylacetate ($\varepsilon = 6.04$) is miscible with the 'less-polar' bis(trifluorosulfonyl)imide and triflate ILs, and only partially miscible with more polar ILs containing carboxylate anions. Brennecke [21] has described miscibility measurements for a series of organic solvents with ILs with complementary results based on bulk properties.

We have shown that, in general, ILs display partitioning properties similar to those of dipolar aprotic solvents, or short chain alcohols. The relationship between octanol/water partitioning and IL/water partitioning [51] (Fig. 3.3-7), despite clear polarity differences between the solvents, allows the solubility or partitioning of

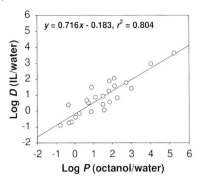

Fig. 3.3-7 Distribution ratios for organic solutes between [BMIM][PF$_6$] and water (neutral pH) correlate with literature partition functions of the solutes between octanol and water (log P).

organic solutes with ILs to be predicted from relative polarities of the materials (using solvatochromatic scales, etc.). Complex organic molecules such as cyclodextrins, glycolipids [48], and antibiotics [52] can be dissolved in ILs; the solubility of these complex molecules increases in the more polar ILs. The interactions are greatest when the ILs have H-bond acceptor capability, for example chloride-containing ILs. The miscibility of ILs with water varies with cation substitution and with anion types; coordination anions generally lead to water soluble IL whereas the presence of large, non-coordinating charge diffuse anions, generates hydrophobic ILs.

From empirical observation, ILs tend to be immiscible with non-polar solvents. Thus ILs can be washed, or contacted with diethyl ether, or hexane to extract non-polar reaction products. On increasing solvent polarity, esters (for example ethyl acetate) exhibit variable solubility with ILs, depending on the nature of the IL. Polar solvents (including chloroform, acetonitrile, and methanol) appear to be totally miscible with all ILs (excepting tetrachloroaluminate ILs and the like, which react). Among notable exceptions, [EMIM]Cl and [BMIM]Cl are insoluble in dry acetone.

We have shown that hydrophobic hexafluorophosphate ILs can be made totally miscible with water, by the addition of alcohols [53,54], the ternary phase diagram for [BMIM][PF$_6$]/water/ethanol (Fig. 3.3-8(a)) shows the large co-miscibility region of the three components. In fact, ethanol forms biphasic mixtures with [BMIM][PF$_6$], [HMIM][PF$_6$], and [OMIM][PF$_6$], the degree of miscibility depending on temperature, and on the water composition of the mixtures (Fig. 3.3-8(b)). In each case, increasing the water content of the IL, increases ethanol solubility. This process has many potential uses for washing and removal of ionic liquids from products and reactors, or catalyst supports and has important implications in designing IL/aqueous two-phase extraction systems.

Aromatic compounds are considerably more soluble in ILs than aliphatics; benzene can be dissolved in [EMIM]Cl/Al$_2$Cl$_6$, [BMIM][PF$_6$], and [EMIM][(CF$_3$SO$_2$)$_2$] at up to a ca. 1:1 ratio, reflecting the importance of CH$\cdots\pi$ interactions and $\pi-\pi$ stacking interactions and formation of liquid clatharate structures. Atwood has

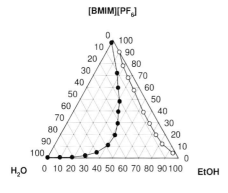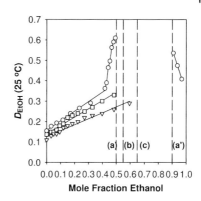

Fig. 3.3-8 Ethanol/water/[BMIM][PF$_6$] ternary phase diagram (a, left) and solute distribution in EtOH/water/IL mixtures (b, right) for [BMIM][PF$_6$] (○), [HMIM][PF$_6$] (□) and [OMIM][PF$_6$] (▽) as a function of initial mole fraction of ethanol in the aqueous phase, measured at 25 °C, from Refs. [53, 54].

shown that the high solubility of aromatics can be attributed to clathrate formation [55] and has described model extraction of toluene from toluene/heptane mixtures using [EMIM][I$_3$] and [BMIM][I$_3$] [56]. The difference in solubility of aromatic and aliphatic hydrocarbons in ILs makes possible the separation of toluene from heptane from naphta cracker feeds [57].

ILs containing 1-alkylisoquinolinium cations [R-ISOQ]$^+$, combined with the bis(perfluoroethylsulfonyl)imide ([N(SO$_2$CF$_2$CF$_3$)$_2$]$^-$, [BETI]$^-$) anion have been reported [58] and were tested for organic partitioning in aqueous/IL two-phase systems. The large, extended aromatic core in the cations of these ILs was anticipated to exhibit a greater affinity for aromatic solutes in IL/aqueous partitioning experiments and it was found that, in particular, the distribution ratio for 1,2,4-trichlorobenzene in [C$_{14}$-ISOQ][BETI] was much greater than in [BMIM][PF$_6$] [5, 59].

For charged or ionizable solutes, a change in the aqueous phase pH resulted in certain ionizable solutes exhibiting pH-dependent partitioning, such that their affinity for the IL decreased upon ionization [6]. Solute ionization effects, as demonstrated for aniline and benzoic acid (Fig. 3.3-9), can modify solubility and partitioning of solutes to an IL by several orders of magnitude.

The pH-dependent partitioning of an ionizable, cationic dye, thymol blue, has also been investigated [6]. The dye partitions preferentially to the IL phase in the neutral, zwitterionic and monoanionic forms (from acidic solution), the partition coefficient to the IL increases with increasing IL hydrophobicity. Under basic conditions, the dye is in the dianionic form and partitions to water (Fig. 3.3-10).

In a study investigating a practical application of ILs to separate organics (hexanoic acid, 1-nonanol, toluene, acetic acid, and cyclohexanone) from water, McFarlane showed that ILs can be effective in the separation of some organic contaminants (toluene, 1-nonanol), but it has limitations for extracting acetic acid which prefers

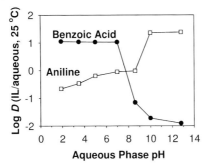

Fig. 3.3-9 Distribution ratios for aniline (pK_b = 9.42) and benzoic acid (pK_a = 4.19) in [BMIM][PF$_6$]/aqueous systems as a function of aqueous phase pH.

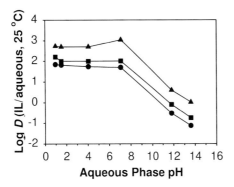

Fig. 3.3-10 pH switchable partitioning of the ionic dye, thymol blue in [BMIM][PF$_6$] (●), [HMIM][PF$_6$] (■), [OMIM][PF$_6$] (▲)/aqueous biphasic systems. From Ref. [6].

the aqueous phase, probably due to the low hydrophobicity of acetate ion, and hexanoic acid which is extracted better at low pH [60], as expected from previous published reports [6].

Also, very low distribution ratios have been reported by Matsumoto for extraction of some organic acids (glycolic, acetic, propionic, lactic, butyric, pyruvic) into [C$_n$MIM][PF$_6$] where n = 4, 6, 8. The low hydrophobicity of the organic acids controls the extraction process [61].

Results of partitioning amino acids in [PF$_6$]$^-$- and [BF$_4$]$^-$-ILs show that aliphatic amino acids partition coefficients are relatively small while aromatic amino acids partition more into ILs and also that more hydrophobic amino acids partition better into the IL [62]. The pH dependence in the distribution of amino acids indicates that, in the range of pH < 1, the number of cationic species decreases with increasing pH and that consequently the partition coefficients decrease as previously reported for other ionizable species [6].

3.3.5
Conclusions

The huge number of possible ILs offers encouraging promise for the possible use of ILs in very different types of separation from metal ions separation through organics separation to chromatographic applications. The challenge in pursuing the successful implementation of ILs in separation techniques is to find or design a suitable IL for a given, specific application. Our contribution aims to show how the first steps have been made in this direction with very promising results. Continued studies to elucidate the driving forces responsible for the success of some initial studies using ILs in separation will contribute to optimizing the processes and implementing more efficient and greener methods in separations.

References

1. R. Sheldon, *Chem. Commun.* **2001**, 2399.
2. W. Keim, P. Wasserscheid, *Angew. Chem. Int. Ed.* **2000**, *39*, 3772.
3. T. Welton, *Chem. Rev.* **1999**, *99*, 2071.
4. J. D. Holbrey, K. R. Seddon, *Clean Prod. Proc.* **1999**, *1*, 233.
5. J. G. Huddleston, A. E. Visser, W. M. Reichert, H. D. Willauer, G. A. Broker, R. D. Rogers, *Green Chem.* **2001**, *3*, 156.
6. A. E. Visser, R. P. Swatloski, R. D. Rogers, *Green Chem.* **2000**, *2*, 1.
7. A. E. Visser, R. P. Swatloski, W. M. Reichert, S. T. Griffin, R. D. Rogers, *Ind. Eng. Chem. Res.* **2000**, *39*, 3596.
8. R. D. Rogers, A. E. Visser, R. P. Swatloski, D. H. Hartman, in *Metal Separation Technologies Beyond 2000: Integrating Novel Chemistry with Processing*, K. C. Liddell, D. J. Chaiko (Eds.), The Minerals, Metals & Materials Society: Warrendale, PA, **1999**, p. 139.
9. S. Dai, Y. H. Ju, C. E. Barnes, *J. Chem Soc., Dalton Trans.* **1999**, 1201.
10. P. Bonhôte, A.-P. Dias, M. Armand, N. Papageorgiou, K. Kalyanasundaram, M. Grätzel, *Inorg. Chem.* **1996**, *35*, 1168.
11. L. Crowhurst, P. R. Mawdsley, J. M. Perez-Arlandis, P. A. Salter, T. Welton, *Phys. Chem. Chem. Phys*, **2003**, *5*, 2790.
12. M. J. Muldoon, C. M. Gordon, I. R. Dunkin, *J. Chem. Soc., Perkin Trans. 2* **2001**, 433.
13. S. N. V. K. Aki, J. F. Brennecke, A. Samanta, *Chem. Commun.* **2001**, 413.
14. D. Behar, C. Gonzalez, P. Neta, *J. Phys. Chem. A* **2001**, *105*, 7607.
15. K. Nakashima, F. Kuboto, T. Maruyama, M. Goto, *Anal. Sci.* **2003**, *19*, 1097.
16. S. Chun, S. Dzyuba, R. A. Bartsch, *Anal. Chem.* **2001**, *73*, 3737.
17. M. P. Jensen, J. A. Dzielawa, P. Rickert, M. L. Dietz, *J. Am. Chem. Soc.* **2002**, *124*, 10664.
18. H. Luo, S. Dai, P. V. Bonnesen, *Anal. Chem.* **2004**, *76*, 2773.
19. Z. Fang, S. Xu, and J. A. Kozinski, *Ind. Eng. Chem. Res.*, **2000**, *39*, 4536.
20. M. Sasaki, Z. Fang, Y. Fukuskima, T. Adschiri, K. Arai, *Ind. Eng. Chem. Res.* **2000**, *39*, 2883.
21. L. A. Blanchard, J. F. Brennecke, *Ind. Eng. Chem. Res.* **2001**, *40*, 287.
22. C. Ohrenberg, W. E. Geiger, *Inorg. Chem.* **2000**, *39*, 2948.
23. C. L. Hussey, *Pure Appl. Chem.* **1988**, *60*, 1763.
24. M. Hasan, I. V. Kozhevnikov, M. R. H. Siddiqui, A. Steiner, N. Winterton, *Inorg. Chem.* **1999**, *38*, 5637.
25. D. Appleby, P. B. Hitchcock, K. R. Seddon, J. E. Turp, J. A. Zora, C. L. Hussey, J. R. Sanders, T. A. Ryan, *J. Chem. Soc., Dalton Trans.* **1990**, *6*, 1879.
26. D. Appleby, R. I. Crisp, P. B. Hitchcock, C. L. Hussey, T. A. Ryan, J. R. Sanders, K.

R. Seddon, J. E. Turp, J. A. Zora, *J. Chem. Soc., Chem. Commun.* **1986**, 483.

27. D. Appleby, C. L. Hussey, K. R. Seddon, J. E. Turp, *Nature* **1986**, *323*, 614.
28. M. Ortwerth, M. J. Wyzlic, R. Baughman, *Acta. Crystallogr.Sect. C* **1998**, *54*, 1594.
29. Y. Chauvin, L. Mussmann, H. Olivier, *Angew. Chem. Int. Ed.* **1996**, *34*, 2698.
30. S. Einloft, F. K. Dietrich, R. F. Desouza, J. Dupont, *Polyhedron* **1996**, *15*, 3257.
31. P. J. Dyson, D. J. Ellis, D. G. Parker, T. Welton, *Chem. Commun.* **1996**, 25.
32. Y. Chauvin, H. Olivier-Bourbigou, *CHEMTECH* **1995**, *25*, 26.
33. M. F. Sellin, P. B. Webb, D. J. Cole-Hamilton, *Chem. Commun.* **2000**, 781.
34. P. Wasserscheid, H. Waffenschmidt, P. Machnitzki, K. W. Kottseiper, O. Stetzler, *Chem. Commun.* **2001**, 451.
35. F. Favre, H. Olivier-Bourbigou, D. Commereuc, L. Saussine, *Chem. Commun.* **2001**, 1360.
36. A. E. Visser, R. P. Swatloski, W. M. Reichert, R. D. Rogers, R. Mayton, S. Sheff, A. Wierzbicki, J. H. Davis Jr., *Chem. Commun.* **2001**, 135.
37. S. V. Ley, C. Ramarao, M. D. Smith, *Chem. Commun.* **2001**, 2278.
38. C. E. Song, E. J. Roh, *Chem. Commun.* **2000**, 837.
39. M. L. Dietz, J. A. Dzielawa, *Chem. Commun.* **2001**, *20*, 2124.
40. A. E. Visser, R. D. Rogers, *J. Solid State Chem.* **2003**, *171*, 109.
41. A. E. Visser, M. P. Jensen, I. Laszak, K. L. Nash, G. R. Choppin, R. D. Rogers, *Inorg. Chem.* **2003**, *42*, 2197.
42. M. P. Jensen, J. Neueifeind, J. V. Beitz, S. Skanthakumar, L. Soderholm, *J. Am. Chem. Soc.*, **2003**, *125*, 15466.
43. M. L. Dietz, J. A. Dzielawa, I. Laszak, B. A. Young, M. P. Jensen, *Green Chem.* **2003**, *5*, 682.
44. H. Luo, S. Dai, P. V. Bonnesen, A. C. Buchanan III, J. D. Holbrey, N. J. Bridges, R. D. Rogers, *Anal. Chem.* **2004**, *76*, 3078.
45. V. A. Cocalia, M. P. Jensen, J. D. Holbrey, S. K. Spear, D. C. Stepinski, R. D. Rogers, *Dalton Trans.* **2005**, 1966.
46. G. W. Parshall, *J. Am. Chem. Soc.* **1972**, *94*, 8716.
47. C. M. Gordon, J. D. Holbrey, A. R. Kennedy, K. R. Seddon, *J. Mater. Chem.* **1998**, *8*, 2627.
48. D. W. Armstrong, L. He, Y. S. Lui, *Anal. Chem.* **1999**, *71*, 3873.
49. F. Favre, H. Olivier-Bourbigou, D. Commereuc, L. Saussine, *Chem. Commun.* **2001**, 1360.
50. A. A. K. Abdul-Sada, P. W. Ambler, P. K. G. Hodgson, K. R. Seddon, N. J. Stewart, Patent No. 9 521 871, **1995**.
51. J. G. Huddleston, H. D. Willauer, R. P. Swatloski, A. E. Visser, R. D. Rogers, *Chem. Commun.* **1998**, 1765.
52. S. G. Cull, J. D. Holbrey, V. Vargas-Mora, K. R. Seddon, G. J. Lye, *Biotechnol. Bioeng.* **2000**, *69*, 227.
53. R. P. Swatloski, A. E. Visser, W. M. Reichert, G. A. Broker, L. M. Farina, J. D. Holbrey, R. D. Rogers, *Chem. Commun.* **2001**, 2070.
54. R. P. Swatloski, A. E. Visser, W. M. Reichert, G. A. Broker, L. M. Farina, J. D. Holbrey, R. D. Rogers, *Green Chem.* **2002**, *4*, 81.
55. J. L. Atwood, Liquid Clathrates, in *Inclusion Compounds*, J. L. Atwood, J. E. D. Davies, D. D. MacNicol (Eds.), Academic Press: London, **1984**, Vol. 1.
56. M. S. Selvan, M. D. McKinley, R. H. Dubois, J. L. Atwood, *J. Chem. Eng. Data.* **2000**, *45*, 841.
57. G. W. Meindersharma, A.J. G. Podt, M. G. Meseguer, A. B. deHaan, Ionic Liquids as Alternatives to Organic Solvents in Liquid-Liquid Extraction of Aromatics, in *Ionic Liquids IV: Fundamentals, Progress, Challenges, and Opportunities*; Rogers, R. D., Seddon, K. R. (Eds.), ACS Symposium Series 902, **2005**, pp. 57–71.
58. A. E. Visser, J. D. Holbrey, R. D. Rogers, *Chem. Commun.* **2001**, 2484.
59. J. G. Huddleston, H. D. Willauer, R. P. Swatloski, A. E. Visser, R. D. Rogers, *Chem. Commun.* **1998**, 1765.
60. J. McFarlane, W. B.Ridenour, H. Luo, R. D. Hunt, D. W. Paoli, *Sep. Sci. Technol.* **2005**, *40*, 1245.
61. M. Matsumoto, K. Mochiduki, K. Fukunishi, K. Kondo, *Sep. Purif. Technol.* **2004**, *40*, 97.
62. J. Wang, Y. Pei, Y. Zhao, Z. Hu, *Green Chem.* **2005**, *7*, 196.

3.4
Gas Solubilities in Ionic Liquids

Jessica L. Anderson, Jennifer L. Anthony, Joan F. Brennecke, and Edward J. Maginn

3.4.1
Introduction

A wide variety of physical properties are important in evaluating ionic liquids (ILs) for potential use in industrial processes. These include pure component properties like density, isothermal compressibility, volume expansivity, viscosity, heat capacity, thermal conductivity, melting point and decomposition temperature. However, a wide variety of *mixture* properties are also important; the most vital of these is the phase behavior of ionic liquids with other compounds. Knowledge of the phase behavior of ionic liquids with gases, liquids, and solids is necessary to assess the feasibility of their use for reactions, separations, and material processing. It is clear that the cation, substituents on the cation, and the anion can be chosen to enhance or suppress the solubility of ionic liquids in other compounds and the solubility of other compounds in the ionic liquids. For instance, increasing the length of the alkyl chain decreases the mutual solubility with water, but some anions (e.g., $[BF_4]^-$) can increase the mutual solubility with water (compared to $[PF_6]^-$, for instance) [1–3]. While many mixture properties and many types of phase behavior are important, here we will focus on the solubility of gases in room temperature ILs.

A primary motivation for understanding gas solubilities in ILs stems from the many successful demonstrations of their use as solvents for reactions [4–6]. Some of these reactions, such as hydrogenations, oxidations, and hydroformylations, involve the reaction of substrates in the ionic liquid solution with permanent and condensable gases. If a gas has limited solubility in the IL, then significant efforts will have to be made to increase interfacial area and enhance mass transfer and/or high pressure operation will be required. This may limit the ability of ILs to compete with conventional solvents, unless there are other significant chemical processing advantages to the IL. Conversely, high solubility or selective solubility of the desired gases might make ILs quite attractive.

A second motivation for understanding gas solubilities in ILs is the possibility of using ILs for the storage of gases. In this way, they could essentially become liquid adsorbents. A novel application of this idea is discussed in the Applications section.

A third motivation for understanding gas solubilities in ILs is the possibility of using ILs to separate gases. Because they are extremely low volatile [7, 8] and would not contaminate the gas stream in even small amounts, ILs have an automatic advantage over conventional absorption solvents for performing gas separations. In addition, their high thermal stability means that they could be used to perform gas separations at higher temperatures than is possible with conventional absorption solvents. Whether used in a conventional absorber arrangement or as a supported liquid membrane, the important physical properties for this application (besides low volatility) are the solubility and diffusivity of the gases of interest in the ILs.

A fourth motivation for studying gas solubilities in ILs is the potential of using compressed gases or supercritical fluids to separate species from an IL mixture. For example, Brennecke et al. have shown that it is possible to recover a wide variety of solutes from ILs using supercritical CO_2 extraction [9]. An advantage of this technology is that the solutes can be removed quantitatively without any cross-contamination of the CO_2 with the IL. Such separations should be possible with a wide variety of other compressed gases, such as C_2H_6, C_2H_4 and SF_6. It has also been shown that separation of ILs from some liquids can be achieved using lower pressure CO_2, which induces liquid/liquid phase separation [10, 11]. Clearly, the phase behavior of the gases with the IL is important for these applications.

Finally, a fifth motivation for exploring gas solubilities in ILs is that they can act as probes of the molecular interactions with the ILs. Information can be discerned on the importance of specific chemical interactions, such as hydrogen bonding, dipole–dipole, dipole–induced dipole, and dispersion forces. Of course, this information can also be determined from the solubility of a series of carefully chosen liquids. However, gases tend to be the smallest size, and, therefore, the simplest molecules with which to probe molecular interactions.

In this chapter, we will first discuss various experimental techniques that can be used to measure gas solubilities and related thermodynamic properties in ILs. Then, we will describe and compare literature data on the solubility of various gases in ILs. Most of these studies have been published since 2000. Finally, we will discuss the impact that gas solubilities in ILs have on the applications described above (reactions, gas storage, gas separations, separation of solutes from ILs) and draw some conclusions.

3.4.2
Experimental Techniques

In this section, we describe some of the various experimental techniques that can be used to measure gas solubilities and related thermodynamic properties.

3.4.2.1 Gas Solubilities and Related Thermodynamic Properties

In general, gas solubilities are measured at constant temperature as a function of pressure. Permanent gases (gases whose critical temperatures are below room temperature) will not condense to form an additional liquid phase no matter how high the applied pressure. However, condensable gases (ones with critical temperatures above room temperature) will condense to form a liquid phase when the vapor pressure is reached. In normal liquids, the solubility of many gases is quite low and can be adequately described at ambient pressure or below by Henry's law. The Henry's law constant is defined as

$$H_1(T, P) \equiv \lim_{x_1 \to 0} \frac{\hat{f}_1^L}{x_1} \tag{3.4-1}$$

where x_1 is the mole fraction of gas in the liquid and f_1^L is the fugacity of the gas (species 1) in the liquid phase. If the gas phase behaves ideally (i.e., the fugacity coefficient is close to 1) then the fugacity is equal to the pressure of gas above the IL sample. This is because there is essentially no IL in the vapor phase due to its nonvolatility. Experimentally, the Henry's law constant can be determined from the limiting slope of the solubility as a function of pressure. A large Henry's law constant indicates low gas solubility and a small Henry's law constant indicates a high gas solubility. One might also choose to express the limiting gas solubility (especially for condensable gases) in terms of an infinite dilution activity coefficient, where the standard state is pure condensed liquid at the temperature of the experiment (P_1^{sat}). In this case, the infinite dilution activity coefficient, γ_1^∞, can be related to the Henry's law constant simply by $\gamma_1^\infty = H_1(T, P) P_1^{sat}(T)$, neglecting the Poynting pressure correction, as it is only important at high pressures.

Also of importance is the effect of temperature on the gas solubility. From this information, one can determine the enthalpy and entropy change experienced by the gas when it changes from the ideal gas state (h_1^{ig} and s_1^{ig}) to the mixed liquid state ($\bar{h}_1$ and $\bar{s}_1$).

$$\Delta h_1 = \bar{h}_1 - h_1^{ig} = R \left(\frac{\partial \ln P}{\partial (1/T)} \right)_{x_1} \tag{3.4-2}$$

$$\Delta s_1 = \bar{s}_1 - s_1^{ig} = -R \left(\frac{\partial \ln P}{\partial \ln T} \right)_{x_1} \tag{3.4-3}$$

Thus, Δh_1 and Δs_1 can be obtained from determining the pressure required to achieve a specified solubility at several different temperatures and constant composition, x_1. In the Henry's law region, Δh_1 and Δs_1 can be found directly from the temperature dependence of the Henry's law constant, as given by the familiar van't Hoff equations:

$$\Delta h_1 = R \left(\frac{\partial \ln H_1}{\partial (1/T)} \right)_P \tag{3.4-4}$$

$$\Delta s_1 = -R \left(\frac{\partial \ln H_1}{\partial \ln T} \right)_P \tag{3.4-5}$$

The enthalpy and entropy of gas dissolution in the ILs provide, respectively, information about the strength of the interaction between the IL and the gas, and the ordering that takes place in the gas/IL mixture.

Of course, a primary concern for any physical property measurement, including gas solubility, is the purity of the sample. Since impurities in ILs have been shown to affect pure component properties such as viscosity [12], one would anticipate that impurities might also affect gas solubilities, at least to some extent. Since ILs are hygroscopic, a common impurity is water. Interestingly, it was found that the solubility of CO_2 in 1-n-butyl-3-methyl bis(trifluoromethylsulfonyl)imide

([BMIM][Tf$_2$N]) did not change significantly from a dry sample (450 ppm water) to one essentially saturated with water (13500 ppm) [13]. However, this is no doubt not a general result, especially for ILs that are totally miscible with water and can absorb significant amounts of water from the atmosphere. There might also be residual impurities, such as chloride, present from the synthesis procedure. In addition, hexafluorophosphate and tetrafluoroborate salts have been shown to decompose, even at room temperature, if they are wet [14]. Various amounts of these decomposition products might explain some of the differences in the literature values of CO_2 solubility in [PF$_6$]$^-$ ILs [13]. As a result, one must always be cognizant of the potential influence of impurities on gas solubilities.

3.4.2.2 The Stoichiometric Technique

The simplest method to measure gas solubilities is by what we will call the stoichiometric technique. It can be done either at constant pressure or with a constant volume of gas. For the constant pressure technique, a given mass of IL is contacted with the gas at a fixed pressure. The liquid is stirred vigorously to enhance mass transfer and allow approach to equilibrium. The total volume of gas delivered to the system (minus the vapor space) is used to determine the solubility. If the experiments are performed at sufficiently high pressure that the ideal gas law does not apply, then accurate equations of state can be employed to convert the volume of gas to moles. For the constant volume technique, a known volume of gas is contacted with the stirred ionic liquid sample. Once equilibrium is reached, the pressure is noted, and the solubility is determined as before. The effect of temperature (and, thus, enthalpies and entropies) can be determined by repeating the experiment at multiple temperatures.

The advantage of the stoichiometric technique is that it is extremely simple. Care has to be taken to remove all gases dissolved in the IL sample initially, but this is easily accomplished because one does not have to worry about volatilization of the IL sample when the sample chamber is evacuated under usual laboratory conditions. The disadvantage of this technique is that it requires relatively large amounts of IL to get accurate measurements for gases that are only sparingly soluble. For instance, at ambient temperature and pressure 10 mL of 1-n-butyl-3-methylimidazolium hexafluorophosphate ([BMIM][PF$_6$]) would take up only 0.2 cm^3 of a gas with a Henry's law constant of 5000 bar. Also, small temperature variations can cause large uncertainties. For instance, for 50 cm^3 of gas, a temperature fluctuation of just 1 °C would cause about a 0.2 cm^3 volume change. For some metal apparatuses of this type, gas adsorption on the metal surfaces can be an additional source of error. Thus, stoichiometric measurements are more accurate for high solubility gases and, in general, require excellent temperature and pressure control and measurement, as well as relatively large samples.

3.4.2.3 The Gravimetric Technique

An alternative technique to the stoichiometric method for measuring gas solubilities has evolved as a result of the development of extremely accurate microbalances. The gravimetric technique involves the measurement of the weight gain of an IL sample when gases are introduced into a sample chamber at a given pressure. There are various commercial apparatuses (e.g. Hiden Analytical, Cahn, Rubotherm) that are well suited for this purpose. The gravimetric technique was originally designed for gas uptake by solids (e.g., zeolites), but it is well-suited for ILs. Even the powerful vacuum ($\sim 10^{-9}$ bar) used to evacuate the system prior to gas introduction does not evaporate most ionic liquid at room temperature.

The main advantage of the gravimetric technique is that it requires a much smaller sample than the stoichiometric technique. In many cases, samples as small as 70 mg are sufficient. Accurate temperature and pressure control and measurement are still required, but gas adsorption on the metal walls of the equipment is no longer a concern because it is only the weight gain of the sample that is measured.

There are two main disadvantages to this technique. First, the sample is placed in a static sample "bucket"; i.e. there is no possibility of stirring. Thus, equilibrium is reached solely by diffusion of the gas into the IL sample. For the more viscous samples this can require equilibration times of as much as several hours. Second, the weight gain must be corrected for the buoyancy of the sample in order to determine the actual gas solubility. While the mass is measured accurately, the density of the sample must also be known accurately for the buoyancy correction. This is a particularly important problem for low solubility gases, where the buoyancy correction is a large percentage of the weight gain. For example, the density of an IL must be known to at least $\pm 0.5\%$ if one wishes to accurately measure the solubility of gases with Henry's law constants greater than 2000 bar. It is, of course, the density of the IL/gas mixture at each state point measured that is needed. For low pressures and low solubility gases, one may be able to use the density of the pure IL at the same temperature. However, for higher solubility gases (like CO_2) or higher pressures, using the density of the pure IL in the buoyancy correction can introduce significant error. A detailed description of a Hiden Analytical (IGA003) microbalance and its use for the measurement of gas solubilities in ILs can be found elsewhere [1].

Quartz crystal microbalances have also been used to measure gas solubilities in ILs [15]. In these experiments a thin layer of IL is spread on a quartz crystal and the mass uptake is measured by the change in vibration frequency of the crystal. An important consideration in these experiments is that a sufficiently large frequency change results from the mass uptake. As a result, this technique is best suited to high solubility gases.

3.4.2.4 Spectroscopic Techniques

Various spectroscopic techniques can also be used to measure gas solubilities in ILs. For instance, Welton and coworkers have used proton NMR spectroscopy [16] to determine the solubility of hydrogen in a series of ILs. Since hydrogen exhibits low solubility in ILs and has such a low molecular weight, it is difficult to measure gravimetrically or by any of the stoichiometric techniques. As a result, it is particularly well suited to determination by spectroscopy. In addition, Kazarian and coworkers have measured CO_2 solubility by infrared spectroscopy [17]. In general, spectroscopic techniques are quite attractive, as long as extinction coefficients do not vary significantly with temperature and pressure, since they can be performed *in situ*.

3.4.2.5 Gas Chromatography

Another method to determine infinite dilution activity coefficients (or the equivalent Henry's law coefficients) is gas chromatography [18, 19]. In this method, the chromatographic column is coated with the liquid solvent (e.g., the IL). The solute (the gas) is introduced with a carrier gas and the retention time of the solute is a measure of the strength of interaction (i.e., the infinite dilution activity coefficient, γ_1^∞) of the solute in the liquid. For the steady state method, γ_1^∞ is given by [18, 19]:

$$F^0(t - t_{\text{ref}})J_3^2 = \frac{RTn_2^1}{\gamma_1^\infty P_1^{\text{sat}}} \qquad (3.4\text{-}6)$$

where F^0 is the flowrate of the carrier gas, t and t_{ref} are the retention times of the solute of interest and a solute that is not retained, respectively, J_3^2 is a correction factor for the pressure drop across the column, R is the gas constant, T is the temperature, n_2^1 is the mass of solvent on the column, and P_1^{sat} is the vapor pressure of the solute. Additional corrections are required for a high pressure, nonideal gas phase. This technique has been used to measure the infinite dilution activity coefficients of a wide variety of liquid solutes in ionic liquids [20–24] and could conceivably be used to get the infinite dilution activity coefficients of condensable gases, as well. This technique would work for condensable gases that are retained by the IL more strongly than the carrier gas (usually helium), which may well be the case for many of the alkanes and alkenes of interest, as will be shown below.

3.4.3 Gas Solubilities

Published studies on the solubilities of gases in ionic liquids (ILs) have increased dramatically since 2000. These studies include common industrial gases such as CO_2, reaction gases such as O_2, as well as other gases of interest, such as light hydrocarbons. The number of ILs synthesized and analyzed has also increased, though the majority of these studies still focus on 1-n-butyl-3-methylimidazolium hexafluorophosphate ([BMIM][PF$_6$]) as a benchmark IL.

The solubility of gases in ILs is extremely important in order to understand and optimize the use of ILs in reactions, separations, and other processes. Although the number of gas solubility studies has increased in recent years, further studies are still needed to ascertain thermophysical relationships between the gases and ILs. These relationships are essential for the future synthesis of ILs, as they provide valuable information on the underlying solvent behavior of ILs. For example, these relationships can be used to tune the ILs for specific purposes [25, 26] and to choose IL functional groups to avoid environmental toxicity [27–29]. An important tool in determining these thermophysical relationships is the enthalpy and entropy of gas dissolution. The enthalpy provides information on the strength of the interactions between the IL and the gas, while the entropy provides information on the ordering that takes place in the gas–IL mixture. This information can complement IL physical property data needed to size equipment and processes for industrial applications.

As mentioned earlier, it is important to pay particular attention to the purity of the ILs and gases in the solubility measurements, as impurities may have a large effect on the measured solubility. Water, for example, has a profound effect on many of the physical properties of ILs [12]. This is important as many ILs are extremely hygroscopic [3,12]. [BMIM][PF_6] can absorb up to 0.16 mole fraction water from the atmosphere [3]. In general, water in the IL causes a decrease in viscosity, while increases in viscosity have been attributed to halide content. Similarly, water and halide content both cause a decrease in density [3, 12]. Thus, impurities can have a profound affect on the transport properties of the IL. As an example, gas solubility in gravimetric solubility measurements is affected by uncertainty in the IL density as propagated by the buoyancy correction. Particular attention should be paid to IL purity and all resulting impurities should be reported for the solubility measurements.

Gas solubility data have been collected both at low pressure, defined here as less than 50 bar, and high pressure, greater than 50 bar. For the low-pressure studies, the solubilities of many gases can be described by Henry's law. Due to the increasing amount of data presented in the literature, Henry's law constants will be used for comparison between ILs to illustrate the differences in gas solubilities. The high pressure data will be compared using the mole fraction of gas at a given pressure and temperature.

The solubilities of both pure and mixed gases are very important for the determination of thermophysical properties. For this section, solubility data will be separated into the following four categories, CO_2, reaction gases (O_2, H_2, CO), other gases (N_2, CH_4, C_2H_6, C_2H_4, SO_2, H_2O, Ar, etc.), and mixed gases.

3.4.3.1 CO_2

CO_2 solubility is of particular importance due to its use in industrial mixtures and its significance for green engineering. CO_2 has one of the highest solubilities in ILs and most solubility measurement techniques can quantify CO_2 sorption.

In the literature there is a variety of CO_2 gas solubility measurements available. The ILs appearing most often include [BMIM][PF_6] and [BMIM][BF_4]. Unfortunately, [BMIM][PF_6] is a particular concern as it degrades to HF in the presence of water, as mentioned previously [14]. Products which result from this degradation can also affect the solubility measurements. The published solubility results include both high and low pressure measurements that utilize numerous apparatus based on gravimetric, stoichiometric, chromatographic, and spectroscopic techniques. Low-pressure solubility results will be presented followed by a discussion of high-pressure CO_2 solubility measurements.

Low-pressure CO_2 solubility

The ability to "tune" IL properties has led to an increase in the number of ILs synthesized in recent years. CO_2 solubilities in a range of ILs are shown in Table 3.4-1. The most commonly used ILs include [BMIM][PF_6] and [BMIM][BF_4]. Figures 3.4-1 and 3.4-2 compare measured Henry's law constants for the binary systems of

CO_2 + [BMIM][PF_6] and CO_2 + [BMIM][BF_4], respectively. Note that the values are plotted on a logarithmic scale. Some of the variation in Fig. 3.4-1 is likely due to degradation products present in the CO_2 + [BMIM][PF_6] system. Nonetheless, the agreement is reasonably good. From Fig. 3.4-2 it is clear that there is good agreement for CO_2 solubility in [BMIM][BF_4] between the data presented by Anthony et al. and Shiflett and coworkers, but that Husson-Borg et al. report somewhat higher solubilities.

There are several trends that appear in the CO_2 solubility data. The first is that the influence of changing the anion is much larger than making changes in the cation. This can be seen from a study of [BMIM][PF_6], [BMIM][BF_4], and [BMIM][Tf_2N] at 25 °C, where the Henry's law constant ranged from 33 bar to 59 bar [30], versus a study increasing the length of the alkyl chain on the cation from [BMIM][Tf_2N], to [1-n-hexyl-3-methylimidazolium][Tf_2N] ([HMIM][Tf_2N]), and [1-n-octyl-3-methylimidazolium][Tf_2N] ([OMIM][Tf_2N]) where the Henry's law constants only decreased from 37 bar to 30 bar at 25 °C [15]. This small effect of changing the substituents on the cation is also seen for ILs with anions other than [Tf_2N]$^-$, such as [PF_6]$^-$ and [BF_4]$^-$ [31]. Secondly, the substitution of a methyl group on the C2 carbon of the imidazolium ring does not largely affect the solubility at low pressure; for example [32], the Henry's law constant at 25 °C of CO_2 in [1-ethyl-3-methylimidazolium][Tf_2N] ([EMIM][Tf_2N]) is 35.6 bar versus 39.6 bar for [1-ethyl-2,3-dimethylimidazolium][Tf_2N] ([EMMIM][Tf_2N]). Lastly, fluorination of the cation and anion increases CO_2 solubility [33], with the anion fluorination having the larger effect. For example, increasing the cation fluorination from [HMIM][Tf_2N] to [$C_6H_4F_9$][Tf_2N] decreases the Henry's law constant from 34.6 bar to 28.5 bar at 25 °C, but increasing the anion fluorination from [HMIM][Tf_2N] to 1-n-hexyl-3-methylimidazolium tris(pentafluoroethyl)trifluorophosphate ([HMIM][eFAP]) decreases the Henry's law constant from 34.6 bar to 21.7 bar at 25 °C [34].

Additionally, new classes of ILs have been developed for particular functions and are generally called task-specific ionic liquids, or TSILs. For CO_2, the TSIL [25] contained an amine functional group on an imidazolium cation which can chemically complex with the CO_2. The CO_2 is sequestered as an ammonium carbamate salt and thus the molar uptake of CO_2 per mole of TSIL can approach the 0.5 theoretical maximum. This per mole uptake of CO_2 by the amine-appended TSIL is comparable to those of standard sequestering amines such as monoethanolamine (MEA). Though the CO_2 absorption is much greater for the TSIL than an IL without the functional group, [BMIM][BF_4], 59 bar [30], versus [1-propylamine-3-butyl-imidazolium] [BF_4], 15 bar [25], the amount of energy required to remove the CO_2 from the IL is much larger. Thus, the TSILs retain some of the disadvantages of conventional amines when used for CO_2 removal in an absorption/desorption processes.

High-pressure CO_2 solubility

High pressure CO_2 gas solubility in ILs is of particular interest for supercritical CO_2 (scCO_2) extractions and biphasic reactions. The challenge in this area is to separate products from reactants, catalysts, or solvents. Published results on these

Table 3.4-1 Low pressure solubility of CO_2 in ILs

Source Reference	Compound Cation	Anion	Henry's law constant (bar)	Temp. (K)	Enthalpy (kJ mol^{-1})	Entropy (J mol^{-1} K^{-1})
34	1-n-butyl-3-methyl-imidazolium	hexafluorophosphate	38.7	283	−16.1	−53.2
			53.4	298		
			81.3	323		
30	1-n-butyl-3-methyl-imidazolium	hexafluorophosphate	38.8	283	−14.3	−47.6
			53.4	298		
			81.3	323		
	1-n-butyl-3-methyl-imidazolium	tetrafluoroborate	41.8	283	−13.9	−45.6
			59.0	298		
			88.6	323		
	1-n-butyl-3-methyl-imidazolium	bis(trifluoromethylsulfonyl)imide	25.3	283	−12.5	−41.3
			33.0	298		
			48.7	323		
	methyl-tri-butyl-ammonium	bis(trifluoromethylsulfonyl)imide	43.5	298		
	methyl-butyl-pyrrolidinium	bis(trifluoromethylsulfonyl)imide	30.2	283	−11.9	−38.7
			38.6	298		
			56.1	323		
15	tri-isobutyl-methyl-phosphonium	tosylate	90.5	298		
	1-propyl-3-methyl-imidazolium	bis(trifluoromethylsulfonyl)imide	37.0	298		
	1-propyl-3-methyl-imidazolium	hexafluorophosphate	52.0	298		
	1-n-butyl-3-methyl-imidazolium	bis(trifluoromethylsulfonyl)imide	37.0	298		
	1-n-hexyl-3-methyl-imidazolium	bis(trifluoromethylsulfonyl)imide	35.0	298		
	1-n-octyl-3-methyl-imidazolium	bis(trifluoromethylsulfonyl)imide	30.0	298		

(*Continued*)

Table 3.4-1 (*Continued*)

Source Reference	Compound Cation	Anion	Henry's law constant (bar)	Temp. (K)	Enthalpy (kJ mol^{-1})	Entropy (J mol^{-1} K^{-1})
	1-methyl-3-(tetradecylfluorooctyl)-imidazolium	bis(trifluoromethylsulfonyl)imide	4.5	298		
	1,4-dibutyl-3-phenyl-imidazolium	bis(trifluoromethylsulfonyl)imide	63.0	298		
	1-n-butyl-3-phenyl-imidazolium	bis(trifluoromethylsulfonyl)imide	180.0	298		
35	1-propyl-3-methyl-imidazolium	hexafluorophosphate	52.0	298		
	1-n-butyl-3-methyl-imidazolium	bis(trifluoromethylsulfonyl)imide	37.0	298		
	1-n-octyl-3-methyl-imidazolium	bis(trifluoromethylsulfonyl)imide	30.0	298		
	1-methyl-3-(tetradecylfluorooctyl)-imidazolium	bis(trifluoromethylsulfonyl)imide	6.0	298		
25	1-propylamine-3-butyl-imidazolium	tetrafluoroborate	14.8	295		
32	1-n-butyl-3-methyl-imidazolium	hexafluorophosphate	38.7	283	−16.1	−53.2
			53.4	298		
			81.3	323		
	1-n-butyl-2,3-dimethyl-imidazolium	hexafluorophosphate	47.3	283	−13	−42.8
			61.8	298		
			88.5	323		
	1-n-butyl-3-methyl-imidazolium	tetrafluoroborate	40.8	283	−15.9	−52.4
			56.5	298		
			88.9	323		
	1-n-butyl-2,3-dimethyl-imidazolium	tetrafluoroborate	45.7	283	−14.5	−47.7
			61.0	298		
			92.2	323		
	1-ethyl-3-methyl-imidazolium	bis(trifluoromethylsulfonyl)imide	25.3	283	−14.2	−46.9
			35.6	298		
			51.5	323		

	1-ethyl-2,3-dimethyl-imidazolium	bis(trifluoromethylsulfonyl)imide	28.6	283	−14.7	−48.7
			39.6	298		
			60.5	323		
36	1-ethyl-3-methyl-imidazolium	bis(trifluoromethylsulfonyl)imide	39.5	303		
	1-ethyl-3-methyl-imidazolium	dicyanamide	79.0	303		
	1-ethyl-3-methyl-imidazolium	triflate	74.0	303		
	1-n-butyl-3-methyl-imidazolium	hexafluorophosphate	59.8	303		
	tri-hexyl-tetradecyl-phosphonium	chloride	30.4	303		
37	1-n-butyl-3-methyl-imidazolium	hexafluorophosphate	60.0	303		
			74.0	313		
			86.0	323		
			95.0	333		
			109.0	343		
38	1-n-hexyl-3-methyl-imidazolium	bis(trifluoromethylsulfonyl)imide	31.6	298		
39	1-n-butyl-3-methyl-imidazolium	tetrafluoroborate	61.0	304	−10.2	−67.8
			62.0	314		
			71.0	324		
			84.0	334		
			87.0	344		
40	1-n-butyl-3-methyl-imidazolium	hexafluorophosphate	42.2	293	−17.24	−79.5
			66.5	313		
			95.0	333		
			125.6	353		
			157.3	373		
			193.2	393		

(*Continued*)

Table 3.4-1 (*Continued*)

Source Reference	Compound Cation	Anion	Henry's law constant (bar)	Temp. (K)	Enthalpy (kJ mol^{-1})	Entropy (J mol^{-1} K^{-1})
33	1-n-hexyl-3-methyl-imidazolium	bis(trifluoromethylsulfonyl)imide	24.2	283		
			31.6	298		
			45.6	323		
	1-methyl-3-(nonafluorohexyl)-imidazolium	bis(trifluoromethylsulfonyl)imide	28.4	298		
			48.5	333		
	1-methyl-3-(tetradecylfluorooctyl)-imidazolium	bis(trifluoromethylsulfonyl)imide	27.3	298		
			44.7	333		
	1-n-hexyl-3-methyl-imidazolium	tris(pentafluoroethyl)trifluorophosphate	25.2	298		
			42.0	333		
	1-n-hexyl-3-methyl-imidazolium	tris(heptafluoropropyl)trifluorophosphate	21.6	298		
			36.0	333		
	1-n-pentyl-3-methyl-imidazolium	tris(nonafluorobutyl)trifluorophosphate	20.2	298		
			32.9	333		
	1-hexyl-3-methylpyridinium	bis(trifluoromethylsulfonyl)imide	25.4	283		
			32.8	298		
			46.2	323		
	1-hexyl-3-methylimidazolium	acesulfamate	113.1	333		
	1-hexyl-3-methylimidazolium	saccharinate	132.2	333		
	(1-methylimidazole)(triethylamine)boronium	bis(trifluoromethylsulfonyl)imide	33.1	298		
41	1-n-butyl-3-methyl-imidazolium	hexafluorophosphate	59.8	303		
	1-ethyl-3-methyl-imidazolium	bis(trifluoromethylsulfonyl)imide	39.5	303		
	1-ethyl-3-methyl-imidazolium	triflate	74.0	303		
	1-ethyl-3-methyl-imidazolium	dicyanamide	79.0	303		
	tri-hexyl-tetradecyl-phosphonium	chloride	30.4	303		

42	1-n-butyl-3-methyl-imidazolium	hexafluorophosphate	42.4	283
			57.1	298
			87.0	323
			126.6	348
	1-n-butyl-3-methyl-imidazolium	tetrafluoroborate	44.6	283
			60.2	298
			93.5	323
			135.1	348
43	poly[1-(4-vinylbenzyl)-3-butyl-imidazolium tetrafluoroborate]		35.6	298
44	poly[1-(4-vinylbenzyl)-3-butyl-imidazolium tetrafluoroborate]		35.6	298
	poly[(1-(4-vinylbenzyl)-3-butyl-imidazolium hexafluorophosphate]		28.7	298
	poly[2-(1-butylimidazolium-3-yl)ethyl-methacrylate tetrafluoroborate]		44.6	298

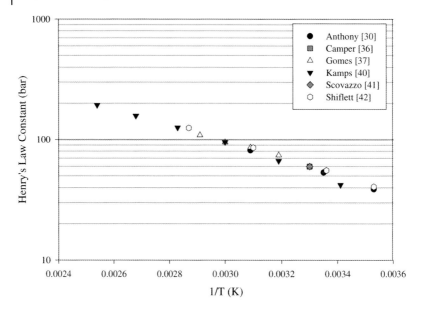

Fig. 3.4-1 Comparison of measured Henry's law constants of the system CO_2 and [BMIM][PF_6].

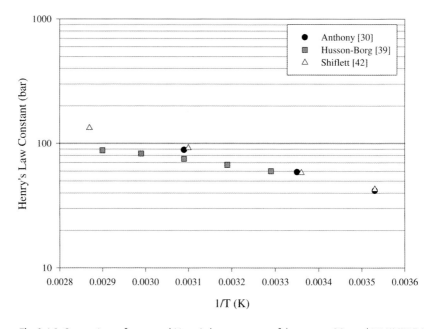

Fig. 3.4-2 Comparison of measured Henry's law constants of the system CO_2 and [BMIM][BF_4].

systems provide fundamental knowledge for developing reactions and separation techniques by utilizing the unique properties of both ILs and scCO$_2$. Table 3.4-2 shows the ILs studied under high pressure CO$_2$, the majority of which are focused on the imidazolium-based cation. Unfortunately, studies on the phase behavior of ILs in the presence of supercritical fluids other than CO$_2$ are scarce.

Supercritical fluid extraction with CO$_2$ has shown that a wide variety of solutes can be extracted from ILs with the solutes being recovered without IL contamination [9, 45, 46]. This is accomplished as CO$_2$ dissolves in the solvent mixture to facilitate extraction, but the IL does not dissolve in CO$_2$, so pure product can be recovered. In addition, IL/CO$_2$ biphasic solutions have been used for a variety of homogeneously catalyzed reactions, as well as for extraction and recovery of organic solutes. Interestingly, the CO$_2$/IL system remains two distinct phases, even under pressures up to 400 bar [46].

Many groups have published phase behavior data which is needed to design extraction and reaction processes necessary for these applications [13, 17, 31, 33, 37, 40, 47–55]. As shown in Table 3.4-2, the imidazolium ILs make up the majority of ILs studied, with very few exceptions. Results from the imidazolium ILs suggest that the solubility of CO$_2$ is strongly dependent on the choice of anion, with the solubility increasing with increasing fluorination on the anion. The cation also affects solubility with increasing alkyl chain length marginally increasing its CO$_2$ solubility. The solubility data indicate that an increase in temperature decreases the CO$_2$ solubility of the IL while an increase in pressure increases the solubility. All of these findings are consistent with the lower-pressure CO$_2$ solubility data.

To compare the CO$_2$ solubility results for different groups, Fig. 3.4-3 shows the mole fraction of CO$_2$ in [BMIM][PF$_6$] at 40 °C. As shown, the results agree reasonably well, though various concentrations of [PF$_6$]$^-$ degradation products are a likely cause for some deviation.

3.4.3.2 Reaction Gases (O$_2$, H$_2$, CO)

Gases such as O$_2$, H$_2$, and CO, are important in oxidations, hydrogenations, hydroformylations, etc. Determining the solubility of these gases is important as the rates of most reactions depend on the concentration of the reacting species. Thus, for reactions involving ILs, one would desire high gas solubilities in the ILs. Unfortunately, there exist very limited data for these gases. H$_2$ in particular is hard to measure due to its low molecular weight.

There are three groups who have studied O$_2$ + IL systems [30, 39, 56]. O$_2$ solubility data are available for [BMIM][PF$_6$] and [BMIM][BF$_4$], as well as a few others. In addition, Kumelan et al. [56], Husson-Borg et al. [39], Urakova et al. [57], and Shah et al. [58] have performed molecular simulations of the O$_2$ + IL system and found good agreement with their individual experimental results. Additional published data [30] from our laboratory show significantly lower O$_2$ solubility in [BMIM][PF$_6$], but these data likely have significantly more uncertainty than originally reported due to neglecting the IL expansion effect of the dissolved gas through the buoyancy correction. A comparison between all reaction gases for [BMIM][PF$_6$] is shown in Fig. 3.4-4.

Table 3.4-2 High-pressure solubility of CO_2 in ILs

Source Reference	Ionic Liquid Cation	Anion	Temp. range (K)	Pressure range (bar)
13	1-n-butyl-3-methyl-imidazolium	dicyanamide	298–333	10–150
	1-n-butyl-3-methyl-imidazolium	nitrate		
	1-n-butyl-3-methyl-imidazolium	tetrafluoroborate		
	1-n-butyl-3-methyl-imidazolium	hexafluorophosphate		
	1-n-butyl-3-methyl-imidazolium	triflate		
	1-n-butyl-3-methyl-imidazolium	bis(trifluoromethylsulfonyl)imide		
	1-n-butyl-3-methyl-imidazolium	methide		
	1-n-hexyl-3-methyl-imidazolium	bis(trifluoromethylsulfonyl)imide		
	1-n-hexyl-2,3-dimethyl-imidazolium	bis(trifluoromethylsulfonyl)imide		
	1-n-octyl-3-methyl-imidazolium	bis(trifluoromethylsulfonyl)imide		
45	1-n-butyl-3-methyl-imidazolium	hexafluorophosphate	313	0–400
46	1-n-butyl-3-methyl-imidazolium	hexafluorophosphate	313–333	1–100
	1-n-octyl-3-methyl-imidazolium	hexafluorophosphate		
	1-n-octyl-3-methyl-imidazolium	tetrafluoroborate		
	1-n-butyl-3-methyl-imidazolium	nitrate		
	1-ethyl-3-methyl-imidazolium	ethylsulfate		
	1-n-butyl-pyridinium	tetrafluoroborate		
9	1-n-butyl-3-methyl-imidazolium	hexafluorophosphate	295	0.98
47	1-n-hexyl-3-methyl-imidazolium	tetrafluoroborate	293–363	5–870
48	trihexyltetradecylphosphonium	chloride	323	104
40	1-n-butyl-3-methyl-imidazolium	hexafluorophosphate	293–393	10–90
17	1-n-butyl-3-methyl-imidazolium	hexafluorophosphate	298–473	1–200
	1-n-butyl-3-methyl-imidazolium	tetrafluoroborate		

49	1-n-butyl-3-methyl-imidazolium	tetrafluoroborate	279–368	5–700
50	1-n-butyl-3-methyl-imidazolium	hexafluorophosphate	313–333	0–125
33	1-methyl-3-(nonafluorohexyl)-imidazolium	bis(trifluoromethylsulfonyl)imide	298–333	10–150
	1-methyl-3-(tetradecylfluorooctyl)-imidazolium	bis(trifluoromethylsulfonyl)imide		
	1-n-hexyl-3-methyl-imidazolium	tris(pentafluoroethyl)trifluorophosphate		
	1-n-pentyl-3-methyl-imidazolium	tris(nonafluorobutyl)trifluorophosphate		
	1-butyl-3-methylimidazolium	trifluoroacetate		
	1-butyl-3-methylimidazolium	pentadecafluorooctanoate		
	1-butyl-nicotinic acid butyl ester	bis(trifluoromethylsulfonyl)imide		
	tetrabutylammonium	docusate		
	PEG-5 codomonium	methylsulfate		
	1-butyl-3-methyl imidazolium	2-(2-methoxyethoxy)ethylsulfate		
	choline	bis(trifluoromethylsulfonyl)imide		
	N,N,N,N-trimethylbutylammonium	bis(trifluoromethylsulfonyl)imide		
51	1-ethyl-3-methyl-imidazolium	hexafluorophosphate	308–366	15–971
52	1-n-hexyl-3-methyl-imidazolium	hexafluorophosphate	298–364	6–946
53	1-n-butyl-3-methyl-imidazolium	hexafluorophosphate	293–364	6–735
31	1-ethyl-3-methyl-imidazolium	hexafluorophosphate	290–370	0–1000
	1-n-butyl-3-methyl-imidazolium	hexafluorophosphate		
	1-n-hexyl-3-methyl-imidazolium	hexafluorophosphate		
54	trihexyl(tetradecyl)-phosphonium	dodecylbenzenesulfonates	305–325	40–90
	trihexyl(tetradecyl)-phosphonium	mesylate		

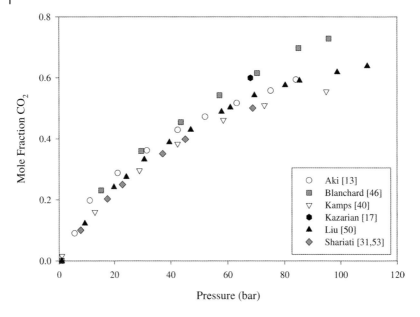

Fig. 3.4-3 Comparison of CO_2 solubility in [BMIM][PF_6] at 40 °C.

Welton and coworkers used NMR spectroscopy to measure hydrogen solubility in a wide variety of ILs, the majority of which are imidazolium-based cations paired with anions such as [PF_6]$^-$, [BF_4]$^-$, and [Tf_2N]$^-$ [16, 59]. Berger et al. [59] reported on the conversion and enantioselectivity of a hydrogenation reaction, as well as kinetic resolution by catalysts immobilized in the ILs. For these reactions, the H_2 concentration in the ionic phase, rather than pressure in the gas phase, was the important kinetic parameter because the reaction occurred in the IL. For the ILs investigated in both studies, Henry's constants ranged from 0.7–7.2 × 10^3 bar.

Carbon monoxide solubility has also been studied for a large number of ILs [30, 60, 61]. CO solubility is similar to or greater than H_2 solubility for the ILs where H_2 solubility data are available. The CO + [BMIM][PF_6] system has also been modeled and compared to a correlation by Hayden and O'Connell [62] which yields very good agreement with the Henry's law constants observed experimentally by Kumelan [60]. However, the available published results differ by a substantial amount, as shown in Fig. 3.4-4. For example, the smallest Henry's law constants found by Kumelan et al. [60] were approximately 70% smaller than those found by Ohlin [61], and approximately an order of magnitude smaller than those found by Anthony et al. [30]. As with H_2, there is little agreement in the literature for these low solubility reactant gases. Further study of these gases is needed to resolve these differences.

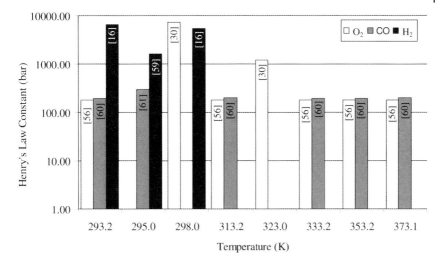

Fig. 3.4-4 Comparison between reaction gas solubilities in [bmim][PF$_6$].

3.4.3.3 Other Gases (N$_2$, Ar, CH$_4$, C$_2$H$_6$, C$_2$H$_4$, H$_2$O, SO$_2$, CHF$_3$, etc.)

There are a number of studies that have investigated the solubilities of other gases and vapors such as light hydrocarbons, water, inert gases, and toxic gases. The availability of such data is still limited, but it is important in furthering the knowledge of structure–property relationships of gas–IL systems. Additionally, the determination of the gas solubilities is important in order to establish gas selectivities for specific applications such as gas separations. The solubility data gained from these studies are vital to understanding how chemical compounds interact with ILs, and for furthering our knowledge of physical data in order to create models which can predict physical properties for ILs.

There are data available for the solubility of inert gases in molten salts. For example, in a 1980 study [63], the solubility of Ar and N$_2$ was determined for LiNO$_3$ and KNO$_3$ at high temperature (263–338 °C for LiNO$_3$ and 335–413 °C for KNO$_3$). In general, N$_2$ and Ar gas solubilities are difficult to measure using many techniques, such as the gravimetric technique, because of their low molecular weight, low solubility, and inert nature. As a result, there is only one study available at present that gives the Henry's law constant for Ar in an IL; the value for Ar in [BMIM][PF$_6$] was reported as 7310 ± 3690 bar at 25 °C [30].

Light hydrocarbons such, as CH$_4$, C$_2$H$_4$, and C$_2$H$_6$, have very limited published gas solubility data [30, 34, 36, 38, 64]. No direct comparison between studies is possible, though close approximations between Camper et al. [36] and Anthony et al. [30] indicate that the results agree well. For example, the Henry's law constant for C$_2$H$_4$ in [BMIM][PF$_6$] has been reported as 180 ± 16 bar at 30 °C [36] and 144 ± 2 bar at 25 °C [30].

Water solubility is of particular interest as many ILs are hygroscopic and can absorb large quantities of water from atmospheric exposure. Measured water solubility values are available from Anthony et al. [1] and Cammarata et al. [65]. Anthony [1] measured the solubility of water vapor in three ILs, [OMIM][BF$_4$], [BMIM][PF$_6$], and [OMIM][PF$_6$], and found the Henry's constants to range between 0.033 and 0.45 bar. Cammarata [65] measured the state of water in ILs with a [BMIM]$^+$ cation and various anions including [PF$_6$]$^-$, [BF$_4$]$^-$, [CF$_3$SO$_3$]$^-$, etc. and found the strength of H-bonding between the anion and the water molecules. Results from this study indicate that [PF$_6$]$^-$ has the weakest H-bonding interactions, while [CF$_3$CO$_2$]$^-$ has the strongest. One implication of these results is that the strength of the H-bonding interaction appears to be in direct correlation with the water solubility in the IL. This is important as the presence of water may have important effects on the solvent properties of ILs, such as density and viscosity.

SO$_2$ has recently received more attention since the need to separate industrial flue gases has become increasingly important. Because this gas is one of the more toxic and difficult to manage gases, its solubility data are limited to one publication with ILs [26]. The IL used in this study could also be classified as a TSIL, as the IL has an attached guanamine functional group specifically intended to improve SO$_2$ solubility. The TSIL ([1,1,3,3-tetramethyl-guanidium][lactate]) absorbed SO$_2$ with a mole ratio of 0.978 at 40 °C when exposed to a simulated flue gas stream composed of 8% SO$_2$ by volume and the balance N$_2$. When exposed to a pure stream of SO$_2$, the mole ratio increased to 1.7. In all cases, the SO$_2$ was removed easily from the IL under vacuum at 40 °C with the IL recycled over many trials showing no loss in SO$_2$ absorbance capacity.

The gas solubility of fluoroform (CHF$_3$) was measured by Peters and coworkers [31, 55]. The initial interest in this gas was for use as a supercritical fluid to extract solutes from an IL. The high-pressure phase behavior of the system CHF$_3$ + [EMIM][PF$_6$] was determined and the results show that the solubility of supercritical CHF$_3$ in [EMIM][PF$_6$] is very high and that the phase behavior is completely different than that of the CO$_2$ + [BMIM][PF$_6$] system. In particular, at high pressures the CHF$_3$ and the IL are completely miscible; this is not the case for CO$_2$.

3.4.3.4 Mixed Gases

It has been shown that H$_2$ solubility increases in the presence of CO$_2$ at high pressures [66]. This study spurred interest in how CO$_2$ affects the solubility of lower solubility gases in ILs. Mixed gas solubilities available in the literature include O$_2$/CO$_2$ and CH$_4$/CO$_2$ in [HMIM][Tf$_2$N] [38]. The presence of CO$_2$ dramatically enhances the O$_2$ and CH$_4$ solubility in ILs with the introduction of only small amounts and low pressures of CO$_2$. The resulting enhancement can be quantified by an enhancement factor (EF), defined as the ratio of the low solubility gas's solubility in the mixed gas system and in the pure gas system, at the same temperature and fugacity. The EFs range from 1.7 to 4.9 for the CO$_2$/O$_2$/[HMIM][Tf$_2$N] system and from 1.4 to 2.6 for the CO$_2$/CH$_4$/[HMIM][Tf$_2$N] system.

Industrial operations will require further experimentation into mixed gas solubilities to gain a better understanding of their properties. In addition to experimental observations, computation simulation and modeling may assist with the combinatorial number of mixed gases.

Table 3.4-3 Enthalpies and entropies for [BMIM][Tf$_2$N] [30]

Gas	Enthalpy (kJ mol^{-1})	Entropy (J mol^{-1} K^{-1})
CO_2	-12.5 ± 0.4	-41.3 ± 1.4
N_2O	-12.5 ± 0.4	-41.5 ± 1.2
C_2H_4	-9.0 ± 2.1	-29.9 ± 6.8
C_2H_6	-9.8 ± 2.9	-32.5 ± 9.5
O_2	20.6 ± 8.2	67.4 ± 26.9

3.4.3.5 Enthalpies and Entropies

As mentioned earlier, the enthalpy and entropy of gas dissolution can yield valuable information about the strength of interactions between the gas and IL and about the degree of ordering present in the gas/IL mixture, respectively. Table 3.4-1 gives the enthalpies and entropies available from the literature for CO_2 + IL systems. In addition to those listed in Table 3.4-1, some enthalpies and entropies exist for other gases, such as O_2 [30, 39], C_2H_4 [30], C_2H_6 [30], and CH_4 [30]. Some of these values, from a study by Anthony et al. [30], are listed in Table 3.4-3 to illustrate the range of interactions and ordering that is possible for different gases in ILs.

Although there are many studies that have measured gas solubility, only a few also report the enthalpy and entropy. A negative enthalpy indicates attractive interactions between the gas and the IL. As shown in Table 3.4-3, CO_2 and N_2O have the strongest enthalpic interactions between the gas and IL. The strength of the enthalpic interactions decreases as $CO_2 \approx N_2O > C_2H_4 > C_2H_6 > CH_4 > O_2$. Note that the value for O_2, although containing a large uncertainty, is actually positive (i.e., repulsion), which means that the solubility of O_2 increases with increasing temperature. It also means that the dissolution of O_2 into the IL is driven by the increase in entropy that results from mixing the two substances, rather than any attraction between the gas and the IL. More measurements of the enthalpy and entropy from gas solubility data will provide further information regarding the interactions between the gas and the IL. This information will be vital to the design of future industrial applications for ILs.

3.4.4
Applications

The solubilities, discussed above, of the various gases in the ionic liquids have important implications for the applications of ILs mentioned in the Introduction. The impact of gas solubilities on reactions, gas storage, gas separations and the use of compressed gases or supercritical fluids to separate solutes from ILs are discussed below.

3.4.4.1 Reactions Involving Gases

In general, the rates of most reactions depend on the concentration of the reacting species with a positive power. Thus, it is generally desirable to have high solubilities of reactants. For reactions in ILs involving gases, this means one would desire high gas solubilities. This is true in all cases except for the instance where the inherent reaction rate is extremely slow; i.e., slower than the natural diffusion of the gas into the IL. Otherwise, one would have to resort to high-pressure operation or vigorous stirring in an attempt to increase the interfacial area and to promote faster mass transfer. In this case, the reaction rate is limited by the rate at which the gas is transferred into the liquid, rather than by the inherent reaction kinetics. This could be a particular challenge when using ILs as solvents, since they tend to be more viscous than conventional solvents [67, 68]. High viscosity means lower diffusivities and more difficulty in attempts to promote adequate mass transfer. It should be noted, though, that the addition of most liquid reactants will likely reduce the viscosity of the solution substantially [2, 3].

Some types of reactions that have been studied in ILs that involve gases are hydrogenations [59, 69–81], oxidations [82, 83] and hydroformylations [69, 84–89]. In addition, some dimerizations and alkylations may involve the dissolution of condensable gases (e.g., ethylene, propylene, isobutene) in the IL solvent [90–94].

Surprisingly, in most of these reaction studies, no effort was made to determine the influence of gas solubility or whether the reaction was, in fact, mass transfer limited. However, due to the low solubility of H_2, O_2, and CO in most ILs, most of the reactions in these studies are likely to have been mass transfer limited. For example, Suarez et al. [71] note that in the absence of stirring the consumption of hydrogen in their hydrogenation reaction is completely stopped. This same group [59] measured Henry's law constants of hydrogen in [BMIM][PF$_6$] and [BMIM][BF$_4$], as discussed above. They correctly note that in comparing reaction results in different ILs, the important physical parameter to hold constant is the gas solubility rather than the gas partial pressure. Moreover, for mass transfer limited reactions, it is not possible to compare results of reaction rates for reactions performed by different research groups. This is because the rates depend solely on the quality of the interphase mass transfer and all the researchers perform the reactions with different mixing conditions.

As mentioned above, the solubilities in ILs of the gases needed for hydrogenation reactions (H_2), oxidation reactions (O_2), and hydroformylation reactions (H_2 and CO) are extremely low. In fact, the solubilities are generally less than in conventional organic solvents [34]. The primary substrate in the hydroformylation reactions is an alkene, the lightest example of which would be ethylene. Since ethylene solubility is reasonably high, the CO and H_2 solubilities are likely to be the limiting factors, rather than the solubility of the alkene. Thus, overall, most ILs do not appear to be good industrial solvents for reactions involving permanent gases such as H_2, O_2, or CO because these reactions would have to be carried out at high pressures or they would be limited by interphase mass transfer. While there are certainly examples of commercial biphasic reactions that are limited by interphase mass transfer [95], that type of reactor configuration is generally not preferred. There would have to be

3.4.4.2 Gas Storage

Even though the solubility of common reactant gases (CO, O_2, H_2) in ILs is generally low, the solubility of some other reactant gases, particularly those used in the microelectronics industry, can be quite high. Air Products [96] has developed a system to use ILs as a storage and delivery medium for highly toxic gases such as AsH_3, PH_3, and BF_3. Currently these gases are provided in cylinders at sub-atmospheric pressure, sometimes using a solid adsorbent, to prevent any possibility of accidental release. ILs have been found to be an effective storage and delivery medium for these gases, essentially serving as a liquid adsorbent. They provide good storage capacity, but the sorption is not strong enough to prevent release of the gases when needed.

3.4.4.3 Gas Separations

Since the solubility of various gases in ILs varies widely, they may be uniquely suited for use as solvents for gas separations [97]. Since they are non-volatile, they cannot evaporate to cause contamination of the gas stream. This is important when selective solvents are used in conventional absorbers, or when they are used in supported liquid membranes. For conventional absorbers, the ability to separate one gas from another depends entirely on the relative solubilities (ratio of Henry's law constants) of the gases. In addition, ILs are particularly promising for supported liquid membranes because they have the potential to be incredibly stable. Supported liquid membranes that incorporate conventional liquids eventually deteriorate because the liquid slowly evaporates. Moreover, this finite evaporation rate limits how thin one can make the membrane. This means that the net flux through the membrane is decreased. These problems could be eliminated with a non-volatile liquid. In the absence of facilitated transport (e.g., complexation of CO_2 with amines to form carbamates), the permeability of gases through supported liquid membranes depends on both their solubility and diffusivity. The flux of one gas relative to the other can be estimated using a simplified solution-diffusion model:

$$\frac{N_A}{N_B} = \frac{H_B \, D_A}{H_A \, D_B} \tag{3.4-7}$$

where N is the flux, H is the Henry's law constant and D is the diffusivity. Thus, the ratio of Henry's law constants is critical in determining the performance of an IL-based supported liquid membrane. The potential of using ILs for gas separations in both absorber and membrane configurations has been discussed by several researchers [35, 98].

The solubility of different gases in the various ILs, as discussed above, suggests that ILs should be excellent candidates for a wide variety of industrially important gas separations. There is also the possibility of doing higher temperature gas separations due to the high thermal stability of the ILs. For supported liquid membranes this would require the use of ceramic or metallic membranes rather than polymeric ones. It may be possible to remove both water vapor and CO_2 from natural gas, in spite of the complication due to enhanced mixed gas solubilities (as discussed in Section 4.3.4) since the sorption capacity of the ILs studied to date for water and CO_2 is significantly greater than CH_4. It should be possible to scrub CO_2 from stack gases composed of N_2 and O_2. Although measurements of H_2S, SO_x, or NO_x solubility in ILs are very limited or nonexistent, it is reasonable to believe that it might be possible to remove these contaminants as well. One should be able to remove CO_2 from syngas. Overall, there appears to be ample opportunity for using ILs for gas separations based upon the widely varying gas solubilities measured thus far.

3.4.4.4 Extraction of Solutes from Ionic Liquids with Compressed Gases or Supercritical Fluids

Brenneke et al. have shown that it is possible to extract a wide variety of solutes from ILs with supercritical CO_2 [9, 45]. The advantage of this technique is that it combines two potentially environmentally benign solvents to perform the reaction and separation steps. Subsequently, several groups have shown that this strategy can be combined with reaction operations [80, 81, 89]. The key to this separation is the phase behavior of ILs with CO_2. Although large amounts of CO_2 dissolve in the IL, no measurable IL dissolves in the CO_2. Conceivably, it should be possible to extract solutes from IL mixtures with other supercritical fluids or compressed gases. For instance, supercritical ethane and propane are excellent solvents for a wide variety of nonpolar and aromatic species. These are not as environmentally benign as CO_2 but may be options if it is shown that the ILs have negligible solubility in them also.

3.4.5
Summary

The solubility of various gases in ionic liquids is extremely important in evaluating ILs as solvents for reactions, separations, and materials processing. There are a number of viable techniques for measuring gas solubilities in ILs, including the stoichiometric method, gravimetric methods, spectroscopy, or even gas chromatography. In general, the measurement of these solubilities is facilitated by the non-volatility of the solvent. For a variety of ILs, researchers have shown very large solubility differences between the relatively high solubility gases (CO_2, C_2H_4, C_2H_6, and CH_4) and the low solubility gases (CO, H_2, O_2, Ar, and N_2). Unfortunately, many of the gases of interest for reactions (H_2, O_2 and CO) are only sparingly soluble in the IL. Nonetheless, a number of toxic gases used in the microelectronics industry are highly soluble in ILs, which can serve as a safe storage and delivery medium. Moreover, the large differences in the solubility of different gases in the IL suggest that ILs may be ideal solvents for performing gas separations. Finally, the wide variety of choice of cations, anions and substituents make the possibility of tailoring ILs for specific gas separations or reactions involving gases an exciting option.

References

1. J. L. Anthony, E. J. Maginn, J. F. Brennecke, *J. Phys. Chem. B*, **2001**, *105*(44), 10942–10949.
2. A. G. Fadeev, M. M. Meagher, *Chem. Commun.*, **2001**(03), 295–296.
3. K. R. Seddon, A. Stark, M. J. Torres, *Pure Appl. Chem.*, **2000**, *72* (12), 2275–2287.
4. J. D. Holbrey, K. R. Seddon, *Clean Prod. Proc.*, **1999**(1), 223–236.
5. T. Welton, *Chem. Rev.*, **1999**, *99*(8), 2071–2083.
6. P. Wasserscheid, W. Keim, *Angew. Chem. Int. Edit.*, **2000**, *39*(21), 3773–3789.
7. C. L. Hussey, *Pure Appl. Chem.*, **1988**, *60*(12), 1763–1772.
8. K. R. Seddon, *Kinet. Catal.*, **1996**, *37*(5), 693–697.
9. L. A. Blanchard, J. F. Brennecke, *Ind. Eng. Chem. Res.*, **2001**, *40*(11), 2550–2550.
10. A. M. Scurto, S. Aki, J. F. Brennecke, *J. Am. Chem. Soc.*, **2002**, *124*(35), 10276–10277.
11. A. M. Scurto, S. Aki, J. F. Brennecke, *Chem. Commun.*, **2003**(5), 572–573.
12. J. A. Widegren, A. Laesecke, J. W. Magee, *Chem. Commun.*, **2005**(12), 1610–1612.
13. S. N. V. K. Aki, B. R. Mellein, E. M. Saurer, J. F. Brennecke, *J. Phys. Chem. B*, **2004**, *108*(52), 20355–20365.
14. R. P. Swatloski, J. D. Holbrey, R. D. Rogers, *Green Chem.*, **2003**, *5*, 361.
15. R. E. Baltus, B. H. Culbertson, S. Dai, H. M. Luo, D. W. DePaoli, *J. Phys. Chem. B*, **2004**, *108*(2), 721–727.
16. P. J. Dyson, G. Laurenczy, C. A. Ohlin, J. Vallance, T. Welton, *Chem. Commun.*, **2003**(19), 2418–2419.
17. S. G. Kazarian, B. J. Briscoe, T. Welton, *Chem. Comm.*, **2000**, 2047–2048.
18. R. J. Laub, R. L. Pecsock, *Physicochemical Applications of Gas Chromatography*, Wiley: New York, **1978**.
19. H. Orbey, S. I. Sandler, *Ind. Eng. Chem. Res.*, **1991**, *30*(8), 2006–2011.
20. A. Heintz, D. V. Kulikov, S. P. Verevkin, *J. Chem. Eng. Data*, **2001**, *46*(6), 1526–1529.
21. A. Heintz, D. V. Kulikov, S. P. Verevkin, *J. Chem. Thermodyn.*, **2002**, *34*(8), 1341–1347.
22. A. Heintz, D. V. Kulikov, S. P. Verevkin, *J. Chem. Eng. Data*, **2002**, *47*(4), 894–899.
23. T. M. Letcher, B. Soko, P. Reddy, N. Deenadayalu, *J. Chem. Eng. Data*, **2003**, *48*(6), 1587–1590.
24. T. M. Letcher, B. Soko, D. Ramjugernath, N. Deenadayalu, A. Nevines, P. K. Naicker, *J. Chem. Eng. Data*, **2003**, *48*(3), 708–711.
25. E. D. Bates, R. D. Mayton, I. Ntai, J. H. Davis, Jr., *J. Am. Chem. Soc.*, **2002**, *124*(6), 926–927.
26. W. Wu, B. Han, H. Gao, Z. Liu, T. Jiang, J. Huang, *Angew. Chem. Int. Ed.*, **2004**, *43*(18), 2415–2417.
27. K. M. Docherty, C. F. Kulpa, *Green Chem.*, **2005**, *7*(4), 185–189.
28. R. J. Bernot, M. A. Brueseke, M. A. Evans-White, G. A. Lamberti, *Environ. Toxicol. Chem.*, **2005**, *24*(1), 87–92.
29. R. J. Bernot, E. E. Kennedy, G. A. Lamberti, *Environ. Toxicol. Chem.*, **2005**, *24*(7), 1759–1765.
30. J. L. Anthony, J. L. Anderson, E. J. Maginn, J. F. Brennecke, *J. Phys. Chem. B*, **2005**, *109*(13), 6366–6374.
31. A. Shariati, C. J. Peters, *J. Supercrit. Fluids*, **2005**, *34*(2), 171–176.
32. C. Cadena, J. L. Anthony, J. K. Shah, T. I. Morrow, J. F. Brennecke, E. J. Maginn, *J. Am. Chem. Soc.*, **2004**, *126*(16), 5300–5308.
33. M. J. Muldoon, S. N. V. K. Aki, J. L. Anderson, J. K. Dixon, J. F. Brennecke, *J. Phys. Chem. B*, **2007**, in press.
34. J. L. Anthony, E. J. Maginn, J. F. Brennecke, *J. Phys. Chem. B*, **2002**, *106*(29), 7315–7320.
35. R. E. Baltus, R. M. Counce, B. H. Culbertson, H. M. Luo, D. W. DePaoli, S. Dai, D. C. Duckworth, *Sep. Sci. Technol.*, **2005**, *40*(1–3), 525–541.
36. D. Camper, P. Scovazzo, C. Koval, R. Noble, *Ind. Eng. Chem. Res.*, **2004**, *43*(12), 3049–3054.
37. M. F. C. Gomes, A. A. H. Padua, *Pure Appl. Chem.*, **2005**, *77*(3), 653–665.
38. D. G. Hert, J. L. Anderson, S. Aki, J. F. Brennecke, *Chem. Commun.*, **2005**(20), 2603–2605.
39. P. Husson-Borg, V. Majer, M. F. C. Gomes, *J. Chem. Eng. Data*, **2003**, *48*(3), 480–485.

40. A. P. S. Kamps, D. Tuma, J. Z. Xia, G. Maurer, *J. Chem. Eng. Data*, **2003**, *48*(3), 746–749.
41. P. Scovazzo, D. Camper, J. Kieft, J. Poshusta, C. Koval, R. Noble, *Ind. Eng. Chem. Res.*, **2004**, *43*(21), 6855–6860.
42. M. B. Shiflett, A. Yokozeki, *Ind. Eng. Chem. Res.*, **2005**, *44*(12), 4453–4464.
43. H. D. Tang, J. B. Tang, S. J. Ding, M. Radosz, Y. Q. Shen, *J. Polym. Sci., Part A: Polym. Chem.*, **2005**, *43*(7), 1432–1443.
44. J. B. Tang, W. L. Sun, H. D. Tang, M. Radosz, Y. Q. Shen, *Macromolecules*, **2005**, *38*(6), 2037–2039.
45. L. A. Blanchard, D. Hancu, E. J. Beckman, J. F. Brennecke, *Nature*, **1999**, *399*, 28–29.
46. L. A. Blanchard, Z. Gu, J. F. Brennecke, *J. Phys. Chem. B*, **2001**, *105*(12), 2437–2444.
47. M. Costantini, V. A. Toussaint, A. Shariati, C. J. Peters, I. Kikic, *J. Chem. Eng. Data*, **2005**, *50*(1), 52–55.
48. J. W. Hutchings, K. L. Fuller, M. P. Heitz, M. M. Hoffmann, *Green Chem.*, **2005**, *7*(6), 475–478.
49. M. C. Kroon, A. Shariati, M. Costantini, J. Spronsenvan, G. J. Witkamp, R. A. Sheldon, C. J. Peters, *J. Chem. Eng. Data*, **2005**, *50*(1), 173–176.
50. Z. M. Liu, W. Z. Wu, B. X. Han, Z. X. Dong, G. Y. Zhao, J. Q. Wang, T. Jiang, G. Y. Yang, *Chem. Eur. J.*, **2003**, *9*(16), 3897–3903.
51. A. Shariati, C. J. Peters, *J. Supercrit. Fluids*, **2004**, *29*(1–2), 43–48.
52. A. Shariati, C. J. Peters, *J. Supercrit. Fluids*, **2004**, *30*(2), 139–144.
53. A. Shariati, K. Gutkowski, C. J. Peters, *Aiche J.*, **2005**, *51*(5), 1532–1540.
54. S. J. Zhang, Y. H. Chen, R. X. F. Ren, Y. Q. Zhang, J. M. Zhang, X. P. Zhan, *J. Chem. Eng. Data*, **2005**, *50*(1), 230–233.
55. A. Shariati, C. J. Peters, *J. Supercrit. Fluids*, **2003**, *25*(2), 109–117.
56. J. Kumelan, A. P. S. Kamps, I. Urukova, D. Tuma, G. Maurer, *J. Chem. Thermodyn.*, **2005**, *37*(6), 595–602.
57. I. Urukova, J. Vorholz, G. Maurer, *J. Phys. Chem. B*, **2005**, *109*(24), 12154–12159.
58. J. K. Shah, E. J. Maginn, *J. Phys. Chem. B*, **2005**, *109*(20), 10395–10405.
59. A. Berger, R. F. de Souza, M. R. Delgado, J. Dupont, *Tetrahedron: Asym.*, **2001**, *12*(13), 1825–1828.
60. J. Kumelan, A. P. S. Kamps, D. Tuma, G. Maurer, *Fluid Phase Equilib.*, **2005**, *228*, 207–211.
61. C. A. Ohlin, P. J. Dyson, G. Laurenczy, *Chem. Commun.*, **2004**(9), 1070–1071.
62. J. G. Hayden, J. P. O'Connell, *Ind. Eng. Chem., Process Des. Dev.*, **1975**, *14*(3), 209–216.
63. W. J. Green, P. E. Field, *J. Phys. Chem.*, **1980**, *84*(23), 3111–3114.
64. J. L. Anthony, J. M. Crosthwaite, D. G. Hert, S. Aki, E. J. Maginn, J. F. Brennecke, Phase equilibria of gases and liquids with 1-n-butyl-3-methylimidazolium tetrafluoroborate, in *Ionic Liquids as Green Solvents: Progress and Prospects*, R. D. Rogers, K. R. Seddon (Eds.), **2003**, ACS Symposium Series 856, pp. 110–120.
65. L. Cammarata, S. G. Kazarian, P. A. Salter, T. Welton, *Phys. Chem. Chem. Phys.*, **2001**(3), 5192–5200.
66. M. Solinas, A. Pfaltz, P. G. Cozzi, W. Leitner, *J. Am. Chem. Soc.*, **2004**, *126*(49), 16142–16147.
67. K. R. Seddon, A. Stark, M. J. Torres, Viscosity and density of 1-alkyl-3-methylimidazolium ionic liquids, in *Clean Solvents*, **2002**, pp. 34–49.
68. P. Bonhote, A. P. Dias, N. Papageorgiou, K. Kalyanasundaram, M. Gratzel, *Inorg. Chem.*, **1996**, *35*(5), 1168–1178.
69. Y. Chauvin, L. Mussmann, H. Olivier, *Angew. Chem.-Int. Edit. Engl.*, **1996**, *34*(23–24), 2698–2700.
70. P. A. Z. Suarez, J. E. L. Dullius, S. Einloft, R. F. DeSouza, J. Dupont, *Polyhedron*, **1996**, *15*(7), 1217–1219.
71. P. A. Z. Suarez, J. E. L. Dullius, S. Einloft, R. F. deSouza, J. Dupont, *Inorg. Chim. Acta*, **1997**, *255*(1), 207–209.
72. A. L. Monteiro, F. K. Zinn, R. F. DeSouza, J. Dupont, *Tetrahedron: Asym.*, **1997**, *8*(2), 177–179.
73. R. T. Carlin, J. Fuller, *Chem. Commun.*, **1997**(15), 1345–1346.
74. T. H. Cho, J. Fuller, R. T. Carlin, *High Temp. Mater. Process*, **1998**, *2*(4), 543558.
75. L. A. Muller, J. Dupont, R. F. de Souza, *Macromol. Rapid Commun.*, **1998**, *19*(8), 409–411.
76. P. J. Dyson, D. J. Ellis, D. G. Parker, T. Welton, *Chem. Commun.*, **1999**(1), 25–26.

77. D. J. Ellis, P. J. Dyson, D. G. Parker, T. Welton, *J. Mol. Catal. A-Chem.*, **1999**, *150*(1–2), 71–75.
78. S. Steines, P. Wasserscheid, B. Driessen-Holscher, *J. Prakt. Chem.-Chem. Ztg.*, **2000**, *342*(4), 348–354.
79. S. Guernik, A. Wolfson, M. Herskowitz, N. Greenspoon, S. Geresh, *Chem. Commun.*, **2001**(22), 2314–2315.
80. R. A. Brown, P. Pollet, E. McKoon, C. A. Eckert, C. L. Liotta, P. G. Jessop, *J. Am. Chem. Soc.*, **2001**, *123*(6), 1254–1255.
81. F. C. Liu, M. B. Abrams, R. T. Baker, W. Tumas, *Chem. Commun.*, **2001**(5), 433–434.
82. J. Howarth, *Tetrahedron Lett.*, **2000**, *41*(34), 6627–6629.
83. L. Gaillon, F. Bedioui, *Chem. Commun.*, **2001**(16), 1458–1459.
84. N. Karodia, S. Guise, C. Newlands, J. A. Andersen, *Chem. Commun.*, **1998**(21), 2341–2342.
85. W. Keim, D. Vogt, H. Waffenschmidt, P. Wasserscheid, *J. Catal.*, **1999**, *186*(2), 481–484.
86. C. C. Brasse, U. Englert, A. Salzer, H. Waffenschmidt, P. Wasserscheid, *Organometallics*, **2000**, *19*(19), 3818–3823.
87. F. Favre, H. Olivier-Bourbigou, D. Commereuc, L. Saussine, *Chem. Commun.*, **2001**(15), 1360–1361.
88. D. J. Brauer, K. W. Kottsieper, C. Liek, O. Stelzer, H. Waffenschmidt, P. Wasserscheid, *J. Organomet. Chem.*, **2001**, *630*(2), 177–184.
89. M. F. Sellin, P. B. Webb, D. J. Cole-Hamilton, *Chem. Commun.*, **2001**(8), 781–782.
90. Y. Chauvin, B. Gilbert, I. Guibard, *Chem. Comm.*, **1990**(23), 1715–1716.
91. Y. Chauvin, S. Einloft, H. Olivier, *Ind. Eng. Chem. Res.*, **1995**, *34*(4), 1149–1155.
92. Y. Chauvin, A. Hirschauer, H. Olivier, *J. Mol. Catal.*, **1994**, *92*(2), 155-165.
93. S. Einloft, F. K. Dietrich, R. F. DeSouza, J. Dupont, *Polyhedron*, **1996**, *15*(19), 3257–3259.
94. Y. Chauvin, H. Olivier, C. N. Wyrvalski, L. C. Simon, R. F. deSouza, *J. Catal.*, **1997**, *165*(2), 275–278.
95. E.G. Kuntz, *Chem. Technol.*, **1987**, *17*(9), 570–575.
96. M. Freemantle, Ionic Liquids Make Splash in Industry, in *Chem. Eng. News*, 83(31), Aug. 1, **2005**.
97. J. F. Brennecke, E. J. Maginn, *Purification of Gas with Liquid Ionic Compounds*, **2003**, U.S. Patent 6, 579, 343.
98. J. L. Anthony, S. N. V. K. Aki, E. J. Maginn, J. F. Brennecke, *Int. J. Environ. Tech. Manag.*, **2004**, *4*(1/2), 105–115.

3.5
Polarity

Tom Welton

It is well known that the choice of solvent can have a dramatic effect upon a chemical reaction [1]. As early as 1862 the ability of solvents to decelerate the reaction of acetic acid with ethanol had been noted [2]. Thirty years later the influence of solvents on reaction equilibria was demonstrated for the first time [3].

Once such effects have been noted it becomes necessary to rationalize the observed results and to classify the solvents. The earliest attempts to do this were by Stobbe, who reviewed the effects of solvents on keto–enol tautomers [4]. Since then many attempts have been made to rationalize solvent effects, some based on observations of chemical reactions, others on physical properties of the solvents and yet others on spectroscopic probes. All of these have their advantages and disadvantages and no one approach can be thought of as exclusively "right". I will organize this chapter by type of measurement and then attempt to summarize the available information at the end.

Most modern discussions of solvent effects rely on the concept of solvent polarity. Qualitative ideas of polarity are based on observations such as "like dissolves like" and are well accepted and understood. However, quantifying polarity has proven to be extraordinarily difficult. Since the macroscopic property polarity is the sum of all possible (non-specific and specific) intermolecular interactions between a solute and the solvent, excluding such interactions leading to definite chemical changes (reactions) of the solute, this is perhaps unsurprising. Hence, it is important that care is taken when discussing the "polarity" of any liquid to ensure that it is clearly understood what is actually being referred to. Attempts to compare polarities measured by different techniques that are sensitive to different properties of the solvent often only lead to confusion.

The most common measure of polarity that is used by chemists in general is that of dielectric constant (ϵ_r). It has been measured for most molecular liquids and is widely available in reference texts, including most commercial catalogues. Although some values for ionic liquids have been derived indirectly (see below), truly direct measurement, which requires a non-conducting medium, is not available for ionic liquids. The usefulness of dielectric constant as a measure of polarity of ionic liquids should also be considered. One of the reasons for the ubiquity of the dielectric constant as a measure of polarity is the large number of solvent phenomena models that rely on considering the solvent to be a continuous dielectric medium. Ionic liquids do not even approximate to continuous dielectric media. They have two different species present, the cations and anions that are highly interconnected. Hence, it is unlikely that the dielectric constant will ever have the privileged status for understanding ionic liquids that it has achieved for molecular solvents. Other methods to determine the polarity of ionic liquids have been used and are the subjects of this chapter. Where appropriate, I have included the literature on higher melting point organic salts.

It should be noted that many of the scales whose use is reported below are based on the effect of the ionic liquids on a single probe molecule. While the response of the probe to the solvent in which it is dissolved is determined by all possible solvent–probe interactions, there is no reason for there to be an equal contribution from all of them for all of probes. Hence, it is important to be as aware as possible of which interactions are likely to have a strong effect on the particular probe used and which have a lesser effect. Since ionic liquids are composed of both anions and cations, either of which may preferentially solvate a particular probe molecule, this problem will be even greater than is usually the case in molecular solvents. It is probably foolhardy to take a scale generated by any one probe as a measure of what is generally understood by the term polarity. However, very useful information can still be gleaned from these studies.

Table 3.5-1 Static dielectric constants for ionic liquids

Ionic liquid	ϵ_r
acetone	20.56
[EMIM][OTf]	15.2
[EMIM][BF$_4$]	12.8
[BMIM][BF$_4$]	11.7
[BMIM][PF$_6$]	11.4
[HMIM][PF$_6$]	8.9
dichloromethane	8.93

3.5.1
Microwave Dielectric Spectroscopy

Microwave dielectric spectroscopy has been used to estimate values for the static dielectric constant of a small number of [RMIM]$^+$ ionic liquids (Table 3.5-1) [5]. Values ranging from 9 to 15 were found, depending on the ionic liquid. This is of the order found for molecular solvents of modest polarity. The dielectric constant was found to decrease as the length of the alkyl chain increased. This is similar to the behavior of homologous series in molecular solvents. The dielectric constant also decreased in the order [OTf]$^-$ > [BF$_4$]$^-$ > [PF$_6$]$^-$.

3.5.2
Chromatographic Measurements

In a series of papers published throughout the 1980s Colin Poole and his coworkers investigated the solvation properties of a wide range of alkylammonium and, to a lesser extent, phosphonium salts. These ionic liquids were used as stationary phases for gas chromatography and the retention of a variety of probe compounds was analyzed using Abraham's solvation parameter model [6]. Since a wide variety of probe solutes were used any problems associated with the use of a single probe, which will inevitably have its own specific chemistry, were removed.

Abraham's model was constructed to describe solute behavior, but it recognizes the intimacy of the solute–solvent relationship and so provides a useful model of solvent properties [7]. The model is based on solvation that occurs in two steps. First, a cavity is generated in the solvent. This process is endoergic, as the self-association of the solvent is overcome. Then the solute is incorporated in the cavity. This step is exoergic as the solvent–solute interactions are formed. By working with G.C., and therefore gaseous solutes, the self-association of the solute can be ignored. The form of the model used in these investigations is [6]:

$$\log K_L = c + r R_2 + s \pi_2^H + a \alpha_2^H + b \beta_2^H + l \log L^{16} \qquad (3.5\text{-}1)$$

Where K_L is the solute gas–liquid partition coefficient, r is the tendency of the solvent to interact through π- and n-electron pairs (Lewis basicity), s the contribution from dipole–dipole and dipole–induced dipole interactions (in molecular solvents), a is the hydrogen bond basicity of the solvent and b is its hydrogen bond acidity, l indicates how well the solvent will separate members of a homologous series and has contributions from solvent cavity formation and dispersion interactions.

Poole applied this model to 38 different organic salts, primarily tetraalkylammonium halides and substituted alkanessulfonates [6]. For the range of salts included in the study, it was found that they were strong hydrogen-bond bases with the precise basicity being a function of the anion. The basicity of the ionic liquid can be diminished by fluorinating the anion. The organic salts had large s-values, which were interpreted as a significant capacity for dipole–dipole and dipole–induced dipole interactions, and it is in the s-value that any Coulombic effect would be expected to show itself.

Some exceptions to this general observation were found, halide and nitrite salts have unusually high hydrogen-bond basicities (as would be expected) and pentacyanopropionide, picrate, triflate and perfluorobezenesulfonate salts had unusually low hydrogen-bond basicities (also as would be expected) but also lower s-values, perhaps due to the weakening of the Coulombic interactions by delocalisation of the charge on the anions.

It was noted that moving from tetrabutylammonium to tetrabutylphosphonium salts with a common anion led to identical solvation properties. This reflects the relatively small differences between these cations.

One interesting point of note is that the component arising from the solvent's ability to form a cavity and its dispersion interactions is unusually high in comparison to non-ionic polar solvents for most of the ionic liquids with poorly associating anions and increases as the cation became bulkier. With the hydrogen-bond base anions, this ability weakens. High values for l are usually associated with nonpolar solvents. This observation agrees with the often-repeated statement that ionic liquids have unusual mixing properties (see Sections 3.3 and 3.4).

Finally, none of the ionic liquids were found to be hydrogen-bond acids [6]. This may simply be because none of the salts had an ion that would be expected to act as a strong hydrogen bond donor. Earlier qualitative measurements on ionic liquid stationary phases of mono-, di- and trialkylammonium salts had suggested that hydrogen bond donation can be important where a potentially acidic proton is available [8–10]. However, more recently, this approach has been applied to a range of substituted imidazolium-based ionic liquids and trialkylammonium-based ionic liquids [11, 12]. In spite of the fact that this study included cations (e.g. tributylammonium) that might be expected to act as hydrogen-bond donors, the ionic liquids did not act as hydrogen-bond donor solvents. Although direct numerical comparison is not possible, these studies concurred with the earlier ones and described the ionic liquids as hydrogen-bond bases that also interact through strong dipolar and dispersion forces. Again, changing the anion of the ionic liquids controlled their hydrogen bond basicity. Those ionic liquids with electron-rich aromatic rings could also interact with solutes via π–π and n–π interactions.

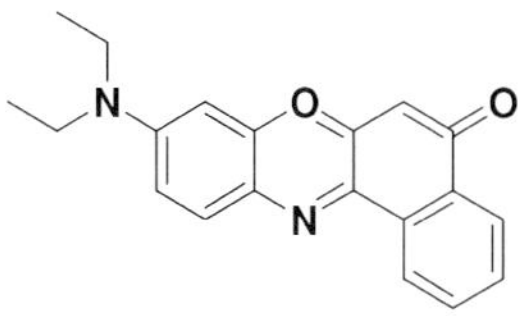

Fig. 3.5-1 Nile Red.

Ionic liquids with highly fluorinated anions can give rise to repulsions for this kind of interaction. The authors used the solvent descriptors to interpret previous chemical and chromatographic observations [12].

3.5.3
Absorption Spectra

The first solvatochromic dye to be used with a number of ionic liquids was Nile Red (Fig. 3.5-1) [13]. The ionic liquids used were composed of $[RMIM]^+$ with $[NO_2]^-$, $[NO_3]^-$, $[BF_4]^-$, $[PF_6]^-$, bis(trifluorosulfonyl)imide ($[N(Tf)_2]^-$) ions. The values of the energy of the electronic transition, E_{NR}, did not vary greatly and fell in the same region as short chain alcohols.

A recent study of the effects of molecular solvents on the E_{NR} value has shown it to be correlated with the Kamlett–Taft $\pi *$ and α scales (see below), and so it is thought to arise from a combination of hydrogen bond donation and dipolarity and polarizability effects [14].

The effect of hydrogen-bond donation to Nile Red can be seen when comparing the E_{NR} values of imidazolium salts protonated on one of the ring nitrogen atoms [15] with the equivalent methyl substituted salt (e.g. E_{NR} for $[BMIM][BF_4]$ = 217.2 kJ mol^{-1} whereas E_{NR} for $[BIM][BF_4]$ = 212.5 kJ mol^{-1}).

The anion also has an effect on the E_{NR} value of the ionic liquids [13]. The E_{NR} values for several 1-ethylimidazolium ionic liquids have been compared to the pK_a of the conjugate acid of the anion of the ionic liquid [15]. This study shows a general trend that more basic anions lead to higher E_{NR} (less polar) values for the ionic liquids. This can largely be attributed to changes in the hydrogen bond donor properties of the ionic liquids. This kind of antagonistic behavior of the ionic liquid ions in contributing to the overall hydrogen bonding properties of the ionic liquids has now been demonstrated on several occasions and is discussed separately below.

The longest wavelength absorption band of Reichardt's dye (2,4,6-triphenylpyridinium-N-4-(2,6-diphenylphenoxide) betaine (Fig. 3.5-2) shows one of the largest solvatochromic shifts known (375 nm between diphenyl ether and water) [16]. It can register effects arising from the solvent dipolarity, hydrogen bonding, and Lewis acidity, with the greatest contribution coming from the hydrogen-bond donor property of the solvent [17]. Its use in the analysis of the polarity of ionic liquids has recently been reviewed [18]. The E_T^N values of a small number of alkylammonium nitrate, thiocyanate, and sulfonate salts [19, 20] have been recorded, as

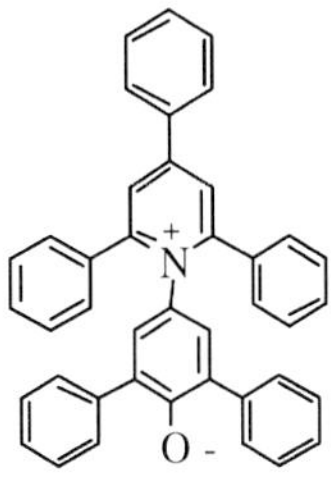

Fig. 3.5-2 Reichardt's dye.

have some substituted imidazolium tetrafluoroborate, hexafluorophosphate, triflate and trifluoromethanesulfonylimide salts [21]. It has been noted that these measurements are highly sensitive to the preparation of the ionic liquids used [18], so it is unfortunate that there are no examples that appear in all studies. The results are also sensitive to reaction conditions, such as the temperature at which the spectra were recorded, so care is needed when comparing the results of different studies.

Table 3.5-2 lists the E_T^N values for the alkylammonium thiocyanates and nitrates and the substituted imidazolium salts. Although measurements have been made on higher melting salts Reichardt's dye is also thermochromic, consequently only measurements made at room temperature are included. It can be seen that the values are dominated by the nature of the cation. For instance, values for monoalkylammonium nitrates and thiocyanates are ca. 0.95–1.01, whereas the two tetraalkylammonium salts have values ca. 0.42–0.46. The substituted imidazolium salts lie between these two extremes, with those with a proton at the 2-position of the ring having higher values than those with this position methylated. This is entirely consistent with the expected hydrogen-bond donor properties of these cations.

The role of the anion is less clear cut. It can be seen that ionic liquids with the same cation but different anions have different E_T^N values. However, the difference in the values for [Et$_4$N]Cl and [Et$_4$N][NO$_3$] is only 0.006, whereas the difference between those for [EtNH$_3$]Cl and [EtNH$_3$][NO$_3$] is 0.318. Less dramatically, the difference in the values for [OMMIM][BF$_4$] and [OMMIM][Tf$_2$N] is only 0.018, whereas the difference between [BMIM][BF$_4$] and [BMIM][Tf$_2$N] is 0.031. Hence, it is clear that the effect of changing the anion depends on the nature of the cation.

Attempts have also been made to separate non-specific effects of the local electric field from hydrogen-bonding effects for a small group of ionic liquids by using the π^* scale of dipolarity/polarizability, the α-scale of hydrogen-bond donor acidity, and the β-scale of hydrogen-bond basicity (Table 3.5-2) [19]. Again, the results are dependent on the experimental details and care should be exercised when comparing results from different groups.

The π^*-values were high for all of the ionic liquids investigated when compared to molecular solvents. The π^*-values result from measuring the ability of the solvent to induce a dipole in the probe solute and should be expected to incorporate the effect of Coulombic interactions from the ions as well as dipole and polarizability effects. This explains the consistently high values for all of the salts in the studies. The values

Table 3.5-2 Solvent polarity measurements of some ionic liquids

Salt	E^N_T	π^*	α	β	Ω
[Pr$_2$NH$_2$][SCN]	1.006	1.16	0.97	0.39	
[sec-BuNH$_3$][SCN]	1.006	1.28	0.91		
water	1.000	1.09	1.17	0.47	0.87
[EtNH$_3$][NO$_3$]	0.954	1.24	0.85	0.46	0.82
[BuNH$_3$][SCN]	0.948	1.23	0.92		
[PrNH$_3$][NO$_3$]	0.923	1.17	0.88	0.52	
[Bu$_3$NH][NO$_3$]	0.802	0.97	0.84		
[BMIM][ClO$_4$]	0.684				0.67
[BMIM][BF$_4$]	0.673	1.09	0.73	0.72	0.66
[BMIM][TfO]	0.667				0.65
[BMIM][PF$_6$]	0.667	0.91	0.77	0.41	0.68
ethanol	0.654	0.54	0.75	0.75	0.72
[BMIM][Tf$_2$N]	0.642				
[BMIM]Cl		1.17	0.41	0.95	
[EtNH$_3$]Cl	0.636				
[OMIM][PF$_6$]	0.633	0.88	0.58	0.46	
[OMIM][Tf$_2$N]	0.630				
[OMIM]Cl		1.09	0.33	0.90	
[Pr$_4$N][CHES][a]	0.62	1.08	0.34	0.80	
[BMIM][CF$_3$CO$_2$]	0.620				
[Bu$_4$N][CHES][a]	0.62	1.01	0.34	0.98	
[Pe$_4$N][CHES][a]	0.58	1.00	0.15	0.91	
[Bu$_4$N][BES][a]	0.53	1.07	0.14	0.81	
[BMMIM][Tf$_2$N]	0.525				
[Bu$_4$N][MOPSO][a]	0.49	1.07	0.03	0.74	
[OMMIM][BF$_4$]	0.543				
[OMMIM][Tf$_2$N]	0.525				
[Et$_4$N][NO$_3$]	0.460				
acetonitrile	0.460	0.75	0.19	0.31	0.69
[Et$_4$N]Cl	0.454				
[Hx$_4$N][PhCO$_2$]	0.420				
diethyl ether	0.117	0.27	0.00	0.47	0.47
cyclohexane	0.009	0.00	0.00	0.00	0.60

[a] CHES is 2-(cyclohexylamino)ethanesulfonate, BES is 2-{bis(2-hydroxoethyl)amino}ethanesulfonate, MOPSO 2-hydroxo-4-morpholinepropanesulfonate.

for quaternary ammonium salts are lower than those of the monoalkylammonium salts. This probably arises from the ability of the charge center on the cation to approach the solute more closely for the monoalkylammonium salts. The values for the imidazolium salts are lower still, reflecting the delocalization of the charge in the cation and the concomitant reduction in Coulombic attraction.

The differences in the hydrogen-bond acidities and basicities between different ionic liquids were far more marked. The α-value is largely determined by the availability of hydrogen-bond donor sites on the cation. Values range from 0.8–0.9 for the monoalkylammonium salts and are slightly lower (0.3–0.8) for the imidazolium salts. In the absence of a hydrogen-bond donor cation values were lower still: in the range 0.1–0.3. It appears that more basic anions give lower values of α with a common cation. This will be discussed at the end of this section (Eqs. (3.5-2) and (3.5-3)).

At first glance the hydrogen-bond basicity β is controlled by the anions, with basicity increasing as the strength of the conjugate acid of the anion decreases. However, while the general trend is clear, the cations do appear to be playing a role. Again this may be a consequence of an antagonistic relationship between the ions in determining this property (see below). It is unfortunate that, to date, no study has used a common anion across a large number of possible cations.

3.5.4
Antagonistic Behavior in Hydrogen Bonding

In several of the studies detailed above the overall ability of the ionic liquid to form a hydrogen bond with a solute molecule appears to come from an antagonistic relationship between its constituent ions. This may be described in terms of two competing equilibria. If we consider hydrogen-bond donation by the cation, first the cation can hydrogen bond to the anion (Eq. (3.5-2)):

$$C^+ + A^- \rightleftharpoons C^+ \cdots A^-$$

$$K'_{eqm} = \frac{[C^+ \cdots A^-]}{[C^+][A^-]} \qquad (3.5\text{-}2)$$

The cation can also hydrogen bond to the solute (Eq. (3.5-3)):

$$C^+ + \text{solute} \rightleftharpoons C^+ \cdots \text{solute}$$

$$K''_{eqm} = \frac{[C^+ \cdots \text{solute}]}{[C^+][\text{solute}]} \qquad (3.5\text{-}3)$$

It can be easily shown that the value of K'' is inversely proportional to the value of K' and that K' is dependent on both the cation and the anion of the ionic liquid. Hence, it is entirely consistent with this model that the difference made by changing the anion should depend on the hydrogen-bond acidity of the cation. A similar argument can be made when considering the ability of the ionic liquid to act as a hydrogen-bond acceptor.

3.5.5
Fluorescence Spectra

A number of workers have attempted to study the polarity of ionic liquids using the fluorescence spectra of polycyclic aromatic hydrocarbons. Of these, the most commonly applied has been that of pyrene [23–26]. The measurements are of the ratio of the intensities of the first and third vibronic bands in the $\pi–\pi^*$ emission spectrum of monomer pyrene (I_1/I_3). The increase in I_1/I_3 values in more polar solvents has been attributed to a reduction in local symmetry [27] but the mechanism for this is poorly understood, although some contribution from solvent hydrogen-bond acidity has been noted [28]. Hence, it is difficult to know what the measurements are telling us about the ionic liquids in anything other than the most general terms. Care is also required when comparing across different studies because it is known that pyrene I_1/I_3 values depend upon the way in which the spectrometer has been set up.

Pyrene I_1/I_3 values measurements have generally placed the ionic liquids in the polarity range of moderately polar solvents. [BMIM][PF$_6$] has a particularly high value (1.84 [25] and 2.08 [24] have been reported). This can be compared to water (1.96), acetonitrile (1.88) and methanol (1.50) [24]. It should be noted that the spectrum of pyrene would be expected to be sensitive to HF, which could well be present in these [PF$_6$]$^-$ ionic liquids, and would lead to artificially high values of I_1/I_3. Other fluorescence probes that have been used give broadly similar results [23–25, 29].

The fluorescence spectrum of 5-N,N,-dimethylamino-1-naphthalenesulfonamide in [BMIM][PF$_6$] has a λ_{max}^{fl} that is higher than dichloromethane, acetonitrile or methanol and comparable to ethylene glycol but has a Stokes shift similar to acetonitrile and lower than the short-chain alcohols used in the study [26]. Of course, the solvation of the probe molecule by the ionic liquid is the same and this clearly indicates that the two values are sensitive to different interactions among those actually occurring. Changes in both λ_{max}^{fl} and the Stokes shift have been associated with a variety of possible interactions and it has not been possible to attribute the apparently differing results to any particular interaction. However, this study does show particularly well the dangers in expecting different polarity probes to give the same answer.

3.5.6
Refractive Index

The refractive index of a medium is the ratio of the speed of light in a vacuum to its speed in the medium and is the square root of the relative permittivity of the medium at that frequency. When measured with visible light, the refractive index is related to the electronic polarizability of the medium. Solvents with high refractive indices, such as aromatic solvents, should be capable of strong dispersion interactions. Unlike the other measures described above, the refractive index is a property of the pure liquid without the perturbation generated by adding a probe species.

Refractive indices for a number of ionic liquids have been reported recently [30]. Increasing the number, length and branching of alkyl chains on the cations increases the refractive index, as does introducing functionality into the chain. Changing the anion of the ionic liquid also affects the refractive index, perhaps with less polarizable anions giving lower values.

3.5.7
EPR Spectroscopy

The ^{14}N hyperfine coupling constant of the EPR spectrum of 4-amino-2,2,6,6-tetramethylpiperidine-1-oxyl, a commercially available stable free radical, has been used to measure the polarity of the a small number of ionic liquids [31]. The values correlate very well with the $E_T(30)$ scale and so arise from a combination of hydrogen-bond donation from the solvent and Coulombic, dipolarity and polarizability effects. EPR also provides information about the tumbling of the probe solute in a solvent by its effect on the spin relaxation times (τ_r). τ_r for the ionic liquids are longer in the molecular solvents studied. This was attributed partly to the higher viscosities of the ionic liquids and partly to a hydrogen-bonding interaction between the cation and the nitroxide functionality of the probe molecule. This technique has the advantage that it does not require that the ionic liquid is transparent at UV and visible wavelengths and may be used for colored ionic liquids.

3.5.8
Chemical Reactions

An alternative avenue to exploring the polarity of a solvent is by investigating its effect on a chemical reaction. Since the purpose of this book is to review the potential application of ionic liquids in synthesis, the effect of ionic liquids on chemical reactions has been treated separately (see Chapter 5, Section 5.1).

3.5.9
Comparison of Polarity Scales

Although a small number of ionic liquids have had their polarity investigated using more than one technique, there has been little attempt to study this systematically. In order to make comparisons between the different solvent polarity scales for ionic liquids there needs to be a sufficient overlap in the ILs used in the studies. Further to this, it is well known that these empirical polarity parameters are easily affected by impurities, so it is necessary to have common samples. So far, there has only been one study of how the same ionic liquids are described by different polarity scales that has used common samples [12, 22]. The Welton and Armstrong groups compared multiple solvation interactions, based on G.C. measurements, to the Kamlet–Taft approach. Eight ionic liquids were directly compared. The non-specific interactions are not directly comparable for the two methods, because they are made up of different contributions, but the hydrogen-bonding effects might be expected to give similar results in both experiments.

The hydrogen-bond basicities of the ionic liquids followed the same trend in both studies, are controlled by the anion and are moderate in value. However, comparisons of hydrogen-bond acidity were, at first sight, contradictory. In the GC study the ionic liquids' hydrogen-bond acidity was found to be dominated by the hydrogen-bond basicity of the anions, with a much lesser contribution from the hydrogen-bond acidity of the cation. Low and even negative values were found for the hydrogen-bond acidity function. The only ionic liquids to display significant hydrogen-bond donor ability where those of the $[N(Tf)_2]^-$ ion. In contrast, the Kamlet–Taft study has strongly emphasized the role of the hydrogen-bond acidity

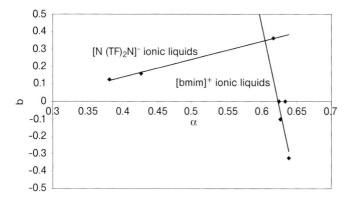

Fig. 3.5-3 Comparison of the hydrogen-bond donor *b* vs. *α* functions for some ionic liquids.

of the cation. A plot (Fig. 3.5-3) of the two functions, α from the Kamlet–Taft study and *b* from the solvation study, reveals a complex relationship.

It can be seen for the $[N(Tf)_2]^-$ ionic liquids, that the hydrogen-bond acidity does indeed vary with cation with $[BMIM]^+$ being the most acidic followed by $[BMPY]^+$ and finally $[BMMIM]^+$ in both studies. However, changing to more basic anions leads to a dramatic drop in the acidity measurements in the solvation study, whereas it has only a limited effect in the Kamlet–Taft experiment. That is, the solvation measurement is anion dominated, whereas the Kamlet–Taft measurement is cation dominated.

A recent theoretical study has provided the likely explanation for these observations [32]. Here it was shown that the hydrogen bond between an imidazolium cation and anionic hydrogen-bond acceptor is dominated by the Coulombic component and is much stronger that that between the cation and a neutral hydrogen-bond acceptor. In the Kamlet–Taft study the hydrogen-bond acceptor is the phenoxide oxygen of Reichardt's dye, whereas all of the probes in the G.C. experiment are neutral molecules. So, the Kamlet–Taft study reveals the interactions of which the ionic liquids are capable in the presence of a strongly and specifically interacting solute and how they are influenced by the nature of the two competing ions, whereas the solvation study reveals the value when the nature of any particular solute has been deliberately down-played and how that is influenced by the nature of the ions.

3.5.10
Conclusions

To date, most of the studies of ionic liquids have used a small set of ionic liquids and have been based on the idea that if the response of a particular probe molecule is like that in some known molecular solvent, then it can be said that the polarities of the ionic liquid and the molecular solvent are the same. This may not necessarily be the case. Only systematic investigations will show whether this is true and it is only when a wide range of ionic liquids have been studied using a wide range of different solvent polarity probes that we will be able to make any truly general statements about the polarity of ionic liquids. Indeed, in our attempts to understand the nature of solvent effects in ionic liquids, we will probably have to refine our notion of polarity itself. However, it is possible to draw some tentative general conclusions.

All ionic liquids are not the same, different combinations of anions and cations lead to solvents with different polarities. No ionic liquids have shown themselves to be "super-polar"; regardless of the method of assessing their polarities, ionic liquids come within the range of molecular solvents. Most general measures of overall polarity place ionic liquids in the range of the short- to medium-chain alcohols.

It becomes more interesting when the solvent properties are broken down into their component parts. Ionic liquids can act as hydrogen-bond acids and/or hydrogen-bond bases or neither. Generally, the hydrogen-bond basicity is determined by the anion and the hydrogen-bond acidity is determined by the cation. There is no obvious unique "ionic effect" to be seen in the available data. So far, no one has attempted to make an ionic liquid of any particular polarity, but this may be possible.

References

1. Reichardt, C., *Solvents and Solvent Effects in Organic Chemistry*, 3rd edn., Wiley-VCH, Weinheim, **2003**.
2. Berthelot, M., Péan de Saint-Giles, L., Ann. Chim. Phys., *3 Ser.* **1862**, *65*, 385; **1862** *66*, 5; **1863**, *68*, 255.
3. Claisen, L., *Liebigs Ann. Chem.* **1896**, *291*, 25; Wislicenus, W., *Liebigs Ann. Chem.* **1896**, *291*, 147; Knorr, L., *Liebigs Ann. Chem.* **1896**, *293*, 70.
4. Stobbe, H., *Liebigs Ann. Chem.* **1903**, *326*, 347.
5. Wakai, C., Oleinikova, A., Ott, M., Weingärtner, *J. Phys. Chem. B*, **2005**, *109*, 17028.
6. Poole, S. K., Poole, C. F., *Analyst* **1995**, *120*, 289 and references therein.
7. Abraham, M. H. *Chem. Soc. Rev.* **1993**, 73.
8. Pacholec, F., Butler, H., Poole, C. F., *Anal. Chem.*, **1982**, *54*, 1938.
9. Poole, C. F., Furton, K. G., Kersten, B. R., *J. Chromatogr. Sci.*, **1986**, *24*, 400.
10. Coddens, M. E., Furton, K. G., Poole, C. F., *J. Chromatogr.*, **1986**, *356*, 59.
11. Armstrong, D. W., He, L., Liu, Y.-S. *Anal. Chem.*, **1999**, *71*, 3873.
12. Anderson, J. L., Ding, J., Welton, T., Armstrong, D. W., *J. Am. Chem. Soc.*, **2002**, *124*, 14247.
13. Carmichael, A. J., Seddon, K. R., *J. Phys. Org. Chem.*, **2000**, *13*, 591.
14. Moog, R. S., Kim, D. D., Oberle, J. J., Ostrowski, S. G., *J. Phys. Chem.*, **2004**, *108*, 9294.
15. Ogihara, W., Aoyama, T., Ohno, H., *Chem. Lett.*, **2004**, *33*, 1414.
16. Reichardt, C., *Chem. Soc. Rev.* **1992**, 147; *Pure Appl. Chem.*, **2004**, *76*, 1903.
17. Taft, R. W., Kamlet, M. J., *J. Am. Chem. Soc.* **1976**, *98*, 2886.
18. C. Reichardt, *Green Chem.*, **2005**, *7*, 339.
19. Poole, S. K., Shetty, P. H., Poole, C. F., *Anal. Chim. Acta* **1989**, *218*, 241.
20. Herfort, I. M., Schneider, H., *Liebigs Ann. Chem.* **1991**, 27.
21. Muldoon, M. J., Gordon, C. M., Dunkin, I. R., *J. Chem. Soc., Perkin Trans. 2* **2001**, 433.
22. Crowhurst, L., Mawdsley, P. R., Perez-Arlandis, J. M., Salter, P. A., Welton, T., *Phys. Chem. Chem. Phys.*, **2003**, *5*, 2790.
23. Street, Jr., K. W., Acree, Jr., W. E., Fetzer, J. C., Shetty, P. H., Poole, C. F., *Appl. Spectrosc.*, **1989**, *43*, 1149.
24. Bonhôte, P., Dias, A.-P., Papageorgiou, N., Kalyanasundaram, K., Grätzel, M., *Inorg. Chem.*, **1996**, *35*, 1168.

25. Baker, S. N., Baker, G. A., Kane M. A., Bright, F. V., *J. Phys. Chem. B*, **2001**, *105*, 9663.
26. Fletcher, K. A., Storey, A., Hendricks, A. E., Pandey S., Pandey S., *Green Chem.*, **2001**, *3*, 210.
27. Dong, D. C., Winnik, M. A., *Can. J. Chem.* **1984**, *62*, 2560.
28. Catalán, J., *J. Org. Chem.*, **1997**, *62*, 8231.
29. Aki, S. N. V. K., Brennecke, J. F., Samanta, A., *Chem. Commun.* **2001**, 413.
30. (a) Przybysz, K., Drzewinska, E., Stanislawska, A., Wysocka-Robak, A., Cieniecka-Roslonkiewicz, A., Foksowicz-Flaczyk, J., Pernak, J., *Ind. Eng. Chem. Res.*, **2005**, *44*, 4599; (b) Huddleston, J. G., Visser, A. E., Reichert, W. M., Willauer, H. D., Broker, G. A., Rogers, R. D., *Green Chem.*, **2001**, *3*, 156; (c) Fredlake, C. P., Crosthwaite, J. M., Hert, D. G., Aki, S. N. V. K., Brennecke, J. F., *J. Chem. Eng. Data*, **2004**, *49*, 954; (d) Awad, W. H., Gilman, J. W., Nyden, M., Harris, R. H., Sutto, T. E., Callahan, J., Trulove, P. C., Delong, H. C., Fox, D. M., *Thermoochim Acta*, **2004**, *409*, 3; (e) Guillett, E., Imbert, D., Scopelliti, R., Bunzli, J. C. G., *Chem. Mater.*, **2004**, *16*, 4063; (f) Anderson, J. L., Armstrong, D. W., *Anal. Chem.*, **2005**, *77*, 6453; (g) Ito, N., Arzhantsev, S., Heitz, M., Maroncelli, M., *J. Phys. Chem. B*, **2004**, *108*, 5771; (h) Poole, C. F., Kersten, B. R., Ho, S. S. J., Coddens, M. E., Furton, M. E., *J. Chromatogr.* **1986**, *352*, 407.
31. Kawai, A., Hidemori, T., Shibuya, K., *Chem. Lett.*, **2004**, *33*, 1464.
32. Hunt, P., Kirchner, B., Welton, T., *Chem Eur. J.*, **2006**, in press.

3.6
Electrochemical Properties of Ionic Liquids

Paul C. Trulove and Robert A. Mantz

The early history of ionic liquids research was dominated by their application as electrochemical solvents. One of the first recognized uses of ionic liquids was as a solvent system for the room-temperature electrodeposition of aluminum [1]. In addition, much of the initial development of ionic liquids was focused on their use as electrolytes for battery and capacitor applications. Until recently, electrochemical studies in the ionic liquids were primarily carried out in the haloaluminate-based systems, and this work has been extensively reviewed [2–9]. Development of non-haloaluminate ionic liquids over the past fifteen years, however, has led to an explosion of research in these systems [10, 11]. Much of the initial interest in these new ionic liquids has been in areas other than electrochemistry. However, this initial slight has been largely corrected, as evidenced by the dramatic growth over the past five years in electrochemically related publications involving non-haloaluminate ionic liquids and the appearance of several good reviews on the subject [12–17].

Ionic liquids possess a variety of properties that make them desirable as solvents for investigating electrochemical processes. They often have wide electrochemical potential windows; they have reasonably good electrical conductivity and solvent transport properties; they have wide liquid ranges and they are able to solvate a wide variety of inorganic, organic, and organometallic species. The liquid ranges of ionic liquids have been discussed in Section 3.1 and the solubility and solvation in Section 3.3. In this section we will deal specifically with the electrochemical properties of ionic liquids (electrochemical window, conductivity, and transport properties). We will discuss the techniques involved in measuring these properties, summarize the relevant literature data, and discuss the effects of ionic liquid components and purity on their electrochemical properties.

In this section we will limit ourselves primarily to discussion of the electrochemical properties of pure ionic liquids. The notable exception to this will be our discussion of the properties of the binary and ternary haloaluminates due to their great historical importance to the development of ionic liquids. Absent from this section will be any discussion of the electrochemical properties of binary and ternary mixtures of non-haloaluminate ionic liquids. These systems are becoming an area of increasing interest due to their potential application to the development of ionic liquid electrolytes for lithium ion batteries. Nonetheless, the far greater complexity of these mixed systems is beyond the scope of this section. For further information on these interesting systems, the reader is directed to the following selected papers [18–22]. Also, notably absent from this section will be discussions of the electrochemical properties of "Brønsted acid–base" ionic liquids. These relatively new systems are normally prepared via the protonation of tertiary amines by Brønsted acids. Many of these ionic liquids have potentially useful properties and are of special interest for applications as proton conductors for hydrogen fuel cells [23–27]. Even so, the electrochemical properties of these systems are highly dependent on the extent of proton transfer and reaction stoichiometry, and as such are beyond the scope of this section.

3.6.1
Electrochemical Potential Windows

A key criterion for selection of a solvent for electrochemical studies is the electrochemical stability of the solvent [28, 29]. This is most clearly manifested by the range of voltages over which the solvent is electrochemically inert. This useful electrochemical potential "window" depends on the oxidative and reductive stability of the solvent. In the case of ionic liquids, the potential window depends primarily on the resistance of the cation to reduction and the resistance of the anion to oxidation (a notable exception to this is in the acidic chloroaluminate ionic liquids where the reduction of the heptachloroaluminate species, $[Al_2Cl_7]^-$, is the limiting cathodic process). In addition, the presence of impurities can play an important role in limiting the potential window of ionic liquids.

The most common method used to determine the potential window of an ionic liquid is cyclic voltammetry (or its digital analogue, cyclic staircase voltammetry). In a three-electrode system, the potential of an inert working electrode is scanned out to successively greater positive (anodic) and negative (cathodic) potentials until background currents rise dramatically due to oxidation and reduction of the ionic liquid, respectively. The oxidative and reductive potential limits are assigned when the background current reaches a threshold value. The electrochemical potential window is the difference between these anodic and cathodic potential limits. Since the choice of the threshold currents is somewhat subjective, the potential limits and corresponding electrochemical window have a significant uncertainty associated with them. Normally this is in the range of ±0.2 V.

It must be noted that impurities in the ionic liquids can have a profound impact on the potential limits and the corresponding electrochemical window. During the synthesis of many of the non-haloaluminate ionic liquids residual halide and water may remain in the final product [30]. Halide ions (Cl^-, Br^-, I^-) are more easily oxidized than the fluorine-containing anions used in most non-haloaluminate ionic liquids. Consequently, the observed anodic potential limit could be appreciably reduced if significant concentrations of halide ions are present.

During the initial development of the non-haloaluminate air and water stable ionic liquids, researchers often ignored potential contamination by water and the corresponding effects on the physical and chemical properties of the ionic liquids. However, as work on these new ionic liquids has progressed it has become apparent that water is an important contaminate to control. Even systems that are commonly referred to as "hydrophobic" will take up water from the atmosphere at levels often greater than 1 mass% [31]. Water can be reduced and oxidized within the electrochemical potential window of many ionic liquids. Consequently, contamination of an ionic liquid with significant amounts of water can decrease both the anodic and cathodic potential limits, and, correspondingly decrease the overall effective electrochemical window [32]. Work by Schröder et al. demonstrated considerable reduction in both the anodic and cathodic limits of several ionic liquids upon the addition of 3% by weight of water [33]. The electrochemical window of 'dry' [BMIM][BF_4] was found to be 4.10 V while that for the ionic liquid with 3% by weight of water was reduced to 1.95 V. In addition to its electrochemistry, water can react with the ionic liquid components (especially anions) to produce products that are electroactive in the electrochemical potential window. This has been well documented in the chloroaluminate ionic liquids where water will react to produce electroactive proton-containing species (e.g., HCl and [HCl_2]$^-$) [4]. In addition, water appears to react with some of the anions commonly used in the non-haloaluminate ionic liquids [34]. For instance, the [PF_6]$^-$ anion is known to react with water to form HF [35, 36].

Glassy carbon (GC), platinum (Pt), and tungsten (W) are the most common working electrodes used to evaluate electrochemical windows in ionic liquids. The choice of the working electrode has some impact on the overall electrochemical window measured. This is due to the effect of the electrode material on the irreversible electrode reactions that take place at the oxidative and reductive limits. For example, W gives a 0.1 to 0.2 V greater oxidative limit for [EMIM]Cl–$AlCl_3$ ionic liquids when compared to Pt due to a greater overpotential for the oxidation of the chloroaluminate anions [37]. GC (and to a lesser extent W) exhibits a large overpotential for the reduction of proton. Under normal circumstances, the electrochemistry of protonic impurities (i.e., water) will not be observed at GC in the ionic liquid electrochemical window. Pt, on the other hand, exhibits generally good electrochemical behavior for proton. Consequently, protonic impurities will give rise to a reduction wave(s) at Pt at potentials positive of the cathodic limit. Interestingly, a comparison of the background electrochemical behavior of an ionic liquid at both Pt and GC working

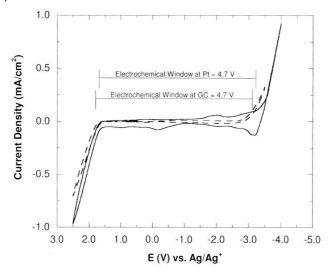

Fig. 3.6-1 The electrochemical window of 76–24 mol% [BMMIM][(CF$_3$SO$_2$)$_2$N]–Li[(CF$_3$SO$_2$)$_2$N] binary melt at (a) platinum working electrode, solid line; (b) glassy carbon working electrode, dashed line. Electrochemical window was set at a threshold of 0.1 mA cm^{-2}. The reference electrode was a silver wire immersed in 0.01 M AgBF$_4$ in [EMIM][BF$_4$] in a compartment separated by a Vicor frit, and the counter electrode was a graphite rod [34].

electrodes can be an excellent qualitative tool for determining whether significant amounts of protonic impurities are present.

Fig. 3.6-1 shows the electrochemical window of a 76–24 mol% [BMMIM][(CF$_3$SO$_2$)$_2$N]–Li[(CF$_3$SO$_2$)$_2$N] ionic liquid at both GC and Pt working electrodes [34]. For the purposes of assessing the electrochemical window, the current threshold for both the anodic and cathodic limits was set at an absolute value of 100 μA cm^{-2}.

As shown in Fig. 3.6-1, GC and Pt exhibit anodic and cathodic potential limits that differ by several tenths of volts. However, somewhat fortuitously, the electrochemical potential windows for both electrodes in this ionic liquid come out to be 4.7 V. What is also apparent from Fig. 3.6-1 is that the GC electrode exhibits no significant background currents until the anodic and cathodic potential limits are reached, while the Pt working electrode shows several significant electrochemical processes prior to the potential limits. This observed difference is most likely due to trace amounts of water in the ionic liquid, which is electrochemically active on Pt but not on GC (*vide supra*).

Tables 3.6-1–3.6-4 contain electrochemical potential windows for a wide variety of ionic liquids. Limited information was available concerning the purity of the ionic liquids listed in Tables 3.6-1–3.6-4, so these electrochemical potential windows must be treated with caution, as it is likely that many of the ionic liquids had residual halides and water present. Ideally, one would prefer to compare anodic and cathodic potential limits instead of the overall ionic liquid electrochemical window, because

Table 3.6-1 The room-temperature electrochemical potential windows for binary and ternary chloroaluminate and related ionic liquids[a]

Ionic liquid System	Cation(s)	Anion(s)	Working electrode[f]	Window (V)	Ref.
Imidazolium (IM)					
60.0–40.0 mol% [EMIM]Cl–AlCl$_3$	[EMIM]$^+$	[AlCl$_4$]$^-$/Cl$^-$	W	2.8	76[b]
50.0–50.0 mol [EMIM]Cl–AlCl$_3$	[EMIM]$^+$	[AlCl$_4$]$^-$	W	4.4	76[b]
45.0–55.0 mol% [EMIM]Cl–AlCl$_3$	[EMIM]$^+$	[Al$_2$Cl$_7$]$^-$/[AlCl$_4$]$^-$	W	2.9	76[b]
45.0–55.0 mol% [EMIM]Cl–AlCl$_3$	[EMIM]$^+$	[Al$_2$Cl$_7$]$^-$/[AlCl$_4$]$^-$	W	2.9	6[b]
60.0–40.0 mol [PMMIM]Cl–AlCl$_3$	[PMMIM]$^+$	[AlCl$_4$]$^-$/Cl$^-$	GC	3.1	34[b]
50.0–50.0 mol% [PMMIM]Cl–AlCl$_3$	[PMMIM]$^+$	[AlCl$_4$]$^-$	GC	4.6	34[b]
40.0–60.0 mol% [PMMIM]Cl–AlCl$_3$	[PMMIM]$^+$	[Al$_2$Cl$_7$]$^-$/[AlCl$_4$]$^-$	GC	2.9	34[b]
45.5–50.0–4.5 mol% [EMIM]Cl–AlCl$_3$–LiCl	[EMIM]$^+$/Li$^+$	[AlCl$_4$]$^-$	W	4.3	41[b,d]
45.5–50.0–4.5 mol% [EMIM]Cl–AlCl$_3$–LiCl	[EMIM]$^+$/Li$^+$	[AlCl$_4$]$^-$	W	4.6	41[b,d,e],
47.6–50.0–2.4 mol% [EMIM]Cl–AlCl$_3$–NaCl	[EMIM]$^+$/Na$^+$	[AlCl$_4$]$^-$	W	4.5	41[b,d]
47.6–50.0–2.4 mol% [EMIM]Cl–AlCl$_3$–NaCl	[EMIM]$^+$/Na$^+$	[AlCl$_4$]$^-$	W	4.6	41[b,d]
45.5–50.0–4.5 mol% [PMMIM]Cl–AlCl$_3$–NaCl	[PMMIM]$^+$/Na$^+$	[AlCl$_4$]$^-$	W	4.6	41[b,d]
45.5–50.0–4.5 mol% [PMMIM]Cl–AlCl$_3$–NaCl	[PMMIM]$^+$/Na$^+$	[AlCl$_4$]$^-$	W	4.7	41[b,d,e]
52.0–48.0 mol% [EMIM]Cl–GaCl$_3$	[EMIM]$^+$	[GaCl$_4$]$^-$/Cl$^-$ [b]	W	2.4	77[b]
50.0–50.0 mol% [EMIM]Cl–GaCl$_3$	[EMIM]$^+$	[GaCl$_4$]$^-$ [b]	W	4.0	77[b]
49.0–51.0 mol% [EMIM]Cl–GaCl$_3$	[EMIM]$^+$	[Ga$_2$Cl$_7$]$^-$/[GaCl$_4$]$^-$ [b]	W	2.2	77[b]

(Continued)

Table 3.6-1 (Continued)

Ionic liquid System	Cation(s)	Anion(s)	Working electrode[f]	Window (V)	Ref.
Pyridinium (P)					
50.0–50.0 mol% [BP]Cl–AlCl$_3$	[BP]$^+$	[AlCl$_4$]$^-$	W	3.6	76[b]
52.0–48.0 mol% [BP]Cl–GaCl$_3$	[BP]$^+$	[GaCl$_4$]$^-$/Cl$^-$ [b]	W	2.2	77[b]
50.0–50.0 mol% [BP]Cl–GaCl$_3$	[BP]$^+$	[GaCl$_4$]$^-$ [b]	W	3.7	77[b]
49.0–51.0 mol% [BP]Cl–GaCl$_3$	[BP]$^+$	[Ga$_2$Cl$_7$]$^-$/[GaCl$_4$]$^-$ [b]	W	2.2	77[b]
Ammonium					
Basic [(CH$_3$)$_2$(C$_2$H$_5$)(C$_2$H$_5$OCH$_2$)N]Cl–AlCl$_3$	[(CH$_3$)$_2$(C$_2$H$_5$)(C$_2$H$_5$OCH$_2$)N]$^+$	[AlCl$_4$]$^-$/Cl$^-$	Pt	3.5	78[b]
Sulfonium					
33.0–67.0 mol% [(CH$_3$)$_3$S]Cl–AlCl$_3$	[(CH$_3$)$_3$S]$^+$	[Al$_2$Cl$_7$]$^-$/[AlCl$_4$]$^-$	GC	2.5	79[b]

[a] Abbreviations for substituents on the imidazolium and pyridinium nitrogens: M = methyl, E = ethyl, P = n-propyl, B = n-butyl.
[b] Voltage window estimated from cyclic voltammograms contained in the reference.
[c] The exact nature of the anions is unknown; anions listed are those that would be expected if the system behaves similar to the chloroaluminates.
[d] Voltage window determined assuming anodic limit of 2.4 V vs. Al/Al(III) reference.
[e] Small amount of [EMIM][HCl]$_2$ added.
[f] Working electrode, Pt = Platinum, GC = Glassy Carbon, W = Tungsten.

different sets of anodic and cathodic limits can give rise to the same value of electrochemical window (see Fig. 3.6-1). However, the lack of a standard reference electrode system within and between ionic liquid systems precludes this possibility. Consequently, significant care must be taken when evaluating the impact of changes in the cation or anion on the overall ionic liquid electrochemical window.

When looking at the data in Tables 3.6-1–3.6-4 it is clear that there is significant variability in magnitude of the electrochemical windows for each type of cation (Fig. 3.6-2).

An important contributor to this variability is the effect of the oxidation of the anion on the anodic limit. However, when we compare cations with similar anions (e.g., $[BF_4]^-$) we can make some broad generalizations as to the relative electrochemical stability of the various cations. The apparent overall trend in the electrochemical stability of the ionic liquid cations follows the order; benzotriazolium < pyridinium < pyrrolinium < imidazolium ≤ pyrazolium ≤ sulfonium ≤ pyrrolidinium ≤ piperidinium ≈ ammonium ≈ morpholinium. Overall, the ionic liquids with four coordinate nitrogens (ammonium, piperidinium, morpholinium) have the largest electrochemical potential windows with most ionic liquids having windows greater than 5 V and some with windows greater than 6 V[1] (for comparison, one of the best non-aqueous electrolyte systems, acetonitrile–tetrabutylammonium hexafluorophosphate, exhibits a potential window of 6.3 V [28]). Because of uncertainties associated with the data listed in Tables 3.6-1–3.6-4, it is impossible to determine what effect changes in the alkyl substituents has on the electrochemical stability of the cation. However, within the group of imidazolium-based ionic liquids there is a clear increase in cation stability when the 2-position on the imidazolium ring is capped by an alkyl substituent, as in $[EMMIM]^+$. It has been proposed that the cathodic limiting reactions of imidazolium cations proceed initially via the reduction of ring protons to molecular hydrogen [38]. Since the 2-position on the imidazolium ring is the most acidic hydrogen [39], it is reasonable to conclude that substitution of an alkyl substituent at that position would result in an improvement in the reductive stability of the imidazolium cation. From the data in Tables 3.6-1–3.6-4, the anion stability towards oxidation appears to follow the order of halides (Cl^-, F^-, Br^-) < chloroaluminates ($[AlCl_4]^-$, $[Al_2Cl_7]^-$) ≤ fluorinated ions ($[PF_6]^-$, $[AsF_6]^-$) ≤ triflate/triflyl ions ($[CF_3SO_2]^-$, $[(CF_3SO_2)_2N]^-$, $[(C_2F_5SO_2)_2N]^-$, $[(CF_3SO_2)_3C]^-$) ≈ fluoroborates ($[BF_4]^-$, $[CF_3BF_3]^-$, $[C_2F_5BF_3]^-$, $[n\text{-}C_3F_7BF_3]^-$, $[n\text{-}C_4F_9BF_3]^-$)

The electrochemical windows exhibited by the chloroaluminates (Table 3.6-1) tend to fall into three ranges that correspond to the types of chloroaluminate ionic liquids; basic, neutral, and acidic. Basic ionic liquids contain an excess of the organic chloride salt (>50 mol%) resulting in the presence of free chloride ion (a Lewis base). This, in turn, significantly restricts the anodic limit of the basic ionic liquids.

---

[1] We must note that there are two imidazolium-based ionic liquids with apparent potential windows greater than 6 V (Table 3.6-2), but since these values are over 1.5 V greater than almost all other imidazolium potential windows they must be treated with some skepticism until additional data are obtained.

Table 3.6-2 The room-temperature electrochemical potential windows for non-haloaluminate imidazolium ionic liquids[a]

Cation	Anion(s)	Working electrode[d]	Window, (V)	Ref.
[MMIM]$^+$	[CF$_3$BF$_3$]$^-$	Pt	4.6	80, 99
[MMIM]$^+$	[(HF)$_{2.3}$F]$^-$	GC	2.4	51[b]
[EMIM]$^+$	[BF$_4$]$^-$	Pt	4.3	44, 81[b]
[EMIM]$^+$	[C(CN)$_3$]$^-$	Pt	2.9	82
[EMIM]$^+$	[C(CN)$_3$]$^-$	GC	3.0	83
[EMIM]$^+$	[CH$_3$CO$_2$]$^-$	Pt	3.6	84[b]
[EMIM]$^+$	[CF$_3$BF$_3$]$^-$	GC	4.6	85
[EMIM]$^+$	[CF$_3$CO$_2$]$^-$	Pt	3.8	86[b]
[EMIM]$^+$	[CF$_3$SO$_2$]$^-$	Pt	4.1	86[b]
[EMIM]$^+$	[CF$_3$SO$_2$]$^-$	Pt	4.3	84[b]
[EMIM]$^+$	[(CF$_3$SO$_2$)$_2$N]$^-$	GC	4.1	53
[EMIM]$^+$	[(CF$_3$SO$_2$)$_2$N]$^-$	Pt	4.5	86[b]
[EMIM]$^+$	[(CF$_3$SO$_2$)$_2$N]$^-$	GC	4.5	97[b]
[EMIM]$^+$	[(C$_2$F$_5$SO$_2$)$_2$N]$^-$	GC	4.1	53
[EMIM]$^+$	[(CF$_3$SO$_2$)(CF$_3$CO)N]$^-$	GC	4.2	98
[EMIM]$^+$	[(CF$_3$SO$_2$)(C$_2$F$_5$SO$_2$)N]$^-$	GC	4.3	96[b]
[EMIM]$^+$	[(HF)$_{2.3}$F]$^-$	Pt	3.1	87[b], 88
[EMIM]$^+$	[(HF)$_{2.3}$F]$^-$	GC	3.0	51, 88
[EMIM]$^+$	[N(CN)$_2$]$^-$	Pt	3.3	82, 89
[EMIM]$^+$	[C$_2$H$_5$SO$_4$]$^-$	GC	4.3	90
[EMIM]$^+$	[NbF$_6$]$^-$	GC	3.6	91[b]
[EMIM]$^+$	[SbF$_6$]$^-$	GC	3.7	91[b]
[EMIM]$^+$	[TaF$_6$]$^-$	Pt	4.5	91, 92[a]
[EMIM]$^+$	[WF$_7$]$^-$	GC	2.2	91[b]
[EMIM]$^+$	[WOF$_6$]$^-$	GC	2.1	93[b]
[BMIM]$^+$	[BF$_4$]$^-$	Pt	4.1	33
[BMIM]$^+$	[(HF)$_{2.3}$F]$^-$	GC	3.1	51
[BMIM]$^+$	[PF$_6$]$^-$	Pt	4.2	33
[PMIM]$^+$	[(HF)$_{2.3}$F]$^-$	GC	3.1	51
[PeMIM]$^+$	[(HF)$_{2.3}$F]$^-$	GC	3.2	51
[HMIM]$^+$	[(HF)$_{2.3}$F]$^-$	GC	3.6	51
[EMMIM]$^+$	[(CF$_3$SO$_2$)$_2$N]$^-$	Pt	4.7	86[c]
[PMMIM]$^+$	[(CF$_3$SO$_2$)$_2$N]$^-$	GC	4.3	53
[PMMIM]$^+$	[(CF$_3$SO$_2$)$_2$N]$^-$	GC	5.2	94
[PMMIM]$^+$	[(CF$_3$SO$_2$)$_3$C]$^-$	GC	5.4	94
[PMMIM]$^+$	[PF$_6$]$^-$	GC	4.3[e]	94[f]
[PMMIM]$^+$	[AsF$_6$]$^-$	GC	4.4[e]	94[f]
[(CH$_3$OC$_2$H$_4$)(CH$_3$)IM]$^+$	[BF$_4$]$^-$	GC	6.4	95
[(CH$_3$OC$_2$H$_4$)(CH$_3$)IM]$^+$	[(CF$_3$SO$_2$)$_2$N]$^-$	GC	6.2	95
[(CH$_3$OC$_2$H$_4$)(CH$_3$)IM]$^+$	[PF$_6$]$^-$	GC	5.4	95

[a] Abbreviations for substituents on the imidazolium (IM) nitrogens: M = methyl, E = ethyl, P = n-propyl, B = n-butyl, Pe = n-pentyl, H = n-hexyl, O = n-octyl, A = allyl.
[b] Voltage window estimated from cyclic voltammograms contained in the reference.
[c] Voltage window may be limited by impurities.
[d] Working electrode, Pt = Platinum, GC = Glassy Carbon, W = Tungsten.
[e] Voltage window at 80 °C. e Voltage window determined assuming cathodic limit of 0.63 V vs. Li/Li$^+$ reference.

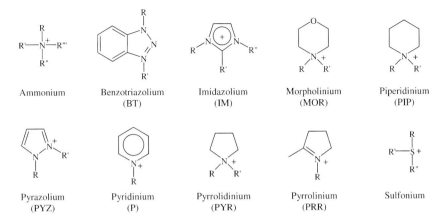

Fig. 3.6-2 Structures for the ionic liquid cations listed in Tables 3.6-1–3.6-10.

Acidic ionic liquids are prepared using an excess of the aluminum chloride (organic chloride < 50 mol%), and contain two chloroaluminate species, $[AlCl_4]^-$ and $[Al_2Cl_7]^-$ (a Lewis acid). Both are anions significantly more stable towards oxidation than the chloride ion. Furthermore, the $[Al_2Cl_7]^-$ is more readily reduced than the organic cation. Thus, the acidic ionic liquids have a limited cathodic range, but an extended anodic potential range. In the special case where the organic chloride salt and aluminum chloride are in equal amounts (50 mol%), these ionic liquids are termed neutral, because they contain only the organic cation and $[AlCl_4]^-$. The neutral chloroaluminate ionic liquids possess the widest electrochemical window, but they are difficult to prepare and maintain at the exact neutral composition. A solution to this problem has been developed through the introduction of a third component to the chloroaluminate ionic liquids, an alkali halide [40]. When an excess of alkali halide (e.g., LiCl, NaCl) is added to an acidic chloroaluminate ionic liquid it dissolves to the extent that it reacts with $[Al_2Cl_7]^-$ ion producing $[AlCl_4]^-$ and the alkali metal cation; this results in an ionic liquid that is essentially neutral, and at that point the alkali halide is no longer soluble. This neutral ionic liquid is "buffered" to both the addition of more $[Al_2Cl_7]^-$ or organic chloride. Consequently, the buffered neutral ionic liquids possess wide, and stable, electrochemical windows. The cathodic limits of the imidazolium-based buffered neutral ionic liquids are not sufficient to obtain reversible alkali metal deposition and stripping. Interestingly, addition of small amounts of proton to the buffered neutral ionic liquids shifts the reduction of the imidazolium cation sufficiently negative that reversible lithium and sodium deposition and stripping can be obtained [38, 41].

3.6.2
Ionic Conductivity

The ionic conductivity of a solvent is of critical importance in its selection for an electrochemical application. There are a variety of DC and AC methods available for the measurement of ionic conductivity. However, in the case of ionic liquids, the vast majority of data in the literature has been collected using one of two AC techniques, the impedance bridge method or the complex impedance method [42, 43]. Both of these methods employ simple two electrode cells to measure the impedance of the ionic liquid (Z). This impedance arises from resistive (R) and capacitive contributions (C), and can be described by the following equation,

$$Z = \sqrt{(1/\omega C)^2 + R^2} \qquad (3.6\text{-}1)$$

where ω is the frequency of the AC modulation. One can see from Eq. (3.6-1) that as the AC frequency increases the capacitive contribution to the impedance becomes vanishingly small and Eq. (3.6-1) reduces to $Z = R$, the resistance of the ionic liquid in the impedance cell. Under these conditions the conductivity (κ) of the ionic liquid may be obtained from the measured resistance using the following equation,

$$\kappa = \frac{l}{AR} \qquad (3.6\text{-}2)$$

where l is the distance between the two electrodes in the impedance cell and A is the area of the electrodes. The term in l/A is often referred to as the cell constant and it is normally determined by measuring the conductivity of a standard solution (usually aqueous KCl).

The impedance bridge method employs the AC version of a Wheatstone bridge (i.e., an impedance bridge) to measure the unknown cell impedance. Impedance measurements are carried out at a relatively high fixed frequency, normally in the range of a few kHz, in order to minimize the impact of the capacitive contribution to the cell impedance. This contribution is often further reduced by increasing the electrode surface area, and correspondingly increasing its capacitance, with a fine deposit of platinum black.

The complex impedance method involves the measurement of the cell impedance at frequencies ranging from a few Hz up to several MHz. The impedance data are collected with standard electrochemical impedance hardware (i.e., potentiostat/impedance analyzer) and are separated out into the real and imaginary components. These data are then graphed in the form of a Nyquist Plot (imaginary vs. real impedance), and the ionic liquid resistance is taken as the point where the data crosses the real axis at high frequency.

In general, there appears to be no significant difference between the data collected by either method. There is some evidence that data collected by the bridge method at lower frequencies may provide an underestimation of the true conductivity [44, 45], but there is no indication that this error is endemic to the impedance bridge method. The instrumentation for the impedance bridge method, although somewhat specialized, is generally less costly than the instrumentation required by the complex impedance method. However, the complex impedance method has gained popularity in recent years, most likely due to the increased availability of electrochemical impedance hardware.

Table 3.6-3 The room-temperature electrochemical potential windows for ionic liquids with ammonium cations

Cation	Anion(s)	Working electrode[c]	Window, (V)	Ref.
$[(C_2H_5)NH_3]^+$	$[(CF_3SO_2)_2N]^-$	GC	3.8	96[a]
$[(C_2H_5)_3NH]^+$	$[(CF_3SO_2)_2N]^-$	GC	4.7	96[a]
$[(C_2H_5)_4N]^+$	$[(CF_3SO_2)(CF_3CO)N]^-$	GC	4.4	96[a]
$[(n-C_3H_7)(CH_3)_3N]^+$	$[(CF_3SO_2)_2N]^-$	GC	5.8	97[a]
$[(n-C_3H_7)(CH_3)_3N]^+$	$[(CF_3SO_2)(CF_3CO)N]^-$	GC	4.7	98
$[(n-C_4H_9)(CH_3)_3N]^+$	$[(CF_3SO_2)_2N]^-$	GC	5.8	99
$[(n-C_6H_{13})(CH_3)_3N]^+$	$[(CF_3SO_2)_2N]^-$	GC	5.6	96[a]
$[(n-C_6H_{13})(CH_3)_3N]^+$	$[(CF_3SO_2)(CF_3CO)N]^-$	GC	4.5	96[a]
$[(n-C_4H_9)(C_2H_5)(CH_3)_2N]^+$	$[CF_3BF_3]^-$	GC	5.7	99
$[(n-C_4H_9)(C_2H_5)(CH_3)_2N]^+$	$[C_2F_5BF_3]^-$	GC	5.6	99
$[(n-C_4H_9)(C_2H_5)(CH_3)_2N]^+$	$[(CF_3SO_2)_2N]^-$	GC	5.7	99
$[(n-C_6H_{13})(C_2H_5)_3N]^+$	$[(CF_3SO_2)_2N]^-$	GC	4.5[b]	45
$[(n-C_6H_{13})(C_4H_9)_3N]^+$	$[(CH_3SO_2)_2N]^-$	GC	5.0	100
$[(n-C_8H_{17})(C_2H_5)_3N]^+$	$[(CF_3SO_2)_2N]^-$	GC	5.0	45
$[(n-C_8H_{17})(C_4H_9)_3N]^+$	$[(CF_3SO_2)_2N]^-$	GC	5.0	45
$[(CH_3OCH_2)(CH_3)_3N]^+$	$[(CF_3SO_2)_2N]^-$	GC	5.2	97[a]
$[(CH_3OC_2H_4)(CH_3)_3N]^+$	$[n-C_3F_7BF_3]^-$	GC	5.0	99
$[(CH_3OC_2H_4)(CH_3)_3N]^+$	$[(CF_3SO_2)_2N]^-$	GC	5.1	99
$[(CH_3OC_2H_4)(C_2H_5)(CH_3)_2N]^+$	$[CF_3BF_3]^-$	GC	5.2	99
$[(CH_3OC_2H_4)(C_2H_5)(CH_3)_2N]^+$	$[(CF_3SO_2)_2N]^-$	GC	5.5	99
$[(CH_3OC_2H_4)(C_2H_5)_2(CH_3)N]^+$	$[BF_4]^-$	GC	5.6	99
$[(CH_3OC_2H_4)(C_2H_5)_2(CH_3)N]^+$	$[BF_4]^-$	Pt	6.0	101
$[(CH_3OC_2H_4)(C_2H_5)_2(CH_3)N]^+$	$[CF_3BF_3]^-$	GC	5.7	99
$[(CH_3OC_2H_4)(C_2H_5)_2(CH_3)N]^+$	$[C_2F_5BF_3]^-$	GC	5.6	99
$[(CH_3OC_2H_4)(C_2H_5)_2(CH_3)N]^+$	$[n-C_3F_7BF_3]^-$	GC	5.6	99
$[(CH_3OC_2H_4)(C_2H_5)_2(CH_3)N]^+$	$[n-C_4F_9BF_3]^-$	GC	5.6	99
$[(CH_3OC_2H_4)(C_2H_5)_2(CH_3)N]^+$	$[(CF_3SO_2)_2N]^-$	GC	5.7	99
$[(CH_3OC_2H_4)(C_2H_5)_2(CH_3)N]^+$	$[(CF_3SO_2)_2N]^-$	Pt	5.7	101
$[(CH_3OC_2H_4)(C_2H_5)_3N]^+$	$[CF_3BF_3]^-$	GC	5.7	99
$[(CH_3OC_2H_4)(C_2H_5)_3N]^+$	$[C_2F_5BF_3]^-$	GC	5.6	99
$[(CH_3OC_2H_4)(C_2H_5)_3N]^+$	$[(CF_3SO_2)_2N]^-$	GC	5.7	99
$[(CH_3OC_2H_4)(C_2H_5)_3N]^+$	$[(CF_3SO_2)_2N]^-$	Pt	5.4	102

[a] Voltage window estimated from cyclic voltammograms contained in the reference.
[b] Voltage window may be limited by impurities.
[c] Working electrode, Pt = Platinum, GC = Glassy Carbon, W = Tungsten.

The conductivity of an electrolyte is a measure of the available charge carriers and their mobility. On the surface one would expect ionic liquids to possess very high conductivities because they are composed entirely of ions. Unfortunately, this is not the case. As a class, ionic liquids possess reasonably good ionic conductivities, comparable to the best non-aqueous solvent/electrolyte systems (~ 10 mS cm^{-1}). However, they are, in general, significantly less conductive than concentrated aque-

Table 3.6-4 The room-temperature electrochemical potential windows for other ionic liquids

Cation	Anion(s)	Working electrode[b]	Window, (V)	Ref.
Benzotriazolium (BT)				
$[(n\text{-}C_4H_9)(CH_3)BT]^+$	$[(CF_3SO_2)_2N]^-$	GC	3.5[c]	103
$[(n\text{-}C_4H_9)(C_2H_5)BT]^+$	$[BF_4]^-$	GC	3.1[d]	104
Morpholinium (MOR)				
$[(C_2H_5)(CH_3)MOR]^+$	$[(CF_3SO_2)_2N]^-$	GC	4.5	105
$[(n\text{-}C_4H_9)(CH_3)MOR]^+$	$[(CF_3SO_2)_2N]^-$	GC	4.5	105
$[(CH_3OC_2H_4)(CH_3)MOR]^+$	$[BF_4]^-$	GC	5.7	95
$[(CH_3OC_2H_4)(CH_3)MOR]^+$	$[Br]^-$	GC	4.0	95
$[(CH_3OC_2H_4)(CH_3)MOR]^+$	$[(CF_3SO_2)_2N]^-$	GC	6.2	95
$[(CH_3OC_2H_4)(CH_3)MOR]^+$	$[PF_6]^-$	GC	5.2	95
Piperidinium (PIP)				
$[(C_2H_5)(CH_3)PIP]^+$	$[(HF)_{2.3}F]^-$	GC	5.1	106
$[(n\text{-}C_3H_7)(CH_3)PIP]^+$	$[(CF_3SO_2)_2N]^-$	GC	5.8	107[a]
$[(n\text{-}C_3H_7)(CH_3)PIP]^+$	$[(CF_3SO_2)(C_2F_5SO_2)N]^-$	GC	5.4	96[a]
$[(n\text{-}C_3H_7)(CH_3)PIP]^+$	$[(HF)_{2.3}F]^-$	GC	5.2	106
$[(n\text{-}C_4H_9)(CH_3)PIP]^+$	$[(HF)_{2.3}F]^-$	GC	5.3	106
$[(CH_3OCH_2)(CH_3)PIP]^+$	$[(CF_3SO_2)_2N]^-$	GC	5.0	96[a]
Pyrazolium (PYZ)				
[1-ethyl-2-methylPYZ]$^+$	$[BF_4]^-$	GC	4.4	108
[1,2-dimethyl-4-fluoroPYZ]$^+$	$[BF_4]^-$	GC	4.1	109
Pyridinium (P)				
$[BP]^+$	$[BF_4]^-$	Pt	3.4	81
Pyrrolidinium (PYR)				
$[(C_2H_5)(CH_3)PYR]^+$	$[(HF)_{2.3}F]^-$	GC	4.7	106
$[(n\text{-}C_3H_7)(CH_3)PYR]^+$	$[(CH_3SO_2)_2N]^-$	GC	4.2	100
$[(n\text{-}C_3H_7)(CH_3)PYR]^+$	$[(HF)_{2.3}F]^-$	GC	5.3	106
$[(n\text{-}C_4H_9)(CH_3)PYR]^+$	$[C_2F_5BF_3]^-$	GC	5.6	85
$[(n\text{-}C_4H_9)(CH_3)PYR]^+$	$[(CH_3SO_2)_2N]^-$	GC	4.0	100
$[(n\text{-}C_4H_9)(CH_3)PYR]^+$	$[(HF)_{2.3}F]^-$	GC	4.7	106
$[(n\text{-}C_4H_9)(CH_3)PYR]^+$	$[(CF_3SO_2)_2N]^-$	GC	5.5	
$[(CH_3OCH_2)(CH_3)PYR]^+$	$[C_2F_5BF_3]^-$	GC	5.3	85
$[(CH_3OC_2H_4)(CH_3)PYR]^+$	$[C_2F_5BF_3]^-$	GC	5.2	85
Pyrrolinium (PRR)				
$[(n\text{-}C_3H_7)(CH_3)PYR]^+$	$[(CF_3SO_2)_2N]^-$	GC	4.0	110
Sulfonium				
$[(C_2H_5)_3S]^+$	$[(CF_3SO_2)_2N]^-$	GC	4.7	111
$[(n\text{-}C_4H_9)_3S]^+$	$[(CF_3SO_2)_2N]^-$	GC	4.8	112

[a] Voltage window estimated from cyclic voltammograms contained in the reference.
[b] Working electrode, Pt = Platinum, GC = Glassy Carbon, W = Tungsten.
[c] Voltage window at 100 °C.
[d] Voltage window at 40 °C.

ous electrolytes. The smaller than expected conductivity of ionic liquids can be attributed to the reduction of available charge carriers due to ion pairing and/or ion aggregation, and to the reduced ion mobility resulting from the large ion size found in many ionic liquids.

The conductivity of ionic liquids often exhibits classical linear Arrhenius behavior above room-temperature. However, as the temperature of these ionic liquids approaches their glass transition temperatures (T_g) the conductivity displays significant negative deviation from linear behavior. The observed temperature-dependent conductivity behavior is consistent with glass-forming liquids, and is often best described using the empirical Vogel–Tammann–Fulcher (VTF) equation,

$$\kappa = AT^{-1/2} \exp[-B/(T-T_o)] \tag{3.6-3}$$

where A and B are constants, and T_o is the temperature at which the conductivity (κ) goes to zero [42]. Examples of Arrhenius plots of temperature-dependent conductivity data for three ionic liquids are shown in (3.6-3). The data in Fig. 3.6-3 are also fit to the VTF equation. As can be seen from these data, the change in conductivity with temperature clearly varies depending on the ionic liquid. For example, the conductivity of the [EMIM][BF$_4$] decreases by a factor of 10 over the 275 to 375 K temperature range while the conductivity of [PMMIM][(CF$_3$SO$_2$)$_2$N] decreases by a factor of 30 over the same range of temperatures (Fig. 3.6-3). The temperature dependence of conductivity for an ionic liquid involves a complex interplay of short- and long-range forces that is strongly impacted by the type and character of the cation and anion. At our current level of understanding it is not possible to accurately predict how the conductivity of a given ionic liquid will vary with temperature. However, much effort is being directed towards this goal [46, 47].

It is now well documented that even small amounts of impurities can significantly impact upon the properties of ionic liquids. For example, the presence of chloride ion has been shown to decrease conductivity while the presence of water results in an increase [30]. Recent work by Widegren et al. [50] has shown that increasing the weight% of water in the ionic liquid [BMIM][(CF$_3$SO$_2$)$_2$N] from 0.001 to 0.09 resulted in a 36% increase in specific conductivity. Furthermore, this increase in conductivity was due almost entirely to the corresponding decrease in ionic liquid viscosity upon addition of water. Because of its ubiquitous nature in laboratory air and its common use as a solvent for the preparation of ionic liquids, water will be present in an ionic liquid unless great care is taken to remove/exclude it. It is clear that in many of the initial studies of ionic liquid conductivity, appropriate precautions were not taken, and as such these data must viewed with care. More recent work has benefited from our greater understanding of the effects of water, and thus the conductivity data, in general, is much less prone to error.

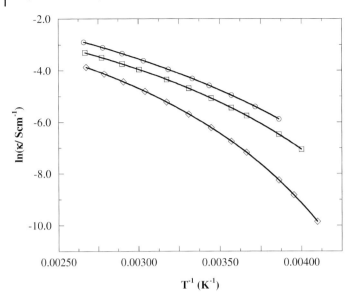

Fig. 3.6-3 Examples of Arrhenius plots of temperature-dependent conductivity for, (○) [EMIM][BF$_4$], (□) [EMIM][(CF$_3$SO$_2$)$_2$N], and (◇) [PMMIM][(CF$_3$SO$_2$)$_2$N]. The solid lines through each set of data represent the best fit of the data to the VTF equation [34, 53, 73].

The room-temperature conductivity data for a wide variety of ionic liquids are listed in Tables 3.6-5–3.6-8.[2]

Due to the growing interest in ionic liquids as electrochemical solvents, there has been a dramatic increase in the amount of conductivity data since the first edition of this volume was published. Only a few years ago it appeared that the room-temperature conductivity of ionic liquids was limited to the range of ~20 mS cm^{-1}. However, the recent development of ionic liquids with the [(HF)$_{2.3}$F]$^-$ anion has resulted in room-temperature ionic liquid conductivities in excess of 100 mS cm^{-1} [51]!

Ionic liquid conductivity appears weakly correlated with the size and type of the cation (Tables 3.6-5–3.6-8). Increasing cation size tends to lead to lower conductivity, most likely due to the lower mobility of the larger cations. Interestingly, the

---

[2] The first edition of this volume included multiple listings of conductivity data for several ionic liquids; these listings represented independent measurements from different researchers and had been included to help emphasize the significant variability in the conductivity data found in the literature (undoubtedly due primarily to the presence of impurities). However, because of the dramatic increase in the amount of conductivity data in the literature over the past few years we have been forced to limit ourselves to one entry for each ionic liquid. In those cases where multiple measurements exist in the literature, we have used our best judgment as to which measurements were the most reliable and/or complete and if necessary listed average values in Tables 3.6-5–3.6-8.

Table 3.6-5 Specific conductivity data for binary haloaluminate ionic liquids.[a]

Ionic liquid system	Cation	Anion(s)	Temperature, (K)	Conductivity (κ), mS cm^{-1}	Molar conductivity (Λ), S cm^2 mol^{-1}	Walden Product ($\Lambda\eta$), Scm^2 mol^{-1} P	Ref.
Imidazolium (IM)							
34.0–66.0 mol% [MMIM]Cl–AlCl$_3$	[MMIM]$^+$	[Al$_2$Cl$_7$]$^-$	298	15.0	4.26	0.721	12[b]
34.0–66.0 mol% [EMIM]Cl–AlCl$_3$	[EMIM]$^+$	[Al$_2$Cl$_7$]$^-$	298	15.0	4.46	0.630	113[b]
50.0–50.0 mol% [EMIM]Cl–AlCl$_3$	[EMIM]$^+$	[AlCl$_4$]$^-$	298	23.0	4.98	0.891	113[b]
60.0–40.0 mol% [EMIM]Cl–AlCl$_3$	[EMIM]$^+$	Cl$^-$, [AlCl$_4$]$^-$	298	6.5	1.22	0.578	113[b]
34.0–66.0 mol% [EMIM]Br–AlBr$_3$	[EMIM]$^+$	[Al$_2$Br$_7$]$^-$	298	5.8	1.89	0.596	113[b,c]
60.0–40.0 mol% [EMIM]Br–AlBr$_3$	[EMIM]$^+$	Br$^-$, [AlBr$_4$]$^-$	298	5.7	1.15	0.767	114[b,c]
40.0–60.0 mol% [PMIM]Cl–AlCl$_3$	[PMIM]$^+$	[AlCl$_4$]$^-$, [Al$_2$Cl$_7$]$^-$	298	11.0	2.94	0.534	113[b]
50.0–50.0 mol% [PMIM]Cl–AlCl$_3$	[PMIM]$^+$	[AlCl$_4$]$^-$	298	12.0	2.79	0.763	113[b]
60.0–40.0 mol% [PMIM]Cl–AlCl$_3$	[PMIM]$^+$	Cl$^-$, [AlCl$_4$]$^-$	298	3.3			113[b]
34.0–66.0 mol% [BMIM]Cl–AlCl$_3$	[BMIM]$^+$	[Al$_2$Cl$_7$]$^-$	298	9.2	3.04	0.585	113[b]
50.0–50.0 mol% [BMIM]Cl–AlCl$_3$	[BMIM]$^+$	[AlCl$_4$]$^-$	298	10.0	2.49	0.674	113[b]
34.0–66.0 mol% [BBIM]Cl–AlCl$_3$	[BBIM]$^+$	[Al$_2$Cl$_7$]$^-$	298	6.0	2.32	0.554	113[b]
50.0–50.0 mol% [BBIM]Cl–AlCl$_3$	[BBIM]$^+$	[AlCl$_4$]$^-$	298	5.0	1.50	0.568	113[b]
Pyridinium (P)							
33.3–66.7 mol% [MP]Cl–AlCl$_3$	[MP]$^+$	[Al$_2$Cl$_7$]$^-$	298	8.1	2.23	0.461	114[b]
33.3–66.7 mol% [EP]Cl–AlCl$_3$	[EP]$^+$	[Al$_2$Cl$_7$]$^-$	298	10.0	2.91	0.513	115[b]
33.3–66.7 mol% [EP]Br–AlBr$_3$	[EP]$^+$	[Al$_2$Cl$_x$Br$_{7-x}$]$^-$	298	8.4			115[b]
33.3–66.7 mol% [EP]Br–AlBr$_3$	[EP]$^+$	[Al$_2$Cl$_x$Br$_{7-x}$]$^-$	298	17.0			115
33.3–66.7 mol% [PP]Cl–AlCl$_3$	[PP]$^+$	[Al$_2$Cl$_7$]$^-$	298	8.0	2.47	0.449	115[b]
33.3–66.7 mol% [BP]Cl–AlCl$_3$	[BP]$^+$	[Al$_2$Cl$_7$]$^-$	298	6.7	2.18	0.458	115[b]

[a] Abbreviations for substituents on the imidazolium and pyridinium nitrogens: M = methyl, E = ethyl, P = n-propyl, B = n-butyl.
[b] Conductivity at 298 K Calculated from least-squares-fitted parameters given in reference.
[c] Conductivity estimated from graphical data provided in the reference.

Table 3.6-6 Specific conductivity data for non-haloaluminate imidazolium (IM)-based ionic liquids[a]

Cation	Anion	Temperature (K)	Conductivity (κ), mS cm^{-1}	Molar conductivity S cm^2 mol^{-1}	Walden product ($\Lambda\eta$), S cm^2 mol^{-1} P	Ref.
[MIM]$^+$	[(CF$_3$SO$_2$)$_2$N]$^-$	293	7.2			116
[MMIM]$^+$	[(CF$_3$SO$_2$)$_2$N]$^-$	293	8.4	2.03	0.894	86
[MMIM]$^+$	[CF$_3$BF$_3$]$^-$	298	15.5			80
[MMIM]$^+$	[(HF)$_{2.3}$F]$^-$	298	110	15.25	0.777	51
[EMIM]$^+$	[BF$_4$]$^-$	298	14	2.24	0.762	44, 118
[EMIM]$^+$	[C(CN)$_3$]$^-$	293	18	3.26	0.587	82, 83
[EMIM]$^+$	[CH$_3$BF$_3$]$^-$	298	9.0	1.51	0.711	119
[EMIM]$^+$	[C$_2$H$_5$BF$_3$]$^-$	298	6.3	1.16	0.833	119
[EMIM]$^+$	[n-C$_3$H$_7$BF$_3$]$^-$	298	5.7	1.14	0.613	119
[EMIM]$^+$	[n-C$_4$H$_9$BF$_3$]$^-$	298	3.2	0.70	0.580	119
[EMIM]$^+$	[n-C$_5$H$_{11}$BF$_3$]$^-$	298	2.7	0.63	0.558	119
[EMIM]$^+$	[CH$_2$CHBF$_3$]$^-$	298	10.5	1.86	0.764	119
[EMIM]$^+$	[CH$_3$CO$_2$]$^-$	293	2.8			86
[EMIM]$^+$	[CH$_3$SO$_3$]$^-$	298	2.7	0.45	0.719	84
[EMIM]$^+$	[CF$_3$BF$_3$]$^-$	298	14.6			80
[EMIM]$^+$	[CF$_3$CO$_2$]$^-$	293	9.6	1.67	0.586	86
[EMIM]$^+$	[C$_3$F$_7$-CO$_2$]$^-$	293	2.7	0.60	0.634	86
[EMIM]$^+$	[CF$_3$SO$_3$]$^-$	298	9.2	1.73	0.741	84
[EMIM]$^+$	[(CF$_3$SO$_2$)$_2$N]$^-$	298	9.2	2.38	0.811	50, 97
[EMIM]$^+$	[(C$_2$F$_5$SO$_2$)$_2$N]$^-$	299	3.4			53
[EMIM]$^+$	[(CF$_3$SO$_2$)(C$_2$F$_5$SO$_2$)N]$^-$	298	4.4	1.21	0.579	96
[EMIM]$^+$	[(CF$_3$SO$_2$)(CF$_3$CO)N]$^-$	298	9.8	2.38	0.595	98
[EMIM]$^+$	[CN$_4$]$^{-[c]}$	298	8.9			120
[EMIM]$^+$	[(HF)$_{2.3}$F]$^-$	298	100	15.5	0.753	51, 87, 88

[EMIM]$^+$	[FeCl$_4$]$^-$	293	18		121
[EMIM]$^+$	[GaCl$_4$]$^-$	293	20		122
[EMIM]$^+$	[N(CN)$_2$]$^-$	293	27	4.43	82
[EMIM]$^+$	[NbF$_6$]$^-$	298	8.5	1.62	91, 92
[EMIM]$^+$	[PF$_6$]$^-$	299	5.2		53
[EMIM]$^+$	[TaF$_6$]$^-$	298	7.1	1.33	91, 92
[EMIM]$^+$	[SbF$_7$]$^-$	298	6.2	1.16	91
[EMIM]$^+$	[WF$_7$]$^-$	298	3.2	0.60	91
[EMIM]$^+$	[WOF$_6$]$^-$	298	3.0	0.57	93
[EMMIM]$^+$	[(CF$_3$SO$_2$)$_2$N]$^-$	293	3.2	0.87	86
[EMM(5)IM]$^+$	[CF$_3$SO$_3$]$^-$	293	6.4	1.32	86
[EMM(5)IM]$^+$	[(CF$_3$SO$_2$)$_2$N]$^-$	293	6.6	1.82	86
[PMIM]$^+$	[BF$_4$]$^-$	298	5.9	1.01	122
[PMIM]$^+$	[CF$_3$BF$_3$]$^-$	298	8.5		80
[PMIM]$^+$	[(HF)$_{2.3}$F]$^-$	298	61	10.45	51
[PMMIM]$^+$	[BF$_4$]$^-$	295	5.9	0.732	118
[PMMIM]$^+$	[PF$_6$]$^-$	308	0.5		53
[PMMIM]$^+$	[(CF$_3$SO$_2$)$_2$N]$^-$	299	3.0		53
[BMIM]$^+$	[BF$_4$]$^-$	298	3.5	0.65	122, 123
[BMIM]$^+$	[CF$_3$BF$_3$]$^-$	298	5.9	1.177	80
[BMIM]$^+$	[(HF)$_{2.3}$F]$^-$	298	33		51
[BMIM]$^+$	[PF$_6$]$^-$	298	1.5	1.223	50, 124
[BMIM]$^+$	[CF$_3$CO$_2$]$^-$	293	3.2	0.487	86
[BMIM]$^+$	[C$_3$F$_7$CO$_2$]$^-$	293	1.0	0.481	86
[BMIM]$^+$	[CF$_3$SO$_3$]$^-$	293	3.7	0.744	86
[BMIM]$^+$	[C$_4$F$_9$SO$_3$]$^-$	293	0.45	0.516	86
[BMIM]$^+$	[(CF$_3$SO$_2$)$_2$N]$^-$	293	3.9	0.595	50, 86
[iBMIM]$^+$	[(CF$_3$SO$_2$)$_2$N]$^-$	293	2.6	0.634	86
[BMMIM]$^+$	[BF$_4$]$^-$	298	0.7		123
[BMMIM]$^+$	[PF$_6$]$^-$	298	0.2		123
[BMMIM]$^+$	[SbF$_6$]$^-$	298	0.4		123
[EEIM]$^+$	[CF$_3$CO$_2$]$^-$	293	7.4	1.41	86

(*Continued*)

Table 3.6-6 (Continued)

Cation	Anion	Temperature (K)	Conductivity (κ), mS cm^{-1}	Molar conductivity S cm^2 mol^{-1}	Walden product ($\Lambda\eta$), S cm^2 mol^{-1} P	Ref.
[EEIM]$^+$	[CF$_3$SO$_3$]$^-$	293	7.5	1.55	0.820	86
[EEIM]$^+$	[(CF$_3$SO$_2$)$_2$N]$^-$	293	8.5	2.37	0.831	86
[EEM(5)IM]$^+$	[(CF$_3$SO$_2$)$_2$N]$^-$	293	6.2	1.82	0.654	86
[BEIM]$^+$	[CF$_3$CO$_2$]$^-$	293	2.5	0.56	0.501	86
[BEIM]$^+$	[CH$_3$SO$_3$]$^-$	298	0.55	0.12		84
[BEIM]$^+$	[CF$_3$SO$_3$]$^-$	298	2.7	0.64		84
[BEIM]$^+$	[C$_4$F$_9$SO$_3$]$^-$	293	0.53	0.17	0.543	86
[BEIM]$^+$	[(CF$_3$SO$_2$)$_2$N]$^-$	293	4.1	1.27	0.608	86
[PeMIM]$^+$	[(HF)$_{2.3}$F]$^-$	298	27	5.61	1.499	51
[HMIM]$^+$	[CF$_3$BF$_3$]$^-$	298	2.8			80
[HMIM]$^+$	[(CF$_3$SO$_2$)$_2$N]$^-$	298	2.2			50
[HMIM]$^+$	[(HF)$_{2.3}$F]$^-$	298	16	3.72	0.959	51
[AAIM]$^+$	[Cl]$^-$	298	0.58			125
[AAIM]$^+$	[Br]$^-$	298	0.74			125
[AAIM]$^+$	[I]$^-$	298	1.21			125
[AAIM]$^+$	[(CF$_3$SO$_2$)$_2$N]$^-$	298	2.48			125
[AMIM]$^+$	[Br]$^-$	298	1.55			125
[AEIM]$^+$	[Br]$^-$	298	1.05			125
[APIM]$^+$	[Br]$^-$	298	0.48			125
[APeIM]$^+$	[Br]$^-$	298	0.17			125
[AOIM]$^+$	[Br]$^-$	298	0.06			125
[AMMIM]$^+$	[BF$_4$]$^-$	298	2.1			123
[(CH$_3$OC$_2$H$_4$)IM]$^+$	[BF$_4$]$^-$	298	4.6			95
[(CH$_3$OC$_2$H$_4$)IM]$^+$	[CF$_3$SO$_3$]$^-$	293	3.6	0.77	0.567	86
[(CH$_3$OC$_2$H$_4$)IM]$^+$	[(CF$_3$SO$_2$)$_2$N]$^-$	293	4.2	1.18	0.639	86, 95
[(CH$_3$OC$_2$H$_4$)IM]$^+$	[PF$_6$]$^-$	298	2.1			95
[CF$_3$CH$_2$MIM]$^+$	[(CF$_3$SO$_2$)$_2$N]$^-$	293	0.98	0.25	0.626	86

[a] Abbreviations for substituents on the imidazolium nitrogens: M = methyl, E = ethyl, P = n-propyl, B = n-butyl, Pe = n-pentyl, H = n-hexyl, O = n-octyl, A = allyl.
[b] I = complex impedance, B = conductivity bridge, U = method unknown (not provided in the reference).
[c] 1,2,3,4-tetrazolium anion.

Table 3.6-7 Specific conductivity data for ionic liquids with ammonium cations

Cation	Anion	Temperature, (K)	Conductivity (κ), mS cm^{-1}	Molar conductivity (Λ), S cm^2 mol^{-1}	Walden product ($\Lambda\eta$), S cm^2 mol^{-1} P	Ref.
Ammonium						
[n-C$_3$H$_7$)(CH$_3$)$_3$N]$^+$	[(CF$_3$SO$_2$)$_2$N]$^-$	298	3.3	0.88	0.631	97
[n-C$_3$H$_7$)(CH$_3$)$_3$N]$^+$	[(CF$_3$SO$_2$)(C$_2$F$_5$SO$_2$)N]$^-$	298	1.2	0.34	0.590	96
[n-C$_3$H$_7$)(CH$_3$)$_3$N]$^+$	[(CF$_3$SO$_2$)(CF$_3$CO)N]$^-$	298	4.3	1.09	0.490	98
[n-C$_3$H$_7$)(C$_2$H$_5$)$_2$(CH$_3$)N]$^+$	[(CF$_3$SO$_2$)$_2$N]$^-$	298	2.2	0.64	0.598	99
[n-C$_4$H$_9$)(C$_2$H$_5$)$_2$(CH$_3$)N]$^+$	[(CF$_3$SO$_2$)$_2$N]$^-$	298	1.6	0.47	0.570	99
[n-C$_4$H$_9$)(C$_2$H$_5$)$_2$(CH$_3$)N]$^+$	[CF$_3$BF$_3$]$^-$	298	2.1	0.50	1.051	99
[n-C$_4$H$_9$)(C$_2$H$_5$)$_2$(CH$_3$)N]$^+$	[C$_2$F$_5$BF$_3$]$^-$	298	2.3	0.61	0.634	99
[n-C$_5$H$_{11}$)$_4$N]$^+$	[(CF$_3$SO$_2$)(C$_2$F$_5$SO$_2$)N]$^-$	298	0.11	0.06	0.372	96
[n-C$_5$H$_{11}$)$_4$N]$^+$	[(CF$_3$SO$_2$)(CF$_3$CO)N]$^-$	298	0.20	0.10	0.382	96
[n-C$_6$H$_{13}$)(C$_2$H$_5$)$_3$N]$^+$	[(CF$_3$SO$_2$)$_2$N]$^-$	298	0.67	0.25	0.411	45
[n-C$_6$H$_{13}$)(C$_4$H$_9$)$_3$N]$^+$	[(CH$_3$SO$_2$)$_2$N]$^-$	298	0.005			100
[n-C$_8$H$_{17}$)(C$_2$H$_5$)$_3$N]$^+$	[(CF$_3$SO$_2$)$_2$N]$^-$	298	0.33	0.13	0.264	45
[n-C$_8$H$_{17}$)(C$_4$H$_9$)$_3$N]$^+$	[(CF$_3$SO$_2$)$_2$N]$^-$	298	0.13	0.07	0.386	45
[CH$_3$OCH$_2$)(CH$_3$)$_3$N]$^+$	[(CF$_3$SO$_2$)$_2$N]$^-$	298	4.7	1.20	0.598	97
[CH$_3$OC$_2$H$_4$)(CH$_3$)$_3$N]$^+$	[n-C$_3$F$_7$BF$_3$]$^-$	298	2.5	0.63	0.478	99
[CH$_3$OC$_2$H$_4$)(C$_2$H$_5$)(CH$_3$)$_2$N]$^+$	[BF$_4$]$^-$	298	1.7	0.31	1.031	99, 126, 127
[CH$_3$OC$_2$H$_4$)(C$_2$H$_5$)(CH$_3$)$_2$N]$^+$	[CF$_3$BF$_3$]$^-$	298	2.5	0.53	0.541	99
[CH$_3$OC$_2$H$_4$)(C$_2$H$_5$)(CH$_3$)$_2$N]$^+$	[C$_2$F$_5$BF$_3$]$^-$	298	3.8	0.91	0.529	99, 127
[CH$_3$OC$_2$H$_4$)(C$_2$H$_5$)(CH$_3$)$_2$N]$^+$	[n-C$_3$F$_7$BF$_3$]$^-$	298	2.6	0.69	0.483	99
[CH$_3$OC$_2$H$_4$)(C$_2$H$_5$)(CH$_3$)$_2$N]$^+$	[n-C$_4$F$_9$BF$_3$]$^-$	298	1.5	0.43	0.442	99
[CH$_3$OC$_2$H$_4$)(C$_2$H$_5$)(CH$_3$)$_2$N]$^+$	[(CF$_3$SO$_2$)$_2$N]$^-$	298	3.1	0.88	0.529	99
[CH$_3$OC$_2$H$_4$)(C$_2$H$_5$)$_2$(CH$_3$)N]$^+$	[BF$_4$]$^-$	298	1.3	0.25	1.076	99, 127

(*Continued*)

Table 3.6-7 (Continued)

Cation	Anion	Temperature, (K)	Conductivity (κ), mS cm^{-1}	Molar conductivity (Λ), S cm^2 mol^{-1}	Walden product ($\Lambda\eta$), S cm^2 mol^{-1} P	Ref.
[(CH$_3$OC$_2$H$_4$)(C$_2$H$_5$)$_2$(CH$_3$)N]$^+$	[CF$_3$BF$_3$]$^-$	298	3.0	0.68	0.734	99
[(CH$_3$OC$_2$H$_4$)(C$_2$H$_5$)$_2$(CH$_3$)N]$^+$	[C$_2$F$_5$BF$_3$]$^-$	298	3.2	0.81	0.553	99, 127
[(CH$_3$OC$_2$H$_4$)(C$_2$H$_5$)$_2$(CH$_3$)N]$^+$	[n-C$_3$F$_7$BF$_3$]$^-$	298	1.9	0.53	0.468	99
[(CH$_3$OC$_2$H$_4$)(C$_2$H$_5$)$_2$(CH$_3$)N]$^+$	[n-C$_4$F$_9$BF$_3$]$^-$	298	1.3	0.40	0.468	99
[(CH$_3$OC$_2$H$_4$)(C$_2$H$_5$)$_2$(CH$_3$)N]$^+$	[(CF$_3$SO$_2$)$_2$N]$^-$	298	2.6	0.78	0.539	99
[(CH$_3$OC$_2$H$_4$)(C$_2$H$_5$)$_3$N]$^+$	[CF$_3$BF$_3$]$^-$	298	2.0	0.49	0.735	99
[(CH$_3$OC$_2$H$_4$)(C$_2$H$_5$)$_3$N]$^+$	[C$_2$F$_5$BF$_3$]$^-$	298	2.4	0.65	0.566	99, 127
[(CH$_3$OC$_2$H$_4$)(C$_2$H$_5$)$_3$N]$^+$	[n-C$_3$F$_7$BF$_3$]$^-$	298	1.8	0.53	0.485	99
[(CH$_3$OC$_2$H$_4$)(C$_2$H$_5$)$_3$N]$^+$	[n-C$_4$F$_9$BF$_3$]$^-$	298	1.1	0.35	0.474	99
[(CH$_3$OC$_2$H$_4$)(C$_2$H$_5$)$_3$N]$^+$	[(CF$_3$SO$_2$)$_2$N]$^-$	298	2.1	0.66	0.562	99, 102
[(CH$_3$OC$_2$H$_4$OC$_2$H$_4$)(C$_2$H$_5$)$_3$N]$^+$	[(CF$_3$SO$_2$)$_2$N]$^-$	298	2.6	0.74	0.467	96

Table 3.6-8 Specific conductivity data for other ionic liquids

Cation	Anion	Temperature, (K)	Conductivity (κ), mS cm^{-1}	Molar conductivity (Λ), S cm^2 mol^{-1}	Walden product ($\Lambda\eta$), S cm^2 mol^{-1} P	Ref.
Morpholinium (MOR)						
[(CH$_3$OC$_2$H$_4$)(CH$_3$)MOR]$^+$	[BF$_4$]$^-$	298	0.087			95
[(CH$_3$OC$_2$H$_4$)(CH$_3$)MOR]$^+$	[(CF$_3$SO$_2$)$_2$N]$^-$	298	0.085			95
[(CH$_3$OC$_2$H$_4$)(CH$_3$)MOR]$^+$	[PF$_6$]$^-$	298	0.00066			95
Piperidinium (PIP)						
[(C$_2$H$_5$)(CH$_3$)PIP]$^+$	[(HF)$_{2.3}$F]$^-$	298	37.2	6.72	1.47	106
[n-C$_3$H$_7$)(CH$_3$)PIP]$^+$	[(CF$_3$SO$_2$)(CF$_3$CO)N]$^-$	298	2.1	0.59	0.580	96
[n-C$_3$H$_7$)(CH$_3$)PIP]$^+$	[(CF$_3$SO$_2$)(C$_2$F$_5$SO$_2$)N]$^-$	298	0.71	0.22	0.626	96
[n-C$_3$H$_7$)(CH$_3$)PIP]$^+$	[(HF)$_{2.3}$F]$^-$	298	23.9	4.67	1.28	106
[n-C$_3$H$_7$)(CH$_3$)PIP]$^+$	[(CF$_3$SO$_2$)$_2$N]$^-$	298	1.51			107
[n-C$_4$H$_9$)(CH$_3$)PIP]$^+$	[(HF)$_{2.3}$F]$^-$	298	12.3	2.62	0.900	106
[(CH$_3$OCH$_2$)(CH$_3$)PIP]$^+$	[(CF$_3$SO$_2$)$_2$N]$^-$	298	2.2	0.64	0.432	96
[(CH$_3$OC$_2$H$_4$)(CH$_3$)PIP]$^+$	[(CF$_3$SO$_2$)$_2$N]$^-$	298	2.7	0.82	0.449	96
[(CH$_3$OC$_2$H$_4$OC$_2$H$_4$)(CH$_3$)PIP]$^+$	[(CF$_3$SO$_2$)$_2$N]$^-$	298	1.7	0.55	0.485	96
Pyridinium (P)						
[BP]$^+$	[BF$_4$]$^-$	298	1.9	0.35	0.358	73
[BP]$^+$	[BF$_4$]$^-$	303	3.0			81
[BP]$^+$	[(CF$_3$SO$_2$)$_2$N]$^-$	298	2.2	0.63	0.359	73[a]
Pyrrolidinium (PRY)						
[(C$_2$H$_5$)(CH$_3$)PYR]$^+$	[(HF)$_{2.3}$F]$^-$	298	74.6	12.5	1.82	106
[n-C$_3$H$_7$)(CH$_3$)PYR]$^+$	[(HF)$_{2.3}$F]$^-$	298	58.1	10.7	1.39	106
[n-C$_3$H$_7$)(CH$_3$)PYR]$^+$	[(CF$_3$SO$_2$)$_2$N]$^-$	298	1.4	0.39	0.248	110
[n-C$_4$H$_9$)(CH$_3$)PYR]$^+$	[(CH$_3$SO$_2$)$_2$N]$^-$	298	0.07	0.02	0.289	100
[n-C$_4$H$_9$)(CH$_3$)PYR]$^+$	[C$_2$F$_5$BF$_3$]$^-$	298	3.5	0.89	0.629	85

(Continued)

Table 3.6-8 (Continued)

Cation	Anion	Temperature, (K)	Conductivity (κ), mS cm^{-1}	Molar conductivity (Λ), S cm^2 mol^{-1}	Walden product ($\Lambda\eta$), S cm^2 mol^{-1} P	Ref.
[(n-C$_4$H$_9$)(CH$_3$)PYR]$^+$	[(HF)$_{2.3}$F]$^-$	298	35.9	7.15	0.429	106
[(n-C$_4$H$_9$)(CH$_3$)PYR]$^+$	[(CF$_3$SO$_2$)$_2$N]$^-$	298	2.2	0.66	0.560	110
[(CH$_3$OCH$_2$)(CH$_3$)PYR]$^+$	[C$_2$F$_5$BF$_3$]$^-$	298	6.8	1.55	0.574	85
[CH$_3$OC$_2$H$_4$)(CH$_3$)PYR]$^+$	[C$_2$F$_5$BF$_3$]$^-$	298	4.5	1.10	0.570	85
[C$_2$H$_5$OC$_2$H$_4$)(CH$_3$)PYR]$^+$	[C$_2$F$_5$BF$_3$]$^-$	298	3.7	0.96	0.470	85
[CH$_3$OC$_2$H$_4$OC$_2$H$_4$)(CH$_3$)PYR]$^+$	[C$_2$F$_5$BF$_3$]$^-$	298	3.0	0.84	0.453	85
Pyrazolium (PYZ)						
1,2-dimethyl-4-fluoro-PYZ	[BF$_4$]$^-$	298	1.3			109
Phosphonium						
[(n-C$_6$H$_{13}$)(CH$_3$)$_3$P]$^+$	[(CF$_3$SO$_2$)$_2$N]$^-$	298	0.92	0.30	0.455	96
[(n-C$_6$H$_{13}$)(CH$_3$)$_3$P]$^+$	[(CF$_3$SO$_2$)(CF$_3$CO)N]$^-$	298	1.4	0.44	0.403	96
Sulfonium						
[(CH$_3$)$_3$S]$^+$	[HBr$_2$]$^-$	298	34	4.62	0.953	127
[(CH$_3$)$_3$S]$^+$	[HBr$_2$]$^-$, [H$_2$Br$_3$]$^-$	298	56	8.41	0.698	128[b]
[(CH$_3$)$_3$S]$^+$	[Al$_2$Cl$_7$]$^-$	298	5.5	1.49	0.586	79, 128
[(CH$_3$)$_3$S]$^+$	[Al$_2$Cl$_6$Br]$^-$	298	4.21	1.12	0.616	128
[(CH$_3$)$_3$S]$^+$	[Al$_2$Br$_7$]$^-$	298	1.44	0.41	0.572	128
[(CH$_3$)$_3$S]$^+$	[(CF$_3$SO$_2$)$_2$N]$^-$	318	8.2	1.85	0.816	112
[(C$_2$H$_5$)$_3$S]$^+$	[(CF$_3$SO$_2$)$_2$N]$^-$	298	7.1	1.94	0.583	112
[(C$_2$H$_5$)$_3$S]$^+$	[(CF$_3$SO$_2$)(C$_2$F$_5$SO$_2$)N]$^-$	298	3.7	1.05	0.579	96
[(C$_2$H$_5$)$_3$S]$^+$	[(CF$_3$SO$_2$)(CF$_3$CO)N]$^-$	298	9.0	2.30	0.645	96
[(n-C$_4$H$_9$)$_3$S]$^+$	[(CF$_3$SO$_2$)$_2$N]$^-$	298	1.4	0.52	0.394	112
Thiazolium						
[1-ethylthiazolium]$^+$	[CF$_3$SO$_3$]$^-$	298	4.2			84

[a] Conductivity at 298 K calculated from VTF parameters given in reference.
[b] Binary composition of 42.0–58.0 mol% [(CH$_3$)$_3$S]Br–HBr.

correlation between the anion type or size and the ionic liquid conductivity is very limited. Other than the higher conductivities observed for ionic liquids with the $[(HF)_{2.3}F]^-$ and $[BF_4]^-$ anions, there appears to be no clear relationship between anion size and conductivity. For example, ionic liquids with large anions such as $[(CF_3SO_2)_2N]^-$ often exhibit higher conductivities than those with smaller anions such as $[CH_3CO_2]^-$.

The conductivity and viscosity of ionic liquids are often combined into what is termed Walden's rule [52],

$$\Lambda \varphi = \text{constant} \tag{3.6-4}$$

where Λ is the molar conductivity of the ionic liquid, and it is given by

$$\Lambda = \kappa M / \rho \tag{3.6-5}$$

where M is the equivalent weight (molecular weight) of the ionic liquid and ρ is the ionic liquid density. Ideally, the Walden Product ($\Lambda \eta$) remains constant for a given ionic liquid, regardless of temperature. The magnitude of the Walden Product for different ionic liquids has been shown to vary inversely with ion size [53, 54]. This inverse relationship between ion size and the magnitude of $\Lambda \eta$ is generally followed for the cations in Tables 3.6-5–3.6-8. The clearest example can be seen for the sulfonium ionic liquids (Table 3.6-8) where increasing cation size from $(CH_3)_3S^+$, $(C_2H_5)_3S^+$, and $(n\text{-}C_4H_9)_3S^+$ results in Walden Products of 0.816, 0.583, and 0.394, respectively. As was the case with conductivity, the size of the anions in Tables 3.6-5–3.6-8 exhibits no clear correlation to the magnitude of the Walden Product.

The relationship between conductivity and viscosity may be viewed through the use of a Walden plot (log Λ versus log $(1/\eta)$) [61]. Plotting the molar conductivity (Λ) instead of the absolute conductivity (κ), to an extent, normalizes the effects of molar concentration and density on the conductivity and, thus, gives a better indication of the number of mobile charge carriers in an ionic liquid. Fig. 3.6-4 shows the Walden Plot for the data in Tables 3.6-5–3.6-8. Data for each of the various types of ionic liquids (haloaluminates, non-haloaluminate imidazoliums, ammoniums, other ionic liquids) were plotted separately on the graph. However, as is clearly shown in Fig. 3.6-4, no difference in the behavior of any of the types of ionic liquids was observed.

As has been observed previously, the vast majority of the ionic liquids fall slightly below the ideal 1:1 Walden line [23, 47, 51, 82]. A linear regression was performed on the data giving a slope of 1.11, an intercept of –0.235, and $R^2 = 0.958$. The fit of the data to the linear relationship is surprisingly good when one considers the wide variety of ionic liquids and the unknown errors in the literature data. The linear behavior in the Walden Plot clearly demonstrates that ionic liquid conductivity and its viscosity are strongly coupled. Furthermore, it shows that all the ionic liquids exhibit normal Walden type conductivity where the conductivity arises from the movement of mobile charge carriers [23, 47].

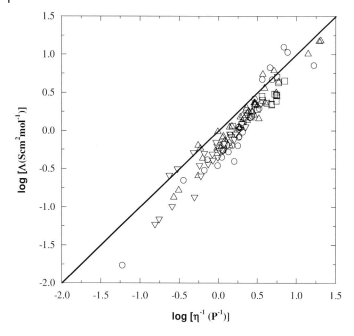

Fig. 3.6-4 The Walden plot of the molar conductivity and viscosity data in Tables 3.6-4–3.6-8. (□) haloaluminate ionic liquids, (△) non-haloaluminate imidazolium ionic liquids, (▽) ammonium ionic liquids, (○) other ionic liquids. The line represents the ideal 1:1 Walden line.

The physical properties of ionic liquids can often be considerably changed through the judicious addition of cosolvents [15, 57–64, 104, 124]. For example, Hussey and coworkers investigated the effect of co-solvents on the physical properties of [EMIM]Cl–AlCl$_3$ ionic liquids [55, 56]. They found significant increases in ionic conductivity upon the addition of a variety of co-solvents. Fig. 3.6-5 displays representative data from this work.

The magnitude of the conductivity increase depends both on the type and amount of the co-solvent (Fig. 3.6-5). The observed effect of co-solvent addition has been explained in terms of the solvation of the constituent ions of the ionic liquid by the co-solvent. This solvation, in turn, reduces ion pairing or ion aggregation in the ionic liquid, resulting in an increase in the number of available charge carriers and an increase in the mobility of these charge carriers. Counteracting this solvating effect is the dilution of the number of free ions as the mole fraction of the co-solvent increases. These counteracting effects help to explain the observed maximum in conductivity for benzene added to a 40.00–60.00 mol% [EMIM]Cl–AlCl$_3$ ionic liquid shown in Fig. 3.6-5.

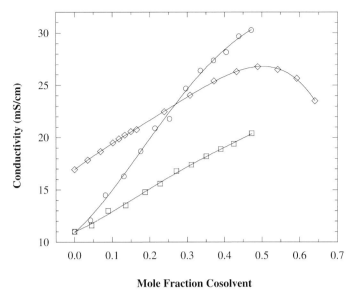

Fig. 3.6-5 Change in the conductivity of [EMIM]Cl–AlCl$_3$ ionic liquids with the mole fraction of co-solvent. (□) benzene or (○) CH$_2$Cl$_2$ added to a 55.56–44.44 mol% [EMIM]Cl–AlCl$_3$ ionic liquid, and (◇) benzene added to a 40.00–60.00 mol% [EMIM]Cl–AlCl$_3$ ionic liquid [55, 56].

3.6.3
Transport Properties

The behavior of ionic liquids as electrolytes is strongly influenced by the transport properties of their ionic constituents [63]. These transport properties relate to the rate of ion movement, and to the manner in which the ions move (as individual ions, ion-pairs, or ion aggregates). Conductivity, for example, depends on the number and mobility of charge carriers. If an ionic liquid is dominated by highly mobile but neutral ion pairs it will have a small number of available charge carriers and thus a low conductivity. The two quantities often used to evaluate the transport properties of electrolytes are ion diffusion coefficients and ion transport numbers. The diffusion coefficient is a measure of the rate of movement of an ion in a solution, and the transport number is a measure of the fraction of charge carried by that ion in the presence of an electric field.

The diffusion coefficients of the constituent ions in ionic liquids have most commonly been measured by either electrochemical or NMR methods. These two methods, in fact, measure slightly different diffusional properties. The electrochemical methods measure the diffusion coefficient of an ion in the presence of a concentration gradient (Fick diffusion) [64], while the NMR methods measure the diffusion coefficient of an ion in the absence of any concentration gradients (self-diffusion) [65]. Fortunately, under most circumstances these two types of diffusion coefficients are roughly equivalent.

There are a number of NMR methods available for evaluating self diffusion coefficients, all of which use the same basic measurement principle [65]. They are all based on application of the spin echo technique under conditions of either a static or a pulsed magnetic field gradient. Essentially, a spin echo pulse sequence is applied to a nucleus on the ion of interest while at the same time a constant or pulsed field gradient is applied to the nucleus. The spin echo of this nucleus is then measured and its attenuation due to the diffusion of the nucleus in the field gradient is used to determine its self-diffusion coefficient. The self diffusion coefficient data for a variety of ionic liquids are given in Table 3.6-9.

Electrochemically generated diffusion coefficients are normally determined from the steady state voltammetric limiting current arising from the reduction or oxidation of the ion of interest. In the case of ionic liquids this requires that the potential of a working electrode be scanned into the cathodic and anodic potential limits in the hopes of obtaining clear limiting current plateaus for the reduction of the cation and the oxidation of the anion, respectively. This process is fraught with difficulty and has met with limited success. The very large limiting currents encountered scanning beyond the normal potential limits lead to significant migration effects, and irreversible electrode reactions (especially for the cation reduction) often foul the working electrode surface. The one successful application of these electrochemical methods in ionic liquids has been in the evaluation of the diffusion coefficient of the chloride ion in basic [EMIM]Cl–$AlCl_3$ ionic liquids [66, 67]. In slightly basic ionic liquids the chloride ion concentration is reasonably low and its oxidation well separated from that of the other anion present ($[AlCl_4]^-$). These diffusion coefficient data are given in Table 3.6-9.

The cation and anion diffusion coefficient data in Table 3.6-9 correlate well with the corresponding ionic conductivity data given in Tables 3.6-5–3.6-8. As expected both the cation and anion diffusion coefficients increase with increasing conductivity. The cation diffusion coefficients for the [EMIM]Cl–$AlCl_3$ ionic liquids decrease significantly as the mol% of $AlCl_3$ decreases below 50%. This "basic" composition regime is characterized by increasing chloride as the mole percent of $AlCl_3$ decreases. The decline in cation diffusion coefficients is consistent with the observation of significant hydrogen bonding between chloride ion and the $[EMIM]^+$ cation [4] that would be expected to lead to reduced cation mobility.

Transport numbers are intended to measure the fraction of the total ionic current carried by an ion in an electrolyte as it migrates under the influence of an applied electric field. In essence, transport numbers are an indication of the relative ability of an ion to carry charge. The classical way to measure transport numbers is to pass a current between two electrodes contained in separate compartments of a two-compartment cell. These two compartments are separated by a barrier that only allows the passage of ions. After a known amount of charge has been passed the composition and/or mass of the electrolytes in the two compartments are analyzed. From these data the fraction of the charge transported by the cation and anion can be calculated. Transport numbers obtained by this method are measured with respect to an external reference point (i.e., the separator) and, therefore, are often referred to as external transport numbers. Two variations of the above method, the

Table 3.6-9 Diffusion coefficients for ionic liquids

Ionic Liquid System	Cation	D_{R^+} (10^{11} m^2 s^{-1})	Anion(s)	D_{X^-} (10^{11} m^2 s^{-1})	Temperature (K)	Method[a]	Ref.
[MMIM][(CF$_3$SO$_2$)$_2$N]	[MMIM]$^+$	5.0	[(CF$_3$SO$_2$)$_2$N]$^-$	2.6	298	PNMR	129[b]
[EMIM][BF$_4$]	[EMIM]$^+$	5.0	[BF$_4$]$^-$	4.2	298	PNMR	73[b]
[EMIM][BF$_4$]	[EMIM]$^+$	3.0	[BF$_4$]$^-$	–	300	PNMR	129
[EMIM][CF$_3$SO$_2$]	[EMIM]$^+$	5	[CF$_3$SO$_2$]$^-$	3	298	FNMR	74[e]
[EMIM][(CF$_3$SO$_2$)$_2$N]	[EMIM]$^+$	5.0	[(CF$_3$SO$_2$)$_2$N]$^-$	3.1	298	PNMR	129[b]
[EMIM][(CF$_3$SO$_2$)$_2$N]	[EMIM]$^+$	5	[(CF$_3$SO$_2$)$_2$N]$^-$	4	298	FNMR	74[e]
[BMIM][BF$_4$]	[BMIM]$^+$	1.5	[BF$_4$]$^-$	1.3	298	PNMR	131[b]
[BMIM][CF$_3$CO$_2$]	[BMIM]$^+$	1.8	[CF$_3$CO$_2$]$^-$	1.4	298	PNMR	133[b]
[BMIM][CF$_3$SO$_3$]	[BMIM]$^+$	1.7	[CF$_3$SO$_3$]$^-$	1.3	298	PNMR	131[b]
[BMIM][(CF$_3$SO$_2$)$_2$N]	[BMIM]$^+$	2.8	[(CF$_3$SO$_2$)$_2$N]$^-$	2.2	298	PNMR	129[b]
[BMIM][(C$_2$F$_5$SO$_2$)$_2$N]	[BMIM]$^+$	1.3	[(C$_2$F$_5$SO$_2$)$_2$N]$^-$	0.86	298	PNMR	131[b]
[BMIM][PF$_6$]	[BMIM]$^+$	0.71	[PF$_6$]$^-$	0.52	298	PNMR	49, 131[b]
[HMIM][(CF$_3$SO$_2$)$_2$N]	[HMIM]$^+$	1.7	[(CF$_3$SO$_2$)$_2$N]$^-$	1.5	298	PNMR	129[b]
[OMIM][(CF$_3$SO$_2$)$_2$N]	[OMIM]$^+$	1.2	[(CF$_3$SO$_2$)$_2$N]$^-$	1.2	298	PNMR	129[b]
[BP][BF$_4$]	[BP]$^+$	0.91	[BF$_4$]$^-$	1.0	298	PNMR	73[b]
[BP][(CF$_3$SO$_2$)$_2$N]	[BP]$^+$	2.4	[(CF$_3$SO$_2$)$_2$N]$^-$	2.0	298	PNMR	73[b]
33.0–67.0 mol% [EMIM]Cl-AlCl$_3$	[EMIM]$^+$	14.4	[Al$_2$Cl$_7$]$^-$/[AlCl$_4$]$^-$	–	303	DNMR	131
50.0–50.0 mol% [EMIM]Cl-AlCl$_3$	[EMIM]$^+$	18	[AlCl$_4$]$^-$	–	258	PNMR	132[c]
50.0–50.0 mol% [EMIM]Cl-AlCl$_3$	[EMIM]$^+$	12.7	[AlCl$_4$]$^-$	–	303	DNMR	132
55.0–45.0 mol% [EMIM]Cl-AlCl$_3$	[EMIM]$^+$	10	Cl$^-$/[AlCl$_4$]$^-$	–	258	PNMR	133[c]
60.0–40.0 mol% [EMIM]Cl-AlCl$_3$	[EMIM]$^+$	4.3	Cl$^-$/[AlCl$_4$]$^-$	–	298	PNMR	133[c]
65.0–35.0 mol% [EMIM]Cl-AlCl$_3$	[EMIM]$^+$	2.2	Cl$^-$/[AlCl$_4$]$^-$	–	298	PNMR	133[c]
70.0–30.0 mol% [EMIM]Cl-AlCl$_3$	[EMIM]$^+$	1.2	Cl$^-$/[AlCl$_4$]$^-$	–	298	PNMR	133[c]
56.0–44.0 mol% [EMIM]Cl-AlCl$_3$	[EMIM]$^+$	–	Cl$^-$/[AlCl$_4$]$^-$	6.1[d]	299	Elec	66
50.5–49.5 mol% [EMIM]Cl-AlCl$_3$	[EMIM]$^+$	–	Cl$^-$/[AlCl$_4$]$^-$	5.7[d]	293	Elec	67
51.0–49.0 mol% [EMIM]Cl-AlCl$_3$	[EMIM]$^+$	–	Cl$^-$/[AlCl$_4$]$^-$	5.3[d]	293	Elec	67
51.5–48.5 mol% [EMIM]Cl-AlCl$_3$	[EMIM]$^+$	–	Cl$^-$/[AlCl$_4$]$^-$	4.6[d]	299	Elec	67

[a] PNMR = Pulse gradient-field spin echo (PGSE) NMR, DNMR = ^{1}H diffusion ordered spectroscopy (DOSY) NMR, Elec = Electrochemistry, FNMR = fringe field NMR.
[b] Diffusion coefficients at 298 K calculated from VTF parameters given in reference.
[c] Only cation diffusion coefficients were determined.
[d] Diffusion coefficient for the chloride ion.
[e] Diffusion coefficients estimated from graphical data provided in the reference.

Moving Boundary method [68] and the Hittorff method [68–71] have been used to measure cation (t_{R+}) and anion (t_{X-}) transport numbers in ionic liquids, and these data are listed in Table 3.6-10.

The measurement of transport numbers by the above electrochemical methods requires a significant amount of experimental effort to generate high quality data [72]. In addition, the methods do not appear applicable to many of the newer non-haloaluminate ionic liquid systems. An interesting alternative to the above method utilizes the NMR generated self-diffusion coefficient data discussed above. If both the cation (D_{R+}) and anion (D_{X-}) self-diffusion coefficients are measured, then both the cation (t_{R+}) and anion (t_{X-}) transport numbers can be determined using the following equations [73, 74];

$$t_{R+} = \frac{D_{R+}}{D_{R+} + D_{X-}} \qquad t_{X-} = \frac{D_{X-}}{D_{R+} + D_{X-}} \qquad (3.6\text{-}6)$$

Transport numbers for several non-haloaluminate ionic liquids generated using ionic liquid self-diffusion coefficients are listed in Table 3.6-10. The interesting question is whether the NMR-generated transport numbers provide the same measure of the fraction of current carried by an ion as the electrochemically-generated transport numbers. The electrochemical experiment measures the relative movement of charge carriers in the presence of an applied field while the NMR experiment measures the movement of all ions, regardless of whether they are free ions or neutral ion pairs. If an ion spends a significant portion of its time as part of a neutral association of ions, will its NMR-generated transport number differ significantly from the electrochemical transport number? Recent work by Watanabe and coworkers [129, 131] has begun to shed some light on this interesting question. When they compared the molar conductivities of a variety of imidazolium ionic liquids measured by impedance methods (Λ_{imp}) with their molar conductivities calculated from NMR self-diffusion data (Λ_{NMR}) they consistently found that $\Lambda_{imp} < \Lambda_{NMR}$. Furthermore, they showed that the magnitude of this difference depended strongly on the length of the alkyl group on the imidazolium cation and on the nature of the anion. They concluded that the larger magnitude of Λ_{NMR} was the result of the movement of neutral ion pairs, and that this ion pairing was dependent on the intermolecular interactions of the cations and anions of the ionic liquids [75]. Thus, this work provides some of the clearest evidence yet of what has long been assumed, namely the poor conductivities of ionic liquids are due to the presence of significant ion pairing.

As one can see from the data in Table 3.6-10, the [EMIM]$^+$ cation carries the major portion of the charge ($t_{R+} = 0.70$) for all the haloaluminate ionic liquids measured. This result is very surprising in view of the large size of the cation compared with that of the extant anions and the fact that the relative fraction of the charge carried by the anions remains essentially constant, even with significant changes in the anion composition occurring with changes in ionic liquid composition. It has been proposed that these observations result from the fact that the smaller anions are more structurally constrained in the charge transport process [69]. However, this explanation seems overly simplistic. The NMR-generated transport numbers in

Table 3.6-10 External ion transport numbers in ionic liquids

Ionic liquid system	Cation	t_{R^+}	Anion(s)	t_{X^-}	Temperature (K)	Method[a]	Ref.
[MMIM][(CF$_3$SO$_2$)$_2$N]	[MMIM]$^+$	0.65	[(CF$_3$SO$_2$)$_2$N]$^-$	0.35	298	PNMR	129[b]
[EMIM][BF$_4$]	[EMIM]$^+$	0.54	[BF$_4$]$^-$	0.46	298	PNMR	73[b]
[EMIM][(CF$_3$SO$_2$)$_2$N]	[EMIM]$^+$	0.64	[(CF$_3$SO$_2$)$_2$N]$^-$	0.36	298	PNMR	73[b]
[EMIM][(CF$_3$SO$_2$)$_2$N]	[EMIM]$^+$	0.62	[(CF$_3$SO$_2$)$_2$N]$^-$	0.38	298	PNMR	129[b]
[BMIM][BF$_4$]	[BMIM]$^+$	0.52	[BF$_4$]$^-$	0.48	298	PNMR	131[b]
[BMIM][CF$_3$CO$_2$]	[BMIM]$^+$	0.56	[CF$_3$CO$_2$]$^-$	0.44	298	PNMR	131[b]
[BMIM][CF$_3$SO$_3$]	[BMIM]$^+$	0.57	[CF$_3$SO$_3$]$^-$	0.43	298	PNMR	131[b]
[BMIM][(CF$_3$SO$_2$)$_2$N]	[BMIM]$^+$	0.56	[(CF$_3$SO$_2$)$_2$N]$^-$	0.44	298	PNMR	129[b]
[BMIM][(C$_2$F$_5$SO$_2$)$_2$N]	[BMIM]$^+$	0.59	[(C$_2$F$_5$SO$_2$)$_2$N]$^-$	0.41	298	PNMR	131[b]
[BMIM][PF$_6$]	[BMIM]$^+$	0.58	[PF$_6$]$^-$	0.42	298	PNMR	49, 131[b]
[HMIM][(CF$_3$SO$_2$)$_2$N]	[HMIM]$^+$	0.52	[(CF$_3$SO$_2$)$_2$N]$^-$	0.48	298	PNMR	129[b]
[OMIM][(CF$_3$SO$_2$)$_2$N]	[OMIM]$^+$	0.50	[(CF$_3$SO$_2$)$_2$N]$^-$	0.50	298	PNMR	129[b]
[BP][BF$_4$]	[BP]$^+$	0.48	[BF$_4$]$^-$	0.52	298	PNMR	73[b]
[BP][(CF$_3$SO$_2$)$_2$N]	[BP]$^+$	0.55	[(CF$_3$SO$_2$)$_2$N]$^-$	0.45	298	PNMR	73[b]
45.0–55.0 mol% [EMIM]Cl–AlCl$_3$	[EMIM]$^+$	0.71	[AlCl$_4$]$^-$	0.23	303	MH	70[c]
			[Al$_2$Cl$_7$]$^-$	0.06	303	MH	70[c]
50.0–50.0 mol% [EMIM]Cl–AlCl$_3$	[EMIM]$^+$	0.70	[AlCl$_4$]$^-$	–	303	MH	68
50.0–50.0 mol% [EMIM]Cl–AlCl$_3$	[EMIM]$^+$	–	[AlCl$_4$]$^-$	0.30	366	MB	68
60.0–40.0 mol% [EMIM]Cl–AlCl$_3$	[EMIM]$^+$	0.71	[AlCl$_4$]$^-$	0.19	298	MH	69[c]
			Cl$^-$	0.10	298	MH	69[c]
70.0–30.0 mol% [EMIM]Cl–AlCl$_3$	[EMIM]$^+$	0.71	[AlCl$_4$]$^-$	0.12	298	MH	69[c]
			Cl$^-$	0.17	298	MH	69[c]
45.0–55.0 mol% [EMIM]Br–AlBr$_3$	[EMIM]$^+$	0.76	[AlBr$_4$]$^-$	0.22	333	MH	70[c]
			[Al$_2$Br$_7$]$^-$	0.02	333	MH	70[c]
50.0–50.0 mol% [EMIM]Br–AlBr$_3$	[EMIM]$^+$	0.76	[AlBr$_4$]$^-$	0.24	333	MH	71
60.0–40.0 mol% [EMIM]Br–AlBr$_3$	[EMIM]$^+$	0.76	[AlBr$_4$]$^-$	0.22	333	MH	71[c]
			Br$^-$	0.02	333	MH	69[c]
70.0–30.0 mol% [EMIM]Br–AlBr$_3$	[EMIM]$^+$	0.76	[AlBr$_4$]$^-$	0.16	333	MH	69[c]
			Br$^-$	0.08	333	MH	69[c]

[a] MB = Moving Boundary, MH = modified Hittorf, PNMR = pulse gradient-field spin echo (PGSE) NMR.
[b] Transport numbers at 298 K determined from self-diffusion data provided in the reference.
[c] Anion transport numbers calculated from formulas provided in the reference.

Table 3.6-10 indicate that, in general, more charge is carried by the cation. However, the relative fraction of this charge is less than that observed in the electrochemical transport data for the haloaluminate ionic liquids. It is unclear at this time if this difference is due to the different anions present in the non-haloaluminate ionic liquids or to differences in the two types of transport number measurements. The apparent greater importance of the cation to the movement of charge demonstrated by the transport numbers (Table 3.6-10) is consistent with the observations made from the diffusion and conductivity data above. Indeed, these data taken in total may indicate that the cation tends to be the majority charge carrier for all ionic liquids, especially the alkylimidazoliums. However, a greater quantity of transport number measurements, performed on a wider variety of ionic liquids, will be needed to ascertain if this is indeed the case.

References

1. (a) F. H. Hurley, T. P. Wier, *J. Electrochem. Soc.*, **1951**, *98*, 203; (b) F. H. Hurley, T. P. Wier, *J. Electrochem. Soc.*, **1951**, *98*, 207.
2. H. L. Chum, R. A. Osteryoung, in *Ionic Liquids*, D. Inman, D. G. Lovering (Eds.), Plenum Press, New York, **1981**, pp. 407–423.
3. R. J. Gale, R. A. Osteryoung, in *Molten Salt Techniques*, Vol. 1, D. G. Lovering, R. J. Gale (Eds.), Plenum Press, New York, **1983**, pp. 55–78.
4. C. L. Hussey, in *Advances in Molten Salt Chemistry*, G. Mamantov, C. B. Mamantov (Eds.), Elsevier, Amsterdam, **1983**, pp. 185–230.
5. G. Mamantov, in *Molten Salt Chemistry*, G. Mamantov, R. Marassi (Eds.), D. Reidel, New York, **1987**, pp. 259–270.
6. R. A. Osteryoung, in *Molten Salt Chemistry*, G. Mamantov, R. Marassi (Eds.), D. Reidel, New York, **1987**, pp. 329–364.
7. C. L. Hussey, *Pure Appl. Chem.*, **1988**, *60*, 1763.
8. C. L. Hussey, in *Chemistry of Nonaqueous Solutions–Current Progress*, G. Mamantov, A. I. Popov (Eds.), VCH, New York, **1994**, pp. 227–275.
9. R. T. Carlin, J. S. Wilkes, in *Chemistry of Nonaqueous Solutions–Current Progress*, G. Mamantov, A. I. Popov (Eds.), VCH, New York, **1994**, pp. 277–306.
10. T. Welton, *Chem. Rev.*, **1999**, *99*, 2071.
11. Y. Ito, T. Nohira, *Electrochim. Acta*, **2000**, *45*, 2611.
12. F. Endres, *ChemPhysChem*, **2002**, *3*, 144.
13. F. Endres, *Z. Phys. Chem.*, **2004**, *218*, 255.
14. M. Koel, *Crit. Rev. Anal. Chem.*, **2005**, *35*, 177.
15. M. C. Buzzeo, R. G. Evans, R. G. Compton, *ChemPhysChem*, **2004**, *5*, 1106.
16. C. Chiappe, D. Pieraccini, *J. Phys. Org. Chem.*, **2005**, *18*, 275.
17. H. Xue, J. M. Shreeve, *Eur. J. Inorg. Chem.*, **2005**, 2573.
18. K. Haymizu, Y. Aihara, H. Nakagawa, T. Nukuda, W. Price, *J. Phys. Chem. B.*, **2004**, *108*, 19527.
19. M. Yoshizawa, A. Narita, H. Ohno, *Aust. J. Chem.*, **2004**, *57*, 139.
20. B. Garcia, S. Lavallee, G. Perron, C. Michot, M. Armand, *Electrochim. Acta*, **2004**, *49*, 4583.
21. I. Nicotera, C. Oliviero, W. A. Henderson, G. B. Appentecchi, S. Passerini, *J. Phys. Chem. B.*, **2005**, *109*, 22814.
22. A. Hayashi, M. Yoshizawa, C. A. Angell, F. Mizuno, T. Minami, M. Tatsumisago, *Electrochem. Solid–State Lett.*, **2003**, *8*, E19.
23. W. Xu, C. A. Angell, *Science*, **2003**, *302*, 422.
24. H. Ohno, in *Electrochemical Aspects of Ionic Liquids*, H. Ohno (Ed.), Wiley, New York, **2005**, Ch. 19.
25. M. A. B. H. Susan, A. Nota, S.

Mitsushima, M. Watanabe, *Chem. Comm.*, **2003**, 938.

26. Z. Du, Z. Li, S. Guo, J. Zhang, L. Zhu, Y. Deng, *J. Phys. Chem. B.*, **2005**, *109*, 19542.
27. A. Noda, Md. A. B. H. Susan, K. Kudo, S. Mitsushima, K. Hayamizu, M. Watanabe, *J. Phys. Chem. B.*, **2003**, *107*, 4024.
28. A. Fry, W. E. Britton, in *Laboratory Techniques in Electroanalytical Chemistry*, 2nd Edn., P. T. Kissinger, W. R. Heineman (Eds.), Marcel Dekker, New York, **1996**, Ch. 15.
29. H. Matsumoto, in *Electrochemical Aspects of Ionic Liquids*, H. Ohno (Ed.), Wiley, New York, **2005**, Ch. 3.
30. K. R. Seddon, A. Stark, M.-J. Torres, *Pure Appl. Chem.*, **2000**, *72*, 2275.
31. C. Hilgers, P. Wasserscheid, in *Ionic Liquids in Synthesis*, P. Wasserscheid, T. Welton (Eds.), Wiley, New York, **2003**, pp. 21–33.
32. B. D. Fitchett, T. N. Knepp, J. C. Conboy, *J. Electrochem. Soc.*, **2004**, *151*, E219.
33. U. Schröder, J. D. Wadhawan, R. G. Compton, F. Marken, P. A. Z. Suarez, C. S. Consorti, R. F. de Souza, J. Dupont, *New J. Chem.*, **2000**, *24*, 1009.
34. T. E. Sutto, H. C. De Long, P. C. Trulove, unpublished results.
35. A. E. Visser, R. P. Swatloski, W. M. Reichert, S. T. Griffin, R. D. Rogers, *Ind. Eng. Chem. Res.*, **2000**, *39*, 3596.
36. J. G. Huddleston, A. E. Visser, W. M. Reichert, H. D. Willauer, G. A. Broker, R. D. Rogers, *Green Chem.*, **2001**, *3*, 156.
37. R. T. Carlin, T. Sullivan, *J. Electrochem. Soc.*, **1992**, *139*, 144.
38. G. E. Gray, J. Winnick, P. A. Kohl, *J. Electrochem. Soc.*, **1996**, *143*, 3820.
39. J. D. Vaugh, A. Mughrabi, E. C. Wu, *J. Org. Chem.*, **1970**, *35*, 1141.
40. T. J. Melton, J. Joyce, J. T. Maloy, J. A. Boon, J. S. Wilkes, *J. Electrochem. Soc.*, **1990**, *137*, 3865.
41. (a) T. L. Riechel, J. S. Wilkes, *J. Electrochem. Soc.*, **1992**, *139*, 977; (b) C. Scordilis-Kelly, R. T. Carlin, *J. Electrochem. Soc.*, **1993**, *140*, 1606; (c) C. Scordilis-Kelly, R. T. Carlin, *J. Electrochem. Soc.*, **1994**, *141*, 873.
42. F. J. Holler, C. G. Enke, in *Laboratory Techniques in Electroanalytical Chemistry*, 2nd Edn., P. T. Kissinger, W. R. Heineman (Eds.), Marcel Dekker, New York, **1996**, Ch. 8.
43. H. Ohno, M. Yoshizawa, T. Mizumo, in *Electrochemical Aspects of Ionic Liquids*, H. Ohno (Ed.), Wiley, New York, **2005**, Ch. 6.
44. (a) J. Fuller, R. T. Carlin, R. A. Osteryoung, *J. Electrochem. Soc.*, **1997**, *144*, 3881; (b) J. Fuller, R. A. Osteryoung, R. T. Carlin, *Abstracts of Papers*, 187th Meeting of The Electrochemical Society, Reno, NV, **1995**, Vol. 95–1, p. 27.
45. (a) J. Sun, M. Forsyth, D. R. MacFarlane, *Molten Salt Forum*, **1998**, *5–6*, 585; (b) J. Sun, M. Forsyth, D. R. MacFarlane, *J. Phys. Chem. B.*, **1998**, *102*, 8858.
46. C. A. Angell, in *Molten Salts: From Fundamentals to Applications*, M. Gaune-Escard (Ed.), Kluwer Academic Publishers, London, **2002**, pp. 305–320.
47. C. A. Angell, W. Xu, M. Yoshizawa, A. Hayashi, J.-P. Belieres, P. Lucas, M. Videa, in *Electrochemical Aspects of Ionic Liquids*, H. Ohno (Ed.), Wiley, New York, **2005**, Ch. 2.
48. J. Sirieix-Plenet, L. Gaillon, P. Letellier, *Talanta*, **2004**, *63*, 979.
49. T. Umecky, M. Kanakubo, Y. Ikushima, *Fluid Phase Equilib.*, **2005**, *228–229*, 329.
50. J. A. Widegren, E. M. Saurer, K. N. Marsh, J. W. Magee, *J. Chem. Thermodynam.*, **2005**, *37*, 569.
51. R. Hagiwara, K. Matsumoto, Y. Nakamori, T. Tsuda, Y. Ito, H. Matsumoto, K. Momota, *J. Electrochem. Soc.*, **2003**, *140*, D195.
52. S. I. Smedley, *The Interpretation of Ionic Conductivity in Liquids*, Plenum, New York, **1980**, Ch. 3.
53. A. B. McEwen, H. L. Ngo, K. LeCompte, J. L. Goldman, *J. Electrochem. Soc.*, **1999**, *146*, 1687.
54. A. M. Elias, M. E. Elias, *Molten Salt Forum*, **1998**, *5–6*, 617.
55. R. L. Perry, K. M. Jones, W. D. Scott, Q. Liao, C. L. Hussey, *J. Chem. Eng. Data*, **1995**, *40*, 615.
56. Q. Liao, C. L. Hussey, *J. Chem. Eng. Data*, **1996**, *41*, 1126.
57. N. Papageorgiou, Y. Athanassov, M.

Armand, P. Bonhôte, H. Pattersson, A. Azam, M. Grätzel, *J. Electrochem. Soc.*, **1996**, *143*, 3099.
58. R. Moy, R.-P. Emmenegger, *Electrochim. Acta*, **1992**, *37*, 1061.
59. J. Robinson, R. C. Bugle, H. L. Chum, D. Koran, R. A. Osteryoung, *J. Am. Chem. Soc.*, **1979**, *101*, 3776.
60. M. Diaw, A. Chagnes, B. Carre, P. Willman, D. Lemordant, *J. Power Sources*, **2005**, *146*, 682.
61. A. Jarosik, S. R. Krajewski, A. Lewandowski, P. Radzimski, *J. Mol. Liq.*, **2006**, *123*, 43.
62. A. Heintz, *J. Chem. Thermodynam.*, **2005**, *37*, 525.
63. M. A.-B. Hasan Susan, A. Noda, M. Watanabe, in *Electrochemical Aspects of Ionic Liquids*, H. Ohno (Ed.), Wiley, New York, **2005**, Ch. 5.
64. A. J. Bard, L. R. Faulkner, *Electrochemical Methods – Fundamentals and Applications*, 2nd Edn., Wiley, New York, **2001**, Ch. 4.
65. P. Stilbs, *Prog. NMR Spectrosc.*, **1987**, *19*, 1.
66. R. T. Carlin, R. A. Osteryoung, *J. Electroanal. Chem.*, **1988**, *252*, 81.
67. L. R. Simonsen, F. M. Donahue, *Electrochim. Acta*, **1990**, *35*, 89.
68. C. J. Dymek, L. A. King, *J. Electrochem. Soc.*, **1985**, *132*, 1375.
69. C. L. Hussey, J. R. Sanders, H. A. Øye, *J. Electrochem. Soc.*, **1985**, *132*, 2156.
70. C. L. Hussey, H. A. Øye, *J. Electrochem. Soc.*, **1984**, *131*, 1623.
71. C. L. Hussey, J. R. Sanders, *J. Electrochem. Soc.*, **1987**, *134*, 1977.
72. C. L. Hussey, in *Molten Salt Chemistry*, G. Mamantov, R. Marassi (Eds.), D. Reidel, New York, **1987**, pp. 141–160.
73. (a) A. Noda, K. Hayamizu, M. Watanabe, *J. Phys. Chem. B*, **2001**, *105*, 4603; (b) A. Noda, M. Watanabe, in *Proceedings of the Twelfth International Symposium on Molten Salts*, P. C. Trulove, H. C. De Long, G. R. Stafford, S. Deki (Eds.), The Electrochemical Society: Pennington NJ, **2000**, Vol. 99–41, pp. 202–208.
74. H. Every, A.G. Bishop, M. Forsyth, D. R. MacFarlane, *Electrochim. Acta*, **2000**, *45*, 1279.
75. S. Tsuzuki, H. Tokuda, K. Hayamizu, M. Watanabe, *J. Phys. Chem. B.*, **2005**, *109*, 16474.
76. M. Lipsztajn, R. A. Osteryoung, *J. Electrochem. Soc.*, **1983**, *130*, 1968.
77. S. P. Wicelinski, R. J. Gale, J. S. Wilkes, *J. Electrochem. Soc.*, **1987**, *134*, 262.
78. J. R. Stuff, S. W. Lander Jr., J. W. Rovang, J. S. Wilkes, *J. Electrochem. Soc.*, **1990**, *137*, 1492.
79. S. D. Jones, G. E. Blomgren, in *Proceedings of the Seventh International Symposium on Molten Salts*, C. L. Hussey, S. N. Flengas, J. S. Wilkes, Y. Ito (Eds.), The Electrochemical Society: Pennington NJ, **1990**, Vol. 90–17, pp. 273–280.
80. Z.-B. Zhou, H. Matsumoto, K. Tatsumi, *Chem. Lett.*, **2004**, *33*, 680.
81. A. Noda, M. Watanabe, *Electrochim. Acta*, **2000**, *45*, 1265.
82. Y. Yoshida, K. Muroi, A. Otsuka, G. Saito, M. Takahashi, T. Yoko, *Inorg. Chem.*, **2004**, *43*, 1458.
83. S. Forsyth, S. R. Batten, Q. Dai, D. R. MacFarlane, *Aust. J. Chem.*, **2004**, *57*, 121.
84. E. I. Cooper, E. J. M. O'Sullivan, in *Proceedings of the Eighth International Symposium on Molten Salts*, R. J. Gale, G. Blomgren (Eds.), The Electrochemical Society: Pennington NJ, **2000**, Vol. 92-16, pp. 386–396.
85. (a) Z.-B. Zhou, H. Matsumoto, K. Tatsumi, *Chem. Lett.*, **2004**, *33*, 1636; (b) Z.-B. Zhou, M. Takeda, M. Ue, *J. Fluorine Chem.*, **2004**, *125*, 471.
86. P. Bonhôte, A.-P. Dias, N. Papageorgiou, K. Kalyanasundaram, M. Grätzel, *Inorg. Chem.*, **1996**, *35*, 1168.
87. (a) R. Hagiwara, T. Hirashige, T. Tsuda, Y. Ito, *J. Fluorine Chem.*, **1999**, *99*, 1; (b) R. Hagiwara, T. Hirashige, T. Tsuda, Y. Ito, *J. Electrochem. Soc.*, **2002**, *149*, D1.
88. The effect of the anion stoichiometry on the electrochemical properties of the $[EMIM]^+$ flourohydregenate ionic liquid is evaluated in R. Hagiwara, Y. Nokamori, K. Matsumoto, Y. Ito, *J. Phys. Chem. B.*, **2005**, *109*, 5445.
89. J. N. Barisci, G. G. Wallace, D. R. MacFarlane, R. H. Baughman, *Electrochem. Commun.*, **2004**, *6*, 22.
90. J. D. Holbrey, W. M. Reichert, R. P.

Swatloski, G. A. Broker, W. R. Pitner, K. R. Seddon, R. D. Rogers, *Green Chem.*, **2002**, *4*, 407.

91. K. Matsumoto, R. Hagiwara, R. Yoshida, Y. Ito, Z. Mazej, P. Benkič, B. Žemva, O. Tamada, H. Yoshino, S. Matsubara, *Dalton Trans.*, **2004**, 144.
92. K. Matsumoto, R. Hagiwara, Y. Ito, *J. Fluorine Chem.*, **2002**, *115*, 133.
93. K. Matsumoto, R. Hagiwara, *J. Fluorine Chem.*, **2005**, *126*, 1095.
94. V. R. Koch, L. A. Dominey, C. Najundiah, M. J. Ondrechen, *J. Electrochem. Soc.*, **1996**, *143*, 798.
95. S.-H. Yeon, K.-S. Kim, S. Choi, H. Lee, H. S. Kim, H. Kim, *Electrochim. Acta*, **2005**, *50*, 5399.
96. H. Matsumoto, H. Sakaebe, K. Tatsumi, *J. Power Sources*, **2005**, *146*, 45.
97. H. Matsumoto, M. Yanagida, K. Tanimoto, M. Nomura, Y. Kitagawa, Y. Miyazaki, *Chem. Lett.*, **2000**, 922.
98. H. Matsumoto, H. Kageyama, Y. Miyazaki, in *Proceedings of the Thirteenth International Symposium on Molten Salts*, P. C. Trulove, H. C. De Long, R. A. Mantz, G. R. Stafford, M. Matsunaga (Eds.), The Electrochemical Society: Pennington NJ, **2002**, Vol. 2002-19, pp. 1057–1065.
99. Z.-B. Zhou, H. Matsumoto, K. Tatsumi, *Chem.-Eur. J.*, **2005**, *11*, 752.
100. J. M. Pringle, J. Golding, K. Baranyani, C. M. Forsyth, G. B. Deacon, J. L. Scott, D. R. MacFarlane, *New J. Chem.*, **2003**, *27*, 1504.
101. T. Sato, G. Masuda, K. Takagi, *Electrochim. Acta*, **2004**, *49*, 3603.
102. T. Sato, T. Maruo, S. Marukane, K. Takagi, *J. Power Sources*, **2004**, *138*, 253.
103. S. Forsyth, D. R. MacFarlane, *J. Mater. Chem.*, **2003**, *13*, 2451.
104. S. Zhang, Y. Hou, W. Huang, Y. Shan, *Electrochim. Acta*, **2005**, *50*, 4097.
105. K.-S. Kim, S. Choi, D. Dembereinyamba, H. Lee, J. Oh, B.-B. Lee, S.-J. Mun, *Chem. Comm.*, **2004**, 828.
106. K. Matsumoto, R. Hagiwara, Y. Ito, *Electrochem. Solid–State Lett.*, **2004**, *7*, E41.
107. H. Sakaebe, H. Matsumoto, *Electrochem. Comm.*, **2003**, *5*, 594.
108. J. Caja, T. D. J. Dunstan, V. Katovic in *Proceedings of the Thirteenth International Symposium on Molten Salts*, P. C. Trulove, H. C. De Long, R. A. Mantz, G. R. Stafford, M. Matsunaga (Eds.), The Electrochemical Society: Pennington NJ, **2002**, Vol. 2002-19, 1014–1023.
109. J. Caja, T. D. J. Dunstan, D. M. Ryan, V. Katovic, in *Proceedings of the Twelfth International Symposium on Molten Salts*, P. C. Trulove, H. C. De Long, G. R. Stafford, S. Deki (Eds.), The Electrochemical Society: Pennington NJ, **2000**, Vol. 99–41, pp. 150–161.
110. D. R. MacFarlane, P. Meakin, J. Sun, N. Amini, M. Forsyth, *J. Phys. Chem. B.*, **1999**, *103*, 4164.
111. J. Sun, D. R. MacFarlane, M. Forsyth, *Electrochim. Acta*, **2003**, *48*, 1707.
112. H. Matsumoto, T. Matsuda, Y. Miyazaki, *Chem. Lett.*, **2000**, 1430.
113. a. J. S. Wilkes, J. A. Levisky, R. A. Wilson, C. L. Hussey, *Inorg. Chem.*, **1982**, *21*, 1263. b. A. A. Fannin Jr., D. A. Floreani, L. A. King, J. S. Landers, B. J. Piersma, D. J. Stech, R. J. Vaughn, J. S. Wilkes, J. L. Williams, *J. Phys. Chem.*, **1984**, *88*, 2614.
114. (a) J. R. Sanders, E. H. Ward, C. L. Hussey, in *Proceedings of the Fifth International Symposium on Molten Salts*, M.-L. Saboungi, K. Johnson, D. S. Newman, D. Inman (Eds.), The Electrochemical Society: Pennington NJ, **1986**, Vol. 86-1, pp. 307–316; (b) J. R. Sanders, E. H. Ward, C. L. Hussey, *J. Electrochem. Soc.*, **1986**, *133*, 325.
115. R. A. Carpio, L. A. King, R. E. Lindstrom, J. C. Nardi, C. L. Hussey, *J. Electrochem. Soc.*, **1979**, *126*, 1644.
116. V. R. Koch, L. L. Miller, R. A. Osteryoung, *J. Am. Chem. Soc.*, **1976**, *98*, 5277.
117. H. Ohno, M. Yoshizawa, *Solid State Ionics*, **2002**, *154–155*, 303.
118. T. E. Sutto, H. C. De Long, P. C. Trulove, in *Progress in Molten Salt Chemistry 1*, R. W. Berg, H. A. Hjuler (Eds.), Elsevier, Amsterdam, **2000**, p. 511.
119. Z.-B. Zhou, H. Matsumoto, K. Tatsumi, *ChemPhysChem*, **2005**, *6*, 1324.
120. W. Ogihara, M. Yoshizawa, H. Ohno, *Chem. Lett.*, **2004**, *33*, 1022.

121. Y. Yoshida, J. Fujii, K. Muroi, A. Otsuka, G. Saito, M. Takahashi, T. Yoko, *Syn. Met.*, **2005**, *153*, 421.
122. T. Nishida, Y. Tashiro, M. Yamamoto, *J. Fluorine Chem.*, **2003**, *120*. 135.
123. P. Kölle, R. Dronskowski, *Eur. J. Inorg. Chem.*, **2004**, 2313.
124. J. Fuller, A. C. Breda, R. T. Carlin, *J. Electroanal. Chem.*, **1998**, *459*, 29.
125. T. Mizumo, E. Marwanta, N. Matsumi, H. Ohno, *Chem. Lett.*, **2004**, *33*, 1360.
126. E. I. Cooper, C. A. Angell, *Solid State Ionics*, **1983**, *9–10*, 617.
127. Z.-B. Zhou, H. Matsumoto, K. Tatsumi, *Chem. Lett.*, **2004**, *33*, 886.
128. M. Ma, K. E. Johnson, in *Proceedings of the Ninth International Symposium on Molten Salts*, C. L. Hussey, D. S. Newman, G. Mamantov, Y. Ito (Eds.), The Electrochemical Society, Pennington NJ, **1994**, Vol. 94-13, pp. 179–186.
129. H. Tokuda, K. Hayamizu, K. Ishii, Md. A. B. H. Susan, M. Watanabe, *J. Phys. Chem. B*, **2005**, *109*, 6103.
130. (a) J.-F. Huang, P.-Y. Chen, I.-W. Sun, S. P. Wang, *Inorg. Chim. Acta*, **2001**, *320*, 7; (b) J.-F. Huang, P.-Y. Chen, I.-W. Sun, S. P. Wang, *Spectrosc. Lett.*, **2001**, *34*, 591.
131. H. Tokuda, K. Hayamizu, K. Ishii, Md. A. B. H. Susan, M. Watanabe, *J. Phys. Chem. B*, **2004**, *108*, 16593.
132. W. R. Carper, G. J. Mains, B. J. Piersma, S. L. Mansfield, C. K. Larive, *J. Phys. Chem.*, **1996**, *100*, 4724.
133. R. A. Mantz, H. C. De Long, R. A. Osteryoung, P. C. Trulove, in *Proceedings of the Twelfth International Symposium on Molten Salts*, P. C. Trulove, H. C. De Long, G. R. Stafford, S. Deki (Eds.), The Electrochemical Society, Pennington NJ, **2000**, Vol. 99-41, pp. 169–176.

4
Molecular Structure and Dynamics

4.1
Order in the Liquid State and Structure

Chris Hardacre

The structure of liquids has been studied for many years. These investigations have, in general, been focused on the arrangements in molecular solvents such as water, *tert*-butanol and simple chlorinated solvents. The field of molten salts and the structures thereof have been much less studied and within this field the study of the structure of room temperature ionic liquids is in its infancy. A variety of techniques have been used to investigate the liquid structure including neutron diffraction, X-ray scattering and extended X-ray absorption fine structure. This chapter summarizes some of the techniques used including practical details and shows examples of where they have been employed previously. The examples given are not meant to be exhaustive and are provided for illustration only. Where possible, examples relating to the more recent air and moisture stable ionic liquids have been included.

4.1.1
Neutron Diffraction

Neutron diffraction is one of the most widely used techniques for studying liquid structure. In the experiment, neutrons are elastically scattered off the nuclei in the sample and are detected at different scattering angles, typically 3° to 40°, for the purpose of measuring intermolecular structure whilst minimizing inelasticity corrections. The resultant scattering profile is then analyzed to provide structural information.

The data taken are normally presented as the total structure factor, $F(Q)$. This is related to the neutron scattering lengths, b_i, the concentrations, c_i, and the partial structure factor $S_{ij}(Q)$ for each pair of atoms i and j in the sample, via Eq. (4.1-1):

$$F(Q) = \sum_{i,j} c_i c_j b_i b_j \left(S_{ij}(Q) - 1 \right) \tag{4.1-1}$$

Ionic Liquids in Synthesis, Second Edition. P. Wasserscheid and T. Welton (Eds.)
Copyright © 2008 WILEY-VCH Verlags GmbH & Co. KGaA, Weinheim
ISBN: 978-3-527-31239-9

where Q is the scattering vector and is dependent on the scattering angle, θ, and the wavelength of the neutrons used, λ.

$$Q = \frac{4\pi \sin \theta}{\lambda} \qquad (4.1\text{-}2)$$

The real space pair distributions, $g_{ij}(r)$ are the inverse Fourier transform of $(S_{ij}(Q)-1)$, that is:

$$g_{ij}(r) = 1 + \frac{1}{2\pi^2 \rho} \int_0^\infty Q^2 \frac{\sin(Qr)}{Qr} \left(S_{ij}(Q) - 1\right) dQ \qquad (4.1\text{-}3)$$

normalized to the atomic density, ρ.

In a neutron diffraction experiment, the quantity measured as a function of angle is the total scattering cross section that consists of two components: (i) neutrons which scatter coherently, that is where phase is conserved and whose signal contains structural information and (ii) incoherently scattered neutrons that result in a background signal. The scattering amplitude is then determined by the concentration, atomic arrangement and neutron scattering lengths of the atoms involved. Since different isotopes have different neutron scattering lengths, it is possible to simplify the analysis of the neutron data simply by isotopic exchange experiments and taking first and second order difference spectra to separate out the partial pair distribution functions. This is clearly set out by Bowron et al. [1] for a mixture of tert-butanol/water and illustrates how isotopic substitution neutron scattering experiments can assist in distinguishing between intermolecular and intramolecular distributions within a sample, both of which would otherwise contribute to a measured diffraction pattern in a complex and often difficult to interpret combination.

4.1.2
Formation of Deuterated Samples

In general, isotopic exchange is both expensive and difficult; however, in the case of many room-temperature ionic liquids the manufacture of deuterated ionic liquids is relatively easily achieved. For example, the general synthesis of 1-alkyl-3-methyl imidazolium salts is shown in Scheme 4.1-1 [2]. This methodology allows maximum

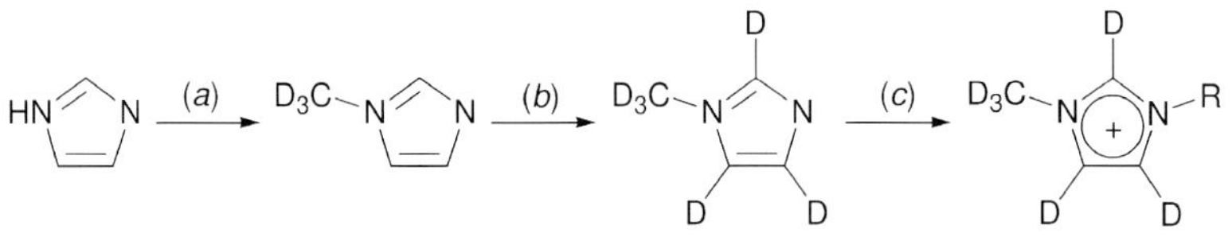

Scheme 4.1-1 Reaction scheme showing a method for deuterating 1-alkyl-3-methyl imidazolium salts with (a) CD_3OD, $RuCl_3/(n\text{-}BuO)_3P$, (b) D_2O, 10% Pd/C and (c) RX (CD_3Cl, C_2D_5I).

flexibility in the deuteration on the imidazolium cation, that is it can be either ring or side chain deuteration or both.

4.1.3
Neutron Sources

The following sources and instruments dominate the studies in the area of liquids and amorphous materials. Although there are a number of sources available each is optimized for a particular class of experiments. The sources can be split into two types: pulsed neutron sources and reactor sources.

4.1.3.1 Pulsed (Spallation) Neutron Sources

One example of a pulsed neutron source is at ISIS, at the Rutherford Appleton Laboratory, UK. This source has the highest flux of any pulsed source in the world, at present, and is therefore one of the most suitable for isotopic substitution work, as this class of experiment tends to be flux limited. At ISIS, two stations are particularly well set up for the examination of liquids

SANDALS station – Small Angle Neutron Diffractometer for Amorphous and Liquid Samples

This station is optimized for making measurements on liquids and glasses that contain light elements such as H, Li, B, C, N, O. This set-up relies on collecting data at small scattering angles and using high energy neutrons. This combination of characteristics has the effect of reducing the corrections necessary for inelastic scattering that otherwise dominate the measured signal and hence complicate the extraction of structural signal information. This instrument is singularly specifically optimized for H–D isotopic substitution experiments.

GEM – GEneral Materials diffractometer

GEM is designed with extremely stable detectors, covering a very large solid angle and is optimized for collecting data at a very high rate. It is a hybrid instrument that can perform both medium-high resolution powder diffraction studies on crystalline systems as well as very accurate total scattering measurements for liquids and glasses. Due to the high stability of the detectors and data acquisition electronics it is suitable for doing isotopic substitution work on systems containing elements with only small differences in the isotope neutron scattering lengths, for example ^{12}C and ^{13}C.

ISIS is only one pulsed source available for the study of liquids. Both the USA, in the form of IPNS (Intense Pulsed Neutron Source) at the Argonne National Lab on the instrument GLAD, and Japan, with the KEK Neutron Scattering Facility (KENS) on instrument Hit II, have facilities for studying liquids similar to SANDALS and GEM but with slightly lower neutron intensity.

4.1.3.2 Reactor Sources

The Institute Laue-Langevin (ILL, Grenoble, France) on instrument D4C is arguably the premier neutron scattering instrument for total scattering studies of liquid and amorphous materials. The neutrons are provided by a reactor source that is very stable and which delivers a very high flux. This makes the ILL ideal for isotopic substitution work for elements with atomic numbers greater than oxygen. For light elements, the inelasticity corrections are a major problem as the instrument collects data over a large angular range and with relatively low energy neutrons. Typically the wavelength of the neutrons used on D4C is 0.7 A which is quite long when compared with a source like ISIS where wavelengths ranging from 0.05 A to 5.0 A can be used on an instrument like SANDALS.

Other reactor sources with instruments like D4C but with much lower flux, and hence longer data collection times, are the Laboratoire Leon Brillouin (LLB, Saclay, France) on the instrument 7C2 and the NFL (Studsvik, Sweden) on the instrument SLAD.

The following website provides links to all the neutron sites in the world: http://www.isis.rl.ac.uk/neutronSites/.

4.1.4
Neutron Cells for Liquid Samples

As with any scattering experiment, the ideal sample holder is one which does not contribute to the signal observed. In neutron scattering experiments, the typical cells used are either vanadium, which scatters neutrons almost completely incoherently, that is with almost no structural components in the measured signal, or a null scattering alloy of TiZr. Vanadium cells react with water and are not ideal for studies of hydrated systems, whereas TiZr is more chemically inert; for example TiZr cells have been used to study supercritical water and alkali metals in liquid ammonia. In addition, TiZr cells capable of performing measurements at high pressure have been constructed. Figure 4.1-1 shows typical sample cells made from Vanadium and TiZr alloy.

Cells used for high-temperature measurements in furnaces often consist of silica sample tubes, supported by thin vanadium sleeves. The key to the analysis is whether it is possible to have a container that scatters in a sufficiently predictable way so that its background contribution can be subtracted. With the current neutron flux available using both pulsed and reactor sources, sample volumes of between 1 and 5 cm^3 are required. Obviously, with increasing flux at the new neutron sources being built in the USA and Japan, the sample sizes will decrease.

4.1.5
Examples

Neutron diffraction has been used extensively to study a range of ionic liquid systems; however, many of these investigations have focused on high temperature

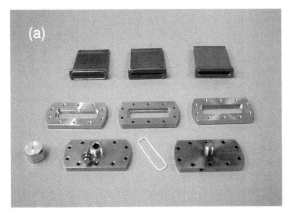

Fig. 4.1-1 Liquid sample cells made from (a) TiZr alloy and (b) vanadium.

materials, for example NaCl studied by Enderby and coworkers [3]. A number of liquid systems with relatively low melting points have been reported and this section summarizes some of the findings of these studies. Many of the salts studied melt above 100 °C, and, therefore, are not room-temperature ionic liquids, but the same principles apply to the study of these materials as to the lower melting point salts.

4.1.5.1 Binary Mixtures

A number of investigations have been focused on alkali haloaluminates, that is mixtures $(MX)_y(AlX_3)_{1-y}$ where M is an alkali metal and X is a halogen (Cl or Br). Blander et al. [4] have used neutron diffraction combined with quantum chemical calculations to investigate the salts formed from KBr and KCl where $y = 0.25$ and 0.33. They showed that for both bromide and chloride salts, $[Al_2X_7]^-$ was the dominant species present, as expected, in full agreement with other spectroscopic techniques such as Raman and infrared [5]. However, unexpectedly, in the case of

Table 4.1-1 Comparison of the neutron scattering cross sections for ^{6}Li and ^{7}Li with all the other atoms (i) present in a binary mixture of LiSCN with AlCl$_3$

Ion pairs	$b_{6_{Li}} b_i$	$B_{7_{Li}} b_i$
Li–S	0.004	−0.005
Li–C	0.009	−0.012
Li–N	0.013	−0.017
Li–Al	0.005	−0.006
Li–Cl	0.041	−0.054

chloride, as the acidity of the melt increases, that is as y decreases, although the proportion of [Al$_3$X$_{10}$]$^-$ ions did increase the change was smaller than that predicted by the stoichiometry. A closer relationship with the stoichiometry was found for bromide. The neutron scattering data showed a strong correlation between the liquid structure and that found in the crystal structure of K[Al$_2$Br$_7$], for example. In both the liquid and crystal, the angle of the Al–Br–Al bridge within the [Al$_2$Br$_7$]$^-$ anion is found to be approximately 109° and both also show that the neighboring structural units pack parallel to each other.

The structure of binary mixtures of AlCl$_3$ and NaCl and LiCl were studied from pure AlCl$_3$ to a 1:1 mixture by Badyal et al. [6]. In the pure AlCl$_3$ liquid, the neutron data indicate that the long-held view that isolated Al$_2$Cl$_6$ dimers make up the structure may not be the true scenario. The structure is reported to be a sparse liquid network made up of polymeric species containing corner-shared tetrahedra. On addition of the alkali halides, the presence of Al–Cl–Al linkages gradually decreases in proportion to the concentration of the halide added. This coincides with the formation of [AlCl$_4$]$^-$ species. The neutron scattering also shows that the long-range order within the liquid decreases as the binary salt mixture is formed, consistent with the gradual breakdown of the polymeric aluminum trichloride structure. Of significance in the 1:1 binary mixture is the high level of charge ordering in the system. For example, in the case of LiCl, features at $r = 6.65$, 9.85 and 12.9 Å in the radial distribution function. These do not correspond to distances in either of the pure components and therefore probably are associated with spacings between [AlCl$_4$]$^-$ units.

Binary mixtures of LiSCN with AlCl$_3$ have also been studied [7]. In this case, the 1:1 mixture is liquid at ambient temperature and therefore provides a system analogous to room temperature ionic liquids. Lee et al. used Li isotope substitution to enable the correlations between Li–X to be isolated. The weighting factors between ^{6}Li and ^{7}Li are positive and negative, respectively, and can be used to distinguish features in the partial radial distribution function. Table 4.1-1 compares the combined neutron scattering cross sections, $b_{Li}b_i$ for both isotopes with each other atom, i, in the liquid. Figure 4.1-2 shows the equivalent total correlation functions for the 1:1 mixture. Clearly the amplitude of the ^{7}Li systems shows negative and reduced features

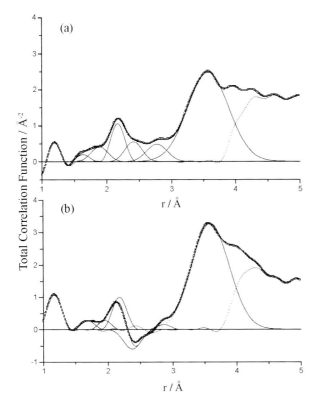

Fig. 4.1-2 Total correlation functions for (a) ^{6}LiSCN/AlCl$_3$ and (b) ^{7}LiSCN/AlCl$_3$. The bold lines are the experimental neutron data (●), the fit (——), the Gaussian functions for each of the atomic pairs used to fit the data (——) and the deviation ( ) used. Reproduced from Ref. [7] with permission.

compared with the ^{6}Li sample and therefore these must be associated with Li–X features.

From the data, the liquid is shown to have tetrahedrally coordinated aluminum with three chlorines and the isocyanate group attached. The neutron data clearly shows nitrogen, as opposed to sulfur, coordination to the aluminum center forming an [AlCl$_3$NCS]$^-$ species which is consistent with a hard base–hard acid interaction compared with the softer sulfur donation. It was also possible to show that a tetrahedral chloride environment is present around the lithium.

Takahashi et al. have studied the structures of AlCl$_3$:[EMIM]Cl mixtures over a range of concentrations from 46 to 67 mol% AlCl$_3$ [8]. Below 50 mol% AlCl$_3$, the neutron data could simply be simulated using the isolated ions, that is [EMIM]$^+$ and [AlCl$_4$]$^-$. Above 50 mol% AlCl$_3$, [Al$_2$Cl$_7$]$^-$ is also known to exist and at 67 mol% AlCl$_3$ becomes the major anion present. Unlike the [AlCl$_4$]$^-$ the [Al$_2$Cl$_7$]$^-$ geometry is changed substantially in the liquid compared with the isolated ions, implying a

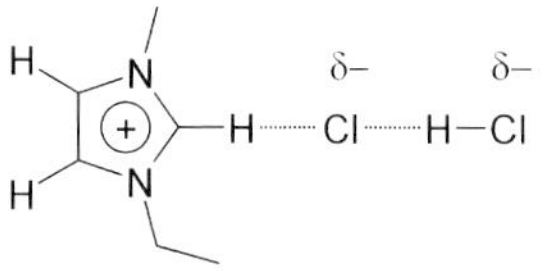

Fig. 4.1-3 Structure proposed for the asymmetric $[HCl_2]^-$ ion bonded to the $[EMIM]^+$ cation in a binary mixture of HCl–[EMIM]Cl. The figure has been redrawn from Ref. [10] with permission.

direct interaction between the imidazolium cation and the anionic species. This is manifested by a decrease in the torsion angle around the central Al–Cl–Al axis from 57.5° to 26.2° and hence a decrease in the Cl–Cl distance across the anion.

Mixtures of HCl and [EMIM]Cl have also been studied [9,10]. By analyzing the first order differences using hydrogen/deuterium substitution, both on the imidazolium ring and the HCl, two intramolecular peaks were observed. These indicated the presence of $[HCl_2]^-$ as an asymmetric species which coupled with analysis of the second order differences, allowed the structure in Fig. 4.1-3 to be proposed.

4.1.5.2 Simple Salts

Bowron et al. [11] have performed neutron diffraction experiments on 1,3-dimethyl imidazolium chloride, [MMIM]Cl in order to model the imidazolium room-temperature ionic liquids. The total structure factors, $F(Q)$, for five 1,3 dimethylimidazolium chloride melts were measured, that is fully protiated, fully deuterated, a 1:1 fully deuterated:fully protiated mixture, ring deuterated only and side chain deuterated only. Figure 4.1-4 shows the probability distribution of chloride around a central imidazolium cation as determined from modeling the neutron data.

From the radial distribution functions, clear charge ordering is seen out to two chloride shells. As well as charge ordering in the system, the specific local structure shows strong interactions between the chloride and the ring hydrogens as well as some interaction between the methyl groups of adjacent imidazolium cations. This is consistent with the crystal structure and implies that the molecular packing and interactions in the first two or three coordination shells is similar in both the crystal and the liquid. The crystal and liquid structures of $[MMIM][PF_6]$ are also closely related [12]. Overall the structures of 1,3-dimethylimidazolium chloride and hexafluorophosphate are very similar, despite the difference in hydrogen bonding ability of the anion, and the $[MMIM][PF_6]$ salt also shows strong ordering in the liquid state, as found in the chloride. Due to the size of the hexafluorophosphate anion compared with the chloride, the structure is expanded and, as a consequence, the spatial distributions of the anions and cations in the hexafluorophosphate salt have an almost mutually exclusive arrangement unlike the arrangement in the chloride salt. The comparison of the $[PF_6]^-$ and $[MMIM]^+$ arrangement around a central $[MMIM]^+$ is shown in Fig. 4.1-4 .

Solute–solvent interactions have been examined by neutron diffraction using mixtures of benzene and $[MMIM][PF_6]$ [13]. The dissolution of benzene is found

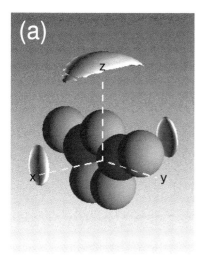
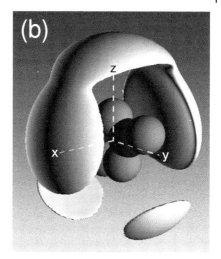
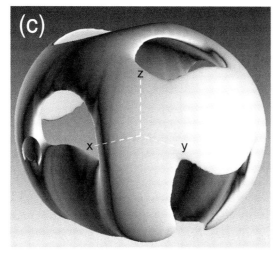

Fig. 4.1-4 Probability distribution of (a) chloride from liquid [MMIM]Cl at 150 °C, (b) hexafluorophosphate anion and (c) the 1,3-dimethyl imidazolium cation from liquid [MMIM][PF$_6$] at 127 °C around a central imidazolium cation, as determined by the EPSR model of the neutron diffraction data.

to increase the cation–cation contacts in the ionic liquid; however, the strong ionic ordering in the liquid is maintained. Furthermore strong ordering of the cations and anions is found around the benzene molecule, with alternating cation–anion layers observed; the first shell anions occupying positions in the plane of the benzene ring, i.e. interacting with the ring hydrogens, whilst the cations interact with the ring electrons, i.e. occupying positions above and below the π-system. Each anion is replaced by approximately 3 benzenes suggesting benzene intercalation into the liquid structure. As with the pure ionic liquids, similar interactions are also found in the solid [14]. The strong interactions of the benzene with the cations and anions result in a homogeneous distribution of the solute throughout the melt with no evidence of micellar formation.

In all these examples, the importance of good simulation and modeling cannot be stressed enough. A variety of methods have been used in this field to simulate the data in the cases studies described above. For example Blander et al. [4] used a semi-empirical molecular orbital method, MNDO, to calculate the geometries of the free haloaluminate ions and used these as a basis for the modeling of the data using the RPSU model [15]. Badyal et al. [6] used reverse Monte Carlo simulations whereas Hardacre and coworkers [11–13] simulated the neutron data from $[MMIM]^+$-based ionic liquids using an empirical potential structure refinement (EPSR) model [16] (for more details about computational modeling of ionic liquids see Section 4.2).

4.1.6
X-ray Diffraction

X-ray diffraction has been used for the study of both simple molten salts and binary mixtures thereof, as well as for liquid crystalline materials. The scattering process is similar to that described above for neutron diffraction with the exception that the scattering of the photons occurs from the electron density and not the nuclei. Therefore the X-ray scattering factor increases with atomic number and the scattering pattern is dominated by the heavy atoms in the sample. Unlike in neutron diffraction, hydrogen, for example, scatters very weakly and its position cannot be determined with any great accuracy.

Compared with the study of the structure of the molten salts, for liquid crystalline materials, full analysis of the scattering profile is not generally performed. In the latter, only the Bragg features are analyzed, that is for a wavelength, λ, incident on the sample at an angle, θ, to its surface normal, the position of the diffraction peaks are determined by Bragg's law, $n\lambda = 2d\sin\theta$. From the angle of diffraction, the periodicity length, d, may be determined.

In both cases, laboratory X-ray sources may be used and the X-ray measurements taken in θ–2θ geometry. For weakly scattering systems synchrotron radiation is helpful.

4.1.6.1 Cells for Liquid Samples
Sample cells include Lindemann/capillary tubes (normally <1 mm in diameter) and aluminum holders. In the latter, thin aluminum windows sandwich the sample in a cylindrical aluminum sample holder. The diffraction from the aluminum in this case is observed and may be used as a calibration standard. For low-temperature materials, the aluminum window can be replaced by the polymer Kapton. Beryllium may also be used [17]. Typically, sample volumes of between 50 and 100 μL are required.

4.1.6.2 Examples
Molten Salts and Binary Mixtures
X-ray diffraction has been performed by Takahashi et al. [18] on 1:1 binary mixtures of $AlCl_3$ with LiCl and NaCl. In agreement with the neutron data obtained by Badyal et al. [6] discussed above, the liquid has a degree of charge ordering with units of four $[AlCl_4]^-$ units surrounding a central $[AlCl_4]^-$ unit at a distance of 6.75 A and 6.98 A for LiCl and NaCl, respectively. Similarly, Igarashi et al. [19] have studied a molten LiF–NaF–KF eutectic mixture. For the ion pairs, Li–F, Na–F and K–F, the nearest neighbor coordination and distances were almost identical to those found in the individual melts of the component salts.

Binary mixtures which melt close to room temperature have also been investigated, namely $AlCl_3$–N-butylpyridinium chloride mixtures. Takahashi et al. [20] have also shown that for the 1:1 composition, $[AlCl_4]^-$ predominates with a tetrahedral environment. At a ratio of 2:1, $[Al_2Cl_7]^-$ becomes the main species. At high temperature, that is above 150 °C, some decomposition to $[AlCl_4]^-$ and Al_2Cl_6 was observed.

High energy X-ray diffraction has been used to examine the liquid structure of binary ionic liquids of 1,3-dialkyl imidazolium fluoride with HF [21]. As found for dimethyl imidazolium salts studied by neutron diffraction (Section 4.1.5.2), the X-ray data showed that in these materials the solid state and liquid structures are closely related. The structures of molten [EMIM]F•HF and [EMIM]F•2.3HF were found to be similar and showed the presence of $[HF_2]^-$ in the liquid, in agreement with the crystalline phase structure. Shodai et al. also studied the structure of liquid $[(CH_3)_4N]F$•nHF ($n = 3$–5). A range of anion structures were found in the liquid, $[(HF)_xF]^-$ ($x = 1$–3); however, structures with higher ratios of HF to F^- ($x = 4$ or 5) were not found in the melt although similar compositions have been reported in the solid state [22].

Liquid Crystals
A wide range of ionic liquids form liquid crystalline phases. This is normally achieved by increasing the amphiphilic character of the cation through substitution with longer, linear alkyl groups. The salts have relatively low melting points, close to room temperature when the alkyl chain length, C_n, is small ($n < 10$), and display liquid crystal mesomorphism when $n > 12$. This section describes some of the results of studies using X-ray diffraction to examine the mesophase and liquid phase. There are many examples of materials which form liquid crystalline phases that have been studied using techniques such as NMR, DSC, single crystal X-ray diffraction and so on which have not been included, for example Refs. [23–26].

Metal-Containing Systems   Many of the systems studied are based on the $[MCl_4]^{2-}$ anion. Neve et al. have studied extensively the formation of liquid crystalline phases on N-alkylpyridinium salts with alkyl chain lengths of $n = 12$–18 with tetrahalometallate anions based upon Pd(II) [27], Cu(II) [28], Co(II), Ni(II), Zn(II) and Cd(II)

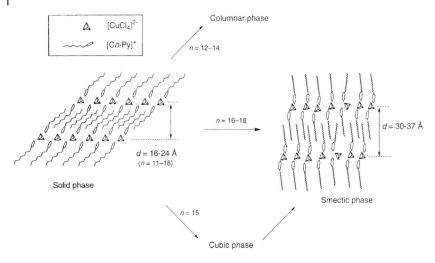

Fig. 4.1-5 Schematic showing the changes in structure of N-alkylpyridinium tetrachlorocuprate salts with varying alkyl chain lengths. Reproduced from Ref. [23] with permission.

[23, 29]. Extensive crystal polymorphism is found which is highly dependent on the metal contained in the salt. For example, in general, the liquid crystalline phases exhibit lamellar-like structures based upon the smectic A structure. For $n = 16$, 18, in the case of $[PdCl_4]^{2-}$ and $[PdBr_4]^{2-}$, this is preceded by an ordered smectic E phase. Cuprate-based pyridinium ionic liquids exhibit a range of structures depending on the alkyl chain length. For $C_{12}-C_{18}$, each solid state structure has a layered periodicity; however, on melting $C_{12}-C_{14}$ exhibit a columnar phase whereas $C_{16}-C_{18}$ simply form a smectic A phase. For $n = 15$, the solid melts into a cubic phase before transforming into the smectic A phase, as seen for longer alkyl chains. Figure 4.1-5 illustrates the changes observed in the latter case.

Similar lamellar structures are formed for 1-alkyl-3-methylimidazolium cations with $[PdCl_4]^{2-}$ when $n > 12$. As with the pyridinium systems, mesomorphic liquid crystal structures are formed based on the smectic A structure [30].

Martin [31] has also shown that ammonium salts display similar behavior. For example [cetyltrimethylammonium]$_2$[ZnCl$_4$] first melts to a S_C-type liquid crystal at 70 °C and then to a S_A-type mesophase at 160 °C. The broad diffraction features observed in the liquid crystalline phases are similar to those in the original crystal phase and show the retention of some of the order originating from the initial crystal on melting, as shown in Fig. 4.1-6.

Needham et al. [32] also used X-ray diffraction to show that, in the case of Mn(II), Cd(II) and Cu(II)-based C_{12}- and C_{14}-ammonium tetrachlorometallate salts, two mechanistic pathways were present on melting to the mesophase. Each pathway was shown to have a minor and major structural transformation. The minor change was thought to be a torsional distortion of the alkyl chains and the major change the melting of the chains to form a disordered layer. The order in which the

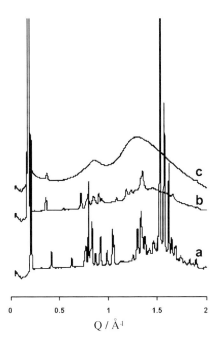

Fig. 4.1-6 Small angle X-ray diffraction data from [cetyltrimethylammonium]$_2$[ZnCl$_4$] at (a) room temperature (solid), (b) 90 °C (S$_C$-phase) and (c) 200 °C (S$_A$-phase). Reproduced from Ref. [25] with permission.

structural changes occur was found to be dependent on the metal and alkyl chain length.

Non-Metal Containing Systems   Non-metal containing salts have been studied extensively. Bradley et al. [33] examined a range of 1-alkyl-3-methylimidazolium-based salts containing chloride, bromide, trifluoromethanesulfonate ([OTf]$^-$), bis(trifluoromethanesulfonyl)amide([Tf$_2$N]$^-$) and [BF$_4$]$^-$ anions. In the mesophase, the X-ray data of these salts were consistent with a smectic A phase with an interlayer spacing of between 22 and 61 Å, which increases uniformly with increasing alkyl chain length, *n*. For a given cation, the mesophase interlayer spacing decreases following the order Cl$^-$ > Br$^-$ > [BF$_4$]$^-$ > [OTf]$^-$ with the bis(trifluoromethanesulfonyl)amide salts not exhibiting any mesophase structure. The anion dependence of the mesophase interlayer spacing is largest for the anions with the greatest ability to form a three-dimensional hydrogen-bonding lattice. As shown in Fig. 4.1-7, on melting to the isotropic liquid, a broad peak is observed in the X-ray scattering data for each salt. This peak indicates that, even within the isotropic liquid phase, some short-range associative structural ordering is still retained. For the halide salts, irreversible changes are observed in the structure on

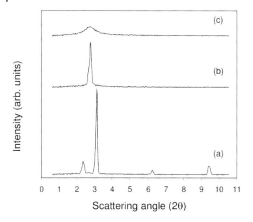

Fig. 4.1-7 Small angle X-ray diffraction data from [C$_{16}$MIM][OTf] at (a) 50 °C, (b) 70 °C and (c) 90 °C, in the crystal, SmA$_2$ and isotropic phases, respectively, on cooling. Reproduced from Ref. [27] with permission.

cooling from the mesophase. The X-ray data show an interlayer spacing of between 50 and 60 Å which collapses to 25–31 Å on cooling. Single-crystal X-ray data have shown that two arrangements of the cation headgroup and the alkyl chain exist which give rise to the different interlayer spacings [34]. De Roche et al. have also investigated the structural behavior of [C$_n$MIM][PF$_6$] ($n = 14,16$) as a function of temperature [35]. Two distinct polymorphs are observed in the solid state associated with increasing disorder of the alkyl chains with some associated anion movement. This is the precursor state to the transition in the S$_A$ mesophase. Hamaguchi and coworkers have reported that similar polymorphic behavior also occurs for short alkyl chain length imidazolium salts. In 1-butyl-3-methyl imidazolium halide salts, the solid state structure is strongly dependent on its thermal history [36]. Two crystal structures are found which are related by rotational isomerism around the α–β carbon bond in the butyl chain. In the liquid state, both structures are found.

Similarly, N-alkylammonium [37] and alkylphosphonium [38] salts form lamellar phases with a smectic bilayer structure. In both cases, the X-ray scattering also showed the isotropic liquid to be not completely disordered, still showing similar features to the mesophase. Buscio et al. [37] showed that in N-alkylammonium chlorides the feature was not only much broader than that observed in the mesophase but increased in width with decreasing chain length.

Other examples include ditholium salts, shown in Fig. 4.1-8 [39]. The scattering data show a range of mesophase behavior is present, as with the metal-containing systems, dependent on alkyl chain length.

For example, for $n = 12$, two transitions within the liquid crystalline region are observed from a nematic columnar phase (N$_{col}$) to a hexagonal columnar lattice (D$_h$) and then finally to a rectangular lattice (D$_r$). The X-ray diffraction data on benzimidazolium salts have also been reported [40], and indicate a switch from a lamellar β

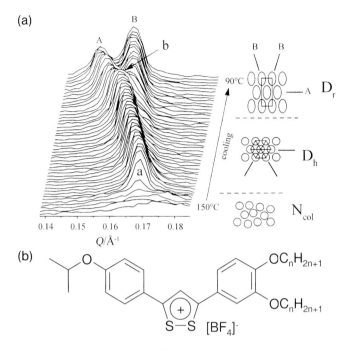

Fig. 4.1-8 (a) Small angle X-ray diffraction data from the ditholium salt shown in (b) for $n = 12$. **A** and **B** correspond to the rectangular lattice vectors shown in D_r and **a** and **b** correspond to the N_{col} to D_h and D_h to D_r phase transitions. Reproduced from Ref. [30] with permission.

phase to the α phase within the liquid crystalline region which is not shown in the differential scanning calorimetry, in some examples. Using X-ray diffraction, Bruce and coworkers have proposed a new structural model for N-alkylpyridinium alkylsulfates [41]. In these liquid crystalline materials, the d spacings obtained are less than the molecular length but are not associated with tilting of the alkyl chains. The new model proposed shows microdomains of interdigitated and non-interdigitated molecules.

Other liquid crystalline materials which have been investigated using X-ray scattering include single and double chained pyridinium [42], N-substituted 4-(5-alkyl-1,3-dioxan-2-yl) pyridinium [43], 1-acetamido-3-alkylimidazolium [44] and 1-alkyl-4-[5-(dodecylsulfanyl)-1,3,4-oxadiazol-2-yl]pyridinium [45] salts. In the former the diffraction allowed an explanation for the differences in single and double substituted salts to be proposed.

In general the X-ray data are used in conjunction with other techniques to obtain as full a picture as possible. For the liquid crystalline materials differential scanning calorimetry (DSC) and polarizing optical microscopy (POM) are conventionally used.

4.1.7
Extended X-ray Absorption Fine Structure Spectroscopy

Extended X-ray absorption fine structure spectroscopy (EXAFS) measures the X-ray absorption as a function of energy and allows the local arrangement of atoms to be elucidated. The absorption results from the excitation of a core electron in an atom. Conventional EXAFS is usually associated with hard X-rays, that is >3–4 keV, to allow measurements to be made outside a vacuum, and requires synchrotron radiation to provide the intensity at the energies involved. At these energies, the core electrons ejected correspond to the 1s (K-edge), 2s (L_I-edge), $2p_{1/2}$(L_{II}-edge) and $2p_{3/2}$(L_{III}-edge) states. As the photon energy is increased past the absorption edge, an oscillatory structure is found, described as the X-ray fine structure. The X-ray fine structure starts at approximately 30 eV past the edge and extends to a range of 1000 eV.

The EXAFS is observed as a modulating change in the absorption coefficient caused by the ejected electron wave backscattering from the surrounding atoms, resulting in interference between ejected and backscattered waves. It is defined as:

$$\chi(k) = \frac{\mu(k) - \mu_0(k)}{\Delta\mu_0} \quad (4.1\text{-}4)$$

where $\chi(k)$ is the EXAFS as a function of the wavenumber of the photoelectron, k, $\mu(k)$ is the measured absorption above the absorption edge, $\mu_0(k)$ is the absorption spectrum without the EXAFS oscillations, that is the background, and $\Delta\mu_0$ is a normalization factor.

The wavenumber is defined at a photon energy, E, above the absorption edge energy, E_0, with respect to the mass of the electron, m_e.

$$k = \sqrt{\frac{2m_e}{\hbar^2}(E - E_0)} \quad (4.1\text{-}5)$$

The EXAFS is related to the wavenumber by:

$$\chi(k) \approx \sum_i \frac{N_i f_i(k)}{kr_i^2} e^{-2\sigma_i^2 k^2} e^{-2r_i/\lambda} \sin[2kr_i + \alpha_i(k)] \quad (4.1\text{-}6)$$

that is, $\chi(k)$ is the sum over N_i backscattering atoms, i, where f_i is the scattering amplitude term characteristic of the atom, σ_i is the Debye–Waller factor associated with the vibration of the atoms, r_i is the distance from the absorbing atom, λ is the mean free path of the photoelectron and α_i is the phase shift of the spherical wave as it scatters from the backscattering atoms. By taking the Fourier transform of the amplitude of the fine structure, that is $\chi(k)$, a real-space radial distribution function of the backscattering atoms around the absorbing atom is produced.

On analysis of the EXAFS data, the local environment around a given absorbing atom can be obtained, that is the type, number and distance of the backscattering atoms. It should be noted that it is not necessary for the surrounding atoms to be formally bonded to the absorbing atom. Typically the distance has an uncertainty of $\pm 1\%$ within a radius of approximately 6 Å. However, the error in the coordination number is strongly dependent on the system studied and can be high. In this regard, comparison with standard materials and the use of EXAFS in conjunction with other techniques to ensure a realistic interpretation of the data is vital.

Since the fine structure observed is only associated with the particular absorption edge being studied, and the energy of the absorption edge is dependent on the element and its oxidation state, the EXAFS examines the local structure around one particular element, and in some cases, an element in a given oxidation state. Therefore by studying more than one absorbing element in the sample, a fuller picture can be obtained.

4.1.7.1 Experimental
Measuring EXAFS Spectra
In general, transmission EXAFS can be used, providing the concentration of the element to be investigated is sufficiently high. The sample is placed between two ionisation chambers, whose signals are proportional to the incident intensity, I_0, and transmitted intensity through the sample, I_t. The transmission of the sample is dependent on the thickness of the sample, x, and the absorption coefficient via a Beer–Lambert relationship:

$$I_t = I_0 e^{-\mu x} \tag{4.1-7}$$

The difference between the $\ln(I_t/I_0)$ before and after the absorption edge, that is the edge jump, should be between 0.1 and 1 to obtain good spectra. This may be calculated using the mass absorption coefficient of a sample:

$$\left(\frac{\mu}{\rho}\right)_{sample} = \sum_i w_i \left(\frac{\mu}{\rho}\right)_i \tag{4.1-8}$$

where ρ is the sample density and the mass weighted average of the mass absorption coefficients of each element in the sample, using weight fractions, w_i.

If the edge jump is too large, the sample should be diluted or the path length decreased. If the edge jump is too small, then addition of more sample is one possibility; however, this is dependent on the matrix in which the sample is studied. For low atomic weight matrices such as carbon-based materials, the path length can be increased without the transmission of the X-rays being affected adversely. In matrices containing high atomic weight elements, such as chlorine, increasing path length will result in a larger edge jump, but it will also decrease the overall transmission of the X-rays. For such samples, fluorescence EXAFS may be performed. In this geometry, the emitted X-rays are measured. Optimally, the sample is placed at 45° to the incident X-rays and the X-ray fluorescence detected at 90° to the direction of the exciting X-rays using, for example, a solid-state detector. The X-ray florescence is proportional to the X-rays absorbed by the sample and therefore can be used to measure the EXAFS oscillations. In general, this technique has a

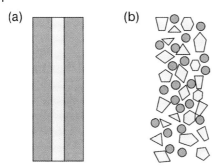

Fig. 4.1-9 Schematic of the sample preparation methods used to study liquid EXAFS (a) thin liquid film sandwich between low atomic weight plates and (b) the liquid (circles) dispersed in a low atomic number matrix (polyhedrons). The figure has been redrawn from Ref. [40] with permission.

poorer signal-to-noise ratio than transmission EXAFS and there are problems with self-absorption effects requiring dilute or thin sample sizes.

Self-absorption occurs when the path length is too large [46] and the X-rays emitted have a significant probability of being absorbed by the remainder of the sample before being detected. This has the consequence of reducing the amplitude of the EXAFS oscillations and producing erroneous results. As the sample becomes more dilute this probability decreases. All the atoms in the sample determine the amount of self-absorption, hence the need for thin samples.

Liquid Set-Ups
There are two major methods in which liquid samples are studied. These are shown in Fig 4.1-9.

These consist of supporting the liquid in an inert low atomic weight matrix such as graphite or boron nitride or sandwiching thin films between low atomic weight plates. The choice of the matrix materials used is a balance between chemical inertness towards the liquid being studied whilst being thermally stable and transparent to the X-rays at the absorption energy. The latter becomes less problematic as the energy of the absorption edge increases. Figure 4.1-10 shows an experimental cell which has been used to measure the EXAFS of ionic liquid samples [47].

Analysis
A number of commercial software packages are available to model the EXAFS data, for example the FEFF program developed by Rehr and coworkers [48], GNXAS developed by Filipponi et al. [49] and EXCURV, developed by Binsted [50]. These analysis packages fit the data to curve wave theory and describe multiple scattering as well as single scattering events. Before analysis, the pre-edge and a smooth post

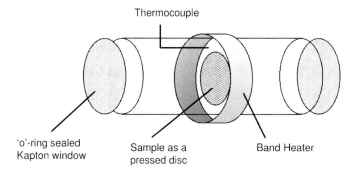

Fig. 4.1-10 Schematic of transmission EXAFS cell. Reproduced from Ref. [36] with permission.

edge background function, that is $\mu_0(k)$, is subtracted from the measured spectra. This is one of the most important procedures and can, if performed poorly, result in loss of amplitude of the EXAFS oscillations or unphysical peaks in the Fourier transform.

Two review articles have recently been published which treat the theory more rigorously and summarise much of the data on general liquid systems [51, 52].

4.1.7.2 Examples
Ionic Liquid Structure
Although EXAFS has only been employed to investigate the structure of high temperature molten salts in detail, one of the first examples of its use examined the liquid structure of two low-temperature tetrabutyl ammonium salts [53]. Crozier et al. studied the liquid structure of [Bu$_4$N][MnBr$_3$] and [Bu$_4$N]$_2$[MnBr$_4$] at temperatures up to 400 K. After taking into account the asymmetry in the atomic distributions of bromine around the manganese and vice versa, Mn–Br bond distances of 2.46 and 2.50 Å were found for [Bu$_4$N][MnBr$_3$] and [Bu$_4$N]$_2$[MnBr$_4$], respectively. The analysis showed a coordination number of 3.14 for the Br around the Mn in the liquid structure of [Bu$_4$N][MnBr$_3$]. This is compared with a coordination number of six in the crystal structure. Similar trigonal structures had not been proposed previously in the liquid state.

Di Cicco and coworkers [54, 55] have examined the structure of molten CuBr using Cu and Br K-edge EXAFS. From the EXAFS data, the Cu–Br bond length distribution was found to be shorter than that derived from neutron data and theoretical models, indicating a more covalent character than previously thought. Similar EXAFS studies on KBr and RbBr are in good agreement with the theory, showing high levels of ionicity [55].

Zn and Rb K-edge EXAFS have also been used to examine the melting of Rb$_2$ZnCl$_4$ compared with the liquid structure of ZnCl$_2$ and RbCl [56]. In molten ZnCl$_2$, the zinc is found to be tetrahedrally coordinated with the tetrahedra linked by corner-sharing chlorines in a weak extended network. In RbCl, significant disorder in the chloride shell around the rubidium is evident and indicates significant movement of the Rb$^+$ and Cl$^-$ in the molten state. In the crystal structure of Rb$_2$ZnCl$_4$, the chlorine coordination number around the Rb is between 8 and 9 whilst the Zn is found in isolated ZnCl$_4$ units. In the molten state, the EXAFS also indicates isolated ZnCl$_4$ units with a chlorine coordination of 7.6 around the Rb. This may be compared with a chlorine coordination of 4.8 in liquid RbCl. The EXAFS clearly shows that the solid and liquid structures of Rb$_2$ZnCl$_4$ are similar and the melt does not rearrange into a simple combination of the component parts.

In the studies described above, the samples were supported in low atomic weight matrices, melted *in situ* and measured in transmission mode. Similarly, second generation ionic liquids have been studied. Carmichael et al. [57] showed that it was possible to support and melt [EMIM]$_2$[NiCl$_4$] and [C$_{14}$MIM]$_2$[NiCl$_4$] in inert matrices such as boron nitride, graphite and lithium fluoride without the EXAFS being affected by the sample matrix used. In these samples, the Ni K-edge EXAFS was investigated between room temperature and 131 °C and even in LiF, where halide exchange was possible, little difference was found in the Ni coordination on melting.

Species Dissolved in Ionic Liquids   A number of systems have been investigated in both chloroaluminate and second generation ionic liquids.

Dent et al. [58] studied the dissolution of [EMIM][MCl$_4$] in [EMIM]Cl–AlCl$_3$ binary mixtures, for M = Mn, Co and Ni, at AlCl$_3$ mole fractions of 0.35 and 0.60 using the M K-edges. Due to the problems associated with the high concentration of chloride, it was not possible to perform transmission experiments and so fluorescence measurements were used. In this case, self-absorption problems were overcome by using a thin film of liquid pressed between two sheets of polythene sealed in a glove box. The coordination of Ni, Co and Mn was found to change from [MCl$_4$]$^{2-}$ to [M(AlCl$_3$)$_4$]$^-$ as the mole fraction of AlCl$_3$ increased. Figure 4.1-11 shows the EXAFS and pseudo-radial distribution functions for M = Co in both the acidic and basic chloroaluminate ionic liquids.

Aluminum coordination was only observed in acidic mixtures, that is at AlCl$_3$ mole fractions greater than 0.5. The latter was surprising given that at 0.60 AlCl$_3$, almost half the anion species are in the form [Al$_2$Cl$_7$]$^-$ yet no coordinating [Al$_2$Cl$_7$] was observed.

Dent et al. [59] also investigated the V K-edge EXAFS for the dissolution of [EMIM][VOCl$_4$] and [NEt$_4$][VO$_2$Cl$_2$] in a basic [EMIM]Cl–AlCl$_3$ and compared the data with that for solid samples. In both cases, the dissolved and solid samples showed similar EXAFS and, for example, no coordination of the chloroaluminate species to the vanadyl oxygen was found.

Due to the decrease in average atomic weight of the medium compared with chloroaluminate systems, second generation ionic liquids may be studied in transmission. Carmichael et al. [57] have shown that solutions of [EMIM]$_2$[NiCl$_4$] in [BMIM][PF$_6$] may be studied by supporting the liquid between two boron nitride discs. The resulting Ni K-edge EXAFS showed a similar local structure to the molten [EMIM]$_2$[NiCl$_4$], described above.

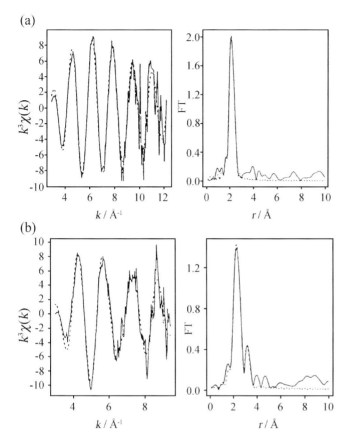

Fig. 4.1-11 The EXAFS data and pseudo-radial distribution functions of Co(II) in (a) basic and (b) acidic chloroaluminate ionic liquid. Reproduced from Ref. [46] with permission.

Baston et al. [60] studied the samples of ionic liquid after the anodization of uranium metal in [EMIM]Cl using the U L_{III}-edge EXAFS to establish both the oxidation state and the speciation of uranium in the ionic liquid. This was part of an ongoing study to replace high-temperature melts, such as LiCl:KCl [61], with ionic liquids. Although it was expected that, when anodized, the uranium would be in the +3 oxidation state, electrochemistry showed that the uranium is actually in a mixture of oxidation states. The EXAFS of the solution showed an edge jump at 17166.6 eV, indicating a mixture of uranium(IV) and uranium(VI). The EXAFS data and pseudo-radial distribution functions for the anodized uranium in [EMIM]Cl are shown in Fig. 4.1-12.

Two peaks were fitted corresponding to a 1:1 mixture of $[UCl_6]^{2-}$ and $[UO_2Cl_4]^{2-}$, in agreement with the position of the edge. Oxidation to uranium(VI) was surprising

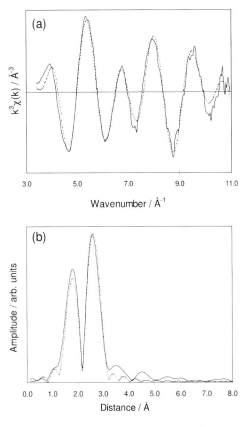

Fig. 4.1-12 The experimental (solid line) and fitted (dashed line) U L(III)-edge (a) EXAFS data and (b) pseudo-radial distribution function following anodization of uranium in [EMIM]Cl. The figure has been redrawn from Ref. [48] with permission.

in this system and may have arisen from the presence of water due to the highly hygroscopic nature of [EMIM]Cl.

Nitrate-based second generation ionic liquids have also been studied for the reprocessing of uranium [62]. In particular, it has been shown that dinuclear dioxouranium(VI) salt containing a bridging oxalate ligand, 1-butyl-3-methylimidazolium μ_4-(O,O,O',O'-ethane-1,2-dioato)-bis{bis(nitrato-O',O)dioxouranate(VI)} ([BMIM]$_2$[{UO$_2$(NO$_3$)$_2$}$_2$(μ-C$_2$O$_4$)]), is formed on oxidative dissolution of uranium(IV) oxide in [BMIM][NO$_3$] using concentrated nitric acid as the oxidising agent. EXAFS was used to examine how the dinuclear dimer was formed. The EXAFS indicated that the oxalate "dinuclear" species remains intact when dissolved in the ionic liquid and the hydrated dioxouranium(VI) cation is not formed. Following oxidative dissolution of UO$_2$ in [BMIM][NO$_3$] in the absence or presence of acetone, similar EXAFS were found in each case. For both solutions, the best fit was obtained with 15:85 mononuclear nitrate:dinuclear oxalate dimer mole ratio present in solution.

EXAFS is also being increasingly used to examine the speciation of metals extracted with ionic liquids from aqueous solutions. Ionic liquids are able to increase the extraction efficiency of actinides compared with commonly used organic solvents. The speciation of aqueous uranium(VI) following extraction in the presence of carbamoylphosphine oxide (CMPO) and tri(n-butyl)phosphate has been investigated. Significant changes in the inner sphere complex on changing the extracting solvent from dodecane to [BMIM][PF$_6$]/[OMIM][NTf$_2$], but not between the ionic liquids, was observed using EXAFS [63]. Two uranyl oxygens, with two coordinating monodentate CMPO molecules (P=O bound) and bidentate nitrate anions i.e. six equatorial oxygens around the uranium, were found in dodecane. In contrast, in both ionic liquids, although the two uranyl oxygens were observed, the number of equatorial oxygens decreased to 4–4.5. The difference is attributed to water molecules complexing to the uranium and also the formation of $[UO_2(NO_3)(CMPO)]^+$ which causes the ionic liquid to cation exchange into the aqueous phase.

An analogous mechanism for cation extraction from aqueous species was also reported by Jensen et al. for Sr^{2+}(aq) studied using EXAFS [64]. Using a representative neutral complexing agent, *cis-syn-cis*-dicyclohexyl-18-crown-6 (DCH18C6), and various counterions ([NO$_3$]$^-$, Cl$^-$, [SO$_4$]$^{2-}$), the speciation was studied in 1-methyl-3-pentylimidazolium bis(trifluoromethylsulfonyl)amide and in octan-1-ol and compared with dissolution of Sr(NO$_3$)$_2$(18C6). Following extraction of Sr^{2+} in the presence of DCH18C6 from an aqueous solution of Sr(NO$_3$)$_2$ into the ionic liquid, the coordination number decreased from 10.4, on dissolving Sr(NO$_3$)$_2$(18C6) in the ionic liquid, to 8.5. In contrast, extraction using octan-1-ol, showed a similar structure as found in Sr(NO$_3$)$_2$(18C6). The difference between the solvents was thought to be due to water displacing the nitrate anion in the case of the ionic liquid. The absence of anion in the inner shell of the strontium cation showed that poor anion extraction occurs using the ionic liquids compared with the organic solvent.

Similar changes are also found for europium(III) complexed to 1-(2-thienyl)-4,4,4-trifluoro-1,3-butanedione (Htta) [65]. In 1-methyl-3-pentylimidazolium bis(trifluoromethylsulfonyl)amide, the EXAFS is consistent with the formation of a $[Eu(tta)_4]^-$ complex. The presence of discrete octahedral anionic species in the ionic liquid is in contrast with organic solvents where neutral species are formed, for example [Eu(tta)$_3$(Htta)] or [Eu(tta)$_3$(H$_2$O)].

In all these studies, EXAFS has illustrated some of the problems associated with using ionic liquids as extraction media [63–65]. In each case, the anions formed caused the ionic liquid to become more hydrophilic by forming a binary ionic liquid, and leaching of the ionic liquid into the aqueous phase occurs.

In none of the above cases has a reaction been performed whilst taking the EXAFS data. Hamill et al. [66] have investigated the Heck reaction, catalyzed by palladium salts and complexes, in room-temperature ionic liquids. On dissolution of palladium ethanoate in [BMIM]$^+$ and *N*-butylpyridinium ([BPy]$^+$)

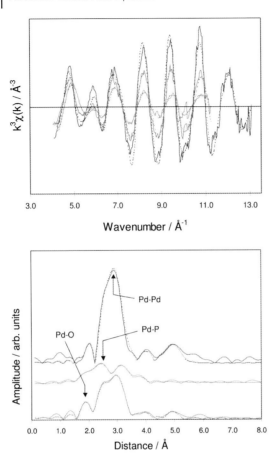

Fig. 4.1-13 Comparison of the experimental (solid line) and fitted (dashed line) (a) EXAFS and (b) pseudo-radial distribution functions from palladium ethanoate in [BMIM][PF$_6$] without and with triphenylphosphine at 80 °C and in the presence of triphenylphosphine and reagents at 50 °C for 20 min. Reproduced from Ref. [50] with permission.

hexafluorophosphate and tetrafluoroborate ionic liquids, and triethyl-hexyl ammonium bis(trifluoromethanesulfonyl)amide, a gradual change from ethanoate coordination to the formation of palladium metal was observed in the Pd K-edge EXAFS as shown in Fig. 4.1-13.

In pyridinium chloride ionic liquids and in 1,2-dimethyl-3-hexyl imidazolium chloride ([HMMIM]Cl), that is where the C(2) position is protected by a methyl group, only [PdCl$_4$]$^{2-}$ was observed; whereas in [HMIM]Cl, the EXAFS showed the formation of a bis-carbene complex. In the presence of triphenylphosphine, Pd–P coordination was observed in all ionic liquids except where the carbene complex was formed. During the Heck reaction, the formation of palladium was found to be quicker than in the absence of reagents. Overall, the EXAFS showed the presence of small palladium clusters approximately 1 nm diameter formed in solution.

4.1.8
X-ray and Neutron Reflectivity

Reflectometry is a useful probe to investigate the structure of multilayers both in self-supporting films and those adsorbed on surfaces [67]. Specular X-ray reflectivity probes the electron density contrast perpendicular to the film. The X-rays irradiate the substrate at a small angle (<5°) to the plane of the sample, are reflected and detected at an equal angle. If a thin film is present on the surface of the substrate, the X-rays may be reflected from the top and the bottom of the film, which gives rise to interference and an oscillatory pattern with changing angle of incidence, known as Kiessig fringes. The pattern obtained is a function of the difference in electron density and roughness at each interface present; rough films give rise to a reduction in the amplitude of the oscillation observed. Analysis of this variation gives information principally about the interfaces, but may also be used to investigate chain layering, for example in metal soaps [68]. Neutron reflectometry is complementary to X-ray reflectometry and probes the neutron scattering length density, and hence composition, profile normal to the surface [69]. As described above for neutron diffraction, by using isotopic substitution, in particular hydrogen and deuterium exchange, study of the arrangements of specific parts of the molecule within a film is possible. In addition using neutrons coupled with isotopic exchange permits the study of less well-ordered structures, i.e. systems that do not give rise to interference fringes or Bragg peaks.

4.1.8.1 Experimental Set-up
Commonly X-ray reflectivity experiments for thin films of liquids, and so on, are performed on silicon single-crystal wafers and the X-rays are reflected off the surface of the wafer [70]. To enable good adhesion, the wafers have to be cleaned, for example in concentrated nitric acid followed by a UV–O_3 treatment, to remove any trace organics. Deposition of the films can then be performed by spin coating from a solution of the salt in a volatile organic solvent. In general, the spin-coated films are too rough to give good reflectivity spectra and the films need to be pre-annealed. X-ray reflectivity measurements may be performed using a laboratory X-ray source as well as synchrotron radiation. Figure 4.1-14 shows a typical cell used for X-ray reflectivity measurements. In neutron reflection measurements, this protocol can also be applied to examine thin supported films. Solid–liquid interfaces are accessed by transmission through a silicon or quartz substrate, commonly, whilst vapor–liquid interfaces may be examined by reflecting from the free liquid surface with the liquid samples contained, usually, in poly(tetrafluoroethylene) troughs. The reflectivity measurements can be performed at either a reactor or spallation neutron source.

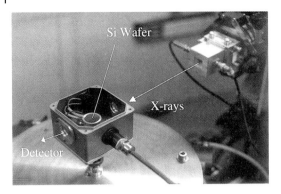

Fig. 4.1-14 A typical cell used for X-ray reflectivity measurements.

4.1.8.2 Examples

Carmichael et al. [71] have used this technique to compare the structure of thin films of [C_{18}MIM][PF_6], [C_{18}MIM][BF_4], [BMIM]$_2$[$PdCl_4$], [C_{12}MIM]$_2$[$PdCl_4$], and [C_{12}MIM][PF_6] to bulk solutions as studied by small angle X-ray scattering. For all the salts studied, Bragg features were clearly visible; however, in most cases, the additional Kiessig fringes were not observed. Figure 4.1-15 shows an example of data collected on a thin film of [C_{18}MIM][PF_6].

The Bragg peaks indicated an ordered local structure within the sample film and the interlayer spacings were reproduced compared with the bulk samples with only minor shifts in layer spacing. The small changes in layer spacing are expected since the thin film structure is not constrained by long-range order effects and hence adopts a slightly different lower energy form. The similarity between the bulk

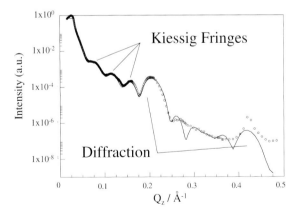

Fig. 4.1-15 Experimental reflectivity data (points) compared with a 5 bilayer model (solid line) for a 156 Å thick [C_{18}MIM][PF_6] film at 298 K. Reproduced from Ref. [54] with permission.

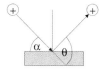

Scattered ions

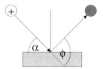

Sputtered (recoiled) ions

Fig. 4.1-16 Schematic of the scattering process showing the scattered ions and the recoiled ions. The figure has been redrawn from Ref. [57] with permission.

samples and the thin film is further exemplified from the modelling of the Kiessig fringes in the case of [C_{18}MIM][PF_6]. This model was comprised of layers of associated 1-ethyl-3-methylimidazolium cation head groups and hexafluorophosphate anions, denoted as the charged region, separated by hydrocarbon chains. Reasonable fits were only obtained with the charged region at both the salt–silicon and salt–air interfaces.

Neutron reflectivity has been used to investigate shorter chain length ionic liquids, [BMIM][BF_4] and [OMIM][PF_6] [72]. As found with the X-ray reflectivity measurements, both ionic liquids are layered with the ionic and alkyl parts of the liquid segregating into a lamellar structure. This ordering is maintained to at least two alkyl lengths. However, the uncertainty in the analysis did not allow determination of whether the charged layer is at the surface or how deep the ordering penetrates in the liquid.

4.1.9
Direct Recoil Spectrometry (DRS)

The surface structure of ionic liquids has been studied using direct recoil spectrometry. In this experiment, a pulsed beam of 2–3 keV inert gas ions is scattered from a liquid surface and the energy and intensity of the scattered and sputtered (recoiled) ions are measured as a function of the incident angle, α, of the ions. Figure 4.1-16 shows a scheme of the process for both the scattered and sputtered ions.

The incident ions cause recoil in the surface atoms. In the studies of ionic liquids, only direct recoil was measured, that is motion in the forward direction. Watson and coworkers [73,74] used time-of-flight analysis with a pulsed ion beam to measure the kinetic energy of the scattered and sputtered ions and therefore determine the mass of the recoiled surface atoms. By relating the measured intensity of the sputtered atoms to the scattering cross section, the surface concentration may be found. The

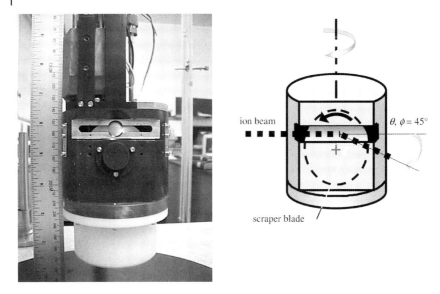

Fig. 4.1-17 The experimental set-up used to generate the thin films of ionic liquid for analysis by direct recoil spectrometry. Reproduced from Ref. [58] with permission.

intensity variation with incident angle also allows the orientation of the atoms on the surface to be elucidated. The scattered and sputtered ions are detected at angles of θ and ϕ, respectively, measured with respect to the incident ion beam. For all the experiments reported to date on ionic liquids, θ and ϕ are equal.

4.1.9.1 Experimental Set-up

Direct recoil spectrometry requires high and ultra-high vacuum conditions for the transport of ions to the sample and to the detector. In this regard the use of ionic liquids, with their corresponding low vapor pressure, is ideal. To prevent contamination of the surface and any surface charging effects, Watson and coworkers used a rotating stainless steel wheel partially submerged in a reservoir holding the liquid sample to continually create a fresh liquid surface. Before analysis, the liquid film passes by a knife-edge, leaving a fresh surface $\sim$ 0.1–0.2 mm thick. Figure 4.1-17 shows the typical sample set-up [75].

4.1.9.2 Examples

A number of ionic liquids have been studied using DRS, namely [OMIM][PF$_6$], [BF$_4$]$^-$, Br$^-$, Cl$^-$; [BMIM][PF$_6$], [BF$_4$]$^-$ and [C$_{12}$MIM][BF$_4$]. The scattering profile as a function of the incident angle for [OMIM][PF$_6$] is shown in Fig. 4.1-18.

In all cases the charged species were found to concentrate at the surface of the liquid under vacuum conditions. Little surface separation of the anions and cations was observed. For the [PF$_6$]$^-$ and [BF$_4$]$^-$ ions, the cation ring was found to

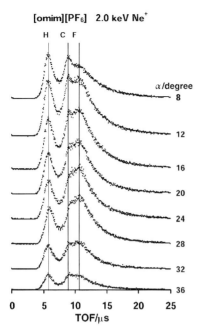

Fig. 4.1-18 Ion intensity as a function of incident angle measured using time of flight direct recoil spectrometry on [OMIM][PF$_6$]. Reproduced from Ref. [57] with permission.

prefer a perpendicular orientation to the surface with the nitrogen atoms closest to the surface. Increasing the alkyl chain length caused the cation to rotate so that the alkyl chain moved into the bulk liquid, i.e. away from the surface, forcing the methyl group closer to the surface. For halide ionic liquids the data were less clear and the cation could be fitted to a number of orientations.

4.1.10
Conclusions

A wide range of structural techniques may be utilised for the study of ionic liquids and dissolved species. Overall, in both high-temperature and low-temperature ionic liquids, as well as for mixtures, a close correlation between the solid structure and the liquid is found. In many cases, significant intermediate order is observed, for example in the form of charge ordering.

Acknowledgements

CH would like to thank James Bowers (University of Exeter), Daniel Bowron (CLRC, Rutherford Appleton Laboratory), Fred Mosselmanns (CLRC, Daresbury

Laboratory), Will Pitner (Queen's University, Belfast), Nick Terrill (CLRC, Daresbury Laboratory) and Philip Watson (University of Oregon) for contributing figures as well as useful discussions and comments in the preparation of this manuscript.

References

1. D. T. Bowron, J. L. Finney, A. K. Soper, *J. Phys. Chem. B* **1998**, *102*, 3551–3563.
2. C. Hardacre, J. D. Holbrey, S. E. J. McMath, *J. Chem. Soc., Chem. Commun.* **2001**, 367–368.
3. F. G. Edwards, J. E. Enderby, R. A. Howe, D. I. Page, *J. Phys. C* **1975**, *8*, 3483–90;
S. Biggin, J. E. Enderby, *J. Phys. C* **1982**, *15*, L305–309.
4. M. Blander, E. Bierwagen, K. G. Calkins, L. A. Curtiss, D. L. Price, M.-L. Saboungi, *J. Chem. Phys.* **1992**, *97*, 2733–2741.
5. S. J. Cyvin, P. Klaeboe, E. Rytter, H. A. Øye, *J. Chem. Phys.* **1970**, *52*, 2776;
J. Hvistendahl, P. Klaeboe, E. Rytter, H. A. Øye, *Inorg. Chem.* **1984**, *23*, 706–715.
6. Y. S. Badyal, D. A. Allen, R. A. Howe, *J. Phys.: Condens Matter* **1994**, *6*, 10193–10220.
7. Y.-C. Lee, D. L. Price, L. A. Curtiss, M. A. Ratner, D. F. Shriver, *J. Chem. Phys.* **2001**, *114*, 4591–4594.
8. S. Takahashi, K. Suzuya, S. Kohara, N. Koura, L. A. Curtiss, M.-L. Saboungi, *Z. Phys. Chem.* **1999**, *209*, 209–221.
9. P. C. Truelove, D. Haworth, R. T. Carlin, A. K. Soper, A. J. G. Ellison, D. L. Price, *Proc. 9th Int. Symp. Molten Salts, San Francisco*, Electrochemical Society, Pennington NJ, **1994**, Vol. 3, pp. 50–57.
10. F. R. Trouw, D. L. Price, *Annu. Rev. Phys. Chem.* **1999**, *50*, 571–601.
11. D. T. Bowron, C. Hardacre, J. D. Holbrey, S. E. J. McMath, A. K. Soper, unpublished results.
12. C. Hardacre, S. E. J. McMath, M. Nieuwenhuyzen, D. T Bowron, A. K Soper, *J. Phys.: Condens. Matter* **2003**, *15*, S159–S166.
13. M. Deetlefs, C. Hardacre, M. Nieuwenhuyzen, O. Sheppard, A. K. Soper, *J. Phys. Chem. B* **2006**, *110*, 12055–12061.
14. J. D. Holbrey, W. M. Reichert, M. Nieuwenhuyzen, O. Sheppard, C. Hardacre, R. D. Rogers, *Chem. Comm.* **2003**, 476.
15. P. A. Egelstaff, D. I. Page, J. G. Powles, *Mol. Phys.* **1971**, *20*, 881; *Mol. Phys.* **1971**, *22*, 994.
16. A. K. Soper, *Chem. Phys.* **1996**, *202*, 295–306; A. K. Soper, *Chem. Phys.* **2000**, *258*, 121–137.
17. F. Vaslow, A. H. Narten, *J. Chem. Phys.* **1973**, *59*, 4949–4954.
18. S. Takahashi, K. Maruoka, N. Koura, H. Ohno, *J. Chem. Phys.* **1986**, *84*, 408–415;
S. Takahashi, T. N. Muneta, N. Koura, H. Ohno, *J. Chem. Soc., Faraday Trans. II* **1985**, *81*, 1107–1115.
19. K. Igarashi, Y. Okamoto, J. Mochinaga, H. Ohno, *J. Chem. Soc., Faraday Trans. I* **1988**, *84*, 4407–4415.
20. S. Takahashi, N. Koura, M. Murase, H. Ohno, *J. Chem. Soc., Faraday Trans. II* **1986**, *82*, 49–60.
21. R. Hagiwara, K. Matsumoto, T. Tsuda, Y. Ito, S. Kohara, K. Suzuya, H. Matsumoto, Y. Miyazaki, *J. Non-Cryst. Solids* **2002**, *312–314*, 414–418; K. Matsumoto, R. Hagiwara, Y. Ito, S. Kohara, K. Suzuya, *Nucl. Instrum. Methods Phys. Res., Sect. B* **2003**, *199*, 29–33.
22. Y. Shodai, S. Kohara, Y. Ohishi, M. Inaba, A. Tasaka, *J. Phys. Chem. A* **2004**, *108*, 1127–1132.
23. C. J. Bowlas, D. W. Bruce, K. R. Seddon, *J. Chem. Soc., Chem. Commun.* **1996**, 1625–1626.
24. R. Kind, S. Plesko, H. Arend, R. Blinc, B. Zeks, J. Seliger, B. Lozar, J. Slak, A. Levstik, C. Filipic, V. Zagar, G. Lahajnar, F. Milia, G. Chapuis, *J. Chem. Phys.* **1979**, *71*, 2118–2130.
25. C. M. Gordon, J. D. Holbrey, A. Kennedy, K. R. Seddon, *J. Mater. Chem.* **1998**, *8*, 2627–2636; J. D. Holbrey, K.R. Seddon, *J. Chem. Soc., Dalton Trans.* **1999**, 2133–2139.

26. P. B. Hitchcock, K. R. Seddon, T. Welton, *J. Chem. Soc., Dalton Trans.* **1993**, 2639–2643.
27. F. Neve, A. Crispini, S. Armentano, O. Francescangeli, *Chem. Mater.* **1998**, *10*, 1904–1913.
28. F. Neve, O. Francescangeli, A. Crispini, J. Charmant, *Chem. Mater.* **2001**, *13*, 2032–2041.
29. F. Neve, O. Francescangeli, A. Crispini, *Inorg. Chim. Acta* **2002**, *338*, 51–58.
30. C. Hardacre, J. D. Holbrey, P. B. McCormac, S. E. J. McMath, M. Nieuwenhuyzen, K. R. Seddon, *J. Mater. Chem.* **2001**, *11*, 346–350.
31. J. D. Martin, ACS Symposium Series on *Industrial Applications of Ionic Liquids*, **2002**, *818* (Ionic Liquids), 413–427.
32. G. F. Needham, R. D. Willett, H. F. Franzen, *J. Phys. Chem.* **1984**, *88*, 674–680.
33. A. E. Bradley, C. Hardacre, J. D. Holbrey, S. Johnston, S. E. J. McMath, M. Nieuwenhuyzen, *Chem. Mater.* **2002**, *14*, 629–635.
34. A. Downard, M. J. Earle, C. Hardacre, S. E. J. McMath, M. Nieuwenhuyzen, S. J. Teat, *Chem. Mater.* **2004**, *16*, 43–48.
35. J. De Roche, C. M. Gordon, C. T. Imrie, M. D. Ingram, A. R. Kennedy, F. Lo Celso, A. Triolo, *Chem. Mater.* **2003**, *15*, 3089–3097.
36. S. Hayashi, R. Ozawa, H. Hamaguchi, *Chem. Lett.* **2003**, *32*, 498–499; S. Saha, S. Hayashi, A. Kobayashi, H. Hamaguchi, *Chem. Lett.* **2003**, *32*, 740–741; H. Katayanagi, S. Hayashi, H. Hamaguchi, K. Nishikawa, *Chem. Phys. Lett.* **2004**, *392*, 460–464.
37. V. Busico, P. Corradini, M. Vacatello, *J. Phys. Chem.* **1982**, *86*, 1033–1034; V. Busico, P. Cernicchlaro, P. Corradini, M. Vacatello, *J. Phys. Chem.* **1983**, *87*, 1631–1635.
38. D. J. Abdallah, A. Robertson, H.-F. Hsu, R. G. Weiss, *J. Am. Chem. Soc.* **2000**, *122*, 3053–3062.
39. F. Artzner, M. Veber, M. Clerc, A.-M. Levelut, *Liq. Cryst.* **1997**, *23*, 27–33.
40. K. M. Lee, C. K. Lee, I. J. B. Lin, *J. Chem. Soc., Chem. Commun.* **1997**, 899–900.
41. C. Cruz, B. Heinrich, A. C. Ribeiro, D. W. Bruce, D. Guillon, *Liq. Cryst.* **2000** *27*, 1625–1631; D. W. Bruce, S. Estdale, D. Guillon, B. Heinrich, *Liq. Cryst.* **1995**, *19*, 301–305.
42. E. J. R. Sudhölter, J. B. F. N. Engberts, W. H. de Jeu, *J. Phys. Chem.* **1982**, *86*, 1908–1913.
43. Y. Haramoto, S. Ujiie, M. Nanasawa, *Liq. Cryst.* **1996**, *21*, 923–925; Y. Haramoto, M. Nanasawa, S. Ujiie, *Liq. Cryst.* **2001**, *28*, 557–560.
44. K-M. Lee, Y-T. Lee, I. J. B. Lin, *J. Mater. Chem.* **2003**, *13*, 1079–1084.
45. D. Haristoy, D. Tsiourvas, *Chem. Mater.* **2003**, *15*, 2079–2083.
46. J. Jaklevic, J. A. Kirby, M. P. Klein, A. S. Robertson, G. S. Brown, P. Eisenberger, *Solid State Commun.* **1977**, *23*, 1679.
47. J. C. Mikkelsen, J. B. Boyce, R. Allen, *Rev. Sci. Instrum.* **1980**, *51*, 388–389.
48. A. L. Ankudinov, B. Ravel, J. J. Rehr, S. D. Conradson, *Phys. Rev. B* **1998**, *58*, 7565–7576; J. J. Rehr, R. C. Albers, *Phys. Rev. B* **1990**, *41*, 8139–8149.
49. A. Filipponi, A. Di Cicco, C. R. Natoli, *Phys. Rev. B* **1995**, *52*, 15122–15134.
50. N. Binsted, EXCURV98: CCLRC Daresbury Laboratory computer program, **1998**.
51. A. Filipponi, *J. Phys.: Condens. Matter* **2001**, *13*, R23–60
52. J. J. Rehr, R. C. Albers, *Rev. Mod. Phys.* **2000**, *72*, 621–654.
53. E. D. Crozier, N. Alberding, B. R. Sundheim, *J. Chem. Phys.* **1983**, *79*, 939–943.
54. A. Di Cicco, M. Minicucci, A. Filipponi, *Phys. Rev. Lett.* **1997**, *78*, 460–463; M. Minicucci, A. Di Cicco, *Phys. Rev. B* **1997**, *56*, 11456–11464.
55. A. Di Cicco, *J. Phys.: Condens. Matter* **1996**, *8*, 9341–9345.
56. L. Hefeng, L. Kunquan, W. Zhonghua, D. Jun, *J. Phys.: Condens. Matter* **1994**, *6*, 3629–3640.
57. A. J. Carmichael, C. Hardacre, J. D. Holbrey, M. Nieuwenhuyzen, K. R. Seddon, *Anal. Chem.* **1999**, *71*, 4572–4574.
58. A. J. Dent, K. R. Seddon, T. Welton, *J. Chem. Soc., Chem. Commun.* **1990**, 315–316.

59. A. J. Dent, A. Lees, R. J. Lewis, T. Welton, *J. Chem. Soc., Dalton Trans.* **1996**, 2787–2792.
60. G. M. N. Baston, A. E. Bradley, T. Gorman, I. Hamblett, C. Hardacre, J. E. Hatter, M. J. F. Healy, B. Hodgson, R. Lewin, K. V. Lovell, G. W. A. Newton, M. Nieuwenhuyzen, W. R. Pitner, D. W. Rooney, D. Sanders, K. R. Seddon, H. E. Simms, R. C. Thied, ACS Symposium Series on *Industrial Applications of Ionic Liquids* **2002**, *818* (Ionic Liquids), 162–177.
61. J. J. Laidler, J. E. Battles, W. E. Miller, J. P. Ackerman, E. L. Carls, *Prog. Nucl. Energy* **1997**, *31*, 131.
62. A. E. Bradley, C. Hardacre, M. Nieuwenhuyzen, W. R. Pitner, D. Sanders, K. R. Seddon, R. C. Thied, *Inorg. Chem.* **2004**, *43*, 2503–2514.
63. A. E. Visser, M. P. Jensen, I. Laszak, K. L. Nash, G. R. Choppin, R. D. Rogers, *Inorg. Chem.* **2003**, *42*, 2197–2199.
64. M. P. Jensen, J. A. Dzielawa, P. Rickert, M. L. Dietz, *J. Am. Chem. Soc.* **2002**, *124*, 10664–10665.
65. M. P. Jensen, J. Neuefeind, J. V. Beitz, S. Skanthakumar, L. Soderholm, *J. Am. Chem. Soc.* **2003**, *125*, 15466–15473.
66. N. A. Hamill, C. Hardacre, S. E. J. McMath, *Green Chem.* **2002**, *4*, 139–142.
67. X.-L. Zhou, S.-H. Chen, *Phys. Rep.* **1995**, *257*, 223–348.
68. U. Englisch, F. Peñacorada, L. Brehmer, U. Pietsch, *Langmuir* **1999**, *15*, 1833–1841.
69. R. K. Thomas, *Annu. Rev. Phys. Chem.* **2004**, *55*, 391–426.
70. M. F. Toney, C. M. Mate, K. A. Leach, D. Pocker, *J. Colloid Interface Sci.* **2000**, *225*, 219–226.
71. A. J. Carmichael, C. Hardacre, J. D. Holbrey, M. Nieuwenhuyzen, K. R. Seddon, *Mol. Phys.* **2001**, *99*, 795–800.
72. J. Bowers, M. C. Vergara-Gutierrez, J. R. P. Webster, *Langmuir* **2004**, *20*, 309–312.
73. T. J. Gannon, G. Law, P. R. Watson, A. J. Carmichael, K. R. Seddon, *Langmuir* **1999**, *15*, 8429–8434; G. Law, P. R. Watson, *Chem. Phys. Lett.* **2001**, *345*, 1–4.
74. G. Law, P. R. Watson, A. J. Carmichael, K. R. Seddon, *Phys. Chem. Chem. Phys.* **2001**, *3*, 2879–2885.
75. M. Tassotto, PhD Thesis, Dept. of Physics, Oregon State University, **2000**.

4.2
Computational Modeling of Ionic Liquids

Patricia A. Hunt, Edward J. Maginn, Ruth M. Lynden-Bell, and Mario G. Del Pópolo

4.2.1
Introduction

The physical characterization and chemical exploration of ionic liquids has progressed significantly over recent years [1–8]. Previously, computational methods have played a relatively minor role in advancing our understanding of ionic liquids. Over the last five years, however, significant developments have been made, new classical potentials have been developed, a wider range of computational tools has been employed, and the number of publications reporting computational results has increased.

This chapter is divided into four key sub-sections. The first is a very brief and basic introduction to the methods used in, and information that can be obtained from, computational studies of ionic liquids. It is primarily aimed at orientating

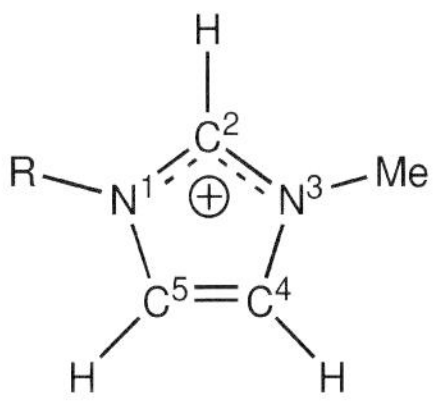

Fig. 4.2-1 1-Alkyl-3-methylimidazolium or [RMIM]$^+$.

those with no background in computational methods. Three sections follow, each concentrating on a particular computational method used in the study ionic liquids: *Ab initio* quantum chemical methods (such as those encompassed by the Hartee–Fock (HF), MP2 and DFT methods), Section 4.2.2; classical molecular dynamics (MD) and Monte Carlo methods, Section 4.2.3; and, finally, *ab initio* molecular dynamics (which are also sometimes referred to as "semi-classical MD" [9] or "first principles MD"), Section 4.2.4. In each case recent advances in both methodology and understanding are presented in more detail.

Many of the ionic liquids that have been examined computationally have been based around the imidazolium cation, Fig. 4.2-1.

Ionic liquids are challenging materials for computational chemists and physicists to study. There are five key problems to be overcome in modeling ionic liquids: (i) the liquid phase, (ii) a high viscosity, (iii) the size of the constituent ions, (iv) the type of interactions that occur, and (v) the electronic structure. Why each of these features is problematic is outlined shortly, after which each of the modern computational methods (quantum chemical, classical MD and MC and *ab initio* MD) is introduced. Each of these methods is capable of dealing with some but not all of the five key problems. For example, quantum chemical methods describe the electronic structure extremely well, but are very expensive and cannot be used to study the large number of molecules (or ions) required to describe the liquid phase. They do however provide valuable insight into the electronic and physical characteristics of the molecules (or ions) that constitute an ionic liquid. Classical MD and MC methods can be used to model a large number of interacting molecules (or ions), and hence portray the liquid state extremely well, however, they are unable to describe the movement of electrons within the system. Thus, in order to obtain a good computational overview of ionic liquids a range of computational methodologies must be employed.

Liquids are difficult to model computationally because the individual molecules (or ions) that make up the liquid are not isolated (as in the gas phase), but are interacting with each other. These interactions are not symmetric and static (as in solids) but are randomized and dynamic. Thus, errors are introduced when attempts are made to extract a "portion" of the liquid for study. This problem can be maneuvered around by taking a large "portion" and by controlling the boundary conditions of the sample. The techniques for extracting a sample and for treating the boundary of a sample are well established within the molecular dynamics methodologies.

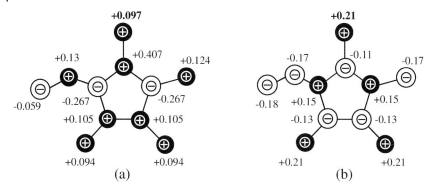

Fig. 4.2-2 Very different charge distributions have been used in classical simulations, (a) Ref. [12] and (b) Ref. [13] for [EMIM]Cl. Hydrogen atoms from the alkyl groups are not shown.

The high viscosity of ionic liquids is also problematic; computational methods that aim to determine thermodynamic properties rely on sampling a large number of "configurations" or snap-shots of the constituent molecules (or ions) in *different* positions or orientations within the liquid [10]. In a viscous liquid this motion is limited and thus it can take a very long time, and be computationally expensive, to build up the required number of "snap-shots" that will produce accurate predictions [11, 12].

Computationally there is always a limit to the number of electrons or atoms that can be modeled. The constituent ions of many ionic liquids, for example imidazolium cations and $[PF_6]^-$ anions, contain a large number of atoms and an even larger number of electrons. Coordinates and velocities have to be calculated for each particle at each step, and thus the size of these systems makes them extremely computationally intensive.

Interactions between the cations and ions of an ionic liquid are complex, they involve not only Coulombic interactions (due to the presence of ions), but Van der Waals interactions (due to the mutual interaction of alkyl groups) and other dipole–dipole derived interactions, all of which can be problematic to model. For example, the aromatic p_π electrons above and below the ring plane of imidazolium cations can interact with the anion's electron cloud, the resulting dispersion forces can be significant and are extremely difficult to describe well. Coulombic interactions in an ionic liquid tend to produce charge ordering, which manifests as an alternation in the sign of the local charge density on moving away from the center of a reference ion. In addition, due to the anisotropy of the ions (see Fig. 4.2-2), alignment of higher multipoles induces orientational correlation in charge densities. This charge correlation attenuates radially away from the charged species, nevertheless the effective forces are very long range and in ionic liquids can extend over 20 Å [12, 14, 15]. To calculate these effects for a liquid made up of ions, especially when there are both charge dense anions and charge diffuse, or polarizable cations (or vice versa) takes significant effort.

Finally, the electronic structure of the constituent ions in ionic liquids is not always easily or well described. Great care must be taken with classical methods because the charge distribution is *input* into the calculation. If the input charge distribution has a large impact on simulation quality, predictions will be erroneous if it is not accurate. More importantly, the magnitude and significance of the errors introduced in this way are not yet well defined for ionic liquids. Quantum chemical methods are used to determine the electronic structure through the electronic wavefunction or density, however the methods for interpretation of these quantities can differ significantly. Take for example the alkyl-substituted imidazolium cation. Different methods have produced very different charge distributions (see Fig. 4.2-2), and thus there is some uncertainty about how to best describe the detailed nature of the cation–anion interactions [13, 16]. This problem has not yet been solved.

The computational effort required to study ionic liquids is extreme. In order to reduce the difficulty of a computation additional approximations can be made. Larger systems can be studied for longer times using classical methods which recover bulk properties at the expense of atomic level electronic properties, while the atomic level electronic properties of smaller systems can be treated accurately by *ab initio* methods at the cost of neglecting bulk properties. Even within each methodology there is a delicate trade off between the complexity of the method and the duration of the calculation. For classical methods this is between the number of atoms and interactions explicitly described versus the duration of the simulation, for example in the choice between using "atomistic" or "united atom" models [12]. For quantum chemical methods, the trade-off is between the amount of electronic correlation recovered versus the time required to obtain convergence in the energy or electron density, for example between using Møller–Plesset (MP2) theory which includes dispersion effects, and density functional theory (DFT) which does not [17].

4.2.1.1 Classical MD and MC

Basic classical molecular dynamics and Monte Carlo methods simulate molecules by fixing the bonding pattern and replacing each atom (or group of atoms) with a potential and a fixed charge. Most current models also include additional terms accounting for bond stretching, angle bending and torsional motion. Interactions between molecules are determined via interactions between these potentials and charges. Some force fields also allow charge centers to move and thus include charge polarization effects. The exact force field used by different groups can vary, and improvements are still being made. Classical calculations are carried out under periodic boundary conditions, a mathematical trick which eliminates the need to describe the distorting effects of a physical boundary; it also has the advantage of reproducing the conditions in the bulk of a material. Charge effects are normally taken care of by Ewald summation [18]. Classical calculations include no quantum mechanics but can be as expensive as *ab initio* methods because many more molecules are simulated (hundreds to thousands rather than tens). To ensure that the model system is in a reliable starting state, it will first be "equilibrated". After equilibration several "production" runs may be carried out. The number of molecules simulated, the duration of both the equilibration and the production runs, and the ability to correctly predict known experimental thermodynamic quantities, are typically used to indicate the quality of a classical calculation [10, 19].

Classical methods are designed to obtain thermodynamic and transport information, for example molar volume, density, viscosity, and surface tension. The effects of pressure and temperature on these properties can also be evaluated, and thus phase transition information such as melting points and glass transition temperatures. If molecular dynamics (in contrast to Monte Carlo) is used, data relating to reorientation of molecules, self-diffusion and residence times are all available. Information can also be obtained from the simulation equations on the contribution made by kinetic, coulombic, intramolecular and dispersion energies to the total potential energy. However, because the charges are fixed and there is no explicit wavefunction included in the classical methods, no electronic information can be obtained.

4.2.1.2 *Ab initio* Quantum Chemical Methods

Ab initio quantum chemical calculations are primarily used to describe the electronic character of individual molecules in the gas phase. Quantum chemical methods can vary widely in their accuracy, depending on the specific approximations taken. Those most applicable to the study of ionic liquids are medium level methods such as DFT and MP2 [20]. Hartree–Fock(HF) level calculations may be carried out as a starting point or to obtain geometries but should be followed by calculations that include some level of electronic correlation. Higher level methods such as Coupled Cluster methods (ie CCSD(T)) are only just accessible, and will not be routine. They do however allow for an estimation of effects hard to recover with the lower level DFT and MP2 methods, such as dispersion, more dynamic correlation, and an estimation of other neglected effects (such as the stabilization afforded by mixing in excited electronic states). The method employed (HF< DFT < MP2 < CCSD(T)) and sophistication of the basis set used are typically used to indicate the quality of an *ab initio* quantum chemical calculation.

Using quantum chemical methods only a small number of molecules can be treated. The isolated system studied by quantum chemical methods cannot recover the liquid environment, the effect of H-bonded extended networks, long-range electrostatic effects, or dynamic effects. However these shortcomings may not be as crucial as originally thought. Recently, it has been suggested that the cation–anion interactions play an important role in determining the solvation structure of ions in ionic liquids [21]. Ionic liquids are also relatively viscous and thus the dynamics is slow, a static model therefore provides a snap-shot of the electronic situation at a given point in time. Quantum chemical methods efficiently and accurately describe short-range interactions and naturally include polarization and charge transfer effects (which are absent in many classical simulations). Moreover, the electronic structure of the ions is well described.

Quantum chemical methods can be used to determine physical properties relating to the electronic wavefunction or density, these include vibrational frequencies and intensities (IR and Raman), dipole moments, polarizabilities, NMR chemical shifts and couplings and estimates for peak positions in electronic spectra [20]. The wavefunction or electronic density can be evaluated directly via population analysis methods that deliver partial charges, bond orders and delocalization parameters. The charges input into many classical simulations are determined using the charge density or an electrostatic potential derived from a quantum chemical calculation. A key advantage of these methods is the insight they give into bonding (for example H-bonding between ions in ionic liquids) and the chemistry of ionic liquids. Energy terms can be evaluated to determine kinetic, electron–electron, and electron–nuclear contributions to the total energy. Bond energies, association energies, pK_as, ionization potentials and electron affinities can also be determined. The energy associated with internal molecular rearrangements, such as torsional rotation in alkyl chains can also be obtained and used to gain chemical insight as well as in the parameterization of force fields.

4.2.1.3 *Ab initio* MD

A group of theoretical methods exists where the electronic wavefuntion is computed, *and* the atomic nuclei are propagated (using classical equations of motion). The Car-Parrinello MD method is one of this type [22–24]. These methods lie between the extremes of the classical and *ab initio* methods, as they include some (quantum) electronic information and some (classical) dynamics information. These methods are called *ab initio* or first principles MD if you come from the classical community and semi-classical MD if you come from the quantum community [9]. *Ab initio* MD methods are far more expensive and cannot simulate as many molecules for as long as the classical simulations, but they are more flexible in that structures are not predetermined and information on the electronic structure is retained. Semi-classical MD can be carried out under periodic boundary conditions and thus the local liquid environment, and any extended bonding network, will be present. These methods hold a great deal of promise for the future study of ionic liquid systems, the first such calculations on ionic liquids were reported in 2005 [21, 25].

4.2.2
Using *Ab Initio* Quantum Chemical Methods to Study Ionic Liquids

4.2.2.1 **Introduction**

Ab initio quantum chemical methods model the detailed electronic interactions within and between molecules. This accuracy is costly and, in general, only a few ions or ion pairs can be computed. There are still relatively few quantum chemical calculations on the gas phase ion pairs that make up ionic liquids. The most accurate methods are too costly to apply to ionic liquid ions, and typically methods such as HF, DFT and MP2 have been employed. A common strategy is to produce a reasonable structure using a hybrid functional DFT method, and then to generate a more accurate wavefunction and energy by performing a single point MP2 computation.

Unlike the HF method, the MP2 method recovers a good portion of the electron correlation in a system. DFT methods are popular because they recover some of the correlation for approximately the cost of a HF computation. Correlation is important for describing aromatic systems such as the imidazolium cation, for accurately recovering dispersion effects, and describing hydrogen bonds well. DFT methods, however, lack a dispersion term, and this is problematic for ionic liquids, as Van der Waals interactions between the alkyl chains of the ions are important. Improvements can be made by the use of coupled cluster or multi-configurational methods, however, these require very large basis sets and are computationally very expensive to perform.

In quantum chemical computations the basis set is very important and, in general, the better the basis set the more accurate the calculation. Cost generally determines the level of basis set achieved. Recent calculations have been performed using basis sets of approximately 6-31+G(d,p) [26–28] or aug-cc-pVDZ [29, 30] quality. Older calculations tend to use slightly less expanded basis sets, for example 6-31G(d). A small number of high accuracy calculations have been carried out and have used better basis sets, such as 6-31++G(d,p)[26–28] and aug-cc-pVTZ [29, 30]. The more accurate quantum chemical calculations will include a zero point energy (ZPE) correction, and a correction for the basis set superposition error (BSSE), typically employing the counterpoise method [31, 32].

Early computational studies of ionic liquids include semi-empirical methods; these have now generally been superseded as advances in computing resources have allowed the routine use of higher level DFT and HF methods. As these methods have not been parameterized for the unusual interactions present in ionic liquids, the level of accuracy is unknown. For example, early attempts to determine the H-bonding in [BMIM]Cl (Fig. 4.2-1) using the AM1 [33] method predicted a very long C^2–H bond (1.56 Å) in favor of a short H–Cl bond (1.40 Å) [34]. The equivalent interaction calculated at the MP2 level predicts the reverse, a strong C^2–H bond (1.11 Å), and weaker H–Cl interaction (2.02 Å).

Early quantum chemical calculations on ionic liquids were focused towards the haloaluminate, and related metal- (Au and Fe) containing melts, these are examined in the following subsection. As the field has developed, this focus has shifted towards imidazolium-based ionic liquids because of their lower melting points and more favorable physical properties. Imidazolium-based ionic liquids are discussed in the third subsection which examines imidazolium cations with small alkyl chains (methyl, ethyl and butyl). The ionic liquids which can be formed from imidazolium cations and small anions such as halides or $[PF_6]^-$ are then discussed, mention is also made of calculations carried out on a few more diverse systems. The electronic structure of the imidazolium-based ionic liquids is the focus of the fourth and final subsection.

4.2.2.2 Acidic Haloaluminate and Related Melts

Early interest in ionic liquids developed from work on low melting point haloaluminate melts, such as $Na[AlCl_4]$. MX–AlX_3 (where M = group 1 cation) melts form $[AlX_4]^-$ or $[Al_2X_7]^-$ depending on the quantity of MX, i.e. the excess of X^- anions in the melt. The earliest computational studies of these species were related to the matrix isolation of $M[AlX_4]^-$ and X = F, Cl, Br [35]. Some of these early investigations are, for their time, of surprisingly good quality. For example, a calculation reported in 1979 on $Li[AlX_4]^-$ was carried out at the HF level using 6-31G on Li and F and (11s7p)/[5s4p] contracted basis set on Al [35]. The semi-empirical MNDO [36] method has been used to investigate the structures and IR spectrum of $[AlCl_4]^-$ and $[Al_2Cl_7]^-$ [37]. By 1992 ab initio HF/6-31G(d) calculations had been performed on $[AlX_4]^-$, $[Al_2X_7]^-$ and $[Al_3X_{10}]^-$ X = Cl or Br, these results were then used in the interpretation of neutron diffraction patterns for the acidic melts. The vibrational spectra of $[AlX_4]^-$ X = F,Cl,Br were also computed [38]. Interest quickly expanded to include the organic cation-based haloaluminates [EMIM]X–AlX_3 which were thought to form $[EMIM][AlX_4]$ or $[EMIM][Al_2X_7]$ in the melt. Semi-empirical methods were used to assist in the interpretation of early vibrational IR spectra, and were referred to in early discussions on the possibility of hydrogen bonding in [EMIM]Cl–$AlCl_3$-based molten salts [34, 37, 39]. A number of configurations for the association of the $[EMIM]^+$ cation and $[AlCl_4]^-$ and $[Al_2Cl_7]^-$ anions were examined [40, 41]. The most stable $[EMIM][AlCl_4]$ structure had one face of the $[AlCl_4]^-$ tetrahedra approaching the imidazolium C^2–H (2.67 A), and three additional H$\cdots$Cl interactions with the first carbon of each alkyl chain (distances ranging from 2.94 to 3.22 A). Mulliken analysis determined the C^2 hydrogen (+0.307e) to be more positive than those at C^4 or C^5 (+0.297e), indicating the higher acidity of this site. The association energy of this structure was calculated to be 284.1 kJ mol^{-1}. The most stable $[BMIM][Al_2Cl_7]$ structure was more complex, five H$\cdots$Cl interactions were identified between four different Cl atoms, the C^2–H$\cdots$Cl interaction was again the strongest (2.81 A). However, the association energy was found to be less than that of $[EMIM][AlCl_4]$, 264.5 kJ mol^{-1} [40, 41].

Chaumont and Wipff have over several years examined the solvation of a number of lanthanide and actinide species in a basic [EMIM]Cl–$AlCl_3$ melt (excess Cl^- anions present), and in $[BMIM][PF_6]$ [42–45]. The ab initio computations were primarily used to assess the accuracy of more detailed classical simulations. For example, MCl_n^{3-n} M = La^{3+}, Eu^{3+} and Yb^{3+} ions [42], EuF_n^{3-n} and $EuF_nCl_{6-n}^{-3}$ [43], UO_2^{2+}, Eu^{3+} and Eu^{2+} cations and their ligated analogs $[EuCl_4(AlCl_4)_2]^{3-}$, $[EuCl_n]^{3-n}$, $[UO_2Cl_2(AlCl_4)_2]^{2-}$, $[UO_2Cl_3(AlCl_4)]^{2-}$ and $[UO_2Cl_4]^{2-}$ [44], and the $[UO_2Cl_2]$, $[UO_2Cl_3]^-$, and $[UO_2Cl_4]^{2-}$ species have all been examined [45]. Calculations were carried out at the HF and B3LYP level with a 6-31+G(d) basis set on the light atoms, Los Alamos relativistic large core pseudo-potentials with associated basis sets on uranium, and a Stuttgart large core pseudo-potential and associated basis set augmented by an additional f-function on the europium [46–50]. Association and ligand exchange energies from ab initio computations and AMBER force field simulations have been extensively and favorably compared. For example, the exchange of one $[AlCl_4]^-$ anion for a Cl^- anion in the gas phase, converting $[UO_2Cl_2(AlCl_4)_2]^{2-}$ into $[UO_2Cl_3(AlCl_4)]^{2-}$ was found to be exothermic by -145.18 kJ mol^{-1} at the B3LYP level, and -194.14 kJ mol^{-1} by classical simulation.

In each case, an extensive, and informative investigation has been carried out on the solvation effects in ionic liquids [44].

Systems related to the haloaluminates by substitution of aluminum for other metals have also been investigated. *Ab initio* gas-phase structures, energies and vibrational frequencies have been calculated for [EMIM]Cl–AuCl$_3$ using Los Alamos pseudo-potentials and associated basis sets on Au and Cl [51, 52]. The [BMIM]Cl–FeCl$_3$ ionic liquids have also been examined [53]. Unrestricted HF and B3LYP calculations have been carried out with all electron basis sets 6-31G(d), Los Alamos pseudo-potentials with associated double zeta basis sets, and CEP-31G (also referred to as SBKJ) pseudo-potentials with associated basis sets [52, 54–56]. *Ab initio* gas-phase structures, energies and vibrational frequencies were calculated for FeCl$_2$, FeCl$_3$, [FeCl$_3$]$^-$, [FeCl$_4$]$^-$, [FeCl$_4$]$^{2-}$ and [Fe$_2$Cl$_7$]$^-$ and used in the analysis of experimental Raman spectra. The UHF/CEP-31G results were found to give a good performance when compared with the all electron UHF/6-31G(d) results [53].

4.2.2.3 Alkyl Imidazolium-based Ionic Liquids
Isolated Imidazolium Cations
Studies of the isolated imidazolium cation have been undertaken primarily to produce parameters for use in force field methods. Shah et al., have calculated the [BMIM]$^+$ cation at the HF/6-31G(d) level, and determined CHelpG [57] charges [58]. De Andrade et al., have calculated the [EMIM]$^+$ and [BMIM] cations at the HF/6-31G(d) level, and determined the RESP charges using AMBER [59]. They also compared the *ab initio* dipole moments and vibrational frequencies with those obtained using force field methods [60].

Several classical simulations have indicated that the shorter alkyl chains (C$_n$H$_{2n+1}$, $n < 5$) are freely rotating at room temperatures [16, 61]. However the torsional parameters required to model these rotations are missing from the AMBER package, and recently *ab initio* calculations have been used to map out the rotational potential energy surfaces in some detail. Liu et al. have studied [RMIM]$^+$ (R = methyl, ethyl) at the MP2/6-31+G(d)//HF/6-31+G(d) level and Lopes et al. have studied the [RMIM]$^+$ (R = ethyl, propyl, butyl) cations using HF/6-31G(d) geometries and single point energy computations at the frozen core MP2 level using a cc-pVTZ basis set with all f functions removed [13, 62]. Recently, the rotation has also been examined at the B3LYP level [17].

The vibrational spectrum of [BMIM]$^+$ has been determined at the HF/6-31+G(d) level [62], and at the B3LYP and MP2 levels and examined for evidence of hydrogen bonding with a Cl$^-$ anion [17].

Imidazolium Cations And Halide Anions
Recently the [RMIM]X (X = Cl, Br, I and R = methyl, ethyl, propyl, butyl) ionic liquids have been investigated in some detail. A crucial problem when studying loosely bound ion pairs is the large number of possible conformations. Two key structural features are dominant, they involve the position of the halide around the imidazolium ring, Fig. 4.2-3(a), and rotation of the alkyl chain, Fig. 4.2-3(b). In both cases

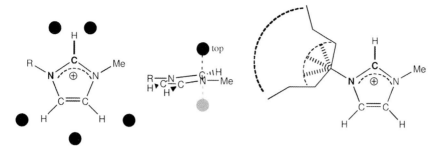

Fig. 4.2-3 (a) Positions of local energy minima for a Cl⁻ anion around an imidazolium cation, (b) volume excluded by alkyl chain rotation.

more than one orientation is stable. Initial studies ignored this problem completely computing only a single ion pair, normally with the halide hydrogen bonding to the most acidic hydrogen atom, that at C^2. However more recent investigations have applied various strategies to determine a number of minima [25,63,64]. The halide anions can take up a number of positions around the imidazolium. Studies have now shown that there are seven primary positions; locations to the "front" of the ring lying between the C^2–H and the alkyl chain or the methyl group, to the "side" of the ring lying between the alkyl chain and C^5–H, or the methyl group and C^4–H and at the "back" of the ring lying between the C^4–H and C^5–H. Additionally the halide can lie above ("top" structure) or below the ring. Depending on the orientation of the alkyl chain the "top" and "bottom" structures may not be degenerate.

Interaction energies have been computed for a range of conformers, $E(\text{interaction}) = E(\text{ion-pair}) - [E(\text{cation}) + E(\text{anion})]$. Chaumont et al., have computed gas-phase interaction energies using the AMBER [59] force field [44]. For [BMIM]Cl with the Cl hydrogen bonding to the C^2–H (with unspecified butyl orientation) or $C^{4/5}$–H, -363.59 and -353.13 kJ mol^{-1}, have been obtained, respectively [44]. This compares to -379.09 kJ mol^{-1} for C^2–H, -344.40 for C^5–H and -341.69 for C^4–H association at the B3LYP/6-31++G(d,p) level [17]. A similar pattern, showing a stronger interaction with C^2–H compared to $C^{4/5}$–H has also been found for [BMIM]Cl, for example, -397.83, -345.58 and -346.54 kJ mol^{-1}, respectively, at the B3LYP/6-31++G(d,p) level [65].

Several groups have recently reported studies examining the relative stability of halide ions in various positions around an imidazolium cation. Turner et al. in 2003 were the first to make a more detailed *ab initio* investigation of these compounds. They attempted a systematic analysis of small [RMIM]$^+$ cations (R = methyl, ethyl, butyl) with a series of halides spanning F$^-$ through to I$^-$ [63]. They placed F$^-$ anions in a grid around each cation and optimized at the HF/STO-3G level. This was followed by a series of calculations of increasing sophistication where each higher level calculation started from a lower level pre-optimized structure, culminating at the MP2/6-31+G(d) level. The calculation of the heavier halide ion pairs was initiated by taking the optimized [RMIM]F structure and replacing the F$^-$ anion with the relevant halide. Relativistic effects, which are expected to be important, were not considered for the I$^-$ and Br$^-$ anion. Hunt and Gould have studied the [BMIM]Cl ion pair in greater detail, locating additional stable structures. They started by placing the Cl$^-$ anion around the cation in "chemically intuitive" positions, structures were pre-optimized at the HF/3-21G level and used to initiate B3LYP/6-31++G(d,p) and then MP2/6-31++G(d,p) level optimizations [64]. However, in some cases, previously reported structures were optimized directly at the B3LYP/6-31++G(d,p) level. Wang et al. followed a similar procedure using the [EMIM]$^+$ and [BMIM]$^+$ cations and Cl$^-$ and Br$^-$ anions, optimizing structures at the B3LYP/6-31G(d) level and carrying out single point energy computations on the optimized structures at the B3LYP/6-31G++(d,p), B3LYP/6-311G+(d,p) and B3LYP/6-311G++(d,p) levels [65]. Bühl et al. have also considered a number of structures for the [MMIM]Cl ion pair using a range of DFT functionals (BP86 [66–68], BLYP [66,69], BPE [70,71] and B3LYP [69,72]) and the MP2 method with a 6-31+G(d,p) basis set. Kossman et al. have considered rings and chains formed from [MMIM]Cl units at the BP86, B3LYP and MP2 level using primarily Ahlrich's TZVP basis set, but selected computations were also performed using Dunning aug-cc-pVDZ and aug-cc-pVTZ basis sets, and Ahlrich's TZVPP basis set [30,73,74].

Turner et al. found that fluorine prefers to covalently bond to the imidazolium at C^2, C^4 and C^5 or to remove a hydrogen atom from these positions and form HF [63]. While the structures with the Cl$^-$ anion associated at the back or to the side of the imidazolium ring are found to be $\approx$30 to 60 kJ mol^{-1} higher in energy than the most stable conformer, the relative energy ordering of the "top" and "front" structures is particularly sensitive to the method of computation [17, 25, 63]. The top structure of [BMIM]Cl is found to be slightly more stable (<1 kJ mol^{-1}) than the front conformer at the CCSD(T)/aug-cc-pVDZ level [17]. DFT methods predict the front conformer to be more stable by $\approx$5 kJ mol^{-1} [17, 25]. In general, lower level methods tend to favor a front over a top conformation [17, 25, 63]. Dispersion effects are very important for the "top" conformer, and so as the sophistication of a calculation increases the "top" conformer becomes increasingly more stable. The energy ordering of the conformers is also affected by the orientation of the alkyl chain in [BMIM]Cl [17]. For the [PMIM]X and [BMIM]X structures the Br$^-$ and I$^-$ anions were most stable lying above the ring [63]. A number of workers have found that optimizing structures at a low level of theory generates minima that do not exist, and misses minima that do exist at higher levels of theory [17, 63, 65]. Overall optimizations carried out at the HF/6-31G(d) level followed by single point calculations at the MP2/6-31++G(d,p) level appear to offer a viable low cost means of obtaining good structures and energies.

Conformational changes within the alkyl chain attached to the imidazolium have also been examined. Two crystal polymorphs of the [BMIM]$^+$ cation can be distinguished by the different orientation of the alkyl chains (and hydrogen-bonding interactions) [75, 76]. Turner et al. found that rotation about C–C bonds of the alkyl chains cost very little energy (for example 0.007 kJ mol^{-1} per 10° rotation in [EMIM]$^+$), except where steric interactions with hydrogen atoms of the imidazolium chain dominated. The smallest frequencies, typically <50 cm^{-1} were associated with alkyl rotation [63]. More recently, the effect of the halide on butyl chain rotational barriers in the [BMIM]Cl ion pair has been investigated by Hunt and Gould [17] Multiple local minima were identified with different torsional angles of the butyl chain; generally these lie within 10 kJ mol^{-1} of each other. The intervening barriers have been found to arise from repulsion between hydrogen atoms on the imidazolium ring and the butyl chain, and steric hindrance with respect to the "substituents" on carbon atoms within the alkyl chain, the substituents being the imidazolium ring on the one hand and the end group of the butyl chain on the other. The presence of the Cl$^-$ anion was found to deepen and flatten the torsional potential energy wells. Moreover, hydrogen atoms within the alkyl chains have been found to participate in H-bonding with the anions. In the butyl structures, weak longer range forces may facilitate bending of the alkyl chain toward the anion. The presence of the anion therefore enhances low energy rotational movement in the alkyl chains (when they are not involved in hydrogen bonding to the anion) [17].

Imidazolium Cations and [PF$_6$]$^-$ Anions

Meng et al. in 2002 used semi-empirical methods to investigate the ionic liquid [BMIM][PF$_6$] [77]. The anion was placed in various positions about the butyl chain, above and below the imidazolium ring, and near the three hydrogen atoms of the imidazolium ring, seven stable ion pairs were identified. A similar process identified only one stable conformer at the HF and DFT levels. The isolated anions and single ion pair were then calculated with a range of basis sets, HF/3-21G(d), HF/6-31G(d), HF/6-31G(d,p), B3LYP/6-31G(d) and B3LYP/6-31G(d,p). Electron correlation for the ion pair was further examined by single point energy calculations at the MP2/6-31G(d) level. It was observed that the "tightness" of the ion pair increased with the complexity of the basis set [77].

The shape of the potential energy surface with respect to proton transfer (i.e. hydrogen bonding) was examined by stepping along proton transfer coordinates and carrying out single point energy calculations with improved basis sets, for example, MP2/6-31+G(d,p)//MP2/6-31G(d,p). The potential curve was found to be highly anharmonic and qualitatively similar for all methods, improving the basis set only lowered the total energy [77]. HF optimized geometries and single point MP2 energy calculations at the B3LYP level with a 6-31G(d,p) basis set were also reported for [BMIM][PF$_6$] and 1-methyl-3-nonylimidazolium hexafluorophosphate at about this time [78].

Talaty et al. in 2004 examined the vibrational spectra of ion pairs composed of [RMIM]$^+$ (R = ethyl, propyl, butyl) cations and [PF$_6$]$^-$ anion both experimentally and at the B3LYP/6-311+G(2d,p) level [79]. The report considers only one type of conformer for each cation, that in which the [PF$_6$]$^-$ anion lies in a "front" position. For each ion pair, six ([EMIM]$^+$) or seven ([PMIM]$^+$ and [BMIM]$^+$) C–H$\cdots$F interactions were found shorter than the Van der Waals distance (2.67 A), in each case three of these interactions involved the C^2–H [79]. Known and systematic errors in the vibrational spectra of the HF, DFT and MP2 methods are sometimes corrected by applying a scale factor, these are obtained by fitting the computed and experimental spectra. A scaling factor of 0.964 has been determined for B3LYP frequencies of [BMIM][PF$_6$], and 0.965 for [PMIM][PF$_6$] and [EMIM][PF$_6$] [79].

Chaumont et al. have computed the association energy of [BMIM][PF$_6$] at the HF/6-31G(d,p) level and determine a BSSE corrected value of -318.82 kJ mol^{-1} [80]. Chaumont et al. have also examined, using primarily classical methods a "humid" [BMIM][PF$_6$] ionic liquid. Nevertheless, a small number of *ab initio* calculations were performed [45]. The association energy of the dimers [PF$_6$]$^-$···H$_2$O and [BMIM]···OH$_2$ (interacting with C$_2$–H) at the B3LYP/6-31+G(d) level was found to be similar, 39.75 kJ mol^{-1}. Simulations, however showed that the preferred order of association is [PF$_6$]$^-$ > [BMIM]$^+$ > H$_2$O indicating that entropic effects dominate in solution [45].

Imidazolium Cations and Other Anions
Gozzo et al. in 2004 reported on calculations carried out at the B3LYP/6-311+G(d,p) level on ionic liquids composed of the [BMIM]$^+$ cation and the anions, [CF$_3$CO$_2$]$^-$, [BF$_4$]$^-$, [PF$_6$]$^-$ and [BPh$_4$]$^-$, comparing the hydrogen-bonding ability of the anions [81]. Only the C^2–H coordinated isomers, with hydrogen bond distances of 1.95, 1.88, 2.01 and 2.43 Å, respectively, were reported. The H-bonding strength of this series of anions was determined using both computational and experimental data; [CF$_3$CO$_2$]$^-$ > [BF$_4$]$^-$ > [PF$_6$]$^-$ > [BPh$_4$]$^-$, indicating that C^2–H...anion distances do not correlate well with interaction strength. Two mixed anion [CF$_3$CO$_2$]$^-$...[BMIM]$^+$...[BF$_4$]$^-$ clusters were examined computationally, however only the conformers with an anion associating with C^2–H and lying between C$^{4/5}$–H were considered. The cluster with [CF$_3$CO$_2$]$^-$, coordinated to the C^2–H was found to be the most stable.

4.2.2.4 The Electronic Structure of Ionic Liquids

The electronic structure of the imidazolium cations is of interest because it impacts on the hydrogen bond acceptor and donor properties of ionic liquids. This in turn relates to the penchant of the solvent to coordinate to, or react with the solvated species. The imidazolium cation is isoelectronic with the "carbene"-like imidazole-2-ylidene. Theoretical calculations on deprotonation of the unsubstituted imidazolium cations determine pK_as of 24.90 and 32.97 for the proton at the C^2 and C$^{4/5}$ positions respectively [82]. The electronic structure of the [BMIM]$^+$ cation is probably best represented by a double bond between C^4–C^5 and a delocalized 3c–4e contribution across N^1–C^2–N^3, however there is also extensive delocalization around the entire ring (Fig. 4.2-1) [64].

Until recently the primary interest in the electronic structure calculations has been to derive partial charges on the imidazolium cation for use in classical simulations [11, 13, 14, 16, 62, 83]. The charge distribution however depends strongly on the method of analysis. The Natural Atomic Orbital (NAO) [84–88] and Mulliken population analysis methods have been applied to the imidazolium cations [16, 64]. A number of methods have been used to determine partial charges from the electrostatic potential; these also show significant variation. There is an especially large discrepancy in the nitrogen atomic charges, which range from ≈-0.4e [12, 16, 89] through essentially neutral [42, 60, 90] to $\approx+0.15$e [11, 13]. The charge on C^2 ranges from -0.11e through essentially neutral to $+0.6$e, and there is significant variation among the charges for the hydrogen atoms at C^2–H, C^4–H and C^5–H. However, the distributed multipole analysis (DMA) [91] used by Hanke et al. is found to be qualitatively consistent with the NAO analysis [12, 64, 89].

Hunt et al. have carried out an analysis of the electronic structure of the [BMIM]Cl ion pair [64]. Natural bond orbital analysis, electron density maps, localized and delocalized molecular orbitals, and the electrostatic potential were used to build up a picture of the electronic structure. Most of the positive charge on the imidazolium cation is carried by the peripheral hydrogen atoms, while all of the carbon atoms, except C^2 which is substantially positive, carry a negative charge. The formally positive nitrogen atoms also carry a significant negative charge.

The Cl$^-$ anion is thus primarily attracted to the positive C^2–H unit, which sits exposed in an area depleted of electron density. The association of the Cl$^-$ anion with the cation is driven by strong Coulombic forces, which ameliorate destabilizing electronic effects; the covalent part of the hydrogen bond involves high energy C^{ring}–H σ^*-orbitals, and adding electrons to an essentially antibonding π-type LUMO. The good hydrogen-bond donor properties of [BMIM]Cl are related to the positive charge on the *entire* C^2–H unit, once this has been eliminated by forming a hydrogen bond, the good hydrogen-bond donor properties of the ionic liquid decrease, as is observed experimentally. The significantly smaller charge on $C^{4/5}$–H units make these sites less attractive, as does repulsion from the local build up in π electron density (due to the double bond between C^4 and C^5). However, these same features also make the imidazolium a better electron donor when coordinated through C^4 and C^5 to a transition metal, as has been observed in recent experiments [92–94]. Interactions governing the "top" conformers are very different from the "in-plane" conformers; here the Cl$^-$ anion interacts with the ring π-electron density.

Solvatochromic experiments have shown [BMIM]$^+$-based ionic liquids to be good hydrogen-bond donors [95], however, complementary GC-based experiments have shown these *same* ionic liquids to be poor hydrogen-bond donors [96]. This discrepancy has been resolved, and traced back to the nature of the probe molecules. The dye used in the solvatochromic experiments is zwitterionic while the probe molecules used in the GC experiment are neutral. Less effective hydrogen bonds are formed with the neutral as opposed to the highly polarized probe molecules. The formation of a hydrogen bond is essentially destabilizing and must be compensated for by the ionic attraction between charged species, without this stabilizing influence only weak hydrogen bonds can form [64].

The unreactive but coordinating environment, and high electrochemical window, of the ionic liquid have been related to the antibonding character of the cation LUMO and closed shell nature of the Cl$^-$ anion. The ability of the ion pair to accept electrons is governed by the system LUMO which is a π^* orbital on the cation and will be destabilized by further electron acceptance, hence the ionic liquid will be unreactive and require a strong external potential before electrons will be accepted. The ability of the ion pair to donate electrons is governed by the HOMO which is the anion HOMO. The Cl$^-$ anion, however, is a closed shell species and unwilling to give up electrons, hence the ionic liquid will be unreactive and will require a strong external potential before electrons can be withdrawn [64].

4.2.3
Atomistic Simulations of Liquids

Atomistic-level simulation of the condensed phase was one of the earliest applications of the digital computer. Metropolis and co-workers at Los Alamos [97] used a stochastic method, "Monte Carlo", to compute the properties of a model fluid in 1953. Alder and Wainwright at Livermore [98, 99], Rahman at Argonne [100] and Verlet at Yeshiva University [101] were among the pioneers in the 1950s and 1960s in the use of a deterministic method called "molecular dynamics" to compute the properties of simple liquids. Most of these early studies were concerned with computing the properties of either idealized "hard" fluids or more realistic but still simple liquids such as argon. A major objective of this work was to compare the results of these computer "experiments" with predictions made from emergent liquid state theories. Since that time, advances in algorithms and techniques, coupled with an explosive growth in computer power, have led to the widespread use of atomistic simulations in the study of condensed phases. It is now common for molecular dynamics and Monte Carlo to be applied to complex systems such as proteins, surfactants, organic mixtures, electrolytes and polymers. While memory and processor speed limited the early simulations to a few hundred atoms, it is now commonplace to see simulations containing millions of atoms. In fact, using 64 000 processors on the new Blue Gene/Light computer, researchers at the Lawrence Livermore National Laboratory [102] have carried out a simulation of 4×10^{10} atoms! As important, the sophistication with which intermolecular interactions can be modeled has improved greatly, and the timescales over which molecular motions can be studied has increased substantially.

The basic idea behind an atomistic-level simulation is quite simple. Given an accurate description of the energetic interactions between a collection of atoms and a set of initial atomic coordinates (and in some cases, velocities), the positions (velocities) of these atoms are advanced subject to a set of thermodynamic constraints. If the positions are advanced stochastically, we call the simulation method *Monte Carlo* or MC [10]. No velocities are required for this technique. If the positions and velocities are advanced deterministically, we call the method *molecular dynamics* or MD [10]. Other methods exist which are part stochastic and part deterministic, but we need not concern ourselves with these details here. The important point is that statistical mechanics tells us that the collection of atomic positions that are obtained from such a simulation, subject to certain conditions, is enough to enable all of the thermophysical properties of the system to be determined. If the velocities are also available (as in an MD simulation), then time-dependent properties may also be computed. If done properly, the numerical method that generates the trajectories provides an *exact* solution for the model system. The degree to which the simulation results agree with experimental data for the "real" system tells us how good the model is. A great deal of attention must therefore be given to the model used to represent the atomic species and their energetic interactions. In the next section, a summary of the types of models used to simulate ionic liquids is provided.

4.2.3.1 Atomistic Potential Models for Ionic Liquid Simulations

From the above discussion, it is apparent that the more accurate the geometric and energetic model one uses in a simulation, the better agreement one should expect between simulated properties and experimental data. The most accurate way of treating the interactions between atoms is through *ab initio* electronic structure calculations. As described elsewhere in this chapter, *ab initio* methods such as Hartree–Fock, density functional theory, Møller–Plesset perturbation theory and coupled cluster approaches are all predictive methods that provide varying levels of accuracy in determining molecular structure and energetics. A major drawback of these approaches, however, is computational cost. The most accurate *ab initio* methods are – even today – too computationally demanding for use with systems bigger than a few ionic liquid ion pairs. This means these techniques cannot be used directly in condensed phase simulations, in which the energetics of hundreds of ion pairs must be computed thousands of times to properly simulate liquid properties. The *ab initio* methods *can* be used in developing approximate classical expressions for the interactions present in these systems. The way in which this is done is the subject of this section. We note in passing that recent work on so-called *ab initio* molecular dynamics such as Car-Parinello MD [22–24] is blurring the distinction between purely *ab initio* methods and the "classical" simulation methods described below. In fact, the first *ab initio* MD simulation study of an ionic liquid was reported recently [21].

The First "Ionic Liquid" Simulations: Alkali Halides

We will use the shorthand term "force field" to mean a set of analytic equations that approximate the energetic interactions among atoms in a system. These interactions take many forms, from strong forces characteristic of covalent bonds to weaker forces representative of Van der Waals interactions. The origins of the current force fields used to simulate room temperature ionic liquids can be found in earlier works on alkali halides. The most widely used functional form is that due to the work of Huggins and Mayer [103]

$$U_{ij}(r_{ij}) = \frac{q_i q_j}{r_{ij}} + B_{ij} \exp(-\alpha_{ij} r_{ij}) - \frac{C_{ij}}{r_{ij}^6} - \frac{D_{ij}}{r_{ij}^8} \qquad (4.2\text{-}1)$$

where i and j can be either a cation or anion and U_{ij} is the potential energy on species i due to species j. It is a pairwise additive potential function containing a Coulombic term where formal charges of $q_i = \pm 1$ are assumed and are fixed at the ion centers, a short-range repulsive term of the Born–Mayer (exponential) form, and attractive Van der Waals terms that represent dipole–dipole and dipole–quadrupole

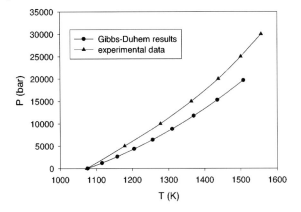

Fig. 4.2-4 Computed [105] solid–liquid coexistence curve for NaCl (Gibbs–Duhem) versus experimental data [111]. Although agreement is good at atmospheric pressure, the slope of the coexistence curve is off and thus the melting point is severely over-estimated at high pressure.

interactions. The Van der Waals terms were parameterized by Mayer [104]. Tosi and Fumi [105] developed parameters for the repulsive part of this function to reproduce the solid phase properties of alkali halides having the rock salt structure. Many authors subsequently used this parameterization (or variations of it) to simulate a wide range of alkali halides in both the solid and molten states. Sangster and Dixon [106] have reviewed much of the early work in this field. The Tosi–Fumi potential still enjoys wide use today. It is generally thought that static properties are reasonably captured with this type of "rigid ion" model, but its ability to accurately model dynamic properties has been questioned due to its neglect of ion-induced polarization. Recent work by Galamba and coworkers [107–109] suggests that shear viscosities and thermal conductivities of molten NaCl and KCl are over-predicted by anywhere from 10–20% with this force field. Even when static properties are captured well at one state point, the Tosi–Fumi force field can yield results that do not agree with experiment at other state points. For example, Fig. 4.2-4 shows the results of melting temperature calculations for NaCl as a function of pressure [110] compared against experimental data. Although the Tosi–Fumi potential reproduces the atmospheric pressure melting point quantitatively, it predicts a significantly higher melting temperature as pressure (and hence density) increases. This is likely to be due to the fact that the model assumes pair-wise interactions, but three-body and higher terms become important as density increases.

What can someone interested in room-temperature ionic liquids learn from the results of these alkali halide simulations? First, despite over thirty years of study and the fact that alkali halide salts have a simpler structure than ionic liquids it is still difficult to quantitatively predict every thermophysical property of these systems with classical simulations, particularly if one insists on using the same force field for every property. Some properties, such as lattice energies and volumetric properties, are modeled very accurately while others, particularly time-dependent properties, are more difficult to predict with simulations. Of course, this is due to the approximate way in which energetics are treated via classical force fields such as Eq. (4.2-1). We should expect similar results for ionic liquid simulations, and that greater accuracy will come with improved treatment of inter- and intramolecular energetics. Second, the simulations have yielded a tremendous amount of quantitative and qualitative information, which has enabled us to understand molten salt systems much better than would have been possible using only experiment or theory. We know that the Coulombic forces lead to order over a much longer length scale than is present in simple liquids [112] and that these forces can lead to the appearance of small voids having lifetimes on the order of ps [112,113]. This is in contrast to normal molecular liquids, where the steeply repulsive part of the potential dominates fluid structure and leads to a more close-packed structure.

Ionic liquid Simulations

Simulating the properties of ionic liquids gives rise to additional challenges not present when simulating molten alkali halides. First, the cations and anions are no longer simple spheres, but are instead multi-atom molecular species. This means that, unless one approximates the ions as rigid, the functional form of Eq. (4.2-1) is inadequate for treating these systems; additional intramolecular terms need to be developed. Second, the large size of the ions means that dynamics are likely to be slow and thus computational costs high. Finally, the general lack of good quality thermophysical data on ionic liquid systems means that it will be difficult to benchmark calculations against experimental data. Despite all this, a tremendous amount of progress has been made in the last few years developing accurate and predictive classical force fields for the simulation of ionic liquids.

All of the classical force fields used in the study of ionic liquids contain elements of the Tosi–Fumi potential model. Like Eq. (4.2-1), these force fields are all pairwise additive, meaning that the energy and force on an atom can be found by summing all its interactions with its neighbors. In addition, the force fields all contain a Coulomb term that accounts for electrostatic interactions, a short-range repulsive term and one or more long-range attractive (dispersion) terms.

Lynden-Bell and coworkers [12] were the first to propose a force field for the class of compounds that can be considered ionic liquids. The form of the potential function they used is

$$U_{ij}(r_{ij}) = \frac{q_i q_j}{r_{ij}} + (A_{ii} A_{jj})^{1/2} \exp(-(B_{ii} + B_{jj})r_{ij}/2) - \frac{(C_{ii} C_{jj})^{1/2}}{r_{ij}^6} \quad (4.2-2)$$

which has the same form as Eq. (4.2-1) but the dipole–quadrupole terms are omitted. The species simulated in this work were dimethylimidazolium chloride ([MMIM][Cl]), 1-ethyl-3-methylimidazolium chloride ([EMIM][Cl]), dimethylimidazolium hexafluorophosphate ([MMIM][PF_6]) and 1-ethyl-3-methylimidazolium hexafluorophosphate ([EMIM][PF_6]). Bond lengths were kept fixed, as were all bond

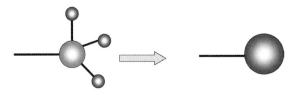

Fig. 4.2-5 The approximation made in converting a methyl group into a single united atom interaction site.

angles except those formed between the N–C–H atoms in the methyl groups. Partial charges q_i were located at each atomic center. The values of these partial charges were determined from *ab initio* calculations of individual cations and anions [12]. Literature values were used for the repulsion–dispersion parameters (A, B, and C).

A common approximation used in liquid phase simulations is to lump carbon atoms and the hydrogen atoms bonded to it into a "united atom", as shown in Fig. 4.2-5. This greatly reduces the number of pairwise interactions that must be computed, but at the sacrifice of some chemical reality.

Lynden-Bell and coworkers examined the validity of this approximation by simulating systems with fully explicit methyl groups (i.e. all hydrogens) as well as united atom models. They found that the united atom approximation leads to a higher density in both the liquid and crystalline states. As a consequence, the dynamics of the united atom system are somewhat slower than for the case of the explicit atom model.

Soon after this initial study, several other groups developed force fields for imidazolium-based ionic liquids. Stassen and coworkers [83] reported a force field for [EMIM][AlCl$_4$] based on the so-called "AMBER" functional form. This functional form is widely used in the simulation of organics and is given by

$$U_{\text{tot}} = \sum_{\text{bonds}} k_b (r - r_0)^2 + \sum_{\text{angles}} k_\theta (\theta - \theta_0)^2 + \sum_{\text{dihedrals}} \frac{V_n}{2} (1 + \cos[n\phi - \gamma])$$
$$+ \sum_{i=1}^{N-1} \sum_{j>1}^{N} \left\{ 4\varepsilon_{ij} \left[\left(\frac{\sigma_{ij}}{r_{ij}}\right)^{12} - \left(\frac{\sigma_{ij}}{r_{ij}}\right)^{6} \right] + \frac{q_i q_j}{r_{ij}} \right\} \quad (4.2\text{-}3)$$

where now we have explicitly written sums over all atom pairs to obtain an expression for the total potential energy U_{tot}. Harmonic terms represent bond stretching and bond angle bending, and a cosine series represents the potential energy of dihedral angle rotation. Repulsion–dispersion interactions are represented with a Lennard-Jones 12-6 type potential, with ε_{ii} representing the depth of the attractive potential for atomic site i, and σ_{ii} representing the collision diameter of atomic site i. Interactions between atoms of a different type are modeled using the Lorentz–Berthelot combining rule

$$\sigma_{ij} = (\sigma_{ii} + \sigma_{jj})/2$$
$$\varepsilon_{ij} = (\varepsilon_{ii}\varepsilon_{jj})^{1/2} \quad (4.2\text{-}4)$$

With the exception of the partial charges, all parameters for the cation were taken directly from the Amber force field [114] using protonated histidine as a surrogate. Anion parameters were taken from other literature force fields [115, 116] and missing intramolecular terms were developed using literature methods [115]. Partial charges on each atom center were determined by carrying out an *ab initio* calculation on a single cation and anion, each of which was forced to have a formal charge of +1 and −1, respectively. Values for the partial charges were then adjusted until the resulting electrostatic potential matched as closely as possible that obtained from the *ab initio* calculations. This group carried out MD simulations for over 100 ps to obtain liquid densities, enthalpies of vaporization and atom–atom pair distribution functions.

At about the same time, Maginn and coworkers published a united atom force field for 1-n-butyl-3-methylimidazolium hexafluorophosphate ([BMIM][PF_6]) [90]. At the time, this was one of the few ionic liquids for which experimental thermophysical properties were available, and so the ability of the force field to reproduce experimental quantities could be assessed. The force field had essentially the same form as that in Eq. (4.2-3). Parameters were developed in a manner similar to that described earlier. The cation and anion were simulated in the gas phase using an *ab initio* method. Bond lengths and angles were held fixed at the values of the minimum energy conformation obtained from the *ab initio* calculations. Partial charges on each atomic center were derived by fitting the first principles electrostatic potential at points selected using the CHELPG procedure [117]. Lennard-Jones and dihedral angle parameters were taken from the "optimized potentials for liquid simulation" (OPLS) force field [118] using analogous compounds as surrogates for the cation. Lennard-Jones parameters for the "united atom" [PF_6]$^-$ were derived via scaling arguments, using a widely-used united atom model for SF_6 as a reference. They used a geometric mean combining rule for unlike parameters. Using an isothermal–isobaric Monte Carlo procedure, various volumetric and structural properties were computed and compared to experimental data. Consistent with the finding of Lynden-Bell and coworkers [12], predicted densities were about 5% higher than experiment, although the trends in density as a function of pressure and temperature change were captured reasonably well.

Force fields for [BMIM][PF_6] that explicitly treat all hydrogens (all-atom models) were developed soon after this by Margulis et al. [14], and Morrow and Maginn [11], while Stassen and coworkers [83] published a force field for the [EMIM]$^+$ and [BMIM]$^+$ cations paired with tetrachloroaluminate and tetrafluoroborate anions. The force fields all have similar functional forms, and parameters were again mainly developed using literature force field parameters for similar compounds and *ab initio* calculations of single ions or ion pairs. In these and later studies, repulsion–dispersion parameters were generally adapted from those available from one of three popular force field databases (Amber [114], OPLS [118] and CHARMM [119]). For [BMIM][PF_6], the added realism of the all-atom model enabled densities to be predicted within 1% of the experimental value [11]. The first indications of restricted dynamics in these systems were also observed [11, 14, 15].

Canongia Lopes and coworkers [13] provided a critical evaluation of the various force fields for imidazolium-based ionic liquids, and proposed a refined force field of the form given in Eq. (4.2-3) for the 1-n-alkyl-3-methylimidazolium cation. Importantly, this force field was designed to be extended to alkyl groups of any length. They also determined parameters for rigid $[PF_6]^-$ and $[NO_3]^-$ anions. Many intramolecular cation parameters were taken directly from the AMBER/OPLS database unless these values were unavailable, in which case new parameters were derived which reproduce the *ab initio* geometry, whereas stretching and bending force constants were taken from the databases for similar heterocyclic compounds. All C–H bonds were treated as rigid. Several new dihedral angle rotation parameters were derived by matching rotational energy profiles to *ab initio* calculations. Partial atomic charges were fitted to electrostatic surface potentials using the CHELPG procedure. The *ab initio* calculations were carried out using a high level of theory (frozen core MP2 with a cc-pVTZ(-f) basis set) and a comparison of the different partial charges obtained in previous works was given [13]. Other extensions and refinements have been made to the imidazolium force fields, including those due to Cadena et al. [120], Margulis [61], Urahata and Ribeiro [16], and Liu et al. [62]. Simulations of imidazolium halide systems have been carried out using these cation force fields along with standard halide anion force fields. Canongia Lopes and Pádua [121] have recently developed a force field for the triflate ($[Tf]^-$) and bis(trifluoromethanesulfonyl)amide ($[Tf_2N]^-$) anions, paying particular attention to the dihedral angle rotation of the $[Tf_2N]^-$ anion. Lee et al. [122] recently developed another force field for the [BMIM] cation, following essentially the same procedures as that outlined above. They also proposed a force field for various fluorinated anions, including $[CF_3COO]^-$, $[C_3F_7COO]^-$, $[CF_3SO_3]^-$ and $[C_4F_9SO_3]^-$.

We note that there has been very recent work devoted to the development of force fields for other types of ionic liquid cations including pyridinium [123] and triazolium [124]. Moreover, an electronically polarizable model has been developed and applied to the simulation of [EMIM][NO_3] [125]. Results from these studies will be discussed in the next section.

Although all of the force fields described above have essentially the same functional form, each has a unique set of parameters that can vary significantly, depending on the assumptions made by the different researchers. Not surprisingly, these force fields often yield different results for properties in the liquid and crystalline phase, although the differences are not as great as might be expected. Some aspects of ionic liquids, such as cation–anion ordering, are observed by all the different studies, regardless of the parameters used in the force fields. In the next section, we review some of the properties that have been computed for pure ionic liquids, assess the strengths and weaknesses of the various force fields and simulation approaches, and discuss some of the insights that these simulations have given us into the behavior of ionic liquids.

4.2.3.2 Atomistic Simulations of Neat Ionic Liquids – Structure and Dynamics

Atomistic simulations provide the positions (and in the case of MD, velocities) of all the atoms in the system consistent with the equilibrium probability distribution

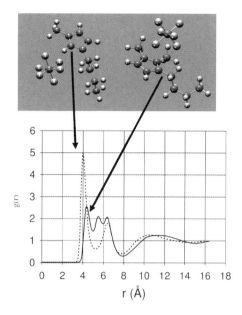

Fig. 4.2-6 RDFs for the $[PF_6]^-$ center of mass with carbon atoms of $[BMMIM]^+$ (solid line, top right structure) and $[BMIM]^+$ (dashed line, top left structure) cations. Note the strong organization of $[PF_6]^-$ about the C^2 ring position in $[BMIM][PF_6]$, but in $[BMMIM][PF_6]$ there is no strong preference for the anion to associate with any of the three carbon sites.

of the statistical mechanical ensemble under investigation. This means that a simulation provides information ranging from the "microscopic" details of molecular organization all the way up to "macroscopic" thermodynamic and dynamic properties. In this section, we summarize what has been learned about these three aspects of ionic liquids from classical atomistic simulations.

Structure

The most common way in which liquid structure is quantified is through a radial distribution function (RDF), commonly given the symbol $g(r)$. This function, which can be obtained experimentally, is a measure of the probability of observing two particular sites at a given distance from each other relative to a random distribution. The function is normalized such that when $g(r) < 1$, there is a "depletion" of these pairs in the system at a distance r, and when $g(r) > 1$, there is a relative enhancement of these types of interactions in the system.

Figure 4.2-6 shows RDFs for the center of mass of $[PF_6]^-$ with the 1-n-butyl-3-methylimidazolium $[BMIM]^+$ and $[BMMIM]^+$ cations. The first sharp peak in $g(r)$ for the $[BMIM][PF_6]$ system is due to the anion localizing near the C_2 carbon of the ring (the carbon between the two nitrogens). This behavior has been observed in nearly all simulations involving 1-alkyl-3-methylimidazolium cations [11, 12, 14, 15,

61, 62, 90, 120]. As described earlier, this is because the hydrogen atom attached to the C_2 carbon of the 1-alkyl-3-methylimidazolium ring is the most acidic site, and thus the negatively charged anion will interact most strongly with this position. This idea has been confirmed experimentally [126] and is consistent with the results of *ab initio* calculations [77, 127]. When the acidic hydrogen is replaced by a methyl group, however, the simulations predict that the anion no longer exhibits a preference for this position, but instead the anion appears to associate equally with all the carbon atoms of the ring [120].

Margulis has shown [61] that as the alkyl chain length R increases from hexyl to dodecyl in [RMIM][PF$_6$], the first and second solvation shells become slightly more ordered for the anion about the C^2 position of the cation. This same physical picture was seen in the 3-dimensional probability distribution plots for anions about cations as shown in Section 4.2.6.

Besides the tendency of anions to localize near acidic regions of the cation, simulations have shown that ionic liquids exhibit much longer-range order than simple molecular liquids. For example, RDFs between the centers of mass of cations and anions do not decay to unity until well after 1.5 nm in most of the simulations of imidazolium-based ionic liquids. Simulations of [EMIM][NO$_3$] have shown that Coulombic interactions are over four times as great as attractive Van der Waals forces, and two orders of magnitude larger than the thermal energy [15]. These strong Coulomb forces lead to charge ordering over at least three coordination shells. This result is consistent with experimental studies, which have also found long-range order in ionic liquids [40, 126, 128].

Dihedral angle distributions along the alkyl chain backbone of [RMIM][PF$_6$] have also been computed [61]. The dihedral angle nearest the ring is found to be nearly always in the *gauche* conformation, while the adjacent angle is nearly always in the *trans* position. All other angles along the chain are more likely to be found in the *trans* position, but some *gauche* conformations are possible.

Simulations carried out by Pádua and coworkers [129] have suggested that some ionic liquid mixtures are not homogeneous, but instead show nanometer-scale separation into continuous and semi-continuous domains of polar and apolar regions. By visualizing the results of MD simulations and using shades of gray for charged regions (atoms in the anion and all atoms directly attached to the cation rings) dark gray and non-polar groups (those atoms beyond the C^2 position in the alkyl chain) light gray, it is possible to see this type of phase segregation. Figure 4.2-7 shows the results from a simulation of 1-hexyl-3-methylimidazolium hexafluorophosphate ([HMIM][PF$_6$]). There is clearly extensive ordering of the alkyl tails and the charged groups into domains that appear to percolate through the entire simulation box. These researchers found that the charged or polar domain formed a tridimensional network of ionic channels, while the uncharged non-polar region formed a discontinuous microphase when the alkyl chain was four carbons long or shorter, and a continuous network for longer chains such as octyl or decyl. A similar result was obtained by Wang and Voth [130], who used a coarse-grained model for the ionic liquid and saw ordering of alkyl chains but not of charged groups.

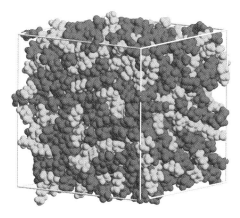

Fig. 4.2-7 Snapshot from an MD simulation of [HMIM][PF$_6$] in which polar regions are shaded dark gray and non-polar regions light gray [129].

Simulations have also been carried out on the crystalline phase of ionic liquids, mainly as a test of the ability of the force fields to reproduce the experimental crystal structure. Starting from the experimental crystal structure, Hanke et al. [12] minimized the potential energy of [MMIM]Cl, [HMIM]Cl, [EMIM]Cl, [MMIM][PF$_6$] and [EMIM][PF$_6$], and compared the resulting structures against experimental data. They found that the crystals distorted very little; the root mean square error for cell edges was on the order of 1%, indicating excellent performance of the force field. Canongia Lopes et al. [13] used their force field to simulate the crystal structures of [MMIM]Cl, [EMIM]Cl, [EMIM][NO$_3$], [EMIM][PF$_6$], and 1-dodecyl-3-methylimidazolium hexafluorophosphate. They also simulated various cations with the triflate and bis(trifluoromethanesulfonyl)amide anion [121]. In all these cases, "flexible cell" isothermal–isobaric MD simulations were conducted for 300 ps, again starting from the experimental structure. Average cell parameters were determined and compared to experimental data from the Cambridge Crystallographic Database. Agreement was generally excellent. In both of these cases, the fact that the force fields yield stable crystal structures that agree with experimental data suggests that much of the physics is being captured for these systems. Indeed, when *ad hoc* force field parameters were substituted for those developed by Canongia Lopes and coworkers [121], it was found that very large distortions in the crystals were observed during MD simulations. We should note that starting simulations of this type from experimental crystal structures is common practice. Reliably predicting crystal structures from first principles using force field-based simulations is still an unsolved problem, even for simple molecules. The fact that these force fields reproduce experimental crystal structures so well for complex ionic liquid molecules is no small achievement.

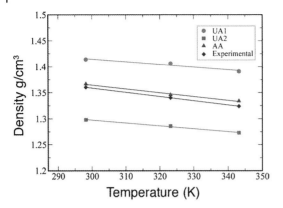

Fig. 4.2-8 Computed liquids densities at 1 bar for [BMIM][PF$_6$].

Thermodynamic Properties

In addition to structural properties, a great deal of effort has gone into modeling thermodynamic properties of pure ionic liquids. One of the most basic thermodynamic properties that must be captured well by a model is the liquid density. As mentioned above, crystalline densities have been determined by conducting simulations starting from an experimental crystal structure. For liquids, however, determining the density involves the generation of a "random" collection of ions and then equilibrating the system at constant pressure and temperature. The volume of the system fluctuates during this process, and the average molar volume is computed. From this, the density of the liquid may be determined. By carrying out a series of simulations at varying temperature and pressure (or by utilizing so-called "fluctuation formulas" of statistical mechanics) one can also determine the volume expansivity, defined as

$$\alpha_P = \frac{1}{V}\left(\frac{\partial V}{\partial T}\right)_P \tag{4.2-5}$$

and the isothermal compressibility, defined as

$$\kappa_T = -\frac{1}{V}\left(\frac{\partial V}{\partial P}\right)_T. \tag{4.2-6}$$

In general, a good force field can match experimental liquid densities within 5% and derivative properties such as the volume expansivity and isothermal compressibility within 10–15% without recourse to parameter fitting. The more details included in the force field, the more accuracy can be expected. For example, Fig. 4.2-8 shows computed densities as a function of temperature for [BMIM][PF$_6$]. UA1 refers to a simple united atom force field developed by Shah et al. [90] that treats [PF$_6$]$^-$ as a sphere, UA2 is a different united atom potential that utilizes a seven-site model for [PF$_6$]$^-$, [58] and AA is an "all-atom" force field that explicitly treats every atom in the system [11].

The experimental data of Gu and Brennecke [131] are shown for comparison. Clearly, the all-atom model matches the data extremely well, especially considering that the model was not adjusted to fit the experimental results. As expected, the simplest united atom model (UA1) overestimates the density by about 5%, but the other united atom model actually underestimates the density by about the same amount. Interestingly, all three models capture the temperature dependence of the density reasonably well. For example, the volume expansivity obtained using the all-atom model is $\alpha_P = 5.5 \times 10^{-4}$ K^{-1} while the experimental value [128] is $\alpha_P = 6.1 \times 10^{-4}$ K^{-1}. Similar good agreement between simulated and experimental densities have been obtained for [EMIM][BF$_4$] [60, 62], [BMIM][BF$_4$] [60, 62, 132], [EMIM][PF$_6$] [13], [BMIM][PF$_6$] [13, 14, 61, 62], [BMIM][NO$_3$] [13], [BMIM][Cl] [13], [HMIM][PF$_6$] [13], [HMIM][Cl] [13], [BMIM][Tf$_2$N] [121], [EMIM][Tf$_2$N] [121], [MMIM][Tf$_2$N] [121], [MPPY][Tf$_2$N] [121], [EMIM][TfO] [121], [MMIM][PF6] [62], [BMMIM][PF$_6$] [61], and [MMIM]Cl [62].

In addition to densities, a range of other thermodynamic properties have been computed for pure ionic liquids including cohesive energy densities/enthalpies of vaporization [11, 62, 83, 90, 132] and heat capacities [123].

Dynamics

The dynamic behavior of ionic liquids is important for both practical and theoretical reasons. From a practical standpoint, bulk transport properties such as the viscosity, self-diffusivity, thermal conductivity and electrical conductivity govern the effectiveness of these liquids in any application. For example, mass transfer of reactants and products is critical to the performance of ionic liquid solvents, and is highly correlated with the self-diffusivity and viscosity. Viscosity also plays a role in the cost of pumping the liquid and its performance as a lubricant. Thermal conductivity is a key parameter for thermal fluid applications, and electrical conductivity is obviously important in electrochemical applications.

From a fundamental standpoint, the dynamics of these systems is also of great interest. Are the dynamics of ionic liquids similar to that of normal molecular liquids, or do the strong Coulombic forces present in these systems give rise to different dynamical behavior? How does the relative size of the ions affect their translational and rotational dynamics? How do the dynamics change as the glass transition temperature is approached? Are these "strong" or "fragile" liquids, in the classification scheme of Angell [133]? Many of these questions are being examined experimentally using various spectroscopic techniques. Atomistic simulations present a particularly powerful method for examining these questions, as they yield extraordinarily detailed information on individual ion motion.

The simulation method of choice for examining the motion of liquids is molecular dynamics. MD gives information in timescales ranging from femtoseconds to several nanoseconds. Dynamical processes that occur over longer time scales than this cannot be accessed with standard simulation methods under current computational limitations.

One property that is particularly amenable for calculation with MD is the self-diffusivity, D_s, which can be calculated via the Einstein relation

$$D_s = \frac{1}{6} \lim_{t \to \infty} \frac{d}{dt} \left\langle [\vec{r}(t) - \vec{r}(0)]^2 \right\rangle \quad (4.2\text{-}7)$$

where the term in pointed brackets is the average mean square displacement (MSD) of all the species in the system. If the MSD is linear with respect to time, the self-diffusivity can be found from the slope of the MSD versus time curve.

Self-diffusivities were first computed for [MMIM][Cl] and [MMIM][PF$_6$] at temperatures ranging from 400–500 K [12]. The MSDs were collected over 15 ps intervals, and the estimated self-diffusivities of the ions were of the order of 1×10^{-8} m^2 s^{-1}. At room temperature, other groups [11, 14, 62] computed the self-diffusivity of [BMIM][PF$_6$] and found it to be of the order of 1×10^{-11} m^2 s^{-1}, which is two orders of magnitude slower than water, but agrees well with experimental studies of ionic liquids. Other studies [16] have found similar values of D_s at 400 K [131] for related systems. Simulations of [EMIM][NO$_3$] at 400 K indicate that the self-diffusivity is 6×10^{-11} m^2 s^{-1} for the anion and 1×10^{-10} m^2 s^{-1} for the cation [15]. A later study by the same group [125] found that the inclusion of polarizability into the model can increase the computed self-diffusivity by as much as a factor of three over that obtained from a fixed charge model. In most cases, it was observed that the cation self-diffusivity was slightly larger than that of the anion, even though the cation is the larger species. This finding is consistent with experimental NMR studies as well [134]. Urahata and Ribeiro explained this trend by examining MD trajectories which indicate that the cation has a preferred translational axis along the direction of the C^2 carbon position in the ring. Small anions, on the other hand, diffuse in a homogeneous manner.

It is important to note that most of the recent MD studies emphasize the importance of carrying out very long simulations to ensure that true diffusive behavior is observed during the simulation. Trajectories that appear to show diffusive (random walk) behavior may in fact still be in the sub-diffusive regime. This can be tested most easily by computing $\beta(t) = d \log(\Delta r^2)/d \log(t)$. At long time, $\beta(t) \to 1$. Figure 4.2-9 shows the MSD as a function of time for the cation 1-n-octyl-3-methylpyridinium bis(trifluoromethylsulfonyl)amide at 298 K [123]. Notice that this trajectory is over 5 ns, and visually appears to be linear with respect to time. Figure 4.2-10 shows how $\beta(t)$ varies with time, however, and it is clear from this plot that the system is still in the sub-diffusive range, even after 5 ns.

This same result has been seen by other researchers [15] and is evidence for "glass-like" dynamics. The simulations suggest that some ionic liquids can exhibit glass-like or super-cooled liquid behavior even above their nominal glass transition temperature. This is likely due to the long-range spatial ordering driven by Coulombic forces between the ions.

Unlike the self-diffusivity, which is a single-molecule property, other transport properties such as the shear viscosity, electrical conductivity and thermal conductivity are collective properties. This makes them much more difficult to compute from a simulation. We are only aware of one direct viscosity calculation for an ionic

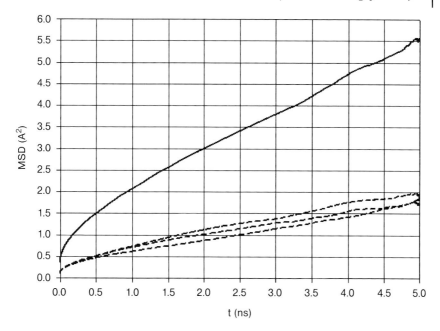

Fig. 4.2-9 MSD versus time for 1-n-octyl-3-methylpyridinium bis(trifluoromethylsulfonyl)amide at 298 K. Solid line is overall MSD and dashed lines are individual Cartesian components [120].

liquid [125], and no comparison was made in this work to experiment. No direct calculations of electrical or thermal conductivity have been reported. There have been attempts to relate these latter three quantities to readily available self-diffusivities by invoking models such as the Nernst–Einstein relation for conductivity [122] or Stokes–Einstein for viscosity [132]. Del Pópolo and Voth examined the electric current autocorrelation function for [EMIM][NO$_3$] at 400 K and found that there was a significant deviation from Nernst–Einstein behavior on the sub-picosecond timescale. They found that oppositely charged ions associate and diffuse together in a collective motion for times of the order of 0.1 ps. This motion contributes to mass transport but not conductivity. Although these associations break up after this time the direct current conductivity would be underestimated under the Nernst–Einstein approximation. There is clearly an area of opportunity for modeling researchers to work on the calculation of collective transport properties.

Simulations have also been used to examine the local dynamics of individual ions in the liquid state. In an early study from Lynden-Bell's group [12], rotational dynamics of the cations and anions of [EMIM][Cl] and [MMIM][PF$_6$] were examined by computing averages of the Legendre polynomials $P_1(\cos\theta(t))$ and $P_2(\cos\theta(t))$ as a function of time, where $\theta(t)$ is the angle moved through by one of three orthogonal vectors associated with an ion. The decorrelation of these order parameters was fitted to an exponential, and rotational relaxation times τ_1 and τ_2 were determined. At

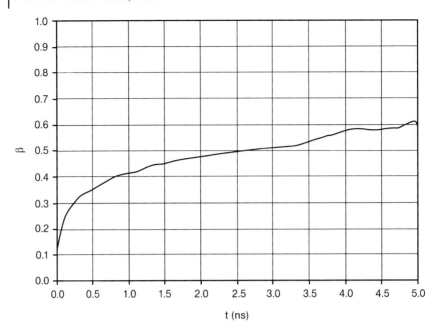

Fig. 4.2-10 Slope of MSD as a function of time for 1-n-octyl-3-methylpyridinium bis(trifluoromethylsulfonyl)amide at 298 K [120]. The fact that β is less than unity indicates sub-diffusive behavior, even after 5 ns.

400 K, the longest rotational time corresponded to an axis parallel to a vector connecting the ring nitrogen atoms. Values of τ_1 along this axis ranged from 47–73 ps for the cation. The $[PF_6]^-$ anion, on the other hand, experienced faster rotation, with $\tau_1 = 4.0$ ps at 400 K. Using a similar technique, Morrow and Maginn found that increasing the alkyl chain length and lowering the temperature to 298 K significantly slows the rotational motion of the cation in [BMIM][PF_6] [11]. They found that the longest rotational relaxation time for the [BMIM]$^+$ cation was $\tau_1 \approx 4$ ns, while the $[PF_6]^-$ anion rotated much faster, with $\tau_1 \approx 29$ ps. They also found that the cations and anions remain associated with one another in "cages", with exchange taking place on a timescale of 2–3 ns. This is much longer than the corresponding exchange times seen in molecular liquids, and is consistent with the slow translational motion observed in diffusion studies. The picture, then, is of associated cations and anions having slow translational and exchange motion. Cations, particularly those with longer alkyl chains, rotate very slowly, while a spherical anion such as $[PF_6]^-$ is free to rotate on a much faster time scale.

Recently, dynamical rotational motion of various pyridinium-based ionic liquids paired with the $[Tf_2N]^-$ anion were examined [124]. In this work, the principal axes of the cations and anions were assigned as the eigenvectors of an atomic mass weighted covariance matrix, defined as

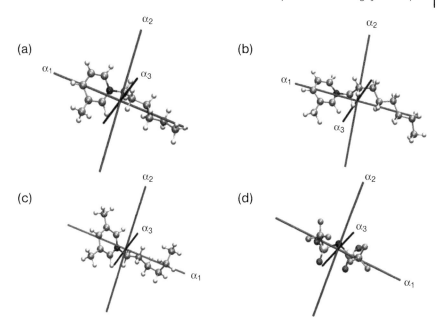

Fig. 4.2-11 Principal axes of (a) 1-n-hexyl-3-methylpyridinium; (b) 1-n-octyl-3-methylpyridinium; (c) 1-n-hexyl-3,5-dimethylpyridinium and (d) [Tf$_2$N]$^-$ ions. The long axis is a_1 the intermediate axis is a_2 and the short axis is a_3.

$$s_{jk} = \frac{\sum_{i=1}^{n} m_i (q_{ij} - \langle q \rangle_j)(q_{ik} - \langle q \rangle_k)}{\sum_{i=1}^{n} m_i} \quad (4.2\text{-}8)$$

where s_{ij} is the weighted covariance of the jth and kth atomic coordinates, n is the total number of atoms in an ion, m_i is the atomic mass of atom i, and q_{ij} represents the three coordinates ($j = x, y, z$) of the ith atom. Figure 4.2-11 shows examples of the principal axes for three cations 1-n-hexyl-3-methylpyridinium, 1-n-octyl-3-methylpyridinium, and 1-n-hexyl-3,5-dimethylpyridinium, paired with the [Tf$_2$N]$^-$ anion. The major axis, a_1, extends along the longest alkyl chain of the cation, as expected. As described above, τ_2 was computed for each axis, and rotational times for the cations ranged from about 2 ns at 348 K to over 12 ns at 298 K. For the anion, rotational times varied from just under 1 ns at 348 K to 4.5 ns at 298 K. These systems clearly exhibit very sluggish dynamics, especially compared to the smaller imidazolium cations.

A number of other studies have subsequently examined the dynamics of individual ions in the liquid state. To cite just two examples, Urahata and Ribeiro [16] and Margulis [61] have computed the vibrational density of states for a number of ionic liquids including [EMIM]Cl, [OMIM]Cl, [BMIM][PF$_6$] and [OMIM][PF$_6$]. The density of states is obtained from a Fourier transform of the velocity autocorrelation functions, and can be compared, at least qualitatively, to experimental IR and Raman spectra. The calculations show a rich range of dynamics, including rapid bond stretching and bending, alkyl chain oscillations, rotations, rattling and librational motion. Anisotropic dynamics was observed in the reorientational dynamics of the imidazolium cations. The flexible alkyl chain allows vibrational motion of the ring, but severely hinders overall rotational motion. The chain itself is capable of adopting many conformations over a relatively short time, however. The rich interplay of dynamics between cation and anion, as well as different regions of the cation itself, lead to some of the interesting dynamical properties observed in ionic liquids.

4.2.4
Simulations of Solutions and Mixtures

One of the important applications of room temperature ionic liquids is as solvents for chemical processes. Many simulations have been carried out to investigate various aspects of solvation of large and small molecules in such liquids. The most important factor determining the chemical potential or solubility of a substance in an ionic liquid is the electrostatic interaction. This was demonstrated by Lynden-Bell and coworkers [135, 136] who showed that the excess chemical potential of water in [MMIM]Cl is much more negative than that of propane. Polar but non-hydrogen-bonding solutes had intermediate chemical potentials. Examining individual contributions of energies and the local structure showed that the main interaction of protic solutes is with anions, and that in [MMIM]Cl each water molecule makes two strong hydrogen bonds to different chloride ions. The importance of the electrostatic interactions between solutes and ions in explaining solvation properties has been confirmed for other systems. For example Maginn et al. [120, 137] confirmed that the reason that CO_2 is more soluble in imidazolium salts than N_2 is also the result of the difference in the local electrostatic fields.

In many cases a polar solute induces local structure in the ionic liquid that depends on both the solute–ion interactions and the ion–ion interactions, which may propagate charge ordering over a considerable distance. An example of this is shown in the solutions of benzene and perfluorobenzene [138]. Immediately above and below the aromatic rings of benzene one finds a region of high concentration of cations, while in the plane of the ring there is a region of high anion concentration. As one goes further from the benzene molecule along the vertical axis successive regions of high concentrations of cations and anions succeed each other. This charge alternation is due to the cation–anion interactions. In solutions of perfluorobenzene the local liquid structure around the aromatic solute is reversed as the molecular quadrupole moment of perfluorobenzene is the opposite to that of benzene itself. The degree of local liquid structure around 1,3,5 trifluorobenzene (which has a very small quadrupole moment) is very small. The energetics of solute–solvent interaction are also small in the latter case, while quite large in the former.

Another example of the same effect was found by Kim and coworkers [139] when they studied the effects of turning on a dipole in a probe diatomic solute dissolved in a range of imidazolium ionic liquids. In the absence of the dipole there was very little local structure of the ionic liquid around the solute, but when charges of $\pm0.5e$ or $\pm1e$ are placed on the two sites, cations and anions cluster closely around the positive and negatively charged ends of the solute. As one moves further away from the solute the angular probability distributions of cations and anions relative to the dumbbell axis show charge oscillations. This work was inspired by the need to understand the response of dye spectra in ionic liquids. Kobrak and Znamenskiy [140, 141] have simulated realistic models of dyes in solution and have examined the solvatochromic shift and the dynamics of the solvation response to excitation. They too deduce that local ion–ion interactions are important. The dynamics of the response to excitation is made up of a fast process, which they find is a collective motion of both anions and cations moving by small amounts, while the slower process involves larger solvent rearrangements.

Ions comprise another class of solutes which have been studied extensively by Wipff and coworkers, who have been particularly interested in lanthanide and uranyl ions and their chloro-complexes [44, 142]. Their studies show that the chloro-complexes are stabilized by solvation in the ionic liquids based on imidazolium cations and the $[PF_6]^-$ anion. The principal interaction of the naked ions is with the anions, but on chlorination the cations move closer to complexes such as $[EuCl_6]^{4-}$ and $[UO_2Cl_4]^{2-}$. In a further study [45] they showed that in an equimolar mixture of water and ionic liquid, water molecules tend to fill the first solvation shell of naked ions in preference to $[PF_6]^-$ ions, and tend to solvate the chloro-complexes in preference to the imidazolium cations.

One important solute is water, which is completely miscible in imidazolium salts with short alkyl side chains and hydrophilic anions such as Cl^-, but is only partially miscible with ionic liquids with longer side chains and less hydrophilic anions. Extraction into an aqueous phase is important for product recovery from an ionic liquid medium. Hanke and Lynden-Bell [143] used simulation to investigate thermodynamic properties and local structure in mixtures of water with [MMIM]Cl and [MMIM][PF$_6$] liquids. They found that the excess energy of solvation was negative for the chloride and positive for the $[PF_6]^-$ liquid, as shown in Fig. 4.2-12 . There is a similar difference in the molar volumes of mixing shown in Fig. 4.2-13 . This is consistent with the perception of the $[PF_6]^-$ anion as being more hydrophobic than the chloride anion.

The simulations showed that water molecules tend to be more strongly associated with anions than with other water molecules. At low concentrations each water molecule is surrounded by ions and isolated from other water molecules. It is only at quite high mole fractions (75%) that significant clusters of water are found.

The question of whether two liquids are miscible or only partially miscible depends on the free energy of mixing. This is difficult to determine from simulations, but the energy or enthalpy of mixing can readily be determined. As the entropy of mixing is almost always positive, it is unlikely that the demixing will occur unless the enthalpy of mixing is positive. The results shown above for mixtures of water and [MMIM]X (X = Cl^- and $[PF_6]^-$) showed that the energy of mixing of water with the chloride liquid was negative, while that with the $[PF_6]^-$ liquid was small and positive. On the other hand 33% and 67% mixtures of benzene, perfluorobenzene and

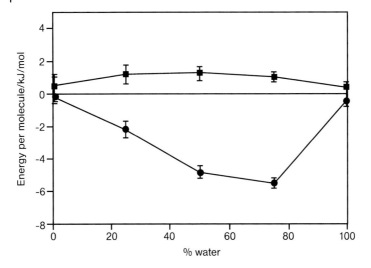

Fig. 4.2-12 Energies of mixing of water with [MMIM]Cl (lower curve) and with [MMIM][PF$_6$] (upper curve). Note that the energy of mixing is favorable for the former and unfavorable for the latter.

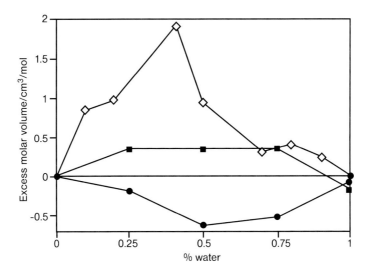

Fig. 4.2-13 Molar volumes of mixing for water with [MMIM]Cl (bottom curve with filled circles) and [MMIM][PF$_6$] (middle curve with filled squares). The open circles show the experimental volumes of mixing for [BMIM][PF$_6$] and water.

1,3,5 trifluorobenzene with [MMIM][PF$_6$] showed negative, near-zero and strongly positive values of the energy of mixing, respectively [89]. This was correlated with the different degrees of interaction of the three substances with the ionic liquid. In particular the 1,3,5-trifluorobenzene has a very small electrostatic field around it and interacts weakly with the ions.

In a recent simulation [45] Wipff and coworkers demonstrated that phase separation towards a water-rich phase and a imidazolium-rich phase began to occur after 10 ns for [OMIM][PF$_6$]/water mixtures but not for [BMIM][PF$_6$]/water.

4.2.5
Simulations of Surfaces

The first simulations of [MMIM]Cl liquid/gas surfaces [144, 145] showed very unusual properties near the surface (see Fig. 4.2-14). Just below the surface there was a region of enhanced density about 5 Å thick. In this region the cations were found to be strongly aligned perpendicular to the surface, suggesting that the high density was the result of closer packing in this layer. Most experimental work has been performed on [BMIM]$^+$ ionic liquids, and shows that the butyl groups point outwards into the vapor phase, supporting the conclusion that there is significant alignment of the cation rings near the surface. Simulations of [BMIM]X surfaces are currently under way for direct comparison with experiments.

Pinilla et al. studied the structure and dynamics of [MMIM]Cl confined between two parallel solid walls [146], this was the first simulation of an ionic-liquid/solid interface. Simulations were performed at various interwall distances. Mass and charge density along the confinement axis revealed a structure of layers parallel to the walls and a corresponding oscillatory profile of electrostatic potential. In particular, the potential drop between a point inside the solid wall and the center of the liquid slab was −0.5 V. Orientational correlation functions indicated that, at the interface, cations orient tilted with respect to the surface but that such orientational order is lost thereafter. A rather singular result was that the ionic diffusion under confinement was faster than in the bulk, at least for the non-corrugated walls used in the model.

4.2.6
Ab initio Simulations of Ionic Liquids

In *ab initio* or first-principles simulations, the potential energy of the system and the forces on the atoms result from the solution of the electronic structure of the system for a fixed nuclear configuration. In the study of solid and liquid phases, the behavior of the electrons is usually described within the framework of density functional theory (DFT) [147, 148]. Once the total energy and atomic forces are known, any traditional simulation method can be used to sample representative nuclear configurations, allowing for the calculation of thermodynamics, dynamics and structural properties.

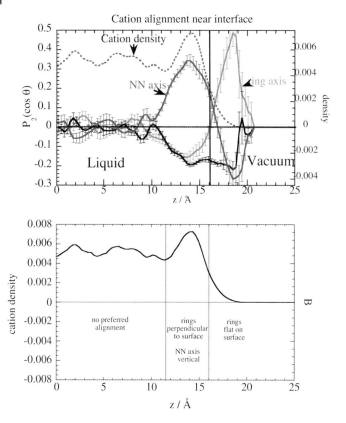

Fig. 4.2-14 The structure of the gas/liquid interface of [MMIM]Cl. The lower graph shows the general structure, while the upper graph shows the evidence for the cation alignment in the form of orientational functions $P_2(\cos\theta)$ for various axes in the ring.

The clear advantage of first principles methods over empirical force fields is in the generality gained on the physical description of the electronic structure of the system. Charge transfer and electronic polarization enter into the picture in a natural way. Accuracy at quantitative levels is also gained, although this factor depends on the approximations made in the quantum mechanical calculation of the electronic density, such as the choice of exchange and correlation functional, type and size of basis sets, etc. *Ab initio* simulations of liquids also pose the challenge of an accurate and efficient sampling, a factor that still remains as the weakness of these methodologies. First-principle simulations are currently restricted to short time scales or properties that converge quickly with the number of configurations generated by the sampling algorithm. A thorough combination of first-principle and empirical or semi-empirical simulations, is always a fruitful approach.

The simplest way of doing *ab initio* molecular dynamics is to converge the electronic structure at every nuclear configuration, calculate the forces on the particles by means of the Hellman–Feynman theorem, and move the atoms as dictated by classical mechanics. Such approach is known as Born–Oppenheimer molecular dynamics and assumes that the system remains always in the electronic ground state. As a more efficient alternative Car and Parrinello proposed in 1985 an extended Lagrangian approach for the propagation of the electronic degrees of freedom [22]. The method, known as Car-Parrinello molecular dynamics, was implemented in the CPMD program [149].

The first *ab initio* simulation of a room temperature molten salt, dimethylimidazolium chloride ([MMIM]Cl), appeared at the beginning of 2005 [21]. The work aimed at providing information on the liquid structure, in order to compare with results from classical force-field simulations and neutron diffraction experiments. Unlike non-associating fluids, in ionic liquids the distribution of ions around certain chemical bonds may depend strongly on the instantaneous electronic structure. Therefore, site–site distribution functions and three-dimensional densities may change when passing from a classical to a quantum mechanical description of the interactions.

For [MMIM]Cl, the level of agreement between the experimental structure and that obtained from classical simulations was good, although some differences were evident in the three-dimensional distribution of Cl^- around $[MMIM]^+$ [12,128]. The origin of such differences was unclear since, on the one hand, three-dimensional distributions are not directly accessible from the experiments (they have to be obtained through the numerical procedure EPSR [128]) and, on the other hand, force fields are normally parameterized to gas-phase *ab initio* calculations of the isolated ions. It is well known that the effective dipole moment of a molecule in a liquid may be significantly different to that in the gas phase.

DFT-based Born–Oppenheimer molecular dynamics simulations were performed with the program SIESTA [150]. In SIESTA the core electrons are replaced by norm-conserving pseudo-potentials and the Kohn–Sham orbitals are expanded in a non-orthogonal basis set of atom-centered orbitals. These orbitals are obtained as a numerical solution of the pseudo-atom with the particularity that they exactly vanish beyond a certain radius. The exchange and correlation functional was the generalized gradient approximation (GGA) of Perdew, Burke and Ernzerhof [70]. The accuracy of the methodology was tested against plane-waves calculations performed with the CPMD code. Two model systems were studied, one of 8 ion pairs and a second of 24 ions. The density of the liquid was set to the experimental value and periodic boundary conditions were used in both cases. Initial configurations for six independent trajectories were generated using known force fields. For the smallest system the total simulation time was 39 ps while for the largest it was just 3.2 ps. This last simulation was used to test the convergence with system size.

The performance of the SIESTA calculation was tested by optimizing the structure of the solid and comparing both with an equivalent plane wave calculation and with the experimental X-ray structure. The intra-molecular structure was well described

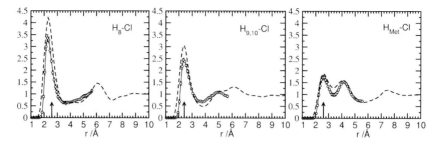

Fig. 4.2-15 Site–site distributions in [MMIM]Cl between Cl$^-$ and the unique hydrogen attached to C^2 of the imidazolium ring (H$_8$), the two equivalent ring hydrogens (H$_{9,10}$) and the methyl hydrogens (H$_{Met}$). Line with circles: system of 8 ions pairs. Dashed line: system of 24 ions. Arrows: solid phase distance. See atomic labeling in Figure 4.2-1.

by the two methods although the position of the anions showed an absolute deviation of 0.17 Å between SIESTA and the experiment, while between the two calculations it was 0.004 Å. The quality of the lattice parameters between the two calculations was reasonable and comparable, although not to the extent of the intra-molecular structure. The disagreement with experiment was attributed to the functional of exchange and correlation (GGA).

In the liquid, a first view of the structure was given by the radial distribution functions calculated from the centers of the ions. The simulations revealed a first peak in the cation–anion distribution at 4.5 Å which was independent of the system size and in good agreement with experiment. On the average each cation was surrounded by 6 anions up to a distance of 6.5 Å. Furthermore, the solvation shell of [MMIM]$^+$ was characterized by anion–cation site–site radial distribution functions and the space density of anions around a central cation.

Figure 4.2-15 shows, for the two model systems, the site–site distributions between Cl$^-$ and the unique hydrogen attached to C^2 of the imidazolium ring (H$_8$), the two equivalent hydrogens of the imidazolium ring (H$_9$ and H$_{10}$), and the methyl hydrogens (H$_{Met}$). These functions showed the largest and more systematic deviations with respect to classical simulations. Clearly, the position of the first peak is independent of the system size. The H$_8$–Cl and H$_{9,10}$–Cl functions showed similar features, a first peak at 2.2 Å and a second broader maximum at 6.0 Å. The first peak indicates localization of Cl$^-$ near the hydrogens while the second is associated with the anions located in the proximity of other ring hydrogens. The coordination number of H$_8$ and H$_{9,10}$ was up to a distance of 3.5 Å.

Figure 4.2-16 shows two views of the spatial distribution of Cl$^-$ around [MMIM]$^+$. The C$_{2v}$ symmetry of the cation was used to symmetrize this function. The isosurfaces plotted correspond to a density of 0.03 Å^{-1} which is six times the average density of anions. Anions appear localized on three regions, each associated with each one of the ring C–H bonds. The highest density corresponds to the Cl$^-$ located in front the unique hydrogen, H$_8$, and was visible even at ten times the average density. For the anions located in the vicinity of the H$_9$ and H$_{10}$, the distribution

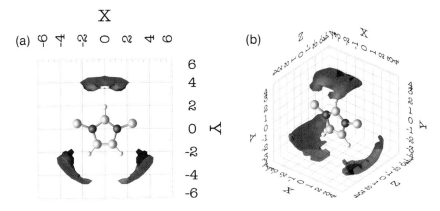

Fig. 4.2-16 Density of Cl$^-$ around [MMIM]$^+$. The iso-surface corresponds to a density level of 0.3 Å^{-1}.

is less symmetric with respect to the C–H bond. Only at very low probability levels were anions observed in positions axial to the methyl groups.

Using a simple geometric criterion, the former Cl–H contacts were interpreted as moderate hydrogen bonds since the C–H distance (1.1 Å) is much shorter than the H–Cl distance (2.2 Å) and the average angles for C^2–H$\cdots$Cl$^-$ and $C^{4,5}$–H$\cdots$Cl$^-$ were 151° and 148°, respectively. Apart from directionality, the stretching frequency of the C^2–H and $C^{4/5}$–H bonds in the liquid was red-shifted by 11% and 9% with respect to those of an isolated cation.

Recently, a second *ab initio* simulation of [MMIM]Cl was reported by Bühl et al. [25]. These authors performed Car-Parrinello molecular dynamics on systems of 25 and 41 ion pairs, considerably increasing the system size with respect to the former study, at the price of much shorter simulation times (5.3 ps in comparison with the 36 ps of Ref. [21]). Plane waves were included up to a cut-off energy of 60 Ry, and the exchange and correlation functional was that of Becke and Perdew [66–68]. The liquid density was fixed to the experimental value and the simulations were initiated from configurations generated by classical simulations.

The structure of the first solvation shell of [MMIM]$^+$, expressed in terms of anion–cation site–site distributions, agreed very well with those reported in Ref. [21]. Again, chlorides appeared located around the imidazolium C–H bonds showing the strongest association with the unique hydrogen. However, it was observed that cation–cation distribution functions did not converge well within the time scale of the simulations. That was expected given from the slow rotational relaxation of [MMIM]$^+$, and of course the observation is also valid for the results of Ref. [21].

By looking at the maximum localized Wannier functions and using different population analysis methods, Bühl et al. suggested that there is a donor–acceptor interaction between the lone pairs of Cl$^-$ and the anti-bonding σ^*_{C-H} orbitals of [MMIM]$^+$. Also the population analysis revealed that the amount of charge transferred from anions to cations ($\sim$0.25e) fluctuated very little during the MD trajectory.

A point to remark when comparing the two available *ab initio* simulations of [MMIM]Cl, concerns a difference in the relative stability and geometry of the most stable gas-phase isomers and their dependence on the parameters used in the calculations (exchange-correlation functional, basis set and pseudo-potentials). The method used in Ref. [21] predicts the isomer with Cl$^-$ localized on top the imidazolium ring to be more stable than the one with the anion on the plane of the ring; while in Ref. [25] the reported trend was the opposite and the energy difference between the two isomers smaller. The origin of such discrepancy is not yet clear. However, the agreement between the solvation structures predicted by the two methods really encourages the use of first-principle simulations to understand certain structural and electronic features of room-temperature molten salts.

4.2.7
Chemical Reactions and Chemical Reactivity

In contrast to inorganic molten salts, the fluidity of ionic liquids at room temperature permits their use as solvents for chemical reactions. Electrostatic properties and charge mobility in ionic liquids can play a distinctive role in chemical reactivity, as compared with neutral solvents. In particular, hydrogen and proton transfer reactions are likely to be sensitive to an ionic environment due to the hydrogen-bond acceptor ability of the anions. Such type of reactions are fundamental in acid-based chemistry and proton transport in solution.

From the methodological point of view, the description of bond breaking and bond formation processes requires the use of first principle simulations or hybrid methods. In the last case quantum mechanics is combined with molecular mechanics (QM/MM). Del Pópolo et al. [151] have studied the solvation structure and transport of an acidic proton in [MMIM]Cl. Calculations were performed with the SIESTA code, using the same technical specifications of Ref. [21] Hydrogen chloride (HCl) was used as model reactant. It was found that the acidic proton remained associated in the molten salt, due to the formation of hydrogen dichloride, [HCl$_2$]$^-$, that appeared as a linear and symmetric molecule in [MMIM]Cl. This can be seen in Fig. 4.2-17, where the distributions of the order parameter, ζ, defined as the difference between the two H–Cl bond lengths and bending angle of [HCl$_2$]$^-$ are shown. The source of stability of the di-anion is the formation of two strong H$\cdots$Cl hydrogen bonds. The first solvation shell of [HCl$_2$]$^-$ is formed by cation hydrogens, both belonging to the imidazolium ring and the methyl groups, but the hydrogen coordination number of each chlorine atom in [HCl$_2$]$^-$ was lower than those of Cl$^-$ in the neat liquid.

Since [HCl$_2$]$^-$ was very stable, no dissociation or spontaneous proton transfer was observed during the course of several dynamical trajectories. By performing umbrella sampling simulations, one of the H$\cdots$Cl bonds was forced to break, ending with the exchange of a Cl$^-$ in [HCl$_2$]$^-$ with a Cl$^-$ belonging to the ionic liquid. Such processes resulted in an effective hopping of the hydrogen atom. The most plausible

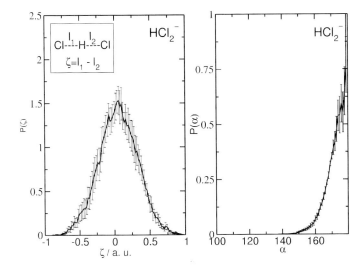

Fig. 4.2-17 Distribution of order parameter (ζ) and bending angle for HCl_2^- in [MMIM]Cl.

mechanism was discussed in terms of structural parameters involving the reacting atoms and their solvation shells.

Acknowledgements

P. Hunt gratefully acknowledges The Royal Society for a University Research Fellowship. E.J.M. thanks Cesar Cadena, David Eike, Timothy Morrow and Jindal K. Shah for their many contributions to this work as well as the US Air Force Office of Scientific Research (F49620-03-1-0212) and the US Department of Energy (DE-FG26-04NT42122) for funding. Agílio Pádua is acknowledged for providing Fig. 4.2-7. M.G.DP. and R.M.LB acknowledge EPSRC (grant GR/S41562). RMLB also thanks the Leverhulme Trust for an Emeritus Fellowship.

References

1. T. Welton, *Coord. Chem. Rev.* **2004**, *248*, 2459.
2. C. Chiappe, D. Pieraccini, *J. Phys. Org. Chem.* **2005**, *18*, 275.
3. T. Welton, P. Wasserscheid, *Ionic Liquids in Synthesis*, VCH-Wiley: Weinheim, **2002**.
4. J. D. Holbrey, K. R. Seddon, *Clean Prod. Proc.* **1999**, *1*, 223.
5. T. Welton, *Chem. Rev.* **1999**, *99*, 2071.
6. P. Wasserscheid, W. Keim, *Angew. Chem. Int. Ed.* **2000**, *39*, 3772.
7. J. Dupont, R. F. de Souza, P. A. Saurez, *Chem. Rev.* **2002**, *102*, 3667.
8. J. S. Wilkes, *J. Mol. Catal. A* **2004**, *214*, 11.
9. G. Worth, P. Hunt, M. Robb, *J. Phys. Chem. A* **2003**, *107*, 621.
10. M. P. Allen, D. J. Tildesley, *Computer Simulation of Liquids*, Oxford University Press: Oxford, **1987**.
11. T. I. Morrow, E. J. Maginn, *J. Phys. Chem. B* **2002**, *106*, 12807.

12. C. G. Hanke, S. L. Price, R. M. Lynden-Bell, *Mol. Phys.* **2001**, *99*, 801.
13. J. Lopes, J. Deschamps, A. Padua, *J. Chem. Phys. B* **2004**, *108*, 2038.
14. C. J. Margulis, H. A. Stern, B. J. Berne, *J. Phys. Chem. B* **2002**, *106*, 12017.
15. M. G. Del Popolo, G. A. Voth, *J. Phys. Chem. B* **2004**, *108*, 1744.
16. S. Urahata, M. Ribeiro, *J. Chem. Phys.* **2004**, *120*, 1855.
17. P. A. Hunt, I. R. Gould, *J. Phys. Chem. A* **2006**, *6*, 2269.
18. T. A. Darden, D. M. York, L. G. Pedersen, *J. Chem. Phys.* **1993**, *98*, 10089.
19. C. G. Hanke, R. M. Lynden-Bell, in *Ionic Liquids in Synthesis*, T. Welton, P. Wasserscheid (Eds.), VCH-Wiley, Weinheim, **2002**, p. 157.
20. F. Jensen, *Introduction to Computational Chemistry*, John Wiley & Sons, Chichester, **1999**.
21. M. G. Del Popolo, R. M. Lynden-Bell, J. Kohanoff, *J. Phys. Chem. B* **2005**, *109*, 5895.
22. R. Car, M. Parrinello, *Phys. Rev. Lett.* **1985**, *55*, 2471.
23. M. E. Tuckerman, M. Parrinello, *J. Chem. Phys.* **1994**, *101*, 1302.
24. D. Marx, J. Hutter, in *Modern Methods and Algorithms in Quantum Chemistry*, J. Grotendorst (Ed.), NIC Series, Juelich, **2000**, Vol. 1, pp. 301–449.
25. M. Bühl, A. Chaumont, R. Schurhammer, G. Wipff, *J. Phys. Chem. B* **2005**, *109*, 18591.
26. M. M. Francl, W. J. Pietro, W. J. Hehre, J. S. Binkley, D. J. DeFrees, J. A. Pople, M. S. Gordon, *J. Chem. Phys.* **1982**, *77*, 3654.
27. P. C. Hariharan, J. A. Pople, *Theor. Chim. Acta* **1973**, *28*, 213.
28. T. Clark, J. Chandrasekhar, G. W. Spitznagel, P. v. R. Schleyer, *J. Comput. Chem.* **1983**, *4*, 294.
29. T. H. Dunning, *J. Chem. Phys.* **1989**, *90*, 1007.
30. D. E. Woon, T. H. Dunning, *J. Chem. Phys.* **1993**, *98*, 1358.
31. S. F. Boys, F. Bernardi, *Mol. Phys.* **1970**, *19*, 553.
32. Z. Meng, A. Dölle, W. R. Carper, *J. Mol. Struct. (Theochem)* **2002**, *585*, 119.
33. M. J. S. Dewar, E. G. Zoebisch, E. F. Healy, J. J. P. Stewart, *J. Am. Chem. Soc.* **1985**, *107*, 3902.
34. K. M. Dieter, C. J. J. Dymek, N. E. Heimer, J. W. Rovang, J. S. Wilkes, *J. Am. Chem. Soc.* **1988**, *110*, 2722.
35. L. A. Curtiss, *Chem. Phys. Lett.* **1979**, 68.
36. M. J. S. Dewar, W. Thiel, *J. Am. Chem. Soc.* **1977**, *99*, 4899.
37. L. P. Davis, C. J. Dymek, J. P. Stewart, H. P. Clark, W. J. Lauderdale, *J. Am. Chem. Soc.* **1985**, *107*, 5041.
38. M. Blander, E. Bierwagen, K. G. Calkins, L. A. Curtiss, S. L. Price, M. Saboungi, *J. Chem. Phys.* **1992**, *97*, 2733.
39. C. J. J. Dymek, J. J. Stewart, *Inorg. Chem.* **1989**, *28*, 1472.
40. S. Takahashi, K. Suzuya, S. Kohara, N. Koura, L. A. Curtiss, M. Saboungi, *Z. Phys. Chem.* **1999**, *209*, 209.
41. S. Takahashi, L. A. Curtiss, D. Gosztola, N. Koura, M. Saboungi, *Inorg. Chem.* **1995**, *34*, 2990.
42. A. Chaumont, G. Wipff, *J. Phys. Chem. B* **2004**, *108*, 3311.
43. A. Chaumont, G. Wipff, *Phys. Chem. Chem. Phys.* **2005**, *7*, 1926.
44. A. Chaumont, G. Wipff, *Chem. Eur. J.* **2004**, *10*, 3919.
45. A. Chaumont, G. Wipff, *Inorg. Chem.* **2004**, *43*, 5891.
46. J. V. Oritz, P. J. Hay, R. L. Martin, *J. Am. Chem. Soc.* **1992**, *114*, 2736.
47. M. Dolg, H. Stoll, A. Savin, H. Preuss, *Theor. Chim. Acta* **1989**, *75*, 173.
48. M. Dolg, H. Stoll, H. Preuss, *Theor. Chim. Acta* **1993**, *85*, 441.
49. T. H. Dunning, P. J. Hay, *Methods of Electronic Structure Theory. Modern Theoretical Chemistry 3*, Plenum Press, New York, **1977**.
50. A. W. Ehlers, M. Böhme, S. Dapprich, A. Gobbi, A. Höllwarth, V. Jonas, K. F. Köhler, R. Stegmann, A. Veldkamp, G. Frenking, *Chem. Phys. Lett.* **1993**, *208*, 111.
51. E. R. Schreiter, J. E. Stevens, M. F. Ortwerth, R. G. Freeman, *Inorg. Chem.* **1999**, *38*, 3935.

52. P. J. Hay, W. R. Wadt, *J. Chem. Phys.* **1985**, *82*, 299.
53. M. S. Sitze, E. R. Schreiter, E. V. Patterson, R. G. Freeman, *Inorg. Chem.* **2001**, *40*, 2298.
54. P. J. Hay, W. R. Wadt, *J. Chem. Phys.* **1985**, *82*, 284.
55. W. J. Stevens, H. Basch, M. Krauss, *J. Chem. Phys.* **1984**, *81*, 6026.
56. T. R. Cundari, W. J. Stevens, *J. Chem. Phys.* **1993**, *98*, 5555.
57. C. M. Breneman, K. B. Wiberg, *J. Comput. Chem.* **1990**, *11*, 361.
58. J. K. Shah, E. J. Maginn, *Fluid Phase Equlib.* **2004**, *222-223*, 195.
59. D. A. Pearlman, D. A. Case, J. W. Caldwell, W. R. Ross, T. E. Cheatham, S. DeBolt, D. Ferguson, G. Seibel, P. Kollman, *Comput. Phys. Commun.* **1995**, *91*, 1.
60. J. de Andrade, E. S. Böes, H. Stassen, *J. Phys. Chem. B* **2002**, *106*, 13344.
61. C. J. Margulis, *Mol. Phys.* **2004**, *102*, 829.
62. Z. Liu, S. Haung, W. Wang, *J. Phys. Chem. B* **2004**, *108*, 12978.
63. E. A. Turner, C. C. Pye, R. D. Singer, *J. Phys. Chem. A* **2003**, *107*, 2277.
64. P. Hunt, B. Kirchner, T. Welton, *Chem. Eur. J.* **2006**, *12*, 6762.
65. Y. Wang, H. Li, S. Han, *J. Chem. Phys.* **2005**, *123*, 174501.
66. A. D. Becke, *Phys. Rev. A* **1988**, *38*, 3098.
67. J. P. Perdew, *Phys. Rev. B* **1986**, *33*, 8822.
68. J. P. Perdew, *Phys. Rev. B* **1986**, *34*, 7406.
69. C. Lee, W. Yang, R. G. Parr, *Phys. Rev. B* **1988**, *37*, 785.
70. J. P. Perdew, K. Burke, M. Ernzerhof, *Phys. Rev. Lett.* **1996**, *77*, 3865.
71. J. P. Perdew, K. Burke, M. Ernzerhof, *Phys. Rev. Lett.* **1997**, *78*, 1396.
72. A. D. Becke, *J. Chem. Phys.* **1993**, *98*, 5648.
73. R. Ahlrichs, M. Bär, M. Häser, H. Horn, C. Kölmel, *Chem. Phys. Lett.* **1989**, *162*, 165–169.
74. S. Kossmann, J. Thar, B. Kirchner, P. Hunt, T. Welton, *J. Chem. Phys* **2006**, accepted.
75. J. D. Holbrey, W. M. Reichert, M. Nieuwenhuyzen, S. Johnston, K. R. Seddon, R. D. Rogers, *Chem. Commun.* **2003**, 1636.
76. S. Saha, S. Hayashi, A. Kobayashi, H. Hamaguchi, *Chem. Lett.* **2003**, *32*, 740.
77. Z. Meng, A. Dölle, W. R. Carper, *J. Mol. Struct.* **2002**, *585*, 119.
78. W. R. Carper, Z. Meng, A. Dölle, *Ionic Liquids in Synthesis*, VCH-Wiley, Weinheim, **2002**.
79. E. R. Talaty, S. Raja, V. J. Storhaug, A. Dölle, W. R. Carper, *J. Phys. Chem. B* **2004**, *108*, 13177.
80. A. Chaumont, E. Engler, G. Wipff, *Inorg. Chem.* **2003**, *42*, 5348.
81. F. C. Gozzo, L. S. Santos, R. Augusti, C. S. Consorti, J. Dupont, M. N. Eberlin, *Chem. Eur. J.* **2004**, *10*, 6187.
82. A. M. Magill, B. F. Yates, *Aust. J. Chem.* **2004**, *57*, 1205.
83. J. de Andrade, E. S. Böes, H. Stassen, *J. Phys. Chem. B* **2002**, *106*, 3546.
84. J. P. Foster, F. Weinhold, *J. Am. Chem. Soc.* **1980**, *102*, 7211.
85. A. E. Reed, F. Weinhold, *J. Chem. Phys.* **1983**, *78*, 4066.
86. A. E. Reed, R. B. Weinstock, F. Weinhold, *J. Chem. Phys.* **1985**, *83*, 735.
87. A. E. Reed, F. Weinhold, *J. Chem. Phys.* **1985**, *83*, 1736.
88. E. Reed, L. A. Curtiss, F. Weinhold, *Chem. Rev.* **1988**, *88*, 899.
89. J. B. Harper, R. M. Lynden-Bell, *Mol. Phys.* **2004**, *102*, 85.
90. J. K. Shah, J. F. Brennecke, E. J. Maginn, *Green Chem.* **2002**, *4*, 112.
91. A. J. Stone, M. Alderton, *Mol. Phys.* **1985**, *56*, 1047.
92. S. Gründemann, A. Kovacevic, M. Albrecht, J. W. Faller, T. H. Crabtree, *Chem. Commun.* **2001**, 2274.
93. A. Kovacevic, S. Gründemann, J. R. Miecznikowski, E. VClot, O. Eisenstein, T. H. Crabtree, *Chem. Commun.* **2002**, 2580.
94. A. R. Chianese, A. Kovacevic, B. M. Zeglis, J. W. Faller, T. H. Crabtree, *Organometallics* **2004**, *23*, 2461.
95. L. Crowhurst, P. R. Mawdsley, J. M. Perez-Arlandis, P. A. Salter, T. Welton, *Phys. Chem. Chem. Phys.* **2003**, *5*, 2790.
96. J. L. Anderson, J. Ding, T. Welton, D. W. Armstrong, *J. Am. Chem. Soc.* **2002**, *124*, 14247.

97. N. Metropolis, A. W. Rosenbluth, M. N. Rosenbluth, A. N. Teller, E. Teller, *J. Chem. Phys.* **1953**, *21*, 1087.
98. B. J. Alder, T. E. Wainwright, *J. Chem. Phys.* **1959**, *31*, 459.
99. B. J. Alder, T. E. Wainwright, *Phys. Rev.* **1962**, *127*, 359.
100. A. Rahman, *Phys. Rev. A* **1964**, *136*, 405.
101. L. Verlet, *Phys. Rev.* **1967**, *159*, 98.
102. http://www.cs.sandia.gov~sjplimp/lammps/bench.html#billion
103. M. L. Huggins, J. E. Mayer, *J. Chem. Phys.* **1933**, *1*, 643.
104. J. E. Mayer, *J. Chem. Phys.* **1933**, *1*, 270.
105. M. P. Tosi, F. G. Fumi, *J. Phys. Chem. Solids* **1964**, *25*, 31.
106. M. J. Sangster, M. Dixon, *Adv. Phys.* **1976**, *25*, 247.
107. N. Galamba, C. A. Nieto de Castro, J. F. Ely, *J. Phys. Chem. B* **2004**, *108*, 3658.
108. N. Galamba, C. A. Nieto de Castro, J. F. Ely, *J. Chem. Phys.* **2004**, *120*, 8676.
109. N. Galamba, C. A. Nieto deCastro, J. F. Ely, *J. Chem. Phys.* **2005**, *122*, 22450.
110. D. M. Eike, J. F. Brennecke, E. J. Maginn, *J. Chem. Phys.* **2005**, *122*, 014115.
111. J. Akella, S. N. Vaidya, G. C. Kennedy, *Phys. Rev.* **1969**, *185*, 1135.
112. L. V. Woodcock, K. Singer, *Trans. Faraday Soc.* **1971**, *67*, 12.
113. J. W. E. Lewis, K. Singer, *J. Chem. Soc., Faraday II* **1975**, *71*, 41.
114. W. D. Cornell, P. Cieplak, C. I. Bayly, I. R. Gould, K. M. Merz, D. M. Ferguson, D. C. Spellmeyer, T. Fox, J. W. Caldwell, P. A. Kollman, *J. Am. Chem. Soc.* **1995**, *117*, 5179.
115. T. Fox, P. A. Kollman, *J. Phys. Chem. B* **1998**, *102*, 8070.
116. S. L. Mayo, B. D. Olafson, W. A. Goddard III, *J. Phys. Chem.* **1990**, *94*, 8897.
117. C. M. Breneman, J. Wiberg, *J. Comput. Chem.* **1990**, *11*, 361.
118. W. L. Jorgensen, J. D. Madura, C. J. Swenson, *J. Am. Chem. Soc.* **1984**, *106*, 813.
119. A. D. MacKerell, D. Bashford, M. Bellott, R. L. Dunbrack, J. D. Evanseck, M. J. Field, S. Fisher, J. Gao, H. Guo, S. Ha, S. Joseph-McCarthy, L. Kuchnir, K. Kuczera, F. T. K. Lau, C. Mattos, S. Michnick, T. Ngo, D. T. Nguyen, B. Prodhom, W. E. Reiher III, B. Roux, M. Schlenkrich, J. C. Smith, R. Stote, J. Straub, M. Watanabe, J. Wiorkiewicz-Kuczera, D. Yin, M. Karplus, *J. Phys. Chem. B* **1998**, *102*, 3586.
120. C. Cadena, J. L. Anthony, J. K. Shah, T. I. Morrow, J. F. Brennecke, E. J. Maginn, *J. Am. Chem. Soc.* **2004**, *126*, 5300.
121. J. N. Canongia Lopes, A. A. H. Pádua, *J. Phys. Chem. B* **2004**, *108*, 16893–16898; erratum, *J. Phys. Chem. B* **2004**, *108*, 1250.
122. S. U. Lee, J. Jung, Y.-K. Han, *Chem. Phys. Lett.* **2005**, *406*, 332–340.
123. C. Cadena, Q. Zhao, R. Q. Snurr, E. J. Maginn, *Phys. Chem. Chem. Phys.* **2006**, *110*, 2821.
124. C. Cadena, E. J. Maginn, unpublished results.
125. T. Yan, C. J. Burnham, M. G. Del Pópolo, G. A. Voth, *J. Phys. Chem. B.* **2004**, *108*, 11877.
126. C. Hardacre, S. E. J. McMath, M. Nieuwenhuyzen, D. T. Bowron, A. K. Soper, *J. Phys.: Condens. Matter* **2003**, *15*, S159.
127. T. I. Morrow, E. J. Maginn, Molecular Structure of Various Ionic Liquids from Gas Phase Ab Initio Calculations, in *Ionic Liquids as Green Solvents*, ACS Symposium Series 856, K. R. Seddon, R. D. Rogers (Eds.), American Chemical Society, Washington, DC, **2003**.
128. C. Hardacre, J. D. Holbrey, S. E. J. McMath, D. T. Bowron, A. K. Soper, *J. Chem. Phys.* **2003**, *118*, 273.
129. A. Pádua, Thermodynamics 2005 Conference, Sesimbra, Portugal, April, **2005**; Congress on Ionic Liquids, Salzburg, Austria, June, **2005**.
130. Y. Wang, G. A. Voth, *J. Am. Chem. Soc.* **2005**, *127*, 12192.
131. Z. Gu, J. F. Brennecke, *J. Chem. Eng. Data* **2002**, *47*, 339.
132. X. Wu, Z. Liu, S. Huang, W. Wang, *Phys. Chem. Chem. Phys.* **2005**, *7*, 2771.
133. W. Xu, E. I. Cooper, C. A. Angell, *J. Phys. Chem. B* **2003**, *107*, 6170.
134. The reported values for the self-diffusivity in Ref. 54 are listed as

having units of cm² s⁻¹ but we assume this is a typographical error. Thus we report here values that are of the order of 1×10^{-11} m² s⁻¹.

135. C. G. Hanke, N. Atamas, R. M. Lynden-Bell, *Green Chem.* **2002**, *4*, 107.
136. R. M. Lynden-Bell, N. A. Atamas, A. Vasilyuk, C. G. Hanke, *Mol. Phys.* **2002**, *100*, 3225.
137. J. K. Shah, E. J. Maginn, *J. Phys. Chem. B* **2005**, *109*, 10395.
138. C. G. Hanke, A. Johansson, J. B. Harper, R. M. Lynden-Bell, *Chem. Phys. Lett.* **2003**, *374*, 85.
139. Y. Shim, M. Y. Choi, H. J. Kim, *J. Chem. Phys.* **2005**, *122*, 044510.
140. V. Znamenskiy, M. N. Kobrak, *J. Phys. Chem. B* **2004**, *108*, 1072.
141. M. N. Kobrak, V. Znamenskiy, *Chem. Phys. Lett.*, **2004**, *395*, 127.
142. A. Chaumont, G. Wipff, *Phys. Chem. Chem. Phys.* **2005**, *5*, 3481.
143. C. Hanke, R. M. Lynden-Bell, *J. Phys. Chem B* **2003**, *107*, 10873.
144. R. M. Lynden-Bell, *Mol. Phys.* **2003**, *101*, 2625.
145. R. M. Lynden-Bell, J. Kohanoff, M. G. Del Pópolo, *Faraday Discuss.* **2005**, *129*, 57.
146. C. Pinilla, M. G. Del Pópolo, R. M. Lynden-Bell, J. Kohanoff, *J. Phys. Chem. B* **2005**, *109*, 17922.
147. R. G. Parr, W. Yang, *Density-Functional Theory of Atoms and Molecules*, Oxford University Press, **1989**.
148. R. Martin, *Electronic Structure: Basic Theory and Practical Methods*, Cambridge University Press, Cambridge, **2004**.
149. CPMD, Copyright IBM Corp **1990–2001**, Copyright MPI für Festkörperforschung Stuttgart 1997–2004. http://www.cpmd.org/.
150. J. M. Soler, E. Artacho, J. Gale, A. Garcia, J. Junquera, P. Ordejon, D. Sanchez-Portal, *J. Phys.: Condens. Matter* **2002**, *14*, 2745. http://www.uam.es/departamentos/ciencias/fismateriac/siesta/
151. M. G. Del Pópolo, J. Kohanoff, R. M. Lynden-Bell, *J. Chem. Phys. B* **2006**, *110*, 8798.

4.3
Translational Diffusion

Joachim Richter, Axel Leuchter, and Günter Palmer

4.3.1
Main Aspects and Terms of Translational Diffusion

Looking at translational diffusion in liquid systems, at least two elementary categories have to be taken into consideration: self-diffusion and mutual diffusion [1, 2]. In a liquid which is in thermodynamic equilibrium and which contains only one chemical species,[1] the particles are in translational motion due to thermal agitation. The term for this motion, which can be characterized as a random walk of the particles, is *self-diffusion*. It can be quantified by observing the molecular displacements

---

[1] *Components* are those substances, the amount of which can be changed independently from others, while *chemical species* mean any particles in the sense of chemistry (atoms, molecules, radicals, ions, electrons) which appear at all in the system [3].

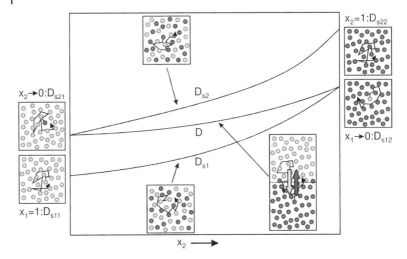

Fig. 4.3-1 Self-diffusion and mutual diffusion in a binary mixture. The self-diffusion coefficients are denoted D_{s1} and D_{s2}, the mutual diffusion coefficient D. At $x_1 = 1$ and $x_2 = 1$ the self-diffusion coefficients of the pure liquids D_{s11} and D_{s22}, respectively, are marked. Extrapolations $x_1 \to 0$ and $x_2 \to 0$ lead to the self-diffusion coefficients D_{s12} and D_{s21}.

of the single particles. The self-diffusion coefficient D_s is introduced by the Einstein relation

$$D_s = \lim_{t \to \infty} \frac{1}{6t} \left\langle |\vec{r}(t) - \vec{r}(0)|^2 \right\rangle \tag{4.3-1}$$

where $\vec{r}(t)$ and $\vec{r}(0)$ denote the locations of a particle at time t and 0, respectively. The brackets indicate that the ensemble average is used.

However, self-diffusion is not limited to one-component systems. As illustrated in Fig. 4.3-1 the random walk of particles of each component in any composition of a multicomponent mixture can be observed.

If a liquid system containing at least two components, is not in thermodynamic equilibrium due to concentration inhomogenities, transport of matter occurs. This process is called *mutual diffusion*. Other synonyms are chemical diffusion, interdiffusion, transport diffusion and, in the case of systems with two components, binary diffusion.

The description of mass transfer requires a separation of the contributions of *convection* and mutual diffusion. While convection means macroscopic motion of complete volume elements, mutual diffusion denotes the macroscopic perceptible relative motion of the individual particles due to concentration gradients. Hence, when measuring mutual diffusion coefficients, one has to avoid convection in the system or, at least has to take it into consideration.

Mutual diffusion is usually described with Fick's first law, here written for a system with two components and one-dimensional diffusion in the z-direction:

$$\vec{J}_i = -D_i \frac{\partial c_i}{\partial z} \quad (i = 1, 2). \tag{4.3-2}$$

Equation (4.3-2) describes the flux density $\vec{J}_i$ (in mol m^{-2} s^{-1}) of component i through a reference plane, caused by the concentration gradient $\partial c_i/\partial z$ (in mol m^{-4}). The factor D_i (in m^2 s^{-1}) is called the diffusion coefficient.

Most mutual diffusion experiments use Fick's second law, which permits the determination of D_i from measurements of the concentration distribution as a function of position and time:

$$\frac{\partial c_i}{\partial t} = D_i \frac{\partial^2 c_i}{\partial z^2} \tag{4.3-3}$$

Solutions for this second-order differential equation are known for a number of initial and boundary conditions [4].

In a system with two components, one finds experimentally the same values for D_1 and D_2, because $\vec{J}_1$ is not independent of $\vec{J}_2$. It follows, that the system is described with only one mutual diffusion coefficient $D = D_1 = D_2$.

In the case of systems containing ionic liquids, components and chemical species have to be differentiated. The system methanol/[BMIM][PF$_6$], for example, consists of two components (methanol and [BMIM][PF$_6$]) but three chemical species (methanol, [BMIM]$^+$ and [PF$_6$]$^-$), assuming that [BMIM][PF$_6$] is completely dissociated. If [BMIM][PF$_6$] is not completely dissociated, one has a fourth species, the undissociated [BMIM][PF$_6$]. From this it follows that the diffusive transport can be described with three and four flux equations, respectively. But the fluxes of [BMIM]$^+$ and [PF$_6$]$^-$ are not independent because of electroneutrality in each volume of the system. Furthermore, the flux of [BMIM][PF$_6$] is not independent of the flux of the ions because of the dissociation equilibrium. Thus, the number of independent fluxes is reduced to one, and the system can be described with only one mutual diffusion coefficient. In addition, one has four self-diffusion coefficients D_s(methanol), D_s([BMIM]$^+$), D_s([PF$_6$]$^-$), and D_s([BMIM][PF$_6$]), so that five diffusion coefficients are necessary to describe the system completely.

4.3.2
Use of Translational Diffusion Coefficients

Following the general trend of looking for a molecular description of the properties of matter, self-diffusion in liquids has become a key quantity for interpreting and modeling transport in liquids [5]. Self-diffusion coefficients can be combined with other data like viscosities, electrical conductivities, densities, etc. to evaluate and improve solvodynamic models like that of the Stokes–Einstein type [6–9]. From temperature-dependent measurements activation energies can be calculated using the Arrhenius or the Vogel–Tamman–Fulcher equation (VTF), to evaluate models which treat the diffusion process similarly to diffusion in the solid state with jump or hole models [1, 2, 7].

From the molecular point of view, the self-diffusion coefficient is more important than the mutual diffusion coefficient, because the different self-diffusion coefficients give a more detailed description of the single chemical species than the mutual diffusion coefficient, which characterizes the system with only one coefficient. Owing to its cooperative nature, a theoretical description of mutual diffusion is expected to be more complex than that of self-diffusion [5]. Moreover, self-diffusion measurements are accessible in pure ionic liquids, while mutual diffusion measurements require mixtures of liquids.

Looking from the applications point of view, mutual diffusion is by far more important than self-diffusion, because the transport of matter plays a major role in many physical and chemical processes, like crystallization, distillation, or extraction. Hence, the knowledge of mutual diffusion coefficients is valuable for modelling and scaling-up of these processes.

The need to predict mutual diffusion coefficients from self-diffusion coefficients often arises, and many efforts have been made to understand and predict mutual diffusion data, for example with approaches like the following extension of the Darken equation [5]:

$$D = (x_2 D_{21} + x_1 D_{12}) \Gamma, \quad \text{with} \quad \Gamma = \frac{d \ln a_1}{d \ln x_1} = \frac{d \ln a_2}{d \ln x_2} \tag{4.3-4}$$

where a_i is the activity of component i. Γ is denoted as the thermodynamic factor. Systems that are near to ideality can be described satisfactorily with Eq. (4.3-4), whereas the equation does not work very well in systems that are far from thermodynamic ideality, even if the self-diffusion coefficients and activities are known. Since systems with ionic liquids show strong intermolecular forces, there is a need to find better predictions of the mutual diffusion coefficients from self-diffusion coefficients.

Since the prediction of mutual diffusion coefficients from self-diffusion coefficients is not accurate enough to be used for modeling chemical processes, complete data sets of mutual and self-diffusion coefficients are necessary and valuable.

4.3.3
Experimental Methods

Nowadays self-diffusion coefficients are almost exclusively measured with NMR methods, using for example the 90–δ–180–δ-echo technique (Stejskal and Tanner sequence) [10–12]. The pulse-echo sequence, which is illustrated in Fig. 4.3-2, can be divided into two periods of time τ. After a 90° radio-frequency (RF) pulse the macroscopic magnetization is rotated from the z-axis into the x–y-plane. A gradient

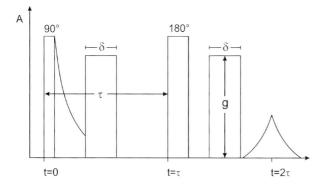

Fig. 4.3-2 Pulse-echo sequence of a NMR experiment for the measurement of self-diffusion coefficients.

pulse of duration δ and magnitude g is applied, so that the spins dephase. After a time τ a 180° RF pulse reverses the spin precession. A second gradient pulse of equal duration δ and magnitude g follows to tag the spins in the same way. If the spins have not changed their position in the sample, the effects of the two applied gradient pulses compensate each other, and all spins refocus. If the spins have moved due to self-diffusion, the effects of the gradient pulses do not compensate and the echo-amplitude is reduced. The decrease of the amplitude A with the applied gradient is proportional to the movement of the spins and is used to calculate the self-diffusion coefficient.

Popular methods for mutual diffusion measurements in fluid systems are the Taylor dispersion method and interferometric methods, such as digital image holography [13, 14].

With digital image holography it is possible to measure mutual diffusion coefficients in systems that are fairly transparent for laser light and whose components have a significant difference in the refractive index. The main idea of this method is to initiate a diffusion process by creating a so-called step-profile between two mixtures of a binary system, which have slightly different concentrations. The change in the step-profile is associated with the change in the optical phase profile, which can be scanned using a coherent laser beam passing perpendicular to the diffusion axis z through the diffusion cell. The state of the diffusion cell at a certain time is stored as a hologram on a CCD camera. The hologram is processed with holograms taken at different times to receive interference patterns which indicate the change in the diffusion cell with time. Using Fick's second law, the diffusion coefficient can be calculated from a single interference pattern. Mutual diffusion coefficients are accessible over the whole composition range of binary mixtures [15].

With electrochemial methods such as chronoamperometry, cyclovoltammetry (CV), or conductivity measurements, the diffusion coefficients of charged chemical species can be estimated in highly diluted solutions [16, 17].

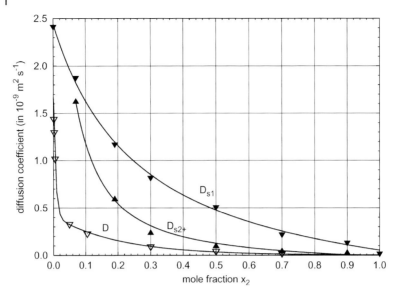

Fig. 4.3-3 Self-diffusion and mutual diffusion coefficients in the system methanol/[BMIM][PF$_6$]. x_2: mole fraction of [BMIM][PF$_6$], D: mutual diffusion coefficient, D_{s1}: self-diffusion coefficient of methanol, D_{s2^+}: self-diffusion coefficient of [BMIM]$^+$.

4.3.4
Results for Ionic Liquids

Typical values of self-diffusion coefficients and mutual diffusion coefficients in aqueous solutions and molten salt systems such as (K,Ag)NO$_3$ are of the order of 10^{-9} m^2 s^{-1}, and the coefficients usually do not vary by more than a factor of 10 over the whole composition range [1,2,15]. From measurements in pure ionic liquids we have learned that their self-diffusion coefficients are only of the order of 10^{-11} m^2 s^{-1}. From this point of view it is interesting to investigate systems of "ordinary" and ionic liquids. Figure 4.3-3 shows the results of first measurements in the system methanol/[BMIM][PF$_6$], which can be seen as a prototype for a system, in which an organic and an ionic liquid are mixed.

Self-diffusion coefficients were measured with the NMR spin-echo method and mutual diffusion coefficients with digital image holography, respectively. As can be seen from Fig. 4.3-3, the diffusion coefficients show the whole bandwidth of diffusion coefficient values from 10^{-9} m^2 s^{-1} on the methanol-rich side down to 10^{-11} m^2 s^{-1} on the [BMIM][PF$_6$]-rich side. The concentration dependence of the diffusion coefficients on the methanol-rich side is extreme and shows that special care and attention should be paid in the dimensioning of chemical processes with ionic liquids.

Since this is just the beginning of investigations into the diffusion behavior and the intermolecular forces in ionic liquid systems, further experimental work has to be done in pure ionic liquids as well as in systems of mixtures of ionic and organic liquids.

References

1. H. J. V. Tyrrell, K. R. Harris, *Diffusion in Liquids*, Butterworths, London, **1984**.
2. E. L. Cussler, *Diffusion - Mass Transfer in Fluid Systems*, Cambridge University Press, Cambridge, **1984**.
3. R. Haase, *Thermodynamics of Irreversible Processes*, Dover Publications, Mineola (NY), **1990**.
4. J. Crank, *The Mathematics of Diffusion*, 2nd edn., Clarendon Press, Oxford, **1975**.
5. H. Weingärtner, in *Diffusion in Condensed Matter*, J. Kärger, P. Heitjans, R. Haberlandt (Eds.), Vieweg, Wiesbaden, **1998**.
6. W. R. Carper, G. J. Mains, B. J. Piersma, S. L. Mansfield, C. K. Larive, *J. Phys. Chem.* **1996**, *100*, 4274.
7. N. Weiden, B. Wittekopf, K. G. Weil, *Ber. Bunsenges. Phys. Chem.* **1990**, *94*, 353.
8. J.-F. Huang, P.-Y. Chen, I. W. Sun, S. P. Wang, *Inorg. Chim. Acta* **2001**, *320*, 7.
9. C. K. Larive, M. Lin, B. J. Piersma, W. R. Carper, *J. Phys. Chem.* **1995**, *99*, 12409.
10. W. S. Price, *Concepts Magn. Reson.* **1997**, *9*, 299.
11. W. S. Price, *Concepts Magn. Reson.* **1998**, *10*, 197.
12. U. Matenaar, J. Richter, M. D. Zeidler, *J. Magn. Reson. A* **1996**, *122*, 72.
13. E. Marquardt, J. Richter, *Opt. Eng.* **1998**, *37*, 1514.
14. E. Marquardt, N. Großer, J. Richter, *Opt. Eng.* **1997**, *36*, 2857.
15. A. Leuchter, J. Richter, *High Temp. Mater. Proc.* **1998**, *2*, 521.
16. C. L. Hussey, I.-W. Sun, S. K. D. Strubinger, P. A. Barnard, *J. Electrochem. Soc.* **1990**, *137*, 2515.
17. R. A. Osteryoung, M. Lipsztajn, *J. Electrochem. Soc.* **1985**, *132*, 1126.

4.4
Molecular Reorientational Dynamics

Andreas Dölle, Phillip G. Wahlbeck, and W. Robert Carper

4.4.1
Introduction

Models for describing liquids should provide us with an understanding of the dynamic behavior of the molecules, and thus of the route of chemical reactions in the liquids. While it is often relatively easy to describe the molecular structure and dynamics of the gaseous or solid state, this is not true for the liquid state. Molecules in liquids can perform vibrations, rotations, and translations. A successful model often used for the description of molecular rotational processes in liquids is the rotational diffusion model, in which it is assumed that the molecules rotate by small angular steps about the molecular rotation axes. One quantity to describe the rotational speed of molecules is the reorientational correlation time τ, which is a measure of the average time elapsed when a molecule has rotated through an angle of the order of 1 radian or approximately 57°. It is indirectly proportional to the velocity of rotational motion.

4.4.2
Experimental Methods

A particularly important and convenient experimental method to obtain information on the reorientational dynamics of molecules is the measurement of longitudinal or spin–lattice relaxation times T_1 of peaks in nuclear magnetic resonance (NMR) spectra [1,2]. These relaxation times describe how fast a nuclear spin system reaches thermal equilibrium after disturbance of the system. Longitudinal relaxation is the relaxation process for the magnetization along the z axis, being parallel to the static magnetic field used in NMR spectroscopy. During this relaxation process, energy is exchanged between the spin system and its environment, the lattice. The measurement of ^{13}C relaxation data [3] has great advantages for the study of the reorientational behavior of organic molecules: For each carbon atom in the molecule, usually one signal is obtained, so that the mobility or flexibility of different molecular segments can be studied. Spin diffusion processes, dipolar ^{13}C–^{13}C interactions and, for ^{13}C nuclei with directly bonded protons, intermolecular interactions can be neglected. The dipolar ^{13}C spin–lattice relaxation rates $1/T_1^{DD}$, which are related to the velocity of the molecular rotational motions (see below), are obtained by measurement of ^{13}C spin–lattice relaxation rates $1/T_1$ and the nuclear Overhauser enhancement (NOE) factors η of the corresponding carbon atoms:

$$\frac{1}{T_1^{DD}} = \frac{\eta}{1.988} \frac{1}{T_1} \qquad (4.4\text{-}1)$$

A simple, but accurate way to determine spin–lattice relaxation rates is the inversion–recovery method [4]. In this experiment, the magnetization is inverted by a 180° radio frequency pulse and relaxes back to thermal equilibrium during a variable delay. The extent to which relaxation is gained by the spin system, is observed after a 90° pulse, which converts the longitudinal magnetization into detectable transverse magnetization. The relaxation times for the different peaks in the NMR spectrum can be obtained by means of a routine for determination of the spin–lattice relaxation time, which is usually implemented in the spectrometer software. When the inversion–recovery pulse sequence is applied under ^1H broadband decoupling conditions, only one signal is observed for each ^{13}C nucleus and the relaxation is governed by only one time constant $1/T_1$. The NOE factors are obtained by comparing signal intensities I_{dec} from ^1H broadband decoupled ^{13}C NMR spectra with those from inverse gated decoupled spectra I_{igdec} with the relation

$$\eta = \frac{I_{dec}}{I_{igdec}} - 1 \qquad (4.4\text{-}2)$$

4.4.3
Theoretical Background

Usually, nuclear relaxation data for the study of reorientational motions of molecules and molecular segments are obtained for non-viscous liquids in the extreme narrowing region where the product of the resonance frequency and the reorientational correlation time is much less than unity [1, 3, 5]. The dipolar ^{13}C spin–lattice relaxation rate of the ^{13}C nucleus i is then directly proportional to the reorientational correlation time τ_i

$$\left(\frac{1}{T_1^{DD}}\right)_{ij} = n_H (2\pi D_{ij})^2 \tau_i \qquad (4.4\text{-}3)$$

with the dipolar coupling constant

$$D_{ij} = \frac{\mu_0}{4\pi} \gamma_C \gamma_H \frac{\hbar}{2\pi} r_{ij}^{-3} \qquad (4.4\text{-}4)$$

where μ_0 is the magnetic permeability of the vacuum, γ_C and γ_H are the magnetogyric ratios of the ^{13}C and ^{1}H nuclei, respectively, $\hbar = h/2\pi$ with the Planck constant h, and r_{ij} is the length of the internuclear vector between the ^{13}C nucleus i and the interacting proton j. For the relaxation of ^{13}C nuclei with n_H directly bonded protons only interaction with these protons has to be taken into account.

Ionic liquids, however, are often quite viscous, and thus the measurements are beyond the extreme narrowing region. Hence, the relaxation rates become frequency dependent. Under these conditions, the equation for the spin–lattice relaxation rate becomes more complex:

$$\left(\frac{1}{T_1^{DD}}\right)_{ij} = \frac{1}{20}(2\pi D_{ij})^2 [J_i(\omega_C - \omega_H) + 3J_i(\omega_C) + 6J_i(\omega_C + \omega_H)] \qquad (4.4\text{-}5)$$

Here, the J_i are the spectral densities with the resonance frequencies ω of the ^{13}C and ^{1}H nuclei, respectively. Now, one has to find an appropriate spectral density to describe the reorientational motions properly [6]. The simplest spectral density commonly used for the interpretation of NMR relaxation data is that introduced by Bloembergen, Purcell and Pound [7].

$$J_{BPP,i}(\omega) = \frac{2\tau_{BPP,i}}{1 + (\omega\tau_{BPP,i})^2} \qquad (4.4\text{-}6)$$

The continuous distribution of correlation times by Cole and Davidson [8] found broad application in the interpretation of relaxation data of viscous liquids and glassy solids. The corresponding spectral density is

$$J_{CD,i}(\omega, \tau_{CD,i}, \beta_i) = \frac{2}{\omega} \frac{\sin(\beta_i \arctan(\omega\tau_{CD,i}))}{(1+(\omega\tau_{CD,i})^2)^{\beta_i/2}} \qquad (4.4\text{-}7)$$

Another way to describe deviations from the simple BPP spectral density is the so-called model-free approach by Lipari and Szabo [9]. This takes into account the reduction of the spectral density usually observed in NMR relaxation experiments.

Although the model-free approach was first applied mainly to the interpretation of relaxation data of macromolecules, it is now also used for fast internal dynamics of small and middle-sized molecules. For very fast internal motions the spectral density is given by

$$J_{LS,i}(\omega) = S_i^2 J_i \qquad (4.4\text{-}8)$$

which simply means a reduction of the BPP or CD spectral density J_i by the generalized order parameter S^2.

The resonance frequencies of the nuclei are given by the accessible magnetic field strengths via the resonance condition. Since the magnets used for NMR spectroscopy usually have only fixed field strengths, the correlation times, that is the rotational dynamics, have to be varied to leave the extreme narrowing regime. One way to vary the correlation times and thus the spectral densities and relaxation data, is to change the temperature. The temperature dependence of the correlation times is often given by an Arrhenius equation:

$$\tau_i = \tau_{A,i} \exp(E_{A,i}/RT) \qquad (4.4\text{-}9)$$

with the gas constant R and the activation energy E_A, which is interpreted in the following as a fit parameter representing a measure of the hindrance of the corresponding reorientational process.

4.4.4
Results for Ionic Liquids

Recently, the measurement of correlation times in molten salts and ionic liquids was reviewed [10] (for more recent references refer to Carper et al. [11]). We have measured the ^{13}C spin–lattice relaxation rates $1/T_1$ and nuclear Overhauser factors η in temperature ranges in and outside the extreme narrowing region for the neat ionic liquid [BMIM][PF$_6$], to observe the temperature dependence of the spectral density. Subsequently, the models for the description of the reorientational dynamics introduced in the theoretical Section 4.4.3 were fitted to the experimental relaxation data. The ^{13}C nuclei of the aliphatic chains can be assumed to relax only via the dipolar mechanism. This is in contrast to the aromatic ^{13}C nuclei, which also relax to an extent via the chemical-shift anisotropy mechanism. The latter mechanism has to be taken into account to fit the models to the experimental relaxation data (refer to Refs. [1] or [3] for more details). Preliminary results are shown in Figs. 4.4-1 and 4.4-2 together with the curves for the fitted functions.

In Table 4.4-1 values for the fit parameters and the reorientational correlation times calculated from the dipolar relaxation rates are given.

The largest correlation times, and thus the slowest reorientational motion, showed the three ^{13}C–^{1}H vectors of the aromatic ring with values between approximately 60 and 70 ps at 357 K, which are values to be expected for viscous liquids like ionic liquids. The activation energies are also in the typical range for viscous liquids. As is seen from Table 4.4-1, the best fit was obtained for a combination of the

Table 4.4-1 Reorientational correlation times τ at 357 K and fit parameters activation energy E_A, Cole–Davidson distribution parameter β, and generalized order parameter S^2

	C2	C4/C5	CH_3(ring)	CH_2(ring)	CH_2–CH_2–CH_2	CH_2–CH_3	CH_2–CH_3
τ_i(357 K)/ps	63	65	1.0	46	26	16	4.7
$E_{A,i}$/[kJ mol^{-1}]	38	37	27	32	26	26	20
β_i	0.46	0.43					
S_i^2	0.65	0.73	0.059	0.37	0.38	0.28	0.075

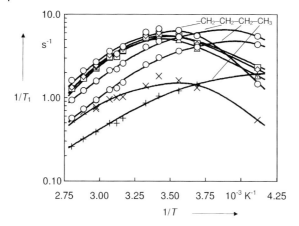

Fig. 4.4-1 ^{13}C relaxation rates $1/T_1$ of [BMIM][PF$_6$] in neat liquid as a function of reciprocal temperature T ($\triangle$: C2, $\square$ and $\diamond$: C4 and C5, X: CH$_3$ (ring), +: CH$_3$ (butyl group), ○: CH$_2$, lines: functions calculated with the fitted parameters).

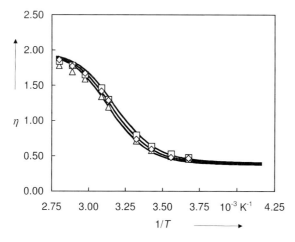

Fig. 4.4-2 {^{1}H}–^{13}C NOE factors η for [BMIM][PF$_6$] in neat liquid as a function of reciprocal temperature T ($\triangle$: C2, $\square$ and $\diamond$: C4 and C5, lines: functions calculated with the fitted parameters).

Cole–Davidson with the Lipari–Szabo spectral density with a distribution parameter β of about 0.45. Cole–Davidson spectral densities are often found for glass-forming liquids. The ring can be taken to be the rigid part of the molecule without internal flexibility, although a generalized order parameter S^2 of less than unity was observed. The value of approximately 0.70 for S^2 is explained by very fast motions like librations in the cage of the surrounding molecules, and vibrations. The reorientational correlation times of the aliphatic ^{13}C nuclei are smaller than those of the aromatic ring, indicating the internal motion of the corresponding molecular segments. The flexibility in the butyl chain increases from the methylene group bound to the rigid and slowly moving imidazolium ring to the methyl group at the end. The other methyl group also exhibited fast motion compared to the rigid aromatic ring. The correlation times of the aliphatic carbons relative to those of the ring and their graduation in the chain are similar to those of alkyl chains in hydrocarbons of comparable size [12]. The experimental ^{13}C spin–lattice relaxation rates for the aliphatic carbons could be fitted by a combination of the Lipari–Szabo with the BPP spectral density. The activation energies and the generalized order parameters decreased from the methylene group bound at the ring to the methyl group at the end of the chain, being again an indication for the increasing flexibility. The methyl groups had the smallest S^2 value of approximately one tenth of the value for the rigid part of the molecule, which is the typical value for fast methyl group rotation.

The very detailed results obtained for the neat ionic liquid [BMIM][PF$_6$] demonstrate clearly the potential of the method to determine the molecular reorientational dynamics in ionic liquids. An even more detailed study is given elsewhere [12a].

4.4.5
Chemical Shift Anisotropy Analysis

At low magnetic fields, it is often reasonable to ignore the contributions from chemical shift anisotropy (CSA). However, modern day investigators are using higher magnetic fields and this often requires an analysis that includes a CSA contribution as the CSA contribution increases with the square of the magnetic field.

$$R_1^{\text{total}} = R_1^{\text{Dipolar}} + R_1^{\text{CSA}} \qquad (4.4\text{-}10)$$

$$R_1^{\text{csa}} = [1/15]\,\gamma_C^2\,H_0^2\,(\Delta\sigma_i)^2\,[1 + (\eta_{\text{csa}}^2/3)]\,J(\omega_C) \qquad (4.4\text{-}11)$$

If one ignores even a small CSA contribution, the error introduced in the calculations may lead to considerable error in the calculation of rotational correlation times and their variation with temperature. In order to determine the contribution of each mechanism (dipolar and CSA) in ^{13}C NMR relaxation studies, a mathematical approach has been devised [13–15] which involves combining several correlation functions followed by a series of iterative steps that lead to the correct rotational correlation times. This results in the determination of the separate dipolar and CSA contributions to the overall relaxation mechanism in ^{13}C NMR relaxation studies. The following section outlines the necessary steps leading to the determination of both dipolar and CSA contributions as given in Eq. (4.4-10).

4.4.6
Stepwise Solution of the Combined Dipolar and NOE Equations

The basic assumption in this analysis [13–15] is that the maximum value of the ^{13}C NOE in Eq. (4.4-12) is determined by the dipolar rotational correlation time obtained

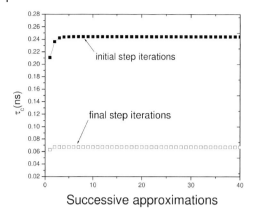

Fig. 4.4-3 Results of successive approximation steps for the calculation of τ_c. Initial iterations obtained from Eq (4.4-5) according to step (1). Final step iterations after cycling through steps (1) through (4) using Eqs. (4.4-5), (4.4-13) and (4.4-1).

from the measured relaxation rate. It is also assumed that dipolar relaxation and chemical shift anisotropy make up the overall relaxation rate.

$$\eta_{i,\max} = \gamma_H \left[6J(\omega_C+\omega_H) - J(\omega_C-\omega_H)\right] / \gamma_C \left[J(\omega_C - \omega_H) + 3J(\omega_C) + 6J(\omega_C + \omega_H)\right] \quad (4.4\text{-}12)$$

The ^{13}C dipolar and chemical shift anisotropy spin-lattice relaxation rates for those carbons bonded to hydrogen may be obtained by iterations of the following steps (1) through (4), followed by step (5).

(1) The experimental T_1s are assumed to be completely dipolar and Eq. (4.4-5) is solved for a pseudo-rotational correlation time. In Eq. (4.4-5), values for ω_C, ω_H, N_H and D_{ij} are known. Experimental values for T_1 are measured. Values of τ_c are calculated by successive approximation steps by setting τ_c on the right-hand side of Eq. (4.4-5) equal to the previously calculated value. The initial value of τ_c is set at 0.01 ns. A constant value of τ_c is usually obtained after ca. 5 successive steps, however the value after 40 successive steps is utilized. An illustration of these values of τ_c is given in Fig. 4.4-3 using data from Antony et al. [14] for the ring C2 carbon resonance at 137 ppm and 30 °C. The value of τ_c after 5, 15 and 40 steps, is 244 ps.

The final step iterations (cycle 5) for the ring C2 carbon of the 1-methyl-3-nonyl-imidazolium cation [14] yield a correlation time of 74 ps vs. an initial step value of 244 ps.

(2) Equations (4.4-5) and (4.4-12) are combined to form Eq. (4.4-13). The experimental T_1s and the pseudo-rotational correlation times, τ_c, from Eq. (4.4-5) are used in Eq. (4.4-13) to calculate $\eta_{\max}$. If these values of $\eta_{\max}$ were greater than 1.988, $\eta_{\max}$ was set equal to 1.988.

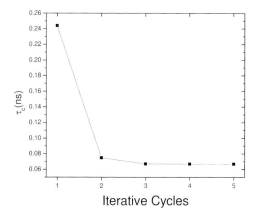

Fig. 4.4-4 Correlation times (τ_c) obtained at the end of each of the first five complete cycles using Eqs. (4.4-5), (4.4-13) and (4.4-1).

$$\eta_{max} = N_H \left[T_1^{DD}/20 \right] (\gamma_H/\gamma_C)(2 D_{ij})^2 [6J_+ - J_-] \qquad (4.4\text{-}13)$$

where $J_+ = \left[2\tau_c/[1 + (\omega_C + \omega_H)^2 \tau_c^2] \right]$ and $J_- = \left[2\tau_c/[1 + (\omega_C - \omega_H)\tau_c^2] \right]$. The η_{max} value is then used to calculate $R_1^{Dipolar}$ from Eq. (4.4-1).

(3) $R_1^{Dipolar}$ is used to calculate a new τ_c as outlined in (1) above. The new τ_c is used as outlined in step (2) above to determine a new η_{max} and the iterative process is repeated until a final self-consistent τ_c and $R_1^{Dipolar}$ are obtained. Figure 4.4-3 contains the final results for each of 5 cycles through steps (1) through (4) using Eqs. (4.4-5), (4.4-13) and (4.4-1). As indicated previously, the correlation time, τ_c, undergoes a major change from 244 to 74 ps. This has the effect of decreasing the relative contribution of the dipolar relaxation rate by ca. 15% and increasing the relative contribution of the chemical shift anisotropy relaxation rate accordingly. The 15% error can easily have major effects on relaxation rate minima and other important physical properties that are used to describe viscous solutions.

Selected values of ^{13}C $R_1^{Dipolar}$ are shown in Fig. 4.4-4 for five iteration steps (cycles). Note that for several ^{13}C resonance frequencies, a considerable change in $R_1^{Dipolar}$ occurred after the first iteration. The $R_1^{Dipolar}$ relaxation rate for ring carbons and methylene carbons adjacent to the imidazolium ring usually undergoes the greatest change as these ^{13}C relaxation rates contain a contribution from R_1^{CSA}. Aliphatic sidechain carbons may or may not contain a relaxation rate contribution from R_1^{CSA} [16].

(4) Finally, the aromatic carbon chemical shift anisotropy (R_1^{CSA}) spin–lattice relaxation rates are determined from Eq. (4.4-10). Equation (4.4-11) is then used to calculate the chemical shift anisotropy ($\Delta\sigma$) for an axially symmetric chemical-shift tensor.

4.4.7
NMR–Viscosity Relationships

Recent studies combine the results for the reorientational dynamics with viscosity data to compare experimental correlation times with correlation times calculated from hydrodynamic models [15]. By combining experimental and theoretical methods, it is possible to make statements about the intermolecular structure and interactions in ionic liquids and about the molecular basis for the specific properties of ionic liquids [15–20].

References

1. T. C. Farrar, E. D. Becker, *Pulse and Fourier Transform NMR. Introduction to Theory and Methods*, Academic Press, New York, **1971**.
2. M. L. Martin, J.-J. Delpuech, G. J. Martin, *Practical NMR Spectroscopy*, Heyden, London, **1980**.
3. J. R. Lyerla, G. C. Levy, *Top. Carbon-13 NMR Spectrosc.* **1972**, *1*, 79.
4. R. L. Vold, J. S. Waugh, M. P. Klein, D. E. Phelps, *J. Chem. Phys.* **1968**, *48*, 3831.
5. A. Abragam, *The Principles of Nuclear Magnetism*. Oxford University Press, Oxford, **1989**.
6. P. A. Beckmann, *Phys. Rep.* **1988**, *171*, 85.
7. a) N. Bloembergen, E. M. Purcell, R. V. Pound, *Phys. Rev.* **1948**, *73*, 679; b) A. Dölle, *J. Phys. Chem. A* **2002**, *106*, 11683.
8. D. W. Davidson, R. H. Cole, *J. Chem. Phys.* **1951**, *19*, 1484.
9. G. Lipari, A. Szabo, *J. Am. Chem. Soc.* **1982**, *104*, 4546.
10. W. R. Carper, Molten Salts, in *Encyclopedia of Nuclear Magnetic Resonance*, D. M. Grant, R. K. Harris (Eds.), John Wiley & Sons, New York, **1995**.
11. C. E. Keller, B. J. Piersma, G. J. Mains, W. R. Carper, *Inorg. Chem.* **1994**, *33*, 5601; C. E. Keller, W. R. Carper, *J. Phys. Chem.* **1994**, *98*, 6865; C. E. Keller, B. J. Piersma, G. J. Mains, W. R. Carper, *Inorg. Chim Acta* **1995**, *230*, 185; C. E. Keller, B. J. Piersma, W. R. Carper, *J. Phys. Chem.* **1995**, *99*, 12998; C. E. Keller, W. R. Carper, *Inorg. Chim. Acta*, **1995**, *238*, 115; C. K. Larive, M. Lin, B. J. Piersma, W. R. Carper, *J. Phys. Chem.* **1995**, *99*, 12409; W. R. Carper, G. J. Mains, B. J. Piersma, S. L. Mansfield, C. K. Larive, *J. Phys. Chem.* **1996**, *100*, 4724; C. K. Larive, M. Lin, B. S. Kinnear, B. J. Piersma, C. E. Keller, W. R. Carper, *J. Phys. Chem. B* **1998**, *102*, 1717.
12. a) J. H. Antony, D. Mentens, A. Dölle, W. R. Carper, P. Wasserscheid, *Chem. Phys. Chem.*, **2003**, *4*, 588; b) P. Gruhlke, A. Dölle, *J. Chem. Soc., Perkin Trans. 2* **1998**, 2159.
13. W. R. Carper, P. G. Wahlbeck, A. Dölle, *J. Phys. Chem. A* **2004**, *108*, 6096.
14. J. H. Antony, A. Dölle, D. Mertens, P. Wasserscheid, W. R. Carper, P. G. Wahlbeck, *J. Phys. Chem. A* **2005**, *109*, 6676.
15. N. E. Heimer, J. S. Wilkes, P. G. Wahlbeck, W. R. Carper, *J. Phys. Chem. A* **2006**, *110*, 868.
16. R. P. Klüner, A. Dölle, *J. Phys. Chem. A* **1997**, *101*, 1657.
17. Z. Meng, A. Dölle, W. R. Carper, *J. Mol. Struct. (THEOCHEM)* **2002**, *585*, 119.
18. J. H. Antony, D. Mertens, T. Breitenstein, A. Dölle, P. Wasserscheid, W. R. Carper, *Pure Appl. Chem.* **2004**, *76*, 255.
19. W. R. Carper, Z. Meng, P. Wasserscheid, A. Dölle, NMR Relaxation Studies and Molecular Modeling of 1-Butyl-3-methyl imidazolium PF_6 [BMIM][PF_6] in *XIII International Symposium on Molten Salts*, P. C. Trulove, H. C. DeLong, R. A. Mantz, (Eds.), *Electrochem. Soc. Proc.*, **2003**, 2002–19, 973.
20. N. Heimer, R. E. Del Sesto, W. R. Carper, *Magn. Reson. Chem.*, **2004**, *42*, 71.

5
Organic Synthesis

5.1
Ionic Liquids in Organic Synthesis: Effects on Rate and Selectivity

Cinzia Chiappe

5.1.1
Introduction

An incredibly large number of papers have appeared in the last few years about organic reactions and catalyzed processes carried out in ionic liquids, demonstrating that these salts can be used with success as solvents for organic reactions. It has often been claimed that they are able to increase both the reactivity and/or selectivity of a number of processes [1]. Unfortunately, little is known about how the use of an ionic liquid can affect the reactivity of solute species.

Chemical reactivity in solution is determined by the ability of the solvent to interact with substrates, intermediates and transition states [2], as well as the intrinsic properties of the reacting solutes. With their unique properties, ionic liquids may induce solvent effects on a wide range of processes. For molecular solvents, the key properties determining the interactions are recorded as "polarity" and they can change the rate and the order of homogeneous chemical reactions, affect the contribution of competing processes, modify the stereochemical behavior and the product distribution. The ability of some ionic liquids to interact with solutes has been investigated [3] using both partition methods and solvatochromic dyes. Through these measurements, it has been shown that they are able to interact with solutes via dipolar and dispersion forces and act as strong hydrogen-bond acceptors. This latter property mainly depends on the anion, while the hydrogen-bond donor ability (when observed, the values of hydrogen-bond acidity depend on the method used for the determination) arises from the cation. In particular, solvatochromic measurements have shown [3c] that all the investigated ionic liquids are characterized by dipolarities (π^*) higher than those of most non-aqueous molecular solvents; depending

---

Ionic Liquids in Synthesis, Second Edition. P. Wasserscheid and T. Welton (Eds.)
Copyright © 2008 WILEY-VCH Verlags GmbH & Co. KGaA, Weinheim
ISBN: 978-3-527-31239-9

Table 5.1-1 Physicochemical properties and solvent parameters for several ionic liquids[a].

	Viscosity (cP)	δ^2 (J cm^{-3})	E_T^N	π^*	α	β
[BMIM][BF$_4$]	233 (303 K)	998	0.67	1.047	0.627	0.376
[BMIM][PF$_6$]	450	912	0.669	1.032	0.634	0.207
[BMIM][SbF$_6$]		992		1.039	0.639	0.146
[BMIM][TfO]		620	0.656	1.006	0.625	0.464
[EMIM][Tf$_2$N]	28					
[BMIM][Tf$_2$N]	52	650	0.644	0.984	0.617	0.243
[BMMIM][BF$_4$][a]			0.576	1.083	0.402	0.363
[BMPY][Tf$_2$N]	85		0.544	0.954	0.427	0.252
[BMMIM][Tf$_2$N]	97.1	586	0.541	1.010	0.381	0.239
[OMIM][PF$_6$]	682		0.633			
[OMIM][Tf$_2$N]	92.7		0.629			
[OMMIM][Tf$_2$N][a]			0.525			
[OMMIM][BF$_4$]			0.543			

[a] [BMIM] = 1-butyl-3-methylimidazolium; [EMIM] = 1-ethyl-3-methylimidazolium; [OMIM] = 1-octyl-3-methylimidazolium; [BMMIM] = 1-butyl-2,3-dimethylimidazolium; [OMMIM] = 1-octyl-2,3-dimethylimidazolium; [BMPY] = 1-butyl-1-methylpyrrolidinium.

on the anion they can have a significant hydrogen-bond basicity (β) and, on the cation, a hydrogen-bond acidity (α) comparable to or lower than that of aniline, although this latter parameter is influenced also by the anion. No measure, however, suggests a "super polar" nature for the ionic liquids investigated (see Chapter 3, Section 3.5).

Ionic liquids have other physicochemical properties that can affect the reactivity of solute species; they are characterized by high viscosities, by elevated values of cohesive pressure and by a high degree of order. These properties are discussed in Chapter 3, but in Table 5.1-1 some physicochemical properties of several widely used ionic liquids are shown. The elevated viscosity may produce a reduction in the rate of diffusion-controlled processes, including the diffusion rates of redox species. The high cohesive pressure values, recently evaluated [4] on the basis of kinetic measurements (see below), show that more energy is necessary to create a hole in ionic liquids, whether to accommodate a solute or to bring all the components to the reaction site. Finally, the strong ion–ion interactions present in the ionic liquids lead to highly ordered three-dimensional supermolecular polymeric networks of anions and cations linked by hydrogen bonds [5] and/or Coulombic interactions [6], producing a completely different environment than found in molecular solvents. Higher energies are necessary to bring solutes together but the formation of the activated complex may be accompanied by a decrease in the order characterizing the overall system, with a consequent increase in the activation entropy.

Moreover, depending upon the cation and anion structure and size, the presence of polar and non-polar domains inside the liquid, which may affect the solvation and diffusion of ions and neutral molecules, can arise [7]. The situation is made even more complicated by the addition of simple molecules, such as water, alcohols and aromatic compounds, which drastically modifies the three-dimensional structure of ionic liquids. In water at infinite dilution, ionic liquid components break away from their pure environment and settle in the water environment, probably as well-separated ions. As the concentration of the ionic liquid increases, its anions and cations begin to interact with each other, directly or mediated by H_2O. At higher concentrations, ionic liquid ions cluster together with a very similar arrangement to that in the pure state [8, 9]. Water and small alcohols are hydrogen bonded strongly with the anion rather than the cation. Conversely, for non-hydrogen-bonding solutes, such as ethers and propane, the interaction is predominantly with the cation. Aromatic compounds (benzene, toluene and xylenes) form clathrate structures in 1-alkyl-3-methylimidazolium salts [10, 11].

The extent of mixing and the distribution of solutes in ionic liquids depend, therefore, on the relative solute–solute and solute–solvent interactions, which can have significant consequences on chemical reactivity and stability. In many ionic liquids, water-sensitive catalysts and chemical reactions are less sensitive to water compared with the situation in organic solvents because water dispersed throughout the ionic liquid cannot act like bulk water.

Finally, although, in general, molecular solvents affect the rates of homogeneous reactions via static or equilibrium solvent effects, dynamic or frictional solvent effects may be present in ionic liquids. The former are traditionally rationalized in terms of the transition-state theory; the solvent modifies the Gibbs energy of activation by differential solvation of the reactants and the activated complex. Here, it is necessary to assume that the required reorientational relaxation of the solvent molecules during the activation process is sufficiently fast and the activated complex is in thermal equilibrium with the solvent. For rapid reactions, in solvents characterized by slow solvation processes, this hypothesis is, however, not necessarily valid and the solvent may affect reaction rates through dynamic frictional effects. In the case of slow solvent relaxation, significant dynamic contributions to the experimentally determined activation parameters, which are completely absent in the conventional transition state theory, can exist and, in the extreme case, solvent reorientation can become the rate-limiting step: in this situation the rate depends on solvent dynamics and will vary with density, internal pressure and viscosity. In ionic liquids the strong anion–cation interactions affect the dynamic nature of solvation. Measured rotation times in ionic liquids are considerably slower than in polar molecular solvents [12, 13] suggesting that, while in polar solvents solvation arises due to the rearrangement of the solvent molecules around an instantaneously created dipole upon absorption of a photon, in ionic liquids the solvation is probably due to the motions of the ions.

It is evident from this brief introduction that ionic liquids may affect reactivity much more than molecular solvents. They are not only mixtures of two species interacting with each other (anion and cation), each able to give specific interactions with the dissolved reagents and/or activated complexes, but the anion–cation interactions make them complex three-dimensional structures able to exercise unusual effects.

This section attempts to show how ionic liquids are able to modify the kinetic and stereochemical course of organic reactions. It will cover, therefore, only a limited number of reactions and even a more limited number of ionic liquids, mainly those reported in Table 5.1-1. Unfortunately, only a few ionic liquids have been used in studies of reactivity and for even fewer are all of the physicochemical properties required to rationalize organic reactivity known. Sometimes, the data reported in the literature have been revised and reinterpreted in the light of more recent findings on the physicochemical properties of the applied ionic liquids. I express regret if I have misunderstood some results or some interpretations. My aim is merely to promote the discussion about ionic liquids' solvent properties in order to increase the understanding of these neoteric solvents and favor the development of more efficient processes.

5.1.2
Ionic Liquid Effects on Reactions Proceeding through Isopolar and Radical Transition States

Traditionally, homogeneous organic reactions are grouped into three classes depending on the character of the activated complex through which the reactions proceed: dipolar, isopolar and free-radical transition state reactions. Reactions occurring through radical or isopolar transition states are generally unaffected, or only marginally affected, by solvent polarity. For these reactions other solvent properties such as cohesive pressure, viscosity and so on may become important.

5.1.2.1 Energy Transfer, Hydrogen Transfer and Electron Transfer Reactions
Despite the fact that reactivity in ionic liquids has been only marginally investigated [1m] a relatively high number of quantitative studies have been carried out on reactions occurring through isopolar or radical transition states. By using a series of representative photochemical reactions, covering a wide range of fundamental processes (energy transfer, hydrogen transfer and electron transfer) it has been shown [14] that the slow molecular diffusion characterizing ionic liquids renders diffusion-controlled processes up to two orders of magnitude slower than in common organic solvents (e.g., acetonitrile, water). However, it is interesting to note that often the kinetic constants determined in ionic liquids for diffusion-controlled processes are at least an order of magnitude higher than the kinetic constants estimated from the ionic liquid's viscosity ($k_{diff} = 8000\ RT/3\eta$) [14b]. Related to this, the kinetic investigation of the exothermic energy transfer from benzophenone triplet ($^3Bp^*$) to naphthalene (N) has shown that this phenomenon can be attributed to the very large solvent-dependent pre-exponential factors, A, characterizing the reaction in ionic liquids that overcompensate for the large E_a values (Scheme 5.1-1).

Moreover, both E_a and A values show a significant dependence on the ionic liquid structure. For any cation the E_a values are 12–15 kJ mol^{-1} higher for the [PF$_6$]$^-$ salts than the [Tf$_2$N]$^-$ and this behavior has been related to the higher extent of ionic cross-linking characterizing the ionic liquids bearing the smaller and more

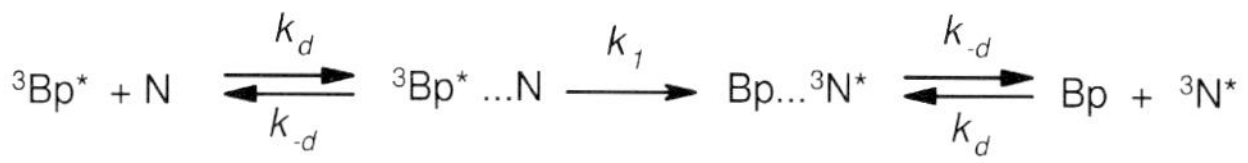

$$^3\text{Bp}^* + \text{N} \underset{k_{-d}}{\overset{k_d}{\rightleftarrows}} {}^3\text{Bp}^*\ldots\text{N} \xrightarrow{k_1} \text{Bp}\ldots{}^3\text{N}^* \underset{k_d}{\overset{k_{-d}}{\rightleftarrows}} \text{Bp} + {}^3\text{N}^*$$

Scheme 5.1-1 Exothermic energy transfer from benzophenone triplet ($^3\text{Bp}^*$) to naphthalene (N).

symmetrical $[PF_6]^-$ anion. On the other hand, as suggested by the authors [14b] this enhanced ionic cross-linking probably results in a greater diffusing mass, and therefore a greater A, whilst creating large cavities for solutes to jump into.

A large and positive entropy of activation, confirming that the solvent ions are freed up on formation of an encounter complex, was also found by the same authors for the diffusive electron-transfer quenching of the excited state of tris(4,4'-bipyridyl)ruthenium, $[Ru(bpy)_3]2+$, by methylviologen, MV^{2+} (Scheme 5.1-2) [15].

The ionic liquid structure, determining viscosity, cohesive pressure and degree of order, therefore, seems to be the main factor affecting the rate of diffusion-controlled reactions, influencing both the enthalpy and the entropy of the process.

Electron and hydrogen transfer processes in ionic liquids have also been widely investigated in the last three to four years through pulse radiolysis [16]. Ionic liquids are excellent media for the generation of radical ions; electrons and holes may be generated in high yield by radiolysis of pure ionic liquids and electrons may be efficiently trapped by both cations and anions. Moreover, the tendency of ionic liquids to supercool, resulting in the formation of more viscous liquids and finally transparent glasses without crystallization, has been used for the generation and spectroscopic characterization of unstable solute radical ions [17], and then, after thermal annealing of the solvent up to ambient temperature, to study the reactivity of these transient species.

Kinetic studies of several electron transfer processes triggered via pulse radiolysis have provided additional information about the properties of these new reaction media. Using this method to generate radicals the rate constants measured in ionic liquids for many diffusion-controlled electron transfer reactions were lower than those measured in water or common organic solvents. Even in this case, however, the authors only partly attributed [16a] the low kinetic constants, found for example for the oxidation of chlorpromazine and Trolox by $CCl_3O_2^\bullet$, to the high viscosity of the ionic liquids used, [BMIM][PF_6] and [BMIM][BF_4]. Considering that these reactions require separation and reassembly of solvent molecules, the effect has been primarily attributed to the high degree of ion-association of the ionic liquids investigated, and it has been suggested that, for these reactions, the rate constants

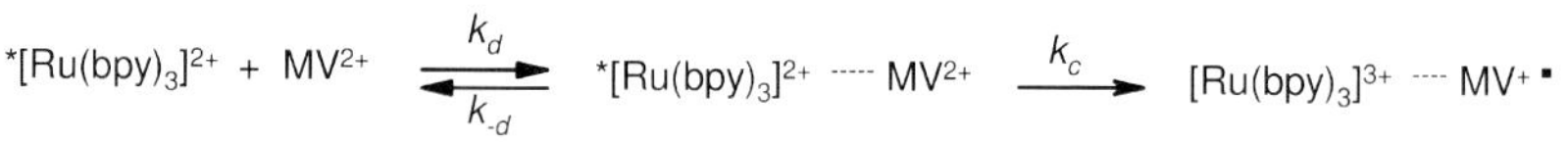

$$^*[Ru(bpy)_3]^{2+} + MV^{2+} \underset{k_{-d}}{\overset{k_d}{\rightleftarrows}} {}^*[Ru(bpy)_3]^{2+}\cdots MV^{2+} \xrightarrow{k_c} [Ru(bpy)_3]^{3+}\cdots MV^{+\bullet}$$

Scheme 5.1-2 Diffusive electron-transfer quenching of the excited state of tris(4,4'-bipyridyl)ruthenium, $[Ru(bpy)_3]2+$, by methylviologen, MV^{2+}.

should be better correlated with the solvent cohesive pressure than with solvent polarity. *"Ionic liquids are highly ordered reaction media, and the activation energies for the reactions carried out in ionic liquids can be high because it is necessary to break the order of the medium to bring all the components to the reaction site"* [16a].

The kinetic studies of pulse radiolysis triggered electron transfer processes in pyridinium- and ammonium-based ionic liquids have confirmed the data arising from laser flash photolysis studies: i.e. that the rate of diffusion-controlled constants measured in ionic liquids are often higher than the rate constants estimated from the viscosity of the IL employed ($k_{\text{diff}} = 8000RT/3\eta$). In this case, two different explanations have been given by the same authors for this experimental result. Initially, since kinetic constants higher than the estimated limits have been found for electron transfer reactions from the N-butylpyridinyl radical (BuPy•) to other substrates (methyl viologen, 4-nitrobenzoic acid and duroquinone, DQ), using an N-butylpyridium IL as solvent, the phenomenon was attributed to an electron hopping through solvent cations ([BuPy]$^+$). However, this behavior was also found for reactions that could not involve such a mechanism, i.e. the previously mentioned quenching of benzophenone triplet by naphthalene [14b], the reaction of solvated electrons with aromatic compounds, and the reaction of pyridinyl radicals (bpy•) with duroquinone (DQ) in ionic liquids having a completely different nature, such as ammonium salts {[Bu$_3$MeN][Tf$_2$N]} (Scheme 5.1-3) [18]. It has therefore been proposed [18] that the viscosity of ionic liquids, which determines the diffusion of whole molecules or ions, does not adequately represent the diffusion of reactants within ionic liquids.

In particular, it has been suggested that the highly ordered structure of these salts may contain voids, and that these voids can accommodate small solute molecules. Furthermore, since the chains present on the cations are flexible they can move more rapidly than the whole cation, permitting a rapid diffusion of solutes from one void to another [18]. The formation of cavities (voids) in ionic liquids has been recently studied via Monte Carlo simulations [19]. Analysis of cavity size distribution functions shows that ionic liquids exhibit a large tendency to form cavities, a property which seems to be correlated to the attractive interactions between ions and, particularly, to the tendency of the ions to associate into ion aggregates.

The presence of voids and the ability of small molecules to move within them have also been recently proposed to explain the reactivity of H• atoms with aromatic solutes in ionic liquids. Indeed, it has been shown [20] that the rate constants for reactions of pyrene and phenanthrene with H• atom are 10 times higher than for the reactions of the corresponding solvated electrons, implying that the diffusion-limited rate constant for the H• atom reaction is higher by one order of magnitude. Furthermore, for both substrates the trend in the ionic liquid examined, [Bu$_3$MeN][Tf$_2$N], is opposite to that observed in water, suggesting that a small

$$\text{BuPy}^\bullet + \text{DQ} \longrightarrow [\text{BuPy}]^+ + \text{DQ}^{\bullet -}$$

Scheme 5.1-3 Reaction of pyridinyl radicals (BuPy•) with duroquinone (DQ).

neutral species, such as the H•atom, can move easily between voids within the ionic liquid, whereas the diffusion of solvated electrons, being charged species, is limited by their interaction with the ionic charges of the medium.

In conclusion, in agreement with other experimental measurements, the kinetic and thermodynamic data arising from the energy transfer, electron transfer and proton abstraction reactions suggest highly ordered structures for ionic liquids, well above their melting points. The highly ordered structure can affect reactivity through entropic effects, although the hypothesis of the presence of voids through which small molecules can move with the help of cation chains is very attractive. Based on this latter hypothesis, ionic liquids could be considered as reaction media characterized by a behavior that lies between that of true solvents and polymeric matrixes [21].

The extensive investigations [16] of Neta et al. have, however, also shown that other factors are able to affect the rate of electron transfer processes, particularly when charged species are involved and the solvent can influence reactivity by changing the energy of solvation of these species. The stability of $Br_2^{•-}$ is indeed much higher in [Bu_3MeN][Tf_2N] than in water and, consequently, the rate constant for oxidation of chlorpromazine by $Br_2^{•-}$ decreases upon changing the solvent from water to an IL [16e]. [Bu_3MeN][Tf_2N] behaves like an aprotic organic solvent, with the energy of solvation of small ions being lower than in water and alcohols.

The concept of ionic liquids as ordered solvents, able to interact with charged species, has also been considered and developed by Hapiot et al. Initially these authors investigated [22] the stability of electrogenerated radical cations in imidazolium- and ammonium-based ionic liquids by cyclic voltammetry and used the current responses to extract the corresponding thermodynamic and kinetic parameters, showing that the use of ionic liquids does not modify the mechanistic pathways of the reactions examined. The electron transfer rates for reactions between aromatic molecules and the electrode decrease by an order of magnitude on going from molecular solvents to ionic liquids, an effect that has been attributed to a higher solvent reorganization during the charge transfer. On the other hand, the kinetics of the chemical reactions triggered by the electron transfer are only slightly affected by the use of ionic liquids. The only notable effect is a decrease in the bimolecular reaction rates, which has been attributed partly to a lowering of the diffusion-controlled kinetic rate constants together with a poorly specified "specific solvation effect" of reactants in these special media characterized by a high-degree of ion association.

More recently, the same authors studying the effects of the ionic liquids [BMIM][Tf_2N], [Me_3BuN][Tf_2N] and [Et_3BuN][Tf_2N] on the cleavage reactivity of aromatic radical anions have given [23] more detailed information about the ability of ionic liquids to affect reactivity through "specific solvation effects" (Scheme 5.1-4). The reactivity of radical anions is considerably modified when changing from acetonitrile to an ionic liquid.

However, the effects mainly depend on the charge delocalization in the radical anion and the nature of the ionic liquid's cation plays a secondary role. When the charge is spread out over the entire molecule, as in the 9-chloroanthracene radical anion, the carbon–halogen bond cleavage accelerates in ionic liquids due

$$ArX + \bar{e} \rightleftarrows ArX^{\bullet-}$$

$$ArX^{\bullet-} \longrightarrow Ar^{\bullet} + X^{-}$$

Scheme 5.1-4 Cleavage reactivity of aromatic radical anions.

to specific interaction (ion pairing) between the cation of the ionic liquid and the small leaving ion. On the contrary, in radical anions where the negative charge is more concentrated on a small area of the molecule, such as in the case of 4-chlorobenzophenone where it is mostly on the oxygen atom of the carbonyl group, a large decrease in the cleavage activity occurs on changing from acetonitrile to ionic liquids. This behavior has been attributed to the fact that, in this latter case, the stabilization of the leaving group still exists but it is largely overcompensated by the stabilization of the radical anion, in agreement with the positive shift of the reduction potential. Ion pair association stabilizes the π^* orbital where the unpaired electron is located and thus decreases its ability to be transferred into the σ^* of the breaking bond. This change in reactivity can be accompanied by a total change of mechanism and the dimerization between two charged radical anions can become favored over the cleavage reaction. It is noteworthy that a strong ion pairing between the imidazolium cation and the dinitrobenzene dianion was previously shown by cyclic voltametric measurements [24].

The question of "ion pairing" in ionic liquids is more complicated and quite different from that encountered in conventional solvents where ion pair association is built between a discrete ion surrounded by solvent molecules and another solvated species. In ionic liquids, there is no molecular solvent to solvate the ionic species. The solvent shell is ionic and an ionic species, dissolved or generated in an ionic liquid, should always be surrounded by the counteranions of the ionic liquid. The formation of clusters between cation(s) and anionic solutes (or between anion(s) and cation solutes) is a process that competes with the "intra-association" between the ions of ionic liquids and depends on their relative strengths.

5.1.2.2 Diels–Alder Reactions

Important information about ionic liquid properties has also been obtained by the study of reactions occurring through isopolar transition states, such as Diels–Alder reactions. Diels–Alder reactions have been widely investigated in molecular solvents and it has been suggested that the hydrogen-bond donating capacity and the cohesive energy density of the solvent are the determining solvent properties able to affect both the kinetic and stereochemical behavior of these reactions. Diels–Alder reactions can indeed be efficiently performed using water as reaction medium [25]; in this solvent the rates and the endo/exo ratios (when they can be observed) are much higher than in common organic solvents.

Recently, ionic liquids have also been used with success as solvents for Diels–Alder reactions. The reactions in ionic liquids are indeed marginally faster than in water but are considerably faster than in diethyl ether. Furthermore, it has

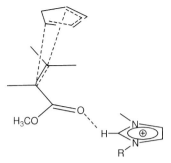

Fig. 5.1-1 Ability of the 1,3-dialkylimidazolium cation to hydrogen bond methyl acrylate in the course of its reaction with cyclopentadiene.

been shown that, as with molecular solvents, the presence of a Lewis acid greatly accelerates the reaction and improves selectivity. The acidity of chloroaluminates [26], or ionic liquids containing $ZnCl_2$ and $SnCl_2$ [27], have been used to this purpose. The molecular origin of how ionic liquids influence this reaction is, however, always a matter of controversy. A solvophobic effect, able to generate an "internal pressure" and to promote the association in a cavity of the solvent, has been initially invoked to explain the kinetic and stereochemical behavior of Diels–Alder reactions carried out in ionic liquids [28, 29]. A more recent study on the reaction of cyclopentadiene with methyl acrylate in several ionic liquids has, however, provided evidence that the ionic liquid's hydrogen-bond donor ability increases reactivity and selectivity [30]. Moreover, the determination of the selectivity in five [BMIM]$^+$-ionic liquids has shown that the nature of the anion also affects the endo/exo ratio. Higher selectivities characterize ionic liquids having the smaller hydrogen-bonding interaction between cation and anion. The *endo* selectivity has therefore been explained considering that the ability of the cation to hydrogen bond methyl acrylate is determined by two competing equilibria (Fig. 5.1-1).

Since both anion (A) and solute (S) can be hydrogen bonded to the cation, the concentration of the bonded methyl acrylate is inversely proportional to the equilibrium constant for the formation of the cation–anion hydrogen-bonded adduct (Scheme 5.1-5).

These data, although they give important information about the hydrogen donor ability of 1,3-dialkylimidazolium ionic liquids, do not exclude the possibility that other factors can influence the rate and selectivity of Diels–Alder reactions. Recently, the kinetic study of reaction between the first excited state of molecular oxygen and

$$[BMIM]^+ + A^- \rightleftharpoons [BMIM]\text{--}A$$

$$[BMIM]^+ + S \rightleftharpoons [BMIM]^+\text{--}S$$

Scheme 5.1-5 Equilibria for the formation of the cation–anion hydrogen-bonded adduct vs. the cation–solute hydrogen-bonded adduct.

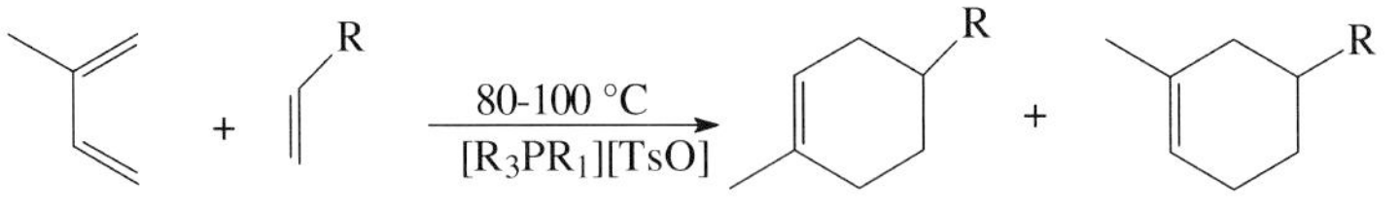

R = CO$_2$CH$_3$	>99:1
R = COCH$_3$	>99:1
R = CN	72:28

Scheme 5.1-6 Diels–Alder reactions of isoprene with methyl acrylate, acrylic acids, but-3-en-2-one and acrylonitrile in phosphonium tosylates.

1,4-dimethylnaphtalene has shown [4] that the variation in reaction kinetic constants in the [BMIM]$^+$-based ionic liquids examined arises from lower activation enthalpies associated with decreased cavitation volumes for the transition states with respect to solvated reagents. The kinetic constants increase with increasing solvent cohesive pressure. Although this reaction does not allow the investigation of the problem of the endo/exo ratio, the authors suggest that in the cases where there are two competing product-forming transition states, that with the most negative $\Delta V^{\#}_{solv}$ should be increasingly favored as the cohesive pressure is increased, if this latter parameter is the dominant controlling factor. In other words, cohesive pressure might also significantly affect the reaction diastereoselectivity. Unfortunately, at the moment only few data of cohesive pressure for ionic liquids have been reported and, therefore, it is not possible to verify the role of this parameter on the selectivity of the previously reported Diels–Alder reactions.

It is finally worth noting that phosphonium tosylates [31], and more recently pyridinium-based ionic liquids [32], have also been used as solvents for the Diels–Alder reactions of isoprene with methyl acrylate, acrylic acids, but-3-en-2-one and acrylonitrile (Scheme 5.1-6).

In phosphonium salts the reactions of isoprene with oxygen-containing dienophiles proceed with high regioselectivity (>99:1), whereas in pyridinium-based ionic liquids selectivity and reactivity depend on the ionic liquid anion ([BF$_4$]$^-$<[CF$_3$COO]$^-$), on dienophile nature and on the reaction time. The high regioselectivity characterizing many of these reactions has been attributed in both cases to the ability of the ionic liquid to coordinate substituents on the dienophile. Unfortunately, in this case the possible role of solvent cohesive pressure cannot be evaluated, since the δ^2 values for pyridinium- and phosphonium-based ionic liquids are, as yet, unknown.

5.1.3
Ionic Liquid Effects on Reactions Proceeding through Dipolar Transition States

Hughes and Ingold, primarily studying substitution and elimination reactions, have extensively investigated the effect of molecular solvents on reactions passing through dipolar-transition states [2]. On the basis of a simple qualitative solvation model, which exclusively considers electrostatic interactions between ions (or dipolar molecules) and solvent molecules in initial and transition states it is generally possible to predict the variations in reaction rates on changing the solvent. Despite the extensive and successful application of this theory, it does contain some inherent limitations. One of these limitations is the assumption that the reactions are controlled by enthalpy changes; the contribution of entropy changes to changes in Gibbs energy of activation are considered negligible. A second limitation concerns the fact that the solvent is treated as a dielectric continuum, able to interact with the solute molecules exclusively through non-specific interactions. Third, the model considers exclusively the static equilibrium transition-state solvation. Fourth, it neglects the changing solvent structure and therefore it ignores the related entropic contributions. In conventional media, interactions inside the solvents are generally small compared to solute–solvent interactions, but surely consideration should be given to solvent association when reactions are carried out in highly structured solvents, such as ionic liquids. It is possible that all of the limitations may be important in the case of ionic liquids, as we will see below.

5.1.3.1 Nucleophilic Substitution Reactions

Nucleophilic substitution reactions in ionic liquids have recently been the subject of both synthetic [33] and kinetics and mechanistic studies [34]. Ionic liquids may be efficient promoting media for nucleophilic displacement reactions and important information about the ionic liquid properties has been obtained from the study of these reactions. The kinetic investigation of bimolecular substitution reactions of the halides on methyl p-nitrobenzenesulfonate, recently carried out by Welton et al. [34] in several ionic media characterized by the same anion, $[Tf_2N]^-$, and different cations ($[BMIM]^+$, $[BMMIM]^+$ and $[BMPY]^+$) or by the same cation, $[BMIM]^+$ and different anions ($[BF_4]^-$, $[PF_6]^-$, $[SbF_6]^-$ and $[Tf_2N]^-$), has shown that the halides' nucleophilicities depend on the ionic liquid structure (Scheme 5.1-7).

In molecular solvents, this reaction can proceed either through discrete anions or through ion pairs. The nucleophilic reactivity of an anion depends significantly on the degree of association with the cation and on the hydrogen-bond donor ability of the solvent. In ionic liquids, halides are coordinated by cations and the presence of "free" anions can be excluded, although coordinated anions are quite different from ion pairs in molecular solvents. In all of the ionic liquids examined the reactivity of chloride, bromide and iodide is lower than in chlorinated solvents, in agreement with the more polar nature of ionic liquids (as suggested by the Kamlet–Taft π^* parameter). Polar solvents should indeed disfavor reactions occurring through more charge diffuse transition states than reagents. The order of halide nucleophilicity is, however, affected by the ionic liquid cation and anion. In $[BMPY][Tf_2N]$,

Scheme 5.1-7 Bimolecular substitution reaction of methyl p-nitrobenzenesulfonate with halides.

the nucleophilicity scale is $Cl^- > Br^- > I^-$, in agreement with the known gas-phase nucleophilicity trend. In [BMIM][Tf$_2$N], the scale is $I^- > Br^- > Cl^-$, indicating that some influence of the IL is acting differentially on the three halides, whereas in [BMMIM][Tf$_2$N] the situation is in some way intermediate, with nucleophilicity changing $Cl^- > I^- > Br^-$. The differences in the kinetic constants on going from one ionic liquid to the next are, however, very small, except where the halide is chloride. Since chloride is the best hydrogen-bond acceptor of the halides, it is highly probable that its nucleophilicity is mainly affected by the ability of the cation to interact through hydrogen bonding with chloride. The halides nucleophilicity is also affected by the anion nature: in [BMIM][PF$_6$] chloride is more nucleophilic than bromide. In [BF$_4$]$^-$ and [SbF$_6$]$^-$ ionic liquids, chloride and bromide have practically the same nucleophilicity, whereas in ionic liquids having [OTf]$^-$ and [Tf$_2$N]$^-$ as anions, the chloride is less nucleophilic than the bromide. Iodide is generally more nucleophilic than bromide, although in [BMIM][Tf$_2$N] they have similar nucleophilicity. It is evident that when the ionic liquid anion is a good hydrogen-bond acceptor (higher β value), it may compete with the halide to form a hydrogen bond with the cation, and an effect on halide nucleophilicity can be expected. However, the correlation between the kinetic constants measured in the different ionic liquids and the β parameter is not complete, suggesting the influence of other solvent effects.

A strong cation–anion association in the ionic liquid can affect not only the ability of the cation to give hydrogen bonding but more generally, it can modify the degree of association of the cation to the nucleophile, affecting both the activation enthalpy and entropy of the process. Evidence in support of this latter hypothesis has been obtained [34] from the examination of the activation parameters. The activation parameters found for the reaction of chloride and bromide in different ionic liquids have been rationalized [34] by considering that, in these bimolecular reactions (S$_N$2), the halide is coordinated to several cations and the enthalpies are generally very similar to those characterizing the reactions via ion pairs in dichloromethane. The activation entropies (often large and negative) are, however, more similar to those found in molecular solvents for free ions (large and negative) than ion pairs (positive). This different behavior between entropy and enthalpy has been explained [34] by considering that the reaction occurs through a pentacoordinate transition state. The activation step is an association process, which should have negative entropy. In dichloromethane, it has a positive value because, when the activated complex is formed, the cation of this ion pair is liberated as a free solvated cation and the leaving group cannot be stabilized by ion pairing (Fig. 5.1-2).

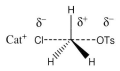

Fig. 5.1-2 Transition state in the bimolecular substitution reaction of methyl p-nitrobenzenesulfonate with chloride.

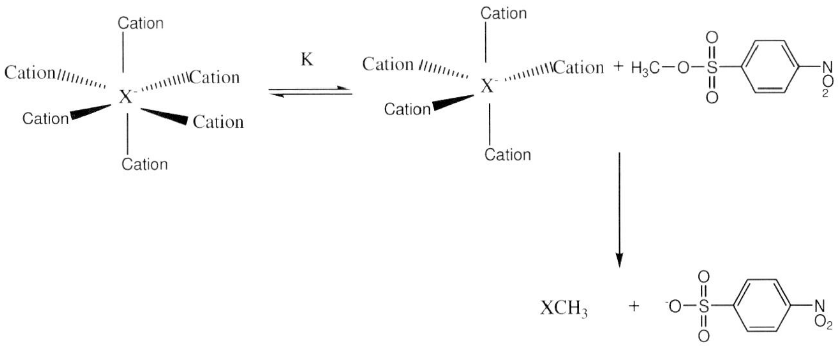

Scheme 5.1-8 Equilibrium between fully coordinated "unavailable" chloride and a one face "available" chloride in the bimolecular substitution reaction of methyl p-nitrobenzenesulfonate with chloride.

In ionic liquids, the leaving group associates instead with the ionic liquid's cation(s). The entropy gained by liberating a cation is cancelled out by the association of another cation with the leaving group. It is, however, interesting to note that the enthalpy and entropy change significantly on changing the ionic liquid's anion. In particular the reactions in [BMIM][BF$_4$] are accompanied by particularly large entropic barriers. This behavior has been interpreted [34c] considering that in the ionic liquid an equilibrium exists between fully coordinated "unavailable" chloride and a one face "available" chloride which can associate with the substrate (Scheme 5.1-8).

This loose association of available chloride with the substrate should represent the ground state. The dissociation of the cations from the halide to a greater or lesser degree compensates for the loss of entropy arising from the S$_N$2 mechanism. If the anion–cation linkages are relatively strong (as in the case of [BMIM][BF$_4$]), the halide is more weakly coordinated and its reaction will be characterized by a lower activation enthalpy associated with a larger activation entropy.

Although ionic liquids have also been used as solvents for alkane fluorination through nucleophilic substitution processes, no quantitative data about this reaction have been reported [35]. The preparative experiments carried out under non-identical conditions cannot be easily compared with each other. However, they seem to suggest that the nucleophilicity and basicity of the fluoride anion in ionic liquids depends on the counteranion and on the presence of cosolvents and water. In the presence of acetonitrile and small amounts of water, the reactions using RbF or CsF proceed much faster than the reaction of KF. In the absence of cosolvents and water, KF practically does not react with alkyl bromides whereas CsF gave mixtures of substitution and elimination products. Moreover, under these conditions a slow degradation of the ionic liquid due to the basicity of CsF has been observed. The different reactivity of the metal fluorides is probably determined by the tightness of the ion pairs, a parameter influenced also by the presence of water in the reaction media. It is, moreover, notable that KF in 1,3-dialkylimidazolium methanesulfonates, in the

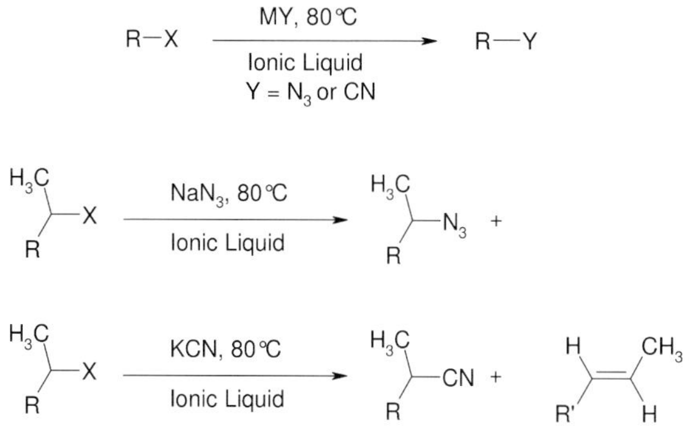

Scheme 5.1-9 Reactions of alkyl halides or tosylates with KCN and NaN$_3$ in different ionic liquids.

absence of cosolvents or added water, has been used successfully as a catalyst for the esterification of carboxylic acids with alkyl halides (generally chlorides) [36]. Under these conditions KF is not able to act as a nucleophile but its interaction with the carboxylic acid through a strong hydrogen bond with the hydroxy proton increases the nucleophilicity of the organic part of the complex and favors the reaction of the acid with the alkyl halide. The ionic liquid behaves in this case as a polar aprotic solvent.

In this research field, the shift of the substitution mechanism from S_N2 to S_N1 and the competition between substitution and elimination have recently been investigated [37]. The reactions of primary, secondary and tertiary halides or tosylates with KCN and NaN$_3$ in [BMIM][PF$_6$], [BMIM][Tf$_2$N] and [HPY][Tf$_2$N] were examined (Scheme 5.1-9). The observed ability of Cl$^-$, Br$^-$, I$^-$, and tosylate to act as leaving groups in the substitution reaction of NaN$_3$ is similar to that reported for the same process in cyclohexane (it exactly corresponds to that calculated for S_N2 reactions in the gas phase), suggesting the absence of strong specific interactions between the examined ionic liquids and the activated complex. The high reactivity of the examined secondary substrates (secondary ≥ primary) in the reaction of NaN$_3$ in ionic liquids disagrees with a pure S_N2 mechanism, although the exclusive formation of substitution products with complete inversion of configuration excluded the S_N1 mechanism for these substrates.

Considering, therefore, the overall reaction order and the product distribution, it has been suggested [37] that the reaction of NaN$_3$ in ionic liquids occurs with a gradual shift of the mechanism from S_N2 to S_N1, depending on the substrate. The reaction should proceed through the rear-side nucleophilic attack of N_3^- with a pure S_N2 mechanism in the case of primary substrates. A pure S_N1 mechanism probably occurs in the case of tertiary substrates, whereas the reaction of secondary halides and tosylates implies "the nucleophilically assisted formation of an ion pair intermediate." Practically, in the case of secondary substrates the bond-making and bond-breaking in the transition state are not two synchronous processes but the latter precedes the former (Scheme 5.1-10).

Evidence in favor of a larger amount of carbocation character in the transition state of secondary substrates in ionic liquids has also been obtained by comparison

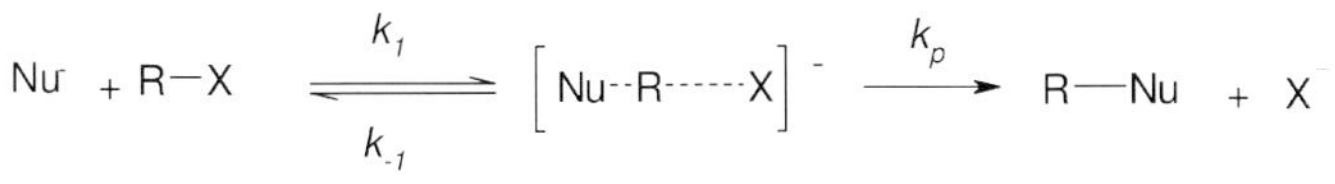

$$\text{Nu}^- + \text{R}-\text{X} \underset{k_{-1}}{\overset{k_1}{\rightleftharpoons}} \left[\text{Nu}\cdots\text{R}\cdots\text{X}\right]^- \overset{k_p}{\longrightarrow} \text{R}-\text{Nu} + \text{X}^-$$

Scheme 5.1-10 Nucleophilically assisted formation of an ion pair intermediate.

of the reaction of 2-bromoheptane with CN^- and N_3^-. The reactivity order of 2-bromoheptane ($N_3^- \gg CN^-$) is opposite to that observed [38] in molecular solvents for displacement reactions on methyl halides (pure S_N2 reaction), but the same as for nucleophilic reactions with carbocations.

Recently, the classical Menschutkin reaction between two neutral reagents has also been investigated in several ionic liquids [39,40]. In the first study by Skrzypczak and Neta, the kinetic constants for the reaction of 1,2-dimethylimidazole with benzylbromide were determined [39] in twelve ionic liquids and a comparable number of molecular solvents (Scheme 5.1-11).

Whereas the rate constants in molecular solvents vary by two orders of magnitude on going from 1-propanol to acetonitrile, the variation in the ionic liquids examined was less than a factor of three and the values are similar to the highest two values in molecular solvents, i.e. in acetonitrile and propylene carbonate. The variations within the ionic liquids indicate a small effect of the anion ($[Tf_2N]^- <$ $[PF_6]^- < [BF_4]^-$) and a moderate effect of cation (adding a methyl at C-2 of the imidazolium ring increases the rate constant slightly) and the observed trend is in agreement with a negative contribution of the ionic liquid's hydrogen-bond donor ability. Whereas the dipolarity/polarizability of ionic liquids stabilizes the transition state of this reaction, the ionic liquid's hydrogen-bond acidity decreases the nucleophilicity of 1,2-dimethylimidazole. The correlation of the rate constants with the solvatochromic parameter $E_T(30)$ is, however, reasonable within each group of similar solvents but very poor when all the solvents are correlated together. Really, for the organic solvents a better correlation has been obtained using a combination of two parameters, π^* (dipolarity/polarizability) and α (hydrogen-bond acidity) but this approach was not tried for ionic liquids, for which only the $E_T(30)$ parameter was known at the time.

In the light of the more recently reported data about the alkylation of tertiary amines (see discussion reported below) [40], I have tried to correlate the rate

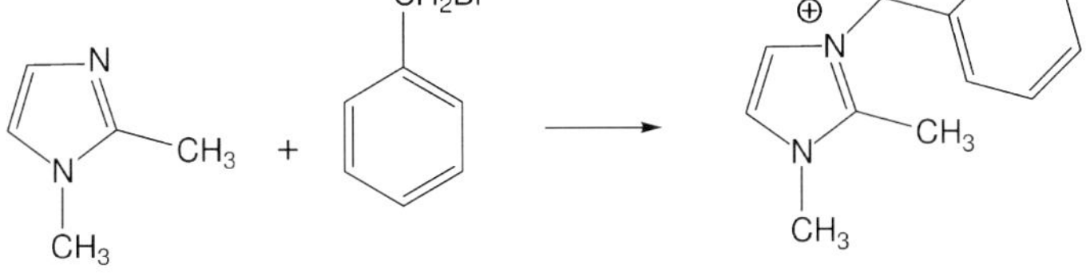

Scheme 5.1-11 Reaction of 1,2-dimethylimidazole with benzyl bromide.

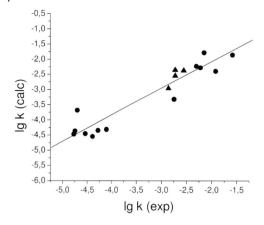

Fig. 5.1-3 Multiparameter correlation of the rate constant for the reaction of 1,2-dimethylimidazole with benzyl bromide with π^*, δ^2 and α solvent parameters. Molecular solvents (•), IL(▲).

constants in ionic liquids with the parameters π^*, α and δ^2. These parameters should take into account not only the dipolarity/polarizability and the hydrogen-bond donor ability but also the anion–cation association. Since all three parameters are known for only a limited number of ionic liquids, the correlation has been verified by also using the data related to the reactions in molecular solvents. The correlation (Fig. 5.1-3) fits Eq. (5.1-1)

$$\log k = -4.27 + 2.13\pi^* - 2.71\alpha + 1.37 \times 10^{-3}\delta^2 \qquad (5.1\text{-}1)$$

with a correlation coefficient $r^2 = 0.93$.

The dependences on π^*, δ and α have quite small uncertainties, respectively ±19%, ±44%, ±15%, showing that all three parameters contribute to the overall reaction rate. Furthermore, the inclusion of the values related to the ionic liquids drastically decreases the uncertainty of the dependence on δ^2 (from ±800% to ±44%) and this parameter, which can be neglected in the case of molecular solvents, becomes important to improve the correlation when ionic liquids are included.

The data arising from the kinetics of the reaction of three n-butylamines (mono, di, tri-substituted amines) with methyl p-nitrobenzene sulfonate in a variety of

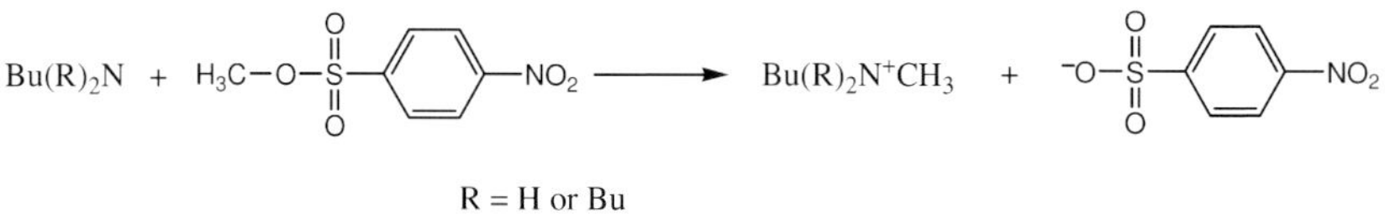

Scheme 5.1-12 Reaction of butylamines with methyl p-nitrobenezene sulfonate in different ionic liquids.

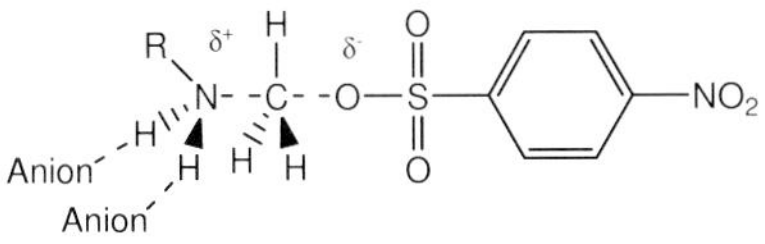

Fig. 5.1-4 Transition state of the reaction of alkylamines with methyl p-nitrobenezene sulfonate.

ionic liquids, acetonitrile and dichloromethane are also in agreement with this analysis (Scheme 5.1-12).

All amines are more nucleophilic in ionic liquids than in the molecular solvents. The Eyring activation parameters reveal that the changes in the activation entropies are mainly responsible, which can be at least partially rationalized on the basis of the differing hydrogen-bonding abilities of ionic liquids, as expressed by the Kamlet–Taft solvent parameters. Reactions of amines in ionic liquids are accelerated by the high polarizability/dipolarity term, π^*, characterizing these solvents (in agreement with the previously reported correlation). Moreover, where the formation of hydrogen bonds between the nucleophile and the anion of the ionic liquid is possible (primary and secondary amines), the reaction is further accelerated by the hydrogen-bonding ability of the ionic liquid (Fig. 5.1-4).

This is unlike the formation of hydrogen bonds between the nucleophile and the cation, possible for all amines, which tends to reduce the reactivity of the nucleophile and should be avoided (in accord with the sign of the α parameter in Eq. (5.1-1)). On the other hand, the kinetic data and the thermodynamic parameters related to the reaction of Bu_3N in [BMPY][OTf] suggest also that, where there is no opportunity for the nucleophile to hydrogen bond to either the anion or cation of the ionic liquid, other effects such as the self-association of the ionic liquid may become important. Related to this feature, it has been suggested [40] that the inability of the reagents to disrupt the Coulombic attraction of the ionic liquid's ions may leave them with a restricted volume in which to move, producing a solvatophobic effect. Furthermore, since the activated complex is a species containing both δ^+ and δ^- moieties, it is capable of interacting with the ionic liquid's ions. In so doing, it may disrupt the interionic interactions and cause the local structure of the ionic liquid to break down, producing an increase in activation enthalpy and entropy ($\Delta S^{\neq}$ becomes less negative) (Fig. 5.1-5).

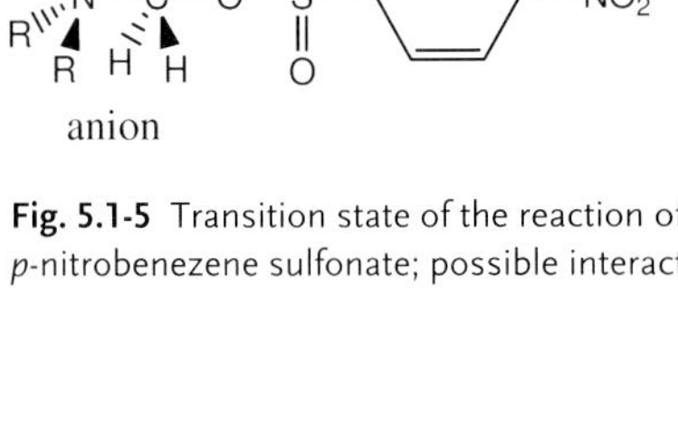

Fig. 5.1-5 Transition state of the reaction of alkylamines with methyl p-nitrobenezene sulfonate; possible interactions in an ionic liquid.

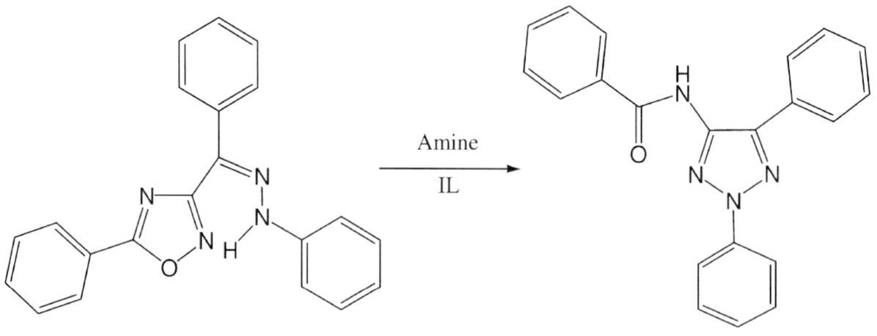

Scheme 5.1-13 Rearrangement of the Z-phenylhydrazone of 3-benzoyl-5-phenyl-1,2,4-oxadiazole in ionic liquids.

In agreement with the behavior of amines, the reactivity of other neutral nucleophiles, such as water and methanol, towards primary and secondary haloalkanes in ionic liquids is generally higher than in molecular solvents and lower amounts of elimination products are formed. Water and methanol are more nucleophilic but less basic in ionic liquids than in molecular solvents [41].

The inability of ionic liquids to interact strongly with neutral nucleophilic species has been considered recently to be the main factor determining the high rate of rearrangement of the Z-phenylhydrazone of 3-benzoyl-5-phenyl-1,2,4-oxadiazole in ionic liquids in comparison to molecular solvents (Scheme 5.1-13) [42].

This intramolecular ring to ring rearrangement, induced by amines, has been investigated in [BMIM][PF$_6$] and [BMIM][BF$_4$]. With the exception of an acid–base equilibrium between the solvent and amine, the data seem to indicate that other interactions (such as substrate–ionic liquid, amine–ionic liquid and amine–amine) have scarce relevance, whereas the high dipolarity/polarizability of ionic liquids stabilizes the rate-determining transition state and so increases the reaction rate.

In comparison to bimolecular nucleophilic substitutions, the monomolecular reaction (S_N1) in ionic liquids has been less extensively investigated. The recent work on the glycosidation of glucopyranosyl fluoride, in the presence of acids has, however, given important information about this kind of reaction (Scheme 5.1-14). The stereoselectivity of the glycosidation of glucopyranosyl fluoride is significantly affected by the ability of the ionic liquid anion to interact with the positively charged intermediate [43].

This reaction, which takes place through an S_N1 type mechanism, involves an oxonium intermediate and, since no epimerization of the generated glycosidic bond

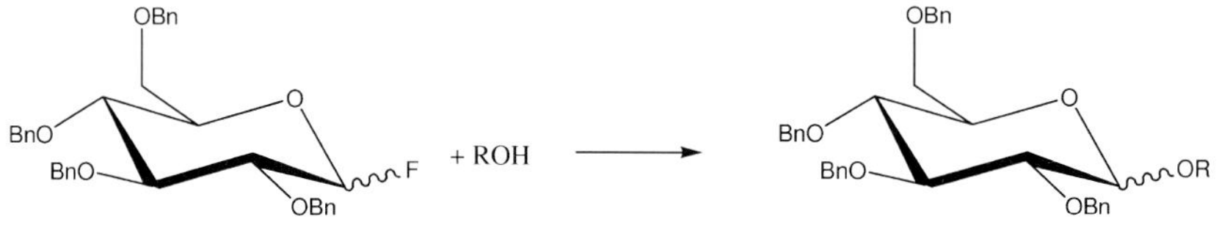

Scheme 5.1-14 Glycosidation of glucopyranosyl fluoride.

Fig. 5.1-6 Ionic intermediates of the S_N1 type glycosidation of glucopyranosyl fluoride in presence of $[Tf_2N]^-$ and $[OTf]^-$ anions, respectively.

occurs during the reaction, the product distribution may be considered to be under kinetic control. In [H−MIM][Tf$_2$N], containing HNTf$_2$, the reaction is characterized by an α-stereoselectivity, attributable to the anomeric effect; the alcohol attacks the oxonium ion preferentially on the α face. However, using [H−MIM][ClO$_4$]/HCl$_4$ or [H−MIM][OTf]/HOTf as solvent, the stereoselectivity inverts (β-stereoselectivity) and this behavior may be rationalized considering a stronger coordination of the perchlorate or trifluoromethanosulfonate with the oxonium intermediate (on the α-face) which drives the alcohol attack on the β-face (Fig. 5.1-6).

It is noteworthy that experiments carried out using acids having a different anion with respect to the ionic liquid show that the β-stereoselectivity is induced exclusively by the anion of the ionic liquid, i.e. by its coordination with the positively charged intermediate. This behavior shows that, independently of the acid determining the formation of the oxonium intermediate, the metathesis reaction is so efficient that the anion of the ionic liquid is always the sole effective counteranion of the oxonium intermediate. The fact that the ionic intermediates undergo a very fast anion exchange is important, not only for the product distribution but also for the kinetic behavior of this kind of reaction. Although in ionic liquids the carbenium ion intermediates are always present as "ion pairs" or "ion aggregates" the rapid exchange of the anion renders the internal return of the primarily formed intermediates unimportant. The kinetic behavior of these species in ionic liquids should be more similar to that characterizing molecular solvents, "solvent separated ion pairs" or "free" ions.

Finally, the data related to the application of ionic liquids for the cleavage of ethers are also notable. Chloroaluminates [44], anhydrous hydrobromic acid in 1-methylimidazole [45] and concentrated hydrobromic acid in [BMIM][BF$_4$] or p-TsOH/[BMIM][Br] in [BMIM][BF$_4$] [46] have been used to regenerate phenols from the corresponding aryl alkyl ethers. In ionic liquids, in the presence of an efficient proton donor (HBr or TsOH), the bromide anion behaves as a strong nucleophile able to cleave ethers. These latter reactions seem, therefore, to indicate that, as opposed to the situation in the absence of acids, the nucleophilicity of the bromide anion in ionic liquids in the presence of proton donors is higher than in molecular solvents (both protic and aprotic). This behavior may be a consequence of a different acidity of HBr in ionic liquids with respect to molecular solvents, associated with the different solvation ability of the two reaction media. In water, HBr is highly dissociated but Br^- is strongly solvated; conversely, in chlorinated solvents Br^- is much less solvated and therefore more nucleophilic, but HBr is practically undissociated. Ionic liquids might represent a good compromise between these two opposite situations.

The increased nucleophilicity of halides ([BMIM]X) in the presence of Brønsted acids has also been used to convert alcohols into alkyl halides under mild conditions. Since the mechanism of this reaction should involve the initial protonation of the OH group followed by nucleophilic displacement, the observed increase in the reaction rate may be attributed to an increased acidity of HBr associated with a higher nucleophilicty of Br^-, together with and enhanced ability of the ionic liquid to aid the charge separation in the transition state [47].

5.1.3.2 Electrophilic Addition Reactions

Electrophilic addition reactions are another class of reactions that have been extensively studied in organic solvents from a mechanistic point of view and bromine addition is one of the most investigated addition reactions. Chiappe et al. have used ionic liquids to synthesize vicinal dihaloalkanes and dihaloalkenes by electrophilic addition of halogens to double and triple bonds (Scheme 5.1-15) [48–50]. Recently, dibromides have also been synthesized [51] in ionic liquids using electrogenerated bromine, whereas bromohydrins have been obtained [52] under two-phase conditions (water/IL) through a vanadium (V) catalyzed oxidation of bromide ions by hydrogen peroxide.

Bromine addition in [BMIM][PF$_6$] and [BMIM][BF$_4$] is a stereospecific *anti*-addition process with dialkyl substituted alkenes, alkyl substituted alkynes and *trans*-stilbenes, whereas *cis*-stilbenes and aryl alkynes give mixtures of *syn*- and *anti*-addition products, although in the case of *cis*-diaryl substituted olefins the *anti*-stereoselectivity is generally higher than in chlorinated solvents. In the case of diaryl substituted olefins, such as stilbenes, it has been shown that stereoselectivity in molecular solvents depends primarily on two factors: (i) the nature of the intermediates; and (ii) the lifetime of the ionic intermediates [53]. Bridged bromiranium

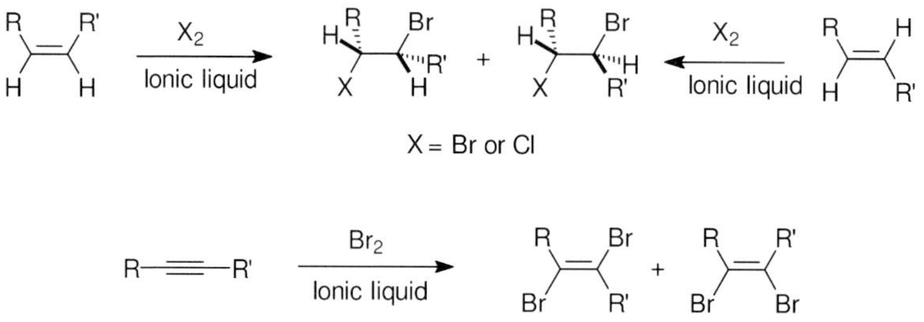

Scheme 5.1-15 Synthesis of vicinal dihaloalkanes and dihaloalkenes by electrophilic addition of halogens to double and triple bonds in various ionic liquids.

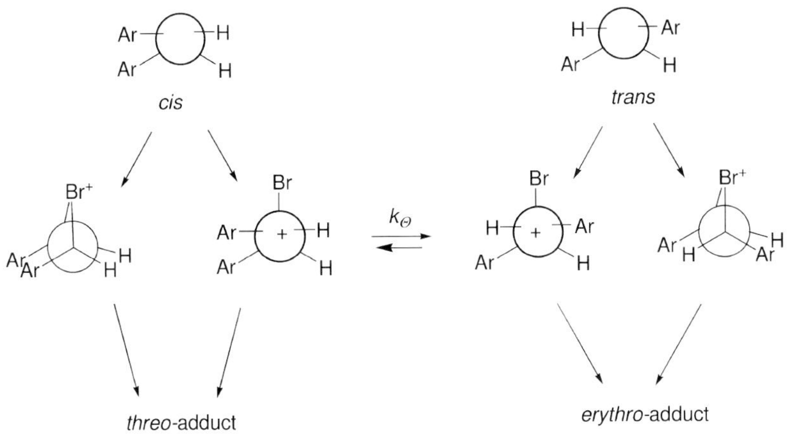

Fig. 5.1-7 Intermediates in the electrophilic addition of bromine to stilbenes.

ions give exclusively *anti*-addition products, whereas open β-bromocarbenium ions generally give mixtures of *syn*- and *anti*-addition products. The nature of the intermediate is determined by the nature of the substituents on the phenyl ring, electron-withdrawing groups favor bridged intermediates, electron-donating groups open β-bromocarbenium ions. Open β-bromocarbenium ions give the *syn*-addition products mainly through an attack *anti* to the counter anion, after rotation around the C–C bond (k_θ) (Fig. 5.1-7) . If the lifetime of the intermediate is sufficiently short, the rotation around the C–C bond is not able to compete with the nucleophilic attack of the counter anion and the *anti*-addition product can be obtained practically exclusively.

In molecular solvents, the nature of the ionic intermediate is not dependent on the properties of the solvent, whereas the lifetimes of the ionic intermediates are. In ionic liquids the nature of the intermediate is probably also not affected by the medium, whereas it does affect the lifetime of the intermediate or the rotation around the C–C bond, favoring the nucleophilic trapping of the first formed bromocarbenium ion.

The electrophilic additions of trihalide species (Br_3^-, ICl_2^- and IBr_2^-) to double and triple bonds have also been investigated. Ionic liquids having a trihalide as their anion have been prepared recently [50] by mixing equimolar amounts of halogens (ICl, or IBr, or Cl_2 or Br_2) to proper 1-alkyl-3-methylimidazolium halides, and the structures of the trihalide ions formed have been investigated by electrospray ionization mass spectroscopy (ESI-MS), UV and NMR. Spectroscopic measurements and reactivity data have shown that liquids bearing $[Br_3]^-$, $[I_3]^-$, $[IBr_2]^-$ and $[ICl_2]^-$ as counteranion are quite stable, whereas those having $[Br_2Cl]^-$, $[I_2Cl]^-$ and $[I_2Br]^-$ as anions are rather unstable, since these latter anions tend to disproportionate into the more stable trihalide species. Unlike protic molecular solvents, which favor dissociation due to the very high energy of solvation of the small Br^- or Cl^- anions, ionic liquids seem to disfavor this process. Their behavior is, therefore, significantly different from alcohols or water.

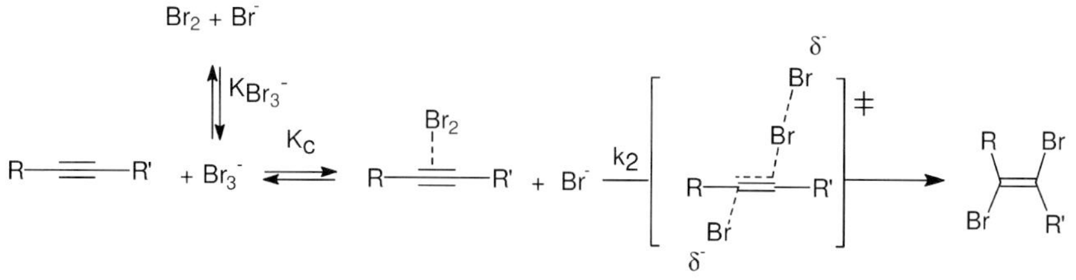

Scheme 5.1-16 Electrophilic addition of Br_3^- to alkynes.

Electrophilic additions of Br_3^- to alkenes and alkynes have been carried out [48–50] both in [BMIM][Br] and in other ionic liquids bearing non-nucleophilic anions (Scheme 5.1-16). The reaction is always completely *anti*-stereospecific, independent of alkene or alkyne structure. It follows a second-order rate law, suggesting a concerted mechanism of the type reported for Br_3^- addition in aprotic molecular solvents, involving a product- and rate-determining nucleophilic attack by bromide on the alkene or alkyne-Br_2 π-complex initially formed.

Since, in chlorinated solvents, Br_3^- addition is positively affected by the ability of the medium to hydrogen bond, the kinetic constants for the addition of Br_3^- to alkynes and alkenes have been measured in several ionic liquids having different hydrogen-bond abilities [49, 54]. Unlike the behavior observed in chlorinated solvents, however, no correlation has been found between the kinetic constants or the activation parameters and the ionic liquid's solvent parameters (π^* or α). The reaction in ionic liquids has, however, some important features with respect to those in molecular solvents. In particular, whereas in chlorinated solvents bond making is not influenced by the solvent hydrogen-bond ability, since the entering bromide ion is present as an ion pair with the tetraalkylammonium cation, in ionic liquids both bond breaking and bond making are affected by this property. Considering, moreover, that in this reaction both events (bond breaking and bond making) probably occur simultaneously in the rate-determining transition state only a very little effect, if any, may be detected.

The kinetic behavior for the addition of ICl_2^- to alkenes and alkynes is quite different. The mixed trihalide ionic liquids, in particular [RMIM][ICl_2] and [RMIM][IBr_2], have been used as excellent iodine-donor reagents for the stereoselective *anti*-iodination of alkenes and alkynes in [BMIM][PF_6] [50]. Very good to almost quantitative yields of vicinal iodochloro or iodobromo adducts were observed for all of the substrates examined. However, kinetic measurements carried out in several ionic media, using [BMIM][ICl_2] as the reactant, have shown [54] that the addition of ICl_2^- is affected much more by the physicochemical properties of the ionic liquid under investigation than the addition of Br_3^-, and in this case a possible role of the hydrogen-bonding donor ability of the ionic liquid has been suggested. The reaction is indeed characterized by a relatively low activation enthalpy, associated with a high activation entropy, when the addition process is performed in [BMIM][Tf_2N] (the ionic liquid among those examined with a relatively higher hydrogen-bonding donor ability). Conversely, when the reaction is performed in liquids characterized

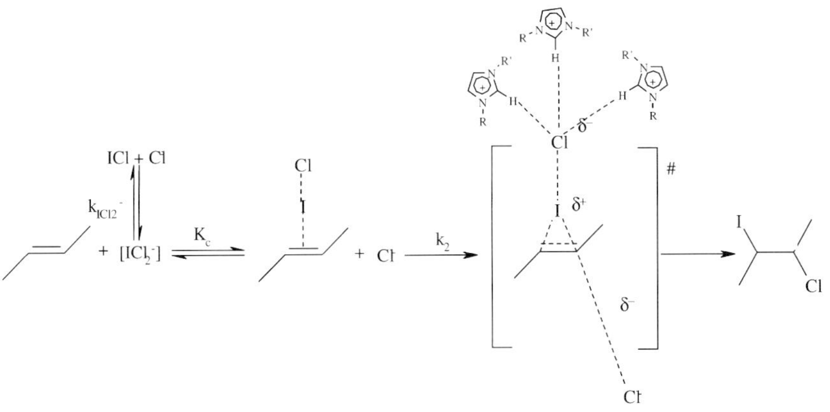

Scheme 5.1-17 Addition of ICl$_2^-$ to alkenes in a 1,3-dialkylimidazolium ionic liquid.

by cations less able to be hydrogen-bond donors, such as [BMMIM][Tf$_2$N] and [BPY][Tf$_2$N], the activation enthalpies increase significantly and activation entropies decrease. It is also interesting to note that a high activation enthalpy, associated with a low activation entropy, has been found in [BMIM][PF$_6$]. This behavior may be rationalized considering the high degree of order characterizing this ionic liquid, which may affect the kinetics of this reaction mainly through entropic effects.

Finally, the different sensitivity of addition processes of the two trihalides (Br$_3^-$ and ICl$_2^-$) to the ionic liquid properties is probably a consequence of the different charge development into the two transition states. The addition of ICl$_2^-$ is most likely characterized by a later transition state, having a greater iodiranium character, and the attack by chloride is not itself the rate-determining step (Scheme 5.1-17). The ability of the ionic liquid to be a hydrogen-bond donor can indeed affect the reaction rate only in the case where bond breaking precedes bond making.

5.1.3.3 Electrophilic Substitution Reactions

Electrophilic aromatic substitutions are reactions of great importance for industry and ionic liquids have been used in these reactions as alternative reaction media, mainly with the aim of reducing the environmental impact of these processes. Only few reactivity data, which might be used to quantitatively estimate the effect of the ionic reaction medium on these reactions, have been collected.

Electrophilic nitrations of aromatics have been performed in several ionic liquids but, as shown [55] by Laali, depending on the nitrating agent and the ionic liquid, nitration of the solvent may compete with the electrophilic substitution of the substrate. This problem has been recently underlined [56] by Lancaster: high yields of aromatic nitration products using HNO$_3$–Ac$_2$O have been obtained mainly in [BMPY][TF$_2$N], which has no aromatic cation. Using this nitrating agent, moreover, reaction of deactivated substrates is highly favored in ionic liquids with respect to molecular solvents. On going from dichloromethane to [BMPY][TF$_2$N] the yield of bromonitrobenzene increases from 0 to 63%. Although no kinetic study has been performed on this reaction, the authors suggested that the increased reactivity might be attributed generically to the increased polarizability/dipolarity of the medium, although they do not exclude the contribution of specific interactions. [BMPY][TF$_2$N] might favor the dissociation of acetyl nitrate to nitronium acetate (the nitronium ion is the nitrating agent) or it could interact more strongly with the Wheland intermediate. A more recent comparison [57] of the behavior of the reaction in [BMPY][Tf$_2$N] with that in [BMPY][OTf] has also shown that the anion affects the reaction; [BMPY][Tf$_2$N] is a better solvent for aromatic nitration. This effect has been attributed to the ease of formation of the nitronium ion in the former solvent, a feature which is consistent with the fact that [BMPY][Tf$_2$N] is a weaker hydrogen-bond acceptor than [BMPY][OTf].

Halogenated aromatic compounds are key intermediates in organic synthesis as they are precursors to a number of organometallic species used for the preparation of highly functionalized molecules. Several electrophilic halogenation reactions have been performed in ionic liquids, although again no systematic mechanistic study has been reported. Regioselective brominations in ionic liquids have been carried out both using NBS [58] or tribromide-based ionic liquids (pentylpyridinium perbromide [59], [BMIM][Br$_3$] and [H−MIM][Br$_3$] [60]). Although products are generally obtained in high yields in the case of tribromide-based liquids a higher efficiency of 3-methylimidazolium tribromide ([H−MIM][Br$_3$]), with respect to the corresponding non-acidic derivative [BMIM][Br$_3$], has been observed [60]. Using [H−MIM][Br$_3$], the halogenation is effective not only with activated compounds, but also with differently substituted benzaldehydes as well as with sterically hindered aromatic compounds. Moreover, it is notable that no phenol is detected from the halogenation of anisole catalysed by [H−MIM][Br$_3$], although 3-methylimidazolium bromohydrogenate ([HMIM][HBr$_2$]), the residue ionic liquid at the end of the reaction, has been used [45] in a 1:2 molar ratio to cleave the ether functionality.

Fluorinated arenes have instead been prepared using N−F fluorine transfer reagents, such as F−TEDA−[BF$_4$] [61]. Substrate selectivity measured in competitive experiments carried out for the electrophilic fluorination of arenes ($k_{mesitylene}$:k_{durene}), using F−TEDA−[BF$_4$] (Selectfluor^{TM}) as the halogenating agent, has suggested [62] that the reaction occurs through a polar mechanism, involving an ionic intermediate (σ-complex) and an ionic transition state in the rate-determining step. Comparison with the reaction in acetonitrile has, furthermore, suggested a slightly greater degree of polar character in the ionic liquid.

The same reagent, F−TEDA, in the presence of iodine in imidazolium and pyridinium ionic liquids has also been used [63] for the regioselective iodination of aromatic compounds. The reaction is *para*-directed when possible. Otherwise, it occurs in the *ortho*-position. In addition, competitive experiments ($k_{mesitylene}$:k_{durene}) suggest a polar mechanism for this process, in agreement with the behavior in molecular solvents.

Fluoroarenes have also been obtained using [EMIM]$^+$ and [BMIM]$^+$ salts with [BF$_4$]$^-$ and [PF$_6$]$^-$ counteranions as solvents for fluorodediazoniation and for *in situ* diazotination with [NO][BF$_4$] and [NO][PF$_6$] [64]. It is generally agreed that in conventional solvents fluorodediazoniation occurs via a tight ion pair. In ionic liquids the diazonium salt is surely coordinated with one or more anions, however, experiments carried out on [ArN$_2$][BF$_4$] in [EMIM][CF$_3$CO$_2$], [EMIM][OTs] and [EMIM][OTf] show that in these media the ionic intermediates behave like "solvent-separated ion pairs." Only products arising from the ionic liquid anion quenching were obtained in all investigated ionic liquids, showing that on mixing metathesis occurs immediately. This behavior is in agreement with the data related to the glycosidation reaction [43], previously discussed.

Finally, it must be noted that the ionic liquid anion may have an active role in this kind of reaction, determining the outcome of some chemical reactions [65]. The reaction of toluene and nitric acid gives a nitrated product in triflate and hydrogensulfate ionic liquids, whereas the monohalogenated product can be obtained in a halide salt. On the other hand, nitric acid acts as an oxidizing agent in methansulfonate ionic liquids to give benzoic acid in quantitative yields. Ionic liquids seem to be able to catalyze these reactions with the only by-product being water.

5.1.4
Conclusions

In this section, it has been highlighted that the use of ionic liquids as solvents for organic reactions is not merely an environmentally friendly alternative, but has considerable additional consequences on reactivity in the form of significant variations in reaction rates and selectivities.

The interactions between cations and anions render the ionic liquids highly ordered three-dimensional supramolecular networks and the possibility for cations or anions to interact with dissolved species (reagents, intermediates or transition states) is a competitive process which depends on the relative strengths of the anion–cation and solute–anion (or cation) interactions. When the solute is able to give interactions that can compete with those present inside the ionic liquid, the latter behaves as a polar solvent. The high polarizability/dipolarity, generally characterizing these solvents, together with the ability of cation and anion to hydrogen bond, determines the solvation of substrates, transition states and intermediates, and therefore their reactivity in ionic liquids. Moreover, since ionic liquids are normally much more ordered solvents than molecular ones, the formation of the activated complex from the reagents may, at least partially, destroy the degree of order of the system, giving significant entropic contributions to the reactions performed in these media.

Other properties of ionic liquids (such as viscosity, degree of order and the cohesive pressure) must be considered carefully, however, when these salts are employed as solvents. The importance of these latter parameters, moreover, increases when the solutes (reagents and transition states) are not able to give competitive interactions with the components of the ionic liquid. In this situation, all the properties which determine the ionic liquid's polarity become unimportant, and the systems, with the extensive association of their cations and anions, behave as non-polar solvents. The high cohesive pressure may favor associative processes, whereas the elevated viscosity may reduce the rate of diffusion-controlled reactions and very fast reactions may become controlled by the solvent re-orientation.

It is, however, notable that the strong attractive interactions between ions, which are at the origin of the tendency of ions to associate in ion aggregates, may determine the formation of cavities inside the IL. The presence of cavities may favor the motion of small, uncharged species and render ionic liquids in this situation much more similar to polymeric matrixes than to true solvents.

In conclusion, ionic liquids probably cannot be considered as polar solvents, nor as non-polar solvents, nor as polymeric matrixes. Their behavior arises from a fine balancing of entropic and enthalpic contributions that involve all of the interactions inside the ionic liquid and all of the possible interactions between cations and/or anions and dissolved species. The ability of ionic liquids to affect reactivity is, therefore, much more complicated to rationalize and more parameters must be considered to explain the "ionic liquid effect" on reaction rate and selectivity than with molecular solvents.

References

1. (a) J. D. Holbrey, K. R. Seddon, *Clean Prod. Proc.* **1999**, *1*, 223–236; (b) M. J. Earle, M. K. R. Seddon, *Pure Appl. Chem.* **2000**, *72*, 1391–1398; (c) T. Welton, *Chem. Rev.* **1999**, *99*, 2071–2083; (d) P. Wasserscheid, M. Keim, *Angew. Chem. Int. Ed.* **2000**, *39*, 3772–3789; (e) R. Sheldon, *Chem. Commun.* **2001**, 2399–2407; (e) H. Olivier-Bourbigou, L. Magna, *J. Mol. Catal. A* **2002**, *182*, 419–437; (f) J. Dupont, R. F. de Souza, P. A. Z. Suarez, *Chem. Rev.* **2002**, *102*, 3667–3692; (g) J. S. Wilkes, *J. Mol. Chem. A* **2004**, *214*, 11–17; (h) C. F. Poole, *J. Chromatogr. A* **2004**, *1037*, 49–82; (i) *Ionic Liquids in Synthesis*, P. Wasserscheid, T. Welton (Eds.), Wiley-VCH, Weinheim, **2003**; (l) T. Welton, *Coord. Chem Rew.* **2004**, *248*, 2459–2477; (m) C. Chiappe, D. Pieraccini, *J. Phys. Org. Chem.* in press.
2. C. Reichard, *Solvents and Solvent Effects in Organic Chemistry*, 3rd edn. VCH, Weinheim, **2003**.
3. (a) J. L. Anderson, J. Ding, T. Welton, D. W. Armstrong, *J. Am. Chem. Soc.* **2002**, *124*, 14247–14254; (b) M. H. Abraham, A. M. Zissimos, J. G. Huddleston, H. D. Willauer, R. D. Rogers, W. E. Acree, *Ind. Eng. Chem. Res.* **2003**, *42*, 413–418; (c) L. Crowhurst, P. R. Mawdsley, J. M. Perez-Arlandis, P. A. Salter, T. Welton, *Phys. Chem. Chem. Phys.* **2003**, *5*, 2790–2794.
4. K. Swiderski, A. McLaen, C. M. Gordon, D. H. Vaughan, *Chem. Commun.* **2004**, 2178–2179.
5. F. C. Gozzo, L. S. Santos, R. Augusti, C. S. Consorti, J. Dupont, M. N. Eberlin, *Chem. Eur. J.* **2004**, *10*, 6187.
6. P. A. Z. Suarez, S. Einloft, J. E. L. Dullius, R. F. de Souza, J. Dupont, *J. Chim. Phys. Phys. Chim. Biol.* **1998**, *95*, 1626.
7. P. Koelle, R. Dronskowski, *Eur. J. Inorg. Chem.* **2004**, *14*, 2313–2320.
8. R. M. Lynden-Bell, N. A. Atamas, V. Vasilyuk, C. G. Hanke, *Mol. Phys.* **2002**, *100*, 3225.
9. H. Katayanagi, K. Nishikawa, H. Shimozaki, K. Miki, P. Westh, Y. Koga, *J. Phys. Chem. B.* **2004**, *108*, 19451–19457.
10. J. D. Holbrey, W. M. Reichert, M. Nieuwenhuyzen, O. Sheppard, C. Hardacre, R. D. Rogers, *Chem. Commun.* **2003**, 476–477.

11. M. Deetlef, C. Hardacre, M. Nieuwenhuyzen, O. Sheppard, A. K. Soper, *J. Phys. Chem. B*. **2005**, *109*, 1593–1598.
12. R. Karmar, A. Samanta *J. Phys. Chem. A* **2002**, *106*, 4447–4452; R. Karmar, A. Samanta, *J. Phys. Chem. A* **2002**, *106*, 6670–6675; R. Karmar, A. Samanta, *J. Phys. Chem. A* **2003**, *107*, 7340–7346.
13. J. A. Ingram, R. S. Moog, N. I. R. Biswas, M. Maroncelli, *J. Phys. Chem. B*. **2003**, *107*, 5926–5932.
14. M. Alvaro, B. Ferrer, H. Garcia, M. Narayana, *Chem Phys. Lett*. **2002**, *362*, 435; M. J. Mildoon, A. J. McLean, C. M. Gordon, I. R. Dunkin, *Chem. Commun*. **2001**, 2364–2365.
15. G. M. Gordon, A. J. McLean, *Chem. Commun*. **2000**, 1395–1396.
16. (a) D. Behar, C. Gonzalez, P. Neta, *J. Phys. Chem. A*, **2001**, *105*, 7607–7614; (b) D. Behar, P. Neta, C. Schlultheisz, *J. Phys. Chem. A* **2002**, *106*, 3139–3147; (c) J. Grodkowski, P. Neta, *J. Phys. Chem. A* **2002**, *106*, 5468–5473; (d) J. Grodkowski, P. Neta, *J. Phys. Chem. A*, **2002**, *106*, 9030–9035; (e) J. Grodkowski, P. Neta, *J. Phys. Chem. A*, **2002**, *106*, 11130–11134.
17. A. Marccinek, J. Zielonka, J. Gębicki, C. M. Gordon I. R. Dunkin, *J. Phys. Chem. A* **2001**, *105*, 9305–9309.
18. A. Skrzypczak, P. Neta, *J. Phys. Chem. A* **2003**, *107*, 7800–7803.
19. F. Bresme, J. Alejendre, *J. Chem. Phys*. **2003**, *118*, 4134–4139.
20. J. Grodkowski, P. Neta, J. F. Wishart, *J. Phys. Chem. A* **2003**, *107*, 9794–9799.
21. D. Y. Chu, J. K. Thomas, *Macromolecules* **1990**, *23*, 2217.
22. C. Lagrost, D. Carrié, M. Vaultier, P. Hapiot, *J. Phys. Chem. A* **2003**, *107*, 745–752.
23. C. Lagrost, S. Gmouh, M. Vaultier, P. Hapiot, *J. Phys. Chem. A* **2004**, *108*, 6175–6182.
24. A. J. Fry, *J. Electroanal. Chem*. **2003**, *546*, 35.
25. J. J. Gajewski, *J. Org. Chem*. **1992**, *57*, 5500–5508.
26. C. W. Lee, *Tetrahedron Lett*. **1999**, *40*, 2461–2464.
27. A. P. Abbott, G. Capper, D. L. Davies, R. K. Rasheed, G. Tambyrajah, *Green Chem*. **2002**, *4*, 24–26.
28. M. J. Early, P. B. McCormac, K. R. Seddon, *Green Chem*. **1999**, *1*, 23–26.
29. J. F. Debreul, J. P. Bazureau, *Tetrahedron Lett*. **2000**, *41*, 7351–7355.
30. A. Aggarwal, N. L. Lancaster, A. R. Sethi, T. Welton, *Green Chem*. **2002**, *4*, 517–520.
31. P. Ludley, N Karodia, *Tetrahedron Lett*. **2001**, *42*, 2011–2014.
32. Y. Xiao, S. Malhotra, *Tetrahedron Lett*. **2004**, *45*, 8339–8342.
33. C. Wheeler, K. N. West, C. L. Liotta, C. A. Eckert, *Chem. Commun*. **2001**, 887–888; N. M. T. Lourenço, C. A. M. Afonso, *Tetrahedron*, **2003**, 789–794; L. Brinchi, R. Germani, G. Savelli, *Tetrahedron Lett*. **2003**, *44*, 2027–2029, Z. M. A. Judeh, H.-Y. Shen, B. C. Chi, L.-C. Feng, S. Selvatoshi, *Tetrahedron Lett*. **2002**, 9381–9384; D. W. Kim, C. E. Song, D. Y. Chi, *J. Org. Chem*. **2003**, *68*, 4281–4285.
34. N. L. Lancaster, T. Welton, G. B. Young, *J. Chem. Soc., Perkin Trans. 2* **2001**, 2267–2270; N. L. Lancaster, P. A. Salter, T. Welton, G. B. Young, *J. Org. Chem*. **2002**, *67*, 8855–8861; N. L. Lancaster, T. Welton, *J. Org. Chem*. **2004**, *69*, 5986–5992.
35. D. W. Kim, C. E. Song, D. Y. Chi, *J. Am. Chem. Soc*. **2002**, *124*, 10278–10279; C. B. Murray, G. Sandford, S. R. Korn, *J. Fluorine Chem*. **2003**, *123*, 81–84; D. W. Kim, Y. S. Choe, D. Y. Chi, *Nuclear Med. Biol*. **2003**, *30*, 345–350.
36. L. Brinchi, R. Germani, G. Savelli, *Tetrahedron Lett*. **2003**, *44*, 6583–6585.
37. C. Chiappe, D. Pieraccini, P. Saullo, *J. Org. Chem*. **2003**, *68*, 6710–6715.
38. R. G. Pearson, H. Sobel, J. Songstand, *J. Am. Chem. Soc*. **1968**, *90*, 319–326; C. D. Ritchie, J. Gandler, *J. Am Chem. Soc*. **1979**, *101*, 7318–7323.
39. A. Skrzypczak, P. Neta, *Int. J. Chem. Kinet*. **2004**, *36*, 253–258.
40. L. Crowhurst, N. L. Lancaster, J. M. P. Arlandis, T. Welton, *J. Am. Chem. Soc*. **2004**, *126*, 11549–11555.
41. D. W. Kim, D. J. Hong, J. W. Seo, H. S. Kim, H. K. Kim, C. E. Song, D. Y. Chi, *J. Org. Chem*. **2004**, *69*, 3186–3189.
42. F. D'Anna, V. Frenna, R. Noto, V. Pace, D. Spinelli, *J. Org. Chem*., **2005**, *70*, 2828–2831.

43. K. Sasaki, S. Matsumura, K. Toshima, *Tetrahedron Lett.* **2004**, *45*, 7043–7047.
44. G. J. Kemperman, T. A. Roeters, P. W. Hilberink, *Eur. J. Org. Chem.* **2003**, 1681–1686.
45. G. Driver, K. E. Johnson, *Green Chem.* **2003**, *5*, 163–169.
46. S. K. Boovanahalli, D. W. Kim, D. Y. Chi, *J. Org. Chem.* **2004**, *69*, 3340–3344.
47. R. X. Ren, J. X. Wu, *Org. Lett.* **2001**, *3*, 3727–3828.
48. C. Chiappe, D. Capraro, V. Conte, D. Pieraccini, *Org. Lett.* **2001**, *3*, 1061–1063.
49. C. Chiappe, V. Conte, D. Pieraccini, *Eur. J. Org. Chem.* **2002**, 2831–2837.
50. O. Bortolini, M. Bottai, C. Chiappe, V. Conte, D. Pieraccini, *Green Chem.* **2002**, *4*, 621–627.
51. G. D. Allen, M. C. Buzzeo, I. G. Davies, C. Villagran, C. Hardacre, R. C. Compton, *J. Phys. Chem. B*, **2004**, *108*, 16322–16327.
52. V. Conte, B. Floris, P. Galloni, A. Silvagni, *Pure Appl. Chem.*, **2005**, *77*, 1571–1581.
53. M. F. Ruasse, G. Lo Moro, B. Galland, R. Bianchini, C. Chiappe, G. Bellucci, *J. Am. Chem. Soc* **1997**, *119*, 12492–12502.
54. C. Chiappe, D. Pieraccini, *J. Org. Chem.* **2004**, *69*, 6059–6064.
55. K. K. Laali, V. J. Gettwert, *J. Org. Chem.* **2001**, *66*, 35–40.
56. N. L. Lancaster, V. Llopis-Mestre, *Chem Commun.* **2003**, 2812–2813.
57. E. Dal, N. L. Lancaster, *Org. Biomol. Chem.* **2005**, *3*, Advance Article.
58. R. Rajagopal, D. V. Jarikote, R. J. Lahoti, T. Daniel, K. V. Srinivasan, *Tetrahedron Lett.* **2003**, *44*, 1815–1817.
59. J. Salasar, D. Romano, *Synlett* **2004**, *7*, 1318–1319.
60. C. Chiappe, E. Leandri, D. Pieraccini, *Chem. Commun.* **2004**, 2536–2537.
61. J. Baudoux, A.-F. Salit, D. Cahard, J.-C. Plaquevent, *Tetrahedron Lett.* **2002**, *43*, 6573–6574.
62. K. K. Laali, G. I. Borodkin, *J. Chem. Soc, Perkin Trans 2* **2002**, 953–957.
63. C. Chiappe, D. Pieraccini, *ARKIV OC* **2002**, 249–255.
64. K.K. Laali, V. J. Gettwert, *J. Fluorine Chem.* **2001**, *107*, 31–34.
65. M. J. Earle, S. P. Katdare, K. R. Seddon, *Org. Lett.* **2004**, *6*, 707–710.

5.2
Stoichiometric Organic Reactions and Acid-catalyzed Reactions in Ionic Liquids

Martyn Earle

The field of reaction chemistry in ionic liquids was initially confined to the use of chloroaluminate(III) ionic liquids. With the development of "neutral" ionic liquids in the mid-1990s, the range of reactions that can be performed has expanded rapidly. In this section, reactions in both chloroaluminate(III) ionic liquids and similar Lewis acidic media are described. In addition, stoichiometric reactions, mostly in neutral ionic liquids, are discussed. Review articles are available by several authors, including Welton [1] (reaction chemistry in ionic liquids), Holbrey [2, 3] (properties and phase behavior), Earle [4] (reaction chemistry in ionic liquids), Rooney [5] (physical properties of ionic liquids), Seddon [6, 7] (chloroaluminate(III) ionic liquids and industrial applications), Wasserscheid [8] (catalysis in ionic liquids), Dupont [9, 10] (catalysis in ionic liquids) and Sheldon [11] (catalysis in ionic liquids).

Ionic liquids have been described as "Designer Solvents" [12]. Simply by making changes to the structure of either the anion, or the cation, or both, properties such as solubility, density, refractive index and viscosity can be adjusted to suit requirements [13, 14]. This degree of control can be of substantial benefit when carrying out solvent extractions or product separations, as the relative solubilities of the ionic and extraction phases can be adjusted to assist with the separation [15]. Separation of the products can also be achieved by other means such as distillation (usually under vacuum), steam distillation and supercritical fluid extraction (CO_2).

To many chemists, performing reactions in ionic liquids may seem daunting and the range of ionic liquids or potential ionic liquids available is very large. However, many have found that performing reactions in ionic liquids is straightforward and practical, compared to similar reactions in conventional organic solvents. This is particularly the case when considering reactions normally carried out in noxious and difficult to remove solvents such as dipolar aprotic solvents like dimethylsulfoxide.

With the growing interest in ionic liquids, reactions were initially performed in various chloroaluminate(III) ionic liquids. Their strong solvating ability was an advantage, but their sensitivity to moisture and strong interactions with certain commonly occurring functional groups limited the scope of reactions in these media. With the discovery of water-stable "neutral" ionic liquids, the range of reactions that can be performed has grown to cover most classes of reactions found in organic chemistry textbooks [16]. The vast majority of the reactions in ionic liquids are now carried out in these ionic liquids.

Chloroaluminate(III) ionic liquids are excellent media for many processes, but suffer from several disadvantages, such as their moisture sensitivity and the difficulty of separation of products containing heteroatoms. Furthermore, these ionic liquids often have to be quenched (usually in water) at the end of a chemical reaction, and are lost in the form of acidic aqueous waste. Hence research is shifting to the investigation of ionic liquids that are stable to water. This allows for straightforward product separation and easier handling. In particular, a number of ionic liquids have been found to be hydrophobic (immiscible with water), but readily dissolve many organic molecules (with the exception of alkanes, some ethers and alkylated aromatic compounds such as toluene). An example of this is the ionic liquid 1-butyl-3-methylimidazolium hexafluorophosphate [BMIM][PF_6] [17], which forms triphasic solutions with alkanes and water [18]. This multiphasic behavior has important implications for clean synthesis and is analogous to the use of fluorous phases in some chemical processes [19]. For example, a reaction can be performed in the ionic liquid, the products separated by distillation or steam stripping, and a by-product extracted with water or an organic solvent.

Most of the chemistry carried out in neutral ionic liquids has centered around the use of [BMIM][PF_6] and [BMIM][BF_4] as solvents. These two ionic liquids were chosen because they are relatively simple to prepare and purify and their properties are well understood. Unfortunately, these two ionic liquids are not ideal solvents for many chemical reactions that are carried out under acidic or basic conditions [20]. This is because $[PF_6]^-$ and $[BF_4]^-$ ions can hydrolyse under either acidic or basic conditions [21] and the $[BMIM]^+$ cation is not stable under basic conditions and can form a highly reactive carbene [22]. Thus ionic liquids based on ammonium and other ions are superseding these ionic liquids [23, 24].

5.2.1
Electrophilic Reactions

Ionic liquids have proven to be excellent media for electrophilic reactions. Many of the first electrophilic reactions were carried out in chloroaluminate(III) salts[1] and other related binary salts. These often proceed smoothly to give products in good yield and selectivity, however, it should be noted that these salts are water sensitive and reaction must be carried out under dry conditions. These salts react with water to give hydrolysed aluminum(III) ionic species and HCl. When a reactant or product contains a heteroatomic functional group, such as a ketone, a strong aluminum(III) chloride adduct is formed. In these cases, this adduct can be difficult to separate from the ionic liquid at the end of a reaction. The isolation of the product often involves destruction of the ionic liquid with water. For products that do not have polar electron-donating functional groups, isolation of the products is straightforward and the ionic liquid can be reused.

The disadvantages associated with the use of chloroaluminate(III) ionic liquids have led various authors to investigate alternative Lewis acidic ionic liquid media such as those based on zinc(II) chloride [25], tin(II) chloride [26] and indium(III) chloride [27]. These ionic liquids are considerably more water stable although much less reactive than the chloroaluminate(III) systems.

5.2.1.1 Friedel-Crafts Reactions

The Friedel-Crafts reaction has been studied in detail by Olah [28, 29]. This reaction can be considered to be a reaction that results in the formation of carbon–carbon bonds that are "catalyzed" by either strong acids or Lewis acids. The term "catalyzed" is not always appropriate since, in many reactions, the so-called catalyst is a stoichiometric reagent.

---

[1] Chloroaluminate(III) salts belong to a class of ionic liquids called binary ionic liquids. The composition of a chloroaluminate ionic liquid is best described by the apparent mole fraction of $AlCl_3$ $\{X(AlCl_3)\}$ present. Ionic liquids with $X(AlCl_3) < 0.5$ contain an excess of Cl^- ions, and are called "basic"; those with $X(AlCl_3) > 0.5$ contain an excess of $[Al_2Cl_7]^-$ ions, and are called "acidic"; melts with $X(AlCl_3) = 0.5$ are called 'neutral'. For example, the binary salt $NaCl–AlCl_3$ ($X = 0.67$) refers to a 1 part NaCl to 2 parts $AlCl_3$ mixture of salts and is described as "acidic".

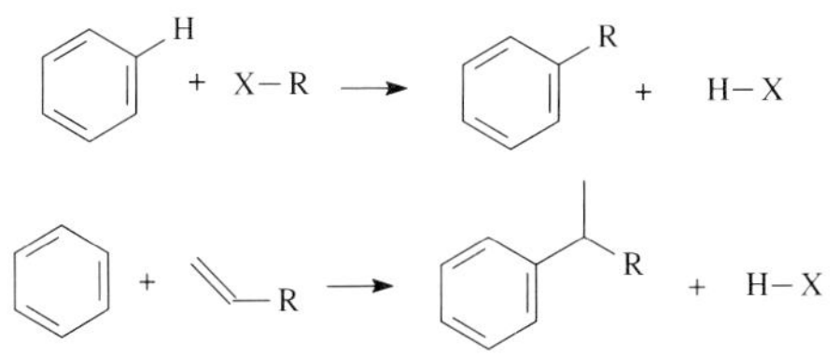

Scheme 5.2-1 The Friedel-Crafts alkylation reaction (R = alkyl, X = leaving group) [29].

Friedel-Crafts alkylation reactions

The Friedel-Crafts alkylation reaction usually involves the interaction of an alkylating agent such as an alkyl halide, alcohol or alkene with an aromatic compound, to form an alkylated aromatic compound (Scheme 5.2-1).

It should be noted that Scheme 5.2-1 denotes idealized Friedel-Crafts alkylation reactions. In practice, there are a number of problems associated with the reaction. These include polyalkylation reactions since the products of a Friedel-Crafts alkylation reaction are often more reactive than the starting material. Also isomerization and rearrangement reactions can occur, which can lead to a large number of products [28, 29]. The mechanism of Friedel-Crafts reactions is not straightforward, and it is possible to propose two or more different mechanisms for a given reaction. Examples of the typical processes occurring in a Friedel-Crafts alkylation reaction are given in Scheme 5.2-2 for the reaction of 1-chloropropane and benzene.

The chemical behavior of acidic chloroaluminate(III) ionic liquids (where $X(AlCl_3) > 0.50$) [6] is that of a powerful Lewis acid. As might be expected, it catalyzes reactions that are conventionally catalyzed by aluminum(III) chloride, without suffering the disadvantage of the low solubility of aluminum(III) chloride in many solvents.

The first examples of alkylation reactions in "molten salts" were reported in the 1950s. Baddeley and Williamson performed a number of intramolecular cyclization

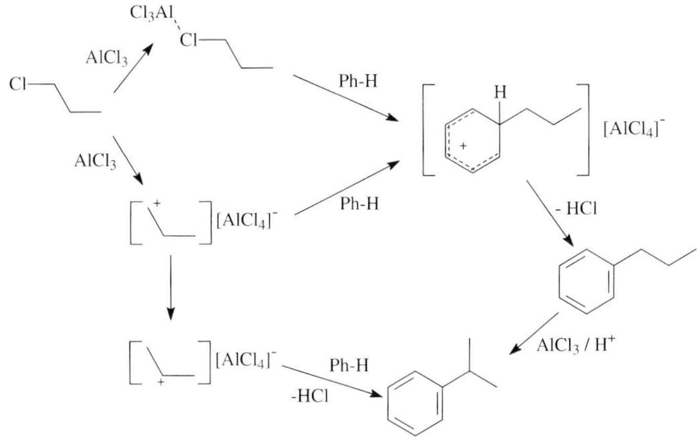

Scheme 5.2-2 The reaction of 1-chloropropane and benzene [29].

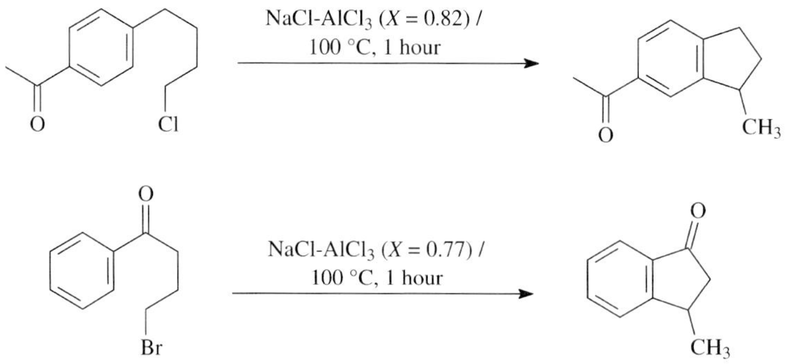

Scheme 5.2-3 The intramolecular cyclization of alkyl chlorides and bromides [30].

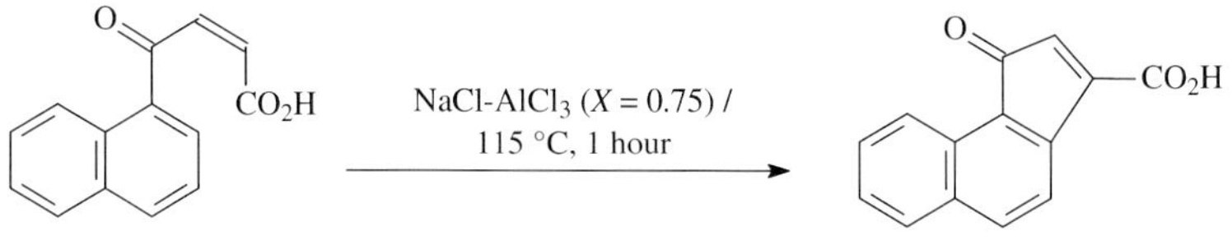

Scheme 5.2-4 The intramolecular cyclization of an alkene in a molten salt [31].

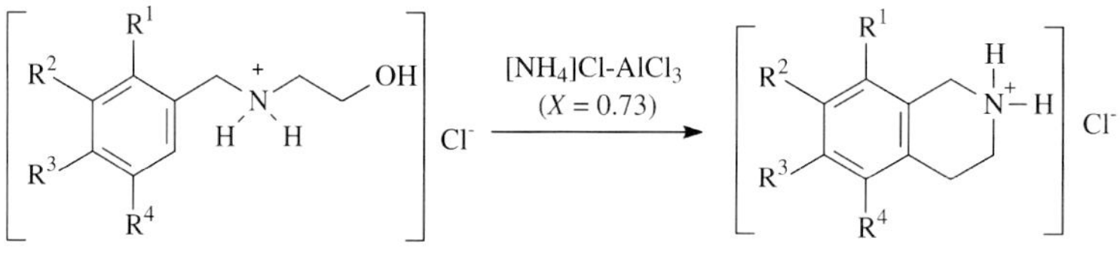

Scheme 5.2-5 The cyclodehydration of N-benzylethanolamine chloride [32].

reactions [30] (Scheme 5.2-3). These were carried out in mixtures of sodium chloride and aluminum chloride. The reactions were run at below the melting point of the salt, and it is presumed that the mixture of reagents acted to lower the melting point.

Baddeley also investigated the cyclization of alkenes in the NaCl–AlCl$_3$ molten salt. An example is given in Scheme 5.2-4 [31].

Mendelson et al. [32] also investigated a number of cyclization reactions. One of these involved the cyclodehydration of N-benzylethanolamine chloride in a molten salt derived from AlCl$_3$ and NH$_4$Cl ($X = 0.73$). This gave rise to the corresponding tetrahydroisoquinoline in 41–80% yield, as shown in Scheme 5.2-5.

Boon et al. investigated the reactions of benzene and toluene in room-temperature ionic liquids based on [EMIM]Cl–AlCl$_3$ mixtures [33] ([EMIM] = 1-ethyl-3-methylimidazolium). The reactions of various alkyl chlorides with benzene in the ionic liquid [EMIM]Cl–AlCl$_3$ ($X = 0.60$ or 0.67) was carried out and the product distribution is given in Table 5.2-1 and Scheme 5.2-6.

Table 5.2-1 The products from the reactions of alkyl chlorides with benzene in [EMIM]Cl–AlCl$_3$ (X = 0.60 or 0.67) [33].

R–Cl	X	R–Cl:C$_6$H$_6$: IL	mono-	di-	tri-	tetra-	penta-	hexa-
methyl[a]	0.67	xs:1:1	1.5	58.5	1.5	26.8	1.4	10.2
ethyl[a]	0.67	xs:1:1	11.5	10.8	33.4	24.4		1.5
n-propyl[b]	0.60	1.25:1.25:1	24.8	19.9	55.3			
n-butyl[b,c]	0.60	1.33:1.33:1	25.0	26.3	48.7			
cyclohexyl	0.60	10:10:1	35.0	30.0	34.4			
benzyl[d]	0.60	0.78:1.17:1	50.0	34.5	15.6			

[a] At reflux temperature of alkyl halide.
[b] Room temperature in dry box.
[c] Only sec-butyl products formed.
[d] Tar formed, only a small amount of alkylated product isolated.

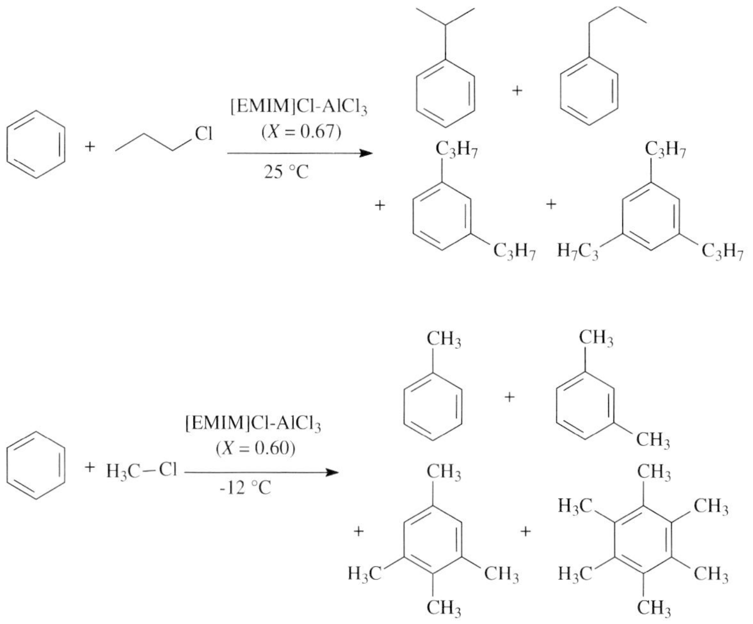

Scheme 5.2-6 The alkylation of benzene with methyl chloride or n-propyl chloride in an ionic liquid [33].

The methylation of benzene with methyl chloride proceeds to give predominantly dimethylbenzene (xylenes) and tetramethylbenzene, with about 10% hexamethylbenzene. In the propylation of benzene with 1-chloropropane, not only does polyalkylation occur, but there is also a considerable degree of isomerization of the *n*-propyl group to the isopropyl isomer (Scheme 5.2-6). In the butylation, complete isomerization of the butyl side chain occurs to give only *sec*-butyl benzenes.

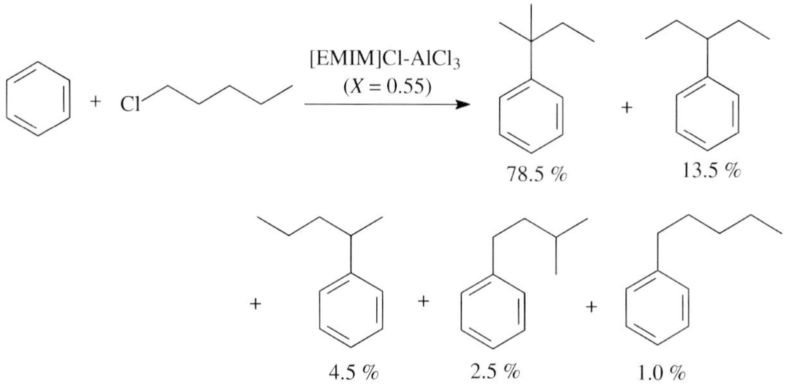

Scheme 5.2-7 The reaction of 1-chloropentane with benzene in [EMIM]Cl–AlCl$_3$ ($X = 0.55$) [34].

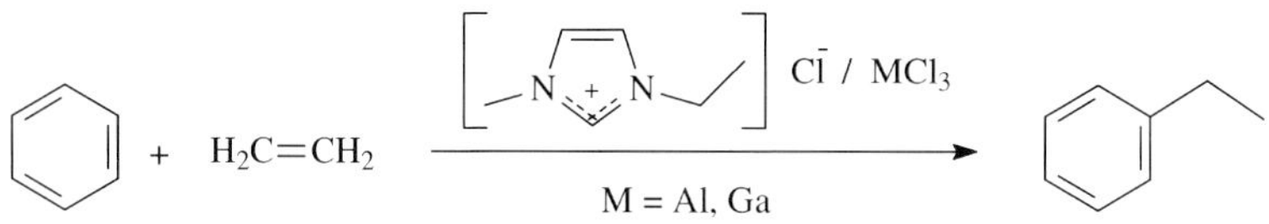

Scheme 5.2-8 The alkylation of aromatic compounds in chloroaluminate(III) or chlorogallate(III) ionic liquids [35].

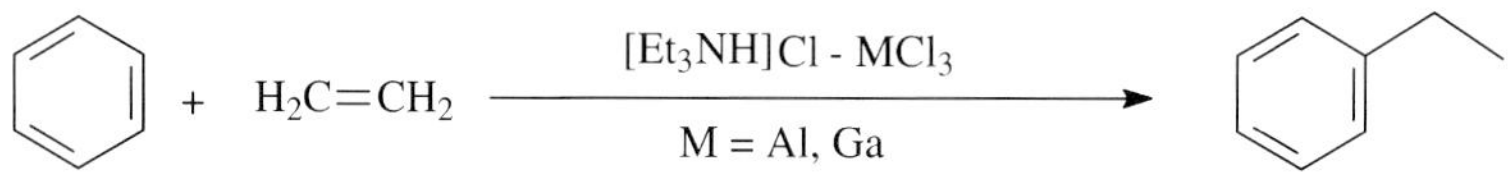

Scheme 5.2-9 The reaction of benzene with ethylene in a triethylammonium ionic liquid [36].

Piersma and Merchant have studied the alkylation of benzene with various chloropentanes in [EMIM]Cl–AlCl$_3$ ($X = 0.55$) [34]. The reaction of 1-chloropentane with benzene gave a mixture of products, with surprisingly only a 1% yield of the unisomerized n-pentylbenzene. The major products of the reaction had all undergone isomerization (Scheme 5.2-7).

The details of two related patents for the alkylation of aromatic compounds with chloroaluminate(III) or chlorogallate(III) ionic liquid catalysts have become available. The first by Seddon and coworkers [35] describes the reaction of ethylene with benzene to give ethylbenzene (Scheme 5.2-8). This is carried out in an acidic ionic liquid based on an imidazolium cation and is claimed for ammonium, phosphonium and pyridinium cations. The anion exemplified in the patent is a chloroaluminate(III) and the claim includes for chlorogallate(III) anions and various mixtures of anions.

The second patent by Wasserscheid and coworkers [36] also describes the reaction of benzene with ethylene in ionic liquids, but exemplifies a different ionic liquid that is suitable for this reaction (Scheme 5.2-9).

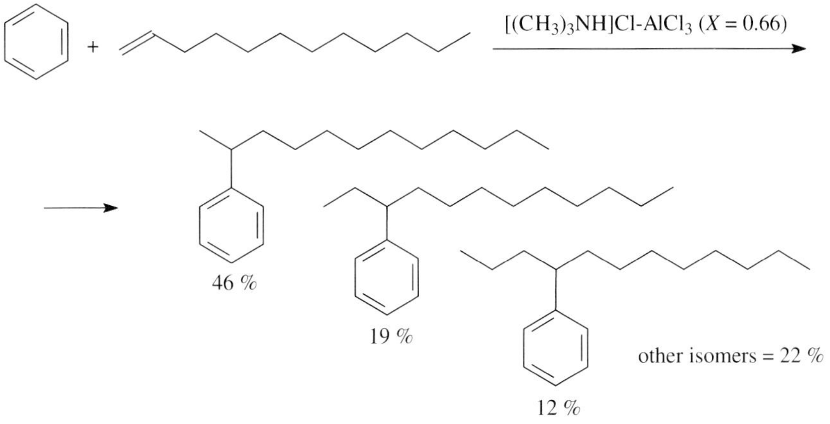

Scheme 5.2-10 The reaction of dodec-1-ene with benzene using an ionic liquid as a catalyst [37].

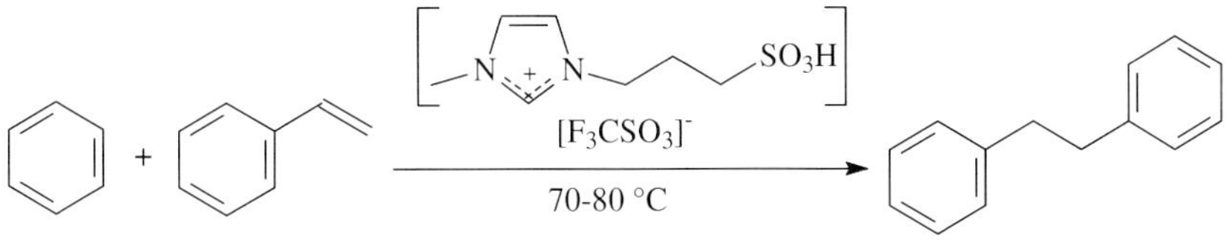

Scheme 5.2-11 The task specific ionic liquid catalyzed alkylation of benzene [42].

The production of linear alkyl benzenes (LABs) is carried out on a large scale for the production of surfactants. The reaction involves the reaction of benzene with a long chain alkene such as dodec-1-ene and often gives a mixture of isomers. Greco et al. have used a chloroaluminate(III) ionic liquid as a catalyst in the preparation of LABs [37] (Scheme 5.2-10).

Johnson and coworkers used two different chloroaluminate(III) ionic liquids [PyH]Cl–AlCl$_3$ or [(C$_2$H$_5$)$_3$S]Br–AlCl$_3$ [38] in the alkylation of benzene with 1-pentene, 2-pentene and 1-octene and obtained the expected monoalkylbenzene when using a high benzene to alkene ratio. Disubstituted and rearrangement products as well as dimerization and oligomerization of the alkene were also found to occur in these reactions [39]. Friedel-Crafts alkylations of 2-methylnaphthalene with long-chain alkenes (mixed C$_{11-12}$ olefins) in acidic [(C$_2$H$_5$)$_3$HN]Cl–AlCl$_3$ ionic liquids modified with HCl gave the expected alkylmethylnaphthalene in up to 90% conversion and up to complete selectivity. The influences of the type and dosage of catalysts, molar ratio of 2-methylnaphthalene to alkenes and solvent to 2-methylnaphthalene, reaction temperature and time were also studied [40]. A task specific ionic liquid, first described by Davis [41] was used in the alkylation of benzene, toluene and xylene [42] with styrene. This is shown in Scheme 5.2-11.

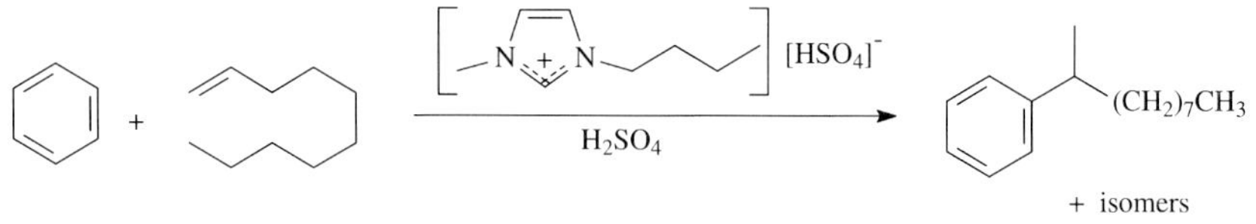

Scheme 5.2-12 The sulfuric acid catalyzed alkylation of benzene in a hydrogensulfate ionic liquid [43].

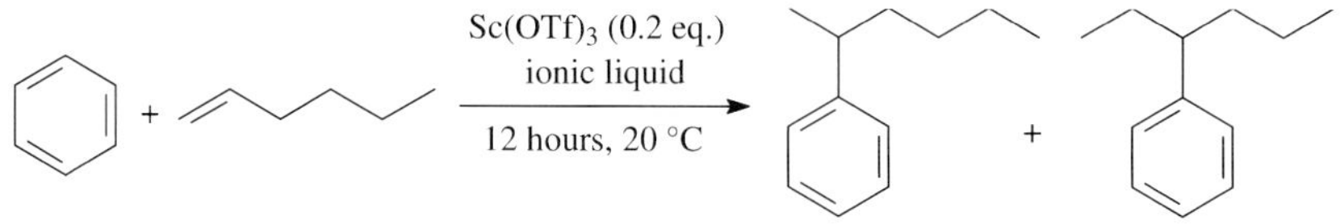

Scheme 5.2-13 The alkylation of benzene with hex-1-ene catalyzed by scandium(III) triflate in ionic liquids [44].

Keim and coworkers have carried out various alkylation reactions of aromatic compounds in ionic liquids substantially free of Lewis acidity [43]. An example is the reaction of benzene with decene in [BMIM][HSO$_4$], using sulfuric acid as the catalyst (Scheme 5.2-12). Keim has also claimed that these acid-ionic liquids systems can be used for esterification reactions.

Other Lewis acids dissolved in ionic liquids have been used for Friedel-Crafts alkylation reactions. Song [44] has reported that scandium(III) triflate in [BMIM][PF$_6$] acts as an alkylation catalyst in the reaction of benzene with hex-1-ene (Scheme 5.2-13).

The ionic liquids that were found to give the expected hexylbenzenes were [BMIM][PF$_6$], [PMIM][PF$_6$], [HMIM][PF$_6$], [EMIM][SbF$_6$] and [BMIM][SbF$_6$]. The reaction did not succeed in the corresponding tetrafluoroborate or trifluoromethanesulfonate ionic liquids. For the successful reactions, conversions of 99% of the hexene to products occurred, with 93–96% of the products being the monoalkylated product. The authors noted that the successful reactions all took place in the hydrophobic ionic liquids. It should be noted that the [PF$_6$]$^-$ and [SbF$_6$]$^-$ ions are less stable to hydrolysis reactions (resulting in the formation of HF) than the [BF$_4$]$^-$ or [OTf]$^-$ ions. The possibility of these reactions being catalyzed by traces of HF cannot be excluded [45]. Metal triflimide salts and bis(trifluoromethanesulfonyl)amide acid (H[Tf$_2$N]) dissolved in ionic liquids such as phosphonium triflimides [46] have also been found to catalyze the alkylation of benzene with alkenes [47].

The alkylation of phenol with *tert*-butyl alcohol was carried out in the ionic liquids [BMIM][PF$_6$] [48], [OMIM][BF$_4$], and [HMIM][BF$_4$]. Comparative studies on the catalytic properties of ionic liquids, H$_3$PO$_4$ and some solid acidic catalysts were carried out under identical reaction conditions, and the solvent effects were studied. The use of ionic liquids was found to enhance the catalytic properties of the catalysts used [49]. However, the claim that [BMIM][PF$_6$] alone can catalyze this reaction is uncertain, and, is probably due to the presence of traces of hydrogen fluoride in the ionic liquid [45, 46]. Dy(OTf)$_3$ dissolved in various [BF$_4$]$^-$ and [PF$_6$]$^-$ ionic

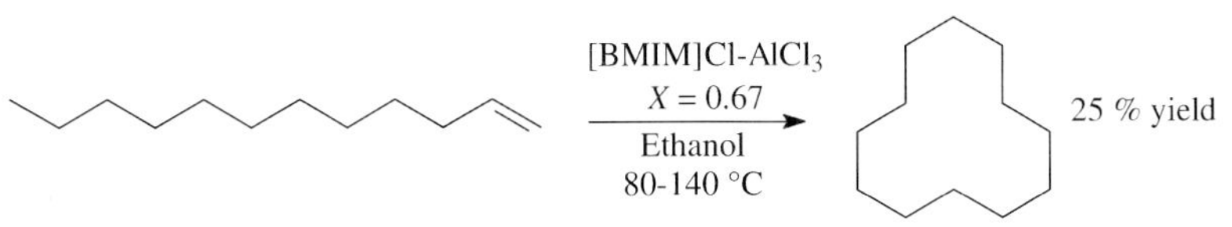

Scheme 5.2-14 The cyclization of dodecene to cyclododecane [55].

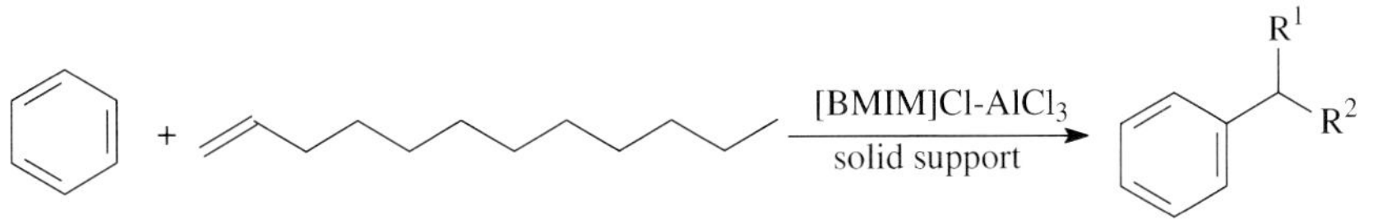

$R^1 = CH_3$, $R^2 = (CH_2)_9CH_3$ 2-phenyldodecane
$R^1 = CH_2CH_3$, $R^2 = (CH_2)_8CH_3$ 3-phenyldodecane
$R^1 = (CH_2)_2CH_3$, $R^2 = (CH_2)_7CH_3$ 4-phenyldodecane
$R^1 = (CH_2)_3CH_3$, $R^2 = (CH_2)_6CH_3$ 5-phenyldodecane
$R^1 = (CH_2)_4CH_3$, $R^2 = (CH_2)_5CH_3$ 6-phenyldodecane

Scheme 5.2-15 The alkylation of benzene with dodecene with an ionic liquid on a solid support [56, 57].

liquids has been used as a catalyst in the alkylation of indoles with ketones to form diindolylmethanes [50]. Reactions are also known to be Brønsted acid catalyzed [51]. A similar alkylation reaction using chloromethyl ether gave rise to a range of diaryl methanes [52] in a range of acidic chloroaluminate ionic liquids. The products were usually immiscible with the ionic liquids and could be easily separated, so that the ionic liquid could be reused.

Alkylation of isobutane with 2-butene using various 1-alkyl-3-methylimidazolium halide–aluminum chloride catalysts has been successfully demonstrated [53]. Among theses ionic liquids, [OMIM]Br–AlCl$_3$ displayed the best performance in terms of activity and selectivity for this reaction. From various parametric studies, such as anion compositions and temperature, optimum catalytic activity was observed at 80 °C and $X = 0.52$. This reaction has also been used to study the Lewis acidity of ionic liquids [54]. An interesting alkylation of an alkene can also be found in the cyclization of dodecene in a [BMIM]Cl–AlCl$_3$ ($X = 0.67$)/ethanol mixture at 6 MPa pressure (Scheme 5.2-14). The authors claim 27% yield and 93% selectivity for the formation of cyclododecane [55].

The alkylation of a number of aromatic compounds by the use of a chloroaluminate(III) ionic liquid on a solid support has been investigated by Hölderich and coworkers [56, 57]. Here the alkylation of aromatic compounds such as benzene, toluene, naphthalene and phenol with dodecene was performed using the ionic liquid [BMIM]Cl–AlCl$_3$ supported on silica, alumina and zirconia. With benzene, monoalkylated dodecylbenzenes were obtained (Scheme 5.2-15).

Table 5.2-2 The product distribution dependence on the catalyst used for the reaction of benzene with dodecene. IL = [BMIM]Cl–AlCl$_3$ ($X = 0.6$), Temperature = 80 °C, with 6 mol% catalyst and benzene to dodecene ratio = 10:1.

Catalyst	2-Phenyl dodecane	3-Phenyl dodecane	4-Phenyl dodecane	5-Phenyl dodecane	6-Phenyl dodecane
AlCl$_3$	46.4	19.4	12.7	12.1	9.5
IL ($X = 0.6$)	36.7	19.0	15.0	15.5	13.8
T 350/IL	42.9	22.8	13.0	11.8	9.4
H-Beta	75.7	19.0	3.8	1.1	0.4
H-Beta/IL	43.9	21.2	12.4	12.0	10.5

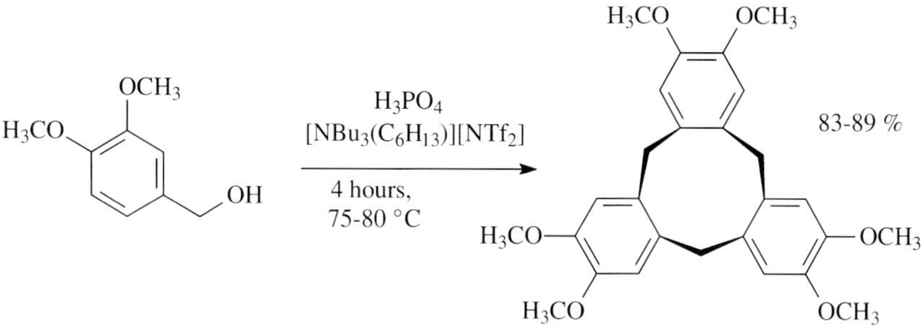

Scheme 5.2-16 The cyclization of 3,4-dimethoxyphenylmethanol in an ionic liquid [58].

The product distribution in the reaction of benzene with dodecene was determined for a number of catalysts (Scheme 5.2-15, Table 5.2-2). As can be seen, the reaction with the zeolite H-Beta gave predominantly the 2-phenyldodecane whereas the reaction in the pure ionic liquid gave a mixture of isomers, with selectivity similar to that of aluminum chloride. The two supported ionic liquid reactions (H-Beta/IL and T 350/IL) gave product distributions again similar to aluminum(III) chloride (T350 is a silica support made by Degussa).

Raston has reported an acid catalyzed Friedel-Crafts reaction [58] where compounds such as 3,4-dimethoxyphenylmethanol were cyclized to cyclotriveratrylene (Scheme 5.2-16). The reactions were carried out in tributylhexylammonium bis(bistrifluoromethanesulfonyl)amide [NBu$_3$(C$_6$H$_{13}$)][Tf$_2$N] using a phosphoric or p-toluenesulfonic acid catalyst. The product was isolated by dissolving the ionic liquid/catalyst in methanol and filtering off the cyclotriveratrylene product as white crystals. Evaporation of the methanol allowed the ionic liquid and catalyst to be regenerated.

The Pictet-Spengler reaction is a form of intermolecular Friedel-Crafts alkylation reaction. An example of this is the ionic liquid catalyzed and Lewis acid catalyzed one-pot Pictet–Spengler reactions of tryptophan methyl ester or tryptamine with aliphatic and aromatic aldehydes [59]. Short reaction times were achieved with the aid of microwave irradiation (Scheme 5.2-17).

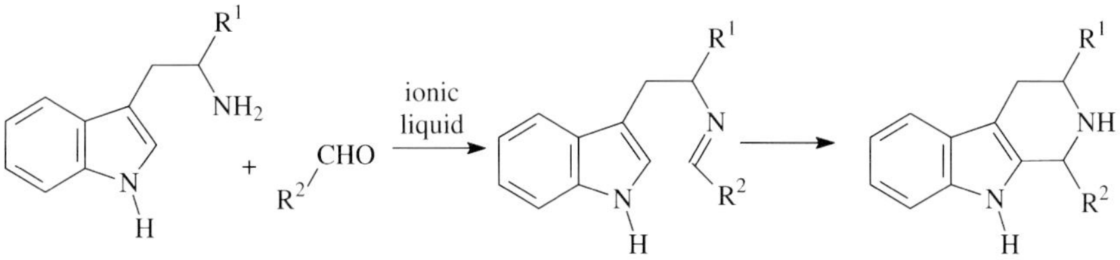

Scheme 5.2-17 The Pictet-Spengler reaction in ionic liquids [59].

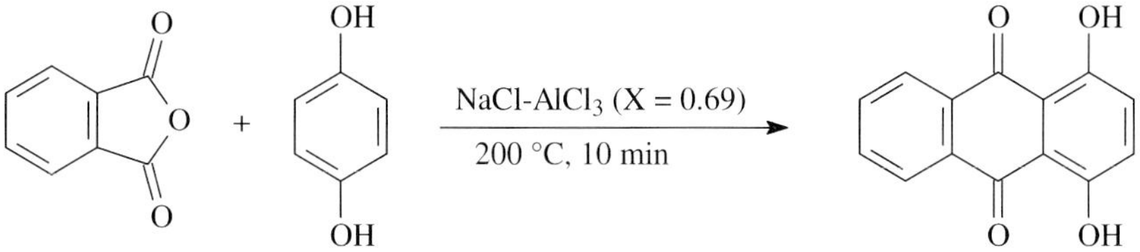

Scheme 5.2-18 The reaction of phthalic anhydride with hydroquinone in NaCl–AlCl$_3$ ($X = 0.69$) [60].

Friedel-Crafts acylation reactions

Friedel-Crafts acylation reactions usually involve the interaction of an aromatic compound with an acyl halide or anhydride in the presence of a catalyst, to form a carbon–carbon bond [28]. As the product of an acylation reaction is less reactive than its starting material, usually monoacylation occurs. The "catalyst" in the reaction is not a true catalyst, as it is often (but not always) required in stoichiometric quantities. For Friedel-Crafts acylation reactions in chloroaluminate(III) ionic liquids or molten salts, the ketone product of an acylation reaction forms a strong complex with the ionic liquid, and separation of the product from the ionic liquid can be extremely difficult. The products are usually isolated by quenching the ionic liquid in water. Current research is moving towards finding genuine catalysts for this reaction, some of which are described in this section.

The first example of a Friedel-Crafts acylation reaction in a molten salt was carried out by Raudnitz and Laube [60]. This involved the reaction of phthalic anhydride with hydroquinone at 200 °C in NaCl–AlCl$_3$ ($X = 0.69$) (Scheme 5.2-18).

Scholl and coworkers [61] performed the acylation of 1-benzoylpyrene with 4-methylbenzoyl chloride in a NaCl–AlCl$_3$ ($X = 0.69$) molten salt (110–120 °C). This gave 1-benzoyl-6-(4-methylbenzoyl)pyrene as the major product (Scheme 5.2-19).

Bruce et al. carried out the cyclization of 4-phenylbutyric acid to tetralone in NaCl–AlCl$_3$ ($X = 0.68$) at 180–200 °C [62]. The reaction of valerolactone with hydroquinone was also performed by Bruce, to give 3-methyl-4,7-dihydroxyindanone using the same ionic liquid and reaction conditions. These reactions are shown in Scheme 5.2-20.

The Fries rearrangement can be considered to be a type of Friedel-Crafts acylation reaction. Two examples of this reaction are given in Scheme 5.2-21. The first is the

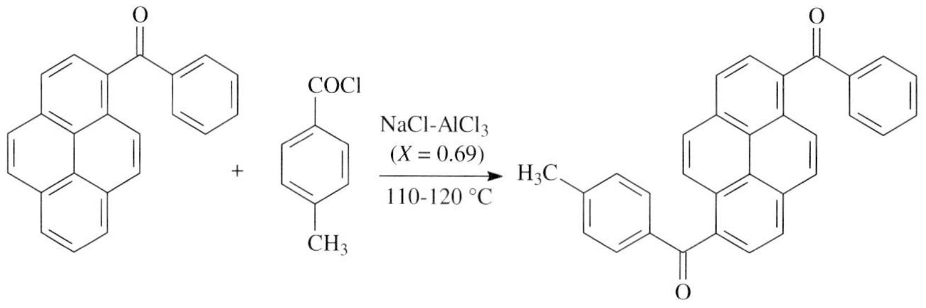

Scheme 5.2-19 The acylation of 1-benzoylpyrene in NaCl–AlCl$_3$ ($X = 0.69$) [61].

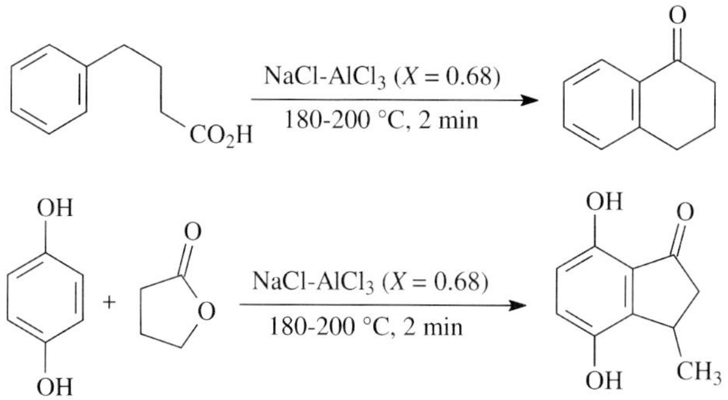

Scheme 5.2-20 The use of NaCl–AlCl$_3$ ($X = 0.68$) in the formation of cyclic ketones [62].

rearrangement of 4,4′-diacetoxybiphenyl to 4,4′-dihydroxy-3,3′-diacetoxybiphenyl in a NaCl–AlCl$_3$ ($X = 0.69$) molten salt [63]. The second example is the rearrangement of phenyl 3-chloropropionate to 2′-hydroxy-3-chloropropiophenone followed by cyclization to an indanone [64].

One of the problems with these NaCl–AlCl$_3$ molten salts is their high melting points and corresponding high reaction temperatures. The high reaction temperatures tend to cause side reactions and decomposition of the products of the reaction. Hence, a number of reactions were carried out under milder conditions, in room temperature ionic liquids. The first example of a Friedel-Crafts acylation in such an ionic liquid was performed by Wilkes and coworkers [33, 69] (Scheme 5.2-22).

The rate of the acetylation reaction was found to be dependent on the concentration of the [Al$_2$Cl$_7$]$^-$ ion and suggested that this ion was acting as the Lewis acid in the reaction. Wilkes goes on to provide evidence that the acylating agent is the acetylium ion [H$_3$CCO]$^+$ [33, 65].

Singer and coworkers have investigated the acylation reactions of ferrocene in ionic liquids made from mixtures of [EMIM]I and aluminum(III) chloride [66, 67]. They found that in the basic or neutral [EMIM]I–AlCl$_3$ ($X < 0.50$) no reaction occurred. In mildly acidic [EMIM]I–AlCl$_3$, the mono-acetylated ferrocene was obtained as the major product. In strongly acidic [EMIM]I–AlCl$_3$ ($X = 0.67$) the diacylated

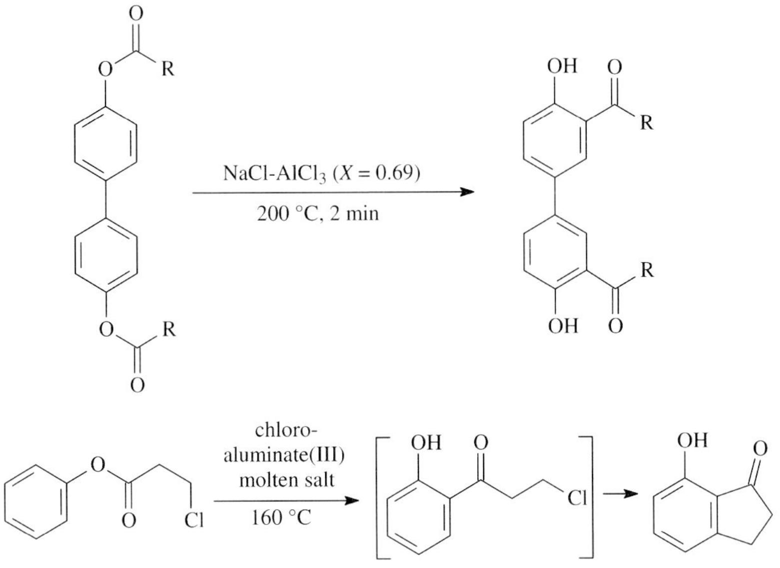

Scheme 5.2-21 The Fries rearrangement in chloroaluminate(III) molten salts [63, 64].

Scheme 5.2-22 The acetylation of benzene in a room temperature ionic liquid [65].

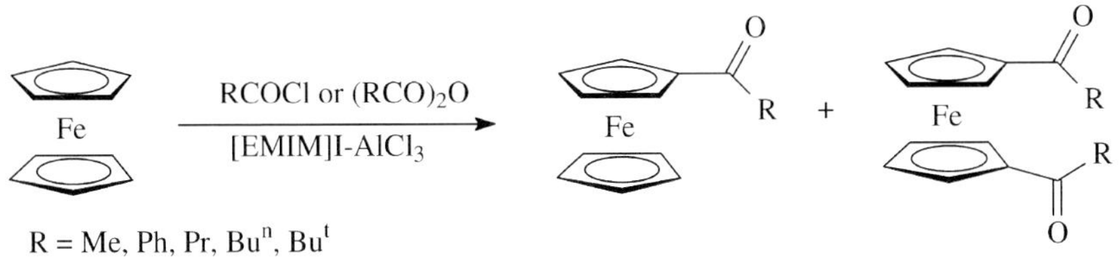

Scheme 5.2-23 The acylation of ferrocene in [EMIM]I–AlCl$_3$ [66, 67].

ferrocene was the major product. Also, when R = alkyl, the diacylated product was usually the major product, but for R = Ph, the monoacylated product was favored (Scheme 5.2-23).

A number of commercially important fragrance molecules have been synthesised by Friedel-Crafts acylation reactions in these ionic liquids. Traseolide® (5-acetyl-1,1,2,6-tetramethyl-3-isopropylindane) (Scheme 5.2-24) has been made in high yield in the ionic liquid [EMIM]Cl–AlCl$_3$ ($X = 0.67$). For the acylation of naphthalene, the ionic liquid gives the highest reported selectivity for the 1-position

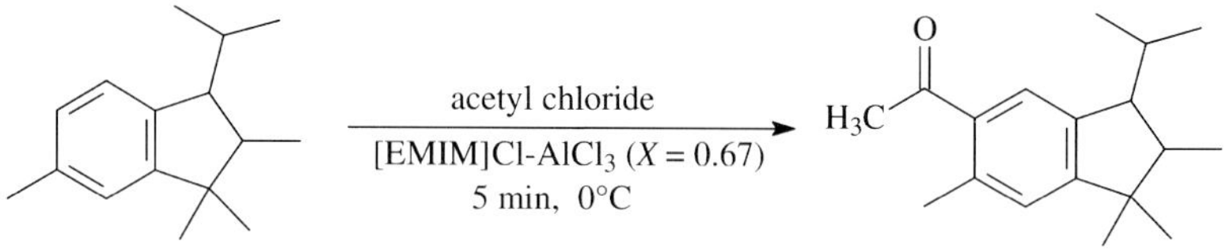

Scheme 5.2-24 The acetylation of 1,1,2,6-tetramethyl-3-isopropylindane in [EMIM]Cl–AlCl$_3$ (X = 0.67) [68].

[68]. The acetylation of anthracene at 0 °C was found to be a reversible reaction. The initial product of the reaction of acetyl chloride (1.1 equivalents) with anthracene is 9-acetylanthracene, formed in 70% yield in less than 5 min. The 9-acetylanthracene was then found to undergo diacetylation reactions, giving the 1,5- and 1,8-diacetylanthracene and anthracene after 24 h (Scheme 5.2-25). This was confirmed by taking a sample of 9-acetyl anthracene and allowing it to isomerize in the ionic liquid. This gave a mixture of anthracene, 1,5-diacetylanthracene and 1,8-diacetylanthracene. It should be noted that a proton source was needed for this reaction to occur, implying an acid catalyzed mechanism (Scheme 5.2-26) [68]. Acylation as well as dealkylation reactions have been performed by Gigante et al. in 1-alkyl-3-methylimidazolium ionic liquid–aluminum(III) chloride mixtures, on methyl dehydroabietate (Fig. 5.2-1) [69]. When contacted with the ionic liquid in the presence of toluene, the isopropyl group was transferred to the toluene. When acetyl chloride was added, acylation took place *ortho*- to the isopropyl group in up to 97% yield. Similar deacylation and transacylation reactions have been observed by Laali and Sarca in the reactions of 2,4,6-trimethylacetophenone and pentamethylaceophenone in various triflate, tetrafluoroborate and hexafluorophosphate ionic liquids [70].

Metal triflimide salts were found to be excellent catalysts for Friedel-Crafts acylation reactions of aromatic compounds [47, 71]. Although these salts will catalyze the reaction by themselves, their reactivity is increased significantly when dissolved in triflimide ionic liquids, and the ionic liquid/catalyst can be recycled [71]. Surprisingly, the metal triflimides with the greatest activity were those of cobalt(II), nickel(II), manganese(II), iron(III) and indium(III) (Scheme 5.2-27).

Wasserscheid et al. found that aluminum chloride dissolves in triflimide ionic liquids to form biphasic solutions. These solutions can be used to promote the Friedel-Crafts acylation reaction, and an interesting variant is the carbonylation of toluene with carbon monoxide (Scheme 5.2-28) [72]. The ionic liquid can be recycled, but the aluminum chloride is lost when the reaction is worked up.

The Friedel-Crafts acylation reaction has also been performed in iron(III) chloride ionic liquids by Seddon and coworkers [73]. An example is the acetylation of benzene (Scheme 5.2-29). The ionic liquids of the type [EMIM]Cl–FeCl$_3$ (0.50 < X < 0.62) are good acylation catalysts, with the added benefit that the ketone product of the reaction can be separated from the ionic liquid by solvent extraction, provided that X is in the range 0.51–0.55. An improvement on the iron(III) binary ionic liquids

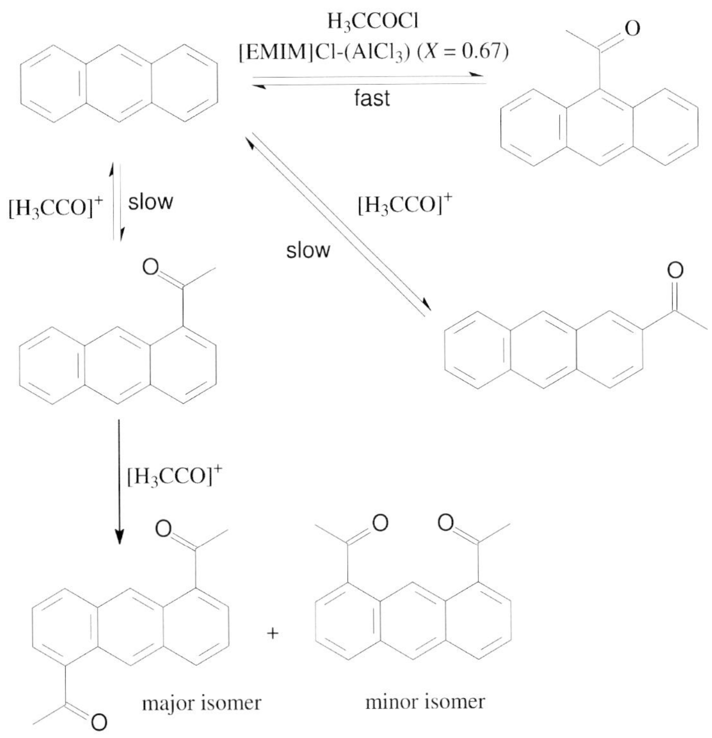

Scheme 5.2-25 The acetylation of anthracene in [EMIM]Cl–AlCl$_3$ ($X = 0.67$) [68].

is the use of indium(III) chloride ionic liquids in the catalytic acylation of aromatic compounds [74]. Although this is less reactive than the iron(III) system, it has the advantage that the ionic liquid is water stable, and the ionic liquid can be recycled by washing the product with water to dissolve the indium ionic liquid, followed by evaporation of the water to regenerate the ionic liquid [75].

The ability of iron(III) chloride to genuinely catalyze Friedel-Crafts acylation reactions has also been recognized by Hölderich and coworkers [76]. By immobilizing the ionic liquid [BMIM]Cl–FeCl$_3$ on a solid support Hölderich was able to acetylate mesitylene, anisole and *m*-xylene with acetyl chloride in excellent yield. The performance of the iron-based ionic liquid was then compared with the corresponding chlorostannate(II) and chloroaluminate(III) ionic liquids. The results are given in Scheme 5.2-30 and Table 5.2-3. As can be seen, the iron catalyst gave superior results to the aluminum or tin-based catalysts. The reactions were also carried out in the gas phase at between 200 and 300 °C. The acetylation reaction was complicated by two side reactions. For example, in the reaction of acetyl chloride with *m*-xylene, the decomposition of acetyl chloride to ketene and the formation of 1-(1-chlorovinyl)-2,4-dimethylbenzene were also found to occur [76].

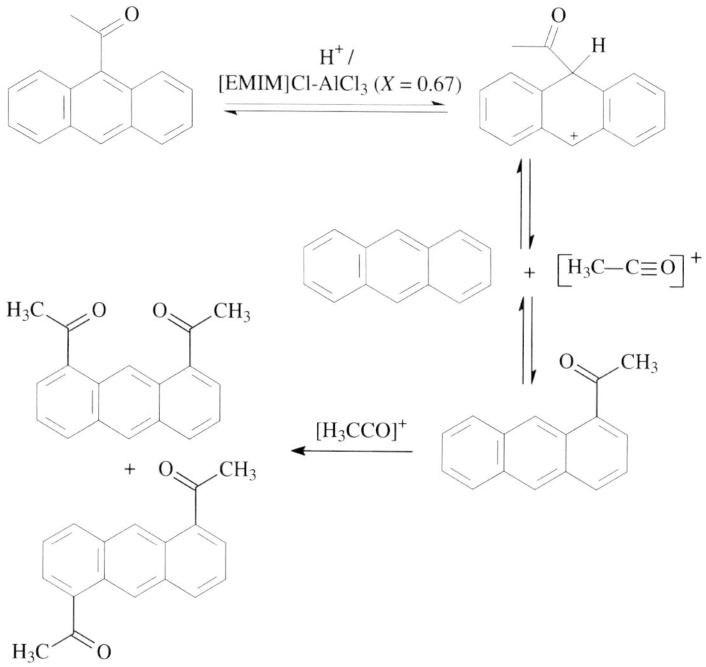

Scheme 5.2-26 Proposed mechanism for the isomerization of 9-acetylanthracene in [EMIM]Cl–AlCl$_3$ ($X = 0.67$) [68].

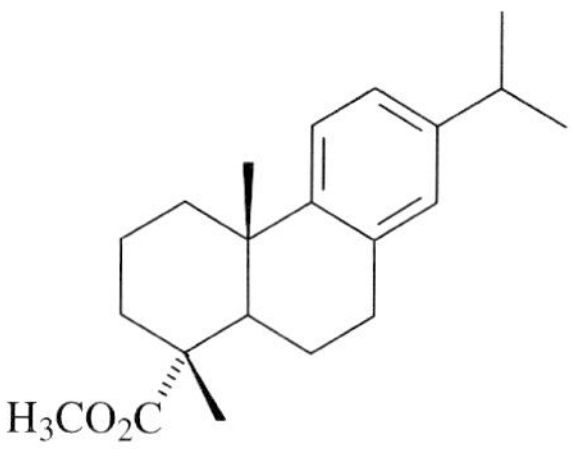

Fig. 5.2-1 The structure of methyl dehydroabietate [69].

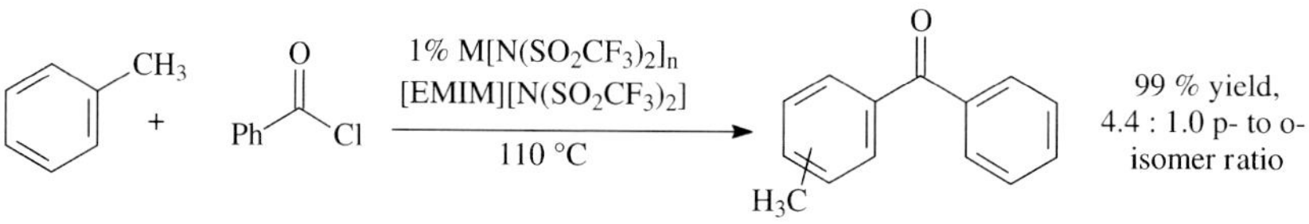

Scheme 5.2-27 Metal triflimides as Friedel-Crafts acylation catalysts [47, 71].

5.2 Stoichiometric Organic Reactions and Acid-catalyzed Reactions in Ionic Liquids

Table 5.2-3 The acylation of aromatics in batch reactions at 100 °C, for 1 h. Ratio of aromatic compound to acetylating agent = 5:1. mes. = mesitylene.

Ionic liquid	Reaction	Molar ratio IL:Ar–H	Conversion (%)	Selectivity (%)
[BMIM]Cl–AlCl$_3$	mes. + AcCl	1:205	68.1	98
[BMIM]Cl–AlCl$_3$	anisole + Ac$_2$O	1:45	8.3	96
[BMIM]Cl–AlCl$_3$	m-xylene + AcCl	1:205	3.5	96
[BMIM]Cl–FeCl$_3$	mes. + AcCl	1:205	94.7	95
[BMIM]Cl–FeCl$_3$	anisole + Ac$_2$O	1:45	100	98
[BMIM]Cl–FeCl$_3$	m-xylene + AcCl	1:205	33.8	79
[BMIM]Cl–SnCl$_2$	anisole + Ac$_2$O	1:45	19.7	94
[BMIM]Cl–SnCl$_2$	m-xylene + AcCl	1:205	3.6	95

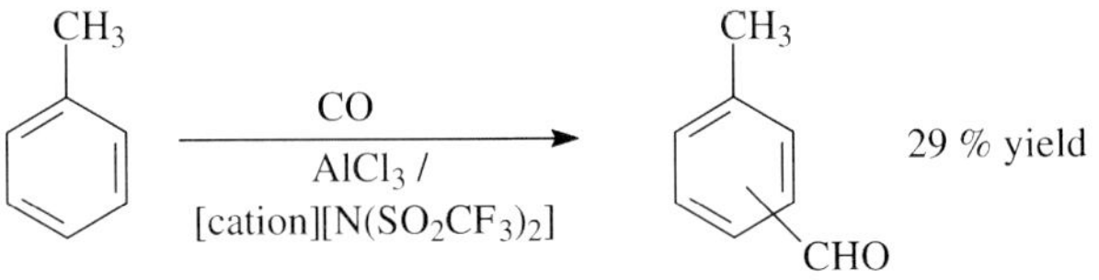

Scheme 5.2-28 The carbonylation of toluene in [cation][Tf$_2$N]–AlCl$_3$ [72].

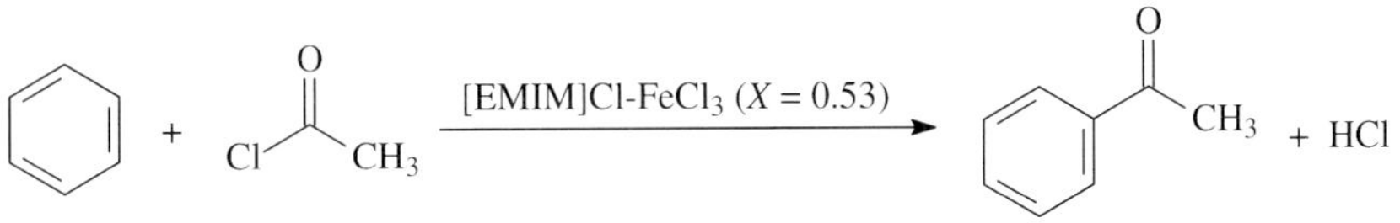

Scheme 5.2-29 The acetylation of benzene in an iron(IIII) chloride-based ionic liquid [73].

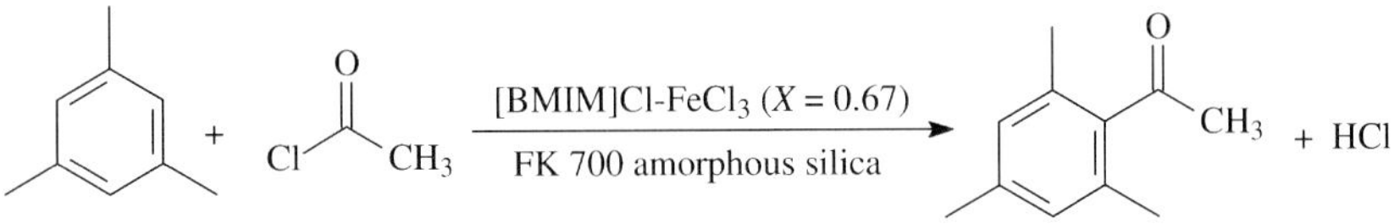

Scheme 5.2-30 The acetylation of aromatics with supported ionic liquids [76] (FK 700 is a type of amorphous silica made by Degussa).

Rebeiro and Khadilkar have investigated the reactions of trichloroalkanes with aromatic compounds. For example, the benzoylation of aromatic compounds in ionic liquids was performed using benzotrichloride, which on aqueous work up gave ketones [77]. Thioamidation of aromatic compounds can be achieved by the reaction of isothiocyanates with aromatic compounds in the presence of chloroaluminate ionic liquids [78].

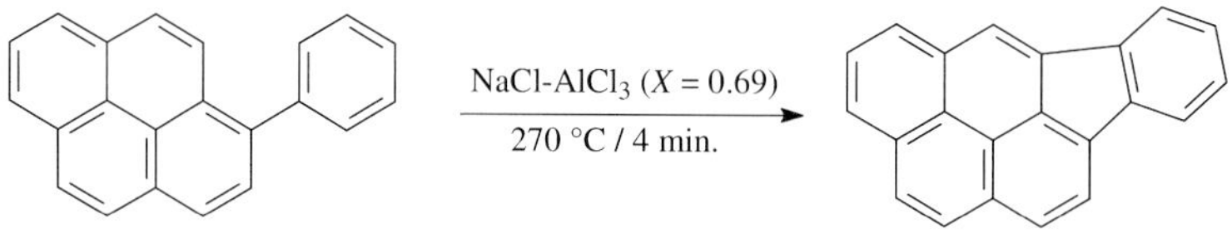

Scheme 5.2-31 The Scholl reaction of 1-phenylpyrene [79].

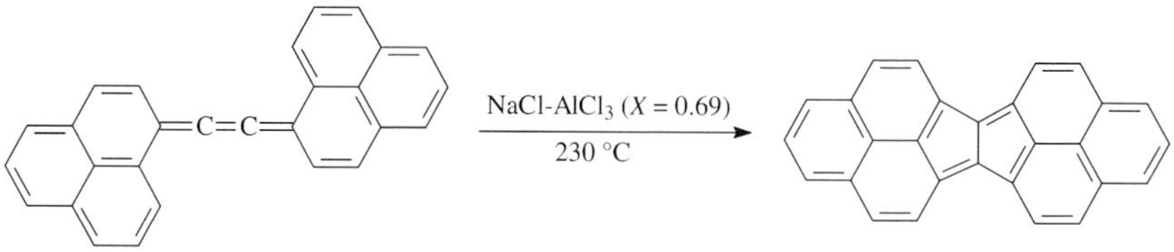

Scheme 5.2-32 The cyclization of an aromatic cumulene in a molten salt [80].

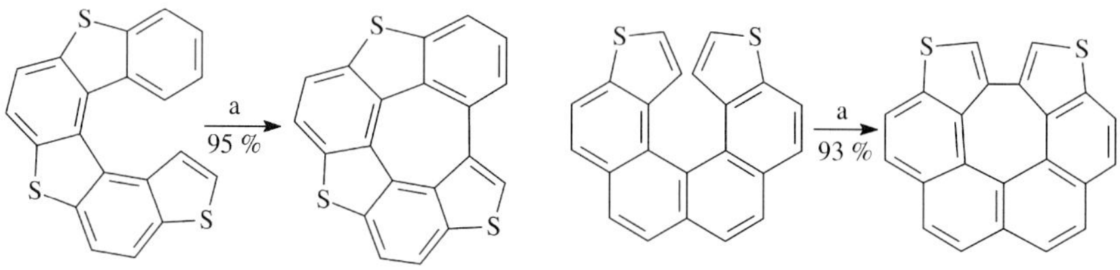

Scheme 5.2-33 The Scholl reactions of two helicines. (a = NaCl–AlCl$_3$ ($X = 0.69$) at 140 °C) [81].

5.2.1.2 Scholl and Related Reactions

One of the first reactions to be carried out in a molten salt (albeit at 270 °C) was the Scholl reaction. This involves the inter- or intra-molecular coupling of two aromatic rings. An example of this reaction is given in Scheme 5.2-31, where 1-phenylpyrene was cyclized to indeno[1,2,3-cd]pyrene [79]. A more elaborate version of the Scholl reaction is shown in Scheme 5.2-32 and involves bicyclization of an aromatic cumulene [80].

Wynberg et al. found that the yields in the cyclization of helicenes could be improved from 10% in an aluminum(III) chloride solution in benzene system to 95% in a NaCl–AlCl$_3$ ($X = 0.69$) molten salt [81]. An example is given in Scheme 5.2-33.

The Scholl reaction involves an overall oxidation of the coupled aromatic rings, yet, there is no obvious oxidizing agent. This leads to the question: what happens to the two hydrogen atoms that are produced in this reaction? It has been suggested that oxygen (air) may act as the oxidant but this currently lacks confirmation [82]. The molten salt NaCl–KCl–AlCl$_3$ (20:20:60) was used in the dimerization of aniline to benzidine (Scheme 5.2-34) [83].

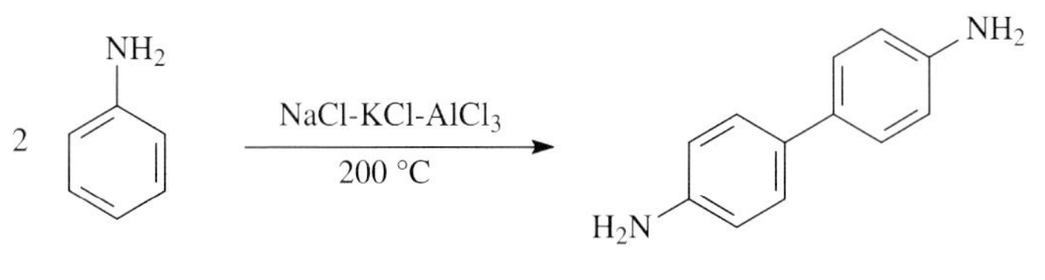

Scheme 5.2-34 The oxidation of aniline to benzidine in a molten salt [83].

$$SbCl_3 \rightleftharpoons [SbCl_2]^+ + Cl^-$$

$$SbCl_3 + AlCl_3 \longrightarrow [SbCl_2]^+ + [AlCl_4]^- \quad \text{acidic}$$

Scheme 5.2-35 The effect of adding aluminum(III) chloride to antimony(III) chloride [84, 85].

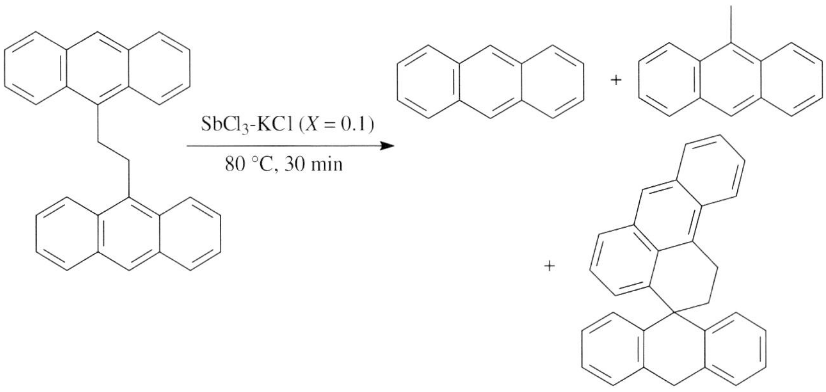

Scheme 5.2-36 The cyclization of 1,2-di-(9-anthryl)ethane in antimony(III) chloride ionic liquids [85].

Buchanan and coworkers studied the behavior of various aromatic compounds in antimony(III) chloride molten salts [84]. These salts can both act as a mild Lewis acid and allow redox reactions to take place. The Lewis acidity of the melt can be tuned by controlling the concentration of $[SbCl_2]^+$ in the melts. Basic melts are formed by adding a few mol% of a chloride donor such as KCl, whereas acidic melts are formed by adding chloride acceptors such as $AlCl_3$ (Scheme 5.2-35).

Examples of reactions that have been carried out in these antimony(III) chloride ionic liquids include the cyclizations of 1,2-di-(9-anthryl)ethane (Scheme 5.2-36) and 1,2-di-(1-naphthyl)ethane (Scheme 5.2-37). A more detailed review of the antimony(III) chloride molten salt chemistry has been published by Pagni [85].

In an attempt to study the behavior and chemistry of coal in ionic liquids, 1,2-diphenylethane was chosen as a molecule to model its chemical reactions. Newman et al. [86] investigated of the behavior of 1,2-diphenylethane in acidic pyridinium chloroaluminate(III) melts ([PyH]Cl–AlCl$_3$). At 40 °C, 1,2-diphenylethane

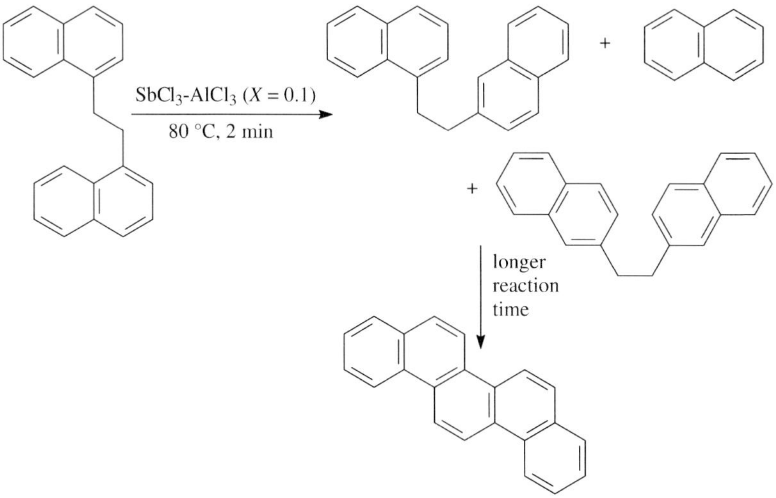

Scheme 5.2-37 The reaction of dinaphthylethane in antimony(III) chloride ionic liquids [85].

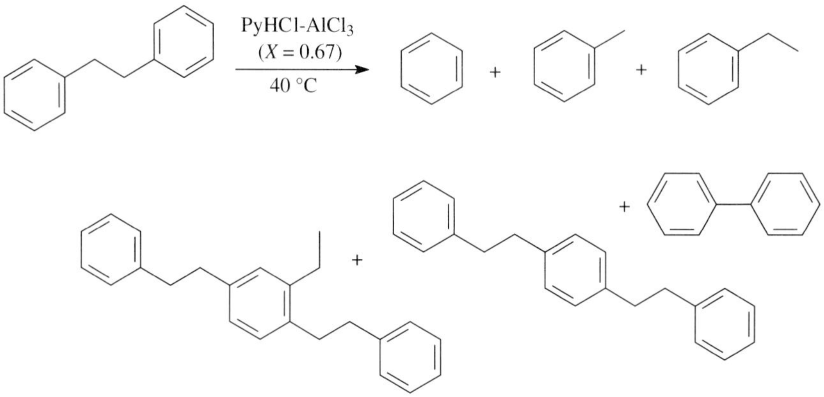

Scheme 5.2-38 The reaction of 1,2-diphenylethane with [PyH]Cl–AlCl$_3$ ($X = 0.67$) [86].

undergoes a series of alkylation and dealkylation reactions to give a mixture of products. Some of the products are shown in Scheme 5.2-38. Newman also investigated the reactions of 1,2-diphenylethane with acylating agents such as acetyl chloride or acetic anhydride in the pyridinium ionic liquid shown [87] and with alcohols such as isopropanol [88].

5.2.1.3 Cracking and Isomerization Reactions

Cracking and isomerization reactions occur readily in acidic chloroaluminate(III) ionic liquids. A remarkable example of this is the reaction of polyethylene, which is converted to a mixture of gaseous alkanes with the formula ($C_n H_{2n+2}$, where

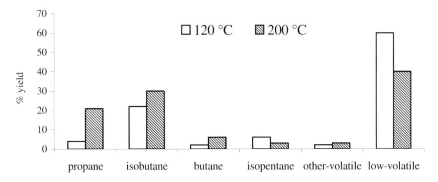

Fig. 5.2-2 The products from the ionic liquid cracking of high-density polyethylene at 120 °C and 200 °C [89].

$n = 3\text{–}5$) and cyclic alkanes with a hydrogen to carbon ratio of less than two (Fig. 5.2-2, Scheme 5.2-39) [89].

The distribution of the products obtained from this reaction depends upon the reaction temperature (Fig. 5.2-2) and differs from other polyethylene recycling reactions in that aromatics and alkenes are not formed in significant concentrations. Another significant difference is that this ionic liquid reaction occurs at temperatures as low as 90 °C, whereas conventional catalytic reactions require much higher temperatures, typically 300–1000 °C [90]. A patent filed under the Secretary of State for Defence (UK) has reported a similar cracking reaction for lower molecular weight hydrocarbons in chloroaluminate(III) ionic liquids [91]. An example is the cracking of hexane to products like propene and isobutene (Scheme 5.2-40). The reaction was

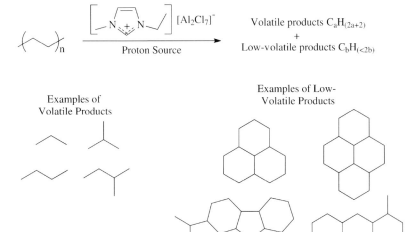

Scheme 5.2-39 The cracking of polyethylene in an ionic liquid [89].

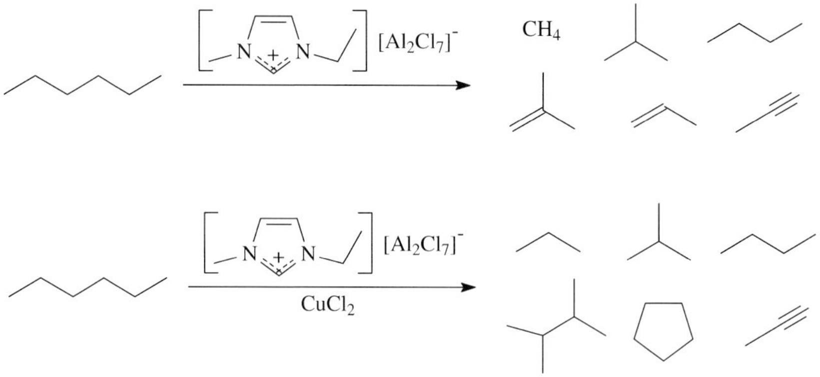

Scheme 5.2-40 The cracking of hexane in [EMIM]Cl–AlCl$_3$ ($X = 0.67$) with and without added copper(II) chloride [91].

also performed with added copper(II) chloride, which gave a significantly different product distribution.

Johnson and coworkers investigated the cracking and isomerization of various alkanes such as nonane, tetradecane and 2-methylpentane in acidic pyridinium chloride–aluminum chloride ionic liquids. Similar product types to the cracking of hexane (above) were observed and after 15 days some polymerization of the cracked products had occurred [91]. A similar reaction occurs with fatty acids (such as stearic acid) or methyl stearate, which undergo isomerization, cracking, dimerization, and oligomerization reactions. This has been used to convert solid stearic acid into the more valuable liquid isostearic acid [92] (Scheme 5.2-41). The isomerization and dimerization of oleic acid and methyl oleate have also been found to occur in chloroaluminate(III) ionic liquids [93].

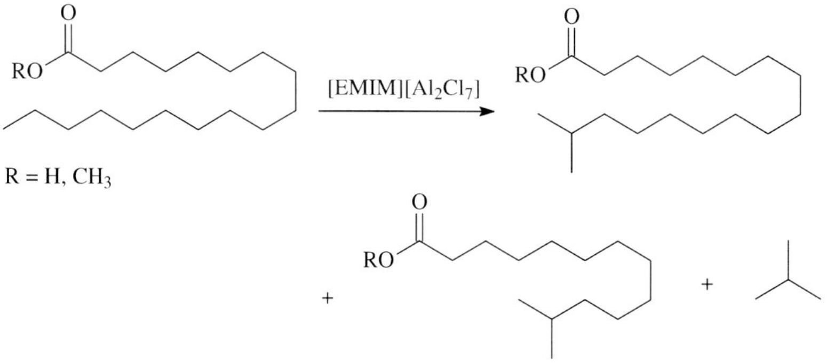

Scheme 5.2-41 The cracking and isomerization of fatty acids and fatty acid methyl esters in chloroaluminate(III) ionic liquids [92, 93].

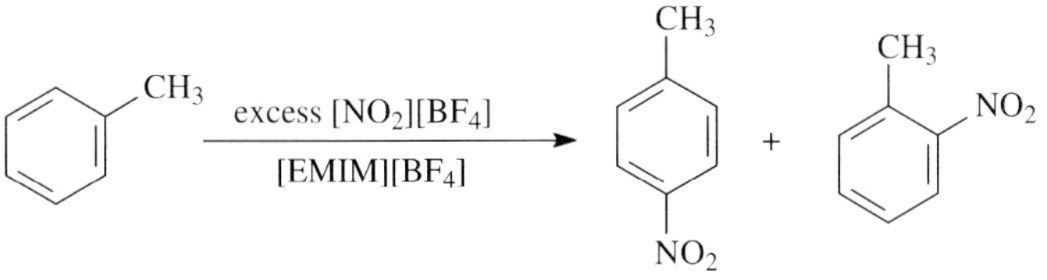

Scheme 5.2-42 The nitration of toluene with [NO$_2$][BF$_4$] in [EMIM][BF$_4$] [95].

5.2.1.4 Electrophilic Nitration Reactions

The first example of an electrophilic nitration in an ionic liquid was performed by Wilkes and coworkers [94]. Here a number of aromatic compounds were nitrated using KNO$_3$ dissolved in chloroaluminate(III) ionic liquids. A number of nitration reactions have also been carried out by Laali et al. [95]. The reactions of nitrates, preformed nitronium salts and alkyl nitrates with aromatic compounds were performed in a wide range of ionic liquids. For example, the reaction of toluene with [NO$_2$][BF$_4$] in [EMIM]Cl, [EMIM][AlCl$_4$], [EMIM][Al$_2$Cl$_7$], [EMIM][BF$_4$], [EMIM][PF$_6$], and [EMIM][OTf] were all performed with varying degrees of success. Of these, the reaction in [EMIM][BF$_4$] (Scheme 5.2-42) gave the best yield (71%, 1.17 o- : p- ratio), but only after the imidazolium ring had undergone nitration (Fig. 5.2-3).

Other methods of nitration that Laali investigated used isoamyl nitrate combined with a Brønsted or Lewis acid in several ionic liquids, with [EMIM][OTf] giving the best yields (69%, 1.0:1.0 o-: p-ratio). In the ionic liquid [HNEt(Pr^i)$_2$] [CF$_3$CO$_2$] (m.p. = 92–93 °C) toluene was nitrated with a mixture of [NH$_4$][NO$_3$] and trifluoroacetic acid (TFAH) (Scheme 5.2-43). This gave ammonium trifluoroacetate [NH$_4$][TFA] as a by-product, which could be removed from the reaction vessel by distillation (sublimation).

The nitration of aromatic compounds with nitric acid in an ionic liquid was shown by Earle et al. [96, 97]. It was found that triflate and triflimide ionic liquids catalyze nitration reactions with nitric acid. This methodology has the advantage that water is the only by-product (Scheme 5.2-44). This process could also be carried out in phosphonium ionic liquids [46]. Acidic ionic liquids such as [EMIM][HSO$_4$] or the Davis-type ionic liquids [41] could also be used but gave lower reaction rates and selectivities [98]. The effect of metal triflates such as Yb(OTf)$_3$ or Cu(OTf)$_2$ dissolved in [N-butyl-N-methylpyrrolidium][triflimide] was investigated by Handy and Egrie [99]. These gave similar yields and selectivities to ionic liquids systems

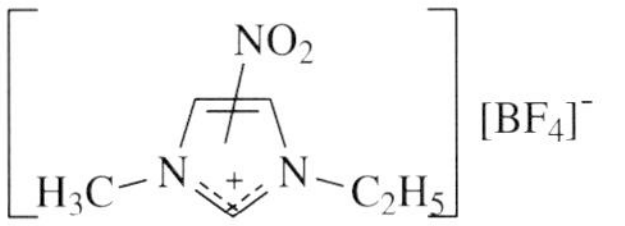

Fig. 5.2-3 The structure of the nitroimidazolium ionic liquid [95].

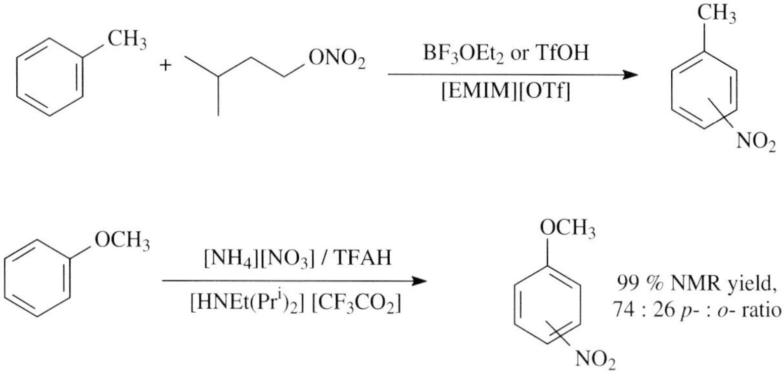

Scheme 5.2-43 Aromatic nitration reactions in ionic liquids [95].

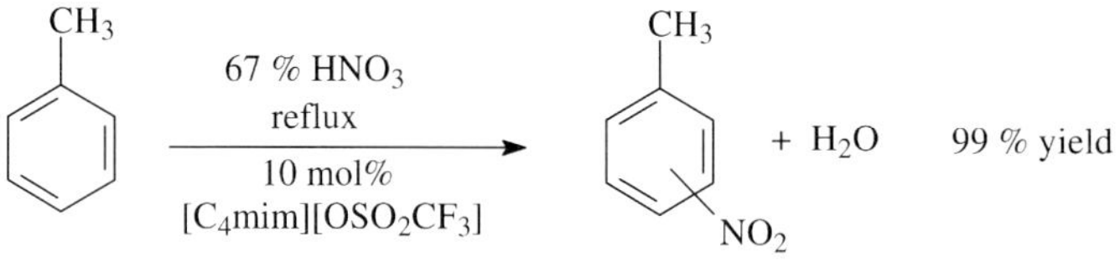

Scheme 5.2-44 The nitration of aromatic compounds in triflate ionic liquids [96, 97].

without the metals present [97]. Another metal-catalyzed nitration is the iron(III) nitrate-catalyzed nitration of phenol in dibutylimidazolium tetrafluoroborate [100]. Addition of dehydrating agents such as acetic anhydride or trifluoroacetic anhydride dramatically improved the reaction rate, but at the expense of producing the corresponding acetic or trifluoroacetic acid as a by-product [101].

The use of ultrasound has been shown to enhance the reaction rate in the nitration of phenols in $[H_5C_2NH_3][NO_3]$ [102] with an $Fe(NO_3)_3$ catalyst [103]. Significant improvements in selectivities and a ten-fold reduction in reaction time were observed compared with similar silent reactions.

5.2.1.5 Electrophilic Halogenation Reactions

Another common reaction is the chlorination of alkenes to give 1,2-dihaloalkanes. Patell et al. have reported that the addition of chlorine to ethene in acidic chloroaluminate(III) ionic liquids gave 1,2-dichloroethane [104]. Under these conditions, the ring of imidazolium ionic liquid is chlorinated. Initially, the chlorination occurs at the 4- and 5- positions of the imidazole ring, and is followed by much slower chlorination at the 2-position. This does not affect the outcome of the alkene chlorination reaction and it was found that the chlorinated imidazolium ionic liquids are excellent catalysts for the reaction. This is shown in Scheme 5.2-45.

Wilkes and coworkers have investigated the chlorination of benzene in both acidic and basic chloroaluminate(III) ionic liquids [65]. In the acidic ionic liquid [EMIM]Cl–AlCl$_3$ (X > 0.5), the chlorination reaction initially gave chlorobenzene,

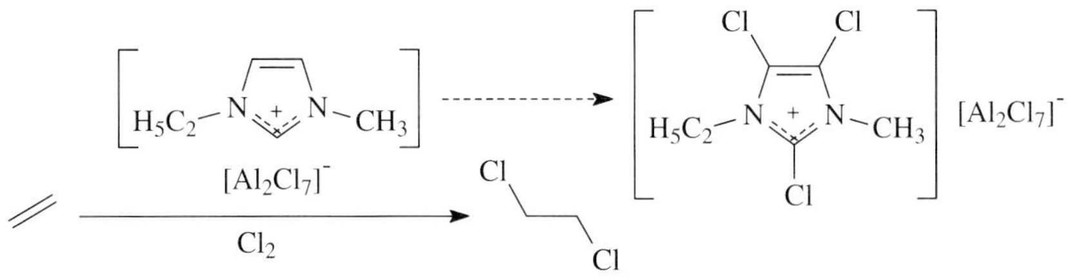

Scheme 5.2-45 The chlorination of ethene to 1,2-dichloroethane [104].

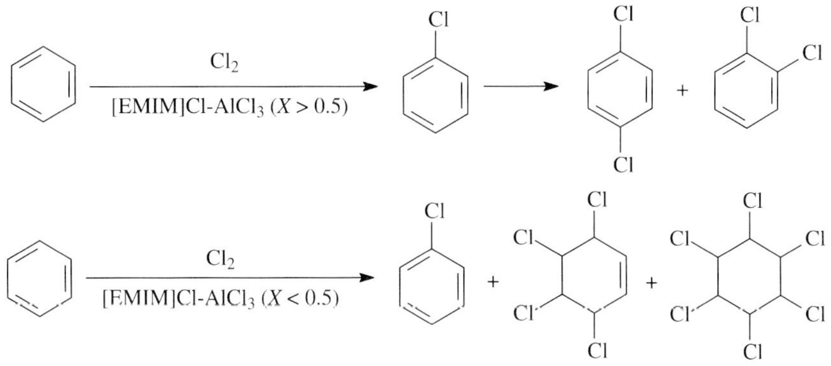

Scheme 5.2-46 The chlorination of benzene in acidic and basic chloroaluminate ionic liquids [65].

which in turn reacts with a second molecule of chlorine to give dichlorobenzenes. In the basic ionic liquid, the reaction is more complex. In addition to the formation of chlorobenzene, addition products of chlorine and benzene are observed. These addition products include various isomers of tetrachlorocyclohexene and hexachlorocyclohexane (Scheme 5.2-46).

With a view to making halogenation reactions more atom efficient and reducing the amount of noxious waste products, an oxidative halogenation reaction was developed by Earle et al. [105]. Following the observation that the reaction of toluene with nitric acid in a halide ionic liquid gave halotoluenes as a mixture of isomers in quantitative yield, it was found that the reaction of hydrogen halide in a nitrate ionic liquid gave the same halotoluenes [97, 105]. The nitrate in this reaction was reduced to nitrogen monoxide (with the corresponding oxidation of halide to hypohalic acid), which in turn could be reoxidised spontaneously with air. The result is that ionic liquids such as [BMIM][NO$_3$] catalyze the oxidative halogenation of arenes with hydrogen halides, using air as the oxidant (Scheme 5.2-47). The only by-product in this reaction is water, making this a very clean reaction [97].

The iodination of aromatic compounds using the electrophilic fluorinating agent 1-(chloromethyl)-1,4-diazabicyclo[2,2,2]octane tetrafluoroborate and iodine was carried out in a range of [BF$_4$]$^-$ and [PF$_6$]$^-$ ionic liquids and generally gave high yields [106]. Trihalide-based ionic liquids have been synthesised and the structure of the

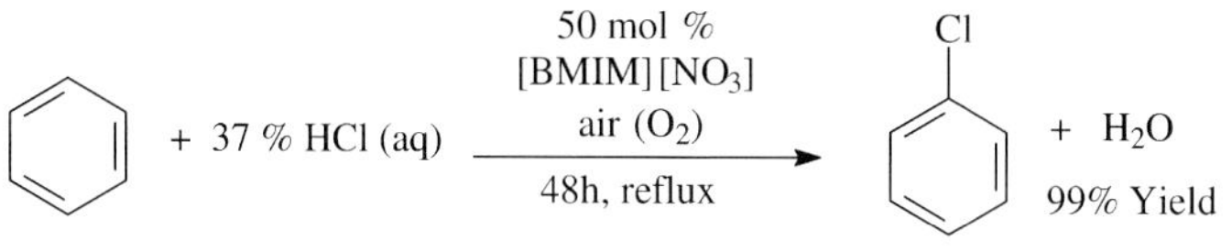

Scheme 5.2-47 The oxidative halogenation of benzene [97, 105].

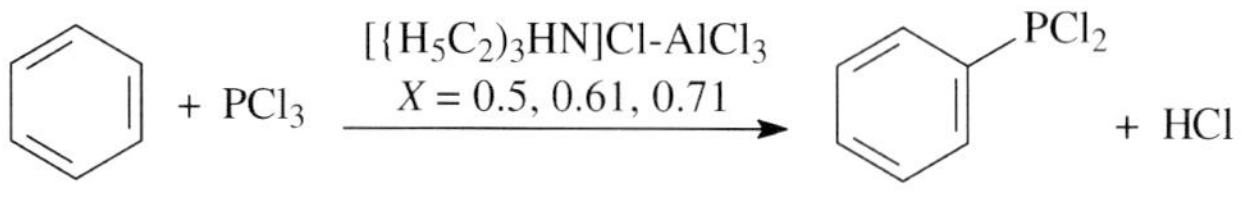

Scheme 5.2-48 The phosphylation of benzene in acidic triethylammonium chloroaluminate(III) ionic liquids [112].

trihalide ions has been investigated by electrospray ionization mass spectroscopy and NMR. They are made by mixing equimolar amount of ICl with [HMIM]Cl and IBr with [BMIM]Br or alternatively Cl_2 or Br_2 with [EMIM]I. These ionic liquids were used as reagent-solvents, or as reagents, by carrying out the reactions in [BMIM][PF_6], in iodobromination and iodochlorination reactions of alkenes and alkynes. Furthermore, the addition of ICl and IBr to [BMIM][PF_6] was investigated. Yields of *vic*-iodochloro or iodobromo adducts from very good to almost quantitative were observed for all the substrates examined [107]. The electrophilic addition of halogens to alkenes and alkynes was also investigated by Chiappe et al., in [BMIM][PF_6] and [BMIM][BF_4] [108, 109].

The bromination of aromatic compounds using *N*-bromosuccinimide (NBS) occurs in high yield in ionic liquids such as dibutylimidazolium tetrafluoroborate [110]. The yields of the monobrominated product were all in the 80% to 98% range. NBS was also used in the Wohl-Ziegler α-bromination of aryl methyl groups [111]. The reaction in [BMIM][PF_6] typically gave a 5 to 10% improvement in the yield, compared with a solvent-free reaction.

5.2.1.6 Electrophilic Phosphylation Reactions

The Friedel–Crafts reaction of PCl_3 and benzene in [(C_2H_5)$_3$HN]Cl–$AlCl_3$ ionic liquids was investigated for the synthesis of dichlorophenylphosphine [112]. A simple product isolation procedure was achieved as the product and ionic liquid formed separate phases at the end of the reaction. The effects of the ionic liquid composition, reactant composition, reaction time and quantity of ionic liquids on this reaction were studied (Scheme 5.2-48).

5.2.1.7 Electrophilic Sulfonation Reactions

The sulfonation of aromatic compounds occurs readily in ionic liquids, with the simplest case being the direct sulfonation of aromatic compounds with sulfur trioxide to give the aryl sulfonic acid [113]. Ionic liquids such as triflate or triflimide ionic liquids were found to enhance the reaction rate. In the reaction of chlorosulfuric acid with aromatic compounds, the reaction in the ionic liquid gave a

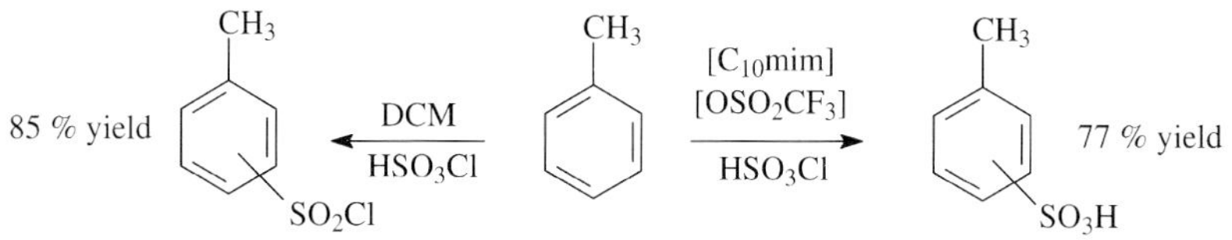

Scheme 5.2-49 The reaction of toluene with chlorosulfuric acid in ionic liquids and dichloromethane [113].

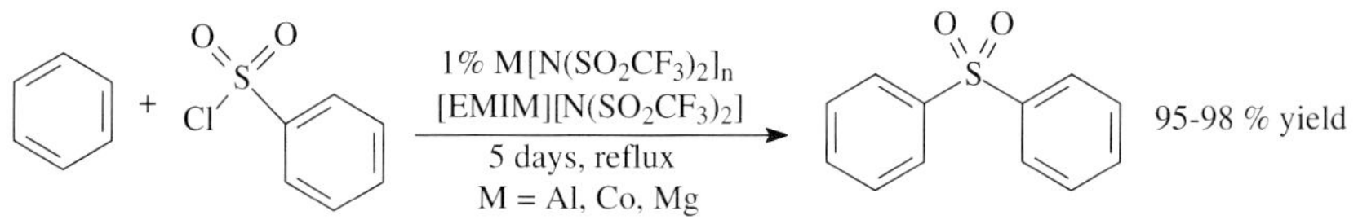

Scheme 5.2-50 The benzenesulfonylation of benzene in an ionic liquid [47].

different product, compared with a similar reaction in dichloromethane. In the molecular solvent DCM the sulfonyl chloride was the major product, whereas in ionic liquid [BMIM][OTf] or [C_{10}MIM][OTf], the sulfonic acid was the major product (Scheme 5.2-49) [113].

In chloroaluminate(III) ionic liquids such as [BMIM]Cl–AlCl$_3$ (X = 0.50–0.67) thionyl chloride forms diarylsulfoxides in 85–96% yield [114]. Sulfuryl chloride (SO$_2$Cl$_2$) shows a different chemical reactivity in [BMIM][Tf$_2$N], where it halogenates the aromatic compound in quantitative yield [114]. By way of contrast, benzenesulfonyl chloride or methanesulfonyl chloride both sulfonylate aromatic compounds such as benzene or toluene in the presence of metal triflimide catalysts [47]. This is shown in Scheme 5.2-50. A similar reaction occurs in chloroaluminate(III) ionic liquids, but these suffer from the disadvantage that the separation of the product from the ionic liquid is difficult, without hydrolysing the ionic liquid [115].

5.2.2
Nucleophilic Reactions

5.2.2.1 Aliphatic Nucleophilic Substitution Reactions
Nucleophilic substitution reactions involving molten salts are well known. A number of examples of molten pyridinium hydrochloride (mp 144 °C) being used in chemical synthesis, dating back to the 1940s, are known. Pyridinium chloride can act as both an acid and as a nucleophilic source of chloride. These properties are exploited in the dealkylation reactions of aromatic ethers [85]. An example involving the reaction of 2-methoxynaphthalene is given in Scheme 5.2-51 and a mechanistic explanation is given in Scheme 5.2-52.

Pyridinium chloride ([PyH]Cl) has also been used in a number of cyclization reactions of aryl ethers [82, 85] (Scheme 5.2-53). Presumably the reaction initially

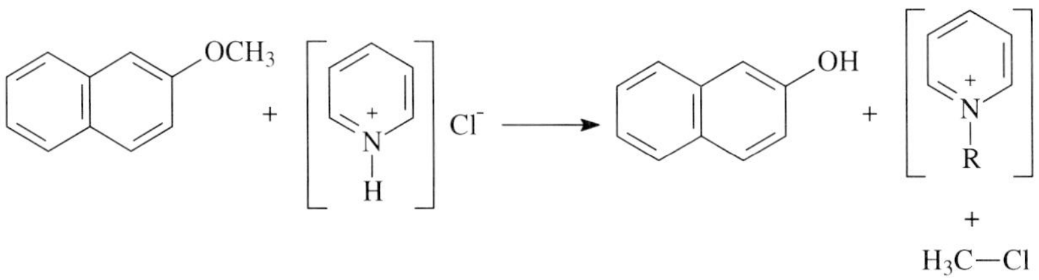

Scheme 5.2-51 The demethylation of 2-methoxynaphthalene to 2-naphthol with pyridinium chloride [16, 82].

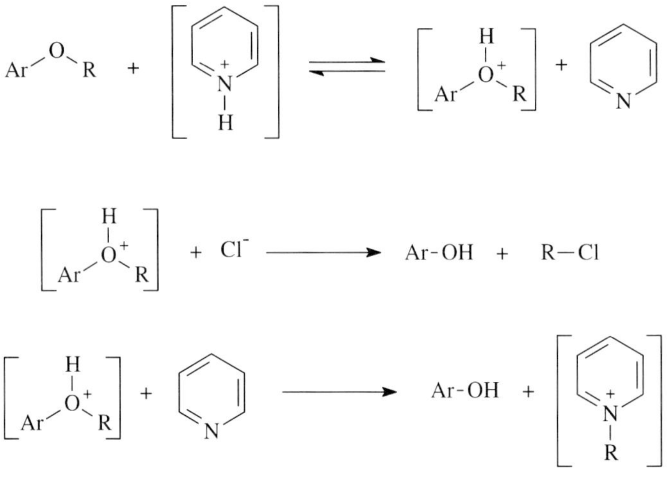

Scheme 5.2-52 A mechanism for the dealkylation of aryl ethers with pyridinium chloride [116].

proceeds by dealkylation of the methyl ether groups to the corresponding phenol. The mechanism of the cyclization is not well understood, but Pagni and Smith have suggested that it proceeds via nucleophilic attack of an Ar–OH or Ar–O⁻ group on the second aromatic ring (in a protonated form) [85].

Tetrabutylammonium fluoride (TBAF) is usually used in the form of the trihydrate or as a solution in tetrahydrofuran (THF). The pure form is difficult to isolate, owing to decomposition to HF, tributylamine and but-1-ene [82, 117] on dehydration. It has been used for a variety of reactions, including as a catalyst for various reactions with silicon compounds [118, 119]. One of its main uses is in the cleavage of silyl ether protecting groups [120]. TBAF has been used as a source of fluoride ions in a number of substitution reactions studied by Cox et al. [121]. Alkyl and acyl halides react with TBAF to give the corresponding alkyl or acyl fluoride in good yield. In the reaction of (R)-2-tosyloctane with TBAF, the product was (S)-2-fluorooctane, confirming an S_N2 type mechanism in the substitution reaction (Scheme 5.2-54).

TBAF has also been used in the preparation of various fluorocarbenes. This involved the photolysis of phenyl- or phenoxyfluorodiazirine, which was in turn

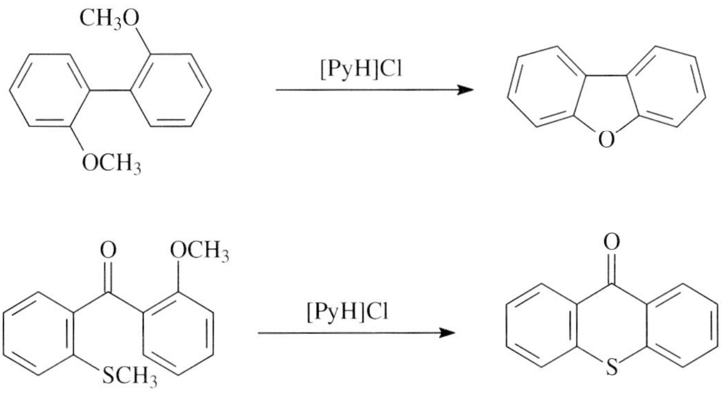

Scheme 5.2-53 Two examples of aryl demethylation reactions followed by cyclization [82, 85].

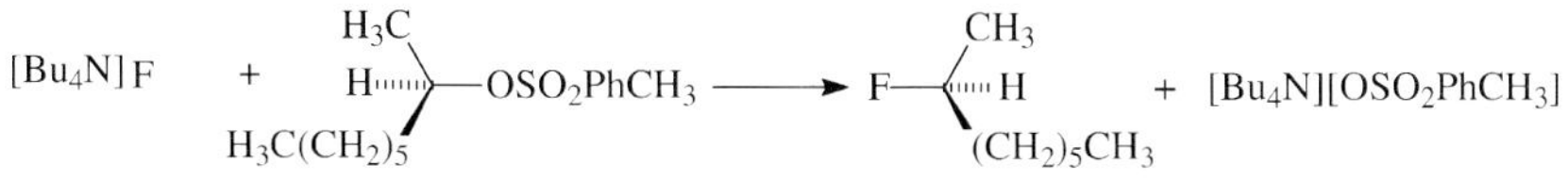

Scheme 5.2-54 The use of TBAF in an S_N2 substitution reaction [82, 121].

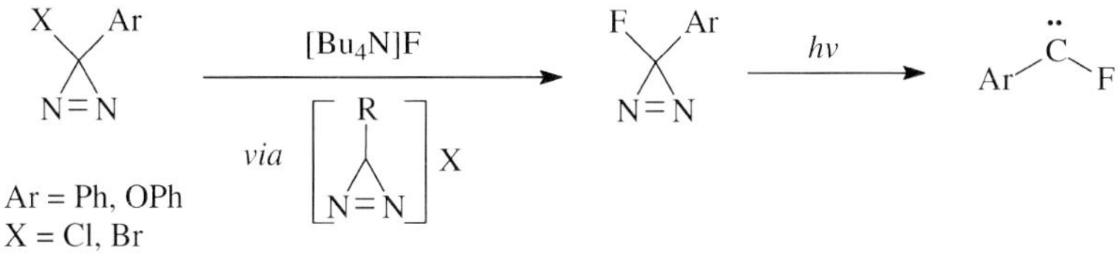

Scheme 5.2-55 The use of TBAF in the preparation of a fluorodiazirine [122, 123].

synthesized from the reaction of TBAF with phenyl- or phenoxyhalodiazirine as in Scheme 5.2-55 [122, 123].

The first attempt at a nucleophilic substitution reaction in an ionic liquid was carried out by Ford and coworkers [124–126]. Here, the rates of reaction of halide ion (in the form of its triethylammonium salt) with methyl tosylate in the molten salt triethylhexylammonium triethylhexylborate were studied (Scheme 5.2-56). This was compared with similar reactions in dimethyl formamide (DMF) and methanol. The reaction rates in the molten salt appeared to be intermediate in rate between methanol and DMF (a dipolar aprotic solvent known to accelerate S_N2 substitution reactions).

The alkylation of sodium 2-naphthoxide with benzyl bromide in tetrabutylammonium and tetrabutylphosphonium halide salts was investigated by Brunet and Badri [127] (Scheme 5.2-57). The yields in this reaction were quantitative and alkylation occurred predominantly on the oxygen atom of the naphthoxide ion (typically 93–97%). The rate of the reaction was slower in the chloride salts due to the benzyl bromide reacting with chloride ion to give the less reactive benzyl chloride.

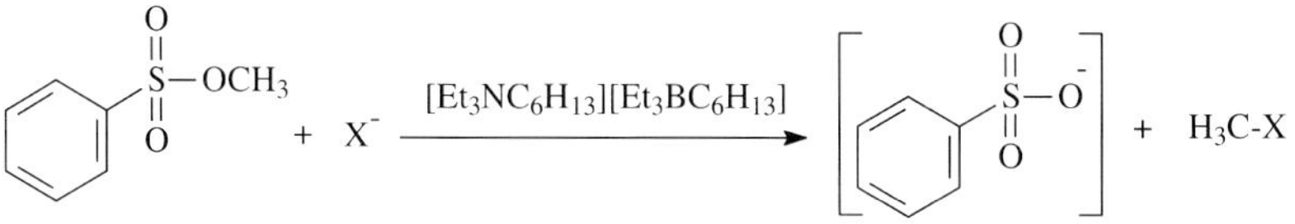

Scheme 5.2-56 The reaction of halide with methyl tosylate in triethylhexylammonium triethylhexylborate [124–126].

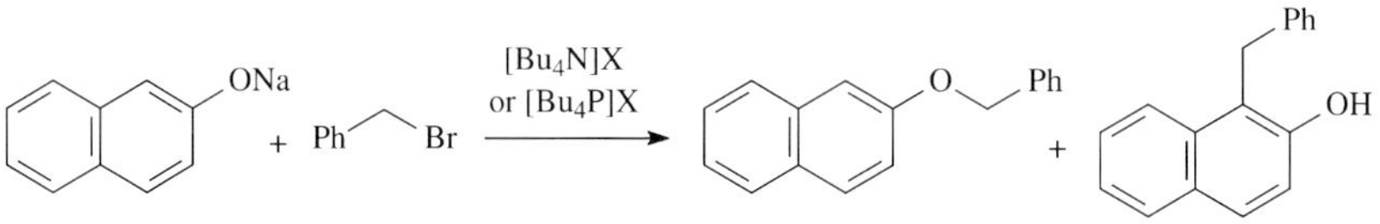

Scheme 5.2-57 The benzylation of sodium 2-naphthoxide with benzyl bromide in ammonium or phosphonium halide salts (X = Cl, Br) [127].

Indole and 2-naphthol undergo alkylation on the nitrogen and oxygen atoms respectively (Scheme 5.2-58), when treated with an alkyl halide and base (usually NaOH or KOH) in [BMIM][PF$_6$] [128]. These reactions occur with similar rates to those carried out in dipolar aprotic solvents such as DMF or DMSO. An advantage of using the room-temperature ionic liquid for this reaction is that the lower reaction temperatures result in higher selectivities for substitution on the oxygen or nitrogen atoms. The by-product (sodium or potassium halide) of the reaction can be extracted with water and the ionic liquid recycled. Indole was also benzylated by reaction with dibenzyl carbonate (DBC) using DABCO as a catalyst in 80% yield in tetrabutylammonium chloride [129]. Similarly, a wide range of halide, [BF$_4$]$^-$ and [PF$_6$]$^-$ ionic liquids were used in the alkylation of benzimidazole in with DBC in 74 to 95% yield [129]. The ionic liquid reactions all gave higher yields when the ionic liquid was present, compared with a similar reaction in acetonitrile. Potassium fluoride has been used as a promoter for the benzylation of phthalimide with benzyl chloride in imidazolium and pyridinium ionic liquids in 91% to 97% yields [130]. Benzyl cinnamate formation from potassium cinnamate and benzyl chloride proceeds considerably faster in ionic liquids than in the dipolar aprotic solvent DMF. The ionic liquid [BMIM][BF$_4$] showed the greatest catalytic capability (97% yield) and the ionic liquid could be recycled and reused [131]. A quantitative study of the nucleophilic displacement reaction of benzoyl chloride with cyanide ion in [BMIM][PF$_6$] was investigated by Eckert and coworkers [132]. The separation of the product 1-phenylacetonitrile from the ionic liquid was achieved by distillation or by extraction with supercritical CO$_2$. 1-Phenylacetonitrile was then treated with KOH in [BMIM][PF$_6$] to generate an anion, which reacted with 1,4-dibromobutane to give 1-cyano-1-phenylcyclopentane (Scheme 5.2-59). This was in turn extracted from the ionic liquid with supercritical CO$_2$. These reactions lead to a build up of KCl or KBr in the ionic liquid, which was removed by washing the ionic liquid with water.

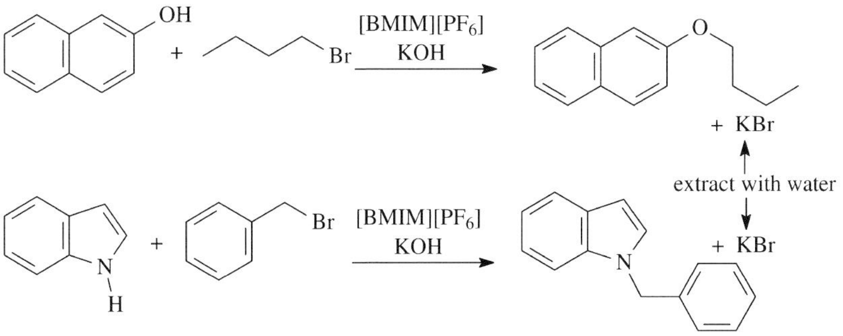

Scheme 5.2-58 Alkylation reactions in [BMIM][PF$_6$] [128].

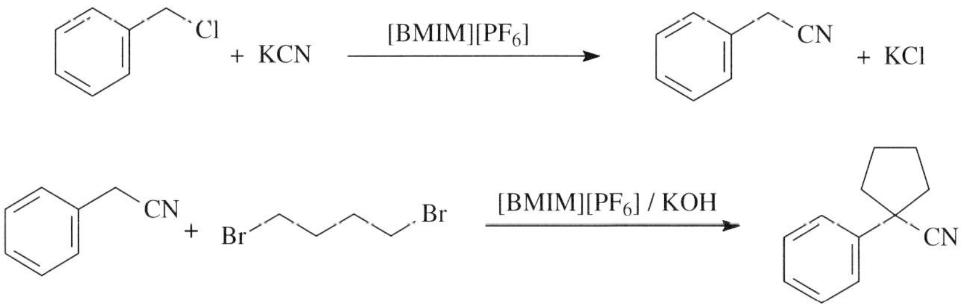

Scheme 5.2-59 The reaction of cyanide with benzyl chloride and its subsequent reaction with 1,4-dibromobutane [132].

Nucleophilic displacement reactions have been carried out under aqueous biphasic conditions with water and the water immiscible ionic liquid [BMIM][PF$_6$]. Nucleophiles such as azide, cyanide, phenoxide and 4-chlorophenoxide all demonstrated good yields in the reaction with a range of alkyl bromides [133]. The addition of a chiral phase transfer catalyst did not result in any observed stereoselectivity. [BMIM][PF$_6$] and [BMIM][BF$_4$] have also been used as solvents in the alkylation of benzotriazole [134] and sodium benzenesulfonates [135]. A range of aromatic alkyl ethers were synthesized in [BMIM] [PF$_6$], [BF$_4$], [OTf], [SbF$_6$] and [O$_2$CCH$_3$] by Chi et al., by the reaction of an alcohol with an alkyl bromide. This was without a base present and therefore required higher reaction temperatures (100 °C) [136]. Chi also reports nucleophilic displacement of alkylmethane sulfonates with halides, cyanide and alkoxides [137]. This methodology has been applied with ^{18}F labeling of molecules [138]. Headley et al. have reported similar substitution reactions of a more functionalized alkyl halide [139]. An example of nucleophilic substitution has been used in the formation of constrained carbanucleosides. This shows that the reaction proceeds through a classic S$_N$2 mechanism, since the stereochemistry at the nucleophilic site has been inverted [140]. The reactions in [BMIM][BF$_4$] gave typical improvements in yield from 15% to 70%. This is shown in Scheme 5.2-60.

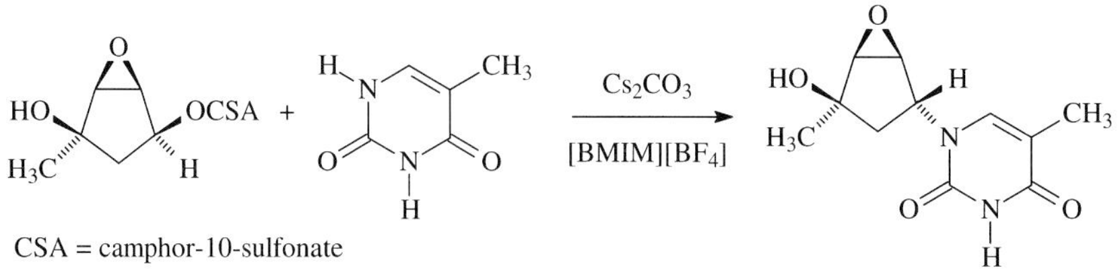

Scheme 5.2-60 The formation of carbanucleosides in [BMIM][BF$_4$] [140].

Welton and coworkers have published a series of papers seeking to understand the kinetics and how the structure of ionic liquids affects nucleophilic substitution reactions. The reactions of imidazolium halides with methyl 4-nitrotosylate in [BMIM][BF$_4$] were examined using UV–visible spectrophotometry [141]. The effect of the structure of the cation on the reaction rate in triflimide ionic liquids [142] and the effect of changing the anion in [BMIM]$^+$ ionic liquids are described [143]. Welton finds that the rates of reactions where the starting materials are charge neutral and the formation of the activated complex involves the development of charges will be accelerated by the use of an ionic liquid solvent. This is an extension of the Hughes–Ingold rule (devised for molecular solvents) to ionic liquids, showing them to be highly polar solvents. Where the formation of hydrogen bonds between a nucleophile and the anion of an ionic liquid is possible, the reaction will be further accelerated [144].

The conversion of alcohols to halides has been described by several authors. Nguyen et al. have used imidazolium halide ionic liquids in the conversion of fatty alcohols to fatty alkyl halides with sodium bromide or sodium iodide [145]. This procedure can be used with direct or microwave heating. Tang describes a similar reaction (the halogenation of diols) using hydrohalic acid and has the advantage that water is the only byproduct [146]. The halogenation of diols is also described by Nguyen [147].

[BMIM][PF$_6$] has been used as a solvent for the nucleophilic addition of lithium halide to epoxides to give a vicinal halohydrin [148]. This nucleophilic addition was also found to work well with amines in [BMIM][BF$_4$] to give β-amino alcohols [149]. These two reactions are shown in Scheme 5.2-61.

A similar reaction is the use of TMS azide and TMS cyanide in the ring opening of epoxides (Scheme 5.2-62) [150]. The authors use chiral catalysts such as Salen-Al or Ti(O-^{i}Pr)$_4$/TADDOL, which give the expected addition product with a modest enantioselectivity. Khodaei et al. have carried out nucleophilic addition and ring opening of epoxides with aromatic amines in the salt tetrabulylammonium bromide, with several bismuth catalysts [151]. These reactions gave excellent yields and improved reaction rates over conventional solvents.

As a demonstration of the complete synthesis of a pharmaceutical in an ionic liquid, Pravadoline was selected as it combines a Friedel-Crafts reaction and a nucleophilic displacement reaction (Scheme 5.2-63) [152]. The alkylation of 2-methylindole with 1-(N-morpholino)-2-chloroethane occurs readily in [BMIM][PF$_6$]

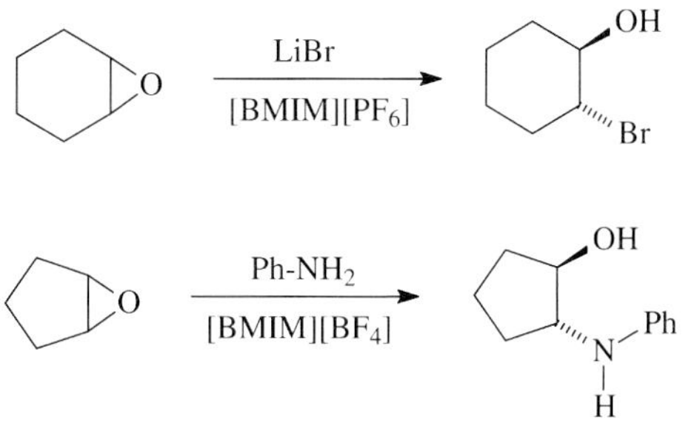

Scheme 5.2-61 The nucleophilic ring opening of epoxides with halides and amines [148, 149].

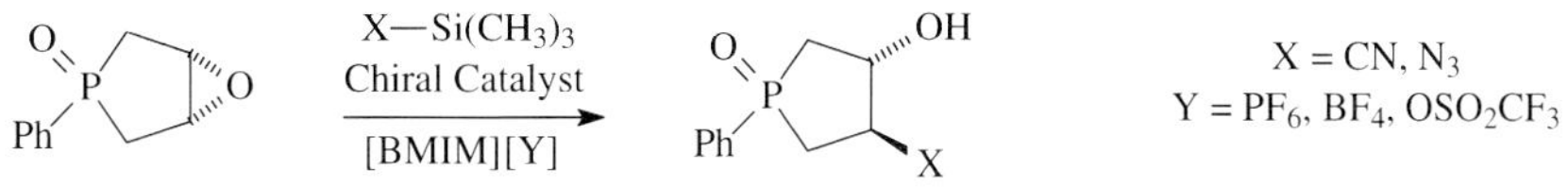

Scheme 5.2-62 The addition of TMS-CN or TMS-N$_3$ to epoxides in ionic liquids [150].

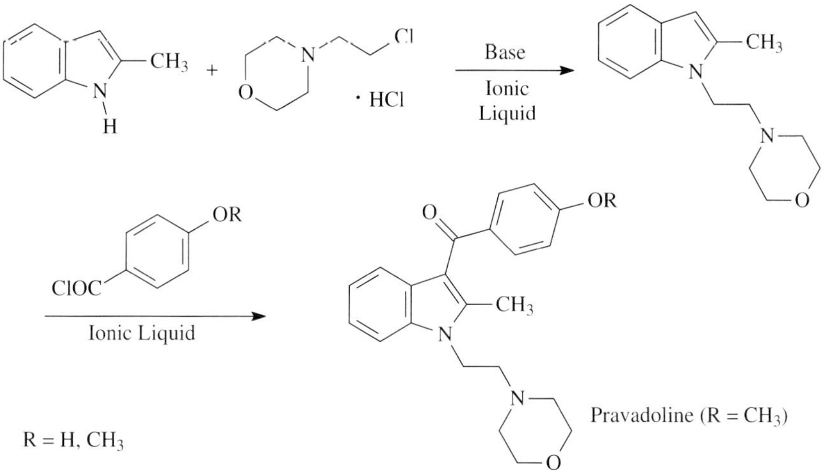

Scheme 5.2-63 The complete synthesis of Pravadoline in [BMIM][PF$_6$] [152].

and [BMMIM][PF$_6$] (BMMIM = 1-butyl-2,3-dimethylimidazolium), in 95–99% yields, respectively, using potassium hydroxide as the base. The Friedel-Crafts acylation step in [BMIM][PF$_6$] at 150 °C occurs in 95% yield and requires no catalyst.

The substitution of trichloroacetimidates with alcohols and glycosides was attempted in [BMIM][PF$_6$] and [EMIM][OTf] [153]. The reactions proceeded to give the appropriate isopropyl or glucoside substitution product in moderate to good yields. These and similar glycosidation reactions are also described by Pakulski [154]. An example is shown in Scheme 5.2-64.

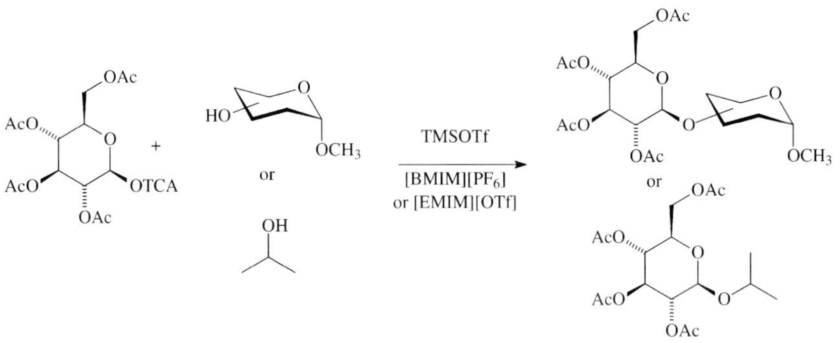

Scheme 5.2-64 The TMSOTf catalyzed substitution of trichloroacetimidate in ionic liquids [154].

Shreeve et al. have synthesized a range of new ionic liquids based on oxazolidine, morpholine and 1,2,4-triazole. These have been found to be good solvents for the Cu(I)-mediated nucleophilic trifluoromethylation of benzyl bromide [155]. Nucleophilic substitution has also been carried out on alkenes in the form of 2-tosyltropone, where the tosyl group was substituted for a chloride in [BMIM][BF$_4$] [156]. Aminopyridines [157] and aminopyrimidines [158] react with α-halo ketones to form imidazopyridines and imidazopyrimidines. These reactions have been investigated in [BF$_4$]$^-$ and [PF$_6$]$^-$ ionic liquids and found to give faster reaction rates than in organic solvents [159].

5.2.2.2 Aromatic Nucleophilic Substitution Reactions

Molten salts have been used for many years in the form of reagents such as fused KOH, pyridinium chloride and tetrabutylammonium fluoride (TBAF) [160]. One of the earliest examples of a molten salt in the literature involves the use of KOH. Examples, dating from 1840 [161] are known. A common use is in the reaction of fused KOH with arene sulfonic acids to produce phenols. Although KOH has a high melting point (410 °C), impurities such as traces of water or carbonates bring the melting point down. An example is given in Scheme 5.2-65 [162].

Diaryliodonium salts undergo a nucleophilic displacement reaction in [BMIM][BF$_4$] with potassium trithiocarbonates to give aryltrithiocarbonates (Scheme 5.2-66) [163]. The reactions in ionic liquids gave significantly higher yields (up to 75%) than in solvents such as THF, DMF, or acetonitrile. Diaryliodonium

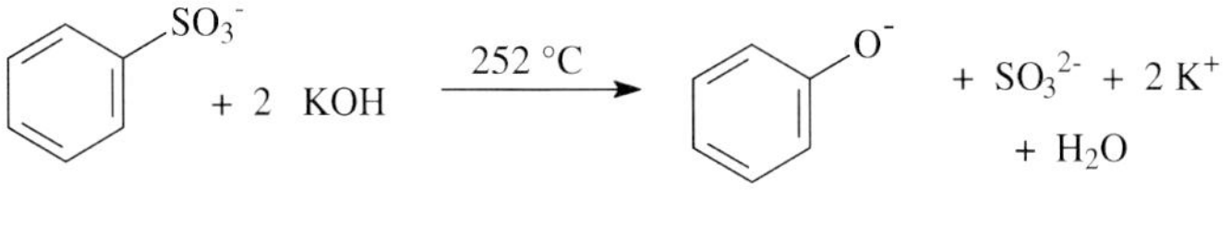

Scheme 5.2-65 The reaction of benzenesulfonates with fused KOH [161, 162].

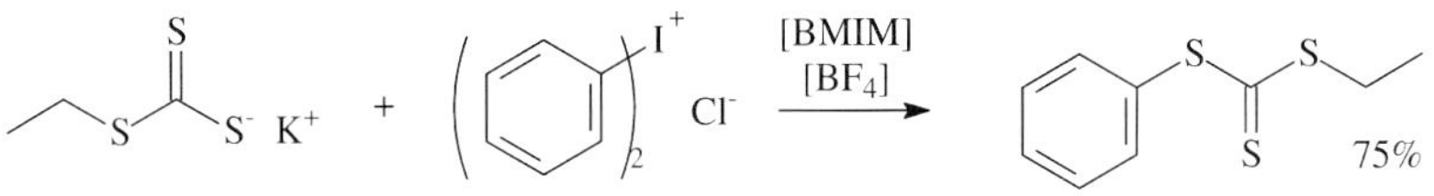

Scheme 5.2-66 The preparation of aryltrithiocarbonates.

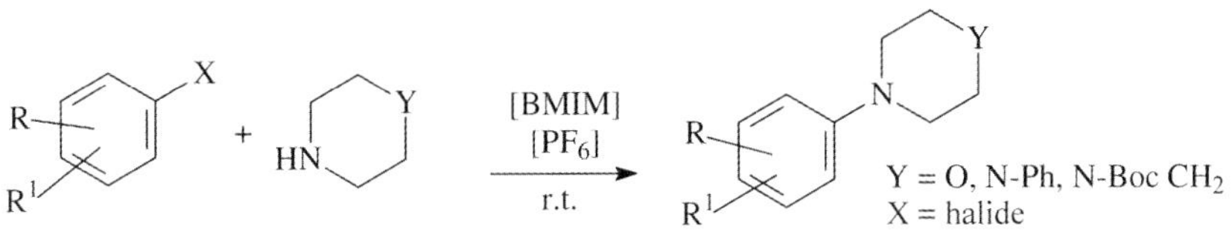

Scheme 5.2-67 The amination of aryl halides in [BMIM][PF$_6$] [165].

salts have also been used to arylate indoles and benzimidazoles in [BMIM][BF$_4$] [164].

The amination of aryl halides has been carried out in [BMIM][BF$_4$] and [BMIM][PF$_6$] by Yadav and coworkers [165]. Conventionally, these reactions are carried out in dipolar aprotic solvents such as DMF or DMSO, and require high reaction temperatures. However, in [BMIM][PF$_6$], these reactions have been shown to proceed at room temperature (Scheme 5.2-67).

A related reaction is the substitution of sulfonate leaving groups with halide ion in 2-substituted tropones [156]. In these reactions, lithium ion is thought by the authors to catalyze the reaction.

5.2.3
Electrocyclic Reactions

5.2.3.1 Diels-Alder Reactions

Lee has used chloroaluminate(III) ionic liquids in the Diels-Alder reaction [166]. The *endo:exo* ratio rose from 5.25 to 19 by changing the composition of the ionic liquid from $X = 0.48$ to $X = 0.51$ (Scheme 5.2-68). The reaction works well giving up to 95% yield, but the moisture sensitivity of these systems is a major disadvantage and the products were recovered by quenching the ionic liquid in water. A by-product of the reactions involving cyclopentadiene is the formation of its dimer, and this dimerization reaction in acidic chloroaluminate(III) ionic liquids was investigated [167]. Unsurprisingly, it was found that the rate of dimerization was a function of the acidity of the ionic liquid. Kumar has also investigated a means of converting an *exo*-selective Diels-Alder reaction to an *endo*-selective reaction, by means of altering the acidity of chloroaluminate(III) ionic liquids [168]. Generally the yields in most of these chemical reactions are too low to be useful, and suffer the usual product isolation problems of chloroaluminate(III) ionic liquids.

Neutral ionic liquids have been found to be excellent solvents for the Diels-Alder reaction. The first example of a Diels-Alder reaction in an ionic liquid was in the

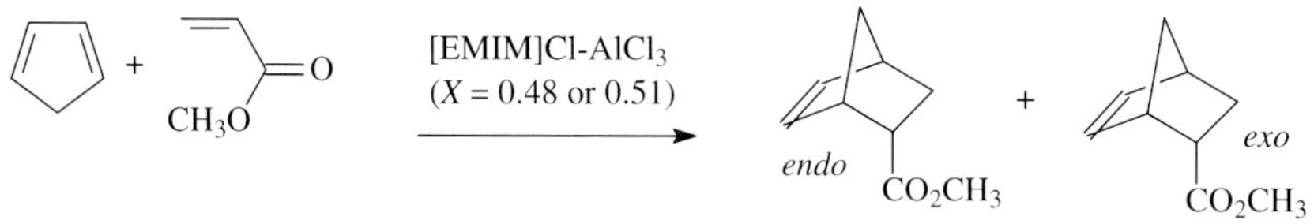

Scheme 5.2-68 The Diels-Alder reaction in a chloroaluminate(III) ionic liquid [166].

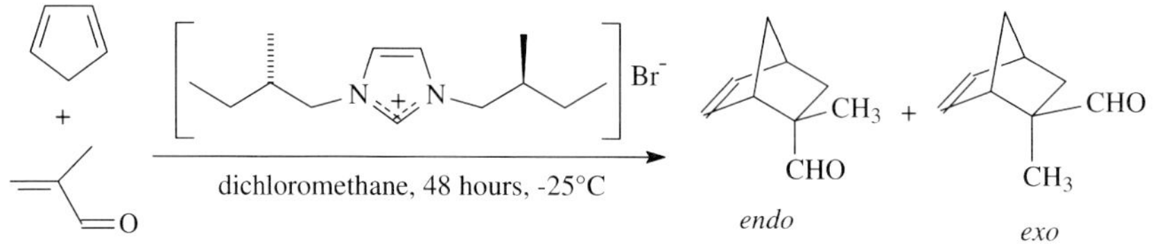

Scheme 5.2-69 Use of a chiral ionic liquid in a Diels-Alder reaction [170].

reaction of methyl acrylate with cyclopentadiene in [EtNH$_3$][NO$_3$] [169], where significant rate enhancement was observed. Howarth et al. investigated the role of chiral imidazolium chloride and trifluoroacetate salts (dissolved in dichloromethane) in the Diels-Alder reaction of cyclopentadiene and either crotonaldehyde or methacrolien [170]. It should be noted that this paper describes one of the first examples of a chiral cationic ionic liquid being used in synthesis (Scheme 5.2-69). The enantioselectivity was found to be <5% in this reaction for both the *endo* (10%) and *exo* (90%) isomers.

A study of the Diels-Alder reaction was carried out by Earle et al. [171]. The rates and selectivities of reactions of ethyl acrylate (EA) with cyclopentadiene (CP) in water, 5 M lithium perchlorate in diethyl ether (5 M LPDE), and [BMIM][PF$_6$] were compared. The reactions in the ionic liquid [BMIM][PF$_6$] are marginally faster than in water, but are both slower than in 5 M LPDE [43, 172] (see Table 5.2-3 and Scheme 5.2-18). It should be noted that these three reactions give up to 98% yield if left for 24 h. The *endo:exo* selectivity in [BMIM][PF$_6$] was similar to 5 M LPDE, and considerably greater than that in water (Table 5.2-4).

Table 5.2-4 Diels-Alder reactions in various solvents [165].

Solvent	Diene	Dieneophile	Product	Time	Yield	a:b ratio
[BMIM][PF$_6$]	CP	EA	1a + 1b	1	36	8.0
5M LPDE	CP	EA	1a + 1b	1	61	8.0
Water	CP	EA	1a + 1b	1	30	3.5
[BMIM][PF$_6$][a]	IP	MVK	2a + 2b	6	98	20
[BMIM][PF$_6$]	IP	MVK	2a + 2b	18	11	4

[a] 5 mol% ZnI$_2$ added, IP = isoprene.

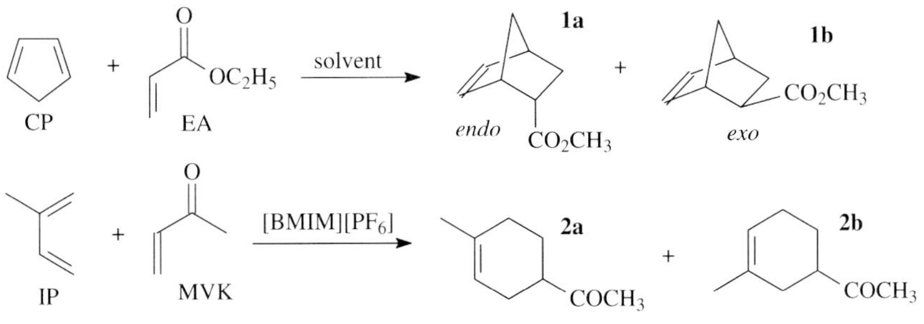

Scheme 5.2-70 The Diels-Alder reactions in Table 5.2-4 [171].

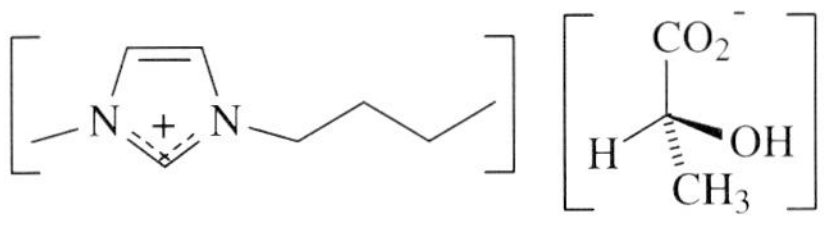

Fig. 5.2-4 An example of a chiral ionic liquid used in the Diels-Alder reaction [43].

In the reaction of isoprene (IP) with methyl vinyl ketone (MVK), the selectivities of the two isomers produced in this reaction can be improved from 4:1 to 20:1 by the addition of a mild Lewis acid such as zinc(II) iodide (5 mol%) to the ionic liquid [BMIM][PF$_6$] (Scheme 5.2-70). One of the key benefits of this is that the ionic liquid and catalyst can be recycled and reused after solvent extraction or direct distillation of the product from the ionic liquid. The reaction was also carried out in the chiral ionic liquid [BMIM][lactate] (Fig. 5.2-4). This was found to give the fastest reaction rates of all the ionic liquids tested, and the lowest *endo:exo* selectivity. The products of the Diels-Alder reaction were found to be racemic and no chiral induction was observed [171].

A similar study was performed by Welton and coworkers, who studied the rate and selectivities of the Diels-Alder reaction of cyclopentadiene with methyl acrylate in a number of neutral ionic liquids [173]. It was found that *endo:exo* ratios decreased slightly as the reaction proceeded, and were dependent on reagent concentration and ionic liquid type. A further study showed that the degree of hydrogen bonding to the ionic liquid of solvent affected the endo:exo ratio, with greater hydrogen bonding giving greater *endo*-selectivity [174].

The use of molten salts based on phosphonium tosylates has also been reported for Diels-Alder reactions [175]. These salts have higher melting points than most ionic liquids in common use and hence the reactions were performed in a sealed tube. The authors claim very high selectivities in the reaction of isoprene with MVK or methyl acrylate. The effect of temperature on the selectivity in phosphonium tosylates gave reduced *endo:exo* ratios at higher temperatures [176]. The Diels-Alder reactions of isoprene with acrylonitrile, acrylic acid and methacrylic acid in pyridinium ionic liquids ([EtPy][BF$_4$] or [EtPy][F$_3$CCO$_2$]) were found to give the expected cyclohexene structures [177]. The authors show that

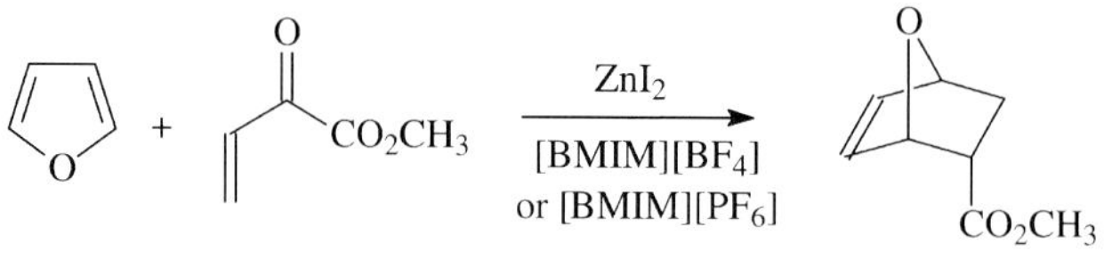

Scheme 5.2-71 The Diels-Alder reaction of furan catalyzed by zinc(II) iodide [180].

far better yields and selectivities are obtained, when compared with a molecular solvent (dichloromethane), and that these ionic liquids give improved performance over phosphonium tosylates. Another class of ionic liquids used in this reaction are binary zinc(II) halide-containing salts. These mildly Lewis acidic binary ionic liquids gave enhanced selectivities over neutral ionic liquids (12:1 endo:exo ratio compared with typically 5:1) [178]. Choline chloride–zinc chloride binary ionic liquids have also been used for these reactions [179]. The Diels-Alder reaction can also be carried out successfully on furan, thiophene and pyrroles in [BMIM][BF$_4$] or [BMIM][PF$_6$] [180]. This is shown in Scheme 5.2-71. Another example of the use of metal salts to enhance the Diels-Alder reaction in ionic liquids is in the addition of small amounts of scandium(III) triflate to [BMIM][OTf] [181]. The reaction in the ionic liquid gave over 10 times the reaction rate (1,4-naphthaquinone + 2,3-dimethylbutadiene) compared to a similar reaction in dichloromethane.

Oh and Meracz looked at the use of chiral auxiliaries to control the stereochemical outcome of the Diels-Alder reaction [182]. The reactions in some cases gave very good enantioselectivities, and complete *endo*-selectivity. The authors also investigated the use of a chiral bis-oxazoline copper(II) complex, which also gave high enantioselectivities (96:4 ratio). These two reactions are shown in Scheme 5.2-72. Another asymmetric Diels-Alder reaction (Scheme 5.2-73) used a chiral platinum(II) complex, and gave 90–93% *ee* in the reaction of cyclopentadiene with an oxazolinoyl α,β-unsaturated ketone [183]. When the reaction was carried out in a molecular solvent such as dichloromethane, the platinum catalyst underwent racemisation, and hence significantly reduced the *ee* of the product. In the ionic liquids investigated, racemization did not occur. Hence the ionic liquids were acting to stabilise the catalyst.

5.2.3.2 Hetero Diels-Alder Reactions

Kitazume and Zulfiqar have investigated the *aza*-Diels-Alder reaction in 1,8-diazabicyclo[5,4,0]-7-undecenium trifluoromethanesulfonate [EtDBU][OTf] [184] (Fig. 5.2-5). This reaction involved the scandium(III) trifluoromethanesulfonate catalyzed reaction of an imine (usually generated *in situ* from an aldehyde and an amine) with a diene. An example of this reaction is given in Scheme 5.2-74. The yields in this reaction were high (80–99%) and it was found that the ionic liquid could be recycled and reused.

Yadav has demonstrated another *aza*-Diels-Alder reaction in the formation of pyrano- and furanoquinolines from anilines, aldehydes and dihydropyran or dihydrofuran in ionic liquids such as [BMIM][BF$_4$] [185]. The reaction involves the

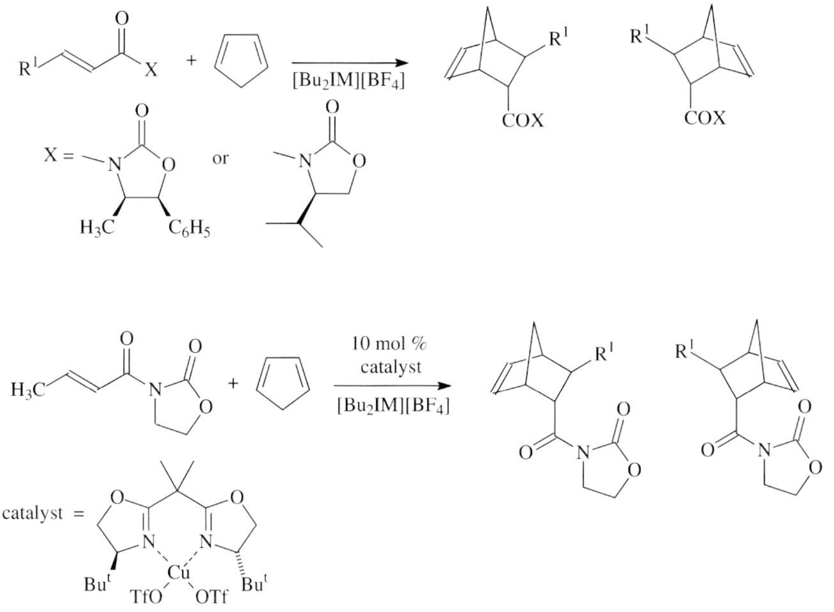

Scheme 5.2-72 The asymmetric Diels-Alder reaction in [Bu₂IM][BF₄] [182].

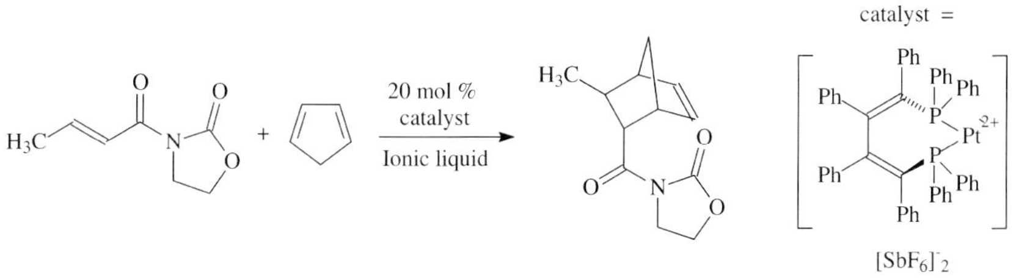

Scheme 5.2-73 The Pt-NUPHOS catalyzed asymmetric Diels-Alder reaction [183].

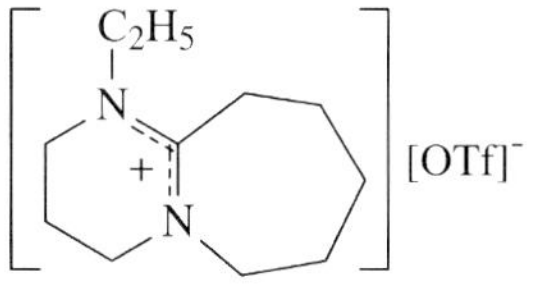

Fig. 5.2-5 The structure of diazabicyclo[5,4,0]-7-undecenium trifluoro methanesulfonate [EtDBU][OTf] [184].

condensation of the aromatic amine with the aldehyde to give an aromatic imine followed by a Diels-Alder reaction, to give the cyclized product in up to 92% yield. An example is shown in Scheme 5.2-75. Yadav also describes a more elaborate hetero Diels-Alder reaction, shown in Scheme 5.2-76. Here an aldehyde undergoes a condensation reaction with the β-dicarbonyl compound followed by a hetero-Diels-Alder reaction in [BMIM][BF₄] [186].

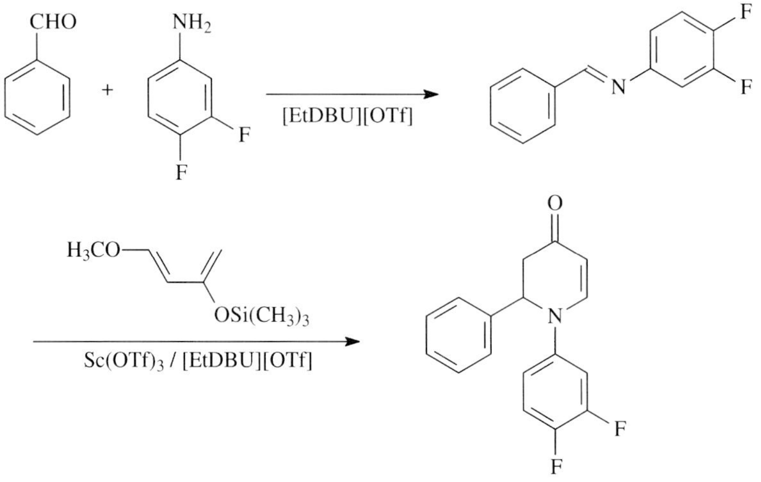

Scheme 5.2-74 The aza-Diels-Alder reaction in an ionic liquid [169].

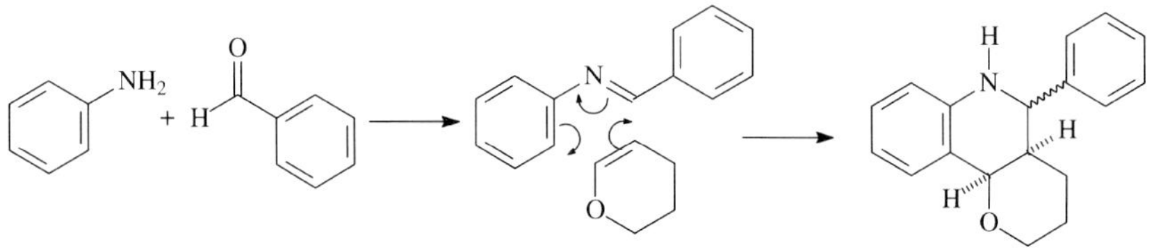

Scheme 5.2-75 The aza-Diels-Alder reaction in ionic liquids [185].

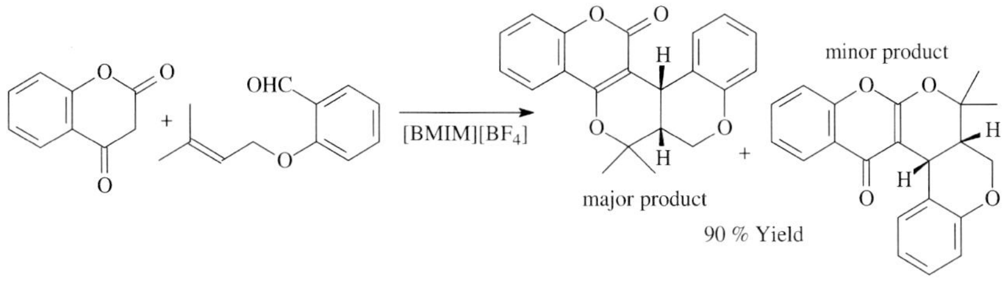

Scheme 5.2-76 The Knoevenagel hetero-Diels-Alder reaction coumarin synthesis [186].

Microwave irradiation has been used to dramatically reduce reaction times in a number of hetero Diels-Alder reactions [187], using [BMIM][PF$_6$] dissolved in dichloroethane to promote the absorption of microwave energy.

5.2.3.3 The Ene Reaction

Kitazume and Zulfiqar have investigated the Claisen rearrangement of several aromatic allyl ethers in ionic liquids, catalyzed by scandium(III) trifluoromethanesulfonate [188]. The reaction initially gave the 2-allylphenol but this reacted further to

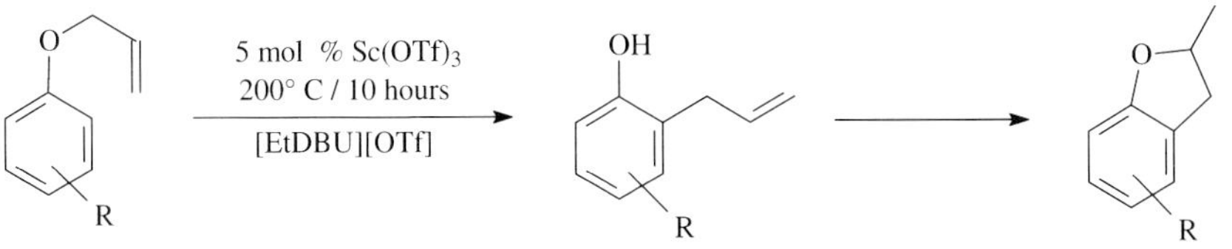

Scheme 5.2-77 The Claisen rearrangement of several phenyl allyl ethers (R = H, 4-CH$_3$, 6-CH$_3$) [188].

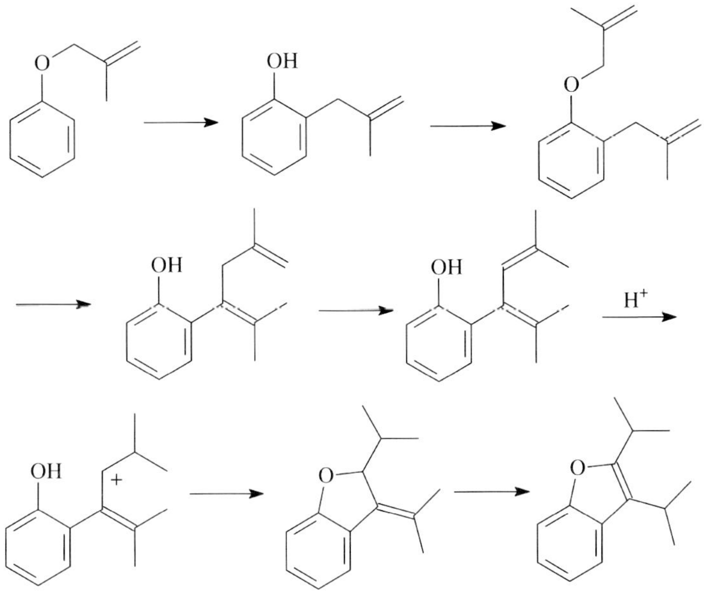

Scheme 5.2-78 Proposed mechanism for the formation of 2,3-diisopropylbenzo[b]furan [188].

give 2-methyl-2,3-dihydrobenzo[b]furan (Scheme 5.2-77). The yields in this reaction were highly dependent on the ionic liquid chosen, with [EtDBU][OTf] giving the best yields (e.g. 91% for R = 6-CH$_3$). Reactions in [BMIM][BF$_4$] and [BMIM][PF$_6$] gave low yields (9–12%).

In order to confirm that 2-allylphenol was indeed an intermediate in the reaction, the authors subjected 2-allylphenol to the same reaction conditions and found that it rearranged to give 2-methyl-2,3-dihydrobenzo[b]furan. In the reaction of 2-methyl-2-propenyl phenyl ether under similar conditions, 2,3-diisopropylbenzo[b]furan was isolated in 15% yield. A plausible mechanistic scheme is given by the authors (Scheme 5.2-78). It involves the Claisen rearrangement of 2-methyl-2-propenyl phenyl ether to 2-(2-methyl-2-propenyl)phenol, followed by a transalkylation of a 2-methylpropenyl group to the phenyl OH group. This undergoes further rearrangements and cyclization to give the 2,3-diisopropylbenzo[b]furan [188].

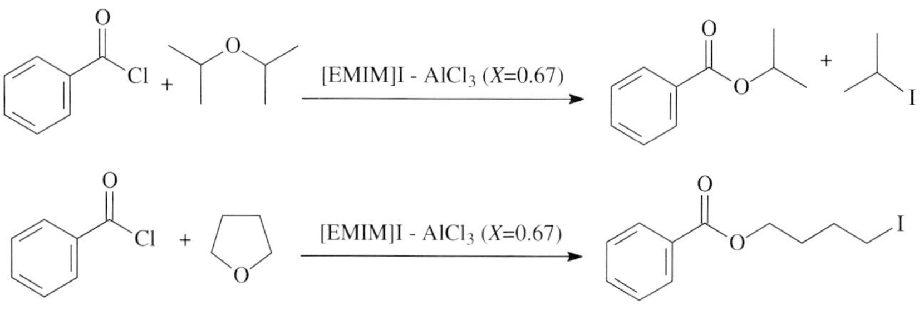

Scheme 5.2-79 The acylative cleavage of ethers in an ionic liquid [190].

5.2.4
Addition Reactions (to C=C and C=O Double Bonds)

5.2.4.1 Esterification Reactions (Addition to C=O)

Esterification reactions readily occur in ionic liquids. A simple example is the reaction of acetic, decanoic and octadecanoic acid with alcohols such as methanol, 1-butanol or 1-octanol. Here, Tang et al. used the ionic liquid [H-MIM][BF$_4$] (this is a simple mixture of 1-methylimidazole and tetrafluoroboric acid) as a solvent and catalyst for the reaction [189]. Singer and coworkers have shown that benzoyl chloride reacts with ethers to give alkyl benzoates [190] in chloroaluminate(III) ionic liquids. This reaction results in the acylative cleavage of ethers, and a number of reactions with cyclic and acyclic ethers were investigated in the ionic liquid [EMIM]I-AlCl$_3$ ($X = 0.67$). Two examples are shown in Scheme 5.2-79.

Esterification reactions can be catalyzed by the ionic liquid 1-butylpyridinium chloride ([BPY]Cl)–AlCl$_3$ ($X = 0.33$) [191, 192]. Deng and coworkers found that higher yields were obtained compared to similar reactions with a sulfuric acid catalyst. A number of primary, secondary and tertiary alcohols were acylated with acetic acid and acetic anhydride in [BMIM][PF$_6$] using metal catalysts [193]. These reactions work well initially, but the catalyst becomes ineffective when the ionic liquid/catalyst is recycled. The authors do not give an explanation for this, however it is known that the [PF$_6$]$^-$ ion dissociates to phosphate and fluoride ions under aqueous conditions and this could lead to catalyst being turned into an inactive fluoride salt [194]. Two examples of the use of Brønsted acidic ionic liquids, where the acid group is in the cation and their use in esterification reactions were published a year later. The first uses [3-(triphenylphosphonium)propanesulfonic acid][tosylate] (Fig. 5.2-6) in the formation of ethyl acetate [195] and the second uses [1-(3-methylimidazolium)butanesulfonic acid] [trifluoromethanesulfonate] (Fig. 5.2-6) in the esterification of acetic acid with alkenes [196].

Ultrasound has been used in the acetylation of a range of alcohols in dibutylimidazolium bromide [197]. Generally a 5 to 10 times enhancement in the reaction rate is observed when compared with the equivalent silent reaction. 1,3-dialkylimidazolium benzoate ionic liquids have been used in the benzoylation of glucose with benzoic anhydride (Scheme 5.2-80) [198]. The reaction was found to

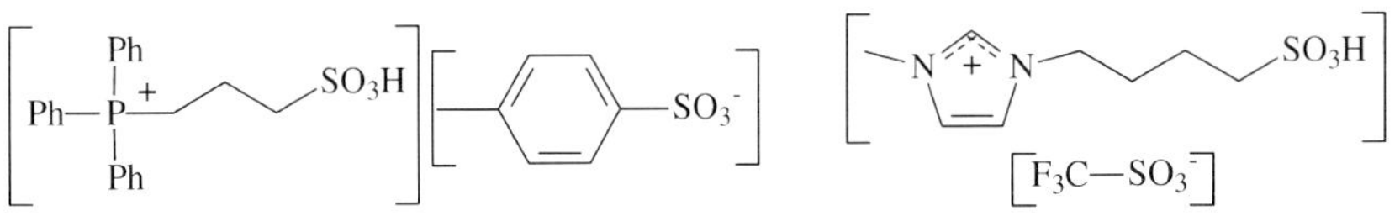

Fig. 5.2-6 Brønsted acidic ionic liquids used in esterification reactions [195, 196].

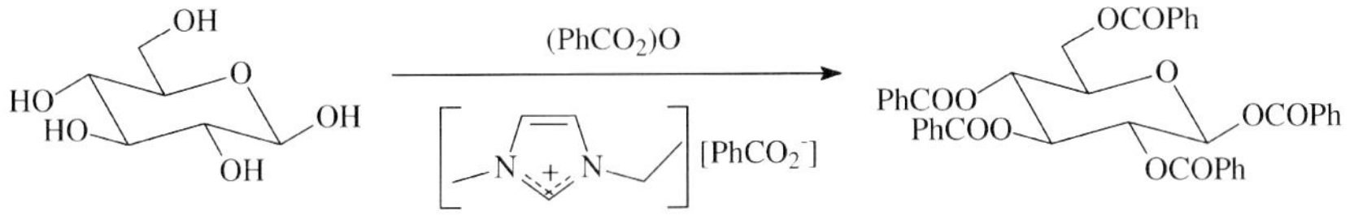

Scheme 5.2-80 Peracylation of β-D-glucose in 1-ethyl-3-methylimidazolium benzoate [198].

work well with anhydrides, but failed when acid chlorides were employed. The authors state that the basicity of the benzoate ion is responsible for promoting the reaction. A similar reaction was described by MacFarlane et al. [199]. Here [BMIM][N(CN)$_2$] was used to promote the acetylation of glucose with acetic anhydride. The dicyanamide ion is mildly basic and was shown to act as a catalyst.

A popular way to form esters and amides is to use a coupling reagent such as dicyclohexylcarbodiimide (DCC). This has been used to form phenolic esters of ferrocenemonocarboxylic acid in [BMIM][BF$_4$] and [BMIM][PF$_6$] in high yield [200]. Unfortunately, the by-product dicyclohexyl urea builds up in the ionic liquid and the authors do not explain how it can be removed from the ionic liquids chosen.

5.2.4.2 Amide Formation Reactions (Addition to C=O)

The first example of a peptide synthesis in an ionic liquid was described by Plaquevent et al.. [BMIM][PF$_6$] was used as a solvent for natural and synthetic amino acids, which were coupled with two ionic coupling agents based on triazoles [201] (Fig. 5.2-7). If no coupling agent or DCC was used, no reaction was observed. Yields as high as 87% were observed and the type of extraction procedure used made a significant difference to the isolated yields.

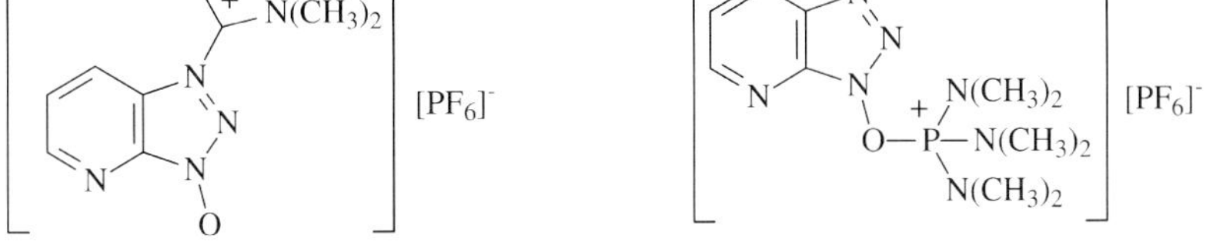

Fig. 5.2-7 The structure of two ionic [PF$_6$]$^-$ peptide coupling agents [201].

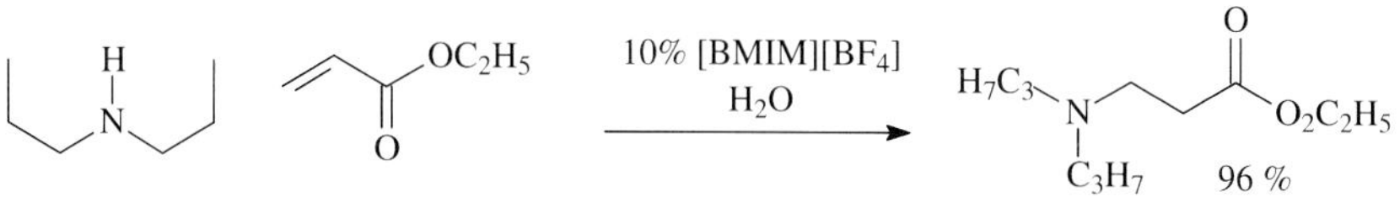

Scheme 5.2-81 The Michael addition of dipropylamine to ethyl acrylate [202].

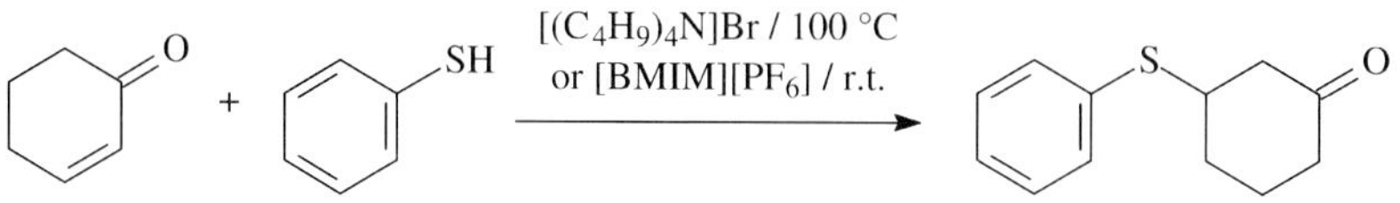

Scheme 5.2-82 The addition of thiols to cyclohexenone [205, 206].

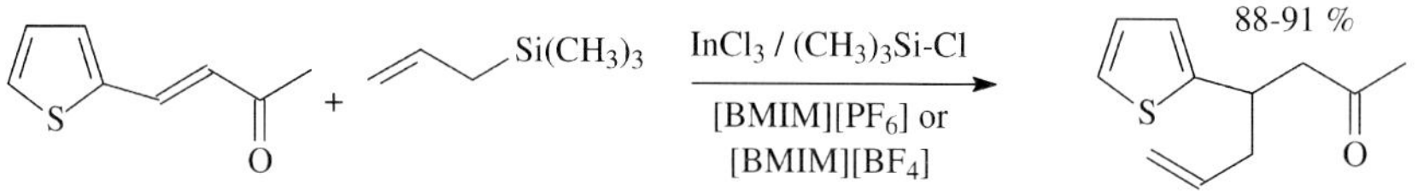

Scheme 5.2-83 The Sakurai reaction in ionic liquids [207].

5.2.4.3 The Michael Reaction (Addition to C=C)

The Michael reaction proceeds very efficiently in many ionic liquids. Traditionally, metal catalysts are used to promote the reaction, however, these catalysts are not necessary in ionic liquids. Xia et al. have found that amines reacts with ethyl or methyl acrylate in aqueous solutions of ionic liquids such as [BMIM][BF$_4$] [202] to give the Michael product in up to 96% yield (Scheme 5.2-81). Similar reactions are also described by Yadav in [BMIM][PF$_6$] [203] where no catalyst is required. Unfortunately neither of these papers give an explanation of why the ionic liquids seem to catalyze the reaction.

In the Michael addition of azide ion to α,β-unsaturated carbonyl compounds, Xia et al. found the reaction worked well with or without an amine base catalyst in up to 95% yield [204]. Two descriptions of the conjugate addition of thiols to alkenes were reported by Ranu [205] and Yadav [206]. Both authors found that thiols such as thiophenol add to cyclohexenone in 92% yield in the ionic liquids [(C$_4$H$_9$)$_4$N]Br or [BMIM][PF$_6$] (Scheme 5.2-82).

An example of attaching a carbon nucleophile to an α,β-unsaturated ketone is the Sakurai reaction. This involves the reaction of allyltrimethylsilane with an α,β-unsaturated ketone to form a δ,ε-unsaturated ketone in the presence of a Lewis acid. Howarth used indium(III) chloride to catalyze this reaction in the ionic liquids [BMIM][BF$_4$] and [BMIM][PF$_6$] [207]. An example of this is shown in Scheme 5.2-83.

The L-proline catalyzed Michael reaction has been used in the synthesis of trifluoromethylated compounds as in Scheme 5.2-84 [208]. The yields in the reaction were in the 44–88% range and no diastereoselectivity was observed in the reaction.

A more detailed study of the diastereoselectivity of the Michael reaction was performed by Hagiwara et al. [209]. In the presence of a catalytic amount of optically

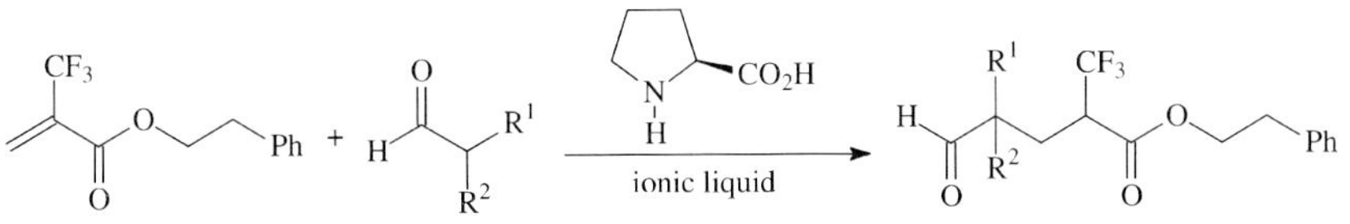

Scheme 5.2-84 The L-proline catalyzed Michael reaction in ionic liquids [208].

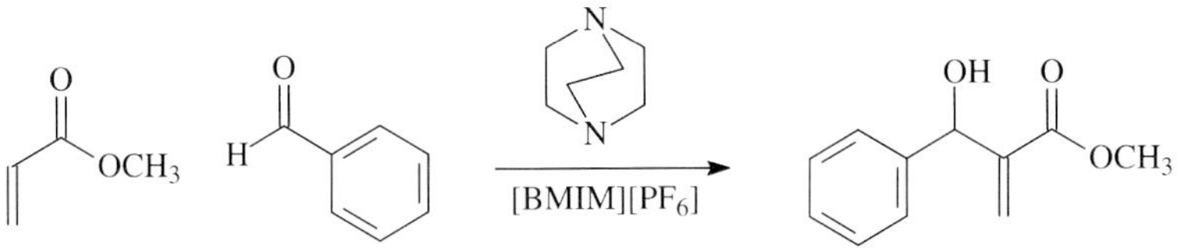

Scheme 5.2-85 The Baylis-Hillman reaction in ionic liquids [210].

active pyrrolidine derivatives, derived from L-proline, the conjugate addition of an unmodified aldehyde to 3-buten-2-one in [BMIM][PF$_6$] was achieved and afforded (2S)-5-keto-aldehyde in up to 59% ee. This reaction was also found to work for enamines derived from aldehydes and for nitroalkenes.

The Baylis-Hillman reaction involves the reaction of an aldehyde with an α,β-unsaturated ketone in the presence of a tertiary amine. This results in the aldehyde adding at the α-position of the carbon–carbon double bond. The reaction is conventionally carried out without a solvent; however this causes problems if the starting materials are solids. Hence the reaction was investigated in ionic liquids such as [BMIM][BF$_4$] and [BMIM][PF$_6$] [210]. The reaction of methyl acrylate with benzaldehyde proceeded 11 to 34 times faster in the ionic liquids than in the solvent acetonitrile. Various Lewis acidic additives had little effect on the reaction, with the exception of lithium perchlorate, which gave a 53 times rate enhancement. This reaction is shown in Scheme 5.2-85.

The same reactions were performed in a range of chloroaluminate(III) ionic liquids by Kumar [211]. By adjusting the aluminum chloride content from $X = 0.45$ to $X = 0.60$, the rate of the reaction was found to increase from 3 to 20 times relative to a reaction in acetonitrile. A problem with this reaction is that the authors quenched the ionic liquid with sodium hydroxide solution which destroys the ionic liquid. This is usually because esters are notoriously difficult the extract from acidic chloroaluminate ionic liquids [73]. The Baylis-Hillman reaction was carried out in a range of neutral ionic liquids to determine which factors affected the reaction rate [212]. Keeping the [BMIM] cation constant, the anions were found to increase the reaction rate in the following order [OTf]$^-$ > [PF$_6$]$^-$ > [N(Tf)$_2$]$^-$ = [OAc]$^-$ > [BF$_4$]$^-$ = [SbF$_6$]$^-$. Scandium(III) and lanthanum(III) triflate were found to enhance the reaction rate, with the lanthanum salt having the greatest effect. The reaction was also carried out in a range of phosphonium ionic liquids and was found to give improved yields compared with [BMIM][PF$_6$] [213]. An explanation as to why the phosphonium ionic liquids gave better yields can be found in a paper

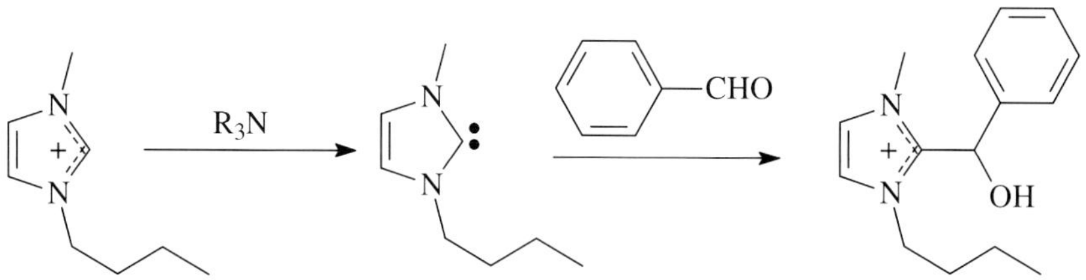

Scheme 5.2-86 The side reaction of imidazolium salts in the presence of a base [214].

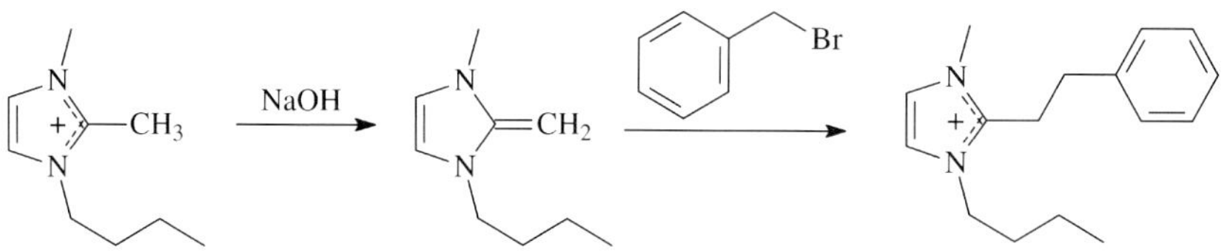

Scheme 5.2-87 The reaction of 2-alkylimidazoles in the presence of base [215].

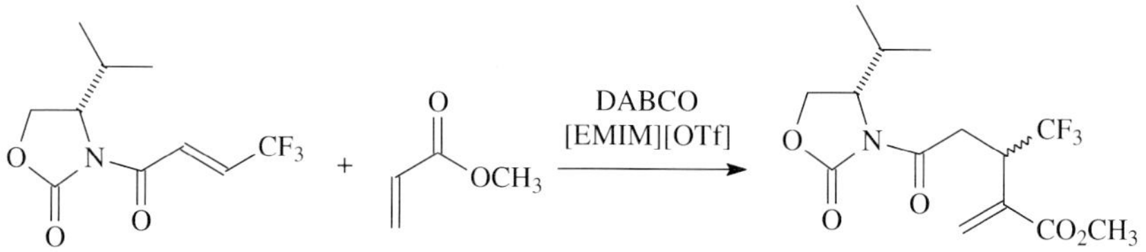

Scheme 5.2-88 A modified Baylis-Hillman reaction with a chiral auxiliary [216].

by Aggarwal et al. [214]. Imidazolium ionic liquids are not stable in the presence of moderate or strong bases. They undergo deprotonation at the 2-position to give a carbene, which can then react with the starting materials. This is shown in Scheme 5.2-86. It should also be mentioned that 1,2,3-trialkylimidazole cations are not stable in the presence of a base. An example of this is in the reaction of 1,2-dimethyl-3-butylimidazolium hexafluorophosphate with benzyl bromide using sodium hydroxide as a base [215]. Here the 2-methyl group reacts with the benzyl bromide to form a new benzylated ionic liquid. A plausible explanation is shown in Scheme 5.2-87. The main conclusion here is that imidazolium ionic liquids are not always suitable for reactions involving bases.

The diastereoselectivity of the Baylis-Hillman reaction was investigated by attaching chiral auxiliaries to α,β-unsaturated ketones [216]. The use of the 4-isopropyloxazoladinone chiral auxiliary in [EMIM][OTf] gave rise to very low diastereomer ratios (45:55) (Scheme 5.2-88). The authors also describe a proline-catalyzed aldol reaction of acetone, butanone, hydroxyacetone and chloroacetone with a chiral imine and obtain up to 9:1 diastereomer ratios.

A potentially very important use of chiral ionic liquids is their use in affecting the stereochemical outcomes of chemical reactions. This was demonstrated in

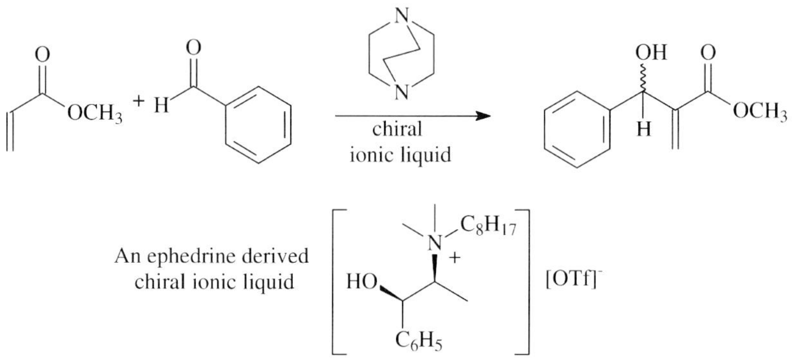

Scheme 5.2-89 The use of a chiral ionic liquid in the Baylis-Hillman reaction [217].

Baylis-Hillman reactions by Loupy et al. [217] (Scheme 5.2-89). Enantioselectivities in the 29–44 % range were obtained by varying the ionic liquid to starting material ratios from 0.5:1 to 3:1. The yields were in the range 65–88 % after 4 days at 30 °C.

5.2.4.4 Methylene Insertion Reactions (Addition to C=O and C=C)

The first methylene insertion reaction in ionic liquids involves the reaction of a trialkyl sulfonium salt with an aldehyde to give an epoxide [4]. Here a sulfide is alkylated with methyl iodide to give a methylsulfonium iodide which adds the methylene group across an aldehyde double bond [218]. The ionic liquid chosen was [BMIM][PF$_6$] and potassium hydroxide was used as a base. However, it is known that [BMIM][PF$_6$] is not stable to KOH over the long term and a base stable ionic liquid would be preferred for this reaction. This is shown in Scheme 5.2-90. A more detailed description of methylene insertion reactions was described by Chandrasekhar et al. who also carried out the addition of trimethylsulfonium iodide to aldehydes in the same ionic liquid, and using the same base [219]. The addition of trimethylsulfonium iodide and trimethylsulfoxonium iodide to α,β-unsaturated ketones also gave rise to addition on the alkene, to give a cyclopropane ring [219]. The addition of ethyldiazoacetate to styrene using a chiral bisoxazoline–copper(II) catalyst was investigated in four ionic liquids [220]. This is shown in Scheme 5.2-91. The reaction gave a roughly 2:1 *trans-* to *cis-* ratio and ee's of the *cis-* and *trans-* isomers ranging from 2 to 94%. The advantage of the ionic liquids reaction compared with one in a molecular solvent is that the bisoxazoline catalyst could be recycled and reused. The effect of halide impurities in this cyclopropanation reaction was also investigated [221].

The first examples of Horner-Wadsworth-Emmons reactions have been given by Kitazume and Tanaka [222]. Here the ionic liquid [EtDBU][OTf], has been used in the synthesis of α-fluoro-α,β-unsaturated esters (Scheme 5.2-92). It was found that when K$_2$CO$_3$ was used as a base, the *E*-isomer was the major product whereas when DBU was used as a base, the *Z*-isomer was the major product. The reaction was

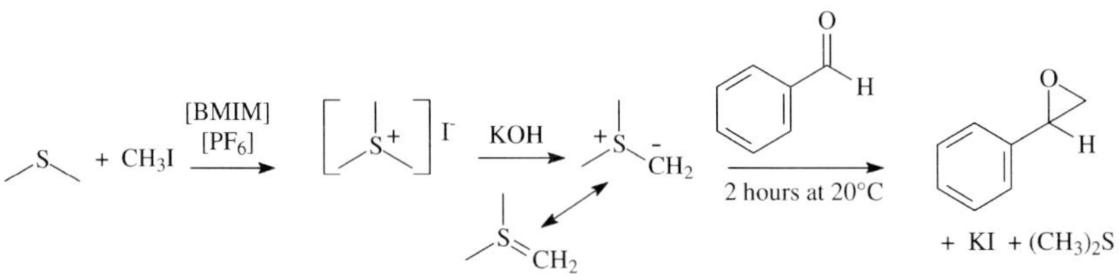

Scheme 5.2-90 The sulfonium iodide prompted methylene addition to aldehydes [218, 219].

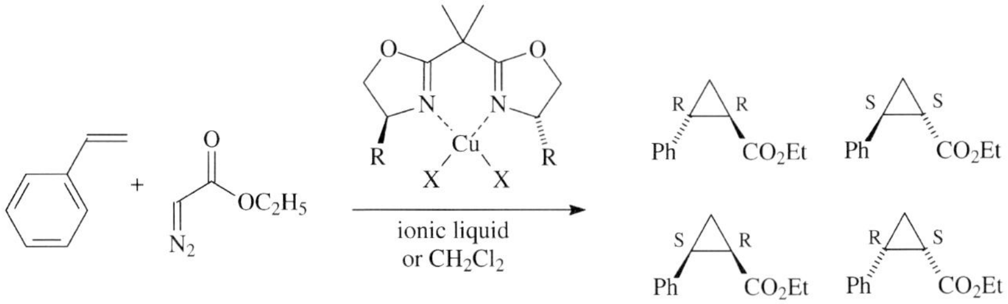

Scheme 5.2-91 The chiral copper(II)-catalyzed reaction of styrene and ethyl diazoacetate [220].

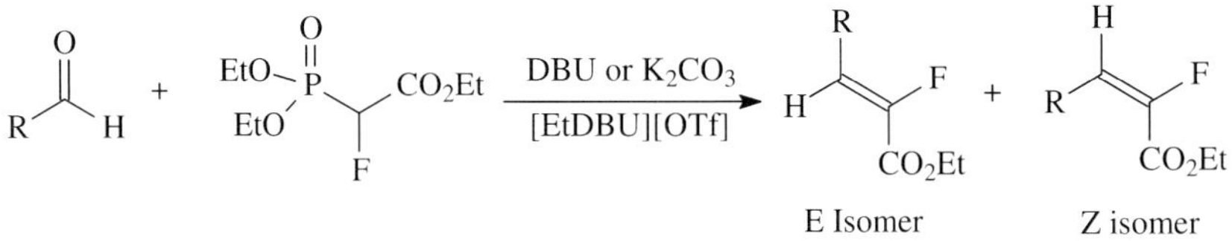

Scheme 5.2-92 The Horner-Wadsworth-Emmons reaction in an ionic liquid [222].

also performed in [EMIM][BF$_4$] and [EMIM][PF$_6$] but gave lower yields that with [EtDBU][OTf] [222].

5.2.4.5 Addition Reactions Involving Organometallic Reagents

The addition of organometallic reagents to carbonyl compounds is an important reaction in organic chemistry, with the Grignard reaction being one example of this. Hence protocols that achieve similar results in ionic liquids are desirable. Gordon and McClusky [223] have reported the formation of homoallylic alcohols, from the addition of allyl stannanes to aldehydes in the ionic liquids [BMIM][BF$_4$] and [BMIM][PF$_6$] (Scheme 5.2-93). It was found that the ionic liquid could be recycled and reused over several reaction cycles. When an imine (formed from the aldehyde) was substituted for the aldehyde, the nitrogen analog was formed. This variation was carried out in [BMIM][BF$_4$] and typically gave 82–93% yields [224].

Kitazume and Kasai have investigated the Reformatsky reaction in three ionic liquids. This involves the reaction of an α-bromoester with zinc to give an α-zinc bromide ester, which in turn reacts with an aldehyde to give an addition product. An example of this reaction is given in Scheme 5.2-94. Moderate to good yields

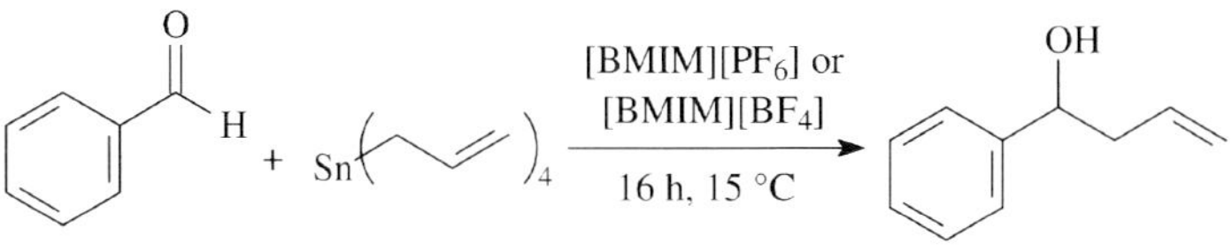

Scheme 5.2-93 Allylation of aldehydes in [BMIM][PF$_6$] or [BMIM][BF$_4$] [223].

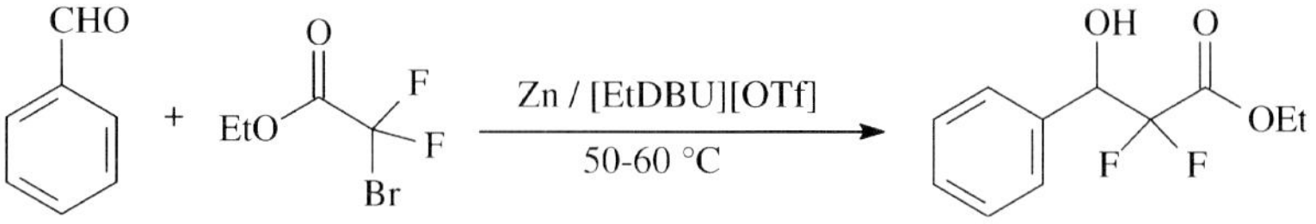

Scheme 5.2-94 The Reformatsky reaction in ionic liquids [225].

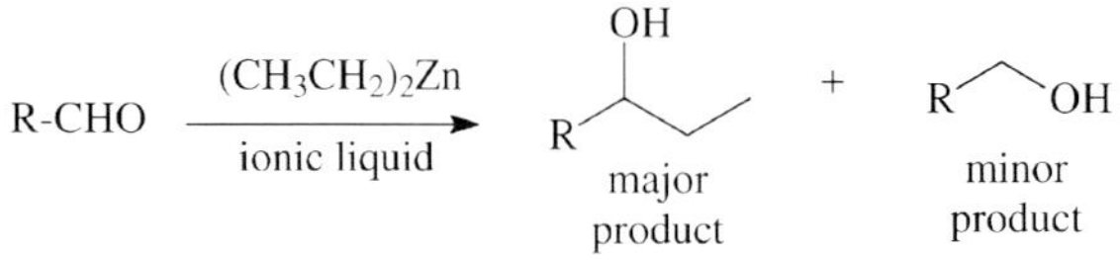

Scheme 5.2-95 The addition of diethyl zinc to aldehydes [226].

(45–95%) were obtained in ionic liquids such as [EtDBU][OTf] for the reaction of ethylbromoacetate or ethylbromodifluoroacetate and benzaldehyde [225].

The ionic liquids, [BMIM]Br, [BMIM][BF$_4$], [BMIM][PF$_6$], [BDMIM][BF$_4$], and [BPY][BF$_4$], were examined as the solvent media for dialkylzinc addition to aldehydes giving the corresponding alcohols. The ionic liquid [BPY][BF$_4$] was found to be the solvent of choice, giving the best yields, and was found to be easily recovered and reused [226] (Scheme 5.2-95). It was found that the imidazolium salts react with diethyl zinc to form a carbene complex of zinc, but the 2-methylimidazolium or pyridinium salts did not react and hence could be recycled.

The reactions of aldehydes with alkynes to give propargyl alcohols are also described in Kitazume and Kasai's paper [226]. Here, various aldehydes such as benzaldehyde or 4-fluorobenzaldehyde were treated with alkynes such as phenylethyne or pent-1-yne in three ionic liquids: [EtDBU][OTf], [BMIM][PF$_6$] and [BMIM][BF$_4$] (Scheme 5.2-96). Both a base (DBU) and Zn(OTf)$_2$ were required for the reaction to be effective. The yields were in the 50–70% range. The best ionic liquid for this reaction depends on the individual reaction.

The enantioselective addition of alkynes to imines using a 2,6-bis(4-phenyloxazolino)pyridine catalyst was carried out by Rosa et al. to give a chiral amino alkyne (Scheme 5.2-97) [227]. The reaction gave similar yields and selectivities to the reaction in toluene, but the catalyst and ionic liquids could be recycled and reused.

McCluskey et al. have also used [BMIM][BF$_4$] as a solvent for the allylation of aldehydes and Weinreb amides [228]. Similar diastereoselectivities and similar or

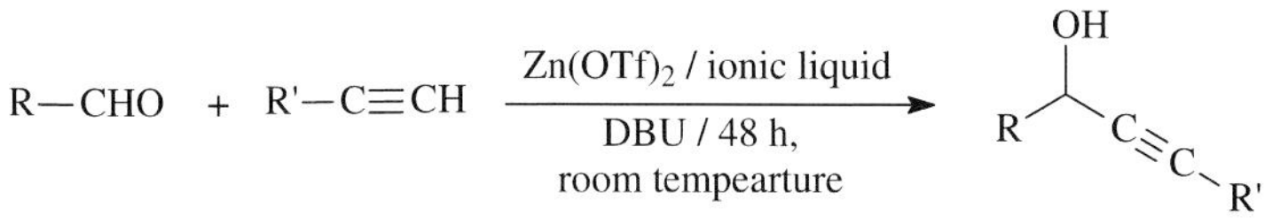

$$R-CHO + R'-C{\equiv}CH \xrightarrow[\text{room tempearture}]{\text{Zn(OTf)}_2 / \text{ionic liquid} \atop \text{DBU / 48 h,}} R\underset{OH}{\overset{}{\underset{|}{C}}}\!\!-C{\equiv}C\!-\!R'$$

Scheme 5.2-96 The zinc triflate-catalyzed coupling of alkynes with aldehydes to give propargyl alcohols in an ionic liquid [226].

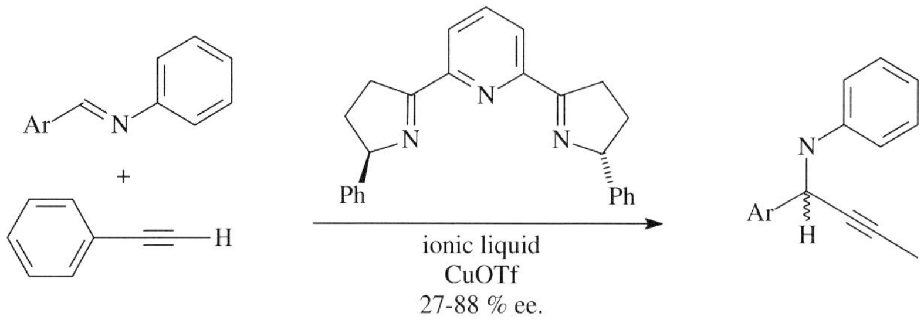

Scheme 5.2-97 The chiral addition of an alkyne to an imine [227]. (ionic liquid, CuOTf, 27–88 % ee.)

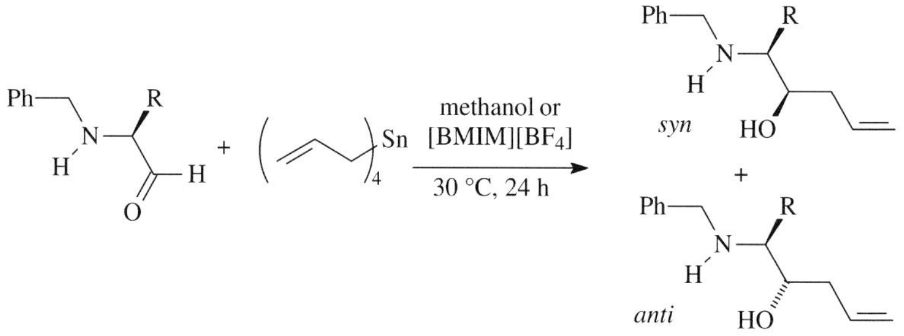

Scheme 5.2-98 The reaction of tetraallylstannane with an aldehyde in methanol or [BMIM][BF$_4$] [228].

slightly lower yields were obtained in this ionic liquid, compared with reactions carried out in methanol (Scheme 5.2-98, Table 5.2-5). The lower yield assigned to the reaction in the ionic liquid is thought to be due to difficulty in extracting the product from the ionic liquid.

Table 5.2-5 The yields and selectivities for the reaction in Scheme 5.98 [228].

R	syn- : anti-	Yield [BMIM][BF$_4$] (%)	Yield methanol (%)	d.e. (%)
CH$_3$	82:18	72	87	64
CH(CH$_3$)$_2$	93:7	70	74	86
PhCH$_2$	93:7	73	82	86

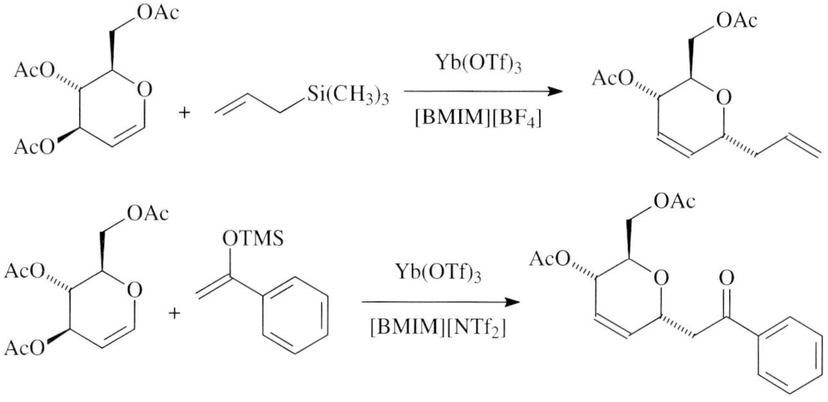

Scheme 5.2-99 The Carbon-Ferrier reaction in [BMIM][BF$_4$] and [BMIM][PF$_6$] [229].

The Carbon-Ferrier reaction involves the addition of an allylsilane or silyl-enol ether to a dihydropyran. This was found to occur in the ionic liquids [BMIM][BF$_4$] and [BMIM][Tf$_2$N] using Yb(OTf)$_3$ as a catalyst [229]. The reactions typically gave 80% yield for the reactions of allyl silanes and 60% yield for the reactions with TMS enol-ethers (Scheme 5.2-99).

The Mukaiyama aldol reaction involves the addition reaction of a TMS-enol ether to an aldehyde. Loh et al. have investigated the reaction of 1-methoxy-2-methyl-1-trimethylsiloxypropene with aliphatic and aromatic aldehydes in chloride, [BF$_4$] and [PF$_6$] ionic liquids. The yields varied considerably and it was found that the chloride ionic liquids gave the best yields (50–74%) [230]. Ruthenium complexes have been used in the addition of allyl alcohols to aldehydes and imines in [BMIM][PF$_6$] [231] (and later in a very similar paper [232]). The addition of a co-catalyst such as indium(III) acetate was found to dramatically improve the yields in some cases and it was found that the ionic liquid/catalyst combination could be recycled. Examples of these reactions are shown in Scheme 5.2-100.

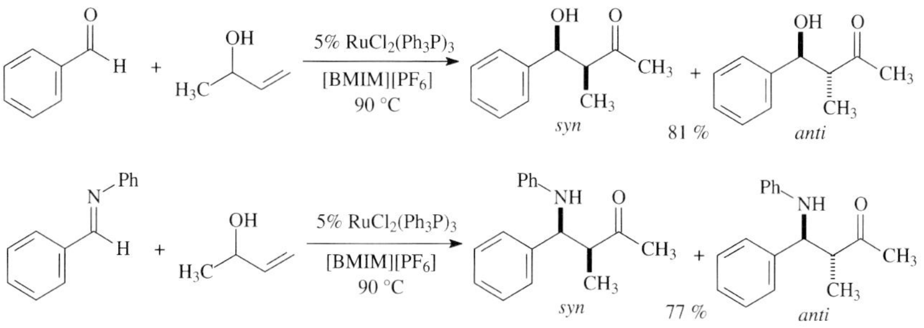

Scheme 5.2-100 The addition of allyl alcohols to aldehydes and imines [231].

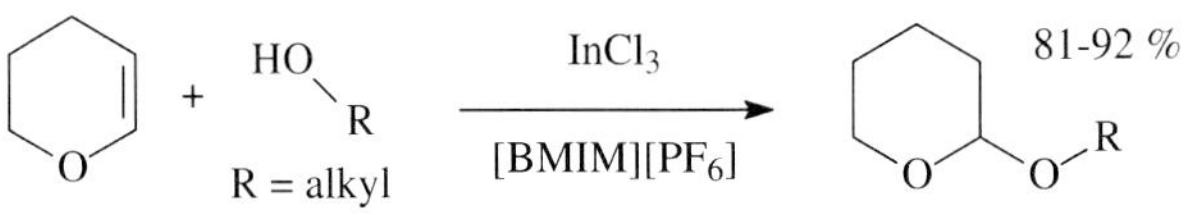

Scheme 5.2-101 The tetrahydropyranylation of alcohols in [BMIM][PF$_6$] [233].

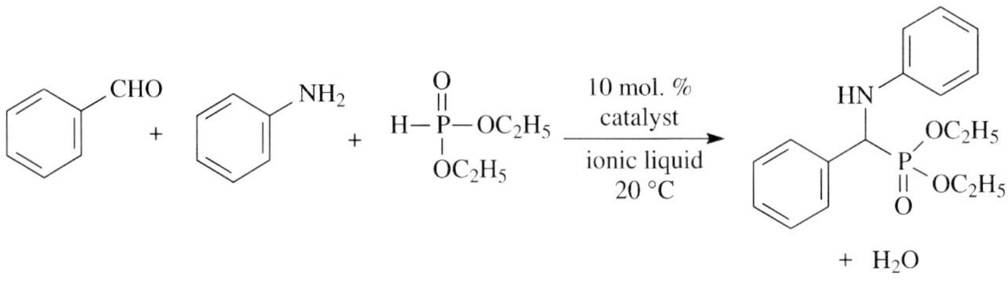

Scheme 5.2-102 The three component reactions of benzaldehyde, aniline and diethyl phosphonate in ionic liquids, catalyzed by lanthanide triflates and indium(III) chloride [235].

5.2.4.6 Miscellaneous Addition Reactions

A common protecting group in organic chemistry is the tetrahydropyranyl group. Yadav found that a 5 mol% solution of indium(III) chloride in [BMIM][PF$_6$] or [BMIM][BF$_4$] catalyzed the addition of alcohols to the alkene in tetrahydropyran (Scheme 5.2-101) [233]. The reaction proceeded at ambient temperatures and in, typically, 90% yield. The authors also claim that the reaction works well in [BMIM][BF$_4$] and to a lesser extent in [BMIM][PF$_6$] without the indium catalyst being present. As this reaction is well known to be catalyzed by traces of acid, and the [BF$_4$]$^-$ and [PF$_6$]$^-$ anions readily give traces of HF, it seems that it is the impurities in the ionic liquid that are promoting the reaction. Indium(III) bromide in [BMIM][PF$_6$] was also found to catalyze the formation of 1,3-dioxanes from alkenes and paraformaldehyde [234].

Lee et al. have investigated the Lewis acid-catalyzed three-component synthesis of α-amino phosphonates [235]. This was carried out in the ionic liquids [BMIM][PF$_6$], [BMIM][OTf], [BMIM][BF$_4$] and [BMIM][SbF$_6$], and the results were compared with a similar reaction carried out in dichloromethane (Scheme 5.2-102). Lee found that the reaction gave good yields (70–99%) in the ionic liquids [BMIM][PF$_6$], [BMIM][OTf] and [BMIM][SbF$_6$] with Lewis acids such as Yb(OTf)$_3$, Sc(OTf)$_3$, Dy(OTf)$_3$, and InCl$_3$. The reaction was also performed in [BMIM][PF$_6$] or dichloromethane using Sm(OTf)$_3$ as the catalyst. The ionic liquid reaction gave a yield of 99% compared with 70% for the reaction in dichloromethane [235]. Amines readily add to aldehydes and ketones to form imines. The addition of amines to ethylacetoacetate or pentane-2,4-dione (acac) results in the formation of an enamine. This was carried out either under solvent-free conditions or in tetrabutylammonium bromide. The reaction in the ionic media generally gave better yields than in the solvent-free reaction [236].

The formation of benzimidazoles, benzoxazoles and benzthiazoles by the addition of 2-(NH$_2$), 2-(OH) and 2-(SH) anilines to acyl chlorides in ionic liquids

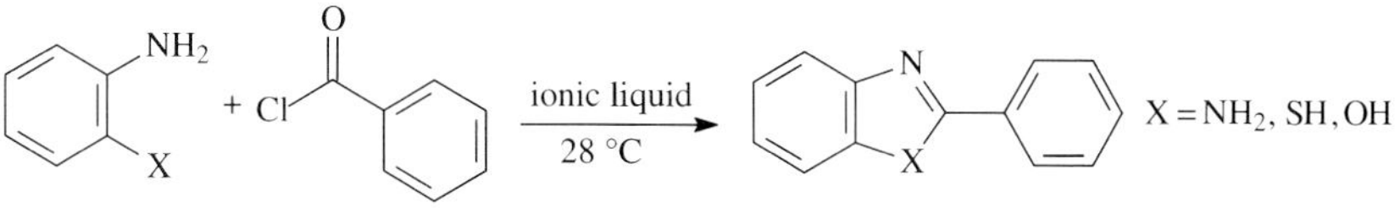

Scheme 5.2-103 The formation of benzimidazoles, benzoxazoles and benzthiazoles in ionic liquids.

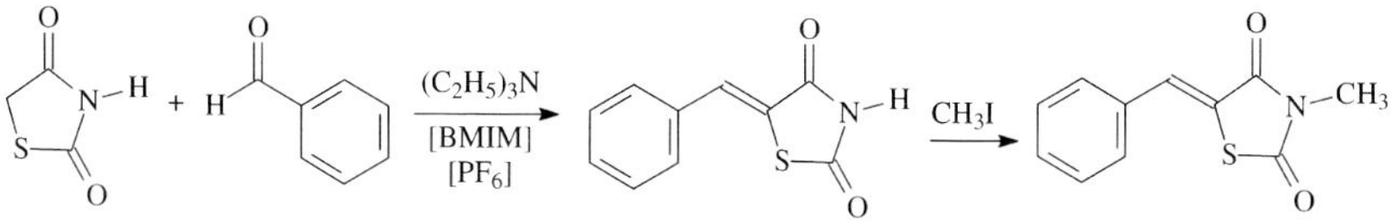

Scheme 5.2-104 The aldol reaction of benzaldehyde with 1,3-thiazolidine-2,4-dione [241].

was studied by Srinivasan et al. [237]. The ionic liquids chosen for the study were 1,3-dibutylimidazolium Cl$^-$, Br$^-$, [BF$_4$]$^-$, [PF$_6$]$^-$ and [ClO$_4$]$^-$ or 1-H-3-butylimidazolium Cl$^-$, Br$^-$, [BF$_4$]$^-$, [PF$_6$]$^-$ and [ClO$_4$]$^-$. The yields in the formation of benzimidazoles were in the 84–95% range and the 1-H-3-butylimidazoles gave slightly faster reaction rates. The yields and reaction times in the formation of benzoxazoles and benzthiazoles were similar. A second report of this reaction using 1-pentylimidazolium bromide as the solvent and an aldehyde instead of an acyl chloride was achieved using microwave heating [238]. Examples are shown in Scheme 5.2-103.

5.2.5
Condensation Reactions

5.2.5.1 General Condensation Reactions

One of the simplest aldol condensations is the reaction of benzaldehyde with acetophenone. Iron(III) chloride hexahydrate in [BMIM][BF$_4$] catalyzes this reaction to give chalcone [239]. This same ionic liquid catalyst combination can also be used in the reaction of benzaldehyde with cyclopentanone and cyclohexanone [240]. The aldol reaction can also be base promoted in ionic liquids such as [BMIM][PF$_6$]. An example is the reaction of benzaldehyde with 1,3-thiazolidine-2,4-dione. The authors follow this reaction with methylation of the imide with methyl iodide (Scheme 5.2-104) [241].

Kitazume et al. have also investigated the use of [EtDBU][OTf] as a medium for the formation of heterocyclic compounds [242]. Compounds such as 2-hydroxymethylaniline readily condense with benzaldehyde to give the corresponding benzoxazine (Scheme 5.2-105). The product of the reaction is readily extracted with solvents such as diethyl ether and the ionic liquid can be recycled and reused. Thioacetals and dithianes formed from the condensation of 2-mercaptoethanol and ethane-1,2-dithiol with aldehydes also form readily in

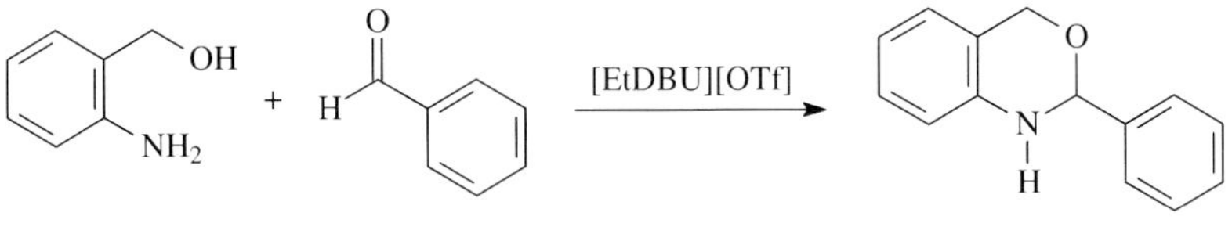

Scheme 5.2-105 The formation of 2-phenylbenzoxazine in [EtDBU][OTf] [242].

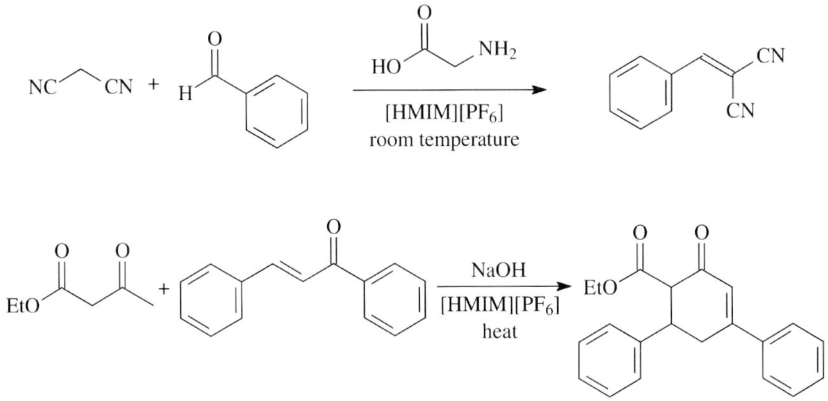

Scheme 5.2-106 The Knoevenagel condensation and Robinson annulation in [HMIM][PF$_6$] [245].

ionic liquids such as [BMIM][BF$_4$] at room temperature [243]. A large range of thioacetalisation reactions were investigated by Patel et al. The ionic liquid chosen was tetrabutylammonium tribromide, which was found to give the desired product when applied in catalytic quantities (0.02 eq.) [244].

Davis and coworkers have carried out the first examples of the Knoevenagel condensation and Robinson annulation reactions [245] in the ionic liquid [HMIM][PF$_6$] (HMIM = 1-hexyl-3-methylimidazolium) (Scheme 5.2-106). The Knoevenagel condensation involved the treatment of propane-1,3-dinitrile with a base (glycine) to generate an anion. This anion adds to benzaldehyde and, following loss of a water molecule, gave 1,1-dicyano-2-phenylethene. The product was separated from the ionic liquid by extraction with toluene. A similar set of Knoevenagel condensation reactions catalyzed by an acetic acid – piperidine mixture in the ionic liquid [BMIM][BF$_4$] were described by Bao et al.. This involved the condensation of diethyl malonate and ethylcyanoacetate with salicylaldehyde [246]. Hydrotalcite (a hydrated magnesium–aluminum carbonate) was used in the reaction in Scheme 5.2-106 and other Knoevenagel reactions in [BMIM][BF$_4$] and [BMIM][PF$_6$]. The reaction was also found to work in the absence of catalyst in 94–100% yield, but gave lower reaction rates [247]. Ethylenediammonium diacetate (EDDA) was used as a catalyst for the Knoevenagel condensation of benzaldehyde with ethylacetoacetate or deithylmalonate in [BMIM][BF$_4$] and [BMIM][PF$_6$] [248]. With 2-hydroxybenzaldehyde, a number of coumarins derivatives were synthesised in

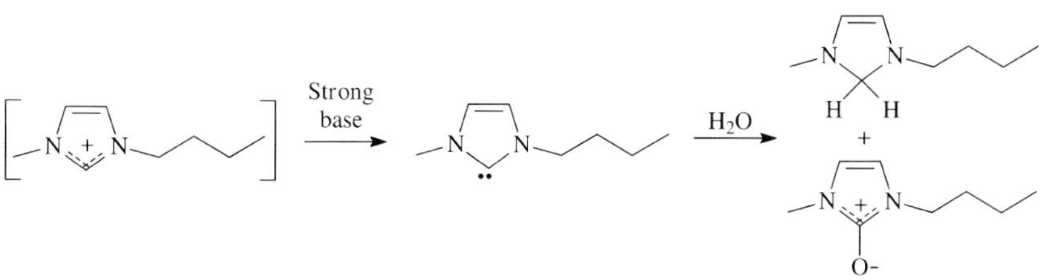

Scheme 5.2-107 The disproportion of imidazolium cations under basic conditions [22, 215].

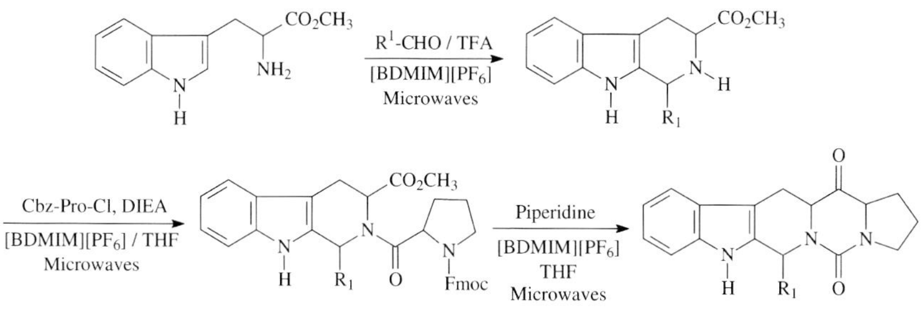

Scheme 5.2-108 The microwave-assisted synthesis of indole alkaloids [251].

90–93% yield. Ionic liquids such as [BPY][NO$_3$], [BMIM][BF$_4$] and [BMIM][PF$_6$] used as microwave absorbers were found to promote the Knoevenagel condensation [249]. The Robinson annulation of ethylacetoacetate and *trans*-chalcone proceeded smoothly to give 6-ethoxycarbonyl-3,5-diphenyl-2-cyclohexenone in 48% yield (Scheme 5.2-106) [245]. The product was separated from the ionic liquid by means of solvent extraction with toluene. In both these reactions, the ionic liquid [HMIM][PF$_6$] was recycled and reused with no lowering of product yield.

The suitability of imidazolium ionic liquids as reaction media for the base-catalyzed Knoevenagel and Claisen reaction was investigated by Garcia et al. [250]. Indirect evidence for the formation of the carbene 1-methyl-3-butylimidazolylidine from [BMIM][PF$_6$] was found and the authors postulate the formation of an imidazolium anion by hydride abstraction from 1,4-cyclohexadiene. This is an antiaromatic species and is previously unknown. A more plausible explanation is the disproportionation of the imidazolium cation via the carbene, shown in Scheme 5.2-107. [22, 215].

A range of microwave-assisted condensation reactions was used in the synthesis of indole alkaloids with pharmaceutical applications. An example is shown in Scheme 5.2-108 [251]. A one-pot synthesis of a pharmaceutically active compound was synthesized by a series of three condensation reactions in a range of [BF$_4$] and [PF$_6$] ionic liquids. This is shown in Scheme 5.2-109 [252].

A range of 1,3-dialkylimidazolium and 1-H-3-alkylimidazolium ionic liquids were used in the condensation of ketones with 2-aminoacetophenones to give quinolines [253] (the Friedlander annulation). The authors discuss how the basicity of the

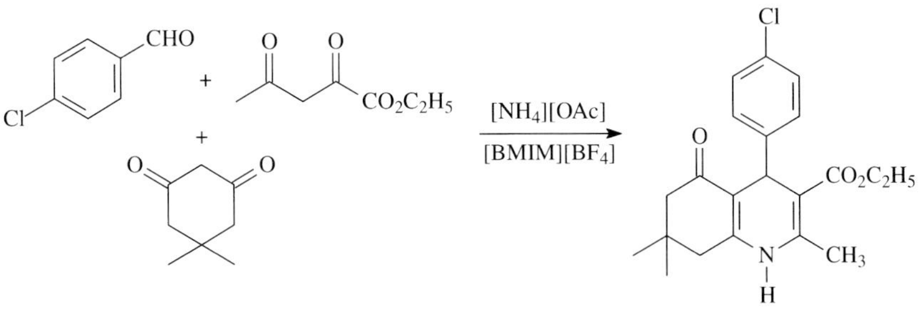

Scheme 5.2-109 The synthesis of polyhydroquinoline derivatives in [BMIM][BF$_4$] [252].

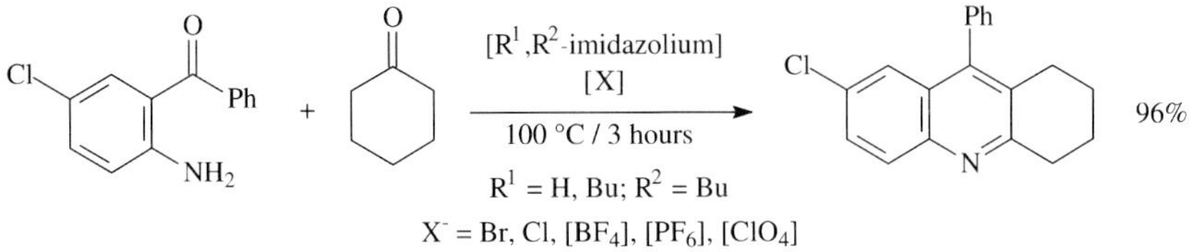

Scheme 5.2-110 The Friedlander annulation in ionic liquids [253].

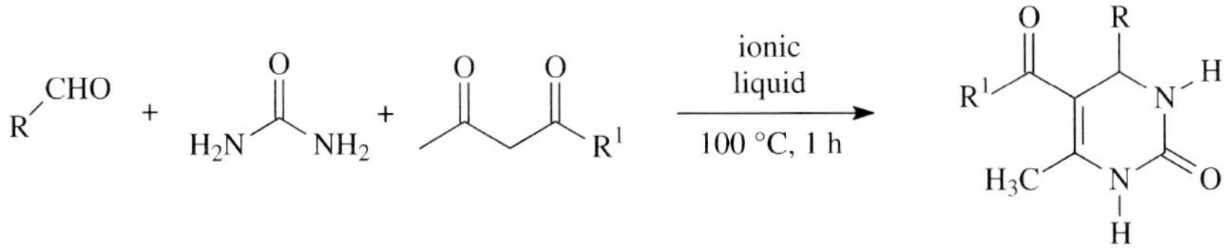

Scheme 5.2-111 The Biginelli reaction in an ionic liquid (R = C$_6$H$_5$, 4-(H$_3$CO)C$_6$H$_4$, 4-Cl-C$_6$H$_4$, 4-(O$_2$N)C$_6$H$_4$, C$_5$H$_{11}$; R^1 = OC$_2$H$_5$, CH$_3$) [254].

anion affects the reaction and an example is shown in Scheme 5.2-110. Deng and Peng have found that certain ionic liquids catalyze the Biginelli reaction [254]. Usually, this reaction is catalyzed by Lewis acids such as InCl$_3$, [Fe(H$_2$O)$_6$]Cl$_3$, BF$_3$.O(C$_2$H$_5$)$_2$, or by acid catalysts such as Nafion-H. The reaction was found to give yields in the 77–99% range for the examples in Scheme 5.2-111 in the ionic liquids [BMIM][PF$_6$] or [BMIM][BF$_4$]. The reaction fails if there is no ionic liquid present or in the presence of tetrabutylammonium chloride. Peng et al. have also developed a method for the formation of triazines in [BMIM][PF$_6$] using KOH as a catalyst [255]. The problem here is that both the [BMIM]$^+$ cation and the [PF$_6$]$^-$ anion are unstable in the long term under the conditions employed in the reaction [215, 216].

1,4-Diketones condense to furans or pyrroles in the presence of a primary amine and an acidic catalyst. Yadav has carried out numerous such reactions in the ionic liquid [BMIM][BF$_4$], using bismuth triflate as a catalyst [256]. The reaction typically gave 80–90% yield and the ionic liquid/catalyst was easily recycled (Scheme 5.2-112). With the addition of a primary amine to a 1,4-diketone, pyrroles can be formed.

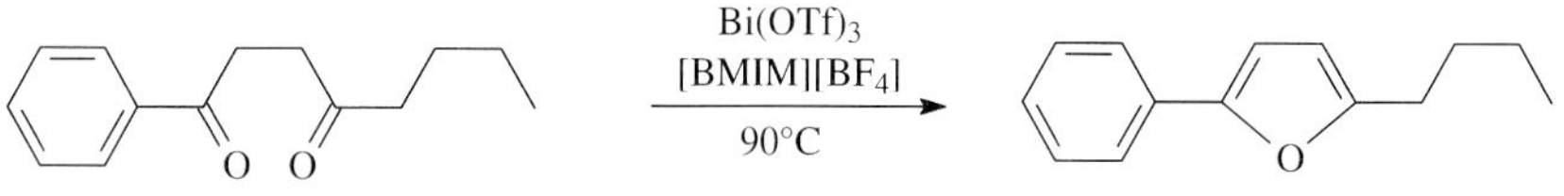

Scheme 5.2-112 The condensation of 1,4-diketones to furans in [BMIM][BF$_4$] [256].

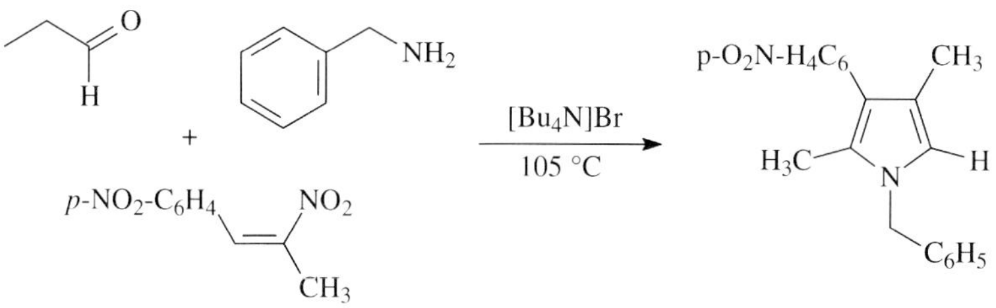

Scheme 5.2-113 The three-component synthesis of pyrroles in [TBA]Br [258].

Yields in the 85–99% range were observed by Yang et al. [257] without the need for a catalyst.

The three-component condensation of an amine, aldehyde and nitroalkene to form substituted pyrroles was carried out in [TBA]Br [258]. No catalyst is required for this reaction and the [TBA]Br molten salt can be recycled. An example is shown in Scheme 5.2-113.

5.2.5.2 The Mannich Reaction

The Mannich reaction involves the condensation of an aldehyde or ketone with an iminium salt usually derived from formaldehyde or other aldehyde and an amine. An example is the use of four acidic ionic liquids in the reaction of benzaldehyde, aniline and acetophenone to give a β-amino ketone as in Scheme 5.2-114 [259].

A Mannich type reaction was used in the three-component synthesis of isoquinolic acids in [BMIM][BF$_4$] or [BMIM][PF$_6$] [260] (Scheme 5.2-115). These reactions are normally catalyzed by Lewis acids such as BF$_3$–OEt$_2$, TiCl$_4$ or SnCl$_4$. The reaction in the ionic liquids does not require a catalyst to be added, although addition of 5 mol% of indium(III) chloride does marginally improve the yield (87–95%).

Another related reaction involves the addition of cyanide to an iminium ion formed from the reaction of an amine with an aldehyde. The use of ionic liquids such as [BMIM][BF$_4$] gave significantly enhanced yields and reaction rates when compared to molecular solvents such as acetonitrile or dichloromethane [261]. This reaction is shown in Scheme 5.2-116.

Potassium alkynyltrifluoroborates react with amines and salicylaldehydes in the presence of benzoic acid to generate highly functionalized amines [262]. Ionic liquids such as [BMIM][BF$_4$] are suitable solvents for the reaction. An example is shown in Scheme 5.2-117.

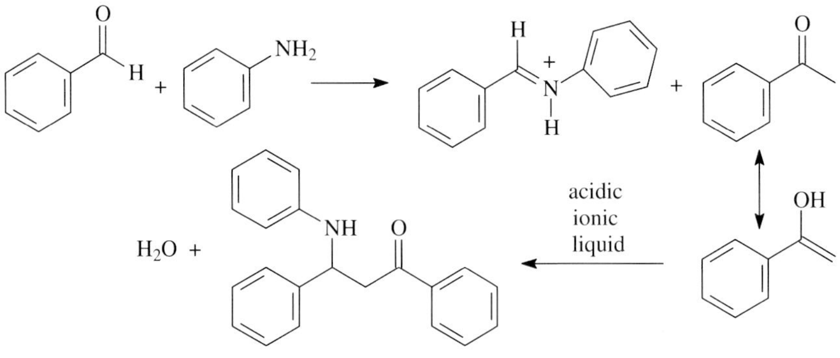

Scheme 5.2-114 The Mannich reaction in acidic ionic liquids [259].

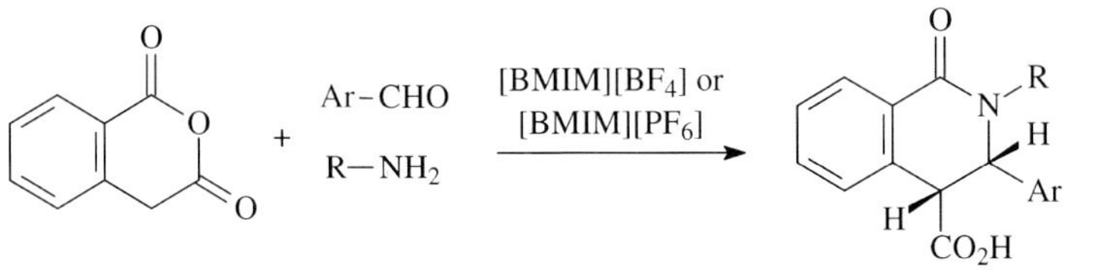

Scheme 5.2-115 The three-component synthesis of *cis*-isoquinolonic acids in ionic liquids [260].

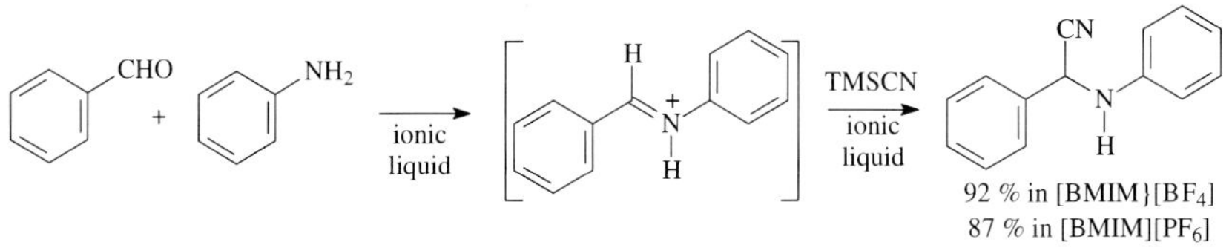

Scheme 5.2-116 The addition of TMS cyanide to an iminium ion formed from the reaction of an amine with an aldehyde [261].

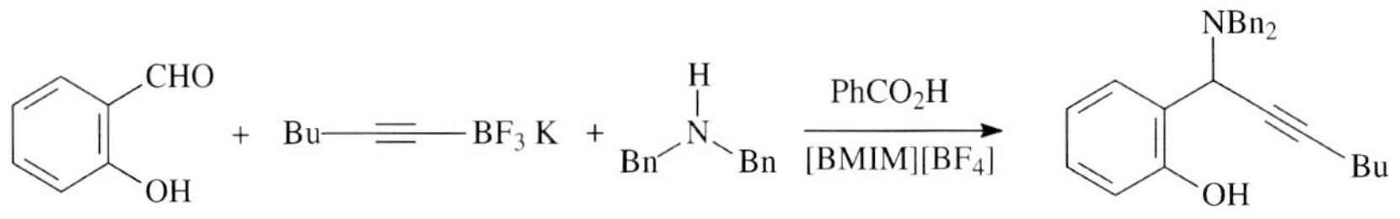

Scheme 5.2-117 The Mannich reaction with alkynyl borates in [BMIM][BF$_4$] [262].

5.2.6
Oxidation Reactions

5.2.6.1 Functional Group Oxidation Reactions

Ionic liquids have been used for a wide variety of oxidations, ranging from the relatively straightforward oxidation of alcohols to ketones, to the more difficult oxidations of alkanes. For clean synthesis, the oxidizing agent should give rise to

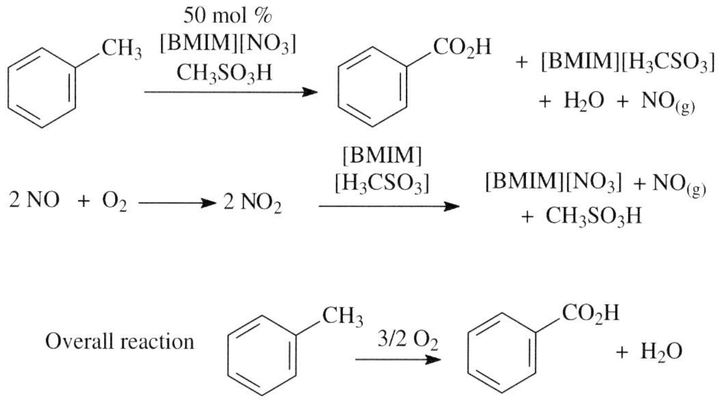

5.2 Stoichiometric Organic Reactions and Acid-catalyzed Reactions in Ionic Liquids

Scheme 5.2-118 The oxidation of toluene to benzoic acid with [BMIM][NO$_3$] [263].

non-toxic by-products and ideally only have water as the by-product. In this respect the ideal oxidizing agents are oxygen (air) or hydrogen peroxide. Earle et al. have developed an oxidation of alkylaromatic compounds to the corresponding carboxylic acid or arylketone in nitrate ionic liquids (Scheme 5.2-118) [263].

Singer and Scammells have investigated the γ-MnO$_2$ oxidation of codeine methyl ether (CME) to thebaine in the ionic liquid [BMIM][BF$_4$] [264]. The ionic liquid was used in different ways and with mixed results (Scheme 5.2-119). For example, the oxidation of CME in the ionic liquid gave 38% yield after 120 h. A similar reaction under biphasic conditions (with diethyl ether) gave a 36% yield of thebaine. This reaction gave a 25% yield of thebaine when carried out in tetrahydrofuran (THF). The authors found that the yield could be increased to 95% by use of sonication of the reaction vessel. The ionic liquid was then used to extract the manganese by-products and impurities from an ethyl acetate solution of the product. A second paper describes the oxidation of codeine to oxycodone, in a two-stage process [265]. The first step is the oxidation of codeine with morphine dehydrogenase (MDA) and NADP (nicotinamide diphosphate) to a codeinone/neopinone mixture, followed by a cobalt(II) acetoacetate-catalyzed hydration to oxycodone. This is shown in Scheme 5.2-120. The ionic liquid chosen is 1-(3-hydroxypropyl)-3-methylimidazolium 2-hydroxyacetate. This shows much stronger hydrogen bonding than ionic liquids such as [BMIM][PF$_6$] and is able to solubilise the opiates under investigation.

A straightforward oxidation of alcohols uses the readily available and cheap oxidant: sodium hypochlorite. A guanidinium ionic liquid was chosen for this reaction because of its greater stability, particularly under basic conditions [266]. Here, the ionic liquid acted as a phase transfer catalyst in the oxidation of benzyl alcohols to the corresponding aldehyde (Scheme 5.2-121). The guanidinium ionic liquid was recovered and found by NMR to be unchanged. Hypervalent iodine reagents were effective in the oxidation of alcohols to ketones in ionic liquids. The reaction simply involves stirring the hypervalent iodine reagent with an alcohol in the ionic liquid at room temperature, followed by extracting the aldehyde or ketone product

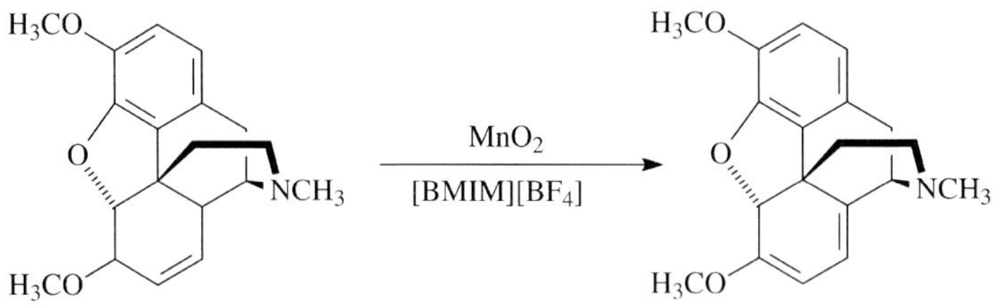

Scheme 5.2-119 The oxidation of CME to thebaine in [BMIM][BF$_4$] [264].

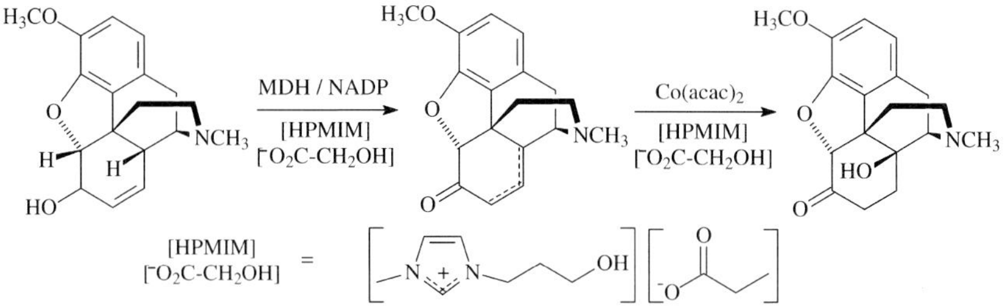

Scheme 5.2-120 The preparation of oxycodone from codeine in an ionic liquid [265].

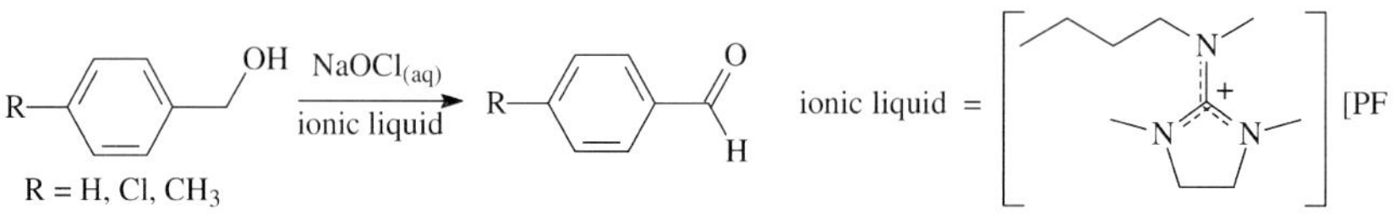

Scheme 5.2-121 The NaOCl oxidation of alcohols to aldehydes [266].

with diethyl ether as in Scheme 5.2-122 [267]. Another way of oxidizing alcohols to aldehydes or ketones is with hydrogen peroxide in [BMIM][BF$_4$] [268]. A catalyst is required and [BMIM][W$_{10}$O$_{23}$] proved effective in this task. The yields are quantitative and the selectivity for the aldehyde (over the carboxylic acid) is >95%.

A more challenging oxidation reaction is the conversion of alkanes to alcohols or ketones. [BMIM][PF$_6$] has been used as part of a solvent (dichloromethane, acetonitrile or dichloroethane are the other components) for manganese porphyrins which catalyze the phenyliodoso diacetate (PhI(OAc)$_2$) oxidation of cyclic alkanes such as cyclohexane, adamantane or tetralin [269]. The reactions gave yields in the 25–55% range, based on the starting alkane. A number of pyridinium ionic liquids were found to be stable to superoxide ion and were used in the functionalization of siloxanes [270]. This involved the electrochemical reduction of dioxygen to superoxide ion. This in turn was used to activate hexamethyldisiloxane and add trimethylsiloxy groups to Ph$_2$Si(OCH$_3$)$_2$ or NC(CH$_2$)$_3$CH$_3$SiCl$_2$.

Yadav et al. have also carried out a range of Baeyer-Villiger oxidations in [BMIM][BF$_4$]. This involves the reaction of a peracid; in this case

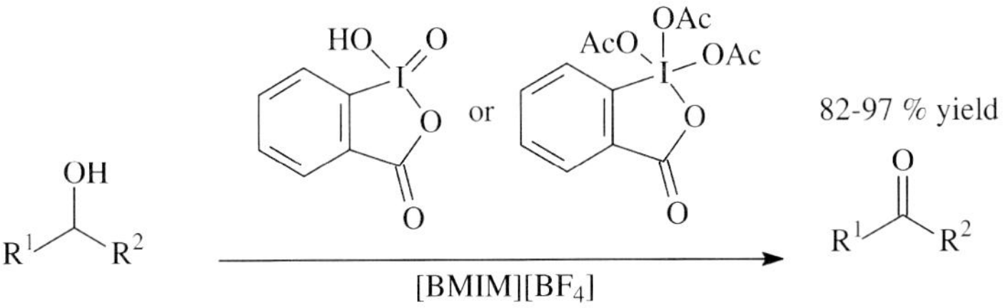

Scheme 5.2-122 Hypervalent iodine oxidations in [BMIM][BF$_4$] [267].

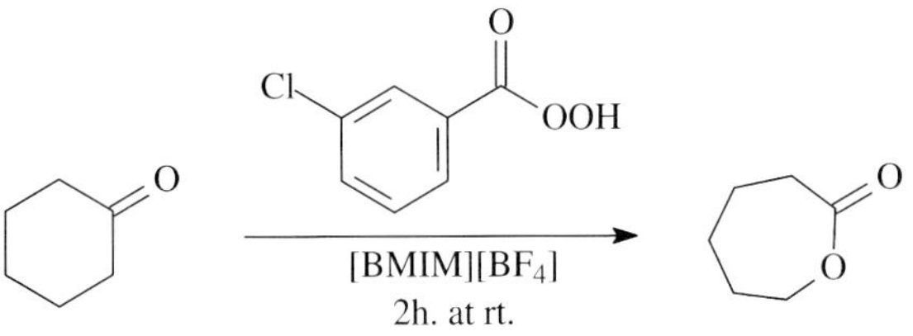

Scheme 5.2-123 The Baeyer-Villiger oxidation in [BMIM][BF$_4$] [271].

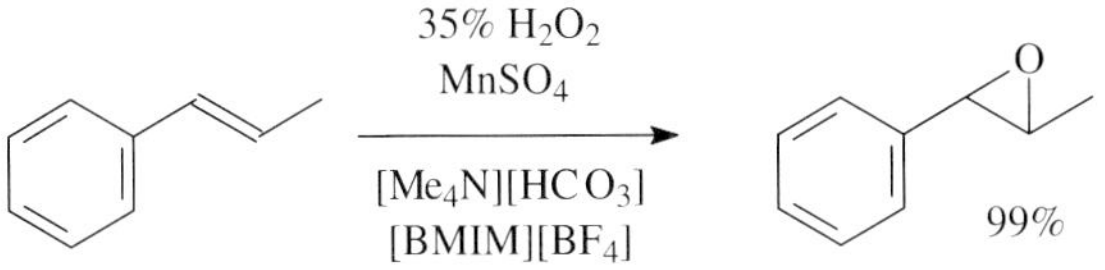

Scheme 5.2-124 The epoxidation of *trans*-1-phenylprop-1-ene with 35% H$_2$O$_2$ [272].

meta-chloroperbenzoic acid with a ketone and results in the insertion of an oxygen atom α to the ketone [271]. An example of this is shown in Scheme 5.2-123.

5.2.6.2 Epoxidation and Related Reactions

Numerous epoxidation reactions have been carried out in ionic liquids. One of the simplest uses hydrogen peroxide in [BMIM][BF$_4$] with tetramethylammonium bicarbonate and manganese(II) sulfate as a catalyst [272]. The reaction was successful (typically 98–99%) various styrenes, cyclooctene, norbornene and pinene and the ionic liquid could be recycled and reused (Scheme 5.2-124).

A series of room-temperature ionic liquids were tested as solvents for dioxomolybdenum(VI) complexes in the catalytic epoxidation of *cis*-cyclooctene, using *tert*-butyl hydroperoxide (TBHP) as the oxygen source [273]. In general, the best results were obtained using the ionic liquid [BMIM][Tf$_2$N]. Upon addition of cyclooctene and TBHP (in decane) to solutions of MoO$_2$X$_2$(*p*-tolyl-(CH$_3$-DAB)) [X = Cl, Me; *p*-tolyl-(CH$_3$-DAB) = *N,N-p*-tolyl-2,3-dimethyl-1,4-diazabutadiene] in [BMIM][Tf$_2$N], biphasic mixtures were obtained. The epoxidation reactions proceeded with 100% selectivity to cyclooctene oxide, but activities were lower than those observed for the same catalysts in the absence of any additional solvent (other than decane). Another

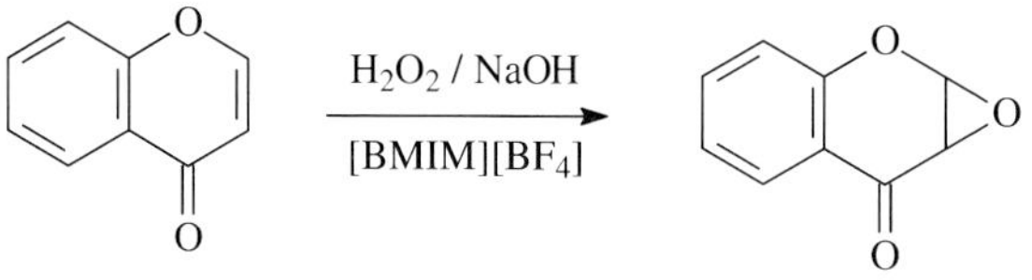

Scheme 5.2-125 Epoxidation reactions in [BMIM][BF$_4$] with alkaline H$_2$O$_2$ [275].

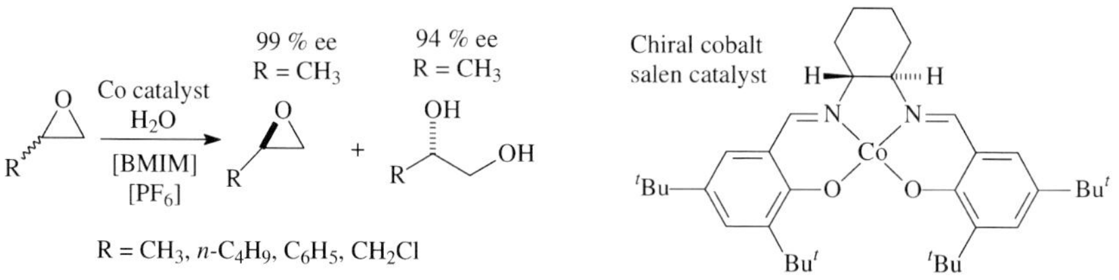

Scheme 5.2-126 The kinetic resolution of racemic epoxides in [BMIM][PF$_6$] or [BMIM][Tf$_2$N] [276].

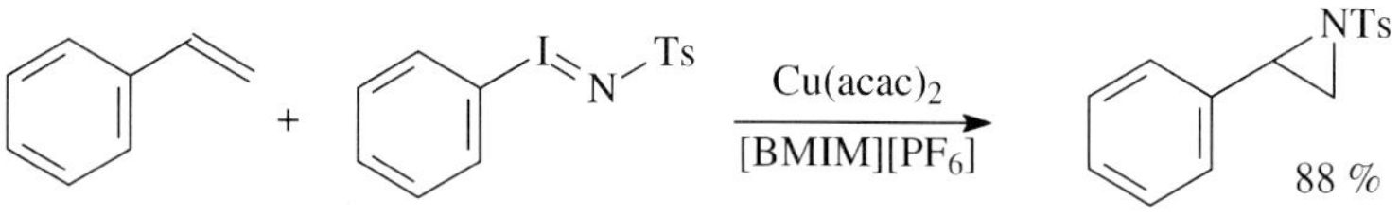

Scheme 5.2-127 The aziridination of alkenes in [BMIM][PF$_6$] [277].

class of metal complexes that promote the epoxidation of alkenes are manganese(III) porphyrins. These complexes in [BMIM][PF$_6$] catalyzed the epoxidation of simple alkenes with hypervalent iodine reagents such as Ph-IO or PhI(OAc)$_2$ [274].

A convenient and efficient procedure for the epoxidation of chromone, isoflavone, and chalcone derivatives using [BMIM][BF$_4$] as solvent and alkaline hydrogen peroxide as oxidant was developed. The reactions proceed in good yields and faster than in conventional solvents. No evidence of formation of compounds derived from the opening of the epoxide ring was observed [275]. An example is shown in Scheme 5.2-125. A very similar alkaline hydrogen peroxide epoxidation in [BMIM][PF$_6$] is also described [276]. In the chiral Co(III)(salen)-catalyzed hydrolytic kinetic resolution of racemic epoxides, in the presence of ionic liquids, a Co(II)(salen) complex is oxidized to catalytically active Co(III)(salen) complex. During the reaction this oxidation state is stabilized against reduction to a Co(II) complex which enables the reuse of the recovered catalyst for consecutive reactions without extra reoxidation. This is shown in Scheme 5.2-126.

The nitrogen equivalent of the epoxidation reaction is the aziridination reaction. Cu(acac)$_2$ was used as a catalyst for the aziridination of alkenes with [N-(p-tolylsulfonyl)imino]phenyliodinane (PhI=NTs) in [BMIM][PF$_6$] [277] and is shown in Scheme 5.2-127.

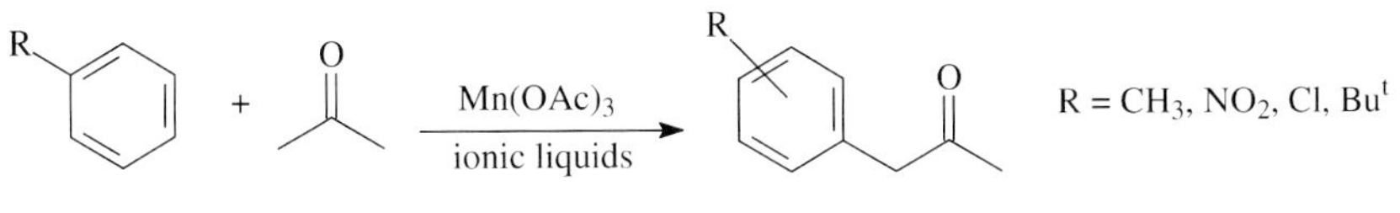

Scheme 5.2-128 The coupling of acetone substituted benzenes in ionic liquids using ultrasound [278].

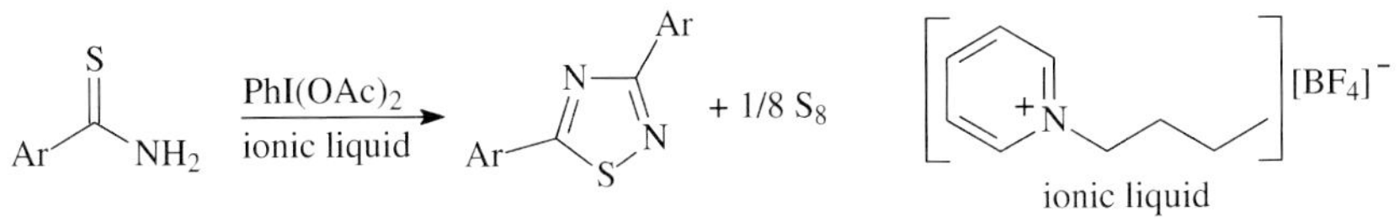

Scheme 5.2-129 The oxidative dimerisation of thioamides [281].

5.2.6.3 Miscellaneous Oxidation Reactions

The oxidative coupling of acetone with monosubstituted benzenes, promoted by manganese(III) acetate in three ionic liquids proceeded using ultrasound irradiation [278]. The reactions in the ionic liquids gave far better yields and improved selectivity for the *para*-isomer than similar reactions in acetic acid. This is shown in Scheme 5.2-128.

The oxidative dimerization of terminal alkynes with oxygen using a TMEDA/CuCl catalyst (the Glaser oxidation) was carried out in [BMIM][PF$_6$] and [BMIM][BF$_4$] [279]. The ionic liquid/catalyst mixture was recyclable (up to 6 times) and typical yields were in the 85–95% range, for 14 separate terminal alkynes. The oxidative coupling of 2-naphthols with iron(III) and copper(II) binary ionic liquids and solutions of iron(III), ruthenium(III) and copper(II) in [BMIM][PF$_6$] were examined [280]. The reactions in the ionic liquids gave a marginal increase in yield over reactions in chlorobenzene or toluene. Phenyliodine(III) diacetate in butylpyridinium tetrafluoroborate promotes the dimerization of thioamides as in Scheme 5.2-129 [281]. This reaction gave 85–93% yields after 15 min at 75 °C. The product was extracted from the ionic liquid and the sulfur was removed by filtration. The yield remained the same when the ionic liquid was reused.

An important class of reactions is the remote functionalization of alkanes. An efficient C–H insertion reaction of H$_2$O$_2$ into hydrocarbons by homogeneous methyltrioxorhenium(VII) (MTO), heterogeneous poly(4-vinylpyridine)/methyl trioxorhenium (PVP/MTO) and microencapsulated polystyrene/methyltrioxorhenium (PS/MTO) systems in ionic liquids, has been developed (Scheme 5.2-130). In some cases higher activity was observed when compared with the same reaction in molecular solvents. The heterogeneous catalysts are stable systems under the reaction conditions and can be recycled for more transformations [282].

The hydroxylation of benzene to give phenol was carried out using a biphasic mixture of hydrophobic ionic liquids such as [OMIM][BF$_4$] or [OMIM][PF$_6$] and water/hydrogen peroxide [283]. A range of transition metal dodecylsulfate catalysts were employed with the iron giving the best results. Conversions of benzene to phenol of 54% were obtained and the ionic liquid/catalyst combination could be

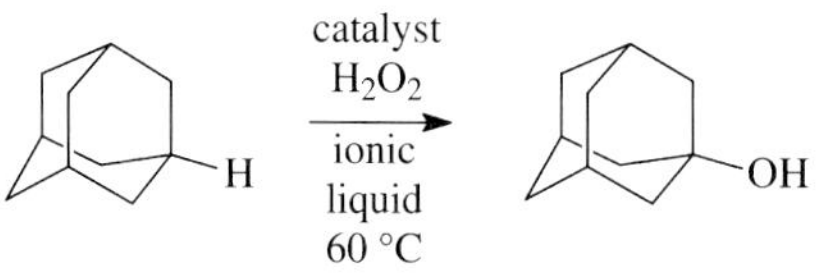

Scheme 5.2-130 The oxidation of adamantane with H_2O_2 [282].

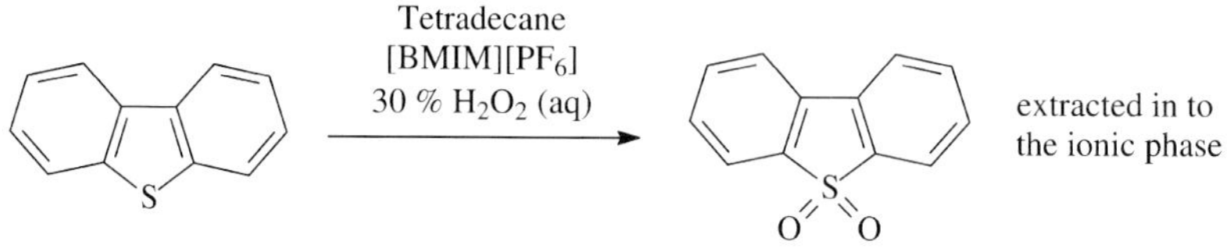

Scheme 5.2-131 The oxidative extraction of dibenzothiophene with [BMIM][PF$_6$] [284].

recycled. The desulfurisation of oils is a reaction of increasing importance as a means of reducing the amount of sulfur dioxide in the environment. Generally this involves oxidising the sulfur compounds to sulfones and extracting them with polar solvents such as dimethylsulfoxide. The oxidation of dibenzothiophene with hydrogen peroxide was carried out in biphasic or triphasic tetradecane/ionic liquid mixture/aqueous hydrogen peroxide [284]. The best desulfurisation process involved extracting dibenzothiophene from tetradecane with [BMIM][PF$_6$] and aqueous H_2O_2 in a triphasic mixture. This is shown in Scheme 5.2-131.

5.2.7
Reduction Reactions

Polycyclic aromatic hydrocarbons dissolve in chloroaluminate(III) ionic liquids to give brightly colored solutions (thought to be due to the protonated aromatic compound [285]). The addition of a reducing agent (such as an electropositive metal and a proton source) results in the selective hydrogenation of the aromatic compound. For example pyrene and anthracene can be reduced to perhydropyrene and perhydroanthracene at ambient temperatures and pressures (Scheme 5.2-132). Interestingly, only the thermodynamically most stable isomer of the product is obtained [286]. This contrasts with catalytic hydrogenation reactions, which require high temperatures and pressures and an expensive platinum oxide catalyst and give rise to an isomeric mixture of products.

Ionic liquids such as [BMIM][BF$_4$] and [EMIM][PF$_6$] have been used in the trialkylborane reduction of aldehydes to alcohols (Scheme 5.2-133) [287]. In the reduction of benzaldehyde with tributylborane, similar yields (90–96%) were obtained for the ionic liquids [EMIM][BF$_4$], [EMIM][PF$_6$], [BMIM][BF$_4$], and [BMIM][PF$_6$]. The effect of electron-releasing and electron-withdrawing groups on the aromatic aldehyde were investigated. In general, electron-withdrawing groups such as halogen give near quantitative yields, but electron-releasing groups such as methoxy reduce the reaction rate and yield.

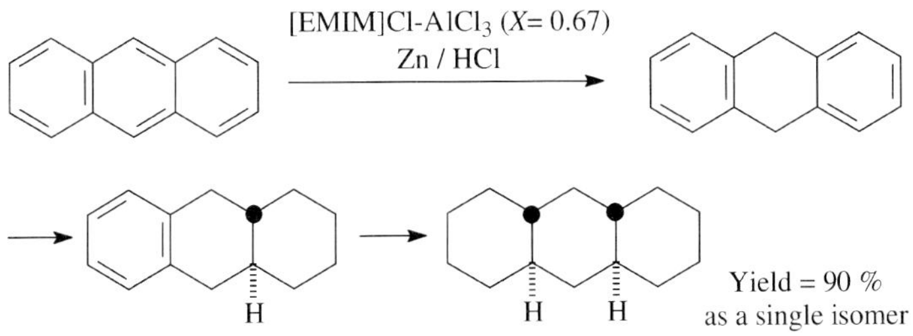

Scheme 5.2-132 The reduction of anthracene to perhydroanthracene [286].

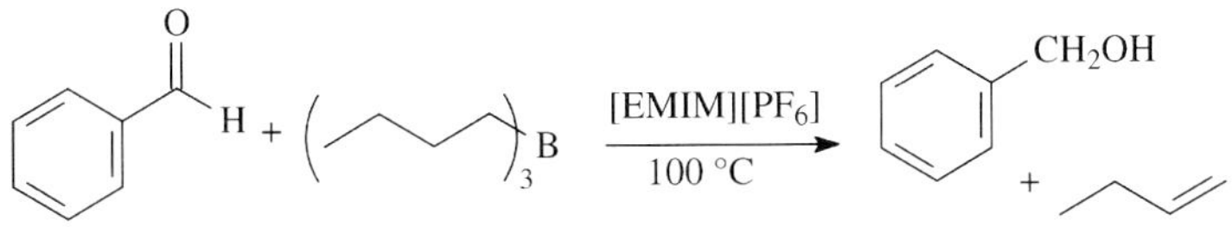

Scheme 5.2-133 The reduction of benzaldehyde in [EMIM][PF$_6$] [287].

This reduction is conventionally affected by reaction of an aldehyde or ketone with sodium borohydride in a solvent bearing an acidic proton such as methanol. In the first reported sodium borohydride reduction Howarth et al. found that the reduction of aldehydes and ketones with NaBH$_4$ in the ionic liquid [BMIM][PF$_6$] can be achieved [288]. The ionic liquid can be recycled, and in some cases the product alcohol may be distilled directly from the ionic liquid eliminating classical organic solvents entirely. In a later example, sodium borohydride was also used to reduce aldehydes and ketones in a range of ionic liquids such as [BMIM][BF$_4$], [HMIM]Br, [EMIM][F$_3$CCO$_2$] [289]. With α,β-unsaturated ketones, the carbonyl was reduced and not the alkene group. Nitro groups readily undergo reduction to amino groups. Usually this is carried out using a tin(II) reagent or zinc in acetic acid. Tin(II) chloride was found to give this reaction in tetrabutylammonium bromide at 90 °C [290]. This procedure gave lower reaction times, but the author does not discuss the recovery of the ionic liquid and the removal of the tin(IV) salts. The reduction of both carbonyl and nitro groups was achieved using a mixture of [BMIM][PF$_6$] and aqueous Na[BF$_4$] [291]. These were biphasic reactions, and the ionic liquid could be recycled and reused. The sodium and borate salt by-products could be removed in the aqueous layer. The photochemical reduction of chlorinated phenols was carried out in [BMIM][PF$_6$] as in Scheme 5.2-134 [292]. Using 253.7 nm irradiation, the authors propose that a chlorine atom is lost and the phenol radical picks up a hydrogen atom from the solvent. The stability of the ionic liquid was investigated and it was found impurities hindered the regeneration of the ionic liquid. The reaction rate diminished to a small extent when the ionic liquid was recycled. The cobalt(I) salen complex has been used in a similar reaction: the debromination of cyclic vic-dibromides in ionic liquids using an electrochemical technique (cyclic voltametry) [293] and dechlorination [294]. The work-up after electrolysis in the ionic liquid

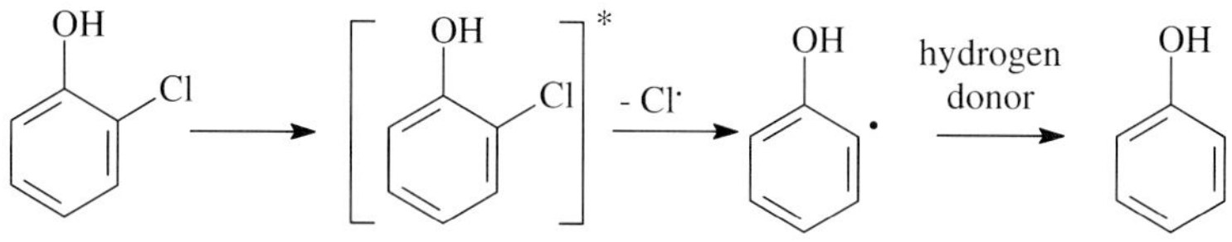

Scheme 5.2-134 The photochemical reduction of 2-chlorophenol to phenol in [BMIM][PF$_6$] [292].

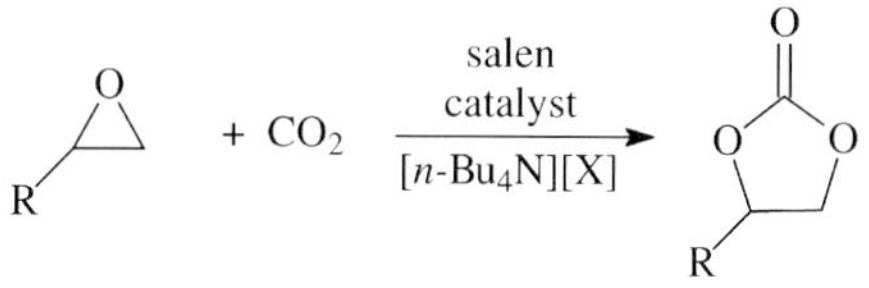

Scheme 5.2-135 The fixation of carbon dioxide in the form of carbonates [296].

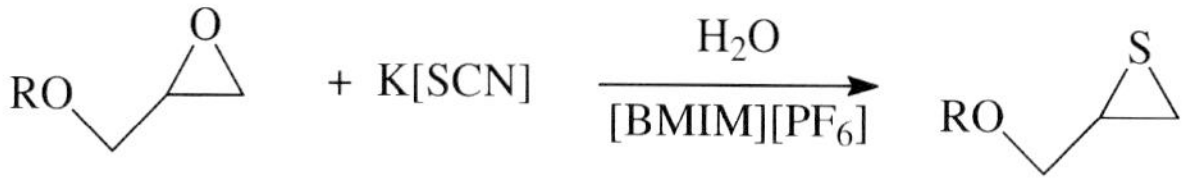

Scheme 5.2-136 The conversion of epoxides to thiiranes in [BMIM][PF$_6$] [297].

proved to be much simpler than that in organic solvents, and the possibility of reuse of the ionic liquid was demonstrated. The related electrochemical dechlorination is also described and explanations are offered for factors affecting this reaction [295].

5.2.8
Miscellaneous Reactions in Ionic Liquids

The chemical fixation of carbon dioxide to cyclic carbonates proceeds effectively under mild conditions using a bifunctional nucleophile–electrophile catalyst system. This was based on tetradentate Schiff-base aluminum complexes [(Salen)AlX] in conjunction with a quaternary ammonium salt [Bu$_4$N][Y]. The electrophilicity of the central Al^{3+} ion and the steric factor of the substituent groups on the aromatic rings of the (Salen)AlX (electrophile), and nucleophilicity and leaving ability of the anion [Y]$^-$ of [n-Bu$_4$N][Y] (nucleophile) had a significant effect on the catalytic activity of the catalyst [296] (Scheme 5.2-135).

A variety of epoxides react with potassium thiocyanate in a [BMIM][PF$_6$]/water mixture (2:1) at room temperature under mild conditions to produce the corresponding thiiranes in high yields. Enhanced rates and improved yields were observed in ionic liquids [297] (Scheme 5.2-136).

The imines formed from aldehydes and primary amines were converted to aziridines with ethyldiazoacetate in [BMIM][PF$_6$] [298]. The catalyst employed was Bi(OTf)$_3$ and the *cis*-isomer was formed predominantly in 75–90% yield (Scheme 5.2-137).

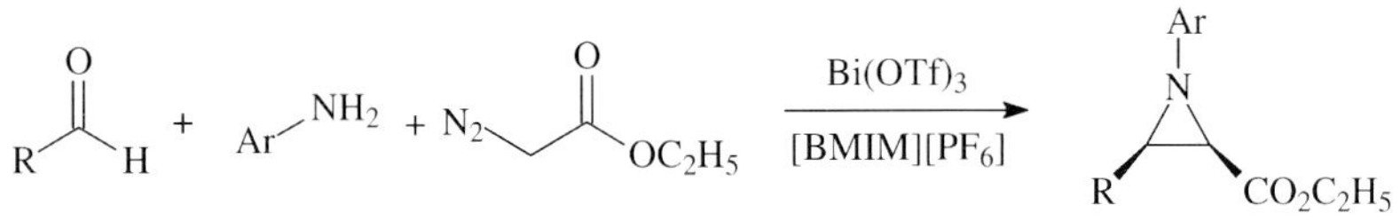

Scheme 5.2-137 The three-component synthesis of aziridines in [BMIM][PF$_6$] [298].

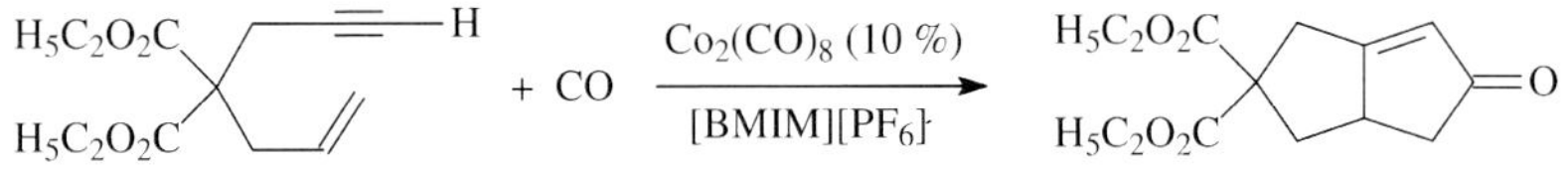

Scheme 5.2-138 The Pauson–Khand annulation in [BMIM][PF$_6$] [300].

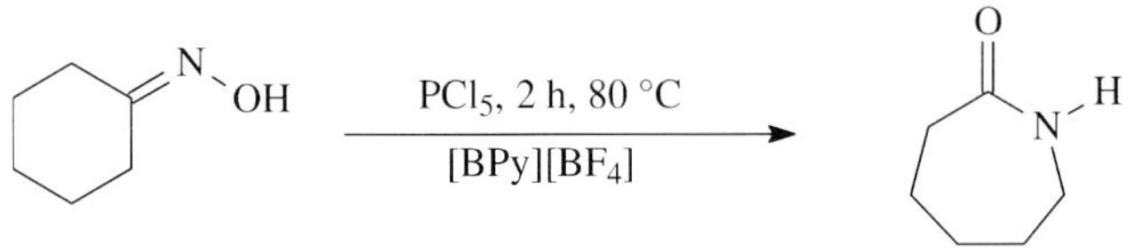

Scheme 5.2-139 The Beckmann rearrangement in ionic liquids [301].

Ionic liquids ([BMIM][PF$_6$]) are suitable media for the Co$_2$(CO)$_8$-catalyzed intramolecular and intermolecular Pauson–Khand annulation, provided that the reaction is carried out under a CO pressure of 10 bar [299]. Two diethyl allyl propargyl malonates were quantitatively converted into the relevant cyclopentenones, whereas heteroatom tethered enynes gave lower yields in their cyclocarbonylation products. An example is shown in Scheme 5.2-138. The reaction also works with the direct addition of ethyldiazoacetate to imines in [BMIM][BF$_4$] and [BMIM][PF$_6$] [300].

Beckmann rearrangements of several ketoximes were performed in the room-temperature ionic liquid based on 1,3-dialkylimidazolium or alkylpyridinium salts containing phosphorus compounds (e.g. PCl$_5$) by Deng and Peng [301] (Scheme 5.2-139). Turnover numbers of up to 6.6 were observed, but the authors did not mention whether the ionic liquid could be reused.

A way of forming cyclic amides is by the reaction of a primary amine with a cyclic anhydride to give an imide. Chen et al. has carried out numerous of these reactions in [BMIM][PF$_6$] to give the cyclic imide in 91–98% yield [302]. Howarth et al. have achieved the coupling of aryl halides with a zinc metal reducing agent in the ionic liquid [BMIM][PF$_6$]. In this reaction, a zero-valent nickel catalyst (or coupling agent) is generated *in situ* from zinc and (Ph$_3$P)$_2$NiCl$_2$ [303] (Scheme 5.2-140).

A novel use of the salt [BMIM][PF$_6$] is to enhance microwave absorption and hence accelerate the rate of a reaction. Ley found that [BMIM][PF$_6$] enhanced the rate of the microwave promoted thionation of amides using a polymer supported thionating agent [304]. Hardacre et al. have developed a protocol for the synthesis of deuterated imidazoles and imidazolium salts [305]. The procedure involves the platinum- or palladium- catalyzed deuterium exchange of 1-methyl-d^3-imidazole with D$_2$O to give 1-methylimidazole-d^6, followed by reaction with a deuterated alkyl halide.

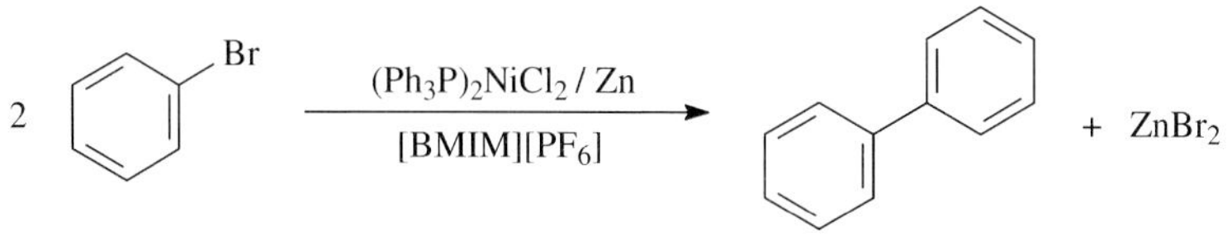

Scheme 5.2-140 The zinc promoted coupling of aryl halides [303].

Finally, a word of caution when using $[BF_4]^-$ and $[PF_6]^-$ ionic liquids – they are not stable and give off HF, particularly when heated in the presence of a proton source or a metal salt [21]. There are many examples of this in this chapter. An example of a HF-catalyzed reaction is: ether formation from alcohols is a classic acid-catalyzed reaction. An ether formation reaction was found to occur in a range of $[BF_4]^-$ ionic liquids, with an example being the addition of methanol to *tert*-butanol to form methyl-*tert*-butyl ether (MTBE) [306]. The author is of the opinion that $[BF_4]^-$ ionic liquids (even hydrophobic ones) can dehydrate alcohols to ether and refers to these ionic liquids as dehydrators. All that is happening here is a simple HF-catalyzed reaction. With many authors not aware of this phenomenon, they resort to all kinds of inappropriate explanations for what is occurring.

References

1. (a) T. Welton, *Chem. Rev.* **1999**, *99*, 2071–2083; (b) T. Welton, *Coord. Chem. Rev.* **2004**, *248*, 2459–2477.
2. J. D. Holbrey. K. R. Seddon, *Clean Prod. Proc.* **1999**, *1*, 223–236.
3. J. S. Wilkes, *J. Mol. Catal. A*, **2004**, *214*, 11–17.
4. M. J. Earle, K. R. Seddon, *Pure Appl. Chem.* **2000**, *72*, 1391–1398.
5. D. W. Rooney, K. R. Seddon, in *The Handbook of Solvents*, G. Wypych (Ed.), ChemTech Publishing, New York, **2001**, pp. 1459–1484.
6. K. R. Seddon, *J. Chem. Tech. Biotech.* **1997**, *68*, 351–356.
7. R. D. Rogers, K. R. Seddon, S. Volkov, in *NATO Science Series II: Mathematics, Physics and Chemistry*, Vol. 92, Kluwer, Dordrecht, **2002**.
8. P. Wasserscheid, W. Keim, *Angew. Chem. Int. Ed.* **2000**, *39*, 3773–3789.
9. J. Dupont, C. S. Consorti, J. Spencer, *J. Braz. Chem. Soc.* **2000**, *11*, 337–344.
10. J. Dupont, R. F. de Souza, P. A. Z. Suarez, *Chem. Rev.* **2002**, *102*, 3667–3692.
11. R. A. Sheldon, *Chem. Commun.*, **2001**, 2399–2407.
12. M. Fremantle, *Chem. Eng. News* **1998** (30th March), *76*, 32–37.
13. C. M. Gordon, J. D. Holbrey, A. R. Kennedy, K. R. Seddon, *J. Mater. Chem.* **1998**, *8*, 2627–2636.
14. K. R. Seddon, A. Stark, M. J. Torres, *Pure Appl. Chem.* **2000**, *72*, 2275–2287.
15. A. E. Visser, R. P. Swatloski, R. D. Rogers, *Green Chem.* **2000**, *2*, 1–4.
16. J. March, *Advanced Organic Chemistry*, 4th edn., Wiley, Chichester, **1992**.
17. J. D. Huddleston, H. D. Willauer, R. P. Swatloski, A. E. Visser, R. D. Rogers, *Chem. Commun.* **1998**, 1765–1766.
18. A. J. Carmichael, M. J. Earle, J. D. Holbrey, P. B. McCormac, K. R. Seddon, *Org. Lett.* **1999**, *1*, 997–1000.
19. L. P. Barthel-Rosa, J. A. Gladysz, *Coord. Chem. Rev.* **1999**, *192*, 587–605.
20. R. D. Rogers, *Chem. Eng. News* **2002**, *80*, 4.
21. R. P. Swatloski, J. D. Holbrey, R. D. Rogers, *Green Chem.* **2003**, *5*, 361–363.

22. M. J. Earle, K. R. Seddon, *World Patent*, 0177081, **2001**.
23. P. Wasserscheid, R. van Hal, A. Bosmann, *Green Chemistry*, **2002**, *4*, 400–404.
24. A. J. Carmichael, M. J. Earle, M. Deetlefs, U. Fröhlich, K. R. Seddon, Ionic liquids: Improved Syntheses and New Products, in *Ionic liquids as Green Solvents, Progress and Prospects*, R. D. Rogers, K. R. Seddon (Eds.), ACS Symposium Series 856, **2003**, pp. 14–31.
25. J. F. Huang, I. W. Sun, *J. Electrochem. Soc.* **2003**, *150*, E299–E306.
26. A. P. Abbott, G. Capper, D. L. Davies, R. H. Rasheed, V. Tambyrajah, *Green Chem.* **2002**, *4*, 24–26.
27. C. Hardacre, B. J. Mcauley, K. R. Seddon, *World Patent*, 03028883, **2003**.
28. G. A. Olah, *Friedel-Crafts and Related Reactions*, Interscience, New York, **1963**.
29. G. A. Olah, *Friedel Crafts Chemistry*, Wiley-Interscience, New York, **1973**.
30. G. Baddeley, R. Williamson, *J. Chem. Soc.* **1953**, 2120–2123.
31. G. Baddeley, G. Holt, S. M. Makar, M. G. Ivinson, *J. Chem. Soc.* **1952**, 3605–3607.
32. W. L. Mendelson, C. B. Spainhour, S. S. Jones, B. L. Lamb, K. L. Wert, *Tetrahedron Lett.* **1980**, *21*, 1393–1396.
33. J. A. Boon, J. A. Levisky, J. L. Pflug, J. S. Wilkes, *J. Org. Chem.* **1986**, *51*, 480–483.
34. B. J. Piersma, M. Merchant, in *Proceedings of the 7th International Symposium on Molten Salts*, C. L. Hussey, S. N. Flengas, J. S. Wilkes, Y. Ito (Eds.), The Electrochemical Society, Pennington, **1990**, pp. 805–821.
35. P. K. G. Hodgson, M. L. M. Morgan, B. Ellis, A. A. K. Abdul-Sada, M. P. Atkins, K. R. Seddon, *US Patent*, 5994602, **1999**.
36. P. Wasserscheid, B. Ellis, H. Fabienne, *World Patent*, 0041809, **2000**.
37. C. C. Greco, S. Fawzy, S. Lieh-Jiun, *US Patent*, 5824832, **1998**.
38. L. Xiao, K. E. Johnson, *Can. J. Chem.* **2004**, *82*, 491–498.
39. X. A. Li, K. E. Johnson, R. G. Treble, *J. Mol. Catal. A* **2004**, *214*, 121–127.
40. Z. Zhao, Z. Li, G. Wang, W. Qiao, L. Cheng, *Appl. Catal. A* **2004**, *262*, 69–73.
41. J. H. Davis, *Chem. Lett.* **2004**, *33*, 1072–1077.
42. K. Qiao, C. Yokoyama, *Chem. Lett.* **2004**, *33*, 472–473.
43. W. Keim, W. Korth, P. Wasserscheid, *World Patent*, 0016902, **2000**.
44. C. E. Song, W. H. Shim, E. J. Roh, J. H. Choi, *Chem. Commun.* **2000**, 1695–1696.
45. M. J. Earle, unpublished results, **1998**.
46. M. J. Earle, A. Ramani, A. J. Robertson, K. R. Seddon, *World Patent*, 03020683, **2003**.
47. M. J. Earle, B. J. Mcauley, A. Ramani, K. R. Seddon, J. M. Thomson, *World Patent*, 02072519, **2002**.
48. H.-Y. Shen, Z. M. A. Judeh, C. B. Ching, *Tetrahedron Lett.* **2003**, *44*, 981–983.
49. H.-Y. Shen, Z. M. A. Judeh, C. B. Ching, Q.-H. Xia, *J. Mol. Catal. A* **2004**, *212*, 301–308.
50. X. Mi, S. Luo, J. He, J.-P. Cheng, *Tetrahedron Lett.* **2004**, *45*, 4567–4570.
51. M. J. Earle, R. A. Fairhurst, H. Heaney, *Tetrahedron Lett.* **1991**, *32*, 6171–6174.
52. K. Qiao, Y.-Q. Deng, *Acta Chim. Sin.* **2003**, *61*, 133–136.
53. K. Yoo, V. V. Namboodiri, R. S. Varma, P. G. Smirniotis, *J. Catal.* **2004**, *222*, 511–519.
54. Y. L. Yang, X. H. Wang, Y. Kou, *Chin. J. Catal.* **2004**, *25*, 60–64.
55. K. Qiao, Y. Deng, *Tetrahedron Lett.*, **2003**, *44*, 2191–2193.
56. C. P. DeCastro, E. Sauvage, M. H. Valkenberg, W. F. Hölderich, *World Patent*, 0132308, **2001**.
57. C. P. DeCastro, E. Sauvage, M. H. Valkenberg, W. F. Hölderich, *J. Catal.* **2000**, *196*, 86–94.
58. J. L. Scott, D. R. MacFarlane, C. L. Raston, C. M. Teoh, *Green Chem.* **2000**, *2*, 123–126.
59. N. Srinivasan, A. Ganesan, *Chem. Commun.* **2003**, 916–917.
60. H. Raudnitz, G. Laube, *Berichte* **1929**, *62*, 509.
61. R. Scholl, K. Meyer, J. Donat, *Chem. Ber.* **1937**, *70*, 2180–2189.
62. D. B. Bruce, A. J. S. Sorrie, R. H. Thomson, *J. Chem. Soc.* **1953**, 2403–2408.
63. G. C. Misra, L. M. Pande, G. C. Joshi, A. K. Misra, *Aust. J. Chem.* **1972**, *25*, 1579–1581.
64. S. Wagatsuma, H. Higuchi, T. Ito, T. Nakano, Y. Naoi, K. Sakai, T. Matsui,

Y. Takahashi, A. Nishi, S. Sano, *Org. Prep. Proced. Int.* **1973**, *5*, 65–70.

65. J. A. Boon, S. W. Lander Jr., J. A. Levisky, J. L. Pflug, L. M. Skrznecki-Cooke, J. S. Wilkes, *Proceedings of the Joint International Symposium on Molten Salts*, 6th edn., **1987**, pp. 979–990.

66. J. K. D. Surette, L. Green, R. D. Singer, *Chem. Commun.* **1996**, 2753–2754.

67. A. Stark, B. L. MacLean, R. D. Singer, *J. Chem. Soc., Dalton Trans.* **1999**, 63–66.

68. C. J. Adams, M. J. Earle, G. Roberts, K. R. Seddon, *Chem. Commun.* **1998**, 2097–2098.

69. C. Baleizao, N. Pires, B. Gigante, M. Joao, M. Curto, *Tetrahedron Lett.* **2004**, *45*, 4375–4377.

70. V. D. Sarca, K. K. Laali, *Green Chem.* **2004**, *6*, 245–248.

71. M. J. Earle, U. Hakala, B. J. McAuley, M. Nieuwenhuyzen, A. Ramani, K. R. Seddon, *Chem. Commun.* **2004**, 1368–1369.

72. N. Brausch, A. Metlen, P. Wasserscheid, *Chem. Commun.* **2004**, 1552–1553.

73. P. N. Davey, M. J. Earle, C. P. Newman, K. R. Seddon, *World Patent* 99 19288, **1999**.

74. M. J. Earle, U. Hakala, C. Hardacre, J. Karkkainen, B. J. McAuley, D. W. Rooney, K. R. Seddon, J. M. Thompson, K. Wähälä, *Chem. Commun.* **2005**, 903–905.

75. C. Hardacre, B. J. McAuley, K. R. Seddon, *World Patent*, 03028883, **2003**.

76. M. H. Valkenberg, C. deCastro, W. F. Hölderich, *App. Catal. A* **2001**, *215*, 185–190.

77. G. L. Rebeiro, B. M. Khadilkar, *Syn. Commun.* **2000**, *30*, 1605–1608.

78. P. U. Naik, S. J. Nara, J. R. Harjani, M. M. Salunkhe, *Can. J. Chem.* **2003**, 1057–1060.

79. P. Studt, *Liebigs Ann. Chem.* **1978**, 528–529.

80. K. Nakasuji, K. Yoshida and I. Murata, *J. Am. Chem. Soc.* **1983**, *105*, 5136–5137.

81. M. B. Groen, H. Schadenberg, H. Wynberg, *J. Org. Chem.* **1971**, *36*, 2797–2809.

82. G. P. Smith, R. M. Pagni in *Molten Salt Chemistry, An Introduction to Selected Applications*, G. Mamantov, R. Marassi (Eds.), D. Reidel Publishing Co., Dordrecht, **1987**, pp. 383–416.

83. H. Imaizumi, S. Sekiguchi, K. Matsui, *Bull. Chem. Soc. Jpn.* **1977**, *50*, 948–952.

84. A. C. Buchanan III, D. M. Chapman, G. P. Smith, *J. Org. Chem.* **1985**, *50*, 1702–1711.

85. R. M. Pagni, in *Advances in Molten Salt Chemistry*, Vol. 6, G. Mamantov, C. B. Mamantov, J. Braunstein (Eds.), Elsevier, Oxford, **1987**, pp. 303–323.

86. D. S. Newman, T. H. Kinstle, G. Thambo, *Proceedings of the Joint International Symposium on Molten Salts*, 6th edn., **1987**, pp. 991–1001.

87. D. S. Newman, T. H. Kinstle, G. Thambo, *J Electrochem. Soc.* **1987**, *134*, C512.

88. D. S. Newman, R. E. Winans, R. L. McBeth, *J Electrochem. Soc.* **1984**, *131*, 1079–1083.

89. C. J. Adams, M. J. Earle, K. R. Seddon, *Green Chem.* **2000**, *2*, 21–24.

90. R. W. J. Westerhout, J. A. M. Kuipers, W. P. M. van Swaaij, *Ind. Eng. Chem. Res.* **1998**, *37*, 841–847.

91. P. N. Barnes, K. A. Grant, K. J. Green, N. D. Lever, World Patent 0040673, **2000**.

92. C. J. Adams, M. J. Earle, J. Hamill, C. M. Lok, G. Roberts, K. R. Seddon, World Patent 9807680, **1998**.

93. C. J. Adams, M. J. Earle, J. Hamill, C. M. Lok, G. Roberts, K. R. Seddon, World Patent 9807679, **1998**.

94. J. A. Boon, S. W. Lander Jr., J. A. Levisky, J. L. Pflug, L. M. Skrznecki-Cooke, J. S. Wilkes, *Proceedings of the Joint International Symposium on Molten Salts*, 6th edn., **1987**, pp. 979–990.

95. K. K. Laali, V. J. Gettwert, *J. Org. Chem.* **2001**, *66*, 35–40.

96. M. J. Earle, S. P. Katdare, K. R. Seddon, *World Patent*, 0230865, **2002**.

97. M. J. Earle, S. P. Katdare, K. R. Seddon, *Org. Lett.* **2004**, *6*, 707–710.

98. K. Qiao, C. Yokoyama, *Chem. Lett.* **2004**, *33*, 808–809.

99. S. T. Handy, C. R. Egrie, Green Synthesis: Aromatic Nitration in Room-Temperature Ionic Liquids, in *Ionic liquids: Industrial applications to Green Chemistry*, R. D. Rogers, K. R.

Seddon (Eds.), ACS Symposium Series 818, **2002**, pp. 134–146.
100. R. Rajagopal, K. V. Srinivasan, *Syn. Commun.* **2003**, *33*, 961–966.
101. N. L. Lancaster, V. Llopis-Mestre, *Chem. Commun.* **2003**, 2812–2813.
102. P. Walden, *Bull. Acad. Imper. Sci.* (St Petersburg), **1914**, 1800.
103. R. Rajagopal, K. V. Srinivasan, *Ultrasonics Sonochem.* **2003**, *10*, 41–43.
104. Y. Patell, N. Winterton, K. R. Seddon, World Patent 0037400, **2000**.
105. M. J. Earle, S. P. Katdare, K. R. Seddon, World Patent, 0230852, **2002**.
106. C. Chiappe and D. Pieraccini, *Arkivoc* **2002**, (*xi*), 249–255.
107. O. Bortolini, M. Bottai, C. Chiappe, V. Conte, D. Pieraccini, *Green Chem.* **2002**, *4*, 621–627.
108. C. Chiappe, D. Capraro, V. Conte, D. Pieraccini, *Org. Lett.* **2001**, *3*, 1061–1063.
109. C. Chiappe, V. Conte, D. Pieraccini, *Eur. J. Org. Chem.* **2002**, 2831–2837.
110. R. Rajagopal, D. V. Jarikote, R. J. Lahoti, T. Daniel, K. V. Srinivasan, *Tetrahedron Lett.* **2003**, *44*, 1815–1817.
111. H. Togo, T. Hirai, *Synlett* **2003**, 702–704.
112. Z.-W. Wang, L.-S. Wang, *Appl. Catal. A* **2004**, *262*, 101–104.
113. M. J. Earle, S. P. Katdare, K. R. Seddon, World Patent, 0230878, **2002**.
114. S. S. Mohile, M. K. Potdar, M. M. Salunkhe, *Tetrahedron Lett.* **2003**, *44*, 1255–1258.
115. S. J. Nara, J. R. Harjani, M. M. Salunkhe, *J. Org. Chem.* **2001**, *66*, 6818–6820.
116. G. P. Smith, R. M. Pagni, in *Molten Salt Chemistry, An Introduction to Selected Applications*, G. Mamantov, R. Marassi (Eds.), D. Reidel Publishing Co., Dordrecht, **1987**, pp. 383–416.
117. R. K. Sharma, J. L. Fry, *J. Org. Chem.* **1983**, *48*, 2112–2114.
118. E. Nakamura, M. Shimizu, I. Kuwajima, J. Sakata, K. Yokoyama, R. Noyori, *J. Org. Chem.* **1983**, *48*, 932–945.
119. G. Majetich, A. Casares, D. Chapman, M. Behnke, *J. Org. Chem.* **1986**, *51*, 1745–1753.
120. T. W. Green, P. G. M. Wuts, *Protective Groups in Organic Synthesis*, Wiley, New York, **1991**.
121. D. P. Cox, J. Terpinski, W. Lawrynowicz, *J. Org. Chem.* **1984**, *49*, 3216–3219.
122. R. A. Moss, W. Lawrynowicz, *J. Org. Chem.* **1984**, *49*, 3828–3830.
123. D. P. Cox, R. A. Moss, J. Terpinski, *J. Am. Chem. Soc.* **1983**, *105*, 6513–6514.
124. W. T. Ford, R. J. Hauri, D. J. Hart, *J. Org, Chem.* **1973**, *38*, 3916–3918.
125. W. T. Ford, R. J. Hauri, *J. Am. Chem. Soc.* **1973**, *95*, 7381–7391.
126. W. T. Ford, *J. Org, Chem.* **1973**, *38*, 3614–3615.
127. M. Badri, J.-J. Brunet, *Tetrahedron Lett.* **1992**, *33*, 4435–4438.
128. M. J. Earle, P. B. McCormac, K. R. Seddon, *Chem. Commun.* **1998**, 2245–2246.
129. W.-C. Shieh, M. Lozanov, O. Repič, *Tetrahedron Lett.* **2003**, *44*, 6943–6945.
130. Y. Hu, Z.-C. Chen, Z.-G. Le, Q.-G. Zheng, *J. Chem. Res.–S* **2004**, 276–278.
131. D.-Q. Xu, S.-P. Luo, B.-Y. Liu, X.-H. Yan, Z.-Y. Xu, *Chin. J. Org. Chem.* **2004**, *24*, 99–102.
132. C. Wheeler, K. N. West, C. L. Liotta, C. A. Eckert, *Chem. Commun.* **2001**, 887–888.
133. N. M. T. Lourenco, C. A. M. Afonso, *Tetrahedron* **2003**, *59*, 789–794.
134. Z.-G. Le, Z.-C. Chen, Y. Hu, Q.-G. Zheng, *J. Chem. Res.-S* **2004**, 344–346.
135. Y. Hu, Z.-C. Chen, Z.-G. Le, Q.-G. Zheng, *J. Chem. Res.-S* **2004**, 267–269.
136. D. W. Kim, D. J. Hong, J. W. Seo, H. S. Kim, H. K. Kim, C. E. Song, D. Y. Chi, *J. Org. Chem.*, **2004**, *69*, 3186–3189.
137. D. W. Kim, C. E. Song, D. Y. Chi, *J. Org. Chem.* **2003**, *68*, 4281–4285.
138. D. W. Kim, Y. S. Choe, D. Y. Chi, *Nucl. Med. Biol.* **2003**, *30*, 345–350.
139. S. R. S. Saibabu Kotti, X. Xu, G. Li, A. D. Headley, *Tetrahedron Lett.* **2004**, *45*, 1427–1431
140. M. L. Paoli, S. Piccini, M. Rodriquez, A. Sega, *J. Org. Chem.* **2004**, *69*, 2881–2883.
141. N. L. Lancaster, T. Welton, G. B. Young, *J. Chem. Soc., Perkin Trans. 2* **2001**, 2267–2270.
142. N. L. Lancaster, P. A. Salter, T. Welton, G. B. Young, *J. Org. Chem.* **2002**, *67*, 8855–8861.
143. N. L. Lancaster, T. Welton, *J. Org. Chem.* **2004**, *69*, 5986–5992.

144. L. Crowhurst, N. L. Lancaster, J. M. P. Arlandis, T. Welton, *J. Am. Chem. Soc.* **2004**, *126*, 11549–11555.
145. H.-P. Nguyen, H. Matondo, M. Baboulène, *Green Chem.* **2003**, *5*, 303–305.
146. H.-H. Wu, J. Sun, F. Yang, J. Tang, M.-Y. He, *Chinese J. Chem.* **2004**, *22*, 619–621.
147. H.-P. Nguyen, P. Kirilov, H. Matondo, M. Baboulène, *J. Mol. Catal. A.* **2004**, *218*, 41–45.
148. J. S. Yadav, B. V. S. Reddy, Ch. S. Reddy, K. Rajasekhar, *Chem. Lett.* **2004**, *33*, 476–477.
149. J. S. Yadav, B. V. S. Reddy, A. K. Basak, A. V. Narsaiah, *Tetrahedron Lett.* **2003**, *44*, 1047–1050.
150. Z. Pakulski, K. M. Pietrusiewicz, *Tetrahedron Asym.* **2004**, *15*, 41–45.
151. M. M. Khodaei, A. R. Khosropour, K. Ghozati, *Tetrahedron Lett.* **2004**, *45*, 3525–3529.
152. M. J. Earle, P. B. McCormac, K. R. Seddon, *Green Chem.* **2000**, *2*, 261–262.
153. L. Poletti, A. Rencurosi, L. Lay, G. Russo, *Synlett* **2003**, 2297–2300.
154. Z. Pakulski, *Synthesis* **2003**, 2074–2076.
155. J. Kim, J. M. Shreeve, *Org. Biomol. Chem.* **2004**, *2*, 2728–2734.
156. M. Cavazza, F. Pietra, *Tetrahedron Lett.* **2004**, *45*, 3633–3634.
157. Y.-Y. Xie, Z.-C. Chen, Q.-G. Zheng, *J. Chem. Res.-S* **2003**, 614–615.
158. D. Q. Xu, B. Y. Liu, Z. Y. Xu, *Chinese Chem. Lett.* **2003**, *14*, 1002–1004.
159. D. Xu, B. Liu, M. Zheng, *J. Chem. Res.-S* **2003**, 645–647.
160. R. M. Pagni, in *Advances in Molten Salt Chemistry*, Vol. 6, G. Mamantov, J. Braunstein (Eds.), Elsevier, Oxford, **1987**, pp. 211–346.
161. M. Fieser, L. Fieser, *Reagents for Organic Synthesis*, Vol. *1*, Wiley-Interscience, New York, **1967**, p. 936, and references therein.
162. C. M. Suter, *The Organic Chemistry of Sulfur, Tetravalent Sulfur Compounds*, Wiley-Interscience, New York, **1944**, p. 420.
163. F.-W. Wang, Z.-C. Chen, Q.-G. Zheng, *J. Chem. Res.-S* **2003**, 810–811.
164. F.-W. Wang, Z.-C. Chen, Q.-G. Zheng, *J. Chem. Res.-R* **2004**, 206–207.
165. J. S. Yadav, B. V. S. Reddy, A. K. Basak, A. Venkat Narsaiah, *Tetrahedron Lett.* **2003**, *44*, 2217–2220.
166. C. W. Lee, *Tetrahedron Lett.* **1999**, *40*, 2461–2462.
167. A. Kumar, S. S. Pawar, *J. Mol. Catal. A* **2004**, *208*, 33–37.
168. A. Kumar, S. S. Pawar, *J. Org. Chem.* **2004**, *69*, 1419–1420.
169. D. A. Jaeger, C. E. Trucker, *Tetrahedron Lett.* **1989**, *30*, 1785–1788.
170. J. Howarth, K. Hanlon, D. Fayne, P. B. McCormac, *Tetrahedron. Lett.* **1997**, *38*, 3097–3099.
171. M. J. Earle, P. B. McCormac, K. R. Seddon, *Green Chem.* **1999**, *1*, 23–25.
172. M. J. Earle, K. R. Seddon, Ionic liquids: Green solvents for the future, in *Clean Solvents: Alternative Media for Chemical Reactions and Processing*, M. Abraham (Ed.), ACS Symposium Series 819, **2002**, pp. 10–25.
173. T. Fisher, A. Sethi, T. Welton, J. Woolf, *Tetrahedron Lett.* **1999**, *40*, 793–795.
174. A. Aggarwal, N. Llewellyn Lancaster, A. R. Sethi, T. Welton, *Green Chem.* **2002**, *4*, 517–520.
175. P. Ludley, N. Karodia, *Tetrahedron Lett.* **2001**, *42*, 2011–2014.
176. P. Ludley, N. Karodia, *Arkivoc* **2002**, (*iii*), 172–175.
177. S. M. Malhotra, Y. Xiao, *Tetrahedron Lett.* **2004**, *45*, 8339–8342.
178. I.-W. Sun, S.-Y. Wu, C.-H. Su, *J. Chinese Chem. Soc.* **2004**, *51*, 367–370.
179. A. P. Abbott, G. Capper, D. L. Davies, R. K. Rasheed, V. Tambyrajah, *Green Chem.* **2002**, *4*, 24–26.
180. I. Hemeon, C. DeAmicis, H. Jenkins, P. Scammells, R. D. Singer, *Synlett* **2002**, 1815–1818.
181. C. E. Song, W. H. Shim, E. J. Roh, S.-G. Lee, J. H. Choi, *Chem. Commun.* **2001**, 1122–1123.
182. I. Meracz, T. Oh, *Tetrahedron Lett.* **2003**, *44*, 6465–6468.
183. S. Doherty, P. Goodrich, C. Hardacre, H.-K. Luo, D. W. Rooney, K. R. Seddon, P. Styring, *Green Chem.* **2004**, *6*, 63–67.
184. F. Zulfiqar, T. Kitazume, *Green Chem.* **2000**, *2*, 137–139.

185. J. S. Yadav, B. V. S. Reddy, J. S. S. Reddy, R. Srinivasa Rao, *Tetrahedron* **2003**, *59*, 1599–1604.
186. J. S. Yadav, B. V. S. Reddy, V. Naveenkumar, R. Srinivasa Rao, K. Nagaiah, *Synthesis* **2004**, 1783–1788.
187. E. Van Der Eycken, P. Appukkuttan, W. De Borggraeve, W. Dehaen, D. Dallinger, C. Oliver Kappe, *J. Org. Chem.* **2002**, *67*, 7904–7907.
188. F. Zulfiqar, T. Kitazume, *Green Chem.* **2000**, *2*, 296–298.
189. H.-P. Zhu, F. Yang, M.-Y. He, *Green Chem.* **2003**, *5*, 38–39.
190. L. Green, I. Hemeon, R. D. Singer, *Tetrahedron Lett.* **2000**, *41*, 1343–1345.
191. Z. Ma, Y. Q. Deng, F. Shi, *Can. Patent* 1247856, **2000**.
192. Y. Q. Deng, F. Shi, J. J. Beng, K. Qiao, *J. Mol. Catal. A* **2001**, *165*, 33–36.
193. S.-G. Lee, J. H. Park, *J. Mol. Catal. A* **2003**, *149*, 49–52.
194. R. D. Rogers, *Chem. Eng. News* **2002**, *80*, 4–5.
195. D. C. Forbes, K. J. Weaver, *J. Mol. Catal. A* **2004**, *214*, 129–132.
196. Y. Gu, F. Shi, Y. Deng, *J. Mol. Catal. A* **2004**, *212*, 71–75.
197. A. R. Gholap, K. Venkatesan, T. Daniel, R. J. Lahoti, K. V. Srinivasan, *Green Chem.* **2003**, *5*, 693–696.
198. S. Murugesan, N. Karst, T. Islam, J. M. Wiencek, R. J. Linhardt, *Synlett* **2003**, 1283–1286.
199. S. A. Forsyth, D. R. MacFarlane, R. J. Thomson, M. von Itzstein, *Chem. Commun.* **2002**, 714–715.
200. C. Imrie, E. R. T. Elago, C. W. McCleland, N. Williams, *Green Chem.* **2002**, *4*, 159–160.
201. H. Vallette, L. Ferron, G. Coquerel, A.-C. Gaumont, J.-C. Plaquevent, *Tetrahedron Lett.* **2004**, *45*, 1617–1619.
202. L.-W. Xu, J.-W. Li, S.-L. Zhou, C.-G. Xia, *New. J. Chem.* **2004**, *28*, 183–184.
203. J. S. Yadav, B. V. S. Reddy, A. K. Basak, A. V. Narsaiah, *Chem. Lett.* **2003**, *32*, 988–989.
204. L.-W. Xu, L. Li, C.-G. Xia, S.-L. Zhou, J.-W. Li, *Tetrahedron Lett.* **2004**, *45*, 1219–1221.
205. B. C. Ranu, S. S. Dey, A. Hajra, *Tetrahedron* **2003**, *59*, 2417–2421.
206. J. S. Yadav, B. V. S. Reddy, G. Baishya, *J. Org. Chem.* **2003**, *68*, 7098–7100.
207. J. Howarth, P. James, J. Dai, *J. Mol. Catal. A* **2004**, *214*, 143–146.
208. A. M. Salaheldin, Z. Yi, T. Kitazume, *J. Fluorine Chem.* **2004**, *125*, 1105–1110.
209. H. Hagiwara, T. Okabe, T. Hoshi, T. Suzuki, *J. Mol. Catal. A* **2004**, *214*, 167–174.
210. J. N. Rosa, C. A. M. Afonso, A. G. Santos, *Tetrahedron* **2001**, *57*, 4189–4193.
211. A. Kumar, S. S. Pawar, *J. Mol. Catal. A* **2004**, *211*, 43–47.
212. E. J. Kim, S. Y. Ko, C. E. Song, *Helv. Chim. Acta* **2003**, *86*, 894–899.
213. C. L. Johnson, R. E. Donkor, W. Nawaz, N. Karodia, *Tetrahedron Lett.* **2004**, *45*, 7359–7361.
214. V. K. Aggarwal, I. Emme, A. Mereu, *Chem. Commun.* **2002**, 1612–1613.
215. M. J. Earle, unpublished results.
216. T. Kitazume, K. Tamura, Z. Jiang, N. Miyake, I. Kawasaki, *J. Fluorine Chem.* **2002**, *115*, 49–53.
217. B. Pérgot, G. Vo-Thanh, D. Gori, A. Loupy, *Tetrahedron Lett.* **2004**, *45*, 6425–6428.
218. P. B. McCormac, K. R. Seddon, *Abs. Papers Am. Chem. Soc.* **1998**, *216*, 300.
219. S. Chandrasekhar, C. Narasihmulu, V. Jagadeshwar, K. V. Reddy, *Tetrahedron Lett.* **2003**, *44*, 3629–3620.
220. J. M. Fraile, J. I. Garcia, C. I. Herrerias, J. A. Mayoral, S. Gmough, M. Vaultier, *Green Chem.* **2004**, *6*, 93–98.
221. D. L. Davies, S. K. Kandola, R. K. Patel, *Tetrahedron Asym.* **2004**, *15*, 77–80.
222. T. Kitazume, G. Tanaka, *J. Fluorine. Chem.* **2000**, *106*, 211–215.
223. C. M. Gordon, A. McClusky, *Chem. Commun.* **1999**, 143–144.
224. J. S. Yadav, B. V. S. Reddy, A. K. Raju, *Synthesis* **2003**, 883–886.
225. T. Kitazume, K. Kasai, *Green Chem.* **2001**, *3*, 30–32.
226. M. C. Law, K.-Y. Wong, T. H. Chan, *Green Chem.* **2004**, *6*, 241–244.
227. J. N. Rosa, A. G. Santos, C. A. M. Afonso, *J. Mol. Catal. A* **2004**, *214*, 161–165.

228. A. McCluskey, J. Garner, D. J. Young, S. Caballero, *Tetrahedron Lett.* **2000**, *41*, 8147–8151.
229. S. Anjaiah, S. Chandrasekhar, R. Grée, *J. Mol. Catal. A* **2004**, *214*, 133–136.
230. S.-L. Chen, S.-J. Ji, T.-P. Loh, *Tetrahedron Lett.* **2004**, *45*, 375–377.
231. X.-F. Yang, M. Wang, R. S. Varma, C.-J. Li, *Org. Lett.* **2003**, *5*, 657–660.
232. X.-F. Yang, M. Wang, R. S. Varma, C.-J. Li, *J. Mol. Catal. A* **2004**, *214*, 147–154.
233. J. S. Yadav, B. V. S. Reddy, D. Gnaneshwar, *New J. Chem.* **2003**, *27*, 202–204.
234. J. S. Yadav, B. V. S. Reddy, G. Bhaishya, *Green Chem.* **2003**, *5*, 264–266.
235. S.-gi Lee, J. H. Park, J. Kang, J. K. Lee, *Chem. Commun.* **2001**, 1698–1699.
236. M. M. Khodaei, A. R. Khosropour, M. Kookhazadeh, *Synlett* **2004**, 1980–1984.
237. R. N. Nadaf, S. A. Siddiqui, T. Daniel, R. J. Lahoti, K. V. Srinivasan, *J. Mol. Catal. A* **2004**, *214*, 155–160.
238. B. C. Ranu, R. Jana, S. S. Dey, *Chem. Lett.* **2004**, *33*, 274–275.
239. X. Zhang, H. Niu, J. Wang, *J. Chem. Res. (S)* **2003**, *32*, 33–35.
240. X. Zhang, X. Fan, H. Niu, J. Wang, *Green Chem.* **2003**, *5*, 267–269.
241. D.-H. Yang, Z.-C. Chen, S.-Y. Chen, Q.-G. Zheng, *Synthesis* **2003**, 1891–1894.
242. T. Kitazume, F. Zulfiqar, G. Tanaka, *Green Chem.* **2000**, *2*, 133–136.
243. J. S. Yadav, B. V. S. Reddy, G. Kondaji, *J. Chem. Res. (S)* **2003**, *32*, 672–673.
244. S. Naik, R. Gopinath, M. Goswami, B. K. Patel, *Org. Biomol. Chem.* **2004**, *2*, 1670–1677.
245. D. W. Morrison, D. C. Forbes, J. H. Davis Jr., *Tetrahedron Lett.* **2001**, *42*, 6053–6055.
246. W. Bao, Z. Wang, Y. Li, *J. Chem. Res. (S)* **2003**, 294–295.
247. F. A. Kahn, J. Dash, R. A. Satapathy, S. K. Upadhyay, *Tetrahedron Lett.* **2004**, *45*, 3055–3058.
248. C. Su, Z.-C. Chen, Q.-G. Zheng, *Synthesis* **2003**, 555–559.
249. X.-M. Xu, Y.-Q. Li, M.-Y. Zhou, Y.-H. Tan, *Chinese J. Org. Chem.* **2004**, *24*, 184–186.
250. P. Formentin, H. Garcia, A. Leyva, *J. Mol. Catal. A* **2004**, *214*, 137–142.
251. Y.-H. Yen, Y.-H. Chu, *Tetrahedron Lett.* **2004**, *45*, 5313–5316.
252. S.-J. Ji, Z.-Q. Jiang, J. Lu, T.-P. Loh, *Synlett* **2004**, 831–835.
253. S. S. Palimkar, S. A. Siddiqui, T. Daniel, R. J. Lahoti, K. V. Srinivasan, *J. Org. Chem.* **2003**, *68*, 9371–9378.
254. J. Peng, Y. Deng, *Tetrahedron Lett.* **2001**, *42*, 5917–5919.
255. Y. Peng, G. Song, *Tetrahedron Lett.* **2004**, *45*, 8137–8140.
256. J. S. Yadav, B. V. S. Reddy, B. Eeshwaraiah, M. K. Gupta, *Tetrahedron Lett.* **2004**, *45*, 5873–5876.
257. B. Wang, Y. Gu, C. Luo, L. Yang, J. Suo, *Tetrahedron Lett.* **2004**, *45*, 3417–3419.
258. B. C. Ranu, S. S. Dey, *Tetrahedron Lett.* **2003**, *44*, 2865–2868.
259. G. Zhao, Tao Jiang, H. Gao, B. Han, J. Huang, D. Sun, *Green Chem.* **2004**, *6*, 75–77.
260. J. S. Yadav, B. V. S. Reddy, K. Saritha Raj, A. R. Prasad, *Tetrahedron* **2003**, *59*, 1805–1809.
261. J. S. Yadav, B. V. S. Reddy, R. Eshwaraiah, M. Srinivas, P. Vishnumurthy, *New J. Chem.* **2003**, *27*, 462–465.
262. G. W. Kabalka, B. Venkataiah, G. Dong, *Tetrahedron Lett.* **2004**, *45*, 729–731.
263. M. J. Earle, S. P. Katdare, *World Patent*, 02/38062 A1, **2002**.
264. R. D. Singer, P. J. Scammells, *Tetrahedron Lett.* **2001**, *42*, 6831–6833.
265. A. J. Walker, N. C. Bruce, *Tetrahedron* **2004**, *60*, 561–568.
266. H. Xie, S. Zhang, H. Duan, *Tetrahedron Lett.* **2004**, *45*, 2012–2105.
267. J. S. Yadav, B. V. S. Reddy, A. K. Basak, A. Venkat Narsaiah, *Tetrahedron* **2004**, *60*, 2131–2135.
268. B. S. Chhikara, S. Tehlan, A. Kumar, *Synlett* **2005**, 64–66.
269. Z. Li, C.-G. Xia, *J. Mol. Catal. A* **2004**, *214*, 95–101.
270. B. Martiz, R. Keyrouz, S. Gmouh, M. Vaultier, V. Jouikov, *Chem. Commun.* **2004**, 674–675.
271. J. S. Yadav, B. V. S. Reddy, A. K. Basak, A. V. Narsaiah, *Chem. Lett.* **2004**, *33*, 248–249.
272. K.-H. Tong, K.-Y. Wong, T. H. Chan, *Org. Lett.* **2003**, *5*, 3423–3425.

273. A. A. Valente, Z. Petrovski, L. C. Branco, C. A. M. Afonso, M. Pillinger, A. D. Lopes, C. C. Romão, C. D. Nunes, I. S. Gonçalves, *Tetrahedron Lett.* **2004**, *45*, 5–11.
274. Z. Li, C.-G. Xin, *Tetrahedron Lett.* **2003**, *44*, 2069–2071.
275. R. Bernini, E. R. Bernini, E. Mincione, A. Coratti, G. Fabrizi, G. Battistuzzi, *Tetrahedron*, **2004**, *60*, 967–971.
276. B. Wang, L.-M. Yang, J.-S. Suo, *Acta Chim. Sin.* **2003**, *61*, 285–290.
277. M. Lakshmi Kantam, V. Neeraja, B. Kavita, Y. Haritha, *Synlett* **2004**, 525–527.
278. Y. H. Zhu, S. Bahmueller, N. S. Hosmane, J. A. Maguire, *Chem. Lett.* **2003**, *32*, 730–731.
279. J. S. Yadav, B. V. S. Reddy, B. Reddy, K. U. Gayathri, A. R. Prasad, *Tetrahedron Lett.* **2003**, *44*, 6493–6496.
280. J. S. Yadav, B. V. S. Reddy, K. U. Gayathri, A. R. Prasad. *New J. Chem.* **2003**, *27*, 1684–1686.
281. M. Yan, Z.-C. Chen, Q.-G. Zheng, *J. Chem. Res. (S)* **2003**, 618–619.
282. G. Bianchini, M. Crucianelli, F. De Angelis, V. Neri, R. Saladino, *Tetrahedron Lett.* **2005**, *46*, 2427–2432.
283. J. Peng, F. Shi, Y. Gu, Y. Deng, *Green Chem.* **2003**, *5*, 224–226.
284. W.-H. Lo, H.-Y. Yang, G.-t. Wei, *Green Chem.* **2003**, *5*, 639–642.
285. G. P. Smith, A. S. Dworkin, R. M. Pagni, S. P. Zingg, *J. Am. Chem. Soc.* **1989**, *111*, 525–530.
286. C. J. Adams, M. J. Earle, K. R. Seddon, *Chem. Commun.* **1999**, 1043–1044.
287. G. W. Kabalka, R. R. Malladi, *Chem. Commun.* **2000**, 2191.
288. J. Howarth, P. James, R. Ryan, *Syn. Commun.* **2001**, *31*, 2935–2938.
289. D. Q. Xu, S. P. Luo, B. Y. Liu, Z. Y. Xu, Y. C. Shen, *Chinese Chem. Lett.* **2004**, 643–645.
290. P. De, *Synlett* **2004**, 1835–1837.
291. H.-M. Luo, Y.-Q. Li, *Chin. J. Chem.* **2005**, *23*, 345–348.
292. Q. Yang, D. D. Dionysiou, *J. Photochem. Photobiol. A: Chem.* **2004**, *165*, 229–240.
293. Y. Shen, T. Tajima, M. Atobe, T. Fuchigami, *Electrochemistry* **2004**, *72*, 849–851.
294. L. Gaillon, F. Bedioui, *J. Mol. Catal. A* **2004**, *214*, 91–94.
295. C. Lagrost, S. Gmouh, M. Vaultier, P. Hapiot, *J. Phys. Chem. A*, **2004**, *108*, 6175–6182.
296. X.-B. Lu, Y.-J. Zhang, B. Liang, X. Li, H. Wang, *J. Mol. Catal. A* **2004**, *210*, 31–34.
297. J. S. Yadav, B. V. S. Reddy, C. S. Reddy, K. Rajasekhar, *J. Org. Chem.*, **2003**, *68*, 2525–2527.
298. J. S. Yadav, B. V. S Reddy, P. N. Reddy, M. S. Rao, *Synthesis* **2003**, 1387–1390.
299. P. Mastrorilli, C. F. Nobile, R. Paolillo, G. P. Suranna, *J. Mol. Catal. A* **2004**, *214*, 103–106.
300. W. Sun, C.-G. Xia, H.-W. Wang, *Tetrahedron Lett.* **2003**, *44*, 2409–2411.
301. P. P. Peng, Y. Q. Deng, *Tetrahedron Lett.* **2001**, *42*, 403–405.
302. Z.-G. Le, Z.-C. Chen, Y. Hu, Q.-G. Zheng, *Synthesis* **2004**, 995–998.
303. J. Howarth, P. James, J. Dai, *Tetrahedron Lett.* **2000**, *41*, 10319–10321.
304. S. V. Ley, A. G. Leach, R. I. Storer, *J. Chem. Soc., Perkin Trans. 1* **2001**, 358–361.
305. C. Hardacre, J. D. Holbrey, S. E. J. McMath, *Chem. Commun.* **2001**, 367–368.
306. F. Shi, H. Xiong, Y. Gu, S. Guo, Y. Deng, *Chem. Commun.* **2003**, 1054–1055.

Ionic Liquids in Synthesis

Edited by
Peter Wasserscheid and
Tom Welton

Further Reading

Endres, F., MacFarlane, D., Abbott, A. (Eds.)

Electrodeposition in Ionic Liquids

2007
ISBN 978-3-527-31565-9

Sheldon, R. A., Arends, I., Hanefeld, U.

Green Chemistry and Catalysis

2007
ISBN 978-3-527-30715-9

Loupy, A. (Ed.)

**Microwaves in Organic Synthesis
Second, Completely Revised and Enlarged Edition**

2006
ISBN 978-3-527-31452-2

Ionic Liquids in Synthesis

Second, Completely Revised and Enlarged Edition

Volume 2

Edited by
Peter Wasserscheid and Tom Welton

WILEY-VCH Verlag GmbH & Co. KGaA

The Editors

Prof. Dr. Peter Wasserscheid
Friedrich-Alexander-Universität
Lehrstuhl für Chemische Reaktionstechnik
Institut für Chemie und Bioingenieurwesen
Egerlandstr. 3
91058 Erlangen
Germany

Prof. Dr. Tom Welton
Imperial College of Science,
Technology and Medicine
Department of Chemistry
South Kensington
London, SW7 2AZ
United Kingdom

1st Edition 2008
   1st Reprint 2008

All books published by Wiley-VCH are carefully produced. Nevertheless, authors, editors, and publisher do not warrant the information contained in these books, including this book, to be free of errors. Readers are advised to keep in mind that statements, data, illustrations, procedural details or other items may inadvertently be inaccurate.

Library of Congress Card No.: applied for

British Library Cataloguing-in-Publication Data
A catalogue record for this book is available from the British Library

Bibliographic information published by the Deutsche Nationalbibliothek
Die Deutsche Nationalbibliothek lists this publication in the Deutsche Nationalbibliografie; detailed bibliographic data are available on the Internet at <http://dnb.d-nb.de.>

© 2008 WILEY-VCH Verlag GmbH & Co. KGaA, Weinheim

All rights reserved (including those of translation into other languages). No part of this book may be reproduced in any form – by photoprinting, microfilm, or any other means – nor transmitted or translated into a machine language without written permission from the publishers. Registered names, trademarks, etc. used in this book, even when not specifically marked as such, are not to be considered unprotected by law.

Composition Aptara, New Delhi, India
Printing Betz-Druck GmbH, Darmstadt
Bookbinding Litges & Dopf GmbH, Heppenheim
Cover Design Adam-Design, Weinheim
Wiley Bicentennial Logo Richard J. Pacifico

Printed in the Federal Republic of Germany
Printed on acid-free paper

ISBN 978-3-527-31239-9

Contents

Preface to the Second Edition *xv*

A Note from the Editors *xix*

Acknowledgements *xix*

List of Contributors *xxi*

Volume 1

1 Introduction *1*
John S. Wilkes, Peter Wasserscheid, and Tom Welton

2 Synthesis and Purification *7*
2.1 Synthesis of Ionic Liquids *7*
Charles M. Gordon and Mark J. Muldoon
2.1.1 Introduction *7*
2.1.2 Quaternization Reactions *9*
2.1.3 Anion-exchange Reactions *13*
2.1.3.1 Lewis Acid-based Ionic Liquids *13*
2.1.3.2 Anion Metathesis *14*
2.1.4 Purification of Ionic Liquids *18*
2.1.5 Improving the Sustainability of Ionic Liquids *20*
2.1.6 Conclusions *23*
2.2 Quality Aspects and Other Questions Related to Commercial Ionic Liquid Production *26*
Markus Wagner and Claus Hilgers
2.2.1 Introduction *26*
2.2.2 Quality Aspects of Commercial Ionic Liquid Production *27*
2.2.2.1 Color *28*
2.2.2.2 Organic Starting Material and Other Volatiles *29*
2.2.2.3 Halide Impurities *30*
2.2.2.4 Protic Impurities *32*

2.2.2.5	Other Ionic Impurities from Incomplete Metathesis Reactions	33
2.2.2.6	Water	33
2.2.3	Upgrading the Quality of Commercial Ionic Liquids	34
2.2.4	Novel, Halide-Free Ionic Liquids	34
2.2.5	Scale-up of Ionic Liquid Synthesis	36
2.2.6	Health, Safety and Environment	37
2.2.7	Corrosion Behavior of Ionic Liquids	41
2.2.8	Recycling of Ionic Liquids	42
2.2.9	Future Price of Ionic Liquids	43
2.3	Synthesis of Task-specific Ionic Liquids	45

James H. Davis, Jr., updated by Peter Wasserscheid

2.3.1	Introduction	45
2.3.2	General Synthetic Strategies	47
2.3.3	Functionalized Cations	48
2.3.4	Functionalized Anions	53
2.3.5	Conclusion	53
3	**Physicochemical Properties**	**57**
3.1	Physicochemical Properties of Ionic Liquids: Melting Points and Phase Diagrams	57

John D. Holbrey and Robin D. Rogers

3.1.1	Introduction	57
3.1.2	Measurement of Liquid Range	59
3.1.2.1	Melting Points	60
3.1.2.2	Upper Limit – Decomposition Temperature	60
3.1.3	Effect of Ion Sizes on Salt Melting Points	62
3.1.3.1	Anion Size	63
3.1.3.2	Mixtures of Anions	64
3.1.3.3	Cation Size	65
3.1.3.4	Cation Symmetry	66
3.1.3.5	Imidazolium Salts	67
3.1.3.6	Imidazolium Substituent Alkyl Chain Length	68
3.1.3.7	Branching	69
3.1.4	Summary	70
3.2	Viscosity and Density of Ionic Liquids	72

Rob A. Mantz and Paul C. Trulove

3.2.1	Viscosity of Ionic Liquids	72
3.2.1.1	Viscosity Measurement Methods	73
3.2.1.2	Ionic Liquid Viscosities	75
3.2.2	Density of Ionic Liquids	86
3.2.2.1	Density Measurement	86
3.2.2.2	Ionic Liquid Densities	86
3.3	Solubility and Solvation in Ionic Liquids	89

Violina A. Cocalia, Ann E. Visser, Robin D. Rogers, and John D. Holbrey

3.3.1	Introduction 89	
3.3.2	Metal Salt Solubility 90	
3.3.2.1	Halometallate Salts 90	
3.3.2.2	Metal Complexes 91	
3.3.3	Extraction and Separations 92	
3.3.4	Organic Compounds 96	
3.3.5	Conclusions 101	
3.4	Gas Solubilities in Ionic Liquids 103	
	Jessica L. Anderson, Jennifer L. Anthony, Joan F. Brennecke, and Edward J. Maginn	
3.4.1	Introduction 103	
3.4.2	Experimental Techniques 104	
3.4.2.1	Gas Solubilities and Related Thermodynamic Properties 104	
3.4.2.2	The Stoichiometric Technique 106	
3.4.2.3	The Gravimetric Technique 107	
3.4.2.4	Spectroscopic Techniques 107	
3.4.2.5	Gas Chromatography 108	
3.4.3	Gas Solubilities 108	
3.4.3.1	CO_2 109	
3.4.3.2	Reaction Gases (O_2, H_2, CO) 117	
3.4.3.3	Other Gases (N_2, Ar, CH_4, C_2H_6, C_2H_4, H_2O, SO_2, CHF_3, etc.) 121	
3.4.3.4	Mixed Gases 122	
3.4.3.5	Enthalpies and Entropies 123	
3.4.4	Applications 123	
3.4.4.1	Reactions Involving Gases 124	
3.4.4.2	Gas Storage 125	
3.4.4.3	Gas Separations 125	
3.4.4.4	Extraction of Solutes from Ionic Liquids with Compressed Gases or Supercritical Fluids 126	
3.4.5	Summary 126	
3.5	Polarity 130	
	Tom Welton	
3.5.1	Microwave Dielectric Spectroscopy 131	
3.5.2	Chromatographic Measurements 131	
3.5.3	Absorption Spectra 133	
3.5.4	Antagonistic Behavior in Hydrogen Bonding 136	
3.5.5	Fluorescence Spectra 137	
3.5.6	Refractive Index 137	
3.5.7	EPR Spectroscopy 138	
3.5.8	Chemical Reactions 138	
3.5.9	Comparison of Polarity Scales 138	
3.5.10	Conclusions 140	
3.6	Electrochemical Properties of Ionic Liquids 141	
	Robert A. Mantz	

3.6.1	Electrochemical Potential Windows	142
3.6.2	Ionic Conductivity	150
3.6.3	Transport Properties	165
4	**Molecular Structure and Dynamics**	**175**
4.1	Order in the Liquid State and Structure	175
	Chris Hardacre	
4.1.1	Neutron Diffraction	175
4.1.2	Formation of Deuterated Samples	176
4.1.3	Neutron Sources	177
4.1.3.1	Pulsed (Spallation) Neutron Sources	177
4.1.3.2	Reactor Sources	178
4.1.4	Neutron Cells for Liquid Samples	178
4.1.5	Examples	178
4.1.5.1	Binary Mixtures	179
4.1.5.2	Simple Salts	182
4.1.6	X-ray Diffraction	184
4.1.6.1	Cells for Liquid Samples	184
4.1.6.2	Examples	185
4.1.7	Extended X-ray Absorption Fine Structure Spectroscopy	190
4.1.7.1	Experimental	191
4.1.7.2	Examples	193
4.1.8	X-ray and Neutron Reflectivity	199
4.1.8.1	Experimental Set-up	199
4.1.8.2	Examples	200
4.1.9	Direct Recoil Spectrometry (DRS)	201
4.1.9.1	Experimental Set-up	202
4.1.9.2	Examples	202
4.1.10	Conclusions	203
4.2	Computational Modeling of Ionic Liquids	206
	Patricia A. Hunt, Edward J. Maginn, Ruth M. Lynden–Bell, and Mario G. Del Pópolo	
4.2.1	Introduction	206
4.2.1.1	Classical MD	209
4.2.1.2	*Ab initio* Quantum Chemical Methods	210
4.2.1.3	*Ab initio* MD	211
4.2.1.4	Using *Ab Initio* Quantum Chemical Methods to Study Ionic Liquids	211
4.2.2.1	Introduction	211
4.2.2.2	Acidic Haloaluminate and Related Melts	212
4.2.2.3	Alkyl Imidazolium-based Ionic Liquids	214
4.2.2.4	The Electronic Structure of Ionic Liquids	218
4.2.3	Atomistic Simulations of Liquids	220
4.2.3.1	Atomistic Potential Models for Ionic Liquid Simulations	221

4.2.3.1	Atomistic Simulations of Neat Ionic Liquids – Structure and Dynamics *226*	
4.2.4	Simulations of Solutions and Mixtures *236*	
4.2.5	Simulations of Surfaces *239*	
4.2.6	*Ab initio* Simulations of Ionic Liquids *239*	
4.2.7	Chemical Reactions and Chemical Reactivity *244*	
4.3	Translational Diffusion *249*	
	Joachim Richter, Axel Leuchter, and Günter Palmer	
4.3.1	Main Aspects and Terms of Translational Diffusion *249*	
4.3.2	Use of Translational Diffusion Coefficients *251*	
4.3.3	Experimental Methods *252*	
4.3.4	Results for Ionic Liquids *254*	
4.4	Molecular Reorientational Dynamics *255*	
	Andreas Dölle, Phillip G. Wahlbeck, and W. Robert Carper	
4.4.1	Introduction *255*	
4.4.2	Experimental Methods *256*	
4.4.3	Theoretical Background *257*	
4.4.4	Results for Ionic Liquids *258*	
4.4.5	Chemical Shift Anisotropy Analysis *261*	
4.4.6	Stepwise Solution of the Combined Dipolar and NOE Equations *261*	
4.4.7	NMR–Viscosity Relationships *264*	
5	**Organic Synthesis** *265*	
5.1	Ionic Liquids in Organic Synthesis: Effects on Rate and Selectivity *265*	
	Cinzia Chiappe	
5.1.1	Introduction *265*	
5.1.2	Ionic Liquid Effects on Reactions Proceeding through Isopolar and Radical Transition States *268*	
5.1.2.1	Energy Transfer, Hydrogen Transfer and Electron Transfer Reactions *268*	
5.1.2.2	Diels–Alder Reactions *272*	
5.1.2.3	Ionic Liquid Effects on Reactions Proceeding through Dipolar Transition States *274*	
5.1.3.1	Nucleophilic Substitution Reactions *275*	
5.1.3.2	Electrophilic Addition Reactions *284*	
5.1.3.3	Electrophilic Substitution Reactions *287*	
5.1.4	Conclusions *289*	
5.2	Stoichiometric Organic Reactions and Acid-catalyzed Reactions in Ionic Liquids *292*	
	Martyn Earle	
5.2.1	Electrophilic Reactions *294*	
5.2.1.1	Friedel-Crafts Reactions *294*	
5.2.1.2	Scholl and Related Reactions *310*	
5.2.1.3	Cracking and Isomerization Reactions *312*	

5.2.1.4	Electrophilic Nitration Reactions *315*
5.2.1.5	Electrophilic Halogenation Reactions *316*
5.2.1.6	Electrophilic Phosphylation Reactions *318*
5.2.1.7	Electrophilic Sulfonation Reactions *318*
5.2.2	Nucleophilic Reactions *319*
5.2.2.1	Aliphatic Nucleophilic Substitution Reactions *319*
5.2.2.2	Aromatic Nucleophilic Substitution Reactions *326*
5.2.3	Electrocyclic Reactions *327*
5.2.3.1	Diels-Alder Reactions *327*
5.2.3.2	Hetero Diels-Alder Reactions *330*
5.2.3.3	The Ene Reaction *332*
5.2.4	Addition Reactions (to C=C and C=O Double Bonds) *334*
5.2.4.1	Esterification Reactions (Addition to C=O) *334*
5.2.4.2	Amide Formation Reactions (Addition to C=O) *335*
5.2.4.3	The Michael Reaction (Addition to C=C) *336*
5.2.4.4	Methylene Insertion Reactions (Addition to C=O and C=C) *339*
5.2.4.5	Addition Reactions Involving Organometallic Reagents *340*
5.2.4.6	Miscellaneous Addition Reactions *344*
5.2.5	Condensation Reactions *345*
5.2.5.1	General Condensation Reactions *345*
5.2.5.2	The Mannich Reaction *349*
5.2.6	Oxidation Reactions *350*
5.2.6.1	Functional Group Oxidation Reactions *350*
5.6.6.2	Epoxidation and Related Reactions *353*
5.2.6.3	Miscellaneous Oxidation Reactions *355*
5.2.7	Reduction Reactions *356*
5.2.8	Miscellaneous Reactions in Ionic Liquids *358*

Volume 2

5.3	Transition Metal Catalysis in Ionic Liquids *369*
	Peter Wasserscheid and Peter Schulz
5.3.1	Concepts, Successful Strategies, and Limiting Factors *372*
5.3.1.1	Why Use Ionic Liquids as Solvents for Transition Metal Catalysis? *372*
5.3.1.2	The Role of the Ionic Liquid *377*
5.3.1.3	Methods for Analysis of Transition Metal Catalysts in Ionic Liquids *383*
5.3.2	Selected Examples of the Application of Ionic Liquids in Transition Metal Catalysis *390*
5.3.2.1	Hydrogenation *390*
5.3.2.2	Oxidation Reactions *405*
5.3.2.3	Hydroformylation *410*
5.3.2.4	Heck Reaction and Other Pd-catalyzed C–C-coupling Reactions *419*
5.3.2.5	Dimerization and Oligomerization Reactions *430*
5.3.2.6	Olefin Metathesis *441*

5.3.2.7	Catalysis with Nanoparticulate Transition Metal Catalysts *444*	
5.3.3	Concluding Remarks: "Low-hanging Fruits" and "High-hanging Fruits"— Which Transition Metal Catalyzed Reaction Should Be Carried Out in an Ionic Liquid? *448*	
5.4	Ionic Liquids in Multiphasic Reactions *464*	
	Hélène Olivier-Bourbigou and Frédéric Favre	
5.4.1	Multiphasic Reactions: General Features, Scope and Limitations *464*	
5.4.2	Multiphasic Catalysis: Limitations and Challenges *465*	
5.4.3	Why Ionic Liquids in Mutiphasic Catalysis? *466*	
5.4.4	Different Technical Solutions to Catalyst Separation through the Use of Ionic Liquids *469*	
5.4.5	Immobilization of Catalysts in Ionic Liquids *473*	
5.4.6	The Scale-up of Ionic Liquid Technology from Laboratory to Continuous Pilot Plant Operation *476*	
5.4.6.1	Dimerization of Alkenes Catalyzed by Ni complexes *477*	
5.4.6.2	Alkylation Reactions *483*	
5.4.6.3	Industrial Use of Ionic Liquids *485*	
5.4.7	Concluding Remarks and Outlook *486*	
5.5	Task-specific Ionic Liquids as New Phases for Supported Organic Synthesis *488*	
	Michel Vaultier, Andreas Kirschning, and Vasundhara Singh	
5.5.1	Introduction *489*	
5.5.2	Synthesis of TSILs *490*	
5.5.2.1	Synthesis of TSILs Bearing a Hydroxy Group *491*	
5.5.2.2	Parallel Synthesis of Functionalized ILs from a Michael-type Reaction *495*	
5.5.2.3	Synthesis of TSILs by Further Functional Group Transformations *496*	
5.5.2.4	Loading of TSIL Supports *500*	
5.5.3	TSILs as Supports for Organic Synthesis *501*	
5.5.3.1	First Generation of TSILs as New Phases for Supported Organic Synthesis *503*	
5.5.3.2	Second Generation of TSILs: The BTSILs *510*	
5.5.3.3	Reactions of Functionalized TSOSs in Molecular Solvents *515*	
5.5.3.4	Lab on a Chip System Using a TSIL as a Soluble Support *523*	
5.5.4	Conclusion *523*	
5.6	Supported Ionic Liquid Phase Catalysts *527*	
	Anders Riisager and Rasmus Fehrmann	
5.6.1	Introduction *527*	
5.6.2	Supported Ionic Liquid Phase Catalysts *527*	
5.6.2.1	Supported Catalysts Containing Ionic Media *527*	
5.6.2.1.1	Process and engineering aspects of supported ionic liquid catalysts *528*	
5.6.2.1.2	Characteristics of ionic liquids on solid supports *529*	
5.6.2.2	Early Work on Supported Molten Salt and Ionic Liquid Catalyst Systems *531*	
5.6.2.2.1	High-temperature supported molten salt catalysts *531*	

5.6.2.2.2	Low-temperature supported catalysts 533	
5.6.2.3	Ionic Liquid Catalysts Supported through Covalent Anchoring 534	
5.6.2.3.1	Supported Lewis acidic chlorometalate catalysts 534	
5.6.2.3.2	Neutral, supported ionic liquid catalysts 537	
5.6.2.4	Ionic Liquid Catalysts Supported through Physisorption or via Electrostatic Interaction 540	
5.6.2.4.1	Supported ionic liquid catalysts (SILC) 540	
5.6.2.4.2	Supported ionic liquid phase (SILP) catalysts incorporating metal complexes 543	
5.6.2.4.3	Supported ionic liquid catalyst systems containing metal nanoparticles 552	
5.6.2.4.4	Supported ionic liquid catalytic membrane systems containing enzymes 554	
5.6.3	Concluding Remarks 555	
5.7	Multiphasic Catalysis Using Ionic Liquids in Combination with Compressed CO_2 558	
	Peter Wasserscheid and Sven Kuhlmann	
5.7.1	Introduction 558	
5.7.2	Catalytic Reaction with Subsequent Product Extraction 560	
5.7.3	Catalytic Reaction with Simultaneous Product Extraction 561	
5.7.4	Catalytic Conversion of CO_2 in an Ionic Liquid/$scCO_2$ Biphasic Mixture 562	
5.7.5	Continuous Reactions in an Ionic Liquid/Compressed CO_2 System 562	
5.7.6	Concluding Remarks and Outlook 567	
6	**Inorganic Synthesis** 570	
6.1	Directed Inorganic and Organometallic Synthesis 569	
	Tom Welton	
6.1.1	Coordination Compounds 569	
6.1.2	Organometallic Compounds 570	
6.1.3	Formation of Oxides 572	
6.1.4	Other Reactions 574	
6.1.5	Outlook 574	
6.2	Inorganic Materials by Electrochemical Methods 575	
	Frank Endres and Sherif Zein El Abedin	
6.2.1	Electrodeposition of Metals and Semiconductors 576	
6.2.1.1	General Considerations 576	
6.2.1.2	Electrochemical Equipment 577	
6.2.1.3	Electrodeposition of Less Noble Elements 578	
6.2.1.4	Electrodeposition of Metals That Can Also Be Obtained From Water 582	
6.2.1.5	Electrodeposition of Semiconductors 585	
6.2.2	Nanoscale Processes at the Electrode/Ionic Liquid Interface 587	
6.2.2.1	General Considerations 587	

6.2.2.2	The Scanning Tunneling Microscope	587
6.2.2.3	Results	589
6.2.3	Summary	604
6.3	Ionic Liquids in Material Synthesis: Functional Nanoparticles and Other Inorganic Nanostructures	609
	Markus Antonietti, Bernd Smarsly, and Yong Zhou	
6.3.1	Introduction	609
6.3.2	Ionic Liquids for the Synthesis of Chemical Nanostructures	609

7	**Polymer Synthesis in Ionic Liquids**	**619**
	David M. Haddleton, Tom Welton, and Adrian J. Carmichael	
7.1	Introduction	619
7.2	Acid-catalyzed Cationic Polymerization and Oligomerization	619
7.3	Free Radical Polymerization	624
7.4	Transition Metal-catalyzed Polymerization	627
7.4.1	Ziegler–Natta Polymerization of Olefins	627
7.4.2	Late Transition Metal-catalyzed Polymerization of Olefins	628
7.4.3	Metathesis Polymerization	630
7.4.4	Living Radical Polymerization	631
7.5	Electrochemical Polymerization	633
7.5.1	Preparation of Conductive Polymers	633
7.6	Polycondensation and Enzymatic Polymerization	634
7.7	Carbene-catalyzed Reactions	635
7.8	Group Transfer Polymerization	636
7.9	Summary	637

8	**Biocatalytic Reactions in Ionic Liquids**	**641**
	Sandra Klembt, Susanne Dreyer, Marrit Eckstein, and Udo Kragl	
8.1	Introduction	641
8.2	Biocatalytic Reactions and Their Special Needs	641
8.3	Examples of Biocatalytic Reactions in Ionic Liquids	644
8.3.1	Whole Cell Systems and Enzymes Other than Lipases in Ionic Liquids	644
8.3.2	Lipases in Ionic Liquids	651
8.4	Stability and Solubility of Enzymes in Ionic Liquids	655
8.5	Special Techniques for Biocatalysis with Ionic Liquids	657
8.6	Conclusions and Outlook	658

9	**Industrial Applications of Ionic Liquids**	**663**
	Matthias Maase	
9.1	Ionic Liquids in Industrial Processes: Re-invention of the Wheel or True Innovation?	663
9.2	Possible Fields of Application	664
9.3	Applications in Chemical Processes	666
9.3.1	Acid Scavenging: The BASIL™ Process	666

9.3.2	Extractive Distillation 669
9.3.3	Chlorination with "Nucleophilic HCl" 670
9.3.4	Cleavage of Ethers 672
9.3.5	Dimerization of Olefins 673
9.3.6	Oligomerization of Olefins 673
9.3.7	Hydrosilylation 674
9.3.8	Fluorination 675
9.4	Applications in Electrochemistry 675
9.4.1	Electroplating of Chromium 675
9.4.2	Electropolishing 676
9.5	Applications as Performance Chemicals and Engineering Fluids 677
9.5.1	Ionic Liquids as Antistatic Additives for Cleaning Fluids 677
9.5.2	Ionic Liquids as Compatibilizers for Pigment Pastes 678
9.5.3	Ionic Liquids for the Storage of Gases 679
9.6	FAQ – Frequently Asked Questions Concerning the Commercial Use of Ionic Liquids 681
9.6.1	How Pure are Ionic Liquids? 681
9.6.2	Is the Color of Ionic Liquids a Problem? 682
9.6.3	How Stable are Ionic Liquids? 682
9.6.4	Are Ionic Liquids Toxic? 683
9.6.5	Are Ionic Liquids Green? 684
9.6.6	How Can Ionic Liquids be Recycled ? 684
9.6.7	How Can Ionic Liquids be Disposed Of? 685
9.6.8	Which is the Right Ionic Liquid? 686

10   Outlook   689

Peter Wasserscheid and Tom Welton

Index   705

Preface to the Second Edition

"And with regard to my actual reporting of the events [. . .], I have made it a principle not to write down the first story that came my way, and not even to be guided by my own general impressions; either I was present myself at the events which I have described or else I heard of them from eye-witnesses whose reports I have checked with as much thoroughness as possible. Not that even so the truth was easy to discover: different eye-witnesses give different accounts of the same events, speaking out of partiality for one side or the other or else from imperfect memories. And it may well be that my history will seem less easy to read because of the absence in it of a romantic element. It will be enough for me, however, if these words of mine are judged useful by those who want to understand clearly the events which happened in the past and which (human nature being what it is) will, at some time or other and in much the same ways, be repeated in the future. My work is not a piece of writing designed to meet the taste of an immediate public, but was done to last for ever."

The History of the Peloponnesian War (Book I, Section 22),
Thucydides (431–413 BC), translated by Rex Warner

Almost five years ago to this day, I wrote the preface to the first edition of this book (which is reproduced herein, meaning I don't have to repeat myself). I was honoured to be asked to do it, and it was an enjoyable task. How often do we, as scientists, get the privilege to write freely about a subject close to our hearts, without a censorious editor's pen being wielded? This is a rite of passage we more normally associate with an arts critic. So when Peter and Tom asked me to write the preface for the second edition, I was again flattered, but did wonder if I could add anything to what I originally wrote.

I was literally shocked when I read my original preface—was this really written only five years ago? How memory distorts with time! The figure illustrating the publication rate, for example—was it only five years ago that we were in awe of the fact that there was a "burgeoning growth of papers in this area"—when the total for 1999 was almost as high as 120! Even the most optimistic of us could not have anticipated how this would look in 2007 (see Figure 1). Approximately two thousand papers on ionic liquids appeared in 2006 (nearly 25% originating in China), bringing the total of published papers to over 6000 (and of these, over 2000 are concerned

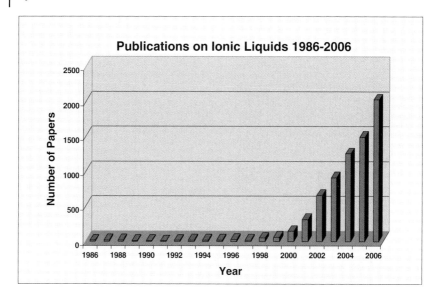

with catalysis)—-and there are also over 700 patents! That is 40 papers appearing per week—more than were being published annually a decade ago. And, on average, a review appears every two to three days. That means there is one review being published for every 20 original papers. If one assumes the garbage factor[1] runs at about 90% (a generous assumption), that means there is a review being published for every two valuable original contributions.

This is a bizarre and surreal situation, which seems more appropriate to a Kurt Vonnegut[2] novel—did buckminsterfullerene and superconductivity have the same problem? And how many papers within this annual flood of reviews say anything critical, useful, or interesting? How many add value to a list of abstracts which can be generated in five minutes using SciFinder or the ISI Web of Knowledge? How many of them can themselves be categorised as garbage? It is the twenty-first century—if a review is just an uncritical list of papers and data, what is its value?

So, am I being cynical and judgemental when I state that 90% of the published literature on ionic liquids adds little or no useful information? The PhD regulations for my University state that a satisfactory thesis must:

(1) Embody the results of research which make a distinct contribution to scholarship and afford evidence of originality as shown by the discovery of new facts, the development of new theory or insight or by the exercise of independent critical powers; and
(2) contain an acceptable amount of original work by the candidate. This work must be of a standard which could be published, either in the form of articles

---

[1] Discussed in the Preface to the First Edition.
[2] Sadly, he died in April 2007.

in appropriate refereed journals or as the basis of a book or research monograph which could meet the standards of an established academic publisher.

Well, clearly (2) is not evidence of (1); examination of the published literature undoubtedly demonstrates that "the results of research which make a distinct contribution to scholarship and afford evidence of originality as shown by the discovery of new facts, the development of new theory" is no longer a criterion for publication in refereed journals. If it was, would we find multiple publication of results from the same authors, or (frighteningly common) publication of work already published elsewhere by another, frequently uncited, group? Would papers on ionic liquids still be appearing where there is no report of the purity or water content of the ionic liquids, where claims of autocatalytic effects from the solvent appear based on reactions carried out in hexafluorophosphate or tetrafluoroborate ionic liquids (which contain HF), where physical properties are reported on impure materials, if the publications were properly refereed? I reject many of the papers which cross my (electronic) desk on these grounds when submitted to the ACS or RSC; months later I will see these papers appear, largely unchanged, in the pages of commercial journals—clearly, you can't keep a bad paper down—publish, and be damned! I have actually heard scientists say "I can't be expected to keep on top of the literature when it is appearing so rapidly." Well, sorry, yes you can—it is your job and duty as a scientist to know the published literature. It has never been easier to keep up-to-date with the literature, but finding and downloading a .pdf file is not the same as reading it!! With 2000 papers appearing in 2006 (and will anyone bet against over 2500 in 2007?), we must exercise our critical faculties to the full; we much teach our students, colleagues and collaborators to look for experimental evidence, not unsubstantiated claims. The field of ionic liquids is vibrant, fascinating, and rewarding, and offers a phenomenal opportunity for new science and technology, but we must guard, as a community, against it getting a reputation (as green chemistry has already gained) for being an area of soft publications by mediocre scientists. And the attacks and carping criticism have started; Murray, in an editorial in the ACS journal *Analytical Chemistry* [*Anal. Chem.*, **2006**, *78*, 2080], rubbished both the areas of ionic liquids and green chemistry; although he later published a mealy-mouthed, insincere apology at the end of a response from Robin Rogers and myself [*Anal. Chem.*, **2006**, *78*, 3480–3481], it is clear that this will not be the last emotive, rather than logical, attack on the field. There are hundreds of outstanding papers being published annually in this area—they must not be tarnished by the hundreds of reports of bad science.

So, having vented my spleen, how do these rhetorical comments relate to this book, which has grown dramatically in size (but, thankfully, not to a size reflecting the growth of the number of publications) since the First Edition? The number of chapters and sections in the Second Edition reflect the broadening of the applications of ionic liquids; wherever a conventional fluid can be used, the option for replacing it with an ionic liquid exists. The present chapters are written from a depth of understanding that did not exist five years ago. Today, there are over a dozen extant industrial processes; in 2002, there were none in the public domain. This has been achieved by ongoing synergistic collaborations between industry and academia, and not by the literally fantastic views expressed recently in an article entitled "Out of the Ivory Tower" [P.L. Short, *Chem. Eng. News*, **2006**, *84* (24th April)] [15–21]. The field has expanded and matured, and so has this Second Edition. The team of expert writers remains impressive—these are authors who are at the top of their field. The chapters radiate the informed writing of specialists; their wisdom is generously shared with us. The editors have performed a Herculean task in bringing this all together in a coherent and smooth account of the whole field as it stands today (although, at the current rate, the total number of papers published will rise above 10000 by 2009). If there is to be a Third Edition, and we will need one, it will have to be in two volumes! So let us hope this book is read by all practitioners of the field—by some for enjoyment, by all for insight and understanding, and by some as a bible. The field continues to expand and intrigue—by the time this book is in print, nearly one thousand more papers will have appeared—this textbook will remain the rock upon which good science will be built. To return to thoughts expressed over two thousand years ago, it will be enough *"if these words [. . .] are judged useful by those who want to understand clearly the events which happened in the past and which (human nature being what it is) will, at some time or other and in much the same ways, be repeated in the future. My work is not a piece of writing designed to meet the taste of an immediate public, but was done to last for ever."*

K.R. Seddon
April, 2007

A Note from the Editors

This book has been arranged in several chapters that have been prepared by different authors, and the reader can expect to find changes in style and emphasis as they go through it. We hope that, in choosing authors who are at the forefront of their particular specialism, this variety is a strength of the book.

In addition to the subjects covered in the first edition we have added five new chapters describing newly emerging areas of interest for ionic liquids in synthesis. The book now ranges from the most fundamental theoretical understanding of ionic liquids through to their industrial applications.

In order to cover the most important advances we allowed the book to double in length. Yet, due to the explosion of interest in the use of ionic liquids in synthesis it has not been possible to be fully comprehensive. Consequently, the book must be didactic with examples from the literature used to illustrate and explain. We hope that no offence is caused to anyone whose work has not been included. None is intended.

Naturally, a multi-authored book has a time gap between the author's submission and the publication which can be different for different contributions. However, this was the same for the first edition of this book and did not harm its acceptance.

Acknowledgements

We would like to sincerely thank everyone who has been involved in the publication of this book. All our authors have done a great job in preparing their chapters and it has been a pleasure to read their contributions as they have come in to us. We are truly grateful for them making our task so painless. We would also like to thank the production team at VCH-Wiley, particularly Dr. Elke Maase, Dr. Rainer Münz and Dr. Tim Kersebohm.

Finally in a project like this, someone must take responsibility for any errors that have crept in. Ultimately we are the editors and this responsibility is ours. So we apologize unreservedly for any mistakes that have found their way into this book.

August 2007

Peter Wasserscheid, Tom Welton

List of Contributors

Jessica L. Anderson
University of Notre Dame
Notre Dame, IN 46556
USA

Jennifer L. Anthony
University of Notre Dame
Notre Dame, IN 46556
USA

Markus Antonietti
Max Planck Institute of Colloids
and Interfaces
Research Campus Golm
14424 Potsdam
Germany

Joan F. Brennecke
University of Notre Dame
Notre Dame, IN 46556
USA

Adrian J. Carmichael
University of Warwick
Dept. of Chemistry
Coventry CV4 7AC
UK

W. Robert Carper
Wichita State University
Dept. of Biophysical and Physical
Chemistry
206 McKinley Hall
1845 Fairmount
Wichita, KS 67260-0051
USA

Cinzia Chiappe
Università di Pisa
Dipartimento di Chimica
Bioorganica e Biofarcia
Via Bonanno Pisano 33
56126 Pisa
Italy

Violina A. Cocalia
Cytec Industries Inc.
Mining Chemicals Department
1937 West Main Street
Stamford, CT 06904
USA

James H. Davis, Jr.
University of South Alabama
Dept. of Chemistry
Mobile, AL 36688-0002
USA

Mario G. Del Pópolo
Queen's University Belfast
Atomistic Simulation Centre
School of Mathematics and Physics
Belfast BT7 1NN
Northern Ireland, UK

Andreas Dölle
RWTH Aachen
Institute of Physical Chemistry
Templergraben 59
52062 Aachen
Germany

Susanne Dreyer
University of Rostock
Department of Chemistry
Albert-Einstein-Str. 3a
18059 Rostock
Germany

Martyn Earle
The Queen's University
School of Chemistry
Stransmills Rd.
Belfast BT9 5AG
Northern Ireland
UK

Marrit Eckstein
RWTH Aachen
Institute for Technical and
Macromolecular Chemistry
Worringerweg 1
52074 Aachen
Germany

Sherif Zein El Abedin
Clausthal University of Technology
Faculty of Natural & Material Sciences
Robert-Koch-Str. 42
38678 Clausthal-Zellerfeld
Germany

Frank Endres
Clausthal University of Technology
Institute of Metallurgy
Robert-Koch-Str. 42
38678 Clausthal-Zellerfeld
Germany

Frédéric Favre
Institut Francais du Pétrole
IFP Lyon
69390 Vernaison
France

Rasmus Fehrmann
Technical University of Denmark
Department of Chemistry
Building 207
2800 Kgs. Lyngby
Denmark

Charles M. Gordon
Pfizer Global Research
and Development
Ramsgate Road
Sandwich
Kent CT13 9NJ
UK

David M. Haddleton
University of Warwick
Dept. of Chemistry
Coventry CV4 7AC
UK

Chris Hardacre
Queen's University Belfast
School of Chemistry and Chemical
Engineering
Stranmillis Road
Belfast BT9 5AG
Northern Ireland
UK

List of Contributors

Claus Hilgers
Solvent Innovation GmbH
Nattermannallee 1
50829 Köln
Germany

John D. Holbrey
Queen's University of Belfast
QUILL, School of Chemistry
and Chemical Engineering
David Keir Building
Stranmillis Road
Belfast BT9 5AG
Northern Ireland
UK

Patricia A. Hunt
Imperial College of Science,
Technology and Medicine
Department of Chemistry
South Kensington
London, SW7 2AY
UK

Andreas Kirschning
University of Hannover
Institute of Organic Chemistry
Schneiderberg 1b
30167 Hannover
Germany

Sandra Klembt
University of Rostock
Dept. of Chemistry
Albert-Einstein-Str. 3a
18059 Rostock
Germany

Udo Kragl
University of Rostock
Dept. of Chemistry
Albert-Einstein-Str. 3a
18059 Rostock
Germany

Axel Leuchter
RWTH Aachen
Institut für Physikalische Chemie
Templergraben 59, Raum 304
52056 Aachen
Germany

Ruth M. Lynden-Bell
University of Cambridge
University Chemical Laboratory
Lensfield Road
Cambridge, CB2 1EW
UK

Matthias Maase
BASF AG
Global New Business Development
Chemical Intermediates for Industrial
Applications
CZ/BS – E 100
67056 Ludwigshafen
Germany

Edward J. Maginn
University of Notre Dame
Notre Dame, IN 46556
USA

Robert A. Mantz
U.S. Army Research Laboratory
2800 Powder Mill Rd
Adelphi, MD 20783-1197
USA

Mark J. Muldoon
Queen's University Belfast
School of Chemistry and Chemical
Engineering
Stranmillis Road
Belfast, BT9 5AG
Northern Ireland
UK

Hélène Olivier-Bourbigou
Institut Francais du Pétrole (IFP)
Division Cinétique et Catalyse
1 / 4 Avenue de Bous Préau
92852 Rueil-Malmaison
France

Günter Palmer
RWTH Aachen
Institut für Physikalische Chemie
Landoltweg 2
52056 Aachen
Germany

Joachim Richter
RWTH Aachen
Institut für Physikalische Chemie
Landoltweg 2
52056 Aachen
Germany

Anders Riisager
Technical University of Denmark
Department of Chemistry
Building 207
2800 Kgs. Lyngby
Denmark

Robin D. Rogers
The University of Alabama
Department of Chemistry
Box 870336
Tuscaloosa, AL 35487-0336
USA

Peter Schulz
Friedrich-Alexander-Universität
Lehrstuhl für Chemische
Reaktionstechnik
Institut für Chemie und
Bioingenieurwesen
Egerlandstr. 3
91058 Erlangen
Germany

Vasundhara Singh
University College of Engineering
Punjabi University
Reader in Chemistry
Department of Basic and Applied
Sciences
Patiala, 147002
India

Bernd Smarsly
Max Planck Institute of Colloids and
Interfaces
Research Campus Golm
14424 Potsdam
Germany

Paul C. Trulove
Centre for Green Manufacturing
Department of Chemistry
United States Naval Academy
572 Holloway Road
Annapolis, MD 21402-5026
USA

Michel Vaultier
Univ. Rennes
Groupe Rech. Physicochim. Struct.
CNRS
35042 Rennes
France

Ann E. Visser
Savannah River National Laboratory
Aiken, SC 29808
USA

Markus Wagner
Solvent Innovation GmbH
Nattermannallee 1
50829 Köln
Germany

Phillip G. Wahlbeck
Wichita State University
Department of Chemistry
Wichita, KS 67260–0051
USA

Peter Wasserscheid
Friedrich-Alexander-Universität
Lehrstuhl für Chemische
Reaktionstechnik
Institut für Chemie und
Bioingenieurwesen
Egerlandstr. 3
91058 Erlangen
Germany

Tom Welton
Imperial College of Science,
Technology and Medicine
Department of Chemistry
South Kensington
London, SW7 2AY
UK

John S. Wilkes
Department of Chemistry
U.S. Air Force Academy
2355 Fairchild Drive
Colorado 80840
USA

Yong Zhou
Max Planck Institute of
Colloids and Interfaces
Research Campus Golm
14424 Potsdam
Germany

5.3
Transition Metal Catalysis in Ionic Liquids

Peter Wasserscheid and Peter Schulz

Many transition metal complexes dissolve readily in ionic liquids, which enables their use as solvents for transition metal catalysis. Sufficient solubility for a wide range of catalyst complexes is an obvious, but not trivial, prerequisite for a versatile solvent for homogenous catalysis. Some of the other approaches to replace traditional volatile organic solvents in transition metal catalysis by "greener" alternatives, such as the use of supercritical CO_2 or perfluorinated solvents, suffer very often from low catalyst solubilities. This limitation is usually overcome by special ligand systems that have to be synthesized prior to the catalytic reaction.

In the case of ionic liquids, a special ligand design is usually not necessary to get catalyst complexes dissolved in the ionic liquid in a sufficiently high concentration. However, it should be mentioned that sometimes the dissolving of a solid, crystalline complex in an (often relatively viscous) ionic liquid can be slow. This is due to restricted mass transfer and can be speeded up by either increasing the exchange surface (ultrasonic bath) or by lowering the ionic liquid's viscosity. The latter is easily achieved by adding small amounts of a volatile organic solvent that dissolves both the catalyst complex and the ionic liquid. As soon as the solution is homogeneous the volatile solvent is then removed *in vacuo*.

Even if no special ligand design is usually required to dissolve transition metal complexes in ionic liquids, the application of an ionic or highly polar ligand is usually required to immobilize the catalyst in the ionic medium. Particularly for continuous catalytic operation, where the ionic catalyst layer is intensively extracted with the non-miscible product phase, it is important to make sure that the amount of catalyst washed from the ionic liquid is extremely low. The full immobilization of the (often quite expensive) transition metal catalyst combined with the possibility of its recycle is usually a crucial criterion for the large-scale use of homogeneous catalysis (for more details see Section 5.4).

The first example of homogeneous transition metal catalysis in an ionic liquid was the platinum catalyzed hydroformylation of ethene in tetraethylammonium trichlorostannate (mp. 78 °C), described by Parshall in 1972 (Scheme 5.3-1(a)) [1]. In 1987, Knifton reported the ruthenium- and cobalt-catalyzed hydroformylation of internal and terminal alkenes in molten [Bu_4P]Br, a salt that falls within the now accepted definition for an ionic liquid (see Scheme 5.3-1(b)) [2]. The first applications of room-temperature ionic liquids in homogeneous transition metal catalysis were described in 1990 by Chauvin et al. and by Wilkes et al.. Wilkes et al. used weakly acidic chloroaluminate melts and studied therein ethylene polymerization with Ziegler-Natta catalysts (Scheme 5.3-1(c)) [3]. Chauvin's group dissolved nickel catalysts in weakly acidic chloroaluminate melts and investigated the resulting ionic catalyst solutions for the dimerization of propene (Scheme 5.3-1(d)) [4].

The potential of ionic liquids as novel media for transition metal catalysis received a substantial boost by the work of Wilkes' group when they described, in 1992, the

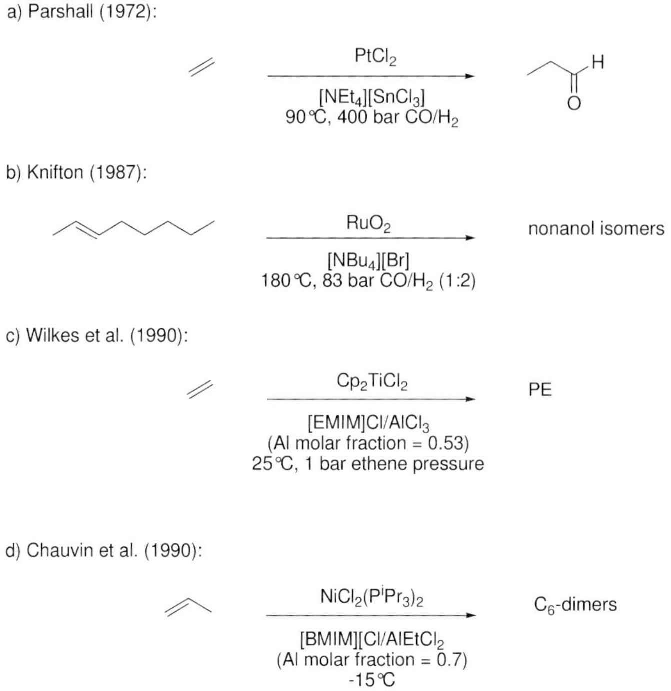

a) Parshall (1972):

$$\text{CH}_2=\text{CH}_2 \xrightarrow[\text{90 °C, 400 bar CO/H}_2]{\text{PtCl}_2,\ [\text{NEt}_4][\text{SnCl}_3]} \text{CH}_3\text{CH}_2\text{CHO}$$

b) Knifton (1987):

$$\text{1-hexene} \xrightarrow[\text{180 °C, 83 bar CO/H}_2\ (1:2)]{\text{RuO}_2,\ [\text{NBu}_4][\text{Br}]} \text{nonanol isomers}$$

c) Wilkes et al. (1990):

$$\text{CH}_2=\text{CH}_2 \xrightarrow[\substack{[\text{EMIM}]\text{Cl}/\text{AlCl}_3 \\ (\text{Al molar fraction} = 0.53) \\ 25\,°\text{C},\ 1\ \text{bar ethene pressure}}]{\text{Cp}_2\text{TiCl}_2} \text{PE}$$

d) Chauvin et al. (1990):

$$\text{propene} \xrightarrow[\substack{[\text{BMIM}][\text{Cl}/\text{AlEtCl}_2] \\ (\text{Al molar fraction} = 0.7) \\ -15\,°\text{C}}]{\text{NiCl}_2(\text{P}^i\text{Pr}_3)_2} \text{C}_6\text{-dimers}$$

Scheme 5.3-1 Early examples of transition metal catalysis in ionic liquids.

synthesis of non-chloroaluminate, room-temperature liquid systems with significantly enhanced stability to hydrolysis, e.g. low melting tetrafluoroborate melts [5]. In contrast to chloroaluminate ionic liquids, these "second generation ionic liquids" offer high tolerance to functional groups, which opens up a much larger range of applications for transition metal catalysis. The first successful catalytic reactions in ionic liquids with tetrafluoroborate ions included the rhodium catalyzed hydrogenation and hydroformylation of olefins [6]. Nowadays, tetrafluoroborate and (the slightly later published [7]) hexafluorophosphate ionic liquids are among the "work horses" for transition metal catalysis in ionic liquids. They combine – as some other ionic liquids with weakly coordinating anions – the properties of relatively polar yet non-coordinating solvents. This special combination makes them extremely suitable solvents for reactions with electrophilic catalysts [8]. Moreover, these ionic liquids are now widely available commercially [9] so that research groups and companies focusing on catalytic applications do not necessarily have to go through all of the synthetic work themselves (for the synthesis of ionic liquids and especially for the quality requirements related to their applications as solvents in homogeneous catalysis see Chapter 2).

However, we have to be aware of a number of limitations when applying tetrafluoroborate and hexafluorophosphate ionic liquids as reaction media in homogeneous catalysis. The major problem is that these anions are sensitive to hydrolysis. The tendency to anion hydrolysis is of course much less pronounced than for chloroaluminate melts (the latter immediately hydrolyze completely to HCl and Al oxides when in contact with sufficient amounts of water). Anion hydrolysis of tetrafluoroborate and hexafluorophosphate ionic liquids has been shown to occur under relatively moderate conditions [10] and well within the time-scale of an ordinary catalytic experiment (over a few hours significant decomposition can be observed in a wet system under slightly elevated temperatures), which makes these effects important for the correct interpretation of the results of even laboratory screening experiments. The hydrolytic formation of HF from the hexafluorophosphate and tetrafluoroborate anion causes the following problems with regard to their use as solvents for transition metal catalysis: (i) loss or partial loss of the ionic liquid solvent; (ii) corrosion problems related to the HF formed; (iii) deactivation of the transition metal catalyst through its irreversible complexation by F^- ions. Consequently, the application of tetrafluoroborate and hexafluorophosphate ionic liquids is effectively restricted to those applications where water-free conditions can be realized at acceptable effort. This pre-condition is certainly fulfilled in cases where the transition metal complex under investigation or the substrates used are water-sensitive themselves, so that the reaction has to be carried out under inert and water-free conditions anyway.

In 1996, Grätzel, Bonhôte and coworkers published the synthesis and properties of ionic liquids with anions containing CF_3 and other fluorinated alkyl groups [11]. These do not show the same sensitivity towards hydrolysis as $[BF_4]^-$ and $[PF_6]^-$-containing systems. In fact, heating [BMIM][$CF_3SO_2)_2N$] with excess water to 100 °C for 24 h did not reveal any hint of anion hydrolysis. The first successful catalytic experiments using these ionic liquid systems have been reported, e.g., the hydrovinylation of styrene catalyzed by a cationic nickel complex in [EMIM][$CF_3SO_2)_2N$] [12]. However, despite the very high stability of these salts to hydrolysis and a number of other very suitable properties (e.g. low viscosity, high thermal stability, easy preparation in halogen-free form due to miscibility-gap with water) the relatively high price of $[(CF_3SO_2)_2N]^-$ and related anions may be a major problem for their practical application in larger quantities (the Li salt is commercially available from both Rhodia and 3M). Moreover, the disposal of spent ionic liquids of this type, e.g., by combustion, is more complicated due to potential HF liberation.

In this context, the use of ionic liquids with halogen-free anions has become more and more popular. In 1998, Andersen et al. described the use of phosphonium tosylates (all with melting points >70 °C) in the Rh-catalyzed hydroformylation of 1-hexene [13]. More recently, much lower melting imidazolium-based alkylsulfate and oligoethersulfate ionic liquids have been prepared [14] and used as solvents in transition metal catalysis [15]. For example, [BMIM][n-$C_8H_{17}SO_4$] (mp = 35 °C) could be used as a catalyst solvent in the rhodium catalyzed hydroformylation of 1-octene. It can be anticipated that the further development of transition metal catalysis in ionic liquids will be driven to a significant extent by the availability of new ionic liquids with different anion systems. In particular, cheap, halogen-free systems that combine weak coordination to electrophilic metal centers and low viscosity with high stability to hydrolysis are highly desirable.

Transition metal catalysis in ionic liquids is a field of great research activity (in particular over the last five years) which has also generated an extensive reviewing practice. Latest examples originated from Welton [16], MacFarlane [17] and Pozzi [18] and compiled the development of the research field in a more or less comprehensive way, though with slightly different emphasis from their specific, different viewpoints. These three up-dated earlier published reviews by Dupont [19], Zhao [20], Haag [21], Dobbs [22], Olivier-Bourbigou [23], Sheldon [24], Gordon [25], Wasserscheid [26], Welton [27] and Seddon [28] on the same topic.

In contrast to most of these reviews, the present section will have a clearly instructive character. It aims to derive general principles from the work published so far in order to give the reader a useful overall understanding of the achievements and remaining challenges in transition metal catalysis using ionic liquids. The chapter will be structured in two major parts. Section 5.3.1 will present general and conceptual aspects, motivations, successful strategies and limiting factors. The latter could not yet be overcome or have been found out in the meanwhile to be intrinsic to the concept of performing transition metal catalysis in ionic liquids.

In Section 5.3.2 we will review selected transition metal catalyzed reactions with the clear aim to further illustrate the principles derived in Section 5.3.1. For this selection a number of criteria were considered: (i) the general and technical relevance of the transition metal catalyzed reaction; (ii) the degree of understanding concerning the ionic liquid's role; and finally (iii) the potential to transfer key results to future, new applications. The (obviously somewhat subjective) selection includes transition metal catalyzed hydrogenation, oxidation, hydroformylation, Pd-catalyzed C–C-coupling and dimerization/oligomerization reactions as well as catalysis involving transition metal nanoparticles. Regarding these selected applications our chapter aims for a sound overview, however comprehensiveness is not guaranteed. The missing work has been left out to avoid information overload. Thus, we would like to apologize at this point for having omitted also a number of sound and important papers for the reason of improved clarity. We do appreciate very much all contributions to advance ionic liquid research with respect to transition metal catalysis!

With the dual approach of Sections 5.3.1 and 5.3.2 we would like to share our experience in particular with readers working in the field of transition metal catalysis, but so far inexperienced in using ionic liquids. Our section aims to provide some guidelines and is intended to encourage these scientists to test and develop ionic liquids as parts of a "tool box" for their future research.

5.3.1
Concepts, Successful Strategies, and Limiting Factors

5.3.1.1 Why Use Ionic Liquids as Solvents for Transition Metal Catalysis?

The non-volatile nature
Probably the most prominent property of an ionic liquid is its extremely low vapor pressure [29]. Transition metal catalysis in ionic liquids can particularly benefit from this on economic, environmental and safety grounds.

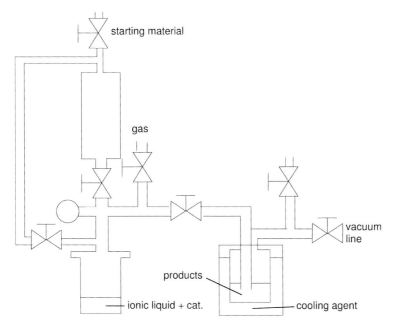

Fig. 5.3-1 Stabilization of an active rhodium catalyst by addition of the ionic liquid [BMIM][PF$_6$] as co-solvent during distillative product isolation – apparatus for distillative product isolation from the ionic catalyst layer.

Obviously, the use of a non-volatile ionic liquid simplifies the distillative work-up of volatile products, especially in comparison to the use of low boiling solvents, where this may save the distillation of the solvent during product isolation. Moreover, common problems related to the formation of azeotropic mixtures of a volatile solvent and the products/by-products formed, are avoided by using a non-volatile ionic liquid. In the Rh-catalyzed hydroformylation of 3-pentenoic acid methyl ester it was even found that the ionic liquid added could stabilize the homogeneous catalyst during the thermal stress of product distillation (Fig. 5.3-1) [30]. This option may be especially attractive technically, due to the fact that the stabilizing effects could already be observed with quite low amounts of ionic liquid added.

As in stoichiometric organic reactions, the application of non-volatile ionic liquids can contribute to the reduction of atmospheric pollution. This is of special relevance for non-continuous reactions where complete recovery of a volatile organic solvent is usually difficult to integrate into the process.

As well as this quite obvious environmental aspect, the switch from a volatile, flammable, organic solvent to an ionic liquid may significantly improve the safety of a given process. This is especially true for oxidation reactions where air or pure oxygen are used as oxidants. Here, the use of common organic solvents is often restricted by the potential formation of explosive mixtures between oxygen and the volatile organic solvent in the gas phase. Therefore, it is anticipated that there

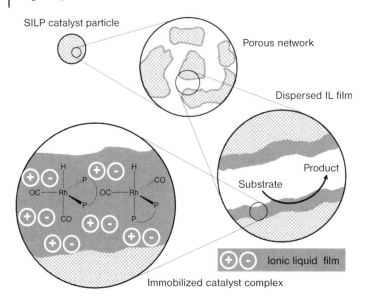

Fig. 5.3-2 Schematic representation of a supported ionic liquid phase (SILP) catalyst exemplified for a typical rhodium hydroformylation catalyst.

will be some technical applications for ionic liquids in the future where solely the advantage of their non-volatile character is used for mainly safety reasons.

Very recently, a new concept to process transition metal catalysis in ionic liquids has been introduced that makes explicit use of the extremely low volatility of ionic liquids. It comprises the immobilization of a transition metal catalyst in an ionic liquid. The resulting solution is confined on the surface of a highly porous solid by various methods such as, e.g. physisorption, tethering, or covalent anchoring of ionic liquid fragments [31]. This preparation results in a solid that contains, in a supported liquid film, a homogeneously dissolved transition metal complex. During catalytic reactions using such a supported ionic liquid phase (SILP) catalyst the feedstock molecules diffuse through the residual pore space of the catalyst, dissolve in the ionic liquid catalyst phase, and react at dissolved transition metal complexes within the thin liquid catalyst film dispersed on the walls of the pores in the support material. The products then diffuse back out of the catalyst phase into the void pore space and further out of the catalyst particle (see Fig. 5.3-2).

The reactants in SILP catalysis are preferentially processed in gaseous form. Processing of solid SILP catalysts in a liquid reaction phase as a slurry requires extremely low solubility of the ionic liquid film in the liquid reaction mixture and affords special constraints upon the mechanical stability of the liquid film. In contrast, for the – by far more attractive – gas-phase applications of SILP catalysts the extremely low volatility of the ionic liquids is the key success factor. It is noteworthy that earlier attempts to apply supported liquid catalysts in continuous gas phase reactions – using organic liquid phases [32] or water [33] as the immobilized liquid phase – resulted in catalyst systems too unstable for technical use due to evaporation of the liquid film with time. In contrast, SILP systems have been shown to be stable in activity and selectivity for several hundreds hours on stream [34]. The excellent performance of SILP catalysts has been demonstrated for hydroformylations [35] (Rh-catalyzed), hydrogenation [36] (Rh-catalyzed), Heck reactions [37] (Pd-catalyzed), hydroaminations [38] (Rh-, Pd- and Zn-catalyzed) and methanol carbonylation [39] (Rh-catalyzed). The concept of using a supported ionic liquid catalyst combines the most attractive features of homogeneous catalysis (such as high specificity and dispersion of the catalyst) with the most attractive features of heterogeneous catalysts (such as large interfacial reaction area and ease of separation of the products from the catalyst). Finally, SILP catalysts allow the application of conventional fixed-bed reactor process technology for homogeneous catalysis. Combined, these features have made the methodology very attractive for future technical applications (for more details about SILP catalysis see Section 5.6).

New opportunities for biphasic catalysis

In comparison to heterogeneous catalyzed reactions, homogeneous catalysis offers several important advantages. The catalyst complex is usually well defined and can be rationally optimized by ligand modification. Every metal center can be active in the reaction. The reaction conditions are usually much milder (T usually <200 °C) and selectivities are often much higher than with heterogeneous catalysts.

Notwithstanding these advantages, the proportion of homogeneous catalyzed reactions in industrial chemistry is still quite low. The main reason for this is the difficulty in separating the homogeneously dissolved catalyst from the products and by-products after the reaction. Since the transition metal complexes used in homogeneous catalysis are usually quite expensive, complete catalyst recovery is crucial in a commercial situation.

Biphasic catalysis in a liquid–liquid system is another powerful approach to combine the advantages of both homogeneous and heterogeneous catalysis. The reaction mixture consists of two immiscible solvents. Only one phase contains the catalyst, allowing easy product separation by simple decantation. The catalyst phase can be recycled without any further treatment. However, the right combination of catalyst, catalyst solvent and product is crucial for the success of biphasic catalysis [40]. The catalyst solvent has to provide excellent solubility for the catalyst complex without competing with the substrate for the free coordination sites at the catalytic center.

Even more attractive is the possibility of optimizing the reaction's activity and selectivity by means of a biphasic reaction mode. This can be realized by *in situ* extraction of catalyst poisons or reaction intermediates from the catalytic layer. However, to benefit from this potential, even more stringent requirements have to be fulfilled by the catalyst solvent since the latter has to provide a specific, very low solubility for the substances that are to be extracted from the catalyst phase under the reaction conditions. Figure 5.3-3 demonstrates this concept, exemplified for an oligomerization reaction. The dimer selectivity of the oligomerization of compound A can be enhanced significantly if the reaction is carried out in biphasic mode using a catalyst solvent with high preferential solubility for A. The A–A formed is then readily extracted from the catalyst phase into the product layer, which reduces the chance for the formation of higher oligomers A–A–A and A–A–A–A.

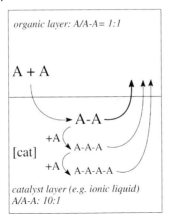

Fig. 5.3-3 Enhanced dimer selectivity in the oligomerization of compound A due to a biphasic reaction mode with a catalyst solvent of high preferential solubility for A.

Keeping all these circumstances in mind, it becomes understandable why the use of traditional solvents for biphasic catalysis (e.g., water or butanediol) has only been able to fulfill this potential in a few specific examples [41]. Whereas, this type of highly specialized liquid–liquid biphasic operation is an ideal field for the application of ionic liquids, mainly due to their exactly tuneable physicochemical properties (see Section 3.3 for more details). Very recently it has been demonstrated that the solubility and miscibility properties of ionic liquids can be varied so widely that even mutually immiscible ionic liquids can be realized [42]. However, applications of these ionic liquid–ionic liquid biphasic systems in catalysis have not yet been described.

Reviewing the so far published literature about transition metal catalysis in ionic liquids it becomes quite clear that by far the greatest part of the work has been carried out in the form of liquid–liquid biphasic catalysis. There are three convincing arguments for the choice of this reaction mode.

1 In this way the catalyst complex immobilized in the ionic liquid can be easily recycled.
2 Liquid–liquid biphasic catalysis is a very efficient way to reuse the relatively expensive ionic liquid itself. Thus – for a commercial application – the ionic liquid may be seen as an investment for the process (in an ideal case) or at least as a "working solution", which means that only a small amount has to be replaced after a certain time of application. In fact, for any kind of recycling process the ionic catalyst solution (transition metal complex + ionic liquid) has to be regarded as an entity. With regards to this aspect liquid–liquid biphasic catalysis with ionic liquids is better regarded as a kind of heterogeneous catalysis using a liquid, non-volatile support material which is a fundamentally different approach compared to conventional homogeneous catalysis in an organic solvent.

3 Most catalytic applications using ionic liquids aim to realize beneficial effects with the lowest possible amount of ionic liquid present in the system. A fully soluble ionic liquid would thus lead to the situation where a very small amount of ionic liquid would be dissolved in a large volume of reaction mixture. In such a scenario the ions of the ionic liquid would be present as either fully dissociated and solvated cations/anions or as solvated ion-pairs depending on the polarity of the reactants. But can this still be regarded as being an ionic liquid? In contrast, an immiscible ionic catalyst solution retains its specific ionic liquid character even if the immiscible, organic reactant/product phase is used in large volumetric excess.

Due to the great importance of liquid–liquid biphasic catalysis for ionic liquids, Section 5.4 will be dedicated to specific aspects related to this mode of reaction, with special emphasis on practical, technical and engineering needs. Section 5.7 will summarize a very interesting recent development for biphasic catalysis with ionic liquids, namely the use of ionic liquid–compressed CO_2 biphasic mixtures in transition metal catalysis.

Activation of a transition metal catalyst in ionic liquids
Apart from the activation of a biphasic reaction by extraction of catalyst poisons, described above, the ionic liquid solvent can activate homogeneously dissolved transition metal complexes by chemical interaction.

In general, it is possible for a chemical interaction between the ionic liquid solvent and a dissolved transition metal complex to be either activating or deactivating. Therefore, understanding these chemical interactions is crucial to benefit from this potential and to avoid deactivation. Everything is traced back to the rather obvious question of how much the presence of a specific ionic liquid influences the electronic and steric properties of the active catalyst complex and – perhaps even more importantly – to what extent the ionic liquid influences the availability of free coordination sites at the catalytic center for the substrates that are supposed to undergo the catalysis. Clearly, an exact knowledge of both the catalytic mechanism in common organic solvents and the chemical properties of the ionic liquid is very helpful in understanding these different effects.

In the following section, the nature of chemical interactions between an ionic liquid and a transition metal catalyst is systematically developed according to the role of the ionic liquid in the different systems.

5.3.1.2 The Role of the Ionic Liquid
Depending on the coordinative properties of the anion and on the degree of the cation's reactivity, the ionic liquid can be regarded as "innocent" solvent, as ligand (or ligand precursor), as co-catalyst or as the catalyst itself.

The ionic liquid as "innocent" solvent
Ionic liquids with weakly coordinating, inert anions (e.g., $[(CF_3SO_2)_2N]^-$, $[BF_4]^-$, $[SbF_6]^-$ or $[PF_6]^-$ under anhydrous conditions) and inert cations (cations that do

not coordinate to the catalyst themselves and that do not form species under the reaction conditions that coordinate to the catalyst) can be considered as more or less "innocent" solvents in transition metal catalysis. In these cases, the role of the ionic liquid is to provide a more or less polar, more or less weakly coordinating medium for the transition metal catalyst that additionally offers special solubility properties for feedstock and products.

However, the chemical inertness of these "innocent" ionic liquids does not mean that the reactivity of a transition metal catalyst dissolved in the ionic liquid is necessarily equal to the reactivity observed in common organic solvents. This becomes understandable from the fact that many organic solvents are applied in catalytic reactions that do not behave as innocent solvents, but show significant coordination to the catalytic center. The reason why these solvents are nevertheless used in catalysis is that some polar or ionic catalyst complexes are not soluble enough in weakly coordinating organic solvents. For example, many cationic transition metal complexes are known to be excellent oligomerization catalysts [43]. However, their usually poor solubility in non-polar solvents often requires a compromise between the solvation and the coordination properties of the solvent, if organic solvents are used. In order to achieve sufficient solubility of the metal complex a solvent of higher polarity is required, that may compete with the substrate for the coordination sites at the catalytic center. Consequently, in these cases, the use of an inert, weekly coordinating ionic liquid (such as a hexafluorophosphate ionic liquid) can result in a clear enhancement of catalytic activity, since these ionic liquids are known to combine high solvation power for polar catalyst complexes (polarity) with weak coordination (nucleophilicity) [44]. It is this combination of properties of the ionic liquids that cannot be realized with water or common organic solvents.

Despite the great potential of these "weakly coordinating" ionic liquids in combination with highly electrophilic, polar or ionic catalyst complexes, it should be noted that the anions of an ionic liquid are more likely to coordinate to the metal center of a dissolved complex than the same anions dissolved in a molecular solvent would do [16]. This point becomes understandable considering a cationic metal center (coordinatively unsaturated with vacant orbitals) dissolved in a molecular solvent. The possibilities are (i) an anion can directly coordinate to the metal center, (ii) the anion and the cation may be well separated in the solvent, with the cation coordinated to molecules of the solvent, or the cation and anion can exist as either (iii) an intimate, or (iv) a solvent separated ion-pair. In a pure ionic liquid there are no molecules available to separate the ions and the cations of the ionic liquid will be repelled by the charge on the metal complex. Hence, only options (i) and (iii) are available. Thus the cationic center will always be closely associated with the anions of the ionic liquid and this interaction will greatly determine its reactivity.

Ionic liquid as solvent and co-catalyst
Ionic liquids formed by the reaction of a halide salt with a Lewis acid (e.g. chloroaluminate or chlorostannate melts) generally act as both solvent and co-catalyst in

transition metal catalysis. The reason for this is that the Lewis acidity or basicity, which is always present (at least latently), results in strong interactions with the catalyst complex. In some cases, the Lewis acidity of an ionic liquid is used to convert the neutral catalyst precursor into the corresponding cationic, catalytically active form. The activation of Cp_2TiCl_2 [45] and $(ligand)_2NiCl_2$ [46] in acidic chloroaluminate melts and the activation of $(PR_3)_2PtCl_2$ in chlorostannate melts [47] are examples of this kind of activation [Eqs. (5.3-1)–(5.3-3)].

$$Cp_2TiCl_2 + [cation][Al_2Cl_7] \rightleftharpoons [Cp_2TiCl][AlCl_4] + [cation][AlCl_4] \qquad (5.3\text{-}1)$$

$$\begin{aligned}&(ligand)_2NiCl_2 + [cation][Al_2Cl_7] + [cation][Al_2EtCl_6] \rightleftharpoons \\ &[(ligand)Ni-CH_2-CH_3][AlCl_4] + 2[cation][AlCl_4] \\ &+ AlCl_3\text{–ligand}\end{aligned} \qquad (5.3\text{-}2)$$

$$\begin{aligned}(PR_3)_2PtCl_2 + [cation][Sn_2Cl_5] &\rightleftharpoons [(PR_3)_2PtCl][SnCl_3] \\ &+ [cation][SnCl_3]\end{aligned} \qquad (5.3\text{-}3)$$

In those cases where the ionic liquid is not directly involved in creating the active catalytic species, a co-catalytic interaction between the ionic liquid solvent and the dissolved transition metal complex may still take place and can result in significant catalyst activation. When a catalyst complex is, for example, dissolved in a slightly acidic ionic liquid some electron-rich parts of the complex (e.g., lone pairs of electrons in the ligand) will interact with the ionic liquid in a way that will usually result in a lower electron density at the catalytic center. If this higher electrophilicity of the catalytic center results in a higher catalytic activity (e.g., as in oligomerization reactions of most olefins) then there is a very good chance of activating the catalyst system in a slightly acidic ionic liquid. In fact, this is the reason why many Ni-catalyzed oligomerization reactions of propene and butene have been carried out very successfully in slightly acidic or buffered chloroaluminate ionic liquids (see Section 5.3.2.5 for more details). It is noteworthy, that this type of co-catalytic influence is well-known in heterogeneous catalysis where, for some reactions, an acidic support activates the dispersed metal catalyst better than do neutral supports. Such comparison justifies regarding acidic ionic liquids in transition metal catalysis as a kind of liquid acidic support for the metal complex dissolved therein.

As one would expect, in those cases where the ionic liquid acts as a co-catalyst, the nature of the ionic liquid becomes very important for the reactivity of the transition metal complex. The chance to optimize the ionic reaction medium by variation of the halide salt, the Lewis acid and the ratio of the two components forming the ionic liquid, opens up enormous potential for optimization. However, the choice of these parameters may be restricted by some possible incompatibilities with the feedstock used. Undesired side reactions caused by the Lewis acidity of the ionic liquid or strong interaction between the Lewis acidic ionic liquid and, for example, some oxygen functionalities in the substrate have to be considered.

Ionic liquid as solvent and ligand/ligand precursor

Both cation and anion of an ionic liquid can act as a ligand or ligand precursor for a transition metal complex dissolved in the ionic liquid.

Anions of the ionic liquid may act, to some degree, as ligands if the catalytic center is cationic, depending on their coordination strength. Indeed, it has been clearly demonstrated that the anion of a cationic transition metal complex is replaced to a large extent by the ionic liquid's anion if they are different [12]. While most of the ionic liquid anions used in catalysis are chosen to interact as weakly as possible with the catalytic center, this situation may change dramatically if the ionic liquid's anion undergoes decomposition reactions. If the hexafluorophosphate anion of an ionic liquid hydrolyses in contact with water, for example, strongly coordinating fluoride ions are liberated that will act as strong ligand and catalyst poison to many transition metal complexes.

With respect to the ionic liquid's cation the situation is quite different, since catalytic reactions with anionic transition metal complexes are not yet very common in ionic liquids. However, the 1,3-dialkylimidazolium cation can act as a ligand precursor for the dissolved transition metal. Its transformation under the reaction conditions into a ligand has been observed in three different ways: (i) formation of metal carbene complexes by oxidative addition of the imidazolium cation; (ii) formation of metal–carbene complexes by deprotonation followed by coordination of the imidazolylidene on the metal center; (iii) dealkylation of the imidazolium cation and formation of a metal imidazole complex. These different ways are displayed in a general form in Scheme 5.3-2.

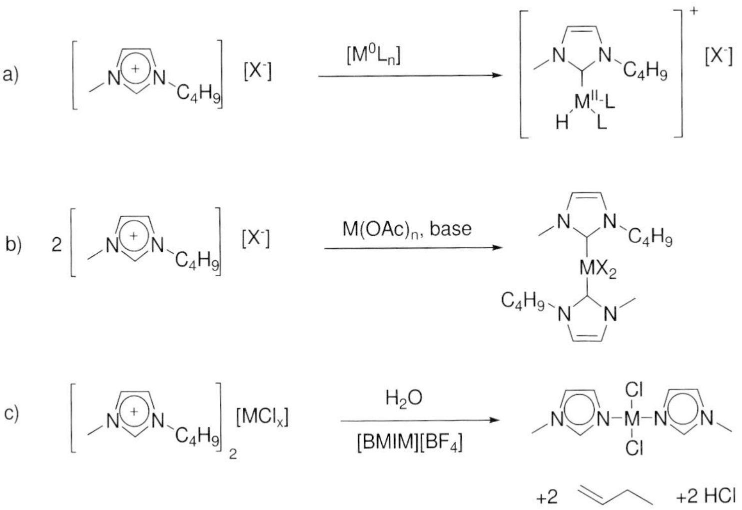

Scheme 5.3-2 Different routes for an *in situ* ligand formation from the ionic liquid's 1,3-dialkylimidazolium cation.

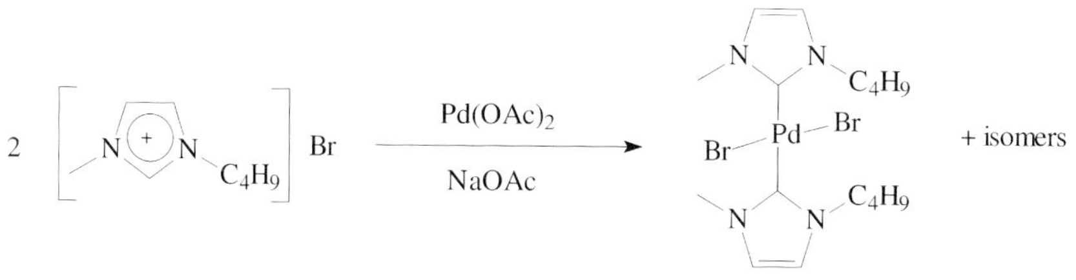

Scheme 5.3-3 Formation of a Pd–carbene complex by deprotonation of the imididazolium cation.

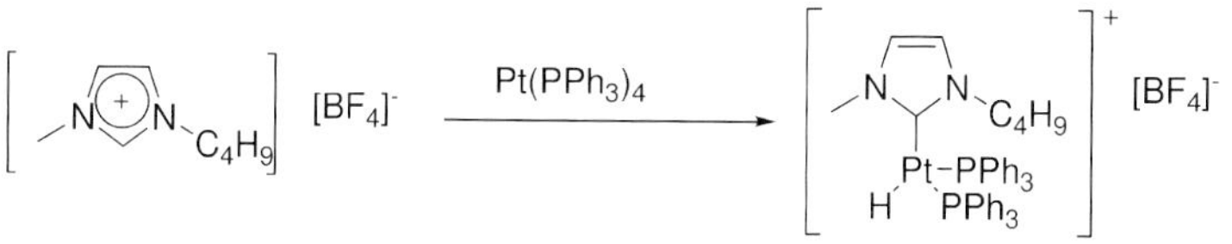

Scheme 5.3-4 Formation of a Pt–carbene complex by oxidative addition of the imidazolium cation.

The second reaction pathway ((b) in Scheme 5.3-2) is based on the well-known, relatively high acidity of the H atom in the 2-position of an 1,3-dialkylimidazolium ion [48]. The latter can be deprotonated (e.g. by basic ligands of the metal complex or by basic reactants) to form a metal–carbene complex. Xiao and coworkers demonstrated that a Pd imidazolylidene complex is formed when Pd(OAc)$_2$ was heated in the presence of [BMIM]Br [49]. The isolated Pd–carbene complex was found to be active and stable in Heck coupling reactions (for more details see Section 5.3.2.4). Welton et al. were later able to characterize an isolated Pd-carbene complex obtained in this way by X-ray spectroscopy [50]. The reaction pathway to form the complex is displayed in Scheme 5.3-3.

However, the formation of the metal–carbene complex was not observed in pure, halide-free [BMIM][BF$_4$], indicating that the formation of carbene depends on the nucleophilicity of the ionic liquid's anion. To avoid the formation of metal–carbene complexes by deprotonation of the imidazolium cation under basic conditions the use of 2-methyl-substituted imidazolium is frequently suggested. However, it should be mentioned here that strong bases can also abstract a proton to form the vinyl imidazolidene species which may also act as a strong ligand to electrophilic metal centers.

Another method of *in situ* metal–carbene complex formation in an ionic liquid is the direct oxidative addition of the imidazolium cation on a metal center in a low oxidation state (see Scheme 5.3-2(a)). Cavell and coworkers have observed oxidative addition by heating 1,3-dimethylimidazolium tetrafluoroborate with Pt(PPh$_3$)$_4$ in refluxing THF [51]. The Pt–carbene complex formed can decompose by reductive elimination. Winterton et al. have also described the formation of a Pt–carbene complex by oxidative addition of the [EMIM]$^+$ cation to PtCl$_2$ in a basic [EMIM]Cl/AlCl$_3$ (free Cl$^-$ ions present) under ethylene pressure [52]. The formation of a Pt–carbene complex by oxidative addition of the imidazolium cation is displayed in Scheme 5.3-4.

Not only platinum forms carbene complexes by oxidative addition of 1,3-dialkylimidazolium salts. Cavell and coworkers also reported the formation of stable carbene complexes of nickel and palladium by reaction with imidazolium ionic liquids [53]. Even in cases where the imidazolium was protected with a methyl group in the 2-position of the imidazolium ring, carbene formation has been observed in the 4- or 5-position in some cases [54].

In the light of these results, it becomes important to question whether a particular catalytic result obtained in a transition metal catalyzed reaction in an imidazolium ionic liquid is in fact caused by a metal–carbene complex formed *in situ*. The following simple experiments can help to check this point out in more detail: (i) variation of ligands in the catalytic system; (ii) application of independently prepared, defined metal–carbene complexes; (iii) investigation of the reaction in pyridinium, tetraalkylammonium, tetraalkylphosphonium or 1,2,3-trialkylimidazolium-based ionic liquids. If the reaction shows significant sensitivity to the use of different ligands, if the application of the independently prepared, defined metal–carbene complex shows a different reactivity than the catalytic system under investigation or if the catalytic result in the ionic liquid with non-carbene-forming cations is similar to the catalytic result in the 1,3-dialkylimidazolium system, then a significant influence of a metal–carbene complex formed *in situ* is unlikely. Of course, even then, the *in situ* formation of a metal–carbene complex cannot be totally excluded, but its lifetime may be very short so that there is not a significant influence on the catalysis.

It is interesting to note that the *in situ* formation of transition metal–carbene complexes has been used to explain both increased catalyst activity and stability as well as catalyst instability and deactivation. While Basset and coworkers attributed the complete deactivation of their Pd catalyst for the telomerization of butadiene and methanol in 1,3-dialkylimidazolium ionic liquids to the formation of highly stable Pd-imidazolylidene from palladium(II) acetate [55], Welton and coworkers ascribed the high activity of their Pd catalyst in the Suzuki cross-coupling reaction to the *in situ* formation of a mixed phosphine/imidazolylidene complex [56]. Both authors may well be right with their respective interpretations. The key to understanding the different behavior is probably the question of whether the ionic liquid as the source of the ligand (which as the solvent is naturally in very large excess vs. the metal complex) still allows the formation of free coordination sites at the active catalytic center. One can speculate that in a mixed phosphine/imidazolylidene system the phosphine ligand sterically prevents complete "carbene saturation" of the palladium center while in the telomerization system the butadiene substrate cannot compete with the carbene ligands for the coordination sites.

Finally, a third way of ligand formation from a 1,3-dialkylimidazolium cation has been described by Dupont and coworkers [57]. They investigated the hydrodimerization/telomerization of 1,3-butadiene with palladium(II) compounds in [BMIM][BF$_4$] and described the activation of the catalyst precursor complex [BMIM]$_2$[PdCl$_4$] via a palladium(IV) compound, which is formed by oxidative addition of the imidazolium with cleavage of the C–N bond of the [BMIM]$^+$ ion leading to dichloro-bis(methylimidazole)-palladium (Scheme 5.3-5). However, this reaction was only observed in the presence of water.

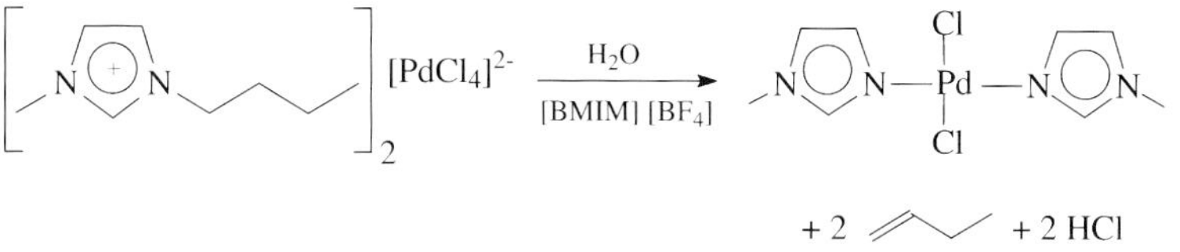

Scheme 5.3-5 Formation of the active Pd catalyst from [BMIM]$_2$[PdCl$_4$] for the hydrodimerization of 1,3-butadiene.

Ionic liquid as solvent and transition metal catalyst
Acidic chloroaluminate ionic liquids have already been described in Section 5.2 as both solvents and catalysts for reactions conventionally catalyzed by AlCl$_3$, e.g. catalytic Friedel-Crafts alkylation [58] or stoichiometric Friedel-Crafts acylation [59]. Similarly, Lewis-acidic transition metal complexes can form complex anions by reaction with organic halide salts. Seddon and coworkers patented, for example, a Friedel-Crafts acylation process based on an acidic chloroferrate ionic liquid catalyst [60].

However, ionic liquids acting as transition metal catalysts are not necessarily based on classical Lewis acids. Dyson et al. published, for example, the ionic liquid [BMIM][Co(CO)$_4$] [61]. The system was obtained as an intense blue–green colored liquid by metathesis reaction of [BMIM]Cl and Na[Co(CO)$_4$]. The liquid was used as a catalyst in the debromination of 2-bromoketones to their corresponding ketones. An ionic liquid with the complex rhodium-carbonyl anion [Rh(CO)$_2$I$_2$]$^-$ was later published by the same authors [62].

Examples in which the cation of the ionic liquid contains the transition metal complex for catalysis have also been published. For example, Forbes and coworkers [63] synthesized a Rh-containing ionic liquid cation by replacing an acetate ligand at the Rh center by a carboxylic acid functionalized imidazolium moiety. The modified dirhodium(II) dimer of this kind was applied as an effective catalyst in the intermolecular cyclopropanation reaction of styrene using ethyl diazoacetate.

In general, the incorporation of the active transition metal catalyst into the cation or anion of an ionic liquid appears to be an attractive concept for applications where a high catalyst concentration is needed. However, the physicochemical properties (in particular melting point and viscosity) of such ionic liquids may be unfavorable in many cases if such an ionic liquid is used in neat form.

5.3.1.3 Methods for Analysis of Transition Metal Catalysts in Ionic Liquids

Earlier in this section we stated that in many respects transition metal catalysis in ionic liquids is better regarded as heterogeneous catalysis on a liquid support than as conventional homogeneous catalysis in an alternative solvent. As in heterogeneous catalysis, support–catalyst interactions are known in ionic liquids and can lead to catalyst activation (see Section 5.3.1.2 for more details). Product separation from an ionic catalyst layer is often easy (at least if the products are not too polar and have a significant vapor pressure) as in classical heterogeneous catalysis. However, mass transfer limitation problems (when the chemical kinetics are fast) and some uncertainty concerning the exact micro-environment around the catalytically active center represent common characteristics for transition metal catalysis in ionic liquids and in heterogeneous catalysis.

However, despite all these similarities we should not forget that it is possible to optimize the catalytically active center in ionic liquids by molecular synthesis and ligand design, which is of course much easier and more efficient than to synthesize highly defined solid surfaces. Another very important difference – that should in particular be treated in this sub-section – is the possibility to analyze the active catalyst in an ionic liquid in a much easier and often more insightful way than this is possible for the surfaces of heterogeneous catalysts. In principle, this important advantage should enable more rational catalyst development and thus much quicker catalyst optimization times.

Unfortunately, most attempts to characterize catalytically active transition metal complexes in ionic liquids have been based on product analysis. There is nothing wrong with the argument that a catalyst is more active because it produces more of the product. However, this is not the type of explanation that can help to develop a more general understanding of what happens to a transition metal complex under catalytic conditions in a certain ionic liquid.

Applying spectroscopic methods to the analysis of active catalysts in ionic liquids is not an easy task. Three aspects illustrate this: (i) As for catalysis in conventional media, the lifetime of the catalytically active species will be very short, which makes it difficult to observe. (ii) In a realistic catalytic scenario the concentration of the catalyst in the ionic liquid will be very low. (iii) The presence and concentration of the substrate will influence the catalyst/ionic liquid interaction. These three aspects alone clearly show that an ionic liquid/substrate/catalyst system is quite complex and may not be easy to study by spectroscopic methods.

In the eighties and early nineties a great deal of effort was made to study transition metal complexes in chloroaluminate ionic liquids (see Section 6.1 for some examples). The investigations at this time generally started with electrochemical studies [64] but included also spectroscopic and complex chemistry experiments [65]. With the development of the first catalytic reactions in ionic liquids, the general research focus turned away from the basic studies of metal complexes dissolved in ionic liquids. Nevertheless, the relatively small number of papers published in the last few years that deal with spectroscopic investigations of transition metal catalysis in ionic liquids clearly demonstrate the value of this kind of analytic work for the overall development of this field.

To underline this point, we aim to present in the following sub-sections a few examples of such successful analytic approaches structured by the different methodologies that were used.

NMR spectroscopy

NMR spectroscopy is a routine method in all synthetic laboratories. *In situ* NMR spectroscopy is, therefore, a natural first choice when it comes to the characterization of catalysts dissolved in ionic liquids. However, this method suffers in particular from the low concentration of the catalyst in the ionic liquid. Moreover, ^{1}H- and ^{13}C-NMR spectroscopic investigations are difficult, since the intense signals of the ionic liquid often make a clear detection of the dissolved catalyst difficult.

Several papers have been published on improving the sensitivity of NMR spectroscopy with regard to the detection of solutes in ionic liquids. Giernoth's group suggested to use the fluorine nuclei of anions like [BF_4]$^-$ or [$(CF_3SO_2)_2$N]$^-$ as an internal ^{19}F lock signal rather then to lock via the deuterium channel using a coaxial tube insert, as the latter practice makes the lock level often a bad indicator for the shim quality [66]. Hardacre and coworkers described the synthesis and application of fully deuterated ionic liquids [67]. An alternative, transition-metal free ring deuteration of imidazolium ionic liquid cations was recently presented by Giernoth [68]. The same author also reported a remarkable method to separate ionic liquid signals from reactant (and potentially also catalyst) signals by making use of the different mobilities of these species in a gradient NMR experiment [69].

Deuteration of the ionic liquid's cation has not only been applied to obtain proton-free ionic liquids for ^{1}H-NMR experiments but also as direct probe for the reactivity of 1,3-dialkylimidazolium based ionic liquids vs. Ir(0) nanoclusters [70]. After addition of D_2 to [BMIM][$(CF_3SO_2)_2$N] in the presence of an Ir(0) nanocluster Finke and coworkers found deuterium incorporation at the 2-H, 4-H, 5-H and 8-H positions of the imidazolium cation while the control experiment in the absence of the Ir(0) cluster showed no D-incorporation. The authors concluded from their ^{2}H-NMR experiments that a sequence of N-heterocylic carbene formation by oxidative addition (see Section 3.1.2 for more details) of the imidazolium cation, H/D scrambling atop the nanocluster surface, followed by the reductive elimination of a C–D bond takes place. From careful kinetic investigations the authors concluded that the coordinatively unsaturated nanocluster surface acts indeed as the true catalyst for the H/D-exchange reaction.

An example of using deuterated reactants for detailed kinetic studies of transition metal catalyzed reactions in ionic liquids was contributed by Abu-Omar and coworkers. They studied the Rh-catalyzed epoxidation of olefins at ambient temperatures using [D_8]styrene and [D_{10}]-cyclohexene [71]. They also applied ^{2}H-NMR experiments of [D_3]-diperoxorhenium, formed *in situ* by reaction of [D_3]-methyltrioxorhenium and urea hydrogen peroxide (UHP) to determine rate constants in single turnover experiments.

An example of the selective introduction of a deuterium probe into the ligand of a transition metal catalyst (followed by *in situ* ^{2}H-NMR spectroscopy) was earlier applied to explain the activation of the square planar Ni-complex (η-4-cycloocten-1-yl)(1,5-diphenyl-2,4-pentandionato-O,O')nickel in slightly acidic chloroaluminate ionic liquids [72]. The D-labeled ligand was prepared according to Scheme 5.3-6 by reacting 1,5-diphenyl-2,4-pentadione with NaH followed by hydrolysis with D_2O. The deuterated ligand was dried and reacted with bis(1,5-cyclooctadiene)nickel [Ni(COD)$_2$].

The ^{2}H-NMR spectra of the deuterated complex obtained in CH_2Cl_2 and in [BMIM]Cl–$AlCl_3$ (1:1.2) are displayed in Fig. 5.3-4.

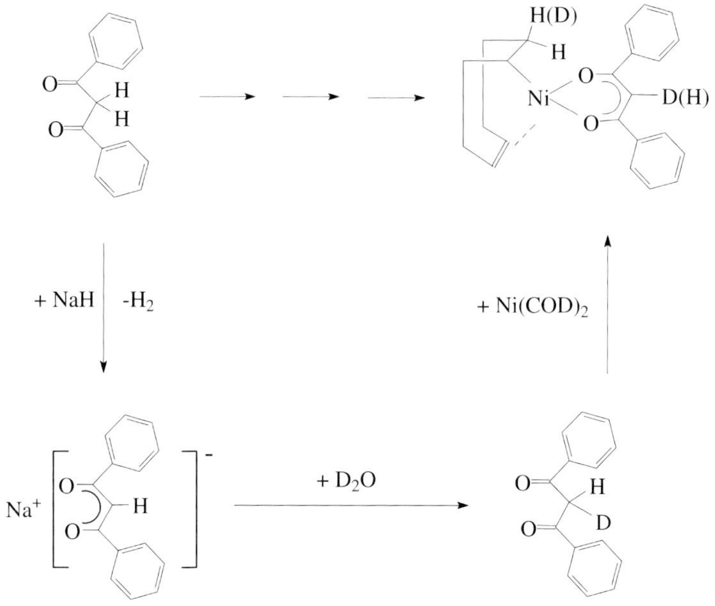

Scheme 5.3-6 Synthesis of a deuterated analogue of the square planar Ni-complex (η-4-cycloocten-1-yl](1,5-diphenyl-2,4-pentandionato-O,O') nickel for ^{2}H-NMR investigations.

While the deuterated complex shows the expected NMR signals in CH_2Cl_2 (two signals from the complex and one signal from the solvent), the ^{2}H-NMR spectrum obtained from the complex in the slightly acidic chloroaluminate ionic liquid shows only one signal indicating that the abstraction of COD is more efficient in the ionic liquid medium. Moreover, the deuterium signal of the acac ligand undergoes a significant downfield shift, suggesting intense electronic interaction between the ligand and the Lewis acidic centers of the melt. These interactions, which should result in an increased electrophilicity of the Ni-center, help to explain the activation of Ni–acac complexes in slightly acidic chloroaluminate ionic liquids.

An alternative way to gain insight into the interactions between the catalyst complex and the ionic liquid is to record changes in the ionic liquid during the addition of the catalyst complex in an indirect manner. This method has been successfully applied by van Eldik and coworkers to understand in more detail the activation of $[(PPh_3)_2PtCl_2]$ in chlorostannate ionic liquids [73]. The authors determined the ^{119}Sn NMR chemical shift of many different melt compositions [BMIM]Cl/$SnCl_2$ and found evidence for the formation of $[Sn_2Cl_5]^-$ anions in the acidic regime of the composition range that plays a major role in the catalyst activation. The change in color from yellow to red during the dissolution of the complex in the ionic liquid could be attributed to the insertion of $SnCl_2$ from the dimeric anion into the Pt–Cl bond.

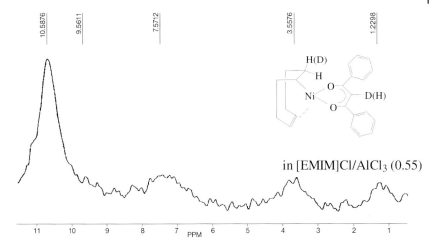

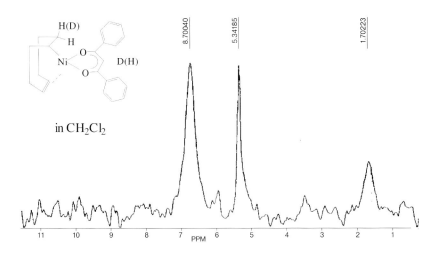

Fig. 5.3-4 ^{2}H-NMR spectra of the deuterated analogue of the square planar Ni-complex (η-4-cycloocten-1-yl) (1,5-diphenyl-2,4-pentandionato-O,O')nickel recorded in CH$_2$Cl$_2$ and in [EMIM]Cl/AlCl$_3$ [X(AlCl$_3$) = 0.55].

UV–Vis spectroscopy

The same authors also studied the kinetics of [(PPh$_3$)$_2$PtCl$_2$] activation in chlorostannate ionic liquids by UV–Vis spectroscopy [73]. Apart from investigations in neat chlorostannate melts they also investigated the activation in [BMIM][(CF$_3$SO$_2$)$_2$N] after addition of [BMIM]Cl/SnCl$_2$ (equimolar composition in tenfold excess to the added Pt(II)-complex). The observed reaction was found to proceed in two steps, the first step could be identified as the nucleophilic substitution of Cl$^-$ by [SnCl$_3$]$^-$ which is the only anion present in this system.

Later van Eldik and coworkers applied UV–Vis spectroscopy to monitor the kinetics of ligand substitution reactions at the complex cation [Pt(II)(2,6-bis(aminomethyl)pyridine)Cl]$^+$ using thiourea and iodide as nucleophiles [74]. They compared the reactions in water, methanol and [BMIM][(CF$_3$SO$_2$)$_2$N] and concluded that with respect to spectral changes, kinetic traces, rate and activation parameters the selected ionic liquid behaves as a "normal", innocent solvent in the ligand substitution reactions under investigation. For the first time the authors reported an activation volume for a ligand substitution reaction in an ionic liquid and correlated this with intrinsic volume changes that result from changes in bond length and bond angles on going to the transition state of the ligand exchange reaction.

An example of monitoring non-steady state kinetics in ionic liquids using UV–Vis spectroscopy was published by Abu-Omar and coworkers [71]. They monitored the methyltrioxorhenium(MTO)-catalyzed olefin epoxidation in [EMIM][BF$_4$] by observing the spectral changes of the Re complexes in the ionic liquid. The decreasing absorbance at 360 nm was attributed to the reactions of both the diperoxorhenium and the monoperoxorhenium complex with the olefinic substrate.

It is noteworthy that the study of transition metal catalysts in ionic liquids by UV–Vis spectroscopy affords, in most cases, colorless ionic liquids. The preparation of colorless ionic liquids is not an easy task for all ionic liquids and several methods have been recently reported by Seddon and coworkers to purify ionic liquids after their synthesis for such analytic purposes [75].

IR spectroscopy

In situ high-pressure infrared spectroscopy has been applied successfully by Dupont, van Leeuwen and coworkers to probe the catalytically active Rh-complex in the 1-octene hydroformylation reaction. The active catalyst formed by reaction of the [Rh(acac)(CO)$_2$] precursor and the ligand 2,7-bissulfonate-4,5-bis(diphenylphosphino)-9,9-dimethylxanthene (sulfoxantphos) in the ionic liquid [BMIM][PF$_6$] [76]. These authors found that the *in situ* IR signals were very similar to the same precursor/ligand mixture in organic solvent, indicating that the same mixture of ee (bis-equatorial) and ea (equatorial-apical) [(diphosphine)Rh(CO)$_2$H] catalyst complex is also the active species in the ionic liquid. Later, these data were used by Riisager and coworkers as a reference to prove the homogeneously dissolved character of the same Rh-complexes in supported ionic liquid phase (SILP) catalytic systems [35].

EPR spectroscopy

EPR spectroscopy has been applied by Kucherov and coworkers to study the nature of the catalytically active species in [W^{n+}(Mo^{n+}) – ionic liquids – olefin] metathesis systems [77]. In this study the key role of isolated paramagnetic complexes of W^{5+} and Mo^{5+} ions in the catalytic metathesis of 1-hexene could be demonstrated. It was shown that the number of EPR-visible isolated paramagnetic Mo^{5+} complexes in the ionic liquid reaches around 90% of the total number of molybdenum ions in the melt. Furthermore it was shown that the kinetics of the metathesis reaction correlates closely with the initial reduction of W^{6+} to W^{5+} species.

Mass spectroscopy
Many catalyst complexes that are suitable for operation and immobilization in ionic liquids are ionic themselves with the charge either located at the metal center or in the ligand sphere. Thus, these complexes are accessible to electrospray ionization mass spectroscopy (ESI-MS). However, it is not a trivial task to investigate an ionic catalyst complex next to the vast excess of ionic liquid ions. Dyson's group reported a combination of ESI-MS with quadrupole ion trap methods that enabled the detection of, e.g. several cationic Rh-complexes in the ionic liquid [BMIM][PF_6] [62]. They could demonstrate that the detection limit of their methodology (catalyst concentrations down to 1×10^{-10} mol l^{-1} were analyzed) is so low that even complexes in the typically very low concentration of a catalytic experiment could be easily characterized, even in the presence of a large excess of ionic liquid ions.

It should be noted here that MS techniques have been used with great success in the recent past to study species dissolved in ionic liquids [78] as well as the aggregate formation of ionic liquid ions in different ion mixtures [79] and in the presence of organic solvents [80], impressively underlining the potential of mass spectrometry for ionic liquid investigation in a more general respect.

Extended X-ray absorbance fine structure (EXAFS) spectroscopy
EXAFS studies have been among the first analytical techniques applied to study metal complexes dissolved in ionic liquids or melts containing metallate anions [65] and more recent examples can also be found in the literature [81]. However, the number of attempts to apply this well established and very successful methodology to the study of metal complexes in direct context with catalytic reactions is relatively small.

Iwasawa and coworkers published, for example, Suzuki cross-coupling reactions using [BMIM]$_2$[NiCl$_4$] as the catalyst precursor [82]. By applying EXAFS analysis they could conclude that the active catalyst obtained after treatment of the precursor with K_3PO_4 or NaO^{t}Bu is indeed the corresponding carbene complex. The analyzed catalytic systems proved to be highly active and selective in the cross-coupling reaction of aryl chlorides with arylboronic acids in the presence of two equivalents of PPh_3. Another example of the successful application of EXAFS studies for the direct study of catalytic active species in ionic liquids has been reported by Dupont and coworkers [83]. These authors investigated the reduction of [Ir(cod)Cl]$_2$ (cod= 1,5-cyclooctadiene) dissolved in different [BMIM] ionic liquids in the presence of 1-decene to yield Ir(0) nanoparticles. Their EXAFS analyses (in combination with *in situ* TEM, SAXS, XPS and XRD of the isolated material) provided evidence for interactions of the ionic liquid with the metal surface and demonstrated the formation of an ionic liquid protective layer surrounding the iridium nanoparticles.

In an earlier publication the same authors had reported the catalytic activity of these Ir nanoparticles in the hydrogenation of several olefinic substrates [84].

XPS spectroscopy
A technique that has only very recently been introduced but promises great potential for the investigation of transition metal catalysts in ionic liquids is X-ray photoelectron spectroscopy (XPS). XPS is a highly surface sensitive method able to analyse quantitatively the elemental composition of an ionic liquid surface vs. ultra-high vacuum (10^{-10} mbar) including the oxidation state of the observed elements. Licence and coworkers pioneered this method for ionic liquids by studying [Pd(OAc)$_2$(PPh$_3$)$_2$] at the surface of the ionic liquid [EMIM][EtOSO$_3$] [85]. They were able to follow spectroscopically the decomposition of the Pd(II) complex to form a Pd(0) species. Later, Maier et al. expanded the methodology by selective detection of photoelectrons emitted under different angles to the normal surface [86]. By this methodology they were able to distinguish the concentration and chemical nature of Pt-complex ions in the top surface layer from those in the layers beneath. Very interestingly, by dissolving the complex [Pt(NH$_3$)$_4$]Cl$_2$ in [EMIM][EtOSO$_3$] a complete surface depletion of Cl$^-$ ions was observed and a surface enrichment of the soft Pt-complex ions was unambiguously detected. These findings may have great implications for multiphasic catalysis with ionic liquids as the possibility of an enrichment of catalyst species may give an important starting point for the interpretation of kinetic data. In more general, XPS spectroscopy of ionic catalyst solutions has been found to be a very powerful analytic tool that makes elegant use of the extremely low volatility of the ionic liquid. The methodology may have a substantial impact on the understanding of liquid surfaces in homogeneous catalysis in general.

Of course, the examples described above are not comprehensive. But it is important to note that the total number of spectroscopic studies to understand better transition metal catalyzed reactions in ionic liquids is still very small – at least compared to the overall number of papers describing catalytic reactions in ionic liquids. It is encouraging however, that the number of sound mechanistic studies is now beginning to increase. Without doubt, much more work of this kind is still needed to better understand the nature of active catalytic species in ionic liquids and to explain some of the observed "ionic liquid" effects on a rational level. It will only be by gaining a thorough understanding of these effects that the true potential of ionic liquids in transition metal catalysis can be realized.

5.3.2
Selected Examples of the Application of Ionic Liquids in Transition Metal Catalysis

5.3.2.1 Hydrogenation
The use of ionic liquids has been successfully studied in many transition metal catalyzed hydrogenation reactions ranging from simple olefin hydrogenation to examples of asymmetric hydrogenation. Almost all applications so far include procedures of multiphase catalysis with the transition metal complex being

immobilized in the ionic liquid by its ionic nature or by means of ionic (or highly polar) ligands.

Hydrogenation, in principle, is very well suited for a biphasic reaction mode using ionic liquids. A large number of known, ionic hydrogenation catalysts are available [87] and the miscibility gap between the saturated reaction products and the ionic liquid is often large so that, in the majority of cases, a biphasic procedure is possible. This allows easy product separation after the reaction with low product solubility in the catalyst phase. Moreover, the formation of heavy, polar side products that would accumulate in the catalyst phase during recycling is not very likely in hydrogenation chemistry.

While the solubility of alkenes in ionic liquids is high enough for most hydrogenation applications (in particular if an appropriate cation design is applied) the hydrogen solubility has been shown to be intrinsically relatively low (see Section 3.3. for more details) [88]. Thus hydrogenation reactions in ionic liquids proceed at low hydrogen concentrations. This is some disadvantage as higher pressures may be required in some cases. However, it is noteworthy that the availability of hydrogen for a hydrogenation reaction results not only from its solubility under equilibrium conditions but reflects as well the ease of its transfer from the gas phase to the melt's surface and into the melt. Since the diffusion of hydrogen into ionic liquids has been found to be relatively fast [89], the latter contribution is of special importance. The high diffusivity of hydrogen results in high hydrogen transfer rates to and into the catalyst layer so that – during a hydrogenation reaction – the consumed hydrogen can be replenished rapidly. This fact explains the high number of successful hydrogenation reactions that have been reported in ionic liquids (see below) despite the usually low hydrogen solubility in these media.

Hydrogenation of olefins

The first transition metal catalyzed hydrogenation reactions in ionic liquids were reported by the groups of de Souza [90] and Chauvin [91] in 1995. De Souza et al. investigated the Rh-catalyzed hydrogenation of cyclohexene in [BMIM][BF$_4$]. Chauvin et al. dissolved the cationic "Osborn complex" [Rh(nbd)(PPh$_3$)$_2$][PF$_6$] (nbd=norbornadiene) in ionic liquids with weakly coordinating anions (e.g., [PF$_6$]$^-$,

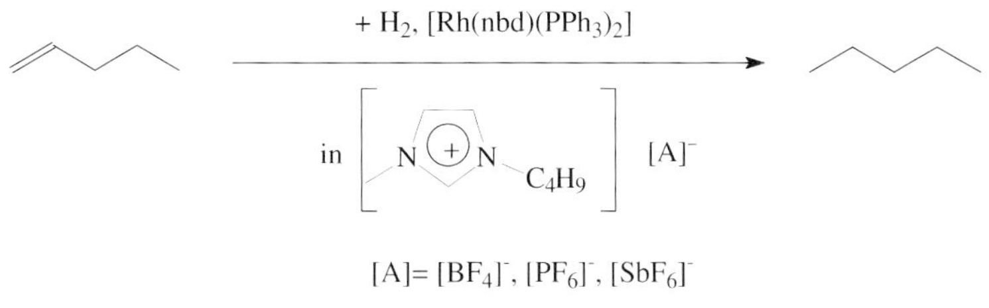

Scheme 5.3-7 Biphasic hydrogenation of 1-pentene with the cationic "Osborn complex" [Rh(nbd)(PPh$_3$)$_2$][PF$_6$] (nbd=norbornadiene) in ionic liquids with weakly coordinating anions.

Table 5.3-1 Rh catalyzed hydrogenation of pent-1-ene.

No.	Solvent	Conversion (%) pent-1-ene	Yield[a] pentane	pent-2-ene	TOF (min^{-1})[b]
1[c]	acetone	99	38	61	0.55
2	[BMIM][SbF$_6$]	96	83	13	2.54
3	[BMIM][PF$_6$]	97	56	41	1.72
4	[BMIM][BF$_4$]	10	5	5	0.15

[a] Cat.: 0.05 mmol; pent-1-ene: 8.4 mmol; solvent: 4 ml, $T = 30\,°C$, $p(H_2) = 0.1$ MPa; $t = 2$ h.
[b] TOF = mol (pentane) per mol (rhodium) and time (min).
[c] 10 ml acetone, 9.2 mmol pent-1-ene.

[BF$_4$]$^-$ and [SbF$_6$]$^-$) and used the ionic catalyst solutions thus obtained according to Scheme 5.3-7 for the biphasic hydrogenation of 1-pentene.

Table 5.3-1 summarizes the most relevant results of this early study. Although the reactants show only limited solubility in the catalyst phase, the rates of hydrogenation in [BMIM][SbF$_6$] are almost five times faster than for the comparable reaction in acetone. However, the reaction was found to be much slower using a hexafluorophosphate ionic liquid. This effect was attributed to the better solubility of pentene in the hexafluoroantimonate ionic liquid. The very poor yield in [BMIM][BF$_4$], however, was due to a high amount of residual Cl$^-$ ions in the ionic liquid leading to catalyst deactivation. At that time the preparation of this tetrafluoroborate ionic liquid in a chloride-free quality was obviously a problem.

All ionic catalyst solutions could be reused repeatedly. The loss of rhodium through leaching into the organic phase was below the detection limit of 0.02%. These results are of general importance for the field of biphasic catalysis since this was the first time that a rhodium catalyst could be "immobilized" in a polar solvent without the use of especially designed ligands. Moreover, Chauvin's group described the selective hydrogenation of cyclohexadiene to cyclohexene by making use of the biphasic reaction system [91]. Since the solubility of cyclohexadiene in [BMIM][SbF$_6$] is about five times higher than the solubility of cyclohexene in the ionic liquid, the latter was obtained in 98% selectivity at 96% conversion.

Following up these pioneering studies several other papers dealing with non-stereoselective hydrogenations of olefins in ionic liquids using liquid–liquid biphasic systems have been published [92–96]. The immobilization of Pd(acac)$_2$ as hydrogenation catalyst in the ionic liquids [BMIM][BF$_4$] and [BMIM][PF$_6$] was reported by Dupont et al. in 2000 [94]. They compared the biphasic hydrogenation of butadiene with the homogeneous system with all reactants being dissolved in CH$_2$Cl$_2$, with the reaction in neat butadiene and with a heterogeneous system using Pd on carbon as catalyst. The paper demonstrates that for 1,3-butadiene hydrogenation the selectivities achieved with Pd(acac)$_2$ dissolved in ionic liquids were similar to those observed under homogeneous conditions and were higher than under heterogeneous conditions (using Pd on carbon as the catalyst) or in neat 1,3-butadiene. The authors extended their investigations to a series of functionalized dienes like sorbic acid, methyl sorbate, 1-nitro-1,3-butadiene, and cyclic dienes.

The group of Dupont also studied the catalytic activity of $RuCl_2(PPh_3)_3$ and $K_3Co(CN)_5$ in [BMIM][BF$_4$] for the hydrogenation of a number of unfunctionalized, unsaturated hydrocarbon compounds [93]. It was found that – with the ruthenium complex as catalyst – the interaction of $RuCl_2(PPh_3)_3$ with the ionic liquid led to a stable, ionic purple solution, and no leaching of the Ru complex could be detected by extraction with hydrocarbon solvents. Turnover frequencies up to 537 h^{-1} were achieved. It is also noteworthy that the hydrogenation in the liquid required less drastic conditions (temperature and hydrogen pressure) compared to the hydrogenation in an aqueous medium using the water-soluble catalyst.

Many low oxidation state transition metal (carbonyl) clusters are salts and can be stabilized in ionic liquids due to their ionic character. Interestingly, Dyson et al. revealed that the activity of certain clusters in the hydrogenation of alkene substrates is up to 3.6-fold faster if these clusters are immobilized in ionic rather than in organic solvents [96]. They evaluated the clusters $[HFe(CO)_{11}]^-$, $[HWOs_3(CO)_{14}]^-$, $[H_3Os_4(CO)_{12}]^-$ and $[Ru_6C(CO)_{16}]^{2-}$ as catalysts/pre-catalysts in the hydrogenation of styrene to ethylbenzene in [BMIM][BF$_4$], octane and methanol respectively. Using $[Ru_6C(CO)_{16}]^{2-}$ as catalyst precursor in [BMIM][BF$_4$], the same research group also obtained good results in the partial reduction of cyclohexadienes to cyclohexene. However, by poisoning the catalytic phase with mercury and by means of high-pressure ^{1}H-NMR spectroscopy experiments they revealed that under the reaction conditions Ru-colloids/Ru-nanoparticles were formed and the latter acted as the catalyst in these reactions. These studies represent therefore examples of heterogeneous catalysis with nanoparticles in ionic liquids and will be discussed in more detail in Section 5.3.2.7. In contrast to the ruthenium-containing cluster, a homogeneous hydride complex was found to be the active catalyst with the osmium-containing clusters being dissolved in ionic liquids [96]. A comparison of the turnover frequencies of the $[H_3Os_4(CO)_{12}]^-$-catalyzed hydrogenation of styrene to ethylbenzene in various ionic liquids is presented in Table 5.3-2.

The turnover frequencies in all the ionic liquids are quite similar although the values are somewhat lower in [BMMIM][PF$_6$], [BMMIM][BF$_4$] and [BMMIM][(CF$_3$SO$_2$)$_2$N] than in the [BMIM]-containing ionic liquids. The highest activity was found for [OMPy][BF$_4$] ([OMPy] = 1-octyl-3-methylpyridinium) being used as the ionic liquid. An explanation for this could be the fact that [BMMIM]-containing ionic liquids possess the highest viscosities and have therefore the lowest mass transfer rates for hydrogen. The experiment with [OMPy][BF$_4$] is special in that only for this ionic liquid does the reaction mixture form a single phase thus preventing any liquid–liquid mass transport resistance. However, this monophasic reaction mode also means that catalyst recycling by simple decantation is not possible in that latter case.

The selective hydrogenation of sorbic acid to *cis*-3-hexenoic acid was demonstrated in ionic liquids by Drießen-Hölscher et al. (Scheme 5.3-8). Based on investigations

Table 5.3-2 Hydrogenation of styrene to ethylbenzene using $[H_3Os_4(CO)_{12}]^-$ as the catalyst precursor in various ionic liquids.

Ionic liquid	TOF (mol mol^{-1} h^{-1})
[BMIM][BF$_4$]	587
[BMIM][PF$_6$]	522
[BMIM][(CF$_3$SO$_2$)$_2$N]	587
[BMMIM][PF$_6$]	392
[BMMIM][BF$_4$]	413
[BMMIM][(CF$_3$SO$_2$)$_2$N]	457
[OMPy][BF$_4$]	718

Conditions: $p(H_2) = 50.7$ bar, 100 °C, 4 h; Cluster concentration 5×10^{-4} M, ionic liquid (1 ml), styrene (1 ml), total reactor volume (30 ml); TOF is the turnover in units of mol (product) mol^{-1} (catalyst) h^{-1}, calculated as an average value over 4 h.

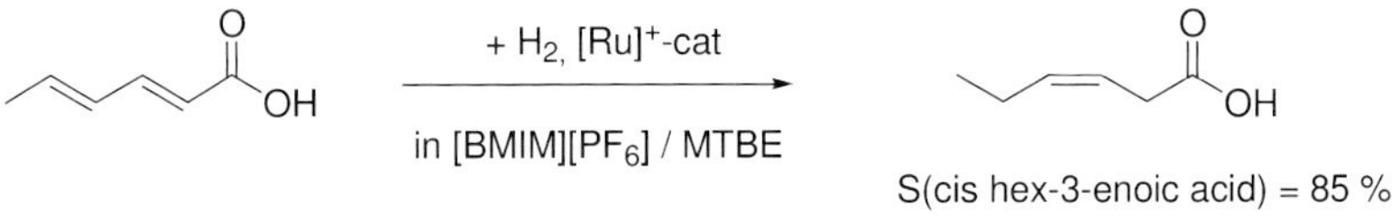

Scheme 5.3-8 Regioselective hydrogenation of sorbic acid in the biphasic system [BMIM][PF$_6$]/MTBE.

in the biphasic system water/n-heptane [97], the ruthenium-catalyzed reaction was studied in the biphasic system [BMIM][PF$_6$]/MTBE [98].

In comparison to an optimized polar organic solvent (e.g. glycol) a more than threefold increase in activity with comparable selectivity to cis-3-hexenoic acid was observed in the ionic liquid. This is explained by a partial deactivation (through complexation) of the catalytic active center in those polar organic solvents that are able to dissolve the cationic Ru-catalyst. In contrast, the ionic liquid [BMIM][PF$_6$] is known to combine high solvation power for ionic metal complexes with relatively weak coordination strength. In this way the catalyst can be dissolved in a "more innocent" environment than is the case if polar organic solvents are used. After the biphasic hydrogenation of sorbic acid, the ionic catalyst solution could be recovered by phase separation and reused four times with no significant loss of selectivity but a decreasing conversion from 74.0 and 77.4%, to 59.6 and 35.8% in the third and fourth run. However, during these recycling experiments the autoclave containing the catalyst phase had to stand overnight between the second and the third run.

Hydrogenation of arenes

The hydrogenation of arenes is industrially important, but so far is dominated by the use of heterogeneous catalysts. Ionic liquids offer, in principle, the chance to use a liquid–liquid biphasic system where the homogeneous catalyst is immobilized and therefore recyclable. Dyson et al. applied ruthenium clusters as catalyst for

Table 5.3-3 Comparative studies of the biphasic hydrogenation reactions of arenes in [BMIM][BF$_4$] and water with [H$_4$Ru$_4$(η^6-C$_6$H$_6$)$_4$][BF$_4$]$_2$ as the catalyst precursor.

Subtrate	Reaction system	Reaction conditions	Conversion (%)	Catalytic turnover[a]/h^{-1}
benzene	ionic liquid	60 atm H$_2$, 90 °C, 2.5 h	91	364
	water	60 atm H$_2$, 90 °C, 2.5 h	88	352
toluene	ionic liquid	60 atm H$_2$, 90 °C, 3 h	72	240
	water	60 atm H$_2$, 90 °C, 3 h	78	261
cumene	ionic liquid	60 atm H$_2$, 90 °C, 2.5 h	34	136
	water	60 atm H$_2$, 90 °C, 2.5 h	31	124

[a] Catalytic turnover is calculated on the assumption that the tetraruthenium catalyst does not break down into monoruthenium fragments, which is entirely consistent with the data.

the hydrogenation of benzene, toluene, cumene, ethylbenzene and chlorobenzene [99,100]. A direct comparison of the two biphasic systems water/organic solvent and ionic liquid/organic solvent showed that the turnover frequencies obtained in the ionic liquid and the aqueous media are similar [99]. The results of hydrogenation in the two biphasic systems are shown in Table 5.3-3. The authors propose, that the catalytical active species is [H$_6$Ru$_4$(η_6-C$_6$H$_6$)$_4$]$^{2+}$ in the ionic liquid-containing system, as was shown for the water-containing system [100]. Again the advantage of the ionic liquid-containing biphasic system was the easy separation of products and the possibility to reuse the catalytic active phase.

Very interesting results were obtained by using [Ru(η^6-p-cymene)(η^2-TRIPHOS)Cl][PF$_6$] in a biphasic system with [BMIM][BF$_4$] as catalyst phase [101]. Hydrogenation of benzene, toluene and ethylbenzene in this ionic liquid proceeded with higher yields and accordingly with higher turnover frequencies than in the monophasic system with CH$_2$Cl$_2$ as solvent. The TOF (yield) of benzene hydrogenation was found to increase from 242 h^{-1} (52% yield) in CH$_2$Cl$_2$ to 476 h^{-1} (82% yield) in the ionic liquid, for toluene from 74 h^{-1} (19% yield) to 205 h^{-1} (42% yield) and for ethylbenzene from 57 h^{-1} (17% yield) to 127 h^{-1} (30% yield), respectively. With allylbenzene as the subtrate the authors observed selective hydrogenation of the benzyl group and the unsaturated alkyl chain remained intact. The hydrogenation yielded allylcyclohexane with a TOF of 329 h^{-1} (84% yield), whereas the system was inactive toward arenes with other alkene substituents such as styrene and 1,3-divinylbenzene.

Hydrogenation of polymers

Transition metal catalyzed hydrogenation in ionic liquids has also been applied to the hydrogenation of polymers. First studies were presented by Dupont's group which investigated the hydrogenation of acrylonitrile–butadiene copolymers [102]. These early studies were later expanded by Rosso and coworkers studying the rhodium catalyzed hydrogenation of polybutadiene (PBD), nitrile–butadiene rubber (NBR) and styrene–butadiene rubber (SBR) in a [BMIM][BF$_4$]/toluene and a

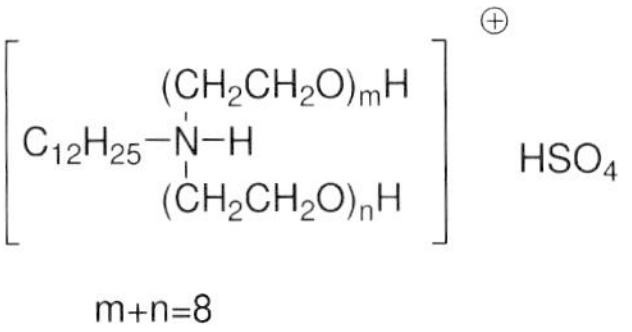

$$\left[\begin{array}{c} (CH_2CH_2O)_mH \\ C_{12}H_{25}-\overset{\oplus}{N}-H \\ (CH_2CH_2O)_nH \end{array} \right] \quad HSO_4^{\ominus}$$

m+n=8

Fig. 5.3-5 Structure of a polyether-modified ammonium ionic liquid as used for the hydrogenation of polystyrene-b-polybutadiene-b- polystyrene (SBS) block copolymer using a Ru/TPPTS catalyst.

[BMIM][BF$_4$]/toluene/water system [103]. The activity of the catalyst followed the trend PBD>NBR>SBR, which corresponds to the order of polymer solubility in the ionic liquid. The degree of hydrogenation (as a percent of total hydrogenation) was 94% for PBD (4 h reaction time), 43% for NBR (4 h reaction time), and 19% for SBR (3 h reaction time).

A polyether modified ammonium salt ionic liquid/organic biphasic system was used by Jiang et al. for the hydrogenation of polystyrene-b-polybutadiene-b-polystyrene (SBS) block copolymer using a Ru/TPPTS complex as catalyst [104]. The ionic liquid is displayed in Fig. 5.3-5.

By addition of triphenylphosphine as promoter ligand, hydrogenation degrees of up to 89% could be achieved. Hydrogenation of the benzene ring and gel formation was not observed. The active catalyst was found to be well immobilized in the ionic liquid and the ionic catalyst phase could be reused three times without significant loss in catalytic activity.

Stereoselective hydrogenation

Since 1995 a number of enantioselective hydrogenation reactions have been described using ionic liquids as the catalyst immobilization phase. In most cases reported so far the role of the ionic liquid was solely to allow facile recycling of the expensive chiral metal complex by liquid–liquid operation. Again, Chauvin et al. pioneered the field describing the hydrogenation of α-acetamido cinnamic acid using [Rh(cod)(−)-(diop)][PF$_6$] as catalyst in a [BMIM][SbF$_6$] melt affording (S)-phenylalanine with 64% enantiomeric excess (ee) [91]. The product was easily and quantitatively separated and the ionic liquid could be recovered. The loss of rhodium was less than 0.02% per run. Dupont and coworkers were able to obtain up to 80% ee in the reaction of 2-arylacrylic acid to (S)-2-phenylpropionic acid with the chiral [RuCl$_2$(S)-BINAP]$_2$NEt$_3$ complex as catalyst in [BMIM][BF$_4$] melts (Scheme 5.3-9) [105]. Both reactions were carried out in two phases with the help of an additional organic solvent (e.g. iPrOH).

In another example de Souza and Dupont studied the asymmetric hydrogenation of α-acetamido cinnamic acid and the kinetic resolution of (±)-methyl-3-hydroxy-2-methylenebutanoate with chiral Rh(I) and Ru(II) complexes in [BMIM][BF$_4$] and [BMIM][PF$_6$] [106]. A special focus of their work was on the influence of H$_2$ pressure on conversion. They determined the hydrogen solubility in the ionic liquid using

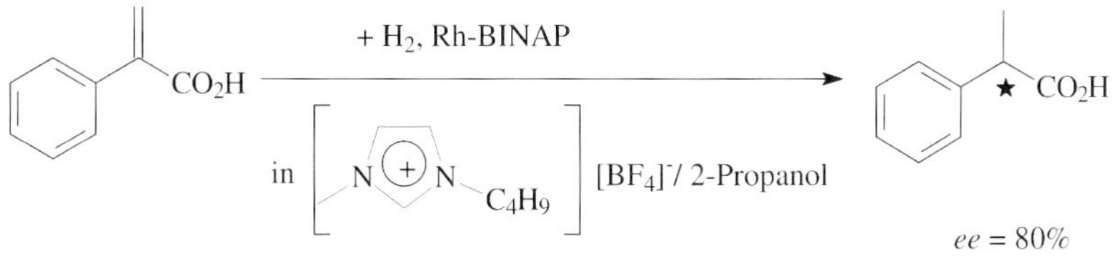

Scheme 5.3-9 Hydrogenation of 2-arylacrylic acid to (S)-2-phenylpropionic acid with the chiral [RuCl$_2$(S)-BINAP]$_2$NEt$_3$ complex as catalyst in [BMIM][BF$_4$].

pressure drop experiments [107]. The solubility values reported are $K = 3.0 \times 10^{-3}$ mol L^{-1} atm^{-1} for H$_2$ in [BMIM][BF$_4$] and 8.8×10^{-4} mol L^{-1} atm^{-1} for H$_2$ in [BMIM][PF$_6$]/H$_2$ at room temperature. These values differ significantly from those determined by Dyson et al. using ^{1}H-NMR [88b]. The reported values suggest that molecular hydrogen is almost four times more soluble in [BMIM][BF$_4$] than it is in [BMIM][PF$_6$] under the same pressure. According to the authors, this difference in solubility leads to the different degrees of conversion observed in their experiments. They reported 73% conversion (93% *ee*) for the reaction in [BMIM][BF$_4$] while only 26% conversion (81% *ee*) was found using [BMIM][PF$_6$] as rection medium (50 bar hydrogen pressure in both experiments).

Geresh et al. applied the chiral rhodium complex [Rh-MeDuPHOS] dissolved in [BMIM][PF$_6$] for the asymmetric hydrogenation of enamides [108] (Scheme 5.3-10). The group focussed on the stabilization of the air-sensitive catalyst in the ionic liquid and described the ionic liquid as being able to protect the air-sensitive complex from attack by atmospheric oxygen. According to the authors this greatly facilitates recycling of the ionic catalyst solution.

The catalytic results were comparable to those of the homogeneous reaction in *i*PrOH and recyclability could be demonstrated over five cycles with constant *ee* though decreasing conversion. Amazingly the catalyst still showed some catalytic activity after storage under atmospheric conditions for 24 h.

In enantioselective hydrogenation substrates can be divided into two classes (Fig. 5.3-6): Class I are substrates which require low H$_2$ pressure to obtain good enantioselectivities while class II substrates require high H$_2$ pressure [109, 110].

In this context the group of Jessop studied the influence of different ionic liquids in the asymmetric hydrogenation of representatives from both classes

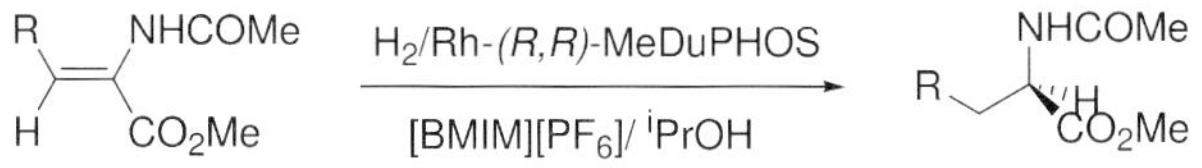

1a. R = H
1b. R = Phenyl

Scheme 5.3-10 Asymmetric hydrogenation of enamides catalyzed by Rh-MeDuPHOS immobilized in [BMIM][PF$_6$].

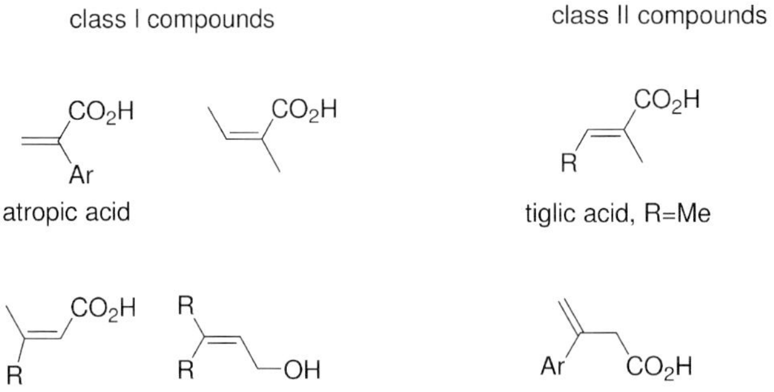

Fig. 5.3-6 Classification of substrates for enantioselective hydrogenation.

[111]. Atropic acid was investigated as an example of class I compounds and the hydrogenation of tiglic acid was selected as an example of class II compounds. For atropic acid a strong dependence of the obtained *ee* on the used solvent was found. The *ee* values varied in the range of 72% to 95% increasing in the following order: MeOH<[EMIM][CF$_3$SO$_3$]<[BMIM][BF$_4$] = [MBPy][BF$_4$]<[BMIM][PF$_6$] = [MMPIM][(CF$_3$SO$_2$)$_2$N)] ([MMPIM]= 1,2-dimethyl-3-propylimidazolium) < [EMIM][(CF$_3$SO$_2$)$_2$N)]. In contrast, enantioselectivities were low for the asymmetric hydrogenation of the class II substrate in ionic liquids (without co-solvent). The best enantioselectivities were obtained using methanol as co-solvent. This effect is presumably due to the reduction of viscosity with methanol addition (i.e. enhanced mass transfer) and increased hydrogen solubility compared to pure ionic liquids. For the reactions in neat ionic liquids, the selectivity was found to depend on the choice of ionic liquid, now increasing in the order: [BMIM][BF$_4$] < [EMIM][O$_3$SCF$_3$] < [BMIM][PF$_6$] = [EMIM][(CF$_3$SO$_2$)$_2$N)] < [MMPIM][(CF$_3$SO$_2$)$_2$N)]. Remarkably these results demonstrate that the effectiveness of these asymmetric hydrogenations in ionic liquids is not only a function of H$_2$ availability. Many solvent parameters including polarity, coordinating ability and hydrophobicity have to be taken into account and add to a complex picture that is still not fully understood.

Attempts to improve the solubility and immobilization of chiral hydrogenation catalysts in ionic liquids were presented by Lee and coworkers [112]. They synthesized a chiral Rh-complex carrying the dicationic bisphosphine ligand depicted in Fig. 5.3-7. Immobilization of the tricationic complex in [BMIM][SbF$_6$] showed better immobilization results in contact with *i*PrOH compared to the non-modified complex Me-BDPMI in the Rh-catalyzed asymmetric hydrogenation of N-acetylphenylethenamine (Scheme 5.3-11). The ionic catalyst solution was reused three times without loss of activity. In the fourth run conversion decreased but high conversions could still be realized by increasing the reaction time.

Finally, a broad screening of ligands and ionic liquids was carried out by Feng et al. for the Rh-catalyzed hydrogenation of enamides [113]. Rhodium-ferrocenyl-diphosphine complexes with taniaphos, josiphos, walphos and mandyphos

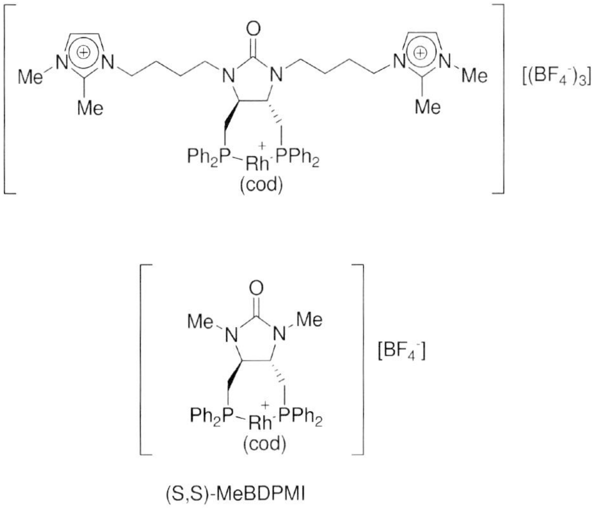

Fig. 5.3-7 Rh-complex with a bisphosphine-containing cation as ligand.

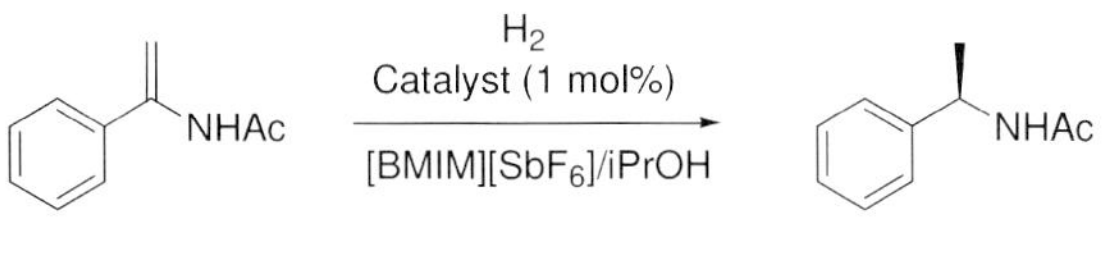

Scheme 5.3-11 Rh-catalyzed asymmetric hydrogenation of N-acetylphenylethenamine.

ligands were identified in this study to be the most effective catalyst complexes (see Fig. 5.3-8).

The screening of the different reaction media proved that the application of an ionic liquid/water mixture – a so-called "wet ionic liquid" – led to improved catalyst recycling compared to the reaction in ionic liquid without co-solvent. This comparative study revealed good results for the hydrogenation of various enamides in wet ionic liquids with respect to conversion, enantioselectivity and catalyst separation. The results obtained with the Rh-taniaphos system in different solvent systems are summarized in Table 5.3-4.

Ketone and imine hydrogenation in ionic liquids

There is only one study on ketone hydrogenation in ionic liquids which applied Rh-complexes as catalysts. Zhu et al. synthesized a new carborane-based room-temperature ionic liquid consisting of an N-n-butylpyridinium cation and the anion $[CB_{11}H_{12}]^-$, and used this ionic liquid as the reaction medium in the asymmetric hydrogenation of unsymmetrical aryl ketones in the presence of the chelating ligand (R)-BINAP and a rhodacarborane catalyst (Table 5.3-5) [114]. Compared to hydrogenation in tetrahydrofuran and in "classical" ionic liquids the best results were indeed achieved with the carborane-based ionic liquid.

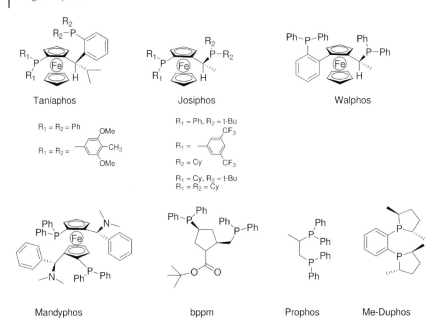

Fig. 5.3-8 Structures of ligands as used for the Rh-catalyzed asymmetric hydrogenation of enamides in ionic liquids.

Lin et al. studied the hydrogenation of β-aryl ketoester using a Ru-BINAP system with different substituents at the 4,4′-position of the BINAP ligand [115]. Best enantioselectivities were achieved with sterically demanding and electron-donating 4,4′-substituents. For example, ee values of 97.2% and 99.5% were obtained for the hydrogenation of ethyl benzoylacetate with R = trimethylsilane (**1**, Fig. 5.3-9), and R = bisphosphonic acid (**2**, Fig. 5.3-9), respectively as substituents. By immobilization of these catalysts in [BMIM][BF$_4$], a slight deterioration in ee values (reduction of ee by 1%) was observed for the trimethylsilane substituted catalyst, while an increase (increase of ee by up to 2.6%) was observed for the biphosphonic acid substituted catalyst. Both catalysts were recycled and reused four times and the ee (conversion) decreased from 97.3% (>98% conversion) to 95.1% (62% conversion) in the case of trimethylsilane as substituent and from 97.5% (98% conversion) to 74.7% (44% conversion) for the biphosphonic acid-substituted catalyst. In both cases no significant leaching was detected. Higher conversions with comparable ee were achieved by using [MMPIM][(CF$_3$SO$_2$)$_2$N] as ionic liquid [116]. These findings can be explained by the absence of the anions [BF$_4$]$^-$ and [PF$_6$]$^-$ (with their potential to liberate the catalyst poison F$^-$ in a hydrolysis reaction) and the substitution of the acidic proton at the 2-position of the imidazolium cation by a methyl group, which safely prevents carbene formation with the transition metal.

Table 5.3-4 Enantioselective hydrogenation of enamides catalyzed by Rh-taniaphos ($R^1 = R^2 =$ Ph) in various ionic liquids/water combinations and in conventional solvents.

No	Reaction medium[a]	Conversion (%)	ee (%)	Catalyst separation[b]	No. of phase

substrate: methyl α-acetamidoacrylate

$$\text{CH}_2=\text{C}(\text{COOCH}_3)(\text{NHCOCH}_3) \xrightarrow[\text{H}_2]{[\text{Rh(nbd)}_2]^+\text{BF}_4^-,\ \text{Taniaphos}} \text{CH}_3-\overset{*}{\text{CH}}(\text{COOCH}_3)(\text{NHCOCH}_3)$$

No	Reaction medium[a]	Conversion (%)	ee (%)	Catalyst separation[b]	No. of phase
1	MeOH-H$_2$O	100	98	–	1
2	MeOH	100	97	–	1
3	i-PrOH	92	95	–	1
4	toluene	47	28	–	1
5	[BMIM]BF$_4$	32	>99	–	1
6	[BMIM]PF$_6$	6	91	–	1
7	[BMIM]BF$_4$/i-PrOH	64	96	+	2
8	[BMIM]PF$_6$/i-PrOH	12	93	+	2
9	[OMIM]BF$_4$/H$_2$O	100	>99	++	2
10	[BMIM]Tf$_2$N/H$_2$O	97	>99	++	2
11	[BMIM]BF$_4$-H$_2$O/toluene	100	>99	++	2

substrate: methyl α-acetamidocinnamate

$$\text{Ph-CH}=\text{C}(\text{COOCH}_3)(\text{NHCOCH}_3) \xrightarrow[\text{H}_2]{[\text{Rh(nbd)}_2]^+\text{BF}_4^-,\ \text{Taniaphos}} \text{Ph-CH}_2-\overset{*}{\text{CH}}(\text{COOCH}_3)(\text{NHCOCH}_3)$$

No	Reaction medium[a]	Conversion (%)	ee (%)	Catalyst separation[b]	No. of phase
20	i-PrOH	100	94	–	1
21	[BMIM]BF$_4$	52	93	–	1
22	[BMIM]BF$_4$/i-PrOH	100	93	+	2
23	[OMIM]BF$_4$/H$_2$O	100	95	++	2
24	[BMIM]BF$_4$-H$_2$O/toluene	100	94	++	2
25	[OMIM]BF$_4$/H$_2$O/toluene	100	95	++	3

[a] ILs: ca. 2 mL, co-solvents: ca. 2–3 mL, S/C = 200, [S] = 0.25 M in co-solvent, room temperature, $p(H_2) = 1$ bar, and $t = 20$ min.
[b] "–" no, "+" good (some leaching), "++" excellent (no leaching).

The same catalyst was used in the asymmetric hydrogenation of β-keto esters in [BMIM][PF$_6$], [BMIM][BF$_4$] and [MMPIM][(CF$_3$SO$_2$)$_2$N] with complete conversions and ee values of up to 99.3% [117].

Other substituted derivatives of BINAP were investigated by Vaultier et al. [118]. The ammonium salt catalysts (**3**, Fig. 5.3-9) and (**4**, Fig. 5.3-9) were prepared *in situ* from the respective bromohydrates and [Ru(η^3-2-methylallyl)$_2$(η^2-COD)] and were immobilized in several ionic liquids. Comparative studies of the hydrogenation of ethyl acetoacetate revealed best results for imidazolium and pyridinium containing ionic liquids. In contrast, no significant ee was observed with the phosphonium

Table 5.3-5 Rh-catalyzed hydrogenation of acetophenone (A) and ethyl benzoylformate (B).[a]

Solvent	Conversion (%)[b]	ee (%)[c]	TOF (h^{-1})[d]
[OMIM][BF$_4$]	100 (A,B)	97.3(A), 99.3(B)	194(A), 201(B)
[BMIM][BF$_6$]	100 (A,B)	97.8(A), 98.2(B)	207(A), 213(B)
[BPy][CB$_{10}$H$_{12}$]	100 (A,B)	99.1(A), 99.5(B)	239(A), 306(B)
tetrahydrofuran	82 (A), 87 (B)	91.3(A), 85.7(B)	96(A), 107(B)

[a] Mol ratio of catalyst/(R)-binap/acetophenone = 1:1.5:1000; reaction conditions: H$_2$ (12 atm), 50 °C, 12 h; [cat.] = 8.1×10^{-4} M.
[b] Determined by GC.
[c] Determined by GC on a Chirasil DEX CB column.
[d] Turnover frequency (=moles of hydrogenation product per mole of Rh per hour) was determined after 3 h.

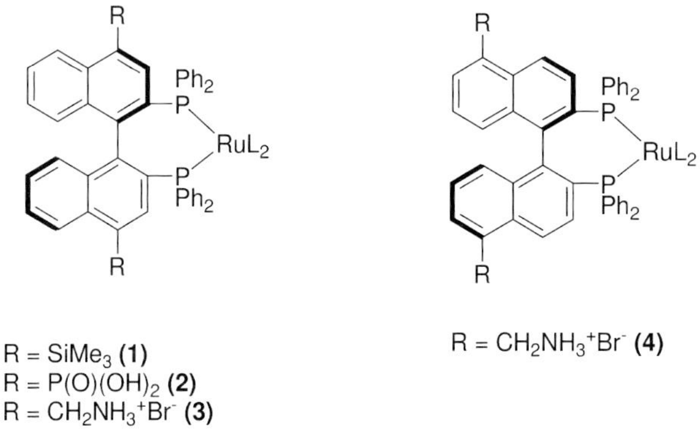

R = SiMe$_3$ (**1**)
R = P(O)(OH)$_2$ (**2**)
R = CH$_2$NH$_3^+$Br$^-$ (**3**)

R = CH$_2$NH$_3^+$Br$^-$ (**4**)

Fig. 5.3-9 Ligands as used in the Ru-catalyzed hydrogenation of β-aryl ketoester in different ionic liquids.

salt. This observation was attributed to problems of solubility and to the ability of complexation for the phosphonium ion. With respect to the selection of anion [BF$_4$]$^-$ appeared superior to [PF$_6$]$^-$ and [(CF$_3$SO$_2$)$_2$N]$^-$.

Ionic liquids have also been applied in transfer hydrogenation. Ohta et al. examined the transfer hydrogenation of acetophenone derivatives with a formic acid–triethylamine azeotropic mixture in the ionic liquids [BMIM][PF$_6$] and [BMIM][BF$_4$] [119]. They compared the TsDPEN-coordinated Ru(II) complexes (**1**, Fig. 5.3-10) with the ionic catalyst which was synthesized from the task specific ionic liquid (**2**, Fig. 5.3-10) in the presence of [RuCl$_2$(benzene)]$_2$.

The enantioselectivities obtained with the catalyst immobilized by the task specific ionic liquid **2** in [BMIM][PF$_6$] were found to be comparable with those of the TsDPEN-coordinated Ru(II) catalyst **1** and reached 93%. Both systems could be recycled five times with only a slight decrease in conversion in cycles 4 and 5 for the TsDPEN-coordinated Ru(II) catalyst **1**. The recycling results are displayed in Table 5.3-6.

Table 5.3-6 Recycling of **1** and **2**-Ru (Fig. 5.3-10) in the asymmetric transfer hydrogenation of acetophenone using the formic acid–triethylamine azeotropic mixture in the ionic liquids [BMIM][PF$_6$].

Cycle	Catalyst 1		Catalyst 2-Ru[a]	
	Conversion (%)[b]	ee (%)[b]	Conversion (%)[b]	ee (%)[b]
1	96	93	98	92
2	99	92	>99	93
3	95	92	99	93
4	88	92	92	93
5	63	93	75	90

Reaction conditions: room temperature, 24 h and S/C = 100.
[a] A mixture of **2** and [RuCl$_2$(benzene)]$_2$ was used.
[b] Determined by capillary GLC analysis using a chiral Cyclodex-B column.

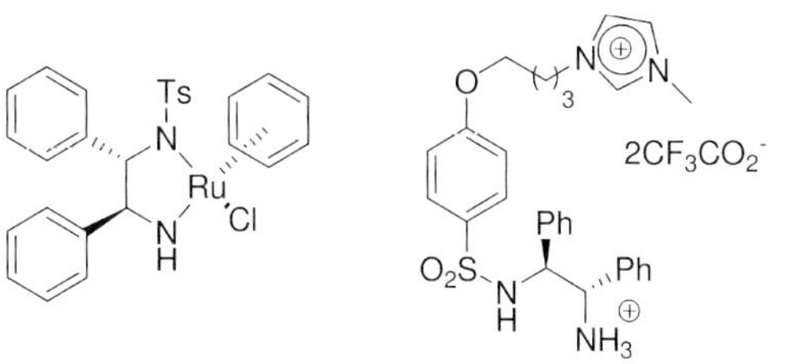

Fig. 5.3-10 Ru-complex and task specific ionic liquid for the transfer hydrogenation of acetophenone derivatives using a formic acid–triethylamine azeotropic mixture.

Very few examples of imine hydrogenation in ionic liquids have been published so far. Giernoth et al. screened eight different ionic liquids (the cations were [BMIM]$^+$, 1-decyl-3-methylimidazolium ([DMIM]$^+$), N-butyl-3-methylpyridinium ([BMPy]$^+$) and N-decyl-3-methylpyridinium ([DMPy]$^+$) combined with the anions [BF$_4$]$^-$ and [(CF$_3$SO$_2$)$_2$N]$^-$) and compared them with toluene as solvent in the hydrogenation of trimethylindolenine with the Ir-XYLIPHOS catalyst system (Scheme 5.3-12) [120].

Besides the immobilization of the catalyst, the authors claim, as benefit of the ionic liquids, a reduction of the reaction time from 23 h in toluene to less than 15 h in [DMIM][BF$_4$] with no loss in selectivity, although the ionic media require slightly higher reaction temperatures. Furthermore, a stabilization of the ionic catalyst solution against atmospheric oxygen is observed. This stabilization effect facilitates the transfer of freshly prepared catalyst to the autoclave and, in general, makes the handling of the ionic liquid/catalyst system much easier.

Summarizing the actual state of the art it can be stated that the application of ionic liquid media for catalytic hydrogenation has been successfully demonstrated

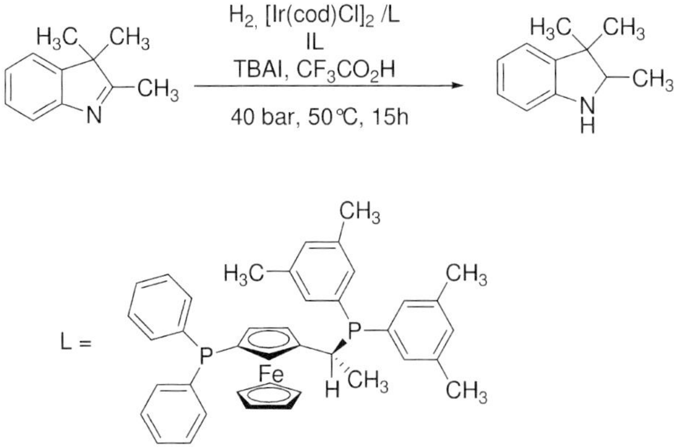

Scheme 5.3-12 Enantioselective hydrogenation of trimethylindolenine using Ir-XYLIPHOS as catalyst.

for many substrates and reactions ranging from simple olefin hydrogenation to examples of asymmetric hydrogenation. Almost all applications so far include procedures of multiphase catalysis with the transition metal complex being immobilized in the ionic liquid. Furthermore it was found by far the most authors that activity and selectivity of known transition metal complexes does not change too much in the ionic environment, indicating weak interaction of the ionic liquid ions and the catalyst complex in most cases. Selectivity optimization based on ligand design – especially important in asymmetric catalysis – has been proven many times to work in the same manner in the ionic liquid environment as in an organic solvent. However, some important differences from the homogeneous reactions in organic solvents have to be taken into account. Hydrogen solubility in ionic liquids is very low compared to most traditional solvents so that hydrogen concentration is low at the catalytic center. Mass transfer of hydrogen to and into the catalyst layer is affected by the viscosity of the ionic liquid (with low viscosities enhancing mass transfer) but has been found to be fast enough in most cases to reach acceptable reaction rates, comparable to those obtained in organic media.

Finally, it should be mentioned that apart from liquid–liquid biphasic mode, catalytic hydrogenation reactions have been carried out with great success in ionic liquid/compressed CO_2 [111, 112] and supported ionic liquid phase (SILP) systems [122]. In both cases the advantages of a molecular defined, homogeneously dissolved hydrogenation catalyst have been combined with very efficient, continuous processing of the catalytic system. Due to the increasing importance of both fields separate sections of this book have been dedicated to both concepts (see Section 5.7 for ionic liquid/compressed CO_2 and Section 5.6 for SILP catalysis) and related hydrogenation examples are described there in more detail.

5.3.2.2 Oxidation Reactions

Catalytic oxidation reactions in ionic liquids have attracted strong interest in recent years. This is not surprising, taking into account the oxidation stability of ionic liquids well known from electrochemical studies [11] and the great commercial importance of oxidation reactions. Moreover, for oxidation reactions with oxygen the non-volatility of the ionic liquid is of real advantage for safety issues. While the application of volatile organic solvents may be restricted by the formation of explosive mixtures in the gas phase this problem does not arise if a non-volatile ionic liquid is used as the solvent. However, along with these very attractive conceptual features of oxidation chemistry in ionic liquids come two principal problems. First, oxygen solubility in most ionic liquids is relatively low [123], meaning that only a small amount of oxygen is available at the catalytic active center for reaction. Second, the polarity of the oxidation product is usually much higher than the polarity of the starting material, making extraction from the ionic liquid sometimes difficult (a convincing solution to this problem is to work in a hydrophobic ionic liquid and to extract the product into water). Since, additionally, almost all preparative oxidations lead to a product of a higher boiling point compared to the feedstock, product recovery by evaporation can be difficult. The field has been recently reviewed in detail by Muzart [124].

First attempts to use ionic liquids in transition metal catalyzed oxidation reactions were described by Howarth et al. in 2000. They oxidized various aromatic aldehydes to the corresponding carboxylic acids using $Ni(acac)_2$ dissolved in [BMIM][PF_6] as the catalyst and oxygen at atmospheric pressure as the oxidant [125]. However, this reaction cannot be considered as a real challenge. Moreover, the catalyst loading used for the described reaction was rather high (3 mol%). The same combination of catalyst, ionic liquid and oxidant has also been used in the synthesis of ethylbenzene hydroperoxide from ethylbenzene [126]. However, in this case it is more accurate to describe the reaction system as a solution of the catalyst and ionic liquid in ethyl benzene. The two principal advantages of the ionic liquid are that it has a greater solubility in the ethylbenzene and the poorly coordinating anion competes less well for the metal center than the previously used tetraalkylammonium halide salts. A more recent attempt to oxidize aldehydes to acids using $MeReO_3$ in 1-butyl-3-methyimidazolium ionic liquids with aqueous H_2O_2 as oxidizing agent has been reported by Bernini et al. [127].

Many examples have been published demonstrating the feasibility of ionic liquid media for the transition metal catalyzed oxidation of alcohols to ketones. An overview is given in Table 5.3-7.

These publications demonstrate that there is no limitation on the applicability of the ionic liquid methodology concerning the sort of alcohol. Nearly all oxidations in ionic liquids given in Table 5.3-7 were tested with primary, secondary, allylic, benzylic alcohols and phenols with good results. Ruthenium with its high number of accessible oxidation states is a very suitable metal for oxidation reactions in ionic liquids (Table 5.3-7, entries 1–3). As an additional benefit it can be used in the form of its perruthenate salt. Thus immobilization and possibly stabilization of the metal is feasible. It should be mentioned that these oxidation reactions have been reported

Table 5.3-7 Oxidation reactions of alcohols in ionic liquids.

Entry	Catalyst	Ionic liquid	Oxidizing agent	Ref.
1	tetra-N-propylammonium-perruthenate (TPAP)	[EMIM][PF$_6$]/CH$_2$Cl$_2$ or [Et$_4$N][Br]/CH$_2$Cl$_2$[a] [BMIM][X][b]	N-methylmorpholine-N-oxide (NMO) or O$_2$	[128, 129]
2	RuCl$_3$ or [RuCl$_2$(PPh$_3$)$_3$]	[Oct$_3$MeN][Cl] or [Me$_4$N][OH]	O$_2$	[130]
3	RuCl$_3$	[BMIM][BF$_4$], [BMIM][PF$_6$], [BMIM][OOCCF$_3$]	O$_2$	[131]
4	[Pd(OAc)$_2$]	[BMIM][BF$_4$]	t-BuOOH	[132]
5	[BMIM]$_4$[W$_{10}$O$_{23}$]	[BMIM][Br]	H$_2$O$_{2(aq)}$	[133]
6	[BMIM]$_3$[PO$_4$(W(O)(O$_2$)$_2$)$_4$]	[BMIM][Br]	H$_2$O$_{2(aq)}$	[134]
7	CuCl + TEMPO	[BMIM][PF$_6$]	O$_2$	[135]
8	[Cu(ClO$_4$)$_2$] + acetamino-TEMPO	[BMPy][PF$_6$]	O$_2$	[136]
9	CuCl$_2$	[BMIM][Cl]/n-Butanol	O$_2$	[137, 138]
10	MnO$_2$	[BMIM][BF$_4$],[c,d] [BMIM][PF$_6$][c]	aerobic	[139, 140]
11	Mn(salen)-complex	[BMIM][PF$_6$]/CH$_2$Cl$_2$	PhI(OAc)$_2$	[141]

[a] Ref. [128].
[b] Ref. [129].
[c] Ref. [139].
[d] Ref. [140].

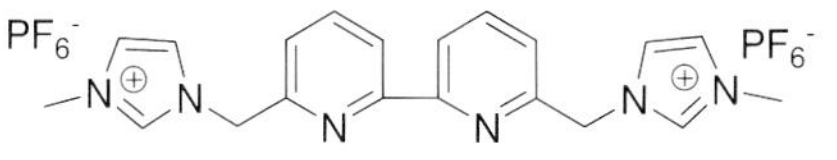

Fig. 5.3-11 Ionic ligand as applied by Wu et al. for the copper catalyzed oxidation of alcohols in [BMIM][PF$_6$].

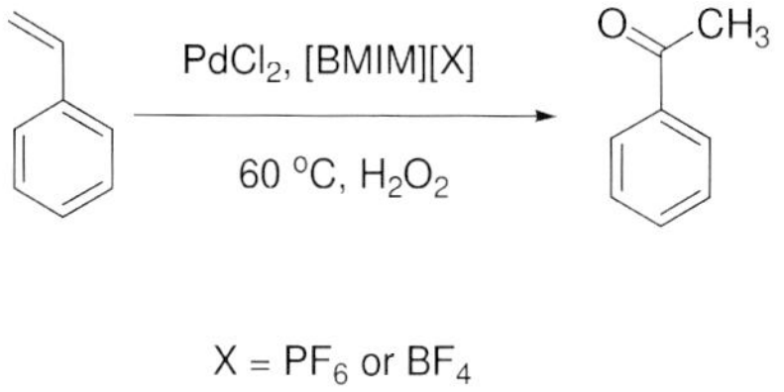

Scheme 5.3-13 The reaction of styrene with H$_2$O$_2$ in the presence of ionic liquids.

to be very sensitive to impurities in the ionic liquid so proper preparation and purification of the ionic medium is necessary [129]. Interesting details concerning the probability of over-oxidation of primary alcohols to acids have been discussed in a number of the published examples (Table 5.3-7, entries 1–3, 7, 8). While it is well known that even traces of water trigger over-oxidation, this consecutive reaction was not found in wet ionic liquids, even addition of water did not lead to significant acid formation. Possibly the strong interaction of the ionic liquid with water makes the latter unavailable for the reaction.

Obviously, all applied ionic liquids were stable against the selected oxidizing agents as no hint of any ionic liquid oxidation is found in the publications. For all these published examples catalyst immobilization and potential recyclability serves as motivation to apply the ionic reaction medium. To improve the immobilization of the catalyst, Wu et al. suggested later a bipyridine ligand carrying two 1-methylimidazolium hexafluorophosphate moieties to be a more appropriate ligand system (Fig. 5.3-11) [142].

The oxidation of alkenes (Wacker oxidation, mainly styrene to acetophenone, Scheme 5.3-13) has been reported to be catalyzed by PdCl$_2$ in the presence of e.g. [BMIM][PF$_6$] [143]. The need for only a small excess (1.15 equiv.) of aqueous H$_2$O$_2$ was demonstrated, which is a significant improvement in H$_2$O$_2$ utilization compared to previously reported methods.

Furthermore thio compounds were oxidized to disulfides with thiols as starting material, using cobalt(II) phtalocyanines in [BMIM][BF$_4$] with oxygen [144]. The synthesis of sulfoxides was reported in ionic liquids using thioethers as starting material, a heterogeneous, mesoporous Ti or Ti/Ge catalyst and hydrogen peroxide [145]. Ionic liquids were also applied as reaction and catalyst immobilization media for the Pt(II)-catalyzed oxidation of ketones [146], the iron(III) porphyrin and phosphotungstic acid catalyzed oxidation of oximes [147] as well as for the Baeyer-Villiger oxidation of cyclic ketones [146, 148].

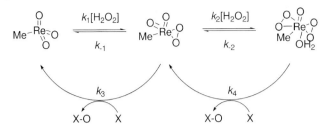

Scheme 5.3-14 Rhenium oxo- and peroxo-species present in the MTO–UHP epoxidation system in ionic liquids.

Since 2000 several publications dealing with epoxidations and dihydroxylations of olefins have been published. The oxidation of alkenes and allylic alcohols using the urea–H_2O_2 adduct (UHP) as oxidant and methyltrioxorhenium (MTO) dissolved in [EMIM][BF_4] as catalyst was described by Abu-Omar et al. [149]. Both MTO and UHP dissolve completely in the ionic liquid. Conversions were found to depend on the reactivity of the olefin and the solubility of the olefinic substrate in the reactive layer. In general, the reaction rates of the epoxidation reaction were found to be comparable to those obtained in classical solvents. Spectroscopic investigations showed that both monoperoxo- and diperoxorhenium species were present in the solution and active in the oxidation chemistry (Scheme 5.3-14, see Section 5.3.1.3 for analytic details).

The same system, with [BMIM][BF_4] as the ionic liquid, was applied for the epoxidation of glycols [150]. Here, no attempt was made to isolate the somewhat unstable epoxide and the reaction was conducted in the presence of dibutylphosphate, which gave the glycosyl phosphate product.

Detailed kinetic investigations by Abu-Omar et al. have elucidated the operation of these rhenium oxo- and peroxo-systems in ionic liquids [71b]. First, it was found that the rate of the oxidation by the diperoxorhenium species (k_4) is greater than that by the monoperoxorhenium species (k_3) for all substrates investigated in all of the ionic liquids applied, whereas in acetonitrile they are approximately the same. For most of the substrates used the k_4 values in [EMIM][BF_4] were similar to those in a 1:1 CH_3CN/H_2O mixture. Investigating different ionic liquids, it was shown that there was no real effect of changing the cation of the ionic liquid, but there was a distinct effect on changing the anion, on both k_3 and k_4. This was attributed to the greater coordinating ability of the nitrate ion compared to [BF_4]$^-$. The kinetics of the formation of the two peroxo species in ionic liquids (k_1 and k_2) have also been studied [151]. The formation of the monoperoxorhenium species was found to be very fast and exact measurement of k_1 was therefore not possible. However, k_2 could be deduced using either UHP or aqueous H_2O_2 (30%) as the peroxide source. In pure [BMIM][NO_3] k_2 was approximately the same as in acetonitrile, but as water was added the rate constant increased. Several water miscible ionic liquids were compared as 9:1 (v/v) ionic liquid/water solutions and k_2 was found to be the same for both peroxide sources in all of them.

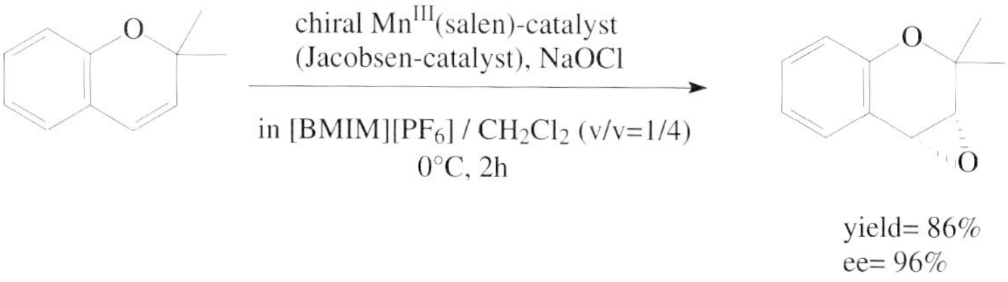

chiral Mn^{III}(salen)-catalyst
(Jacobsen-catalyst), NaOCl

in [BMIM][PF$_6$] / CH$_2$Cl$_2$ (v/v=1/4)
0°C, 2h

yield= 86%
ee= 96%

Scheme 5.3-15 Mn-catalyzed asymmetric epoxidation in a [BMIM][PF$_6$]/CH$_2$Cl$_2$ (v/v = 1/4) solvent mixture.

Song and Roh investigated the epoxidation of e.g. 2,2,-dimethylchromene with a chiral Mn^{III}(salen) complex (Jacobsen catalyst) in a mixture of [BMIM][PF$_6$] and CH$_2$Cl$_2$ (1:4 v/v) using NaOCl as the oxidant (Scheme 5.3-15) [152].

Even if the reaction medium consisted mainly of CH$_2$Cl$_2$ the authors described a clear enhancement of the catalyst activity by the addition of the ionic liquid. In the presence of the ionic liquid a 86% conversion of 2,2-dimethylchromene was observed after 2 h. Without the ionic liquid the same conversion was only obtained after 6 h. In both cases the enantiomeric excess was as high as 96%. Moreover, the ionic catalyst solution could be reused several times after product extraction. However, after five recycles, the conversion dropped from 83% to 53%, which was explained, according to the authors, by a slow degradation process of the Mn^{III} complex.

For non-activated olefins like styrol, cyclohexene and cyclooctene Chauhan et al. investigated iron(III)porpyhrine systems in combination with H$_2$O$_2$ as a suitable epoxidation system [153]. The metalloporphyrines were used as model catalysts for cytochrome P450 and were immobilized in [BMIM]Br. The yields were found to depend on the applied olefin and varied from 42% for cyclohexene, 74% for styrene, to 81% for cyclooctene. Recycling was possible for the epoxidation of styrene in 5 runs under biphasic reaction conditions and the yield decreased from 74% for the first run to 62% for the fifth run.

For the oxidation of alkanes Li et al. used iodobenzene diacetate [PhI(OAc)$_2$] as oxygen source. An electron-deficient manganeseporphyrin catalyst was immobilized in [BMIM][PF$_6$] and tested in liquid–liquid biphasic reaction mode with CH$_2$Cl$_2$ as organic phase [154]. They found the catalyst more active in the [BMIM][PF$_6$]/CH$_2$Cl$_2$ system than in neat CH$_2$Cl$_2$. The increase in activity was attributed by them to the higher polarity of the ionic liquid.

As a further oxygen source *tert*-butyl hydroperoxide (TBHP) was applied in combination with dioxomolybdenum(VI) complexes immobilized in ionic liquid for the epoxidation of *cis*-cyclooctene [155]. Again, a strong dependence of reactivity, selectivity and the ability for recycling conditions on the chosen ionic liquid and solvent was described.

Finally, it should be mentioned that Yamaguchi et al. used the SILP concept to immobilize an oxidation catalyst [156]. They found that peroxotungstate [\{W(=O)(O$_2$)$_2$(H$_2$O)\}$_2$(μ-O)]$^{2-}$ can be immobilized on a dihydroimidazolium-based ionic liquid-modified SiO$_2$ surface to get an efficient heterogeneous epoxidation system with H_2O_2 as oxidant. The system showed the same activity as the corresponding homogeneous analogue. The catalyst was reused three times without any loss in catalytic activity and selectivity (99% yield, >99% selectivity). More details about catalytic SILP systems are to be found in Section 5.6.

5.3.2.3 Hydroformylation

In hydroformylation, biphasic catalysis is a well-established method for effective catalyst separation and recycling. In the case of Rh-catalyzed hydroformylation reactions this principle is technically realized in the Ruhrchemie-Rhône-Poulenc-process, where water is used as the catalyst phase [157]. Unfortunately, this process is limited to C_2–C_5-olefins due to the low water solubility of higher olefins. Nevertheless, the hydroformylation of many higher olefins is of commercial interest. One example is the hydroformylation of 1-octene for the selective synthesis of linear nonanal. The latter can be obtained in high selectivity by application of special ligand systems around the catalytic center. However, the additional costs related to these ligands make it even more economically attractive to develop new methods for an efficient catalyst separation and recycling. In this context, biphasic catalysis using an ionic liquid as catalyst layer is a highly promising approach.

As early as 1972 Parshall described the platinum-catalyzed hydroformylation of ethene in tetraethylammonium trichlorostannate melts, $[NEt_4][SnCl_3]$ [1]. The ionic liquid used for these investigations has a melting point of 78 °C. Later, the platinum-catalyzed hydroformylation in the chlorostannate room-temperature ionic liquid $[BMIM]Cl/SnCl_2$ was studied in the author's group anew. The hydroformylation of 1-octene was carried out with remarkable n/iso- selectivities (Scheme 5.3-16) [158].

Despite the limited solubility of 1-octene in the ionic catalyst phase, a remarkable activity of the platinum catalyst was achieved (turnover frequency (TOF) = $126\,h^{-1}$). However, the system has to be carefully optimized to avoid significant formation of hydrogenated by-product. Detailed studies to identify the best reaction conditions revealed that in the chlorostannate ionic liquid $[BMIM]Cl/SnCl_2$ $[X(SnCl_2) = 0.55]$ the highest ratio of hydroformylation to hydrogenation is found at high syngas pressure and low temperature. At 80 °C and 90 bar CO/H_2 pressure more than

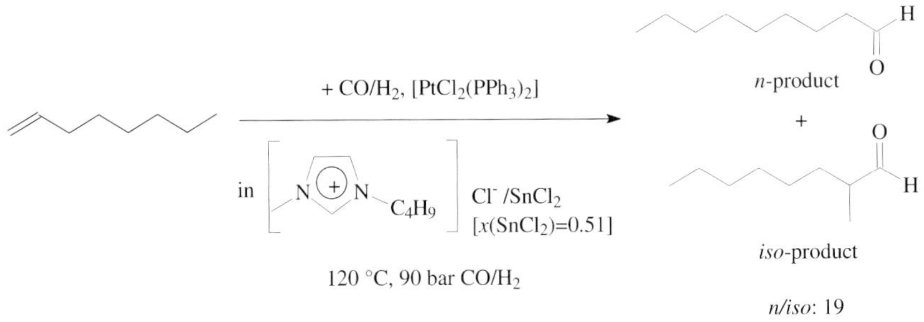

Scheme 5.3-16 Biphasic, Pt-catalyzed hydroformylation of 1-octene using a slightly acidic [BMIM]Cl/SnCl$_2$ ionic liquid as catalyst layer.

90% of all products are n-nonanal and iso-nonanal, the ratio between these two hydroformylation products being as high as 98.6:1.4 (n/iso = 70.4) [158].

Moreover, these experiments revealed some unique properties of the chlorostannate ionic liquids. In contrast to other known ionic liquids, the chlorostannate system combines a certain Lewis-acidity with high compatibility to functional groups. The first led, in the hydroformylation of 1-octene, to the activation of $(PPh_3)_2PtCl_2$ by a Lewis acid–base reaction with the acidic ionic liquid medium. The high compatibility to functional groups is demonstrated by the catalytic reaction in the presence of CO and hydroformylation products.

Later, van Eldik et al. studied in more detail the kinetics of the formation of the active hydroformylation catalysts *cis*-$[Pt(PPh_3)_2Cl(SnCl_3)]$ and *cis*-$[Pt(PPh_3)_2(SnCl_3)_2]$ from the precursor *cis*-$[Pt(PPh_3)_2Cl_2]$ in the presence of $SnCl_2$ in different imidazolium-based chlorostannate ionic liquids (for analytic details see Section 5.3.1.3) [73].

Ruthenium- and cobalt-catalyzed hydroformylation of internal and terminal alkenes in molten $[PBu_4]Br$ was reported by Knifton as early as 1987 [2]. He described a stabilization of the active ruthenium-carbonyl complex by the ionic medium. An increased catalyst lifetime at low synthesis gas pressures and higher temperatures was observed.

First investigations of the rhodium-catalyzed hydroformylation in room-temperature liquid molten salts were published by Chauvin et al. in 1995 [6, 159]. The hydroformylation of 1-pentene with the neutral catalyst system $[Rh(CO)_2(acac)]$/triarylphosphine was carried out in a biphasic reaction using [BMIM][PF_6] as the ionic liquid. However, with none of the ligands tested was it possible to combine high activity, complete retention of the catalyst in the ionic liquid and high selectivity for the desired linear hydroformylation product at this time. The use of PPh_3 resulted in significant leaching of the Rh catalyst out of the ionic liquid layer. In this case, the catalyst is active in both phases, which makes a clear interpretation of solvent effects on the reactivity difficult. The catalyst leaching could be suppressed by the application of sulfonated triaryl phosphine ligands, but a major decrease in catalytic activity was found with these ligands (TOF = 59 h^{-1} with tppms compared to 333 h^{-1} with PPh_3). Moreover, all of the ligands used in Chauvin's work showed poor selectivity to the desired linear hydroformylation product (n/iso-ratio between 2 and 4). Obviously, the Rh-catalyzed, biphasic hydroformylation of higher olefins in ionic liquids requires the use of ligand systems that are specifically designed for this application. Thus, these early results stimulated the research for other immobilizing, ionic ligand systems that provide good catalyst immobilization without deactivation of the catalyst.

Ligand optimization studies for hydroformylation in ionic liquids
A first ligand system especially designed for the use in ionic liquids was described in 2000 by Salzer et al. [160]. Cationic ligands with a cobaltocenium backbone were successfully used in the biphasic, Rh-catalyzed hydroformylation of 1-octene. 1,1′-Bis(diphenylphosphino)cobaltocenium hexafluorophosphate (cdpp) proved to

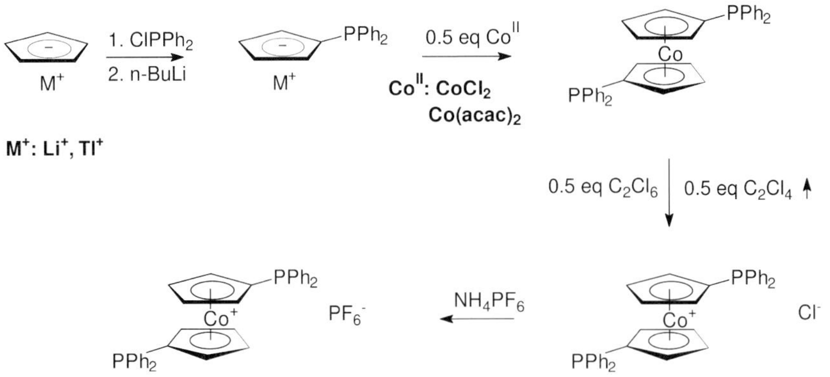

Scheme 5.3-17 Synthesis of 1,1′-Bis(diphenylphosphino)cobaltocenium hexafluorophosphate.

be an especially promising ligand. The compound can be synthesized according to Scheme 5.3-17 by mild oxidation of 1,1′-bis(diphenylphosphino)cobaltocene with C_2Cl_6 and anion exchange with $[NH_4][PF_6]$ in acetone. (for detailed ligand synthesis see Ref. [160]).

The results obtained in the biphasic hydroformylation of 1-octene are presented in Table 5.3-8. In order to evaluate the properties of the ionic diphosphine ligand with a cobaltocenium backbone, the results with the cdpp ligand are compared with those obtained with PPh_3, two common neutral bidentate ligands and with Natppts as a standard anionic ligand [160].

It is noteworthy that a clear enhancement of the selectivity to the linear hydroformylation product is observed only with cdpp (Table 5.3-8, entry e). With all other ligands, the n/iso-ratios are in the range 2 to 4. While this is in accordance with known results in the case of PPh_3 (entry a) and dppe (entry c) (in comparison to the monophasic hydroformylation [161]) and also with reported results in the case of

Table 5.3-8 Comparison of different phosphine ligands in the Rh-catalyzed hydroformylation of 1-octene in [BMIM][PF$_6$].

Entry	Ligand	TOF/h^{-1}	n/iso	S (n-ald)[a] (%)
a	PPh$_3$	426	2.6	72
b	tppts	98	2.6	72
c	dppe	35	3.0	75
d	dppf	828	3.8	79
e	cdpp	810	16.2	94

Conditions: ligand/Rh: 2:1, CO/H$_2$ =1:1, $t = 1$ h, $T = 100$ cC, $p = 10$ bar, 1-octene/Rh = 1000, 5 mL [BMIM][PF$_6$]; dppe: bis(diphenylphosphinoethane); dppf: 1,1′-bis(diphenylphosphino)ferrocene.
[a] S (n-ald) = selectivity to n-nonanal in the product.

Natppts (entry b; in comparison to the biphasic hydroformylation of 1-pentene in [BMIM][PF$_6$] [6]), it is more remarkable for the bidentate metallocene ligand dppf.

Taking into account the high structural similarity of dppf and cdpp, their different influence on the reaction's selectivity has to be attributed to electronic effects. The electron density at the phosphorus atoms is significantly lower in the case of cdpp due to the electron-withdrawing effect of the formal cobalt(III) central atom in the ligand. This interpretation is supported by former work from Casey et al. [162] and Duwell et al. [163]. These groups described positive effects of ligands with electron-poor phosphorus atoms in selective hydroformylation reactions, which they attribute to their ability to allow back-bonding from the catalytically active metal atom. It has to be pointed out that with the phosphinocobaltocenium ligand cdpp the reaction takes place almost exclusively in the ionic liquid phase (almost clear and colorless organic layer, less than 0.5% Rh in the organic layer). An easy catalyst separation by decantation was possible. Moreover, it was found that the recovered ionic catalyst solution could be reused at least one more time with the same activity and selectivity as in the original run [160].

Cationic phosphine ligands containing guanidiniumphenyl moieties were originally developed to make use of their pronounced solubility in water [164, 165]. They were shown to form active catalytic systems in Pd mediated C–C coupling reactions between aryl iodides and alkynes (Castro-Stephens-Sonogashira reaction) [166] and Rh catalyzed hydroformylation of olefins in aqueous two-phase systems [167].

The modification of neutral phosphine ligands with cationic phenylguanidinium groups proved to be a very powerful tool to immobilize Rh-complexes in ionic liquids such as e.g. [BMIM][PF$_6$] [168]. The guanidinium-modified triphenylphosphine ligand was prepared according to Scheme 5.3-18 by anion exchange with [NH$_4$][PF$_6$] in aqueous solution from the corresponding iodide salt. The latter can be prepared as previously described by Stelzer et al. [165].

In contrast to the use of PPh$_3$ as the ligand, the reaction takes place solely in the ionic liquid layer when the guanidinium-modified triphenylphosphine is applied. In the first catalytic run the hydroformylation activity was found to be somewhat lower than with PPh$_3$ (probably due to the fact that some of the activity observed with PPh$_3$

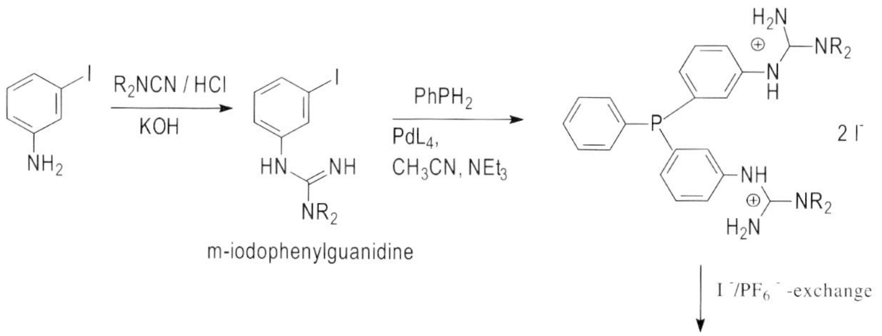

Scheme 5.3-18 Synthesis of a guanidinium-modified triphenylphosphine ligand.

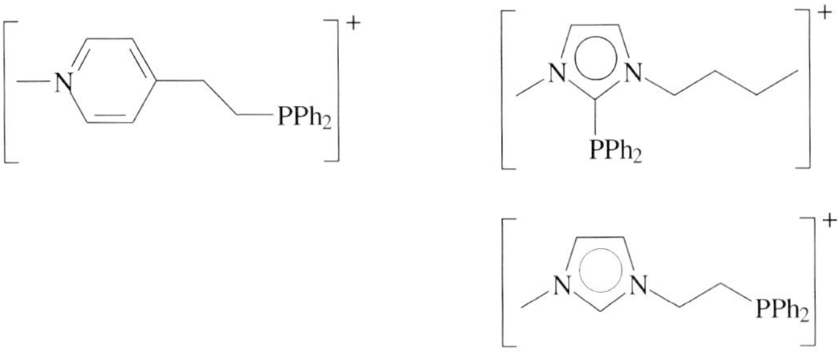

Fig. 5.3-12 Cationic diphenylphosphine ligands as used in the biphasic, Rh-catalyzed hydroformylation of 1-octene in e.g. [BMIM][PF$_6$].

takes place in the organic layer). However, due to the excellent immobilization of the Rh catalyst with the guanidinium modified ligand [leaching is <0.07% per run according to ICP analysis (detection limit)], the catalytic activity does not drop over the first ten recycling runs. For the recycling runs the organic layer was decanted after each run (under normal atmosphere) and the ionic catalyst layer remained in the autoclave for the next hydroformylation experiment. Already after five recycling runs, the overall catalytic activity obtained with the ionic catalyst solution containing the guanidinium-modified ligand is higher than can be realized with the simple PPh$_3$ ligand. With both ligands the n/iso-ratio of the hydroformylation products is in the expected range of 1.7–2.8.

Alternative methods to immobilize monodentate phosphine ligands by attaching them to ionic groups with high similarity to the ionic liquid's cation have also been reported. Both pyridinium-modified phosphine ligands [169] and imidazolium-modified phosphine ligands [170, 171] have been synthesized and applied in Rh-catalyzed hydroformylation (see Fig. 5.3-12). While the presence of the ionic group led to better immobilization of the Rh catalyst in the ionic liquid in all cases no outstanding reactivity or selectivity were observed with these ligands. This is not really surprising since all these ligands are electronically and sterically closely related to PPh$_3$.

Further development was aimed at adopting this immobilization concept to a ligand structure that promises better regioselectivity in the hydroformylation reaction. It is well-known that bidentate phosphine ligands with large P–metal–P bite angles form highly regioselective hydroformylation catalysts [172]. Here, xanthene type ligands (P–metal–P∼110°) developed by van Leeuwen's group proved to be especially suitable allowing, for example, an overall selectivity of 98% towards the desired linear aldehyde in 1-octene hydroformylation [173, 174].

While unmodified xanthene ligands (Fig. 5.3-13(a)) show highly preferential solubility in the organic phase in the biphasic mixture 1-octene/[BMIM][PF$_6$], even at room temperature, the application of the guanidinium-modified xanthene ligand (Fig. 5.3-13(b)) resulted in excellent immobilization of the Rh catalyst in the ionic liquid.

The guanidinium-modified ligand is synthesized by reacting the xanthenediphosphine [175] with iodophenylguanidine in a Pd(0)-catalyzed coupling reaction. The

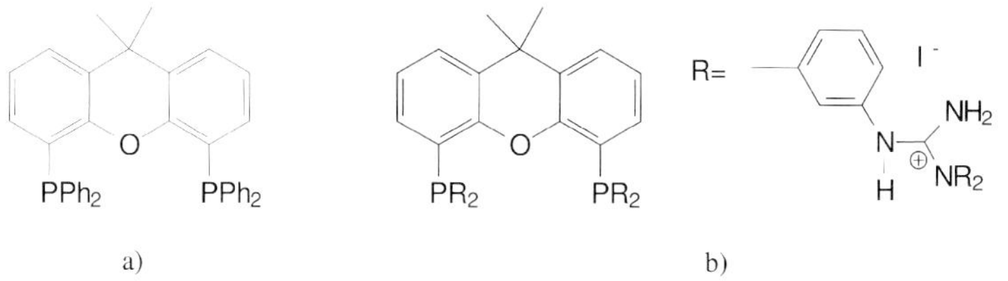

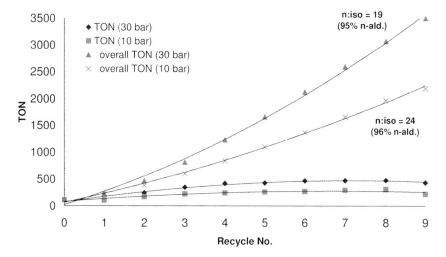

Fig. 5.3-13 Unmodified (a) and guanidinium-modified xanthene ligand (b) as used in the biphasic, Rh-catalyzed hydroformylation of 1-octene.

Fig. 5.3-14 Recycling experiments – Rh-catalyzed, biphasic 1-octene hydroformylation in [BMIM][PF$_6$] using a guanidinium-modified diphosphine ligand with a xanthene backbone.

ligand was tested in the Rh-catalyzed hydroformylation in ten consecutive recycling runs. The results are presented in Fig. 5.3-14. It is noteworthy that the catalytic activity increases during the first runs, achieving a stable level only after the fourth recycling run. This behavior is attributed to a certain catalyst pre-forming time but also to impurities of iodoaromate in the ligand used. Probably, these are slowly washed out of the catalyst layer during the first catalytic runs.

After ten consecutive runs the overall turnover number reached 3500 mol 1-octene converted/mol Rh-catalyst. In agreement with these recycling experiments, no Rh could be detected in the product layer by AAS or ICP, indicating a leaching of less then 0.07%. In all experiments very good selectivities for the linear aldehyde were obtained, thus proving that the attachment of the guanidinium moiety to the xanthene backbone does not influence its known positive effect on the regioselectivity of the reaction. Thus, these results demonstrate that the modification of known phosphine ligands with guanidinium groups is a simple and very efficient method to fully immobilize transition metal complexes in ionic liquids.

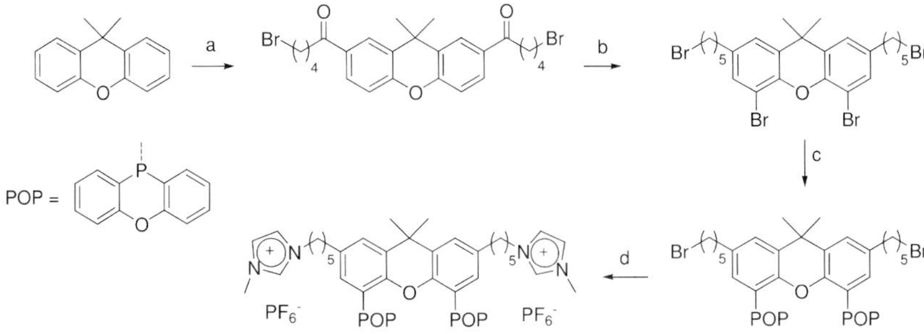

Scheme 5.3-19 Synthesis of a dicationic phenoxaphosphino-modified xantphos type ligand as used in the hydroformylation of 1-octene in [BMIM][PF$_6$].

An imidazolium-based dicationic phenoxaphosphino-modified xantphos type ligand was applied by van Leeuwen et al. for the hydroformylation of 1-octene in [BMIM][PF$_6$] [176]. The ligand was prepared in a six-step synthesis according to Scheme 5.3-19.

The ligand provided excellent immobilization of the catalyst system with no Rh or P leaching detectable in the organic product phase and no loss in catalyst activity or selectivity observable in seven recycling experiments. In contrast, a catalyst preformation period was observed during the first four cycles, after which the catalyst activity reached a constant TOF level of 110 h^{-1}. Additionally, the ionic catalyst solution was found to be pretty robust as it could be stored under air for more than 14 days without loss of activity. This study was extended recently by the same group still using dicationic phenoxaphosphino-modified xantphos ligands for the Rh-catalyzed hydroformylation of 1-octene in [BMIM][PF$_6$] [177]. They carried out detailed investigations on stirring speed and catalyst concentration variation. When lowering the rhodium concentration from 6.4 mmol l^{-1} to 1.7 mmol l^{-1} a dramatic increase in the TOF together with a slight increase in regioselectivity was observed with TOFs up to 7400 h^{-1} and product rations n-nonanal/iso-nonanl as high as 64 being reported. The results indicated that the overall reaction rate of the system was still influenced by mass transport issues, even at a stirring rate of 1600 min^{-1}.

Apart from charged phospine ligands, phosphite ligands have also been explored in ionic liquids. Since phosphite ligands are usually unstable in aqueous media this adds, apart from the much better solubility of higher olefins in ionic liquids, another important advantage to biphasic hydroformylation using ionic liquids in comparison to the, well-known, biphasic reaction in water. The group of Olivier-Bourbigou has shown, for example, that phosphite ligands can be used in the Rh-catalyzed hydroformylation of 1-hexene in various imidazolium and pyrrolidinium ionic liquids with better selectivity to linear aldehyde compared to the well-known phosphine systems [169]. An earlier study by Keim et al. used a bulky phosphite ligand to promote the selective Rh-catalyzed hydroformylation of methyl-3-pentenoate in [BMIM][PF$_6$] [30].

Ionic liquid optimization studies for hydroformylation in ionic liquids

Apart from all these attempts to improve the immobilization of the phosphine ligands in the ionic medium, other research activities were directed to optimize the structure of the ionic liquid medium for specific hydroformylation applications. Olivier-Bourbigou and coworkers investigated the hydroformylation of 1-hexene in a variety of ionic liquids with imidazolium and pyrrrolidinium cations and a range of different anions [169]. Applying [Rh(CO)$_2$(acac)] with four equivalents of the charged phosphine TPPMS as the catalyst system they measured the turnover frequency of the catalyst in the different ionic liquids and found for different [BF$_4$]$^-$, [PF$_6$]$^-$, [CF$_3$SO$_3$]$^-$, [CF$_3$CO$_2$]$^-$ melts the TOFs to be dependent on the solubility of the 1-hexene in the respective ionic liquid.

An important step towards the technical applicability of ionic liquids in industrial hydroformylation is the successful replacement of hexafluorophosphate (and other halogen-containing) ionic liquids by some cheap and halogen-free ionic liquids. Rh-catalyzed hydroformylation offers a specific potential in this respect as the reaction is not too sensitive to slightly coordinating anions such as alkyl- and aryl sulfonates and alkylsulfates. The first practical attempts were made by Andersen et al. [13]. They investigated the Rh-catalyzed hydroformylation of 1-hexene in high melting phosphonium tosylate salts, such as butyltriphenylphosphonium tosylate (mp = 116–117 °C). Obviously, the high melting point of the salts used makes the processing of the reaction difficult, even if the authors describe easy product isolation by pouring the product off the solid catalyst medium at room temperature.

Later, Wasserscheid et al. introduced much lower melting 1,3-dialkylimidazolium benzenesulfonate, tosylate and octylsulfate ionic liquids to biphasic, Rh-catalyzed hydroformylation of 1-octene [15]. The catalyst activity obtained with these systems was in all cases equal or even higher than that with the commonly used [BMIM][PF$_6$]. Taking into account the much lower costs of the ionic medium, the better hydrolysis stability and the wider disposal options related to, for example, an octylsulfate ionic liquid in comparison to [BMIM][PF$_6$], the technical relevance of this anion switch is very obvious.

In order to eliminate the possibility for *in situ* carbene formation Raubenheimer et al. synthesized 1-alkyl-2,3-dimethylimidazolium triflate ionic liquids and applied these as solvents in the rhodium catalyzed hydroformylation of 1-hexene and 1-dodecene [178]. Both, the classical Wilkinson type complex [RhCl(TPP)$_3$] and the chiral, stereochemically pure complex (−)-(η^4-cycloocta-1,5-diene)-(2-menthyl-4,7-dimethylindenyl)rhodium(I) were applied. The Wilkinson catalyst showed low selectivity towards n-aldehydes whereas the chiral catalyst formed branched aldehydes predominantly. Hydrogenation was significant with up to 44% alkanes being formed and also a significant activity for olefin isomerization was observed. Additionally, hydroformylation was found to be slower in the ionic liquid than in toluene. Some of the findings were attributed by the authors to the lower gas solubility in the ionic liquid and the slower diffusion of the reactive gases H$_2$ and CO into the ionic medium.

Highly viscous triazine-based ionic liquids have been recently applied for the rhodium catalyzed hydroformylation of 1-octene [179]. Hydroformylation was carried out at 69 bar syngas pressure and total 1-octene conversion was achieved within 14 to 19 h. Only the two isomers n-nonanal and 2-methyloctanal were obtained as products and no hydrogenation or isomerization of the feedstock was reported. The catalyst was shown to be recyclable in principle with only slight decrease in activity and selectivity over two cycles. Rh leaching was determined by ICP to be between 0.40 and 26.5 ppm.

The hydroformylation of 1-tetradecene in novel ionic liquids consisting of ammonium salts with polyether-tail has been reported by Jin et al. [180]. The highly viscous salts were synthesized by protonation of a polyether-tailored amine with p-toluenesulfonic acid. The Rh-catalyzed hydroformylation reactions were carried out at 105 °C and 50 bar syngas pressure in a biphasic system using heptane as extraction phase. Severe Rh leaching was observed when using TPP as ligand, similar to results reported earlier by Chauvin [6b]. The ligands TPPTS and octylpolyethyleneglycol-phenyl-phosphite were applied in order to achieve better immobilization of the Rh in the ionic liquid phase. The Rh leaching could be suppressed in this way without loss of activity. However, the selectivity to the linear pentadecanal was extremely low in both cases with only 17 and 19% of the desired product being formed respectively.

Process related studies
Process related studies of transition metal catalyzed hydroformylation in ionic liquids mainly focused on a better utilization of the amount of ionic liquid and thus on a reduction of the amount of ionic liquids required. A key aspect in this context is to enhance the solubility of the reactants CO, H_2 and olefin in the ionic liquid and to enhance the mass transfer between the gas phase/organic fluid and the ionic liquid phase.

The solubilities of H_2 [181] and CO [182] in many different ionic liquids have been measured in this context by Dyson and coworkers using high-pressure ^{1}H- or ^{13}C-NMR spectroscopy. CO solubility at ambient conditions (Henry's law constants) were found to be in the range of 8.05–0.67 × 10^2 MPa which were in the same range than those measured for H_2 in the same ionic liquids (about 6.6–0.7 × 10^2 MPa). In order to establish the effect of the CO solubility on the rate of hydroformylation reactions, the same authors studied the [RhH(CO)(PPh$_3$)$_3$] catalyzed hydroformylation of 5-hexen-2-one comparing catalyst activities in the different ionic liquids with the determined gas solubilities.

As expected from the known kinetics of hydroformylation the authors found that increasing hydrogen solubility in the ionic liquid (either induced by increased hydrogen pressure or by higher hydrogen solubility) increased the reaction rate while higher CO concentration slowed the hydroformylation down.

A well established way in multiphasic catalysis to improve the substrate solubility in the catalyst phase is the addition of appropriate co-solvents. This approach has been studied for the hydroformylation of 1-octene with a [Rh(CO)$_2$(acac)]-(sulfoxantphos) catalyst system in [BMIM][PF$_6$] by Dupont et al. [183]. They found that the selectivity to n-nonanal was best in pure [BMIM][PF$_6$] compared to the same ionic liquid with e.g. added co-solvent toluene. They reported catalyst activities which were very similar for the reaction in the pure ionic liquid and the ionic liquid/toluene system.

Another interesting development to improve the processibility of the Rh-catalyzed hydroformylation reaction in ionic liquids is to work in the presence of supercritical CO_2 as extraction phase. The use of $scCO_2$ as the mobile phase in such a continuous multiphasic operation brings along a number of very attractive features such as very elegant product removal from the catalyst phase, reduction of the ionic liquid's melting point [184] and viscosity [185] (and thus diffusion rate [186]). Most important for the hydroformylation application may be the fact that the relatively large amount of CO_2 that dissolves in the ionic liquid (see Section 3.3. for details) acts as a very powerful cosolvent to increase hydrogen solubility in the ionic liquid [187]. The continuous hydroformylation of 1-octene in the biphasic system [BMIM][PF_6]/$scCO_2$ has been studied in detail by Cole-Hamilton et al. [188, 189]. Together with other examples of transition metal catalysis in ionic liquid/$scCO_2$ systems this specific example will be described in detail in Section 5.7.

Finally, SILP catalysis has been introduced in the last three years using mainly hydroformylation as the model reaction to develop the technology [190]. Due to the general importance of SILP catalysis a separate section is devoted to this topic (see Section 5.6) in which all details of the studied SILP-hydroformylation systems can be found.

5.3.2.4 Heck Reaction and Other Pd-catalyzed C–C-coupling Reactions
Early studies and reactions in [NBu$_4$]Br
The Heck reaction and other related transformations for selective C–C-couplings are receiving a great deal of attention among synthetic chemists, due to their versatility for fine chemical synthesis. However, these reactions suffer, in many cases, from the instability of the Pd catalysts used, leading to high catalyst consumption and difficult processing. In recent years, many different versions of the Heck reaction have been developed using ionic liquids as the reaction medium and these have already been extensively reviewed from different perspectives [191].

The use of ionic liquids as reaction media for the palladium-catalyzed Heck reaction was first described by Kaufmann et al. in 1996 [192]. The reaction of bromobenzene with acrylic acid butyl ester to *trans*-cinnamic acid butyl ester succeeded in high yield in molten tetraalkylammonium and tetraalkylphosphonium bromide salts, without addition of phosphine ligands (Scheme 5.3-20).

The authors describe a stabilizing effect of the ionic liquid on the palladium catalyst. In almost all reactions no precipitation of elemental palladium was observed, even at complete conversion of the aromatic halide. The reaction products were isolated by distillation from the non-volatile ionic liquid.

Extensive studies of the Heck reaction in low melting salts have been presented by Hermann and Böhm [193]. Their results indicate that the application of ionic solvents show clear advantages in comparison to commonly used organic solvents (e.g., DMF), especially for the conversion of the commercially interesting chloroarenes. With almost all catalyst systems tested an additional activation and stabilization was

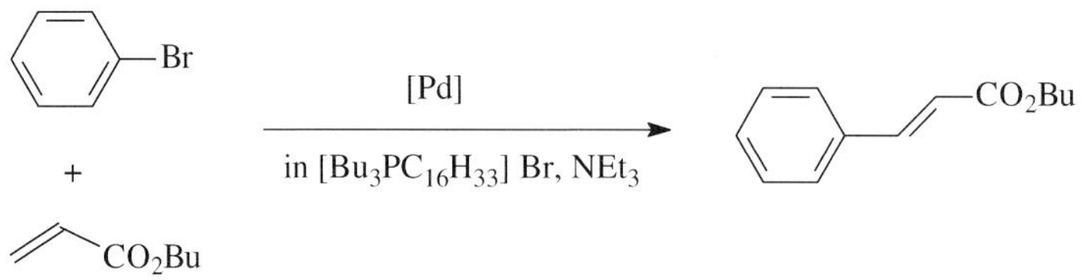

Scheme 5.3-20 Pd-catalyzed Heck reaction of acrylic acid butyl ester with bromobenzene carried out in a phosphonium bromide salt.

observed. Molten [NBu$_4$]Br (mp = 103 °C) proved to be a specially suitable reaction medium among the ionic solvent systems investigated. In the reaction of bromobenzene with styrene using diiodo-bis(1,3-dimethylimidazolin-2-ylidene)-palladium(II) as catalyst the yield of stilbene could be increased from 20% (DMF) to over 99% ([NBu$_4$]Br) under otherwise identical conditions. Again, a distillative product separation from the non-volatile ionic catalyst solution was possible. The latter could be re-used up to 13 times without significant drop in activity. Additional advantages of the new solvent concept arise from the excellent solubility of all reacting molecules in the ionic solvent and the possibility of using cheap inorganic bases.

This work was followed up by other research groups using different substrates and other Pd-precursor/ligand combinations in molten [NBu$_4$]Br for Heck coupling. Muzart et al. described the coupling of aryl iodides and bromides with allylic alcohols to the corresponding β-arylated carbonyl compounds [194]. Calò et al. reported the Heck coupling of substituted acrylates with bromobenzene catalyzed by Pd-benzothiazole carbene complexes in molten [NBu$_4$]Br [195]. The same solvent was found to be essential in investigations carried out by Buchmeiser et al. aiming for the Pd-catalyzed Heck coupling of aryl chlorides and for the amination of aryl bromides [196]. Later, Handy and coworkers demonstrated that the addition of tetrabutylammonium salts to [BMIM][BF$_4$] also had a beneficial effect for the reaction of methyl acrylate with iodobenzene. Depending on the anion of the tetrabutylammonium salt, the yield was improved from 53% (no additional salt) to 96% ([Bu$_4$N]Br), 90% ([Bu$_4$N]Cl), and 89% (([Bu$_4$N]I) respectively [197]. Similar findings were earlier described by Jeffery for Heck reactions in a strictly anhydrous organic medium, in a water–organic solvent mixture or in water alone, where it was believed that the salts were acting as phase-transfer catalysts [198]. Zou et al. reported Heck reactions of aryl halides with butyl acrylate and Suzuki reaction of p-tolylboronic acid with iodobenzene in high-melting-point [Bu$_4$N][BF$_4$], N,N-dialkylpyrrolidiniums and N,N-dialkylpiperidiniums melts with additional water or toluene to form biphasic mixtures [199]. They found a significant melting point suppression of the ionic liquid caused by water or toluene. In this way a ligand-less biphasic reaction system using PdCl$_2$ as catalyst could be realized at 90 °C. While one of the obvious objectives of the use of ionic liquids in Heck reactions is to increase the solubility of the organic starting materials in the catalyst solution, the ability to generate stable Heck catalysts without ligand addition suggests that the ionic liquid environment acts as a catalyst stabilizing medium in the system. Tetraalkylammo-

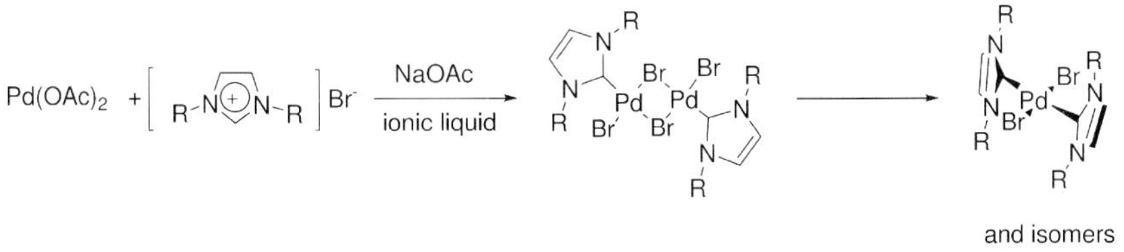

Scheme 5.3-21 The formation of [PdBr$_2$(BMimy)$_2$] in [BMIM]Br.

nium salts have also been used to stabilize nanoparticles/colloids when used as a solute in molecular solvents [200]. It was suggested by Herrmann that such effects may also play a role when [Bu$_4$N]Br was used as the ionic liquid [201].

In situ *carbene complex and Pd nanoparticle formation*
N-heterocyclic carbene (NHC) ligands have been used in volatile organic solvents to prepare catalysts for a wide variety of reactions, many palladium catalyzed [202]. The use of imidazolium-based ionic liquids in the Pd-catalyzed Heck reaction always carries the possibility of an *in situ* formation of Pd–carbene complexes (for more details see Section 5.3.1.2). The formation of the latter under the reaction conditions of the Heck reaction was confirmed by the investigations of Xiao et al. [49]. They described a significantly enhanced reactivity of the Heck reaction in [BMIM]Br in comparison to the same reaction in [BMIM][BF$_4$] and explained this difference with the fact that only in the bromide melt was the formation of Pd–carbene complexes observed. While 'palladium black' was observed to precipitate during reactions performed in [BMIM][BF$_4$], Xiao et al. could isolate the complexes [PdBr$_2$(BMimy)$_2$] and [Pd$_2$(μ-Br)$_2$Br$_2$(BMimy)$_2$] (BMimy = 1-butyl-3-methylimidazolylidene) from the [BMIM]Br solution (see Scheme 5.3-21). Synthesis of Pd-imidazolylidene complexes from imidazolium halide and Pd(OAc)$_2$ has been reported to proceed particularly easily [203].

However, when the authors dissolved independently prepared [PdBr$_2$(BMimy)$_2$] in [BMIM][BF$_4$] the yield of the reaction was the same as that when palladium acetate was used. This suggests that, if the [PdBr$_2$(BMimy)$_2$] complex is responsible for the improved yields in [BMIM]Br, it must rapidly transform in [BMIM][BF$_4$].

Indication for a possible transformation pathway was given by an EXAFS study of palladium acetate dissolved in [BMIM][BF$_4$], [BMIM][PF$_6$], [BPy][BF$_4$] and [BPy][PF$_6$] showing that a gradual change to palladium metal occurred [204]. In contrast, the use of [HMIM]Cl led to the formation of [PdCl$_2$(HMimy)$_2$] which is entirely consistent with Xiao's observations above [49]. Interestingly, with [HMMIM]Cl (HMMIM = 1-hexyl-2,3-dimethylimidazolium) where the blocked C-2 position prevents carbene complex formation, [PdCl$_4$]$^{2-}$ is formed rather than the metal. This is clearly the result of the relative coordinating strengths of the Cl$^-$ anion in comparison to [BF$_4$]$^-$ and [PF$_6$]$^-$. In this study palladium acetate itself was never found to be stable in solution in the ionic liquid. When the experiment was repeated in the presence of PPh$_3$, [HMMIM]Cl yielded [PdCl$_2$(PPh$_3$)$_2$] while [HMIM]Cl still gave

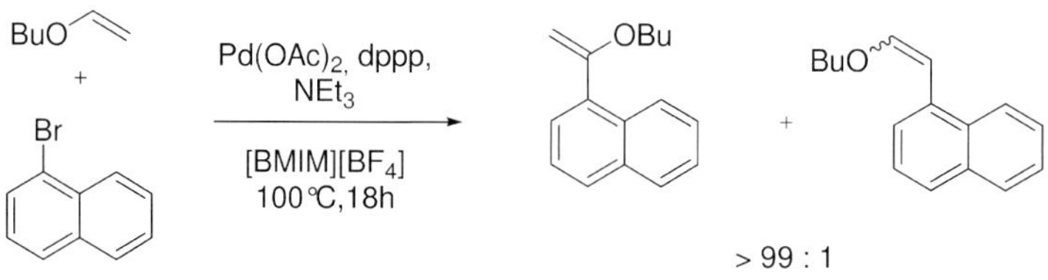

Scheme 5.3-22 Pd-catalyzed, regioselective arylation of butyl vinyl ether in an [BMIM][BF$_4$] ionic liquid.

[PdCl$_2$(HMimy)$_2$]. However, when the [BF$_4$]$^-$ and [PF$_6$]$^-$ ionic liquids were studied under these conditions, nanoparticulate palladium (0.8–1.6 nm diameter) was seen.

The situation might be different if bidentate phosphine ligands were used. This is at least indicated by another study by Xiao's group. They investigated the regioselective arylation of butyl vinyl ether using again Pd(OAc)$_2$ as catalyst precursor in [BMIM][BF$_4$] but the ligand in use was 1,3-bis(diphenylphosphino)propane (dppp) (Scheme 5.3-22) [205].

They compared the results in ionic liquids with those obtained in four conventional organic solvents. Interestingly, no formation of palladium black was observed in the ionic liquid while this was always the case with the organic solvents. Furthermore, the reaction in the ionic liquid proceeded with very high selectivity to the α-arylated compound whereas variable mixtures of the α- and β-isomers were obtained in the organic solvents DMF, DMSO, toluene and acetonitrile.

In recent times, many Heck coupling reactions have been carried out using Pd nanoparticles that were previously prepared for this purpose in ionic liquids. These examples will be presented in Section 5.3.2.7 which is entirely devoted to nanoparticulate catalysis in ionic liquids.

Heck reactions with Pd-complexes carrying carbenes or ionic ligands

Besides their *in situ* formation, carbene ligands have also been synthesized deliberately as alternatives to phosphine ligands for application in the Pd-catalyzed Heck reaction. For example, Shreeve et al. synthesized a NHC bearing catalyst (Scheme 5.3-23) for Heck and Suzuki coupling reactions, which was immobilized in [BMIM][PF$_6$] [206]. This air and moisture stable catalyst was tested in several reactions with good yields in up to six recycling experiments. For the Heck cross-coupling reaction of aryl halides with n-butyl acrylate (Scheme 5.3-23(a)) the yields were in the range of 68–94%, dependent on R. The catalyst dissolved in ionic liquid could be recycled up to six times. For Heck cross-coupling reactions of aryl halides with styrene (Scheme 5.3-23(b)) and for Suzuki coupling reactions (Scheme 5.3-23(c)) yields around 90% were realized and up to five recycling runs of the ionic catalyst solution could be demonstrated.

A particularly successful concept for catalyst immobilization in ionic liquids is the use of ionic ligands. Some examples of Pd-complexes carrying ionic ligands that have been applied in Heck reactions are listed in Table 5.3-9. Entries 1 and

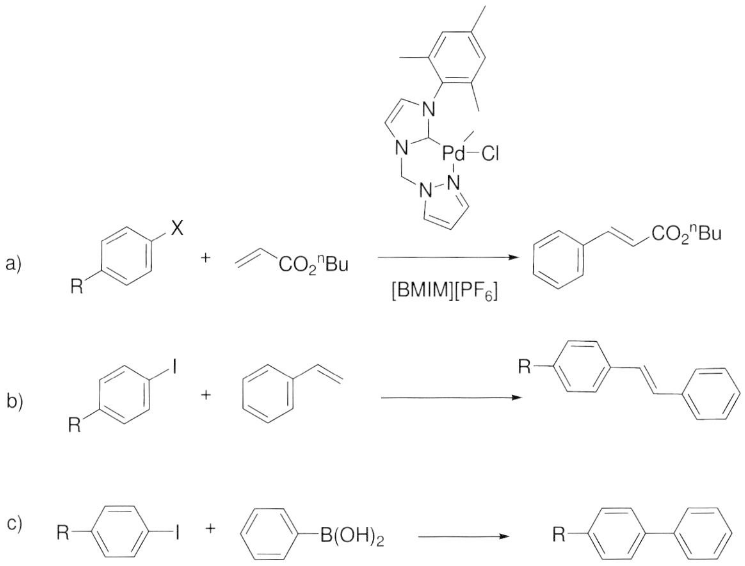

Scheme 5.3-23 Pd–carbene catalyst as synthesized and applied for Heck and Suzuki coupling reactions in [BMIM][PF$_6$] by Shreeve and coworkers.

3 gave high yields and good recyclability results with up to 11 cycles. The oxime carbapalladacycle (entry 2) gave low yields. This observation was ascribed to the possibility of carbene formation at the imidazolium functional group under the basic conditions necessary for the reaction. This and a rather low ligand stability may explain the observed, significant palladium leaching. Significant leaching was also found when supporting this imidazolium-modified carbapalladacycle catalyst on Al/MCM-41 and using the resulting solid as a SILP Heck catalyst.

Ligand-less Heck reaction in ionic liquids

A special focus of modern research in Pd-catalyzed Heck-coupling reactions is to investigate the ligand-free reaction in ionic liquids using Pd(OAc)$_2$, PdCl$_2$ or Pd/C as catalyst precursors [210]. In these examples the role of the ionic liquid is as either solvent, ligand or both. Of special interest for these studies is the regioselectivity of the Heck products when electron-rich olefins, such as acyclic enol ethers, silanes, and enol amides are applied as substrates. Possible products are the branched olefin (α, Scheme 5.3-24) and the linear olefin (β, Scheme 5.3-24).

The formation of the different products can be explained by two reaction pathways, with a non-polar transition state leading to the linear olefin (Scheme 5.3-25, Path A) and a polar transition state to branched olefins (Scheme 5.3-25, Path B). Hallberg and coworkers demonstrated that the regiocontrol in the arylation of electron-rich olefins such as enol ethers is governed by a range of parameters, including, among others, the electronic properties of the aromatic ring and the choice

Table 5.3.9 Examples of ionic liquid-modified ligands for Heck[a], Suzuki[b] and Sonogashira[c] reactions.

Entry	Catalyst	Dissolved in	No. of recycling cycles	Ref.
1[a]			11	[207]
2[a,b]		[BMIM][PF$_6$] or SILP (Al/MCM-41)	5	[208]
3[a,b,c]		Y=PF$_6$ or NTf$_2$ R= H or Me R'=various	9	[209]

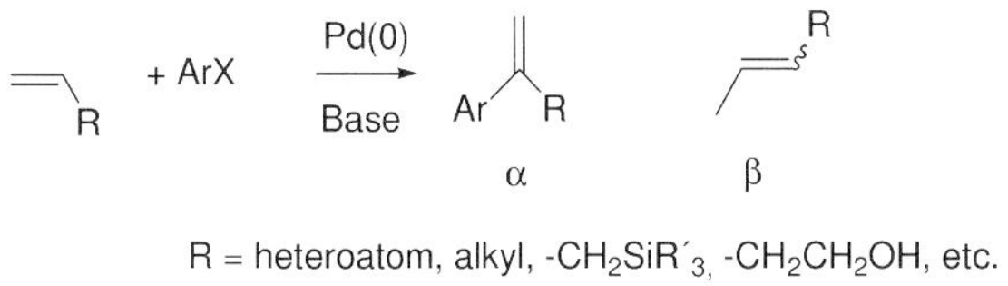
+ ArX
Pd(0)
Base
Ar
R
α
β
R = heteroatom, alkyl, -CH2SiR´3, -CH2CH2OH, etc.

Scheme 5.3-24 Regioselectivity in Heck reactions with electron-rich olefins.

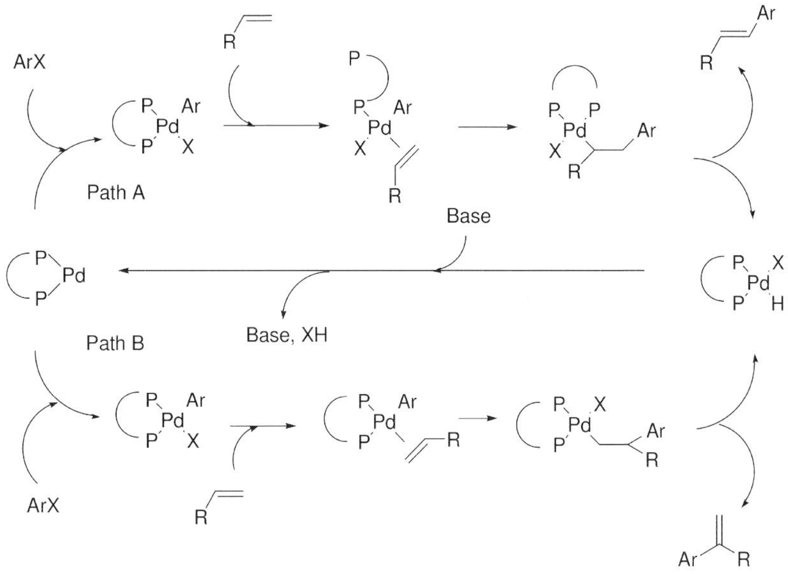
ArX
Path A
Base
Base, XH
Path B
ArX
Ar
R
Pd

Scheme 5.3-25 Regioselectivity of the Heck reaction.

of ligands and halide additives [211]. From earlier work it is known that addition of stochiometric quantities of thallium or silver salt, replacement of arylating halids by triflate and use of bidentate ligands promote the predominant formation of the branched product [212].

Xiao et al. [213] recently reported that the intermolecular Heck arylation with several electron-rich olefins can be effected with both aryl bromides and iodides in a highly regioselective manner by the use of ionic liquids. Regioselectivity towards the branched α-olefin was up to 99/1 and there was no need for additional halide scavenger. In this case the ionic liquid seems to serves not only as a quite polar solvent but also as halide scavenger.

Application-oriented Heck studies in ionic liquids

Seddon's group described the option of carrying out Heck reactions in ionic liquids that do not completely mix with water. These authors studied different Heck reactions in the triphasic system [BMIM][PF_6]/water/hexane [214]. While the [BMIM]$_2$[$PdCl_4$] catalyst remained in the ionic liquid, the products dissolved in the organic layer. The salt formed as a by-product of the reaction ([H-base]X) was extracted into the aqueous phase.

An example of a continuous Heck reaction using a microflow system was reported by Liu and coworkers [215]. Their reaction was catalyzed by a [Pd(PPh$_3$)Cl$_2$(BMimy)] carbene complex, which was immobilized in the ionic liquid [BMIM][(CF$_3$SO$_2$)$_2$N]. The coupling product, butyl cinnamate, was produced in an overall yield of 80% (115.3 g, 10 g h^{-1}), the ionic liquid containing Pd catalyst was continuously recycled. Unfortunately, the authors do not comment on the Pd leaching into the organic phase during their continuous operation.

A highly thermally stable (up to 280 °C), Heck active catalyst system was obtained by Enders and coworkers through covalent anchoring of a N-heterocyclic carbene palladium/ionic liquid matrix on a silica surface [216]. Pd-NHC complexes were generated *in situ* in an imidazolium-type ionic liquid matrix (pre-functionalized with a trimethoxysilylpropyl group) grafted on a silica surface. With this system aryl iodides and bromides were rapidly (2–24 h) converted to the corresponding Heck products with excellent yields (81–99%) depending on the sort of olefin substrate. The catalyst was reused four times without loss of activity, giving a total TON of 36 600. Investigation of the catalyst after reaction by TEM and EDX showed again the formation of Pd nanoparticles (10–40 nm), which were obviously immobilized on the silica surface in an excellent manner as no leaching of palladium in the organic phase was observed.

A non-covalent immobilization of Heck catalyst on silica (SILP concept) has been realized by Hagiwara et al. [217]. They used a silica surface, supported with Pd(OAc)$_2$ dissolved in [BMIM][PF$_6$]. This catalyst was applied to the Mizoroki–Heck reaction of aryl halides with acrylate without a ligand in n-dodecane as solvent. It was six times reused and the overall TON reached 68 400 (for more details see Section 5.6).

Other Pd-catalyzed C–C-coupling reactions

Closely related to Heck coupling, the Suzuki cross-coupling reaction (the coupling reaction of a halogenoarene with an arylboronic acid or ester) [218] has also been successfully performed in ionic liquids. Welton *et al.* used Pd(PPh$_3$)$_4$ as catalyst in [BMIM][BF$_4$] to convert bromo- and iodoarenes with phenylboronic acid (Scheme 5.3-26) [219].

The best results were achieved by pre-heating the aryl halide in the ionic liquid with the Pd-complex to 110 °C. The arylboronic acid and Na$_2$CO$_3$ were later

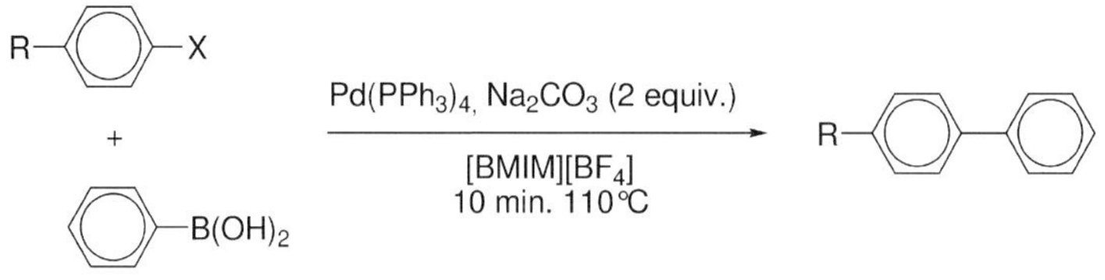

Scheme 5.3-26 Pd-catalyzed Suzuki cross-coupling reaction in a [BMIM][BF$_4$] ionic liquid.

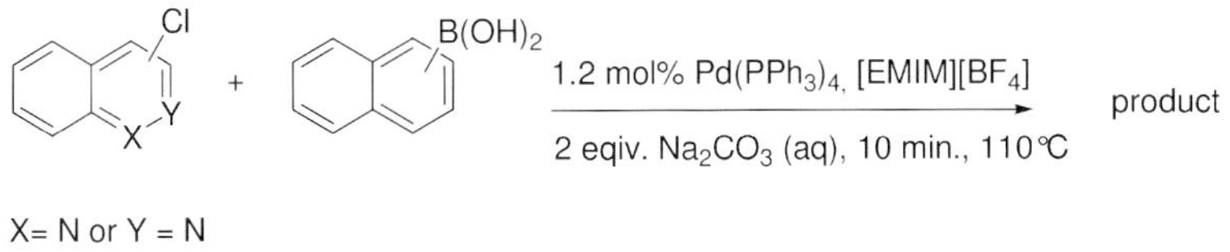

Scheme 5.3-27 Suzuki cross-couplings of heterocyclic chloroarenes with naphthaleneboronic acids in [EMIM][BF$_4$].

X = N or Y = N

Reaction: heterocyclic chloroarene + naphthaleneboronic acid, 1.2 mol% Pd(PPh$_3$)$_4$, [EMIM][BF$_4$], 2 eqiv. Na$_2$CO$_3$ (aq), 10 min., 110 °C → product

added to start the reaction. Several advantages over the reaction performed under the conventional Suzuki conditions were described. The reaction showed significantly enhanced activity in the ionic liquid (TOF = 455 h^{-1} in [BMIM][BF$_4$] in comparison to 5 h^{-1} under conventional Suzuki conditions). The formation of the homo-coupling aryl by-product was suppressed. Moreover, the ionic catalyst layer could be reused after extraction of the products with ether and removal of the by-products NaHCO$_3$ and NaXB(OH)$_2$ with excess water. Over three further reaction cycles no deactivation was observed using this protocol.

Later, the catalyst complex [Pd(PPh$_3$)$_4$] dissolved in [EMIM][BF$_4$] was successfully applied for the coupling of heterocyclic chloroarenes with naphthaleneboronic acids (Scheme 5.3-27) [220]. Products were obtained in higher yields (43–81%), enhanced reaction rate and improved selectivity compared to conventional organic solvent.

In the light of the discussion on the potential role of Pd–carbene complexes in coupling reactions in ionic liquids it is of particular interest to note that an addition of halide ions to the reaction solution was reported to be necessary to form stable Suzuki coupling systems [219]. Investigations by Welton and coworkers in which the catalytically active solutions were analyzed in more detail revealed indeed the presence of a mixed phosphine–imidazolylidene palladium complex (Fig. 5.3-15) [221]. The cation 1-butyl-2-phenyl-3-methylimidazolium could be identified in the reaction mixture [222]. This ion is likely to form by reductive elimination from a palladium species containing both a phenyl ring and a BMimy ligand thus confirming the involvement of the related Pd–carbene complex in the catalytic process.

Interestingly, Suzuki coupling in ionic liquids has also been reported in the absence of any phosphine ligand. When [(CH$_3$CN)$_2$PdCl$_2$] was dissolved in

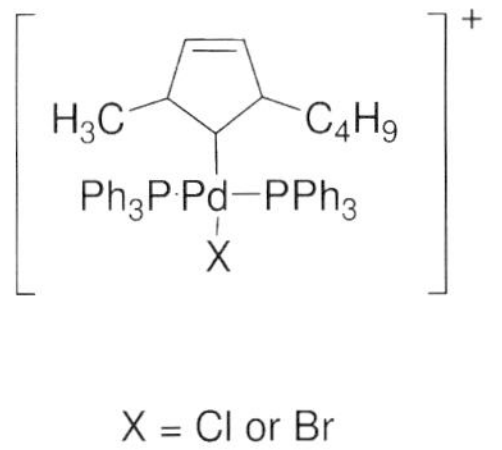

X = Cl or Br

Fig. 5.3-15 Phosphine-imidazolylidene palladium ion [(BMimy)(PPh$_3$)$_2$PdX]$^+$ as identified from the reaction mixture of a Suzuki reaction in [BMIM][BF$_4$].

preferentially [BMPy][(CF$_3$SO$_2$)$_2$N] or [BMMIM][BF$_4$] together with four equivalents of an imidazol derivative (best results gave 1-phenylimidazole and 1-methylbenzimidazole) and this mixture was heated, the Suzuki reaction proceeded with a TOF of up to 221 h^{-1}. To learn more about this sort of catalysis Welton and coworkers added aqueous Na$_2$CO$_3$ solution as a base to the catalytic mixture and analyzed the resulting mixture by ESI-MS. Among several unidentified palladium-complexes they could identify the carbene complex [(MIM)$_2$Pd(BMimy)Cl]$^+$ which is likely to be the catalytic active species [223]. Remarkably, no carbene-complex was built by heating [(CH$_3$CN)$_2$PdCl$_2$] in the ionic liquid itself without additional aqueous Na$_2$CO$_3$ solution.

A certain pre-treatment was also found to be necessary for the Ni-catalyzed Suzuki coupling. Iwasawa's group found that the addition of anhydrous K$_3$PO$_4$ improved the reactivity of [BMIM][NiCl$_4$] significantly and they suggested that by addition of this weak base the catalytically active nickel carbene complex can form. With no addition of phosphine ligand the reaction of 4-chlorotoluene and phenylboronic acid gave a yield of 23%, which could be improved up to 96% after addition of two equivalents of PPh$_3$ [82]. This catalytic system was also immobilized on silica. The immobilization made the pretreatment rather complicated (treatment with NaOtBu for 30 min at room temperature followed by addition of PPh$_3$) and the pre-treatment had to be repeated for every single reuse. Nevertheless, yields up to 93% for the first two runs could be realized with this immobilized system which decreased to 69% and 64% yield in the following runs.

Further examples of phosphine-free Pd-catalyzed Suzuki coupling were reported by Shreeve et al. [224]. They applied the monoquaternary 2,2'-bisimidazolium-based ionic liquid shown above in Table 5.3-9 (entry 1) as ionic liquid and ligand and reported good results for the transformation of phenylboronic acid with chlorobenzene derivatives. The system remained active for 14 consecutive runs with yields of 80–90%, depending on the electron-donating or -withdrawing properties of the substituents at the chlorobenzene substrate.

The Suzuki reaction has also been carried out in ionic liquids using ultrasound irradiation as an energy source [225]. For the reaction in [BMIM][BF$_4$] significant *homo*-coupling of phenylboronic acid was reported when the reaction was performed in air. In this case, decomposition of the Pd-complex prevented repetitive catalytic runs. However, when the imidazolylidene complex [PdX$_2$(BBimy)$_2$] was independently prepared and used as the source of palladium, comparable conversions to Pd(OAc)$_2$ were observed. However, in this case no decomposition to Pd metal or *homo*-coupling was observed in the experiment under otherwise identical conditions. The complex [PdX$_2$(BBimy)$_2$] could be quantitatively recovered from the catalytic mixture and used in three repetitive catalytic runs for the reaction of 4-bromoanisole with phenylboronic acid with only a slight decrease in activity. In contrast to ultrasound-irradiated Heck reactions with the same complex under similar conditions [226] no evidence of Pd nanoparticle formation could be found in the Suzuki reactions.

To achieve a better Pd-immobilization in the ionic liquid, Dyson et al. synthesized and applied *N*-butyronitrile pyridinium ionic liquids (Scheme 5.3-28) [227]. Indeed,

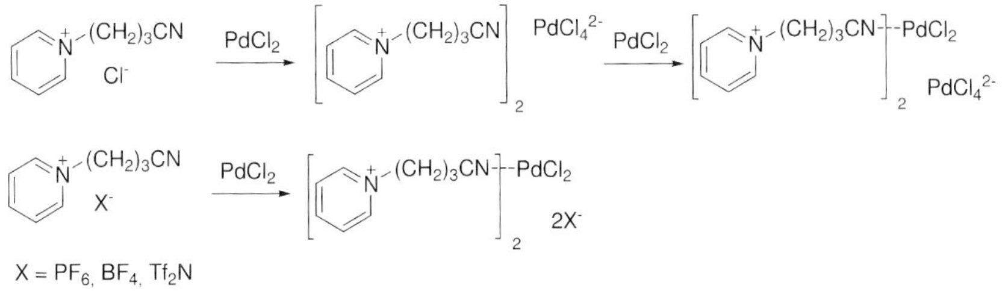

Scheme 5.3-28 PdCl$_2$ immobilization in N-butyronitrile functionalized pyridinium ionic liquids.

the leaching of palladium into the organic layer was significantly less than in the non-functionalized pyridinium system. The method led also to the isolation of the nitrile-stabilized palladium complex that was found to be air stable and did not decompose on washing with water or alcohols at room temperature. However, decomposition was observed when contacting the complex with water or alcohols over prolonged periods of time. The catalyst/ionic liquid system was evaluated in the Suzuki coupling reaction of iodobenzene and phenylboronic acid and in the Stille coupling reaction of iodobenzene and phenyltributylstannane. Catalyst immobilization in the nitrile-functionalized ionic liquid enabled recycling for 9 runs without significant loss of activity (all yield >80%). In contrast, immobilization in the non-functionalized ionic liquid [BPy][(CF$_3$SO$_2$)$_2$N] led to complete deactivation after 6 runs. In contrast to the Suzuki coupling reaction, significant nanoparticle formation was observed for the Stille coupling. Transmission electron microscopy (TEM) analysis of the particles formed *in situ* in the Stille reaction reveals that the nitrile-functionalized ionic liquid exerts a superior nanoparticle-stabilizing effect compared to the non-functionalized ionic liquid.

A number of other Stille coupling reactions have also been reported by Handy et al. [228]. Using [PdCl$_2$(PhCN)$_2$]/Ph$_3$As/CuI in [BMIM][BF$_4$] good yields and good catalyst recyclability (up to five times) were reported for the reaction of α-iodoenones with vinyl and aryl stannanes (Scheme 5.3-29). However, the reported reaction rates were significantly lower than those obtained in N-methylpyrrolidinone (NMP). Additional examples of Stille coupling reactions in different ionic liquids have been more recently reported by Chiappe's group [229].

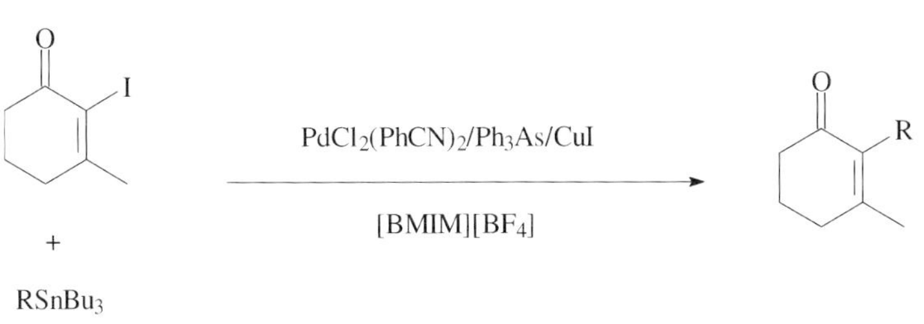

Scheme 5.3-29 Pd-catalyzed Stille coupling of α-iodoenones with vinyl and aryl stannanes in [BMIM][BF$_4$].

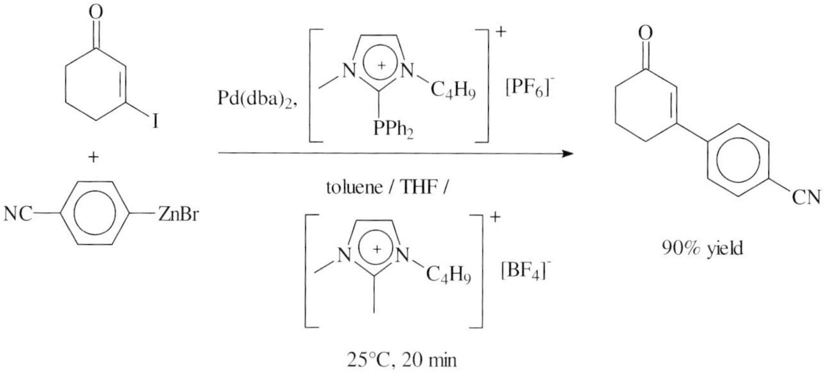

Scheme 5.3-30 Pd-catalyzed cross-coupling of organozinc compounds (Negishi cross-coupling) in [BMMIM][BF$_4$].

Also Pd-catalyzed Negishi cross-coupling reactions have been described in ionic liquids. Knochel and coworkers investigated the reaction between organometallic zinc compounds and aryl iodide in [BMMIM][BF$_4$] using an ionic phosphine ligand. Scheme 5.3-30 illustrates the reaction for the formation of a 3-substituted cyclohexenone from 3-iodo-cyclohex-2-en-1-one [170]. The reaction was carried out in an ionic liquid/toluene biphasic system, which allowed easy product recovery from the catalyst by decantation. However, attempts to recycle the ionic catalyst phase resulted in significant catalyst deactivation, after only the third recycle.

Finally, it should be mentioned that Pd-catalyzed allylations [230, 231], allylic alkylation [231] and substitution [232] reactions as well as Trost-Tsuji-coupling [233] have also been carried out in different ionic liquids with some success.

In summary it can be stated that low melting salts and ionic liquids are very interesting reaction media for Pd-catalyzed coupling reactions. However, the role of the ionic liquid is variable and may be very complex. Ionic liquids have been found to act as solvents, ligands, ligand precursors and dispergents. In many cases more than one of these functions was taken by the ionic reaction medium. 1,3-dialkylimidazolium ionic liquids, especially, are often not inert under the reaction conditions. Moreover, the formation of Pd nanoparticles has always to be taken into account as a possible explanation for the observed results. These particles seem to be stabilized in the ionic liquids (see Sections 5.3.2.7 and 6.3 for more details) and the resulting mixtures have shown excellent catalytic performance in at least some cases.

5.3.2.5 Dimerization and Oligomerization Reactions

Biphasic catalysis is not a new concept for catalytic oligomerization chemistry. On the contrary, the oligomerization of ethylene was the first commercial example of a biphasic, transition metal catalyzed reaction. The process is known under the name "Shell higher olefins process (SHOP)" and the first patents originate from as early as the late sixties [234]. While the SHOP uses 1,4-butanediol as the catalyst phase, it turned out in the following years of research that many highly attractive catalyst systems for dimerization and oligomerization are not compatible with polar organic solvents or water. This is because high electrophilicity of the metal center is a key prerequisite for catalytic activity in oligomerization. The higher the electrophilicity of the metal center the higher usually is its catalytic activity but the lower, at the same time, is its compatibility with polar organic solvents or water. Consequently, many cationic transition metal complexes are excellent oligomerization catalysts [43] but none of these systems could be realized in a biphasic reaction mode using water or polar organic solvents as the catalyst phase.

Chloroaluminate-based dimerization and oligomerization systems
One example of a technically important oligomerization reaction that could not be carried out in a liquid–liquid biphasic mode prior to the invention of chloroaluminate ionic liquids is the Ni-catalyzed dimerization of propene and/or butenes. This reaction was intensively studied in the sixties [235] and commercialized later as the "Dimersol-process" by the Institut Français du Pétrole (IFP). In this reaction the active catalytic species forms *in situ* by reaction of a Ni(II)-source and an aluminum alkyl co-catalyst. In the conventional process the reaction takes place in a monophasic reaction mode in an organic solvent or – technically preferred – in the alkene feedstock. After reaction the catalyst is destroyed by addition of an aqueous solution of a base and the precipitated Ni salt is filtered off and has to be disposed of. 25 Dimersol units are currently in operation producing octane booster for gasoline with a total processing capacity of 3.4 Mtons per year. Taking into account the significant consumption of nickel and aluminum alkyls related to the monophasic Dimersol process it is not surprising that IFP research teams were looking for new solvent concepts to allow a biphasic version of the Dimersol chemistry. Chloroaluminate ionic liquids proved to be highly attractive in this respect.

As early as 1990, Chauvin and his coworkers from IFP published their first results on the biphasic, Ni-catalyzed dimerization of propene in ionic liquids of the type [BMIM]Cl/AlCl$_3$/AlEtCl$_2$ [4]. In the following years the nickel-catalyzed oligomerization of short-chain alkenes in chloroaluminate melts became one of the best investigated applications of transition metal catalysts in ionic liquids to the present.

Due to its importance, some basic principles of the Ni-catalyzed dimerization of propene in chloroaluminate ionic liquids should be presented here. Table 5.3-10 displays some reported examples, which have been selected to explain the most important aspects of oligomerization chemistry in chloroaluminate ionic liquids [236].

The Ni-catalyzed oligomerization of olefins in ionic liquids requires a careful choice of the ionic liquid's acidity. In basic melts (Table 5.3-10, entry (a)) no dimerization activity is observed. Here, the basic chloride ions prevent the formation of free coordination sites on the nickel catalyst. In acidic chloroaluminate melts, an oligomerization reaction takes place even in the absence of a nickel catalyst (entry (b)). However, no dimers, but a mix of different oligomers, are formed by cationic oligomerization. Superacidic protons and the reactivity of the acidic anions $[Al_2Cl_7]^-$ and $[Al_3Cl_{10}]^-$ may account for this reactivity.

Table 5.3-10 Selected results of the Ni-catalyzed propene dimerization in chloroaluminate ionic liquids.

	Ionic liquid	Composition of the ionic liquid (molar ratio)	Ni-complex	Activity kg/g(Ni) h^{-1}	Product DMB/M2P/nH[a]
(a)	[BMIM]Cl/AlCl$_3$	1/0.8	NiBr$_2$L$_2$[b]	0	
(b)	[BMIM]Cl/AlCl$_3$	1/1.5		c	
(c)	[BMIM]Cl/AlEtCl$_2$	1/1.2	NiCl$_2$	2.5	5/74/21
(d)	[BMIM]Cl/AlEtCl$_2$	1/1.2	NiCl$_2$(iPr$_3$P)$_2$	2.5	74/24/2
(e)	[BMIM]Cl/AlCl$_3$/AlEtCl$_2$	1/1.2/0.1	NiCl$_2$(Pr$_3$P)$_2$	12.5	83/15/2

T = −15 °C;
[a] DMB = dimethylbutenes, M2P = methylpentenes, nH = n-hexenes,
[b] L = 2-methylallyl;
[c] highly viscous oligomers from cationic oligomerization were obtained.

The addition of aluminum alkyl compounds is known to suppress this undesired cationic oligomerization activity. In the presence of $NiCl_2$ as catalyst precursor, the ionic catalyst solution is formed and shows high activity for the dimerization (entry (c)). Without added phosphine ligands, a product distribution is obtained with no particular selectivity. With added phosphine ligand, the distribution of regioisomers in the C6-fraction is influenced by the steric and electronic properties of the ligand used in the same way as known from the catalytic system in organic solvents [235] (entry (d)). At longer reaction times, a decrease in the selectivity to highly branched products is observed. It has been postulated that a competing reaction of the basic phosphine ligand with the hard Lewis acid $AlCl_3$ takes place. This assumption is supported by the observation that the addition of a soft competing base such as tetramethylbenzene can prevent the loss in selectivity.

Unfortunately, investigations with ionic liquids containing high amounts of $AlEtCl_2$ reveal several limitations: The reductive effect of the aluminum alkyl affects the temperature stability of the nickel catalyst. At very high aluminum alkyl concentrations, a precipitation of black metallic nickel is observed, even at room temperature.

Based on these results the Institut Français du Pétrole (IFP) has developed a biphasic version of their established monophasic "Dimersol" process, which is offered for licensing under the name "Difasol" process [237]. The "Difasol" process uses slightly acidic chloroaluminate ionic liquids with small amounts of aluminum alkyls as the solvent for the catalytic nickel center. In comparison to the established "Dimersol" process, the new biphasic ionic liquid process drastically reduces the consumption of Ni catalyst and aluminum alkyls. Additional advantages arise from the good performance obtained with highly diluted feedstocks and the significantly improved dimer selectivity of the "Difasol" process (for more detailed information see Section 5.4).

Wasserscheid and coworkers studied the selective, Ni-catalyzed, biphasic dimerization of 1-butene to linear octenes. Therefore, a catalytic system that is well-known for its ability to form linear dimers from 1-butene in conventional organic solvents [238] – namely the square planar Ni-complex (η-4-cycloocten-1-yl)(1,1,1,5,5,5,-hexafluoro-2,4-pentandionato-O,O')nickel, [(H-COD)Ni(hfacac)] – was used in chloroaluminate ionic liquids.

For this specific task, ionic liquids containing aluminum alkyls proved to be unsuitable, due to their strong isomerization activity [241]. Since, mechanistically, only the linkage of two 1-butene molecules can lead to the formation of linear octenes, isomerization activity of the solvent inhibits the formation of the desired product. Therefore, slightly acidic chloroaluminate melts that enable selective nickel catalysis without the addition of aluminum alkyls have been developed [239]. It was found that an acidic chloroaluminate ionic liquid buffered with small amounts of weak organic bases provides a solvent which allows a selective, biphasic reaction using nickel [(H-COD)Ni(hfacac)].

The function of the base is to trap any free acidic species in the melt which might initiate cationic side reactions. A suitable base has to fulfill a number of requirements. Its basicity has to be in the appropriate range to provide enough

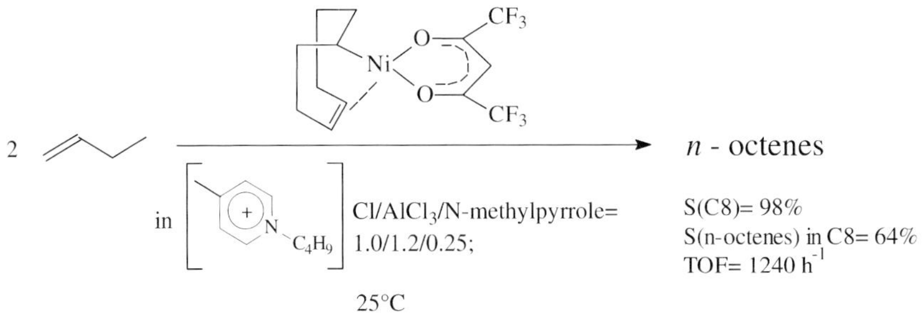

Scheme 5.3-31 Ni-catalyzed, biphasic, linear dimerization in a slightly acidic, buffered chloroaluminate ionic liquid.

reactivity to eliminate all free acidic species in the melt. At the same time, it has to be non-coordinating with respect to the catalytically active Ni center. Another important feature is a very high solubility in the ionic liquid. During the reaction, the base has to remain in the ionic catalyst layer, even under intense extraction of the ionic liquid by the organic layer. Finally, the base has to be inert to the 1-butene feedstock and the oligomerization products.

The use of pyrrole and N-methylpyrrole was found to be preferred. By the addition of N-methylpyrrole, all cationic side reactions could be effectively suppressed and only dimerization products produced by Ni catalysis were obtained. In this case the dimer selectivity was as high as 98%. Scheme 5.3-31 shows the catalytic system that allowed the first successful application of [(H-COD)Ni(hfacac)] in the biphasic linear dimerization of 1-butene.

The comparison of the dimerization of 1-butene with [(H-COD)Ni(hfacac)] in chloroaluminate ionic liquids with the identical reaction in toluene is quite instructive. First, the reaction in the ionic liquid solvent is biphasic with no detectable catalyst leaching, enabling easy catalyst separation and recycling. While [(H-COD)Ni(hfacac)] requires an activation temperature of 50 °C in toluene, the reaction proceeds in the ionic liquid even at −10 °C. This indicates that the catalyst activation, which is believed to be the formation of the active Ni-hydride complex, proceeds much more efficiently in the chloroaluminate solvent (for more details on mechanistic studies see Section 5.3.1.3). Furthermore, the product selectivities obtained in both solvents reveal significantly higher dimer selectivities in the biphasic case. This can be understood by considering the fact that the C8-product is much less soluble in the ionic liquid than the butene feedstock (by about a factor 4). During the reaction, a fast extraction of the C8-product into the organic layer takes place, thus preventing consecutive C12-formation. The linear selectivity is high in both solvents, however, somewhat lower in the ionic liquid solvent.

To create reliable data on the lifetime and overall activity of the ionic catalyst system, a loop reactor was constructed and the reaction was carried out in continuous mode [240]. The most important results of these studies are presented in

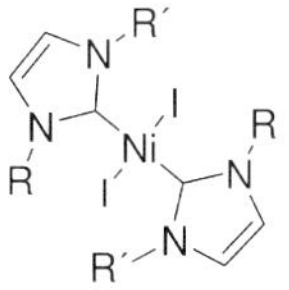

1 R=Me, R′ = ⁿPr
2 R=Me, R′ = ⁿBu
3 R=Me, R′ = ⁱPr
4 R= ⁱPr, R′ = ⁱPr

Fig. 5.3-16 Ni–carbene complexes as used in the dimerization of propene and 1-butene in chloroaluminate ionic liquids.

Section 5.4, together with much more detailed information about the processing of biphasic reactions with an ionic liquid catalyst phase.

[Ni(MeCN)$_6$][BF$_4$]$_2$ dissolved in the slightly acidic chloroaluminate system [BMIM]Cl/AlCl$_3$/AlEtCl$_2$ (ratio = 1:1.2:0.25) has also been used for the dimerization of butenes [241]. The reaction showed high activity even at 10 °C and under atmospheric pressure with a turnover frequency of 6840 h^{-1} and a productivity of 6 kg oligomer per gram Ni per hour. The distribution of the butene dimers obtained (typically 39 ± 1% dimethylhexenes, 56 ± 2% monomethylheptenes and 6 ± 1% n-octenes) was reported to be independent of the addition of phosphine ligands. Moreover, the product mix was independent of feedstock, with both 1-butene and 2-butene yielding the same dimer distribution, with only 6% of the linear product. This clearly indicates that the catalytic system used here is not only an active oligomerization catalyst but is also highly active for isomerization.

In our group nickel carbene complexes (Fig. 5.3-16) dissolved in slightly acidic, buffered chloroaluminate ionic liquids have been applied for the dimerization of propene and 1-butene [242].

Interestingly these complexes showed high activity without addition of alkyl aluminum compounds in the ionic liquid while they are almost inactive in toluene. These results are interpretable in terms of catalyst stabilization by the imidazolium-based ionic liquid. Reductive elimination of imidazolium is also possible as in toluene as in the ionic liquid but in the ionic liquid, a rapid reoxidation via addition of the solvent imidazolium cation seems possible and may prevent the formation of Ni^0 deposits associated with catalyst deactivation. The carbene complex with R = n-Bu showed the highest activity with a dimer yield of 70.2% (TOF = 7020 h^{-1}). The preferred product of the nickel–carbene catalyzed reaction is methylpentene. Additional phosphine ligand had no significant influence on the distribution of the products in this case.

Catalytic systems based on chloroaluminate ionic liquids have also been used for the selective dimerization of ethene to butenes [243]. Dupont et al. dissolved [Ni(MeCN)$_6$][BF$_4$]$_2$ in the slightly acidic chloroaluminate system [BMIM]Cl/AlCl$_3$/AlEtCl$_2$ (ratio = 1:1.2:0.25) and obtained 100% butenes at −10 °C and 18 bar ethylene pressure (TOF = 1731 h^{-1}). Unfortunately, the more valuable 1-butene was not produced selectively but a mixture of all linear butene isomers was obtained. The active nickel catalyst is formed *in situ* in the ionic liquid by reaction of the precursor with the alkylaluminate(III) species in the ionic liquid.

Later, de Souza et al. compared the reactivity of [Ni(MeCN)$_6$][BF$_4$]$_2$ (activated with AlEt$_2$Cl or AlEt$_3$ as cocatalysts) in the homogeneous phase, immobilized in ionic liquids, and heterogenized in zeolite [244]. They found that in the homogeneous phase the Ni catalyst showed high activity (TOF = 2451 h^{-1}) and 97% selectivity to C4 products (with 30% 1-butene in C4) under mild conditions was obtained. Under biphasic conditions, when the catalyst was immobilized in [BMIM]Cl/AlCl$_3$/AlEtCl$_2$, ethylene dimerization occurred with 83% selectivity to 1-butene. Finally, immobilization of the catalyst in zeolite NaX gave 78% selectivity to 1-butene.

Further comparison of the homogenous and biphasic conditions for the oligomerization of ethene was carried out in chloroaluminate systems by de Souza and coworkers using nickel-1,2-diiminophosphorane complexes. They found, that the catalyst had a very much higher activity for ethylene oligomerization when immobilized in a [BMIM]Cl/AlCl$_3$/AlEt$_2$Cl-system than in a homogeneous (chlorobenzene) phase [245]. The catalyst could be easily recycled in the chloroaluminate system with increasing catalyst activity from the first to the third cycle. Possible reasons for this surprising behavior are discussed, such as displacement of the diiminophosphorane ligand leading to new highly active catalytic species. The selectivity to 1-olefins was also improved in the ionic liquid but was still very low due to significant isomerization activity of the catalytic system in use.

Biphasic ethylene oligomerization reactions have also been described in chloroaluminate ionic liquids using catalytic metals other than nickel. Olivier-Bourbigou et al. dissolved the tungsten complex [Cl$_2$W=NPh(PMe$_3$)$_3$] in a slightly acidic [BMIM]Cl/AlCl$_3$ system and used this ionic catalyst solution in ethylene oligomerization without adding a co-catalyst [246]. At 60 °C and 40 bar a product distribution containing 81% butenes, 18% hexenes and 1% higher oligomers was obtained with good activity (TOF = 1280 h^{-1}). However, the selectivity for the more valuable 1-olefins was found to be relatively low (65%).

The selective, chromium catalyzed trimerization of ethylene to 1-hexene was disclosed in a patent by SASOL [247]. In this work a biphasic reaction system was explored using alkylchloroaluminate ionic liquids as the catalyst phase.

Dimerization and oligomerization of olefins in non-chloroaluminate systems

Biphasic oligomerization with ionic liquids is not restricted to chloroaluminate systems. Especially in those cases where the, at least, latent acidity or basicity of the chloroaluminate causes problems, neutral ionic liquids with weakly coordinating anions can be used with great success.

As already mentioned above, the Ni-catalyzed oligomerization of ethylene in chloroaluminate ionic liquids was found to be characterized by high oligomerization and high isomerization activity. The latter leads to a rapid consecutive transformation of the 1-olefins formed into mixtures of far less valuable internal olefins. Higher 1-olefins (HAOs) are an important group of industrial chemicals that find a variety of end uses. Depending on their chain length, they are components of plastics (C$_4$–C$_6$ HAOs in copolymerization), plasticizers (C$_6$–C$_{10}$ HAOs through

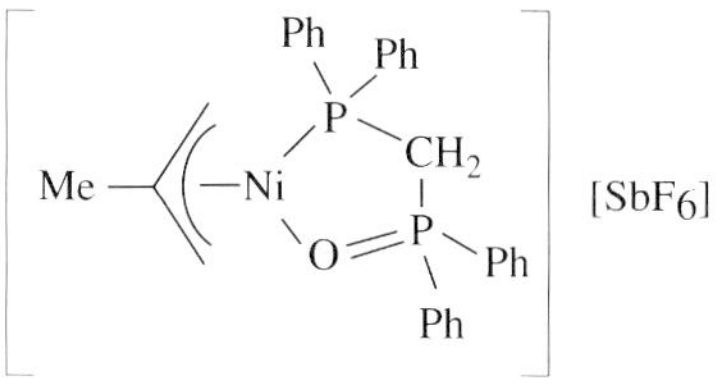

Fig. 5.3-17 The cationic Ni-complexes [(mall)Ni(dppmo)] [SbF$_6$] as used for the biphasic oligomerization of ethylene to 1-olefins in e.g. [BMIM][PF$_6$].

hydroformylation), lubricants (C$_{10}$–C$_{12}$ HAOs through oligomerization) and surfactants (C$_{12}$–C$_{16}$ HAOs through arylation/sulfonation).

Besides the neutral nickel/phosphine complexes used in the SHOP, cationic Ni-complexes such as [(mall)Ni(dppmo)][SbF$_6$] (see Fig. 5.3-17) have attracted some attention as highly selective and highly active catalysts for ethylene oligomerization to HAOs [248].

However, all attempts to carry out a biphasic ethylene oligomerization with this cationic catalyst using traditional organic solvents, such as 1,4-butandiol (which is used in the SHOP) resulted in almost complete catalyst deactivation by the solvent. This reflects the much higher electrophilicity of the cationic complex [(mall)Ni(dppmo)] [SbF$_6$] in comparison to the neutral Ni-complexes used in the SHOP.

Wasserscheid et al. demonstrated that the use of hexafluorophosphate ionic liquids allows, for the first time, a selective, biphasic oligomerization of ethylene to 1-olefins with the cationic Ni-complex [(mall)Ni(dppmo)][SbF$_6$] (Scheme 5.3-32) [44, 249].

Obviously, the ionic liquid's ability to dissolve the ionic catalyst complex, combined with a low solvent nucleophilicity, opens up the possibility for biphasic processing. Furthermore, it was found that the biphasic reaction mode led, in this specific reaction, to improved catalytic activity, selectivity and to enhanced catalyst lifetime.

The higher activity of the catalyst [(mall)Ni(dppmo)][SbF$_6$] in [BMIM][PF$_6$] (TOF = 25425 h^{-1}) vs. the reaction under identical conditions in CH$_2$Cl$_2$ (TOF = 7591 h^{-1}) can be explained by the fast extraction of products and side products out of the catalyst layer into the organic phase. A high concentration of internal olefins (from oligomerization and consecutive isomerization) at the catalyst is known to reduce catalytic activity due to the formation of quite stable Ni–olefin complexes.

The selectivity of the ethylene oligomerization reaction is clearly influenced by the biphasic reaction mode. The oligomers were found to be much shorter in the biphasic system, due to restricted ethylene availability at the catalytic center dissolved in the ionic liquid. This behavior correlates to the ethylene solubility in the different solvents under the reaction conditions. The ethylene solubility in 10 ml CH$_2$Cl$_2$ was determined to be 6.51 g at 25 °C/50 bar vs. only 1.1 g ethylene dissolved in [BMIM][PF$_6$] under identical conditions. Since the rate of ethylene insertion

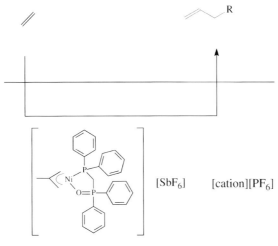

Scheme 5.3-32 Biphasic ethylene oligomerization with cationic Ni-complexes in a [BMIM][PF$_6$] ionic liquid.

is dependent on the ethylene concentration at the catalyst, but the rate of β-H-elimination is not, it becomes understandable that low ethylene availability at the catalytic active center favors the formation of shorter oligomers. In good agreement with this, a shift of the oligomer distribution was observed if the ionic liquid's cation was modified with longer alkyl chains. With increasing alkyl chain length of the ionic liquid's cation the oligomer distribution obtained became gradually broader following the higher ethylene solubility in these ionic liquids. However, all biphasic oligomerization experiments still show much narrower oligomer distributions than in the case of the monophasic reaction in CH_2Cl_2 (under identical conditions).

As well as the oligomer distribution, the selectivity to 1-olefins is of great technical relevance. Despite the much higher catalytic activity this selectivity was even slightly higher in [BMIM][PF$_6$] compared to CH_2Cl_2. The overall 1-hexene selectivity in C6-products was found to be 88.5% in [BMIM][PF$_6$] vs. 85.0% in CH_2Cl_2. Interestingly, less of the internal hexenes (formed by consecutive isomerization of 1-olefins) are obtained in the case of biphasic oligomerization using the ionic liquid solvent. This is explained by the much lower solubility of the higher oligomerization products in the catalyst solvent [BMIM][PF$_6$]. Since the 1-olefins formed are quickly extracted into the organic layer, consecutive isomerization of these products at the Ni center is suppressed in comparison to the monophasic reaction in CH_2Cl_2.

It is noteworthy, that the best results could only be obtained with very pure ionic liquids and by using an optimized reactor set-up. The content of halide ions and water in the ionic liquid was found to be a crucial parameter since both impurities

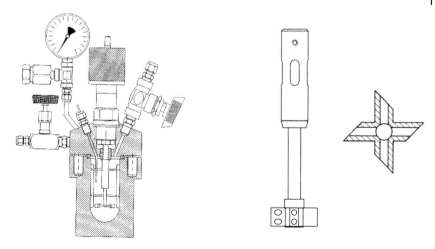

Fig. 5.3-18 150 ml autoclave with special stirrer design to maximize ethylene intake into an ionic liquid catalyst layer as used for the Ni-catalyzed liquid–liquid biphasic ethylene oligomerization in [BMIM][PF$_6$].

poison the cationic catalyst. Furthermore, the catalytic results proved to be highly dependent on all modifications that influence the mass transfer rate of ethylene into the ionic catalyst layer. A 150 ml autoclave with baffles stirred from the top with a special gas entrainment stirrer (see Fig. 5.3-18) gave far better results than a standard autoclave stirred with a magnetic stirrer bar.

Finally, it could be demonstrated that the ionic catalyst solution can, in principle, be recycled. By repetitive use of the ionic catalyst solution an overall activity of 61 106 mol ethylene converted per mol catalyst could be realized after two recycle runs.

An example of a biphasic, Ni-catalyzed co-dimerization in ionic liquids with weakly coordinating anions was described by Leitner et al. [12]. The hydrovinylation of styrene was investigated in the biphasic system ionic liquid/compressed CO_2 using a chiral Ni catalyst. Since it was found that this reaction benefits particularly from this unusual biphasic solvent system more details about this specific application will be given in Section 5.7.

Dupont and coworkers studied the Pd-catalyzed dimerization [250] and the Fe-catalyzed cyclo-dimerization [251] of butadiene in non-chloroaluminate ionic liquids. The biphasic dimerization of butadiene is an attractive research goal since the products formed, 1,3,5-octatriene and 1,3,6-octatriene, are sensitive towards undesired polymerization, so that separation by distillation is usually not possible. These octatrienes are of some commercial relevance as intermediates for the synthesis of fragrances, plastizisers and adhesives. By using PdCl$_2$ with two equivalents of the ligand PPh$_3$ dissolved in [BMIM][PF$_6$], [BMIM][BF$_4$] or [BMIM][CF$_3$SO$_3$], it was possible to obtain the octatrienes with 100% selectivity (after 13% conversion) (Scheme 5.3-33) [250]. The turnover frequency (TOF) was in the range of 50 mol butadiene

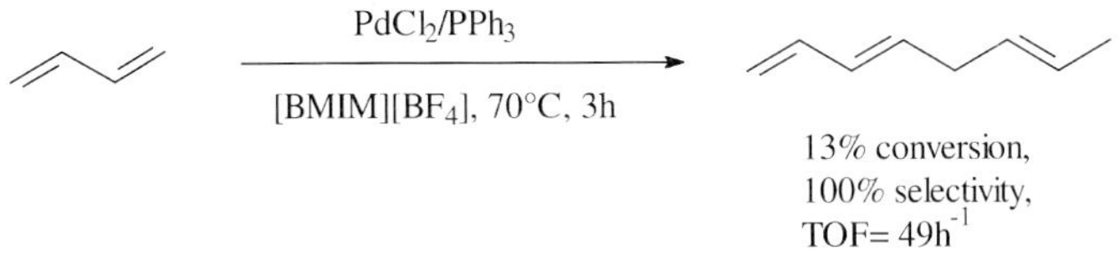

Scheme 5.3-33 Biphasic, Pd-catalyzed dimerization of butadiene in [BMIM][BF$_4$].

Reaction: butadiene → octatriene, PdCl$_2$/PPh$_3$, [BMIM][BF$_4$], 70 °C, 3 h. 13% conversion, 100% selectivity, TOF = 49 h^{-1}.

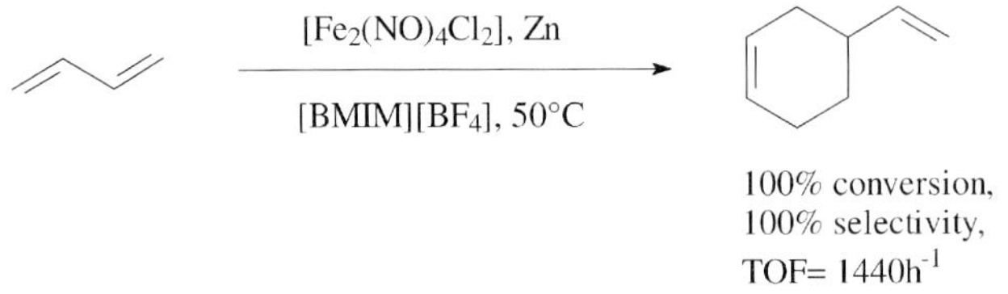

Scheme 5.3-34 Biphasic, Fe-catalyzed cyclotrimerization of butadiene in [BMIM][BF$_4$].

Reaction: butadiene → 4-vinylcyclohexene, [Fe$_2$(NO)$_4$Cl$_2$], Zn, [BMIM][BF$_4$], 50 °C. 100% conversion, 100% selectivity, TOF = 1440 h^{-1}.

converted per mol catalyst per hour, which represents a substantial increase in catalyst activity in comparison to the same reaction under identical conditions (70 °C, 3 h, butadiene/Pd = 1250) in THF (TOF = 6 h^{-1}).

The cyclo-dimerization of 1,3-butadiene was carried out in [BMIM][BF$_4$] and [BMIM][PF$_6$] using an *in situ* iron catalyst system. The catalyst was prepared by reduction of [Fe$_2$(NO)$_4$Cl$_2$] with metallic zinc in the ionic liquid. At 50 °C, the reaction proceeded in [BMIM][BF$_4$] to give full conversion of 1,3-butadiene and 4-vinylcyclohexene was formed in 100% selectivity. The observed catalytic activity corresponded to a turnover frequency of at least 1440 h^{-1} (Scheme 5.3-34).

The authors correlate the observed catalytic activity to the solubility of the 1,3-butadiene feedstock in the ionic liquid, which was found to be two times higher in the tetrafluoroborate ionic liquid vs. the corresponding hexafluorophosphate system. It is noteworthy that the same reaction in a monophasic systems using toluene as the solvent was found to be significantly less active (TOF = 240 h^{-1}).

Dimerization of functionalized olefins

The dimerization of functionalized olefins is of general technical importance. For example, the dimerization of methylacrylate (MA) to Δ^2-dihydrodimethylmuconate (DHM) leads to a highly interesting intermediate which can be transformed to both fine chemicals (such as cyclopentenones) and adipic acid. A continuous, liquid–liquid biphasic version of this Pd catalyzed reaction was realized by Tkatchenko et al. for the first time using [BMIM][BF$_4$]/toluene as reaction system (Scheme 5.3-35) [252].

While the monophasic reaction in organic solvents is known to suffer from product inhibition, the continuous reaction in the liquid–liquid biphasic system allowed to overcome this limitation by *in situ* product extraction from the ionic catalyst phase. To avoid metal leaching out of the ionic liquid phase, the catalyst

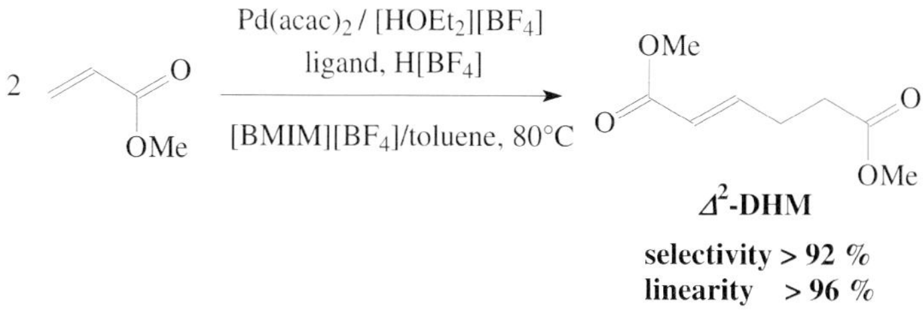

Scheme 5.3-35 Pd-catalyzed dimerization of methylacrylate in the biphasic system [BMIM][BF$_4$]/toluene.

Scheme 5.3-36 *In situ* generated ammonium phosphine ligand as used for the continuous Pd-catalyzed dimerization of methylacrylate in the biphasic system [BMIM][BF$_4$]/toluene.

Pd(acac)$_2$ was immobilized with an ionic ligand, which was generated *in situ* by protonation of 2-(dibutylphosphino)-*N,N*-dimethylethaneamine (Scheme 5.3-36).

In this way a continuous dimerization was realized over 50 h time-on-stream producing Δ^2-dihydrodimethylmuconate in an overall TON of more than 4000 (product selectivity >90%). This clearly demonstrates the potential of the biphasic ionic liquid system to bring the Pd-catalyzed MA dimerization closer to a technical realization.

By using *sc*CO$_2$ instead of toluene as second phase, Pd-leaching could be avoided, even without additional ammoniumphosphine ligand and the TOF and selectivity was increased to 195 h^{-1} and >98% respectively [253]. In another publication Tkachenko and coworkers demonstrated that the selective tail-to-tail dimerization of methyl acrylate can also be carried out with very good results in protonated *N*-butyl-imidazole, [H-BIM][BF$_4$], giving a first example of the versatility of such simple acid–base ionic liquids in continuous catalytic processes [254].

5.3.2.6 Olefin Metathesis
Ru-catalyzed ring-closing metathesis (RCM)
The ability of ionic liquids to form miscibility gaps with many organic liquids has been the starting point for attempts to realize efficient catalyst recycling in liquid–liquid biphasic metathesis reactions. The field was pioneered by Bayer AG [255] and by Buijsman and coworkers [256]. The latter group reported, in 2001, that a solution of Grubbs Ru-catalyst complex (first generation) in [BMIM][PF$_6$] promoted the ring-closing metathesis (RCM) of a number of different dienes for at least three catalytic cycles. However, catalytic performance decreased significantly over recycling which was attributed by the authors to catalyst leaching into the product phase.

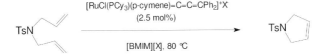

Scheme 5.3-37 Ring-closing metathesis of diallyltosylamide catalyzed by a Grubbs first generation complex in [BMIM][PF$_6$].

Slightly later, the ring-closing metathesis of diallylamides in 1-butyl-3-methylimidazolium salts using ruthenium allenylidene salts as catalyst precursors was reported by Olivier-Bourbigou and Dixneuf [257]. They realized a strong influence of the ionic liquid's anion on the catalytic performance with triflate ionic liquids showing the best performance. These observations are in good agreement with the known counter-ion sensitivity of the cationic ruthenium allenylidene complexes and suggest that effective anion exchange takes place in the ionic reaction medium. The RCM reaction of diallyltosylamide was reported to proceed in full conversion after 2.5 h at 80 °C with 97% selectivity for the closed-ring product (Scheme 5.3-37).

However, they observed slow decomposition of the catalyst in the ionic liquid, behavior that was also observed in organic solvents. This decomposition prevented successful recycling of the ionic catalyst solution and significant loss of catalytic activity was already observed in the third cycle.

A Grubbs Ru-catalyst with a cationic, imidazolium based tag (Fig. 5.3-19) was prepared and evaluated by Mauduit and coworkers [258]. Again the test reaction was the RCM reaction of diallyltosylamide.

They reported full conversion in 45 min (60 °C) with no cross-metathesis product being formed. Product extraction could be easily achieved with toluene and the ionic liquid catalyst solution was recycled nine times with conversions still exceeding 95% in the tenth cycle. The superior performance of the imidazolium-tagged catalyst in the recycling protocol was demonstrated by baseline runs with Grubbs and Hoveyda's catalyst in the same ionic liquid under identical conditions, both showing strong deactivation already in the first recycle. The obviously quite high catalyst stability of the imidazolium-tagged Ru-complex in [BMIM][PF$_6$] could be additionally demonstrated by the fact that the ionic catalyst solution could be stored for several months without loss of catalytic activity. A very similar strategy of

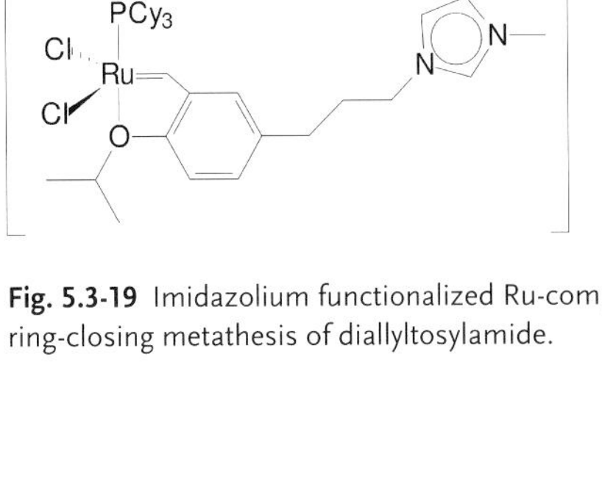

Fig. 5.3-19 Imidazolium functionalized Ru-complex as used for the ring-closing metathesis of diallyltosylamide.

attaching a slightly different imidazolium tag to a Grubbs first generation catalyst was explored by Yao et al. using a broader set of substrates [259].

The same authors later expanded the concept and additionaly provided an imidazolium-tagged Howeyda-Grubbs ruthenium carbene catalyst for the RCM reaction [260]. The resulting system proved to be highly active for the conversion of di-, tri- and tetrasubstituted diene and enyne substrates. In the catalyst solvent system [BMIM][PF$_6$]–CH$_2$Cl$_2$ (volume ratios: 1:1 to 1:9) the catalyst could be recycled 17 times with only very slight loss in activity. Also in this work it was demonstrated that the imidazolium tag is essential to obtain a stable and recycleable catalyst.

The use of methylene chloride as cosolvent has to be seen as a drawback with regard to the "greenness" of the reaction protocol, even so the authors claim that its use minimizes the amount of ionic liquid volume and thus ionic liquid cost. The critical aspect of using methylene chloride was avoided by Mauduit et al. by using a [BMIM][PF$_6$]/toluene and also a [BMIM][(CF$_3$SO$_2$)$_2$N]/toluene biphasic reaction system together with an imidazolium-tagged N-heterocyclic carbene ruthenium complex as the catalyst [261]. A high level of recyclability and reactivity could be achieved in the RCM reaction of several substrates including tri-substituted and oxygen-containing representatives. Only very low residual levels of ruthenium (1–22 ppm) were detected in the products.

Finally, it should be mentioned here that Kiddle and coworkers have successfully carried out several RCM reactions in [BMIM][BF$_4$] under microwave irradiation [262].

Ru-catalyzed self-cross-metathesis

The self-cross metathesis (CM) of 1-octene to form 7-tetradecene catalyzed by several different Ru-carbene complexes (0.02 mol%) has been investigated by Williams and coworkers [263]. While the first generation Grubbs catalyst underwent rapid deactivation in both the applied ionic liquids (several imidazolium tetrafluoroborate, hexafluorophosphate and bis(trifluormethylsulfonyl)amide salts), the second generation Grubbs catalyst showed significantly less secondary metathesis products in the ionic liquid compared to the reaction under identical conditions in organic solvents. Thus, in many cases very high selectivities (>98%) towards 7-tetradecene could be obtained in the ionic liquid media, even at very high 1-octene conversions (>95%).

Another comparative study of the CM reaction catalyzed by a second generation Grubbs carbene complex (without imidazolium tag) has been recently presented by Tang and coworkers [264]. They compared the catalytic transformation of styrene in [BMIM][BF$_4$] and [BMIM][PF$_6$] with the same reaction in CH$_2$Cl$_2$ (3 h, 45 °C) and obtained comparable yields of the CM product in all solvents in the first cycle. In the case of the two ionic liquids the product was extracted with diethyl ether and the remaining ionic catalyst solution could be recycled four times with only a small drop in catalytic activity (product yield 75% in the fourth cycle compared to 85% in the first cycle in [BMIM][PF$_6$]). The scope of the study also included the successful self-cross-metathesis of *p*-methylstyrene, *p*-chlorostyrene, α-methylstyrene and acrylonitrile in [BMIM][PF$_6$] while some ether and amine functionalized olefins failed to react even after prolonged reaction times (12 h, 50 °C).

Ru-catalyzed ring-opening polymerization
Ring-opening metathesis polymerization (ROMP) has also been successfully carried out in ionic liquids. Dixneuf and coworkers studied the biphasic ROMP reaction of norbornene in a 1-butyl-2,3-dimetylimidazolium hexafluorophosphate/toluene mixture using a ruthenium allenylidene pre-catalyst [265]. While the applied ionic liquid medium immobilized the cationic Ru-complex, the polymer formed during the reaction could be isolated in the toluene phase. The ionic catalyst solution was recycled successfully, however, reloading of the ionic liquid with a new portion of the Ru catalyst was necessary after 5 cycles to keep polymer yields high (for more details on polymerization reactions in ionic liquids see Chapter 7).

Tungsten and molybdenum catalyzed metathesis reactions
Apart from Ru-catalyzed metathesis in ionic liquids two examples of tungsten catalyzed metathesis reactions have also been reported. Vasnev and coworkers studied the metathesis of 1-hexene catalyzed by WCl$_6$ in tetrafluoroborate ionic liquids [266]. In the reaction sequence under investigation, 1-hexene first isomerized to 2-hexene which then formed 4-octene and ethylene by cross-metathesis. The yields of the metathesis product were demonstrated to increase with increasing reaction temperature and with addition of a tin-containing promoter. 4-octene was obtained in selectivies up to 97% (at 25% 1-hexene conversion) when tetrabutyl tin was applied as the promoter in this reaction.

5.3.2.7 Catalysis with Nanoparticulate Transition Metal Catalysts
The scientific interest in catalysis by transition metal nanoparticles has seen a dramatic increase in recent years and significant progress has been made in improving selectivity, efficiency and recyclability of the catalytic systems [267]. Usually nanoparticulate catalysts are prepared from a metal salt, a reducing agent and a stabilizer and are supported on oxides, charcoal or zeolites.

Ionic liquids are quite unique media for the synthesis of nanoparticles. This perception has rapidly developed over the last three years and Section 6.3 of this book is devoted to the synthetic aspects of particle preparation in ionic liquids. Thus ionic liquids represent both innovative liquid support materials and stabilizers for catalytic reactions using transition metal nanoparticles.

In the context of this chapter we aim to illustrate the state-of-the-art in this rapidly progressing field of ionic liquid catalysis, exemplified for selective hydrogenation and Heck reactions. Other applications of nanoparticulate catalyst systems have been reported in hydrosilylation reactions [268], Suzuki [269] and Stille coupling [270].

Hydrogenation reactions catalyzed by nanoparticulate transition metal catalysts

Catalytic hydrogenation reactions have to date been explored using nanoparticles of palladium, platinum, ruthenium, iridium and rhodium.

Pd nanoparticles (2–5 nm) have been obtained by Han and coworkers by reducing a solution of palladium acetate in [BMIM][PF$_6$] with H$_2$ [271]. Phenanthroline was added to the reaction mixture to stabilize the nanoparticles. The suspension of the particles was applied to catalyse the hydrogenation of alkenes, and could be recycled several times without loss of activity. However, when the phenanthroline was omitted, the system severely lost its reactivity on the first recycle, presumably due to particle agglomeration. This gives an indication that the ionic liquid alone is not able to stabilize the nanoparticles in this system.

Han's group also reported the immobilization of Pd nanoparticles onto molecular sieves using 1,1,3,3-tetramethylguanidinium lactate as the ionic liquid [272]. In this case excellent synergistic effects between the nanoparticles, the ionic liquid and the molecular sieve support were reported, leading to a very active hydrogenation catalyst for the hydrogenation of 1-hexene and cyclohexene (TOF up to 66 min^{-1} for 1-hexene hydrogenation).

Dupont's group later investigated the selective hydrogenation of 1,3-butadiene using Pd nanoparticles (particle size: 4.9 ± 0.8 nm) obtained from reduction of Pd(acac)$_2$ in [BMIM][PF$_6$] or [BMIM][BF$_4$] [273]. Selectivities to 1-butene of 72% could be achieved in this system at 99% butadiene conversion. The high selectivity to the partial hydrogenation product was attributed to the four times higher solubility of butadiene in the applied ionic liquid compared to the desired butenes. The lack of isomerization activity in the system was interpreted by the authors as an indication for a Pd-cluster surface reaction rather than a homogeneous reaction caused by Pd-complexes leached from the surface.

In this context we recall the fact that the 1,3-dialkylimidazolium ion is not innocent vs. Pd(0) and the ionic liquid's cation may convert into a carbene ligand, especially when catalysis is carried out at higher temperatures. Carbene ligands formed in this way can either be bound to the nanoparticle surface or form mononuclear carbene complexes with metal atoms leached from the nanoparticle's surface [267b].

Dupont and coworkers obtained Pt nanoparticles (2–2.5 nm) in [BMIM][PF$_6$] by reducing dissolved Pt$_2$(dba)$_3$ with H$_2$ (4 atm) at 75 °C [274]. The formed nanoparticles catalyzed the hydrogenation of both alkenes and arenes under the same, relatively mild conditions. However, the ionic liquid suspension gave a less active catalytic system (lower TOF) compared to the same particles under solventless conditions or in acetone. The generated Pt(0) nanoparticles were found to be quite stable and could be re-used, as solid, or re-dispersed in [BMIM][PF$_6$] several times with little loss of catalytic activity.

Ru nanoparticles have also been prepared in ionic liquids and used for catalytic hydrogenation reactions. Dupont's group described the reduction of RuO$_2$ with hydrogen in different ionic liquids with the [BMIM] cation [275]. The Ru nanoparticles were characterized by TEM and XRD and were 2.0–2.5 nm in diameter with a narrow size distribution. The authors demonstrated that the particles dispersed in the ionic liquid were less prone to oxidation compared to isolated nanoparticles.

The Ru nanoparticles in [BMIM][PF$_6$] proved to form a very stable catalytic system for the hydrogenation of 1-hexene with turnover numbers of 110000 (based on total Ru) or 320000 (based on Ru surface atoms) being obtained over 17 recycles of the catalyst system. These authors also investigated the reduction of RuO$_2$ dissolved in [BMIM][PF$_6$] with Na[BH$_4$] and also obtained a highly active catalytic system of ionic liquid supported Ru nanoparticles which was successfully applied in the hydrogenation of olefins under mild conditions [276]. In another publication, Dupont's group described an optimized benzene hydrogenation protocol using Ru nanoparticles in ionic liquids to obtain the partial hydrogenation product cyclohexene [277]. Making use of the preferential solubility of benzene compared to cyclohexene in the ionic liquid [BMIM][PF$_6$] these authors were able to realize selectivities to cyclohexene of up to 39%, however this high selectivity was only realized at low benzene conversion with only a 2% yield of cyclohexene.

A solid supported form of Ru nanoparticles in ionic liquids was again introduced by Han and coworkers [278]. They applied 1,1,3,3-tetramethylguanidinium([TGA])-exchanged Montmorillonite – naturally occurring, negatively charged two-dimensional silicate sheets separated by interlayers of [TGA] ions – as a supporting and stabilizing medium for the Ru nanoparticles. The favorable synergistic effects of the [TMG]$^+$ ions, the Montmorillonite support and the Ru particles led to a catalytic material exhibiting high activity in benzene hydrogenation (TOF = 4000 h^{-1}) and excellent stability.

Finally, a very interesting approach has been suggested by the same group to apply an ionic liquid to the preparation of Ru nanoparticles of defined size on a mesoporous silica support [279]. The authors made use of the ionic liquid [TGA][lactate] to customize particles obtained by reduction of dissolved RuCl$_3$ on the silica support. However, prior to catalysis, the ionic liquid film was removed thermally by heating to 220 °C for 3 h. In this way a quite active catalyst for the hydrogenation of benzene to cyclohexane (TOFs up to 83 h^{-1} at 10 bar H$_2$ pressure) was obtained. Such an approach obviously combines successfully features of SILP catalysis (see Section 5.6 for details) with the use of catalytic nanoparticles.

Iridium and rhodium nanoparticles have also been used as hydrogenation catalysts, suspended in [BMIM][PF$_6$] [280]. The nanoparticles were precipitated from solutions of RhCl$_3$·3H$_2$O or [Ir(COD)Cl]$_2$ by applying H$_2$ (4 atm) at 75 °C, giving particles of mean diameter 2.3 nm and 2.1 nm, respectively. The nanoparticles were used to hydrogenate benzene, but in both cases were less reactive than the same nanoparticles under either solventless conditions or in acetone. The iridium particles remained unchanged by reaction, whereas the rhodium particles clearly agglomerated. An approach to prevent this particle agglomeration was presented later by Kou and coworkers [281]. They applied ionic liquid–soluble ionic copolymers containing imidazolium moieties to successfully stabilize Rh nanoparticles in [BMIM][BF$_4$] and thus reached unprecedented catalytic lifetime and activity (total turnover number of 20000 obtained in four recycles) of the particles in benzene hydrogenation under relatively harsh conditions (75 °C, 40 bar H$_2$). Ir nanoparticles have also been applied to catalyze the hydrogenation of various ketones to the corresponding alcohols [84]. Particles of 2.3 ± 0.4 nm were obtained by reduction of [Ir(cod)Cl]$_2$ in [BMIM][PF$_6$] and showed a turnover frequency of 208 h^{-1} in the reduction of cyclohexanone to cyclohexanol (75 °C; 6 bar H$_2$).

It should be noted, that the examples of nanoparticulate hydrogenation catalysis in ionic liquids presented here do not represent the full list of successful, published applications. Several other groups have recently contributed details on the formation, stabilization and immobilization of nanoparticles in ionic liquids and have described the catalytic activity and reaction kinetics of these systems in different hydrogenation reactions [282].

Heck reaction catalyzed by nanoparticulate catalysts

An early account of the role of Pd nanoparticles in the Heck reaction was published by Srinivasan and coworkers in 2001 [283]. They carried out Heck reactions of different aryl iodides with acrylates and styrene at ambient temperature and observed a significant rate enhancement by combining ultrasonic irradiation with the use of the ionic liquid 1,3-dibutylimidazolium tetrafluoroborate as the reaction medium. Under the sonochemical conditions the formation of Pd–biscarbene complexes was observed and these later transformed into Pd nanoparticles.

In a study dealing with the Heck arylation of 2-methyl-prop-2-en-1-ol and 1-*tert*-butyl-4-iodobenzene to obtain the fragrance β-Lilial, Seddon and coworkers have studied several homogeneous and heterogeneous Pd catalysts in bromide and bis(trifluoromethylsulfon)amide ionic liquids (Scheme 5.3-38) [284]. Induction periods of the catalyst – for both Pd/C and for [Pd(btz)$_2$I$_2$] – were observed and associated by the authors with the formation of Pd nanoparticles in solution.

Calo et al. reported that, in the presence of tetrabutylammonium acetate as base, a palladium complex with benzothiazolidene as ligands as well as palladium acetate is transformed to Pd nanoparticles which catalyze regio- and stereo-specific Heck reactions [285]. The same authors have also applied the complex [PdI$_2$(C$_1$btz)$_2$] (C$_1$btz = methylbenzthiazolylidene, Fig. 5.3-20) as a source of palladium in [Bu$_4$N]Br, with no added ligands [286]. In this case, they have identified the formation of

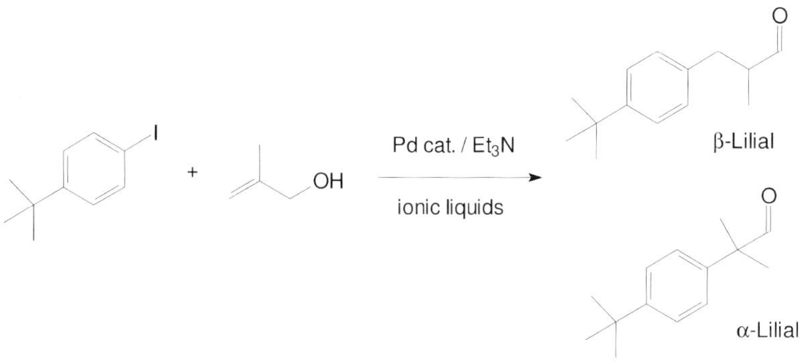

Scheme 5.3-38 Heck arylation of 2-methyl-prop-2-en-1-ol and 1-*tert*-butyl-4-iodobenzene for the synthesis of β-Lilial.

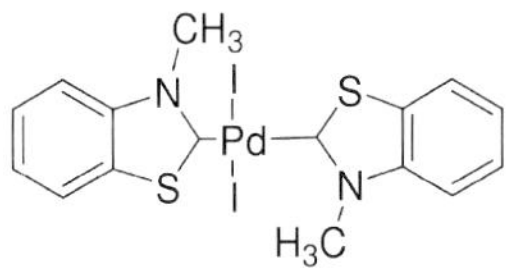

Fig. 5.3-20 Bis(methylbenzthiazolylidene)palladium diiodide.

nanoparticles of 2–6 nm diameter which are responsible for the catalytic activity in the reaction of cinnamates with aryl halides to give β-aryl-substituted cinnamic esters. The presence of tetrabutylammonium acetate was found to be crucial for the formation of the nanoparticles.

Calo and coworkers have also studied the regioselectivity of Pd-nanoparticle catalyzed Heck arylations using aryl bromide and the 1,1-disubstituted olefins butyl methylacrylate and α-methylstyrene as the reactants in tetrabutylammonium bromide as the reaction medium and with tetrabutylammonium acetate as the base [287].

Dupont et al. immobilized Pd(0) nanoparticles in [BMIM][PF_6] and observed for the coupling of aryl halides with butyl acrylate that the size of the particles increased over reaction [288]. They also described significant metal leaching (up to 34%) from the ionic phase to the organic phase at low substrate conversions, which dropped to 5–8% leaching at higher conversions. These results strongly suggest that the Pd(0) nanoparticles served as reservoir for a homogeneous catalytic active species in this case.

An immobilized version of a Heck reaction catalyzed by Pd nanoparticles has been very recently described by Karimi and Enders [289]. The nanoparticles were obtained as a result of the covalent anchoring of a N-heterocyclic carbene palladium/ionic liquid matrix on a silica surface and their nature was confirmed by TEM coupled with EDX analysis. The catalyst showed high thermal stability (up to 280 °C) and could be recycled four times for the reaction of bromobenzene with methylacrylate achieving a total turnover number of 36600. After carrying out a hot filtration process, the authors could not detect any Pd in the filtrate. The filtrate also showed no further reaction progress. From these findings the authors concluded that the reaction was, in their case, indeed catalyzed by the heterogeneous Pd particles and not from monomolecular Pd-complexes leached from the surface.

5.3.3
Concluding Remarks: "Low-hanging Fruits" and "High-hanging Fruits" – Which Transition Metal Catalyzed Reaction Should Be Carried Out in an Ionic Liquid?

Obviously, there are many good reasons to study ionic liquids as alternative solvents in transition metal catalyzed reactions. Besides the engineering advantage of their extremely low volatility, the investigation of new biphasic reactions with an ionic catalyst phase is of special interest. The possibility of adjusting solubility properties by different cation/anion combinations allows a systematic optimization of the biphasic reaction (e.g., with regard to product selectivity). Attractive options to improve selectivity in multiphase reactions derive from the preferential solubility of only one reactant in the catalyst solvent or from the *in situ* extraction of reaction intermediates out of the catalyst layer. Moreover, the application of an ionic liquid catalyst layer enables a biphasic reaction mode in many cases where this is not possible using water or polar organic solvents (e.g., due to incompatibility with the catalyst or problems with substrate solubility).

However, the concept of using ionic liquids in transition metal catalysis brings along some general features that have to do with the general properties of ionic liquids. In the last three years we have learned a lot more about ionic liquids and we have realized that despite some properties of ionic liquids being tuneable over a wide range, others are more or less intrinsic to the approach. In the light of such a property profile it may be a valuable exercise to evaluate in general terms different applications of transition metal catalysis with regard to their fit to this property profile. Such an evaluation may help to estimate the research effort required and the time to success that has to be expected. It also may give an indication at which point additional fundamental research is necessary to make an envisaged transition metal-mediated reaction or process a success.

What are the criteria that have to be considered in such an evaluation process? In the following paragraphs we have tried to condense our more than ten years of experience in transition metal catalysis in ionic liquids into a five-point check list that is hopefully helpful in this context.

Ionic liquid stability under process conditions

A first very important and far from trivial point is ionic liquid stability under the conditions required for the catalytic application under investigation. This point becomes especially interesting given the fact that in far the most cases the process development will aim for an extensive ionic liquid recycling, which naturally implies special requests for long term stability issues.

Ionic liquid stability is known to be a function of temperature (for details see Section 3.1) but the presence of nucleophiles/bases and the water content also have to be considered. There is no doubt that, under the conditions of a catalytic reaction, temperature stability issues are more complicated than under the conditions of a TGA experiment. The presence of the catalyst complex, the reactants and impurities in the system may well influence the thermal stability of the ionic liquid. Basic and nucleophilic counter-ions, reactants and metal complexes may not only lead to deprotonation of 1,3-dialkylimidazolium ions (to form carbene moieties that will undergo further consecutive reactions) but will also promote thermal dealkylation of the ionic liquid's cation. If basic reaction conditions are required for the catalysis only tetraalkylphosphonium ions can be recommended as the ionic liquid's cation at this point in time. Tetraalkylphosphonium cations have been recently shown to display reasonabe stability, even under strongly basic conditions [290]. In contrast, all nitrogen-based cations suffer to some extent from either carbene formation, Hofmann elimination or rapid dealkylation (with alkyl transfer onto the nucleophilic anion).

Water is likely to be present in all practically relevant catalytic applications unless extreme precautions are taken or the system is self-drying (e.g., due to the fact that strong Lewis-acids or metal alkyls are used as co-catalysts). Water will influence the ionic liquid's thermal stability significantly if any part of the ionic liquid is prone to hydrolysis. Apart from the well-known hydrolysis lability of tetrafluoroborates and hexafluorophosphates, water will thus also affect the stability of ester functionalities in the ionic liquid, e.g. the stability of alkyl sulfate anions. The presence of Brønsted acidity in the reaction system will further promote this kind of thermally induced hydrolysis reaction. Additionally, in strong Lewis-acidic ionic liquids care has to be taken to avoid incompatibilities between oxygen and nitrogen functionalities in the reactants or impurities and the ionic liquid's Lewis acidic group (usually a complex anion). It is for example obvious that the Pd-catalyzed dimerization of methylacrylate cannot be carried out in acidic chloroaluminate ionic liquids since the ionic liquid's anion would decompose in an irreversible reaction with the substrate methylacrylate.

Taking all these aspects into account and trying to come to a general view of the thermal stability of ionic liquids in catalytic processes we can safely state today that the chance to identify a sufficient thermally stable ionic liquid for transition metal catalysis is quite realistic if the temperature does not exceed 180 °C. This, by the way, is also the temperature limit encountered for the stability of most organometallic transition metal complexes used in homogeneous catalysis. The latter consideration somehow relativizes the restrictions imposed by the limited thermal stability of many ionic liquids for the specific application of transition metal catalysis.

Catalyst solubility in the ionic liquid without deactivation
In many cases, catalyst solubility is not a strong concern for transition metal catalysis in ionic liquids. Most transition metal complexes are polarizable enough to be well soluble in most ionic liquids. However, it is important to consider the ability of the ionic liquid to interact with the coordination sites of the metal complexes involved in the catalytic cycle.

In a general (and generalizing) view on this issue it can be stated that suitably selected ionic liquids are very likely to form catalytically active ionic catalyst solutions with a given transition metal catalyst if the latter is neither extremely electrophilic (acidic) nor extremely nucleophilic (basic). While extremely electrophilic catalyst complexes are likely to coordinate strongly even with those anions of the ionic liquid solvent which are generally regarded as weakly coordinating, extremely nucleophilic catalytic centers are likely to react with the ionic liquid's cation. Carbene complex formation by oxidative addition as well as dealkylation of the cation are possible deactivation pathways of the catalyst in such a case.

Feedstock concentration in the ionic catalyst solution
Usually, the rate of a catalytic reaction depends on the concentration of the reactants in a positive reaction order. Thus, the rate of the catalytic reaction increases if a higher concentration of the feedstock is available in the catalyst phase. As mentioned earlier (see Section 5.3.1.2) the available feedstock concentration depends,

in multiphasic catalysis, on both the thermodynamic equilibrium solubility of the reactant in the ionic liquid and on the mass transfer rate of the reactant into the ionic reaction medium.

To evaluate *a priori* whether mass transfer or solubility issues will seriously affect the probability for success in multiphasic ionic liquid catalysis the following considerations may be valuable.

It should be checked first whether solubility data for the reactants of the envisaged reactions in any ionic liquid are available. Special attention should be given to the solubility of gaseous reactants. These solubilities might be strongly dependent on temperature and pressure (see Section 3.4 for details). If the solubility of one reactant appears very low the ionic liquid's structure should be selected so as to maximize the solubility of this reactant.

Potential limitations by mass transfer issues become more likely if the reactivity of the transition metal complex in the ionic liquid is high (thus quickly converting all reactant entering the catalytic phase), the size of the reactant is large and the viscosity of the ionic liquid under the reaction conditions is high (diffusion coefficients become small for the diffusion of big molecules into highly viscous media). Therefore the application of low viscosity ionic liquids is particularly advisable if the catalyst complex dissolved in the ionic liquid is likely to be highly active.

Product isolation

Since ionic liquids have extremely low vapor pressures, one of the usual methods of product isolation, evaporation of the solvent, is not an option. The product isolation techniques available are distillation/sublimation of the product from the ionic liquid, precipitation of the product and extraction of the product into another solvent. In the latter case it is often imperative to have the product soluble in an extracting solvent but the catalyst still well immobilized in the ionic liquid, so that the catalyst remains in the ionic liquid phase. This prevents the need for subsequent purification steps to remove the catalyst from the product. In many situations, e.g. in cases where solid reactants have only limited solubility in the ionic liquid, it is preferable to have the extracting solvent already present during the reaction.

In an attempt to predict whether product isolation will be trivial or will be a major problem for an envisaged application of transition metal catalysis in ionic liquids Fig. 5.3-21 has been designed. Here, we have grouped the different established techniques of product isolation from ionic liquids in the order of increasing operating effort. Obviously, the classification is strongly related to specific product properties making certain groups of products much more easily recoverable from ionic catalyst solutions than others.

From these considerations it is quite clear that applications that require a complicated and costly separation technique (near the top in Fig. 5.3-21) need to justify the use of the ionic liquid technique with great advantages in the reaction step. Only for quite exciting and somehow unique chemistries found in ionic liquids may the application of specific membrane separation techniques [291] or even the development of fundamentally new seperation techniques be justified.

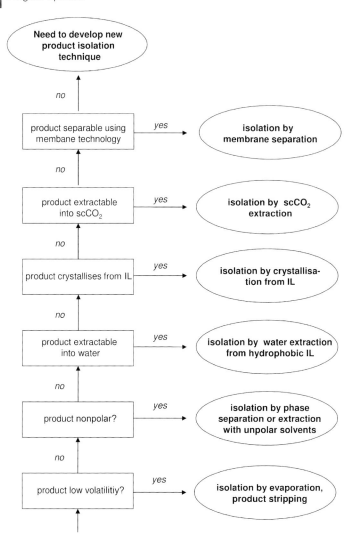

Fig. 5.3-21 Work-up options for transition metal catalysis in ionic liquids.

Recyclability and processibility

Let us first make a very general but nevertheless very true and important statement: One should never try to immobilize and recycle a bad catalyst. That in particular is true for catalyst immobilization in ionic liquids.

"Bad" may mean in some cases that the catalyst does not meet the activity or selectivity criteria to justify any recycling effort in the frame of a process development scheme. "Bad" may mean in other cases that a very selective and also initially quite active catalyst shows poor long-term stability. Obviously, one can easily waste a lot of time in immobilizing an intrinsically unstable catalyst complex in an ionic liquid. However, this work will always result in the frustrating recovery of a catalytically inactive ionic liquid containing the catalyst degradation products. Exceptions to this general rule may arise if the ionic liquid shows significant stabilizing effect for the active catalytic species. Without doubt, it is much more advisable in such a scenario to focus development resources on exploring potential stabilizing ionic liquid – catalyst interactions rather than trying to fight metal leaching into the organic phase.

Once a very stable ionic catalyst solution that shows all required selectivity and production rate characteristics has been identified, metal leaching into the product phase is indeed the next issue that has to be addressed. Excellent advances have been made in recent years in this field. In many applications it was possible to suppress catalyst leaching down to ppb levels using ionic ligands attached to the catalytic metal (see this section and Section 5.4 for numerous examples). Another strategy that very effectively avoids leaching problems is to isolate the reaction products from the ionic catalyst solution via the gas phase. This approach has been very effectively realized in the SILP catalysis technique (see Section 5.6 for details) and builds on the extremely low volatility of transition metal complexes dissolved in ionic liquids.

When it finally comes to continuous processing of transition metal catalysis in ionic liquid–organic biphasic reaction mode, some additional aspects have to be taken into account. First is the ease of phase separation that will determine the size of the separator unit and thus indirectly the ionic liquid hold-up required. Another very important aspect is the build-up of side-products or feedstock impurities in the ionic catalyst phase. Side-products and impurities that are likely to build up in the ionic liquid are relatively polar in nature and this brings along a significant risk of unfavorable interactions with the transition metal catalyst complex. Apart from this, all build-up of undesired components in the ionic liquid will also affect the ionic liquid's physicochemical properties. Therefore, a continuous build-up of components in the ionic catalyst phase that is not restricted by thermodynamic limits (e.g. solubility limits) will always require an extensive purge of the ionic catalyst solution.

It was our intention to discuss in this last and concluding part of our section the *a priori* evaluation of the chances and risks of an envisaged research project in transition metal catalysis using ionic liquids. If all five critical issues that have been discussed above can be answered in favor of the ionic liquid concept, a successful and rapid development is probable and existing separation technologies, known ionic liquids and established reactor concepts are likely to work. It can be expected in these cases that the specific properties of ionic liquids will fit the specific application very convincingly and, therefore, application of the ionic liquid technology might well be superior compared to the state of the art. In these cases, it can be anticipated that successful and very product focused research is possible without the need for further development of the ionic liquid methodology itself.

In contrast, additional and more or less fundamental work is necessary if one or more of the named aspects are unfavorable to the envisaged concept. Optimized ionic liquid structures, new transition metal complexes and/or a different separation concept might have to be developed in these cases. We do not want to discourage anybody from trying to apply ionic liquids in those reactions also but at least a significantly lower success rate and a significantly longer time to success has to be expected. But, high hanging fruits are usually sweetest and harvesting one of them may justify in the end all the extra effort that was necessary to get there.

References

1. G. W. Parshall, *J. Am. Chem. Soc.* **1972**, *94*, 8716–8719.
2. J. F. Knifton, *J. Mol. Catal.* **1987**, *43*, 65–78.
3. R. T. Carlin, J. S. Wilkes, *J. Mol. Catal.* **1990**, *63*(2), 125–129.
4. Y. Chauvin, B. Gilbert, I. Guibard, *J. Chem. Soc., Chem. Commun.* **1990**, 1715–1716.
5. J. S. Wilkes, M. J. Zaworotko, *J. Chem. Soc., Chem. Commun.* **1992**, 965–967.
6. (a) P. A. Z. Suarez, J. E. L. Dullius, S. Einloft, R. F. de Souza, J. Dupont, *Polyhedron* **1996**, *15*, 1217–1219; (b) Y. Chauvin, L. Mußmann, H. Olivier, *Angew. Chem., Int. Ed. Engl.* **1995**, *34*, 1149–1155.
7. J. Fuller, R. T. Carlin, H. C. de Long, D. Haworth, *J. Chem. Soc., Chem. Commun.* **1994**, 299–300.
8. P. Wasserscheid, C. M. Gordon, C. Hilgers, M. J. Maldoon, I. R. Dunkin, *Chem. Commun.* **2001**, 1186–1187.
9. A list of commercial suppliers of tetrafluoroborate and hexafluorophosphate salts includes (with no guarantee for completeness): Solvent Innovation GmbH (www.solvent-innovation.com), BASF (www.basionics.de), Merck (www.merck.de), Bioniqs Ltd. (www.bioniqs.com), Iolitec GmbH & Co. KG (www.iolitec.de), Kanto Chemical CO., Inc. (www.kanto.co.jp), Sachem Inc. (www.sacheminc.com); Fluka (www.fluka.com), Acros Organics (www.acros.be).
10. R. P. Swatloski, J. D. Holbrey, R. D. Rogers, *Green Chem.* **2003**, *5*, 361–363.
11. P. Bonhôte, A.-P. Dias, N. Papageorgiou, K. Kalyanasundaram, M. Grätzel, *Inorg. Chem.* **1996**, *35*, 1168–1178.
12. A. Bösmann, G. Francio, E. Janssen, M. Solinas, W. Leitner, P. Wasserscheid, *Angew. Chem., Int. Ed. Engl.* **2001**, *40*, 2697–2699.
13. N. Karodia, S. Guise, C. Newlands, J.-A. Andersen, *Chem. Commun.* **1998**, 2341–2342.
14. (a) P. Wasserscheid, S. Himmler, S. Hörmann, P. S. Schulz, *Green Chem.* **2006**, *8*(10), 887–894; (b) P. Wasserscheid, R. van Hal Roy, A. Bösmann, *Proc. Electrochem. Soc.*, **2002**, *2002–19* (Molten Salts XIII), 146–154 [*Chem. Abstr.* **2003**, 684180].
15. P. Wasserscheid, R. van Hal, A. Bösmann, *Green Chem.* **2002**, *4*, 400–404.
16. T. Welton, *Coord. Chem. Rev.* **2004**, *248*, 2459–2477.
17. S. A- Forsyth, J. M. Pringle, D. R. MacFarlane, *Aust. J. Chem.*, **2004**, *57*, 113–119
18. G. Pozzi, I. Shepperson, *Coord. Chem. Rev.* **2003**, *242*, 115–124.
19. J. Dupont, R. F. De Souza, P. A. Z. Suarez, *Chem. Rev.* **2002**, *102*, 3667–3692.
20. D. Zhao, M. Wu, Y. Kou, E. Min, *Catal. Today*, **2002**, *74*, 157–189.
21. C. C. Tzschucke, C. Markert, W. Bannwarth, S. Roller, A. Hebel, R. Haag, *Angew. Chem., Int. Ed Engl.* **2002**, *41*, 3964–4000.
22. A. P. Dobbs, M. R. J. Kimberley, *Fluorine Chem.* **2002**, *118*, 3–17.
23. H. Olivier-Bourbigou, L. Magna, *J. Mol. Catal. A: Chemical* **2002**, *2484*, 1–19.
24. R. Sheldon, *Chem. Commun.* **2001**, 2399–2407.
25. C. M. Gordon, *Appl. Catal. A: Gen.* **2001**, *222*, 101–117.
26. P. Wasserscheid, W. Keim, *Angew. Chem., Int. Ed. Engl.* **2000**, *39*, 3772–3789.
27. (a) T. Welton, *Chem. Rev.* **1999**, *99*, 2071–2083; (b) P. J. Dyson, D. J. Ellis, T. Welton, *Platinum Metals Review*, **1998**, *42*, 135–140.
28. J. D. Holbrey, K. R. Seddon, *Clean Prod. Process.* **1999**, *1*, 223–226.
29. (a) M. J. Earle, J. Esperanca, M. A. Gilea,

J. N. C. Lopes, L. P. N. Rebelo, J. W. Magee, K. R. Seddon, J. A. Widegren, *Nature* 2006 *439*, 831–834 (b) P. Wasserscheid, *Nature* 2006, *439*(7078), 797.

30. W. Keim, D. Vogt, H. Waffenschmidt, P. Wasserscheid, *J. Catal.* 1999, *186*, 481–486.
31. (a) C. P. Mehnert, *Chem. Eur. J.* 2005, *11*, 50–56; (b) A. Riisager, R. Fehrmann, M. Haumann, P. Wasserscheid, *Eur. J. Inorg. Chem.* 2006, 695–706.
32. J. Hjortkjaer, B. Heinrich, M. Capka, *Appl. Organomet. Chem.* 1990, *4*(4), 369–374.
33. (a) J. P. Arhancet, M. E. Davis, J. S. Merola, B. E. Hanson, *Nature* 1989, *339*, 454–455; (b) M. E. Davis, *CHEMTECH* 1992, 498–502; (c) M. E. Davis, J. P. Arhancet, B. E. Hanson, US Patent, US4994427, **1991**, Virginia Tech, USA; (d) M. E. Davis, J. P. Arhancet, B. E. Hanson, US Patent, US4947003, **1990**, Virginia Tech, USA.
34. A. Riisager, R. Fehrmann, M. Haumann, B. S. K. Gorle, P. Wasserscheid, *Ind. Engl. Chem. Res.* 2005, *44*(26), 9853–9859.
35. (a) C. P. Mehnert, R. A. Cook, N. C. Dispenziere, M. Afeworki, *J. Am. Chem. Soc.* 2002, *124*, 12932–12933; (b) A. Riisager, R. Fehrmann, S. Flicker, R. van Hal, M. Haumann, P. Wasserscheid, *Angew. Chem., Int. Ed. Engl.* 2005, *44*, 815–819.
36. (a) C. P. Mehnert, E. J. Mozeleski, R. A. Cook, *Chem. Commun.* 2002, 3010–3011; (b) A. Wolfson, I. F. J. Vankelecom, P. A. Jacobs, *Tetrahedron Lett.* 2003, *44*, 1195–1198.
37. H. Hagiwara, Y. Sugawara, K. Isobe, T. Hoshi, T. Suzuki, *Org. Lett.* 2004, *6*, 2325–2328.
38. S. Breitenlechner, M. Fleck, T. E. Müller, A. Suppan, *J. Mol. Catal. A: Chem.* 2004, *214*, 175–179.
39. A. Riisager, B. Jørgensen, P. Wasserscheid, R. Fehrmann, *Chem. Commun.* 2006, 994–996.
40. B. Drießen-Hölscher, P. Wasserscheid, W. Keim, *CATTECH*, 1998, June, 47–52.
41. (a) T. Prinz, W. Keim, B. Drießen-Hölscher, *Angew. Chem., Int. Ed. Engl. Engl.* 1996, *35*, 1708–1710; (b) C. Dobler, G. Mehltretter, M. Beller, *Angew. Chem., Int. Ed. Engl.* 1999, *38*, 3026–3028.
42. A. Arce, M. J. Earle, S. P. Katdare, H. Rodriguez, K. R. Seddon, *Chem. Commun.* 2006, 2548–2550.
43. (a) R. B. A. Pardy, I. Tkatschenko, *J. Chem. Soc., Chem. Commun.* 1981, 49–50; (b) J. R. Ascenso, M. A. A. F. de, C. T. Carrando, A. R. Dias, P. T. Gomes, M. F. M. Piadade, C. C. Romao, A. Revillon, I. Tkatschenko, *Polyhedron* 1989, *8*, 2449–2457; (c) P. Grenouillet, D. Neibecker, I. Tkatschenko, *J. Organomet. Chem.* 1983, *243*, 213–22; (d) J.-P. Gehrke, R. Taube, E. Balbolov, K. Kurtev; *J. Organomet. Chem.* 1986, *304*, C4–C6.
44. P. Wasserscheid, C. M. Gordon, C. Hilgers, M. J. Maldoon, I. R. Dunkin, *Chem. Commun.* 2001, 1186–1187.
45. R. T. Carlin, R. A. Osteryoung, *J. Mol. Catal.* 1990, *63*, 125–129.
46. Y. Chauvin, S. Einloft, H. Olivier, *Ind. Engl. Chem. Res.* 1995, *34*, 1149–1155
47. (a) H. Waffenschmidt, P. Wasserscheid, *J. Mol. Catal.* 2001, *164*, 61–66; (b) P. Illner, A. Zahl, R. Puchta, N. van Eikema Hommes, P. Wasserscheid, R. van Eldik, *J. Organomet. Chem.* 2005, *690*, 3567–3576.
48. (a) A. J. Arduengo, R. L. Harlow, M. Kline, *J. Am. Chem. Soc.* 1991, *113*, 361–363; (b) A. J. Arduengo, H. V. R. Dias, R. L. Harlow, *J. Am. Chem. Soc.* 1992, *114*, 5530–5534; (c) G. T. Cheek, J. A. Spencer, *9th Int. Symp. on Molten salts*, C. L. Hussey, D. S. Newman, G. Mamantov, Y. Ito (Eds.), The Electrochemical Society, New York, **1994**, pp. 426–432; (d) W. A. Herrmann, M. Elison, J. Fischer, C. Koecher, G. R. J. Artus, *Angew. Chem., Int. Ed. Engl.* 1995, *34*, 2371–2374; (e) D. Bourissou, O. Guerret, F. P. Gabbaï, G. Bertrand, *Chem. Rev.* 2000, *100*, 39–91.
49. L. Xu, W. Chen, J. Xiao, *Organometallics* 2000, *19*, 1123–1127.
50. C. J. Mathews, P. J. Smith, T. Welton, A. J. P. White, *Organometallics*, 2001, *20*(18), 3848–3850.
51. D. S. McGuinness, K. J. Cavell, B. F.

Yates, *Chem. Commun.* **2001**, 355–356.
52. M. Hasan, I. V. Kozhevnikov, M. R. H. Siddiqui, C. Fermoni, A. Steiner, N. Winterton, *Inorg. Chem.* **2001**, *40*(4), 795–800.
53. (a) N. D. Clement, K. J. Cavell, *Angew. Chem., Int. Ed. Engl.* **2004**, *43*, 3845–3847; (b) N. D. Clement, K. J. Cavell, C. Jones, C. J. Elsevier, *Angew. Chem., Int. Ed.* **2004**, *43*, 1277–1279.
54. D. Bacciu, K. J. Cavell, I. A. Fallis, L. Ooi, *Angew. Chem., Int. Ed. Engl.* **2005**, *44*, 5282–5284.
55. L. Magna, Y. Chauvin, G. P. Niccolai, Jean-Marie Basset, *Organometallics*, **2003**, *22*, 4418–4425.
56. F. McLachlan, C. J. Mathews, P. J. Smith, T. Welton, *Organometallics*, **2003**, *22*, 5350–5337.
57. J. E. L. Dullius, P. A. Z. Suarez, S. Einloft, R. F. de Souza, J. Dupont, J. Fischer, A. D. Cian, *Organometallics* **1998**, *17*, 815–819.
58. J. A. Boon, J. A. Levisky, J. L. Pflug, J. S. Wilkes, *J. Org. Chem.* **1986**, 480–486.
59. (a) M. J. Earle, K. R. Seddon, C. J. Adams, G. Roberts, *Chem. Commun.* **1998**, 2097–2098; (b) A. Stark, B. L. MacLean, R. D. Singer, *J. Chem. Soc., Dalton. Trans.* **1999**, 63–66.
60. P. N. Davey, C. P. Newman, K. R. Seddon, M. J. Earle, World Patent, WO 9919288 (to Quest International B.V., Neth.), **1999** [*Chem Abstr.* **1999**, *130*, 281871].
61. R. J. C. Brown, P. J. Dyson, D. J. Ellis, T. Welton, *Chem. Commun.* **2001**, 1862–1863.
62. P. J. Dyson, J. S. McIndoe, D. Zhao, *Chem. Commun.* **2003**, 508–509.
63. D. C. Forbes, S. A. Patrawala, K. L. T. Tran, *Organometallics* **2006**, *25*, 2693–2695.
64. (a) T. B. Scheffler, C. L. Hussey, K. R. Seddon, C. M. Kear, P. D. Armitage, *Inorg. Chem.* **1983**, *22*, 2099–2100; (b) T. M. Laher, C. L. Hussey, *Inorg. Chem.* **1983**, *22*, 3247–3251; (c) T. B. Scheffler, C. L. Hussey, *Inorg. Chem.* **1984**, *23*, 1926–1932; (d) P. B. Hitchcock, T. J. Mohammed, K. R. Seddon, J. A. Zora, C. L. Hussey, E. H. Ward, *Inorg. Chim. Acta* **1986**, *113*, L25–L26.
65. (a) D. Appleby, C. L. Hussey, K. R. Seddon, J. E. Turp, *Nature* **1986**, *323*, 614–616; (b) A. J. Dent, K. R. Seddon, T. Welton, *J. Chem. Soc. Chem. Commun.* **1990**, 315–316.
66. R. Giernoth, D. Bankmann, N. Schlörer, *Green Chem.* **2005**, *7*, 279–282.
67. C. Hardacre, S. E. J. McMath, J. D. Holbrey, *Chem. Comm.* **2001**, 367–368.
68. R. Giernoth, D. Bankmann, *Tetrahedron Lett.* **2006**, *47*(25), 4293–4296.
69. R. Giernoth, D. Bankmann, *Europ. J. Org. Chem.* **2005**, *21*, 4529–4532.
70. L. S. Ott, M. L. Cline, M. Deetlefs, K. R. Seddon, R. G. Finke, *J. Am. Soc.* **2005**, *127*, 5758–5789.
71. (a) A. Durazo, M. M. Abu-Omar, *Chem. Commun.* **2002**, 66–67; (b) G. S. Owens, A. Durazo, M. M. Abu-Omar, *Chem. Eur. J.* **2002**, *8*, 3053–3059.
72. P. Wasserscheid, PhD Thesis, RWTH Aachen, Aachen, **1998**.
73. P. Illner, A. Zahl, R. Puchta, N. van Eikema Hommes, P. Wasserscheid, R. van Eldik, *J. Organomet. Chem.* **2005**, *690*, 3567–3576.
74. C. F. Weber, R. Puchta, N. J. R. Van Eikema Hommes, P. Wasserscheid, R. Van Eldik, *Angew. Chem., Int. Ed. Engl.* **2005**, *44*, 6033–6038.
75. M. J. Earle, C. M. Gordon, N. V. Plechkova, K. R. Seddon, T. Welton, *Anal. Chem.* **2007**, *79*(2), 758–764.
76. S. M. Silva, R. P. J. Bronger, Z. Freixa, J. Dipont, P. W. N. M. Van Leeuwen, *New J. Chem.* **2003**, *27*, 1294–1296.
77. A. V. Kucherov, A. V. Vasnev, A. A. Greish, L. M. Kustov, *J. Mol Catal. A: Chem.* **2005**, *237*, 165–171.
78. (a) M. Zabet-Moghaddam, R. Krüger, E. Heinzle, A. Tholey, *J. Mass Spectrom.* **2004**, *39*(12), 1484–1505; (b) D. Zhao, *Aust. J. Chem.* **2004**, *57*(5), 509.
79. R. Bini, O. Bortolini, C. Chiappe, D. Pieraccini, T. Siciliano, *J. Phys. Chem. B* **2007**, *111*, 598–604.
80. S. Dobritz, W. Ruth, U. Kragl, *Adv. Syn. Catal.* **2005**, *347*(9), 1273–1279.
81. A. E. Bradley, C. Hardacre, M. Nieuwenhuyzen, W. R. Pitner, D. Sanders, K. R. Seddon, R. C. Thied, *Inorg. Chem* **2004**, *43*, 2503–2514.

82. C. Zhong, T. Sazaki, M. Tada, Y. Iwasawa, *J. Catal.* **2006**, *242*, 357–364.
83. G. S. Fonseca, G. Machado, S. R. Teixera, G. H. Fecher, J. Morais, M. C. M.m Alves, J. Dupont, *J. Colloid Interface Sci.* **2006**, *301*, 193–204.
84. G. S. Fonseca, J. D. Scholten, J. Dupont, *Synlett* **2004**, 1525–1528.
85. E. F. Smith, I. J. Villar Garcia, D. Briggs, P. Licence, *Chem. Commun.* **2005**, 5633–5635.
86. F. Maier, J. M. Gottfried, J. Rossa, D. Gerhard, P. S. Schulz, W. Schwieger, P. Wasserscheid, H.-P. Steinrück, *Angew. Chem., Int. Ed. Engl.* **2006**, *45*(46), 7778–7780.
87. P. A. Chaloner, M. A. Esteruelas, F. Joó, L. A. Oro, *Homogeneous Hydrogenation*, Kluwer Academic Publisher, Dordrecht, **1994**.
88. (a) J. L. Anthony, E. J. Maginn, J. F. Brennecke, *J. Phys. Chem. B* **2002**, *106*, 7315–7320; (b) P. J. Dyson, G. Laurenczy, C. A. Ohlin, J. Vallance, T. Welton, *Chem. Commun.* **2003**, 2418–2419; (c) A. Berger, R. F. de Souza, M. R. Delgado, J. Dupont, *Tetrahedron: Asymmetry* **2001**, *12*, 1825–1828.
89. D. Morgan, L. Ferrguson, P. Scovazuzo, *Ind. Engl. Chem. Res.* **2005**, *44*, 4815–4823.
90. P. A. Z. Suarez, J. E. L. Dullius, S. Einloft, R. F. de Souza, J. Dupont, *Polyhedron* **1996**, *15*, 1217–1219.
91. Y. Chauvin, L. Mußmann, H. Olivier, *Angew. Chem., Int. Ed. Engl.* **1995**, *107*, 2941–2943.
92. P. A. Z. Suarez, J. E. L. Dullius, S. Einloft, R. F. de Souza, J. Dupont, *Polyhedron* **1996**, *15*, 1217–1219.
93. P. A. Z. Suarez, J. E. L. Dullius, S. Einloft, R. F. de Souza, J. Dupont, *Inorg. Chim. Acta* **1997**, *255*, 207–209.
94. J. Dupont, P. A. Z. Suarez, A. P. Umpierre, R. F. De Souza, *J. Braz. Chem. Soc.* **2000**, *11*(3), 293–297.
95. L. M. Rossi, G. Machado, P. F. P. Fichtner, S. R. Teixeira, J. Dupont, *Catal. Lett.* **2004**, *92*, 149–155.
96. D. Zhao, P. J. Dyson, G. Laurenczy, J. S. McIndoe, *J. Mol. Catal. A: Chem.* **2004**, *214*, 19–25.
97. (a) B. Drießen-Hölscher, J. Heinen, *J. Organomet. Chem.* **1998**, *570*, 141–146; (b) J. Heinen, M. S. Tupayachi, B. Drießen-Hölscher, *Catal. Today* **1999**, *48*, 273–278.
98. S. Steines, B. Drießen-Hölscher, P. Wasserscheid, *J. Prakt. Chem.* **2000**, *342*, 348–354.
99. P. J. Dyson, D. J. Ellis, D. G. Parker, T. Welton, *Chem. Commun.* **1999**, 25–26.
100. P. J. Dyson, D. J. Ellis, W. Henderson, G. Laurenczy, *Adv. Synth. Catal.* **2003**, *345*, 216–221.
101. C. J. Boxwell, P. J. Dyson, D. J. Ellis, T. Welton, *J. Am. Chem. Soc.* **2002**, *124*, 9334–9335.
102. L. A. Müller, J. Dupont, R. F. de Souza, *Macromol. Rapid. Commun.* **1998**, *19*, 409–411.
103. S. MacLeod, R. J. Rosso, *Adv. Synth. Catal.* **2003**, *345*, 568–571.
104. L. Wei, J. Y. Jiang, Y. H. Wang, Z. L. Jin, *J. Mol. Catal. A: Chem.* **2004**, *221*, 47–50.
105. A. L. Monteiro, F. K. Zinn, R. F. de Souza, J. Dupont, *Tetrahedron: Asym.* **1997**, *2*, 177–179.
106. A. Berger, R. F. De Souza, M. R. Delgado, J. Dupont, *Tetrahedron: Asym.* **2001**, *12*(13), 1825–1828.
107. A. Deimling, B. M. Karandikar, Y. T. Shah, N. L. Carr, *Chem. Eng. J.* **1984**, *29*(3), 127–140.
108. S. Guernik, A. Wolfson, M. Herskowitz, N. Greenspoon, S. Geresh, *Chem. Commun.* **2001**, 2314–2315.
109. R. Noyori, *Asymmetric Catalysis in Organic Synthesis*, John Wiley and Sons, New York, **1994**.
110. Y. Sun, R. N. Landau, J. Wang, C. LeBlond, D. G. Blackmond, *J. Am. Chem. Soc.* **1996**, *118*, 1348–1353.
111. P. G. Jessop, R. R. Stanley, R. A. Brown, C. A. Eckert, C. L. Liotta, T. T. Ngo, P. Pollet, *Green Chem.* **2003**, *5*, 123–128.
112. S. G. Lee, Y. J. Zhang, J. Y. Piao, H. Yoon, C. E. Song, J. H. Choi, J. Hong, *Chem. Commun.* **2003**, 2624–2625.
113. B. Pugin, M. Studer, E. Kuesters, G. Sedelmeier, X. Feng, *Adv. Synth. Catal.* **2004**, *346*, 1481–1486.
114. Y. Zhu, K. Carpenter, C. C. Bun, S. Bahnmueller, C. P. Ke, V. S. Srid, L. W. Kee, M. F. Hawthorne, *Angew. Chem.,*

115. A. Hu, L. Ngo Helen, W. Lin, *Angew. Chem., Int. Ed. Engl.* **2004**, *43*, 2501–2504.
116. H. L. Ngo, A. G. Hu, W. B. Lin, *Tetrahedron Lett.* **2005**, *46*, 595–597.
117. H. L. Ngo, A. Hu, W. Lin, *Chem. Commun.* **2003**, 1912–1913.
118. M. Berthod, J.-M. Joerger, G. Mignani, M. Vaultier, M. Lemaire, *Tetrahedron: Asym.* **2004**, *15*, 2219–2221.
119. I. Kawasaki, K. Tsunoda, T. Tsuji, T. Yamaguchi, H. Shibuta, N. Uchida, M. Yamashita, S. Ohta, *Chem. Commun.* **2005**, 2134–2136.
120. R. Giernoth, M. S. Krumm, *Adv. Synth. Catal.* **2004**, *346*, 989–992.
121. (a) R. A. Brown, P. Pollet, E. McKoon, C. A. Eckert, C. L. Liotta, P. G. Jessop, *J. Am. Chem. Soc.* **2001**, *123*, 1254–1255; (b) F. Liu, M. B. Abrams, R. T. Baker, W. Tumas, *Chem. Commun.* **2001**, 433–434; (c) M. Solinas, A. Pfaltz, P. G. Cozzi, W. Leitner, *J. Am. Chem. Soc.* **2004**, *126*, 16142–16147.
122. (a) C. P. Mehnert, *Chem.-Eur. J.* **2004**, *11*, 50–56; (b) C. P. Mehnert, E. J. Mozeleski, R. A. Cook, *Chem. Commun.* **2002**, 3010–3011; (c) A. Wolfson, I. F. J. Vankelecom, P. A. Jacobs, *Tetrahedron Lett.* **2003**, *44*, 1195–1198.
123. (a) J. Jacquemin, P. Husson, V. Majer, M. F. C. Gomes, *Fluid Phase Equilib.* **2006**, *240*, 87–95; (b) J. Jacquemin, M. F. Costa Gomes, P. Husson, V. Majer, *J. Chem. Thermodyn.* **2006**, *38*, 490–502; (c) J. Kumelan, A. Perez-Salado Kamps, I. Urukova, D. Tuma, G. Maurer, *J. Chem. Thermodyn.* **2005**, *37*, 595–602; (d) J. L. Anthony, J. L. Anderson, E. J. Maginn, J. F. Brennecke, *J. Phys. Chem. B* **2005**, *109*, 6366–6374.
124. J. Muzart, *Adv. Synth. Catal.* **2006**, *348*, 275–295.
125. J. Howarth, *Tetrahedron Lett.* **2000**, *41*(34), 6627–6629.
126. R. Alcántara, L. Canoira, P. Guilherme-Joao, P. Pérez-Mendo, *Appl. Catal. A*, **2001**, *218*, 269–279.
127. R. Bernini, A. Coratti, G. Provenzano, G. Fabrizi, D. Tofani, *Tetrahedron* **2005**, *61*, 1821–1825.
128. S. V. Ley, C. Ramarao, M. D. Smith, *Chem. Commun.* **2001**, 2278–2279.
129. V. Farmer, T. Welton, *Green Chem.* **2002**, *4*, 97–102.
130. A. Wolfson, S. Wuyts, D. E. De Vos, I. F. J. Vankelecom, P. A. Jacobs, *Tetrahedron Lett.* **2002**, *43*, 8107–8110.
131. R. F. de Souza, J. Dupont, J. E. D. L. Dullius, *J. Braz. Chem. Soc.* **2006**, *17*, 48–52.
132. K. R. Seddon, A. Stark, *Green Chem.* **2002**, *4*, 119–123.
133. B. S. Chhikara, S. T Ehlan, A. Kumar, *Synlett* **2005**, 63–66.
134. B. S. Chhikara, R. Chandra, V. Tandon, *J. Catal.* **2005**, *230*, 436–439.
135. I. A. Ansari, R. Gree, *Org. Lett.* **2002**, *4*, 1507–1509.
136. N. Jiang, A. J. Ragauskas, *Org. Lett.* **2005**, *7*, 3689–3692.
137. H. Sun, K. Harms, J. Sundermeyer, *J. Am. Chem. Soc.* **2004**, *126*, 9550–9551
138. H. Sun, X. Li, J. Sundermeyer, *J. Mol. Catal. A: Chem.* **2005**, *240*, 119–122.
139. I. Hemeon, N. W. Barnett, N. Gathergood, P. J. Scammells, R. D. Singer, *Aust. J. Chem.* **2004**, *57*, 125–128.
140. W. L. Bao, Q. Wang, Y. F. Zheng, *Chin. Chem. Lett.* **2004**, *15*, 1029–1032.
141. J. W. Li, W. Sun, L. W. Xu, C. G. Xia, H. W. Wang, *Chin. Chem. Lett.* **2004**, *15*, 1437–1440.
142. X.-E. Wu, L. Ma, M.-X. Ding, L.-X. Gao, *Chem. Lett.* **2005**, *34*, 312–313.
143. V. V. Namboodiri, R. S. Varma, E. Sahle-Demessie, U. R. Pillai, *Green Chem.*, **2002**, *4*, 170–173.
144. S. M. S. Chauhan, A. Kumar, K. A. Srinivas, *Chem. Commun.* **2003**, 2348–2349.
145. (a) V. Cimpeanu, V. I. Parvulescu, P. Amoros, D. Beltran, J. M. Thompson, C. Hardacre, *Chem. Eur. J.* **2004**, *10*, 4640–4646; (b) V. Cimpeanu, A. N. Parvulescu, V. I. Parvulescu, D. T. On, S. Kaliaguine, J. M. Thompson, C. Hardacre, *J. Catal* **2005**, *232*, 60–67; (c) V. Cimpeanu, C. Hardacre, V. I. Parvulescu, J. M. Thompson, *Green Chem.* **2005**, *7*, 326–332.
146. V. Conte, B. Floris, P. Galloni, V. Mirruzzo, A. Scarso, D. Sordi, G. Strukul, *Green Chem.* **2005**, *7*, 262–266.
147. N. Jain, A. Kumar, S. M. S. Chauhan,

Tetrahedron Lett. **2005**, *46*, 2599–2602.
148. R. Bernini, A. Coratti, G. Fabrizi, A. Goggiamani, *Tetrahedron Lett.* **2003**, *44*, 8991–8994.
149. G. S. Owens, M. M. Abu-Omar, *Chem. Commun.* **2000**, 1165–1166.
150. G. Soldaini, F. Cardona, A. Goti, *Tetrahedron Lett.* **2003**, *44*, 5589–5592.
151. G. S. Owens, M. M. Abu-Omar, *J. Mol. Cat. A: Chem.* **2002**, *187*, 215–225.
152. C. E. Song, E. J. Roh, *Chem. Commun.* **2000**, 837–838.
153. K. A. Srinivas, A. Kumar, S. M. S. Chauhan, *Chem. Commun.* **2002**, 2456–2457.
154. Z. Li, C. G. Xia, C. Z. Xu, *Tetrahedron Lett.* **2003**, *44*, 9229.
155. A. A. Valente, Z. Petrovski, L. C. Branco, C. A. M. Afonso, M. Pillinger, A. D. Lopes, C. C. Romao, C. D. Nunes, I. S. Goncalves, *J. Mol. Catal. A: Chem.* **2004**, *218*, 5–11.
156. K. Yamaguchi, C. Yoshida, S. Uchida, N. Mizuno, *J. Am. Chem. Soc.* **2005**, *127*, 530–531.
157. (a) E. G. Kuntz, E. Kuntz, German Patent, DE 2627354 (to Rhone-Poulenc S. A., Fr.), **1976**, [*Chem. Abstr.* **1977**, *87*, 101944]; (b) E. G. Kuntz, *CHEMTECH* **1987**, *17*, 570–575; (c) B. Cornils, W. A. Herrmann, *Aqueous-Phase Organometallic Catalysis*, Wiley-VCH, Weinheim, **1998**.
158. P. Wasserscheid, H. Waffenschmidt, *J. Mol. Catal. A: Chem.* **2000**, *164*(1–2), 61–67.
159. Y. Chauvin, H. Olivier, L. Mußmann, Y. Chauvin, European Pateent, EP 9106535 (to Institut Francais Du Petrole, Fr.), **1997** [*Chem. Abstr.* **1997**, *127*, 65507].
160. C. C. Brasse, U. Englert, A. Salzer, H. Waffenschmidt, P. Wasserscheid, *Organometallics* **2000**, *19*(19), 3818–3823.
161. J. D. Unruh, R. Christenson, *J. Mol. Catal.* **1982**, *14*, 19–34.
162. C. P. Casey, E. L. Paulsen, E. W. Beuttenmueller, B. R. Proft, L. M. Petrovich, B. A. Matter, D. R. Powell, *J. Am. Chem. Soc.* **1997**, *119*, 11817–11825.
163. W. R. Moser, C. J. Papile, D. A. Brannon, R. A. Duwell, *J. Mol. Catal.* **1987**, *41*, 271–292.
164. A. Heßler, O. Stelzer, H. Dibowski, K. Worm, F. P. Schmidtchen, *J. Org. Chem.* **1997**, *62*, 2362–2369.
165. P. Machnitzki, M. Teppner, K. Wenz, O. Stelzer, E. J. Herdtweck, *Organomet. Chem.* **2000**, *602*, 158–169.
166. H. Dibowski, F. P. Schmidtchen, *Angew. Chem., Int. Ed. Engl.* **1998**, *37*, 476–478.
167. O. Stelzer, F. P. Schmidtchen, A. Heßler, M. Tepper, H. Dibowski, H. Bahrmann, M. Riedel, German Patent, DE 19701245 (to Celanese G.m.b.H., Germany), **1998** [*Chem. Abstr.* **1998**, *129*, 149094].
168. P. Wasserscheid, H. Waffenschmidt, P. Machnitzki, K. Kottsieper, O. Stelzer, *Chem. Commun.* **2001**, 451–452.
169. F. Favre, H. Olivier-Bourbigou, D. Commereuc, L. Saussine, *Chem. Commun.* **2001**, 1360–1361.
170. J. Sirieix, M. Ossberger, B. Betzemeier, P. Knochel, *Synlett.* **2000**, 1613–1615.
171. (a) K. W. Kottsieper, O. Stelzer, P. Wasserscheid, *J. Mol. Catal. A: Chem.* **2001**, *175*(1–2), 285–288; (b) D. J. Brauer, K. W. Kottsieper, C. Liek, O. Stelzer, H. Waffenschmidt, P. Wasserscheid, *J. Organomet. Chem.* **2001**, *630*(2), 177–184.
172. C. P. Casey, G. T. Whiteker, M. G. Melville, L. M. Petrovich, L. J. A. Gavey, D. R. J. Powell, *J. Am. Chem. Soc.* **1992**, *114*, 5535–5543.
173. M. Kranenburg, Y. E. M. van der Burgt, P. C. J. Kamer, P. W. N. M van Leeuwen, K. Goubitz, J. Fraanje, *Organometallics* **1995**, *14*, 3081–3089.
174. P. W. N. M. van Leeuwen, P. C. J. Kamer, J. N. H. Reek, P. Dierkes, *Chem. Rev.* **2000**, *100*, 2741–2770.
175. P. Dierkes, S. Ramdeehul, L. Barloy, A. De Cian, J. Fischer, P. C. J. Kamer, P. W. N. M. van Leeuwen, *Angew. Chem., Int. Ed. Engl.* **1998**, *37*, 3116–3118.
176. R. P. J. Bronger, S. M. Silva, P. C. J. Kamer, P. W. N. M. van Leeuwen, *Chem. Commun.* **2002**, 3044–3045.
177. R. P. J. Bronger, S. M. Silva, P. C. J. Kamer, P. W. N. M. van Leeuwen, *Chem. Commun.* **2002**, 3044–3045.
178. O. Stenzel, H. G. Raubenheimer, C. Esterhuysen, *J. Chem. Soc., Dalton Trans.* **2002**, 1132–1138.

179. B. A. Omotowa, J. M. Shreeve, *Organometallics* **2004**, *23*, 783–791.
180. F. Z. Kong, J. Y. Jiang, Z. L. Jin, *Catal. Lett.* **2004**, *96*, 63–65.
181. J. P. Dyson, G. Laurenczy, C. A. Ohlin, J. Vallance, T. Welton, *Chem. Comm.* **2003**, 2418–2419.
182. C. A. Ohlin, P. J. Dyson, G. Laurenczy, *Chem. Comm.* **2004**, 1070–1071.
183. J. Dupont, S. M. Silva, R. F. de Souza, *Catal. Lett.*, **2001**, *77*, 131.
184. A. M. Scurto, W. Leitner, *Chem. Comm.* **2006**, 3681–3683.
185. (a) S. Garcia, N. M. T. Lourenco, D. Lousa, A. F. Sequeira, P. Mimoso, J. M. S. Cabral, C. A. M. Afonso, S. Barreiros, *Green Chem.* **2004**, *6*, 466–470; (b) Z. Liu, W. Wu, B. Han, Z. Dong, G. Zhao, J. Wang, T. Jiang, G. Yang, *Chem. Eur. J.* **2003**, *9*, 3897–3903.
186. D. Camper, C. Becker, C. Koval, R. Noble, *Ind. Engl. Chem. Res.* **2006**, *45*, 445–450.
187. M. Solinas, A. Pfaltz, P. G. Cozzi, W. Leitner, *J. Am. Chem. Soc.* **2004**, *126*, 16142–16147.
188. M. F. Sellin, P. B. Webb, D. J. Cole-Hamilton, *Chem. Comm.* **2001**, 781–782.
189. D. J. Cole-Hamilton, *Science*, **2003**, *299*, 1702–1706.
190. (a) A. Riisager, R. Fehrmann, S. Flicker, R. van Hal, M. Haumann, P. Wasserscheid, *Angew. Chem., Int. Ed. Engl.* **2005**, *44*, 815–819; (b) A. Riisager, R. Fehrmann, M. Haumann, B. S. K. Gorle, P. Wasserscheid, *Ind. Engl. Chem. Res.* **2005**, *44*, 9853–9859; (c) A. Riisager, R. Fehrmann, M. Haumann, P. Wasserscheid, *Eur. J. Inorg. Chem.* **2006**, 695–706; (d) A. Riisager, R. Fehrmann, M. Haumann, P. Wasserscheid, *Top. Catal.* **2006**, *40*, 91–102.
191. (a) J. Dupont, R. F. de Souza, P. A. Z. Suarez, *Chem. Rev.* **2002**, *102*, 3667–3691; (b) T. Welton, P. J. Smith, *Adv. Organomet. Chem.* **2004**, *51*, 251–284; (c) F. Alonso, I. P. Beletskaya, M. Yus, *Tetrahedron* **2005**, *61*, 11771–11835.
192. D. E. Kaufmann, M. Nouroozian, H. Henze, *Synlett.* **1996**, 1091–1092.
193. (a) W. A. Herrmann, V. P. W. Böhm, *J. Organomet. Chem.* **1999**, *572*, 141–145; (b) V. P. W. Böhm, W. A. Hermann, *Chem. Eur. J.* **2000**, *6*, 1017–1025.
194. S. Bouquillon, B. Ganchegui, B. Estrine, F. Henin, J. Muzart, *J. Organomet. Chem.* **2001**, *634*, 153–156.
195. (a) V. Calò, A. Nacci, L. Lopez, A. Napola, *Tetrahedron Lett.* **2001**, *42*, 4701–4703; (b) V. Calo, A. Nacci, A. Monopoli, L. Lopez, A. di Cosmo, *Tetrahedron*, **2001**, *57*, 6071–6077.
196. J. Silberg, T. Schareina, R. Kempe, K. Wurst, M. R. Buchmeiser, *J. Organomet. Chem.* **2001**, *622*, 6–18.
197. S. T. Handy, M. Okello, *Tetrahedron Lett.* **2003**, *44*, 8395–8397.
198. for a review of Jeffery's results, see: T. Jeffery, *Tetrahedron* **1996**, *52*, 10113–10130.
199. G. Zou, Z. Wang, J. Zhu, J. Tang, M. Y. He, *J. Mol. Catal. A: Chem.* **2003**, *206*, 193–198.
200. H. Bönnemann, R. Brinkmann, R. Köppler, P. Neiteler, J. Richter, *Adv. Mater.* **1992**, *4*, 804–806.
201. (a) W. A. Herrmann, V. P. W. Böhm, *J. Organomet. Chem.* **1999**, *572*, 141–145; (b) V. P. W. Böhm, W. A. Herrmann, *Chem. Eur. J.*, **2000**, *6*, 1017–1025.
202. (a) W. A. Herrmann, *Angew. Chem., Int. Ed. Engl.*, **2002**, *41*, 1290–1309 and references therin; (b) W. A. Herrmann, T. Weskamp, V. P. W. Böhm, *Adv. Organomet. Chem.*, **2001**, *48*, 1–69.
203. W. A. Herrmann, M. Elison, J. Fischer, C. Kocher, G. R. J. Artus, *Chem. Eur J.*, **1996**, *2*, 772–780.
204. N. Hamill, C. Hardacre, S. E. J. McMath, *Green Chem.*, **2002**, *4*, 139–142.
205. L. Xu, W. Chen, J. Ross, J. Xiao, *Org. Lett.* **2001**, *3(2)*, 295–297.
206. R. Wang, B. Twamley, J. N. M. Shreeve, *J. Org. Chem.* **2006**, *71*, 426–429.
207. J.-C. Xiao, B. Twamley, J. M. Shreeve, *Org. Lett.* **2004**, *6*, 3845–3847.
208. A. Corma, H. Garcia, A. Leyva, *Tetrahedron* **2004**, *60*, 8553–8560.
209. R. Wang, M. P. Melissa, J. M. Shreeve, *Org. Biomol. Chem.* **2006**, *4*, 1878–1886.
210. (a) G. Zou, W. Huang, Y. Xiao, J. Tang, *New J. Chem.* **2006**, *30*, 803–809; (b) R. Wang, M. Piekarski Melissa, M. Shreeve Jeanne, *Org. Biomol. Chem.* **2006**, *4*,

1878–1886; (c) H. Hagiwara, Y. Sugawara, T. Hoshi, T. Suzuki, *Chem. Commun.* **2005**, 2942–2944; (d) B. Park Soon, H. Alper, *Org. Lett.* **2003**, *5*, 3209–3212; (e) M. Choudary Boyapati, S. Madhi, S. Chowdari Naidu, L. Kantam Mannepalli, B. Sreedhar, *J. Am. Chem. Soc.* **2002**, *124*, 14127–14136; (f) H. Hagiwara, Y. Shimizu, T. Hoshi, T. Suzuki, M. Ando, K. Ohkubo, C. Yokoyama, *Tetrahedron Lett.* **2001**, *42*(26), 4349–4351.

211. (a) G. D. Davis Jr., A. Hallberg, *Chem. Rev.* **1989**, *89*, 1433–1445; (b) M. Larhed, C. M. Andersson, A. Hallberg, *Tetrahedron* **1994**, *50*, 285–304; (c) M. Larhed, C. M. Andersson, A. Hallberg, *Acta Chem. Scand.* **1993**, *47*, 212–217; (d) C. M. Andersson, J. Larsson, A. Hallberg, *J. Org. Chem.* **1990**, *55*, 5757–5761.

212. W. Cabri, I. Candiani, A. Bedeschi, *J. Org. Chem.* **1993**, *58*, 7421–7426.

213. (a) Z. Hyder, J. Mo, J. Xiao, *Adv. Syn. Catal.* **2006**, *348*, 1699–1704; (b) J. Mo, S. Liu, J. Xiao, *Tetrahedron* **2005**, *61*, 9902–9907; (c) J. Mo, L. Xu, J. Xiao, *J. Am. Chem. Soc.* **2005**, *127*, 751–760.

214. A. J. Carmichael, M. J. Earle, J. D. Holbrey, P. B. McCormac, K. R. Seddon, *Org. Lett.* **1999**, *1*, 997–1000.

215. S. Liu, T. Fukuyama, M. Sato, I. Ryu, *Org. Process. Res. Dev.* **2004**, *8*, 477–481.

216. B. Karimi, D. Enders, *Org. Lett.* **2006**, *8*, 1237–1240.

217. H. Hagiwara, Y. Sugawara, K. Isobe, T. Hoshi, T. Suzuki, *Org. Lett.* **2004**, *6*, 2325.

218. A. Suzuki, *J. Organomet. Chem.*, **1999**, *576*, 147–168.

219. C. J. Mathews, P. J. Smith, T. Welton, *Chem. Comm.* **2000**, 1249–1250.

220. C.-H. Yang, C.-C. Tai, Y.-T. Huang, I. W. Sun, *Tetrahedron* **2005**, *61*, 4857–4861.

221. C. J. Mathews, P. J. Smith, T. Welton, A. J. P. White, D. J. Williams, *Organometallics*, **2001**, *20*, 3848–3850.

222. C. J. Mathews, P. J. Smith, T. Welton, *Organometallics* **2003**, 5350–5357.

223. C. J. Mathews, P. J. Smith, T. Welton, *J. Mol. Catal. A: Chem.* **2004**, *214*, 27–32.

224. J.-C. Xiao, J. N. M. Shreeve, *J. Org. Chem.* **2005**, *70*, 3072–3078.

225. R. Rajagopal, D. V. Jarikote, K. V. Srinivasan, *Chem. Commun.* **2002**, 616–617.

226. R. R. Dashmukh, R. Rajagopal, K. V. Srinivasan, *Chem. Commun.*, **2001**, 1544–1545.

227. D. Zhao, Z. Fei, J. Geldbach Tilmann, R. Scopelliti, J. Dyson Paul, *J. Am. Chem. Soc.* **2004**, *126*, 15876–15882.

228. S. T. Handy, X. Zhang, *Org. Lett.* **2001**, *3*(2), 233–236.

229. C. Chiappe, G. Imperato, E. Napolitano, D. Pieraccini, *Green Chem.* **2004**, *6*, 33–36.

230. W. Chen, L. Xu, C. Chatterton, J. Xiao, *Chem. Commun.*, **1999**, 1247–1248.

231. J. Ross, W. Chen, L. Xu, J. Xiao, *Organometallics* **2001**, *20*, 138–142.

232. S. Toma, B. Gotov, I. Kmentova, E. Solcaniova, *Green Chem.* **2000**, *2*, 149–151.

233. (a) W. Chen, L. Xu, C. Chatterton, J. Xiao, *Chem. Commun.* **1999**, 1247–1248; (b) C. de Bellefon, E. Pollet, P. Grenouillet, *J. Mol. Catal.* **1999**, *145*, 121–126.

234. P. W. Glockner, W. Keim, R. F. Mason, R. S. Bauer, (Shell Internationale Research Maatschappij N. V.) Application: DE, **1971**, 27 pp.

235. G. Wilke, B. Bogdanovic, P. Hardt, P. Heimbach, W. Keim, M. Kröner, W. Oberkirch, K. Tanaka, E. Steinrücke, D. Walter, H. Zimmermann, *Angew. Chem., Int. Ed. Engl.* **1966**, *5*, 151–154.

236. (a) Y. Chauvin, B. Gilbert, I. Guibard, *J. Chem. Soc. Chem. Commun.* **1990**, 1715–1716; (b) Y. Chauvin, S. Einloft, H. Olivier, *Ind. Engl. Chem. Res.* **1995**, *34*, 1149–1155; (c) Y. Chauvin, S. Einloft, H. Olivier, French Patent, FR 93/11,381 (to Institut Francais Du Petrole, Fr.), **1996** [*Chem. Abstr.* **1995**, *123*, 144896c].

237. (a) M. Freemantle, *Chem. Eng. News* **1998**, *76*(13), 32–37; (b) E. Burridge, ECN Chemscope **1999**, May, 27–28; (c) H. Olivier, *J. Mol. Catal. A: Chem.* **1999**, *146*(1–2), 285–289; (d) H. Olivier-Bourbigou, P. Travers, J. A. Chodorge, *Petroleum Technology Quarterly*, Autumn **1999**, 141–149.

238. W. Keim, B. Hoffmann, R. Lodewick, M.

Peukert, G. Schmitt, J. Fleischhauer, U. Meier, *J. Mol. Catal.* **1979**, *6*, 79–97.

239. (a) B. Ellis, W. Keim, P. Wasserscheid, *Chem. Commun.* **1999**, 337–338; (b) P. Wasserscheid, W. Keim, World Patent, WO 9847616 (to BP Chemicals), **1997** [*Chem. Abstr.* **1998**, *129*, 332457].

240. (a) M. Eichmann, PhD Thesis, RWTH-Aachen, **1999**; (b) P. Wasserscheid, M. Eichmann, *Catal. Today* **2001**, *66*(2–4), 309–316.

241. (a) Y. Chauvin, H. Olivier, C. N. Wyrvalski, L. C. Simon, R. F. de Souza, *J. Catal.* **1997**, *165*, 275–278; (b) L. C. Simon, J. Dupont, R. F. de Souza, *J. Mol. Catal.* **1998**, *175*, 215–220.

242. D. S. McGuinness, W. Mueller, P. Wasserscheid, K. J. Cavell, B. W. Skelton, A. H. White, U. Englert, *Organometallics* **2002**, *21*, 175.

243. S. Einloft, F. K. Dietrich, R. F. de Souza, J. Dupont, *Polyhedron* **1996**, *19*, 3257–3259.

244. M. Oberson de Souza, R. F. de Souza, *Curr. Top. Catal.* **2002**, *3*, 267–273.

245. K. Bernardo-Gusmao, L. F. Trevisan Queiroz, R. F. de Souza, F. Leca, C. Loup, R. Reau, *J. Catal.* **2003**, *219*, 59–62.

246. H. Olivier, P. Laurent-Gérot, *J. Mol. Catal. A: Chem.* **1999**, *148*, 43–48.

247. J. T. Dixon, J. J. C. Grove, A. Ranwell, World Patent, WO 0138270 (to Sasol Technology Ltd.), [*Chem. Abstr.* **2001**, *135*, 7150].

248. (a) I. Brassat, PhD Thesis, RWTH Aachen, **1998**; (b) I. Brassat, W. Keim, S. Killat, M. Möthrath, P. Mastrorilli, C. Nobile, G. J. Suranna, *Mol. Catal. A: Chem.* **2000**, *43*, 41–58; (c) I. Brassat, U. Englert, W. Keim, D. P. Keitel, S. Killat, G. P. Suranna, R. Wang, *Inorg. Chim. Acta* **1998**, *280*, 150–162.

249. P. Wasserscheid, C. Hilgers, W. Keim, *J. Mol. Catal. A: Chem.* **2004**, *214*, 83–90.

250. S. M. Silva, P. A. Z. Suarez, R. F. de Souza, J. Dupont, *Polymer Bull.* **1998**, *41*, 401–405.

251. R. A. Ligabue, R. F. de Souza, J. Dupont, *J. Mol. Catal. A: Chem.* **2001**, *169*, 11–17.

252. (a) J. Zimmermann, P. Wasserscheid, I. Tkatchenko, S. Stutzmann, *Chem. Commun.* **2002**, 760–761; (b) J. Zimmermann, I. Tkatchenko, P. Wasserscheid, *Adv. Synth. Catal.* **2003**, *345*, 402–409.

253. D. Ballivet-Tkatchenko, M. Picquet, M. Solinas, G. Francio, P. Wasserscheid, W. Leitner, *Green Chem.* **2003**, *5*, 232–235.

254. (a) M. Picquet, S. Stutzmann, I. Tkatchenko, I. Tommasi, J. Zimmermann, P. Wasserscheid, *Green Chem.* **2003**, *5*, 153–162; (b) M. Picquet, I. Tkatchenko, I. Tommasi, P. Wasserscheid, J. Zimmermann, *Adv. Synth. Catal.* **2003**, *345*(8), 959–962; (c) M. Picquet, D. Poinsot, S. Stutzmann, I. Tkatchenko, I. Tommasi, P. Wasserscheid, J. Zimmermann, *Top. Catal.* **2004**, *29*(3–4), 139–143.

255. C. Gurtler, M. Jautelat, European Patent, EP 1035093 (to Bayer A.-G., Germany), **2000** [*Chem. Abstr.* **2000**, *133*, 237853].

256. R. C. Buijsman, E. van Vuuren, J. G. Sterrenburg, *Org. Lett.* **2001**, *3*(23), 3785–3787.

257. D. Semeril, H. Olivier-Bourbigou, C. Bruneau, P. H. Dixneuf, *Chem. Commun.* **2002**, 146–147.

258. N. Audic, H. Clavier, M. Mauduit, J.-C. Guillemin, *J. Am. Chem. Soc.* **2003**, *125*, 9248–9249.

259. Q. Yao, Z. Zhang, *Angew. Chem., Int. Ed. Engl.* **2003**, *42*, 3395–3398.

260. Q. Yao, M. Sheets, *J. Organomet. Chem.* **2005**, 3577–3584.

261. (a) H. Clavier, N. Audic, M. Mauduit, J.-C. Guillemin, *Chem. Commun.* **2004**, 2282–2283; (b) H. Clavier, N. Audic, J.-C. Guillemin, M. Mauduit, *J. Organomet. Chem.* **2005**, *690*, 3585–3599.

262. K. G. Mayo, E. H. Nerhoof, J. J. Kiddle, *Org. Lett.* **2002**, *4*, 1567–1570.

263. D. B. G. Williams, M. Ajam, A. Ranwell, *Organometallics* **2006**, *25*(12), 3088–3090.

264. X. Ding, X. Lv, B. Hui, Z. Chen, M. Xiao, B. Guo, W. Tang, *Tetrahedron Lett.* **2006**, 2921–2924.

265. S. Csihony, C. Fischmeister, C. Bruneau, I. Horwarth, P. Dixneuf, *New J. Chem.* **2002**, *28*(11), 1667–1670.

266. A. V. Vasnev, A. A. Greish, L. M. Kustov, *Russian Chem. Bull., Int. Ed.* **2004**, *53*(10), 2187–2191.

267. (a) M. Moreno-Mañas, R. Pleixats, *Acc. Chem. Res.*, **2003**, *36*, 638–643; (b) D. Astruc, F. Lu, J. R. Aranaez, *Angew.*

Chem., Int. Ed. Engl. **2005**, *44*, 7852–7872.
268. T. Geldbach, D. Zhao, N. C. Castillo, G. Laurenczy, B. Weyershausen, P. J. Dyson, *J. Am. Chem. Soc.* **2006**, *128*(30), 9773–9780.
269. V. Calo, A. Nacci, A. Monopoli, F. Montingelli, *J. Org. Chem.* **2005**, *70*(15), 6040–6044.
270. C. Chiappe, D. Pieraccini, D. Zhao, Z. Fei, P. J. Dyson, *Adv. Synth. Catal.* **2006**, *348*(1 + 2), 68–74.
271. J. Huang, T. Jiang, B. Han, H. Gao, Y. Chang, G. Zhao, W. Wu, *Chem. Commun.* **2003**, 1654–1655.
272. J. Huang, T. Jiang, H. Gao, B. Han, Z. Liu, W. Wu, Y. Chanf, G. Zhao, *Angew. Chem., Int. Ed. Engl.* **2004**, *43*, 1397–1399.
273. A. P. Umpierre, G. Machado, G. H. Fecher, J. Morais, J. Dupont, *Adv. Synth. Catal.* **2005**, *347*, 1404–1412.
274. C. W. Scheeren, G. Machado, J. Dupont, P. F. P. Fichtner, S. R. Texeira, *Inorg. Chem.* **2003**, *42*, 4738–4742.
275. L. M. Rossi, G. Machado, P. F. P. Fichtner, S. R. Teixera, J. Dupont, *Catal. Lett.* **2004**, *92*(3–4), 149–155.
276. L. M. Rossi, J. Dupont, G. Mchado, P. F. P. Fichtner, C. Radtke, I. J. R. Baumvol, S. R. Teixeira, *J. Braz. Chem. Soc.* **2004**, *15*(6), 904–910.
277. E. T. Sileira, A. P. Umpierre, L. M. Rossi, G. Machado, J. Morais, G. V. Soares, I. J. R. Baumvol, S. R. Teixera, P. F. P. Fichtner, J. Dupont, *Chem. Eur. J.* **2004**, *10*. 3734–3740.
278. A. Miao, Z. Liu, B. Han, J. Huang, Z. Sun, J. Zhang, T. Jiang, *Angew. Chem., Int. Ed. Engl.* **2006**, *45*, 266–269.
279. J. Huang, T. Jiang, B. Han, W. Wu, Z. Liu, Z. Xie, J. Zhang, *Catal. Lett.* **2005**, *103*(1–2), 59–62.
280. (a) J. Dupont, G. S. Fonseca, A. P. Umpierre, P. F. P. Fichtner, S. R. Teixeira, *J. Am. Chem. Soc.* **2002**, *124*, 4228–4229; (b) G. S. Fonseca, A. P. Umpierre, P. F. P. Fichtner, S. R. Teixeira, J. Dupont, *Chem. Eur. J.* **2003**, *9*, 3263–3269.
281. X. Mu, J. Meng, Z. Li, Y. Kou, *J. Am. Chem. Soc.* **2005**, *127*, 9694–9695.
282. (a) K. Anderson, S. Cortinas Fernandez, C. Hardacre, P. C. Marr, *Inorg. Chem. Commun.* **2004**, *7*, 73–76; (b) R. Tatumi, H. Fujihara, *Chem. Commun.* **2005**, 83–84; (c) V. Mévellec, B. Leger, M. Mauduit, A. Roucoux, *Chem. Commun.* **2005**, 2838–2839; (d) G. S. Fonseca, E. T. Silveira, M. A. Gelesky, J. Dupont, *Adv. Synth. Catal.* **2005**, *347*, 847–853; (e) G. S. Fonseca, J. B. Domingos, F. Nome, J. Dupont, *J. Mol. Catal. A: Chem.* **2006**, *248*, 10–16.
283. R. R. Deshmaukh, R. Rajagopal, K. V. Srinivasan, *Chem. Commun.* **2001**, 1544–1555.
284. S. A. Forsyth, H. Q. Nimal Gunaratne, C. Hardacre, A. McKeown, D. W. Rooney, K. R. Seddon, *J. Mol. Catal A: Chem.* **2005**, *231*, 61–66.
285. V. Calo, A. Nacci, A. Monopoli, *J. Mol. Catal. A: Chem.* **2004**, *214*, 45–56.
286. V. Calò, A. Nacci, A. Monopoli, S. Laera, N. Cioffi, *J. Org. Chem.* **2003**, *68*, 2929–2933.
287. V. Calo, A. Nacci, A. Monopoli, A. Detomaso, P. Iliade, *Organometallics* **2003**, *22*, 4193–4197.
288. C. C. Cassol, A. P. Umpierre, G. Machado, S. I. Wolke, J. Dupont, *J. Am. Chem. Soc.* **2005**, *127*, 3298–3299.
289. B. Karimi, D. Enders, *Org. Lett.* **2006**, *8*(6), 1237–1240.
290. (a) M. J. Earle, U. Frohlich, S. Huq, S. Katdare, R. M. Lukasik, E. Bogel, N. V. Plechkova, K. R. Seddon, World Patent, WO 2006072785 (to Queen's University of Belfast) [*Chem. Abstr.* **2006**, *145*, 145001]; (b) M. J. Earle, K. R. Seddon, S. Forsyth, U. Frohlich, N. Gunaratne, S. Katdare, World Patent, WO 2006072775 (to Queen's University of Belfast) [*Chem. Abstr.* **2006**, *145*, 145000].
291. P. Wasserscheid, U. Kragl, J. Kröckel, World Patent, WO 2003039719 (to Solvent Innovation, GmbH, Cologne) [*Chem. Abstr.* **2003**, *138*, 387435].

5.4
Ionic Liquids in Multiphasic Reactions

Hélène Olivier-Bourbigou and Frédéric Favre

5.4.1
Multiphasic Reactions: General Features, Scope and Limitations

While the solubility of organometallic complexes in common organic solvents appears to be an advantage in terms of site availability and tunability and reaction selectivity and activity, it is a major drawback in terms of catalyst separation and recycling. The quest for new catalyst immobilization or recovery strategies to facilitate its reuse is unceasing. Immobilization of the catalyst on a solid support has been largely studied. Except for Ziegler-Natta and metallocene-type polymerization processes, in which the catalyst is not recycled due to its high activity, this technology has not been developed industrially, mainly because of problems of catalyst leaching and deactivation. One most successful approach to close the advantage/disadvantage gap between homogeneous and heterogeneous catalysis is multiphasic catalysis [1]. In its simplest version, there are only two liquid phases (biphasic catalysis or two-phase catalysis). The catalyst is dissolved in one phase (generally a polar phase) while the products and the substrates are found in the other. The catalyst can be separated by decantation and recycled under mild conditions.

It is important to make the distinction between the multiphasic catalysis concept and transfer-assisted organometallic reactions or phase tranfer catalysis (PTC). In this latter approach, a catalytic amount of quaternary ammonium salt $[Q]^+[X]^-$ is present in an aqueous phase. The catalyst lipophilic cation $[Q]^+$ transports the reactant's anion Y^- to the organic phase, as an ion-pair, and the chemical reaction occurs in the organic phase of the organic/aqueous two-phase mixture [2].

The use of multiphasic catalysis has proven its potential in important industrial processes. In 1977, the first large-scale commercial catalytic process to benefit from two-phase liquid/liquid technology was the Shell Higher Olefin Process (SHOP) for oligomerization of ethene into α-olefins, catalyzed by nickel complexes dissolved in diols such as 1,4-butanediol. Subsequently, the advancement in two-phase homogeneous catalysis has been demonstrated by the biphasic aqueous hydroformylation as an economically competitive large-scale process. The first commercial oxo plant, developed by Ruhrchemie-Rhône-Poulenc for the production of butyraldehyde from propene, came on stream in 1984. This is an example of a gas–liquid–liquid multiphasic system in which the homogeneous rhodium-based catalyst is immobilized in a water phase by coordination to the hydrophilic trisulfonated triphenylphosphine ligand (TPPTS) [3]. The catalyst separation is more effective and simpler than in classical rhodium processes, but separation of by-products from the catalyst is also an important issue.

Since then, water has emerged as a useful solvent for organometallic catalysis. In addition to the hydroformylation reactions, several other industrial processes employing homogeneous catalysis have been converted to aqueous-phase procedures [4].

5.4.2
Multiphasic Catalysis: Limitations and Challenges

Multiphasic (biphasic) catalysis relies on the transfer of organic substrates into the catalyst phase or on catalysis at the phase boundary. Most organic substrates do not have sufficient solubility in the catalyst phase (particularly in water) to give practical reaction rates in catalytic applications. Therefore, although the use of aqueous-biphasic catalysis has proven its potential in important industrial processes, the current applications of this technique remain limited: first to catalysts that are stable in the presence of water, and secondly to substrates that have significant water solubility. Many studies have focused on improving the affinities between the two liquid aqueous/organic phases, either through increasing the liphophilic character of the catalyst phase or even by immobilizing the catalyst on a support. For example, rapid stirring, emulsification, and sonication have been used to increase the interfacial area. The addition of co-solvents to the aqueous phase has been investigated extensively as a means of improving the solubility of higher olefinic substrates in the catalyst-containing phase. Application of detergents or micellar processes promoting the substrate transfer to the interface or the addition of co-ligands such as PPh_3 – or even ligands with an amphiphilic character, or modified cyclodextrins – also play rate-enhancing roles. The development of supported aqueous-phase catalysis (SAPC) which involves the dissolution of an aqueous-phase complex on a thin layer of water adhering to a silica surface, opens the way to the reactivity of hydrophobic substrates. Although all these techniques can change the solubility of organic substrates in the aqueous phase or favor the concentration of the active center at the interface, they can also cause leaching of a proportion of the catalyst into the organic phase.

The major advantage of the use of two-phase catalysis is the easy separation of the catalyst and product phases. However, the co-miscibility of the product and catalyst phases can be problematic. An example is given by the biphasic aqueous hydroformylation of ethene to propanal. First, the propanal formed contains water, which has to be removed by distillation. This is difficult due to formation of azeotropic mixtures. Secondly, a significant proportion of rhodium catalyst is extracted from the reactor with the products, which prevents its efficient recovery. Nevertheless the reaction of ethene itself in the water-based Rh-TPPTS system is fast. It is the high solubility of water in the propanal that prevents the application of the aqueous biphasic process [5].

To overcome these limitations, there has been much investigation of novel methods, one of them focused on the search for alternative solvents [6,7]. Table 5.4-1 lists different approaches of biphasic catalysis, with some of their respective advantages and limitations.

Although already well-known, perfluorinated solvents have only quite recently proved their utility in many organic and catalyzed reactions. The main advantage of these solvents is that their miscibility with organic products can be tuned by varying the temperature. Fluorous-phase catalysis makes possible the association of homogeneous phase catalysis (thus avoiding problems of mass-transfer limitations) and a biphasic separation of the catalyst and reaction mixture [8]. However, these solvents are still relatively expensive and require costly, specially designed, ligands to keep the catalyst in the fluorous phase during the separation. In addition, a significant amount of perfluorinated solvent can remain dissolved in the organic phase, and contamination of the products can occur. To date, there are no industrial developments of this technology, due to the lack of competitiveness.

Supercritical carbon dioxide ($ScCO_2$) has also emerged as a highly promising reaction medium. In combination with homogeneous catalysis, its benefits could be the potential increase in the reaction rates (absence of gas–liquid phase boundary, high diffusion rates) and selectivities, and also lack of toxicity [9]. In combination with water, it has been used in a biphasic system to perform the hydrogenation of cinnamaldehyde. Gas–liquid–liquid mass transfer limitations were ruled out, due to the very high solubility of reactant gas in $ScCO_2$ [10]. Although elegant, this approach still appears relatively expensive, especially for the bulk chemical industry. Furthermore, the low solubility of interesting substrates may hamper the commercialization of $ScCO_2$ in the fine chemical industry. A very recent and highly interesting development is the combination of an ionic liquid catalyst phase and a product phase containing $scCO_2$. This approach is presented in more detail in Section 5.7.

Further progress in multiphasic catalysis will rely on the development of alternative techniques that allow the reactivity of a broader range of substrates, the efficient separation of the products, and recovery of the catalyst, while remaining economically viable.

5.4.3
Why Ionic Liquids in Mutiphasic Catalysis?

Notwithstanding their very low vapor pressure, their good thermal stability (for thermal decomposition temperatures of several ionic liquids see Refs. [11, 12]) and their wide operating range, the key property of ionic liquids is the potential to tune their physical and chemical properties by varying the nature of the anions and cations. An illustration of their versatility is given by their exceptional solubility characteristics, which make them good candidates for multiphasic reactions (see Section 3.3). Their miscibility with water, for example, depends not only on the hydrophobicity of the cation, but also on the nature of the anion and on the temperature.

N,N'-Dialkylimidazolium cations are of particular interest because they generally give low melting salts, are more thermally stable than their tetraalkylammonium

Table 5.4-1 Advantages and limitations of different approaches for multiphasic "homogeneous" catalysis

Catalyst phase	Product phase	Advantages	Limitations
Water (+co-solvent)	Organic liquid	• Easy product separation and catalyst recycling • Lower cost of chemical processes • Lack of toxicity of water	• Low reaction rate for water poorly miscible substrates • Mass transfer limits rate of reaction • Treatment of spent water
Polar solvent	Organic liquid	• Solvent effect	• Use of volatile organic solvent • Co-miscibility of the two phases
Fluorinated organic solvent	Organic liquid	• Temperature dependence of the miscibility of fluorinated phase with organic solvents	• Solvent and ligand costs • Product contamination
Water	Supercritical fluids (e.g. CO_2)	• Organic co-solvent not needed • High miscibility of CO_2 with gas	• Poor solvating ability of supercritical fluids • High investment and operating costs
Ionic liquid	Organic liquid	• Tunability of the solubility characteristics of the ionic liquids • Solvent effect	• Ionic liquid costs • Disposal of spent ionic liquids
Ionic liquid	Supercritical fluids (e.g. CO_2)	• Organic co-solvent not needed • Tunability of the solubility characteristics of the ionic liquids • Presence of CO_2 reduces ionic liquid's viscosity	• Ionic liquid costs • High pressure apparatus needed

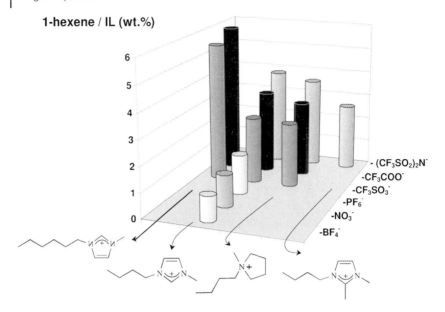

Fig. 5.4-1 Solubility of 1-hexene in different ionic liquids as a function of the nature of the anions and cations.

analogs, and have a wide spectrum of physicochemical properties available. For the same [BMIM]$^+$ cation, the [BF$_4$]$^-$, [CF$_3$SO$_3$]$^-$, [CF$_3$CO$_2$]$^-$, [NO$_3$]$^-$, and halide salts display a complete miscibility with water at 25 °C. On cooling the [BMIM][BF$_4$]/water solution to 4 °C, however, a water-rich phase separates. Similarly, changing the [BMIM]$^+$ cation to the longer chain, more hydrophobic [HMIM]$^+$ (1-hexyl-3-methylimidazolium) cation affords a BF$_4^-$ salt that shows low co-miscibility with water at room temperature. On the other hand, the [BMIM][PF$_6$], [BMIM][SbF$_6$], [BMIM][NTf$_2$] (NTf$_2$=N(CF$_3$SO$_2$)$_2$) and [BMIM][BR$_4$] ionic liquids show a very low miscibility with water, but the shorter, symmetrically substituted, [MMIM][PF$_6$] salt becomes water soluble. One might therefore expect that modification of the alkyl substituents of the imidazolium ring could produce different and very tunable ionic liquid properties.

The influence of the nature of cations and anions on the solubility characteristics of the resulting salts with organic substrates is also discussed in Section 3.3. It has been shown (Fig. 5.4-1) that increasing the length of the alkyl chain on the imidazolium cation can increase the solubility of 1-hexene, but so can tuning the nature of the anion.

A comparison of the solubility of α-olefins with increasing numbers of carbon atoms in water and in [BMIM][BF$_4$] (Fig. 5.4-2), shows that olefins are at least 100 times more soluble in ionic liquids than in water.

Addition of co-solvents can also change the co-miscibility characteristics of ionic liquids. As an example, the hydrophobic [BMIM][PF$_6$] salt can be completely dissolved in an aqueous-ethanol mixture containing between 0.5 and 0.9 mole fraction

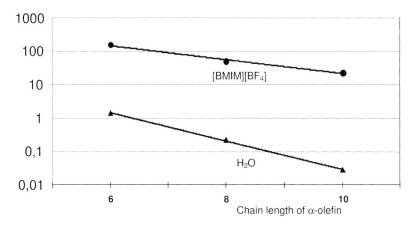

Fig. 5.4-2 Comparison of the solubility of α-olefins with different numbers of carbon atoms in water and in [BMIM][BF$_4$].

of ethanol, whereas the ionic liquid itself is only partially miscible with pure water or pure ethanol [13]. The mixing of different salts can also result in systems with modified properties (e.g. conductivity, melting point).

One of the key factors controlling the reaction rate in multiphasic processes (for reactions taking place in the bulk catalyst phase) is the reactant solubility in the catalyst phase. Thanks to their tunable solubility characteristics, the use of ionic liquids as catalyst solvents can be a solution to the extension of aqueous two-phase catalysis to organic substrates presenting a lack of solubility in water, and also to moisture sensitive reactants and catalysts. With the different examples presented below, we will report how ionic liquids can have advantageous effects on reaction rate and selectivity of homogeneous catalyzed reactions.

5.4.4
Different Technical Solutions to Catalyst Separation through the Use of Ionic Liquids

In general, homogeneous catalysis suffers from complicated and expensive catalyst separation from the products. Homogeneous catalysts are very often unstable at the high temperatures necessary for the distillation of high boiling products. Multiphasic catalysis makes possible the separation of products under mild conditions. Different technologies to separate the products, and to recycle the catalytic system when using ionic liquids as one of these phases, have been proposed (Table 5.4-2).

The simplest case (Table 5.4-2, case (a)) is when the ionic liquid is able to dissolve the catalyst, and displays a partial solubility with the substrates and a poor solubility with the reaction products. Under these conditions, the product upper phase, also containing the unconverted reactants, is removed by simple phase decantation,

Table 5.4-2 Different technologies for multiphasic reactions making use of ionic liquids

Lower phase (during the reaction)	Upper phase (during the reaction)	Mode of separation catalyst phase/products	Ref.
Ionic liquid + catalyst	Organic liquid (Products + unreacted substrates)	a) Decantation (liquid–liquid)	[50]
		b) Filtration of the ionic liquid on cooling	[14, 15]
Ionic liquid + catalyst + part of the products	Organic liquid (part of the products + unreacted substrates)	c) Product extraction with an organic co-solvent immiscible with the ionic liquid	[18]
or		d) Distillation	[16]
Ionic liquid + catalyst + products	No upper phase	e) Separation after addition of a co-sovent miscible with the ionic liquid, immiscible with the products	
Ionic liquid + catalyst	Products + unreacted substrates + CO_2	f) Extraction with $ScCO_2$	[19]
Supported ionic liquid + catalyst	Organic liquid (products) or gas	g) Phase separation	[21]

and the ionic liquid containing the catalyst is recycled. This can be illustrated by transition-metal catalyzed olefin transformations to non-polar hydrocarbon products such as olefin oligomerization, hydrogenation, isomerization, metathesis and acidic olefin alkylation with isobutane. Transition-metal catalysts can also be immobilized in ionic liquids with melting points just above room temperature (Table 5.4-2, case (b)). The reaction occurs in a two-phase liquid–liquid system. By cooling the reaction mixture, the products can be separated by filtration from the "solid" catalyst medium, which can then be recycled. The advantages of this technique have been demonstrated for the hydrogenation of 1-hexene catalyzed by ruthenium-phosphine complexes in [BMIM][Cl]/$ZnCl_2$ [14] and for the hydroformylation of 1-hexene in the high melting phosphonium tosylate ionic liquids [15].

Because of the low vapor pressure of ionic liquids, product distillation without azeotrope formation can reasonably be anticipated if the products are not too high boiling. An example is given by the hydroformylation of methyl-3-pentenoate in [BMIM][PF_6] with catalysis by a homogeneous Rh-phosphite system. In the absence of ionic liquid, deactivation of the catalyst is observed. Through the use of the [BMIM][PF_6] salt, the catalyst is stabilized and can be successfully reused after distillation of the products [16]. Nevertheless, this separation technique remains demanding in energy, and the eventual accumulation of high-boiling by-products in the non-volatile ionic liquid phase can be a problem.

When the products are partially or totally miscible in the ionic phase, separation is much more complicated (Table 5.4-2, cases (c)–(e)). One advantageous option can be to perform the reaction in one single phase, thus avoiding diffusional limitation, and to separate the products in a further step by extraction. Such technology has already been demonstrated for aqueous biphasic systems. This is the case for the palladium-catalyzed telomerization of butadiene with water, developed by Kuraray, which uses a sulfolane/water mixture as the solvent [17]. The products are soluble in water, which is also the nucleophile. The high-boiling by-products are extracted with a solvent (such as hexane) that is is immiscible in the polar phase. This method has the advantage that (i) the catalyst and the products can be separated without heating them, so that thermal deactivation is avoided, and (ii) the extraction is achieved for all the compounds, so that the accumulation of catalyst poisons and high boiling by-products is minimal. This technology can be applied when ionic liquids are used as the catalyst polar phase (Fig. 5.4-3).

A co-solvent that is poorly miscible with ionic liquids but highly miscible with the products can be added in the separation section (after the reaction) to facilitate the product separation. The Pd-mediated Heck coupling of aryl halides or benzoic anhydride with alkenes, for example, can be performed in [BMIM][PF_6], the products being extracted with cyclohexane. In this case, water can also be used as an extraction solvent, to remove the salt by-products formed in the reaction [18]. From a practical point of view, the addition of a co-solvent can result in cross-contamination, and it has to be separated from the products in a supplementary step (distillation). More interestingly, unreacted organic reactants themselves (if they have a non-polar character) can be recycled to the separation step and can be used as the extractant co-solvent [19].

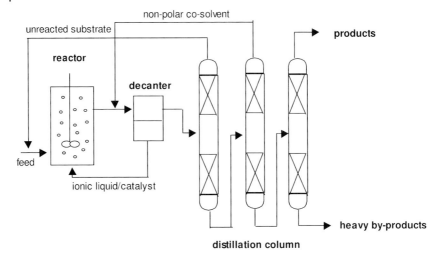

Fig. 5.4-3 Example of extraction method for product separation from ionic liquid/catalyst reaction mixture.

When water-miscible ionic liquids are used as solvents, and when the products are partly or totally soluble in these ionic liquids, the addition of polar solvents, such as water, in a separation step after the reaction can make the ionic liquid more hydrophilic and facilitate the separation of the products from the ionic liquid/water mixture (Table 5.4-2, case (e)). This concept has been developed by Union Carbide for the higher alkene hydroformylation catalyzed by Rh-sulfonated phosphine ligand in an N-methylpyrrolidone (NMP)/water system. Thanks to the presence of NMP, the reaction is performed in one homogeneous phase. After the reaction, water is added in a mixer followed by efficient phase separation in a settler. One advantage of this process is its flexibility and good performance with respect to the olefin carbon number.

The combination of ionic liquids with supercritical carbon dioxide is an attractive approach as both these solvents present complementary properties (volatility, polarity scale, etc.). Compressed CO_2 dissolves quite well in ionic liquids, but ionic liquids do not dissolve in CO_2. It decreases the viscosity of ionic liquids, thus facilitating mass transfer during catalysis. The separation of the products can be effective in a solvent-free form and the CO_2 can be recycled by recompressing it back into the reactor. Continuous flow catalytic systems based on the combination of these two solvents have been reported [19]. This concept is developed in more detail in Section 5.7.

Membrane techniques have already been combined with two-phase liquid catalysis. The main function of this method is to perform fine separation of undesirable constituents from the catalytic system after phase decantation has already performed the coarse separation of the catalyst from the products. This technique can be applied to ionic liquid systems as a promising approach for the selective removal of volatile solutes from ionic liquids [20].

Ionic liquids have been demonstrated to be effective membrane materials for gas separation when supported within a porous polymer support. However, supported ionic liquid membranes offer another versatile approach by which to perform two-phase catalysis. This technology combines some of the advantages of an ionic liquid as catalyst solvent with the ruggedness of the ionic liquid–polymer gels. Transition-metal complexes based on palladium or rhodium have been incorporated into a gas-permeable polymer gel composed of [BMIM][PF_6] and poly(vinylidene fluoride)–hexafluoropropylene copolymer and then were used to investigate the hydrogenation of propene [21].

5.4.5
Immobilization of Catalysts in Ionic Liquids

Two main methodologies have been developed for the use of ionic liquids in catalytic or organic reactions. In the first, the ionic liquid is both the catalyst and the reaction solvent. An example is acid-catalyzed reactions in which Lewis acidic ionic liquids such as acidic chloroaluminates are both active catalytic species and solvents of carbenium ions. In this case, dissolution of the inorganic Lewis acid (e.g. $AlCl_3$) in the organic phase is not observed. In the second approach, discussed in Section 5.3, the ionic liquid acts as a "liquid support" of homogeneous catalysts. In this technology, the catalyst (in general a transition-metal complex) is immobilized in the ionic phase and the products form the upper phase, as described in Section 5.4.4. To achieve the development of such an approach in a continuous process, the key point is to immobilize and stabilize the catalyst in the ionic liquid in the presence of an organic second phase with minimum loss of metal. Two approaches have been investigated:

1 The active species is known to be ionic in organic conventional solvents.
2 The active species is characterized as a non-charged complex.

In the first case, one may expect that the catalyst should remain ionic and be retained without modification in the ionic liquid. Different successful examples have been reported, such as hydrogenation reactions catalyzed by the cationic [HRh(PPh_3)$_2$(diene)][PF_6] complexes [22] and aromatic hydrogenation catalyzed by the [$H_4Ru_4(C_6H_6)_4$][BF_4]$_2$ cluster [23]. In the presence of hydrogen, this latter complex probably forms the [$H_6Ru_4(C_6H_6)_4$][BF_4] complex, which acts as the effective arene hydrogenation catalyst. Another example is olefin dimerization catalyzed by the cationic [HNi(olefin)][A] (A is a chloroaluminate anion) complexes. These species can be formed by *in situ* alkylation of a nickel (II) salt with an acidic alkylchloroaluminate ionic liquid acting both as the solvent and as the co-catalyst [24]. The cationic [(methallyl)Ni($Ph_2PCH_2PPh_2$(O))][SbF_6] complex proved to be stable and active for ethene oligomerization in [PF_6]$^-$-based ionic liquids without the addition of Lewis acid. The high electrophilicity of the Ni center, which is responsible for the activity of the catalyst, is probably not altered by the ionic solvent [25]. In the Suzuki reaction, the active species in [BMIM][BF$_4$] is believed to be the tricoordinated [Pd(PPh$_3$)$_2$(Ar)][X] complex that forms after oxidative addition of the aryl halide to [Pd^0(PPh$_3$)$_4$] [26]. Because of their low nucleophilicity, ionic liquids do not compete with the unsaturated organic substrate for coordination to the electrophilic active metal center. The different recycle experiments demonstrate the stability of these organometallic complexes in ionic liquids.

Not only cationic, but also anionic, species can be retained without addition of specially designed ligands. The anionic active [HPt(SnCl$_3$)$_4$]$^{3-}$ complex has been isolated from the [NEt$_4$][SnCl$_3$] solvent after hydrogenation of ethylene [27]. The PtCl$_2$ precursor used in this reaction is stabilized by the ionic salt (liquid at the reaction temperature) since no metal deposition occurs at 160 °C and 100 bar. The catalytic solution can be used repeatedly without apparent loss of catalytic activity.

In the second case, in which the active catalytic species is assumed to be uncharged, leaching of the transition metal in the organic phase can be limited by the use of functionalized ligands. As the triumph of aqueous biphasic catalysis follows the laborious work involved in the development of water-soluble ligands, recent investigations have focused on the synthesis of new ligands with tailor-made structures for highly active and selective two-phase catalysts and for good solubility in the ionic liquid phase [28]. These ligands are mainly phosphorus ligands with appropriate modifications (Scheme 5.4-1).

Polar groups such as the cationic phenylguanidinium **1–3** [29, 30], imidazolium and pyridinium groups **4** and **5** [31], and the 2-imidazolyl groups **6** and **7** [32] have been reported. A cobaltocinium salt bearing phosphine donors **8** [33] has also been described. Phosphites are well known ligands in homogeneous Rh-catalyzed hydroformylation affording enhanced reaction rates and regioselectivities. Since they are unstable towards hydrolysis, examples of their use in aqueous biphasic catalysis are rare. Ionic liquids offer suitable alternative solvents compatible with phosphites **9** [29].

To date, these functionalized ligands have been investigated on the laboratory scale, in batch operation to immobilize the rhodium catalyst in hydroformylation. Good rhodium retention results were obtained after several recycles. However, optimized ligand/metal ratio and leaching and decomposition rates, which can result in the formation of inactive catalyst, are not known for these ligands and testing in continuous mode is required. As a reference, in the Ruhrchemie-Rhône Poulenc process, the losses of rhodium are $<10^{-9}$ g Rh per kg n-butyraldehyde.

Certain amines, when linked to TPPTS, form ionic solvents liquid at quite low temperature. Bahrman [34] used these ionic liquids as both ligands and solvents for the Rh catalyst for the hydroformylation of alkenes. In this otherwise interesting approach, however, the ligand/rhodium ratio, which influences the selectivity of the reaction, is difficult to control.

As well as phosphorus ligands, heterocyclic carbene ligands **10** have proven to be interesting donor ligands for stabilization of transition metal complexes (especially palladium) in ionic liquids. The imidazolium cation is usually presumed to

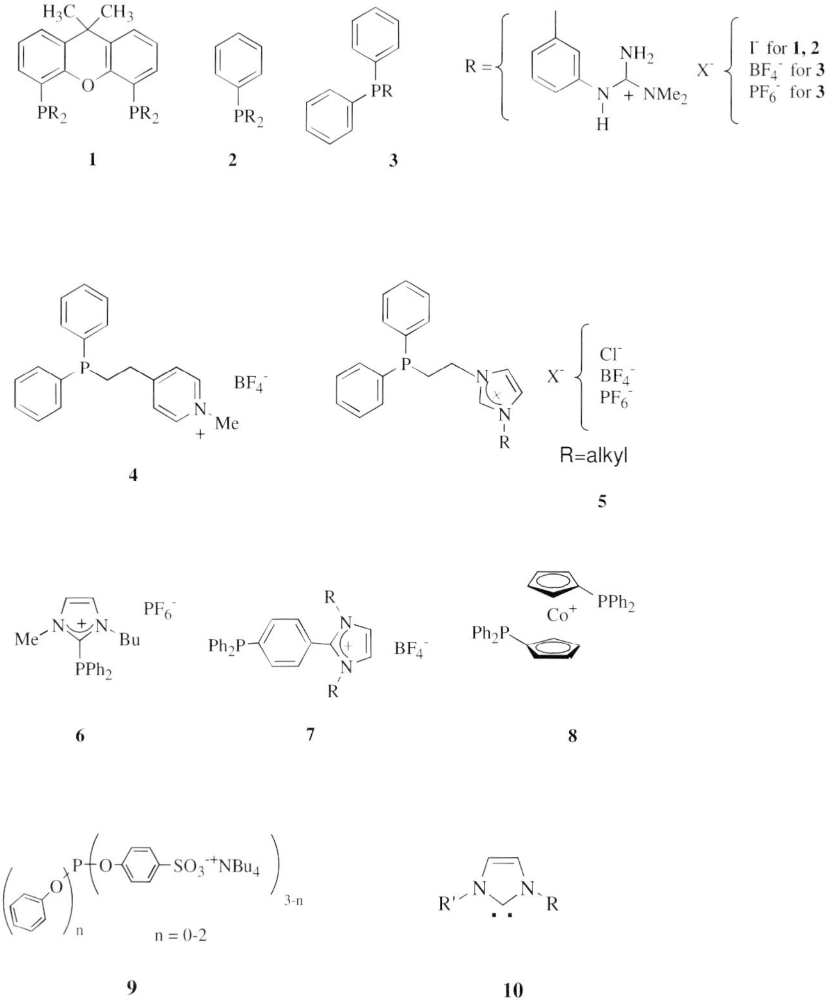

Scheme 5.4-1 Ligands 1–10.

be a simple inert component of the solvent system. However, the proton on the carbon atom at position 2 in the imidazolium is acidic and this carbon atom can be deprotonated by, for example, basic ligands of the metal complex, to form carbenes (Scheme 5.4-2).

The ease of formation of the carbene depends on the nucleophilicity of the anion associated with the imidazolium. For example when $Pd(OAc)_2$ is heated in the presence of [BMIM][Br], the formation of a mixture of Pd imidazolylidene complexes occurs. Palladium complexes have been shown to be active and stable catalysts for Heck and other C–C coupling reactions [35]. The highest activity and stability of palladium is observed in the ionic liquid [BMIM][Br]. Carbene complexes

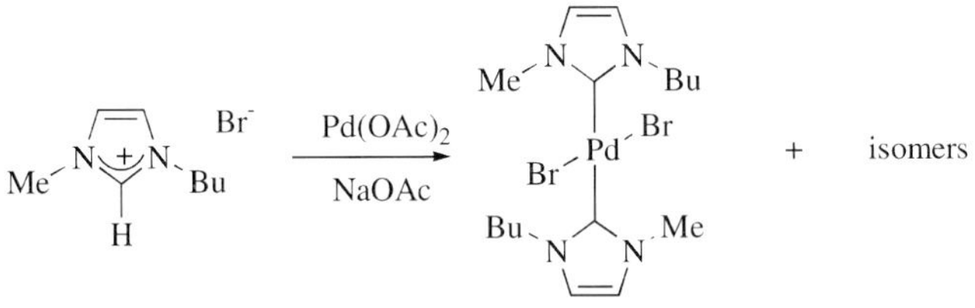

Scheme 5.4-2 Formation of carbene complexes by dialkylimidazolium salt deprotonation.

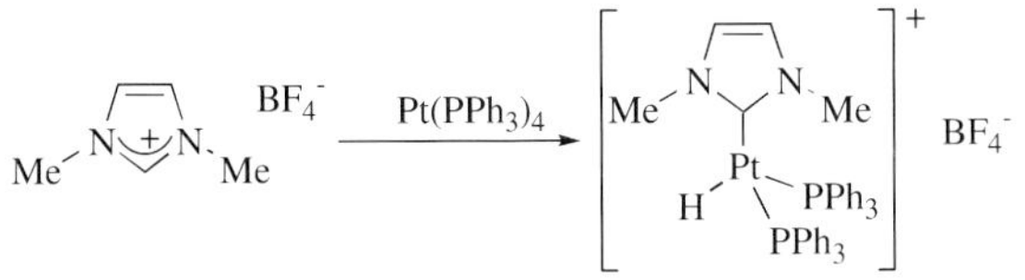

Scheme 5.4-3 Formation of carbene complexes by oxidative addition to Pt(0).

can be formed not only by deprotonation of the imidazolium cation but also by direct oxidative addition to metal(0) (Scheme 5.4-3). These heterocyclic carbene ligands can be functionalized with polar groups in order to increase their affinity for ionic liquids. While their donor properties can be compared to those of donor phosphines, they have the advantage over phosphines of being stable toward oxidation.

5.4.6
The Scale-up of Ionic Liquid Technology from Laboratory to Continuous Pilot Plant Operation

The increasing number of applications that make use of ionic liquids as solvents or catalysts for organic and catalytic reactions emphasizes their key advantages over organic solvents and their complementarity with respect to water or other "green" solvents. For scaling up to large-scale production, however, kinetic models are very often required and have to be developed for an optimum reactor design. In this type of multiphasic (biphasic) catalysis, one important parameter is the location of the reaction: does the reaction take place in the bulk of the liquid, at the interface or simultaneously at both sites? For a reaction in the bulk of the liquid (e.g. in the ionic liquid), the liquid (and/or gaseous) reactants would first have to dissolve in the catalyst solution phase before the start of the chemical reaction. The reaction rate would therefore be determined by the concentration of the reactants in the catalyst phase. It is important to be able to identify mass transfer limitations that occur when the reaction rate is higher than the mass tranfer velocity. In some cases the existence of mass transfer limitations can be used advantageously to control the exothermicity of reactions. For example, a reduction in stirring can be a means to decrease the reaction rate without having to destroy the catalyst. In single phase homogeneous reactions, catalyst poisons (such as CO or CO_2) are sometimes deliberately injected in the reactor to stop the reaction.

In the aqueous biphasic hydroformylation reaction, the site of the reaction has often been discussed (and contested) and is dependent on reaction conditions (temperature, partial pressure of gas, stirring, use of additives) and reaction partners (type of alkene) [36, 37]. It has been suggested that the positive effects of co-solvents indicate that the bulk of the aqueous liquid phase is the reaction site. By contrast, the addition of surfactants or other surface- or micelle-active compounds accelerates the reaction, which apparently indicates that the reaction occurs at the interfacial layer.

Therefore, important parameters such as phase transfer phenomena (i.e. solubility of the reactants in the ionic liquid phase), volume ratio of the different phases, efficiency of mixing so as to provide maximum liquid–liquid interfacial area, are key factors in determining and controlling reaction rates and kinetics. Kinetic models have been developed for aqueous biphasic systems and are continuously refined to improve agreement with experimental results. These models might be transferable to biphasic catalysis with ionic liquids, but more data concerning the solubility of liquids (and gas) in these new solvents and the existence of phase equilibria in the presence of organic upper phases have still to be accumulated (see Sections 3.3 and 3.4).

The influence of the concentration of hydrogen in [BMIM][PF_6] and [BMIM][BF_4] on the asymmetric hydrogenation of α-acetamidocinnamic acid catalyzed by rhodium complexes bearing a chiral ligand has been investigated. Hydrogen was found to be four times more soluble in the [BF_4]$^-$-based salt than in the [PF_6]$^-$-based one, under the same pressure. This difference in molecular hydrogen concentration in the ionic phase (rather than pressure in the gas phase) has been correlated with the remarkable effect on the conversion and enantioselectivity of the reaction [38].

In the rhodium-catalyzed hydroformylation of 1-hexene, it has been demonstrated that there is a correlation between the solubility of 1-hexene in ionic liquids and reaction rates (Fig. 5.4-4) [29].

However, information concerning the characteristics of these systems under the conditions of a continuous process is still very limited. From a practical point of view, the concept of ionic liquid multiphasic catalysis can be applicable only if the resultant catalytic lifetimes and the elution losses of catalytic components into the organic or extractant layer containing products are within commercially acceptable ranges. To illustrate these points, two examples of applications run on continuous pilot operation will be described: (i) biphasic dimerization of olefins catalyzed by nickel complexes in chloroaluminates, and (ii) the biphasic alkylation of aromatic hydrocarbons with olefins and light olefin alkylation with isobutane, catalyzed by acidic chloroaluminates.

5.4.6.1 Dimerization of Alkenes Catalyzed by Ni complexes

The Institut Français du Pétrole has developed and commercialized a process, named Dimersol X, based on a homogeneous catalyst, which selectively produces dimers from butenes. The low-branching octenes produced are good starting materials for isononanol production. This process is catalyzed by a system based on a nickel(II) salt, soluble in a paraffinic hydrocarbon, activated with an

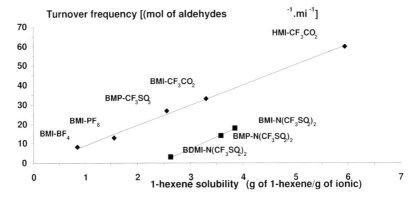

Fig. 5.4-4 Turnover frequency of Rh-catalyzed hydroformylation as a function of 1-hexene solubility in the ionic liquids. Reaction conditions: Rh(CO)$_2$(acac) 0.075 mmol, 1-hexene/Rh = 800, TPPTS/Rh = 4, heptane as internal standard, CO/H$_2$ = 1 (molar ratio), P = 2 MPa, T = 80 °C, TOF determined at 25% conversion of 1-hexene. [BMP] = N,N-butylmethylpyrrolidinium; [BMMIM] = 1-butyl-2,3-dimethylimidazolium.

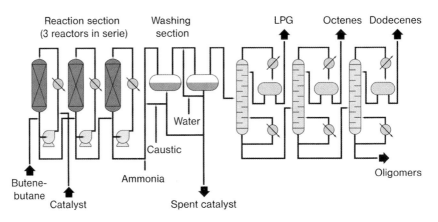

Fig. 5.4-5 Dimersol process.

alkylaluminum chloride co-catalyst directly inside the dimerization reactor. The reaction is second order in monomer concentration and first order in catalyst concentration. The butene conversion level is highly dependent on its initial concentration in the feed. For 70 wt.% butene concentration in the feed, commercial Dimersol X technology can achieve 80% conversion of butenes with 85% octene selectivity. A process flow diagram is shown in Fig. 5.4-5.

The reaction takes place at low temperature (40–60 °C) in a series of well-mixed reactors (two or more, up to four). The pressure is chosen to maintain all reactants and products in the liquid phase (no gas phase). Mixing and heat removal are

Table 5.4-3 Solubilities of 1-butene and n-butane in the acidic mixture composed of 1-butyl-3-methyl imidazolium chloride: aluminum chloride: ethylaluminum dichloride (1:1.22:0.06 molar ratio) as a function of temperature and under atmospheric pressure

Temperature (°C)	Solubility of 1-butene (wt.%)[a]	Solubility of butane (wt.%)
10	4.5	2
20	2	1

[a] Isomerization of 1-butene into 2-butene is observed.

ensured by an external circulation loop over a heat exchanger system. The two components of the catalytic system are injected separately into the circulation loop with precise flow control. The residence time in the reactor can be between 5 and 10 h. At the outlet of the reaction section, the catalyst present in the effluent is chemically neutralized and the products are washed with caustic soda and water to remove the deactivated catalyst. The catalyst components are not recycled. The product effluent is finally distilled to remove unreacted olefins and inert hydrocarbons, which can be used as such (LPG), or sent back to the cracker. The octenes are finally separated from the heavier oligomers by another distillation column. The Dimersol catalytic system is sensitive to impurities, such as polyunsaturated hydrocarbons or polar compounds that can strongly coordinate to the nickel metal center or react with the alkylaluminum chloride co-catalyst. A light pre-treatment is usually sufficient to remove such impurities.

Despite all the advantages of this process, one main limitation is the continuous catalyst carry-over by the products, with the need to deactivate it and dispose of wastes. One way to optimize catalyst consumption and waste disposal is to operate the reaction in a biphasic system. The first difficulty was to choose a "good" solvent. N,N'-Dialkylimidazolium chloroaluminate ionic liquids proved to be the best candidates. They are liquid at the reaction temperature, butenes are reasonably soluble in them (Table 5.4-3), and they are poorly miscible with the products (Table 5.4-2, case (a)). The chloroaluminate efficiently dissolves and stabilizes the nickel catalyst in the ionic medium without the addition of special ligand. The ionic liquid plays the role of both catalyst solvent and co-catalyst. Its Lewis acidity can be adjusted to get the best performance. The catalytically active nickel complex is generated directly in the ionic liquid by reaction of a commercialized nickel(II) salt, as used in the Dimersol process, with an alkylaluminum chloride derivative.

The biphasic system has been evaluated in terms of activity, selectivity, recyclability and lifetime of the ionic liquid, in a continuous flow pilot operation. A representative industrial feed (Raffinate II), composed of 70 wt.% butenes (27% of which is 1-butene) and 1.5 wt.% isobutene (the remainder being n-butane and isobutane) enters continuously into the well mixed reactor containing the ionic liquid and the nickel catalyst. Injection of fresh catalyst components can be made to compensate for the detrimental effects of random impurities present in the feed and for a slight catalyst carryover by the organic phase. The reactor is operated full of liquid. The effluent (a mixture of the two liquid phases) leaves the reactor through an overflow and is transferred to a phase separator. The separation of the ionic liquid (density around 1200 g L^{-1}) and the oligomers occurs rapidly and completely (favored by the difference in densities). The ionic liquid and the catalyst are recycled to the reactor. A continuous run has been carried out over a period of 5500 h with an industrial Raffinate-2 feed. Throughout the whole duration of the run, the conversion of butenes was more than 70 wt.% and the octene selectivity around 95 wt.%. This is about 10 wt.% more than the selectivity already obtained with the monophasic homogeneous Dimersol X. With the biphasic system, octene selectivity remained higher than 90%, even for 80 wt.% butene conversion. This higher selectivity is a direct consequence of the low solubility of octenes in the ionic liquid. In the monophasic Dimersol system, at high monomer conversion, the high concentration of dimers induces increasing production of trimers and tetramers through a consecutive mechanism. In the biphasic system, the low solubility of octenes in the ionic phase containing the catalyst reduces consecutive reactions and enhances octene selectivity. After the 5500 h of running, the test was deliberately stopped. The addition of fresh ionic liquid was not required during the test, which demonstrates the stability of the ionic liquid under the reaction conditions. Relative to the homogeneous Dimersol process, the nickel consumption was decreased by a factor of 10.

Despite the utmost importance of physical limitations such as solubility and mixing efficiency of the two phases, an apparent first-order reaction rate relative to the olefin monomer was determined experimentally. It has also been observed that an increase in the nickel concentration in the ionic phase results in an increase in the olefin conversion.

In the homogeneous Dimersol process, the olefin conversion is highly dependent on the initial concentration of monomer in the feed, which limits the applicability of the process. The biphasic system is able to overcome this limitation and promotes the dimerization of feeds poorly concentrated in olefinic monomer.

The ratio of the ionic liquid to the organic phase present in the reactor also plays an important role. A too high level of ionic liquid results in much longer decantation time and causes lower dimer selectivity.

Based on these results a new biphasic process named Difasol has been developed (Fig. 5.4-6). Because of the solubility of the catalyst in the ionic phase and the poor miscibility of the products, the Difasol unit is essentially reduced to a continuous stirred tank reactor followed by a phase separator. As the mixing of the two phases plays an important role in determining the reaction rate, a mechanically stirred type reactor rather than a loop reactor was chosen. To combine efficient decantation and a reasonable size for the settler in the process design, it has been proposed that the separation of the two phases be performed in two distinct setting zones arranged in parallel [39]. The first settler has a moderate residence time and separates the ionic phase which is returned to the reactor while the organic phase is circulated, via a pump-around through a heat exchanger. The other settler has a longer residence time and sends product phase to the neutralization section. The Difasol reaction section and settling sections can ideally be integrated as a finishing reaction section

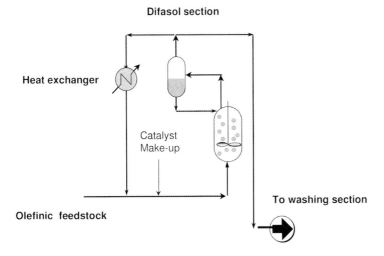

Fig. 5.4-6 Difasol reaction section.

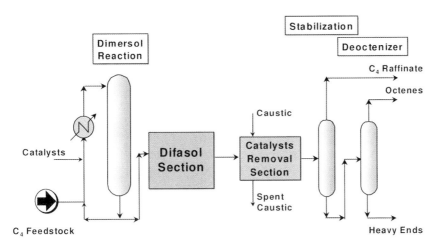

Fig. 5.4-7 Process scheme integrating Dimersol and Difasol.

after a first homogeneous Dimersol reactor (Fig. 5.4-7). The high Difasol efficiency on diluted feed, such as unreacted butenes exiting the Dimersol reactor, allows such a combination. This first homogeneous step offers the possibility to purify the feed of trace impurities very efficiently. Another interesting approach to remove the impurities from the feed consists of circulating the feed to be treated and the ionic liquid already used in the dimerization section, as a counter-current [40].

The Dimersol-Difasol arrangement ensures more efficient overall catalyst utilization and an increase in the yield of octenes by about 10 wt.% (Table 5.4-4). Table 5.4-5 shows a simplified mass balance comparison for the Dimersol process and

Table 5.4-4 Comparison of performance between homogeneous Dimersol process and [Dimersol + Difasol] arrangement with a feed containing 79 wt.% butenes

	Dimersol	Dimersol + Difasol
Octenes yield	0.68	0.75
Relative nickel consumption	100	70

Table 5.4-5 Unit performances: Butenes conversion and oligomer selectivity with a feed containing 79 wt.% butenes

	Dimersol unit	[Dimersol + Difasol] arrangement		
		Dimersol reactor	Difasol reactor	Global arrangement
Conversion (%)	80	50	75	81
Selectivity (%)	85	93	91	92
Relative chemical consumption (including IL) per ton octenes	Base: 100			82

Table 5.4-6 CAPEX and OPEX comparison for the two arrangements Dimersol and [Dimersol + Difasol]

Unit configuration	Dimersol	[Dimersol + Difasol]
Total investment (a)	Base: 100	95
Octenes yield (%) (b)	68	75
relative CAPEX per ton of octenes ((a)/(b))	1.5	1.3
Total utilities and chemicals (c)	Base: 100	92
relative OPEX per ton of octenes ((c)/(b))	1.5	1.2

Basis: C4 Feed 160 000 t a^{-1} C4 cut. Feed 79 wt.% butenes.

the [Dimersol + Difasol] combination. In the [Dimersol + Difasol] arrangement, the catalyst consumption in the Dimersol reactor is reduced to induce a low intermediate conversion. The Difasol reactor converts the remaining butenes. Both butene conversion and octene selectivity are clearly improved using the [Dimersol + Difasol] arrangement. This arrangement features very interesting CAPEX and OPEX savings per ton of octenes produced (Table 5.4-6).

The ionic liquid-based Difasol technology improves further the performance of the classical Dimersol for lightly branched octenes production. Since the catalyst is concentrated and operates in the ionic phase and also probably at the phase boundary, the reaction volume is much lower in the biphasic technology compared to the conventional single-phase Dimersol process, in which the catalyst concentration in the reactor is very low [41].

A similar dimerization catalytic system has been investigated [42] in a continuous flow loop reactor in order to study the stability of the ionic liquid solution. The catalyst used is the organometallic nickel(II) complex (Hcod)Ni(hfacac) (Hcod=cyclooct-4-ene-1-yl and hfacac=1,1,1,5,5,5-hexafluoro-2,4-pentanedionato-O,O′) and the ionic liquid is an acidic chloroaluminate based on the acidic mixture of 1-butyl-4-methylpyridinium chloride and aluminum chloride. No alkylaluminum is added, but an organic Lewis base is added to buffer the acidity of the medium. The ionic catalyst solution is introduced in the reactor loop at the beginning of the reaction and the loop is filled with the reactants (total volume 160 mL). The feed enters continuously in the loop and the products are continuously separated in a settler. The overall activity is 18.000 (TON). The selectivity to dimers is in the 98% range and the selectivity to linear octenes is 52%.

5.4.6.2 Alkylation Reactions

BP Chemicals studied the use of chloroaluminates as acidic catalysts and solvents for aromatic alkylation [43]. At present, the $AlCl_3$ existing technology (based on "red oil" catalyst) is still used industrially, but continues to suffer from poor catalyst separation and recycle [44]. The aim of the work was to evaluate the $AlCl_3$-based ionic liquids, with the emphasis placed on the development of a clean and recyclable system for the production of ethylbenzene (benzene/ethene alkylation) and synthetic lubricants (alkylation of benzene with 1-decene). The production of linear alkyl benzene (LAB) has also been developed by Akzo [45]. The ethylbenzene experiments were run by BP in a pilot loop reactor similar to that described for the dimerization (Fig. 5.4-8).

Ionic liquids operate in true biphasic mode. While the recovery and recyclability of the ionic liquid was found to be more efficient than with the conventional $AlCl_3$ catalyst (red oil), the selectivity to the monoalkylated aromatic hydrocarbon was lower. In this gas–liquid–liquid reaction, the solubility of the reactants in the ionic phase (e.g. the ratio benzene/ethene in the ionic phase) and the mixing of the phases were probably critical. This is an example where the engineering aspects are of the upmost importance.

The use of acidic chloroaluminates as alternative liquid acid catalysts for the alkylation of light olefins with isobutane, for the production of high octane number gasoline blending components, is also a challenge. This reaction has been performed in a continuous flow pilot plant operation at IFP [46] in a reactor vessel similar to that used for dimerization. The feed, a mixture of olefin and isobutane, is pumped continuously into the well-stirred reactor containing the ionic liquid catalyst. In the case of ethene, which is less reactive than butene, [pyridinium]Cl]/$AlCl_3$ (1:2 molar ratio) ionic liquid proved to be the best candidate (Table 5.4-7).

The reaction can be run at room temperature and provides good quality alkylate (dimethylbutanes are the major products) over a period of 300 h. When butenes are used instead of ethene, lower temperature and a fine-tuning of the acidity of the

Table 5.4-7 Alkylation of ethylene and 2-butene with isobutane. Semi-continuous pilot-plant results

Operating conditions/nature of the olefin	Ethene		2-Butene	
Nature of ionic liquid	[Pyridinium, HCl]/AlCl$_3$ (1:2 molar ratio)		[BMIM][Cl]/AlCl$_3$	
Olefin content in the feed (wt.%)	14–20		12–14	
VVH[a] (h^{-1})	0.2		0.35–0.45	
Temperature (°C)	25		5	
Test duration (h)	354		520	
Olefin conversion (wt.%)	60–90		>98	
Production (g products/g ionic liquid)	121		172	
Product distribution (wt.%)				
	i-C$_6$[b]	75–90	Light ends[e]:	5–10
	i-C$_8$[c]	10–17	i-C$_8$[c]	80–90 (>90% TMP)
	C$_8^+$[d]	<5	C$_8^+$[d]	5–10
MON[f]	90–94		90–95	
RON[f]	98–101		95–98	

[a] volume of olefin/(volume of ionic liquid hour).
[b] i-C$_6$ = 2,2- and 2,3-dimethylbutanes.
[c] i-C8 = isooctanes.
[d] C$_8^+$ = hydrocarbon products having more than eight carbon atoms.
[e] Light ends = hydrocarbon products with fewer than eight carbon atoms.
[f] RON = research octane number, MON = motor octane number.

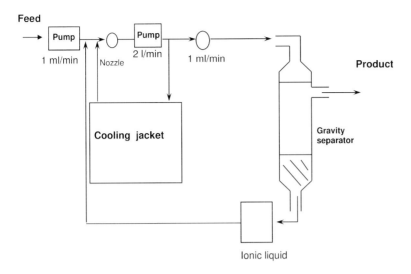

Fig. 5.4-8 Loop reactor as used in aromatic hydrocarbon alkylation experiments.

ionic liquid are required to avoid cracking reactions and heavy by-product formation. Continuous butene alkylation has been performed for more than 500 h with no loss of activity and stable selectivity. A high level of mixing is essential for a high selectivity and thus for a good quality alkylate. These applications are promising, but efforts are still needed to compete with the existing effective processes based on hydrofluoric and sulfuric acids.

5.4.6.3 Industrial Use of Ionic Liquids

What can drive the switch from homogeneous existing procesess to novel biphasic ionic liquid technologies? One major point is probably a higher cost-effectiveness. This can result from improved reaction rate and selectivity, associated with more efficient catalyst recovery and better environmental compatibility. The cost of ionic liquids can, of course, be a limiting factor to their development. However, this cost has to be weighed against that of current chemicals or catalysts. If the ionic liquid can be recycled and if its lifetime proves to be long enough, then its initial price is probably not the critical point. In Difasol technology, for example, ionic liquid cost, expressed with respect to the octenes produced, is lower than that of catalyst components.

The manufacture of ionic liquids on an industrial scale is also to be considered. [BMIM]Cl is already a commercial ionic liquid that has been produced by BASF on a ton scale [47]. Chloroaluminate laboratory preparations proved to be easily extrapolated to large scale. These chloroaluminate salts are corrosive liquids in the presence of protons. When exposed to moisture, they produce hydrochloric acid, similarly to aluminum chloride. However, this can be avoided by the addition of some proton scavenger such as alkylaluminum derivatives. In Difasol technology, for example, carbon-steel reactors can be used with no corrosion problem. The purity of ionic liquids is a key parameter especially when they are used as solvents for transition-metal complexes (see Section 5.3). The presence of impurities arising from their mode of preparation can change their physical and chemical properties. Even trace amounts of impurities (e.g. Lewis bases, water, chloride anion) can poison the active catalyst, due to its generally low concentration in the solvent. The control of ionic liquid quality is thus of the utmost importance.

As relatively new compounds, only limited research has been carried out to evaluate the biological effects of ionic liquids (see Section 2.2 and Chapter 10). The topical effect of [EMIM][Cl]/$AlCl_3$ melts and its organic compound [EMIM][Cl], on the integument of laboratory rat has been investigated. The study reports that [EMIM][Cl] is not in itself responsible for tissue damage, however, the chloroaluminate salt can induce tissue irritation, inflammation and necrosis due to the presence of aluminum chloride. However, treatments for aluminum chloride and hydrochloric acid are well-documented. Fortunately, a lot of different studies on the toxicity and ecotoxicity of ionic liquids are now under way [48]. Few data [49] relating to the disposal of used ionic liquids are available. In Difasol technology, the used ionic liquid is taken out of the production system and the reactor is refilled with fresh catalyst solution.

5.4.7
Concluding Remarks and Outlook

Compared to classical processes involving thermal separation, biphasic techniques offer simplified process schemes and no thermal stress for the organometallic catalyst. The concept requires that the catalyst and the product phases separate rapidly, to achieve a practical approach to the recovery and recycling of the catalyst. Owing to their tuneable solubility characteristics, ionic liquids prove to be good candidates for multiphasic techniques. They extend the applications of aqueous-biphasic systems to a broader range of organic hydrophobic substrates and water-sensitive catalysts (see Section 5.3 and Ref. [50]).

To be applied industrially, performance must be superior to that of the existing catalytic systems (activity, regioselectivity and recyclability). The use of ionic liquid biphasic technology for nickel-catalyzed olefin dimerization proved to be successful and this system has been developed and is now proposed for commercialization. However, much effort remains if the concept is to be extended to non-chloroaluminate ionic liquids. In particular, the true potential of ionic liquids (and mixtures containing ionic liquids) could be achievable if an even more substantial body of thermophysical and thermodynamic properties were amassed in order that the best medium for a given reaction could be chosen. As far as industrial applications are concerned, the easy scale-up of two-phase catalysis can be illustrated by the first oxo commercial unit with an initial capacity of 100 000 tons extrapolated by a factor of 1:24 000 (batch-wise laboratory development to production reactor) after a development period of 2 years [4].

References

1. (a) B. Cornils, W. A. Herrmann, *Applied Homogeneous Catalysis with Organometallic Compounds*, B. Cornils, W. A. Herrmann (Eds.), Wiley-VCH, Weinheim, **2000**, p. 575; (b) *Multiphase Homogeneous Catalysis*, B. Cornils, W. A. Herrmann, I. T. Horvath, W. Leitner, S. Mecking, H. Olivier-Bourbigou, D. Vogt (Eds.), Wiley-VCH, Weinheim, **2005**.
2. V. E. Dehmlow, in: *Aqueous-Phase Organometallic Catalysis: Concept and Applications*, B. Cornils, A. W. Herrmann (Eds.), Wiley-VCH, Weinheim, **1998**, p. 207.
3. (a) B. Cornils, E. G. Kuntz, in *Aqueous-Phase Organometallic Catalysis: Concept and Applications*, B. Cornils, A. W. Herrmann (Eds.), Wiley-VCH, Weinheim, **1998**, p. 271; (b) B. Cornils, E. G. Kuntz, in *Multiphase Homogeneous Catalysis*, B. Cornils, W. A. Herrmann, I. T. Horvath, W. Leitner, S. Mecking, H. Olivier-Bourbigou, D. Vogt (Eds.), Wiley-VCH, Weinheim, **2005**, Vol. 1, p. 148.
4. B. Cornils, *Org. Process Res. Dev.* **1998**, *2*, 121.
5. J. Herwig, R. Fischer, in *Rhodium-catalyzed Hydroformylation*, in *Catalysis by Metal Complexes*, Vol. 22, P. W. N. M. van Leeuwen, C. Claver (Eds.) Kluwer Academic, the Netherlands, **2000**, Vol. *22*, p. 189.
6. *Modern Solvents in Organic Synthesis*, P. Knochel (Ed.), *Top. Curr. Chem.* **1999**, *206*.
7. J. N. Reek, P. C. J. Kamer, P. W. N. M. van Leeuwen, in *Rhodium-catalyzed Hydroformylation*, in *Catalysis by Metal Complexes*, Vol. 22, P.W. N. M. van Leeuwen, C. Claver (Eds.), Kluwer Academic, the Netherlands, **2000**, p. 253.
8. (a) I. T. Horvath, *Acc. Chem. Res.* **1998**, *31*, 641; (b) I. T. Horvath, in *Multiphase Homogeneous Catalysis*, B. Cornils, W. A. Herrmann, I. T. Horvath, W. Leitner, S. Mecking, H. Olivier-Bourbigou, D. Vogt (Eds.), Wiley-VCH, Weinheim, **2005**, Vol. 1, p. 339.
9. (a) P. G. Jessop, T. Ikariya, R. Noyori, *Chem. Rev.* **1999**, *99*, 475; (b) W. Leitner, in *Multiphase Homogeneous Catalysis*, B. Cornils, W. A. Herrmann, I. T. Horvath, W. Leitner, S. Mecking, H. Olivier-Bourbigou, D. Vogt (Eds.), Wiley-VCH, Weinheim, **2005**, Vol. 2, p. 605.
10. B. M. Bhanage, M. Shirai, M. Arai, Y. Ikushima, *Chem. Commun.* **1999**, 1277.
11. J. G. Huddleston, A. E. Visser, W. M. Reichert, H. D. Willauer, G. A. Broker, R. D. Rogers, *Green Chem.* **2001**, *3*, 156.
12. H. L. Ngo, K. LeCompte, L. Hargens, A. B. McEwen, *Thermochim. Acta* **2000**, *97*, 357–358.
13. R. P. Sawtloski, A. E. Visser, M. W. Reichert, G. A. Broker, L. M. Farina, J. D. Holbrey, R. D. Rogers, *Chem. Commun.* **2001**, 2070.
14. J. Dupont, P. A. Z. Suarez, A. P. Umpierre, R. F. de Souza, *Catal. Lett.* **2001**, *73*, 211.
15. N. Karodia, S. Guise, G. Newlands, J.-A. Andersen, *Chem. Commun.* **1998**, 2341.
16. W. Keim, D. Vogt, H. Waffenschmidt, P. Wasserscheid, *J. Catal.* **1999**, *186*, 481.
17. N. Yoshimura, in *Aqueous Phase Organometallic Catalysis, Concept and Applications*, Wiley-VCH, Weinheim, **1998**, p. 408.
18. A. J. Carmichael, M. J. Earle, J. D. Holbrey, P. B. McCormac, K. R. Seddon, *Org. Lett.* **1999**, *1*, 997.
19. M. Freemantle, *Chem. Eng. News* **2001**, 41.
20. T. Schäfer, C. A. Rodrigues, A. M. A. Carlos, J. G. Crespo, *Chem. Commun.* **2001**, 1622.
21. R. T. Carlin, J. Fuller, *Chem Commun.* **1997**, 1345.
22. Y. Chauvin, L. Mussmann, H. Olivier, *Angew. Chem. Int. Ed. Eng.* **1995**, *34*, 2698.
23. P. J. Dyson, D. J. Ellis, D. G. Parker, T. Welton, *Chem. Commun.* **1999**, 25.
24. Y. Chauvin, S. Einloft, H. Olivier, *Ind. Eng. Chem. Res.* **1995**, *34*, 1149.
25. P. Wasserscheid, C. M. Gordon, C. Hilgers, M. J. Muldoon, I. R. Dunkin, *Chem. Commun.* **2001**, 1186.
26. C. J. Mathews, P. J. Smith, T. Welton, *Chem. Commun.* **2000**, 1249.
27. G. W. Parshall, *J. Am. Chem. Soc.* **1972**, *94*, 8716.

28. Z. Fei, J. Geldbach, D. Zhao, P. J. Dyson, *Chem. Eur. J.*, **2006**, *12*, 2122.
29. F. Favre, H. Olivier-Bourbigou, D. Commereuc, L. Saussine, *Chem. Commun.* **2001**, 1360.
30. P. Wasserscheid, H. Waffenschmidt, P. Machnitzki, K. W. Kottsieper, O. Stelzer, *Chem. Commun.* **2001**, 451.
31. D. J. Brauer, K. W. Kottsieper, C. Liek, O. Stelzer, H. Waffenschmidt, P. Wasserscheid, *J. Oganomet. Chem.* **2001**, *630*, 177.
32. K. W. Kottsieper, O. Stelzer, P. Wasserscheid, *J. Mol. Catal.* **2001**, *175*, 285.
33. C. C. Brasse, U. Englert, A. Salzer, H. Waffenschmidt, P. Wasserscheid, *Organometallics* **2000**, *19*, 3818.
34. Celanese Chemicals Europe (H. Bahrmann, H. Bohnen) Ger. Offen., DE 19 919 494 (**2000**).
35. (a) L. Xu, W. Chen, J. Xiao, *Organometallics* **2000**, *19*, 1123; (b) C. Mathews, P. J. Smith, T. Welton, A. J. P. White, D. J. Williams, *Organometallics*, **2001**, *20*, 3848.
36. O. Wachsen, K. Himmler, B. Cornils, *Catal. Today* **1998**, *42*, 373.
37. Y. Zhang, Z.-S. Mao, J. Chen, *Ind. Eng. Chem. Res.* **2001**, *40*, 4496.
38. A. Berger, R. F. de Souza, M. R. Delgado, J. Dupont, *Tetrahedron:Asym.* **2001**, *12*, 1825.
39. Institut Français du Pétrole (C. Bronner, A. Forestière, F. Hugues) US 6 203 712 B1 (**2001**).
40. Institut Français du Pétrole (H. Olivier, D. Commereuc, A. Forestière, F. Hugues) US 6284937 (**2001**).
41. F. Favre, A. Forestière, F. Hugues, H. Olivier-Bourbigou, J. A. Chodorge, *Oil Gas Eur. Mag.* **2005**, *2*, 83.
42. P. Wasserscheid, M. Eichmann, *Catal. Today* **2001**, *66*, 309.
43. BP Chemicals Limited, (B. Ellis, F. Hubert, P. Wasserscheid) PCT Int., WO 00/41809 (**2000**).
44. S. E. Knipling, *Petroleum Technology Quarterly* **2001**, Autumn, 123.
45. Akzo Nobel (C. P. M. Lacroix, F. H. M. Dekker, A. G. Talma, J. W. F. Seetz) WO 98/03454 (**1998**).
46. Y. Chauvin, A. Hirschauer, H. Olivier, *J. Mol. Catal.* **1994**, *92*, 155.
47. M. Maase, *Plenary Lecture, Conference on Ionic Liquids*, Salzburg **2005**.
48. B. Jastorff, K. Mölter, P. Behrend, U. Botti-Weber, J. Filser, A. Heimers, B. Ondruschka, J. Ranke, M. Schäfer, H. Schröder, A. Stark, P. Stepnowski, F. Stock, R. Störmann, S. Stolte, U. Welz-Biermann, S. Ziegert, J. Thöming, *Green Chem.* **2005**, *7*, 362.
49. British Nuclear Fuels (A. J. Jeapes, R. C. Thied, K. R. Seddon, W. R. Pitner, D. W. Rooney, J. E. Hatter, T. Welton) PCT Int. WO 01/15175 (**2001**).
50. (a) H. Olivier, *J. Mol. Catal. A: Chem.* **2002**, *419*, 182; P. Wasserscheid, W. Keim, *Angew. Chem. Int. Ed.* **2000**, *39*, 3772; (c) R. Sheldon, *Chem. Commun.* **2001**, 2399; (d) Z. Zhang, *Adv. Catal.* **2006**, 153–237; (e) T. Welton, *Coord. Chem. Rev.* **2004**, *248*, 2459.

5.5
Task-specific Ionic Liquids as New Phases for Supported Organic Synthesis

Michel Vaultier, Andreas Kirschning, and Vasundhara Singh

Here we provide a comprehensive review of new task-specific ionic liquids (TSILs) and binary task-specific ionic liquids (BTSILs) as functional ionic liquids designed as alternate soluble supports for supported organic synthesis (SPOS) and combinatorial chemistry. The applications of these supports are based mainly on imidazolium, ammonium and pyridinium salts. The versatility of the supports has been

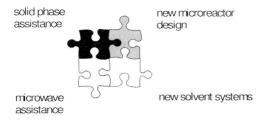

Fig. 5.5-1 Enabling technologies for organic synthesis.

demonstrated for the synthesis of small libraries of heterocyclic compounds, multicomponent reactions, coupling reactions, peptide synthesis, sulfonamide preparations and multistep synthesis of sulfonylated tyrosine derivatives, to mention a few. The use of both ionic liquids and molecular solvents as reaction media is possible. Support recycling, high loading, reaction monitoring by HPLC, ^{1}HNMR and GC/MS are some of the significant advantages of this method. Further, a lab on a chip system using task-specific ionic liquids which is based on using an ionic liquid droplet as an e-microreactor has been developed for solution phase synthesis.

5.5.1
Introduction

Organic synthesis seems to have become so very advanced that basically every molecular target, however structurally complex it may be, can be addressed. However, many developments from research laboratories lack practicality with respect to scale-up, easy and rapid work-up and product isolation, as well as recyclability of precious catalysts. There is a quest to combine new synthetic methodologies with new techniques termed "enabling technologies for organic synthesis" which has been devised to facilitate organic synthesis [1] and hence allow more rapid incorporation into industrial processes [2]. Typical enabling technologies are microwave assistance [3], new solvent systems such as ionic liquids [4], continuous flow reactors [5] and immobilization of chemically active species such as reagents and homogeneous catalysts [6], all of which have recently seen widespread application in research laboratories. Truly new synthetic technology platforms, however, will not be based on the individual use of these new techniques but will require the integration of two or more of these enabling techniques (Fig. 5.5-1).

Solid phase peptide synthesis, as described by Merrifield in 1963 [7], has opened the way for supported organic synthesis as a widely employed technique facilitating the preparation of a large number of compounds. These techniques lie behind combinatorial chemistry and library synthesis. Solid supports such as cross-linked polystyrene beads have been widely used [8]. The success of this approach relies mostly on the ease of product purification and isolation, usually by simple operations such as washing and filtration, which makes automation possible. At the same time, the heterogeneous nature of these polymers gives rise to a number of problems, including nonlinear reaction kinetics, solvation problems, difficult access to reaction sites, slow reaction rates, difficulties in monitoring reactions, just to mention a few. Another severe drawback is the low loading of these resins, typically in the range of 1 mmol g^{-1} or less. These limitations have led to alternative methodologies tending to restore homogeneous reaction conditions. Thus, soluble polymer supports such as polyethylene glycols (PEGs), polyvinyl alcohol, soluble polystyrene [9] and dentritic polyglycerols [10] have attracted much attention in recent years. Since these are soluble in commonly used molecular solvents, reactions can be carried out under solution phase conditions. Purification of the functionalized support, usually by precipitation and filtration, is still an easy process. Nevertheless, some limitations still exist including a low loading capacity (usually $\ll$ mmol g^{-1}), aqueous solubility, insolubility in ether solvents, retention of solvent traces, and interference of the PEG framework with certain reagents eventually leading to degradation. Fluorous solvents and tags have also been successfully used along these lines [11].

Interest in ionic liquids is growing exponentially owing to their many interesting properties [12]. In the above context, TSILs are of particular importance as they combine new solvent systems with features of solid phase chemistry. Task specific ionic liquids have been defined as ionic liquids bearing a functional group covalently tethered either to the cation or the anion or both. This concept was introduced by J. Davis Jr. and Wierbicki [13] in 2000 following the demonstration that a thiazolium IL could interact specifically with a solute and function both as a solvent and a catalyst for the benzoin condensation [14]. The idea was that the incorporation of one or several functional groups into the ions of an ionic liquid should confer special properties or reactivities upon them such as the capacity to behave as reagents, catalysts or new reaction media [15]. This extends to binary ionic liquids (BILs) resulting from the addition of ionic solutes such as functional onium salts which do not have to be liquids at temperature below 100 °C but introduce functional groups thus conferring some special property to the liquid. These binary task specific ionic liquids (BTSILs) therefore constitute another family of TSILs. These concepts proved to be extremely fruitful and are under active investigation in a number of research groups and a large part of the work in this area has been reviewed by Davis [16]. The idea of specifically designing either simple TSILs or BTSILs in order to use them as phases for SPOS arose in the groups of Bazureau [17] and Vaultier [18] in early 2000. Several groups are now involved in this area. TSILs have been demonstrated to be very efficient phases for SPOS.

5.5.2
Synthesis of TSILs

TSILs used as soluble supports usually contain functionalized cations. There are several ways to incorporate functionality into an ionic structure. The most widely used method relies on the alkylation of a nucleophile such as a tertiary amine,

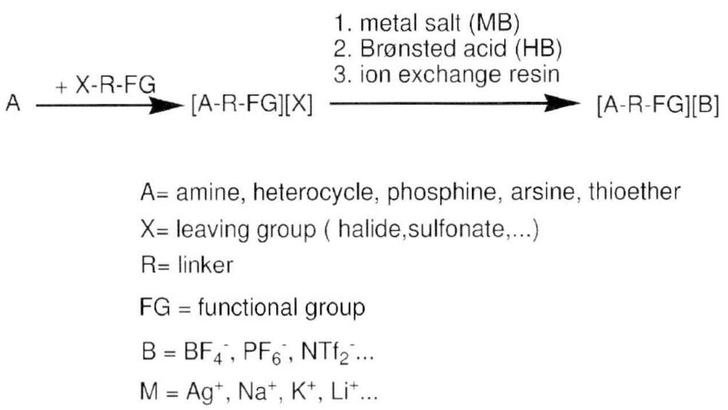

5.5 Task-specific Ionic Liquids as New Phases for Supported Organic Synthesis

$$A \xrightarrow{+\ X\text{-}R\text{-}FG} [A\text{-}R\text{-}FG][X] \xrightarrow[\text{3. ion exchange resin}]{\text{1. metal salt (MB)} \atop \text{2. Brønsted acid (HB)}} [A\text{-}R\text{-}FG][B]$$

A= amine, heterocycle, phosphine, arsine, thioether
X= leaving group (halide, sulfonate,...)
R= linker
FG = functional group
B = BF_4^-, PF_6^-, NTf_2^-...
M = Ag^+, Na^+, K^+, Li^+...

Scheme 5.5-1

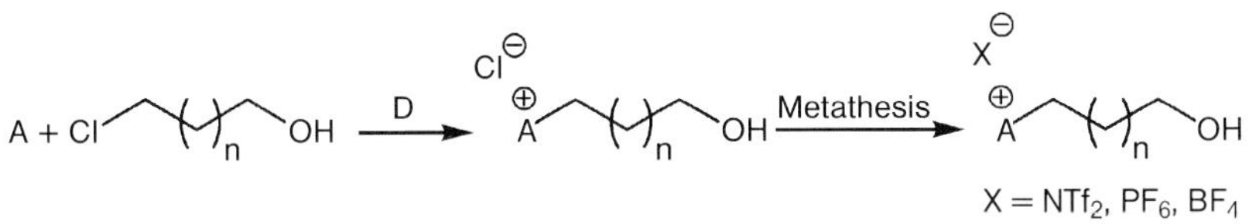

$$A + Cl^- \longrightarrow \underset{n}{\diagdown\!\diagup}\text{OH} \xrightarrow{D} Cl^\ominus\ A^\oplus\ \underset{n}{\diagdown\!\diagup}\text{OH} \xrightarrow{\text{Metathesis}} X^\ominus\ A^\oplus\ \underset{n}{\diagdown\!\diagup}\text{OH}$$

X = NTf_2, PF_6, BF_4

Scheme 5.5-2

heterocycle or a phosphine by a bifunctional organic molecule typically bearing an organic halide at one end and the desired functional group at the other, provided that the nucleophilic reagent does not interfere with the functional group. Once the initial onium salt is isolated and purified, anion metathesis is performed to pair the functional cation with an appropriate anion leading to the TSIL. The most widely used techniques for this exchange of anions are the reaction of the onium halides with a silver, sodium, potassium or lithium salt in an appropriate solvent, such as water or acetonitrile, or with a concentrated aqueous solution of the acid, provided that the desired TSIL is only poorly, or not at all, miscible with water (Scheme 5.5-1) [16, 19].

5.5.2.1 Synthesis of TSILs Bearing a Hydroxy Group

The hydroxy group is one of the most useful functions for anchoring via esterification or alkylation, to mention but two possibilities. The methodology described in Scheme 5.5-1 has been widely used for the preparation of several ω-hydroxylated onium salts according to Scheme 5.5-2. Simply heating a mixture of the nucleophile and the ω-chloroalcohol in acetonitrile [18] or without added solvent under microwave irradiation [20] led to the onium chlorides in very good yields. Anion metathesis with aqueous $H[PF_6]$ or $H[BF]_4$ [18] or $K[PF]_6$ or $[NH_4][BF_4]$ [19] in acetonitrile gave the desired TSILs. Triflimides have been obtained by exchange with $Li[NTf_2]$ in water and isolated by simple decantation. Some examples are given in Table 5.5-1. Based on this sequence, so-called poly(ethylene glycol) ionic liquid phases based on imidazolium cations have been prepared in good yields [20].

Table 5.5-1 Synthesis of ω-hydroxylated onium salts

Entry	A	n	X^-	yield (%)	Ref.
1	MIM[a,b]	0	Cl^-	98	[17]
2	MIM[c]	0	$[BF_4]^-$	85	[17]
3	MIM[b,c]	1	Cl^-	94	[18, 20]
4	MIM[c]	1	$[PF_6]^-$	90	[19]
5	MIM[c]	1	$[NTf_2]^-$	90	[18]
6	NMe_3[c]	0	$[NTf_2]^-$	90	[18]
7	NMe_3[d]	1	Cl^-	82	[18]
8	NMe_3[e]	1	$[PF_6]^-$	67	[18]
9	NMe_3[f]	1	$[BF_4]^-$	82	[18]
10	NMe_3[c]	1	$[NTf_2]^-$	86	[18]
11	NMe_3[g]	2	Cl^-	94	[21]
12	NMe_3[c]	2	$[NTf_2]^-$	86	[21]
13	NMe_3[h]	3	Cl^-	62	[21]
14	NMe_3[c]	3	$[NTf_2]^-$	93	[21]
15	PBu_3[c]	1	Cl^-	72	[18]
16	PBu_3[c]	1	$[NTf_2]^-$	90	[18]
17	 pyridinium[i]	1	Cl^-	73	[18]
18	 pyridinium[c]	1	$[NTf_2]^-$	90	[18]

[a] mp 86–88 °C.
[b] MW heating has been used.
[c] liquid at RT.
[d] mp 158–160 °C.
[e] mp 124–126 °C.
[f] mp 110–112 °C.
[g] mp 118–120 °C.
[h] mp 178–180 °C.
[i] mp 68–70 °C.

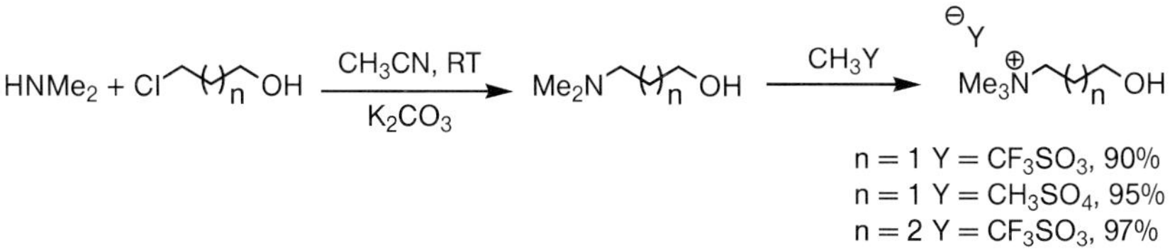

$$HNMe_2 + Cl\text{-}(CH_2)_n\text{-}OH \xrightarrow[K_2CO_3]{CH_3CN, RT} Me_2N\text{-}(CH_2)_n\text{-}OH \xrightarrow{CH_3Y} Me_3N^+\text{-}(CH_2)_n\text{-}OH \;\; Y^-$$

n = 1 Y = CF_3SO_3, 90%
n = 1 Y = CH_3SO_4, 95%
n = 2 Y = CF_3SO_3, 97%

Scheme 5.5-3

The alkylation step with functionalized alkylating agents is sometimes troublesome, leading to impure salts. Since it is not easy to purify liquid or even solid organic salts, it could be advantageous to synthesize the ω-amino alcohols first, which, in principle, are easy to get pure, and then alkylate with simple alkylating agents. This also brings other advantages: first, by using simple alkyl sulfonates or sulfates, halogen-free ionic liquids can be obtained and second, these ionic liquids can be prepared *in situ* under water-free conditions. This is illustrated by the following examples (Scheme 5.5-3) [21].

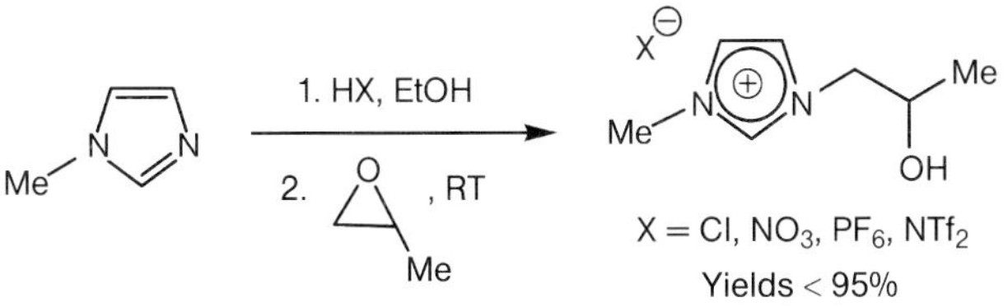

Scheme 5.5-4

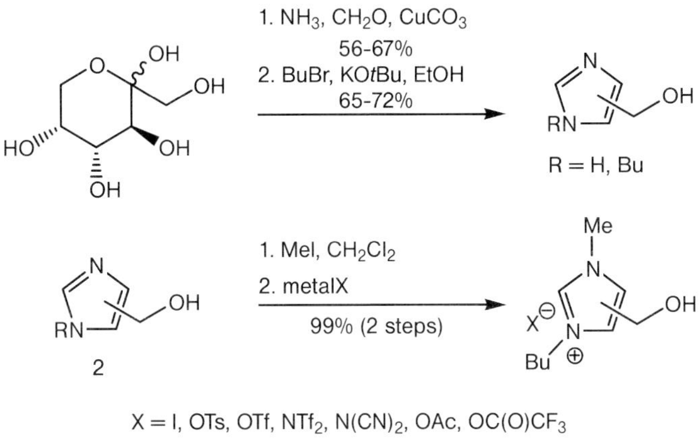

Scheme 5.5-5

Several other methodologies have been employed for the synthesis of hydroxylated TSILs. 2-Hydroxypropyl-functionalized imidazolium salts have been prepared in excellent yields by the reaction of protonated 1-methylimidazole with propylene oxide, the acid providing the anionic component of the resultant ionic liquid (Scheme 5.5-4) [22].

Fructose can be readily transformed into hydroxymethylene-imidazole according to a reaction developed by Trotter and Darby [23]. Alkylation leads to imidazolium cations (Scheme 5.5-5) [24].

Chiral β-hydroxylimidazolium salts containing one chiral carbon in the α-position to the imidazolium ring have been synthesized from natural α-aminoacids (Scheme 5.5-6) [25].

Ionic liquids derived from the alkaloid ephedrine or (S)-valinol have been readily obtained on a kilogram scale in a three-step synthesis: a Leuckart-Wallach reaction followed by alkylation with Me$_2$SO$_4$ and ion exchange in aqueous solution (Scheme 5.5-7) [26].

In principle, any ω-functionalized alkylating agent could be used for the quaternization of a tertiary amine or phosphine, provided that the alkylation reaction is chemoselective and the alkylating agent stable under the alkylation conditions. Several examples, selected for their potential to give further functionalized ionic liquids and therefore serve as soluble supports, are reported below including: ionic

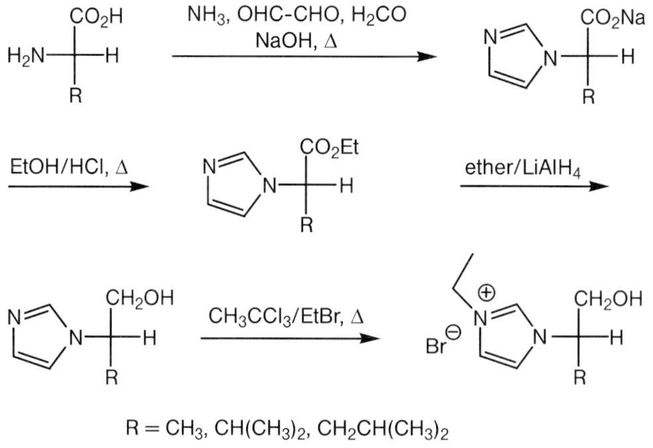

Scheme 5.5-6

R = CH$_3$, CH(CH$_3$)$_2$, CH$_2$CH(CH$_3$)$_2$

Scheme 5.5-7

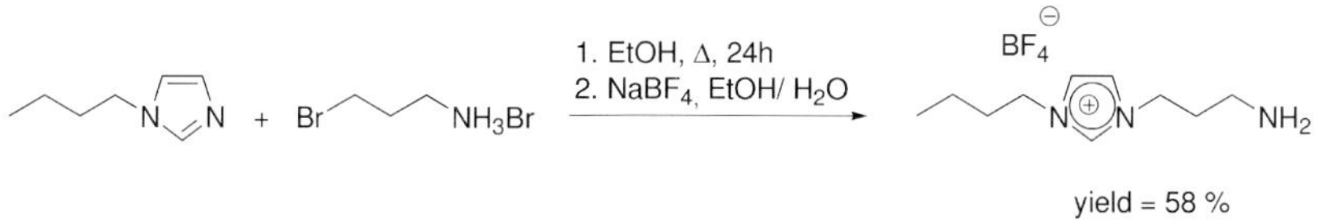

Scheme 5.5-8

yield = 58 %

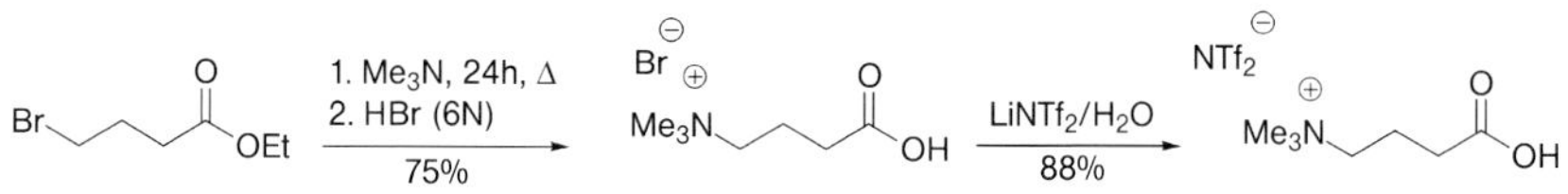

Scheme 5.5-9

liquids bearing a primary amino group (Scheme 5.5-8) [27], a carboxylic function (Scheme 5.5-9) [21] or a bromine (Scheme 5.5-10) [21].

Several ω-bromoalkyltrimethylammonium salts have been prepared in good yields according to Scheme 5.5-10 (Table 5.5-2). These salts are well suited for further transformations into other TSILs bearing an amino group, for example.

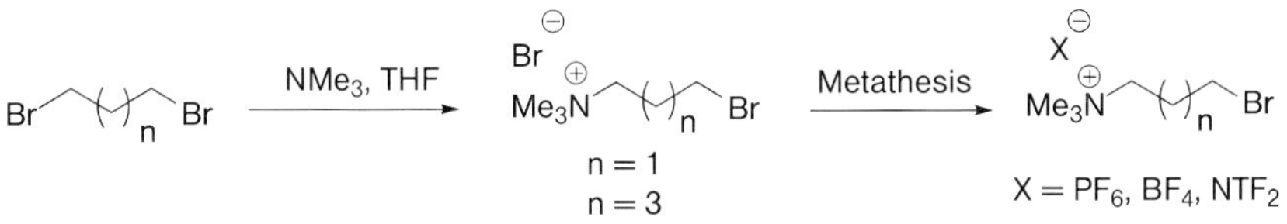

Scheme 5.5-10

Table 5.5-2 Synthesis of ω-bromoalkyl trimethylammonium salts [18, 21].

Entry	n	X⁻	Yield (%)	Mp (°C)
1	1	Br⁻	99	212–214
2	1	[PF$_6$]⁻	88	144–146
3	1	[NTf$_2$]⁻	86	liquid
4	3	Br⁻	96	138–140
5	3	[NTf$_2$]⁻	84	liquid
6	3	[PF$_6$]⁻	97	138–140
7	3	[BF$_4$]⁻	78	liquid

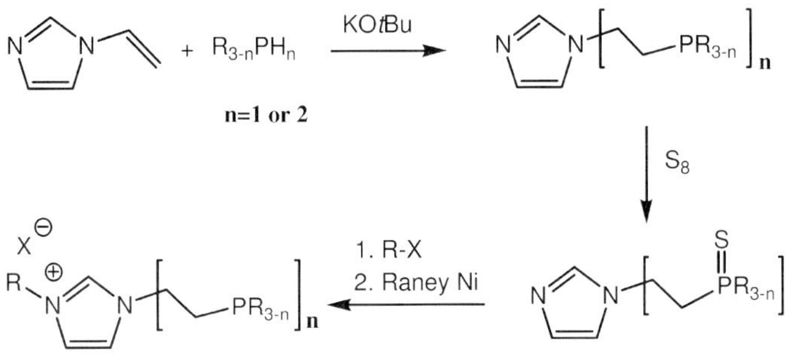

Scheme 5.5-11

A roundabout strategy may be necessary in a number of cases where the alkylation step is not compatible with the electrophilic reagent used, thus making protective groups necessary. A good example of this kind of methodology is the synthesis of imidazolium substituted mono or bis-phosphines from 1-vinylimidazole (Scheme 5.5-11) [28].

5.5.2.2 Parallel Synthesis of Functionalized ILs from a Michael-type Reaction

A very efficient alternative method for the synthesis of functionalized ionic liquids in a simple two-step process has been proposed by Wasserscheid et al. [29]. In this approach, the imidazole or the required nucleophile is protonated by the acid whose anion will be incorporated into the final ionic liquid. A Michael acceptor,

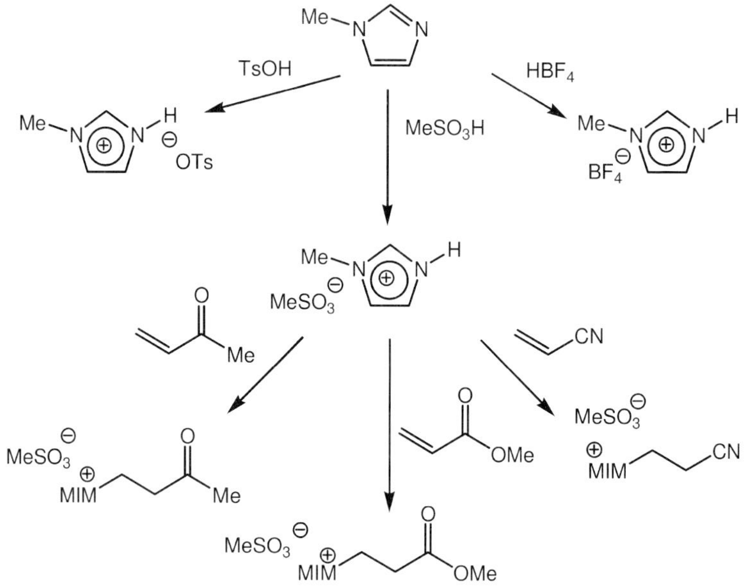

Scheme 5.5-12

which inserts into the N–H or Nu–H bond, is added to the resultant salt. This approach is widely applicable, leading to TSILs in excellent yields (Scheme 5.5-12). It is notable that there is no need for extraction, filtration or further anion metathesis steps. Since no by-product is formed, the two-step sequence proceeds with perfect atom economy. This methodology allows for the rapid parallel synthesis of libraries of TSILs in a combinatorial fashion (this has been realized with three amines, three carboxylic acids and five Michael acceptors). Another interesting feature of this new synthesis is its potential for large-scale production of cheap industrial ionic liquids since several amines or nitrogen heterocycles, acids and Michael acceptors are produced on a large scale. Substituted acrylates such as methyl methacrylate or cinnamate were not as reactive, leading to incomplete reactions or no reaction at all. Thermal stability might be a concern since the Michael-type addition reaction was found to be reversible at higher temperature. This drawback could be circumvented by further functional group transformations.

5.5.2.3 Synthesis of TSILs by Further Functional Group Transformations

Several functional group transformations that lead to new TSILs starting from hydroxyl-, carboxylic acid, hydroxyl-, amino- or halo-substituted TSILs are possible. Primary, secondary and tertiary amines have been prepared in very good yields from the corresponding ω-bromo derivatives by nucleophilic substitution according to Scheme 5.5-13 [21]. Primary and secondary amines can be further used to graft other functional groups and, therefore, act as soluble supports, as we will see later. Tertiary amines can serve as supported bases or ligands.

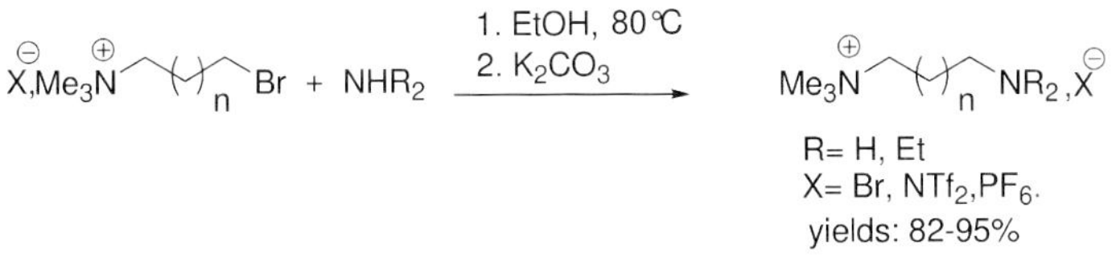

5.5 Task-specific Ionic Liquids as New Phases for Supported Organic Synthesis

$$X,Me_3N^{+}\overset{}{\underset{n}{\frown}}Br + NHR_2 \xrightarrow[\text{2. }K_2CO_3]{\text{1. EtOH, 80°C}} Me_3N^{+}\overset{}{\underset{n}{\frown}}NR_2, X^{-}$$

R= H, Et
X= Br, NTf$_2$, PF$_6$.
yields: 82–95%

Scheme 5.5-13

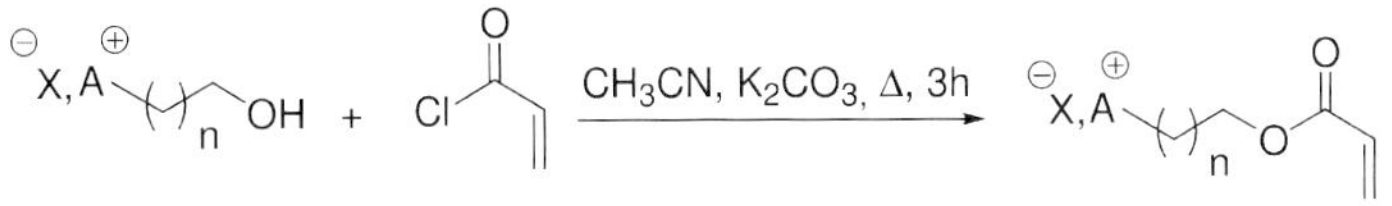

$$X,A^{-}\overset{}{\underset{n}{\frown}}OH + Cl\overset{O}{\underset{}{\frown}}\overset{}{\underset{}{=}} \xrightarrow{CH_3CN, K_2CO_3, \Delta, 3h} X,A^{-}\overset{}{\underset{n}{\frown}}O\overset{O}{\underset{}{\frown}}\overset{}{\underset{}{=}}$$

Scheme 5.5-14

Table 5.5-3 Synthesis of TSILs bearing an acrylic ester [21]

Entry	A (or A+)	n	X	Yield (%)[a]
1	NMe$_3$	1	[NTf$_2$]$^-$	90
2	NMe$_3$	2	[NTf$_2$]$^-$	98
3	NMe$_3$	2	Br	98
4	NMe$_3$	2	[BF$_4$]$^-$	93
5	NMe$_3$	2	[PF$_6$]$^-$	90[b]
6	NMe$_3$	3	[NTf$_2$]$^-$	77
7	NMe$_3$	5	[NTf$_2$]$^-$	78
8	PBu$_3$	2	[NTf$_2$]$^-$	77
9	 pyridyl	2	[NTf$_2$]$^-$	80
10	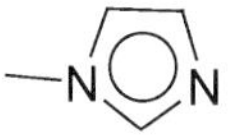 N-methylimidazolyl	2	[NTf$_2$]$^-$	85
11 (Ref. [30])	 N-methylimidazolyl	1	[PF$_6$]$^-$	98
12 (Ref. [31])	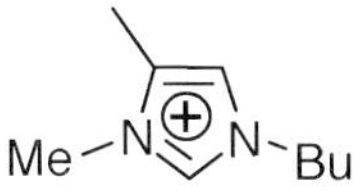 Me–N⊕N–Bu (imidazolium)	0	[NTf$_2$]$^-$	96

[a] Oils at RT. Yields are for isolated pure compounds.
[b] mp = 70–72 °C.

The ester link has been extensively used to graft new functional molecules. Acrylic esters linked to onium cations via a spacer have been obtained in excellent yields and purity by simple acylation with acryloyl chloride or esterification with acrylic acid of the ω-hydroxyl TSILs under standard conditions (Scheme 5.5-14).

Powdered potassium carbonate proved to be an efficient base to remove HCl from the reaction mixture. Some examples are reported in Table 5.5-3.

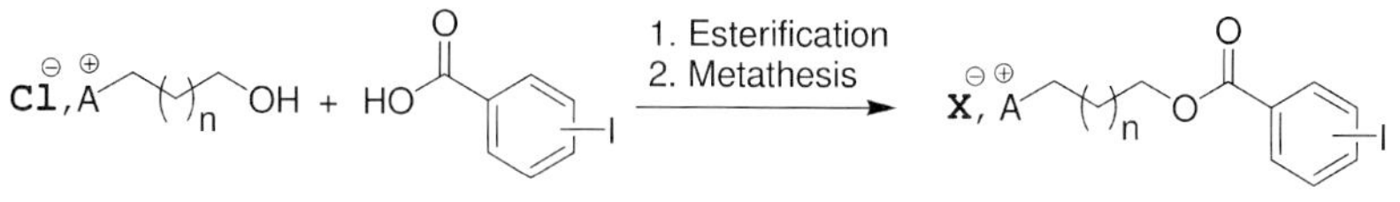

Scheme 5.5-15

Table 5.5-4 Synthesis of 4-iodobenzoic esters

Entry	A	Nn	X	Yield (%)	mp (°C)
1	NMe3	0	Cl	93	258–260
2	NMe3	0	[BF4]	90	200–202
3	MIM	0	Br	92	228–230
4	MIM	0	[BF4]	85[b,33]	146–148
5	NMe3	1	Cl	95	218–220
6	NMe3	1	[BF4]	83	190–192
7	NMe3	1	[NTf2]	95[b,21]	48–50
8	NMe3	4	[BF4]	87	170–172
9	NMe3	1	[NTf2]	88	78–80

[a] Yields are for isolated pure salts.
[b] The 3-iodo benzoic ester has also been prepared.

The acrylic moiety has been grafted onto more complicated ionic liquid phases called poly(ethyleneglycol)ionic liquid phases (PEG$_n$-ILPs). 3- and 4-Iodobenzoic esters have also been grafted by an esterification of the corresponding benzoic acid, performed in the presence of the coupling mixture DCC/DMAP. Excellent results were obtained provided that the dicyclohexylurea formed as a by-product is properly removed (Scheme 5.5-15). It is worth noting that all these salts are high melting compounds that are easy to purify by recrystallisation, and they cannot be called [32] ionic liquids. Nevertheless, they have proved to be very useful as part of BTSILs and should be included here [21].

Several examples are reported in Table 5.5-4. These esters are well suited for Heck and Suzuki-Miyaura coupling reactions.

Sophisticated functionalized ionic liquids aimed at supporting 2-hydroxybenzaldehyde have been prepared by esterification of 1-(2-hydroxyethyl)-3-methylimidazolium tetrafluoborate (Scheme 5.5-16) [17].

Natural α-aminoacids can also be grafted on to ionic liquids by simple esterification as illustrated in (Scheme 5.5-17) [34].

The esters derived from acetylacetic acid have been realised either from ethylacetoacetate by transesterification with 1-(2-hydroxyethyl)-3-methylimidazolium tetrafluoborate under MW irradiation [35] or by reaction of hydroxyalkyltrimethylammonium salts with a ketene dimer (Scheme 5.5-18) [36].

Two new TSILs bearing 2-hydroxybenzylamine based on an imidazolium substructure and used for the extraction of americium ions have been reported by Ouadi et al. [37] (Scheme 5.5-19).

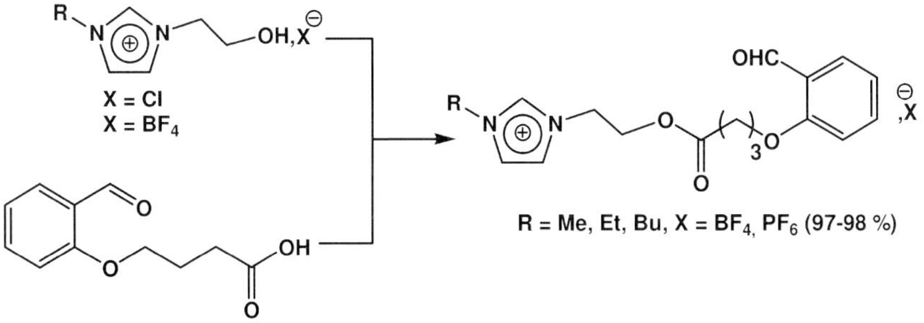

5.5 Task-specific Ionic Liquids as New Phases for Supported Organic Synthesis

Scheme 5.5-16

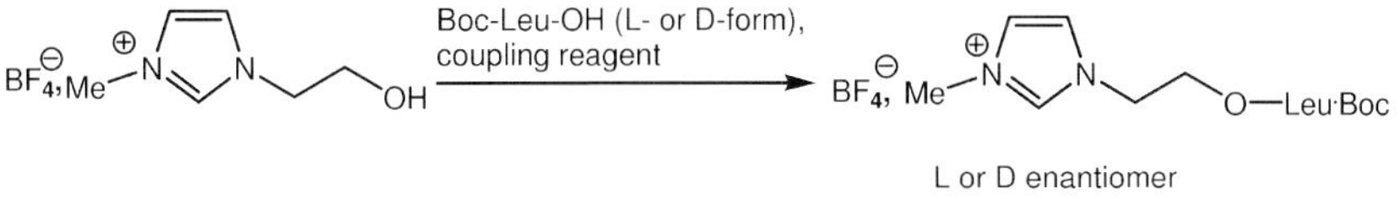

R = Me, Et, Bu, X = BF$_4$, PF$_6$ (97-98 %)

Scheme 5.5-17

Boc-Leu-OH (L- or D-form), coupling reagent

L or D enantiomer

Scheme 5.5-18

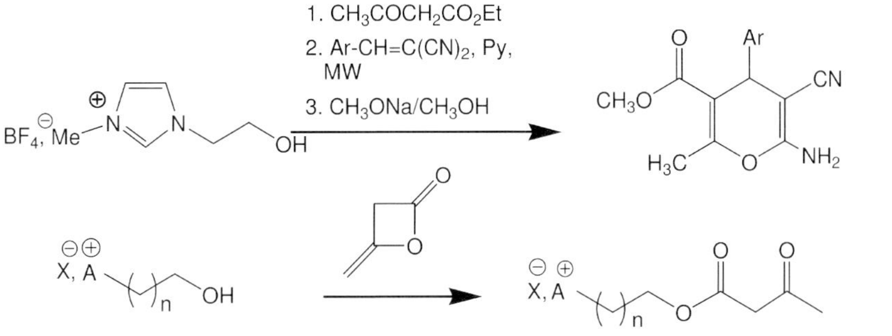

1. CH$_3$COCH$_2$CO$_2$Et
2. Ar-CH=C(CN)$_2$, Py, MW
3. CH$_3$ONa/CH$_3$OH

Li et al. [38] have synthesized a series of N-ester and N-carboxyl appended pyridinium task specific ionic liquids and have found them to be immiscible with acetone and dichloromethane, unlike the imidazolium ionic liquids, and to have improved thermal stability up to 300 °C.

The ether link has also been used to give ionic liquids bearing Wang-type linkers via the Williamson alkylation [39, 40] (Scheme 5.5-20).

From what is reported above, it can be seen that many of the examples of so-called TSILs are either high melting solids, or highly viscous oils or waxes. This leads logically to the extension of the concept of TSILs to that of task-specific onium salts, TSOSs, free from the restriction of being liquid below 100 °C [21]. TSOSs can be useful as soluble supports for solution phase organic synthesis. There are three possibilities to be considered. First, if the TSOSs are liquid at or near room

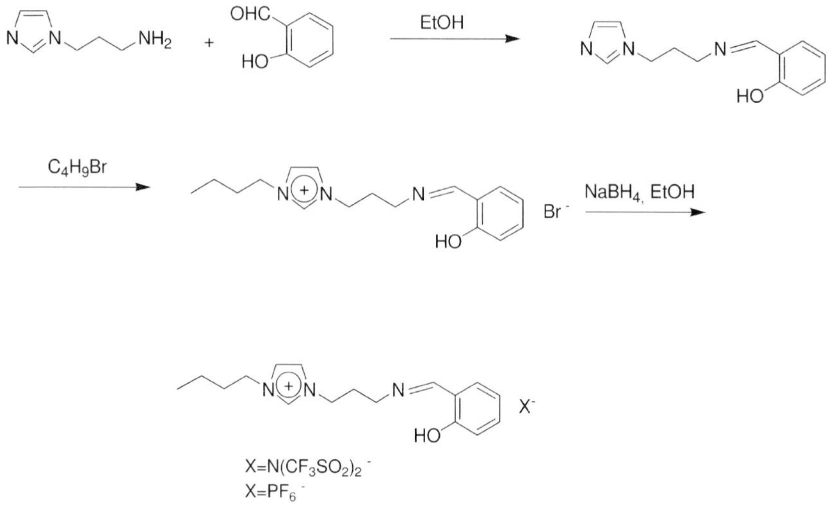

Scheme 5.5-19

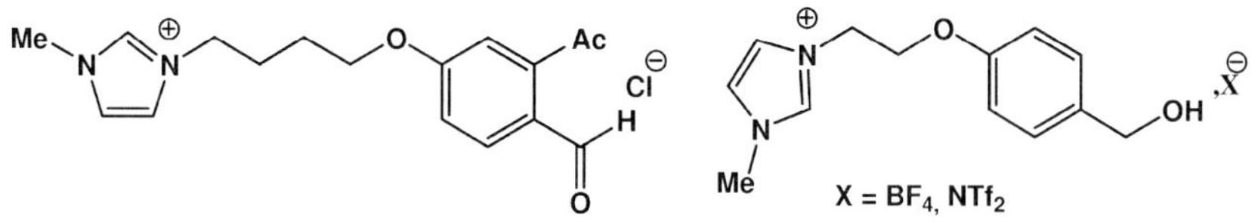

Scheme 5.5-20

temperature, they belong to the first generation of TSILs which rapidly become very viscous or waxes or solids as their functionalization increases and, therefore, become no longer easy to handle. Second, TSOSs can dissolve, whatever their physical state is, i.e. liquid or waxy or crystalline solid, in a non-functionalized ionic liquid which plays the role of a solvent giving BTSILs. The third possibility is to dissolve TSOSs in a molecular solvent and use them as classical soluble supports. These three possibilities have been illustrated in the literature and widen the area of SPOS. It can be safely anticipated that this area is going to expand very rapidly, owing to the interest in these new soluble supports.

5.5.2.4 Loading of TSIL Supports

One of the main limitations of traditional solid or soluble supports is their low loading capacity which is related to the molecular weight of these polymers. Loadings of over 1 mmol g^{-1} of polymer are rarely reached. Also, these polymers are used diluted either as suspensions or in solution in some molecular solvent. Then, we should consider the loading of the suspension or the solution that will be used for the reaction to be carried out. Most of the time low loadings such as 0.1 mmol g^{-1},

Table 5.5-5 Loading capacity of the onium salts [A–(CH$_2$)$_n$OH][X]

A	n	X	Molecular weight	Loading capacity (mmol g^{-1})
MIM[a]	2	Cl	146.5	6.82
MIM	2	[NTf$_2$]$^-$	391	2.56
Me$_3$N	2	Cl	139.5	7.17
Me$_3$N	2	[NTf$_2$]$^-$	384	2.60
Me$_3$N	3	Cl	153.5	6.51
Me$_3$N	3	[BF$_4$]$^-$	205	4.87
Me$_3$N	3	[PF$_6$]$^-$	263	3.80
Me$_3$N	4	Cl	167.5	5.97
Me$_3$N	4	[OTf]$^-$	283	3.53
Me$_3$N	4	[NTf$_2$]$^-$	398	2.51
Me$_3$N	6	Cl	181.5	5.50
Me$_3$N	6	[NTf$_2$]$^-$	426	2.34
—		MPEG 5000	5000	0.2

[a] MIM = methylimidazolium.

or less, of suspension or solution are common. TSOSs and TSILs are low molecular weight soluble supports, as illustrated by the examples given in Table 5.5-5. It can be seen that chlorides have a high loading capacity, up to more than 7 mmol g^{-1}, owing to their low molecular weight, which compares to the loading capacity of MPEG 5000 which is 0.2 mmol g^{-1}. It is also worth noting that a one molar solution of MPEG 5000 contains 5000 mg of polymer per ml of solution, which is not realistic as compared to 139.5 mg for choline chloride.

The loading capacity of the TSOSs and TSILs can be easily increased by using polyfunctional cations. For example, an ammonium cation can be substituted by at most four functional substituents. These kind of multifunctional salts have been prepared [18, 21] and their loading capacity is reported in Table 5.5-6. This loading capacity could even be increased by dendrimerization if necessary [10].

5.5.3
TSILs as Supports for Organic Synthesis

Combinatorial chemistry is a powerful tool for the generation of libraries of compounds for the screening of functional molecules. Solid-phase organic synthesis (SPOS) is an efficient method for high-throughput synthesis which exhibits several shortcomings, such as the heterogeneous nature of the reaction and difficulties in reaction monitoring. By using soluble polymer supports, the familiar reaction conditions of organic chemistry in solution are reinstated, and yet product purification is still facilitated. However, the low loading capacity is the main limitation. So, a search for alternative soluble supports for high-throughput organic synthesis is necessary. In recent years, there has been more attention focused on the TSILs as soluble supports. RTILs have a number of interesting physical and chemical

5 Organic Synthesis

Table 5.5-6 Loading capacity of polyfunctional TSILs [21]

$$X, Me_{4-y}N^{+}\text{—}\!\!\!\sim\!\!\!\text{—OH}_y$$

Entry	y	X	Molecular weight	Loading capacity mmol g^{-1} (mf g^{-1})
1	2	Cl	197.75	5.05 (10.11)
2	2	[NTf$_2$]	442.25	2.26 (4.54)
3	3	Cl	241.80	4.13 (12.4)
4	3	[NTf$_2$]	486.30	2.05 (6.17)
5	4	Cl	285.85	3.50 (14)
6	4	[NTf$_2$]	530.35	1.88/7.51

^{a}mf g^{-1} represents the number of millifunctions/g of support.

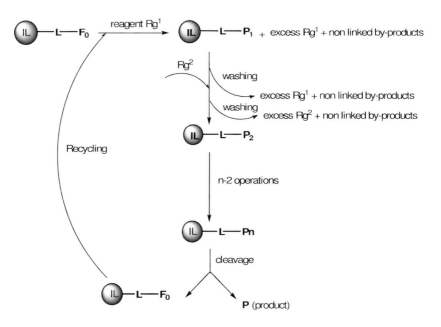

Fig. 5.5-2 General concept of ionic liquid supported synthesis.

properties which have been discussed in other chapters. Some of them are essential for the development of new soluble supports including thermal and chemical stabilities, insignificant vapor pressure, non-flammability, tuneable solubility. Supporting catalysts in RTILs in order to immobilise them in one liquid phase of a biphasic system has led to interesting developments over the last few years [39–41]. Less effort has been spent on supported non-metallic reagents [42]. The general concept of ionic liquid supported synthesis is shown in Fig. 5.5-2.

The TSIL used as a support is drawn like a solid support, i.e. the function F_0 used as the anchor point is linked to the ionic head via a linker which can be more or less sophisticated. Results from the literature show that using sophisticated linkers, such as those described for solid phase supported synthesis is not necessary and moreover could bring additional problems. Grafting of the first reagent R_1 leads to a new TSIL where F_0 has been transformed into P_1 which is mixed with an excess of reagent used for the first chemical transformation and eventually by-products. This excess reagent and non-linked by-products can be removed by washing with an appropriate solvent, or by stripping under high vacuum and heating, or by other separation techniques. The resulting purified TSIL can then be further functionalized. It is notable that it is usually very easy to follow the reactions by standard spectroscopic techniques, such as high-resolution NMR, FTIR, UV etc. and/or by chromatographic methods such as HPLC or TLC. This is a definite advantage over the classical Merrifield-type resins or polymeric soluble supports. After several transformations, the desired product P is liberated from the support by cleavage and can be separated from the starting TSIL in classical ways such as extraction with a non-miscible solvent, bulb to bulb distillation under high vacuum, column chromatography etc. Then, the starting TSIL can be recycled. At the moment, two distinct kinds of TSILs have been developed and reported in the literature: a first generation of pure functionalized ionic liquids and a second one which is based on binary functionalized ionic liquids or BTSILs.

5.5.3.1 First Generation of TSILs as New Phases for Supported Organic Synthesis

TSILs derived from 1-(2-hydroxyethyl)-3-methylimidazolium tetrafluoroborate or hexafluorophosphate have been used in Knoevenagel reactions, 1,3-dipolar cycloadditions [17] synthesis of 4-thiazolidinones [43] and 2-thioxotetrahydropyrimidin-4-(1H)-ones [30] under focused microwave irradiation. Under these conditions, Knoevenagel reactions with malonate derivatives catalyzed by 2% piperidine are over after a short reaction time (15–60 min) at 80 °C. Washing of the TSIL obtained with AcOEt or Et_2O was followed by cleavage with sodium methanolate in methanol. After removal of solvent *in vacuo*, the expected compounds were extracted with CH_2Cl_2 and isolated in 87–98% yields without the need for silica gel chromatography. The starting TSIL could eventually be recovered and reused in another cycle of the synthesis (Scheme 5.5-21).

Benzaldimines have been obtained from the same supported benzaldehyde and various alkylamines in short reaction times (20 min) and good yields. A regioselective 1,3-dipolar cycloaddition with an imidate derived from dimethylaminomalonate led to a cycloadduct, treatment of which with 10% NaOMe in MeOH resulted in the diethyl 2-imidazoline-4,4-dicarboxylate in 84% yield. These two examples show clearly the interest and the simplicity of this methodology. Product isolation in a good state of purity necessitates only an extraction, because non-linked side-products have been removed by simple washing of the TSILs. No chromatography was necessary. Standard analytical methods (NMR, TLC) can be used to monitor reaction progress. This methodology has been extended to the parallel synthesis of some heterocycles. Condensation of α-mercaptocarboxylic acids with the supported

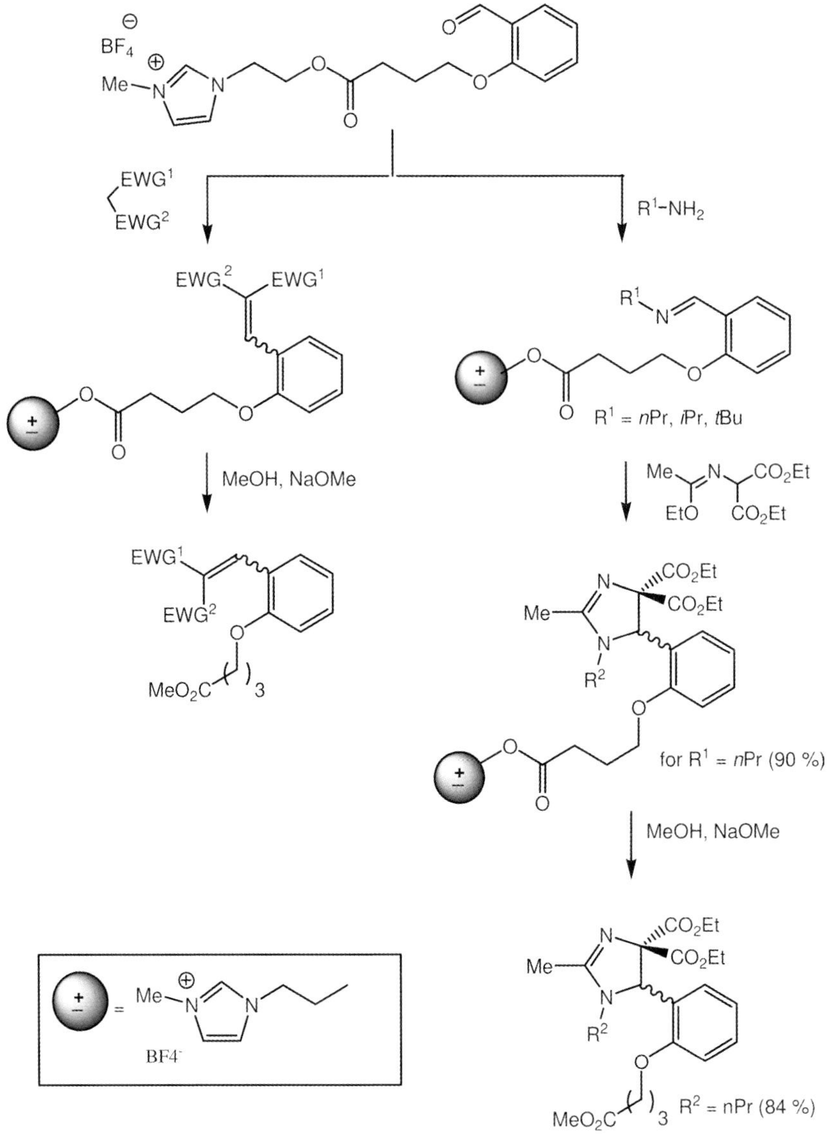

Scheme 5.5-21

imines gave 4-thiazolidinones in 12–86% conversion. Ester aminolysis with primary or secondary amines catalysed by ^{t}BuOK under MW heating led to 4-thiazolidinones in modest overall yields ranging from 25 to 61% after purification by extraction and flash chromatography (Scheme 5.5-22) [41]

MW heating of supported thioureas ω-substituted by an ester function used as a grafting element to the imidazolium moiety in the presence of diethylamine

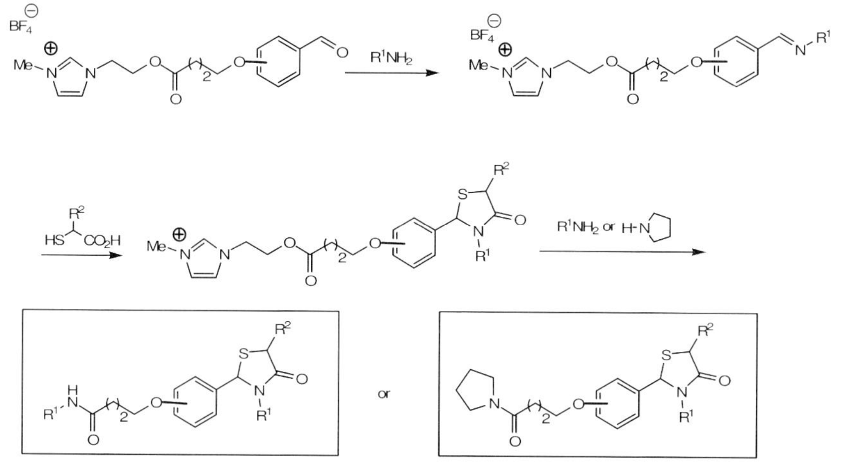

Scheme 5.5-22

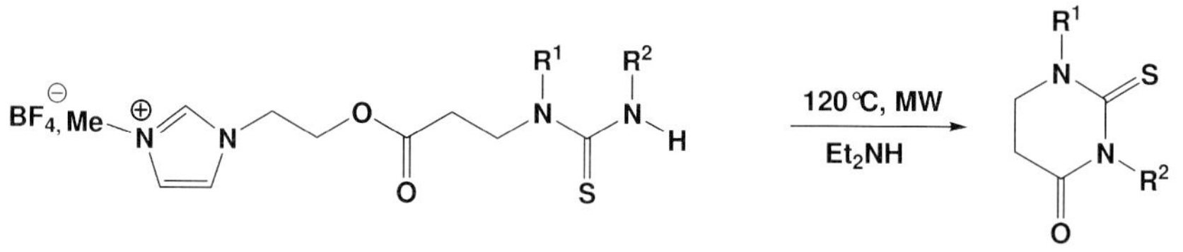

Scheme 5.5-23

led to free 2-thioxotetrahydropyrimidin-4-(1*H*)-ones by an intramolecular cyclisation–cleavage sequence in good yields (67–85%) after flash chromatography (Scheme 5.5-23) [30].

Reaction of IL-bound acetoacetate with excess arylidenemalononitriles under MW irradiation gave bound 4-*H* pyranes. Excess reagent could be removed by simple washing. Cleavage by sodium methoxide in methanol (monitored by TLC) gave methyl-6-amino-5-cyano-4-aryl-2-methyl-4*H*-pyran-3-carboxylates in good yields (85–91%) and purities (85–99%) without further purification (Scheme 5.5-24). The recovered IL after cleavage can be simply reused in a second run. According to the authors, but not described, the methodology is compatible with automatic manipulation [35].

Supported liquid acrylic esters have been prepared from hydroxylated imidazolium-based TSILs and used neat in (4+2) Diels-Alder cycloadditions. First, Handy et al. [31] used a fructose-derived ionic liquid to support acrylic acid and performed the Diels-Alder cycloaddition with several dienes including cyclopenta- and cyclohexadienes and butadiene derivatives at 120 °C for 12 h in the presence of hydroquinone (Scheme 5.5-25).

Excess diene and other volatiles were removed *in vacuo* and the supported cycloadducts washed with cyclohexane. Transesterification with MeOH or EtOH

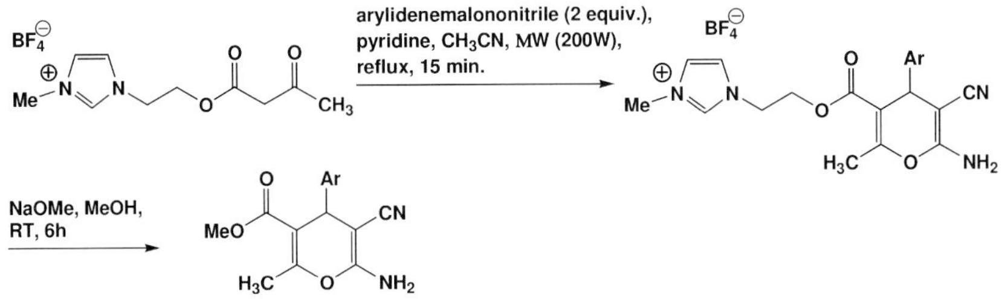

Scheme 5.5-24

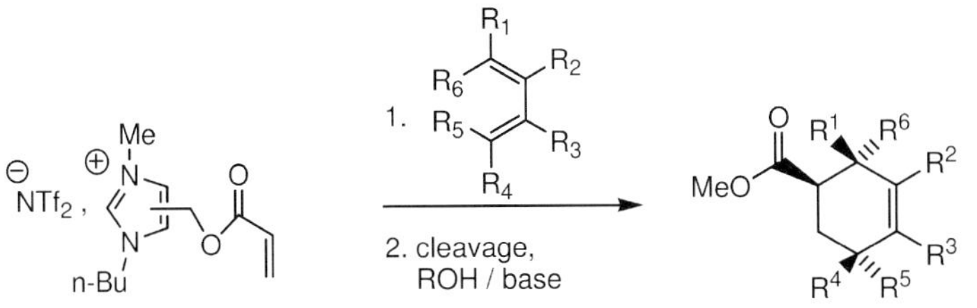

Scheme 5.5-25

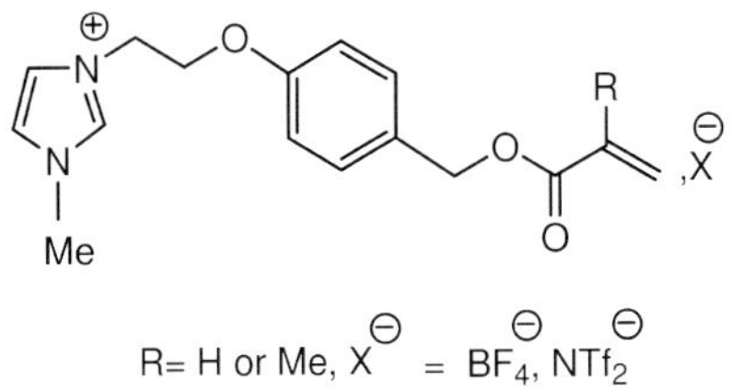

Scheme 5.5-26

under basic conditions appeared to be troublesome, because of the ease of deprotonation of the imidazolium cation at the 2-position and led to a dark brown viscous liquid. The most effective transformation was the cyanide-mediated transesterification in methanol leading to the methyl esters of the corresponding cycloaddition adducts in good overall yields (46–63% for the 3-step sequence). Ethyl and benzyl esters also could be obtained using this methodology. The ionic support was readily recycled. More complicated linkers have been used by Grée et al. [38] to support acrylic esters on imidazolium derived ionic liquids via a Wang- type linker (Scheme 5.5-26).

Again, good results were obtained, but nothing was said about the recycling protocol of the IL support. The use of a complex linker does not seem to be a limiting addition, unlike the former system. These acrylic esters have also been used as partners in several other reactions such as 1–4 additions of nucleophiles such as pyrrolidine or thiophenol in the presence of Et_3N, Heck coupling and dihydroxylation of the obtained cinnamic esters. Methyl esters could be isolated in

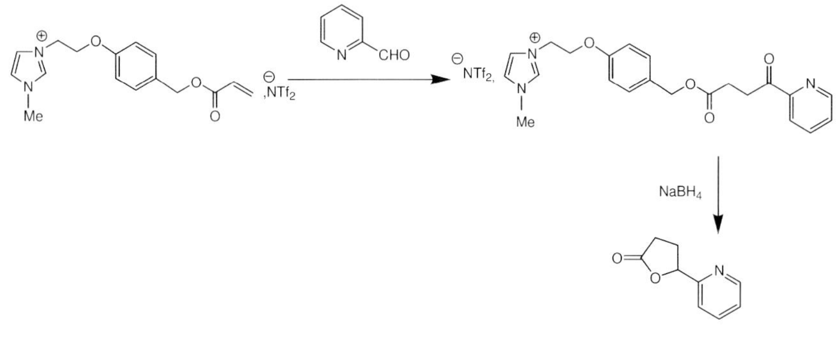

Scheme 5.5-27

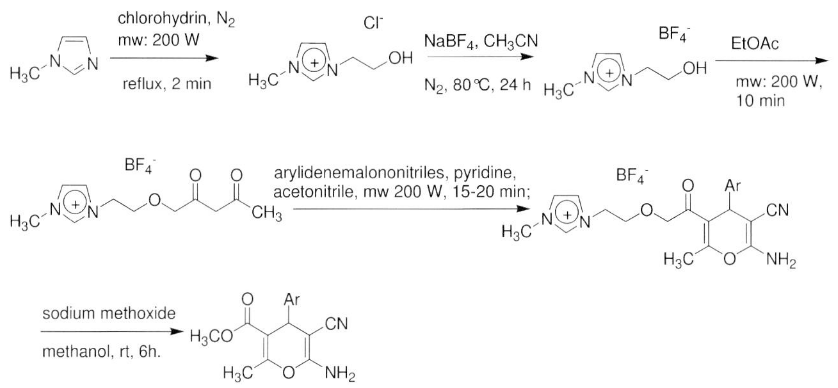

Scheme 5.5-28

good yields after transesterification with methanol under basic conditions. Finally, the Stetter reaction of pyridine-2-carboxaldehyde led to the expected 1,4-dicarbonyl compound (87% yield) which after reduction with sodium borohydride afforded a lactone by a cyclorelease cleavage from the support (50% yield) (Scheme 5.5-27).

The combined effect of microwave heating and functional ionic liquid as soluble support has been demonstrated for the synthesis of methyl 6-amino-5-cyano-4-aryl-2-methyl-4H-pyran-3-carboxylate [44], which are polyfunctionalised 4H-pyrans with pharmacological activity. The ionic liquid bound acetoacetate was treated with various substituted arylidenemalononitriles to give supported 4H-pyran derivatives. After cleavage, the target compounds were obtained in good yields (93–97%) and high purities (99%) without chromatographic purification (Scheme 5.5-28).

A microwave dielectric heating assisted TSIL phase synthesis [45] of 1,4-dihydropyridines, 3,4-dihydropyrimidin-2(1H)-ones, pyridines and polyhydroquinolines using a TSIL as a soluble support was described. The efficiency of the ionic liquid phase organic synthesis (IoLiPOS) methodology was demonstrated by using a one-pot three-component condensation. The structure of the intermediates

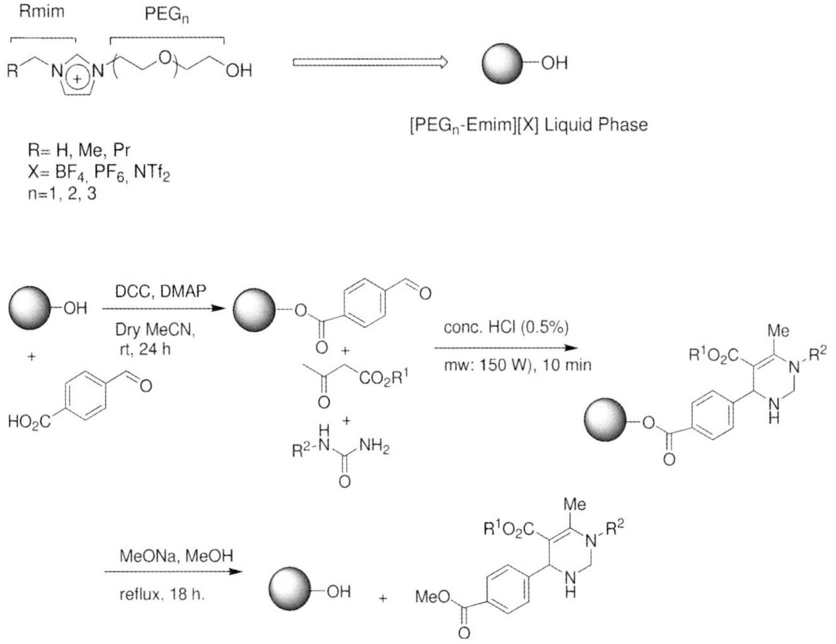

Scheme 5.5-29

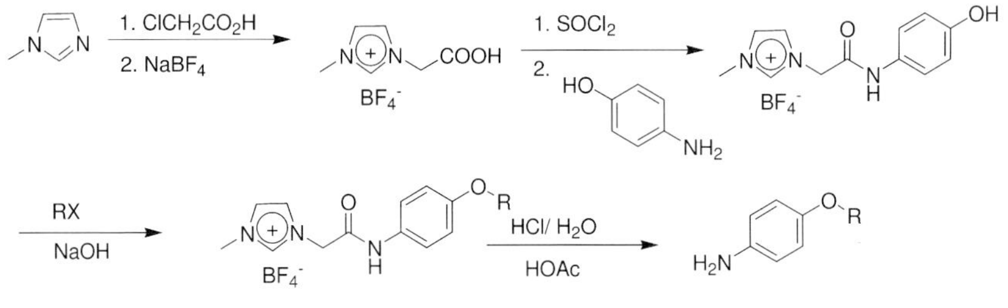

Scheme 5.5-30

in each step was verified routinely by spectroscopic analysis and, after cleavage the target compounds were obtained in good yields and high purities (Scheme 5.5-29).

Song et al. [46] have used a carboxyl-functionalized ionic liquid as soluble support to synthesize a small library of 4-aminophenyl ethers via Williamson reaction and extracting with ethyl acetate in good yields (75–80%) and purities (99%). The recovered ionic liquid support was reused several times with consistent loading capacity (Scheme 5.5-30).

Wang et al. [47] have reported a novel and efficient protocol for ionic liquid supported synthesis of oligosaccharides, which have biological and therapeutic importance. This methodology is remarkable for improved phase separation from organic solvent or aqueous phase. The ionic liquid supported oligosaccharides were

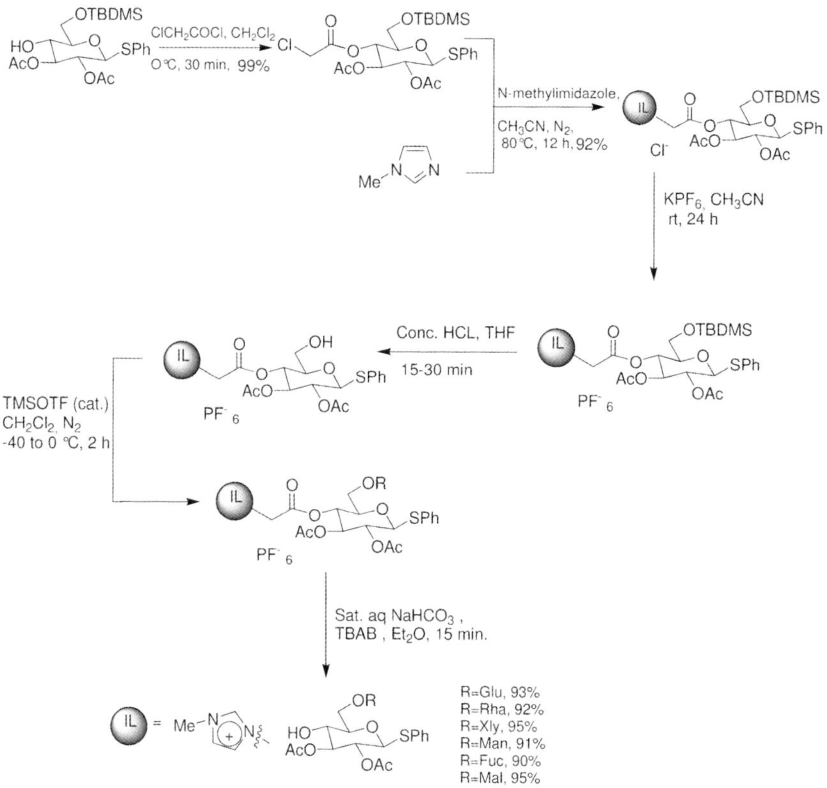

Scheme 5.5-31

soluble in polar organic solvents such as acetone, chloroform, methanol, and are not soluble in hexane and ether, which made it possible to remove the excess reagent and impurities by simple washing with ether or hexane, providing an attractive alternative to the clasical solid- and fluorous-phase synthesis, combining the advantage of performing homogeneous chemistry on a relatively large scale while avoiding large excesses of reagents (Scheme 5.5-31).

A microwave-assisted liquid phase Gewald synthesis [48] of 2-aminothiophenes was developed using the TSIL, 1-(2-hydroxyethyl)-3-methylimidazolium tetrafluoroborate, [2-(OH)EMIM][BF$_4$] as soluble support. This new synthetic method is simple and efficient, and the products are obtained in good to excellent yields with high purities, without the need for chromatographic purification.

This hydroxy functionalized ionic liquid [49], [2-(OH)EMIM][BF$_4$] was prepared under microwave irradiation via methylimidazolium, chlorohydrin and sodium tetrafluoroborate as materials. A microwave assisted solution phase parallel synthesis of 3,4-dihydropyridones was developed using [2-(OH)EMIM][BF$_4$] as a soluble support. First, the [2-(OH)EMIM][BF$_4$] was reacted with ethyl acetoacetate to obtain IL-bound acetoacetate, then this intermediate was reacted with Meldrum's acid, ammonium acetate and aromatic aldehyde. Finally 3,4-dihydropyridone derivatives were obtained in good yields and high purities without chromatographic purification after mild cleavage with sodium methoxide in methanol at room temperature. The target compounds had a purity of 95–98% and the yield was 83–93%. The recovered functionalized ionic liquid could be recycled at least six times without losing its activity. It was clear that the solution phase synthesis using functionalized ionic liquids as supports could be used in the combinatorial synthesis of 3,4-dihydropyridone derivatives in high efficiency.

In summary, these TSILs offer several advantages over the usual supports:

- Higher loading capacity is achieved due to the lower molecular weight of the functionalized ionic liquid.
- Since the intermediates can be purified by simple extraction, the desired products can be obtained in high yields and purities without further chromatographic purification.
- This methodology can be applied to parallel synthesis and small libraries have been prepared.
- In some cases, depending on the treatment, the recovery of the ionic liquid after cleavage is possible and it can be reused in another cycle without losing its activity.

At present the key challenges in using this type of support are the limited stability of the imidazolium cation in the presence of base and the viscosity of these TSILs, since the more they are substituted the more viscous they become. Therefore, it is difficult to handle and stir them, necessitating the use of MW for heating, for example.

5.5.3.2 Second Generation of TSILs: The BTSILs

The question of viscosity can be solved by dissolving a functionalized salt (i.e. TSOS) in a non-functionalized ionic liquid used as a solvent. The viscosity of these binary mixtures is not sensitive to further functionalization of the TSOS dissolved in the solvent. Moreover, it is possible to use TSOSs which are high melting onium salts. Furthermore, these salts can have a cation and (or) an anion different from the solvent. The only necessity is that the salt itself must be reasonably soluble in the ionic liquid used as the solvent [21]. This has widened the range of accessible and useful salts to any functionalized onium salt available by simple organic synthesis and, therefore, very much broadened the scope of the utilization of TSILs as soluble supports. The limited stability of the imidazolium cation in the presence of base could also be largely solved by using tertiary ammonium or phosphonium cations either for the IL solvent or the TSOSs. These cations also offer a better chemical stability towards basic, reducing, oxidizing and nucleophilic reagents. Many different kinds of reactions have been looked at using BTSILs as soluble supports, including Diels-Alder cycloadditions, C–C bond formation by coupling reactions (Heck, Suzuki-Miyaura, Sonogashira), multicomponent reactions, and peptide synthesis [18,21,44–60].

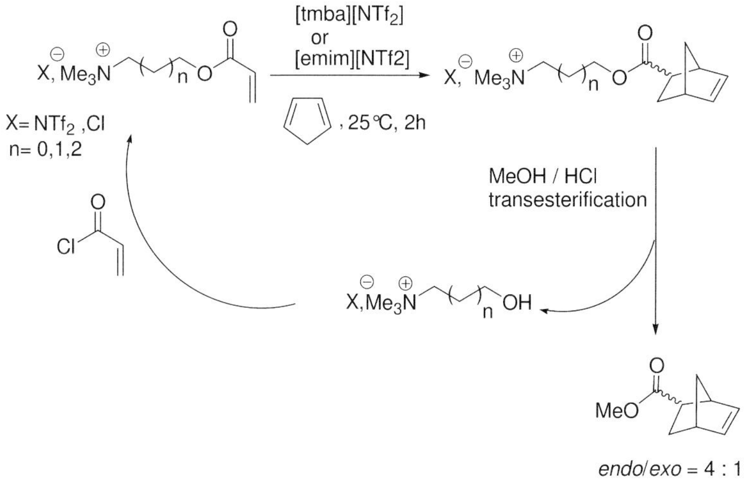

5.5 Task-specific Ionic Liquids as New Phases for Supported Organic Synthesis | 511

Scheme 5.5-32

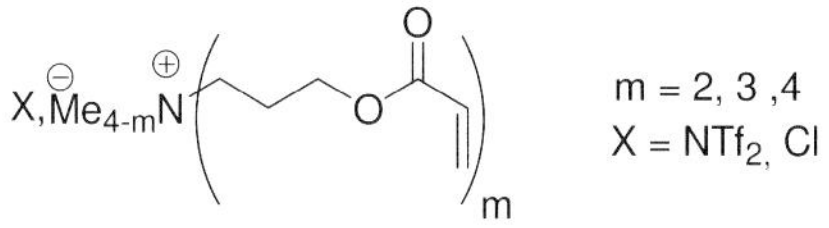

Scheme 5.5-33

The reaction of supported acrylic esters as 0.85 M solutions in an ionic liquid solvent such as n-butyltrimethylammonium or ethylmethylimidazolium triflimide ([BuMe$_3$N][NTf$_2$] or [EMIM][NTf$_2$]) with an excess of cyclopentadiene at room temperature goes to completion within 2 h. After removal of the excess diene by washing with ether, the cycloadducts are cleaved from the support by transesterification in boiling methanol in the presence of traces of HCl. Extraction with ether or bulb to bulb distillation under vacuum allows the isolation of the methyl esters of the adducts in excellent yields (80–90%) and purities (Scheme 5.5-32). No influence of either the nature of the solvent or the supporting salts on the isolated yields of the norbornene derivatives or the *endo/exo* selectivity was observed [11, 21].

No leaching of the supports was observed during the extraction operations and the solutions could be recycled and reused without problems. Other 1,3-dienes gave similarly good results. The productivity of the BTSILs could be increased by using the polyfunctional supporting salts carrying 2, 3 or 4 acrylic groups (Scheme 5.5-33). Isolated yields of the adducts and the *endo/exo*- ratio after transesterification were the same.

These reactions could be easily monitored by routine NMR-spectroscopy (see Fig. 5.5-2) at the concentrations used to perform the reaction, i. e. 0.85 mol l^{-1} and the crude reaction mixture was analyzed by capillary GC [51].

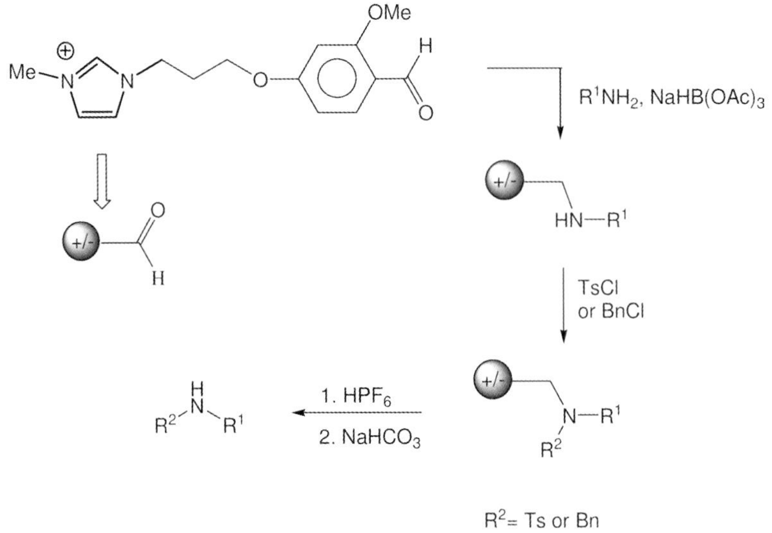

Scheme 5.5-34

An *N*-methylimidazolium chloride-based ionic support bearing an aldehyde, analogous to the known AMEBA [52] solid support [53] readily dissolves in an ionic liquid solvent such as n-butylimidazolium hexafluorophosphate ([BMIM][PF$_6$]). It has been used to prepare a set of diverse sulfonamides and amides according to (Scheme 5.5-34) [39] with isolated yields ranging from 26 to 54% and purities from 55 to 95%, as determined by NMR spectroscopy, comparable to those obtained from solid support chemistry [53]. Reactions were monitored by HPLC. No leakage of the ionically supported substrate into the aqueous or organic phase occurred.

This methodology has been applied to the multistep synthesis of a sulfonylated tyrosine derivative, a highly potent analog of the antithrombic drug tirofiban according to Scheme 5.5-30. The target compound was obtained in 11% yield over 5 steps after purification by preparative HPLC (Scheme 5.5-35).

Multicomponent reactions are an increasingly important class of reactions because they combine simplicity, atom economy and efficiency in terms of both yields and the introduction of molecular diversity [52]. These reactions have been described on both solid and soluble supports, as well as giving an easy entry into libraries of compounds. BTSILs are appropriate soluble supports for such reactions. This is illustrated by the tetrahydroquinoline synthesis developed in solution and on solid supports [52]. The reaction of an aniline supported on an ammonium salt in solution of [BuMe$_3$N][NTf$_2$] with benzaldehydes and electron-rich olefins such as styrene, cyclopentadiene and indene in the presence of a trace of TFA, led quantitatively and rapidly (20 to 60 min) to the corresponding tetrahydroquinolines at room temperature according to (Scheme 5.5-36) [21, 51].

After washing the crude reaction mixture to eliminate the excess reagents and impurities, followed by transesterification, methyl esters of a series of the desired

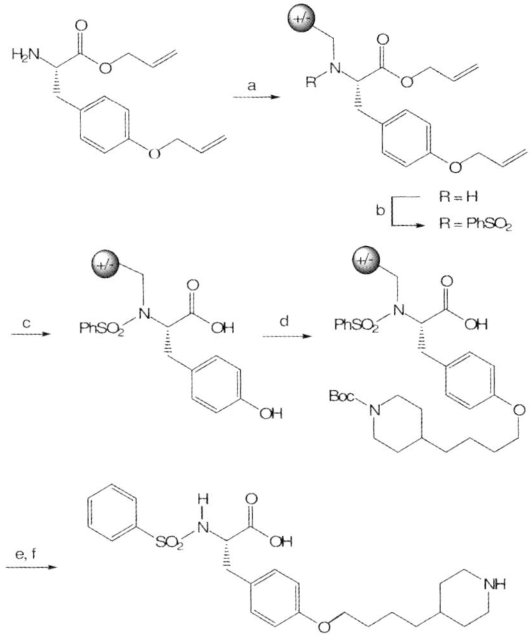

5.5 Task-specific Ionic Liquids as New Phases for Supported Organic Synthesis | 513

All reactions are performed in [BMIM][PF$_6$]. (a) NaBH(OAc)$_3$, HOAc, 16h; (b) PhSO$_2$Cl, Et$_3$N, 1h; (c) Pd(PPh$_3$)$_4$, pyrrolidine, 90°C, 45 min; (d) NaH, then N-Boc-4-(4-iodobutyl)-piperidine, 16h; (e) 2% HPF$_6$, 10 min; (f) Et$_3$N, then extraction with H$_2$O.

Scheme 5.5-35

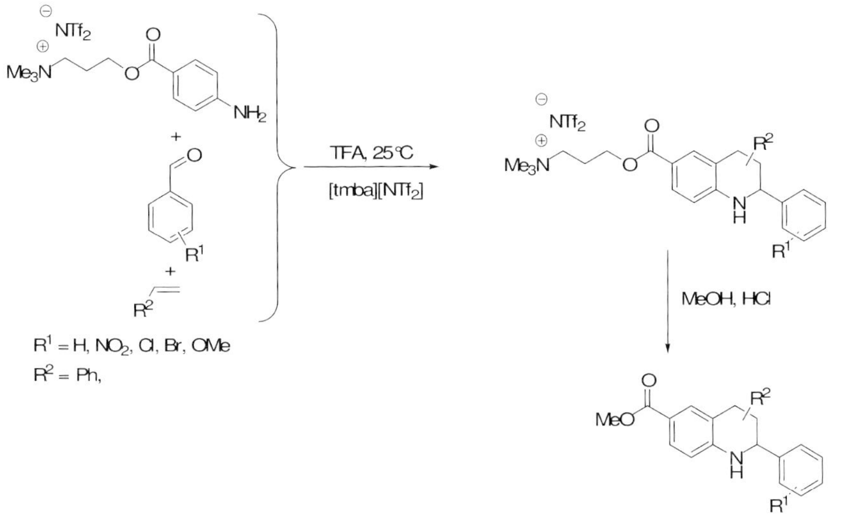

R^1 = H, NO$_2$, Cl, Br, OMe
R^2 = Ph,

Scheme 5.5-36

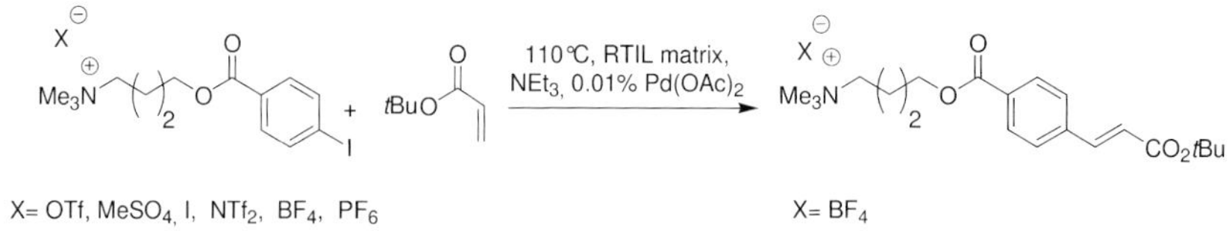

Scheme 5.5-37

tetrahydroquinolines could be isolated in yields ranging from 65 to 90% with excellent purity.

Ionic liquids have been used extensively to perform transition metal catalyzed reactions [56]. Among these, Heck, Suzuki-Miyaura and Sonogashira C–C coupling reactions are among the most synthetically useful ones and have been studied in detail in RTILs. Supporting one of the components of the reaction on a BTSIL is an idea which comes immediately to mind. The catalyst itself can be supported via supported ligands but this will not be addressed in this chapter. All other components have been supported on BTSILs. The interest in this methodology resides in the possibility of washing out excess reagents and by-products resulting from side reactions such as reduction, not linked to the support, or the possibility of performing parallel synthesis or any other combinatorial technique.

Heck coupling has been studied in detail [18, 21]. It has been shown that, for any given catalyst, every parameter of the reaction, such as the nature of the base, of the solvent (cation and anion), of the supporting TSOS (cation and anion) and temperature has its own influence on the reaction. The solvent effect is best illustrated with the reaction of a supported 4-iodobenzoic ester on trimethylammonium tetrafluoroborate TSOS at 110 °C in the presence of 100 ppm of $Pd(OAc)_2$ solubilized in different RTILs as solvents, as illustrated in Scheme 5.5-37. [BMIM][NTf$_2$], [BMIM][PF$_6$], [BMIM][BF$_4$], [BMIM][OTf], [BuMe$_3$N][NTf$_2$] and [(OH)PrMe$_3$N][NTf$_2$] have been used as solvent. Conversions were monitored by HPLC. Under these conditions, no reaction was observed in imidazolium matrices but conversions were complete in trimethylammonium ILs. The anion of the supporting salt is also important. After 2 h of reaction under the mentioned conditions in [BuMe$_3$N][NTf$_2$] as a solvent, 100% conversion was observed with [BF$_4$]$^-$, [PF$_6$]$^-$, [NTf$_2$]$^-$ and [OTf]$^-$ whereas 33 and 69% conversions were observed for I$^-$ and [CH$_3$SO$_4$]$^-$, respectively. It is worth noting that with methyl sulfate, a cheap anion, the reaction goes to completion after 5 h. This methodology has been applied to the parallel synthesis of a small library of 9 supported cinnamic esters in a [BuMe$_3$N][NTf$_2$] solvent according to (Scheme 5.5-38). Reactions were monitored by NMR spectroscopy and were completed within 2 h. The E-configurated olefins were obtained exclusively.

Mixing these 9 cinnamic esters and performing the transesterification with MeOH, EtOH, PrOH and BuOH led to 4 libraries of 9 different esters. The GC trace of the methyl ester library showed an excellent state of purity of the crude reaction mixture and therefore the efficiency of the methodology [21].

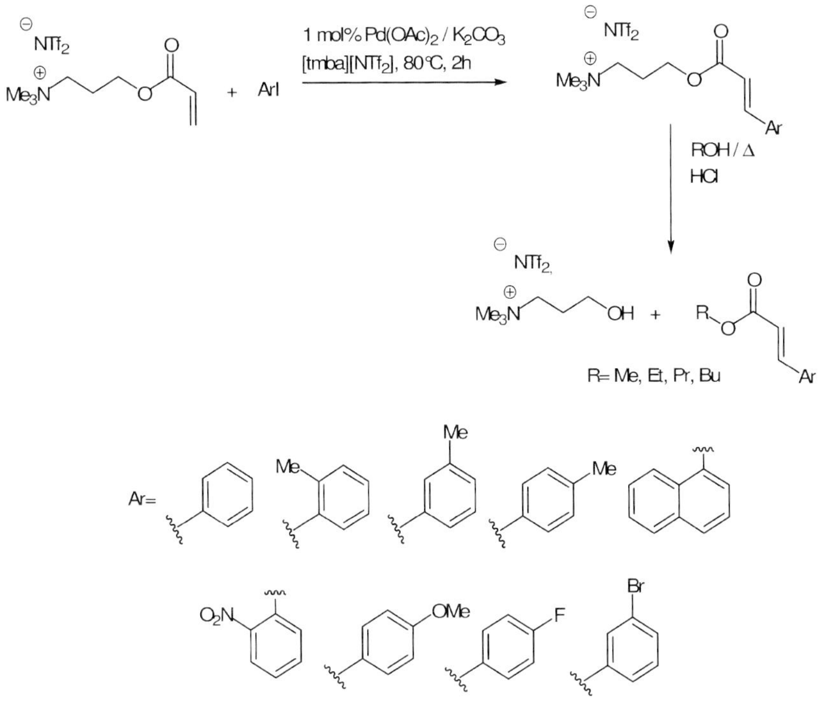

Scheme 5.5-38

Suzuki-Miyaura coupling, involving the reaction of aryl boronic derivatives with aryl halides in the presence of a palladium catalyst, is a very powerful method for the preparation of biaryls. Solutions of aryl bromides or iodides supported on onium salts via an ester link in an RTIL lead to BTSILs that can be engaged in the Suzuki-Miyaura coupling reaction with aryl boronic derivatives. Like the Heck reaction, this coupling is also sensitive to the variations in different parameters. The possibility to synthesize libraries of biarylic derivatives by parallel synthesis is illustrated by (Scheme 5.5-39) [51].

The coupling reactions went to completion within 12 h at room temperature in the 3-hydroxypropyltrimethylammonium triflimide ([(OH)PrMe$_3$N][NTf$_2$]) used as the solvent in the presence of powdered potassium carbonate base. Under these conditions, 1% or less homocoupling occurred. The corresponding symmetrical biphenyl was washed out before transesterification with methanol had been conducted, which gave the biphenyl esters in yields ranging from 90 to 95% and in excellent purities.

5.5.3.3 Reactions of Functionalized TSOSs in Molecular Solvents

The interest in using solutions of TSOSs in RTILs (i.e., BTSILs) as reaction media for supported organic chemistry has been reported above. These TSOSs are

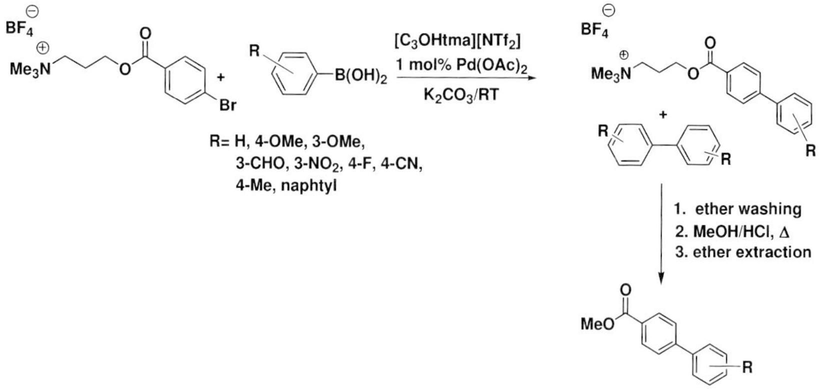

Scheme 5.5-39

usually prepared in molecular solvents such as dichloromethane, DMF, acetonitrile, to mention just a few. Recently, many chemical transformations of onium salts have been reported in solution in these solvents, although the authors claim that it is ionic liquid supported organic synthesis it would be more accurate to use the expression "task-specific onium salt supported organic synthesis" or "TSOS supported organic synthesis". This idea that TSOSs can be used as soluble supports for organic synthesis in molecular solvents has been patented [21]. Indeed, the use of solutions of TSILs or TSOSs in molecular solvents offers a wide and complementary broadening of the use of BTSILs.

Coupling reactions are an excellent field for testing new soluble supports. For example, a small library of stilbenes was easily obtained by the reaction of a supported styrene on a functionalized trimethylammonium tetrafluoborate with a mixture of 7 aryl iodides in DMF at 100 °C for 2 h. The reaction was monitored by HPLC and, after completion, the DMF was evaporated under vacuum and the residue washed several times with ether and water. Transesterification of this mixture with acidic methanol led, after evaporation and extraction with ether, to a library of 7 stilbenes in an excellent state of purity as determined by GC/MS. No impurity and particularly no starting styrene could be detected, showing that the reaction was complete (Scheme 5.5-40) [21, 51].

DMF has also been used as a solvent for Suzuki-Miyaura coupling. A supported 4-bromobenzoic ester in the presence of potassium carbonate was reacted with an array of aryl boronic acids in parallel. After 5 h at 80 °C, the reaction mixture was filtered, DMF evaporated and the products washed with ether, recrystallized and collected by filtration (Scheme 5.5-41) [51].

The results are displayed in Table 5.5-7 and show the efficiency of these systems.

Chan et al. illustrated the use of imidazolium tetrafluoborates functionalized with 3- and 4-iodobenzoic esters in Suzuki-Miyaura cross-coupling reactions [57]. These esters were prepared by esterification of 1-(2-hydroxyethyl)-3-methylimidazolium tetrafluoboratetetrafluoborate in a mixture of acetonitrile and dichloromethane

Table 5.5-7 Suzuki-Miyaura coupling in DMF

Entry	X	R	Yield (%)[a]	mp (°C)
1	[BF$_4$]$^-$	–C$_6$H$_5$	87	204–206
2	[BF$_4$]$^-$	–C$_6$H$_4$-OMe (o, m, p)	82	212–214
3	[NTf$_2$]$^-$	–C$_6$H$_4$-OMe (o, m, p)	73	nd[b]
4	[NTf$_2$]$^-$	–C$_6$H$_4$-CHO	65	138–140
5	[BF$_4$]$^-$	–C$_6$H$_4$-NO$_2$ (o, m, p)	75	146–148
6	[BF$_4$]$^-$	–C$_6$H$_4$-F (o, m, p)	82	132–134
7	[BF$_4$]$^-$	–C$_6$H$_4$-CN (o, m, p)	70	164–166

Table 5.5-7 Continued

Entry	X	R	Yield (%)	mp (°C)
8	[BF$_4$]$^-$	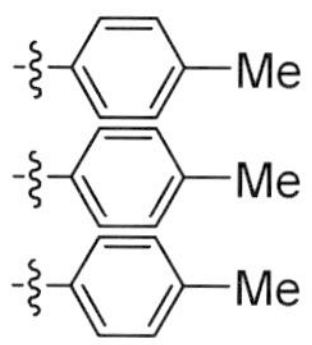 aryl-Me (three isomers shown)	82	198–200
9	[BF$_4$]$^-$	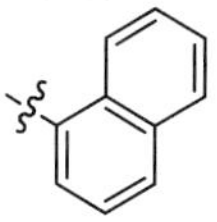 naphthyl	73	oil

^{a}Yields are for isolated pure recristallised compounds.
^{b}nd: not determined.

and were obtained, after the usual work-up, as solids melting at 113–115 °C for the 4-iodo and 76–79 °C for the 3-iodo derivatives (Scheme 5.5-42) [57].

Then, these iodobenzoates were submitted to Suzuki-Miyaura coupling with aryl boronic acids in water as the solvent in the presence of palladium acetate as the catalyst and cesium fluoride as the base. Some homocoupling products were washed out before cleavage by transesterification, extraction and purification of the biaryl products, which were obtained in yields ranging from 55 to 83% for the two steps. This methodology allows for the isolation of the intermediate supported biaryls as salts in a very good state of purity by simple evaporation of water followed by ether washing, which removed the excess of the starting boronic acid and the homocoupling product, leaving behind the supported coupled products.

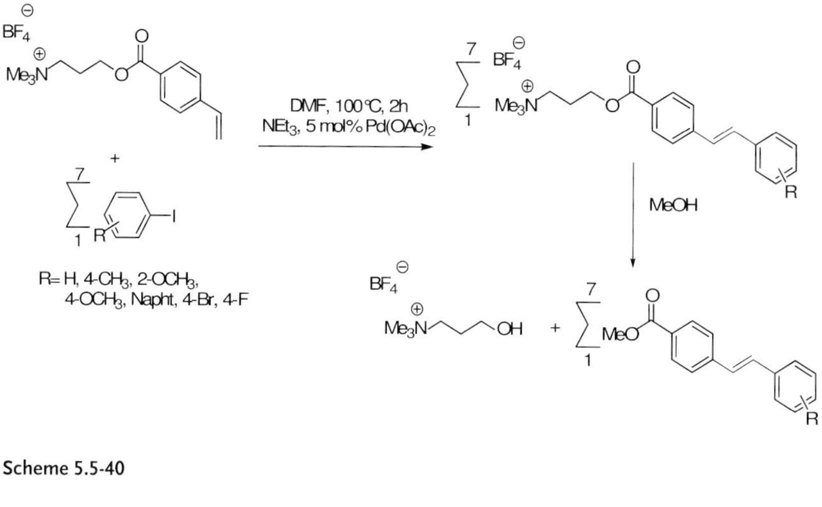

Scheme 5.5-40

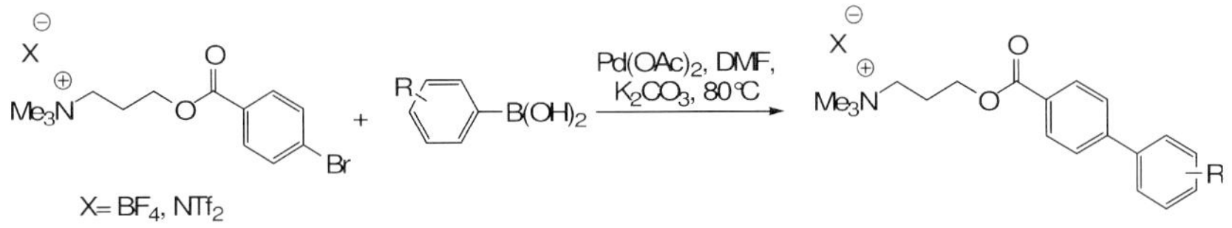

5.5 Task-specific Ionic Liquids as New Phases for Supported Organic Synthesis | 519

Scheme 5.5-41

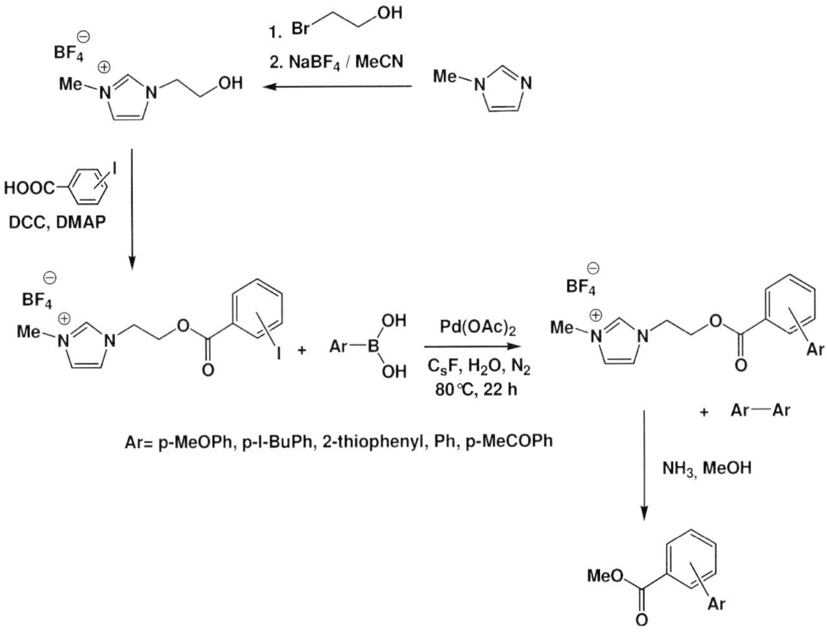

Scheme 5.5-42

Ar= p-MeOPh, p-I-BuPh, 2-thiophenyl, Ph, p-MeCOPh

The Sonogashira coupling reaction also illustrates the possibilities of this methodology. A small library of six supported aryl acetylenes was obtained according to Scheme 5.5-38. The supported propionic acid on the triethylammonium triflimide was esterified in MeCN in 90% yield. Coupling was realized in a parallel manner in MeCN, leading quantitatively to the aryl acetylenes in 20 min at room temperature (Scheme 5.5-43). The results are summarized in Table 5.5-8 [21, 51].

This kind of methodology has also been used for the preparation of thioureas (Scheme 5.5-44) [30].

1,4-addition of primary amines on supported acrylic esters has been performed in acetonitrile followed by the reaction of the secondary amines obtained with isothiocyanates to yield thioureas. Isolation of these compounds only requires solvent evaporation, washing with ether and drying under vacuum. Good yields were obtained.

Isothiocyanato esters appended to (polyethylene glycol)-ionic liquid phases have been prepared by nucleophilic substitution with potassium thiocyanate in refluxing

Table 5.5-8 Sonogashira coupling in CH$_3$CN at RT

Entry	Aryl iodide	Yield (%)[a]
1	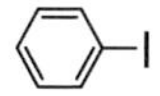 Ph–I	77
2	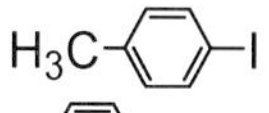 H$_3$C–C$_6$H$_4$–I	89
3	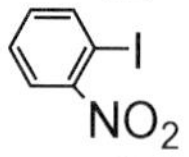 2-NO$_2$–C$_6$H$_4$–I	81
4	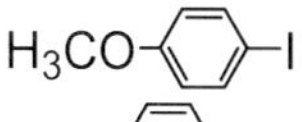 H$_3$CO–C$_6$H$_4$–I	92
5	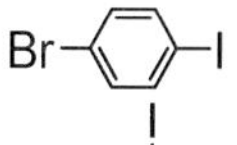 Br–C$_6$H$_4$–I	80
6	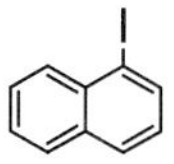 1-Naphthyl–I	86

[a] Yields refer to isolated pure products.

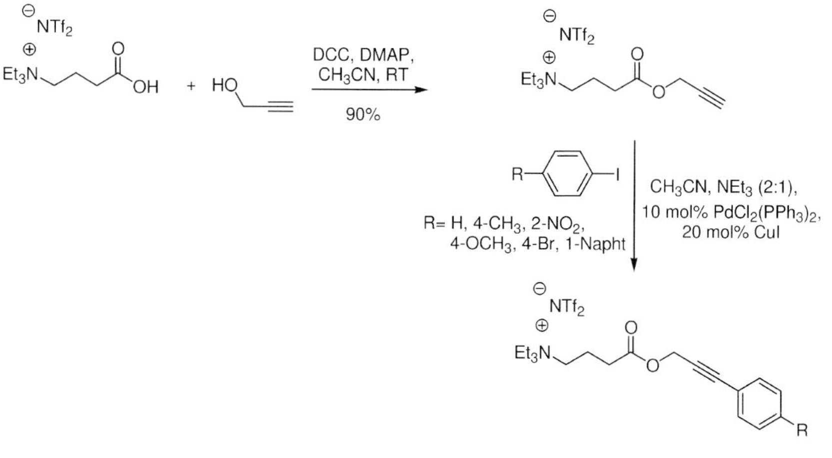

Scheme 5.5-43

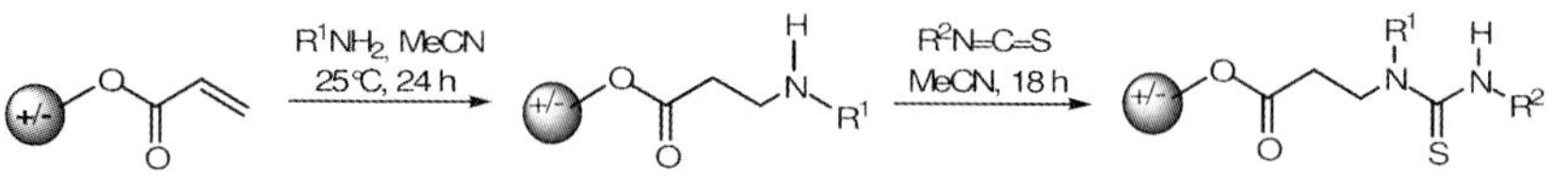

Scheme 5.5-44

anhydrous MeCN and isolated in very good yields after elimination of the solvent under vacuum, washing and further drying under vacuum (Scheme 5.5-45) [50].

Strategies for the synthesis of peptides supported on TSOS in both directions (C to N and N to C), and convergent as well, have been demonstrated [50]. Reverse peptide

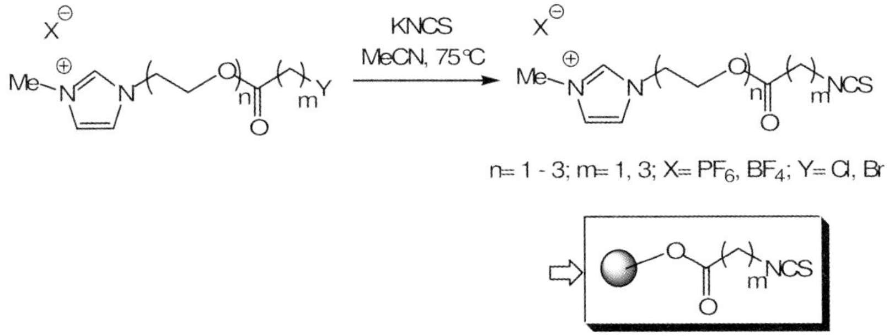

Scheme 5.5-45

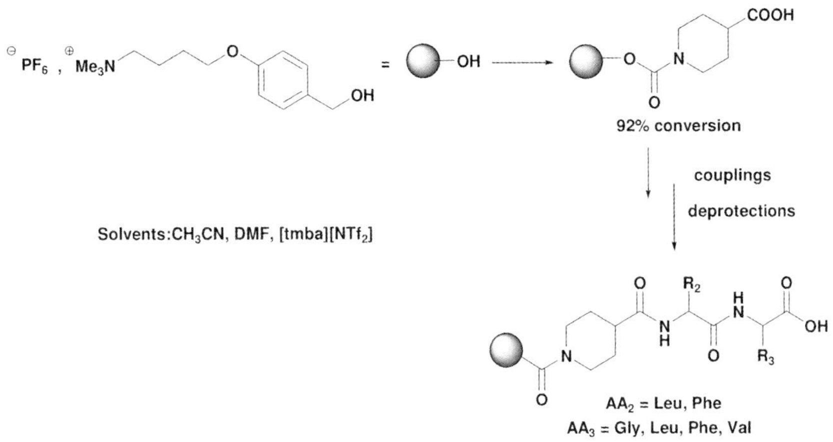

Scheme 5.5-46

synthesis to obtain C-terminally modified peptides has been validated by using a TSOS derived from a Wang-type linker. The first aminoacid, isonipecotic acid, was anchored via a mixed carbonate derived from p-nitrophenyl chloroformate (92%). The following step is a peptide coupling, which can be done using HOBt/DCC (or DIC) as reagents and only 1.5 equivalents of amino acid. The conversion is quantitative when the matrix is MeCN or [BuMe$_3$N][NTf$_2$]. Gly-OMe, Ala-OMe, Val-OMe, Leu-OMe, Phe-OMe were coupled as second aminoacids with yields of isolated supported peptides in the range of 85%. Methyl ester function is saponified with Me$_3$SiOK and completed within 2 h. A second peptidic coupling with either Gly-OMe, Leu-OMe, Phe-OMe or Val-OMe is realized to give supported tripeptides. It is then possible to cleave the peptides from the support with TFA or TMSBr (Scheme 5.5-46).

Direct peptide synthesis supported on TSOS has also been developed using a Fmoc strategy with a linker functionalized by a benzhydril alcohol. Grafting of the first aminoacid is followed by a metathesis of the chloride or bromide anion

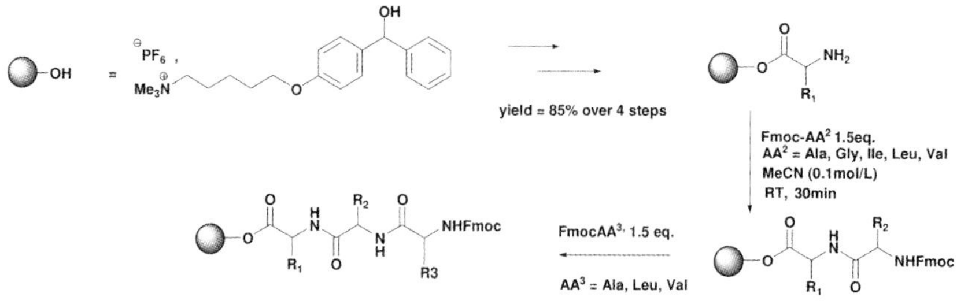

522 | 5 Organic Synthesis

Scheme 5.5-47

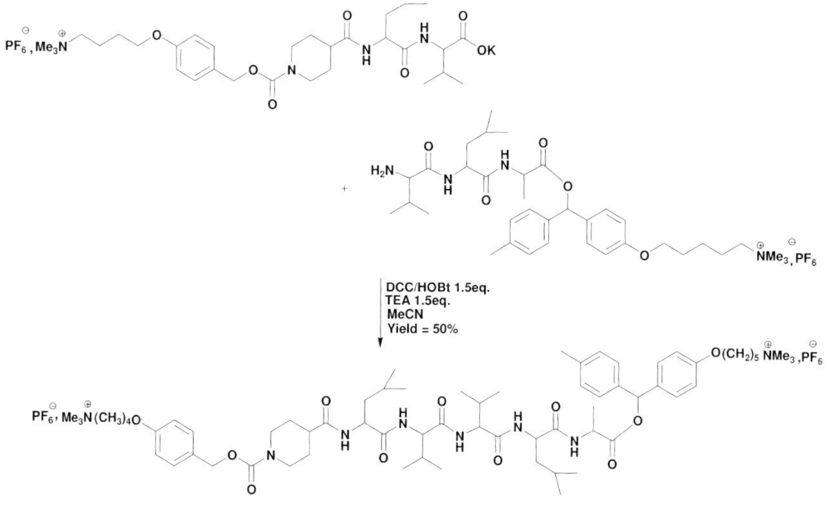

Scheme 5.5-48

to hexafluorophosphate and Fmoc deprotection. The supported aminoacids are isolated with an 85% yield over four steps. Ala, Phe, Gly, Ile, Leu and Val have been supported in this manner. A second aminoacid i.e. Fmoc-Ala, Fmoc-Gly, Fmoc-Ile, Fmoc-Leu and Fmoc-Val, is coupled following the same procedure and work-up. Either HBTU or DCC/HOBt can be used as coupling reagents. After Fmoc deprotection, the supported pure dipeptides are isolated with yields of around 85%. Further coupling under the same conditions leads to supported tripeptides with good yields. These peptides can be cleaved easily from the ionic support (Scheme 5.5-47).

Finally, convergent peptide synthesis under homogeneous conditions has been demonstrated: two supported peptides prepared by inverse and direct peptide synthesis have been successfully coupled to obtain a supported hexapeptide as a bis-ammonium salt (Scheme 5.5-48) [50].

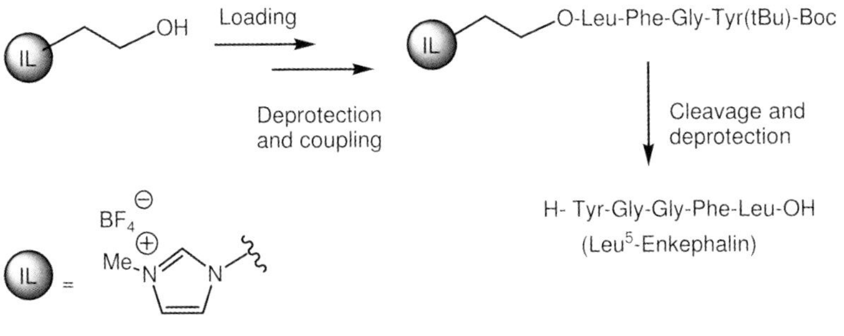

Scheme 5.5-49

The synthesis of a bioactive pentapeptide, Leu^5-enkephalin, in good yield and reasonable purity was developed on functionalized imidazolium salts in a MeCN/CH$_2$Cl$_2$ solvent mixture (Scheme 5.5-49) [59].

This approach, named ILSPS for ionic liquid supported peptide synthesis, offers potential advantages as well as limitations in comparison to the existing methods of peptide synthesis. One important advantage is that, at each step, the intermediate can be purified easily by solvent washings. High loading is an important feature of these supports due to their low molecular weight. This is interesting for large-scale preparation, as is their low cost. Furthermore, the structure and purity of each intermediate can be verified by routine spectroscopic methods and HPLC.

5.5.3.4 Lab on a Chip System Using a TSIL as a Soluble Support

A powerful tool to synthesize easily minute amounts of organic compounds on demand by using both ionic liquids droplets as microreactors and electrowetting as a fluidic motor has been described. These droplets can be moved, divided and combined on an open digital microfluidic lab-on-a-chip system [60]. This has been demonstrated with BTSILs used as reaction media and supports, properly functionalized to perform the Grieco's multicomponent synthesis of tetrahydroquinolines (Fig. 5.5-3). It is assumed that this original concept should impact many areas, notably combinatorial chemistry, parallel synthesis, optimization of protocols, synthesis of dangerous products and embedded chemistry in a portable device.

5.5.4
Conclusion

Although the chemistry of TSILs, BTSILs and TSOSs is still in its infancy, it can be safely stated that, generally speaking, they should have a bright future. Almost any functionalization of these salts is possible, therefore opening the way to task-specific materials and molecules. Numerous developments are predictable, not only in this area of supported chemistry, but in several other fields of chemistry including combinatorial chemistry, physical organic chemistry, microreactors and material sciences. It clearly appears from this review that TSILs, BTSILs and TSOSs are a

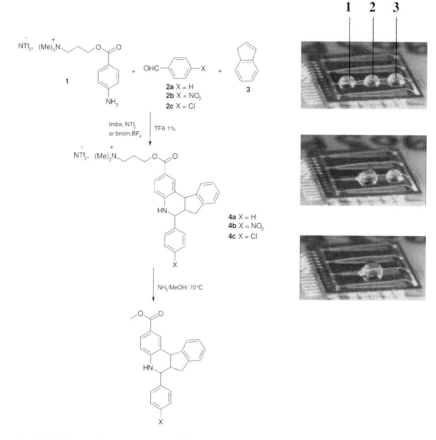

Fig. 5.5-3 Schematic representation of Grieco's reaction in a droplet microreactor. Photographs of droplet displacement during the Grieco's reaction are shown on the right of this figure. The droplet containing **1** was allowed to converge towards the droplet containing **2** and TFA. After mixing, the obtained droplet was directed towards the droplet containing **3**, thus leading quantitatively to adducts **4** which could be cleaved to the corresponding methyl esters by transesterification.

valuable addition to the arsenal of tools available for supported chemistry and in particular in the field of solution phase chemistry. Several advantages of these new supports have been pointed out in this chapter including simple synthesis, solution phase reaction conditions, flexibility, diversity of feasible chemistry, stability, high loading, easy monitoring of reactions, to mention but a few. Problems arise from availability from commercial sources, viscosity, chemical stability which depends very much on both the cations and the anions to be used, and sensitivity to moisture which requires special precautions. No doubt solutions will be found to all these problems.

References

1. Enabling technologies in synthesis found a larger forum at the Gordon Research Conference: Facilitating Organic Synthesis, 7–11.3. **2004**, Ventura Beach, CA, U.S.A..
2. H.-U. Blaser, Siegrist, H. M. Steiner, *Fine Chemicals through Heterogeneous Catalysis*, Wiley-VCH, Weinheim, **2001**, pp. 389; K. U. Schöning, N. End, *Top. Curr. Chem.* **2004**, *242*, 241–273 and 273–319; H.-U. Blaser, A. Indolese, F. Naud, U. Nettekoven, A. Schnyder, *Adv. Synth. Catal.* **2004**, *346*, 1583–1599.
3. Reviews on microwave assisted synthesis: T. Cablewski, A. F. Faux, C. R. Strauss, *J. Org. Chem.* **1994**, *59*, 3408–3412; *Microwaves in Organic Synthesis*, Loupy, A. (Ed) Wiley-VCH, Weinheim, **2002**; C. O. Kappe, *Curr. Opin. Chem. Biol.* **2002**, *6*, 314–320; B. Desai, C. O. Kappe, in *Top. Curr. Chem.* **2004**, *242*, 177–208; C. O. Kappe, *Angew. Chem. Int. Ed.* **2004**, *43*, 6250.
4. P. Wasserscheid, T. Welton (Eds.), *Ionic Liquids in Synthesis*, Wiley-VCH, Weinheim, **2003**.
5. Review on continuous flow reactors in organic chemistry: G. Jas, A. Kirschning, *Chem. Eur. J.* **2003**, *9*, 5708–5723.
6. Reviews on immobilized catalysts: Immobilized Catalysts, A. Kirschning (Ed.), *Top. Curr. Chem.* **2004**, *242*; W. Solodenko, T. Frenzel, A. Kirschning, in *Polymeric Materials in Organic Synthesis and Catalysis*, M. R. Buchmeiser (Ed.), Wiley-VCH, Weinheim, **2003**, pp. 201–240; B. Clapham, T. S. Reger, K. D. Janda, *Tetrahedron* **2001**, *57*, 4637–4662; Recoverable Catalysts and Reagents, J. A. Gladysz (Ed.), *Chem. Rev.* **2002**, *102*, 3215–3216; C. A. McNamara, M. J. Dixon, M. Bradley, *Chem. Rev.* **2002**, *102*, 3275–3300.
7. R. B. Merrifield, *J. Am. Chem. Soc.* **1963**, *85*, 2149–2154.
8. K. C. Nicolaou, R. Hanko, W. Hartwig, *Handbook of Combinatorial Chemistry*, Wiley-VCH, Weinheim, **2002**, Vol. 1 and 2.
9. J. D. Gravert, K. D. Janda, *Chem. Rev.* **1997**, *97*, 489–550; P. H. Toy, K. D. Janda, *Acc. Chem. Res.* **2000**, *33*, 546–554; T. J. Dickerson, N. N. Reed, K. D. Janda, *Chem Rev.* **2002**, *102*, 3325–3343.
10. S. Roller, C. Siegers, R. Haag, *Tetrahedron* **2004**, *60*, 8711–8720 and references cited therein.
11. I. T. Horvath, *Acc. Chem. Res.* **1998**, *31*, 641–650; D. P. Curran, *Angew. Chem. Int. Ed.* **1998**, *37*, 1175–1196.
12. K. E. Gutowski, G. A. Broker, H. D. Willauer, J. G. Swatloski, J. D. Holbrey, R. D. Rogers, *J. Am. Chem. Soc.* **2003**, *125*, 6632–6633.
13. A. Wierbicki, J. H. Davis Jr., in *Proceedings of the Symposium on Advances in Solvent Selection and Substitution for Extraction*, AIChe, New York, **2000**.
14. J. H. Davis Jr., K. J. Forester, *Tetrahedron Lett.* **1999**, *40*, 1621–1623.
15. S. W. Dzyuba, R. A. Bartsch, *Tetrahedron Lett.* **2002**, *43*, 4657–4660.
16. J. H. Davis Jr., *Chem. Lett.* **2004**, *33*, 1072–1077.
17. J. Fraga-Dubrueil, J. P. Bazureau, *Tetrahedron Lett.* **2001**, *42*, 6097–6100.
18. M. Vaultier, S. Gmouh, French Patent, FR 2 845 084, Application No. 2002-11910, WO 2004029004.
19. P. Bonhote, P. A. P. Dias, M. Armand, N. Papageorgiou, K. Kalyanasundaram, M. Graetzel, *Inorg. Chem.* **1996**, *35*, 1168–1178; J. H. Davis, C. M. Gordon, C. Hilgers, P. Wasserscheid, in *Ionic Liquids in Synthesis*, P. Wasserscheid, T. Welton (Eds.), Wiley-VCH, Weinheim, **2003**, Ch. 2.
20. J. Fraga-Dubrueil, M. H. Famelart, J. P. Bazureau, *Org. Process. Res. Dev.* **2002**, *6*, 374–378.
21. M. Vaultier, S. Gmouh, F. Hassine, French Patent FR 2 857 360, FR Application No. 2003-8413, WO2005005345.
22. J. Dolbrey, M. B. Turner, W. M. Reichert, R. D. Rogers, *Green Chem.* **2003**, *5*, 731–736.
23. J. Trotter, W. Darby, *Organic Synthesis*, Wiley, New York, **1955**, Collect. Vol. III, p. 460.
24. S. T. Handy, M. Okello, G. Dickenson, *Org. Lett.* **2003**, *5*, 2513–2515.

25. W. Bao, Z. Wang, Y. Li, *J. Org. Chem.* **2003**, *68*, 591–593.
26. P. Wasserscheid, A. Bösmann, C. Bolm, *Chem. Commun.* **2002**, 200–201.
27. D. Bates, R. D. Mayton, I. Ntai, J. H. Davis, Jr., *J. Am. Chem. Soc.* **2002**, *124*, 926–927.
28. K.W. Kottsieper, O. Stelzer, P. Wasserscheid, *J. Mol. Catal.* **2001**, *175*, 285–288.
29. P. Wasserscheid, B. Drießen-Hölscher, R. van Hal, H.C. Steffens, J. Zimmermann, *Chem. Commun.* **2003**, 2038–2039.
30. J. J. Hakou, J. Van den Eynde, J. Hamelin, J. P. Bazureau, *Tetrahedron* **2004**, *60*, 3745–3753.
31. S. T. Handy, M. Okello, *Tetrahedron Lett.* **2003**, *44*, 8399–8402.
32. J. S. Wilkes, in *Ionic Liquids in Synthesis*, P. Wasserscheid, T. Welton (Eds.), Wiley-VCH, Weinheim, **2003**, Ch. 1.
33. W. Miao, T. H. Chan, *Org. Lett.* **2003**, *5*, 5003–5005.
34. W. Miao, T. H. Chan, *J.Org. Chem.* **2005**, *70*, 3251–3255.
35. F. Yi, Y. Peng, G. Song, *Tetrahedron Lett.* **2005**, *46*, 3931–3933.
36. D. Dhar, M. Vaultier, unpublished results.
37. A. Ouadi, B. Gadenne, P. Hesemann, J.E. Moreau, I. Billard, C. Gaillard, S. Mekki, G. Moutiers, *Chem. Eur. J.* **2006**, *12*, 3074–3081.
38. Li Xue-Hui, Zhang Lei, Wang Le fu, Tang Ying Biao, *Wuli Huaxue Xuebao*, **2006**, *22* (4), 430–435. Li Xue-Hui, Jiang Yan Bin, Zhang Lei, Li Rong, *Wuli Huaxue Xuebao* **2006**, *22* (6), 747–751.
39. M. de Kort, A. W. Tuin, S. Kuiper, H. S. Overkleeft, G. A. van der Marel, R. C. Buisjsman, *Tetrahedron Lett.* **2004**, *45*, 2171–2175.
40. S. Anjaiaha, S. Chandrasekhar, R. Grée, *Tetrahedron Lett.* **2004**, *45*, 571–573.
41. C. C. Brasse, U. Englert, A. Salzer, H. Waffenschmidt, P. Wasserscheid, *Organometallics* **2000**, *19*, 3818; P. Wasserscheid, H. Waffenschmidt, P. Machnitzki, K. Kottsieper, O. Selzer, *Chem. Commun.* **2001**, 451–452; C. Baeizao, B. Gigante, H. Garcia, A. Corma, *Tetrahedron Lett.* **2003**, *44*, 6813–6816; N. Audic, H. Clavier, M. Mauduit, J. C. Guillemin, *J. Am. Chem. Soc.* **2003**, *125*, 9248–9249; T. J. Geldbach, P. J. Dyson, *J. Am. Chem. Soc.* **2004**, *126*, 8114–8115.
42. S. T. Handy, M. Okello, *J. Org. Chem.* **2005**, *70*, 2874–2877.
43. J. Fraga-Dubreuil, J. P. Bazureau, *Tetrahedron* **2003**, *59*, 6121–6130.
44. F. Yi, Y. Peng, G. Song, *Tetrahedron Lett.* **2005**, *46*, 3931–3933.
45. J. C. Legeay, J. J. V. Eynde, J. P. Bazureau, *Tetrahedron* **2005**, *61*, 12386.
46. Y. Peng, F. Yi, G. Song, Y. Zhang, *Monatsh. Chem.*, **2005**, *136*, 1751–1755.
47. J. Y. Huang, M. Lei, Y. G. Wang, *Tetrahedron Lett.* **2006**, *47*, 3047.
48. Y. Hu, P. Wei, H. Huang, S. Q. Han, P. K. Ouyang, *Heterocycles*, **2006**, *68*, 375–380.
49. F. P. Yi, Y. Q. Peng, G. H. Song, *Huaxue Xuebao* **2006**, *64*, 1145–1150.
50. C. Roche, A. Commerçon, M. Vaultier, French Patent, FR 2 882 057, date 2006/08/18, Application No. FR 2005-1557, date 2005/02/16.
51. F. Hassine, Ph.D.thesis, University Hassan II, Casablanca, Marocco, May 2005.
52. For acid sensitive methoxy benzaldehyde.
53. A. M. Fivush, T. M. Wilson, *Tetrahedron Lett.* **1997**, *38*, 7151–7154.
54. J. Zhu, H. Bienayme, *Multicomponent Reactions*, Wiley-VCH, Weinheim, **2005**; I. Ugi, C. Steinbrückner, *Chem. Ber.* **1961**, *94*, 734: Kolb, B. Beck, A. Dömling, *Tetrahedron Lett.* **2002**, *43*, 6897; A. Fayol, J. Zhu, *Org. Lett.* **2005**, *7*, 239.
55. P. Grieco, A. Bahas, *Tetrahedron Lett.* **1988**, *29*, 5855–5858; A. Kiselyov, R. W. Armstrong, *Tetrahedron Lett.* **1998**, *54*, 5089–5096.
56. T. Welton, P. J. Smith, *Adv. Organomet. Chem.* **2004**, *51*, 251–284; V. Calo, A. Nacci, A. J. Monopoli, *J. Mol. Catal. A: Chem.* **2004**, *214*, 45–56; P. Wasserscheid, in *Ionic Liquids in Synthesis*, P. Wasserscheid, T. Welton (Eds.) Wiley-VCH, Weinheim, **2003**, pp. 213–257; J. Dupont, R. F. de Souza, P. A. Z. Suarez, *Chem. Rev.* **2002**, *102*, 3667–3691.
57. W. Miao, T. H. Chan, *Org. Lett.* **2003**, *5*, 5003–5005.
58. H. Hakkou, J. J. Vanden Eynde, J. Hamelin, J. P. Bazureau, *Synthesis*, **2004**, 1793–1798.

59. W. Miao, T. H. Chan, *J. Org. Chem.* **2005**, 70, 3251–3255.
60. P. Dubois, G. Marchand, Y. Fouillet, J. Berthier, T. Douki, F. Hassine, S. Gmouh, M. Vaultier, *Anal. Chem.* **2006**, 78, 4909–4917. French Patent FR2 872 715, WO 2006018560.

5.6
Supported Ionic Liquid Phase Catalysts

Anders Riisager and Rasmus Fehrmann

5.6.1
Introduction

An important concept in catalysis which attracts increasing attention involves the transformation of homogeneous catalytic reactions into heterogeneous processes where the catalyst can be easily separated from the reaction mixture, allowing reuse and continuous flow operation [1]. In the literature several general reviews have dealt with the subject [2–5] along with special surveys focusing on issues such as multiphase catalytic approaches [6–10], organometallic catalysis [11], enantioselective catalysis [12], oxidation catalysis [13], and Lewis acid catalysis [14, 15].

In Section 5.3 it was demonstrated with many examples that ionic liquids are indeed a very attractive class of solvents for catalysis in liquid–liquid biphasic operation (for some selected reviews see Refs. [16–20]). In this section, we will focus on a different way to apply ionic liquids in catalysis, namely the use of an ionic liquid catalyst phase supported on a solid carrier, a technology that has become known as "supported ionic liquid phase (SILP)" catalysis. In comparison to the conventional liquid–liquid biphasic catalysis in ionic liquid–organic liquid mixtures, the concept of SILP-catalysis combines well-defined catalyst complexes, nonvolatile ionic liquids, and porous solid supports in a manner that offers a very efficient use of the ionic liquid catalyst phase, since it is dispersed as a thin film on the surface of the high-area support. Recently, the initial applications using such supported ionic liquid catalysts have been briefly summarized [21]. In contrast to this report, where the applications were distinguished by the choice of support material, the compilation here will divide the applications using the supported ionic liquid catalysts into sections according to the nature of the interaction between the ionic liquid catalyst phase and the support.

5.6.2
Supported Ionic Liquid Phase Catalysts

5.6.2.1 Supported Catalysts Containing Ionic Media
The idea to immobilize catalytically active salts or salt solutions on a support to obtain a solid heterogeneous-like catalyst system, maintaining the selectivity and

efficiency of the homogeneous catalyst, is far from new. The principle of the concept was already noted in 1968 [22] and has, for example, later been realized for organic reactions using the methodology of gas–liquid phase-transfer catalysis (GL-PTC) [23]. In GL-PTC, organic reactions are conducted by a continuous-flow procedure in which the organic, gaseous reactants are converted by passing the gas through a liquid phase-transfer catalyst supported on a porous, solid material (typically organic polymers or inorganic oxides with low surface acidity). Usually, the liquid phase-transfer catalyst is composed of relatively stable and non-volatile, low melting tetraalkyl phosphonium salts (e.g. [Bu$_4$P]Br with a melting point of 103 °C) with or without the presence of additional co-catalysts, providing activation of anions to carry out reactions such as halogen exchange, forming alkyl halides from alcohols, transesterification and isomerization [24, 25].

In a more fundamental perspective the concept of confining catalytically active salts on supports may be regarded as a branch of the common concept of supported liquid phase (SLP) catalysts, where a non- or low-volatile catalyst solution of, for example, pyrosulfates [26], hydrophilic polymer glycols [27–29], phthalates [22] or water (i.e. supported aqueous phase, SAP) [30–33] are deposited on a high-surface area porous support.

In the literature terms such as supported molten salt (SMS) catalysts, supported ionic liquid catalysts (SILC) and supported ionic liquid-phase (SILP) catalysts, have been used somewhat indiscriminately to describe catalyst systems containing a catalytic ionic phase. In this section we will use the terms "molten salt" or "ionic liquid" to indicate the melting point of the fluid phase in the systems. Furthermore, we will distinguish between the terms "SILC" and "SILP." "SILP" is used when the ionic liquid is performing mainly as an immobilizing solvent for the catalytic components. "SILC" is used in cases where the ionic liquid itself, ionic liquid ions or ionic liquid-like fragments are behaving as the catalytic species.

5.6.2.1.1 Process and engineering aspects of supported ionic liquid catalysts

During catalytic reactions using supported ionic liquid-type catalysts gaseous or vapor-phase reactants diffuse through the residual pore space of the catalyst, dissolve in the liquid catalyst phase, and react at catalyst sites within the thin liquid catalyst film dispersed on the walls of the pores in the support material, as illustrated in Fig. 5.6-1. The products then diffuse back out of the catalyst phase into the void pore space and further out of the catalyst particle.

In general, the catalytic performance of SLP-type catalysts depends considerably on the amount of the liquid solvent in the supported catalyst, that is liquid loading (α, defined as the ratio between the liquid volume and support pore volume). Thus, on the one hand the amount of liquid must be large enough to constitute a thin film on the solid, while on the other hand, still low enough to avoid pore plugging that would restrict the diffusion of the reactants. As a consequence, the catalyst layer is often 20 Å or less, corresponding to a diffusion layer of only a few molecules thick. This makes diffusion problems (which are regularly found for reactions in bulk liquids, and especially with the relatively viscous ionic liquids) very unlikely. In addition, the distribution of the liquid inside the porous support is also of

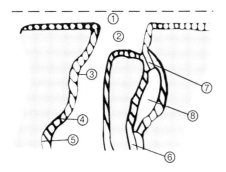

①	Boundary layer
②	Pore space
③	Gas-liquid-interface
④	Liquid film
⑤	Liquid-solid-interface
⑥	Filled pore
⑦	Blocking of pores
⑧	Blocked pore space

Fig. 5.6-1 The principle of a supported liquid phase (SLP) catalyst [34].

great importance and may generally be determined by several interacting factors including, for example, pore radii distribution of the support, the capillary forces, the wetting characteristics of the liquid and possible ionic liquid–support interactions [34]. Moreover, percolation theory may be used to model the liquid distribution in SLP catalysts [35]. More details of the behavior and catalytic performance of SLP-type catalysts with respect to liquid distribution and diffusion can be found in the literature (e.g. Refs. [22, 26, 36–40]).

The concept of using supported ionic liquid catalysts combines the most attractive features of homogeneous catalysis such as the uniform nature of the catalytic centers, high specificity and selectivity of the catalyst, with the most attractive features of heterogeneous catalysts, such as large interfacial reaction area, ease of separation of products and catalyst, high system stability and the potential to use very simple processing such as fixed-bed reactor technology. This quite unique combination has made the methodology very attractive for many applications, as will be illustrated in the following sections.

It should, however, be pointed out that the SILP and SILC techniques are only straightforward to apply for continuous gas-phase reactions, where the advantage of using a liquid with very low vapor pressure can be fully exploited. In contrast, the application of the technology in liquid-phase reactions may be very restricted, since even a minor solubility of the ionic liquid in the feedstock/product mixture will remove the catalyst from the surface (due to the very small amount of ionic liquid on the support). Even worse, the film of the immobilized catalyst phase can physically be removed from the support by mechanical forces, for example, the convective liquid flow, even in the case of complete immiscibility.

5.6.2.1.2 Characteristics of ionic liquids on solid supports

In contrast to the increasing information on the bulk properties of ionic liquids, very little is known about the interfacial structure of ionic liquids with other phases. Examinations of gas–ionic liquids interfaces of 1,3-dialkylimidazolium-based liquids using direct recoil spectroscopy [41] or surface vibrational spectroscopy [42–44] have provided some general indicative information about the surface composition and orientation of the imidazolium cations with respect to the interface. Here, it was found that the preferred orientation of the imidazolium cations at the interface could be either perpendicular or parallel (with some tilting and twist of the ring structure with respect to the interface) depending on factors such as water in the ionic liquid, the length of the alkyl chains in the imidazolium cations, and anion interactions. Moreover, the examinations performed with direct recoil spectroscopy suggested that the surface of the ionic liquids is ionic in nature and equally occupied by the anions and cations with no segregation [44].

With respect to the interfacial structure of ionic liquids with solids, the molecular layering and local order in thin liquid crystalline films of [RMIM][PF$_6$] ionic liquids with long alkyl chains (e.g., R = dodecyl and octadecyl) on solid silicon supports have been studied using X-ray reflectivity [45]. Here, thin films of the crystalline ionic liquids with a thickness of 100–210 A were deposited on polished and cleaned silicon wafers by initial spin-coating of a solution of the ionic liquids in methanol followed by heating the samples into the isotropic liquid state and crystallization.

In all cases, the initial spin-coated films showed a featureless smooth decay profile in the X-ray reflectivity data, indicating that the ionic liquid was not deposited in a regular, ordered manner (i.e. rough films). On melting, however, thin homogeneous, isotropic liquid films were observed which, upon further cooling to crystallization, displayed sharp Bragg diffraction peaks in the reflectivity profiles, indicative of an ordered structure. Furthermore, modeling of data obtained from an ordered film of, for example, 1-methyl-3-octadecylimidazolium hexafluorophosphate, suggested, that a layered ionic liquid structure adopted on the silicon consisted of a multilayer stack of interdigitated amphiphillic bilayers with charged layers at the silicon and air interfaces, that is, the surface of the crystalline liquid is predominately ionic in character. This result is in accordance with analogous results obtained from measurements of gas–liquid interfaces by direct recoil spectroscopy [44].

In another study which used a support more relevant to supported ionic liquid catalysts, sum-frequency vibrational spectroscopy (SFVS) was used to study the structure of ionic liquid cations present at the interface between a silica support and imidazolium ionic liquids [RMIM][X], with R = hexyl, octyl or decyl, and X = $(CF_3(CF_2)_nSO_2)_2N$ with $n = 0$ or 1 [46]. The measurements were performed by introducing the ionic liquids in a thoroughly cleaned fused silica prism.

For all the examined imidazolium-based ionic liquids, the ring structure of the planar imidazolium cation (based on the HCCH-ring dipole orientation) was found to be positioned in a tilted, vertical position at an angle (θ) of 16 to 32° with respect to the silica surface normal (see Fig. 5.6-2). This is in contrast to the parallel orientation found in studies of similar ionic liquids at the air–ionic liquid interface [41].

Furthermore, the orientation of the alkyl chain (i.e. hexyl, octyl or decyl chain) on the imidazolium cation was pointing in a nearly perpendicular orientation away from the silica surface with a high degree of conformational order and few gauche defects. In Table 5.6-1 some of the structural parameters determined for the examined ionic liquids are summarized.

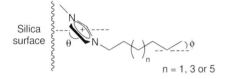

Fig. 5.6-2 Schematic illustration of the orientation of imidazolium cations at the surface of silica determined by sum-frequency vibrational spectroscopy [46].

Table 5.6-1 Tilt angles for the terminal methyl and imidazolium ring determined from the SFVS intensity data [46]

	Alkyl-CH$_3$ tilt angle, ø (°)	Alkyl chain tilt[a] (°)	HCCH tilt angle, θ (°)
[HMIM][(CF$_3$SO$_2$)$_2$N]	36 ± 4	1 ± 4	30 ± 4
[OMIM][(CF$_3$SO$_2$)$_2$N]	38 ± 8	3 ± 8	21 ± 8
[decylMIM][(CF$_3$SO$_2$)$_2$N]	39 ± 6	4 ± 6	32 ± 5
[HMIM][(CF$_3$CF$_2$SO$_2$)$_2$N]	25 ± 5	10 ± 5	16 ± 6
[OMIM][(CF$_3$CF$_2$SO$_2$)$_2$N]	32 ± 4	3 ± 4	17 ± 5
[decylMIM][(CF$_3$CF$_2$SO$_2$)$_2$N]	42 ± 10	7 ± 10	20 ± 4

a Assuming an all-trans chain conformation.

In addition to the interfacial structure of the imidazolium cation, the organization of water at the surface was also determined using water-equilibrated ionic liquids. Here, nearly identical ice-like and "water-like" features were observed for both the water–silica surface and the water-containing ionic liquid–silica surface. Moreover, the water molecules associated with the ionic liquids at the silica interface were hydrogen-bonded in small, structurally defined configurations with the bis(perfluoroalkylsulfonyl)imide anions, consistent with previous infrared spectroscopic studies of similar materials [47].

In summary, the few initial studies conducted so far on the interfacial structure between ionic liquids and solid surfaces clearly suggest – in a more general context – that confinement of ionic liquids on solid surfaces definitely induces some distinctive, structural control of the molecular layering of the ionic liquid and of the ionic liquid distribution (i.e. wetting ability). However, further investigations are surely needed to elucidate these effects in the context of supported ionic liquid catalysis in more detail.

5.6.2.2 Early Work on Supported Molten Salt and Ionic Liquid Catalyst Systems
5.6.2.2.1 High-temperature supported molten salt catalysts

Development and applications of supported ionic salt catalyst systems can generally be divided into periods, which are closely related to the liquid temperature range of the ionic salts used (Fig. 5.6-3). Since low-melting ionic liquids have only been prepared relatively recently, most applications using these supported molten salt

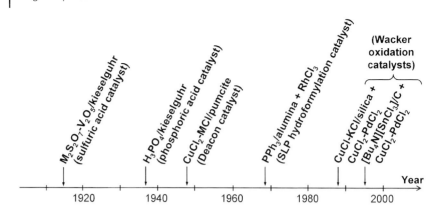

Fig. 5.6-3 Development of supported ionic salt catalyst systems.

catalysts have been restricted to temperatures higher than 200 °C and hence to catalyst systems which are stable at these temperatures.

The first supported molten salt catalyst systems date from 1914, where BASF filed a patent on a silica-supported V_2O_5-alkali pyrosulfate sulfur dioxide oxidation catalyst [48], which even today – as a slightly modified catalyst system – is still the preferred catalyst for sulfuric acid production [49]. However, it took many years to realize in the 1940s [50, 51], that the catalyst system actually was a molten salt SLP-type system which is best described by a mixture of vanadium alkali sulfate/hydrogensulfate/pyrosulfate complexes at reaction conditions in the temperature range 400–600 °C with the vanadium complexes playing a key role in the catalytic reaction [49].

In the period from the late 1930s to the late 1970s several other molten salt catalyst systems (mostly with salts melting in the range 200–500 °C) have been prepared and applied for various organic reactions, for example, condensations, cracking reactions, isomerizations, halogenations and oxidations (for more details see Ref. [52] and references cited therein). In many cases these supported molten salt catalysts proved to be highly reactive but also quite corrosive catalyst phases. Some of the most important supported catalyst systems developed during this period include the industrial Deacon catalyst [26] and the related oxychlorination catalyst systems [52] used for the oxidation of hydrogen chloride and the oxidative chlorination of unsaturated hydrocarbons (e.g. ethylene), respectively. Generally the catalysts are composed of a supported molten salt mixture of $CuCl_2$–$CuCl$/alkali metal chlorides/rare earth chlorides often promoted with other transition metal chlorides (e.g. $ZnCl_2$ and $FeCl_3$) for the oxychlorination reactions. Also the solid silica- or kiselguhr-supported phosphoric acid catalyst used for petrochemical reactions such as, for example, olefin oligomerization and alkylations, was developed during these years after their initial invention in the 1930s [26]. In 1970, systems of supported cupper oxide and mixed CuO/M-oxide (M = Cr, Co, Mn or V) was reported as hydrocarbon oxidation catalysts [53]. The systems were generated by thermal treatment of an eutectic mixture of alkali metal salts containing copper salts (e.g. $CuSO_4$–$NaHSO_4$–$KHSO_4$, melting point <160 °C), on high-surface area alumina or silica. Noticeably, only the calcinated catalysts were applied for oxidation reactions and not the supported liquid salt pre-catalysts themselves. A few years later, however, the use of analogous supported eutectic salt catalyst systems containing alkali metal halides and transition metal halides such as, for example, KCl–LiCl–$MnCl_2$ supported on alumina, were applied by Monsanto [54] for the formation of C_2–C_4 alkenes by gas-phase dehydrogenation of the corresponding alkanes at temperatures above the temperature of fusion of the eutectics of the systems.

5.6.2.2.2 Low-temperature supported catalysts

An early and possibly the first example of a supported catalyst consisting of a relatively low-melting salt comprising a transition metal species was reported for aerial partial oxidation of ethylene to acetaldehyde (Wacker oxidation) [55]. The catalyst contained $CuCl_2$–$PdCl_2$ dissolved in an eutectic CuCl/KCl melt (65:35 mol%) on a porous silica gel support, and proved stable during 150 h continuous reaction (195 °C, 16 bar), providing a product selectivity for acetaldehyde of 95%. In contrast, an analogous water–ethylene glycol (50:50 vol.%) catalyst system deactivated fast during reaction due to evaporation of the volatile liquid solvent, clearly indicating the importance of using a low-volatile reaction medium for the continuous process. Hence, although the melting point of the supported molten salt (melting point 150 °C) was higher than defined for an ionic liquid, the potential of such supported catalyst systems was clearly documented at this early stage.

In a later work, both the CuCl/KCl molten salt Wacker oxidation system and a $[Bu_4N][SnCl_3]$ system (melting point 60 °C) was applied to the electrocatalytic generation of acetaldehyde from ethanol by co-generation of electricity in a fuel cell [56]. In the cell set-up, porous carbon electrodes supported with an ionic liquid catalyst electrolyte were separated by a proton conducting membrane (Fig. 5.6-4), and current efficiency and product selectivity up to 87% and 83%, respectively, were reported at 90 °C.

In addition to the Wacker oxidation catalysts, supported eutectic molten salt CuCl/KCl-based catalyst systems have also been examined for other processes including, for example, production of synthesis gas from methanol for the use as on-board hydrogen production in vehicles [57] and quantitative combustion of chlorinated hydrocarbons to CO_x and HCl/Cl_2 at ambient pressure (200–500 °C) with silica-based systems [58, 59].

Silica-supported catalyst systems comprised of tetra-n-butylammonium chloride and $PdCl_2$ with $CoCl_2$/$CuCl_2$ promotors (melting points ca. 60 °C) have also been used for hydrodechlorination of chloroform with hydrogen at 90–150 °C [60]. As one reason for the observed catalyst deactivation the authors propose the thermal ionic liquid decomposition ($[Bu_4N]Cl$, $T_d = 170$ °C), which seems very likely since tetraalkyl ammonium salts are known to undergo dealkylation under the applied conditions [61, 62].

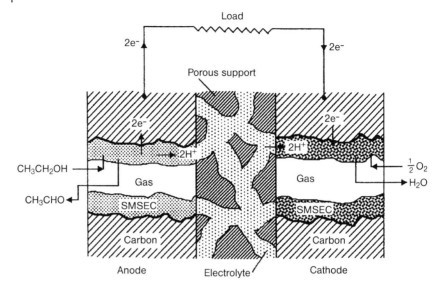

Fig. 5.6-4 Schematic illustration of a supported ionic liquid fuel cell containing the Wacker oxidation system (SMSEC: supported molten salt electro-catalyst) for co-generation of acetaldehyde and electricity from ethanol [56].

5.6.2.3 Ionic Liquid Catalysts Supported through Covalent Anchoring
5.6.2.3.1 Supported Lewis acidic chlorometalate catalysts

Supported Lewis acidic ionic liquid catalysts were first introduced for selective low-temperature alkylations of olefins to generate hydrocarbon fuels in two-phase systems [63–65]. Later, related systems were developed for Friedel-Crafts alkylations in liquid- and gas-phase reactions [66, 67] and for gas-phase disproportionation of C_5-feedstocks [68]. Generally, the catalysts were prepared by confining pre-formed, catalytically active ionic liquids on supports (e.g. silica) by impregnation. In a typical Friedel-Crafts catalyst preparation a previously dried support was treated with a chloroaluminate ionic liquid until its appearance changed from dry to wet powder [69, 70]. Then the mixture was stirred for an extended period of time before the excess ionic liquid was removed via Soxhlet extraction and vacuum dried. Mostly 1,3-dialkylimidazolium or pyridinium chloroaluminate ionic liquids were used, which induced reactions between the chloroaluminate anions and the support surface OH-groups, resulting in covalent anchoring of anions to the support accompanied by HCl liberation ((Scheme 5.6-1), top) [67, 69]. In principle, this method allows the cation to be widely varied, while the anion should have a high affinity for oxygen to remain surface bonded. Otherwise, leaching can be expected in liquid phase reactions, as has been demonstrated in acylation reactions using analogous silica- and charcoal-supported SILC materials containing less Lewis acidic and less oxophilic chloroferrate anion [71].

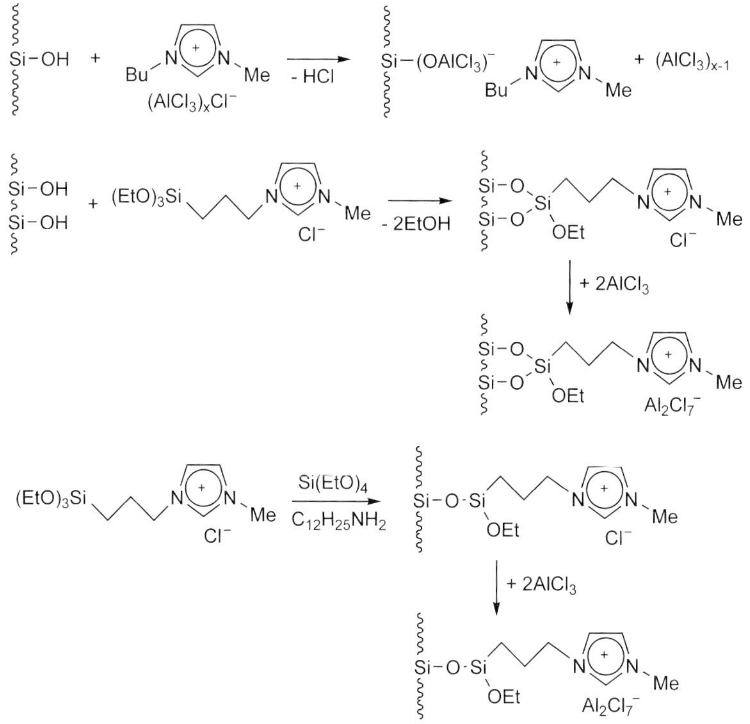

Scheme 5.6-1 Preparation of supported Lewis acidic chloroaluminate ionic liquids catalysts by impregnation with [BMIM][(AlCl$_3$)$_x$Cl] (top), by grafting of 1-(triethoxysilylpropyl)-3-methylimidazolium chloride followed by AlCl$_3$ addition (middle), and by sol–gel formation followed by AlCl$_3$ addition (bottom) [73].

Depending on the characteristics of the support materials different amounts of ionic liquid could be maintained on the surface following the order SiO$_2$ > Al$_2$O$_3$ > TiO$_2$ > ZrO$_2$ after thorough extraction. As expected, low surface area and low surface OH group density in the latter supports result in poor loadings compared to silica and alumina. More importantly, pore damage was observed by X-ray diffraction in some structured supports, apparently by the hydrochloric acid formed in the catalyst preparation method. The possible effect on the catalytic performance of the catalysts in the presence of residual HCl from the preparation was not documented, but it can be assumed to be beneficial for the examined Friedel-Crafts reaction. Moreover, it was shown by leaching studies and examination of reaction filtrates that the ionic liquid catalyst loss was negligible. Consequently, the observed catalyst deactivation in recycling experiments was ascribed to moisture-induced catalyst degradation and the possible adsorption of reactants and by-products [67].

In an alternative synthetic approach, the formation of hydrochloric acid was avoided. The supported Lewis acidic ionic liquids were prepared by an immobilization route involving the confinement of the ionic liquid's cation [73]. Using this method, the cation of, e.g. a chloride ionic liquid modified with an alkoxysilyl group, was covalently attached (i.e. grafted) to the surface. Then the chloride anion was transformed by $AlCl_3$ to the Lewis acidic anion $[Al_2Cl_7]^-$ ((Scheme 5.6-1), middle). Soxhlet extraction of the supported ionic liquid catalyst after treatment with the aluminum chloride removed most of the unreacted reagent, and XRD analysis confirmed far less damage to structured supports than previously observed for the technique releasing HCl during synthesis.

Comparison of results obtained from the alkylation of benzene with dodecene with the resulting catalysts made by the two immobilization methods revealed, in particular, excellent activity and selectivity for the materials obtained by cation grafting. Furthermore, these initial – but highly promising – results have been accompanied by an increased industrial interest for commercialization of these grafted ionic liquid catalyst systems. Hence, Johnson Matthey Catalysts announced (ICC Paris, July 2004) the commercialization of such grafted ionic liquid alkylation/acylation catalysts in late 2004, through their associated chemical company Alfa Aesar [74].

Alternatively, a sol–gel synthetic strategy has been reported for the preparation of a closely related type of supported Lewis acidic ionic liquid catalyst material (Scheme 5.6-1, bottom), which subsequently was used for Friedel-Crafts alkylations of benzene with dodecene [72]. Preparation first involved the formation of MCM-41 silica support material containing cationic imidazolium groups bonded to surface silicon atoms, by hydrothermal reaction between 1-(triethoxysilyl)propyl-3-methylimidazolium chloride and a mixture of tetraethylorthosilicate (TEOS) and dodecylamin. After removal of the amine template by Soxhlet extraction, $AlCl_3$ was introduced, leading to the Lewis acidic sol–gel catalyst. Using a comparable approach, ordered mesoporous silicas with hexagonal or lamellar structures incorporating similar covalently bound ionic liquid species have also been synthesized afterwards [75].

Besides the already mentioned acidic aluminum chloride catalysts, alternative Friedel-Crafts catalysts such as supported acidic tin catalysts have also been developed. The tin-based catalysts were prepared by a method which closely resembled the already mentioned two-step grafting method devised for the aluminum chloride catalyst. Here, $SnCl_4$ was anchored on silica materials modified with tetraalkylammonium chloride moieties obtained for example, from reaction with [3-(trimethoxysilyl)propyl]octadecyldimethylammonium chloride, thereafter, reaction of the Lewis acid with the chloride moieties leads to formation of pentacoordinated anionic tin species forming catalytically active complexes (i.e. $[R_4N][SnCl_5]$ species), associated with the surface. The supported tin catalysts were employed for condensation reactions of olefins with aldehydes forming unsaturated alcohols (Prins condensation, Scheme 5.6-2) [76].

The condensation reactions between formaldehyde and isobutene using the acidic supported tin catalysts generally resulted in high selectivity (88–93%) for formation of the preferred product 3-methyl-3-butene-1-ol (MBOH). In addition, the catalysts were resistant to leaching of the active tin species showing no significant

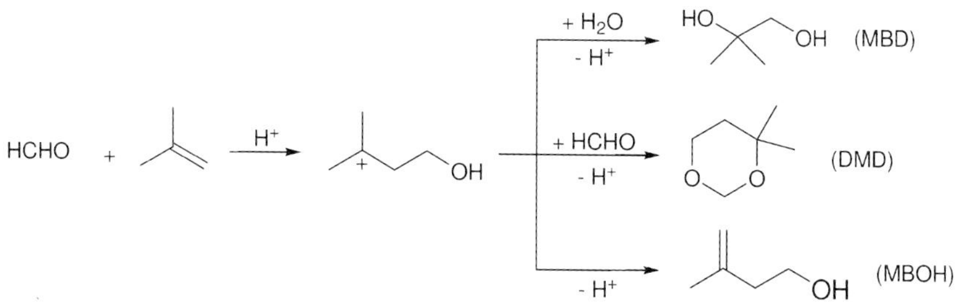

Scheme 5.6-2 Prins condensation of isobutene with formaldehyde (MBD: 3-methylbutane-1,3-diol, DMD: 4,4-dimethyl-1,3-diol, MBOH: 3-methyl-3-buten-1-ol catalyzed by tin(IV) chloride species [76].

decrease in activity in two consecutive runs and no activity of the product filtrate. In comparison, catalysts prepared by immobilizing $SnCl_4$ directly on silica or by immobilizing preformed [(propyl)$_4$N][SnCl$_5$] on silica gave lower product selectivity. In addition, there was considerable leaching of tin species from the $SnCl_4$-silica catalysts.

5.6.2.3.2 Neutral, supported ionic liquid catalysts

A non-acidic supported palladium-based ionic liquid catalyst has also been prepared using the approach of attaching an ionic liquid cation to a support, followed by impregnation with a palladium complex. The catalyst was made by initial grafting of 1-(trimethoxysilylpropyl)-3-methylimidazolium chloride onto pre-dried silica followed by impregnation with $PdCl_2$ [77] (Scheme 5.6-3). For comparison, silica-supported $PdCl_2$ catalyst and the homogeneous [BMIM]$_2$[PdCl$_4$] catalyst were also prepared.

The catalyst could be reused in three consecutive runs for Suzuki coupling reactions between derivates of bromobenzenes and phenylboronic acid in *m*-xylene in the presence of different bases (e.g. K_2CO_3), producing biphenyls in almost unchanged yields of 89–94%. The comparable yield using the supported $PdCl_2$ catalyst was only 81%.

Structural parameter derivation from curve-fitting analysis of EXAFS spectra of the supported $PdCl_2$ catalyst and the attached Pd catalyst, respectively, established the Pd–Cl distances in the catalysts to be almost the same, but with the latter having a higher palladium coordination number close to four (Table 5.6-2). This confirmed the quite uniform preparation of the attached catalyst on the support with mainly $[PdCl_4]^{2-}$ species present.

In another approach, heterogeneous olefin epoxidation catalysts were prepared by electrostatic attachment of binuclear peroxotungstate complexes onto an imidazolium-grafted silica support by ion-exchange [78]. The catalyst was made by initial grafting of the imidazolium-cations onto a pre-dried silica support by treatment with 1-octyl-3-(3-triethoxysilylpropyl)-imidazolium hexafluorophosphate.

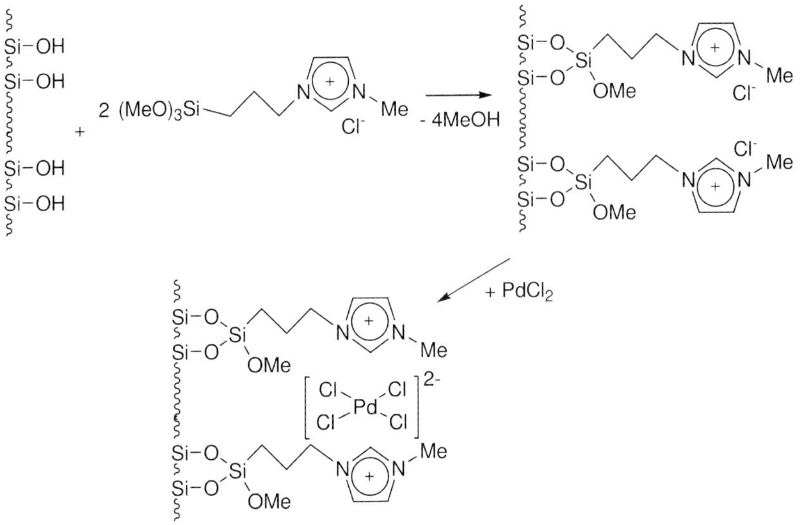

Scheme 5.6-3 Preparation of grafted supported ionic liquid palladium catalyst [78].

Table 5.6-2 Structural parameters derived from EXAFS spectra of the Pd catalysts [78]

Catalyst	Shell	Coordination number	R(Å)	$\sigma^2(Å^2)$
Supported (PdCl$_2$)	Pd–Cl	2.7 ± 0.2	2.32 ± 0.02	5 (± 2) × 10^{-3}
Attached (PdCl$_4^-$)	Pd–Cl	3.5 ± 0.1	2.31 ± 0.01	2.1 (± 0.3) × 10^{-3}

Analysis of the modified support by elemental analysis, IR and MAS-NMR confirmed the anchoring of the ionic liquid fragments and revealed the presence of 0.85 ionic liquid fragments per nm^2 on the support, and a number of reacted silanol groups corresponding to 54% of the original content. Afterwards, ion exchange of the [PF$_6$]$^-$ ions with an aqueous solution of the potassium salt of a binuclear peroxotungstate complex containing [{W(=O)(O$_2$)$_2$(H$_2$O)}$_2$(μ-O)]$^{2-}$ anions afforded the supported catalyst, which contained the complex anion in its original structure (determined by Raman spectroscopy). Notably, the BET surface area of the support increased by 10% after the ion exchange (from 184 m^2 g^{-1} to 203 m^2 g^{-1}).

The supported peroxotungstate catalyst proved to be very efficient for the epoxidation of various olefins using H$_2$O$_2$ (Fig. 5.6-5), with a catalytic activity comparable with a corresponding homogeneous tetraalkylammonium analog under the same conditions. The catalyst could be easily recovered by filtration and reused at least three times for the epoxidation of cyclooctene without loss of catalytic performance (99% yield, >99% selectivity). In accordance with this result no tungsten species

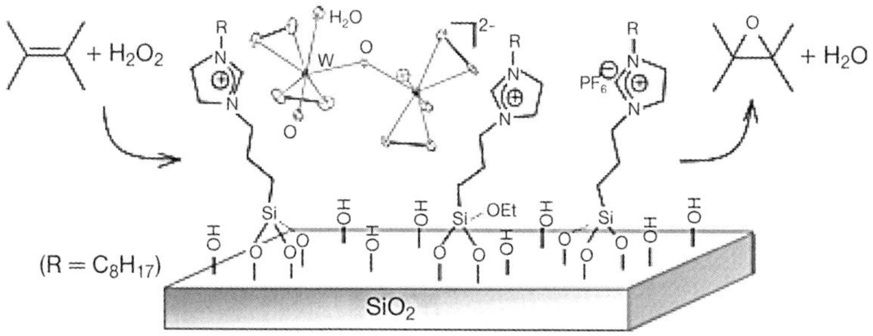

Fig. 5.6-5 Olefin epoxidation with H_2O_2 using a supported ionic liquid catalyst containing an imidazolium-grafted silica support with electrostatic attached $[\{W(=O)(O_2)_2(H_2O)\}_2(\mu\text{-}O)]^{2-}$ complex anions [79].

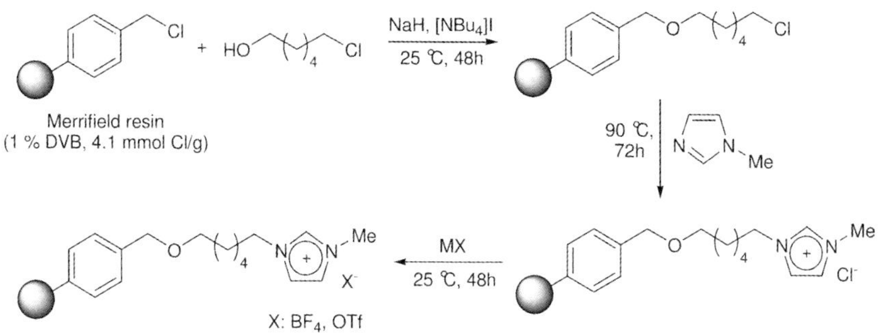

Scheme 5.6-4 Preparation of polymer-supported imidazolium salt catalysts, where DVB = divinylbenzene and MX = $NaBF_4$ or $KOTf$ [80].

were detected in the filtrate after reaction by ICP-AES analysis (detection limit of 16 ppb).

Even though most of the supported ionic liquid catalysts prepared thus far have been based on silica or other oxide supports, a few catalysts have been reported where other support materials have been employed. One example involves a polymer-supported ionic liquid catalyst system prepared by covalent anchoring of an imidazolium compound via a linker chain to a polystyrene support [79]. Using a multi-step synthetic strategy the polymeric support (e.g. Merrifield resin among others) was modified with 1-hexyl-3-methylimidazolium cations (Scheme 5.6-4) and investigated for nucleophilic substitution reactions including fluorinations with alkali-metal fluorides of haloalkanes and sulfonylalkanes (e.g. mesylates, tosylates and triflates).

The polymer-supported catalysts proved to be reusable, efficient catalysts for particular fluorination of various substrates giving excellent yields (>90%), and much higher activity than obtained for systems containing the same amount of ionic liquid in a system applying free ionic liquid. This observation was explained by a significantly increased nucleophilicity of the metal salt in the supported system compared with that obtained in conventional reaction systems, and proved to some extent to be cation dependent.

5.6.2.4 Ionic Liquid Catalysts Supported through Physisorption or via Electrostatic Interaction

A second and somewhat simpler approach that can be applied to obtain supported ionic liquid catalyst systems involves the treatment of a solid, porous carrier material by a substantial amount of a catalytically active ionic liquid, allowing the reaction to take place in the dispersed phase. In these systems the ionic liquid phase can itself act as the catalytically active component or it may contain other dissolved compounds or reagents, for example, transition metal complexes, which function as the catalytically active species (i.e. generating SILP catalysts). Importantly, the ionic liquid catalyst phase in these SILP catalyst systems are confined to the carrier surface only by weak van der Waals interactions and capillary forces interacting in the pores of the support. In special cases electrostatic attachment of the ionic liquid phase may also be applied. Usually, the catalysts are prepared by traditional impregnation techniques, where a volatile solvent is used initially to reduce viscosity for the impregnation process and is finally removed by evaporation leaving the ionic catalyst solution dispersed on the support.

Preparation of SILP catalyst systems may, however, also include initial coverage of the support by covalently attached ionic liquid fragments corresponding to a monolayer-coverage to avoid some detrimental surface interaction. However, this additional treatment is often not necessary when the ionic liquids used are non-acidic and do not undergo reactions with the support surface. Therefore, catalysts containing neutral ionic liquids with relatively inert ions are usually applied for SILP catalyst applications as they are easier to prepare and to handle than the corresponding acidic ionic liquids (e.g. due to the sensitivity to hydrolysis of the latter). However, in this context it should not be forgotten that the frequently applied 1,3-dialkylimidazolium cations may undergo carbene formation with basic supports or undesired metal–carbene formation under certain circumstances [80]. On the anion side, a general trend to avoid halogen-containing ions can be observed [81], since the presence of fluorine may be especially problematic, even if hydrolysis is not an issue. This is due to restricted disposal options and due to the elevated price of the hydrolysis stable fluorinated anions (e.g. $[(CF_3SO_2)_2N]^-$, $[CF_3SO_3]^-$, $[C_4F_9SO_3]^-$). Consequently, SILP catalysts employing halogen-free ionic liquids have been reported recently, as shown later.

5.6.2.4.1 Supported ionic liquid catalysts (SILC)

In the simplest methodology using supported ionic liquid catalysts based on a non-bonding ionic liquid attachment to the support, the ionic liquid phase behaves both as reaction medium and catalyst.

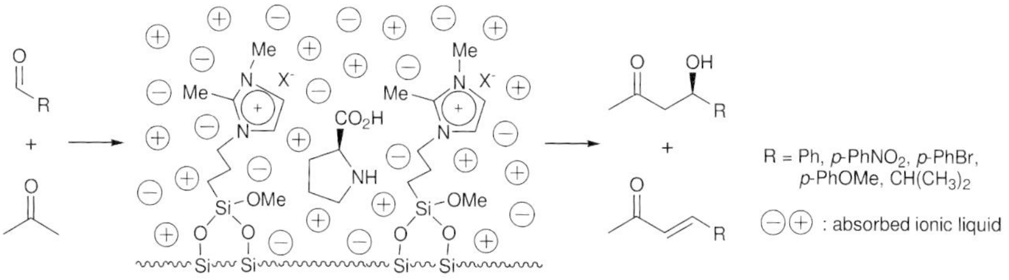

Fig. 5.6-6 L-proline catalyzed aldol formation using supported ionic liquid phase asymmetric catalysis [83].

An approach using this method was applied to prepare a magnesium oxide supported choline hydroxide ionic liquid catalyst (choline hydroxide: (2-hydroxyethyl)trimethylimidazolium hydroxide) by impregnation [82]. The catalyst was used as a solid, basic catalyst for liquid-phase aldol condensation reactions of various carbonyl compounds forming products with interest for the pharmacological, flavor and fragrance industries. In several reactions a better catalyst performance (i.e. activity and product selectivity) was obtained using this SILC than the analogous homogeneous catalyzed reactions using choline or NaOH as catalyst, clearly demonstrating the potential of such catalysts for these C–C bonding reactions. However, no results from consecutive reactions or results from leaching studies have been reported for the catalyst at this stage.

In a related study, L-proline catalyzed asymmetric aldol condensation between aldehydes and ketones was examined using SILP asymmetric catalysts [83] (Fig. 5.6-6). In this approach, the catalysts were prepared by impregnation of [BMIM][PF$_6$] ionic liquid containing L-proline directly onto a silica gel support, or by treatment of an additional amount of the ionic liquid containing L-proline onto a silica gel support pre-modified with a monolayer of ionic liquid attached to the carrier surface via grafting of the 1-(trimethoxysilylpropyl)-3-methylimidazolium cation (as shown earlier in Scheme 5.6-1, middle).

The catalytic performance of the SILC prepared by the different methods and containing various anions was compared with other L-proline supported catalysts and ionic liquid catalyst solutions containing L-proline for reactions between, for example, acetone and benzaldehyde. The results indicated that the surface of the silica gel must be functionalized with covalently attached ionic liquid in order to achieve yields and high ee's (up to 64% ee), comparable with those obtained using analogous homogeneous (60% ee) or bulk ionic liquid systems (71–76% ee) containing L-proline. Moreover, it was realized that the additional phase of ionic liquid increased the yield, with the ionic liquid anion playing an important but not well understood role. Furthermore, the supported catalyst system containing an additional ionic liquid catalyst phase could be reused in three consecutive reactions with highly reproducible results. In another approach, a silica gel supported ionic liquid catalyst was prepared by the sol–gel technique from tetraethyl silicate in the presence of various ionic liquids including, for example, [RMIM][X] (R = ethyl, butyl and decyl; X = [BF$_4$]$^-$ and [PF$_6$]$^-$) [84]. The supported catalysts (containing

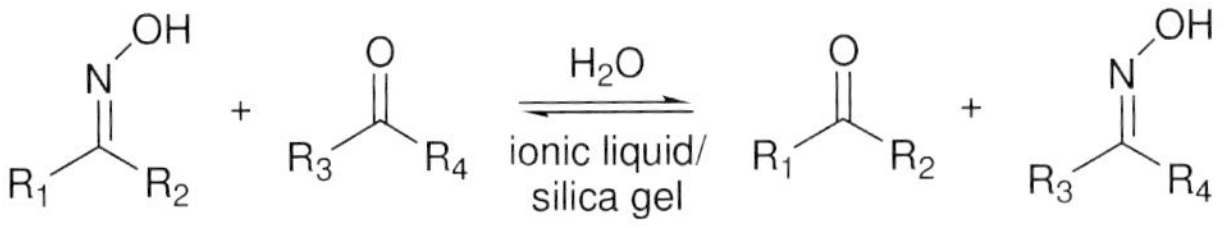

Scheme 5.6-5 Supported ionic liquid catalyzed oxime transformation [84].

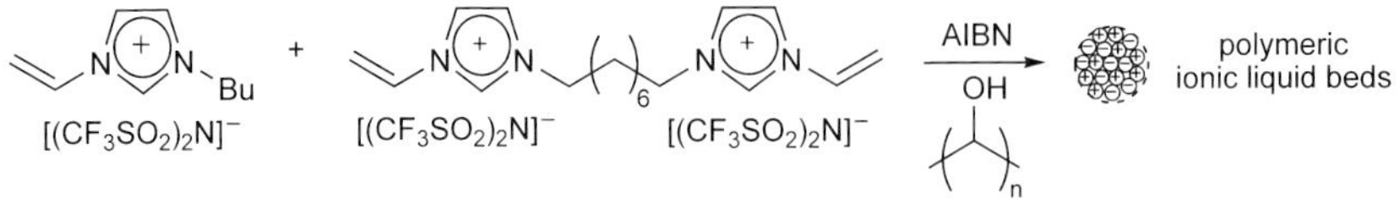

Scheme 5.6-6 Preparation of polymerized supported ionic liquid beads [86].

20–25 wt.% of the ionic liquids) were used to catalyze oxime transformations of various alkyl and aryl compounds (e.g. acetone and cyclohexanone oxime) in aqueous solvents by simultaneous C=N and C=O bond exchange (Scheme 5.6-5), and proved to give 10–20 times higher TONs than reactions performed using bulk ionic liquids as catalysts, which indicates some synergism between the ionic liquid and the support. Furthermore, it was realized that the presence of water was a key factor for obtaining high catalytic efficiency, which was explained by initial hydroxylamine formation from the reaction of water with the originally present oxime compound. Also here, no leaching studies were reported.

Also, in a closely related study, a similar sol–gel synthetic pathway was used to prepare silica gel supported ionic liquid deoximation catalysts from carboxylic acid-functionalized ionic liquids [85]. Here various aryl and alkyl oximes were converted into the corresponding oxo compounds with high conversions (up to 94%) and excellent selectivity (>99%) in aqueous acetone at room-temperature by co-production of 2-propanone oxime. TONs of up to 200 h^{-1} obtained with the SILC were about four times higher than the TONs obtained using pure ionic liquids as catalysts.

Using a quite different approach, polymeric beads of supported ionic liquid palladium catalysts comprised of polymerized ionic liquid monomers and palladium complexes have been synthesized using traditional suspension polymerization methods [86]. Here, polymeric ionic liquid beads were made from polymerization of 1-butyl-3-vinylimidazolium bis(trifluoromethyl sulfonyl)imide and poly(vinylalcohol) by heating with AIBN (2,2'-azobis(2-methylpropionitrile)) in the presence of 1,1'-bis[1,8-octyl)-3-vinylimidazolium bis(trifluoromethyl sulfonyl)imide as cross-linker (Scheme 5.6-6). The ionic liquid support beads proved to be thermally stable up 250 °C which is significantly higher than conventional vinyl resins.

After impregnation of the polymeric support material with palladium(II) tri-(o-tolyl)phosphine complexes the resulting catalyst material was tested in the Heck coupling of bromobenzene with butyl acrylate in DMF to yield 75% of E-3-phenylacrylic acid butyl ester product. Analysis (ICP-MS) of the filtrate after the reaction revealed, however, leaching of 1–2 ppm palladium metal from the catalyst. Additionally, the $[(CF_3SO_2)_2N]^-$ anion could readily be ion exchanged with other anions of catalytic interest (e.g. $[CoCl_4]^-$) by passing a solution of 1-decyl-3-methyl imidazolium tetrachlorocobaltate through a column of the polymeric beads, thus indicating the potential for rapid catalyst modifications in these systems.

5.6.2.4.2 Supported ionic liquid phase (SILP) catalysts incorporating metal complexes

When a substantial amount of an ionic liquid is immobilized on a support, the formation of a film of free ionic liquid on the carrier may act as an inert reaction phase to dissolve various homogeneous catalysts [87]. This further implies that although the resulting material appears as a solid, the active species dissolved in the ionic liquid phase on the support still comprises the attractive features of a dissolved homogeneous catalyst, for example, high specificity and uniform nature of the catalytically active sites.

In the first reported example using this approach, supported hydroformylation catalysts were prepared by impregnation of a surface-modified silica gel containing covalently anchored ionic liquid fragments with ionic liquid solutions of the precursor [Rh(acetylacetonate)(CO)$_2$] and the trisulfonated triphenylphosphine ligand as either trisodium salt (TPPTS) or as (1-butyl-3-methylimidazolium) salt (TPPTI) (ligand/Rh ratio of 10) [88]. The ligand TPPTI was found to dissolve in both [BMIM][BF$_4$] and [BMIM][PF$_6$], while the corresponding sodium salt only dissolved in [BMIM][BF$_4$].

The initial preparation of the catalyst involved modification of a pre-dried silica gel support by treatment of the support in an immobilizing step with functionalized ionic liquids, such as 1-butyl-3-(3-triethoxysilylpropyl)imidazolium tetrafluoroborate or hexafluorophosphate. Analysis of the surface coverage revealed an average of 0.4 ionic liquid fragments per nm^2, corresponding to the involvement of approximately 35% of all the hydroxy groups of the pre-treated silica gel. Treatment of the obtained support with additional ionic liquid resulted in a free film of ionic liquid on the support corresponding to an ionic liquid phase loading of 25 wt.%. After solvent removal under reduced pressure, a slightly yellow-colored powder was obtained.

The prepared catalysts were investigated for the hydroformylation of 1-hexene to produce heptanal in a batch-wise, liquid phase reaction (Table 5.6-3), and all results obtained were compared to the identical reaction in the liquid–liquid biphasic process design.

A comparison between the supported [BMIM][BF$_4$]-based catalyst and the biphasic ionic liquid reaction showed that the supported system exhibited a slightly enhanced activity (TOF of 65 min^{-1} versus 23 min^{-1}) due to the higher concentration of rhodium complexes at the interface and the generally larger interface of the supported system. As expected for the applied mono-phosphine ligand, low selectivities (n/iso-heptanal ratios) of 2.4 to 2.5 were obtained. Importantly, at high aldehyde concentrations the ionic liquid [BMIM][BF$_4$] was found to partially dissolve in the organic phase leading to a considerable rhodium loss of up to 2.1 mol%. The

Table 5.6-3 Evaluation of the hydroformylation reaction of 1-hexene to form n,iso-heptanal using supported ionic liquid phase catalysis (SILP), biphasic catalysis and homogeneous catalysis [88]

Entry	Condition[a]/ligand	Solvent	Time(min)	Yield (%)	n/iso	TOF[b](min^{-1})
1	SILP/TPPTI	[BMIM][BF$_4$]	300	33	2.4	65
2	SILP/TPPTS	[BMIM][BF$_4$]	240	40	2.4	56
3	SILP/TPPTI	[BMIM][PF$_6$]	270	46	2.4	60
4	SILP/no ligand	[BMIM][PF$_6$]	180	85	0.4	190
5	biphasic/TPPTI	[BMIM][BF$_4$]	230	58	2.2	23
6	biphasic/TPPTI	[BMIM][PF$_6$]	180	70	2.5	22
7	biphasic/TPPTS	H$_2$O	360	11	23	2.4
8	homogenous/PPh$_3$	toluene	120	95	2.6	400

[a] Reaction conditions: All runs were conducted at 100 °C with a Rh/ligand ratio of 1:10, SILP runs were evaluated in a 70 ml autoclave at 1500 psi (100 bar) and biphasic and homogeneous catalyst systems were evaluated in a 300 ml autoclave at 600 psi (40 bar).
[b] TOF defined as molaldehyde per molrhodium per minute (full reaction time).

metal loss could be somewhat suppressed at lower aldehyde concentrations and higher ligand excess in single runs. However, pronounced catalyst deactivation was found, even at lower conversion, during recycling of the catalyst independent of the presilylation of the support [89].

Hydroformylation of 1-hexene has also later been performed using a MCM-41 supported Rh-TPPTS catalyst based on the ionic liquid 1,1,3,3-tetramethylguanidinium lactate (TMGL) [90]. Here the catalysts exhibited practically unchanged performance in twelve consecutive runs providing about 50% conversion and also low n/iso ratios of ca. 2.5.

In a continuing effort to explore the full technical potential of supported ionic liquids the concept has been extended to continuous flow processes. The reactions studied were Rh-phosphine catalyzed hydroformylation of propene and 1-octene using more technically attractive continuous flow fixed-bed reaction designs [91–93] (more detailed description of the reaction set-ups may be found in Refs. [94, 95] for gas-phase reactions and in Ref. [96] for liquid-phase reactions).

The supported ionic liquid phase catalyst systems were primarily prepared by immobilizing [Rh(acetylacetonate)(CO)$_2$] and the applied phosphine ligands in a film of either [BMIM][PF$_6$] or [BMIM][n-C$_8$H$_{17}$OSO$_3$] ionic liquid on silica gel (either dried or partly dehydroxylated by thermal treatment), by using impregnation with degassed solutions of methanol or water–dichloromethane. This procedure allowed the generation of catalysts with varying ligand/rhodium ratios and ionic liquid loadings. Further addition of silica gel resulted in a catalyst rhodium metal loading of 0.2 wt.%. Three different phosphine ligands were used for the preparation of the supported catalyst, all modified with charged groups to increase the ionic liquid solubility (Fig. 5.6-7).

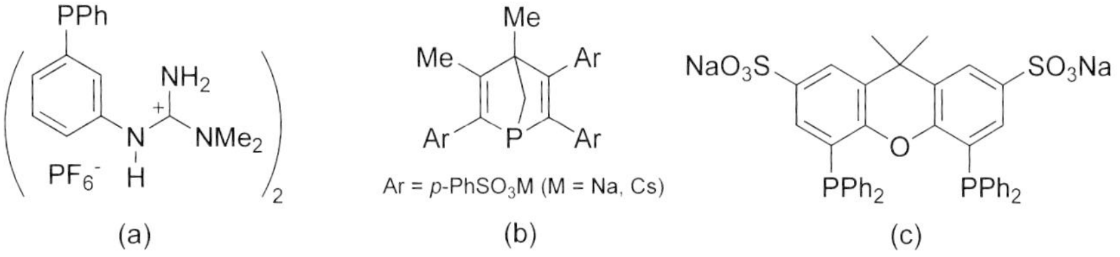

Fig. 5.6-7 The charged phosphine ligands bis(m-phenylguanidinium)phenylphosphine hexafluorophosphate, guanidinium (a), tricesium 3,4-dimethyl-2,5,6-tris(p-sulfonatophenyl)-1-phosphanorbornadiene, NORBOS (b) and disodium 4,5-bis(diphenylphosphino)-2,7-bis(sulfonato)-9,9-dimethylxanthene, sulfoxantphos (c), examined for supported ionic liquid phase continuous Rh-catalyzed propene hydroformylation [91–93].

In an application for continuous biphasic, liquid–liquid 1-octene hydroformylation using a SILP Rh-NORBOS catalyst containing [BMIM][PF$_6$], a steady catalyst performance (TOF of 44 h^{-1} and n/iso ratio of 2.6) was achieved after 3–4 h of reaction. Furthermore, no leaching of rhodium metal could be detected by ICP-AES analysis of outlet samples, at least after this relatively short reaction time.

In more comprehensive studies, the catalytic performance of the silica-supported catalyst based on the different ligands was found to be essentially uninfluenced by the type of ionic liquid in the gas-phase propene hydroformylation reactions, as no significant differences were observed with [BMIM][PF$_6$] or [BMIM][n-C$_8$H$_{17}$OSO$_3$] being the ionic liquids applied (except for the catalyst preformation which appeared to depend on the solubility of the ligand and precursor). Furthermore, this was also shown to be the case in a study performed with analogous catalysts based on alternative oxide supports such as titania, alumina and zirconia [97]. In Table 5.6-4 the most relevant results for the hydroformylation of propene with the different catalyst systems based on the ionic liquid [BMIM][PF$_6$] and pre-dried silica support are compiled.

The performance of the catalyst in 5 h reactions was found to be drastically influenced by the catalyst composition with respect to ligand/rhodium ratio and ionic liquid loading. When using catalysts with low ligand to rhodium ratios, only low selectivity for the linear product was obtained (entries 1 to 3 and 8 to 10). This was ascribed, in part, to the presence of most of the active rhodium species as ligand-free, less selective complexes under these conditions. Additionally, an increased ionic liquid loading led to a decrease in activity, which indicated that the catalysts were operating under certain mass-transfer limitations. In accordance with this, the highest activities were initially observed for the ionic liquid free catalysts (entries 1, 6, 8 and 11).

The use of catalyst systems containing the bidentate phosphine ligand sulfoxantphos proved particular interesting, as excellent n/iso ratios of 23 (i.e. linear product selectivities up to 96%) were attained with these systems (entries 12 to 14).

Table 5.6-4 Continuous, gas-phase Rh-phosphine catalyzed propene hydroformylation using silica gel-based supported ionic liquid phase catalysts containing [BMIM][PF$_6$][a] [91] and [92]

Entry	Ligand	L/Rh[b]	Ionic liquid loading wt. %	α[c]	TOF[d] (h^{-1})	n/iso[e]	Linearity (%)
1	guanidinium	2.9	–	0.00	55.5	0.9	47.4
2	guanidinium	2.9	108.4	0.78	20.6	0.9	47.4
3	guanidinium	2.9	138.9	1.00	16.8	1.0	50.0
4	NORBOS	11.3	6.9	0.05	88.4	2.0	66.7
5	NORBOS	11.3	20.8	0.15	79.4	1.3	56.5
6	NORBOS	21.4	–	0.00	45.8	2.8	73.8
7	NORBOS	21.3	6.9	0.05	28.2	2.6	72.2
8	sulfoxantphos	2.5	–	0.00	37.4	1.7	63.0
9	sulfoxantphos	2.4	23.6	0.17	1.5	1.8	64.0
10	sulfoxantphos	2.5	68.1	0.49	5.1	2.0	66.4
11	sulfoxantphos	10.2	–	0.00	40.8	16.9	94.4
12	sulfoxantphos	10.0	25.0	0.18	34.9	22.6	95.8
13	sulfoxantphos	10.0	72.2	0.52	25.4	22.0	95.8
14	sulfoxantphos	20.0	27.8	0.20	16.7	23.7	96.0
15[e]	–	–	105.6	0.76	–	–	–

[a] Reaction conditions: Propene:CO:H$_2$ = 1:1:1, $T = 100\,°C$, $p = 5$ bar for guanidinium and NORBOS, $p = 10$ bar for sulfoxantphos, reaction time = 5 h. Supported catalysts: 0.2 wt. % rhodium metal, dried silica gel (110 °C, 24 h, in vacuo), [BMIM][PF$_6$] used for impregnation.
[b] Molar ligand to metal ratio.
[c] Ratio of ionic liquid volume to support pore volume.
[d] TOF defined as mol aldehyd per mol rhodium per hour.
[e] Support loaded with ionic liquid only.

Remarkably, at a low ligand/rhodium ratio of 2.5 (entries 9 and 10), the selectivity of the Rh-sulfoxantphos system was significantly lower and comparable to the best selectivity obtained with the monodentate phosphines. This results was in contrast to analogous experiments performed in ionic liquid–liquid biphasic mode where high selectivities were obtained [81, 98], and suggested some ligand loss due to reactions between the ligand and the solid surface.

Importantly, it was also realized in these initial studies that the catalysts deactivated in prolonged use with simultaneous decrease in catalytic activity and selectivity, independent of the type of ionic liquid, ionic liquid loading α, and the ligand/rhodium ratio of the system. However, additional studies demonstrated a suitable thermal pre-treatment of the silica support to be necessary to obtain stable SILP catalysts. This technique was applied to prepare Rh-sulfoxantphos catalysts with [BMIM][n-C$_8$H$_{17}$OSO$_3$] and similar systems without ionic liquid. Both systems were compared in 60 h continuous propene hydroformylation reactions using similar conditions as previously reported (Table 5.6-4, entries 11 and 12, Fig. 5.6-8) [93].

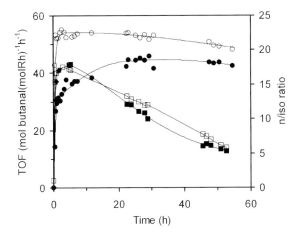

Fig. 5.6-8 Continuous hydroformylation of propene with silica Rh-sulfoxantphos/[BMIM][n-$C_8H_{17}OSO_3$] catalysts (ligand/rhodium = 10) having ionic liquid loadings of $\alpha = 0$ (□, ■) and $\alpha = 0.1$ (○, ●). Activity (closed symbols) and selectivity (n/iso ratio, open symbols) [93].

In the reaction using the Rh-sulfoxantphos catalyst without ionic liquid, the activity and selectivity decreased sharply after 5–10 h on stream, while the SILP system ($\alpha = 0.1$) reached its maximum activity only after 30 h and maintained this level stable up to 60 h (i.e. TON > 2400) along with a high selectivity corresponding to a n/iso ratio of 21–23. It can also be deduced that the apparently negative ionic liquid effect on the catalyst activity measured in the short time reactions can be explained by a delayed formation of the catalytically active species.

Besides the importance of the ionic liquid solvent, it was also shown that a relatively large excess of sulfoxantphos ligand is required to obtain a stable catalyst system, even in reactions using a thermally pre-treated support. This aspect was further recognized by solid-state ^{29}Si and ^{31}P MAS-NMR measurements to be directly related to an irreversible reaction of the ligand with the acidic silanol surface groups during catalysis. Moreover, FT-IR measurements on the supported catalysts under synthesis gas at catalytically relevant conditions, verified the complex formation in the supported ionic liquid phase to be similar to that observed for Rh-sulfoxantphos and analogous xanthene ligand-based rhodium complexes in organic and ionic liquid solvents (Table 5.6-5) [99]. This proved that the studied reactions were indeed homogeneously catalyzed, as expected. Finally, the FT-IR examinations established the catalyst instability to be associated with degradation of the co-existing isomeric [HRh(CO)$_2$(sulfoxantphos)] complexes, which in turn is correlated to the accessible amount of ligand available for complex formation. Thus, the prerequisites for obtaining an active, highly selective, and durable SILP hydroformylation catalyst, were shown to involve both the presence of ionic liquid solvent and a relatively large excess of bisphosphine ligand to compensate for some detrimental surface reactions.

Table 5.6-5 Comparison of infrared ν(CO) bands in [HRh(CO)$_2$(L)] complexes in different systems [93]

	Solvent	ea-isomer ν(CO) (cm^{-1})	ee-isomer ν(CO) (cm^{-1})
SILP Rh-sulfoxantphos	[BMIM][n-C$_8$H$_{17}$OSO$_3$]	1994, 1948	2035(w), 1964
[HRh(CO)$_2$(sulfoxantphos)][a]	[BMIM][PF$_6$]	1985, 1935	2032, 1967
[HRh(CO)$_2$(xantphos)][a]	benzene	1991, 1941	2036, 1969
[HRh(CO)$_2$(thixantphos)][a]	cyclohexane	1999, 1953	2040, 1977

[a] Ref. [99].

The first reported SILP hydrogenation catalyst was prepared by confining a [BMIM][PF$_6$] ionic liquid phase containing [Rh(norbornadiene)(PPh$_3$)$_2$][PF$_6$] complexes onto the carrier by impregnation (contained 25 wt.% ionic liquid, corresponding to an average ionic liquid catalyst layer of 6 Å) [100]. This catalyst was used for liquid-phase hydrogenations of, e.g. 1-hexene and compared to analogous reactions performed in bulk ionic liquid or organic media with the same Rh-complex. A superior activity of the SILP catalysts was revealed with up to 100 times higher TOF under similar reaction conditions. This enhanced activity was attributed to its higher interfacial complex concentration. Moreover, the SILP catalyst showed excellent durability and was reused for 18 consecutive runs without any significant loss of activity and without indication of rhodium metal leaching (the level remained below the detection limit of 33 ppb).

In a different approach, supported ionic liquid poly(vinylidene fluoride) membranes containing an active phase of [Rh(norbornadiene)(PPh$_3$)$_2$][PF$_6$] complex dissolved in different ionic liquids with fluorinated anions (e.g. [RMIM][X] (R = ethyl, butyl and X = [BF$_4$]$^-$, [PF$_6$]$^-$, [(CF$_3$SO$_2$)$_2$N]$^-$ and [CF$_3$SO$_3$]$^-$), were employed to examine the hydrogenation of propene and ethane [101]. To accomplish the hydrogenation reactions, the olefin was maintained at atmospheric pressure on the feed side of the membrane, while a flow of hydrogen gas at the same pressure was swept over the permeate side to remove hydrogenated products when these had passed through the catalytic ionic liquid bulk phase of the membrane (Fig. 5.6-9).

The maximum hydrogenation rates for the catalytic membrane materials based on the different ionic liquids followed the order [EMIM][(CF$_3$SO$_2$)$_2$N)] > [EMIM][CF$_3$SO$_3$] > [EMIM][BF$_4$] > [BMIM][PF$_6$], which proved to be different from the relative gas solubilities estimated from gas permeability studies in the liquids. However, the order of reactivity was in line with results previously found for some of the ionic liquids in analogous biphasic liquid–liquid hydrogenation of 1-pentene [102], which showed the importance of the coordination strength of the ionic liquid anions on reaction rates.

Supported ionic liquid phase hydrogenation catalysts based on a polymeric poly(allyldimethylammonium chloride) support have also been prepared by attachment of [BMIM][PF$_6$] solutions of [RhCl(PPh$_3$)$_3$] (Wilkinson

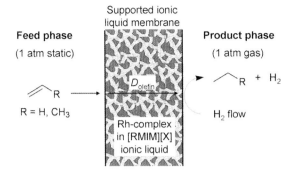

Fig. 5.6-9 Schematic illustration of olefin hydrogenation with a supported ionic liquid poly(vinylidene fluoride) membrane containing [Rh(norbornadiene)(PPh$_3$)$_2$][PF$_6$] complexes.

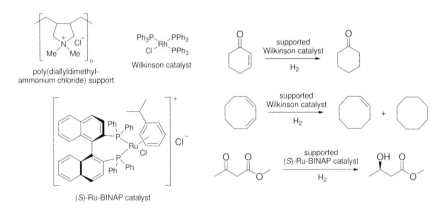

Fig. 5.6-10 Hydrogenation reactions using poly(diallyldimethylammonium chloride) supported ionic liquid phase catalysts [103].

catalyst) and [RuCl(S-BINAP)]Cl (chloro-[(S)-(-)-2,2′-bis(diphenylphosphino)-1,1′-biphenyl](isopropylbenzene)-ruthenium(II)-chloride), respectively [103]. The supported catalysts were applied for liquid-phase hydrogenations of 2-cyclohexen-1-one and 1,3-cyclooctadiene and asymmetric hydrogenation of methyl acetoacetate (Fig. 5.6-10).

For all reactions studied, the activity of the supported catalysts was higher than for the similar biphasic ionic liquid system, which was ascribed to improved mass transfer between the substrates and the ionic liquid phase. In addition, the observed product selectivities of 64–87% and enantioselectivity of 97% for the SILP-Ru-(S)-BINAP catalyzed reaction equalled those of the homogeneous reference reactions. No indication of rhodium metal leeching was found by AAS analysis of the reaction filtrate.

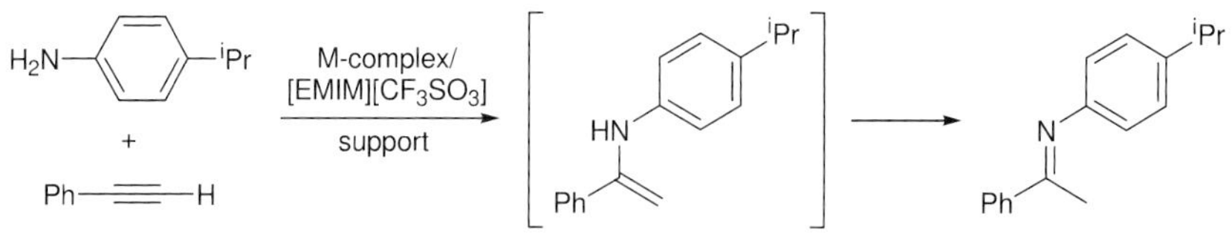

Scheme 5.6-7 Synthesis of (4-isopropylphenyl)-(1-phenylethylidene)amine by hydroamination of phenylacetylene with 4-isopropylamine using supported ionic liquid phase metal catalysts [104].

Catalytic hydroamination is another type of reaction, that recently has been tested using the concept of SILP catalysis containing late transition metal complexes [104]. In this study, hydroamination of phenylacetylene was carried out by direct addition of 4-isopropylamine (Scheme 5.6-7) using different supported catalysts obtained by shock freezing (followed by freeze-drying) a mixture of [EMIM][CF$_3$SO$_3$]-dichloromethane solutions of cationic rhodium(I), palladium(II), cupper(I) and zinc(II), respectively, containing a macroporous, diatomé earth support (Chromosorb P, pre-modified by dichlorodimethylsilane treatment, i.e. silylated) in liquid nitrogen.

For the SILP catalyst containing the complexes [Rh(DPPF)(2,5-norbornadiene)][ClO$_4$], [Pd(DPPF)][CF$_3$SO$_3$]$_2$ and Zn[CF$_3$SO$_3$]$_2$ (DPPF: 1,1'-bis-(diphenylphosphino)ferrocene), the rate of reaction was found to be 2–6 times higher than in the corresponding homogeneous catalyzed reactions. In contrast, for the SILP-[Cu$_2$(C$_6$H$_5$CH$_3$)][CF$_3$SO$_3$]$_2$ catalyst a significantly lower catalytic activity was observed than for the corresponding homogeneous system, possibly due to competing coordination of the ionic liquid and the substrate for the copper(I) center. Notably, product selectivity (85–100%) was also significantly increased for all four supported catalysts compared to the homogeneous reactions.

Pd-complexes have also been impregnated on an amorphous silica support with the aid of a solution containing [BMIM][PF$_6$] dissolved in tetrahydrofuran and these systems were applied as highly efficient catalysts for promoting Mizoroki-Heck coupling reactions between various aryl halides and cyclohexyl acrylate in alkanes without the presence of additional ligand (Scheme 5.6-8) [105].

With optimized reaction conditions using a supported Pd(acac)$_2$-containing catalyst, a TOF of 8000 h^{-1} was reached in the reaction of iodobenzene with cyclohexyl acrylate in dodecane using tributylamine as base. Moreover, the catalyst could readily be reused up to six times giving yields of 89–98% without taking any precautions (e.g. avoidance of air), if surface deposited tri-butylammonium iodide salt was removed by washing with aqueous sodium hydroxide at an intermediate stage during the consecutive reactions. ICP-analysis of the organic product layer after reaction revealed only depletion of trace amounts of palladium metal (less than 0.24%) from the supported ionic liquid catalyst phase.

In another approach, an oxime carbapalladacycle ionic liquid catalyst was attached to an aluminosilicate support (Al/MCM-41, Si/Al 13) via electrostatic interaction by

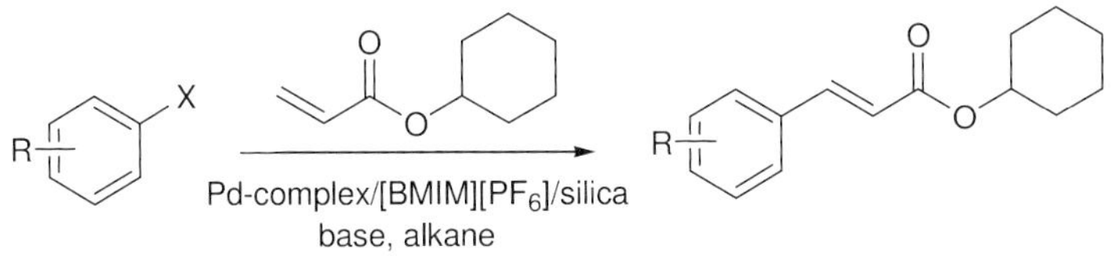

X = I, Br, 2-CHO/4-MeO/5-OH
R = H, 4-CH₃, 4-Br, 4-I, 4-Ac, 4-NO₂, 4-MeO

Scheme 5.6-8 Pd-catalyzed Mizoroki–Heck coupling reactions between aryl halides and cyclohexyl acrylate in alkanes using silica/[BMIM][PF₆] supported ionic liquid phase catalysts [105].

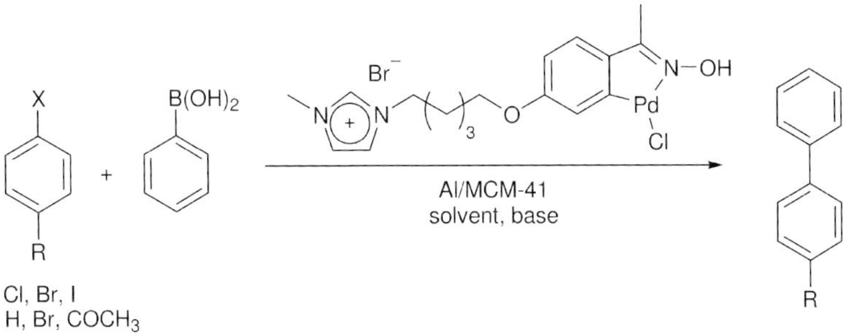

X = Cl, Br, I
R = H, Br, COCH₃

Scheme 5.6-9 Base-promoted Suzuki cross-coupling of halobenzenes and phenylboronic acid using Al/MCM-41-supported oxime carbapalladacycle ionic liquid catalyst [106].

impregnation, and tested for base-promoted Suzuki cross coupling of halobenzenes and phenylboronic acid in toluene and dimethylformamide, respectively (Scheme 5.6-9) [106].

The performance of the supported palladium ionic liquid catalyst was found to be very dependent on the organic solvent, revealing higher activity (comparable to analogous homogeneous catalyzed reactions) and higher product selectivity towards Suzuki-coupled products compared to homo-coupled products in a polar solvent such as DMF (no homo-coupled products observed) than in toluene (1:2 to 1:5 distribution of products). Also, using the catalyst in DMF, coupling of less reactive bromo- and chlorobenzenes proved possible. Unfortunately, it was also shown that the catalyst could only be reused for a few consecutive runs with similar activity (resulting in 86–90% conversion) and selectivity, due to the partial instability of the oxime carbapalladacycle complex under the reaction conditions.

In a different approach based on membrane technology, oligomerization of ethene has also been examined using a poly(ethersulfone)-supported ionic liquid membrane containing [EMIM]Cl–AlCl₃ ionic liquids with or without the presence of [NiCl₂{P(cyclohexyl)₃}₂] dimerization catalyst and dichloroethylaluminate as an acid scavenging co-catalyst [107].

Table 5.6-6 Pd-catalyzed hydrogenation reactions, with nanoparticles actively supported on molecular sieves with a TMGL layer [109]

Entry	Olefins	Olefin/Pd molar ratio	T (°C)	t (h)	Conv. (%)	TOF (min^{-1})
1[a]	cyclohexene	12000	20	10	100	20.0
2	cyclohexene	12000	40	7	100	28.3
3[b]	cyclohexadiene	12000	20	3	98	65.3
4	1-hexene	12000	20	3	100	66.7
5[c]	1-hexene	120	20	6	100	0.30
6[d]	cyclohexene	500	40	5	100	1.67
7[e]	cyclohexene	12000	20	10	100	20.0
8[f]	cyclohexene	12000	20	10	100	20.0
9[g]	cyclohexene	12000	20	10	100	20.0
10[h]	cyclohexene	12000	20	10	100	20.0

[a] The product was cyclohexane.
[b] The product was cyclohexene; no cyclohexane was detected.
[c] Based on data from G. Schmid, *Chem. Rev.*, **1992**, *92*, 1709 and G. Schmid, M. Harm, *J. Am. Chem. Soc.*, **1993**, *115*, 2047, using a heterogeneous catalyst with phenanthroline-protected Pd nanoparticles.
[d] The reaction was carried out using Pd nanoparticles in 1-butyl-3-methylimidazolium hexafluorophosphate as the catalyst, see Ref. [110].
[e-h] The catalyst was used a second, third, fourth, and fifth time, respectively, after the experiment described in entry 1.

5.6.2.4.3 Supported ionic liquid catalyst systems containing metal nanoparticles

Ionic liquids often induce low interfacial tensions when combined with other phases [108]. This characteristic results in high nucleation rates that can be applied to generate metallic nanoparticles that are further stabilized to a high extent, preventing particle aggregation. This very interesting property has recently been used for the first time to prepare a molecular sieve (Al:Si = 1:1) SILP catalyst containing Pd nanoparticles (diameter of most particles 1–2 nm as determined by TEM) which were stabilized by the ionic liquid 1,1,3,3-tetramethylguanidinium lactate (TMGL) [109]. Here, the alkaline nature and the ability of the guanidinium ions to coordinate metal particles were combined to make a stable catalytic material containing 20 wt.% (corresponding to an average supported ionic liquid layer of 0.4 nm) of a strongly absorbed ionic liquid phase. Notably, the palladium nanoparticles were larger than the average thickness of the ionic liquid layer, thus leaving part of the particles less covered by the ionic liquid.

The supported nanoparticle catalyst system was used for solvent-free hydrogenation reactions of cyclohexene, 1-hexene, and 1,3-dicyclohexadiene, respectively, and compared with a similar biphasic ionic liquid system and with a heterogeneous supported nanocatalyst (Table 5.6-6).

The supported ionic liquid nano-catalysts showed unprecedented activity (entries 1 to 4) and high selectivity for the hydrogenation of cyclohexadiene to cyclohexene (entry 3). This was attributed to a much stronger absorption of the hexadiene to the Pd particles than the cyclohexene, resulting in preferential hydrogenation of the

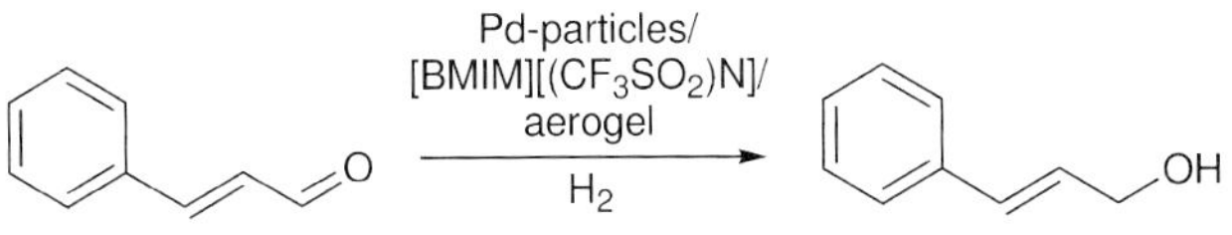

Scheme 5.6-10 2,4-hydrogenation of cinnamaldehyde using aerogel-supported ionic liquid catalyst material containing palladium nanoparticles [112].

alkadiene compared to the alkene. Additionally, the SILP catalyst could be reused for another four runs without any loss of activity (entries 7 to 10), and without palladium metal leaching to the organic products (determined by AAS) or metal particle aggregation into larger particles on the molecular sieves (determined by TEM). Thus, the combination of palladium nanoparticles, ionic liquid and molecular sieves revealed an excellent synergistic effect to enhance the activity and durability of the catalyst.

Also, using a closely related approach, an aerogel-supported catalyst material containing palladium nanoparticles was made by an ionic liquid mediated synthesis, which previously had proved to be a convenient and effective way of preparing silica aero-gel materials [111]. The synthesis involved initial formation of palladium(0) colloids (approximate diameter 1 nm) from Pd(acetate)$_2$ and triphenylphosphine in [BMIM][(CF$_3$SO$_2$)$_2$N] followed by incubation with tetraethylorthosilicate (TEOS) and formic acid. After gelation and aging, most of the ionic liquid solvent was, however, removed by Soxhlet extraction with acetonitrile, resulting in an apparently dry monolithic catalyst material containing highly dispersed Pd nanoparticles [112]. The catalyst was tested for liquid-phase hydrogenation of an α,β-unsaturated aldehyde (cinnamaldehyde) into the corresponding unsataturated alcohol (Scheme 5.6-10), and for Mizoroki-Heck coupling of iodobenzene with butyl acrylate in DMF.

In these reactions the catalyst proved very active (TOF up to 1307 h^{-1}) with very high selectivity for the desired products. On recycling, however, significantly lower activities were obtained due to severe Pd metal leaching (6.8% loss during reuse in the Heck reaction). However, TEM examination of the catalyst post reaction established the metal particles to preserve the original sizes with no sign of aggregation, clearly emphasizing the stabilizing effect of the ionic liquid on the catalyst material.

Supported catalyst materials containing metal particles have also been prepared by immobilizing ionic liquids in membrane materials. These systems have been tested for propene hydrogenation to propane in an asymmetric membrane arrangement [113, 114]. In the study, supported ionic liquid-polymer gel composite membrane materials composed of various air-stable ionic liquids, for example, [RMIM][X] (R = ethyl, butyl and X = [BF$_4$]$^-$, [PF$_6$]$^-$, [(CF$_3$SO$_2$)$_2$N]$^-$), and poly(vinylidene fluoride)-hexafluoropropylene copolymers were prepared by incorporation of a heterogeneous catalyst containing palladium metal (10 wt.%) on activated carbon. To accomplish the hydrogenation reactions, the ionic liquid membrane material was

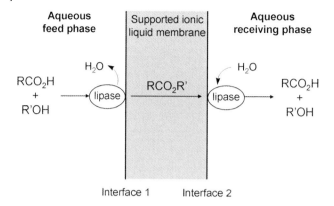

Fig. 5.6-11 Lipase facilitated separation of carboxylic acids using a supported ionic liquid membrane [118].

loaded into a membrane reactor set-up with the supported palladium metal particles on the permeate side of the membrane reactor. In this configuration, the palladium was activated by a hydrogen flow over the permeate side of the membrane, and the propene feed gas hydrogenated after it passed through the bulk of the membrane and reached the ionic liquid catalyst layer on the permeate side. Using [EMIM][(CF$_3$SO$_2$)$_2$N] as the catalyst phase a constant propene conversion of 73% could be achieved after 2 h of reaction.

5.6.2.4.4 Supported ionic liquid catalytic membrane systems containing enzymes

Supported liquid membranes (SLM) containing ionic liquids have been used successfully for selective transport of, e.g. alcohols, ketones, amines [115, 116] and aromatic hydrocarbons [117], according to the concentration gradient of the substance and its solubility in the ionic liquid.

Further development of the SLM-ionic liquid methodology has been coupled with lipase-catalyzed esterification and ester-hydrolysis reactions in the feed gas (interface 1) and receiving phase (interface 2), respectively, to facilitate selective transport of various organic acids with aryl groups (via their esters) from aqueous solutions by utilizing different substrate specificity of lipases in a SLM system (Fig. 5.6-11) [118]. In the enzymatic SLM systems a poly(propene) membrane with water-immiscible [RMIM][X] (R = butyl, hexyl, octyl and X = [PF$_6$]$^-$ and [(CF$_3$SO$_2$)$_2$N]$^-$) ionic liquid phases were used.

Using the same enzyme-SLM coupled principle as above, optical resolution of racemic ibuprofen (2-(4-isobutylphenyl)propionic acid) has also been achieved by utilizing the enantioselectivity of lipases for selective transport of (S)-ibuprofen [119]. For example, when using the poly(propene)/[BMIM][(CF$_3$SO$_2$)$_2$N] membrane system containing lipases originating from *Candida rugosa* and *Porsine pancreas* for esterification and hydrolysis in the feed gas phase and receiving phase, respectively, a maximum optical resolution ratio (i.e. ratio between initial permeate flux of (S)- and (R)-ibuprofen) of 3.9 and an enantiomeric excess (*ee*) of 75% of the (S)-isomer in the feed phase was obtained after 48 h of operation. In perspective, this result clearly demonstrates the potential for conducting enantioselective separations with enzyme-facilitated SLM systems based on ionic liquids.

5.6.3
Concluding Remarks

This section summarizes the recent progress in supported ionic liquid catalysis, demonstrating synthetic applications where the ionic liquid can play its role as either "innocent" solvent (SILP) or the catalyst itself (SILC), depending on the specific cation/anion combination and the reaction under investigation.

Supported ionic liquid catalysis combines well-defined catalyst species, non-volatile ionic liquids, and porous solid supports in a manner which offers advantages over regular biphasic ionic liquid–organic liquid systems, e.g. substantially reduced amounts of catalyst and ionic liquid, higher turnovers, no loss of organic solvent and no catalyst leaching. Especially, the advantage of non-volatility of ionic liquids compared to traditional supported catalysts comprising organic solvents with low vapor pressure (SLP) or water (SAP) on supports, which are clearly limited by solvent evaporation, makes it possible to perform long-term stable, gas-phase, continuous processes using homogeneous, molecularly defined catalyst complexes in continuous fixed-bed reaction technology.

References

1. D. J. Cole-Hamilton, *Science* **2003**, *299*, 1702–1706, and references therein.
2. J. C. Bailar Jr., *Catal. Rev.* **1974**, *10*, 17–36.
3. A. Choplin, F. Quignard, *Coord. Chem. Rev.* **1998**, *178–180*, 1679–1702.
4. M. H. Valkenberg, W. F. Hölderich, *Catal. Rev., Sci. Eng.* **2002**, *44*, 321–374.
5. F. Quignard, A. Choplin, in *Comprehensive Coordination Chemistry II*, Vol. 9, J. A. McCleverty, T. J. Meyer (Eds.), Elsevier, Amsterdam, **2004**, pp. 445–470.
6. B. Driessen-Hölscher, *Adv. Catal.* **1998**, *42*, 473–505.
7. *Applied Homogeneous Catalysis with Organometallic Compounds*, 1st edn., B. Cornils, W. A. Herrmann (Eds.), VCH-Wiley, Weinheim, Germany, **1996**.
8. W. A. Herrmann, C. W. Kohlpaintner, *Angew. Chem. Int. Ed. Engl.* **1993**, *32*, 1524–1544.
9. E. de Wolf, G. van Koten, B. J. Deelman, *Chem. Soc. Rev.* **1999**, *28*, 37–41.
10. C. C. Tzschucke, C. Markert, W. Bannwarth, S. Roller, A. Hebel, R. Haag, *Angew. Chem. Int. Ed.* **2002**, *41*, 3964–4000.
11. C. Copéret, M. Chabanas, R. P. Saint-Arroman, J.-M. Basset, *Angew. Chem. Int. Ed.* **2003**, *42*, 156–181.
12. P. McMorn, G. J. Hutchuings, *Chem. Soc. Rev.* **2004**, *33*, 108–122.
13. D. E. De Vos, B. F. Sels, P. A. Jacobs, *Adv. Catal.* **2001**, *46*, 1–87.
14. A. Corma, H. García, *Chem. Rev.* **2002**, *102*, 3837–3892.
15. A. Corma, H. García, *Chem. Rev.* **2003**, *103*, 4307–4365.
16. P. Wasserscheid, W. Keim, *Angew. Chem. Int. Ed. Engl.* **2000**, *39*, 3772–3789.
17. C. M. Gordon, *Appl. Catal. A: General* **2001**, *222*, 101–117.
18. D. Zhao, M. Wu, Y. Kou, E. Min, *Catal. Today* **2002**, *74*, 157–189.
19. H. Olivier-Bourbigou, L. Magna, *J. Mol. Catal. A: Chem.* **2002**, *182–183*, 419–437.

20. J. Dupont, R. F. de Souza, P. A. Z. Suarez, *Chem. Rev.* **2002**, *102*, 3667–3692.
21. C. P. Mehnert, *Chem. Eur. J.* **2005**, *11*, 50–56.
22. P. R. Rony, *Chem. Eng. Sci.* **1968**, *23*, 1021–1034.
23. M. Cinouini, S. Colonna, H. Molinari, F. Montanari, P. Tundo, *J. Chem. Soc., Chem. Commun.* **1976**, 394–396.
24. P. Tundo, G. Moraglio, F. Trotta, *Ind. Eng. Chem. Res.* **1989**, *28*, 881–890.
25. P. Tundo, *Gazz. Chim. Ital.* **1990**, *120*, 69–76.
26. J. Villadsen, H. Livbjerg, *Catal. Rev., Sci. Eng.* **1978**, *17*, 203–272.
27. G. J. K. Acres, G. C. Bond, B. J. Cooper, J. A. Dawson, *J. Catal.* **1966**, *6*, 139.
28. K. T. Wan, M. E. Davis, *Nature* **1994**, *370*, 449–450.
29. M. J. Naughton, R. S. Drago, *J. Catal.* **1995**, *155*, 383–389.
30. J. P. Arhancet, M. E. Davis, J. S. Merola, B. E. Hanson, *Nature* **1989**, *339*, 454–455.
31. M. E. Davis, *CHEMTECH* **1992**, 498–502.
32. M. E. Davis, J. P. Arhancet, B. E. Hanson, *US Patent* 4 994 427, **1991**, Virginia Tech, USA.
33. M. E. Davis, J. P. Arhancet, B. E. Hanson, *US Patent* 4 947 003, **1990**, Virginia Tech, USA.
34. A. Beckmann, F. J. Keil, *Chem. Eng. Sci.* **2003**, *58*, 841–847.
35. F. J. Keil, H.-P. Detering, A. Klein, *Per. Polytech., Chem. Eng.* **1994**, *38*, 99–103.
36. R. Datta, R. G. Rinker, *J. Catal.* **1985**, *95*, 181–192.
37. H. Livbjerg, T. S. Christensen, T. T. Hansen, J. Villadsen, *Sâdhanâ* **1987**, *10*, 185–216.
38. J. Haugaard, H. Livbjerg, *Chem. Eng. Sci.* **1998**, *53*, 2941–2948.
39. L. A. Abramova, S. P. Baranov, A. A. Dulov, *Appl. Catal. A: General* **2000**, *193*, 243–250.
40. L. A. Abramova, S. P. Baranov, A. A. Dulov, *Appl. Catal. A: General* **2000**, *193*, 251–256.
41. S. Baldelli, *J. Phys. Chem. B.* **2003**, *107*, 6148–6152.
42. T. J. Gannon, G. Law, P. R. Watson, A. J. Carmichael, K. R. Seddon, *Langmuir* **1999**, *15*, 8429–8434.
43. G. Law, P. R. Watson, *Chem. Phys. Lett.* **2001**, *345*, 1–4.
44. G. Law, P. R. Watson, A. J. Carmichael, K. R. Seddon, *Phys. Chem. Chem. Phys.* **2001**, *3*, 2879–2885.
45. A. J. Carmichael, C. Hardacre, J. D. Holbrey, M. Nieuwenhuyzen, K. R. Seddon, *Mol. Phys.* **2001**, *99*, 795–800.
46. B. D. Fitchett, J. C. Conboy, *J. Phys. Chem. B.* **2004**, *108*, 20255–20262.
47. L. Cammarata, S. G. Kazarian, P. A. Salter, T. Welton, *Phys. Chem. Chem. Phys.* **2001**, *3*, 5192–5200.
48. E. Blum, *Swiss Patent*, CH71 326 **1914**, BASF, Germany.
49. O. B. Lapina, B. S. Balzhinimaev, S. Boghosian, K. M. Eriksen, R. Fehrmann, *Catal. Today* **1999**, *51*, 469–479.
50. J. H. Frazer, W. J. Kirkpatrick, *J. Am. Chem. Soc.* **1940**, *62*, 1659–1660.
51. H. F. A. Topsøe, A. Nielsen, *Trans. Danish Acad. Tech. Sci.* **1948**, *1*, 3–17.
52. C. N. Kenney, *Catal. Rev., Sci. Eng.* **1975**, *11*, 197–224.
53. *GB Patent*, GB1 351 802 **1970**, Societe Cooperative Metachimie, Belgium.
54. D. B. Fox, E. H. Lee, *CHEMTECH* **1973**, *3*, 186–189.
55. V. Rao, R. Datta, *J. Catal.* **1988**, *114*, 377–387.
56. S. Malhotra, R. Datta, *Proc. Electrochem. Soc.* **1994**, *94*, 773–780.
57. A. D. Schmitz, D. P. Eyman, *Energy Fuels* **1994**, *8*, 729–740.
58. R. M. Lago, M. L. H. Green, S. C. Tsang, M. Odlyha, *Appl. Catal. B: Environ.* **1996**, *8*, 107–121.
59. M. L. H. Green, R. M. Lago, S. C. Tsang, *J. Chem. Soc., Chem. Commun.* **1995**, 365.
60. X. Wu, Y. A. Letuchy, D. P. Eyman, *J. Catal.* **1996**, *161*, 164–177.
61. M. R. R. Prasad, K. Krishnan, K. N. Ninan, V. N. Krishnamurthy, *Thermochim. Acta* **1997**, *297*, 207–210.
62. M. Amirnasr, Md. K. Nazeeruddin, M. Grätzel, *Thermochim. Acta* **2000**, *348*, 105–114.
63. E. Benazzi, A. Hirschauer, J. F. Joly, H. Olivier, J. Y. Bernhard, *Eur. Patent*, 0553

009 **1993**, Institut Français du Pétrole, France.
64. E. Benazzi, H. Olivier, Y. Chauvin, J. F. Joly, A. Hirschauer, *Abstr. Pap. Am. Chem. Soc.* **1996**, *212*, 45.
65. E. Benazzi, Y. Chauvin, A. Hirschauer, N. Ferrer, H. Olivier, J. Y. Bernhard, *US Patent* 5 693 585 **1997**, Institut Français du Pétrole, France.
66. F. G. Sherif, L.-J. Shyu, *World Patent*, 9903 163 **1999**, Akzo Nobel Inc., USA.
67. C. deCastro, E. Sauvage, M. H. Valkenberg, W. F. Hölderich, *J. Catal.* **2000**, *196*, 86–94.
68. B. B. Randolph, E. L. Sughrue, G. W. Dodwell, *US Patent Application*, 20 050 033 102 **2005**, Hitchcock Fish & Dollar, Richmond, USA.
69. W. F. Hölderich, H. H. Wagner, M. H. Valkenberg, *Spec. Publ. R. Soc. Chem* **2001**, *266*, 76–93.
70. M. H. Valkenberg, C. deCastro, W. F. Hölderich, *Stud. Surf. Sci. Catal.* **2001**, *135*, 4629–4636.
71. M. H. Valkenberg, C. deCastro, W. F. Hölderich, *Appl. Catal. A: General* **2001**, *215*, 185–190.
72. M. H. Valkenberg, C. deCastro, W. F. Hölderich, *Green Chem.* **2002**, *4*, 88–93.
73. M. H. Valkenberg, C. deCastro, W. F. Hölderich, *Top. Catal.* **2001**, *14*, 139–144.
74. http://www.synetix.com/technical/news.htm, **2004**, Johnson Matthey Catalysts.
75. B. Gadenne, P. Hesemann, J. J. E. Moreau, *Chem. Commun.* **2004**, 1768–1769.
76. T. M. Jyothi, M. L. Kaliya, M. V. Landau, *Angew. Chem. Int. Ed.* **2004**, *40*, 2881–2883.
77. T. Kume, T. Sasaki, Y. Iwasawa, *Photon Factory Act. Rep.* **2003**, *20*, 28.
78. K. Yamaguchi, C. Yoshida, S. Uchida, N. Mizuno, *J. Am. Chem. Soc.* **2005**, *127*, 530–531.
79. D. W. Kim, D. Y. Chi, *Angew. Chem. Int. Ed.* **2004**, *43*, 483–485.
80. J. Dupont, J. Spencer, *Angew. Chem. Int. Ed.* **2004**, *43*, 5296–5297.
81. P. Wasserscheid, M. Haumann, Catalyst Recycling Using Ionic Liquids, in *Catalyst Separation, Recovery and Recycling*, D. J. Cole-Hamilton, R. P. Tooze (Eds.), Catalysis by Metal Complexes Series, vol. 30, **2006**, Chapter 7.
82. S. Abelló, F. Medina, X. Rodríguez, Y. Cesteros, P. Salagre, J. E. Sueiras, D. Tichit, B. Coq, *Chem. Commun.* **2004**, 1096–1097.
83. M. Gruttadauria, S. Riela, P. L. Meo, F. D'Anna, R. Noto, *Tetrahedron Lett.* **2004**, *45*, 6113–6116.
84. D. Li, F. Shi, Y. Deng, *Tetrahedron Lett.* **2004**, *45*, 6791–6794.
85. D. Li, F. Shi, S. Guo, Y. Deng, *Tetrahedron Lett.* **2004**, *45*, 265–268.
86. M. J. Muldoon, C. M. Gordon, *J. Polym. Sci. A: Polym. Chem.* **2004**, *42*, 3865–3869.
87. C. P. Mehnert, R. A. Cook, *US Patent*, 6 673 737 **2004**, ExxonMobil Research and Engineering Company, USA.
88. C. P. Mehnert, R. A. Cook, N. C. Dispenziere, M. Afeworki, *J. Am. Chem. Soc.* **2002**, *124*, 12932–12933.
89. C. P. Mehnert, R. A. Cook, E. J. Mozeleski, N. C. Dispenziere, M. Afeworki, *Supported ionic liquid catalysis for hydroformylation and hydrogenation reactions*, **2003**, 226th ACS National Meeting, New York, USA.
90. Y. Yang, H. Lin, C. Deng, J. She, Y. Yuan, *Chem. Lett.* **2005**, *34*, 220–221.
91. A. Riisager, P. Wasserscheid, R. van Hal, R. Fehrmann, *J. Catal.* **2003**, *219*, 252–255.
92. A. Riisager, K. M. Eriksen, P. Wasserscheid, R. Fehrmann, *Catal. Lett.* **2003**, *90*, 149–153.
93. A. Riisager, R. Fehrmann, S. Flicker, R. van Hal, M. Haumann, P. Wasserscheid, *Angew. Chem. Int. Ed.* **2005**, *44*, 815–819.
94. B. Heinrich, Y. Chen, J. Hjortkjær, *J. Mol. Chem.* **1993**, *80*, 365–375.
95. A. Riisager, K. M. Eriksen, J. Hjortkjær, R. Fehrmann, *J. Mol. Catal. A: Chem.* **2003**, *193*, 259–272.
96. J. Hjortkjær, Y. Chen, B. Heinrich, *Appl. Catal.* **1991**, *67*, 269–278.
97. A. Riisager, R. Fehrmann, P. Wasserscheid, R. van Hal, Supported Ionic Liquid Phase Catalysis-Heterogenization of Homogeneous Rhodium Phosphite Catalysts in *Ionic*

558 | 5 Organic Synthesis

Liquids IIIB: Transformations and Processes, R. Rogers, K. R. Seddon (Eds.), American Chemical Society, Washington DC, 2005, pp. 334–349.
98. H. Bohnen, J. Herwig, D. Hoff, R. van Hal, P. Wasserscheid, Eur. Patent, 1 400 504-A1 2003, Celanese Chemical Europe GmbH, Germany.
99. S. M. Silva, R. P. J. Bronger, Z. Freixa, J. Dupont, P. W. N. M. van Leeuwen, New J. Chem. 2003, 27, 1294–1296.
100. C. P. Mehnert, E. J. Mozeleski, R. A. Cook, Chem. Commun. 2002, 3010–3011.
101. T. H. Cho, J. Fuller, R. T. Carlin, High Temp. Mater. Process. 1998, 2, 543–558.
102. Y. Chauvin, L. Mussmann, H. Olivier, Angew. Chem. Int. Ed. Engl. 1995, 34, 2698–2700.
103. A. Wolfson, I. F. J. Vankelecom, P. A. Jacobs, Tetrahedron Lett. 2003, 44, 1195–1198.
104. S. Breitenlechner, M. Fleck, T. E. Müller, A. Suppan, J. Mol. Catal. A: Chem. 2004, 214, 175–179.
105. H. Hagiwara, Y. Sugawara, K. Isobe, T. Hoshi, T. Suzuki, Org. Lett. 2004, 6, 2325–2328.
106. A. Corma, H. García, A. Leyva, Tetrahedron 2004, 60, 8553–8560.
107. R. T. Carlin, T. H. Cho, J. Fuller, Proc. Electrochem. Soc. 1998, 98, 180–185.
108. M. Antonietti, D. Kuang, B. Smarsly, Y. Zhou, Angew. Chem. Int. Ed. 2004, 43, 2–6.
109. J. Huang, T. Jiang, H. Gao, B. Han, Z. Liu, W. Wu, Y. Chang, G. Zhao, Angew. Chem. Int. Ed. 2004, 43, 1397–1399.
110. J. Huang, T. Jiang, B. Han, H. Gao, Y. Chang, G. Zhao, W. Wu, Chem. Commun. 2003, 1654–1655.
111. S. Dai, Y. H. Ju, H. J. Gao, J. S. Lin, S. J. Pennecook, C. E. Barnes, Chem. Commun. 2000, 243–244.
112. K. Anderson, S. C. Fernández, C. Hardacre, P. C. Marr, Inorg. Chem. Commun. 2004, 7, 73–76.
113. R. T. Carlin, J. Fuller, Chem. Commun. 1997, 1345–1346.
114. R. T. Carlin, T. H. Cho, J. Fuller, Proc. Electrochem. Soc. 2000, 99, 20–26.
115. L. C. Branco, J. G. Crespo, C. A. M. Afonso, Chem. Eur. J. 2002, 8, 3865–3871.
116. L. C. Branco, J. G. Crespo, C. A. M. Afonso, Angew. Chem. Int. Ed. 2002, 41, 2771–2773.
117. M. Matsumoto, Y. Inomoto, K. Kondo, J. Membrane Sci. 2005, 246, 77–81.
118. E. Miyako, T. Maryyama, N. Kamiya, M. Goto, Biotech. Lett. 2003, 25, 805–808.
119. E. Miyako, T. Maruyama, N. Kamiya, M. Goto, Chem. Commun. 2003, 2926–2927.

5.7
Multiphasic Catalysis Using Ionic Liquids in Combination with Compressed CO_2

Peter Wasserscheid and Sven Kuhlmann

5.7.1
Introduction

Ionic liquids are often considered as promising solvents for "clean processes" and "green chemistry" mainly due to their non-volatile character [1, 2]. These two catchwords represent current efforts to drastically reduce the amounts of side and coupling products, as well as the solvent and catalyst consumption in chemical processes. As another "green solvent" concept for chemical reactions the replacement of volatile organic solvents by supercritical CO_2 ($scCO_2$) is frequently discussed [3].

$scCO_2$ combines an environmentally benign character (non-toxic, non-flammable) with favorable physicochemical properties for chemical synthesis. Catalyst separation schemes have been developed on the basis of the tuneable phase behavior of $scCO_2$ (CESS process) [4].

However, ionic liquids and $scCO_2$ are not competing concepts for the same applications. While ionic liquids can be considered as replacements for polar organic solvents, the use of $scCO_2$ can cover those applications where non-polar solvents are usually used.

With regard to homogeneous transition metal catalyzed reactions, both media show complementary strengths and weaknesses. While ionic liquids are known to be excellent solvents for many transition metal catalysts (see Section 5.3), the solubility of most transition metal complexes in $scCO_2$ is poor. Usually, a special ligand design (e.g. phosphine ligands with fluorous "ponytails" [3]) is required to allow sufficient catalyst concentration in the supercritical medium. However, the product isolation from the solvent is always very easy in the case of $scCO_2$, while the product isolation from an ionic catalyst solution can become more and more complicated, depending on the solubility of the product in the ionic liquid and on the product's boiling point.

In cases where product solubility in the ionic liquid and the product's boiling point are high, the extraction of the product from the ionic liquid with an additional organic solvent is frequently proposed. This approach often suffers from some catalyst losses (due to small mutual solubility) and causes additional steps in the work-up. Moreover, the use of an additional, volatile extraction solvent may nullify the "green solvent" motivation to use ionic liquids as non-volatile solvents.

Beckman, Brennecke and their research groups first realized that the combination of $scCO_2$ and an ionic liquid can offer special advantages. They observed that, although $scCO_2$ is surprisingly soluble in some ionic liquids, the reverse is not the case, with no detectable ionic liquid solubilization in the CO_2 phase. Based on these results they described a method to remove naphthalene quantitatively from the ionic liquid [BMIM][PF_6] by extraction with $scCO_2$ [5]. Subsequent work by Brennecke's team has applied the same procedure to extract a large variety of different solutes from ionic liquids, without observing any ionic liquid contamination in the isolated substances [6]).

Research efforts aimed at quantifying the solubility of CO_2 in ionic liquids revealed a significant influence of the ionic liquid's water content on the CO_2 solubility. While water-saturated [BMIM][PF_6] (up to 2.3 wt.% water) has a CO_2 solubility of only 0.13 mol fraction, 0.54 mol fraction CO_2 dissolves in dry [BMIM][PF_6] (about 0.15 wt.% water) at 57 bar and 40 °C [7]. Kazarian et al. used ATR-IR to determine the CO_2 solubility in [BMIM][PF_6] and [BMIM][BF_4]. They reported a solubility of 0.6 mol fraction CO_2 in [BMIM][PF_6] at 68 bar and 40 °C [8]. More recent research activities by Brennecke et al. have been directed towards the elucidation of the solvent strength [9] and the phase behavior of $scCO_2$/IL mixtures [10] and the determination of the solubility of various gases in ionic liquids. These studies have revealed that the solubility of CO_2 in ionic liquids is mainly dependent on the anion structure. While [PF_6] and [BF_4]-based ILs show similar solubility, an increased solubility can be achieved using [BTA]-type ILs [11]. This behavior is largely independent of the cation structure (see also Section 3.4).

The same authors also described the solubility of other gases such as oxygen, nitrous oxide, ethylene, ethane etc. in ionic liquids [12] and the beneficial influence of added CO_2 making oxygen and methane more soluble in ionic liquids [13]. Interestingly, Brennecke was able to show that oxygen solubility could be enhanced by a factor of 5.7 using [HMIM][BTA] and relatively low pressure (5.8 bar total pressure). Leitner and coworkers reported a dramatic increase in hydrogen solubility in the ionic liquid [EMIM][BTA] by pressurizing the ionic liquid with CO_2 [14].

Other interesting and practically very relevant features of ionic liquid/scCO$_2$ biphasic systems have also been described. Han's group quantified for the first time the effect of compressed CO_2 on the viscosity of [BMIM][PF$_6$]. They concluded that compressed CO_2 can indeed be effectively applied to reduce the viscosity of ionic liquids, e.g. to increase mass transfer rates. Similar effects were qualitatively observed before by other groups for many different ionic liquids. In addition, Leitner and Scurto reported very recently that the presence of compressed CO_2 leads to a very large melting point suppression of many ammonium and phosphonium salts [15]. Thus, a simple ammonium salt, such as [NBu$_4$][BF$_4$] (mp 156 °C) could be applied in catalytic reactions as an ionic liquid (melting point down to 55 °C) when the salt was contacted with 150 bar of CO_2.

These publications demonstrate impressively that adding compressed CO_2 to an ionic liquid can significantly improve important physicochemical properties of the system. This offers a general tool to expand the application range of ionic liquids into more demanding process conditions.

5.7.2
Catalytic Reaction with Subsequent Product Extraction

The first application including a catalytic reaction in an ionic liquid and a subsequent extraction step with scCO$_2$ was reported by Jessop et al. in 2001 [16]. They described two different, asymmetric hydrogenation reactions using [Ru(OAc)$_2$(tolBINAP)] as the catalyst dissolved in the ionic liquid [BMIM][PF$_6$]. In the asymmetric hydrogenation of tiglic acid (Scheme 5.7-1), the reaction was carried out in a [BMIM][PF$_6$]/water biphasic mixture with excellent yield and selectivity. When the reaction was

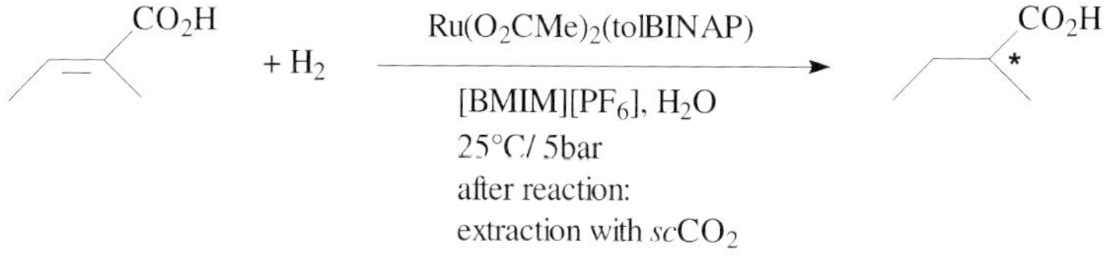

Scheme 5.7-1 Asymmetric, Ru-catalyzed hydrogenation of tiglic acid in [BMIM][PF$_6$] followed by extraction with scCO$_2$.

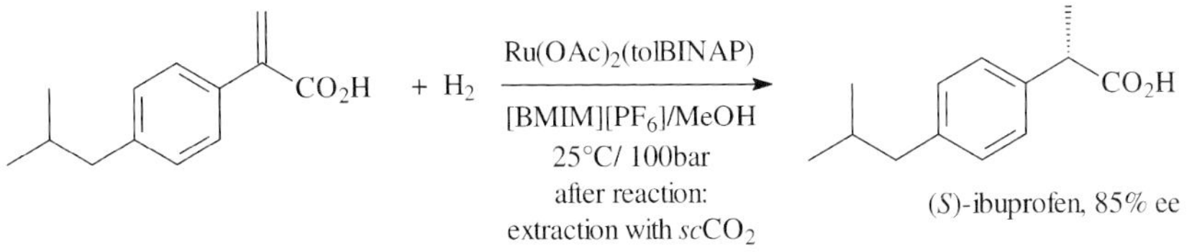

Scheme 5.7-2 Synthesis of ibuprofen by asymmetric, Ru-catalyzed hydrogenation in [BMIM][PF$_6$] with product isolation by subsequent extraction with scCO$_2$.

completed, the product was isolated by scCO$_2$ extraction without contamination by either catalyst or ionic liquid.

Similarly, the asymmetric hydrogenation of isobutylatropic acid to the anti-inflammatory drug ibuprofen has been carried out (Scheme 5.7-2). Here, the reaction was carried out in a [BMIM][PF$_6$]/MeOH mixture, again followed by a product extraction using scCO$_2$ (for more details on these hydrogenation reactions see Section 5.3).

Nunes da Ponte, Afonso and coworkers showed that asymmetric dihydroxylation of olefins could also be carried out successfully in [OMIM][PF$_6$], leading to taxol fragments, which were reported to have antitumor activity [17]. Product recovery by extraction with diethyl ether as well as scCO$_2$ was attempted with the latter showing better yield and lower osmium content. In both cases recycling of the catalyst IL phase was achieved for six consecutive times without loss of activity. The same authors reported that the asymmetric Sharpless dihydroxylation could also be applied to various other substrates (e.g. 1-hexene, styrene etc,) using the above-mentioned technique [18].

5.7.3
Catalytic Reaction with Simultaneous Product Extraction

Baker, Tumas and coworkers reported catalytic hydrogenation reactions in a biphasic reaction mixture consisting of the ionic liquid [BMIM][PF$_6$] and scCO$_2$ [19]. In the hydrogenation of 1-decene with Wilkinson's catalyst [RhCl(PPh$_3$)$_3$] at 50 °C and 48 bar H$_2$ (total pressure 207 bar) a conversion of 98% after 1 h was reported, corresponding to a turnover frequency (TOF) of 410 h^{-1}. Under identical conditions the hydrogenation of cylohexene proceeded with 82% conversion after 2 h (TOF = 220 h^{-1}). The isolated ionic catalyst solution could be recycled in consecutive batch experiments up to four times. However, the fact that a biphasic hydrogenation of 1-decene can be successfully achieved is not a special benefit of the unconventional biphasic system [BMIM][PF$_6$]/scCO$_2$. In fact, no reactivity advantage with using scCO$_2$ in place of a more common alkane solvent for such a biphasic system can be concluded from the reported results.

Later, Ballivet-Tkatchanko and coworkers compared the palladium catalyzed dimerization of methyl acrylate in ionic liquids (monophasic) and IL/scCO$_2$ (biphasic) [20]. They found that both reactions exhibited similar selectivities for tail-to-tail dihydromuconate (>98%) as well as TOF/TON (95 h^{-1} after 1 h/220 h^{-1} after 3 h). The latter was found to be a good indicator for the feasibility of a continuous mode of operation, as one might have expected a detrimental influence from the substrate being mainly dissolved in the CO_2 phase.

More recently, Leitner and coworkers studied the enantioselective hydrogenation of N-(1-phenylethylidene)aniline catalyzed by cationic iridium complexes with chiral phosphinooxazoline ligands in various ionic liquid/scCO$_2$ systems [14]. They proved for this specific system a plethora of beneficial effects. While the presence of the scCO$_2$ enabled the hydrogenation by high hydrogen availability (caused by both enhanced solubility and enhanced mass transfer rate due to lower ionic liquid viscosity), the presence of the ionic liquid led to an activation and stabilization of the cationic catalyst complex. Seven repetitive runs without significant catalyst deactivation could be realized.

5.7.4
Catalytic Conversion of CO_2 in an Ionic Liquid/scCO$_2$ Biphasic Mixture

Parallel to their findings concerning hydrogenation in IL/scCO2 (see Section 5.7.3), Baker, Tumas and coworkers also described the [RuCl$_2$(dppe)]-catalyzed (dppe=Ph$_2$P–(CH$_2$)$_2$–PPh$_2$) hydrogenation of CO_2 in the presence of dialkylamines to obtain N,N-dialkylformamides [19]. The reaction of di-n-propylamine in the system [BMIM][PF$_6$]/scCO$_2$ resulted in complete amine conversion, obtaining in high selectivity the desired N,N-di-n-propylformamide. The latter showed very high solubility in the ionic liquid phase and complete product isolation by extraction with scCO$_2$ proved to be difficult. However, product extraction with scCO$_2$ became possible once the ionic catalyst solution had become completely saturated with the product.

Other examples of CO_2 as a C1-building block include its reaction with propylene oxide as described by Kawanami and Ikushima [21]. They showed that the formation of cyclic carbonates in scCO$_2$ and [OMIM][BF$_4$] (the IL acts as the catalyst) can be achieved with various substrates in very high yield and selectivity (in some cases >99%) and up to 77 times faster than previously reported. Arai and coworkers have also demonstrated that epoxidation of olefinic substrates (i.e. styrene) with subsequent formation of the cyclic carbonate could be carried out in a one-pot synthesis using IL/scCO$_2$. Yields of up to 33% could be achieved in tetrabutylammonium bromide with *tert*-butyl hydroperoxide as the oxidizing agent [22]. The same authors have also reviewed the feasibility of using scCO$_2$ for the formation of cyclic carbonates [23].

5.7.5
Continuous Reactions in an Ionic Liquid/Compressed CO_2 System

Cole-Hamilton and coworkers demonstrated for the first time a flow apparatus for a continuous catalytic reaction using the biphasic system [BMIM][PF$_6$]/scCO$_2$ [24].

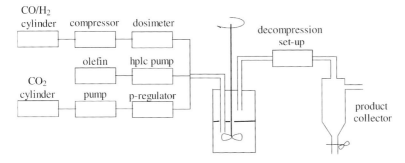

Fig. 5.7-1 Continuous flow apparatus as used for the hydroformylation of 1-octene in the biphasic system [BMIM][PF$_6$]/scCO$_2$.

They investigated the continuous Rh-catalyzed hydroformylation of 1-octene over periods up to 33 h using the ionic phosphine ligand [PMIM]$_2$[PhP(C$_6$H$_4$SO$_3$)$_2$] ([PMIM] = 1-n-propyl-3-methylimidazolium). No catalyst decomposition was observed during the reaction time and Rh leaching into the scCO$_2$/product stream was less than 1 ppm. The selectivity for the linear hydroformylation product was found to be stable over the reaction time (n/iso = 3.1).

During the continuous reaction, alkene, CO, H$_2$ and CO$_2$ were fed separately into the reactor which contained the ionic liquid catalyst solution. The products and non-converted feedstock were removed from the ionic liquid still dissolved in scCO$_2$. After decompression, the liquid product was collected and analyzed.

Obviously, the motivation for performing this hydroformylation reaction in a continuous flow reactor arose from some problems during the catalyst recycling when the same reaction was first carried out in repetitive batch mode. In the latter case, Cole-Hamilton et al. observed a continuous drop of the product's n/iso-ratio from 3.7 to 2.5 over the first nine runs. Moreover, the isomerization activity of the system increased during the batchwise recycling experiments and after the ninth run Rh leaching became significant. From ^{31}P-NMR investigations the authors concluded that ligand oxidation due to contamination of the systems with air (during the reactor openings for recycling) led to the formation of [RhH(CO)$_4$] as the active catalytic species. This is known to show more isomerization activity and lower n/iso ratio than the phosphine-modified catalyst system. Moreover, [RhH(CO)$_4$] is more soluble in scCO$_2$, which explains the observed leaching of rhodium into the organic layer.

All these problems related to the batchwise catalyst recycling could be convincingly overcome by applying the above described continuous operation mode. The authors concluded that the scCO$_2$–ionic liquid biphasic system provides a method for continuous flow homogeneous catalysis with integrated separation of the products from the catalyst and from the reaction solvent. Most interestingly, this unusual, continuous biphasic reaction mode enables the quantitative separation of relatively high boiling products from the ionic catalyst solution under mild temperature conditions and without using an additional organic extraction solvent. More

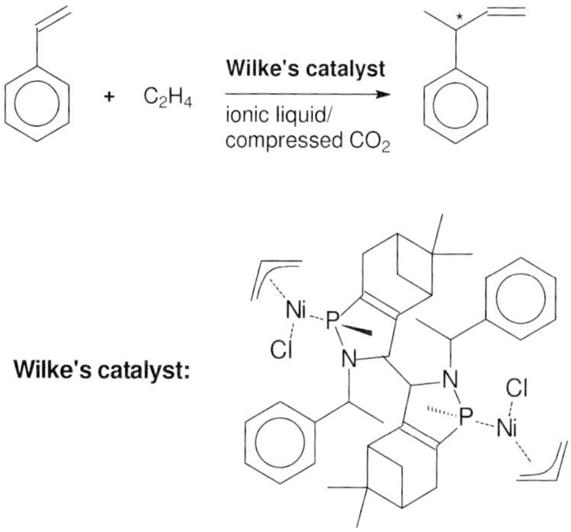

Scheme 5.7-3 The enantioselective hydrovinylation of styrene using Wilke's catalyst.

recently, Cole-Hamilton and coworkers were also able to show for the first time that high-boiling substrates (up to 1-dodecene) could be hydroformylated in a continuous operation using the methodology described above (IL = [BMIM][PF$_6$]). In the corresponding aqueous biphasic catalysis, alkene chain-length is limited to C$_6$ by solubility and thus mass-transfer effects. The same phenomenon was observed in the IL/scCO$_2$ reaction mode (TOF = 5–10 h^{-1} vs. 500–700 h^{-1} for aqueous biphasic propene hydroformyliation), but to a lesser extent. Moreover, by increasing the alkyl chain-length in the imidazolium cation from butyl to octyl the reaction rate could be doubled, and by exchanging the [PF$_6$]$^-$ anion to [BTA]$^-$ a further increase of more than 20% could be achieved. Thus, the conversion in the continuous reactor could be pushed to over 80% and, by adjusting the residence time, an optimized turnover frequency of up to 500 h^{-1} could be achieved. Rhodium leaching is well below 1 ppm and the long-term stability of the catalyst could be proven over a period of 80 h [25].

Slightly later and independently from Cole-Hamilton's pioneering work, the author's group demonstrated, in collaboration with Leitner et al., that the combination of a suitable ionic liquid and compressed CO$_2$ can offer much greater potential for homogeneous transition metal catalysis than being just a new protocol for easy product isolation and catalyst recycling. In the Ni-catalyzed hydrovinylation of styrene it was possible to activate, tune and immobilize the well-known Wilke complex by use of this unusual biphasic system (Scheme 5.7-3). Obviously, this reaction benefits from this special solvent combination in a new and highly promising manner.

Hydrovinylation is the transition metal catalyzed co-dimerization of alkenes with ethene yielding 3-substituted 1-butenes [26]. This powerful carbon–carbon bond forming reaction can be achieved with high enantioselectivity using Wilke's complex as a catalyst precursor [27]. In conventional solvents, this pre-catalyst needs to be activated with a chloride abstracting agent, e.g. Et$_3$Al$_2$Cl$_3$. Leitner et al. reported the use of Wilke's complex in compressed CO_2 (under liquid and under supercritical conditions) after activation with alkali salts of weekly-coordinating anions such as Na[BARF] ($[BARF]^- = [(3,5\text{-}CF_3)_2C_6H_3)_4B]^-$) [28].

At first, the reaction was investigated in batch mode using different ionic liquids with weakly coordinating anions as the catalyst medium and compressed CO_2 as a simultaneous extraction solvent. These experiments revealed that the activation of Wilke's catalyst by the ionic liquid medium is obviously very dependent on the nature of the ionic liquid's anion. Comparison of the results in different ionic liquids with $[EMIM]^+$ as the common cation showed that the catalyst's activity decreases in the order $[BARF]^- > [Al\{OC(CF_3)_2Ph\}_4]^- > [(CF_3SO_2)_2N]^- > [BF_4]^-$. This trend is consistent with the estimated nucleophilicity/coordination strength of the anions.

Interestingly, the specific environment of the ionic solvent system appears to activate the chiral Ni catalyst beyond a simple anion exchange reaction. This becomes obvious by the fact that even the addition of a 100-fold excess of $Li[(CF_3SO_2)_2N]$ or $Na[BF_4]$ in pure, compressed CO_2 lead at best to a moderate activation of Wilke's complex in comparison to the reaction in ionic liquids with the corresponding counter ion (e.g. 24.4% styrene conversion with 100-fold excess of $Li[(CF_3SO_2)_2N]$ in comparison to 69.9% conversion in $[EMIM][(CF_3SO_2)_2N]$ under otherwise identical conditions).

In the biphasic batch reaction the best reaction conditions were assessed for the system $[EMIM][(CF_3SO_2)_2N]$/compressed CO_2. It was found that increasing the partial pressure of ethylene and decreasing the temperature helped to suppress the concurrent side reactions (isomerization and oligomerization). 58% conversion of styrene (styrene/Ni = 1000/1) was achieved after 1 h under 40 bar of ethylene at 0 °C whereby 3-phenyl-1-butene was detected as the only product with 71% ee of the (R)-isomer.

However, attempts to reuse the ionic catalyst solution in consecutive batch experiments failed. While the products could be readily isolated after reaction by extraction with $scCO_2$, the active nickel species deactivated rapidly within three to four batchwise cycles. The fact that no such deactivation was observed in later experiments using the continuous flow apparatus described below (see Fig. 5.7-2) clearly indicates the deactivation of the chiral Ni catalyst being mainly related to the instability of the active species in the absence of substrate.

In the continuous hydrovinylation experiments, the ionic catalyst solution was placed into the reactor R where it was in intimate contact with the continuous reaction phase entering from the bottom (no stirring was used in these experiments). The reaction phase was made up in the mixer from a pulsed flow of ethylene and a continuous flow of styrene and compressed CO_2.

Figure 5.7-3 shows the results of a lifetime study for Wilke's catalyst dissolved, activated and immobilized in the system $[EMIM][(CF_3SO_2)_2N]$/compressed CO_2. Over a period of over 61 h, the active catalyst showed a remarkably stable activity while the enantioselectivity dropped only slightly over this period. These results clearly indicate – at least for the hydrovinylation of styrene with Wilke's catalyst – that an ionic liquid catalyst solution can show excellent catalytic performance under continuous product extraction with compressed CO_2.

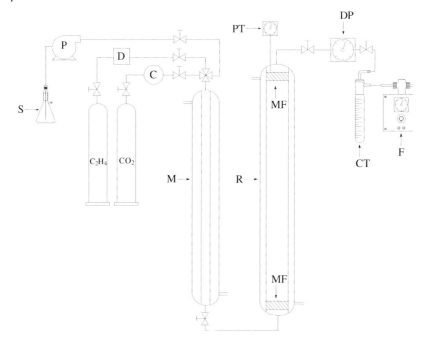

Fig. 5.7-2 Schematic view of the continuous flow apparatus used for the enantioselective hydrovinylation of styrene in the biphasic system [EMIM][(CF$_3$SO$_2$)$_2$N]. The parts are labeled as follows (alphabetically): C, Compressor; CT, cold trap; D, dosimeter; DP, depressurizer; F, flowmeter; M, mixer; MF, metal filter; P, HPLC pump; PT, pressure transducer and thermocouple; R, reactor; S, styrene.

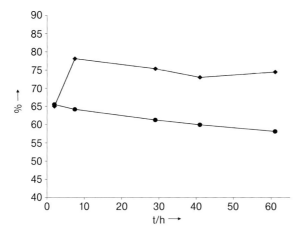

Fig. 5.7-3 Lifetime study of Wilke's catalyst in the hydrovinylation of styrene activated and immobilized in the system [EMIM][(CF$_3$SO$_2$)$_2$N]/compressed CO$_2$ (● ee; ◆ conversion).

In the context of continuous-flow operations using $scCO_2$/ILs, it has to be mentioned that this concept has also been successfully applied to biocatalyzed reactions. Among others, Reetz et al. were able to demonstrate that kinetic resolution of alcohols is possible by lipase-catalyzed alcohol esterification in ILs and subsequent continuous extraction of the ester derivatives with $scCO_2$ [29]. For further information, see Chapter 8.

5.7.6
Concluding Remarks and Outlook

The combination of ionic liquids and compressed CO_2 – which are on the extreme ends of the volatility and polarity scale – offers a new intriguing immobilization technique for homogeneous catalysis.

In comparison to catalytic reactions in compressed CO_2 alone, many transition metal complexes are much more soluble in ionic liquids without the need for special ligands. Moreover, the ionic liquid catalyst phase provides the potential to activate and tune the organometallic catalyst. Furthermore, product separation from the catalyst is now possible without exposing the catalyst to a variation of temperature, pressure or substrate concentration.

In contrast to the use of pure ionic liquid, the presence of compressed CO_2 greatly decreases the viscosity of the ionic catalyst solution, thus facilitating mass transfer during the catalytic reaction. Moreover, high boiling products with some solubility in the ionic liquid phase can now be removed without the use of an additional organic solvent. Finally, the use of compressed CO_2 as the mobile phase enables the use of a reactor design that is very similar to a classical fixed bed reactor [30]. Thus, the combination of ionic liquids and compressed CO_2 provides a new highly attractive approach that benefits from the advantages of both homogeneous and heterogeneous catalysis. Moreover, this approach promises to overcome some of the well-known limitations of conventional biphasic catalysis (catalyst immobilization, feedstock solubility in the catalytic phase, solvent cross-contamination, mass transfer limitation). In particular, the combination of non-volatile ionic liquids with non-hazardous CO_2 offers fascinating new possibilities to design environmentally benign processes.

References

1. M. Freemantle, *Chem. Eng. News* **1998**, 76(13), 32.
2. (a) M. Freemantle, *Chem. Eng. News* **1999**, 77(1), 23; (b) D. Bradley, *Chem. Ind.* **1999**, 86; (c) M. Freemantle, *Chem. Eng. News* **2000**, 78(20), 37.
3. (a) P. G. Jessop, W. Leitner (Eds.). *Chemical Synthesis Using Supercritical Fluids*, Wiley-VCH, Weinheim, **1999**; (b) For reviews, see P. G. Jessop, T. Ikariya, R. Noyori, *Science*, **1995**, 269, 1065–1069; P. G. Jessop, T. Ikariya, R. Noyori, *Chem. Rev.* **1999**, 99, 475–493; M. Poliakoff, S. M. Howdle, S. G. Kazarian, *Angew. Chem. Int. Ed. Engl.* **1995**, 34, 1275–1295.

4. (a) G. Franciò, K. Wittmann, W. Leitner, *J. Organomet. Chem.* **2001**, *621*, 130–142; (b) S. Kainz, A. Brinkmann, W. Leitner, A. Pfaltz, *J. Am. Chem. Soc.* **1999**, *121*, 6421–6429; (c) D. Koch, W. Leitner, *J. Am. Chem. Soc.* **1998**, *120*, 13398–13404; (d) W. Leitner, *Pure Appl. Chem.* **2004**, *76*, 635–644.
5. L. A. Blanchard, D. Hancu, E. J. Beckman, J. F. Brennecke, *Nature* **1999**, *299*, 28–29.
6. L. A. Blanchard, J. F. Brennecke, *Ind. Eng. Chem. Res.* **2001**, *40*, 287–292.
7. L. A. Blanchard, Z. Gu, J. F. Brennecke, *J. Phys. Chem. B* **2001**, *105*, 2437.
8. S. G. Kazarian, B. J. Biscoe, T. Welton, *Chem. Commun.* **2000**, 2047–2048.
9. C. P. Fredlake, M. J. Muldoon, S. N. V. K. Aki, T. Welton, J. F. Brennecke, *Phys. Chem. Chem. Phys.* **2004**, *6*, 3280–3285.
10. S. N. V. K. Aki, B. R. Mellein, E. M. Saurer, J. F. Brennecke, *J. Phys. Chem. B* **2004**, *108*, 20355–20365.
11. C. Cadena, J. L. Anthony, J. K. Shah, T. I. Morrow, J. F. Brennecke, E. J. Maginn, *J. Am. Chem. Soc.* **2004**, *126*, 5300–5308.
12. J. L. Anthony, J. L. Anderson, E. J. Maginn, J. F. Brennecke, *J. Phys. Chem. B* **2005**, *109*, 6366–6374.
13. D. G. Hert, J. L. Anderson, S. N. V. K. Aki, J. F. Brennecke, *Chem. Commun.* **2005**, *20*, 2603–2605.
14. M. Solinas, A. Pfaltz, P. G. Cozzi, W. Leitner, *J. Am. Chem. Soc.* **2004**, *126*, 16142–16147.
15. A. M. Scurto, W. Leitner, *Chem. Commun.* **2006**, 3681–3683.
16. R. A. Brown, P. Pollett, E. McKoon, C. A. Eckert, C. L. Liotta, P. G. Jessop, *J. Am. Chem. Soc.* **2001**, *123*, 1254–1255.
17. A. Serbanovic, L. C. Branco, M. N. da Ponte, C. A. M. Afonso, *J. Organomet. Chem.* **2005**, *690*, 3600–3608.
18. L. C. Branco, A. Serbanovic, M. Numes da Ponte, C. A. M. Afonso, *Chem. Commun*, **2005**, 107–109.
19. F. Liu, M. B. Abrams, R. T. Baker, W. Tumas, *Chem. Commun.* **2001**, 433–434.
20. D. Ballivet-Tkatchenko, M. Picquet, M. Solinas, G. Franciò, P. Wasserscheid, W. Leitner, *Green Chem.* **2003**, *5*, 232–235.
21. H. Kawanami, A. Sasaki, K. Matsui, Y. Ikushima, *Chem. Commun.* **2003**, 896–897.
22. J. Sun, S. Fujita, B. M. Bhanage, M. Arai, *Catal. Commun.* **2004**, *5*, 83–87.
23. J. Sun, S. Fujita, M. Arai, *J. Organomet. Chem.* **2005**, *690*, 3490–3497.
24. (a) M. F. Sellin, P. B. Webb, D. J. Cole-Hamilton, *Chem. Commun.* **2001**, 781–782; (b) D. J. Cole-Hamilton, M. F. Sellin, P. B. Webb, World Patent, WO 0202218 (to the University of St. Andrews) **2002** [*Chem. Abstr.* **2002**, *136*, 104215].
25. P. B. Webb, M. F. Sellin, T. E. Kunene, S. Williamson, A. M. Z. Slawin, D. J. Cole-Hamilton, *J. Am. Chem. Soc.* **2003**, *125*, 15577–15588.
26. For reviews, see (a) P. W. Jolly, G. Wilke, *Applied Homogenous Catalysis with Organic Compounds 2*, B. Cornils, W. A. Herrman (Eds.), Wiley-VCH, **1996**, pp. 1024–1048; (b) T. V. RajanBabu, N. Nomura, J. Jin, B. Radetich, H. Park, M. Nandi, *Chem. Eur. J.* **1999**, *5*, 1963–1968.
27. G. Wilke, J. Monkiewicz, H. Kuhn, German Patent, DE 3618169 (to Studiengesellschaft Kohle m.b.H., Germany), **1987** [*Chem. Abstr.* **1988**, *109*, P6735].
28. A. Wegner, W. Leitner, *Chem. Commun.* **1999**, 1583–1584.
29. (a) M. Reetz, W. Wiesenhöfer, G. Francio, W. Leitner, *Adv. Synth. Catal.* **2003**, *345*, 1221–1228; (b) M. T. Reetz, W. Wiesenhöfer, G. Franciò, W. Leitner, *Chem. Commun.* **2002**, 992–993.
30. For use of CO_2 as a mobile phase in classical heterogeneous catalysis, see: W. K. Gray, F. R. Smail, M. G. Hitzler, S. K. Ross, M. Poliakoff, *J. Am. Chem. Soc.* **1999**, *121*, 10711–10718.

6
Inorganic Synthesis

6.1
Directed Inorganic and Organometallic Synthesis

Tom Welton

Although a great deal of excitement has surrounded the use of ionic liquids as solvents for organic synthesis, the rational synthesis of inorganic and organometallic compounds in ionic liquids has remained largely unexplored. However, there are a few promising examples beginning to appear.

6.1.1
Coordination Compounds

Some halogenometallate species have been observed that have formed spontaneously during spectro-electrochemical studies in ionic liquids. For example $[MoCl_6]^{2-}$, which is hydrolyzed in water, coordinated by solvent in polar molecular solvents and has salts that are insoluble in non-polar solvents, can only be observed in basic $\{X(AlCl_3) < 0.5\}$ chloroaluminate ionic liquids [1]. However, this work has been directed at the measurement of electrochemical data, rather than exploiting the ionic liquids as solvents for synthesis [2]. It has also been shown that the tetrachloroaluminate ion will act as a bidentate ligand in acidic $\{X(AlCl_3) > 0.5\}$ chloroaluminate ionic liquids, forming $[M(AlCl_4)_3]^-$ ions [3]. Again, this was as a result of the spontaneous formation of the complexes, rather than a deliberate attempt to synthesize them.

The only reports of directed synthesis of coordination complexes in ionic liquids are of oxo-exchange chemistry. Exposure of chloroaluminate ionic liquids to water leads to the formation of a variety of aluminum oxo- and hydroxo-containing species [4]. Dissolution of more oxophilic metals than aluminum will generate metal oxohalide species. Hussey et al. have used phosgene ($COCl_2$) to deoxochlorinate $[NbOCl_5]^{2-}$ [5] (Scheme 6.1-1).

Ionic Liquids in Synthesis, Second Edition. P. Wasserscheid and T. Welton (Eds.)
Copyright © 2008 WILEY-VCH Verlags GmbH & Co. KGaA, Weinheim
ISBN: 978-3-527-31239-9

$$[NbOCl_5]^{2-} \underset{O^{2-}}{\overset{COCl_2}{\rightleftarrows}} [NbCl_6]^-$$

Scheme 6.1-1 Nb(V) oxo-exchange chemistry in a basic [EMIM]Cl–AlCl$_3$ ionic liquid.

$$[VO_2Cl_2]^- \underset{PhIO}{\overset{triphosgene}{\rightleftarrows}} [VOCl_4]^{2-} \underset{O_2}{\overset{triphosgene}{\rightleftarrows}} [VCl_6]^{3-}$$

Scheme 6.1-2 Vanadium oxo-exchange chemistry in a basic [EMIM]Cl–AlCl$_3$ ionic liquid.

Triphosgene {bis(trichloromethyl)carbonate} has been used to deoxochlorinate [VOCl$_4$]$^{2-}$ to [VCl$_6$]$^{3-}$ and [VO$_2$Cl$_2$]$^-$ to [VOCl$_4$]$^{2-}$ [6] (Scheme 6.1-2). In both these cases the deoxochlorination was accompanied by spontaneous reduction of the initial products.

More recently, it was found that [MPP]$_2$[Yb(Tf$_2$N)$_4$] {where MPP is 1-methyl-1-propylpyrrolidinium}could be crystallized from a solution of YbI$_2$ in [MPP][Tf$_2$N] [7]. However, when NdI$_3$ was added to the same ionic liquid, [MPP]$_3$[NdI$_6$] precipitated [8]. In both of these examples a metathesis reaction is occurring, e.g.:

$$2NdI_3 + 5[MMP][Tf_2N] \rightarrow [MPP]_3[NdI_6] + [MPP]_2[Nd(Tf_2N)_5]$$

The subtle change of the ionic liquid to [BMP][Tf$_2$N] leads to the precipitation of [BMP][NdI$_6$][Tf$_2$N] [8]. If, one day, this behavior can be controlled, this could lead to a method of preparing a range of interesting structures.

6.1.2
Organometallic Compounds

With the current excitement that is being generated by the, so-called, N-heterocyclic carbenes (NHCs, imidazolylidenes) [9], reports of imidazolium-based ionic liquids being used to prepare metal imidazolylidene complexes are starting to appear [10]. The first came from Xiao et al., who prepared bis(imidazolylidene)palladium(II) dibromide in [BMIM]Br [11]. All four possible conformers were formed (Scheme 6.1-3).

In the presence of triphenylphosphine and four equivalents of chloride, bis(triphenylphosphine)(1-butyl-3-methylimidazolylidene)palladium(II) chloride is formed (Scheme 6.1-4) [12]. Whereas in [EMIM]Cl–AlCl$_3$ (1.3:1), Pt(C$_2$H$_4$)(EMIMY)Cl$_2$ (where EMIMY is 1-ethyl-3-methylimidazol-2-ylidene) can be formed in the presence of ethane.

It has been shown that NHC palladium (or nickel or platinum) hydride complexes can be formed by the oxidative addition of imidazolium cations of ionic liquids to palladium(0) {or Ni(0) or Pt(0)} species [13]. It has been shown that it is not only the C2 of the imidazolium ring that can bond to metals, but also the other ring carbons [14].

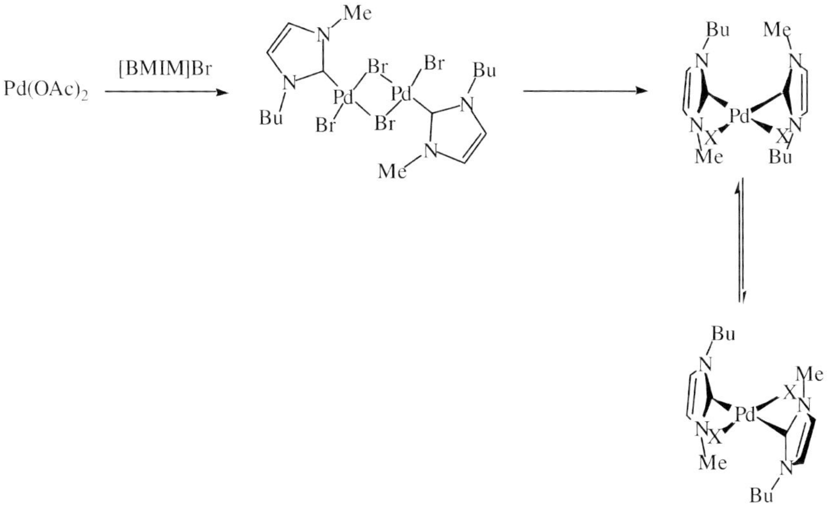

Scheme 6.1-3 The formation of bis(1-butyl-3-methylimidazolylidene) palladium(II) dibromide in [BMIM]Br.

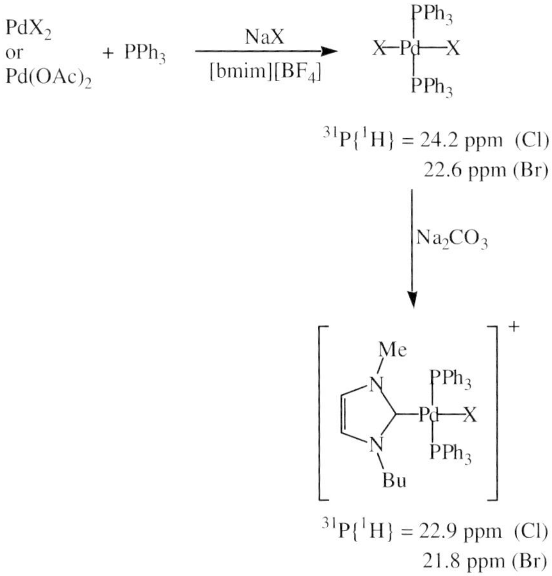

Scheme 6.1-4 The formation of bis(triphenylphosphine)(1-butyl-3-methyl imidazolylidene)palladium(II) chloride in [BMIM][BF$_4$].

^{31}P{^{1}H} = 24.2 ppm (Cl)
22.6 ppm (Br)

^{31}P{^{1}H} = 22.9 ppm (Cl)
21.8 ppm (Br)

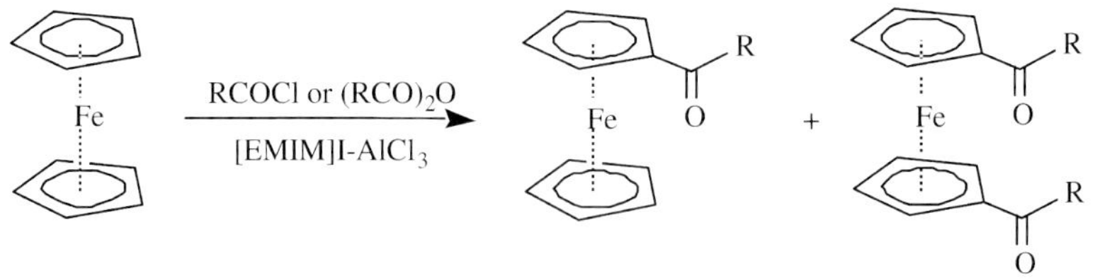

Scheme 6.1-5 The acylation of ferrocene in [EMIM]I–AlCl$_3$.

In another reaction of the cation of the ionic liquid, Dyson et al. [15] have reacted 1,3-dipropynylimidazolium and 1,3-dipentynylimidazolium ions with Co$_2$(CO)$_8$. One of the products formed, 3-dipentynylimidazolium bis(dicobalthexacarbonyl) tetrafluoroborate, was an ionic liquid itself.

Singer and coworkers have investigated the acylation reactions of ferrocene in ionic liquids made from mixtures of [EMIM]I and aluminum(III) chloride [16, 17]. The ionic liquid acts as both solvent and source of the Friedel-Crafts catalyst. In mildly acidic $\{X(AlCl_3) > 0.5\}$ [EMIM]I–AlCl$_3$, the monoacylated ferrocene was obtained as the major product. In strongly acidic [EMIM]I–AlCl$_3$ $\{X(AlCl_3) = 0.67\}$ the diacylated ferrocene was the major product. When R = alkyl, the diacylated product was usually the major product, but for R = Ph, the monoacylated product was favored (Scheme 6.1-5).

In another study that relies on chloroaluminate chemistry, the Fisher-Hafner ligand exchange reactions of ferrocene were investigated [18]. Again the acidic ionic liquids acted as a combination of solvent and catalyst. In these reactions it was necessary to add [BMIM][HCl$_2$] as a proton source, to generate the cyclopentadiene leaving group (Scheme 6.1-6).

The strong halide abstracting properties of acidic $\{X(AlCl_3) = 0.67\}$[BMIM] Cl–AlCl$_3$ have been also used for the synthesis of the "piano stool" complexes [Mn(CO)$_3$(η^6-arene)]$^+$ [19] (Scheme 6.1-7).

In all of the above cases using chloroaluminate ionic liquids the products were isolated by the destruction of the ionic liquids by addition to water.

6.1.3
Formation of Oxides

Recently, there has been some attention given to the preparation of oxides from solutions in ionic liquids. The first of these was the formation of a silica aerogel in [EMIM][Tf$_2$N] {Tf$_2$N= bis(trifluoromethylsulfonyl)imide} [20] (Scheme 6.1-8). Formic acid was added to tetramethylorthosilicate in the ionic liquid, yielding a gel that cured over a period of three weeks. Here, it was the non-volatile nature of the ionic liquid which prevented the loss of solvent during the curing process that was exploited. The ionic liquid was retrieved from the aerogel by extraction with acetonitrile.

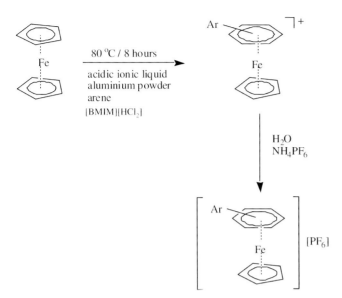

Scheme 6.1-6 Arene exchange reactions of ferrocene in [BMIM]Cl–AlCl$_3$.

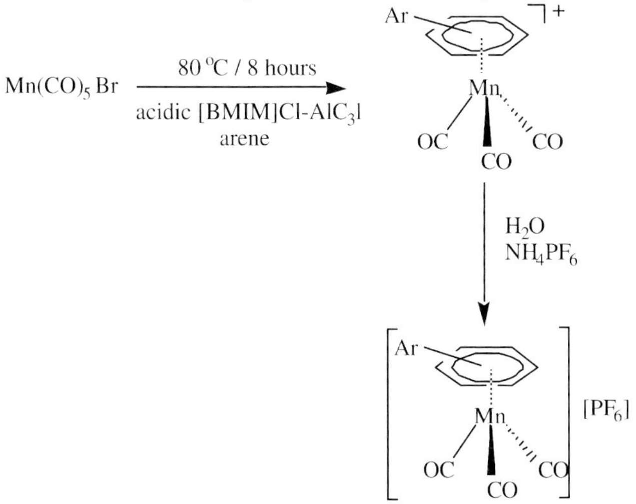

Mn(CO)$_5$Br $\xrightarrow[\text{arene}]{\text{80 °C / 8 hours} \atop \text{acidic [BMIM]Cl-AlCl}_3\text{l}}$ [Mn(CO)$_3$(Ar)]

↓ H$_2$O, NH$_4$PF$_6$

[Mn(CO)$_3$(η^6-arene)][PF$_6$]

Scheme 6.1-7 The synthesis of [Mn(CO)$_3$(η^6-arene)]$^+$ "piano stool" complexes in [bmim]Cl–AlCl$_3$.

2 HC(O)OH + (CH$_3$O)$_4$Si $\xrightarrow{\text{[EMIM][Tf}_2\text{N]}}$ SiO$_2$ + 2 CH$_3$OH + 2 HC(O)OCH$_3$

Scheme 6.1-8 The formation of a SiO$_2$ aerogel in [EMIM][Tf$_2$N].

Subsequently a number of preparations of silica have been reported [21]. It has been shown that it is possible to control the morphology of the deposited silica by changing the ionic liquid (for more details see Section 6.3) [22].

TiO_2 has been prepared by the careful hydrolysis of $TiCl_4$ in [BMIM][BF_4] [23]. The TiO_2 precipitated from the solution and, as with the silica preparation described above, acetonitrile was used to extract any of the ionic liquid contaminating the product (for more details see Section 6.3).

6.1.4
Other Reactions

The deposition of metals has been observed in a number of catalytic processes in ionic liquids and this will be discussed in Section 6.3. This has encouraged the exploration of the possibility of the chemical deposition of metals from solutions in ionic liquids. For instance, Ir(0) nanoparticles 2.3 ± 0.4 nm in diameter have been prepared by the reduction of [Ir(cod)Cl]$_2$ in [BMIM][PF_6] with H_2 [24]. In a particularly elegant experiment, palladium nanoparticles were first formed by the thermal decomposition of palladium(II) acetate in the presence of triphenylphosphine (to give a regular particle size of 1 nm) in [BMIM][Tf$_2$N]. A silica aerogel support was then formed around the nanoparticles by adding (EtO)$_4$Si and formic acid to the mixture [25].

The reaction of Cu(NO$_3$)$_2$·3H$_2$O with 1,3-bis(4-pyridyl)propane (bpp) in [BMIM][BF_4] at 140 °C for 3 days yielded orange crystals of a Cu(I) coordination polymer, [Cu(bpp)][BF_4] [26]. The reduction of Cu(II) to Cu(I) in these reactions without the need to add a reducing agent has been noted in the past. This was the first time that a solvothermal synthesis had been conducted in an ionic liquid. The technique was used again in the synthesis of [NH$_3$CH$_2$CH$_2$NH$_3$]$_2$[B$_2$P$_4$O$_{14}$(OH)] from a mixture of boric acid, phosphoric acid and 1,2-diaminoethane in [BMIM]Cl that was heated at 110 °C for 6 days. The ionic liquid was removed by washing the crystals formed with water.

6.1.5
Outlook

In spite of recent developments, there is no doubt that inorganic and organometallic synthesis is still lagging behind organic synthesis in ionic liquids. This is not due to lack of importance. For instance, if ionic liquids are to find use in biphasic catalysis, the point will arrive when the ionic liquid layer can no longer be recycled. It is only by understanding the chemistry of the dissolved catalysts, deliberately prepared to be difficult to remove, that they can be transformed into materials that can be extracted from the ionic liquids.

Ionic liquids hold as much promise for inorganic and organometallic synthesis as they do for organic synthesis. Their lack of vapor pressure has already been exploited [20] as have their interesting solubility properties. The field can only be expected to accelerate from its slow beginnings.

References

1. T. B. Scheffler, C. L. Hussey, K. R. Seddon, C. M. Kear, P. D. Armitage, *Inorg. Chem.* **1984**, *23*, 1926.
2. C. L. Hussey, *Pure Appl. Chem.* **1988**, *60*, 1763.
3. A. J. Dent, K. R. Seddon, T. Welton, *J. Chem. Soc., Chem. Commun.* **1990**, 315.
4. T. Welton, *Chem. Rev.* **1999**, *99*, 2071.
5. I. W. Sun, E. H. Ward, C. L. Hussey, *Inorg. Chem.* **1987**, *26*, 4309.
6. A. J. Dent, A. Lees, R. J. Lewis, T. Welton, *J. Chem. Soc, Dalton Trans.* **1996**, 2787.
7. A.-V. Mudring, A. Babai, S. Arenz, R. Giernoth, *Angew. Chem. Int. Ed.* **2005**, *44*, 5485.
8. A. Babai, A.-V. Mudring, *Inorg. Chem.* **2006**, *45*, 4874.
9. (a) W. A. Herrmann, C. Kocher, *Angew. Chem. Int. Ed.* **1997**, *36*, 2163; (b) D. Bourissou, O. Guerret, F. P. Gabbai, G. Bertrand, *Chem. Rev.* **2000**, *100*, 39.
10. T. Welton, P. J. Smith, *Adv. Organomet. Chem.* **2004**, *51*, 251.
11. L. Xu, W. Chen, J. Xiao, *Organometallics* **2000**, *19*, 1123.
12. F. Mclachlan, C. J. Mathews, P. J. Smith, T. Welton, *Organometallics* **2003**, *22*, 5350.
13. (a) N. D. Clement, K. J. Cavell, C. Jones, C. J. Elsevier, *Angew. Chem. Int. Ed.* **2004**, *43*, 1277; (b) D. S. McGuinness, K. J. Cavell, B. F. Yates, B. W. Skelton, A. H. White, *J. Am. Chem. Soc.* **2001**, *123*, 8317.
14. A. R. Chianese, A. Kovacevic, B. M. Zeglis, J. W. Faller, R. H. Crabtree, *Organometallics* **2004**, *23*, 2461.
15. Z. F. Fei, D. B. Zhao, R. Scopelliti, P. J. Dyson, *Organometallics* **2004**, *23*, 1622.
16. J. K. D. Surette, L. Green, R. D. Singer, *Chem. Commun.* **1996**, 2753.
17. A. Stark, B. L. MacLean, R. D. Singer, *J. Chem. Soc, Dalton Trans.* **1999**, 63.
18. P. J. Dyson, M. C. Grossel, N. Srinivasan, T. Vine, T. Welton, D. J. Williams, A. J. P. White, T. Zigras, *J. Chem. Soc., Dalton Trans.* **1997**, 3465.
19. D. Crofts, P. J. Dyson, K. M. Sanderson, N. Srinivasan, T. Welton, *J. Organomet. Chem.* **1999**, *573*, 292
20. S. Dai, Y. H. Ju, H. J. Gao, J. S. Lin, S. J. Pennycook, C. E. Barnes, *Chem. Commun.* **2000**, 243.
21. (a) Y. Zhou, J. H. Schattka, M. Antonietti, *Adv. Mater.* **2003**, *15*, 1452; (b) D. M. Li, F. Shi, Y. Q. Deng, *Tetrahedron Lett.* **2004**, *45*, 265; Y. Zhou, J. H. Schattka, M. Antonietti, *Nano Lett.* **2004**, *4*, 477; (c) B. Gadenne, P. Hesemann, J. J. E. Moreau, *Chem. Commun.* **2004**, 1768.
22. B. G. Trewyn, C. M. Whitman, V. S. Y. Lin, *Nano Lett.* **2004**, *4*, 2139.
23. Y. Zhou, M. Antonietti, *J. Am. Chem. Soc,* **2003**, *125*, 14960.
24. G. S. Fonseca, J. D. Scholten, J. Dupont, *Synlett* **2004**, 1525.
25. K. Anderson, S. C. Fernandez, C. Hardacre, P.C. Marr, *Inorg. Chem. Commun.* **2004**, *7*, 73.
26. K. Jin, X. Huang, L. Pang, J. Li, A. Appel, Scot Wherland, *Chem. Commun.* **2002**, 2872.

6.2
Inorganic Materials by Electrochemical Methods

Frank Endres and Sherif Zein El Abedin

In Section 6.2.1 we will report on the literature on the electrodeposition of metals and semiconductors from ionic liquids and briefly introduce basic considerations for electrochemical experiments. Section 6.2.2 describes new results elucidating the nature of the electrode/ionic liquids interface. A short introduction to *in situ* scanning tunneling microscopy (STM) will also be given. For a more detailed overview on the electrodeposition in ionic liquids we would like to refer to "Electrodeposition in Ionic Liquids", ed. by F. Endres, A. Abbot and D. R. MacFarlane, Wiley-VCH, ISBN 978-3-527-31565-9.

6.2.1
Electrodeposition of Metals and Semiconductors

6.2.1.1 General Considerations

Electrodeposition is one of the main fields of electrochemistry, both in technical processes and in basic research. In principle, all metals and semiconductors can be obtained by electrolysis of the respective salts in aqueous or organic solutions, molten salts or ionic liquids. As well as electrowinning of the elements electrocoating of materials for corrosion protection is an important field in industry and in fundamental research. Especially on the nanometer scale, much work has been done in the last 20 years with the help of the scanning tunneling microscope. Insight was obtained into how the initial stages of metal deposition influence the bulk growth. Furthermore, the role of brighteners that are added to solutions to make shiny deposits could be understood: they adsorb at growing clusters and force the metal to grow layer by layer instead of in the form of clusters [1]. However, aqueous solutions are disadvantageous to the electrodeposition of less noble elements because of their limited electrochemical windows. For light, refractory and rare earth metals water fails as a solvent because hydrogen evolves before the metal deposits. For such purposes ionic liquids are ideal solvents because they have – depending on their composition – wide electrochemical windows combined with a good solubility for most of the metal salts and semiconductor precursors [2]. Many technical processes, such as electrowinning of the rare earth and refractory metals, Mg, Al and several others, are performed in high-temperature molten salts. In contrast to these systems that are highly corrosive, leading to difficulty in finding materials that withstand chemical attack by the melts, the design of electrochemical cells is much easier for low melting point ionic liquids. They combine, more or less, the advantages of the classical molten salts and those of aqueous media: due to their wide electrochemical windows several metals and alloys, such as Al, La and some others, that are accessible from high-temperature molten salts, can be deposited at room temperature. Furthermore, metals that can be obtained from aqueous media, can in most cases also be deposited from ionic liquids, often with superior quality as hydrogen evolution does not occur. Pd is a good example as deposits from aqueous solution can contain varying amounts of hydrogen that can make the deposits rather brittle. From ionic liquids, however, shining even nanosized bulk Pd deposits can easily be obtained. These features and their good ionic conductivities of between 10^{-3} and 10^{-1} $(\Omega\, cm)^{-1}$, make ionic liquids interesting media for electrodeposition.

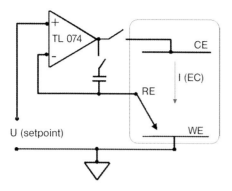

Fig. 6.2-1 Simplified circuit of a potentiostat with working electrode (WE) on ground. Reference electrode (RE) and potentiostatic setpoint are fed to the inverting and non-inverting input of an operational amplifier. The counter electrode (CE) is connected to the output of the operational amplifier. (EC): electrochemical current.

6.2.1.2 Electrochemical Equipment

Any redox couple has a defined electrode potential on the electrochemical potential scale. For example, in aqueous solutions a silver wire immersed in a solution containing Ag^+ ions with activity 1 has a potential of $+799$ mV vs. the normal hydrogen electrode. At more positive potentials an Ag electrode would dissolve, at more negative values Ag would deposit from the ions. If one applies a certain voltage between two electrodes, an electrochemical reaction may occur, depending on the applied voltage. If one wants to know the processes involved, e.g. during electrodeposition, it is necessary to know exactly the electrode potential vs. a reference electrode. It can be measured via a third electrode that is immersed in the solution. This is the classical three- electrode set-up with the working electrode of interest (WE), the reference electrode (RE) and the counter electrodes (CE). However, any current $I(EC)$ that flows through the cell will influence the electrode potentials so that a stable value would not easily be obtained. A potentiostat enables the potential of the working electrode to be controlled precisely with respect to a reference electrode. It always applies the desired value U(setpoint) to the working electrode, any desired changes are usually applied within microseconds. A simplified set-up based on an operational amplifier (OPA), where the working electrode is put to earth, is presented in Fig. 6.2-1.

The reference electrode (RE) is connected to the inverting input of an operational amplifier (here: Texas Instruments TL 074), the setpoint is applied between earth and the non-inverting input of the operational amplifier. For electronic reasons:

$$U(CE) = F[U(RE) - U(\text{setpoint})] \qquad (6.2\text{-}1)$$

where F is the amplification factor.

As F has typical values from ca. 10^6–10^7 it follows that

$$U(\text{RE}) = U(\text{setpoint}) \qquad (6.2\text{-}2)$$

Because of the high amplification factor good electrode contacts are required as any fluctuation would lead to strong oscillations of the output of the OPA. To prevent such problems a capacitor of roughly 1 μF can be inserted between the reference and the counter electrode thus damping such oscillations. U(setpoint) can be a constant voltage or any externally generated signal. In cyclic voltammetry, for example, a linearly varying potential is applied between an upper and a lower limit. If electrode reactions occur in the applied potential range, a current flows and is plotted vs. the electrode potential, thus giving the cyclic voltammograms (CV). Depending on the systems the currents are limited by kinetics or transport limitation, so peak currents are observed and can be evaluated to give insight into the electrochemical processes [3].

6.2.1.3 Electrodeposition of Less Noble Elements

Aluminum electrodeposition

The Al electrodeposition from chloroaluminate ionic liquids was investigated on different substrates by several authors with classical electrochemical methods such as cyclic voltammetry, potential step experiments and *ex situ* techniques [4–7]. In all cases, Al deposition was only observed in the acidic regime and the quality of the deposits was reported to be superior to those obtained from organic solutions. The deposition on substrates such as glassy carbon, tungsten and platinum is preceded by a nucleation step and the deposition is electrochemically quasi-reversible, i.e. it is not solely diffusion controlled, the charge transfer reaction also plays an important role. On Pt, however, from the published electrochemical data there are some hints of underpotential phenomena. The bulk deposits of Al are rather granular and the current density has an influence on the size of the clusters, with a tendency to smaller crystals with higher current densities. If dry toluene [4] or benzene [8] is added to the liquid, mirror bright deposits have been reported. It is likely that the organic molecules play the role of brighteners. Such effects have been known for a long time and organic molecules such as crystal violet are widely used in aqueous electroplating processes to deposit shiny layers of Cu, Ag and others [9]. The miscibility of the chloroaluminates with toluene, xylene and other organic solvents has the further advantage that the liquid can easily be washed away from the samples after the electrodeposition has been performed, so that clean substrates can be prepared as easily as from aqueous solutions.

Although these chloroaluminate-based liquids will most likely not replace high-temperature molten salts for Al electrowinning purposes, they could become important in the electroplating of Al and several Al alloys. Recently, Endres et al. succeeded, using special electrochemical techniques and special bath compositions, in preparing high quality deposits of nanocrystalline metals such as Al with grain sizes down to only several nanometers [10]. Such nanocrystalline deposits are interesting as coatings for corrosion protection, for example. Furthermore, they could show for the first time that nanocrystalline aluminum, with an average crystallite size of about 34 nm, can be electrodeposited without any additives from the water and air stable ionic liquid 1-butyl-1-methylpyrrolidinium bis(trifluoromethylsulfonyl)amide containing $AlCl_3$ [11]. The advantages of this ionic liquid over chloroaluminate ionic liquids, which were intensively investigated in the past, are that it is water and air stable, easy to dry and purify as well as it is easy to handle. Presumably, it behaves as a surfactant, thus leading to Al nanocrystals.

Electrodeposition of less noble elements and aluminum alloys
In technical processes elements such as alkali-, alkaline earth-, refractory- or rare earth- metals are obtained by high-temperature molten salt electrolysis [12, 13]. In many cases eutectics of alkali halides are used as solvents. In the cases of La, Ce and others the metal halides can be directly electrolyzed. The temperatures vary from about 450 °C to more than 1000 °C. On the one hand, these are pretty difficult experimental conditions; on the other, in many cases a high electronic conductivity is observed as soon as metal is deposited. In liquid NaCl, for example, Na dissolves easily and a nonmetal/metal transition is observed with rising Na content [14]. As a consequence, the current efficiency during electrolysis can reduce enormously due to a partial electronic short circuit. Consequently, it would be interesting to apply low-melting ionic liquids for electrowinning or electroplating of such elements. Previously, only a few metals have been deposited in elemental form from low melting ionic liquids. In most cases the chloroaluminate systems were employed and some aluminum alloys with interesting properties have been reported in the literature.

Sodium and lithium Both sodium [15] and lithium [16] electrodeposition was successful in neutral chloroaluminate ionic liquids that contained protons. These elements are interesting for Na- or Li-based secondary batteries, where the metals would serve directly as the anode material. The electrodeposition is not possible in basic or acidic chloroaluminates, only proton-rich NaCl or LiCl buffered neutral chloroaluminate liquids were feasible. The protons enlarged the electrochemical window towards the cathodic regime so that the alkali metal electrodeposition became possible. For Na the proton source was dissolved HCl that was introduced via the gas phase or via 1-ethyl-3-methylimidazolium hydrogen dichloride. Triethanolamine hydrogen dichloride was employed as the proton source for Li electrodeposition. For both alkali metals, reversible deposition and stripping were reported on tungsten and stainless steel substrates, respectively.

Gallium Elemental gallium can be electrodeposited from both chloroaluminate [17] and chlorogallate [18] ionic liquids. In the latter case 1-ethyl-3-methylimidazolium chloride was mixed with $GaCl_3$, thus giving a highly corrosive ionic liquid that was studied for GaAs thin film electrodeposition. In the chloroaluminates Ga can be deposited from Lewis acidic systems. It was found that the electroreduction from

Ga(III) leads first to Ga(I), then upon further reduction the elemental Ga forms from Ga(I). On glassy carbon the electrodeposition involves instantaneous three-dimensional nucleation with diffusion-controlled growth of the nuclei. No alloying with Al was reported if deposition of Ga was performed in the Ga(I) diffusion regime. Reproducible electrodeposition of Ga is a promising route to binary and ternary compound semiconductors. A controlled electrodeposition of GaX quantum dots (X = P, As, Sb) would be very attractive for nanotechnology.

Iron The electrodeposition of iron was investigated in neutral and acidic chloroaluminates [19, 20]. Although iron can be deposited even from certain aqueous solutions, ionic liquids give the advantage of deposition of a high quality product in elemental form without side reactions such as hydrogen evolution or oxidation by water if the potential control is switched off. For nanotechnology especially this is an interesting feature as iron is a magnetic material. It was reported that elemental Fe can be reversibly deposited on several substrates such as tungsten or glassy carbon in acidic chloroaluminate liquids, even though the electrode potential for its electrodeposition is very close to the Al deposition potential. The reduction can be performed from $FeCl_3$ that is reduced to $FeCl_2$ prior to the deposition of the element, or directly from $FeCl_2$. The fact that the electrode potentials for Al and Fe deposition are close together makes it possible to deposit Fe–Al alloys with interesting structural properties.

Aluminum alloys with iron, cobalt, nickel, copper and silver The bulk deposition of alloys of Al with Fe, Co, Ni, Cu and Ag was investigated with electrochemical and *ex situ* analytical techniques by Carlin et al. [21]. The alloys were prepared under near steady state diffusion-controlled conditions. For $CoAl_x$, $FeAl_x$ and $CuAl_x$ compositions with $x \approx 1$ were obtained; for $NiAl_x$ kinetic phenomena complicated a reliable analysis. In the case of $AgAl_x$ the authors report that an analysis was precluded by a dendritic growth of the deposits. All the alloy systems showed complex electrodissolution and the nature of the oxidation process is different for the alloys produced in specific potential regimes. However, one has to keep in mind that classical electrochemistry and ex situ analysis give mainly integral information on the deposits. Nanometer resolution in ex situ methods is not yet a straightforward standard procedure. Nevertheless, although the alloy deposition is obviously complicated, the results are quite interesting for the electrodeposition of thin alloy films, as alloys of Al with Fe, Ni or Co could give magnetic nanostructures that are perhaps more stable than the respective elements.

Aluminum alloys with niobium and tantalum Nb and Ta can be obtained in elemental form from high-temperature molten salts. Nb and Ta are widely used as coatings for corrosion protection as they form – like Al – thin oxide layers that protect the underlying material from attack. In technical processes several high-temperature molten salts are employed for electrocoating and the morphology of the deposit is strongly influenced by the composition of the baths. Some attempts have been carried out to deposit Nb and Ta from ionic liquids [22, 23]. Koura et al. [22] focussed on the electrodeposition of AlNb$_x$ alloys from room-temperature ionic liquids that contain both AlCl$_3$ and chlorides of Nb. The authors reported, at temperatures between 90 and 140 °C, Nb contents of up to 29 wt.% in the deposits. Cheek et al. [23] employed chloroaluminate liquids at room temperature and AlNb$_x$ films could only be obtained if NbCl$_5$ was pre-reduced in a chemical reaction. The authors report that Nb powder is the most effective reducing agent for this purpose. Recently, Endres et al. have reported the first results on the electrodeposition of elemental, crystalline tantalum in the ionic liquid 1-butyl-1-methylpyrrolidinium bis(trifluoromethylsulfonyl)amide ([BMP][Tf$_2$N]) [24, 25]. The use of high-temperature molten salts of alkali metal halides as electrolytes was hitherto the only possible way to electrodeposit Ta. Nevertheless, they could electrodeposit 500 nm thick crystalline tantalum layers in the ionic liquid [BMP][Tf$_2$N] containing TaF$_5$ at 150–200 °C. By addition of LiF to the electrolyte, the quality and the adherence of the electrodeposit were found to be improved. XRD patterns of the electrodeposit clearly show the characteristic patterns of crystalline tantalum.

Aluminum alloys with titanium Titanium is an interesting material for corrosion protection and lightweight construction. Hitherto it could only be deposited in high quality from high-temperature molten salts although attempts were made to deposit it from organic solutions and even aqueous media. In general, the bulk electrodeposition of Ti is complicated because traces of water form immediately, passivating oxide layers on Ti. In ionic liquids – obviously – the deposition of the element has not yet been successful either. However it was reported [26] that AlTi$_x$ alloys can be obtained from chloroaluminates. The corrosion resistance of the layers was reported to be superior to pure Al and seems to become even better with increasing Ti content. However, Ti was not deposited in elemental form without co-deposition of Al. In chloroaluminates Al is nobler than Ti therefore, at room temperature, only codeposits and alloys can be obtained. Furthermore kinetic effects play a role.

Aluminum alloys with chromium The electrodeposition of Cr has been investigated in acidic chloroaluminates by Ali et al. [27]. They report that the Cr content in the AlCr$_x$ deposit can vary from 0 to 94 mol%, depending on the deposition parameters. The deposit consists both of Cr-rich and Al-rich solid solutions as well as intermetallic compounds. An interesting feature of these deposits is their high-temperature oxidation resistance; the layers seem to withstand temperatures up to 800 °C, so that coatings with such an alloy could be interesting for applications in engines.

Lanthanum and aluminum–lanthanum alloys It was reported, that La can be electrodeposited from chloroaluminate liquids [28]. Whereas from the pure liquid only AlLa$_x$ alloys can be obtained, the addition of excess LiCl and small quantities of thionylchloride (SOCl$_2$) to a LaCl$_3$-saturated melt allows the deposition of elemental La. However, the electrodissolution seems to be somewhat kinetically hindered.

This result could perhaps be interesting for coating purposes as elemental La can normally only be deposited in high-temperature molten salts, which requires much more difficult experimental or technical conditions. Furthermore La and Ce electrodeposition would be important as their oxides have interesting catalytic activity, for instance as oxidation catalysts. A controlled deposition of thin metal layers followed by a selective oxidation could perhaps produce catalytically active thin layers that would be of interest for fuel cells or waste gas treatment.

6.2.1.4 Electrodeposition of Metals That Can Also Be Obtained From Water

As mentioned above, most of the metals that can be deposited from aqueous solutions can also be obtained from ionic liquids. One could raise the question as to whether this makes sense, as aqueous solutions are much easier to handle. However, there are two properties of the ionic liquids that are superior to aqueous solutions: on the one hand their electrochemical windows are much wider so that side reactions during electrodeposition can easily be prevented. Whereas Palladium, for example, gives brittle deposits in aqueous media due to hydrogen evolution and dissolution of hydrogen in the metal, in ionic liquids shiny nanosized deposits can be obtained [10]. On the other hand the temperature can be varied over a wide range, in some cases over an interval of up to 400 °C. Variation of the temperature has, in general, a strong effect on the kinetics of the deposition and on the surface as well as on the interface mobility of the deposits. Although there are only a few studies on temperature variation upon electrodeposition in ionic liquids, this is an attractive research field as the gap to the classical high-temperature molten salts could be closed.

Indium and antimony
The electrodeposition of In on glassy carbon, tungsten and nickel has been reported by Liu et al. [29]. In basic chloroaluminates elemental indium is formed via a three-electron reduction step from the $[InCl_5]^{2-}$ complex, but In(I) species are also reported in the literature [30]. The overpotential deposition involves progressive three-dimensional nucleation on nickel and on a finite number of active sites on carbon and tungsten. In electrodeposition was reported not to occur from acidic liquids but liquids based on $InCl_3$ and organic salts were successfully used to deposit InSb [30, 31]. Sb electrodeposition on tungsten, platinum and glassy carbon has been reported by Osteryoung et al. [32, 33]. The metal can be deposited from acidic liquids but, in part, irreversible behavior is observed. Sun et al. showed that InSb can be electrodeposited in the Lewis basic 1-ethyl-3-methylimidazolium chloride/tetrafluoroborate ionic liquid [34]. InSb is a direct semiconductor, and quantum dots of InSb that were made under UHV conditions have already been successfully studied for LASER applications [35]. Quantum dots are widely investigated nowadays and are a rapidly growing research field. Definite electrodeposition from ionic liquids would be an important contribution.

Tellurium and cadmium
Te electrodeposition has been reported in basic chloroaluminates [36] where the element is formed from the $[TeCl_6]^{2-}$ complex in a four-electron reduction step.

Furthermore, metallic Te can be reduced to Te^{2-} species. On glassy carbon the electrodeposition of the element involves three-dimensional nucleation. A systematic study of the electrodeposition in different ionic liquids would be of interest because – as with InSb – a defined co-deposition with cadmium could produce the direct semiconductor CdTe. Although this semiconductor can be deposited from aqueous solutions in a layer by layer process [37], the variation of the temperature over a wide range would be interesting as the grain sizes and the kinetics of the reaction would be influenced.

Cd electrodeposition has been reported by Noel et al. and by Chen et al. [38, 39]. $CdCl_2$ was used to buffer neutral chloroaluminate liquids and the element could be deposited [38]. Chen et al. used a basic 1-ethyl-3-methylimidazolium chloride/tetrafluoroborate ionic liquid to deposit Cd successfully [39]. It is formed on platinum, tungsten and glassy carbon from $CdCl_4^{2-}$ in a quasi-reversible two-electron reduction process. This result is promising as Te could perhaps also be deposited from such an ionic liquid thus giving a system for direct CdTe electrodeposition.

Copper and silver
The electrodeposition of Cu has been widely investigated in chloroaluminate ionic liquids. It has been reported from acidic liquids only. Furthermore, some interesting deviations compared to the behavior in aqueous solutions have been observed. If $CuCl_2$ is added to an acidic liquid, Cu(II) undergoes two one-electron reduction steps on glassy carbon and on tungsten [40, 41]. In the first step Cu(I) is formed, in the second step the metal deposits. At high overvoltages for the deposition an alloying with Al begins [42]. The electrodeposition of Cu from a tetrafluoroborate ionic liquid was investigated by Chen et al. [43] on polycrystalline tungsten, platinum and on glassy carbon. On Pt, UPD phenomena were reported, whereas on tungsten and glassy carbon only OPD was apparent. *Ex situ* analysis proved that the deposit was solely composed of copper.

The electrodeposition of Ag from ionic liquids has also been intensively investigated [44–46]. As in the case of Cu, this metal was only deposited from acidic chloroaluminates. The deposition occurs in one step from Ag(I). On glassy carbon and tungsten three-dimensional nucleation was reported [44]. Furthermore, it was reported that Ag can also be deposited in a one-electron step from tetrafluoroborate ionic liquids [46]. However, the charge transfer reaction seems to play an important role in this medium and the deposition is not as reversible as in the chloroaluminate systems, a common phenomenon in air and water stable ionic liquids.

Nickel and cobalt
Electrodeposition of nickel and cobalt has been investigated intensively in aqueous solutions. Both metals are interesting for nanotechnology as magnetic nanostructures can be formed in aqueous solutions [47]. However, the bulk electrodeposition is accompanied by a massive hydrogen evolution. Both elements can also be deposited from acidic chloroaluminate liquids [48, 49]. Cobalt and zinc–cobalt alloys

can be deposited in the ionic liquid zinc chloride/1-ethyl-3-methylimidazolium chloride (40–60 mol%) containing Co(II) at 80 °C [50]. Abbott et al. have reported the synthesis and characterization of new, moisture stable, Lewis acidic ionic liquids made from metal chlorides and quaternary ammonium salts that are commercially available [51]. They have shown that mixtures of choline (2-hydroxyethyltrimethylammonium) chloride [(Me$_3$NC$_2$H$_4$OH)Cl] and MCl$_2$ (M = Ni, Zn, Sn) give conducting and viscous liquids at or around room temperature. These ionic liquids are easy to prepare, water and air insensitive and their low costs enables their use in large-scale applications. From these ionic liquids Ni can be electrodeposited.

Palladium and gold
Palladium electrodeposition is of special interest for catalysis and for nanotechnology. It was reported that it can be deposited from basic chloroaluminate liquids, in the acidic regime the low solubility of PdCl$_2$ and passivation phenomena complicate the deposition [52]. However, thick Pd layers are difficult to obtain from basic chloroaluminates. With different melt compositions and special electrochemical techniques at temperatures up to 100 °C Endres et al. succeeded in depositing mirror bright and thick nanocrystalline palladium coatings [10]. Sun et al. have reported that Pd–Au [53] and Pd–Ag [54] alloys can be electrodeposited in a basic 1-ethyl-3-methylimidazolium chloride/ tetrafluroborate ionic liquid.

Gold electrodeposition has been reported from chloroaluminate-based liquids and from a liquid made of an organic salt and AuCl$_3$ [55, 56]. Although gold can be electrodeposited in high quality from aqueous solutions the latter result is interesting, especially with respect to the deposition of unusual alloys of gold with less noble elements.

Zinc and tin
The electrodeposition of Zn was investigated in acidic chloroaluminate liquids on gold, platinum, tungsten and glassy carbon [57]. On the metals underpotential deposition phenomena were seen, whereas on glassy carbon only three-dimensional bulk deposition was observed. At higher overvoltages co-deposition with Al was reported. As Zn is widely used in the automobile industry for corrosion protection a co-deposition with Al could also be interesting for certain applications. An aprotic electrodeposition of zinc would be highly welcome in the steel industry as in aqueous solutions hydrogen embrittlement of steel is a problem with certain modern steel qualities. It was shown that Lewis acidic ZnCl$_2$–[EMIM]Cl ionic liquids (in which the molar amount of ZnCl$_2$ is higher than 33 mol%) are potentially useful for the electrodeposition of zinc and zinc-containing alloys [58–60].

Tin has been electrodeposited from basic and acidic chloroaluminate liquids on platinum, gold and glassy carbon [61]. On Au the deposition starts in the UPD regime and from the electrochemical data one monolayer was reported. Furthermore, there seems to be some evidence for alloying between Sn and Au. On glassy carbon three-dimensional growth of Sn occurs.

Chromium

Electrodeposition of chromium in a mixture of choline chloride and chromium(III) chloride hexahydrate has recently been reported [62]. A dark green, viscous liquid can be formed by mixing choline chloride with chromium(III) chloride hexahydrate and the physical properties are characteristic for an ionic liquid. The eutectic composition is found to be 1:2 choline chloride/chromium chloride. Chromium can be electrodeposited efficiently from this ionic liquid to yield a crack-free deposit [62]. Addition of LiCl to the ionic liquid choline chloride–$CrCl_3 \cdot 6H_2O$ allowed the deposition of nanocrystalline black chromium films [63]. The use of this ionic liquid might offer an environmentally friendly process for electrodeposition of chromium instead of the current chromic acid-based baths (see Chapter 9 for details concerning the industrial implementation of this technology).

6.2.1.5 Electrodeposition of Semiconductors

Many studies on semiconductor electrodeposition have been performed in the past in different solutions such as aqueous media, organic solutions, molten salts, as well as a few ionic liquids. A good overview on the topic has been presented by Pandey et al. [64]. However, industrial standard procedures have not yet been established. As well as bulk deposits of semiconductors for photovoltaic applications, thin layers or quantum dots would be of great interest in basic research and in nanotechnology. Hitherto, most of the basic studies on semiconductor formation and their characterizations have been performed under ultrahigh vacuum (UHV) conditions. Molecular beam epitaxy is a widely employed method for such purposes. Furthermore, in technical processes chemical or physical vapor deposition are still methods of great importance. Although high quality deposits can be obtained, such processes are cost-intensive and the layers are consequently expensive. A simple and cheaper electrodeposition would surely be of commercial interest. The work of Stickney [37] has shown that electrochemical atomic layer epitaxy (ECALE) in aqueous media is a suitable method to deposit compound semiconductors with qualities comparable to those made by vacuum techniques. In special electrochemical polarization routines, the elements of a compound semiconductor are successively deposited, layer by layer. Unfortunately direct electrodeposition of CdTe, CdSe and others is difficult in aqueous solutions, and at room temperature in many cases the elements are co-deposited in varying amounts together with the desired semiconductor. Variation of the temperature can strongly affect the quality of the electrodeposits [65]. In general, a direct deposition of a compound semiconductor would be interesting, as it would be less time consuming than the elegant ECALE process. Although there are only a few articles on semiconductor electrodeposition from ionic liquids, these media are interesting for such studies for several reasons. They have low vapor pressures and as a consequence – depending on the system – the temperature can be varied over several hundred degrees, so that kinetic barriers in compound formation can be overcome. Furthermore – due to the wide electrochemical windows – compounds such as GaAs, that are hardly accessible from aqueous solutions, can be obtained. Also, for the GaSb, InSb, InP and ternary compound semiconductor electrodeposition ionic liquids could be interesting, especially if higher temperatures are applied.

As well as compound semiconductors elemental semiconductors can be obtained from ionic liquids. Si and Ge are widely used as wafer material for different electronic applications; furthermore junctions of n- and p-doped Si are still interesting for photovoltaic applications. A controlled electrodeposition of both elements and their mixtures would surely also be interesting for nanotechnology as Ge quantum dots made under UHV conditions show interesting photoluminescence.

GaAs

The direct electrodeposition of GaAs from ionic liquids has been studied mainly by two groups. Wicelinski et al. [66] used an acidic chloroaluminate melt at 35–40 °C to co-deposit Ga and As. However, it was reported that Al underpotential deposition occurs on Ga. Carpenter et al. employed an ionic liquid that was based on $GaCl_3$ to which $AsCl_3$ was added [18, 67]. Unfortunately, in these studies the quality of the deposits was not convincing and both pure arsenic and gallium could be found in the deposits. Nevertheless, these studies have to be regarded as the first steps in the electrodeposition of Ga-based semiconductors. Furthermore, thermal annealing could improve the quality of the deposits.

InSb

The principle of InSb electrodeposition is the same as for GaAs. An ionic liquid based on $InCl_3$ is formed to which $SbCl_3$ is added [30, 31]. At 45 °C InSb can be directly electrodeposited but elemental In and Sb are also reported to deposit. The ratio of In/Sb depends strongly on the deposition potential. Notwithstanding some problems, the authors of these studies are optimistic that ionic liquids based on $GaCl_3$ and $InCl_3$ may also be useful in depositing ternary compound semiconductors such as AlGaAs, InGaSb and others. Sun et al. showed that InSb can be electrodeposited in the Lewis basic 1-ethyl-3-methylimidazolium chloride/tetrafluoroborate ionic liquid containing In(III) and Sb(III) [34].

ZnTe

The electrodeposition of ZnTe was reported by Lin et al. [68]. They prepared a liquid that contained $ZnCl_2$ and [EMIM]Cl in a molar ratio of 40/60. Propylene carbonate was used as the cosolvent to obtain melting points near room temperature. 8-Quinolinol was added to shift the reduction potential for Te to more negative values. Under certain potentiostatic conditions a stoichiometric deposition could be obtained. After thermal annealing the bandgap was determined by absorption spectroscopy to be 2.3 eV, in excellent agreement with ZnTe made by other methods. This study shows convincingly, that wide bandgap semiconductors can be made from ionic liquids.

Germanium

In situ STM studies on Ge electrodeposition on gold from an ionic liquid were performed by the Endres group [69, 70]. In the first studies they used dry [BMIM][PF_6]

as a solvent and dissolved GeI_4 in an estimated concentration of 0.1–1 mmol l^{-1}; the substrate was Au(111). This ionic liquid has, in its dry state, an electrochemical window of more than 4 V on gold and the bulk deposition of Ge started several hundreds of mV positive of the solvent decomposition. Furthermore, distinct underpotential phenomena were observed. We will give some insight into the nanoscale processes at the electrode surface in Section 6.2.2.3.

Silicon
Using the ionic liquid 1-butyl-1-methylpyrrolidinium bis(trifluoromethylsulfonyl)-imide ([BMP][Tf$_2$N]) Endres et al. have shown, for the first time, that silicon can be electrodeposited at room temperature [71]. By means of scanning tunneling spectroscopy, they could prove that the electrodeposited layer obtained in [BMP][Tf$_2$N] containing $SiCl_4$ is elemental, semiconducting silicon. A silicon layer with a thickness of about 100 nm exhibits a band gap of 1.0 ± 0.2 eV, which is shown by *in situ* current/voltage tunneling spectroscopy, indicating that semiconducting silicon had been electrodeposited.

6.2.2
Nanoscale Processes at the Electrode/Ionic Liquid Interface

6.2.2.1 General Considerations
Almost 10 years ago we started in situ STM studies on electrochemical phase formation from ionic liquids. There was no knowledge of the local processes of phase formation in ionic liquids but due to wide electrochemical windows these systems give access to elements that cannot be obtained from aqueous solutions. In the rapidly growing field of nanotechnology, where semiconductor nanostructures will play an important role, we see a great chance for electrodeposition of nanostructures from ionic liquids. It is known that germanium quantum dots on silicon made by molecular beam epitaxy under UHV conditions show an interesting photoluminescence, around 1 eV [72]. Furthermore, LASERS based on compound semiconductor quantum dots such as InSb have been discussed in the literature [35]. An electrochemical routine would be preferable to a UHV process if comparable results could be obtained. For this reason, the electrochemical processes and the factors that influence the deposition and the stability of the structures have to be understood on the nanometer scale.

6.2.2.2 The Scanning Tunneling Microscope
The main technique that is employed for electrochemical *in situ* studies on the nanometer scale is the STM that was invented in 1982 by Binnig and Rohrer [73] and combined a little later with a potentiostat to allow electrochemical experiments [74]. The principle of its operation is remarkably simple; a typical circuit is shown in Fig. 6.2-2.

The left side is essentially identical to the potentiostat circuit presented in Section 6.2.1.2. The right side is the pre-amplifier of the STM. An atomically sharp metal tip

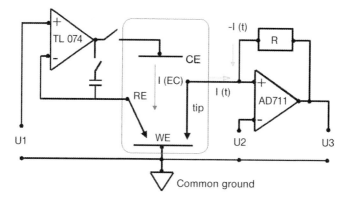

Fig. 6.2-2 Simplified circuit of electrochemical STM set-up. In addition to the potentiostat (see Fig. 6.2-1) an STM preamplifier is added to which the tip is connected. U1: potentiostatic setpoint, U2: tunneling voltage, I(t): tunneling current, U3 = −R I(t).

is located roughly 1 nm over an electronically conductive substrate, here the working electrode of interest (WE). This is carried out by computer control with the help of step motors and micrometer screws as well as piezoelectric elements, to which the tip is connected. If a potential U2 is applied between tip and sample (typically 5–500 mV) a tunneling current I(t) flows that has typical values between 0.1 and 10 nA, depending on the distance. This current is transformed into a voltage, $U3 = -RI(t)$, that can be further processed. The tunneling current is strongly dependent on the distance (d) and is a function of the density of states of tip D(tip) and sample D(sample). For a simple case one obtains in a first approximation [75]:

$$I(\text{tunnel}) = U(\text{bias}) \times D(\text{tip}) \times D(\text{sample}) \times \exp[-\text{const.} \times d] \quad (6.2\text{-}3)$$

Due to this strong distance dependence local height changes can, in principle, be detected in the picometer range. There are two modes of operation: in the "constant height mode" the tip is scanned over the surface at a constant height and the local changes in the tunneling current are acquired. In the "constant current mode" the distance between the tip and sample is kept constant by an electronic feedback control. It works such that U3 is amplified and finally fed back via an adding amplifier to the piezo control. It is also clear that the STM tip acts as an electrode in the electrochemical cell. As soon as a voltage is applied to the tip, a current can flow. Such Faradaic currents (for example the deposition of metal, hydrogen or oxygen evolution) can easily reach some hundreds of nanoamperes. Macroscopically this is negligible but as the tunneling currents are only some nanoamperes, the active sites of the tip, except for the very end, have to be insulated with paint or glass. Thus, the faradaic currents can be reduced down to the picoampere range making stable tunneling conditions under electrochemical conditions possible.

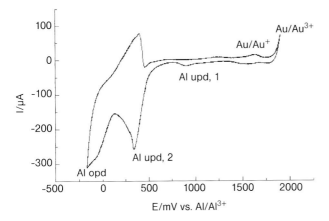

Fig. 6.2-3 Cyclic voltammogram of acid [BMIM]Cl/AlCl$_3$ on Au(111): at electrode potentials > +1.2 V vs. Al/AlCl$_3$ Au oxidation begins, UPD processes are observed at about +900 and +400 mV before bulk deposition of Al starts (from Ref. [76]).

6.2.2.3 Results

Aluminum electrodeposition on Au(111)

The processes during electrodeposition of aluminum were investigated on the nanometer scale for the first time in [76]. As already pointed out in the sections above, Al is an important metal for various applications in technical processes. In order to get insight into the growth of the metal and to understand the initial stages of the phase formation we performed in situ STM experiments under electrochemical conditions during electroreduction of AlCl$_3$ in an acid chloroaluminate ionic liquid composed of AlCl$_3$ and [BMIM]Cl. This liquid is extremely corrosive when water is present, so we had to build our own STM heads that allow measurements under inert gas conditions. The cyclic voltammogram on Au(111) exhibits several UPD and one OPD process, as can be seen in Fig. 6.2-3.

At electrode potentials > +1200 mV vs. Al/AlCl$_3$ gold oxidation starts first at the steps between different terraces, at higher potentials pits are formed that lead rapidly to a complete disintegration of the substrate.

We could identify the following electrode processes at the surfaces during electrodeposition: upon a potential step from $E > +1000$ mV to $E = +950$ mV vs. Al/AlCl$_3$ two-dimensional islands formed irreversibly on the surface. Their height is 250 ± 20 pm, indicative of Au(111). We attributed this observation to the formation of Au–Al compounds followed by an expelling of surplus Au atoms to the surface. At an electrode potential of +400 mV small Al islands with an averaged height of 230 ± 20 pm start growing (Fig. 6.2-4).

At +100 mV vs. Al/AlCl$_3$ clusters of up to 1 nm in height form. If a potential step to +1100 mV vs. Al/AlCl$_3$ is performed, the clusters dissolve immediately, but both holes as well as gold islands of up to two monolayers in height remain on the

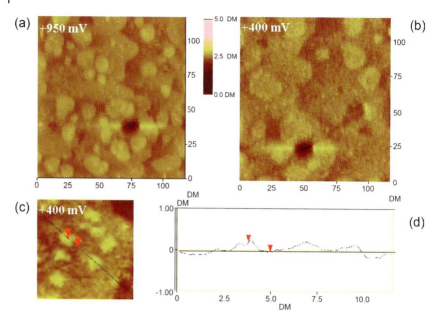

Fig. 6.2-4 Underpotential phenomena during Al reduction in acid [BMIM]Cl/AlCl$_3$ on Au(111): (a) at +950 mV vs. Al/AlCl$_3$ 2D islands with a height of 250 ± 20 pm form; (b) at +400 mV 2D Al islands with an average height of 230 ± 20 pm are reversibly deposited; (c) the islands are shown with a higher resolution; (d) typical height profile.

surface. It is likely that a strong alloying between Au and Al took place both in the surface and in the deposited clusters (Fig. 6.2-5).

In the overpotential deposition regime we observed that nanosized Al was deposited in the initial stages. Furthermore a certain transfer of Al from the scanning tip to the Al-covered substrate was observed. Accidentally we succeeded in an indirect tip-induced nanostructuring of Al on growing Al (Fig. 6.2-6).

These results are quite interesting. The initial stages of Al deposition lead to nanosized deposits. Indeed, based on the STM studies, we succeeded in making bulk deposits of nanosized Al with special bath compositions and special electrochemical techniques [10], also in air and in water-stable ionic liquids [11]. The preliminary results on tip-induced nanostructuring show that nanosized modifications of electrodes by less noble elements are possible in ionic liquids thus giving access to new structures that cannot be made in aqueous media.

Copper electrodeposition on Au(111)

Copper is an important metal and has been widely investigated in electrodeposition studies from aqueous solutions. Numerous publications exist in the literature on this topic. Furthermore, technical processes have been established in aqueous solutions to make Cu interconnects on microchips. In general, the quality of the

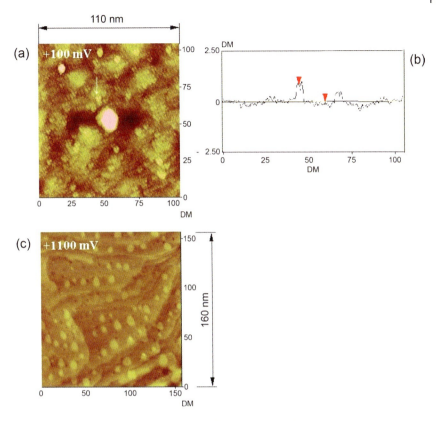

Fig. 6.2-5 Underpotential phenomena during Al reduction in acid [BMIM]Cl/AlCl$_3$ on Au(111): (a) at +100 mV vs. Al/AlCl$_3$ nanoclusters with heights up to 1 nm form; (b) a typical height profile is shown. (c) Upon a potential step to +1100 mV vs. Al/AlCl$_3$ the clusters dissolve immediately and leave holes in the surfaces as well as small Au islands; an alloying between Al and Au is very likely.

deposits is strongly influenced by the bath compositions. On the nanometer scale one finds in the underpotential deposition regime different superstructures if different counter-ions are used in the solutions. A co-adsorption between the metal atoms and the anions has been reported. Before the bulk deposition begins one Cu monolayer forms on Au(111) in the underpotential regime [77].

We performed for the first time in situ STM experiments in an acid chloroaluminate ionic liquid composed of AlCl$_3$ and [BMIM]Cl both with CuCl and CuCl$_2$ as copper sources [78]. The motivation was based mainly on two facts: in the chloroaluminates copper is deposited from Cu(I), and Cu$^+$ can be regarded as a naked cation because there is no solvation shell, unlike in aqueous media. As a consequence we expected distinct deviations from the behavior in aqueous solutions. The cyclic voltammogram on Au(111) exhibits three UPD processes followed by

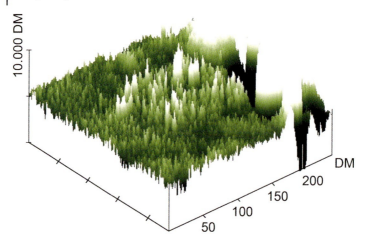

Fig. 6.2-6 The initial stages of Al overpotential deposition lead to nanosized deposits. A jump to contact transfer of Al from the scanning tip to the growing Al was observed (from Ref. [76]).

three-dimensional Cu growth in the OPD regime. At potentials >+1000 mV vs. Cu/Cu$^+$ gold oxidation starts first at the steps, at higher electrode potentials bulk oxidation of gold begins (Fig. 6.2-7).

Figure 6.2-8 (a) and (b) were acquired at +500 mV and at +450 mV vs. Cu/Cu$^+$ and show that at +450 mV vs. Cu/Cu$^+$ monolayer high Cu clusters nucleate at the steps between different Au terraces. Thus the couple of shoulders in the cyclic voltammogram is correlated to this surface process.

If the electrode potential is further reduced to +350 mV, a hexagonal superstructure with a periodicity of 2.4 ± 0.2 nm is observed. With respect to the interatomic distances in the Au(111) structure at the surface this corresponds – within the error limits – to an 8 × 8 superstructure (see Fig. 6.2-9).

The integrated charge would correspond to 0.7 ± 0.1 Cu monolayers, thus either a less closely packed Cu layer or an anion co-adsorption that can both lead to a Moiré superstructure are probed. In the investigated solution [Al$_2$Cl$_7$]$^-$ is the predominant anion. At +200 mV vs. Cu/Cu$^+$ the superstructure disappears and a completely closed Cu monolayer is observed, with a charge corresponding to 1.0 ± 0.1 Cu monolayers.

In contrast to aqueous solutions we observed the growth of a second 200 ± 20 pm high monolayer at +50 mV, together with clusters of heights up to 1 nm (Fig. 6.2-10).

This result was quite surprising as a second Cu monolayer has not yet been reported in aqueous solutions. Furthermore, clusters of up to 1 nm in height have not yet been reported in the UPD regime in aqueous solutions. Here, it cannot be excluded completely that the clusters contain small amounts of Al. Such an underpotential alloying could enormously stabilize the clusters, especially as several stoichiometric Al–Cu compounds are known.

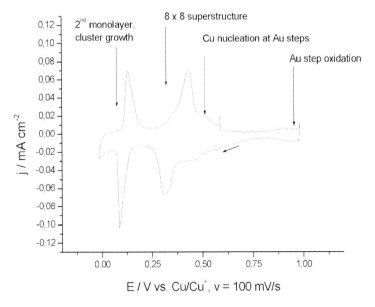

Fig. 6.2-7 Cyclic voltammogram of CuCl in acid [BMIM]Cl/AlCl$_3$ on Au(111): 3 UPD processes are observed that are correlated to decoration of Au steps by copper, formation of an 8 × 8 superstructure followed by a Cu monolayer. Before the bulk deposition a second monolayer grows together with clusters.

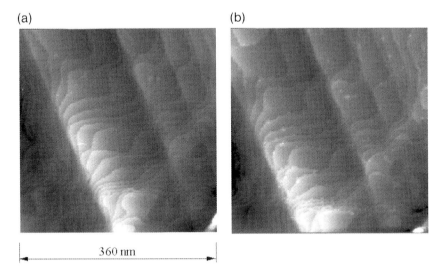

Fig. 6.2-8 Underpotential phenomena during Cu reduction in acid [BMIM]Cl/AlCl$_3$ on Au(111): a potential step from +500 mV vs. Cu/Cu$^+$ (a) to +450 mV leads to the growth of small Cu islands at the steps of the gold terraces (b) (from Ref. [78]).

6 Inorganic Synthesis

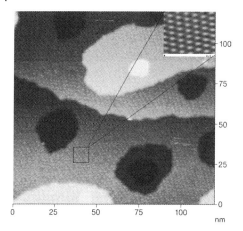

Fig. 6.2-9 Underpotential phenomena during Cu reduction in acid [BMIM]Cl/AlCl$_3$ on Au(111): at +350 mV an 8 × 8 superstructure is observed, the integrated charge would correspond to 0.7 ± 0.1 Cu monolayers (from Ref. [78]).

In the OPD regime, finally, the Cu bulk phase starts growing.

Tantalum Tantalum has unique properties that make it useful for many applications, from electronics to mechanical and chemical systems. Many efforts have been made to develop an electroplating process for the electrodeposition of Ta. High-temperature molten salts were found to be efficient baths for the electrodeposition of refractory metals. To the best of our knowledge, until now no successful attempts have been made for Ta electrodeposition at room temperature or even at low temperature in ionic liquids. We present here the first results of tantalum electrodeposition in the air and water stable ionic liquid 1-butyl-1-methyl-pyrrolidinium bis(tri-fluoromethylsulfonyl)amide ([BMP][Tf$_2$N]).

Figure 6.2-11 shows the cyclic voltammogram of [BMP][Tf$_2$N] containing 0.5 M TaF$_5$ on Au(111) at room temperature. As shown, two reduction processes are recorded in the forward scan. The first starts at a potential of −0.5 V with a peak at −0.75 V, it might be correlated to the electrolytic reduction of Ta(V) to Ta(III). The second process starts at a potential of −1.5 V and is accompanied by the formation of a black deposit on the electrode surface. This can be attributed to the reduction of Ta(III) to Ta metal simultaneously with the formation of insoluble tantalum compounds on the electrode surface. The anodic peak recorded on the backward scan is due to the dissolution of the electrodeposit which, however, does not seem to be complete. At $E > 1.5$ V the anodic current increases as a result of gold dissolution. The deposit obtained is less adherent and can be easily removed by washing with acetone.

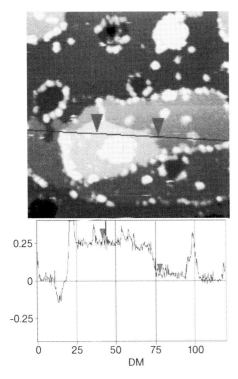

Fig. 6.2-10 Underpotential phenomena during Cu reduction in acid [BMIM]Cl/AlCl$_3$ on Au(111): at +50 mV a second monolayer with a height of 200 ± 20 pm grows together with a pronounced deposition of clusters that contain Cu and perhaps also a little amount of Al (from Ref. [78]).

We also performed the electrodeposition of Ta at different temperatures up to 200 °C. It was found that the mechanical quality and the adherence of the electrodeposits improve at 200 °C.

Moreover, the quality and the adherence of the electrodeposit were found to be improved upon addition of LiF to the electrolyte. The SEM micrograph of the Ta electrodeposit (Fig. 6.2-12(a)) recorded potentiostatically at −1.8 V in [BMP][Tf$_2$N] containing 0.25 M TaF$_5$ and 0.25 M LiF on a Pt electrode at 200 °C for 1 h shows a smooth, coherent and dense layer. XRD patterns of the electrodeposit clearly show the characteristic patterns of crystalline tantalum, as shown in Fig. 6.2-12(b).

In situ STM measurements under potentiostatic conditions can give valuable information on the electrodeposition of Ta in the employed ionic liquid [BMP][Tf$_2$N]. The STM picture of Fig. 6.2-13(a) shows a typical surface of gold Au(111) in the ionic liquid [BMP][Tf$_2$N] containing 0.5 M TaF$_5$ at open circuit potential. As seen, the surface is characterized by terraces with average step heights of about 250 pm, typical for Au(111). By applying a potential of −1.25 V (vs. Pt) (see Fig. 6.2-11), the

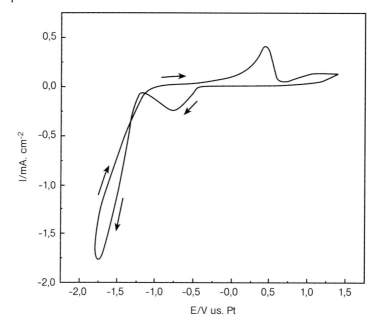

Fig. 6.2-11 Cyclic voltammogram of 0.5 M TaF$_5$ in [BMP][Tf$_2$N] on Au(111) at room temperature. Scan rate 10 mV s^{-1} (from Ref. [24]).

nature of the surface changes, as seen in the STM picture of Fig. 6.2-13(b). A rough layer of Ta is formed rapidly and some triangularly shaped islands with heights of several nanometers grow above the deposited layer (Fig. 6.2-13(b)). With time, these islands grow vertically and laterally and finally merge together to form a thick layer.

The three-dimensional STM picture of Fig. 6.2-14(a) shows the topography of the electrodeposit, with a thickness of about 300 nm. In order to investigate if the *in situ* deposit is metallic or not, current/voltage tunneling spectroscopy was carried out. A typical *in situ* tunneling spectrum of the 300 nm thick layer of the electrodeposit at different positions is shown in Fig. 6.2-14(b). As seen, the *I–U* spectrum exhibits metallic behavior with an exponential rise in the current revealing that the electrodeposited layer might be elemental Ta. From this, together with the *ex situ* measurements we can conclude that the reduction of TaF$_5$ in [BMP][Tf$_2$N] leads to an at least 500 nm thick layer of metallic tantalum.

Germanium electrodeposition on Au(111)

As a third example of our *in situ* STM results we would like to draw attention to the electrodeposition of Germanium [69, 70, 79]. Germanium is an elemental semiconductor with a band gap of 0.68 eV at room temperature in the microcrystalline phase. Furthermore, in contrast to metals, its crystal structure is determined by the tetrahedral symmetry of the Ge atoms, so the diamond structure is thermodynamically the most stable one. Germanium can hardly be obtained in aqueous solutions

(a)

(b)
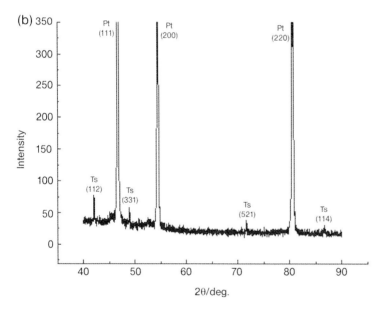

Fig. 6.2-12 (a) SEM micrograph of the electrodeposit formed potentiostatically on Pt in [BMP][Tf$_2$N] containing 0.25 M TaF$_5$ and 0.25 LiF at a potential of -1.8 V for 1 h at 200 °C; (b) XRD patterns of the deposited layer obtained potentiostatically on Pt in [BMP][Tf$_2$N] containing 0.25 M TaF$_5$ and 0.25 LiF at a potential of -1.8 V for 1 h at 200 °C.

598 | 6 Inorganic Synthesis

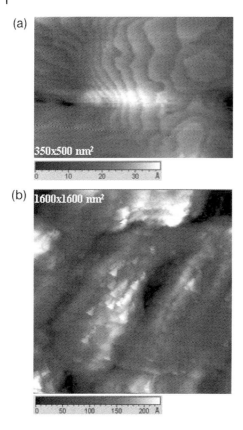

Fig. 6.2-13 (a) *In situ* STM picture of Au(111) in [BMP][Tf$_2$N] containing 0.5 M TaF$_5$ at the open circuit potential (−0.2 V); (b) *In situ* STM picture of the electrodeposit obtained at a potential of −1.25 V.

as its deposition potential is very close to hydrogen evolution. However, the ionic liquid [BMIM][PF$_6$] (and other ones) can be prepared easily with water amounts below 5 ppm and is therefore ideally suited for such electrodeposition studies. The pure liquid shows only capacitive behavior on Au(111), as can be seen in the cyclic voltammogram (Fig. 6.2-15), that was acquired with a scan rate of 1 mV s^{-1} under inert gas.

If GeI$_4$ is added in an estimated concentration between 0.1 and 1 mmol l^{-1} several processes are observed in the cyclic voltammogram (Fig. 6.2-16).

The electrode process at −500 mV on this potential scale is correlated to the growth of 250 ± 20 pm high islands. They grow immediately upon a potential step from the open circuit potential to −500 mV (arrow in Fig. 6.2-17).

They form a monolayer that is rich in defects. A similar result is obtained if the electrodeposition is performed from GeCl$_4$. There, also, 250 ± 20 pm high islands are observed as the first structures on the electrode surface. They can be oxidized

(a)

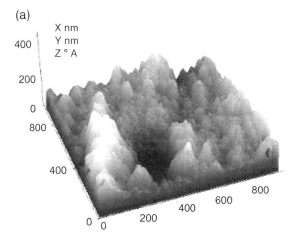

(b)

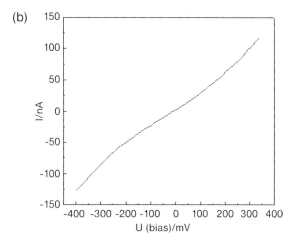

Fig. 6.2-14 (a) *In situ* 3D STM picture of about 300 nm thick layer of the electrodeposit; (b) *In situ* I–U tunneling spectrum of the electrodeposit.

reversibly and disappear completely from the surface. With GeI_4 the oxidation is more complicated, because the electrode potential for the gold step oxidation is too close to that of the island electrodissolution so that the two processes can hardly be distinguished. The gold step oxidation occurs already at +10 mV vs. the former open circuit potential, at +485 mV the oxidation of iodide to iodine starts.

In the reductive regime a strong, apparently irreversible, reduction peak is observed, located at −1510 mV vs. the quasi-reference electrode used in this system. With the *in situ* STM we saw a certain influence of the tip on the electrodeposition process. Therefore the tip was retracted, the electrode potential was set to −2000 mV,

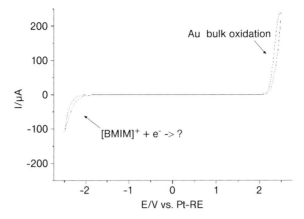

Fig. 6.2-15 Cyclic voltammogram of dry [BMIM][PF$_6$] on Au(111): between the anodic and the cathodic limits only capacitive currents flow: an electrochemical window of a little more than 4 V is obtained (from Ref. [69]).

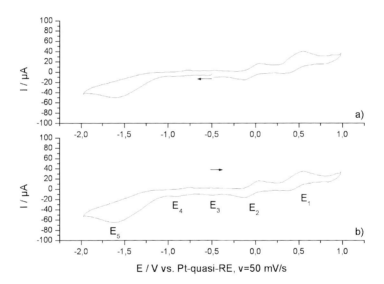

Fig. 6.2-16 Cyclic voltammogram of 0.1–1 mmol l^{-1} GeI$_4$ on gold in dry [BMIM][PF$_6$], starting at -500 mV towards cathodic (a) and anodic (b) regime. Two quasireversible (E_1 and E_2) and two apparently irreversible (E_4 and E_5) diffusion-controlled processes are observed. E_3 is correlated with the growing of 2D islands on the surface, E_4 and E_5 with the electrodeposition of germanium, E_2 with gold step oxidation and E_1 likely with the iodine/iodide couple. Surface area: 0.5 cm^2 (from Ref. [69]).

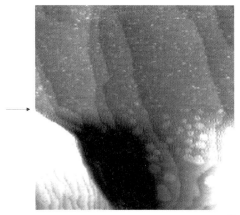

200 × 200 nm²

Fig. 6.2-17 UPD phenomena of Ge on Au(111) in dry [BMIM][PF$_6$]: upon a potential step from the open circuit potential to −500 mV vs. the Pt quasi-reference electrode 2D islands with an average height of 250 ± 20 pm start growing (from Ref. [69]).

and after 2 h the tip was reapproached. The surface topography that we obtained is presented in Fig. 6.2-18.

The surface consists of terraces with a height of 330 ± 30 pm. Within the limits of error, this is the value that has to be expected for Ge(111) bilayers. Furthermore, we could observe that the first electrodeposition leads to a less ordered surface structure with nanoclusters, which transforms on the time scale of about 1 h into a layered structure. With GeBr$_4$ a transformation of clusters into such a layered surface was only partly seen, with GeCl$_4$ this transformation could not be observed. The oxidation of the deposited germanium is also a complicated process. We found that mainly chemical oxidation by GeI$_4$ takes place, together with some electrooxidation. It is likely that kinetic factors play a dominant role.

If the germanium layers are partly oxidized by a short potential step to −1500 mV, random wormlike nanostructures form that heal in a complex process if the electrode potential is set back to more negative values (Fig. 6.2-19).

As well as electrodissolution and electrodeposition periphery and surface diffusion play an important role.

Unfortunately only thin films of about 20 nm in thickness could be obtained with GeI$_4$. Owing to experimental limitations, an *ex situ* analysis was difficult, but XPS showed clearly that elemental Ge was obtained. However, thick layers of Ge in the micrometer regime can be obtained using GeBr$_4$ and GeCl$_4$ as sources of Ge. The surface structure of a roughly 5 μm thick Ge film on gold is shown in Fig. 6.2-20. An interesting feature of the deposits is that they seem to consist of wires that have grown from the surface towards the solution. A higher resolution image shows that these structures seem to consist of coherent nanoparticles. In a wider sense these structures can be regarded as coherent nanosized wires.

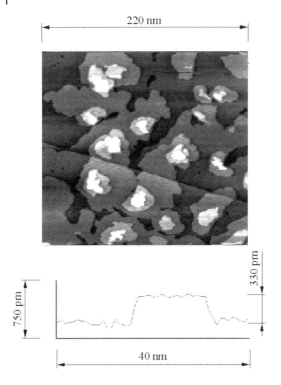

Fig. 6.2-18 Bulk deposition of Ge on Au(111) in dry [BMIM][PF$_6$]: at −2000 mV terraces with an average height of 330 ± 30 pm are obtained, indicative of Ge(111) bilayers.

Silicon

Silicon is one of the most important semiconductors as it is the basis of any computer chip. There have been several attempts in the past to electrodeposit silicon in organic solvents [80–82], but the authors report a disturbing effect of water, which can hardly be avoided in organic solvents. There have also been studies on the electrodeposition of silicon in high-temperature molten salts [83]. It was reported that silicon can also be electrodeposited in a low-temperature molten salt [84]. In this study the authors employed 1-ethyl-3-methylimidazolium hexafluorosilicate, and at 90 °C they could deposit a thin layer of silicon. However, this film reacted with water to form SiO$_2$, so that evidence of whether the deposited silicon species was elemental or even semiconducting is missing. Recently, we have shown that silicon can be well electrodeposited on the nanoscale in the room-temperature ionic liquid 1-butyl-1-methylpyrrolidinium bis(trifluoromethylsulfonyl)amide saturated with SiCl$_4$ [71]. This liquid exhibits on highly oriented pyrolytic graphite (HOPG) an electrochemical window of 4 V, which is limited in the anodic regime by the degradation of HOPG, in the cathodic regime by the irreversible reduction of the organic cation (Fig. 6.2-21).

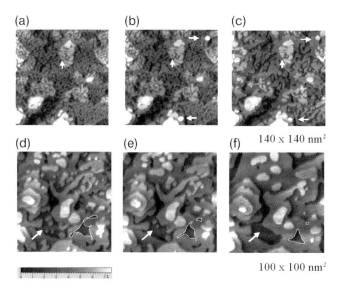

Fig. 6.2-19 Wormlike nanostructures in Ge(111) can be obtained by partial oxidation, they heal in a complex process comprising electrodeposition/electrodissolution and periphery diffusion. Ring-like defects transform to points as predicted by Grayson's theorem (vertical arrow, a: 0 min, b: 8 min, c: 20 min – time with reference to (a), electrodeposition of clusters occurs (horizontal arrows). Furthermore the clusters can dissolve (arrows in d–f) and pinch-off phenomena are observed (manually surrounded structures, d: 0 min, e: 12 min, f: 52 min – time with reference to (d)).

If the $SiCl_4$-saturated ionic liquid is investigated, a strong reduction current sets in at an electrode potential which is 600 mV positive of the cathodic decomposition limit of the liquid. After having passed the lower switching potential the anodic scan crosses the cathodic one at −2000 mV vs. Fc/Fc^+ which is typical for nucleation. Approaching an electrode potential of +400 mV vs. Fc/Fc^+ a strong oxidation current starts, which is in part correlated to the $SiCl_4$ reduction process beginning at −1600 mV vs. Fc/Fc^+ and in part correlated to HOPG oxidation. As with $SiCl_4$ in the liquid a similar oxidation behavior is observed if the scan is started towards positive potentials (Fig. 6.2-21).

Figure 6.2-22 shows a high-resolution SEM picture of an electrodeposited silicon layer on a gold substrate. As can be seen, the deposit contains small crystallites with sizes between 50 and 200 nm. The deposit can keep its dark appearance even under air. The EDX analysis showed only gold from the substrate and silicon, but no detectable chlorine. This proves that elemental silicon was electrodeposited.

Figure 6.2-23(a) shows the STM picture of an about 100 nm thick silicon layer that was electrodeposited at −1600 mV vs. Fc/Fc^+, probed under potential control with the *in situ* STM. Its surface is smooth on the nanometer scale and its topography is similar to that of a germanium layer of comparable thickness [79]. Figure 6.2-23(b)

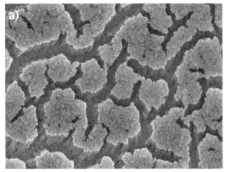

Fig. 6.2-20 The SEM picture of an approximately 5 μm thick Ge film on gold (obtained at −500 mV vs. Ge) shows that in a wider sense it is composed of coherent nanosized wires that seem to be built up of individual nanoclusters; (b) shows such a site at higher resolution.

shows an *in situ* current/voltage tunneling spectrum of HOPG (curve 1) and of the 100 nm thick silicon layer (curve 2). The spectra are of the same quality all over the surface. Whereas the tunneling spectrum of HOPG is – as expected – metallic, for the silicon deposit a typical band gap is observed. An evaluation of the band gap gives a value of 1.0 ± 0.2 eV. This value is quite similar to the value that we observed for hydrogen terminated n-doped Si(111) in an ionic liquid [85]. The value of microcrystalline silicon in the bulk phase at room temperature is 1.1 eV. In the light of these results, it can be concluded that elemental, intrinsic semiconducting silicon was electrodeposited from the employed ionic liquid.

6.2.3
Summary

We have presented in this section an insight into the electrodeposition of metals, alloys and semiconductors from ionic liquids. As well as environmental

6.2 Inorganic Materials by Electrochemical Methods | 605

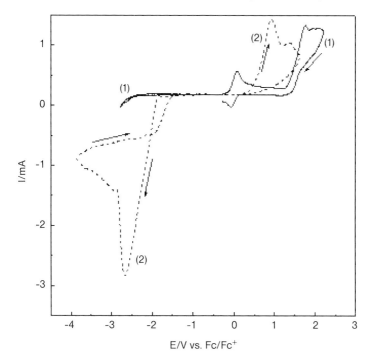

Fig. 6.2-21 (1) Electrochemical window of [BMP][Tf$_2$N] on HOPG with the ferrocene/ferrocinium (Fc/Fc$^+$) couple; (2) cyclic voltammogram of SiCl$_4$ saturated in the same ionic liquid. Scan rate each: 10 mV s^{-1} (from Ref. [71]).

Fig. 6.2-22 SEM micrograph of electrodeposited silicon, made potentiostatically at −2.7 V vs. Fc/Fc$^+$.

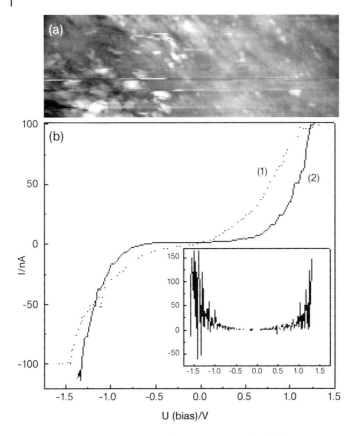

Fig. 6.2-23 (a) *In situ* STM picture of an about 100 nm thick film (600 nm × 200 nm); (b) *in situ* current/voltage tunneling spectra of HOPG (curve 1) and of the silicon electrodeposit (curve 2) on HOPG.

considerations these media have the great advantage of giving access to elements that cannot be obtained from aqueous solutions. Not only could technical procedures profit, but an interesting insight into the nanoscale processes during electrodeposition of elements such as germanium, silicon and others is also possible. Semiconductor nanostructures, especially, will be important in future nanotechnologies. Ionic liquids give access to a great variety of elements and compounds. We believe, therefore, that electrodeposition in ionic liquids is an important contribution to nanotechnology. Perhaps it will be possible in the future to establish nanoelectrochemical processes, for instance to make nanochips. In any case, many more studies will be necessary to understand the deposition processes on the nanometer scale.

References

1. *Nanoscale Probes of the Solid/Liquid Interface, NATO ASI series E 288*, A. A. Gewirth, H. Siegenthaler (Eds.), Kluwer Academic Publishers, Dordrecht, **1995**.
2. *Chemistry of Nonaqueous Solutions: Current Progress*, G. Mamantov, A. I. Popov, VCH, Weinheim, **1994**.
3. A. J. Bard, L. R. Faulkner, *Electrochemical Methods: Fundamentals and Applications*, 2nd edn., John Wiley & Sons, New York, **2001**.
4. S. Takahashi, K. Akimoto, I. Saeki, *Hyomen Gijutsu* **1989**, *40*(1), 134.
5. P. K. Lai, M. Skyllas-Kazacos, *J. Electroanal. Chem.* **1988**, *248*(2), 431.
6. Y. Zhao, T. J. VanderNoot, *Electrochim. Acta*, **1997**, *42* (11), 1639.
7. M. R. Ali, A. Nishikata, T. Tsuru, *Indian J. Chem. Technol.* **1999**, 6 (6), 317.
8. Q. Liao, W. R. Pitner, G. Stewart, C. L. Hussey, *J. Electrochem. Soc.* **1997**, *144* (3), 936.
9. J. Fischer, US patent 2 828 252, **1958**.
10. M. Bukowski, F. Endres, H. Natter, R. Hempelmann, patent pending DE- 101 08 893.0-24 (29.08.2002).
11. S. Zein El Abedin, E. M. Moustafa, R. Hempelmann, H. Natter, F. Endres, *Electrochem. Commun.* **2005**, *7*, 1116.
12. D. Wei, M. Okido, *Curr. Top. Electrochem.* **1997**, *5*, 21.
13. I. Galasiu, R. Galasiu, J. Thonstad, *Nonaqueous Electrochem.* **1999**, 461.
14. W. Freyland, *Z. Phys. Chem.* **1994**, *184*, 139.
15. G. E. Gray, P. A. Kohl, J. Winnick, *J. Electrochem. Soc.* **1995**, *142* (11), 3636.
16. B. J. Piersma, *Proc. Electrochem. Soc.* **1994**, *94*(13), 415.
17. P.-Y. Chen, Y.-F. Lin, I.-W. Sun, *J. Electrochem. Soc.* **1999**, *146* (9), 3290.
18. M. W. Verbrugge, M. K. Carpenter, *AIChE J.* **1990**, *36* (7), 1097.
19. C. Nanjundiah, K. Shimizu, R. A. Osteryoung, *J. Electrochem. Soc.* **1982**, *11*, 2474.
20. M. Lipsztajn, R. A. Osteryoung, *Inorg. Chem.* **1985**, *24*, 716.
21. R. T. Carlin, H. C. De Long, J. Fuller, P. C. Trulove, *J. Electrochem. Soc.* **1998**, *145* (5), 1598.
22. N. Koura, T. Kato, E. Yumoto, *Hyomen Gijutsu* **1994**, *45*(8), 805.
23. G. T. Cheek, H. C. De Long, P. C. Trulove, *Proc. Electrochem. Soc.* **2000**, 99-41, 527.
24. S. Zein El Abedin, H. K. Farag, E. M. Moustafa, U. Welz-Biermann, F. Endres, *Phys. Chem. Chem. Phys.* **2005**, *7*, 2333.
25. S. Zein El Abedin, U. Welz-Biermann, F. Endres, *Electrochem. Commun.* **2005**, *7*, 941.
26. N. Guo, J. Guo, S. Xiong, *Fushi Kexue Yu Fanghu Jishu* **1998**, *10*(5), 290.
27. M. R. Ali, A. Nishikata, T. Tsuru, *Electrochim. Acta* **1997**, *42*(15), 2347.
28. T. Tsuda, Y. Ito, *Proc. Electrochem. Soc.* **2000**, 99–41, 100.
29. J. S-Y. Liu, I.-W. Sun, *J. Electrochem. Soc.* **1997**, *144* (1), 140.
30. M. K. Carpenter, M. W. Verbrugge, *J. Mater. Res.* **1994**, 9 (10), 2584.
31. M. K. Carpenter, M. W. Verbrugge, US Patent 92-926103, **1993**.
32. M. Lipsztjan, R. A. Osteryoung, *Inorg. Chem.* **1985**, *24* (21), 3492.
33. D. A. Habboush, R. A. Osteryoung, *Inorg. Chem.* **1984**, *23* (12), 1726.
34. M.-H. Yang, M.-C. Yang, I.-W. Sun, *J. Electrochem. Soc.* **2003**, *150*, C544.
35. A. F. Tsatsulnikov, S. V. Ivanov, P. S. Kopev, I.L. Krestnikov, A. K. Kryganovskii, N. N. Ledentsov, M. V. Maximov, B. Y. Meltser, P.V. Nekludov, A. A. Suvorova, A. N. Titkov, B. V. Volovik, M. Grundmann, D. Bimberg, Z. I. Alferov, A. F. Ioffe, *Microelectron. Eng.* **1998**, 43/44, 85.
36. E. G.-S. Jeng, I.-W. Sun, *J. Electrochem. Soc.* **1997**, *144*(7), 2369.
37. B. Gregory, J. L. Stickney, *J. Electroanal. Chem.* **1994**, *365*, 87.
38. M. A. M. Noel, R. A. Osteryoung, *J. Electroanal. Chem.* **1996**, *293*, 139.
39. P.-Y. Chen, I.-W. Sun, *Electrochim. Acta* **2000**, *45* (19), 3163.
40. C. L. Hussey, L. A. King, R. A. Carpio, *J. Electrochem. Soc.* **1979**, *126* (6), 1029.
41. C. Nanjundiah, R. A. Osteryoung, *J. Electrochem. Soc.* **1983**, *130* (6), 1312.
42. B. J. Tierney, W. R. Pitner, J. A. Mitchell, C. L. Hussey, *J. Electrochem. Soc.* **1998**, *145*(9), 3110.

43. P.-Y. Chen, I.-W. Sun, *Proc. Electrochem. Soc.* **1998**, *98–11*, 55.
44. X.-H. Xu, C. L. Hussey, *J. Electrochem. Soc.* **1992**, *139*, 1295.
45. F. Endres, W. Freyland, *J. Phys. Chem. B* **1998**, *102*, 10229.
46. Y. Katayama, S. Dan, T. Miura, T. Kishi, *J. Electrochem. Soc.* **2001**, *148* (2), C102.
47. W. Schindler, D. Hofmann, J. Kirschner, *J. Electrochem. Soc.* **2001**, *148*(2), C124–C130.
48. J. A. Mitchell, W. R. Pitner, C. L. Hussey, G. R. Stafford, *J. Electrochem. Soc.* **1996**, *143*(11), 3448.
49. W. R. Pitner, C. L. Hussey, G. R. Stafford, *J. Electrochem. Soc.* **1996**, *143* (1), 130.
50. P.-Y Chen, I.-W Sun, *Electrochim. Acta* **2001**, *46*, 1169.
51. A. P. Abbott, G. Capper, D. L. Davies, H. L. Munro, R. K. Rasheed, V. Tambyrajah, *Chem. Comm.* **2001**, 1010.
52. H. C. De Long, J. S. Wilkes, R. T. Carlin, *J. Electrochem. Soc.* **1994**, *141*(4), 1000.
53. C.-C Tai, F.-Y Su, I.-W. Sun, *J. Electrochem. Soc.* **2004**, *151*, C811.
54. C.-C Tai, F.-Y Su, I.-W. Sun, *Electrochim. Acta* **2005**, *50*, 5504.
55. X.-H. Xu, C.L. Hussey, *Proc. Electrochem. Soc.* **1992**, *16*, 445.
56. E. R. Schreiter, J. E. Stevens, M. F. Ortwerth, R. G. Freeman, *Inorg. Chem.* **1999**, *38* (17), 3935.
57. W. R. Pitner, C. L. Hussey, *J. Electrochem. Soc.* **1997**, *144* (9), 3095.
58. J. F. Haung, I-W. Sun, *Adv. Funct. Mater.* **2005**, *15*, 989.
59. J. F. Haung and I.-W. Sun, *Chem. Mater.* **2004**, *16*, 1829.
60. H. Y. Hsu, C. C. Yang, *Z. Naturforsch. B* **2003**, *58*, 1055.
61. X.-H. Xu, C. L. Hussey, *J. Electrochem. Soc.* **1993**, *140*, 618.
62. A. P. Abbott, G. Capper, D. L. Davies, R. K. Rasheed, *Chem. Eur. J.* **2004**, *10*, 3769.
63. A. P. Abbott, G. Capper, D. L. Davies, R. K. Rasheed, J. Archer, C. John, *Trans. Inst. Met. Finish.* **2004**, *82*, 14.
64. R. K. Pandey, S. N. Sahu, S. Chandra, *Handbook of Semiconductor Electrodeposition*, Marcel Dekker, New York, **1996**.
65. A. Raza, R. Engelken, B. Kemp, A. Siddiqui, O. Mustafa, *Proc. Arkansas Acad. Sci.* **1995**, *49*, 143.
66. S. P. Wicelinski, R. J. Gale, *Proc. Electrochem. Soc.* **1987**, *134*, 262.
67. M. K. Carpenter, M. W. Verbrugge, *J. Electrochem. Soc.* **1987**, *87* (7), 591.
68. M.-C. Lin, P.-Y. Chen, I.-W. Sun, *J. Electrochem. Soc.* **2001**, *148*(10), C653.
69. F. Endres, C. Schrodt, *Phys. Chem. Chem. Phys.* **2000**, *24*, 5517.
70. F. Endres, *Phys. Chem. Chem. Phys.* **2001**, *3*(15), 3165.
71. S. Zein El Abedin, N. Boressinko, F. Endres, *Electrochem. Commun.* **2004**, *6*, 510.
72. O. Leifeld, A. Beyer, E. Müller, D. Grützmacher, K. Kern, *Thin Solid Films* **2000**, *380*, 176.
73. G. Binnig, H. Rohrer, *Helv. Phys. Acta* **1982**, *55*, 726.
74. R. Sonnenfeld, P. K. Hansma, *Science* **1986**, *232*, 211.
75. *Scanning Tunneling Microscopy and Spectroscopy: Theory, Techniques and Applications*, D.A. Bonnell (Ed.), John Wiley & Sons, New York, **2000**.
76. C. A. Zell, F. Endres, W. Freyland, *Phys. Chem. Chem. Phys.* **1999**, *1*, 697.
77. T. Will, M. Dietterle, D. M. Kolb in *Nanoscale Probes of the Solid/Liquid Interface, NATO ASI series E 288*, A. A. Gewirth, H. Siegenthaler (Eds.), Kluwer Academic Publishers, Dordrecht, **1995**, pp. 137–162.
78. F. Endres, A. Schweizer, *Phys. Chem. Chem. Phys.* **2000**, *2*, 5455.
79. F. Endres, S. Zein El Abedin, *Phys. Chem. Chem. Phys.* **2002**, *4*, 1640.
80. A. K. Agrawal, A. E. Austin, *J. Electrochem. Soc.* **1981**, *128*, 2292.
81. J. Gobet, H. Tannenberger, *J. Electrochem. Soc.* **1986**, *133*, C322.
82. J. Gobet, H. Tannenberger, *J. Electrochem. Soc.* **1988**, *135*, 109.
83. T. Matsuda, S. Nakamura, K. Ide, K. Nyudo, S. J. Yae, Y. Nakato, *Chem. Lett.* **1996**, *7*, 569.
84. Y. Katayama, M. Yokomizo, T. Miura, T. Kishi, *Electrochemistry* **2001**, *69*, 834.
85. W. Freyland, C. A. Zell, S. Zein El Abedin, F. Endres, *Electrochim. Acta* **2003**, *48* 3053.

6.3
Ionic Liquids in Material Synthesis: Functional Nanoparticles and Other Inorganic Nanostructures

Markus Antonietti, Bernd Smarsly, and Yong Zhou

6.3.1
Introduction

Ionic liquids (ILs) are organic salts with low melting points, sometimes as low as −96 °C [1]. They are attractive for many fields of chemistry and industry, due to their potential as "green" recyclable alternatives to the traditional organic solvents [2]. They possess a wide liquid range, in some cases in excess of 400 °C. Their very favorable properties, such as high polarity, negligible vapor pressure, high ionic conductivity, and thermal stability, make them effective in catalysis [3, 4], as inert solvents in electrochemistry [5], for polymer synthesis [6, 7], and for transferring enzymatic reactions to non-aqueous media [8]. The most extensively studied ILs are composed of the organic 1-alkyl-3-methylimidazolium ion with a variety of counter-ions. Newer systems include species with additional functionality, e.g. long chain amphiphilic ILs with both lyotropic [9] and thermotropic liquid crystallinity [10].

Apart from the extensive use of ILs in catalysis and organic/inorganic synthesis [11, 12] (see also Chapter 5 and Section 6.1) their advantages for materials chemistry, and especially for the synthesis of novel nanostructures, have been only gradually realized in recent years. This contribution will actualize a recent review [13] and aims to introduce this specific field of ionic liquid applications which undoubtedly promises great potential for future research and development.

6.3.2
Ionic Liquids for the Synthesis of Chemical Nanostructures

ILs have been used in the electrosynthesis of nanostructures; the preparation of various metallic nanoparticles, such as palladium [14], iridium [15], or semiconductor nanoparticles, such as stable Ge-nanoclusters [16], have been described. In some cases, wire-like structures were obtained, for instance Ti nanowires formed on graphite by electroreduction, as described by Freyland et al. [17]. In all these cases, mainly the large electrochemical window and the relatively high polarity of ILs are exploited.

As ILs can reach electrochemical windows of more than 4 V, depending on the cation/anion combination, they have also been used as solvents for the electrodeposition of high quality nanocrystalline films of a wide range of metals, alloys and semiconductors, especially a whole range of elements that cannot be electrodeposited from aqueous solution, such as the light and refractory metals. The related work has been well reviewed [18].

Recently, the electrodeposition of calamitic Fe-nanostructures from the ionic melt $AlCl_3$–1-butyl-3-methylimidazolium chloride ($AlCl_3$/[BMIM]Cl) on Au(111) and their characterization by *in situ* electrochemical STM microscopy have been described [19].

Very fine and stable noble metal nanoparticles (Ir(0) and Ru(0), 2.0–2.5 nm in diameter) can, of course, also be synthesized in ILs by chemical reduction [20]. Metal nanoparticles are extraordinarily stable in ILs as a solvent/stabilizer-system, as indicated by the really extraordinarily high turn-over numbers in catalytic hydrogenation.

These applications employ, at first sight, just one of the advantages of this very special class of solvents (the large electrochemical window), but one can further consider other aspects such as:

- Although polar, ILs can have low interfacial tensions, which in addition seem to adapt to the other phase (e.g. 1-butyl-3-methylimidazolium tetrafluoroborate has an interfacial tension of only ca. 38 mN m^{-1} against air [21]). Thus, nucleation rates can be extraordinarily high, and very small nanostructures can be generated that only weakly undergo Ostwald-ripening.
- Low interface energies for larger objects can be translated into good stabilization or solvation of molecular species. Obviously, the IL structure adapts to many species, as it can provide hydrophobic regions and a high directional polarizability that can be put parallel and perpendicular to the dissolved species. In simple words: materials synthesis is like making a reaction in pure ligand.
- The high thermal stability and low volatility enables one to carry out reactions well beyond 100 °C in non-pressurized devices.
- They allow inorganic synthesis from very polar starting materials under ambient conditions and under water-poor conditions. Thus hydroxide or oxyhydrate formation and the coupled generation of amorphous species can be suppressed, as low amounts of water drive the mass balance to completely condensed systems, which are usually directly crystalline.
- The most important advantage however is an unconventional and very rare property which cannot be sufficiently underlined: ILs form extended H-bonded systems in the liquid state [22] and are therefore highly structured [23, 24]. ILs are therefore "supramolecular" solvents. This fact was just recently nicely supported by neutron reflectometry measurements on surfaces of 1-alkyl-3-methylimidazolium-based ionic liquids where strong self-structuring with a characteristic length scale of about 4 nm was observed [25]. Presumably not generally known, solvent structuring is the molecular basis of most molecular recognition and self-organization processes, with water being the most prominent example.[1] This special quality can be used as the "entropic driver" for spontaneous, well-defined and extended ordering of nanoscale structures.

---

[1] The search for such solvents is a classic goal in colloid chemistry. Although much less pronounced, formamide and water-free hydrazine are also self-structured and allow, for instance, micelle formation of surfactants.

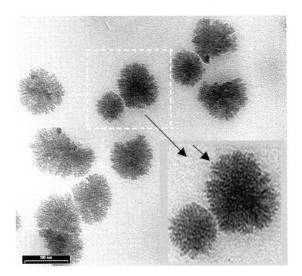

Fig. 6.3-1 Sponge-like anatase with high surface area, as synthesized by the hydrolysis of TiCl$_4$ in ILs.

The first work on inorganic sol–gel reactions focused on the formation of silica aerogels, and it turned out that such aerogels can be dried without a supercritical drying procedure [26]. This again indicates a very low interfacial tension of the binary system coupled with low capillary forces. It is even more interesting to make crystallizable species by sol–gel reactions in water-poor reaction media, as exemplarily pursued by Zhou et al. [27]. They hydrolyzed titanium tetrachloride in 1-butyl-3-methylimidazolium tetrafluoroborate with some reaction water (water-poor conditions), whereupon 2–3 nm sized anatase powders with surface areas of 554 m^{2}g^{-1}, assembling to larger, spherical sponge-like superstructures, were obtained in a low-temperature synthesis (at 80 °C). These experiments look simple, but illustrate the advantages of ILs in several ways. First, sol–gel reactions in water usually provide amorphous titania, which has to be calcined above 350 °C to give the desired crystalline anatase. This usually leads to the avoidance of direct employment of anatase in organic systems. Also, the nucleation rate of titania from the amorphous solid state is rather low (usually, particles of ca. 20 nm in size are obtained). The IL solvent therefore not only provides direct synthesis of crystalline species under ambient conditions, but also increases the nucleation rate by more than a factor of 1000, again using its low interfacial energy and adaptability. Only this enables the delicacy of the resulting structures. The structures obtained are depicted in Fig. 6.3-1. These results have also been recently supported by a second research group [28].

These anatase sponge structures combine the convenient handling of larger spheres with a very high surface area and narrow pore size distribution and are expected to have potential in solar energy conversion, catalysis and optoelectronic devices, e.g. for the potential one-shot synthesis of dye-sensitized titania solar cells. IL-based quasi-solid-state electrolytes have just recently been employed for such regenerative photoelectrochemical cells and yielded a solar-to-electricity efficiency of 6.6% at illumination of air mass of 1.5 (AM 1.5, 100 mW cm^{-2}) and >7% at lower light intensities. In those experiments, the ILs were just used as a solvent, but nanostructure synthesis still took place by classical means [29, 30]. It has also been reported that the use of ILs for π-conjugated polymer-based electrochemical devices has enhanced the electrochemically cycled lifetimes of the π-conjugated polymers up to 1 million cycles without failure, and fast cycle switching speeds of 100 ms were found [31]. Just using the non-miscibility of the ILs with anhydrous toluene, Nakashima et al. reported the synthesis of a different TiO_2 morphology in ILs, hollow microspheres via a so-called interfacial sol–gel reaction [32].

The strong surface binding of ILs onto various nanoparticles was employed by Itoh et al. who showed that the hydrophilic and hydrophobic properties of gold nanoparticles can be tuned by anion exchange of the IL moiety [33]. In that work, the efficient hydrophilic to hydrophobic transfer of the gold nanoparticles formed from aqueous 3.3′-[disulfanylbis(hexane-1,6-diyl)]bis(1-methyl-1H-imidazol-3-ium)dichloride media to water-immiscible 1-hexyl-3-methylimidazolium hexafluorophosphate can be achieved via addition of $H[PF_6]$ to the aqueous solution to exchange Cl^-. The system has potential for the recycling of noble metal nanoparticles by solvent-induced immiscibility after a catalytic process. Wei et al. recently reported on similar phase-transferred gold nanoparticles and gold nanorods using an IL system [34]. CdTe nanoparticles were transferred from water to an IL by Nakashima et al., thus greatly increasing the photoluminescence activity [35].

The presence of a secondary thiol functionality in an IL resulted in a one-pot, size-selective synthesis of gold and platinum nanoparticles [36]. The diameters could be controlled to be between 2.0 and 3.5 nm, depending on the number of thiol groups. The size distribution in those cases was so small that spontaneous ordering into cubic arrays was observed. In an independent paper, similar thiol-functionalized ILs were used to generate IL-stabilized Au nanoparticles, which showed an extraordinary stability in aqueous dispersions against proteins, high salt concentrations, or the ionic liquid itself [37].

Backed by the same advantages of ILs as a reaction medium, i.e. strong surface binding and the high-temperature operating window, microwave-assisted synthesis of single-crystalline tellurium nanorods and nanowires has been reported recently [38]. Monodisperse chromium nanoparticles could be made by thermolysis of a Fischer carbene complex [39].

Solvent self-structuring and supramolecular effects become important when higher concentrations of inorganic reactant are applied. Classical ILs, such as 1-butyl-3-methylimidazolium tetrafluoroborate give nicely nanostructured inorganic gels, as shown for silica [40], with a sponge like, bicontinuous phase with a characteristic length of 5 nm. Based on NMR and Raman spectrometry, this was explained by the fact that the IL molecules spontaneously form a double layer by binding to silica. A model of the resulting structure, as revealed from silica nanocasting, is presented in Fig. 6.3-2.

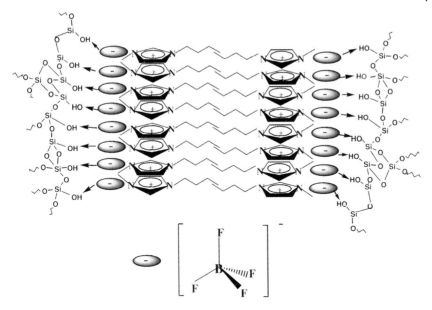

Fig. 6.3-2 Schematic presentation of the self-organization of 1-butyl-3-methylimidazolium tetrafluoroborate, as revealed by silica nanocasting. Hydrophobic interactions and hydrogen bonding between the imidazolium rings and the counterions drive this structure formation.

This bilayer formation sounds unusual for such a small molecule, but just reflects the very strong tendency of the ILs to form extended H-bonded networks, in this case an undulating two-dimensional structure. There is a good chance that the liquid structures of ILs and their mixtures with other solvents are organized in similar fashion.

The self-organization of ILs can be supported (better: simplified and made less multistable) by using amphiphilic species with a longer hydrophobic tail. Again, due to a combination of H-bonded networks and polarity contrast (amphiphilicity), very well organized lyotropic phases are obtained for both the pure ILs and their mixtures with water, oils, and reactants. Lee et al. have independently successfully employed such ILs as templates for the synthesis of periodic mesoporous organosilicas, PMOs [41]. This tolerance of self-organization against monomer choice and loading is again very unusual and, even in water, is only found for some special surfactants forming three-component microemulsions.

Using sol–gel synthesis again, it is possible to employ these oriented phases for material synthesis, in the special case using 1-hexadecyl-3-methylimidazolium chloride, 1_{16}. The condensation products of these oriented phases give quite perfect textures, as shown in Fig. 6.3-3 for a silica made from a lamellar amphiphilic IL mesophase [42].

614 | 6 Inorganic Synthesis

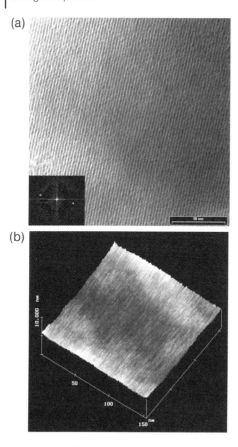

Fig. 6.3-3 (a) TEM image of 1_{16}-templated porous silica prepared in a sol–gel reaction of 1 g of 1_{16} with 1 g of silica at a temperature of 40 °C. The scale bar is 50 nm. Inset: 2d Fourier transform of the picture.
(b) AFM picture of a surface of the same material.

Following the reasoning given above, these phases always interact strongly with surfaces and usually align perpendicular to the substrate. This is the opposite to water systems where a parallel orientation is preferred, and can be explained by the very strong polarizability of the supramolecular IL structure along the aligned H-bonded networks, i.e. perpendicular to the main molecular axis. It is indeed observed that the lamellae or sheets are exactly parallel over wide areas (see Fourier transform), and that even at the surface a perfect homeotropic organization prevails (see AFM). The real structure is presumably not only lamellar but must also contain pillars or bridges, as the described structure is stable throughout the removal of the IL and does not collapse. The silicas made this way essentially have the structural characteristics of clay minerals but there is the freedom to adjust the chemical composition to the requirements of use. It can be assumed that similar

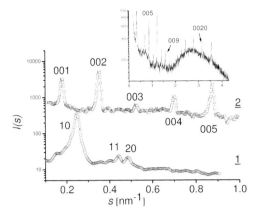

Fig. 6.3-4 XRD curves for two mesostructured IL (1_{16})/silica materials varying in the water content of the precursor solution ($s = 2/\lambda \sin \theta$) with a balance of template to silica of about 1:1. **1**: two-dimensional hexagonal mesostructure obtained with an abundance of water. **2**: Lamellar mesostructure obtained using only stoichiometric amounts of water, keeping the natural H-network intact. This structure was depicted in Fig. 6.3-3. The inset shows the higher-order reflections of this species which go up to the 22nd order.

orientation effects also exist in the non-amphiphilic ILs, which might be the reason why they function as lubricants with an extraordinary performance [43], beyond any effect predicted from the molecular structure. In a recent paper, Cooper et al. have extended this work to the synthesis of zeolitic crystalline materials[44]. Here again, the IL takes the roles of both a template and a solvent, and a whole set of new zeotypes, SIZ-1–SIZ-4, could be made.

The role of extra water in such recipes has turned out to be complex and again a question of supramolecular chemistry. It seems to be safe to state that its structure and chemical reactivity is far from that of bulk water, as it is tightly bound and activated in the H-bonded system of the IL. Therefore, reactions with water take place quite rapidly. On the other hand, water as a solvating ligand seems to be excluded due to the IL-binding, as for instance deduced from the absence of so-called "solvent pores" [30]. This is a quite unique situation for colloid chemistry and material synthesis.

Water, however, modifies the patterns of self-organization, and this is why the structural outcome of such reactions depends greatly on the water content. The peculiar self-aggregation behavior of the IL/water system is seen by comparing two sol–gel derived IL–silica hybrid materials (using 1_{16}), prepared with varying amounts of water (Fig. 6.3-4), but the same ratio of IL to silica.

While sample 2 was obtained under "water-poor conditions", sample 1 was made with a tenfold excess of water. It is seen in X-ray scattering patterns that this difference in the water content strongly affects the self-organization behavior: Sample 1 corresponds to a 2D hexagonal mesophase, while sample 2 corresponds to a lamellar structure with a long period of $d = 5.6$ nm. The latter sample can be indexed up to the 22nd interference order, an exceptional perfection for self-organized mesostructures on the nanometer scale. Note that the sample is not crystalline on the atomic scale, as indicated by the typical halo in the wide angle. This high order is also reflected in the reaction products made from such phases, as indicated in Fig. 6.3-3 or seen in a polarized optical microscopy image, with textures spanning several millimeters.

These results demonstrate that ILs can also be applied in water-rich media, where they "only" play the role of a classical surfactant, however with a very strong tendency towards self-organization with high order.

Simultaneous application of polymer latexes organized by sedimentation aggregation and amphiphilic ILs in water-rich media as templates for porous silica led to structures with organized bimodal porosity where both typical textures coexisted [45]. The resulting supermicroporous inverse opals were discussed as optical sensor elements, where reflection contrast depends critically on the absorption of spurious amounts of organic molecules. Furthermore, ILs allow the preparation of hierarchical pore morphologies with two types of mesopores. Using 1_{16} and certain block copolymers as templates (producing mesopores of ca. 2.6 nm and 13 nm, respectively), a highly mesoporous silica could be obtained with the IL mesopores being located between the 13 nm mesopores [46, 47]. This mesopore hierarchy has to be regarded as remarkable as ionic surfactants and block copolymers in water usually either form compound ("mixed") micelles, or the two types of micelles form separated lyotropic phases. By contrast, it was demonstrated that IL and block copolymer micelles self-assemble into an alloy-like hierarchical mesopore morphology, which is probably due to the aforementioned interaction properties of ILs.

To conclude, we expect that ILs will find, besides organometallic synthesis, catalysis and electrochemistry, a further rich field of application in the synthesis of nanostructured solids, either to make nano-objects (e.g. particles and fibers) with very special and otherwise non-addressable properties or for the design of nanopores and nanochannels in solids. It was reasoned that it is the quite singular combination of energetic adaptability towards other molecules and phases plus the strong H-bonded driven solvent structure which makes ILs a potential key tool in the realization of a new generation of chemical nanostructures.

References

1. K. R. Seddon, A. Stark, M. J. Torres, *Pure Appl. Chem.* **2000**, *72*, 2275.
2. T. Welton, *Chem. Rev.* **1999**, *99*, 2071.
3. J. Dupont, R. R. de Suoza, P. A. Z. Suarez, *Chem.Rev.* **2002**, *102*, 2667.
4. R. Sheldon, *Chem. Comm*, **2001**, *23*, 2399.
5. J. Fuller J, R. T. Carkin, R. A. Osteryoung, *J. Electrochem. Soc.* **1997**, *144*, 3881.
6. Y. L. Zhao, J. M. Zhang, J. Jiang, C. F. Chen, F. Xi, *J. Polym. Sci. A* **2001**, *40*, 3360.
7. P. Kubisa, *Prog. Polym. Sci.* **2001**, *29*, 3.
8. F. Van Rantwijk, R. M. Lau, R. A. Sheldon, *Trends Biotechnol.* **2003**, *21*, 131.
9. T. A. Bleasdale, G. J. T. Tiddy, E. Wyn-Jones, *J. Phys. Chem.* **1991**, *95*, 5385.
10. F. Neve, O. Francescangeli, A. Crispini, *Inorg. Chim. Acta* **2002**, *338*, 51.

11. C. K. Lee, H. W. Huang, I. J. B. Lin, *Chem. Commun.* **2000**, 1911.
12. P. J. Dyson, *Transition Met. Chem.* **2002**, *27*, 353.
13. M. Antonietti, D. B. Kuang, B. Smarsly, Y. Zhou, *Angew. Chem. Int. Ed.* **2004**, *43*, 4988.
14. R. R. Deshmukh, R. Rajagopal, K. V. Srinivasan, *Chem. Commun.* **2001**, 1544.
15. J. Dupont, G. S. Fonseca, A. P. Umpierre, P. F. P. Fichtner, S. R. Teixeira, *J. Am. Chem. Soc.* **2002**, *124*, 4228.
16. (a) F. Endres, S. Z. Abedin, *Chem.Commun.* **2002**, *8*, 892; (b) S. Z. Abedin, N. Borissenko, F. Endres, *Electrochem. Commun.* **2004**, *6*, 510.
17. I. Mukhopadhyay, W. Freyland, *Langmuir* **2003**, *19*, 1951.
18. F. Endres, *Chemphyschem.* **2002**, *3*, 144.
19. C. L. Aravinda, W. Freyland, *Chem. Commun.* **2004**, *23*, 2754.
20. G. S. Fonseca, A. P. Umpierre, P. F. P. Fichtner, S. R. Teixera, J. Dupont, *Chem. Eur. J.* **2003**, *9*, 3263.
21. As measured with a du Nuoy tensiometer from Krüss, this value seems to depend only weakly on water content; for ILs in water, see: J. Bowers, C. P. Butts, P. J. Martin, M. C. Vergara-Gutierrez, R. K. Heenan, *Langmuir* **2004**, *20*, 2191.
22. A. Elaiwi, S. B. Hitchcock, K. R. Seddon, N. Srinivasan, Y. M. Tan, T. Welton, J. A. Zora, *J. Chem. Soc.*, Dalton Trans. **1995**, *21*, 3467.
23. A. Mele, C. D. Tran, S. H. D. Lacerda, *Angew. Chem. Int. Ed.* **2003**, *42*, 4364.
24. S. Saha, S. Hayashi, A. Kobayashi, H. Hamaguchi, *Chem.Lett.* **2003**, *32*, 740.
25. J. Bowers, M. C. Vergara-Gutierrez, J. R. P. Webster, *Langmuir* **2004**, *20*, 309.
26. S. Dai, Y. H. Ju, H. J. Gao, J. S. Lin, S. J. Pennycock, C. E. Barnes, *Chem. Commun.* **2000**, *3*, 243.
27. Y. Zhou, M. Antonietti, *J. Am. Chem. Soc.* **2003**, *125*, 14960.
28. K. Yoo, H. Choi, D. D. Dionysiou, *Chem. Commun.* **2004**, *17*, 2000.
29. (a) P. Wang, S. M. Zakeeruddin, P. Comte, I. Exnar, M. Gratzel, *J. Am. Chem. Soc.* **2003**, *125*, 1166; (b) P. Wang, S. M. Zekeeruddin, J. E. Moser, M. Grätzel, *J. Phys.Chem. B.* **2003**, *107*, 13280.
30. P. Wang, S. M. Zakeeruddin, R. Humphry-Baker, M. Grätzel, *Chem. Mater.* **2004**, *16*, 2694.
31. W. Lu, A. G. Fadeev, B. H. Qi, E. Smela, B. R. Mattes, J. Ding, G. M. Spinks, J. Mazurkiewicz, D. Z. Zhou, G. G. Wallace, D. R. MacFarlane, S. A. Forsyth, M. Forsyth *Science* **2002**, *297*, 983.
32. T. Nakashima, N. Kimizuka, *J. Am. Chem. Soc.* **2003**, *125*, 6386.
33. H. Itoh, K. Naka, Y. Chujo, *J. Am. Chem. Soc.* **2004**, *126*, 3026.
34. G. T. Wei, Z. S. Yang, C. Y. Lee, H. Y. Yang, C. R. C. Wang *J. Am. Chem. Soc.* **2004**, *126*, 5036.
35. T. Nakashima, T. Kawai, *Chem. Commun.* **2005**, *12*, 1643–1645.
36. K. S. Kim, D. Demberelnyamba, H. Lee, *Langmuir* **2004**, *20*, 556.
37. R. Tatumi, H. Fujihara, *Chem. Comm.* **2005**, *1*, 83.
38. Y. J. Zhu, W. W. Wang, R. J. Qi, X. L. Hu, *Angew. Chem. Int. Ed.* **2004**, *43*, 1410.
39. S. U. Son, Y. J. Jang, K. Y. Yoon, C. H. An, Y. Hwang, J. G. Park, H. J. Noh, J. Y. Kim, J. H. Park, T. Hyeon, *Chem. Comm.* **2005**, *1*, 86.
40. Y. Zhou, M. Antonietti, *NanoLett.* **2004**, *4*, 477.
41. B. Lee, H. M. Luo, C. Y. Yuan, J. S. Lin, S. Dai, *Chem. Commun.* **2004**, *2*, 240.
42. Y. Zhou, M. Antonietti, *Chem. Mater.* **2004**, *16*, 544.
43. H. Z. Wang, Q. M. Lu, C. F. Ye, W. M. Liu, Z. J. Cui, *Wear* **2004**, *256*, 44.
44. E. R. Cooper, C. D. Andrews, P. S. Wheatley, P. B. Webb, P. Wormald, R. E. Morris, *Nature* **2004**, *430*, 1012.
45. Y. Zhou, M. Antonietti, *Chem. Commun.* **2003**, *20*, 2564.
46. D. Kuang, T. Brezesinski, B. Smarsly, *J. Am. Chem. Soc.* **2004**, *126*, 10534.
47. T. Brezesinski, C. Erpen, K. I. Iimura, *Chem. Mater.* **2005**, *17*, 1683.

7
Polymer Synthesis in Ionic Liquids

David M. Haddleton, Tom Welton, and Adrian J. Carmichael

7.1
Introduction

Polymers are essential to modern society. They are found in every household as plastics, fibers, coatings, detergents, adhesives etc. So, it is not surprising that the use of ionic liquids as solvents for polymerization reactions is now being extensively explored. This is the subject of this chapter and reviews in the literature have already covered aspects of this field in some detail [1].

7.2
Acid-catalyzed Cationic Polymerization and Oligomerization

Strong Brønsted acids that have non-nucleophilic anions (e.g. $HClO_4$ and CF_3CO_2H) are capable of initiating cationic polymerization with vinyl monomers that contain an electron-donating group adjacent to a carbon–carbon double bond, (e.g. vinyl ethers, isobutylene, styrene and dienes). Lewis acids are also used as initiators in cationic polymerization with the formation of high molecular weight polymers. These include metal halides (e.g. $AlCl_3$, BF_3 and $SbCl_5$), organometallic species (e.g. $EtAlCl_2$) and oxyhalides (e.g. $POCl_3$). Lewis acids are often used in the presence of a proton source (e.g. H_2O, HCl and MeOH) or a carbocation source (e.g. ^tBuCl) which leads to an acceleration in the rate of polymerization [2]. There have been a number of studies of the use of ionic liquids in this role.

Studies on the dimerization and hydrogenation of olefins with transition metal catalysts in acidic chloroaluminate (III) ionic liquids report the formation of higher molecular weight fractions consistent with cationic initiation [3–6]. These studies ascribed the occurrence of the undesired side reaction to both Lewis acid and proton-catalyzed polymerization routes. Studies have shown that when protons, from HCl as the source, are dissolved at ordinary temperatures and pressures in the

Ionic Liquids in Synthesis, Second Edition. P. Wasserscheid and T. Welton (Eds.)
Copyright © 2008 WILEY-VCH Verlags GmbH & Co. KGaA, Weinheim
ISBN: 978-3-527-31239-9

acidic[1] ionic liquid [EMIM]Cl–AlCl$_3$ {X(AlCl$_3$) = 0.55} they are superacidic with a strength similar to a liquid HF–Lewis acid mixture [7]. The precise Brønsted acidity observed depends on proton concentration and on ionic liquid composition. The ambient-temperature chloroaluminate(III) ionic liquids are extremely sensitive to moisture, reacting exothermically to give chlorooxoaluminate(III) species and generating HCl. Since moisture is ever-present, even in the most carefully managed systems, chloroaluminate(III) ionic liquids generally possess superacidic protons. In addition, acidic chloroaluminate(III) ionic liquids contain Lewis acid species [8], for example, [Al$_2$Cl$_7$]$^-$, so it is unsurprising that with the combination of these factors acidic chloroaluminate(III) ionic liquids catalyze the cationic oligomerization and polymerization of olefins. Attempts to avoid these side reactions led to the preparation of alkylchloroaluminate(III) ionic liquids and buffered chloroaluminate(III) ionic liquids [3, 4, 6].

Of course, it is more desirable to form high molecular weight polymers. Using the ability of the chloroaluminate(III) ionic liquids to catalyze cationic polymerization reactions was patented by Ambler et al. of BP Chemicals Ltd in 1993 [9]. They used acidic [EMIM]Cl–AlCl$_3$ {X(AlCl$_3$) = 0.67} for the oligomerization of butene to give products that find application as lubricants. The oligomerization can be carried out by bubbling butene through the ionic liquid. The product forms a separate layer that floats upon the ionic liquid and can be isolated by a simple process. Alternatively, the oligomerization is carried out by injecting the ionic liquid into a vessel charged with butene. After a suitable settling period the poly(butene) is isolated similarly. Chain transfer to impurities, ionic liquid, monomer and polymer terminate the propagation reaction, resulting in the low-mass products. Hence, the products of these reactions are liquid oligomers as opposed to polymers.

The synthesis of higher molecular weight polymers via cationic polymerization requires the formation of charged centers that live for long enough to propagate without chain transfer or termination. For this to occur stabilization of the propagating species by solvation is generally required. In addition, low temperatures are usually employed in an attempt to reduce side reactions that destroy the propagating centers. Use of a pure isobutene feedstock gives poly(isobutene) with properties that depend upon the reaction temperature. As the temperature is reduced the molecular weight of the product is reported to increase dramatically as a result of the rates of the side reactions and the rate of polymerization being reduced, Table 7-1 [9].

Ionic liquid catalyzed polymerization of butene is not limited to the use of pure alkene feedstocks, which can be relatively expensive. More usefully, the technology can be applied to mixtures of butenes, for example, the low value hydrocarbon feedstocks raffinate I and raffinate II have been used. The raffinate feedstocks are principally C4 hydrocarbon mixtures rich in butenes. When these feedstocks are

---

[1] The quantity of AlCl$_3$ present in the ionic liquid determines the physical and chemical properties of the liquid. When the mole fraction, X(AlCl$_3$), is below 0.5 the liquids are referred to as basic. When X(AlCl$_3$) is above 0.5 the liquids are referred to as acidic and at an X(AlCl$_3$) of precisely 0.5 they are referred to as neutral.

Table 7-1 Polymerization of isobutene using the acidic ionic liquid [EMIM]Cl–AlCl$_3$ ($X(AlCl_3) = 0.67$) [29]

Reaction temperature (°C)	Yield (% w/w)	Molecular weight of product (g mol^{-1})
−23	26	100 000[a]
0	75	3000 and 400[b]

[a] Polystyrene equivalents.
[b] Bimodal.

polymerized using acidic chloroaluminate(III) ionic liquids polymeric/oligomeric products are obtained which have higher molecular weights than those obtained by conventional processes, even though higher reaction temperatures are used. With the ionic liquid catalyzed process, although isobutene conversion is much higher than the n-butene conversion, the polymers produced have a much higher incorporation of n-butenes than would be possible with conventional cationic polymerization processes, Table 7-2 [9].

The ionic liquid process has a number of advantages over traditional cationic polymerization processes, for example, the Cosden process which employs a liquid phase aluminium(III) chloride catalyst to polymerize butene feedstocks [10]. The separation and removal of the product from the ionic liquid phase as the reaction proceeds allows the polymer to be obtained simply and in a highly pure state. Indeed, the polymer contains so little of the ionic liquid that an aqueous wash step can be dispensed with. This separation also means that further reaction (e.g. isomerization) of the polymer's unsaturated ω-terminus is minimized. In addition to obtaining the desired product easily, the ionic liquid is not destroyed by any aqueous washing procedure so it can be reused in subsequent polymerization reactions, resulting in a reduction in operating costs. The ionic liquid technology does not require a massive capital investment and is reported to be easily retrofitted to existing Cosden process plants.

Table 7-2 Polymerization of Raffinate I using the acidic ionic liquid [EMIM]Cl–AlCl$_3$ ($X(AlCl_3) = 0.67$): conversion of the individual components [29]

Olefin fraction	Raffinate I feedstock Concentration (% w/w)	Reacted (% w/w)
isobutene	46	91
1-butene	25	47
trans-2-butene	8	34
cis-2-butene	3	37

Further development of this original work which used [EMIM]Cl–AlCl$_3$ {X(AlCl$_3$) = 0.67} as the ionic liquid has found that by replacing the ethyl group attached to the imidazolium ring with alkyl groups of increasing length (e.g. octyl, dodecyl and octadecyl) the catalytic activity of the ionic liquid towards the oligomerization of the olefins increases. Thus, the longer the alkyl chain the greater the degree of polymerization achieved [11]. This is most probably the result of the increased solubility of the polymers in the ionic liquids with the longer alkyl chains and provides an additional method for altering the product distribution. Increased polymer yield with the raffinate I feedstock was achieved by the use of an [EMIM]Cl–AlCl$_3$ ionic liquid that contained a small portion of the quaternary ammonium salt [NEt$_4$]Cl. The ternary ionic liquid [NEt$_4$]Cl–[EMIM]Cl–AlCl$_3$ with the molar ratio 0.08:0.25:0.67 used under the same reaction conditions as the binary ionic liquid [EMIM]Cl–AlCl$_3$ {X(AlCl$_3$) = 0.67} produced ∼70% of a polymer/oligomer mixture as opposed to the ∼40% polymer/oligomer produced with the original binary system. Both systems produce oligomers with M_n = 1000 g mol^{-1} [12]. These examples demonstrate the ability to tune the ionic liquid's properties by changing the ancillary substituents. This allows the solvent to be adapted to the needs of the reaction, as opposed to altering the reaction to the needs of the solvent.

This technology has been utilized by BP Chemicals for the production of lubricating oils with well-defined characteristics (e.g. pour point and viscosity index). It is used in conjunction with a mixture of olefins (i.e. different isomers and different chain length olefins) to produce lubricating oils of higher viscosity than is obtainable by conventional catalysis [13]. Unichema Chemie BV have applied these principles to more complex monomers, e.g. unsaturated fatty acids, to create a mixture of products [14].

Apart from one mention in the original patent of the synthesis of a high molecular weight poly(isobutene), Table 7-1 [9], the remaining work has largely been concerned with the preparation of lower weight oligomers. In 2000, Symyx Technologies Inc. protected a method for the production of high molecular weight poly(iso-olefin)s without the use of very low temperatures [15]. Symyx used the [EMIM]Cl–AlCl$_3$ ionic liquid to produce poly(isobutene)s with weight average molecular weights (M_w) in excess of 100 000 g mol^{-1} which are of use in the automotive industry due to their low oxygen permeability and mechanical resilience, Table 7-3. The table shows that at temperatures as high as –40 °C polymers with molecular weights higher than half a million are obtained. As expected, when the temperature is increased the molecular weight decreases. In all cases the yield is less than 50%. By running the reaction under biphasic conditions, reducing the concentration of isobutene and adding ethylaluminum(III) dichloride the reaction yield becomes quantitative, Table 7-4. This shows that, in addition to using temperature to control the molecular weight of the product, control can also be achieved through the quantity of ethylaluminum(III) dichloride added to the reaction, that is the more alkylaluminum(III) that is added the lower the molecular weight of the product. It might be expected that the ethylaluminum(III) dichloride would act as a proton scavenger which should stop the polymerization but it seems that it acts as either a strong Lewis acid and/or an alkylating agent promoting polymerization.

Table 7-3 Polymerization of isobutene to high molecular weight poly(isobutene)s using the ionic liquid [EMIM]Cl–AlCl$_3$ [35]

Quantity of ionic liquid (μl)	Quantity of isobutene (μl)	Temperature (°C)	Yield (%)	M_w (g mol^{-1})
10	483	−40	38	526 000
10	483	−30	33	302 000
10	483	−20	45	128 000

For the results reported in both Table 7-3 and Table 7-4 the only detail reported concerning the ionic liquid is that it was [EMIM]Cl–AlCl$_3$. No details are forthcoming about the aluminum(III) chloride content. Simpler and cheaper chloroaluminate(III) ionic liquids prepared using cations derived from the reaction of a simple amine and hydrochloric acid (e.g. Me$_3$N·HCl and Bu$_2$NH·HCl) have been successfully used in the polymerization of isobutene and styrene [16]. Although these ionic liquids have much higher melting points than their imidazolium analogues they are liquid at temperatures suitable for their use in the preparation of low molecular weight oligomers (i.e. 1000 to 4000 g mol^{-1}). This reduces one of the barriers to exploitation of the technology, the relatively high expense of the imidazolium halide salts.

Vijayaraghavan and MacFarlane [17] reported the cationic polymerization of styrene using AlCl$_3$ as a catalyst in N-butyl-N-methylpyrrolidinium bis(trifluoromethanesulfonyl)amide. The products were compared with the polymers formed in a traditional organic solvent (dichloromethane, DCM). The ionic liquid reaction produced lower molecular weights and polydispersity compared to the reaction in DCM. This was attributed to differences in the concentration of the active initiator species in each case. They also reported the use of a novel Brønsted acid catalyst, bis(oxalato)boric acid (HBOB), in the cationic polymerization of styrene in the ionic liquid. Reaction temperatures of 60 °C were required

Table 7-4 Polymerization of isobutene to high molecular weight poly(isobutene)s using the ionic liquid [EMIM]Cl–AlCl$_3$ under biphasic conditions [35]

Quantity of ionic liquid (μl)	Quantity of hexane (μl)	Quantity of isobutene (μl)	Quantity of EtAlCl$_2$ (μl)	Yield (%)	M_w (g mol^{-1})
50	321	25	11	100	276 000
50	310	25	23	100	235 000
50	298	25	34	100	186 000

Conditions: temperature = −30 °C; [EtAlCl$_2$] = 1 M solution in hexane.

for this initiator to work. Again, the molecular weights and polydispersity of the reactions in ionic liquids were lower than in DCM, but this time the yields were very much higher (>90% rather than 15%). Biedron and Kubisa [18] used a 1-phenetyl chloride/$TiCl_4$ initiating system in [BMIM][PF_6]. Although the polymerization proceeded to a high conversion, chain transfer was significant, leading to a lack of control over the molecular weight and polydispersity. The same group have reported the cationic ring-opening polymerization of 3-ethyl-3-hydroxymethyloxetane (EOX) to a branched multihydroxy polyether in nearly quantitative conversion using a boron trifluoride initiator in [BMIM][BF_4] [19]. The molecular weights of the polymers were in the same range as those obtained in bulk polymerization or polymerization in organic solvents, but the formation of aggregates was considerably reduced.

Microwave-assisted living cationic ring-opening showed an enhanced polymerization rate in comparison to the reaction in common organic solvents; the ionic liquid was efficiently recovered and reused in new reaction cycles, completely avoiding the use of volatile organic compounds [20].

7.3
Free Radical Polymerization

Free radical polymerization is a key method used by the polymer chemist to produce a wide range of polymers [21]. It is used for the addition polymerization of vinyl monomers including styrene, vinyl acetate, tetrafluoroethylene, methacrylates, acrylates, (meth)acrylonitrile and (meth)acrylamides, etc. in bulk, solution and aqueous processes. The chemistry is relatively easy to exploit and is tolerant to many functional groups and impurities. Consequently it is the most studied polymerization technique in ionic liquids.

The first use of ionic liquids in free radical addition polymerization was as an extension to the doping of polymers with simple electrolytes for the preparation of ion-conducting polymers. Several groups have prepared polymers suitable for doping with ambient-temperature ionic liquids with the aim of producing polymer electrolytes of high ionic conductance. Many of the polymers prepared are related to the ionic liquids employed, for example, poly(1-butyl-4-vinylpyridinium bromide) and poly(1-ethyl-3-vinylimidazolium bis(trifluoromethanesulfonyl)imide [22–25].

Noda and Watanabe [26] reported a simple synthetic procedure for the free radical polymerization of vinyl monomers to give conducting polymer electrolyte films. Direct polymerization in the ionic liquid gives transparent, mechanically strong and highly conductive polymer electrolyte films. This was the first time that ambient-temperature ionic liquids had been used as a medium for free radical polymerization of vinyl monomers. The ionic liquids [EMIM][BF_4] and [NBPY][BF_4] (where NBPY is N-butylpyridinium) were used with equimolar amounts of suitable monomers and the polymerization was initiated by prolonged heating (12 h at 80 °C) using benzoyl peroxide. Suitable monomers for this purpose are ones that dissolve in the ionic liquid to give transparent homogeneous solutions, Table 7-5, with unsuitable monomers phase separating and therefore not being subjected to polymerization.

Table 7-5 Compatibility of the ionic liquids [EMIM][BF$_4$] and [NBPY][BF$_4$] with monomers and their polymers [42]

	[EMIM][BF$_4$]		[NBPY][BF$_4$]	
	Monomer	Polymer	Monomer	Polymer
MMA	X	–	O	X
Acrylonitrile	O	X	O	X
Vinyl acetate	O	no reaction	O	no reaction
Styrene	X	–	X	–
2-hydoxyethyl methacrylate	O	△	O	△

Legend: O, transparent homogenous solution; X, phase separated; △, translucent gel.

Of all the monomers found to give transparent homogeneous solutions only vinyl acetate failed to undergo polymerization. In all of the other polymerizations, with the exception of 2-hydoxyethyl methacrylate (HEMA), the polymer was insoluble in the ionic liquid and phase separated. The compatibility of HEMA with the ionic liquids resulted in its use for the preparation of polymer electrolyte films, which were found to be highly conductive. For film formation, the reaction mixtures were simply spread between glass plates and heated; no degassing procedures were carried out. Analysis of the films found that the amount of unreacted monomer was negligible, indicating fast polymerization. In all of the reactions reported by Noda and Watanabe no characterization of the polymers or indeed analysis of the polymerization reactions was carried out [26].

They followed this by changing to the 1-ethyl-3-methyl imidazolium bis(trifluoromethane sulfonyl)imide ionic liquid [27]. This allowed them to polymerize methyl methacrylate (MMA) in the presence of a small amount of a crosslinker to give self-standing, flexible, and transparent films. The glass transition temperatures of the gels decreased with increasing mole fraction of [EMIM][Tf$_2$N] and behaved as a completely compatible binary system of poly(methyl methacrylate) (PMMA) and [EMIM][Tf$_2$N].

Others workers have looked into the kinetics and the types of polymers formed by the free radical polymerization reactions of vinyl monomers using ambient-temperature ionic liquids as the solvent [28, 29]. The free radical polymerization of methyl methacrylate (MMA) in [BMIM][PF$_6$] initiated by 2,2'-azobisisobutyronitrile (AIBN) at 60 °C proceeds rapidly, causing a large increase in viscosity that hampers efficient stirring of the reaction mixture. The polymerization reactions produce poly(methyl methacrylate) (PMMA) with very high molecular weights, Table 7-6 [29]. The rate constants of propagation and termination of MMA in [BMIM][PF$_6$] were measured using the pulsed laser polymerization technique across a range of temperatures and Arrhenius parameters for the rate of propagation were calculated. The decrease in activation energy leads to large increases in the rate of propagation. In addition, the rate of termination decreases by an order of magnitude as the ionic liquid concentration is increased to 60% v/v. The increase in propagation rate

Table 7-6 Free radical polymerization of MMA in the ionic liquid [BMIM]PF$_6$ [44]

Reaction medium	[AIBN] (w/v%)	Conversion (%)	M_n (g mol^{-1})	PDi
[BMIM][PF$_6$]	1	25	669 000	1.75
[BMIM][PF$_6$]	2	27	600 000	1.88
[BMIM][PF$_6$]	4	36	416 000	2.22
[BMIM][PF$_6$]	8	56	240 000	2.59
toluene	1	3	58 300	1.98

Conditions: temperature = 60 °C; time = 20 min; 20% v/v monomer in ionic liquid.

was attributed to the increased polarity of the medium, while the decrease in the termination rate was attributed to its increased viscosity [29]. Copolymerization data for styrene and MMA by charge transfer polymerization with hydroquinone have been reported with a tendency for alternating copolymerization [30].

The effects of increasing the concentration of initiator (i.e. increased conversion, decreased M_n and broader PDi) and reducing the reaction temperature (i.e. decreased conversion, increased M_n and narrower PDi) for the polymerizations in ambient-temperature ionic liquids are the same as observed in conventional solvents. Mays et al. reported similar results and, in addition, used ^{13}C NMR to investigate the stereochemistry of the PMMA produced in [BMIM][PF$_6$]. They found that the stereochemistry is almost identical to that for PMMA produced by free radical polymerization in conventional solvents [28]. The homopolymerization and copolymerization of several other monomers are also reported. Similar to what was found by Noda and Watanabe, in many cases the polymer was not soluble in the ionic liquid and thus phase separated [28, 29]. Free radical polymerization of n-butyl methacrylate in ionic liquids based on imidazolium, pyridinium, and alkylammonium salts as solvents was investigated with a systematic variation of the length of the alkyl substituents on the cations, and employing different anions such as tetrafluoroborate, hexafluorophosphate, tosylate, triflate, alkyl sulfates and dimethyl phosphate [31].

Mays et al. have investigated the formation of block co-polymers in ionic liquids [32]. They noted that for conventional free radical copolymerization of styrene and methyl methacrylate reactivities were significantly different from those obtained in conventional organic solvents or in the bulk, offering the possibility of forming new monomer sequences [33]. Haddleton et al. applied the reversible addition-fragmentation chain transfer (RAFT) methodology to the polymerization of acrylates and methacrylates [34]. This is a living radical polymerization and as such leads to polymers with narrow polydispersities, so overcoming this drawback of conventional radical polymerization. The rates of reaction were shown to be greater in the ionic liquids than in toluene, the conventionally used solvent. Attempts to polymerize styrene were prevented by the poor solubility of low molecular weight poly(styrene) in the ionic liquids, leading to precipitation and termination of the polymerization process.

7.4
Transition Metal-catalyzed Polymerization

Ionic liquids have been widely used as solvents for transition metal-catalyzed reactions (see Chapter 5). They can act simply as the solvent or sometimes as a co-catalyst or catalyst activator. They are often used in biphasic systems, with the catalyst retained in the ionic liquid phase and the products separated in an organic solvent phase.

7.4.1
Ziegler–Natta Polymerization of Olefins

Ziegler–Natta polymerization is used extensively for the polymerization of simple olefins (e.g. ethene, propene and 1-butene) and is the focus of much academic attention, as even small improvements to a commercial process operated on this scale can be important. Ziegler–Natta catalyst systems, which in general are early transition metal compounds used in conjunction with alkylaluminum compounds, lend themselves to study in the chloroaluminate(III) ionic liquids, especially those with an acidic composition.

During studies into the behavior of titanium(IV) chloride in chloroaluminate(III) ionic liquids Carlin et al. carried out a brief study to investigate whether Ziegler–Natta polymerization was possible in an ionic liquid [35]. They dissolved $TiCl_4$ and $EtAlCl_2$ in [EMIM]Cl–$AlCl_3$ ($X(AlCl_3) = 0.52$) and bubbled ethylene through for several minutes. After quenching, poly(ethene) with a melting point of 120–130 °C was isolated in very low yield, thus demonstrating that Ziegler–Natta polymerization works in these liquids, albeit not very well. Polymerizations of α-olefins in acidic [BMIM]Cl–$AlCl_3$ with added ethylaluminum dichloride $TiCl_4$ produced branched, atactic polymers with narrow monomodal polydispersities as waxes or oils in high yields. Molar masses varied from $M_w = 650$ to 1620 g mol^{-1} ($M_n = 440$–970 g mol^{-1}) regardless of the monomer used. This was in contrast to the same reaction in toluene, which afforded mainly linear poly(ethene) with a broader polydispersity and a higher molar mass.

The same ionic liquid was employed to give higher yields of poly(ethene) using bis(η-cyclopentadienyl)titanium(IV) dichloride in conjunction with $Me_3Al_2Cl_3$ as co-catalyst [36]. However, the catalytic activities are still low when compared to other homogeneous systems and may be attributed to, among other things, low solubility of ethene in the ionic liquids or the presence of alkylimidazole impurities which coordinate and block the active titanium sites. In chloroaluminate(III) ionic liquids of a basic composition no catalysis is observed which was ascribed to the formation of the inactive $[Ti(\eta\text{-}C_5H_5)_2Cl_3]^-$ species. In comparison, the zirconium and hafnium analogues, $[Zr(\eta\text{-}C_5H_5)_2Cl_2]$ and $[Hf(\eta\text{-}C_5H_5)_2Cl_2]$, showed no catalytic activity towards the polymerization of ethene in either acidic or basic ionic liquids. This is presumably due to the presence of stronger M–Cl bonds that preclude the formation of a catalytically active species.

7.4.2
Late Transition Metal-catalyzed Polymerization of Olefins

The surge in development of late transition metal polymerization catalysts has been due, in part, to the need for systems that can copolymerize ethene, and related monomers, with polar comonomers under mild conditions. Late transition metals have a lower oxophilicity relative to early transition metals and therefore a higher tolerance for a wider ranger of functional groups (e.g. –COOR and –COOH groups) [37]. A nickel complex **1** has been used (Fig. 7-1) for the homopolymerization of ethene in an ambient-temperature ionic liquid [38]. **1** was used under mild biphasic conditions with the ternary ionic liquid [BMIM]Cl–AlCl$_3$–EtAlCl$_2$ (1.0:1.0:0.32, $X(AlCl_3) = 0.57$) and toluene, producing poly(ethene) which was easily isolated from the reaction mixture by decanting the upper toluene layer. This permitted the ionic liquid and **1** to be recycled for use in further polymerizations. However, before reuse trimethylaluminum(III) was added to overcome the loss of free alkylaluminum species into the separated organic phase [38]. The characteristics of the isolated poly(ethene) depend upon several reaction conditions. On increasing the reaction temperature from –10 to +10 °C the melting point decreases from 123 to 85 °C due to a greater amount of chain branching and also results in a decrease in the M_w from 388 000 to 280 000 g mol^{-1}. Reusing the catalyst/ionic liquid solution also has an effect, with subsequent reactions giving a progressive shift from crystalline to amorphous polymer, with a period that gives rise to bimodal product distributions. This change is due to the changing composition of the ionic liquid as fresh co-catalyst is added after each polymerization run, which leads to the formation of different active species.

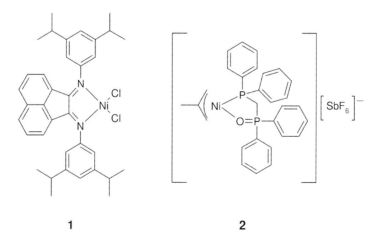

Fig. 7-1 Nickel catalysts used for the polymerization and oligomerization of ethylene in ambient-temperature ionic liquids [48, 49].

A related study used the air stable ionic liquids [RMIM][PF$_6$] (R = butyl–decyl) as solvents for the oligomerization of ethene to higher α-olefins [39]. The reaction uses the cationic nickel complex **2** (Fig. 7-1) under biphasic conditions to give oligomers of up to nine repeat units with better selectivity and reactivity than in conventional solvents. Modifications of the ionic liquids' cation gave access to a further tuning of the product distribution. Recycling of the catalyst/ionic liquid solution was possible with little change in selectivity and only a small drop in activity.

Poly(phenylethene) has been prepared in high yield in [BMIM][BF$_4$] and [BUPY][BF$_4$] ionic liquids using rhodium(I) catalysts with Et$_3$N as a co-catalyst [40]. It was shown that the catalysts could be recycled without significant loss of activity. Subsequently [41], it was shown that the amine co-catalyst was unnecessary in these ionic liquids. It was not clear whether this was due to the inherent basicity of the [BF$_4$]$^-$ ion or the presence of F$^-$. When a range of ionic liquids were used for this polymerization the yield and molecular weight of the polymer were found to depend on the ionic liquid used [42]. The use of alcohol co-solvents was also found to increase the molecular weights of the products.

Polyketones formed by the palladium-catalyzed alternating copolymerization of styrene and CO have become the subject of a great deal of attention [43]. Almost concurrently, two papers appeared describing this reaction in ionic liquids [44, 45]. Seddon et al. used [Pd(bipy)$_2$][PF$_6$] in a variety of ionic liquids (Scheme 7-1) [44]. The complex was active in methanol [ca. 17.4 kg (g Pd)$^{-1}$], but the product polyketone was contaminated with palladium. In the ionic liquids the activity of the catalyst was dependent on the nature of the anion, decreasing in the order [N(SO$_2$CF$_3$)$_2$]$^-$ > [PF$_6$]$^-$ > [BF$_4$]$^-$ for both pyridinium- and imidazolium-based liquids. Increasing the length of the alkyl chain of the pyridinium ionic liquids [RPY][N(SO$_2$CF$_3$)$_2$] (R = butyl–decyl) was reported to afford higher yields, although some palladium leaching was noted for the longer chain salts. However, in [HexPy][NTf$_2$] no palladium decomposition and/or leaching was detected and repetitive catalytic runs were possible with no loss of activity after four runs. When [Pd(OAc)$_2$(bipy)] was used as the catalyst [44], a 10:1 mixture of [HexPY][NTf$_2$]/methanol was needed to observe any reaction. This was believed to be because the methanol reacts with L$_n$Pd^{II} and CO to form [LPdC(O)OCH$_3$]$^+$, which initiates the copolymerization. Bromide contamination (0.5% w/w) of [HexPY][N(SO$_2$CF$_3$)$_2$] was reported to completely stop the copolymerization. Repeat catalytic runs resulted in progressively lower activities because of palladium precipitation.

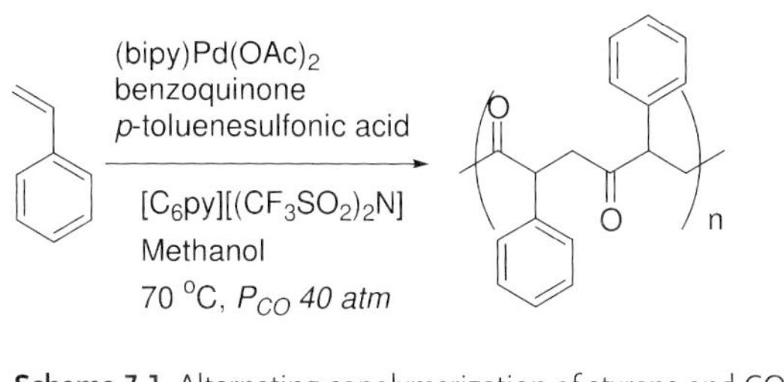

Scheme 7-1 Alternating copolymerization of styrene and CO in [HexPY][NTf$_2$].

7.4.3
Metathesis Polymerization

Metathesis polymerization has become an important tool for polymer synthesis and has even attracted a Nobel Prize [46–49]. Acyclic diene molecules are capable of undergoing intramolecular and intermolecular reactions in the presence of appropriate transition metal catalysts, for example, molybdenum alkylidene and ruthenium carbene complexes. The intramolecular reaction, called ring-closing olefin metathesis (RCM), leads to cyclic compounds and the intermolecular reaction, called acyclic diene metathesis (ADMET) polymerization, leads to oligomers and polymers. Altering the dilution of the reaction mixture can to some extent control the intrinsic competition between RCM and ADMET.

Gürtler and Jautelat of Bayer AG have protected methods that use chloroaluminate(III) ionic liquids as solvents for both cyclization and polymerization reactions of acyclic dienes [50]. They employed the neutral ionic liquid [EMIM]Cl–AlCl$_3$ (X(AlCl$_3$) = 0.5) to immobilize a ruthenium carbene complex for biphasic ADMET polymerization of an acyclic diene ester, Fig. 7-2. The reaction is an equilibrium process, therefore, removal of ethylene drives the equilibrium to products. The reaction proceeds readily at ambient temperatures producing mostly polymeric materials but also ∼10% dimeric material.

Cyclic olefins can undergo ring-opening metathesis polymerization (ROMP). The first ROMP to be carried out in an ionic liquid was of norbornene in a biphasic system consisting [DMMIM][PF$_6$] and toluene, with a cationic ruthenium allenylidene precatalyst, [(p-cymene)RuCl(PCy$_3$)(=C=C=CPh$_2$)][OTf] [51]. The ionic nature of the catalyst allowed it to be effectively retained in the ionic liquid layer while the toluene-soluble polymer was separated. Both the catalyst and the ionic liquid were reused several times. It is interesting to note that this mode of operation was more effective than one in which the reaction in the ionic liquid was first allowed to go to completion and then was washed with toluene, which gave a greatly diminished yield for a second cycle. When the better known Grubbs-type catalysts were used under similar conditions both catalyst deactivation and extraction into the

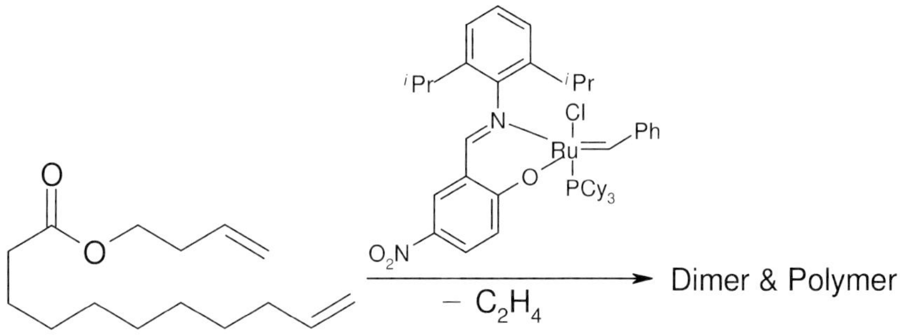

Fig. 7-2 Acyclic diene metathesis polymerization (ADMET) reaction carried out in neutral ionic liquid [EMIM]Cl–AlCl$_3$ (X(AlCl$_3$) = 0.5) [52].

organic phase were found to occur. Vygodskii has been very active in this area, reporting ROMP of many functional norbornenes, e.g., *exo-, endo*-5-norbornene-2-carbonitrile in a variety of ionic liquids with a series of ruthenium-based catalysts, e.g. $RuCl_2Py_2(IMesH_2)(CHPh)$. ROMP of norbornene derivatives in ionic solvents proceeded with high speed and offered access to high molecular weight polymers (M_w up to 1 500 000 g mol^{-1}) in high yields [52].

7.4.4
Living Radical Polymerization

As discussed in Section 7.3 conventional free radical polymerization is a widely used technique that is relatively easy to employ. However, it does have its limitations. It is often difficult to obtain predetermined polymer architectures with precise and narrow molecular weight distributions. Transition metal-mediated living radical polymerization is a powerful method that has been developed to overcome these limitations [53, 54]. It permits the synthesis of polymers with varied architectures (e.g. blocks, stars and combs) and with predetermined end groups (e.g. rotaxanes, biomolecules and dyes).

A potential limitation to commercialization of this technology is that relatively high levels of catalyst are often required. Indeed, it is usual that one mole equivalent is required for each growing polymer chain to achieve acceptable rates of polymerization. The catalysts generally co-precipitate with the polymer and need to be removed post production. This is wasteful and expensive and a number of approaches have been reported to overcome this obstacle including supported catalysts [55], fluorous biphase reactions [56] and also the use of ionic liquids [29, 57, 58]. It was found that copper(I) bromide in conjunction with *N*-propyl-2-pyridylmethanimine as ligand catalyses the living radical polymerization of MMA in the neutral ionic liquid [BMIM][PF$_6$]. The reaction progresses in a manner consistent with a living polymerization, that is, good first-order kinetic behavior and evolution of number average molecular weight (M_n) with time were observed, and a final product with low M_n and PDi values was obtained [57]. Polymerization in the ionic liquid proceeded much more rapidly than in conventional organic solvents, indeed, polymerization occurred at 30 °C in [BMIM][PF$_6$] at a rate comparable to that found in toluene at 90 °C.

The cationic nature of the copper(I) catalyst means that it is immobilized in the ionic liquid. This permits the PMMA product to be obtained, with negligible copper contamination, by a simple extraction procedure using toluene as the solvent (a solvent in which the ionic liquid is not miscible). The ionic liquid/catalyst solution was subsequently reused.

The technique of copper(I) bromide-mediated living radical polymerization is compatible with other ambient-temperature ionic liquids. It proceeds smoothly in hexyl- and octyl-3-methylimidazolium hexafluorophosphate and tetrafluoroborate ionic liquids. However, using [BMIM][BF$_4$] for the polymerization of MMA generates a product with a bimodal product distribution. Figure 7-3 shows this trace together with a trace from a similar reaction carried out in [BMIM][PF$_6$] [29]. The mass distribution for [BMIM][PF$_6$] shows a single narrow low molecular weight

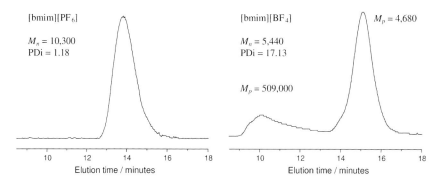

Fig. 7-3 SEC traces for the Cu(I)Br-mediated living radical polymerization of MMA in the ionic liquids [BMIM][X] (X = PF$_6$ or BF$_4$) [44].

peak consistent with living radical polymerization, whereas, the mass distribution for [BMIM][BF$_4$] shows a similar peak and also an additional peak that is broad and at high molecular weight. This high molecular weight peak is consistent with the results observed for conventional free radical polymerization in ionic liquids, as discussed in Section 7.3. This anomalous result can be rationalized in terms of the synthetic method used to prepare the ionic liquids. Of all the ionic liquids used [BMIM][BF$_4$] is the only one in the study that was miscible with water, therefore, it was the only one not subjected to an aqueous work-up and so was contaminated with halide salts [59]. The halide salts might poison the catalyst with subsequent polymerization proceeding via two different mechanisms. Alternatively, it might be that under living polymerization conditions the terminal halide atom on the propagating polymer chain does not fully separate from the polymer during propagation, creating a "caged-radical" which undergoes propagation. Under appropriate conditions separation occurs, leading to irreversible homolytic fission and the production of free radicals. Conventional free radical polymerization ensues in competition with the atom transfer mechanism, giving high conversion and high mass polymer alongside the low mass polymer from the living mechanism. This implies that the rate of termination in conventional radical propagation is drastically reduced, maybe by coordination with the cation or anion from the ionic liquid that also prevents recombination with the halide atom.

Atom-transfer radical polymerization (ATRP) of acrylates has been investigated in [BMIM][PF$_6$] [60, 61]. The solubility of the monomer in the ionic liquid chosen depends very much upon the substituent on the monomer. Homogeneous polymerization of methyl acrylate gave living polymerization with narrow polydispersity polymers and good molecular weight control. Higher order acrylates gave heterogeneous reactions with the catalyst remaining in the ionic liquid phase. Although deviations from living polymerization behavior were observed, butyl acrylate showed controlled polymerization. The same group are currently extending this work and also looking at cationic vinyl polymerization and various ring-opening polymerization reactions.

Similar reactions have been attempted in [BMIM][BF$_4$] and [C$_{12}$MIM][BF$_4$] [62]. While the reaction in [C$_{12}$MIM][BF$_4$] proceeded smoothly to give PMMA and the ionic liquid was reused after purification it failed to do so in [BMIM][BF$_4$]. This was attributed to the lack of solubility of the PMMA product in this ionic liquid.

ATRP has also been used for the polymerization of ionic liquid-like, vinylbenzyl-substituted imidazolium salt, monomers to give ionic polymers [63]. The reaction rates and ultimate conversions were dependent upon the solvent used and reaction conditions (concentrations of reagents/catalysts, temperature etc.). Given that polar aprotic solvents, such as acetonitrile and DMF were preferred for the reactions and each led to different rates, it would be interesting to have seen the effect of using a non-polymerizable ionic liquid as the solvent for the reaction.

Kubisa et al. [64] have been exploring the use of chiral ionic liquids in polymer synthesis. Using ionic liquids with a chiral substituent on the imidazolium ring for the ATRP of methyl acrylate gave a small but definite effect on polymer tacticity, with more isotactic polymer formed than in simple [BMIM][PF$_6$]. They also found that the use of ionic liquids led to fewer side reactions. Ionic liquids have been used as solvents in biphasic ATRP to facilitate the separation of the products from the catalysts [65].

Percec [66] has demonstrated the effect of the 1-butyl-3-methylimidazolium hexafluorophosphate ionic liquid on the living radical polymerization of MMA initiated with arenesulfonyl chlorides and catalyzed by the self-regulated Cu$_2$O/2,2'-bipyridine catalyst. A dramatic acceleration of the polymerization was observed with an initiation efficiency of 100%, giving polymer with molecular weight distribution of 1.1 and perfect bifunctional chain-ends.

7.5
Electrochemical Polymerization

7.5.1
Preparation of Conductive Polymers

As conducting media, often with large electrochemical windows, ionic liquids clearly have potential as solvents for electrosynthesis. One of the few areas in which this has been exploited is in the electrochemical synthesis of conducting polymers.

Electronically conducting polymers have a number of potential applications including as coatings for semiconductors [67], in electrocatalysis [68] and as charge storage materials [69]. Of these poly(*para*-phenylene) (PPP), which is the simplest of the poly(arene) classes, possesses properties that include excellent thermal stability, high coke number and good optical and electrophysical characteristics [70]. For PPP to be utilized in devices and advanced materials it should have a high relative molecular mass (M_r), a homogenous structure and good submolecular packing.

The acidic ionic liquid [NBPY]Cl–AlCl$_3$ (X(AlCl$_3$) = 0.67) has been used as a solvent for the electrochemical polymerization of benzene to PPP as conductive films which were flexible and transparent. The films were prepared with very high relative molecular masses with degrees of polymerization up to 200 [71, 72]. The electrochemical polymerization of benzene to PPP has not been carried out exclusively in [NBPY]Cl–AlCl$_3$ ionic liquids. Other reports describe the use of [NBPY]Cl–AlCl$_2$(OEt), [NCTPY]Cl–AlCl$_3$ (where NCTPY is N-cetylpyridinium) and [EMIM]Cl–AlCl$_3$ with the best results observed in the traditional aluminum(III) chloride ionic liquids [73–75].

The electrochemical oxidation of fluorene in [EMIM]Cl–AlCl$_3$ ionic liquids of an acidic or neutral composition gives poly(fluorene) films that are more stable and have a less complicated electrochemical behavior than those prepared in acetonitrile, the usual solvent. Basic ionic liquids cannot be used, as chloride ions are more easily oxidized than fluorine [76]. A number of aromatic compounds that contain heteroatoms, e.g. pyrrole, aniline and thiophene can also be oxidised electrochemically in chloroaluminate(III) ionic liquids to give polymer films [77–80]. In ionic liquids of an acidic composition the electrochemical polymerization of the nitrogen- and sulfur-containing compounds is either more difficult or not possible at all due to the formation of adducts with AlCl$_3$ [81]. Any interactions between benzene and AlCl$_3$ are not significant enough to influence its polymerization to PPP [75].

The chemical synthesis of PPP has also been carried out in ambient-temperature ionic liquids. The oxidative dehydropolycondensation of benzene was carried out in the acidic ionic liquid [NBPY]Cl–AlCl$_3$ ($X(AlCl_3) = 0.67$) using CuCl$_2$ as the catalyst [71, 82]. This gave PPP with relative molecular masses considerably higher than in conventional solvents and M_r could be tuned by varying the benzene concentration. The high M_r values observed were attributed to the greater solubility of PPP in the ionic liquid which permits a greater degree of polymerization before phase separation terminates the reaction.

7.6
Polycondensation and Enzymatic Polymerization

Ionic liquids could be suitable media for many forms of direct polycondensation due to the high temperatures often employed in this type of polymerization. The formation of aromatic polyamides from the polycondensation of diamines and diacid chlorides has been successfully completed in various ionic liquids [83]. The molecular weights of the polymers formed were found to depend on the solubility of the polymer in the particular ionic liquid being used, with greater solubility leading to higher molecular weights. Vygodskii and coworkers have reported the synthesis of a range of condensation polymers including polyamides [84], polyimides [85] and polyesters. The synthesis of poly(glycolic acid) (PGA) by polyesterification of glycolic acid has been studied using 1,3-dialkylimidazolium salts, as reaction media. Low PGA yields were obtained by the direct polyesterification of glycolic acid at 200–240 °C, ascribed to monomer evaporation during the reaction [86]. Caprolactone has also been polymerized by microwave-assisted ring-opening polymerization in the presence of 1-butyl-3-methylimidazolium tetrafluoroborate ionic liquid using zinc oxide as a catalyst is investigated [87], and interfacial polymerization to polyureas [88].

This has been extended to the use of enzymes in ring-opening polymerization of lactones and related monomers which is of increasing interest and, not surprisingly, ionic liquids have started to be featured. This follows from the use of enzymatic reactions for small molecule transformations [89]. Nara et al. investigated the lipase-catalyzed polycondensation of diethylene octane-1,8-dicarboxylate and 1,4-butanediol in [BMIM][PF$_6$] [90]. At ambient temperature low molecular weight oligomers (M_n) of ca. 2000 g mol^{-1} were obtained after 7 days, the molecular weight increasing to 4500 g mol^{-1} at 60 °C. The low polydispersity of 1.03–1.26 reported was attributed to the insolubility of the polymer in the IL once it exceeds a certain molecular weight. In a second report Uyama et al. investigated both lipase-catalyzed ring-opening polymerization (ROP) of ϵ-caprolactone (CL) and the polycondensation of diethyl adipate and diethyl sebacate, respectively, with 1,4-butanediol [91]. The ROP in [BMIM][PF$_6$] resulted in oligomers ($M_n < 850$ g mol^{-1}) over 3 days, whilst $M_n = 4200$ g mol^{-1} was obtained after 7 days. The molecular weights of the polycondensates were less than 1500 g mol^{-1} in both [BMIM][PF$_6$] and [BMIM][BF$_4$]. Heise has reported the enzymatic synthesis of polyesters in [BMIM][Tf$_2$N], [BMIM][PF$_6$] and [BMIM][BF$_4$]. For ϵ-caprolactone they found that [BMIM][PF$_6$] and [BMIM][BF$_4$] resulted in an inhomogeneous reaction mixture upon polymerization, leading to polymerization characteristics similar to bulk polymerization. In contrast, for [BMIM][Tf$_2$N] characteristics similar to polymerization in toluene were observed. Molecular weights of 7000–9500 g mol^{-1} were obtained. For the polycondensation of dimethyl adipate and dimethyl sebacate with 1,4-butanol the low volatility of the ionic liquids allowed the reactions to proceed in an open vessel at temperatures close to the boiling point of the condensation by-product. Molecular weights up to 5400 g mol^{-1} were obtained. This, in combination with the tuneable solvent hydrophilicity of ionic liquids, could offer an advantage in the polymerization of highly polar monomers with low solubility in organic solvents. The authors concluded that the tuneable solvent hydrophilicity of ILs, might offer an advantage in the polymerization of highly polar monomers with low solubility in organic solvents such as sorbitol. Given the recent interest in the use of enzyme catalysis in ionic liquids, it is likely that this area of research will become increasingly important.

High molecular weight aromatic poly(1,3,4-oxadiazole)s (PODS) were prepared by a "one-pot" procedure from dicarboxylic acid and a hydrazine salt (sulfate, phosphate) or dicarboxylic acid dihydrazides in a mixture of ionic liquid and triphenyl phosphate, used both as a solvent and condensing agent. The polycyclization occurred at a sufficiently low temperature of 210 °C to give film-forming PODS [92].

7.7
Carbene-catalyzed Reactions

Dialkylimidazolium-based ionic liquids can act as sources of imidazolylidenes, often referred to as N-heterocyclic carbenes (NHCs), which can act as nucleophilic catalysts [93]. Hedrick et al. have used ionic liquids as both the source of the NHC

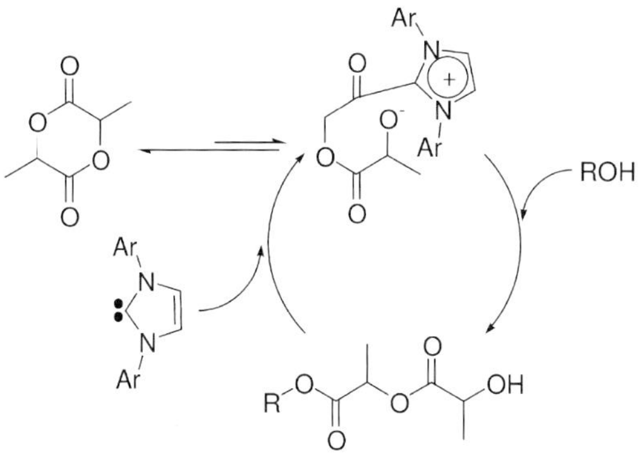

Scheme 7-2 The NHC catalysed ring-opening polymerization of lactide.

and as one of the solvents in a biphasic ring-opening polymerization of lactide and lactones. [EMIM][BF$_4$] with THF provided an excellent medium for the reaction (Scheme 7-2), which was initiated by the addition of a small amount of potassium *tert*-butoxide, which deprotonates the imidazolium ring to generate the NHC. The now neutral NHC transfers to the THF layer, where the reaction occurs. The reaction is terminated by addition of a protic ammonium salt, which regenerates the imidazolium cation by proton transfer.

The NHC complex (1-ethyl-3-methylimidazol-2-ylidene)silver(I) chloride is an ionic liquid, and was found to catalyze the ring-opening polymerization of lactide at elevated temperatures to give narrowly dispersed polylactide of predictable molecular weight [94]. Here, the ionic liquid is acting as a source of the NHC by the thermal decomposition of the silver imidazolylidene complex cation.

7.8
Group Transfer Polymerization

Group transfer polymerization (GTP) of MMA has been found to produce polymers in [BMPY][NTf$_2$] with a reduced polydispersity compared to that in molecular solvents [95]. Again the solubility of the product in the ionic liquid seemed to be the limiting factor in the molecular weight of the polymer formed. The GTP was initiated by a silyl ketene acetyl and [Bu$_4$N]Br was used as a nucleophilic catalyst (although the reaction did proceed in the absence of the catalyst). The role of the nucleophile is to attack the silicon and generate an anionic species that is the actual propagating species. This species is in equilibrium with the neutral, non-propagating species. The potential of ionic liquids to favor the ionic species over the neutral one should mean that they have great potential to be exploited for this reaction. Exposure of the reaction mixture to air did not lead to deactivation of the system.

7.9
Summary

The use of ionic liquids for polymerizations is still in its infancy, but is growing rapidly. There have now been a number of different types of polymerization tried in ionic liquids, but for many of these there are only one or two examples in the literature. However, ionic liquid technology has already brought a number of benefits to polymer synthesis. For example, the application of chloroaluminate(III) ionic liquids as both solvent and catalyst for the cationic polymerization of olefins has generated a system that not only produces cleaner polymers than traditional processes but permits the recovery and reuse of the ionic liquid solvent/catalyst. Ionic liquids have allowed the preparation of high molecular weight conducting polymers such as poly(*para*-phenylene), and have been useful for the immobilization of transition metal polymerization catalysts and their separation from products, thus offering a potential solution to a problem that prevents the commercialization of transition metal-mediated living radical polymerization. Free radical polymerization proceeds at a much faster rate and there is already evidence that both the rate of propagation and rate of termination are affected.

The use of neutral ionic liquids for free radical polymerization highlights one of their problems: their relatively high viscosity. The viscosity of the reaction mixture has a significant effect on the outcome of polymerization reactions and these liquids can have viscosities much higher than conventional organic solvents. The free radical polymerization of MMA in [BMIM][PF$_6$] generates polymers with high molecular weights which, when combined with the reduced fluidity of the ionic liquid, causes the reaction mixture to set after a very short time. This problem can be avoided in polymerization reactions if phase separation of the product occurs, for example, as with the free radical polymerization of MMA in [NBPY][BF$_4$]; or if the reaction is operated under biphasic conditions, for example, the reported ADMET polymerization of an acyclic diene ester.

As well as viscosity, other factors to be aware of include the purity of the ionic liquids. The presence of residual halide ions in neutral ionic liquids can poison transition metal catalysts and different levels of proton impurities in chloroaluminate(III) ionic liquids can alter the product distribution of the reaction. The reduced temperatures required for many polymerization reactions in ionic liquids, together with the reduced solubility of oxygen in ionic liquids compared to conventional solvents, means that two of the most common quenching methods are reduced in effectiveness. When detailed studies are being carried out, in particular kinetic studies, it is necessary to completely stop further reaction so that accurate data are obtained.

One of the common observations across several of the techniques used is that the precipitation of the polymer product from the ionic liquid terminates the reaction. The advantage of this is that it leads to narrower polydispersities than are often found. However, it can lead to lower molecular weights than other methodologies. With the synthetic flexibility that ionic liquids have it may prove possible to manipulate the solubility of polymers in the ionic liquids in a sufficiently controlled manner that they can be used as a means to prepare polymers of desired molecular weights.

References

1. (a) P. Kubisa, *Prog. Polym. Sci.* **2004**, *29*, 3–12; (b) H. Pinkowska, *Polymer* **2006**, *51*, 836; (c) N. Winterton, *J. Mater. Chem.* **2006**, *16*, 4281.
2. G. Odian, *Principles of Polymerization*, Wiley, New York, **1991**.
3. Y. Chauvin, B. Gilbert, I. Guibard, *J. Chem. Soc., Chem. Commun.* **1990**, 1715.
4. B. Ellis, W. Keim, P. Wasserscheid, *J. Chem. Soc., Chem. Commun.* **1999**, 337.
5. P. A. Suarez, J. E. L. Dullius, S. Einloft, R. F. de Souza, J. Dupont, *Polyhedron* **1996**, *15*, 1217.
6. P. Wasserscheid, W. Keim, World Patent, **1998**, WO 98/47616.
7. G. P. Smith, A. S. Dworkin, R. M. Pagni, S. P. Zingg, *J. Am. Chem. Soc.* **1989**, *111*, 5075.
8. J. A. Boon, J. A. Levisky, J. L. Pflug, J. S. Wilkes, *J. Org. Chem.* **1986**, *51*, 480.
9. P. W. Ambler, P. K. G. Hodgson, N. J. Stewart, European Patent, **1993**, EP 0558187.
10. K. Weissermel, H.-J. Arpe, *Industrial Organic Chemistry*, VCH, Weinheim, **1997**.
11. A. A. K. Abdul-Sada, P. W. Ambler, P. K. G. Hodgson, K. R. Seddon, N. J. Stewart, World Patent, **1995**, WO 95/21871.
12. A. A. K. Abdul-Sada, K. R. Seddon, N. J. Stewart, World Patent, **1995**, WO 95/21872.
13. P. M. Atkins, M. R. Smith, B. Ellis, European Patent, **1997**, EP 0791643.
14. G. Roberts, C. M. Lok, C. J. Adams, K. R. Seddon, M. J. Earle, J. Hamill, World Patent, **1998**, WO 98/07679.
15. V. Murphy, World Patent, **2000**, WO 00/32658.
16. F. G. Sherif, L. J. Shyu, C. P. M. Lacroix, A. G. Talma, US Patent, **1998**, US 5731101.
17. R. Vijayaraghavan, D. R. MacFarlane, *Chem. Commun.* **2004**, 700.
18. T. Biedron, P. Kubisa, *J. Polym. Sci. A,* **2004**, *42*, 3230.
19. T. Biedron, M. Bednarek, P. Kubisa, *Macromol. Rapid Commun.* **2004**, *25*, 878.
20. C. Guerrero-Sanchez, R. Hoogenboom, U. S. Schubert, *Chem. Commun.* **2006**, 3797.
21. G. Moad, D. H. Solomon, *The Chemistry of Free Radical Polymerization*, Pergamon, Oxford, **1995**.
22. M. Watanabe, S. Yamada, N. Ogata, *Electrochim. Acta* **1995**, *40*, 2285.
23. J. Fuller, A. C. Breda, R. T. Carlin, *J. Electrochem. Soc.* **1997**, *144*, L67.
24. M. Hirao, K. Ito-Akita, H. Ohno, *Polym. Adv. Technol.* **2000**, *11*, 534.
25. H. Ohno, *Electrochim. Acta* **2001**, *46*, 1407.
26. A. Noda, M. Watanabe, *Electrochim. Acta* **2000**, *45*, 1265.
27. M. A. Susan, T. Kaneko, A. Noda, M. Watanabe, *J. Am. Chem. Soc.* **2005**, *127*, **4976**.
28. H. Zhang, L. Bu, M. Li, K. Hong, J. W. Mays, R. D. Rogers, *Chem. Commun.* **2002**, 1368.
29. (a) S. Harrisson, S. R. Mackenzie, D. M. Haddleton, *Chem. Commun.* **2002**, 2850; (b) S. Harrisson, S. R. Mackenzie, D. M. Haddleton, *Macromolecules* **2003**, *36*, 5072.
30. R. Vijayaraghavan, D. R. MacFarlane, *Eur. Polym. J.* **2006**, *42*, 2736.
31. V. Strehmel, A. Laschewsky, H. Wetzel, E. Gornitz, *Macromolecules* **2006**, *39*, 923.
32. H. W. Zhang, K. L. Hong, J. W. Mays, *Macromolecules* **2002**, *35*, 5738.
33. H. W. Zhang, K. L. Hong, M. Jablonsky, J. W. Mays, *Chem. Commun.* **2003**, 1356.
34. S. Perrier, T. P. Davis, A. J. Carmichael, D. M. Haddleton, *Eur. Poly. J.* **2003**, *39*, 417.
35. R. T. Carlin, R. A. Osteryoung, J. S. Wilkes, J. Rovang, *Inorg. Chem.* **1990**, *29*, 3003.

36. R. T. Carlin, J. S. Wilkes, *J. Mol. Catal.* **1990**, *63*, 125.
37. S. D. Ittel, L. K. Johnson, M. Brookhart, *Chem. Rev.* **2000**, *100*, 1169.
38. M. F. Pinheiro, R. S. Mauler, R. F. de Souza, *Macromol. Rapid Commun.* **2001**, *22*, 425.
39. (a) P. Wasserscheid, C. M. Gordon, C. Hilgers, M. J. Muldoon, I. R. Dunkin, *J. Chem. Soc., Chem. Commun.* **2001**, 1186; (b) P. Wasserscheid, C. Hilgers, W. Keim, *J. Mol. Catal. A* **2004**, *214*, 83.
40. P. Mastrorilli, C. F. Nobile, V. Gallo, G. P. Suranna, G. Farinola, *J. Mol. Catal. A* **2002**, *184*, 73.
41. P. Mastrorilli, C. F. Nobile, V. Gallo, G. P. Suranna, R. Giannandrea, *Stud. Surf. Sci. Catal.* **2003**, *145*, 535.
42. A. M. Trzeciak, J. J. Ziolkowski, *Appl. Organomet. Chem.* **2004**, *18*, 124.
43. A. Sommazzi, F. Garbassi, *Prog. Polym. Sci.* **1997**, *22*, 1547.
44. C. Hardacre, J. D. Holbrey, S. P. Katdare, K. R. Seddon, *Green Chem.* **2002**, *4*, 143.
45. M. A. Klingshirn, G. A. Broker, J. D. Holbrey, K. H. Shaughnessy, R. D. Rogers, *Chem. Commun.* **2002**, 1394.
46. A. Fürstner, *Angew. Chem., Int. Ed. Engl.* **2000**, *39*, 3012.
47. M. R. Buchmeiser, *Chem. Rev.* **2000**, *100*, 1565.
48. R. H. Grubbs, *Handbook of Metathesis*, 1st edn., R. H. Grubbs (Ed.), Wiley-VCH, Weinheim, Germany, **2003**.
49. R. R. Schrock, *Chem. Rev.* **2002**, *102*, 14.
50. C. Gürtler, M. Jautclat, European Patent, **2000**, EP 1035093.
51. S. Csihony, C. Fischmeister, C. Bruneau, I. T. Horvath, P. H. Dixneuf, *New J. Chem.* **2002**, *26*, 1667.
52. Y. S. Vygodskii, A. S. Shaplov, E. I. Lozinskaya, O. A. Filippov, E. S. Shubina, R. Bandari, M. R. Buchmeiser, *Macromolecules* **2006**, *39*, 7821.
53. K. Matyjaszewski, *J. Macromol. Sci., Pure Appl. Chem.* **1997**, *10*, 1785.
54. K. Matyjaszewski, J. Xia, *Chem. Rev.* **2001**, *101*, 2921.
55. D. M. Haddleton, D. Kukulj, A. P. Radigue, *J. Chem. Soc., Chem. Commun.* **1999**, 99.
56. D. M. Haddleton, S. G. Jackson, S. A. F. Bon, *J. Am. Chem. Soc.* **2000**, *122*, 1542.
57. A. J. Carmichael, D. M. Haddleton, S. A. F. Bon, K. R. Seddon, *J. Chem. Soc., Chem. Commun.* **2000**, 1237.
58. V. Percec, C. Grigoras, *J. Polym. Sci. A* **2005**, *43*, 5609.
59. K. R. Seddon, A. Stark, M. J. Torres, *Pure Appl. Chem.* **2000**, *12*, 2275.
60. T. Biedron, P. Kubisa, *Macromol. Rapid Commun.* **2001**, *22*, 1237.
61. H. Y. Ma, X. H. Wan, X. F. Chen, Q. F. Zhou, *J. Polym. Sci. A Polym. Chem.* **2003**, *41*, 143.
62. H. Y. Ma, X. H. Wan, X. F. Chen, Q. F. Zhou, *Polymer* **2003**, *44*, 5311.
63. H. D. Tang, J. B. Tang, S. J. Ding, M. Radosz, Y. Q. Shen, *J. Polym. Sci. A Polym. Chem.* **2005**, *43*, 1432.
64. (a) T. Biedron, P. Kubisa, *Polym. Int.* **2003**, *52*, 1584; (b) T. Biedron, P. Kubisa, *J. Polym. Sci. A*, **2005**, *43*, 3454; (c) T. Biedron, P. Kubisa, *J. Polym. Sci. A Polym. Chem.* **2005**, *43*, 3454.
65. (a) S. J. Ding, M. Radosz, Y. Q. Shen, *Macromolecules* **2005**, *38*, 5921; (b) S. J. Ding, H. D. Tang, M. Radosz, Y. Q. Shen, *J. Polym. Sci. A-Polym. Chem.* **2004**, *42*, 5794; (c) *Prog. Polym. Sci.*, **2004**, *29*, 1053.
66. V. Percec, C. Grigoras, *J. Polym. Sci. A Polym. Chem.* **2005**, *43*, 5609.
67. A. J. Frank, K. Honda, *J. Phys. Chem.* **1982**, *86*, 1933.
68. R. A. Bull, F. R. Fran, A. J. Bard, *J. Electrochem. Soc.* **1983**, *130*, 1636.
69. B. J. Feldman, P. Burgmayer, R. W. Murray, *J. Am. Chem. Soc.* **1985**, *107*, 872.
70. P. Kovacic, M. B. Jones, *Chem. Rev.* **1987**, *87*, 357.
71. V. M. Kobryanskii, S. A. Arnautov, *Synth. Met.* **1993**, *55*, 1371.
72. S. A. Arnautov, V. M. Kobryanskii, *Macromol. Chem. Phys.* **2000**, *201*, 809.
73. S. A. Arnautov, *Synth. Met.* **1997**, *84*, 295.
74. D. C. Trivedi, *J. Chem. Soc., Chem. Commun.* **1989**, 544.
75. L. M. Goldenberg, R. A. Osteryoung, *Synth. Met.* **1994**, *64*, 63.
76. L. Janiszewska, R. A. Osteryoung, *J. Electrochem. Soc.* **1988**, *135*, 116.
77. P. G. Pickup, R. A. Osteryoung, *J. Am. Chem. Soc.* **1984**, *106*, 2294.
78. J. Tang, R. A. Osteryoung, *Synth. Met.* **1991**, *45*, 1.

79. L. Janiszewska, R. A. Osteryoung, *J. Electrochem. Soc.* **1987**, *134*, 2787.
80. R. T. Carlin, R. A. Osteryoung, *J. Electrochem. Soc.* **1994**, *141*, 1709.
81. T. A. Zawodzinski, L. Janiszewska, R. A. Osteryoung, *J. Electroanal. Chem.* **1988**, *255*, 111.
82. V. M. Kobryanskii, S. A. Arnautov, *J. Chem. Soc., Chem. Commun.* **1992**, 727.
83. Y. S. Vygodskii, E. I. Lozinskaya, A. S. Shaplov, *Macromol. Rapid Commun.* **2002**, *23*, 676.
84. E.I. Lozinskaya, A.S. Shaplov, Y.S. Vygodskii, *Eur. Polym. J.* **2004**, *40*, 2065.
85. Y.S. Vygodskii, E.I. Lozinskaya, A.S. Shaplov, *Macromol. Rapid Commun.* **2002**, *23*, 676.
86. S. Dali, H. Lefebvre, R. El Gharbi, A. Fradet, *J.Polym. Sci. A-Polym. Chem.* **2006**, *44*, 3025.
87. L. Q. Liao, L. J. Liu, C. Zhang, S. Q. Gong, *Macromol. Rapid Commun.* **2006**, *27*, 2060.
88. L. Zhu, C. Y. Huang, Y. H. Patel, J. Wu, S. V. Malhotra, *Macromol. Rapid Commun.* **2006**, *27*, 1306.
89. (a) F. van Rantwijk, R. M. Lau, R. A. Sheldon, *Curr. Opin. Biotechnol.* **2002**, *13*, 565; (b) S. Park, R. J. Kazlauskas, *Curr. Opin. Biotechnol.* **2003**, *14*, 432.
90. S. J. Nara, J. R. Harjani, M. M. Salunkhe, A. T. Mane, P. P. Wadgaonkar, *Tetrahedron Lett.* **2003**, *44*, 1371.
91. H. Uyama, T. Takamoto, S. Kobayashi, *Polym J.* **2002**, *34*, 94.
92. E. I. Lozinskaya, A. S. Shaplov, M. V. Kotseruba, L. I. Komarova, K. A. Lyssenko, M. Y. Antipin, D. G. Golovanov, Y. S. Vygodskii, *J. Polym. Sci. A. Polym. Chem.* **2006**, *380*, 44.
93. (a) E. F. Conner, G. W. Nyce, A. Mock, J. L. Hedrick, *J. Am. Chem. Soc.* **2002**, *124*, 914; (b) G. W. Nyce, J. A. Lamboy, E. F. Conner, R. M. Waymouth, J. L. Hedrick, *Org. Lett.* **2002**, *4*, 3587; (c) G. A. Grasa, R. M. Kissling, S. P. Nolan, *Org. Lett.* **2002**, *4*, 3583; (d) G. A. Grasa, T. Guveli, R. Singh, S. P. Nolan, *J. Org. Chem.* **2003**, *68*, 2812.
94. A. C. Sentman, S. Csihony, R. M. Waymouth, J. L. Hedrick, *J. Org. Chem.* **2005**, *70*, 2391.
95. R. Vijayaraghavan, D. R. MacFarlane, *Chem. Commun.* **2005**, 1149.

8
Biocatalytic Reactions in Ionic Liquids

Sandra Klembt, Susanne Dreyer, Marrit Eckstein, and Udo Kragl

8.1
Introduction

Biocatalytic reactions and production processes have been established as useful tools for several decades. The Reichstein process for the oxidation of D-sorbitol to L-sorbose using whole microorganisms, which is still in use, was introduced as early as 1934 [1]. Several years ago, BASF introduced a lipase-catalyzed process for the kinetic resolution of chiral amines [2]. Throughout the history of biocatalysis, alternative reaction conditions have been investigated to overcome problems such as substrate solubility, selectivity, yield or catalyst stability. Some progress has been made by the use of organic solvents [3, 4], the addition of high salt concentrations [5], the use of microemulsions [6] or supercritical fluids [7]. Recently the methods of gene technology, e.g. site-directed mutagenesis and directed evolution, have added new and powerful tools for the development of improved biocatalysts [8, 9]. Thus, it was unsurprising, that researchers in the field of biocatalysis focused on ionic liquids as novel solvents to find new solutions to known problems.

In this chapter we try to summarize the work in this field reported so far. First, we give a short introduction to the different forms of biocatalytic reactions, highlighting some special properties of biocatalysts.

8.2
Biocatalytic Reactions and Their Special Needs

Biotechnological processes can be divided into fermentation processes and biotransformations. In a fermentation process products are formed as primary or secondary metabolites by microorganisms or higher cells from components in the fermentation broth. Product examples are amino acids, vitamins or antibiotics such as penicillin or cephalosporine. In these cases co-solvents are sometimes used for *in situ* product extraction. The term biotransformation or biocatalysis is used for processes where a starting material (precursor) is converted into the desired product

Ionic Liquids in Synthesis, Second Edition. P. Wasserscheid and T. Welton (Eds.)
Copyright © 2008 WILEY-VCH Verlags GmbH & Co. KGaA, Weinheim
ISBN: 978-3-527-31239-9

in just one step. This can be done either by using whole cells or by using (partly) purified enzymes. Product examples range from bulk chemicals (e.g. acrylamide) to fine chemicals and chiral synthons (e.g. chiral amines or alcohols). There are several books and reviews dealing with the use of biotransformations either on a laboratory or an industrial scale [1, 10–13].

Nature has optimized its biocatalysts to perform best in an aqueous environment, at neutral pH, temperatures below 40 °C and at low osmotic pressure. Sometimes, these conditions are contrary to the requirements of the chemist or process engineer to optimize a reaction with respect to space-time yield or high product concentration in order to facilitate downstream processing. Furthermore, enzymes and whole cells are often inhibited by products or substrates. This might be overcome by using continuously operated stirred tank reactors, fed-batch reactors or reactors with *in situ* product removal [14, 15]. To increase the solubility of substrates and/or products the addition of organic solvents is a common practice [16].

Generally, there are three ways to use organic solvents or ionic liquids in a biocatalytic process:

1. as a pure solvent,
2. as a co-solvent in aqueous systems or
3. in a biphasic system.

When either the organic solvent or the ionic liquid is used as a pure solvent, the control of the water content, or rather the water activity, is of crucial importance as a minimum amount is necessary to maintain the enzyme activity. For ionic liquids, the same methods can be used to operate a reaction at constant water activity as those established for organic solvents [17, 70, 72]. As pure solvents and in biphasic systems [BMIM][PF$_6$] or [BMIM][(CF$_3$SO$_2$)$_2$N], for example, are used. Water-miscible ionic liquids can be used in monophasic systems, e.g. [BMIM][BF$_4$] or [MMIM][MeSO$_4$].

It should be noted, that, despite the success of the application of conventional organic solvents, there is no general rule as to which a solvent is "enzyme friendly". To a certain extent, the log P concept, based on the distribution coefficient between water and octanol, can be used as a guideline [18]. In general, solvents with a log $P > 3$ such as xylene (3.1) or hexane (3.9) are less deactivating than those with a log $P < 3$ such as ethanol (−0.24). Kaar et al. measured log P values for three different imidazolium-based ionic liquids (hexafluorophosphate, acetate, nitrate). Values range from −2.39 to −2.90. This would indicate that ionic liquids are highly hydrophilic and would inactivate enzymes. Certainly the hydrophilicity of the co-solvent is important as it allows interaction and breaking of hydrogen bonds that are stabilizing the tertiary structure of the protein. Such interactions are very likely to occur with ionic liquids as well. Surprisingly, enzymes and even whole cells are active in various ionic liquids, as will be shown in Section 8.3. However, the polarities of ionic liquids have been investigated by different groups [19–23]. The polarities of different ionic liquids such as [BMIM][PF$_6$] or [EMIM][(CF$_3$SO$_2$)$_2$N] are similar to those of polar solvents such as ethanol or *N*-methylformamide. On Reichardt's normalized polarity scale, ranging from 0 for tetramethylsilane to 1 for water, ionic liquids have polarities between 0.3 and 1. Toluene (0.1) and MTBE (0.35) are less polar [21, 22, 24] (for more details on the polarity of ionic liquids see Chapter 3.5). Both of these solvents are commonly used as water-immiscible solvents in enzyme catalysis. When used with whole cells organic solvents often damage the cell membrane.

When starting our first experiments using available ionic liquids in screening programs to identify suitable systems we encountered several difficulties such as pH shifts or precipitation. More generally, the following aspects should be taken into account when ionic liquids are used with biocatalysts:

- In some cases impurities in the ionic liquids lead to dramatic pH shifts causing enzyme inactivation. Sometimes this may be overcome simply by titration or the use of higher buffer concentrations. In other cases purification of the ionic liquid or an improved synthesis may be necessary.
- Enzymatic reactions are often performed in aqueous buffer solution; addition of increasing amounts of ionic liquids sometimes causes precipitates of unknown composition.
- To maintain enzymatic activity a minimal amount of water has to be present, best described by the water activity. However, water present in the reaction system may cause hydrolysis of some ionic liquids.
- Some enzymes require metal ions, such as cobalt, manganese or zinc, for their activity; if these are removed by the ionic liquid by complexation enzyme inactivation may occur.
- The ions of the liquid or impurities in the ionic liquid may act as reversible or irreversible enzyme inhibitors.
- For kinetic investigations and activity measurements either photometric assays or—due to the higher complexity of the reactants converted by biocatalysts—HPLC methods are often used. The ionic liquid itself or impurities may interfere with the analytical method.
- Unlike conventional organic solvents most of the research groups prepare the ionic liquids themselves. This may be the reason why, with nominally the same ionic liquid, different results are sometimes obtained. Park and Kazlauskas performed a washing procedure with aqueous sodium carbonate and found improved reaction rates, but this might also be related to a more precisely defined water content/water activity in the reaction system [22].

As with organic solvents, proteins are not soluble in most of the ionic liquids when they are used as pure solvent (examples of the solubility of enzymes in ionic liquids can be found in Section 8.4). As a result the enzyme is either applied as immobilized enzyme coupled to a support or as a suspension in its native form. For production processes the majority of enzymes are used as immobilized catalysts in order to facilitate handling and to improve their operational stability [25–27]. As a support either inorganic materials, such as porous glass, or different organic polymers are used [28]. These heterogeneous catalyst particles are subject to internal and external mass transport limitations which are strongly influenced by the viscosity of the

Table 8-1 Whole cell systems in ionic liquids

Biocatalyst	Ionic liquid[a]	Reaction system	Ref.
Whole cells			
Rhodococcus R312	1	Biotransformation of 1,3-dicyanobenzene	32
bakers yeast	1	Reduction of ketones	33
Sporomusa termitida	1	Sodium caffeate reduction	36
Lactobacillus kefir	2	IL-effect on cell membrane	35

[a] 1—1-butyl-3-methylimidazolium hexafluorophosphate, 2—1-butyl-3-methylimidazolium bis(trifluoromethylsulfonyl)imide.

reaction medium. For [BMIM][(CF$_3$SO$_2$)$_2$N] a dynamic viscosity of 52 mPa s has been reported at 20 °C [19]. For comparison, MTBE has a viscosity of only 0.34 mPa s. The viscosity can be reduced to a large extent by increasing the temperature or adding small amounts of an organic solvent [29] (for more information on viscosity of ionic liquids see Section 3.2). This important aspect of the use of ionic liquids in biocatalysis should be the subject of further studies.

8.3
Examples of Biocatalytic Reactions in Ionic Liquids

Due to their special properties and possible advantages ionic liquids have proved to be interesting solvents for biocatalytic reactions for solving some of the problems stated above. After initial trials using ethylammonium nitrate in salt-water mixtures more than 15 years ago [30], results using ionic liquids as a pure solvent, as a co-solvent or for biphasic systems have been reported recently. The reaction systems are summarized in Tables 8-1 to 8-3. In Table 8-1 examples of biocatalytic systems concerning whole cells are listed. Enzyme systems, except for lipases, are shown in Table 8-2. A separate overview of investigated lipase systems in ionic liquids is given in Table 8-3. Some of the entries will be discussed in more detail in the following sections.

8.3.1
Whole Cell Systems and Enzymes Other than Lipases in Ionic Liquids

In 1984 Magnuson et al. investigated the influence of ethylammonium nitrate/water-mixtures on enzyme activity and stability [30]. At low [H$_3$NEt][NO$_3$] concentrations an increased activity of alkaline phosphatase was reported. The same ionic liquid was used by Flowers and coworkers who found improved protein refolding after denaturation [31].

Compared to enzyme catalysis much less information on the use of ionic liquids with whole cell systems has been reported [32, 33]. In most cases [BMIM][PF$_6$] was used in a two-phase system as substrate reservoir and/or for *in situ* removal of the

Table 8-2 Enzymes other than lipases in ionic liquids

Biocatalyst	Ionic liquid[a]	Reaction system	Ref.
Alkaline phosphatase from E.coli	30	Enzyme activity and stability assayed by hydrolysis of p-nitrophenol phosphate	30
Esterases	1	Transesterification of 1-phenylethanol	48
Pyranose oxidase from Peniophora sp		Enzyme screening using MALDI-ToF	87
Hen egg white lysozyme	30	Protein renaturation	31
Thermolysine	1	Synthesis of Z-aspartame	37
β-galactosidase subtilisin	3	Hydrolytic activity	43
	19	Synthesis of N-acetyllactosamine	44
β-galactosidase from Bacillus circulans	18,19	Synthesis of lactose by reverse hydrolysis	46
Petptide amidase	18,19	Amidation of H-Ala-Phe-OH	46
Protease α-chymotrypsin	12	Stability investigations	41
	1,6	Transesterification of N-acetyl-L-phenylalanine ethyl ester	38
	1,3,12,13,31	Transesterification of N-acetyl-L-tyrosin ethyl ester	39
Papain	3	Resolution of racemic amino acid derivatives	42
Formate dehydrogenase	19	Regeneration of NADH	50
Morphine dehydrogenase	33	Synthesis of the drug oxycodone	79,80
Alcohol dehydrogenase	2	Enantioselective reduction of 2-octanone in a biphasic system	51
Hydrolase purified cress epoxid hydrolase	1–3	Hydrolysis of trans-β-methylstyrene oxide	47
Hydroxynitrile lyases from Prunus amygdalus/ Manihot esculenta	3,7,8,9	Transcyanation of acetone cyanohydrin, thermostability	53
Prunus amygdalus/ Hevea brasiliensis	3,8,11	Cyanohydrin reaction of chain aldehydes	52
Horseradish peroxidase	3	Enzyme immobilization in IL	54

[a] 1—1-butyl-3-methylimidazolium hexafluorophosphate, 2—1-butyl-3-methylimidazolium bis(triflouromethylsulfonyl)imide, 3—1-butyl-3-methylimidazolium tetrafluoroborate, 6—1-octyl-3-methylimidazolium hexafluorophosphate, 7—1-hexyl-3-methylimidazolium tetrafluoroborate, 8—1-pentyl-3-methylimidazolium tetrafluoroborate, 9—1-propyl-3-methylimidazolium tetrafluoroborate, 10—1-ethyl-3-methylimidazolium hexafluorophosphate, 11—1-ethyl-3-methylimidazolium tetrafluoroborate, 12—1-ethyl-3-methylimidazolium bis(triflouromethylsulfonyl)imide, 13—1-ethyl-3-methylimidazolium tetrafluoroborate, 18—1-butyl-3-methylimidazolium methylsulfate, 19—1,3-dimethylimidazolium methylsulfate, 30—ethylammonium nitrate, 31—methyltrioctylammonium bis(triflouromethylsulfonyl)imide, 33—1-(3-hydroxy-propyl)-3-methylimidazolium glycolate.

product formed, thereby increasing the catalyst productivity. Scheme 8-1 shows the reduction of ketones with baker's yeast in the system [BMIM][PF$_6$]/water.

Similarly, the recovery of n-butanol from a fermentation broth has been investigated by in situ extraction using [OMIM][PF$_6$] [34].

Table 8-3 Lipases in ionic liquids

Biocatalyst	Ionic liquid[a]	Reaction system	Ref.
Candida antarctica lipase B	2	Acylation of octan-1-ol and kinetic resolution of 1-phenylethanol using IL/scCO$_2$	81
	2,12	Synthesis of butyl butyrate and kinetic resolution of 1-phenylethanol using IL/scCO$_2$	82
	34,35	Transesterification in IL/scCO$_2$-systems	83
	1,2,12,31	Kinetic resolution of glycidol (2,3-epoxy-1-propanol) in IL/scCO$_2$-systems	84
	1,3	Alcoholysis, amminolysis, perhydrolysis	56
	1,10	Kinetic resolution of sec. alcohols	62
	1,2,12,13	Synthesis of butyl butyrate by transesterification	63
	1–5	Kinetic resolution of allylic alcohols	64
	several	Kinetic resolution of (R,S)-1-phenylethanol, acylation of β-glucose	22
	14	Synthesis of fatty acid esters	58
	1,3	Synthesis of glucose fatty acid esters	59
	1	Water activity control, transesterification	72
Pseudomonias cepacia lipase	1,10	Kinetic resolution of sec. alcohols	62
	1–5	Kinetic resolution of allylic alcohols	64
	several	Kinetic resolution of (R,S)-1-phenylethanol, acylation of β-glucose	22
	1,3	Formation of 4,6-di-O-acetyl-D-glucal, alcoholysis of decanol	57
	1	Kinetic resolution of diols	67
	32	Chiral IL, kinetic resolution of 1-(4-methoxy-phenyl)ethanol	68,69
Pseudomonias sp. lipase	2	Kinetic resolution of (R,S)-1-phenylethanol	44, 71
Candida rugosa lipase	1–5	Kinetic resolution of allylic alcohols	64
	1	Hydrolysing Naproxen methyl ester	65
	1,6	Water activity control, esterification	73
	1,2,6,15	Enantioselective transport of (S)-ibuprofen	85
Porcine liver	1–5	Kinetic resolution of allylic alcohols	64
Porcine pancreas	1,2,6,15	Enantioselective transport of (S)-ibuprofen	85
Screening of 8 lipases and 2 esterases	10 different	Kinetic resolution of (R,S)-1-phenylethanol	61
3 lipases	1,3	Synthesis of simple esters	43

[a] 1—1-butyl-3-methylimidazolium hexafluorophosphate, 2—1-butyl-3-methylimidazolium bis(triflouromethylsulfonyl)imide, 3—1-butyl-3-methylimidazolium tetrafluoroborate, 4—1-butyl-3-methylimidazolium trifluoromethanesulfonate, 5—1-butyl-3-methylimidazolium hexafluoroantimonate, 6—1-octyl-3-methylimidazolium hexafluorophosphate, 7—1-hexyl-3-methylimidazolium tetrafluoroborate, 10—1-ethyl-3-methylimidazolium hexafluorophosphate, 12—1-ethyl-3-methylimidazolium bis(triflouromethylsulfonyl)imide, 13—1-ethyl-3-methylimidazolium tetrafluoroborate, 14—1-(1-methylbutyl)-3-methylimidazolium tetrafluoroborate, 15—1-hexyl-3-methylimidazolium hexafluorophosphate, 31—methyltrioctylammonium bis(triflouromethylsulfonyl)imide, 32—(-)-N-ethylnicotinium bis(triflouromethylsulfonyl)-imide, 34—(3-cyanopropyl)-trimethylammonium bis(triflouromethylsulfonyl)imide, 35—butyltrimethylammonium bis(triflouromethylsulfonyl)imide.

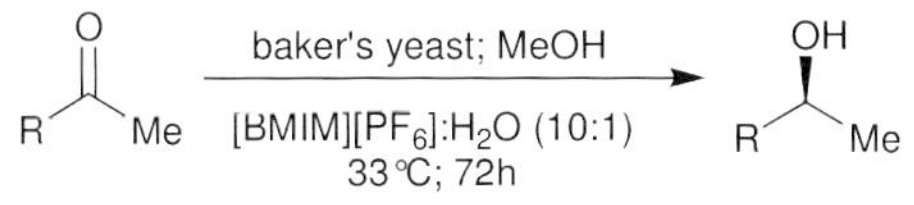

R-C(=O)-Me →[baker's yeast; MeOH / [BMIM][PF$_6$]:H$_2$O (10:1) / 33 °C; 72h] R-CH(OH)-Me

R = -C$_4$H$_9$ (yield 22%, ee$_S$ 95%)
R = -CH$_2$-COOEt (yield 75%, ee$_S$ 84%)   [33]

Scheme 8-1

Pfründer et al. succeeded in applying [BMIM][(CF$_3$SO$_2$)$_2$N] as a substrate reservoir and *in situ* extracting agent for the whole cell catalyzed synthesis of fine chemicals. Choosing the asymmetric reduction of 4-chloroacetophenone to (R)-1-(4-chlorophenyl)ethanol catalyzed by *Lactobacillus kefir* as a model reaction system, they found that [BMIM][(CF$_3$SO$_2$)$_2$N] exhibited very good solvent properties without destructive effects on the cell membrane of *Lactobacillus kefir*. In organic solvents the degree of damage increases with decreasing log *P* values. Ionic liquids that even have negative log *P* values seem to increase membrane integrity. As mentioned earlier, the suitability of an ionic liquid for whole cell catalysis cannot be deduced only from log *P* values and biocompatibility alone [35].

Lenourry et al. doubt, in agreement with general opinion, that [BMIM][PF$_6$] is a suitable ionic liquid for whole cell biotransformations due to its instability, viscosity and the toxicity of its hydrolysis products. Their reduction of caffeate using *Sporomusa termitida* did not give satisfactory results [36].

In the first publication describing the preparative use of an enzymatic reaction in ionic liquids, Erbeldinger et al. reported the use of the protease thermolysin for the synthesis of the dipeptide Z-aspartame [37]. The reaction rates were comparable to those found in conventional organic solvents such as ethyl acetate. Additionally, the enzyme stability was increased in the ionic liquid. The ionic liquid was recycled several times after the removal of non-converted substrates by extraction with water and product precipitation.

The protease α-chymotrypsin was used for transesterification reactions by three groups [38–40]. N-Acetyl-L-phenylalanine ethyl ester or N-acetyl-L-tyrosine ethyl ester were transformed into the corresponding propyl esters (Scheme 8-2).

The first group to investigate this system was Laszlo and Compton. They used [OMIM][PF$_6$] and [BMIM][PF$_6$] and compared the results with other organic solvents such as acetonitrile or hexane [38]. Their investigation was focused on the influence of the water content on enzyme activity as well as on the ratio of transesterification and hydrolysis. They found that, as with polar organic solvents, a certain amount of water is necessary to maintain enzymatic activity. For both ionic liquids and organic solvents, the rates are of the same order of magnitude. Data on the recycling of the enzyme or its stability were not given.

Iborra and coworkers examined the transesterification of N-acetyl-L-tyrosine ethyl ester in different ionic liquids and compared their stabilizing effect relative to that found with 1-propanol as solvent [39]. Despite the fact that in the ionic liquids tested the enzyme activity reached only 10 to 50% of the value in 1-propanol, the

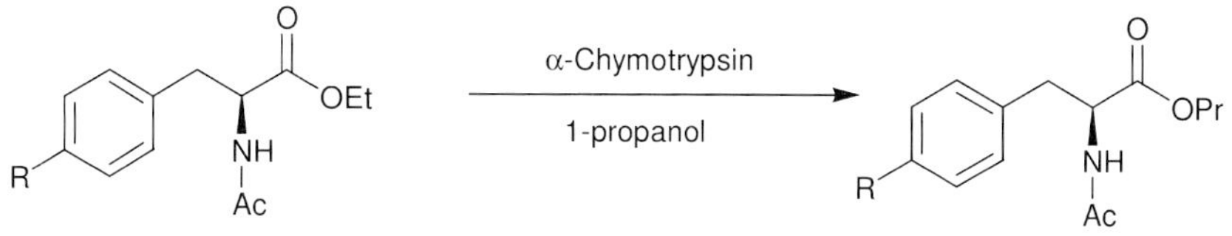

[38]: ionic liquid (up to 1.0% v/v H_2O); 40 °C

R= H

N-acetyl-L-phenylalanine
ethyl ester

[39]: ionic liquid (2% v/v H_2O); 50 °C

R= OH

N-acetyl-L-tyrosine
ethyl ester

Scheme 8-2

increased stability led to higher final product concentrations. In both studies fixed water contents were used.

Eckstein et al. demonstrated that in ionic liquids the enzyme is active at lower water activities than in organic solvents [40].

Furthermore, the stability of α-chymotrypsin in [EMIM][$(CF_3SO_2)_2N$] was studied by De Diego et al. Results were compared to those obtained in 1-propanol, a deactivating medium, and an aqueous solution of sorbitol, an enzyme-stabilizing medium. Using fluorescence and circular dichroism studies they showed that of the solvents used only the ionic liquid was able to stabilize the enzyme via the formation of a flexible and more compact 3D structure [41].

The enzymatic resolution of racemic amino acid derivatives in [BMIM][BF_4] by papain was tested by Lou et al. Higher hydrolytic activity and enantioselectivity concerning asymmetric hydrolysis of D,L-p-hydroxyphenylglycine methyl ester was achieved in ionic liquid/buffer-solution compared to organic solvent solutions. Nearly no hydrolytic activity was observed in pure ionic liquid [42].

Husum et al. found the hydrolytic activities of the protease subtilisin and of β-galactosidase from *E. coli* in a 50% aqueous solution of the water miscible ionic liquid [BMIM][BF_4] to be comparable to those in 50% aqueous solutions of ethanol or acetonitrile [43].

The transfer galactosylation with β-galactosidase from *Bacillus circulans* for the synthesis of N-acetyllactosamine starting from lactose and N-acetylglucosamine [44] has been studied. When performing the reaction in an aqueous system the problem of this approach lies in the secondary hydrolysis of the product by the same enzyme. As a consequence, yields are less than 30%, and it is important to separate enzyme and product when the maximum yield is obtained. By addition of 25% v/v of [MMIM][$MeSO_4$] as a water-miscible co-solvent the water activity of the medium is decreased and, consequently, the secondary hydrolysis of the formed product is effectively suppressed, resulting in doubling the yield to almost 60%! Kinetic studies demonstrated that the enzyme activity is not influenced by the presence of the ionic liquid. The enzyme is stable under the conditions employed, allowing its repeated use after filtration with a commercially available ultrafiltration membrane [45].

References [38, 39, 44] describe so-called kinetically controlled syntheses starting from activated substrates such as ethyl esters or lactose. For two reaction systems it could be demonstrated that ionic liquids can also be useful in a thermodynamically controlled synthesis starting with single components [46]. In both cases, as with the results presented in Ref. [37], the ionic liquids are used with addition of less than 1% water, which is necessary to maintain the enzyme activity. The yields observed are similar or better than those obtained with conventional organic solvents.

Chiappe et al. investigated the applicability of soluble epoxide hydrolases (sHE) as biocatalyst in ionic liquids for the hydrolysis of trans-β-methylstyrene oxide giving the corresponding (1S, 2R)-*erythro*-1-phenylpropane-1,2-diol as the main product. Purified cress sEH showed almost the same activity in ionic liquids as in buffer solution. Attempts to carry out the same process in *tert*-butanol and MTBE were unsuccessful. It was found that the nature of the anion had only a moderate effect on the reaction rate. The conversion was only slightly slower in the more hydrophilic [BMIM][BF_4] than in the more hydrophobic [BMIM][PF_6] and [BMIM][$(CF_3SO_2)_2N$]. Reactivity data also show that the viscosity has only a minor effect. Product enantioselectivity was maintained or slightly enhanced in all three ionic liquids compared to the buffer solution. The highest enantiomeric ratios were found in [BMIM][PF_6]. With the exception of [BMIM][BF_4], ionic liquids could be reused at least four times without loss in enzyme activity or selectivity. Spontaneous oxirane hydrolysis was suppressed in ionic liquids [47].

Persson et al. investigated the stability, activity, and enantioselectivity of two esterases in ionic liquids compared to organic solvents. Specific activity was measured in the transesterification of 1-phenylethanol with vinyl acetate in [BMIM][PF_6]. The lyophilized powder was active in hexane but not in the ionic liquid, whereas the immobilized enzyme was active in both solvents. To estimate the influence of the specific reaction medium applied, different organic solvents and ionic liquids were tested, with hexane showing the best results. Attempts to correlate solvent polarity (Reichardt's dye index) and enzyme activity or enantioselectivity were not successful. An increase in enzyme stability up to 30-fold was observed in ionic liquid. This can be explained by electrostatic interactions between ionic liquid and protein, resulting in a more rigid protein. A higher kinetic barrier needs to be overcome to unfold the enzyme in ionic liquids as compared to suspended enzymes in organic solvents [48].

Oxidoreductases are of special interest for the enantioselective reduction of prochiral ketones [49]. Formate dehydrogenase from *Candida boidinii* was found to be stable and active in mixtures of [MMIM][$MeSO_4$] with buffer. The application of alcohol dehydrogenases for enzyme-catalyzed reactions in the presence of water-miscible ionic liquids could make use of another advantage of these solvents: they increase the solubility of hydrophobic compounds in aqueous systems. By addition of 40% v/v of [MMIM][MeSO$_4$] to water the solubility of e.g. acetophenone is increased from 20 to 200 mmol L^{-1}. Trying to take advantage of this phenomenon we studied the effect of the addition of several water-miscible ionic liquids on the activity of yeast alcohol dehydrogenase (YADH) and pig liver alcohol dehydrogenase (PL-ADH). Unfortunately, a decrease in enzyme activity was observed for both enzymes. In the presence of 20% (v/v) [MMIM][MeSO$_4$] the activity of PL-ADH for the oxidation of ethanol was reduced to 22% and YADH showed no activity at all. Best results could be obtained in the presence of 20% [BMIM][CF$_3$SO$_3$]: residual activities were 8% for YADH and 41% for PL-ADH respectively. Due to the decrease in enzyme activity, further studies on the reduction of hydrophobic substrates in the presence of the investigated ionic liquids were set aside [50].

In contrast, the alcohol dehydrogenase from *Lactobacillus brevis* could successfully be applied in a biphasic system containing buffer and the water immiscible ionic liquid [BMIM][(CF$_3$SO$_2$)$_2$N]. By running this system for the highly enantioselective reduction of 2-octanone advantages of a two-phase system could be exploited. Within this reaction the enzyme requires the cofactor NADPH. Since the cofactor is expensive it was regenerated by a substrate-coupled approach, whereby 2-propanol is oxidized to acetone yielding the reduced cofactor. In the investigated reduction the aqueous phase contains the enzyme and the cofactor, whereas the substrate 2-octanone is dissolved in the organic phase. The partitioning behavior of substrates and products during the reaction is of great importance for the reaction thermodynamics. In a two-phase system consisting of buffer and MTBE the partition coefficients of 2-propanol and acetone are approximately 1. When replacing MTBE by [BMIM][(CF$_3$SO$_2$)$_2$N] the values change significantly. While 2-propanol remains preferentially in the aqueous phase ($m = 0.4$), acetone is removed by the ionic liquid ($m = 2.0$). This partition behavior of the cosubstrate and the coproduct will lead to a positive shift in the thermodynamics of cofactor regeneration, which is the rate limiting factor of the reaction. 2-Propanol is permanently available for regeneration in the aqueous phase, while acetone is continuously removed by the ionic liquid. In addition, acetone is known as an inhibitor for alcohol dehydrogenase. Extraction of acetone from the buffer phase reduces this inhibiting effect. Investigating the first 180 min of the reduction in the biphasic system, the reaction is much faster in the presence of the ionic liquid, leading to a conversion of 88% compared to only 61% in the presence of MTBE (Fig. 8-1) [51]

Promising results using hydroxynitrile lyases were achieved by Gaisberger et al. and Lou et al. in tetrafluoroborate-based ionic liquids, especially when using the enzyme from *Prunus amygdalus*. Gaisberger et al. used the lyases for reactions with long chain aldehydes. Lou et al. discovered increased thermal enzyme stability, improved enantioselectivity and enzymatic activity for a transcyanation reaction at low ionic liquid contents in buffer solution [52, 53]

As mentioned earlier, the way of immobilizing enzymes plays a key role in enzyme stability, solubility and reusability. In addition, mass transport limitations can affect reaction rates drastically. Liu et al. immobilized horseradish peroxidase (HRP) in a [BMIM][BF$_4$] based sol-gel matrix. Sol–gel-derived silica glasses are very popular for encapsulation of biomolecules. Compared to results in the conventional

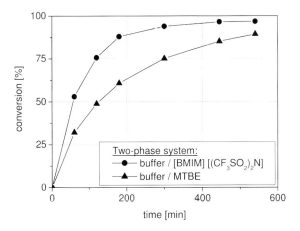

Fig. 8-1 Reduction of 2-octanone to (R)-2-octanol. Substrate-coupled NADPH regeneration [51].

silica matrix without ionic liquid, dramatically enhanced activity (30-fold higher) and excellent thermal stability of HRP immobilized within the ionic liquid-based matrix were obtained. The entrapped enzyme also showed a significantly improved thermal stability. It is assumed that the ionic liquid acts as a template for the formation of the mesoporous matrix and thus improves the mass transfer. As a stabilizer the ionic liquid protects the immobilized HRP from being inactivated by alcohol formed in the sol-gel process [54].

8.3.2
Lipases in Ionic Liquids

The majority of enzymes reported so far to be active in ionic liquids are lipases, the "work horses" of biocatalysis [11]. Designed by nature to work at aqueous/organic interfaces for the cleavage of fats and oils, making the cleavage products accessible as nutrients, lipases in general tolerate and are active in pure organic solvents. This concept has been pioneered by Klibanov and coworkers [17, 55].

The work from Sheldon and coworkers was the second publication demonstrating the potential use of enzymes in ionic liquids and the first on lipases [56]. They compared the reactivity of *Candida antarctica* lipase in ionic liquids such as [BMIM][PF_6] and [BMIM][BF_4] with the reactivity in conventional organic solvents. In all cases the reaction rates were similar for all the reactions investigated: alcoholysis, ammoniolysis and perhydrolysis.

One of the novel properties exhibited by enzymes in ionic liquids, as compared to aqueous solution, is their altered enzyme specificity, such as enhanced regioselectivity and enantioselectivity. Examples of increased regioselectivity in lipase-catalyzed reactions are given in the following section.

Nara et al. investigated the formation of 4,6-di-O-acetyl-D-glucal by *Pseudomonas cepacia* and the dependence of the reaction rate and regioselectivity on the solvent

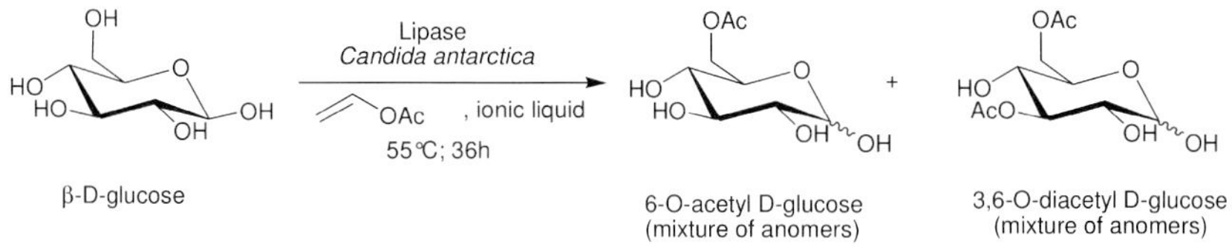

example: 1-Methoxyethyl-3-methylimidazolium ([MOEMIM]) [BF$_4$] dissolves ~5mg/ml glucose at 55°C
yield 99%; selectivity: 93% 6-O-acetyl D-glucose [22]

Scheme 8-3

used. For hydrolysis, the best regioselectivity was achieved in a buffer/[BMIM][PF$_6$] two-phase system with 84% of the monoacylated product in the acylation mixture. Results were compared to THF/buffer and [BMIM][BF$_4$]/buffer systems. Alternatively, the alcoholysis using decanol in the same solvents was investigated. In this case the extent of product formation was slightly higher in THF than in [BMIM][PF$_6$]. In both reaction systems, the use of [BMIM][BF$_4$] did not show satisfactory results. Results in THF and [BMIM][PF$_6$] could be proved in preparative experiments. Attempts to recycle the lipase-ionic liquid systems resulted in a decrease in yield during three runs for hydrolysis from 80% to 74%. In alcoholysis the yields only decreased marginally from 44% to 41% [57].

83% conversion and 65% yield of ascorbyl oleate were obtained for the esterification studied by Park et al. The synthesis of this fatty acid ester of L-ascorbic acid catalyzed by *Candida arctica* lipase B was highly regioselective. The reaction was carried out in a specially synthesized ionic liquid, [PentMIM][BF$_4$], to increase solubility of the acid. Due to low solubility the ester produced inhibits the reaction by precipitation on the surface of the immobilized lipase. To avoid precipitation, hexane or porous propylene, which are more hydrophobic than the resin, were added [58].

One particular feature of ionic liquids is their ability to act as solvents for hydrophobic compounds and hydrophilic compounds, such as carbohydrates. Park and Kazlauskas reported the regioselective acylation of glucose with 99% yield and 93% selectivity in [MOEMIM][BF$_4$] (where MOE = CH$_3$OCH$_2$CH$_2$). Values are much higher than in organic solvents commonly used for this purpose [22] (Scheme 8-3)

These excellent solvation properties were also used by Ganske et al. They were able to synthesize glucose fatty acid esters directly from glucose and fatty acid ester or free fatty acid, respectively. With their mixture of ionic liquid ([BMIM][BF$_4$] or [BMIM][PF$_6$]) and *tert*-butanol the group found a system containing the sugar, the acid or acid ester, and the active enzyme *Candida arctica* lipase B at the same time [59, 60].

Another advantage for the performance of lipase-catalyzed reactions in ionic liquids as compared to the reaction in organic solvents is an enhancement of enantioselectivity.

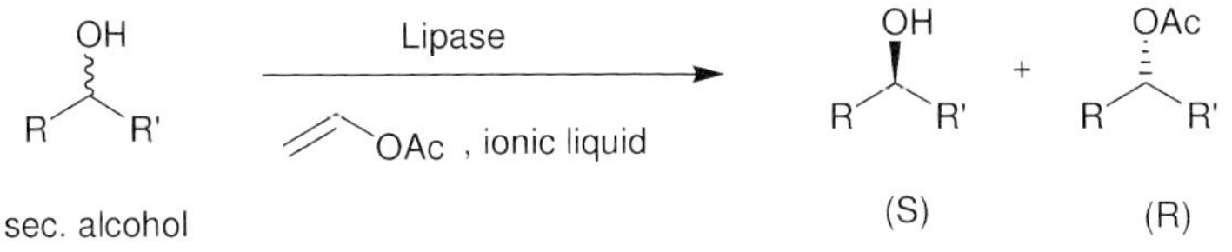

Scheme 8-4

In our group the kinetic resolution of 1-phenylethanol was investigated for a set of eight different lipases and two esterases in ten ionic liquids with MTBE as the reference [61].

Lipases and esterases are often used for the kinetic resolution of racemates either by hydrolysis, esterification or transesterification of suitable precursors. (Scheme 8-4) illustrates the principle for the resolution of secondary alcohols by esterification with vinyl acetate

In the transesterification of 1-phenylethanol with vinyl acetate no activity for esterases was observed. For the lipases from *Pseudomonas sp. and Alcaligenes sp.* an improved enantioselectivity was observed in [BMIM][(CF$_3$SO$_2$)$_2$N] when compared with MTBE as solvent. Best results were obtained for *Candida antarctica* lipase B in [BMIM][CF$_3$SO$_3$], [BMIM][(CF$_3$SO$_2$)$_2$N] and [OMIM][PF$_6$]. Contrary to the observations of other groups almost no activity was observed in [BMIM][BF$_4$] and [BMIM][PF$_6$]. This might be due to the quality of the ionic liquids we used at that time. Other groups investigating the same system observed good activities in these ionic liquids [22, 62–64]. Park and Kazlauskas even demonstrated the influence of additional washing steps upon the enzyme activity [22]. All groups reported excellent enantioselectivities. In addition to our own work several groups reported the repeated use of the lipase after the work-up procedure. In all cases the remaining substrates and formed products were extracted either by using ether or hexane. As a consequence of the conditions employed there was a slight reduction in enzyme activity after each cycle.

Xin et al. used water-saturated [BMIM][PF$_6$] to overcome a number of problems when hydrolyzing Naproxen methyl ester by *Candida rugosa* lipase during the synthesis of the drug Naproxen ((S)-(+)-2-(6-methoxy-2-naphtyl) propionic acid). Results were compared with those achieved in isooctane. Isooctane maintains lipase activity but has a very low water activity, which is necessary for good conversion and enantioselectivity. More polar solvents, such as e.g. toluene, may be employed to overcome the low water capacity problem, however, at the price of decreasing enzyme activity and stability. Again in [BMIM][PF$_6$] lipase activity could be maintained compared to reactions in isooctane, even when repeatedly used. This was attributed to the higher solubility of methanol in the ionic liquid than in isooctane. In addition, the equilibrium constant could be decreased and the enantioselectivity could be enhanced [65].

The often increased enantioselectivity of enzymes in ionic liquids was also explored by Kamal and Chouhan for the synthesis of enantiomerically pure 1,2-diols in ionic liquids. The kinetic resolution of the diol was catalyzed by immobilized lipase from *Pseudomonas cepacia* in [BMIM][PF$_6$] [66]. Kamal and Chouhan also reported a lipase-mediated resolution of chlorohydrin and its application in the synthesis of the optically active catalytic agent NPS-2143. NPS-2143 can start the stimulation of new bone growth during the treatment of osteoporosis. Lipase from *Pseudomonas cepacia* immobilized on diatomite gave the best results out of a screening of six lipases for the transesterification of chlorohydrin. Testing different solvents [BMIM][PF$_6$] provided best conversions and enantioselectivities of the used ionic liquids comparable to that received in toluene and diisopropyl ether [67].

As a new medium for enantioselective catalysis Kitazume prepared the chiral ionic liquid [(−)-N-ethylnicotinium][(CF$_3$SO$_2$)$_2$N] from (*S*)-nicotine. This ionic liquid was examined as the solvent for the kinetic resolution of 1-(4-methoxyphenyl)ethanol mediated by *Pseudomonas cepacia* lipase. Only low enantioselectivities were obtained in this new medium [68, 69]

As already mentioned, control of water content is of great importance in enzyme catalysis. Studies on *Pseudomonas sp.* lipase have also revealed a strong influence of the water content of the reaction medium [70]. In order to compare the enzyme activity and selectivity as a function of the water present in solvents of different polarities, it is necessary to use the water activity a_w in these solvents. We used the method of water activity equilibration over saturated salt solutions [71] and could demonstrate that, in contrast to MTBE, which is commonly used for this type of reaction, the enantioselectivity of the lipase is less influenced either by the water content or the temperature when the reaction is performed in [BMIM][(CF$_3$SO$_2$)$_2$N].

Berberich et al. used salt hydrate pairs to control water activity in [BMIM][PF$_6$]. The results were in good agreement with that obtained for water activity control using saturated salt solutions. The advantage of pre-equilibration is that the contact of the enzyme with the used salt and thus enzyme deactivation can be avoided. On the other hand it is only applicable for initial rate measurements. This disadvantage can be overcome by controlling water activity with salt hydrate pairs. Berberich et al. measured initial rate - water activities for the transesterification reaction of methyl methacrylate with 2-ethylhexanol in either hexane or [BMIM][PF$_6$]. Both reaction systems gave similar profiles [72].

Another approach concerning the influence of water content during lipase catalysis was chosen by Ulbert et al. They investigated the esterification of different 2-substituted-propionic acids with 1-butanol in three different ionic liquids, THF, toluene, and *n*-hexane, catalyzed by *Candida rugosa* lipase, with respect to enantioselectivity and enzyme recycling. In order to study the clear effect of solvents on enzyme activity, reaction rate data obtained at the same water activity (same hydration of the enzyme) must be compared. Therefore, they determined the water content for optimal hydration of the lipase, giving either maximum yield or maximum enantioselectivity, in all of the media studied. After adjusting the optimum water concentration for the medium, the reaction was followed up to 10% conversion, so that continuous water activity control could be avoided. Based on data obtained up to 10% conversion, reaction rates and enantioselectivities were calculated. Both the enzyme activity and the enantioselectivity as a function of water concentration varied along bell-shaped curves. The maxima of these curves determine the water concentrations where either the highest E-value ($c_w(E)$) can be obtained or reaction proceeds with the highest reaction rate (c_w(yield)). In each solvent, $c_w(E)$ tends to be lower than c_w(yield). This observation was explained by the fact that lower hydration levels enhance the rigidity of enzymes and thus improve the enantioselectivity. The highest yield could be reached in n-hexane and [BMIM][PF$_6$]. Remarkably higher enantioselectivities were obtained in [BMIM][PF$_6$] and [OMIM][PF$_6$] compared to those in organic solvents. When comparing log P values with yield and enantioselectivity for the different solvents no correlation between solvent hydrophobicity and activity/enantioselectivity of the lipase could be found. Long term experiments were carried out with continuous water adjustment by pervaporation. In enzyme recycling experiments lipase retained 92% of its original activity in [BMIM][PF$_6$] and 95% in [OMIM][PF$_6$] after five cycles with nearly constant enantioselectivity (n-hexane: 55%, toluene: 50%) [73].

8.4
Stability and Solubility of Enzymes in Ionic Liquids

The major drawback for conducting enzymatic reactions in ionic liquids is the relatively low activity of enzymes suspended in ionic liquids. There are only a few examples so far with dissolved enzymes in ionic liquids with low water content. In many cases the protein has been modified. Some ionic liquids such as [Et$_3$MeN][MeSO$_4$] are known to dissolve enzymes in the presence of a small amount of water, others even in the absence of water [37, 74]. However, most enzymes dissolved in ionic liquids show reduced catalytic activity, presumably due to changes in their conformational state. To overcome this limitation, several strategies have evolved in order to enhance enzymatic activity in ionic liquids, including the addition of a small amount of water to the ionic liquids [38, 40] or the immobilization of enzymes on solid supports [23, 38]. Furthermore, polyethylene glycol (PEG) is known to provide high stability and dispersibility of enzymes in ionic liquids. Several research groups attempted to improve the solubility and activity of enzymes in ionic liquids by covalent modification with linear PEG chains [38, 74–76]. For example, PEG-lipase complexes prepared by Maruyama et al. [76] were suspended in ionic liquids and exhibited improved activity for the alcoholysis reaction between vinyl acetate and 2-phenyl-1-propanol in comparison to common organic solvents (n-hexane, isooctane and dimethylsulfoxide). Furthermore, the enantioselectivity for the R-isomer of the alcoholysis between racemic 1-phenylethanol and vinyl acetate catalyzed by the PEG-lipase complexes (lipases from *Pseudomonas cepacia* and *Pseudomonas fluorescens*) could be increased from E values of 17 and 37 in n-hexane to E values of 120 and 280 in [BMIM] [(CF$_3$SO$_2$)$_2$N] (Scheme 8-5).

Nakashima et al. succeeded in dissolving Subtilisin Carlsberg in common ionic liquids without the addition of water by conjugation of the enzyme with comb-shaped polyethylene glycol PM$_{13}$. These PM$_{13}$–Sub complexes could be solubilized in a variety of ionic liquids including [BMIM][PF$_6$], [EMIM] [(CF$_3$SO$_2$)$_2$N], [C$_2$(OH)MIM] [(CF$_3$SO$_2$)$_2$N] and [C$_2$OCH$_2$MIM] [(CF$_3$SO$_2$)$_2$N] at protein concentrations of at least 1 mg ml^{-1}. Investigations of the enzymatic activity of

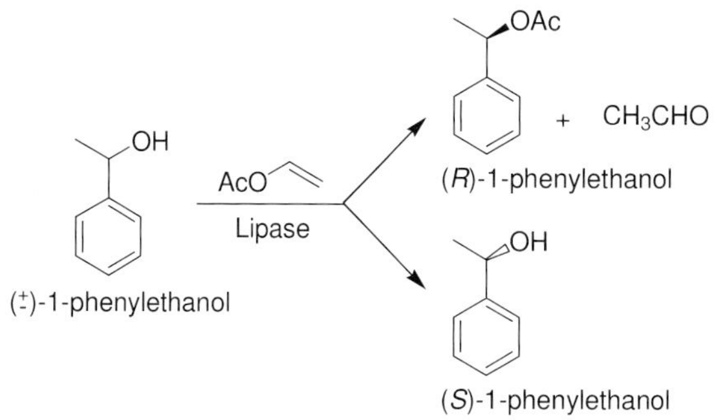

Scheme 8-5 Alcoholysis between 1-phenylethanol and vinyl acetate [76].

PM$_{13}$–Sub dissolved in [EMIM] [(CF$_3$SO$_2$)$_2$N] for the transesterification of N-acetyl-L-phenylalanine ethyl ester with 1-butanol further revealed an increase in the reaction rate (306 nmol min^{-1} (mg enzyme)$^{-1}$) in comparison to performing the reaction in the organic solvent toluene (93 nmol min^{-1} (mg enzyme)$^{-1}$). This approach offers the potential to obtain high enzymatic activity in pure ionic liquids without immobilization of enzyme or addition of water [77].

Another strategy to enhance the solubility and activity of enzymes in ionic liquids can be found in the design of the ionic liquids themselves. In 2005 Fujita et al. reported the development of a family of biocompatible ionic liquids for the solubilization and stabilization of proteins [78]. These ionic liquids consist of potentially biocompatible anions such as dicyanamide, saccharinate or dihydrogen phosphate and cations including the well known 1-butyl-1-methylpyrrolidinium cation and the cation choline. Cytochrome c from horse heart was chosen as a model protein to investigate the solubilizing and stabilizing effects of these ionic liquids. Fujita et al. found that up to 3 mM Cytochrome c could be dissolved in 1-butyl-1-methyl pyrrolidinium dihydrogen phosphate after adding 10–20 wt.% water to the ionic liquid. Moreover, the secondary structure is retained in the solubilized protein and thermal stability is dramatically increased. Because the proteins of potential pharmaceutical use often exhibit a low stability *in vitro*, the ionic liquids described by Fujita et al. could help to overcome these problems and allow a widespread use of some protein therapeutics.

Walker and Bruce reported a combined biological and chemical catalysis. An overall improvement of synthesis steps of the drug oxycodone could be achieved in the hydroxylated ionic liquid 1-(3-hydroxypropyl)-3-methylimidazolium glycolate. The ionic liquid was found to permit both the important intermediate codeinone/neopinone equilibrium and the hydration of the neopinone double bond under bis(acetylacetonato)cobalt(II) catalysis without additional extraction steps [79]. The mentioned combination of a moderately hydrophilic cation with a hydrophobic anion was found to be the most capable of dissolving both morphine dehydrogenase and its associated nicotinamide cofactor whilst permitting the retention of considerable catalytic activity against codeine [80].

8.5
Special Techniques for Biocatalysis with Ionic Liquids

Ionic liquids are always associated with green chemical processing. Usually only the ionic liquid as reaction medium is taken into account and not the recyclability and recoverability of products and catalysts, which is often carried out using common organic solvents. Supercritical CO_2 (scCO_2) has been described as an excellent solvent for the transport of hydrophobic compounds out of ionic liquids. It is highly soluble in ionic liquids, whereas the ionic liquid is not soluble in scCO_2. In an appropriate reactor for phase separation, free or immobilized enzymes can be dissolved or suspended in the ionic liquid phase (catalytic phase) while substrates and/or products reside largely in the supercritical phase (extractive phase). scCO_2 itself has a deactivating effect on enzymes but in ionic liquid/scCO_2 biphasic systems enzymes remain stable even when activity is lower than without the supercritical phase. Reetz et al. used this behavior when acylating octan-1-ol by vinyl acetate using lipase B from *Candida antarctica* in [BMIM][$(CF_3SO_2)_2$N]. After achieving complete conversion, high yields and no loss of enzyme activity after three repeated runs, they developed a continuous process for the kinetic resolution of 1-phenylethanol. Again, the activity of the enzyme was fully retained and the enantiomeric discrimination was high [81]. Using the same enzyme, Lozano et al. tested the synthesis of butyl butyrate from vinyl butyrate and butan-1-ol, and the kinetic resolution of *rac*-1-phenylethanol by transesterification with vinyl propionate in [EMIM][$(CF_3SO_2)_2$N]/scCO_2 and [BMIM][$(CF_3SO_2)_2$N]/ scCO_2 systems. The synthesis of butyl butyrate proceeded with high selectivity. The specific activity of the CALB-[BMIM][$(CF_3SO_2)_2$N-system with scCO_2 was 10-fold lower than that observed without scCO_2. High enantioselectivities and the enzyme protective effect of ionic liquids could also be observed for the kinetic resolution reaction [82]. By testing transesterification reactions using short-chain alkyl esters and, again, *Candida anatarctica* lipase B, Lozano et al. could also show the suitability of quaternary ammonium-based ionic liquids as enzyme stabilizing media under supercritical conditions. They pointed out that the two key criteria for designing such a two-phase process are not only the enzymatic stabilisation by the ionic liquid (protection against the scCO_2), but also the consideration of mass transfer phenomena between the scCO_2 and the highly viscous ionic liquid phase [83].

The same group reported the lipase-catalyzed kinetic resolution of glycidol (2,3-epoxy-1-propanol) by transesterification with vinyl alkyl esters in four different ionic liquids. All three assayed lipases increased their synthetic activity by using ionic liquids as the reaction medium compared to toluene, and exhibited high selectivities. Unfortunately, enantioselectivity could not be improved in the ionic liquids; this seems to be a specific property of each individual biocatalyst. Two of the three tested enzymes were able to catalyse the resolution in ionic liquid/scCO_2, although the corresponding activities were clearly lower than in ionic liquid without scCO_2. Again, the enantioselectivity remained unchanged in supercritical conditions with respect to the ionic liquid media. The authors attribute the observed activity loss in the biphasic system to limitations in the mass-transfer across the ionic liquid

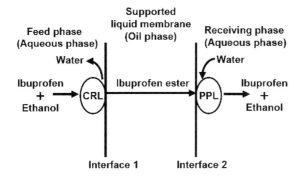

Scheme 8-6 Schematic diagram of enantioselective transport of (S)-ibuprofen through a lipase-facilitated support liquid membrane.

layer around the biocatalyst, rather than to enzyme deactivation phenomena. The dissolution of CO_2 greatly decreases the polarity of the ionic liquids, a phenomenon which clearly affects negatively the efficiency of the transport of polar compounds, such as glycidol, from the extractive phase to the catalytic phase [84].

A new and promising way of combining reaction and extraction are supported liquid membranes (SLM). Miyako et al. used SLMs on the basis of different water-immiscible ionic liquids and aliphatic hydrocarbons to achieve pure (S)-ibuprofen together with an enzymatic resolution step (Scheme 8-6). Only the (S)-enantiomer is esterified by the enzyme attached to the feed interface and able to enter the membrane. The ionic-liquid phase allows the selective transport of the more hydrophobic ibuprofen methyl ester from the aqueous-feed phase to the receiving phase. At the interface of the receiving phase, the (S)-ibuprofen methyl ester is hydrolysed to (S)-ibuprofen by another lipase. As (S)-ibuprofen is more hydrophilic it is not transported back through the supported ionic-liquid phase. Best results were achieved with [BMIM][$(CF_3SO_2)_2N$] as liquid membrane phase and the enzyme combination mentioned in Table 8-3 [85, 86].

Bungert et al. showed the applicability of MALDI-ToF MS with ionic liquid matrices as a tool for the quantitative screening of enzymes. As model system, ten variants of pyranose oxidase from the white rot fungi *Peniophora sp.* converting glucose to 2-keto-D-glucose in buffer solution were used. The results obtained by the faster MALDI-ToF procedure (including less preparation steps) were in good agreement with that achieved by additional HPLC measurements [87].

8.6
Conclusions and Outlook

The results reported clearly demonstrate the potential of ionic liquids as solvents for biotransformations. The variations possible for tailor-made solvents may have a similar impact as the pioneering work of Klibanov for the use of enzymes in pure organic solvents [55]. Further studies are necessary to reveal the reasons for the effects observed, such as better stability, selectivity or suppression of side reactions. Due to their ionic nature, ionic liquids may interact with charged groups of the enzyme, either in the active site or at its periphery, causing changes in the enzyme's structure. In order to use ionic liquids in biocatalytic reactions in some cases special properties or purities are required e.g. in order to avoid changes in the pH of the reaction medium.

When ionic liquids are used as replacements for organic solvents in processes with non-volatile products, downstream processing may become complicated. This might be true for many biotransformations where the better selectivity of the biocatalyst is used to transform more complex molecules. In such cases product isolation can be done by extraction with supercritical CO_2, for example [88]. Recently, membrane processes such as pervaporation and nanofiltration have been used. For less volatile compounds such as phenylethanol the use of pervaporation has been reported by Crespo and coworkers [89–91]. A separation process based on nanofiltration [92] has been developed that is especially well suited for isolation of non-volatile compounds such as carbohydrates or charged compounds. It may also be used for an easy recovery and/or purification of ionic liquids.

However, there is still a long way to go before ionic liquids become routinely used in biocatalysis. From today's perspective the following issues will have to be addressed in much more detail in future research.

- Demonstration of the stability and recyclability over prolonged periods of time under reaction conditions.
- Investigation of mass transport limitations for biocatalysts immobilized on heterogeneous supports.
- Development of suitable methods for product isolation if these are of limited or no volatility.

References

1. A. Liese, K. Seelbach, C. Wandrey, *Industrial Biotransformations*, Wiley-VCH, Weinheim, **2000**.
2. F. Balkenhohl, K. Ditrich, B. Hauer, W. Ladner, *J. Prakt. Chem./Chem.-Ztg.* **1997**, *339*, 381.
3. G. Carrea, S. Riva, *Angew. Chem.-Int. Ed.* **2000**, *39*, 2226.
4. J. M. S. Cabral, M. R. Aires-Barros, H. Pinheiro, D. M. F. Prazeres, *J. Biotechnol.* **1997**, *59*, 133.
5. Y. L. Khmelnitsky, S. H. Welch, D. S. Clark, J. S. Dordick, *J. Am. Chem. Soc.* **1994**, *116*, 2647.
6. B. Orlich, R. Schomaecker, *Biotechnol. Bioeng.* **1999**, *65*, 357.
7. T. Hartmann, E. Schwabe, T. Scheper, D. Combes, in *Stereoselective Biocatalysis*, 1st edn., R. N. Patel (Ed.), Marcel Dekker, New York, **2000**, p. 799.
8. U. T. Bornscheuer, M. Pohl, *Curr. Opin. Chem. Biol.* **2001**, *5*, 137.
9. F. H. Arnold, *Nature* **2001**, *409*, 253.
10. K. Faber, *Biotransformations in Organic Chemistry*, Springer, Berlin, **2004**.
11. U. T. Bornscheuer, R. J. Kazlauskas, *Hydrolases in Organic Synthesis*, Wiley-VCH, Weinheim, **2005**.

12. U. T. Bornscheuer, K. Buchholz, *Eng. Life Sci.* **2005**, *5*, 309.
13. A. Schmid, J. S. Dordick, B. Hauer, A. Kieners, M. Wubbolts, B. Witholt, *Nature* **2001**, *409*, 258.
14. J. E. Bailey, D. F. Ollis, *Biochemical Engineering Fundamentals*, McGraw-Hill, New York, **1986**.
15. U. Kragl, A. Liese, in *Encyclopedia of Bioprocess Technology*, C. Flickinger, S. W. Drew (Eds.), Wiley-VCH, New York, **2000**.
16. A. M. Klibanov, *Nature* **2001**, *409*, 241.
17. R. H. Valivety, P. J. Halling, A. R. Macrae, *Biochim. Biophys. Acta*, **1992**, *1118*, 218.
18. C. Laane, S. Boeren, R. Hilhorst, C. Veeger, Elsevier, Amsterdam, **1987**.
19. P. Bonhote, A. P. Dias, N. Papageorgiou, K. Kalyanasundaram, M. Grätzl, *Inorg. Chem.* **1996**, *35*, 1168.
20. A. J. Carmichael, K. R. Seddon, *J. Phys. Org. Chem.* **2000**, *13*, 591.
21. S. Aki, J. F. Brennecke, A. Samanta, *Chem. Commun.* **2001**, *413*,
22. S. Park, R. J. Kazlauskas, *J. Org. Chem.* **2001**, *66*, 8395.
23. J. L. Kaar, A. M. Jesionowski, J. A. Berberich, R. Moulton, A. J. Russell, *J. Am. Chem. Soc.* **2003**, *125*, 4125.
24. C. Reichardt, *Green Chem.* **2005**, *7*, 339.
25. U. Kragl, L. Greiner, C. Wandrey, Wiley, New York, **2000**.
26. W. Tischer, V. Kasche, *Trends Biotechnol.* **1999**, *17*, 326.
27. E. Katchalski-Katzir, D. M. Kraemer, *J. Mol. Catal. B* **2000**, *10*, 157.
28. W. Keim, B. Drießen-Hölscher, Wiley-VCH, Weinheim, **1997**.
29. P. Wasserscheid, W. Keim, *Angew. Chem.-Int. Ed.* **2000**, *39*, 3773.
30. D. K. Magnuson, J. W. Bodley, D. F. Evans, *J. Solution Chem.* **1984**, *13*, 583.
31. C. A. Summers, R. A. Flowers, *Protein Sci.* **2000**, *9*, 2001.
32. S. G. Cull, J. D. Holbrey, V. Vargas-Mora, K. R. Seddon, G. J. Lye, *Biotechnol. Bioeng.* **2000**, *69*, 227.
33. J. Howarth, P. James, J. F. Dai, *Tetrahedron Lett.* **2001**, *42*, 7517.
34. A. G. Fadeev, M. M. Meagher, *Chem. Commun.* **2001**, 295.
35. H. Pfruender, M. Amidjojo, U. Kragl, D. Weuster-Botz, *Angew. Chem.-Int. Ed.* **2004**, *43*, 4529.
36. A. Lenourry, J. Gardiner, G. Stephens, *Biotechnol. Lett.* **2005**, *27*, 161.
37. M. Erbeldinger, A. J. Mesiano, A. J. Russell, *Biotechnol. Prog.* **2000**, *16*, 1129.
38. J. A. Laszlo, D. L. Compton, *Biotechnol. Bioeng.* **2001**, *75*, 181.
39. P. Lozano, T. de Diego, J. P. Guegan, M. Vaultier, J. L. Iborra, *Biotechnol. Bioeng.* **2001**, *75*, 563.
40. M. Eckstein, M. Sesing, U. Kragl, P. Adlercreutz, *Biotechnol. Lett.* **2002**, *24*, 867.
41. T. De Diego, P. Lozano, S. Gmouh, M. Vaultier, J. L. Iborra, *Biotechnol. Bioeng.* **2004**, *88*, 916.
42. W. Y. Lou, M. H. Zong, H. Wu, *Biocatal. Biotransform.* **2004**, *22*, 171.
43. T. L. Husum, C. T. Jorgensen, M. W. Christensen, O. Kirk, *Biocatal. Biotransform.* **2001**, *19*, 331.
44. U. Kragl, N. Kaftzik, S. H. Schofer, M. Eckstein, P. Wasserscheid, C. Hilgers, *Chim. Oggi-Chem. Today* **2001**, *19*, 22.
45. N. Kaftzik, P. Wasserscheid, U. Kragl, *Org. Process Res. Dev.* **2002**, *6*, 553.
46. N. Kaftzik, S. Neumann, M. R. Kula, U. Kragl, in *Ionic Liquids as Green Solvents: Progress and Prospects*, Vol. 856, **2003**, pp. 206.
47. C. Chiappe, E. Leandri, S. Lucchesi, D. Pieraccini, B. D. Hammock, C. Morisseau, *J. Mol. Catal. B-Enzym.* **2004**, *27*, 243.
48. M. Persson, U. T. Bornscheuer, *J. Mol. Catal. B-Enzym.* **2003**, *22*, 21.
49. M.-R. Kula, U. Kragl, in *Stereoselective Biocatalysis*, 1st edn., R. N. Patel (Ed.), Marcel Dekker, Inc., New York, **2000**, p. 839.
50. N. Kaftzik, unpublished results.
51. M. Eckstein, M. Villela, A. Liese, U. Kragl, *Chem. Commun.* **2004**, 1084.
52. R. P. Gaisberger, M. H. Fechter, H. Griengl, *Tetrahedron-Asym.* **2004**, *15*, 2959.
53. W. Y. Lou, R. Xu, M. H. Zong, *Biotechnol. Lett.* **2005**, *27*, 1387.
54. Y. Liu, M. J. Wang, J. Li, Z. Y. Li, P. He, H. T. Liu, J. H. Li, *Chem. Commun.* **2005**, 1778.
55. A. M. Klibanov, *CHEMTECH* **1986**, *16*, 354.

56. R. M. Lau, F. van Rantwijk, K. R. Seddon, R. A. Sheldon, *Org. Lett.* **2000**, *2*, 4189.
57. S. J. Nara, S. S. Mohile, J. R. Harjani, P. U. Naik, M. M. Salunkhe, *J. Mol. Catal. B-Enzym.* **2004**, *28*, 39.
58. S. Park, F. Viklund, K. Hult, R. J. Kazlauskas, *Green Chem.* **2003**, *5*, 715.
59. F. Ganske, U. T. Bornscheuer, *Org. Lett.* **2005**, *7*, 3097.
60. F. Ganske, U. T. Bornscheuer, *J. Mol. Catal. B-Enzym.* **2005**, *36*, 40.
61. S. H. Schofer, N. Kaftzik, P. Wasserscheid, U. Kragl, *Chem. Commun.* **2001**, 425.
62. K. W. Kim, B. Song, M. Y. Choi, M. J. Kim, *Org. Lett.* **2001**, *3*, 1507.
63. P. Lozano, T. De Diego, D. Carrie, M. Vaultier, J. L. Iborra, *Biotechnol. Lett.* **2001**, *23*, 1529.
64. T. Itoh, E. Akasaki, K. Kudo, S. Shirakami, *Chem. Lett.* **2001**, 262.
65. J. Y. Xin, Y. J. Zhao, Y. G. Shi, C. G. Xia, S. B. Li, *World J. Microbiol. Biotechnol.* **2005**, *21*, 193.
66. A. Kamal, G. Chouhan, *Tetrahedron Lett.* **2004**, *45*, 8801.
67. A. Kamal, G. Chouhan, *Tetrahedron-Asym.* **2005**, *16*, 2784.
68. C. Baudequin, D. Bregeon, J. Levillain, F. Guillen, J. C. Plaquevent, A. C. Gaumont, *Tetrahedron-Asym.* **2005**, *16*, 3921.
69. T. Kitazume, US patent, U.S. 0,031,875,2001.
70. M. Eckstein, P. Wasserscheid, U. Kragl, *Biotechnol. Lett.* **2002**, *24*, 763.
71. H. L. Goderis, G. Ampe, M. P. Feyten, B. L. Fouwe, W. M. Guffens, S. M. Vancauwenbergh, P. P. Tobback, *Biotechnol. Bioeng.* **1987**, *30*, 258.
72. J. A. Berberich, J. L. Kaar, A. J. Russell, *Biotechnol. Prog.* **2003**, *19*, 1029.
73. O. Ulbert, T. Frater, K. Belafi-Bako, L. Gubicza, *J. Mol. Catal. B-Enzym.* **2004**, *31*, 39.
74. M. B. Turner, S. K. Spear, J. G. Huddleston, J. D. Holbrey, R. D. Rogers, *Green Chem.* **2003**, *5*, 443.
75. T. Maruyama, S. Nagasawa, M. Goto, *Biotechnol. Lett.* **2002**, *24*, 1341.
76. T. Maruyama, H. Yamamura, T. Kotani, N. Kamiya, M. Goto, *Org. Biomol. Chem.* **2004**, *2*, 1239.
77. K. Nakashima, T. Maruyama, N. Kamiya, M. Goto, *Chem. Commun.* **2005**, 4297.
78. K. Fujita, D. R. MacFarlane, M. Forsyth, *Chem. Commun.* **2005**, 4804.
79. A. J. Walker, N. C. Bruce, *Tetrahedron* **2004**, *60*, 561.
80. A. J. Walker, N. C. Bruce, *Chem. Commun.* **2004**, 2570.
81. M. T. Reetz, W. Wiesenhofer, G. Francio, W. Leitner, *Chem. Commun.* **2002**, 992.
82. P. Lozano, T. de Diego, D. Carrie, M. Vaultier, J. L. Iborra, *Chem. Commun.* **2002**, 692.
83. P. Lozano, T. de Diego, S. Gmouh, M. Vaultier, J. L. Iborra, *Biotechnol. Prog.* **2004**, *20*, 661.
84. P. Lozano, T. De Diego, D. Carrie, M. Vaultier, J. L. Iborra, *J. Mol. Catal. A-Chem.* **2004**, *214*, 113.
85. E. Miyako, T. Maruyama, N. Kamiya, M. Goto, *Chem. Commun.* **2003**, 2926.
86. C. A. M. Afonso, J. G. Crespo, *Angew. Chem.-Int. Ed.* **2004**, *43*, 5293.
87. D. Bungert, S. Bastian, D. M. Heckmann-Pohl, F. Giffhorn, E. Heinzle, A. Tholey, *Biotechnol. Lett.* **2004**, *26*, 1025.
88. L. A. Blanchard, J. F. Brennecke, *Ind. Eng. Chem. Res.* **2001**, *40*, 287.
89. T. Schafer, C. M. Rodrigues, C. A. M. Afonso, J. G. Crespo, *Chem. Commun.* **2001**, 1622.
90. L. Gubicza, N. Nemestothy, T. Frater, K. Belafi-Bako, *Green Chem.* **2003**, *5*, 236.
91. P. Izak, N. M. M. Mateus, C. A. M. Afonso, J. G. Crespo, *Sep. Purif. Technol.* **2005**, *41*, 141.
92. J. Kröckel, U. Kragl, *Chem. Eng. Technol.* **2003**, *26*, 1166.

9
Industrial Applications of Ionic Liquids

Matthias Maase

9.1
Ionic Liquids in Industrial Processes: Re-invention of the Wheel or True Innovation?

Low melting salts in industrial processes are not new. There is a whole range of examples in which salts that are liquid at low temperatures happened to appear in chemical plants. In some cases the process developers might not even have been aware of it. These salts were primarily chosen for their specific function as a catalyst, supporting salt or reactive intermediate. Being a liquid in addition to this function was certainly welcome, but in most cases not a strict requirement. Among the list of examples is the use of "red oil" as a Friedel-Crafts catalyst in the ethylation of benzene [1]. "Red oil" is a low melting salt, formed by a carbocation and a tetrachloroaluminate anion. However, this technology has since been replaced by a zeolite-based process.

A second example is the chlorination of acid with phosgene (Scheme 9-1) which is performed in low melting liquid Vilsmeier salts [2].

The catalyst is actually a mixture of low melting salts. It is liquid at the reaction temperature of 60 °C and forms a separate phase which is immiscible with the liquid product phase. No additional solvent is needed. In the case of higher carboxylic acids the corresponding acid chlorides can be separated as a liquid phase and used directly without any further purification. This solvent-free process comprising the use of a low melting salt offered a range of advantages:

- no need for phosgene excess
- off-gas free of phosgene
- higher space–time yields
- no distillation of the product required. This avoids thermal stress to a sensitive product and provides higher yields.

This process was established by BASF in 1990.

In 1996, Eastman established a process for the rearrangement of vinyloxirane to dihydrofuran (Scheme 9-2)[3]. This reaction needs the simultaneous activation by a Lewis base, being a phosphonium iodide, and by Lewis acid being an organotin

Ionic Liquids in Synthesis, Second Edition. P. Wasserscheid and T. Welton (Eds.)
Copyright © 2008 WILEY-VCH Verlags GmbH & Co. KGaA, Weinheim
ISBN: 978-3-527-31239-9

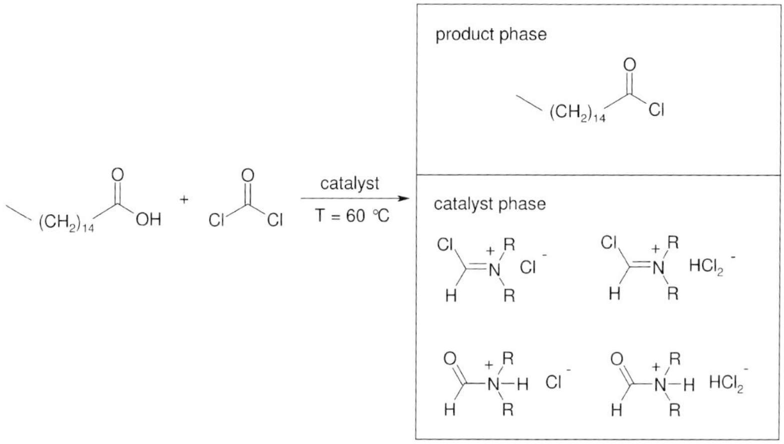

Scheme 9-1

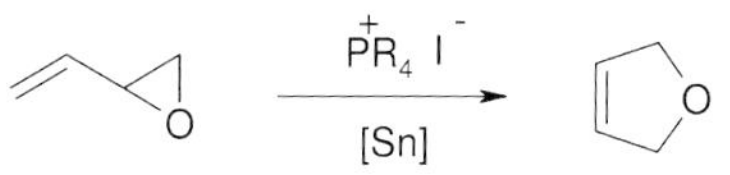

Scheme 9-2

compound. The phosphonium iodide that Eastman used was a low melting salt that also provided a liquid catalyst phase. However, due to a lack of demand for the product, this process was stopped recently.

These examples show that low melting salts have already been used in industrial plants in the past. Now, what about the use of *"ionic liquids"* in industrial processes? Is this a true innovation? How are they different from *"low melting salts"*? Of course in terms of chemistry there is no difference between the *"low melting salts"* that have been described earlier and *"ionic liquids."* In today's definition a Vilsmeier catalyst would be an ionic liquid: it has a rather low melting point and it consists totally of ions. The difference is in the mind set of chemists, engineers and process developers. Today people are aware of the unique properties of ionic liquids and they exploit them on purpose and in a dedicated way to solve an existing problem. Ionic liquid applications are clearly not a re-invention of the wheel. All the processes that will be described in the following are making use of ionic liquids by design. It is exciting and encouraging to see how elegant, how diverse and how innovative the companies have been in designing their specific ionic liquid-based solutions.

9.2
Possible Fields of Application

Until now ionic liquids have received most attention for electrochemical and chemical research [4]. This is also reflected by Table 9-1 of industrial applications that have been reported to date.

However, three entries in Table 9-1 represent applications already outside the classic chemical use as solvents or process chemicals. In these cases ionic liquids are

Table 9-1 Reported industrial applications of ionic liquids

Company	Process	IL is acting as:	Scale
BASF	acid scavenging	auxiliary	commercial
	extractive distillation	extractant	pilot
	chlorination	solvent	commercial
IFP	olefin dimerization	solvent	pilot
Degussa	hydrosilylation	solvent	pilot
	compatibilizer	performance additive	commercial
Arkema	fluorination	solvent	pilot
Chevron Phillips	olefin oligomerization	catalyst	pilot
Scionix	electroplating (Cr)	electrolyte	pilot
Eli Lilly	cleavage of ether	catalyst/reagent	pilot
Air Products	storage of gases	liquid support	pilot
Iolitec/Wandres	cleaning fluid	performance additive	commercial
Linde	gas compression	liquid piston	pilot

used as performance chemicals. There is an additional application cluster (Fig. 9-1) that solely makes use of the physical properties of the materials rather than their chemical behavior. Ionic liquids as so-called engineering fluids or new materials can be a functional part of devices, equipment and machinery that will be used in a variety of industries, such as automotive, air- and spacecraft, textile, electronics, machinery, energy, etc. The most remarkable application in this context has been reported by Linde AG [5]. The so-called "ionic compressor" uses ionic liquids as a "liquid piston" allowing an almost isothermal high pressure compression of gases. Major

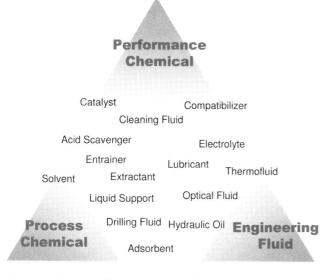

Fig. 9-1 Application fields of ionic liquids. © BASF Aktiengesellschaft 2006.

advantages of this new technology are about ten times longer service times (i.e. operation time without maintenance) and 20% less energy demand. Moreover, the apparatus is less complex, since it contains only 8 rather than some 500 moving parts.

Another excellent example of ionic liquids as new intelligent materials is the use of ionic liquids for gas storage. Storage of toxic gases under high pressure in metal cylinders is often unacceptable due to the possibility of leakage or rupture of the cylinder. Air Products has developed an entirely new technology to solve this problem [6]. Ionic liquids are used as liquid support for the storage of gases. To increase the affinity of the gas for the ionic liquid, a Lewis acid-Lewis base interaction is utilized. This allows for high gas storage capacities without the need for pressure.

For all these kinds of new applications the chemical industry as provider of ionic liquids can act as a driving force that fosters ionic liquid-based innovations in many of its customer industries.

9.3
Applications in Chemical Processes

9.3.1
Acid Scavenging: The BASIL™ Process

> The ionic liquid is acting as an **auxiliary**.
> Specific benefits of the ionic liquid are:
>
> - no handling of solids
> - better heat transfer
> - higher chemical yield
> - higher space–time yield
> - lower investment cost
> - higher sustainability of the process.

In 2002 BASF established the first dedicated industrial-scale ionic liquid-based process [7]. The so-called BASIL™ process (BASIL = **B**iphasic **A**cid **S**cavenging utilizing **I**onic **L**iquids) is used for the synthesis of alkoxyphenylphosphines, which are important raw materials in the production of BASF's Lucirines® (Scheme 9-3), substances that are used as photoinitiators to cure coatings and printing inks by exposure to UV light.

HCl is formed during the synthesis of diethoxyphenylphosphine (Scheme 9-4).

Scavenging with a tertiary amine results in a thick, non-stirrable slurry (Fig. 9-2). These problems significantly lower the yield and capacity of the process. In order to provide a minimum of mixing and heat transfer of the exothermic reaction a solvent usually has to be added.

How can ionic liquids (dis)solve the problem? If an acid has to be scavenged with a base, the formation of a salt cannot be avoided, but why not form a liquid salt instead? The idea of BASF researchers was to use an ionic

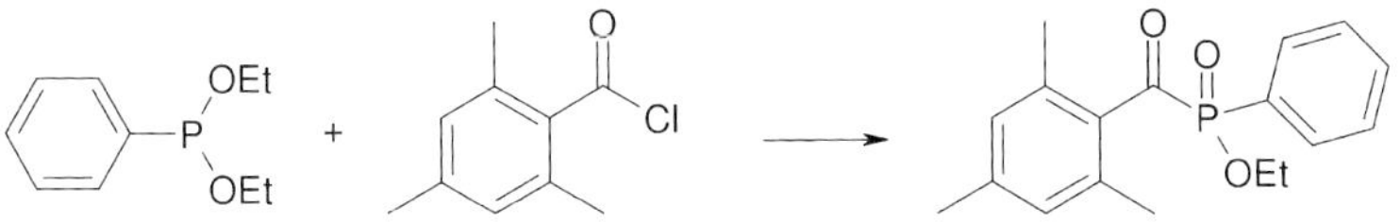

Scheme 9-3

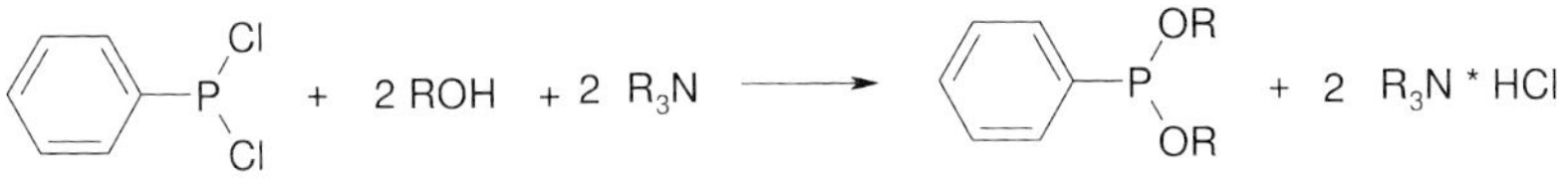

Scheme 9-4

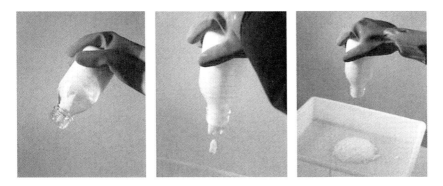

Fig. 9-2 The slurry that is formed when a tertiary amine is used as an acid scavenger. © BASF Aktiengesellschaft 2003.

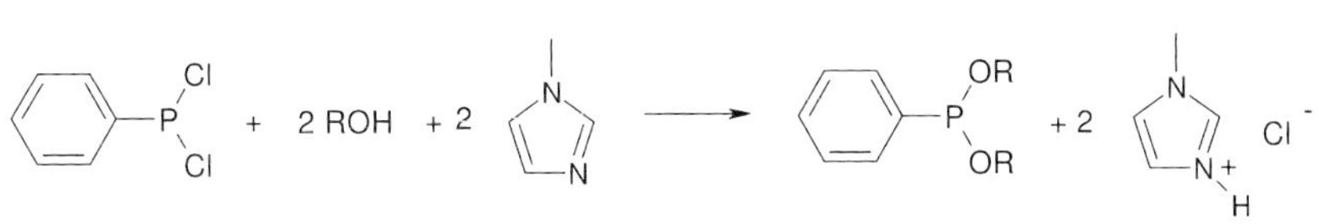

Scheme 9-5

liquid precursor as an acid scavenger. The concept turned out to be successful with the BASF product 1-methylimidazole. After the reaction with HCl an ionic liquid is formed: 1-methyl-imidazolium chloride, which has a melting point of about 75 °C (Scheme 9-5).

During the reaction two clear liquid phases occur (Fig. 9-3) that can easily be separated. The upper phase is the pure product – solvent is no longer needed – the lower phase the pure ionic liquid. 1-methyl-imidazolium chloride as an ionic liquid has a great advantage over the classical dialkylated systems: it can be switched

Fig. 9-3 The BASIL™ process. After the reaction two clear liquid phases are obtained - the upper being the pure product the lower being the ionic liquid [HMIM]Cl. © BASF Aktiengesellschaft 2002.

on and off just by protonation and deprotonation. This is crucial if recycling and purification of the ionic liquids is required.

Further investigations revealed that BASIL™ is not restricted to phosphorylation chemistry but is a general solution to all kinds of acid scavenging [5]. Acylations and silylations have been investigated successfully as well as an elimination reaction. BASIL™ is also applicable to extractive acid removal from organic phases, for example for the purpose of purification. 1-Methylimidazole is doing a perfect job by scavenging the acid. Looking at it more closely, one discovers that 1-methylimidazole also helps in setting the acid free. In other words: it acts as a nucleophilic catalyst [8]. BASF discovered that the phosphorylation reaction is complete in less than a second. Having eliminated the formation of solids and having increased the reaction rate, new reactor concepts were possible. BASF was now able to do the same reaction that was previously done in a large vessel in a little jet reactor, the size of a thumb (Fig. 9-4). In so doing, the productivity of the process has been increased by a factor of 8×10^4 to 690 000 kg m^{-3} h^{-1}. At the end of 2004, BASF successfully started a dedicated BASIL™ plant using this jet stream reactor technology.

Only recently an eco-efficiency analysis has shown that the BASIL™ technology is by far more sustainable than the process using tertiary amines [9].

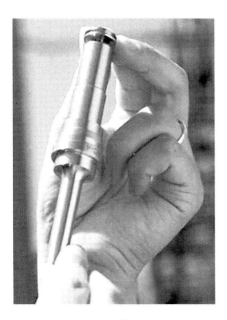

Fig. 9-4 Today the BASIL™ process is run in a small jet reactor that has a capacity of 690 000 kg m^{-3} h^{-1}. © BASF Aktiengesellschaft 2002.

9.3.2
Extractive Distillation

> The ionic liquid is acting as an **entrainer**.
> Specific benefits of the ionic liquid are:
>
> - breaking of azeotropes
> - less energy consumption
> - less equipment (distillation columns)
> - lower investment

In many cases, the formation of azeotropes does not allow the separation of two or more compounds by a simple distillation. Very well known azeotropes with an industrial relevance are, for example, water/ethanol or water/THF. Sometimes the azeotrope can be broken by addition of another compound. These compounds are called entrainers. It was found that ionic liquids work as entrainers for a whole range of azeotropic systems [10]. Very high separation factors can be achieved, especially if water is part of the azeotropic mixture. Ionic liquids are usually hygroscopic materials with a strong affinity to water. Obviously, the interactions between ionic liquid and water are much stronger than those between water and the other component of the azeotrope. The ionic liquids literally grabs the water and releases the second compound, which can be distilled off as a pure material. Figure 9.5 displays the

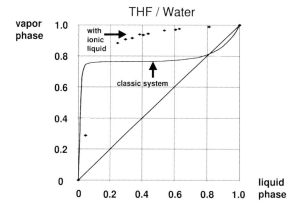

Fig. 9-5 Equilibrium phase diagram for the system THF/water. The solid line shows the classic azeotropic mixture. The dots indicate how the azeotrope has been broken after addition of the ionic liquid. The amount of THF in the vapor phase is always greater than in the liquid phase. © BASF 2004.

classic vapor liquid diagram for THF/water. The dots indicate the change after the addition of ionic liquid.

To afford a sufficient separation the amount of ionic liquid added has to be in the range of 30 to 50 wt.%. This sounds like a large percentage, but is an even smaller quantity than that usually needed with a classic entrainer like dimethylformamide (DMF). In all cases the entrainer has of course to be recycled. Figure 9.6 shows a generic flow chart for the ionic liquids-based process.

The advantage of an ionic liquid over a classic entrainer is that ionic liquids have negligible vapor pressure. This means that the entrainer itself has not to be distilled hence a second separation column is not necessary and a lot of energy can be saved. A benchmark calculation has revealed a saving potential of ca. 37% for energy cost and 22% for the investment. BASF has run an extractive distillation process in a pilot plant continously for 3 months. Although the ionic liquid faced a severe thermal treatment of about 250 °C in the recycling step its performance remained unchanged, without the need for a purge. This again underlines how thermally stable some ionic liquids are and that extremely high recycling rates are possible without a decrease in performance.

9.3.3
Chlorination with "Nucleophilic HCl"

The ionic liquid is acting as a **solvent**.
Specific benefits of the ionic liquid are:

- HCl substitutes for phosgene
- high selectivity at high conversion

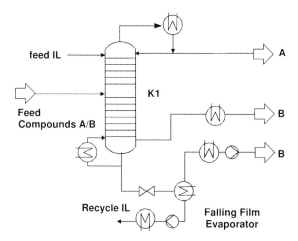

Fig. 9-6 Generic flow chart for the ionic liquid-based process. The ionic liquid is acting as an extractant, washing down compound B. Compound A is released and can be distilled as a pure compound at the head of the column. B is removed at the bottom. The ionic liquid is recycled into the process after having removed residual B in a falling film evaporator.
© BASF 2004.

Industrial chlorinating agents for alcohols are phosgene, $SOCl_2$, PCl_3 and PCl_5. Phosgene is a cheap raw material and usually gives excellent yields. However, the handling of phosgene requires enormous safety efforts. In principle, the chlorination of alcohols can also be achieved by reaction with HCl gas. Unfortunately, in the case of diols, conversion is usually less than 100% and the reaction stops after formation of cyclic or open-chain ethers (Scheme 9-6).

Surprisingly the reactivity profile completely changes when the reaction is performed in an ionic liquid. Obviously either the nucleophilicity of HCl or the nucleofugicity of the leaving group (water) is drastically increased. Apparently the ether side products are cleaved, allowing further reaction to the desired bischlorinated product. This type of chemistry might be similar to what Eli Lilly has described for the cleavage of aromatic methoxy ethers in an pyridinium hydrochloride melt [11]. Figure 9-7 shows how selectivities improve when the chlorination of butanediol is performed in an ionic liquid rather than in an organic solvent [12].

Butanediol is completely soluble in the ionic liquid. When the reaction proceeds a second organic phase occurs which consists of the reaction product

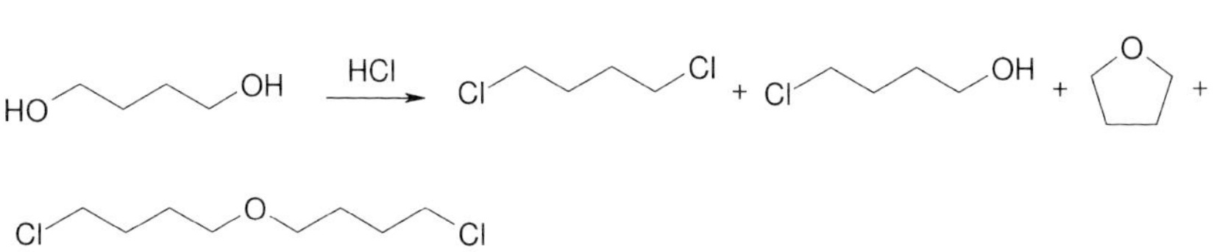

Scheme 9-6

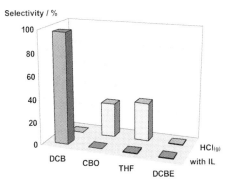

Fig. 9-7 Higher selectivities for the product 1,4-dichlorobutane (DCB) are achieved if the chlorination of the butanediol with HCl gas is performed in an ionic liquid rather than in the pure alcohol. Side products are usually ethers (tetrahydrofuran, THF, dichlorobutyl ether, DCBE) or the monochlorinated product 1-chlorobutane-4-ol (CBO). © BASF 2004.

1,4-dichlorobutane. After complete conversion the organic phase is just separated off. Water which is formed during the reaction is distilled from the remaining ionic liquid which can be used for the next run without any further work-up. HCl is by far the cheapest chlorinating agent. It is widely available, can be handled in standard equipment and is fully consumed in the reaction. With $SOCl_2$ and phosgene only one chlorine atom can be utilized for the chlorination reaction. SO_2 and HCl, and CO_2 and HCl are formed, respectively, as gaseous by-products. The heavy evolution of gaseous by-products can limit the space-time yield in large-scale production.

9.3.4
Cleavage of Ethers

> Ionic liquid is reacting as a **catalyst/reagent**
> Specific benefits of the ionic liquid are:
>
> - HCl can be used as a cheap cleaving agent of aromatic methoxy groups
> - lower cost

Eli Lilly has published the de-methylation of an aromatic methoxy group in ionic liquids [9]. Usually the cleavage of aromatic methoxy ethers requires very harsh conditions and inconvenient reagents, such as BBr_3 [13] or HBr in boiling acetic acid [14]. 4-Hydroxyphenylbutyric acid has been produced by Eli Lilly on a 190 L pilot scale with yields of >94% by reacting 4-methoxyphenylbutyric acid for 2 h at 210 °C in the ionic liquid pyridinium hydrochloride ([PyH]Cl) (Scheme 9-7).

Even if [PyH]Cl with a melting point of 153 °C [15] is somewhat outside the current definition of ionic liquids (T_m <100 °C) this example shows how reactivity,

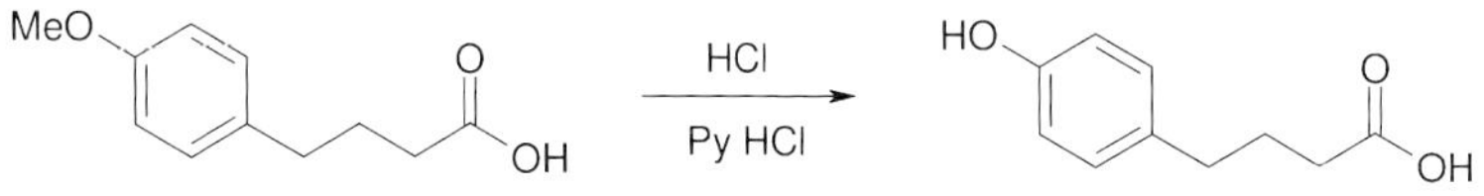

MeO—⟨aryl⟩—CH₂CH₂CH₂C(=O)OH →(HCl, Py HCl)→ HO—⟨aryl⟩—CH₂CH₂CH₂C(=O)OH

Scheme 9-7

in this case of HCl, can be dramatically increased in the ionic environment. This is in-line with BASF's observations that ether by-products are easily cleaved by HCl gas in ionic liquids during the chlorination of alcohols [12].

9.3.5
Dimerization of Olefins

> Ionic liquid is acting as a **solvent**.
> Specific benefits of the ionic liquid are:
>
> - higher catalyst activity
> - higher catalyst stability
> - higher selectivity

The Institut Francaise du Petrol (IFP) was the first to develop an ionic liquid-based process to a pilot scale. This is the DIFASOL technology which can be used for the dimerization of olefins. This process is described in detail in Section 5.4 of this book.

9.3.6
Oligomerization of Olefins

> Ionic liquid is acting as a **catalyst**.
> Specific benefit of the ionic liquid is:
>
> - provides product with a unique viscosity profile useful for application as a lubricant.

Chevron Phillips has invented a process utilizing an acidic ionic liquid for olefin oligomerization [?]. The process produces synthetic lubricant base oil which is used in a variety of lubricants including gear oils, greases and automotive engine lubricants. In 2004, the pilot process produced commercial quantities of over 450 tons of polyalphaolefins. Synthetic lubricants are of considerable interest due to tightening lubricant industry specifications. Polyalphaolefins (PAOs) represent the largest sector of the synthetic lubricants market, which is split between low (2–

10 cSt at 100 °C) and high viscosity ranges. PAOs are carefully designed synthetic oils that are highly branched isoparaffins, oligomerized from alpha olefins. Typically, the alpha olefin of choice is 1-decene.

CPChem's Synfluid® PAO 25 cSt, which is produced with the ionic liquid catalyst has a viscosity at 100 °C, which is unique to the industry. Prior to this product, a 25 cSt PAO was only available commercially by blending higher and lower viscosity PAOs. The specific applications for this product are typically gear oils, greases and various industrial lubricants.

9.3.7
Hydrosilylation

> Ionic liquid is acting as a **catalyst**.
> Specific benefits of the ionic liquid are:
>
> - recovery and reusability of the catalyst phase
> - shorter reaction times due to higher catalyst loadings
> - improvement in product quality.

The hydrosilylation reaction is a widely used method for the synthesis of organo-modified silanes and siloxanes. The addition of Si–H to C–C double bonds is usually catalyzed by homogenous or colloidal Pt catalysts. The major drawback of this reaction is that the catalyst cannot easily be removed from the product after completion of the reaction. To lower the losses of precious metal catalysts, the amount of catalyst is usually decreased, but this decreases reaction speed. Degussa has managed to develop an elegant biphasic approach to run the hydrosilylation reaction [17] (Scheme 9-8).

The catalyst is now dispersed in the ionic liquid phase, from which the pure product separates as a new liquid phase that can be easily decanted after the reaction. The ionic liquid catalyst phase is still active and can be reused. It is important that standard hydrosilylation catalysts can be used without further modification. Degussa has been running this process on a pilot scale achieving conversions of >99%. In most cases the detectable platinum content of the products was <1 ppm.

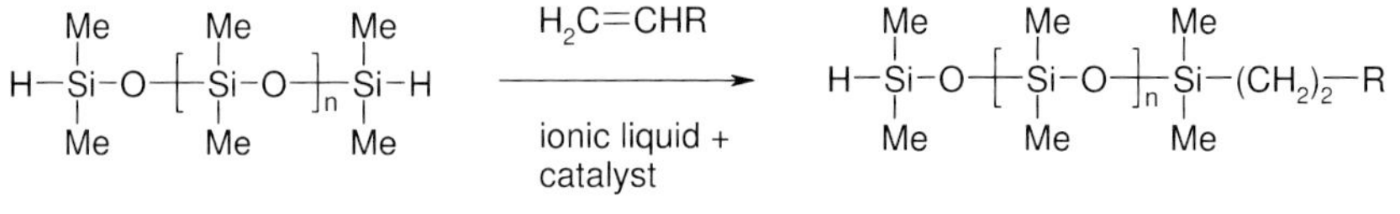

Scheme 9-8

9.3.8
Fluorination

> Ionic liquid is acting as a **catalyst**.
> Specific benefits of the ionic liquid are:
>
> - higher catalyst activity
> - higher stability of the catalyst to reductive deactivation
> - avoidance of chlorine co-feed
> - higher selectivities towards perfluorinated products

Fluorinated hydrocarbons are used as refrigerants in the air-conditioning and refrigeration industry. Chlorofluorocarbons are being phased out according to the Montreal protocol due to their ozone depletion potential. This is mainly attributed to the chlorine content. Hence chlorofluorocarbons will be replaced by chlorine-free hydrofluorocarbons (HFCs). HFCs can be manufactured from chlorinated hydrocarbons by reacting them with HF in order to achieve a chlorine/fluorine substitution. The state-of-the-art catalyst for this reaction is $SbCl_5$. However, this catalyst suffers from a reductive deactivation with the formation of Sb(III) species. To overcome the loss of catalyst, chlorine is co-fed for re-oxidation of Sb(III) to Sb(V). This is technically feasible, but consumes chlorine as a raw material and leads to back-chlorination of the fluorinated products, and hence to significantly lower yields. Arkema has demonstrated at a pilot stage that imidazolium-based ionic liquids with a $[SbF_6]^-$ ion can eliminate the deactivation problem [18] (Fig. 9-8). The pilot reaction has been running for more than 400 h, exceeding selectivities of 99.5%, without any noticeable deactivation.

9.4
Applications in Electrochemistry

9.4.1
Electroplating of Chromium

> Ionic liquid is acting as an **electrolyte**.
> Specific benefits of the ionic liquid are:
>
> - usage of less toxic Cr(III) salts rather than highly toxic Cr(VI) as raw material for chromium plating
> - reduced power consumption

The deposition and dissolution of metals is a multi-billion dollar industry that is almost totally based on aqueous acids and alkalis. The use of water limits the metals

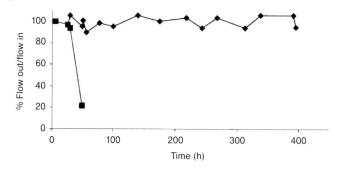

Fig. 9-8 Liquid-phase fluorination of trichloroethylene (TCE), continuously run in the absence of Cl2 feed (◆ $Q^+Cl^- + 2\,SbF_5$; ■ $SbCl_5$) [186].

that can be deposited and produces large volumes of aqueous waste. The processes are often hindered by low current efficiencies.

One of the largest sectors of the electroplating market is that of chromium deposition. The major disadvantage of the current process of chromium plating is that it requires the use of chromic acid-based electrolytes comprising hexavalent chromium. The toxicity and carcinogeneity associated with Cr(VI) has resulted in wide-ranging environmental legislation in the USA and Europe to reduce its use. Other disadvantages of the existing technology are economic, such as the low current efficiency for the reduction of Cr(VI) in acid media. Furthermore, the difference in overpotential between chromium and hydrogen reduction results in the evolution of hydrogen gas, which can lead to embrittlement in the substrate, thus reducing quality and yield.

Scionix, a joint venture company between the University of Leicester and Whyte Chemicals Ltd, has developed an ionic liquid containing Cr(III) salts [19], which are significantly less toxic than the carcinogenic Cr(VI) species. The process also operates with >90% current efficiency, resulting in significantly reduced power consumption and making the ionic liquid technology a more environmentally benign form of plating [20]. Moreover, since these are not aqueous solutions, there is negligible hydrogen evolution. Hence essentially crack-free, highly corrosion resistant deposits are obtained. The process has operated on a 50 L pilot scale and is currently being developed by a number of industrial partners.

9.4.2
Electropolishing

Ionic liquid is acting as an **electrolyte**.
Specific benefits of the ionic liquid are:
- high current efficiency
- improved surface finish
- improved corrosion resistance

Electropolishing of stainless steel is an effective way of increasing corrosion resistance and decreasing wear, in addition to the obvious esthetic benefits. Current electropolishing technology primarily uses sulfuric and phosphoric acids mixtures. These are naturally corrosive, harmful to work with and must be neutralized before disposal. Acid-based electropolishing is an inherently inefficient process: only 10–20% of the energy supplied is utilized for metal dissolution. The scale of this activity worldwide represents a significant environmental concern.

Scionix has developed an alternative concept to forming eutectic-based ionic liquids which is to complex the anion of choline chloride with a hydrogen-bonding compound rather than a metal halide [21, 22]. The ionic liquids allow electropolishing with high current efficiency (>80%), improved surface finish and improved corrosion resistance [23].

Scionix is currently developing a commercially viable medium-to-large-scale electropolishing plant using these novel electrolytes in collaboration with UK-based Anopol Ltd. The new process significantly reduces the total volume of effluent and its toxicity. When the metal, usually iron, is electrochemically oxidized it complexes with the hydrogen-bond donor and the insoluble complex precipitates to the base of the cell. The cell liquid can be filtered periodically and the metal complex collected. This allows for suitable recycling of both the ionic liquid and the metal, preventing environmental emissions. The process is currently running in a 1200 L demonstrator module in Birmingham, UK (Fig. 9-9).

9.5
Applications as Performance Chemicals and Engineering Fluids

9.5.1
Ionic Liquids as Antistatic Additives for Cleaning Fluids

> Ionic liquid is acting as **cleaning agent, antistatic**.
> Specific benefit of the ionic liquid is:
>
> - enabling electric conductivity without formation of solid residues.

The cleaning of high value surfaces, e.g. in the automotive, furniture or electronic industry, is a challenging problem, in particular if small charged particles have to be removed. Wandres Micro-Cleaning GmbH has developed a cleaning-system that uses moistened instead of dry filaments (Fig. 9-10).

The liquid film (water) is brought onto the filament with the help of very small, micrometer-sized water droplets. These droplets are generated in a Venturi nozzle (Fig. 9-10). To avoid electrostatic charging of the surface, a supporting salt is usually added that facilitates electrical conductivity. If sodium chloride is used as a support-

Fig. 9-9 1200 L demonstrator module for electropolishing. © Scionix 2005.

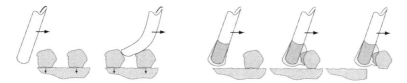

Fig. 9-10 Removal of dust particles by the filament of a brush. The removal is more efficient if the filament is coated with a liquid film. © Iolitec 2005.

ing salt a solid is precipitated in the nozzle leading to encrusting and blocking of the system (Fig. 9-11).

Ionic liquids can offer a unique solution to this problem, since as *liquid* salts they can provide electrical conductivity without precipitation of a solid, hence without formation of deposits. The ionic liquid solution to this problem was developed and established by Iolitec in cooperation with Wandres Micro-Cleaning GmbH.

9.5.2
Ionic Liquids as Compatibilizers for Pigment Pastes

> Ionic liquid is acting as a **compatibilizer**.
> Specific benefits of the ionic liquid are:
>
> - stabilize pigments in pigment pastes
> - provide truly universal water-based pigment pastes suitable for water- *and* solvent-based paints and coatings.

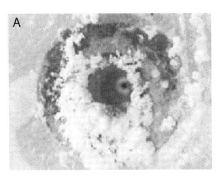 

Fig. 9-11 Venturi nozzle for the generation of small water droplets. If the water contains 1 wt.% of sodium chloride then after 10 h encrustation occurs. By using an ionic liquid as supporting salt, solid is no longer deposited. © Iolitec 2005.

Up to now only very little attention has been paid to ionic liquids as performance chemicals. Degussa has managed to develop an application of ionic liquids as so-called compatibilizers for pigment pastes [24].

Paints and coatings can be tinted by adding a small amount of a color concentrate – the pigment paste – to a white base paint. The advantage is clear. One can derive more or less any colour from the same base paint formulation, just by adding the right mix of pigments. To achieve a homogenous and stable coloring, the paste needs some additives which, for example, prevent the pigments from sedimentation or flocculation. For environmental reasons, it is preferable to use water-based pigment pastes. However, the water-based systems cannot be used universally for both water- and solvent-based paints. This problem has been solved by addition of ionic liquids to the pigment paste. Now the pigments are stable in both water-based paints and in solvent-based ones. The stability of the pigments can be demonstrated by the "rub-out" test (Figs. 9-12 and Fig. 9-13).

9.5.3
Ionic Liquids for the Storage of Gases

> Ionic liquid is acting as a **liquid support**.
> Specific benefits of the ionic liquid are:
>
> - storage of hazardous gases possible without pressure
> - Higher safety.

The semiconductor manufacturing industry uses a number of hazardous specialty gases such as phosphine (PH_3), arsine (AsH_3) and boron trifluoride (BF_3) for doping, etching, and thin-film deposition. These gases are highly toxic and py-

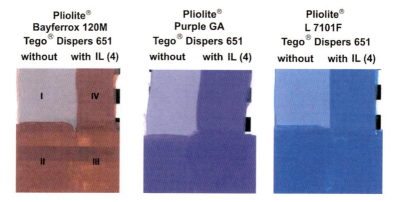

Fig. 9-12 Rub-out test: draw downs of Pliolite® tinted with universal colorants based on Tego® Dispers 651 adjusted with ionic liquid Tego® Dispers 662C. © Degussa 2005.

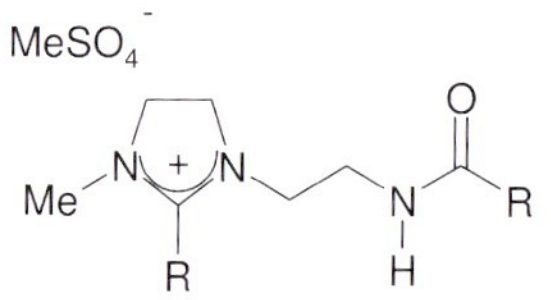

Fig. 9-13 Tego® Dispers 662C.

rophoric. Therefore storage and handling is challenging and requires enormous safety efforts. For example, storage of toxic gases under high pressure in metal cyclinders is often unacceptable due to the possibility of leakage or rupture of the cylinder. To overcome the risks of high-pressure storage, these gases are often stored under low pressure by adsorption on solid supports such as zeolites [25].

Disadvantages of this technology include: low capacities, delivery limitations and low thermal conductivity. Air Products has developed an entirely new technology to solve this problem [6]. Ionic liquids are used as a liquid support for the storage of gases. To increase the affinity of the gas to the ionic liquid a Lewis acid-Lewis base interaction is utilized. This allows high gas storage capacities without the need for pressure. PH_3 for example is a Lewis base. Accordingly, for its absorption a Lewis acidic ionic liquid such as [BMIM][Cu_2Cl_3] is used. BF_3 in turn is a Lewis acid. In this case a Lewis basic ionic liquid like [BMIM][BF_4] is used as liquid support (Scheme 9-9).

The customer can release the gas from the pressure-less cylinder either by applying vacuum or by increasing the temperature. The storage capacities for PH_3 and BF_3 in different supports are summarized in Table 9-2.

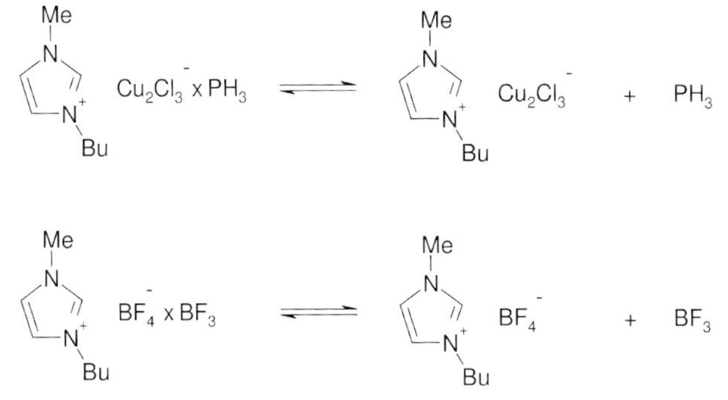

Scheme 9-9

Table 9-2 Storage capacities for PH_3 and BF_3 in different supports

Gas	Support	Temperature	Load (mol L^{-1})	Reversibility (%) (pressure (Torr))	Working capacity (mol L^{-1})
PH_3	[BMIM][Al$_2$Cl$_7$]	RT	1.8	89 (20–760)	1.6
PH_3	[BMIM][AlCl$_4$]		0.06		
PH_3	[BMIM][Cu$_2$Cl$_3$]	RT	7.6	84 (736)	6.4
PH_3	[BMIM][CuCl$_2$]	RT	1.4		
PH_3	CH$_3$SO$_3$H	RT	0.86	75 (20–514)	0.6
PH_3	CF$_3$CO$_2$H	RT	5.3	0 (20–721)	0
PH_3	TiCl$_4$	12 °C	13.8	41 (44–428)	5.6
PH_3	zeolite 5A			66 (20–710)	1.9
BF_3	[BMIM][BF$_4$]	RT	5.2	70 (20–724)	3.6
BF_3	tetraglyme		12.3		0

9.6
FAQ – Frequently Asked Questions Concerning the Commercial Use of Ionic Liquids

9.6.1
How Pure are Ionic Liquids?

While the purity of ionic liquids has already been discussed from the synthetic perspective in Section 2.2, the following section aims to highlight this aspect from the viewpoint of an industrial company aiming to apply ionic liquids in its processes and products. What exactly is purity? Most people would define purity by the actual content of the desired compound expressed in weight percent. Already, this is not an easy thing to do with ionic liquids. As salts they intrinsically consist of two compounds, a cation and an anion. For example a sample of [EMIM]Cl might contain 5 wt.% of [EMIM][HSO$_4$] as impurity. This means the ionic liquid is 100 wt.% pure in terms of the cation [EMIM]$^+$ and 95 wt.% pure in terms of the anion chloride.

Weight percent is only one of many possible definitions of purity. A catalysis chemist would define purity as being free of any coordination species, like halides that deactivate the metal catalyst by formation of stable complex compounds. An electrochemist would define purity as having no oxidizable impurities that reduce the electrochemical window. An engineer might prefer not to have impurities that affect the viscosity and finally the end-user will define purity as being free of residual potentially toxic alkylating agents. These examples show that the targeted application alone can define what purity means. In the end, ionic liquids do not provide purity, but performance.

9.6.2
Is the Color of Ionic Liquids a Problem?

Colored materials are quite often perceived as being impure. In fact, most ionic liquids themselves are colorless liquids. However, they tend to become colored, especially during prolonged thermal treatment. The good news is that the color stays persistently in the ionic liquid and usually cannot be extracted in organic products or solvents. Until today nobody has managed to isolate the colorant because the quantitites are just too low. It is assumed that in the case of imidazolium-based ionic liquids oligomers of the imidazole or even radical ions might cause the color. For commercial applications the color is usually not a problem, since the actual reaction products are not affected. For example BASF has been running a pilot plant for an extractive distillation process for 3 months continuously. The color of the ionic liquid turned black, but the performance stayed constant without the need for any purge. During the whole run the product was colorless, clear and completely within the required specification. For this application it would have been a waste of money to start with a colorless and, hence higher price, ionic liquid. Moreover, a colorless ionic liquid would have turned black under the required high temperatures anyway.

9.6.3
How Stable are Ionic Liquids?

In most cases ionic liquids show remarkably high thermal stabilities of $>200\,°C$. A decomposition pathway of the usually very stable imidazolium-based ionic liquid is the back alkylation of the anion (Scheme 9-10).

The temperature at which this reaction occurs most likely depends on the nucleophilicity of the anion. It turned out that onset measurements from DSC alone are

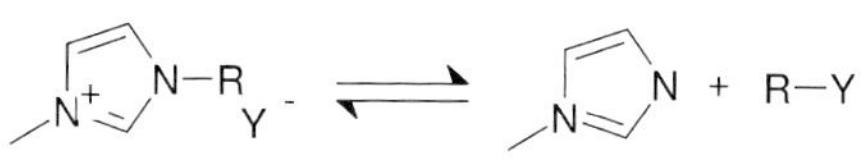

Scheme 9-10

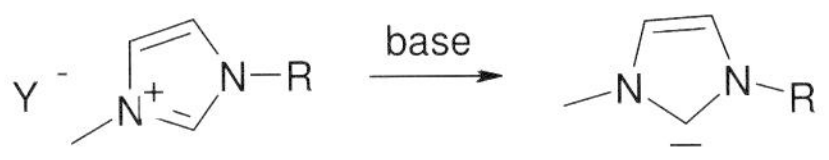

Scheme 9-11

not suitable to determine the thermal stability. A more valuable indication is provided by TGA analysis which shows the loss of weight due to the distillation of the volatile alkylating agents or other decomposition products. For [EMIM][CH$_3$COO] the temperature at which 10% loss of weight is observed is, for example, 215 °C [26]. [EMIM][CH$_3$SO$_3$] or [EMIM][EtOSO$_3$] show higher thermal stabilities, with corresponding decomposition temperatures of 330 °C [26]. Under basic conditions imidazolium-based ionic liquids tend to form carbenes (Scheme 9-11), which can undergo further decomposition, such as irreversible disproportionation.

9.6.4
Are Ionic Liquids Toxic?

There is no way to answer this question in a general way. Since ionic liquids can consist of so many chemically different types of cations and anions this question has to be answered case by case. The "magic ionic liquid" which meets all requirements in toxicity, ecotoxicity, stability and performance simply does not exist. Up to now there is still very limited information available regarding a full toxicological profile of ionic liquids. However, recently some valuable data have been published [27]. These data were collected for the notification process of the corresponding ionic liquids and represent examples from the most common classes of cations: EMIM, BMIM and ammonium (Table 9-3).

Table 9-3

	BMIM Cl[a]	EMIM EtOSO$_3$[b]	MTEOA MeOSO$_3$[c]
Acute oral toxicity	toxic	not harmful	not harmful
Skin irritation	irritant	non-irritant	non-irritant
Eye irritant	irritant	non-irritant	non-irritant
Sensitization	non-sensitizing	non-sensitizing	non-sensitizing
Mutagenicity	non-mutagenic	non-mutagenic	non-mutagenic
Biological degradability	not readily degradable	not readily degradable	readily biodegradable
Toxicity to daphniae	autelytoxic	acutely not harmful	acutely not harmful
Toxicity to fish	acutely not harmful	—	acutely not harmful

[a] BMIM Cl = 1-Butyl-3-methylimidazolium chloride.
[b] EMIM EtOSO3 = 1-Ethyl-3-methylimidazolium ethylsulfate.
[c] MTEOA MeOSO3 = Tris-(2-hydroxyethyl)-methylammonium methylsulfate.

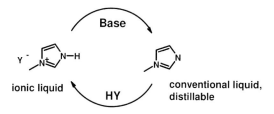

Scheme 9-12

These examples show that ionic liquids can be toxic as well as rather harmless and biodegradable. However, the harmless ionic liquids might not necessarily also provide the best performance. In the end, a balanced decision between toxicological and performance properties has to be made to gain the best fit to the targeted application. Some ionic liquids might only be suitable for being handled by skilled personnel in chemical plants, others for a use closer to the end-user will have to be, of course non-toxic, and readily biodegradable.

9.6.5
Are Ionic Liquids Green?

Besides their technical and economical potential, ionic liquids are often claimed to be green solvents or green materials. The "greenness" has often been justified with an important property being that ionic liquids have negligible vapor pressure, hence are not released to the environment by evaporation. Certainly, there might be applications where exactly this property can help to make a particular process greener than a corresponding process with organic solvents. However, this is not necessarily, and not generally, the case. It might be misleading to assume that non-volatility alone makes a material "green." A more precise approach would evaluate the whole process "from cradle to grave." This obviously includes the manufacture of raw materials as well as the final product. For example, the energy consumption in every step has to be considered as well as emissions. This can be achieved by performing a so-called eco-efficiency analysis, in order to evaluate which process from a set of possible alternatives is the most sustainable one. This analysis has been done for the ionic liquid-based BASIL™ process, which showed it to be more sustainable with regard to both economics and environment [7].

9.6.6
How Can Ionic Liquids be Recycled ?

Recycling of ionic liquids is easy, if protonated cations are used. In this case the ionic liquids can be switched off by deprotonation (Scheme 9-12). The resulting amine or imidazole is a conventional neutral molecule that can be distilled for recycling or purification purposes.

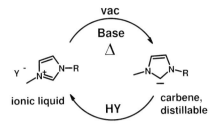

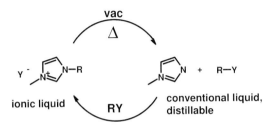

Scheme 9-13

It is more difficult with alkylated cations. Apart from purification or recycling by a liquid-liquid extraction two principal "distillation" methods have been reported. The first is the formation of distillable carbenes [28] the second is the back-alkylation of the anion [29] (Scheme 9-13).

Imidazolium cations can be deprotonated by bases to form neutral carbenes. These carbenes are surprisingly stable and can be distilled. The ionic liquid can be recycled by further reaction of the carbene with an acid.

The controlled decomposition reaction allows for a recycling or purification process of the ionic liquid [29]. In this case the ionic liquid is thermally cleaved. The neutral imidazole and alkylating agent are distilled, collected and re-reacted.

However, in some cases it is even possible to distill the entire ionic liquid without decomposition. [HMIM]Cl, for example, can be distilled at 150 °C at 0.5 mbar [30]. This has recently been demonstrated to be true also for fully alkylated cations [31]. These materials require higher temperatures of about 300 °C at 6 mbar.

9.6.7
How Can Ionic Liquids be Disposed Of?

Ionic liquids are frequently reported as being non-flammable. This is only true up to the temperatures at which decomposition takes place. Some ionic liquids already start to decompose at 120 °C, others are stable up to nearly 400 °C. Upon decomposition neutral and volatile molecules are formed which of course can burn. This explains why ionic liquids do indeed have flash points, even if they are usually much higher than 100 °C. However, ionic liquids can easily be disposed of by incineration, which is usually done at temperatures above 600 °C. At these very high temperatures even the toughest organic ionic liquids will give up.

9.6.8
Which is the Right Ionic Liquid?

The answer is, in principle, easy: the right ionic liquid has the best fit to the requirements of the targeted application. But how can the "best fit" be described? The following criteria might be a guideline to address this question:

- Best fit to required physical parameters (viscosity, melting point etc.)
- Best fit to thermal stability requirements (RT, < 150 °C, >150°C, >300°C etc.)
- Best fit to chemical properties (inert, catalytically active, shifting equilibria etc.)
- Best price/performance ratio ("simple" ionic liquid or highly functionalized etc.)
- Best fit to toxicology requirements (non-toxic, biodegradable etc.)
- Best fit to notification requirements (EINECS listed or new compound)

Regarding prices for ionic liquids in commercial large-scale applications, it is not recommended to consider current prices for lab-scale quantities in order to prefer one liquid over another. Even if commonly stated in the literature, it is not true that, for example, "imidazolium" is generally speaking the most expensive cation. If the required quantities grow to a certain level imidazolium, pyridinium, ammonium and phosphonium will, more or less, end up at a very similar level of cost. For a good choice it is much more important to consider the performance of the relevant cations. Surprisingly the anions are quite often ignored in price discussions. However, they can influence the bulk-scale price level even more significantly. Availability of ionic liquids in bulk quantities is not an issue. They can be purchased from several different manufacturers today.

References

1. *Industrielle Organische Chemie*, K. Weissermel, H.-J. Arpe (Eds.), Wiley-VCH, Weinheim, 5th edn., **1998**, pp. 369–373.
2. EP 367 050 (**1989**), to BASF; DE 40 28 774 (**1990**), to BASF; EP 452 806 (**1991**), to BASF.
3. Stephen Falling (Eastman), oral presentation, COIL meeting, 19–22. July **2005**, Salzburg.
4. (a) J. S. Wilkes, J. A. Levisky, R. A. Wilson, C. L. Hussey, *Inorg. Chem.* **1982**, *21*, 1263; (b) M. Freemantle, *Chem. Eng. News* **1998**, *76*, 32; (c) T. Welton, *Chem. Rev.* **1999**, *99*, 2071; (d) P. Wasserscheid, W. Keim, *Angew. Chem., Int. Ed.* **2000**, *39*, 3772.
5. M. Kömpf, *Linde Technology*, January 2006, pp. 24–29; http://www.linde.com/international/web/linde/likelindeeng.nsf/docbyalias/nav_technology.
6. US 04/0206241, to Air Products.
7. WO 03/062171; WO 03/062251; WO 05/061416 (BASF AG); M. Maase, *Chem. unserer Zeit*, **2004**, 434; M. Freemantle, *Chem. Eng. News 81*, 9; R. D. Rogers, K. R. Seddon, *Nature Mater.* **2003**, *2*, 363; K. R. Seddon, *Science* **2003**, *302*, 792–793.
8. J. Chojnowski, M. Cypryk, W. Fortuniak, *Heteroat. Chem.* **1991**, *2*, 63–70.

9. M. Maase (BASF), oral presentation, COIL meeting, 19–22. July **2005**, Salzburg. www.oeea.de.
10. WO 02/074718, to BASF; C. Jork, M. Seiler, Y.-A. Beste, W. Arlt, *J. Chem. Eng. Data* **2004**, *49*(4), 852; WO 05/016484, to BASF.
11. C. R. Schmid, C. A. Beck, J. S. Cronin, M. A. Staszak, *Org. Process Res. Dev.* **2004**, *8*, 670.
12. WO 05/026089, to BASF.
13. (a) J. F. W. McOmie, M. L. Watts, D. E. West, *Tetrahedron* **1968**, *24*, 2289; (b) Boron Tribromide, in *Electronic Encyclopedia of Reagents for Organic Synthesis*, L. Paquette (Ed.), John Wiley & Sons, New York.
14. (a) D. Papa, E. Schwenk, H. Hankin, *J. Am. Chem. Soc.* **1947**, *69*, 3018; (b) C. S. Yi, L. C. Martinelli, C. D. Blanton, Jr., *J. Org. Chem.* **1978**, *43*, 405.
15. D. Mootz, J. Hocken, *Z. Naturforsch.* **1989**, *44*(10), 1239.
16. U.S. Patent 6,395,948, to Chevron Phillips.
17. (a) B. Weyerhausen, K. Hell, U. Hesse, *Green Chem.* **2005**, *5*, 283; (b) EP 1382630, to Goldschmidt.
18. (a) WO 01/81353, to Arkema; (b) P. Bonnet in B. Comils et al. (eds) "Multiphase Homogeneous Catalysis", Wiley-VCH, Weinheim, **2005**, pp. 535–542.
19. A. P. Abbott, G. Capper, D. L. Davies, R. K. Rasheed, *Chem. Eur. J.* **2004**, *10*, 3769–3773.
20. A. P. Abbott, G. Capper, D. L. Davies, R. K. Rasheed, J. Archer, C. John, *Trans. Inst. Met. Fin.* **2004**, *82*, 14–16.
21. A. P. Abbott, G. Capper, D. L. Davies, R. K. Rasheed, V. Tambyrajah, *Chem. Commun.* **2003**, 70–71.
22. A. P. Abbott, D. Boothby, G. Capper, D. L. Davies, R. K. Rasheed, *J. Am. Chem. Soc.* **2004**, *126*, 9142–9147.
23. A. P. Abbott, G. Capper, B. G. Swain and D. A. Wheeler, *Trans. Inst. Met. Fin.* **2005**, *83*, 51–53.
24. B. Weyerhausen, K. Lehmann, *Green Chem.* **2005**, *1*, 15–19; EP 1566413 (Goldschmidt)
25. US 4744221, to Olin Corp..
26. Measurements by BASF for different BASIONICS^{TM}.
27. M. Maase (BASF), oral presentation, COIL meeting, 19–22. July **2005**, Salzburg; BASF Material Safety Data Sheets; www.basionics.com
28. WO 01/77081, to QUILL; WO 05/019183, to BASF.
29. WO 01/15175, to British Nuclear Fuels plc; DE 10002420, to BASF.
30. WO 2005068404, to BASF AG.
31. M. J. Earle, J. M. S. S. Esperanca, M. A. Gilea, J. N. Canongia Lopes, L. P. N. Rebelo, J. W. Magee, K. R. Seddon, J. A. Widegren, *Nature* **2006**, *439*, 831.

10
Outlook

Peter Wasserscheid and Tom Welton

It was our intention to provide in the nine previous chapters the essential information for a modern understanding of the nature of ionic liquids as well as a review of all different synthetic applications that have benefited from ionic liquid technology so far. In the last five years – since the first edition of this book appeared – the field of ionic liquid chemistry has advanced greatly. Much insight has been added, many new promising applications have appeared but also some of the paradigms have had to be modified.

Let us review a little the predictions for the future that we made in the Outlook chapter of the first edition in Spring 2002. In the first edition of this book we tried to answer the questions we were most commonly asked when telling people about ionic liquids for the first time. Let us compare the old answers and prophesies in 2002 with today's situation. One question at the time was:

What is going to be the first area of broad, commercial ionic liquid application? In 2002, this question had to be answered in very vague and general terms. Apart from some information on a humidity sensor commercialized by Novasina S.A. there was no published example of any ionic liquid process, application or device. As Chapter 9 very clearly illustrates, this situation has fundamentally changed only four years later. The list of published examples is quite impressive and many ionic liquid experts in industry and in academia know about more examples that have not been made accessible to the general public for different reasons. Apart from this significant number of existing technical applications, an even greater variety of very promising development projects of different economic impact are in the pipeline of both academic groups and industrial companies around the world. To give only one example of potentially very high relevance we would like to mention here the joint research efforts by Robin Rogers' group/University of Alabama together with BASF to commercialize a new cellulose processing principle based on ionic liquids to replace the existing CS_2 technology [1]. If the so far presented advantages of the new ionic liquid-based technology hold true we will soon see a number of new biopolymer treatment plants with an ionic liquid hold-up in the range of several thousand tons per plant! There is no doubt that such a development will be a major influence in the field of ionic liquid chemistry in general.

Ionic Liquids in Synthesis, Second Edition. P. Wasserscheid and T. Welton (Eds.)
Copyright © 2008 WILEY-VCH Verlags GmbH & Co. KGaA, Weinheim
ISBN: 978-3-527-31239-9

In 2002 we predicted that non-synthetic applications would have a great chance to be among the first technical ionic liquid applications. This assumption has held true as Chapter 9 clearly proves. Non-synthetic applications are particularly attractive due to their often much shorter development times. Usually, the improvement over existing technology is only based on one or very few specific properties of the ionic liquid, whereas, for most synthetic applications a complex mixture of physicochemical properties in dynamic mixtures has to be considered. This point is well illustrated by the fact that all liquid–liquid biphasic catalysis involves both a reaction and an extraction step. Hence, the ionic liquid catalyst solution has to fulfil at the same time all of the requirements to work as a superior reaction medium as well as its role as a suitable extraction medium. The result is a significantly more complex set of material requirements which prolongs the specific ionic liquid development and testing times.

From this reasoning we can expect also for the years to come that non-synthetic applications will have a significant share of newly arising technical ionic liquid applications.

Table 10-1 gives a selection of the most relevant non-synthetic applications of ionic liquids that are discussed, developed or already technically applied today (for more details concerning the applied examples see Chapter 9). All applications can be grouped into three areas, electrochemical, analytical and engineering applications.

It is very remarkable that most of the non-synthetic applications rely on only a few key success factors and those are very similar in each group. While electrochemical applications benefit mostly from the wide electrochemical window of ionic liquids, analytical applications often profit from the special solubility properties of ionic liquids. The steeply growing area of engineering fluid applications includes material processing with ionic liquids, separation technologies and applications in process machinery and plant applications. All these applications make use – in a more or less pronounced manner – of the negligible vapor pressure of ionic liquids.

How does one identify a promising non-synthetic application for ionic liquid technology? Our answer to this question four years ago was also very vague. We made some general comments that researchers on synthetic and non-synthetic applications should collaborate to broaden the range of general physicochemical properties of ionic liquids and speculated that such better data ranges would lead to the identification of new non-synthetic applications.

Today, much more data on the physicochemical properties of pure ionic liquids are available. We have learned which ionic liquid properties are well tuneable and which properties are more or less intrinsic to the group. Even a prediction of the properties of mixtures of organic substances and ionic liquids with respect to solubility, miscibility and evaporation properties has become possible to some extent. The Cosmotherm methodology – a molecular modelling-based approach [20] – has, for example, already shown some promise in this respect. Today this method is used by a number of groups for the prediction of trends in the structural optimization of ionic liquids, in particular for extraction [21] and extractive distillation [22], with some success. However, more work is necessary to further develop and refine such methods for them to become fully reliable, quantitative tools for all types of ionic

Table 10-1 Non-synthetic application of ionic liquids – selected examples, key success factors and references

Application	Key success factors	Selected refs.
electrochemical applications		
ionic liquids as electrolyte in batteries	wide electrochemical window	2
ionic liquids in supercapacitors	wide electrochemical window	3
electrodeposition of metals from ionic liquids	wide electrochemical window, solvation power	4
dye-sensitized solar cells	negligible volatility, conductivity	5
analytic applications		
ionic liquids as active component in sensors	conductivity, negligible volatility	6
ionic liquid as matrix for mass spectroscopy	solvation power, negligible volatility	7
ionic liquid as stationary phase for gas chromatography	tuneable solvation, negligible volatility	8
"engineering liquids"		
- materials processing		
ionic liquids in biomass processing	special solvation power	1
ionic liquids as plasticizers	negligible volatility, plasticizing properties	9
ionic liquid as antistatics	conductivity, negligible vapor pressure	10
ionic liquids as dispersents, surfactant or detergent	surface activity	11
ionic liquids as liquid crystals	liquid crystalline behavior	12
- separation technologies		
ionic liquids as extraction media:		
ionic liquid/aqueous	hydrophobicity of some ionic liquids, solvation properties	13
ionic liquid/organic	lipophobicity, solvation properties	14
ionic liquids for extractive distillation	solvation properties, negligible volatility	15
ionic liquids in gas separation and gas storage systems	solvation properties, negligible volatility	16
- process machinery		
ionic liquid as lubricant	surface activity, lubricative properties, non-flammability	17
ionic liquids as hydraulic fluid and as working fluids in pumps and compressors	very low compressibility, special gas solubility properties	18
Ionic liquids as heat transfer or heat storage medium	high heat capacity, good heat conductivity	19

liquid properties. Given the extremely impressive activities that have developed in the last four years around the computational and theoretical treatment of ionic liquids and ionic liquid mixtures (see Section 4.2 for details) we expect great advances in this field in the up-coming years.

Is a lack of understanding the major limitation for the development of ionic liquid methodology?

Unfortunately, the answer is still, yes! The fact that we have in hand much more reliable physicochemical data on pure, binary ionic liquids today does not hide the fact that we still have to greatly improve our fundamental understanding of ionic liquids to make full use of their potential. This concerns the detailed investigation of the nature of interactions between the ions of an ionic liquid (and here not only binary ionic liquids should be considered!), on the one hand, and of the nature of interactions between ionic liquids and dissolved substances or surfaces of solids, on the other.

We are still very much at the beginning in our understanding of the order in the liquid state, the structure and the dynamics of ionic liquids. Even less developed is our insight into the specific solvation properties of ionic liquids versus molecular compounds of different polarity, dissolved ions, electrons, metal complexes, biological systems or transition states of reactions. Also, the specific interactions of ionic liquids with the surfaces of metals (e.g. under electrochemical control), organic or inorganic solids (e.g. polymers, nanoparticles or highly porous supports) as well as with the surfaces of poorly miscible or immiscible liquids are certainly of great relevance. All these properties are expected to be strongly dependent on the specific combination of ion structures forming the ionic liquid and an interesting and quite unexplored question is how much we can vary certain features by mixing binary ionic liquids or by introducing new functional groups into task-specific ionic liquids.

By raising the question *"Is there a "universal" ionic liquid at the present state of development?"* we discussed in the first edition of this book the perspective for the technical production of ionic liquids with regard to the future number of so-called "bulk" ionic liquids. With the term "bulk ionic liquid" we named those ionic liquids that would find use in large scale applications or that would form valuable intermediates for other ionic liquids so that such ionic liquids would be produced on a large scale (at least at a ton scale) and would be available on this scale for less than 50 €/kg.

Indeed, we have seen in the last four years a certain concentration on a few ionic liquids that are now envisaged for large-scale applications and in some cases larger scale production has already taken place. A very important issue in understanding this development is the necessity for a legal registration process for these ionic liquids. In Europe, for example, ionic liquids that are produced in quantities greater than 100 kg per year have to undergo a number of toxicological and ecotoxicological tests involving costs of about 50 000 € per ionic liquid and about six months of testing time. For the distribution of production quantities greater than one ton, additional, and even more sophisticated, tests are required.

To the best of our knowledge only [EMIM][EtOSO$_3$] (jointly by Solvent Innovation GmbH and Degussa AG) and [BMIM]Cl (by BASF AG) have been registered in this way to date. Consequently, only these two ionic liquids – and a couple of low melting ammonium salts that have been used industrially for many years and are treated as "existing substances" (e.g. Solvent Innovation's AMMOENG family [23])– can be distributed and applied at this time in amounts larger than 100 kg per year.

It is likely that in the years to come more ionic liquids will be registered, however, this process will be purely application driven. That means that the next ionic liquids for registration will be the candidates that have proven the best performance/cost ratio in an up-coming larger-scale industrial application. As a consequence we anticipate for the next couple of years the registration of quite basic and low-cost ionic liquids rather than the registration of task-specific or highly functionalized ionic liquids. As the cost for the synthesis of more sophisticated ionic liquids will be always an issue, the likeliness of their use in large volumes is significantly lower.

In this context it is very important to note that "new" ionic liquids cannot be considered from a legal perspective as mixtures or parts of already registered cation/anion combinations. The toxicity or ecotoxicity of [EMIM]Cl can, for example, not be deduced for registration purposes from a combination of the data obtained for [EMIM][EtOSO$_3$] and [BMIM]Cl. This further emphasizes the need to focus the resources for ionic liquid registration on the candidates with the highest probability for large-scale applications.

Having said all this we would like to point out very clearly that we consider the constant quest for new, highly functionalized or task-specific ionic liquid structures as a very important part of the development of the whole ionic liquid research area. These may still have many, economically highly attractive applications, even if the total volume of their use is expected to be much smaller. These task-specific ionic liquids are likely to form the key functional materials in devices like sensors or chromatographic columns or the "magic ingredient" in product formulations, like electrochemical baths or lubricants.

It should also be noted that not all functionalized ionic liquids are necessarily very expensive e.g. readily available, functionalized compounds, such as amino acids, may be applied as starting materials for their synthesis. Therefore, it is not excluded that functionalized ionic liquids of this kind will also make the transition to bulk ionic liquids over time. An example that impressively illustrates this possibility is the successful introduction of choline-based ionic liquids by Abbott's group [24] and Scionix for the electrodeposition of Cr-layers (for more details see Chapter 9).

Which type of reaction should be studied in an ionic liquid, and what reaction can be carried out in an ionic liquid that is not possible in organic solvents or water?

At this point we would like to re-emphasize the statement that we made four years ago. Not all chemistry makes sense in ionic liquids!

Why? The concept of working in liquid salts has some general implications that– regardless of the great tuneability of some ionic liquid properties – define their reasonable application areas for synthesis. In the conclusion of Section 5.3 this has been demonstrated in detail under the title "Low hanging fruits and high hanging fruits" for the specific example of transition metal catalysis. Here we would like to generalize the most important aspects. To perform successful synthesis in ionic liquids the latter should (i) be stable under the envisaged or necessary reaction conditions (e.g. vs. acidity, basicity, water, elevated temperature); (ii) should be able to dissolve all relevant parts of the reaction mixture (reactants, catalyst, co-catalysts, products) in the desired quantities and reactivities; (iii) should provide options to isolate the reaction product and, finally, (iv) should be easily processable and recyclable for long-term operation or repetitive use.

The last aspect is especially important if the ionic liquid approach does not change the chemistry of the synthetic application fundamentally, so that the recycling of the ionic liquid can be expected to be the key factor to justify its higher price compared to traditional organic solvents.

Of course it is always more attractive to perform reactions in ionic liquids that are impossible to carry out in other media. But how does one identify these? With greater knowledge of ionic liquids and their specific, often unique, property profiles, this question has become easier to answer. It is quite likely that all kinds of unique chemistry in ionic liquids will somehow build on their unique properties. Therefore, it is still by far the most promising way to start from a detailed understanding of the special properties of each ionic liquid material that is considered for application.

For the last five years, our impression is that by far more unique applications of ionic liquids have been invented by making cleverer use of quite obvious ionic liquid features (such as, e.g. their negligible volatility under ambient conditions, unusual feedstock/product solubility properties and/or unusual miscibility properties) than by any fancy new and highly sophisticated insight into e.g. the structural organization of the ionic liquid itself or the distinct ability of ionic liquids to stabilize unusual transition states. Here, we still see many possibilities for new, more courageous concepts with great potential for surprising and very unusual reactivity. Honestly, we were quite wrong about our prophesies in 2002 that the combination of wide electrochemical window of ionic liquids with their ability to serve as solvents for transition metal catalysts would lead, shortly, to many new and unique applications. The number of convincing examples in this challenging field is still very low, unfortunately. We continue to believe in the power of this concept and the few papers that have appeared have demonstrated it in principle [25]. But it has also become clear that the combination of complex ionic liquids with complex electrochemistry and a complex, unique reaction, lead to a research field in which 'high hanging fruits' prevail. More collaboration between electrochemists, reaction engineers and synthetic chemists may be necessary to achieve successes more quickly and more often.

So far we have just commented on our own prophesies made in the Outlook of the first edition of this book. At this point we feel quite satisfied by the fact that we were not completely wrong at the time.

But what about new questions – questions that came up in the last five years?

It is right that in the last five years a couple of new questions have come up which are of interest for the future development of the ionic liquid field in total. To limit the scope of this new Outlook we decided to deal with four of these questions in closer detail, where we would like to share with the reader our personal opinions.

Definitions

Much debate has been generated around definitions of ionic liquids in the recent past. Recently, a discussion has started among ionic liquid experts about how flexibly the term "ionic liquids" should be used. The generally accepted definition of an ionic liquid is that, when pure, it consists entirely of ions and melts below 100 °C.

First, it should be recognized that there are many definitions used in science that when pushed to their limits either fail or lead to misunderstandings. Even very well accepted definitions such as donor, acceptor, amphoteric and inert do not have rigid boundaries. Of course this is so; nature does not recognize the pigeon holes that we create for its phenomena. The reason that we have definitions is that they can be helpful for us to organize our knowledge and thoughts and they are only helpful when they are just that – *helpful*. Their purpose is certainly not to create artificial boundaries and conflict. So, with that in mind we tiptoe into the debate.

Is it really helpful to distinguish between ionic liquids and molten salts?

The term ionic liquid has been around for a very long time. In 1981 a book titled "Ionic Liquids" [26] was published. the vast majority of which covered high-temperature systems. However, by the time that it came for one of us to write a review in 1999 [27] it had become common place for the term to imply that the salt was low-melting and that implication was made in writing for the first time.

It is absolutely true to say that there is nothing special about 100 °C, it just happens to be the boiling point of water. There is certainly not going to be any great difference between materials that melt just below or just above this temperature. Indeed, important understanding of low-melting systems can be found in the literature of high-temperature systems, particularly because this part of the field has a long history. So, any definition that relates to the melting point of the salt cannot be about its chemical or physical nature beyond the one fact. In that case should it be abandoned?

There are real practical differences that arise when working at higher or lower temperatures. For instance, for those people who are interested in synthetic organic chemistry and catalysis the thermal instability of their reagents, products and catalysts is a major issue. At temperatures much above a couple of hundred degrees or so, most of the compounds of interest to them will decompose. Consequently, the thermal operating range of their solvents is an important issue. So, having a descriptor that recognizes this is *helpful*, at least to them.

Of, course there is something special about room temperature. It is the temperature at which one finds rooms. Again there is a practical, but no scientific, distinction between liquids that melt below or above room temperature. Here, thankfully, the melting temperature is made explicit in the commonly used names, *room-temperature ionic liquid and ambient-temperature ionic liquid*.

Ionic liquid, molten salt, liquid salt, and fused salt have all been used to describe salts in the liquid phase. Those who are interested in low melting salts have commonly preferred to use the name ionic liquid. This has become established and it would probably be both impossible and undesirable to reverse the trend. However, if they believe it to be anything other than a useful moniker they will miss out on the knowledge, expertise and understanding of their colleagues working in higher-temperature systems.

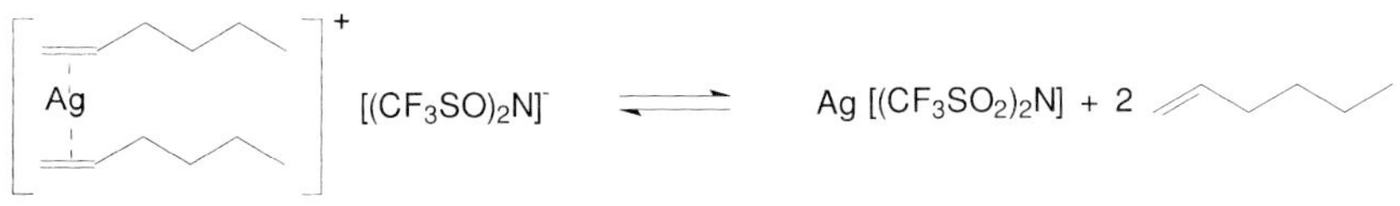

[Scheme showing Ag+ coordinated with two alkene ligands in brackets with + charge, paired with [(CF₃SO)₂N]⁻, in equilibrium with Ag[(CF₃SO₂)₂N] + 2 alkene]

Scheme 10-1 Ag$^+$-based ionic liquid.

If the debate that had grown up around this issue were just a matter of dinner-party entertainment it would probably be amusing. However, there have been times when it has probably been damaging to the field. It should stop.

Is it helpful to distinguish between solvents that are entirely composed of ions when pure and those that are in equilibrium with molecular forms?

At first sight, this may seem like a very simple question with a simple answer, yes. The problem is, of course, that many organic and inorganic ions are in chemical equilibrium with neutral species. Depending on the position of this equilibrium – and the latter is a function of temperature and pressure – even a pure "ionic liquid" may contain significant amounts of neutral molecules. Of course, this will greatly influence all properties of the substance. Volatility, viscosity, chemical reactivity etc. will greatly differ from the hypothetical mixture of the individual ions if free molecular species such as amines, phosphines, Bronsted-acids or acid esters form as neutral molecules in an equilibrium reaction under the conditions of the ionic liquid application.

The most common cations used for ionic liquids to date, e.g., tetraalkylammonium, tetralkylphosphonium, 1,3-dialkylimidazolium, 1,2,3-trialklyimidazolium and 1-alkylpyridinium ionic liquids all have negligible concentrations of neutral molecules in the pure ionic liquid, certainly under ambient conditions. However, even ionic liquids with these archetypical cations may contain significant amounts of neutral molecules if the chosen anion is characterized by a significant nucleophilicity and they are being used at elevated temperatures.

Of course, the intrinsic presence of neutral molecules in "pure" ionic liquids becomes more and more relevant if the equilibrium constant is less and less in favor of the ionic compound. For the N-protonated 1-alkylimidazolium salts the amount of neutral, unprotonated alkylimidazole is obviously a function of the anion's basicity. Here, the wide range of potential anion basicities has provoked the question as to whether these systems should at all be considered as ionic liquids. However, some convincing applications of these substances in catalysis [28] and organic synthesis [29] (see BASIL process in Chapter 9) and physical measurements [30] have led to the now widely accepted view that at least combinations of 1-alkylimidazoles with strong acids can be regarded as ionic liquids.

A similar discussion as for protonated ionic liquids can be expected for the complex cation systems described by Dai and coworkers (see Scheme 10-1) [31].

Given the complex formation constants of metal-ion organic complexes, such ionic liquids will never consist uniquely of ions but will still contain some amount of uncomplexed organic ligand and free metal ion. Of course, in some cases the metal ligand complex will be very strong and then the amount of free ligand will be negligible. But what amount of neutral molecule can be considered as a reasonable limit value in order to still call the system an ionic liquid?

Generalizing Dai's ionic liquid concept of complexed inorganic cations, another question regarding the ionic liquid definition becomes relevant. Many inorganic ions form very stable hydrate coordination spheres. Should we still call a fully hydrated cation/anion system an ionic liquid? One could argue that the water in the first coordination shell may be so strongly bound that just bigger, stable ions are formed. Thus the system could still be regarded as consisting entirely of ions. However, if we accept this, where is the limit value of water content from whence we have to consider the system a solution of ions in an aqueous environment?

As one possible way to define such a limit value, one could consider the ion concentration in the system in mol of ions per volume as a suitable criterion. After all, we have no problem in calling water, ethanol and ammonia molecular liquids, in spite of the fact that they all have autoionization reactions that lead to measurable concentrations of ions at room temperature. However, a comparison of the ion concentration between a saturated KCl solution in water and the typical ionic liquid [BMIM][PF$_6$] immediately shows the unsuitability of such an approach. While a saturated KCl solution in water at 22 °C has an ion concentration of 4.6 mol l^{-1}, the ion concentration of [BMIM][PF$_6$] is only 0.0054 mol l^{-1} at the same temperature [32]. This means that a general ion concentration at which a system can be considered to be an ionic liquid certainly does not exist. Again, we find ourselves in a position where exploring the limits of a definition leads to it breaking down.

Beyond the discussion of ionic liquids when pure, in most situations in which ionic liquids are used some other molecular species is also present. In all organic reactions or liquid–liquid biphasic catalysis a certain concentration of organic molecules will be naturally present in the ionic liquid as reactant or product. Up to what amount of neutral molecules dissolved in the original ionic liquid can we still consider the system as an ionic liquid? It seems reasonable to say that while the ionic liquid is in excess the mixture can be described as a solution *in* the ionic liquid, but once the molecular liquid is in excess it becomes a solution *of* the ionic liquid in the molecular solvent. While such a highly diluted system can no longer be reasonably taken as an ionic liquid, the system may still display some effects originating from the ionic liquid's ions present in the system (such as activation of solutes, solvatation of ionic species, etc.). This leads to another question that goes beyond the discussion about a general ionic liquid definition. At what critical concentration of the ionic liquid's ions can we expect to observe ionic liquid-like behavior and ionic liquid effects (whatever this means)?

A few, selected examples from the literature illustrate this aspect further. Seddon et al. described, for example, the great influence of relatively small amounts of molecular solvents on the physicochemical properties of ionic liquids [33]. Song et al. reported significant activation of an Mn(salen) complex in a solution consisting of 20 vol% of ionic liquid in CH_2Cl_2 versus pure CH_2Cl_2 [34]. Wasserscheid and coworkers found that the strength of diasteriomeric interactions between a chiral ionic liquid's ion and a chiral substrate or counter-ion is strongly dependent on the concentration of the substrate in the ionic liquid [35].

Of course, all these concentration effects are highly dependent on the nature of the substrate dissolved in the ionic liquid as well as on the nature of the ionic liquid's cation and anion. Given the enormous possibility to vary these, and the highly dynamic character of reacting systems, it becomes clear that a detailed understanding of the role of the ionic liquid in reaction mixtures is still far from complete.

In the end it is probably most useful to define ionic liquids as systems that contain only negligible amounts of neutral molecules under the conditions in which they are being employed.

Have we underestimated in the past the volatility of ionic liquids?

Ionic liquids have in the past been widely reported and treated as non-volatile substances. Their non-volatility has even at times been used as part of the definition of an ionic liquid. Recently, Rebelo's group (in collaboration with Seddon's and Magee's groups) has demonstrated that selected families of aprotic ionic liquids can be distilled without decomposition [36]. These authors reported that certain ionic liquid structures, known for their very high thermal stability (in particular bis(trifluoromethylsulfonyl)amide salts) can be distilled under relatively mild conditions, e.g. distillation of 1-hexyl-3-methylimidazolium bis(trifluoromethylsulfonyl)amide has been described to occur at 170 °C and 0.07 mbar. Even thermal separation of two ionic liquids has been demonstrated. These findings expanded significantly on earlier publications discussing the probability of thermal decomposition of ionic liquids (e.g. by transalkylation [37] or by proton transfer [38]) and a re-formation of the ionic liquid from the decomposition products in the colder parts of the applied apparatus.

At first sight, these findings seem to be in contradiction to many previously reported applications of ionic liquids which successfully profit from their assumed non-volatility. Examples include the use of ionic liquids in extractive distillation processes (including long-term stability tests in pilot plants over three months at 175 °C/50 mbar without significant loss of ionic liquid [39]) or the application of ionic liquids in the presence of ultra-high vacuum conditions for XPS analysis (here the ionic liquid was exposed for up to several days in a vacuum of 10^{-10} mbar without visible decomposition or evaporation [40]). However, a detailed analysis reveals that these applications have been examined using other ionic liquid structures and/or conditions significantly different from those applied in the study by Rebelo et al. [36].

As a consequence of these findings, the vapor pressure of an ionic liquid can no longer be considered to be zero. As, from a physicochemical perspective, all matter (even a stone!) has some vapor pressure, this former perception was questionable anyway. However, Rebelo's study also showed that the vapor pressure of ionic liquids is indeed negligible at near-ambient conditions. For applications under harsh temperature and high vacuum conditions, however, the vapor pressure of the applied ionic liquids will have some relevance and should be known. Thus, we expect that in the future the vapor pressure of ionic liquid structures will be determined experimentally and that these data will just add to the property and application profile of each specific ionic liquid.

In this context it should be briefly noted that the discovery of the volatility of some ionic liquids under harsh conditions is not the only shift in paradigms that has been observed in ionic liquid research over the last five years. For example, it has been claimed that no ionic liquid would mix at all with scCO$_2$ [41]. However, this statement is certainly not true in its most general sense. In the same way as some ionic liquid structures have been found to dissolve significantly in a nonpolar solvent like hexane (the ionic liquid's cation carrying extended alkyl chains, of course) these same ionic liquids will also show some solubility in scCO$_2$.

Another, particularly restricting example of a recent paradigm change concerns the thermal stability of ionic liquids. We know today for certain that thermal stabilities of ionic liquids have been seriously overestimated in the past. Reported stabilities of over 400 °C in early papers [42] do not reflect long-term stabilities but have been obtained in TGA-onset measurements with steep heating ramps. In more recent times, the temperature limits for long-term stability have been found to be more than 100 °C lower than these earlier reported data for the same ionic liquids [43].

But there are also shifts in paradigms that have clearly opened the field and will certainly inspire a lot of great science in the future. For example, ionic liquids have been perceived in the past to be notoriously unstable towards strong bases due to carbene formation, Hofmann-elimination or dealkylation [44]. Recent work by Clyburne's group [45] and QUILL/Belfast [46] has impressively demonstrated, however, that some ionic liquid structures can indeed be quite base stable and these ionic liquids have been successfully applied in organic reactions in contact with strongly basic reagents.

These examples stand as representatives for a larger number of scientific achievements of the last five years that have to some extent reconditioned our view on ionic liquids. The fact that some of these findings seem to implicate quite fundamental changes in our general perception of ionic liquids just reflect – in our eyes – the scientifically still quite immature state of the whole field. With such a restricted data set, it is particularly dangerous to make statements of the kind "ionic liquids are …". The only thing that can be known for sure at this stage is that ionic liquids are a diverse set of materials that have wide-ranging properties and that, without doubt, there is still plenty of room for surprises in the future.

Are ionic liquids really green?

As can be seen above, perhaps the better question is: are there green ionic liquids? There is no doubt that the negligible vapor pressure of ionic liquids at ambient conditions is a strong green asset in comparison to volatile organic solvents. Atmospheric pollution by ionic liquids is very unlikely, even if the liquids are handled in open containers. Moreover, the vapor pressure of organic reactants and products is reduced when dissolved in ionic liquids, also reducing their potential for atmospheric pollution. In addition, a reduced volatility of solvents and reactants adds to the operational safety of a given process, as mixtures of volatile organics and air are always linked with explosion hazards. Moreover, there is no doubt that the ionic liquids concept builds very strongly on the idea of solvent and solvent/catalyst recycling. It has been demonstrated in many examples (see Chapter 5 for details) that in this way the application of ionic liquids can result in a significantly reduced environmental impact.

Based on these facts ionic liquids have been promoted for quite some time successfully under the banner of "Green Chemistry". This happened in spite of little being known at the time about the toxicity and ecotoxicity of ionic liquids. In the last five years this view, however, has been seriously questioned by three arising aspects.

1. Some ionic liquids have been found to release highly toxic and very corrosive degradation products when undergoing hydrolytic decomposition. This discovery was especially harmful for the green image of ionic liquids as it addressed, in particular, the main work horses of ionic liquid research of the time, namely the hexafluorophosphate and tetrafluoroborate systems which indeed release HF on contact with water [47]. Today, we know that many ionic liquids are very stable to hydrolysis. Moreover, we have developed many ionic liquids which are completely halide-free [48] so that the release of HCl or HF can certainly be avoided in both hydrolysis and combustion processes.
2. Some ionic liquids have been reported to exhibit high toxicity and high ecotoxicity (for an excellent review see Ref. [49]). Indeed, it has been quite clearly demonstrated that the ionic nature does not prevent ionic liquids from being toxic. However, it is also evident from today's research results that neither the ionic nature nor the structural elements that are needed to design low melting salts necessarily lead to toxicity. Consequently, there are highly toxic ionic liquids *and* ionic liquids with no toxicological and ecotoxicological risk, depending on the chemical structure of their individual ions. For example, a number of hydrophilic ionic liquids have been shown to be non-toxic and [EMIM][EtOSO$_3$] has even passed the registration process without a need to be labeled as harmful. It is remarkable that the toxicity of ionic liquids can often be related to the extent of their lipophilicity with more lipophilic cations and anions resulting in higher ionic liquid toxicity. While it is encouraging to see that we are beginning to understand first structure-toxicity relations for ionic liquids, the fact that [BMIM]Cl had to be labelled as "toxic" as a result of its registration process was not helpful for the green image of ionic liquids.
3. Environmental assessment tools have been introduced to support decision-making processes towards the development of sustainable chemical products and processes. When applying these tools to solvents, aspects like environmental persistence, health and safety hazards, and spatial range as well as energy consumption in a full lifecycle analysis have to be taken into account. In a study by Hungebühler and coworkers [50] 13 organic solvents have been compared using these tools under the pre-requisite that the performance of all solvents in a hypothetical application would be identical. Not surprisingly, in this study methanol turned out to be the most benign solvent as almost all energy spent to produce methanol can be recovered in the combustion process for its disposal. Obviously, such a lifecycle assessment favors very basic solvent concepts with liquids that are accessible, with very few refinement or synthesis steps, from raw materials such as crude oil or natural gas. As a consequence, ionic liquids fail to come very favorably out of such an assessment, as was pointed out by Prof. Hungerbühler in his presentation at the "Green Solvents for Engineering" conference in Friedrichshafen/Germany in October 2006. This is understandable due to the fact that it requires many more synthetic steps to produce an

imidazolium salt than to produce any of the organic solvents commonly applied on a large industrial scale.

From all three aspects we have to conclude that ionic liquids are not *intrinsically* green. So, why are ionic liquids still discussed in the context of "Green Chemistry" (and we strongly believe that this is correct!)? Quite simply: Ionic liquids are a class of fascinating, new liquid materials with unique combinations of properties. By making use of these unique properties, more efficient and greener processes and devices can be realized. As a consequence of this statement the "greenness" of an ionic liquid will be always defined by its performance in a specific application – just as for any other performance chemical.

Once we have accepted that the "greenness" of an ionic liquid is dominated by its performance, it is no longer a problem that replacing a traditional volatile, organic solvent by an ionic liquid does not always lead to a greener process. We can also understand that even a toxic ionic liquid may be the "greenest" choice if it performs in an outstanding manner in a given application. Finally, from this viewpoint, we can also deal with the results of the lifecycle assessment studies for ionic liquids. An evaluation of the "greenness" of an ionic liquid that does not take into account its performance in a specific application is like evaluating the "greenness" of a car without taking into account driving performance and fuel consumption as performance criteria. In such a case the car with the most basic design will always win, of course, as it has the best energy lifecycle with regard to the environmental impact of its production and its disposal. Ionic liquids are like highly sophisticated modern cars optimized for maximum fuel economy when applied in the right manner. Compared to methanol and toluene ionic liquids will always be more expensive and more difficult to produce and to dispose of. As a result of this consideration, the extra investment to use an ionic liquid has always to be largely overcompensated by a superior ionic liquid performance to realize Green Chemistry with ionic liquids.

With this Outlook we have tried to comment on some topical questions about ionic liquids. Of course, we know that it is extremely difficult and probably foolish to try to predict the future, but we found the discussions that were initiated by the Outlook of the first edition of this book so fruitful and so important for our own work that we did not want to miss the opportunity to write an Outlook of this kind again.

There is no doubt that the field of ionic liquids has developed at breath-taking speed over the last few years. In this time, a good part of the way to make ionic liquids a standard tool for the synthetic chemist has already been traveled. Yes, there are still open questions and many unrevealed aspects, but this is a good part of the excitement of working with ionic liquids. The field is still very young and being part of it is challenging, competitive, highly interdisciplinary, very fundamental and – at the same time – also very applied. We are convinced that we will very soon see many new ionic liquid applications. We also expect a number of fundamentally important ground-breaking developments enabled by the use of ionic liquids in the years to come. Let us continue to explore ionic liquids in the intensity that has developed over the last three years. The best is still to come!

References

1. R. P. Swatloski, R. D. Rogers, J. D. Holbrey, U.S. 10/256,521; PCT/US02/31404; WO 03/029329 A2, (to BASF AG), **2003**.
2. (a) S. Tobishima, *Electrochemistry* **2002**, *70*(3), 198–202; (b) A. Webber, G. E. Blomgren, *Advances in Lithium-Ion Batteries* **2002**, 185–232; (c) J. S. Wilkes, *NATO Science Series, II: Mathematics, Physics and Chemistry* **2003**, *92* (*Green Industrial Applications of Ionic Liquids*), pp. 295–320; (d) M. Galinski, A. Lewandowski, I. Stepniak, *Electrochim. Acta* **2006**, *51*(26), 5567–5580.
3. (a) J. D. Stenger-Smith, C. K. Webber, N. Anderson, A. P. Chafin, K. Zong, J. R. Reynolds, *J. Electrochem. Soc.* **2002**, *149*(8), A973–A977; (b) A. Balducci, U. Bardi, S. Caporali, M. Mastragostino, F. Soavi, *Electrochem. Commun.* **2004**, *6*(6), 566–570; (c) E. Frackowiak, G. Lota, J. Pernak, *Appl. Phys. Lett.* **2005**, *86*(16), 164104/1–164104/3.
4. (a) F. Endres, *Phys. Chem. Chem. Phys.* **2001**, *3*, 3165–3174; (b) F. Endres, *ChemPhysChem* **2002**, *3*(2), 144–154; (c) F. Endres, S. Z. El Abedin, *Phys. Chem. Chem. Phys.* **2006**, *8*(18), 2101–2116; (d) S. Z. El Abedin, F. Endres, *ChemPhysChem* **2006**, *7*(1), 58–61.
5. (a) P. Wang, S. M. Zakeeruddin, I. Exnar, M. Graetzel, *Chem. Commun.* **2002**, 2972–2973; (b) H. Matsumoto, T. Matsuda, *Electrochemistry* **2002**, *70*(3), 190–194; (c) V. Jovanovski, E. Stathatos, B. Orel, P. Lianos, *Thin Solid Films* **2006**, *511–512*, 634–637; (d) T. Oda, S. Tanaka, S. Hayase, *Jpn. J. Appl. Phys., Part 1* **2006**, *45*(4A), 2780–2787; (e) F. Mazille, Z. Fei, D. Kuang, D. Zhao, S. M. Zakeeruddin, M. Graetzel, P. J. Dyson, *Inorg. Chem.* **2006**, *45*(4), 1585–1590.
6. (a) S. Dai, C.-Y. Yuan, B. Lee, C. Liang, R. D. Makote, H. Luo, *Proc.-Electrochem. Soc.* **2002**, *2002-19*(*Molten Salts XIII*), 295–304; (b) S. Pandey, *Anal. Chim. Acta* **2006**, *556*(1), 38–45.
7. (a) Armstrong, L.-K. Zhang, L. He, M. L. Gross, *Anal. Chem.* **2001**, *73*(15), 3679–3686; (b) M. Koel, *Crit. Rev. Anal. Chem.* **2005**, *35*(3), 177–192.
8. (a) D. W. Armstrong, L. He, Y.-S. Liu, *Anal. Chem.* **1999**, *71*(17), 3873–3876; (b) A. Berthod, L. He, D. W. Armstrong, *Chromatographia* **2001**, *53*(1/2), 63–68; (c) J. Liu, J. A. Jonsson, G. Jiang, *Trends Anal. Chem.* **2005**, *24*(1), 20–27.
9. (a) M. P. Scott, M. Rahman, C. S. Brazel, *Eur. Polym. J.* **2003**, *39*(10), 1947–1953; (b) M. Rahman, C. S. Brazel, *Polym. Deg. Stabil.* **2006**, *91*(12), 3371–3382.
10. J. Pernak, A. Czepukowicz, R. Pozniak, *Ind. Eng. Chem. Res.* **2001**, *40*, 2379–2383.
11. (a) D. Fu, R. J. Card, WO 32711 (to Schlumberger Technology Corporation, USA) **2000** [*Chem. Abstr.* **2000**, *133*, 32553; (b) B. Weyershausen, K. Lehmann, *Green Chem.* **2005**, *7*, 15–19.
12. (a) C. M. Gordon, J. D. Holbrey, A. R. Kennedy, K. R. Seddon, *J. Mater. Chem.* **1998**, *8*(12), 2627–2736; (b) J. D. Holbrey, K. R. Seddon, *J. Chem. Soc., Dalton Trans.* **1999**, *13*, 2133–2140; (c) T. Mukai, M. Yoshio, T. Kato, Takashi, H. Ohno, *Chem. Lett.* **2004**, *33*(12), 1630–1631; (d) J. Baudoux, P. Judeinstein, D. Cahard, J.-C. Plaquevent, *Tetrahedron Lett.* **2005**, *46*(7), 1137–1140.
13. (a) J. G. Huddleston, R. D. Rogers, *Chem. Commun.* **1998**, 1765–1766; (b) J. D. Holbrey, K. R. Seddon, *J. Chem. Soc., Dalton Trans.* **1999**, *13*, 2133–2140; (c) A. E. Visser, R. P. Swatloski, M. W. Reichert, S. T. Griffin, R. D. Rogers, *Ind. Eng. Res.* **2000**, *39*(10), 3596–3604; (d) A. E. Visser, R. P. Swatloski, W. M. Reichert, J. H. Davis, Jr., R. D. Rogers, R. Mayton, S. Sheff, A. Wierzbicki, *Chem. Commun.* **2001**, 135–136; (e) S. H. Park, D. Demberelnyamba, S. H. Jang, M. W. Byun, *Chem. Lett.* **2006**, *35*(9), 1024–1025; (f) M. L. Dietz, *Sep. Sci. Technol.* **2006**, *41*(10), 2047–2063; (g) A. Ouadi, B. Gadenne, P. Hesemann, J. J. E. Moreau, I. Billard, C. Gaillard, S. Mekki, G. Moutiers, *Chem. Eur. J.* **2006**, *12*(11), 3074–3081.
14. (a) A. Bösmann, L. Datsevitch, A. Jess, A. Lauter, C. Schmitz, P. Wasserscheid, *Chem. Commun.* **2001**, 2494–2495; (b) J. Esser, P. Wasserscheid, A. Jess, *Green Chem.* **2004**, *6*(7), 316–322; (c) Y. Nie,

C. Li, A. Sun, H. Meng, Z. Wang, *Energy Fuels* **2006**, *20*(5), 2083–2087.

15. (a) C. Jork, M. Seiler, Y.-A. Beste, W. Arlt, *J. Chem. Eng. Data* **2004**, *49*, 852–857; (b) M. Seiler, C. Jork, A. Kavarnou, W. Arlt, R. Hirsch, Rolf. *AIChE J.* **2004**, *50*(10), 2439–2454; (c) C. Jork, C. Kristen, D. Pieraccini, A. Stark, C. Chiappe, Y.-A. Beste, W. Arlt, *J. Chem. Thermodyn.* **2005**, *37*(6), 537–558.

16. (a) A. Yokozeki, M. B. Appl. *Energy* **2006**, *84*(3), 351–361; (b) D. Camper, J. Bara, C. Koval, R. Noble, *Ind.Eng. Chem. Res.* **2006**, *45*(18), 6279–6283; (c) Q. Gan, D. Rooney, M. Xue, G. Thompson, Y. Zou, *J. Membrane Sci.* **2006**, *280*(1+2), 948–956; (d) D. J. Tempel, P. B. Henderson, J. R. Brzozowski, R. M. Pearlstein, D. Garg, U.S. Pat. Appl. Publ. 06/060818 (to Air Products and Chemicals Inc.) **2006**.

17. (a) C. Ye, W. Liu, Y. Chen, L. Yu, *Chem. Commun.* **2001**, 2244–2245; (b) W. Liu, C. Ye, Q. Gong, H. Wang, P. Wang, *Tribology Lett.* **2002**, *13*(2), 81–85; (c) H. Zhao, *Chem. Engin. Commun.* **2006**, *193*(12), 1660–1677; (d) J. Qu, J. J. Truhan, S. Dai, H. Luo, P. J. Blau, *Tribology Lett.* **2006**, *22*(3), 207–214; (e) H. Kamimura, T. Kubo, I. Minami, S. Mori, *Tribology Int.* **2007**, *40*(4), 620–625.

18. (a) C. Hilgers, M. Uerdingen, M. Wagner, P. Wasserscheid, E. Schlücker, WO 06/087333 (to Solvent Innovation GmbH), **2006**; (b) R. Adler, H. Mayer, DE 102005026916, (to Linde AG), **2006**; (c) A. Bösmann, T. J. S. Schubert, DE 102004033021 (to Iolitec GmbH).

19. (a) M. E. van Valkenburg, R. L. Vaughn, M. Williams, J. S. Wilkes, *Proc.-Electrochem. Soc.*, **2002**, *19*, 112–123; (b) G. Olbert, T. Mattke, M. Fiene, O. Huttenloch, U. Hammon, DE 10316418 (to BASF AG), **2004**; (c) M. E. van Valkenburg, R. L. Vaughn, M. Williams, J. S. Wilkes, *Thermochim. Acta* **2005**, *425*(1–2), 181–188.

20. F. Eckert, A. Klamt, *AIChE J.* **2002**, *48*(2), 369–385.

21. C. Jork, C. Kristen, D. Pieraccini, A. Stark, C. Chiappe, Y.-A. Beste, W. Arlt, *J. Chem. Thermodyn.* **2005**, *37*(6), 537–558.

22. Z. Lei, W. Arlt, P. Wasserscheid, *Fluid Phase Equilib.* **2006**, *241*(1–2), 290–299.

23. for more information see www.solvent-innovation.com

24. (a) A. P. Abbott, G. Capper, D. L. Davies, H. L. Munro, R. K. Rasheed, V. Tambyrajah, *Chem. Commun.* **2001**, 2010–2011;
(b) A. P. Abbott, G. Capper, D. L. Davies, R. K. Rasheed, V. Tambyrajah, *Green Chem.* **2002**, *4*, 24–26; (c) R. Calderon Morales, V. Tambyrajah, P. R. Jenkins, D. L. Davies, A. P. Abbott, *Chem. Commun.* **2004**, 158–159.

25. L. Gaillon, F. Bedioui, *Chem. Commun.* **2001**, 1458–1459.

26. D. Inman, *Ionic Liquids*, Plenum Press, New York, **1981**.

27. T. Welton, *Chem. Rev.*, **1999**, *99*, 2071–2084.

28. M. Picquet, I. Tkatchenko, I. Tommasi, P. Wasserscheid, J. Zimmermann, *Adv. Syn. Catal.* **2003**, *345*(8), 959–962.

29. (a) WO 03/062171 (to BASF AG); (b) WO 03/062251 (to BASF AG); (c) WO 05/061416 (to BASF AG); (d) M. Freemantle, *Chem. Eng. News* **2003**, *81*, 9; (e) R. D. Rogers, K. R. Seddon, *Nature Mater.* **2003**, *2*, 363; (f) K. R. Seddon, *Science*, **2003**, *302*, 792–793.

30. M. Yoshizawa, W. Xu, C. A. Angell, *J. Am. Chem. Soc.*, **2003**, *125*, 15411–15419.

31. J. F. Huang, H. M. Luo, S. Dai, *J. Electrochem. Soc.* **2006**, *153*, J9–J13.

32. Calculation based on density data taken from: J. Fuller, A. C. Breda, R. T. Carlin, *J. Electroanal. Chem.* **1998**, *459*, 29–34.

33. K. R. Seddon, A. Stark, M.-J. Torres, *Pure Appl. Chem.* **2000**, *72*, 2275–2287.

34. E. C. Song, E. J. Roh, *Chem. Commun.* **2000**, 837–38.

35. (a) P. Wasserscheid, A. Bösmann, C. Bolm, *Chem. Commun.* **2002**, 200–201; (b) S. Schulz, A. Bösmann, N. Müller, P. Wasserscheid, *Angew. Chem., Int. Ed.* **2007**, in press.

36. M. J. Earle, J. Esperanca, M. A. Gilea, J. N. C. Lopes, L. P. N. Rebelo, J. W. Magee, K. R. Seddon, J. A. Widegren, *Nature* **2006** *439*, 831–834.

37. M. Maase, World Pat., WO 05/068404 (to BASF AG, Germany), **2005**.

38. (a) M. Maase, K. Massonne, World Pat., WO 05/019183 (to BASF AG, Germany) **2005**; (b) M. Volland, V. Seitz, M. Maase,

M. Flores, R. Papp, K. Massonne, V. Stegmann, K. Halbritter, R. Noe, M. Bartsch, W. Siegel, M. Becker, O. Huttenloch, World Pat., WO 03/062251 (to BASF AG, Germany) **2003**; (c) M. J. Earle, K. R. Seddon, World Pat., WO 01/077081 (to QUILL) **2001**.

39. Y. Beste, M. Eggersmann, H. Schoenmakers, *Chem. Ing. Technik* **2005**, *77*(11), 1800–1808. (2005).
40. (a) E. F. Smith, I. J. Villar Garcia, D. Briggs, P. Licence, *Chem. Commun.* **2005**, 5633–5635; (b) J. M. Gottfried, F. Maier, J. Rossa, D. Gerhard, P. S. Schulz, P. Wasserscheid, H.-P. Steinrück, *Z. Phys. Chem.* **2006**, *220*, 1439–1453; (c) F. Maier, J. M. Gottfried, J. Rossa, D. Gerhard, P. S. Schulz, W. Schwieger, P. Wasserscheid, H.-P. Steinrück, *Angew. Chem., Int. Ed.* **2006**, *45*(46), 7778–7780.
41. L. A. Blanchard, D. Hancu, E. J. Beckman, J. F. Brennecke, *Nature* **1999**, *299*, 28–29.
42. e.g. P. Bonhôte, A.-P. Dias, N. Papageourgiou, K. Kalyanasundaram, M. Grätzel, *Inorg. Chem.* **1996**, *35*, 1168–1178.
43. K. J. Baranyai, G. B. Deacon, D. R. MacFarlane, J. M. Pringle, J. L. Scott, *Aust. J. Chem.* **2004**, *57*, 145–147.
44. M. J. Earle in P. Wasserscheid, T. Welton, *Ionic Liquids in Synthesis*, VCH-Wiley, Weinheim, **2002**, pp. 174–213.
45. T. Ramnial, D. D. Ino, J. A. C. Clyburne, *Chem. Commun.* **2005**, 325–327.
46. M. J. Earle, U. Frohlich, S. Huq, S. Katdare, R. M. Lukasik, E. Bogel, N. V. Plechkova, K. R. Seddon, WO 06/072785 (to QUILL/Belfast).
47. R. P. Swatloski, J. D. Holbrey, R. D. Rogers, *Green Chem.* **2003**, *5*(4), 361–363.
48. P. Wasserscheid, R. van Hal, A. Bösmann, *Green Chem.* **2002**, *4*, 400–404; S. Himmler, S. Hörmann, R. van Hal, P. S. Schulz, P. Wasserscheid, *Green Chem.* **2006**, *8*(10), 887–894.
49. B. Jastorff, K. Mölter, P. Behrend, U. Botti-Weber, J. Filser, A. Heimers, B. Ondruschka, J. Ranke, M. Schäfer, H. Schröder, A. Stark, P. Stepnowski, F. Stock, R. Störmann, S. Stolte, U. Welz-Biermann, S. Ziegert, J. Thöming, *Green Chem.* **2005**, *7*, 362–372.
50. S. Hellweg, U. Fischer, M. Scheringer, K. Hungerbühler, *Green Chem.* **2004**, *6*, 418–427.

Index

a
absorption spectrum 190
acesulfamate 22
1-acetamido-3-alkylimidazolium salt 188
acetonitrile-tetrabutylammonium hexafluorophosphate 147
5-acetyl-1,1,2,6-tetramethyl-3-isopropylindane 305
acetylation 305–306
N-acetylglucosamine 648
acid scavenging 666ff.
addition 334ff.
– *anti* 284
– electrophilic 284ff.
– organometallic reagent 340–344
– *syn* 284ff.
AIBN, see 2,2'-azobis(2-methylpropionitrile)
aliphatic nucleophilic substitution 319–326
alkali halide 221f.–223
alkene 476
alkoxysulfate anion 23
alkyl chain length 67–69
N-alkylammonium salt 188
alkylation 300
1-alkyl-4-[5-(dodecylsulfanyl)-1,3,4-oxadiazol-2-yl]pyridinium salt 189
4-(5-alkyl-1,3-dioxan-2-yl) pyridinium
– N-substituted 189
1-alkyl-3-methylimidazolium [RMIM]$^+$ 8ff., 227–228, 267, 285
– halide 11, 17, 285
– hexafluorophosphate 228, 612, 626, 631
– trihalide 286
alkylchloroaluminate 620
1-alkylimidazole 9ff.
– C-substituted 11
1,3-alkylimidazolium 7f.
1-alkylisoquinolinium cation [R-ISOQ]$^+$ 99
alkylphosphonium salt 188
1-alkylpyridinium
– cation 4, 13
– ionic liquid 696
alkylsulfate anion 23
allylation 429
allyltrimethylsilane 336
aluminum
– alkyl co-catalyst 431
– alloy 579ff.
– chromium 581
– cobalt 580
– copper 580
– electrodeposition 578ff.
– iron 580
– lanthanum 581–582
– nickel 580
– niobium 580
– salen 358
– silver 580
– tantalum 580
– titanium 581
aluminumchloride 85ff., 149, 296ff., 301ff., 306, 314, 318f., 334, 478–479, 483, 486, 536
aluminum sodium chloride 18
AMBER functional form 224
ambient-temperature ionic liquid 695
AMEBA 512
amide formation 335
amination 327
amine
– primary 281
– secondary 281
2-aminoacetophenone 347
aminoacid 521
– Fmoc 521
1-(3-aminopropyl)imidazole 49f.
aminopyridine 326

aminopyrimidine 326
AMMOENG™ 23
ammonium trifluoroacetate [TFA] 315
aniline 310
anion
   – cation association 280ff.
   – functionalized 53
   – hydrolysis 370
   – metathesis 14, 491, 496
   – mixture 64
   – size 63
   – solute interaction 289
anion-exchange reaction 13
anthracene 306
antimony 582, 675
   – (III) chloride 311–312
   – electrodeposition 582
antistatic additive 677
application
   – electrochemistry 675–677
   – industrial 663ff.
   – non-synthetic 690
arene 394–395
aromatic compound 309
aromatic nucleophilic substitution 326–327
arsine 679
aryl halide 327
arylboronic acid 426
aryldiazonium
   – tetrafluoroborate 300
arylidenemalononitrile 505
aryltrithiocarbonate 326
atom-transfer radical polymerization (ATRP) 632f.
atomistic potential model 221
atomistic simulation 220ff.
azeotrope 669ff.
aziridination 354
2,2′-azobis(2-methylpropionitrile) (AIBN) 625

b
Baeyer-Villiger oxidation 352, 407
[BARF]⁻ 565
BASF lucirines® 666
BASIL™ (biphasic acid scavenging utilizing ionic liquids) 666ff.
Baylis-Hillman reaction 337–338
beckmann rearrangement 359
benzaldehyde 341
benzene 297ff., 318
   – methylation 297
benzimidazole 344

1H-benzotriazole 42
benzoxyzole 344
1-benzoyl-6-(4-methylbenzoyl)pyrene 303
3-benzoyl-5-phenyl-1,2,4-oxadiazole 282
benzoylation 309, 334
1-benzoylpyrene 303
benzthiazole 344
benzyl chloride 322
benzylation 322
bicyclization 310
Biginelli reaction 348
(R)-BINAP 399
(S)-BINAP 548
binary mixture 179ff.
bioaccumulation 38
biocatalysis 641ff., 657
biological activity 38
biotransformation 642
biphasic acid scavenging utilizing ionic liquids, *see* BASIL™
bis(acetylacetonato)cobalt(II) catalysis 656
1,1′-bis-(diphenylphosphino)-cobaltocene 411
1,1′-bis-(diphenylphosphino)-ferrocene (DPPF) 550
1,1′-bis[1,8-octyl]-3-vinylimidazolium bis[(trifluoromethyl)sulfonyl]amide 542
bis(oxalate)boric acid (HBOB) 623
bis-oxazoline copper(II) complex 330ff.
bis(perfluoroalkylsulfonyl)imide anion 531
bis(perfluoroethylsulfonyl)imide anion [BETI]⁻ 99
2,6-bis(4-phenyloxazolino)pyridine 341
1,3-bis(4-pyridyl)propane (bpp) 574
bis(trifluoromethylsulfonyl)amide BTA, NTF, [Tf$_2$N]⁻, TFSA, TFSI 33ff., 46f., 79, 105, 111–114, 119, 197, 232, 572, 587
bis(trifluorosulfonyl)imide [N(Tf)$_2$]⁻ 133
bistrifluoromethanesulfonylimidic acid (H[NTf$_2$]) 300
2,7-bisulfonate-4,5-bis(diphenylphosphino)-9,9-dimethylxanthene 388
BMIMy (1-butyl-3-methylimidazolylidene) 421
Born Oppenheimer MD 241
boron trifluoride 679
branching 69
bromination 318
ω-bromoalkyltrimethylammonium salt 494
β-bromocarbenium ion 285
N-bromosuccinimide (NBS) 318
Brønsted acid 334, 450
BTSIL, *see* ionic liquid

[BUPY] tetrafluoroborate   629
3-buten-2-one   337
tert-butyl alcohol   300
tert-butyl hydroperoxide (TBHP)   353
1-*n*-butyl-2,3-dimethylimidazolium [BMMIM]$^+$ 139, 227, 325
— bis(trifluoromethylsulfonyl)imide   144, 275ff., 392
— bis(trifluoromethylsulfonyl)imide/ lithium bis(trifluoromethylsulfonyl)imide   144
— hexafluorophosphate   325, 392
— tetrafluoroborate   392, 429
1-*n*-butyl-3-methylimidazolium [BMIM]$^+$ 139, 227, 288, 294, 341, 388, 467
— bis(trifluoromethylsulfonyl)imide   33, 106, 154, 275ff., 315ff., 370, 384ff., 467, 552f., 635, 647ff.
— Br$_3$ 288
— bromide   13, 286, 380
— chloride   8ff., 36, 98, 215ff., 385, 692ff.
— cobalt tetracarbonyl   382
— μ4-(O,O,O',O'-ethane-1,2-dioato)-bis{bis(nitrato-O',O) dioxouranate(VI)}   196
— hexafluoroantimonate   275, 300, 391ff., 467
— hexafluorophosphate   32ff., 93ff., 106ff., 124, 202, 213ff., 239, 251, 275ff., 293, 322ff., 340ff., 352ff., 391ff., 412ff., 436ff., 467ff., 513, 540ff., 558ff., 574, 624ff., 642ff., 697
— hydrogensulfate   300
— lactate   329
— nickel chloride   388
— octylsulfate   21f., 370, 543ff.
— tetrachloronickelate   427
— tetrachloropalladate   200, 424, 536
— tetrafluoroborate   17ff., 34, 109f., 124, 201, 275ff., 322, 341ff., 352ff., 380f., 392ff., 407ff., 425ff., 444, 467, 513, 542, 558, 612, 629ff.
— electrochemical potential window   143
— tin chloride   385f., 409
— trifluoromethanesulfonate   319ff., 438ff., 513, 650
— trihalide   286
1-*n*-butyl-3-methylimidazolium [BMIM] chloride
— aluminumchloride   301, 319, 610, 627
— ironchloride   307
— zinc chloride   470

1-butyl-3-methylimidazolylidene [BMimy]   420
N-butyl-3-methylpyridinium [BMPy]$^+$ 402
N-butyl-N-methylpyrrolidinium [BMPY]$^+$ 139, 275, 287f.
— bis(trifluoromethylsulfonyl)imide   287f., 623
— trifluoromethanesulfonate   281ff.
1-butyl-3-methylpyrrolidinium [BMP]   656
— bis(trifluoromethylsulfonyl)imide   581ff.
N-butyl-N-methylpyrrolidium bis(trifluoromethylsulfonyl)imide   315
1-butyl-2-phenyl-3-methylimidazolium   426
1-butyl-3-(3-triethoxysilylpropyl)imidazolium
— hexafluorophosphate   542
— tetrafluoroborate   542
butylation   297
n-butylimidazolium hexafluorophosphate   511
1-butylpyridinium [BPY]$^+$
— bis(trifluoromethylsulfonyl)imide   287
1-butylpyridinium chloride   334
— aluminumchloride   334
N-butylpyridinium [BuPy]$^+$, [BPy]$^+$, [NBPY]   197, 270, 624ff.
— bis(trifluoromethylsulfonyl)imide   428
— chloride aluminum chloride   634
— nitrate   346
— tetrafluoroborate   341, 624ff.
N-butylpyridinyl radical BuPy·   270
n-butyltrimethylammonium bis(trifluoromethylsulfonyl)imide   510ff.

c
cadmium   582
— electrodeposition   582
CALB-[BMIM][(CF$_3$SO$_2$)$_2$N]-system   657
Candida antarctica   651ff.
Candida rugosa   654
— lipase   654
capillary tube   184
ε-caprolactone (CL)   635
Car-Parinello MD   221, 241ff.
carbamoylmethylphosphine oxide (CMPO)   93
carbamoylphosphine oxide (CMPO)   197
carbene
— catalyzed reaction   636
— complex   420
— N-heterocyclic (NHC)   633

carbohydrate   652
Carbon-Ferrier reaction   343
carbon dioxide   109ff.
   – high-pressure solubility   110, 117
   – low-pressure solubility   109–110
   – supercritical (scCO$_2$)   13, 43, 90, 110, 117, 419, 441, 466, 559–562, 657
carbon oxide   117
Castro-Stephens-Sonogashira reaction   413
catalysis   559ff.
   – biphasic   374ff.
   – multiphasic   465ff., 558ff.
   – transition metal   369ff.
catalyst   672ff.
   – Heck   422
   – immobilization   472
   – low temperature supported   532
   – neutral supported ionic liquid   536
   – separation   468
   – silica-supported vanadium(V)-oxide-alkali pyrosulfate sulphur dioxide oxidation catalyst   531
   – solubility   449
   – supported ionic liquid (SILC)   527ff.
   – supported ionic liquid phase (SILP)   445, 526ff., 539ff.
cation
   – functionalized   48
   – size   65
   – symmetry   66
cell
   – liquid sample   184
   – whole cell system   644f.
CESS process   559
N-cetylpyridinium chloride   634
   – aluminumchloride   634
chemical reaction   244
chemical reactivity   244
chemical shift anisotrophy (CSA) analysis   261
chloride
   – aluminum   85ff., 149, 296ff., 301ff., 306, 314, 318f., 334, 478–479, 483, 486, 536
   – hydrogen bonding   215
   – iron   214
   – uranyl   213
chlorination   670
chloroaluminate   213, 283, 293ff., 483
   – dimerization   430
   – ionic liquid   327, 619
   – oligomerization   430
chlorometalate catalyst   533

1-(chloromethyl)-1,4-diazabicyclo[2,2,2]octane tetrafluoroborate   317
chlorostannate   378ff.
1-(1-chlorovinyl)-2,4-dimethylbenzene   307
chlorpromazine   269
choline (2-hydroxy-ethyltrimethylammonium)   656
   – chloride   584f.
   – hydroxide   540
chromatographic measurement   131
chromium   585, 675
   – (III)-chloride hexahydrate   585
   – (VI)   675f.
   – electrodeposition   585, 675
cinnamaldehyde   553
cinnamic ester   506
cleaning fluid   677
CMPO, see carbamoylmethylphosphine oxide or carbamoylphosphine oxide
cobalt
   – (II) acetoacetate   351
   – bis(acetylacetonato)cobalt(II) catalysis   656
   – 1,1′-bis-(diphenylphosphino)-cobaltocene   412
   – (I) chloride   532
   – (I) salen complex   358
   – (II) salen complex   355
   – (II) triflimide   306
cobaltocinium   411, 474
   – electrodeposition   583–584
   – tetracarbonyl   382
codeine methyl ether (CME)   351
compatibilizer   678
computational modelling   206ff.
condensation reaction   345
coordination compound   569
copper
   – (II)   355
   – (II) acetylacetonate   354
   – Au(111)   590ff.
   – bis-oxazoline copper(II) complex   330ff.
   – (I) chloride   532
   – (II) chloride   314, 532
   – (II) complex   330
   – electrodeposition   583
   – oxide   531
   – trifluoromethanesulfonate   300
corrosion   41
coumarin   346
coupled cluster method   210
covalent anchoring   533

cracking reaction 312ff.
crotonaldehyde 328
cumulene 310
cuprate-based pyridinium ionic liquid 186
cyanex-272 95
cis-cyclooctene 353, 408
cyclopentadiene (CP) 328
cytochrome *c* 656

d
DBU 314
DCH18C6
  (*cis-syn-cis*-dicyclohexyl-18-crown-6) 197
Deacon catalyst 531
dealkylation 319
1-decyl-3-methylimidazolium $[C_{10}MIM]^+$,
  $[DMIM]^+$ 403
    – trifluoromethanesulfonate 319
    – tetrafluoroborate 403
N-decyl-3-methylpyridinium $[DMPy]^+$ 402
density functional theory (DFT) 239
    – Born Oppenheimer MD 241
density measurement 86
designer solvent 293
1,2-di-(9-anthryl)ethane 311
di-(2-ethylhexyl)phosphoric acid (HDEHP) 95
1,2-di-(1-naphthyl)ethane 311
4,4′-diacetoxybiphenyl 304
1,3-dialkylimidazolium 38, 347, 359, 380
    – ionic liquid 696
N,N'-dialkylimidazolium chloroaluminate
  ionic liquid 479
1,3-dialkylmethylimidazolium cation 14
    – mono-functionalized 52
N,N-dialkylpiperidinium 420
N,N-dialkylpyrrolidinium 420
diastereoselectivity 336ff.
1,8-diazabicyclo[5,4,0]-7-undecene 9
1,8-diazabicyclo[5,4,0]-7-undecenium
  [EtDBU]
    – trifluoromethanesulfonate 330ff.
diazotination
    – *in situ* 288
dibenzyl carbonate (DBC) 322
1,3-dibutylimidazolium tetrafluoroborate 447
2-(dibutylphosphino)-
  N,N-dimethylethaneamine 440
dichlorodioxovanadate 194
dichloroethylaluminate 550
dicyclohexylcarbodiimide (DCC) 335

Diels Alder reaction 272ff., 327ff.
    – aza 330
    – cycloaddition 505ff.
    – hetero 331f.
Difasol process 433, 480, 673
differential scanning calorimetry (DSC) 189
diffusion 165f.
    – coefficient 167, 251ff.
    – mutual 250ff.
    – translational 251
digital image holography 253
Δ^2-dihydrodimethylmuconate (DHM) 440
4,4′-dihydroxy-3,3′-diacetoxybiphenyl 304
2,3-diisopropylbenzo[*b*]furan 333
1,4-diketone 348
dimerization 430ff., 476
    – aniline 310
    – functionalized olefin 439
    – olefin 673
Dimersol process 431ff., 480ff.
dimethyl formamide (DMF) 321
dimethyl phosphate 626
1,2-dimethyl-3-butylimidazolium
  hexafluorophosphate 338
1,2-dimethyl-3-hexyl imidazolium chloride
  [HMMIM]Cl 198
1,2-dimethyl-3-propyl-imidazolium [MMPIM] 398ff.
5-N,N-dimethylamino-1-
  naphthalenesulfonamide 137
dimethylbenzene 297
2,2-dimethylchromene 409
dimethylimidazolium [MMIM] 467
1,4-dimethylnaphthalene 274
[DMMIM] tetrafluoroborate 630
dioctylsulfosuccinate (docusate) anion 22
dioxomolybdenum(VI) complex 353
dioxouranium(VI) salt 196, 213, 237
1,3-dipolar cycloaddition 48
dipolar relaxation 262
1,2-diphenylethane 312
dipolarity 279ff.
direct recoil spectrometry (DRS) 201
disposal 685
distillation
    – extractive 669
distributed multipole analysis (DMA) 219
3,3′-[disulfanylbis(hexane-1,6-diyl)]bis(1-
  methyl-1H-imidazol-3-ium)dichloride 612
dodecyl-3-methylimidazolium 633
dppe (bis(diphenyl-phosphin)ethan) 561

DPPF, see
  1,1′-bis-(diphenylphosphino)-ferrocene
duroquinone (DQ)   270
Dy   300, 344
  – trifluoromethanesulfonate   300, 344

e
ECOENG™   23
  – 41M   29
  – 212   35ff.
electrochemical atomic layer epitaxy (ECALE)   585
electrochemical method   253
electrochemical potential window   142ff.
electrochemical property   141ff.
electrochemistry
  – application   675–677
electrocyclic reaction   327ff.
electrode/ionic liquid interface   587
electrodeposition   576ff.
  – aluminum   578
  – antimony   582
  – gallium   579
  – indium   582
  – ionic liquid   582ff.
  – iron   580
  – lithium   579
  – metal   582–587
  – semiconductor   585ff.
  – sodium   579
electrolyte   675f.
electron transfer reaction   268
electronic structure   218
electrophilic addition   284
electrophilic halogenation   316
electrophilic nitration   315
electrophilic phosphorylation   318
electrophilic reaction   294
electrophilic substitution   287
electrophilic sulfonation   318
electroplating   675
electropolishing   676f.
electrospray ionization mass spectroscopy (ESI-MS)   285, 389
electrostatic interaction   540
enantioselectivity   328ff., 554ff., 654
endo-selective reaction   327ff.
ene reaction   332
energy transfer   268
enthalpy   123
entrainer   669
entropy   123
  – driver   610

environment   37
enzyme
  – solubility   655
  – stability   655
ephedrine   493
epoxidation   353, 388
epoxide hydrolase
  – soluble (sHE)   649
EPR (electron paramagnetic resonance) spectroscopy   138, 387f.
Escherichia coli   648
ESI-MS, see electrospray ionization mass spectroscopy
esterase   653
esterification   334
ether
  – cleavage   672
  – conjugate   5f.
1-ethyl-2,3-dimethylimidazolium [EMMIM]$^+$ 149
  – bis(trifluoromethylsulfonyl)imide   110
  – hexafluorophosphate   122
1-ethyl-2-hydroxo-3-methylimidazolium tetrafluoroborate   508
1-ethyl-3-methylimidazol-2-ylidene (EMIMY)   570, 636
1-ethyl-3-methylimidazolium [EMIM]$^+$ 4, 194, 288, 563
  – acetate   19
  – aluminumchloride   85, 213
  – antimony   582
  – bis(trifluoromethylsulfonyl)imide   110, 197, 370, 392, 510, 553ff., 572, 642
  – bromide   22
  – cation   58, 170
  – chloride   10ff., 85, 194, 213, 229ff., 316, 587, 627
  – ethyl sulfate, see also ECOENG™   212, 389, 692ff.
  – fluoride   185
  – hexafluoroantimonate   300
  – hexafluorophosphate   223ff., 315, 356
  – hexafluorosilicate   602
  – indium   582
  – nitrate   226ff.
  – [OTs]   289
  – tetrachloroaluminate   224
  – tetrachloroaurate   91
  – tetrachloronickel   194
  – tetrachlorooxovanadate   194
  – tetrafluoroborate   31f., 153, 316, 355, 440, 626ff.

– trifluoroacetat   289, 357
– trifluoromethanesulfonate   289, 315ff., 330, 397
1-ethyl-3-methylimidazolium chloride [EMIM]Cl   98, 223
– aluminumchloride   85ff., 149, 296ff., 301ff., 306, 314, 318f., 334, 478–479, 483, 486, 536
– ironchloride   306
1-ethyl-3-methylimidazolium iodide [EMIM]I   304
– aluminumchloride   304
ethylacrylate (EA)   328
ethylaluminum(III) dichloride   619ff.
ethylammonium nitrate   316
ethylbromoacetate   341
ethylbromodifluoroacetate   341
ethylenediammonium diacetate (EDDA)   346
europium   213
– hexachloroeuropate $[EuCl_6]^{4-}$ 213
– 1-(2-thienyl)-4,4,4-trifluoro-1,3-butadione (Htta)   197
extended X-ray absorption fine structure (EXAFS)   93, 190, 389
– analysis   192
– spectrum   191
extraction   92
– solutes from ionic liquid   126

f
F-TEDA   288
F-TEDA-BF4   288
ferrocene
– 1,1′-bis-(diphenylphosphino)-ferrocene (DPPF)   549
– [Pd(1,1′-bis-(diphenylphosphino)-ferrocene)] trifluoromethanesulfonate   549
– [Rh(1,1′-bis-(diphenylphosphino)-ferrocene)(2,5-norbornadiene)] perchlorate   549
Fick's first law   251
Fick's second law   251ff.
fluorescence spectrum   137
fluorination   675
– electrophilic   288
α-fluoro-α,β-unsaturated ester   339
fluoroarene   288
Fmoc   521
force field   222ff.
– Tosi-Fumi   222
Friedel-Crafts reaction   294ff., 534ff.

– acylation   303ff.
– alkylation   295ff., 534
Friedlander annulation   347
functional group transformation   496
furanoquinoline   330

g
β-galactosidase   648
galactosylation   648
gallium
– arsenide   586
– electrodeposition   579
gas   117ff.
– Henry's law constant   121
– mixed   122
– separation   125
– storage   125
gas chromatography (GC)   108
gas solubility   103ff.
– high-pressure carbon dioxide   117
– low-pressure carbon dioxide   110
– measurement   106ff.
general materials (GEM) diffractometer   177
generalized gradient approximation (GGA)   241
germanium   586
– Au(111)   587
– bromide   601
– chloride   601
– electrodeposition   586
– Ge(111)   601
Glaser oxidation   355
glass transition temperature   153
glucose   334
glycosidation   325
gold
– Au(111)   589ff.
– electrodeposition   584
gravimetric gas solubility measurement   107
green chemistry   699ff.
green solvent   684
Grubbs-type catalyst   630

h
haloaluminate
– acidic   212
halometallate salt   90
Hartree-Fock (HF) level calculation   210
Heck reaction   21, 419ff., 447, 509
– arylation   447f.
– coupling   506
– ligand-less   422
– nanoparticulate catalyst   447
Henry' law constant   105ff.

N-heterocyclic carbene (NHC)   570
hexafluorophosphate [PF$_6$]   33ff., 98ff., 106ff., 122ff., 148, 202, 213ff., 239, 251, 275ff., 293, 322ff., 340ff., 352ff., 368f., 390ff., 412ff., 436ff., 467ff., 513, 540ff., 558ff., 574, 626ff., 642ff., 697
[HexPy] bis(trifluoromethylsulfonyl)imide   629f.
1-n-hexyl-3,5-dimethylpyridinium   235
1-n-hexyl-3-methylimidazolium [HMIM]$^+$ 110, 528
   – bis(trifluoromethylsulfonyl)imide   110, 283, 559
   – chloride   667
   – hexafluorophosphate   98, 228, 300, 346f.
   – perchlorate   283
   – tetrafluoroborate   300, 440
   – tribromide   288
   – trifluoromethanesulfonate   283
   – tris(pentafluoroethyl) trifluorophosphate [eFAP]   110
1-n-hexyl-3-methylpyridinium   235
higher 1-olefin (HAO)   436
highly oriented pyrolytic graphite (HOPG)   602ff.
Htta, *see* 1-(2-thienyl)-4,4,4-trifluoro-1,3-butadione
hydroamination   375
hydroformylation   368ff., 410ff., 471
hydrogen   117
hydrogen bond
   – acidity   279
   – bonding   136
   – donor ability   280
hydrogen transfer   268
hydrogenation   372, 390ff., 444
   – enantioselective   562
hydrosilylation   674
2-hydroxy-ethyltrimethylammonium   656
2-hydroxyethyl methacrylate (HEMA)   625
1-(2-hydroxyethyl)-3-methylimidazolium tetrafluoroborate   497, 515
2-hydroxymethylaniline   345
3-hydroxypropyltrimethylammonium bis(trifluoromethylsulfonyl)imide   514

i
ibuprofen   554, 658
ICP-MS, *see* mass spectrometry
ILSPS (ionic liquid supported peptide synthesis)   523
imidazolium   322, 473
   – cation   214ff.
   – hexafluorophosphate   217
   – salt   28, 67
imidazolium halide   52
   – anion   214ff.
   – 1,3-bis-functionalized   52
2-imidazolyl   474
imidazolylidene   570
imine   399
indeno[1,2,3-cd]pyrene   310
indium   582
   – (III) acetate   343
   – (III) bromide   344
   – (III) chloride   294, 336ff.
   – (III) triflimide   306
   – antimonide   586
   – electrodeposition   582
inorganic material   575ff.
   – electrochemical method   575ff.
inorganic synthesis   569
intense pulsed neutron source (IPNS)   177
iodination
   – regioselective   288
iodobenzene diacetate [PhI(OAc)$_2$]   352, 408
4-iodobenzoic ester   498
ion chromatography   31
ion pair
   – solvent-separated   289
ion pairing   272
ion size   62
ion transport number   169
ionic complex   91
ionic compound   91
ionic conductivity   150
ionic liquid (IL)   1
   – alkali halide   221
   – alkyl imidazolium-based   214
   – amino acid-based   22
   – ammonium cation   151ff.
   – atomistic simulation   220ff.
   – binary haloaluminate   157
   – binary task-specific (BTSIL)   488, 510ff.
   – bioaccumulation   38
   – biocatalysis   641ff.
   – biological activity   38
   – cation   150
   – co-catalyst   377f.
   – color   28, 682
   – commercial production   26ff.
   – computational modelling   206ff.
   – corrosion   41
   – covalent anchoring   534

– cuprate-based pyridinium   186
– decomposition   60
– density   86ff.
– diffusion coefficient   167
– disposal   685
– dynamics   175ff., 231ff.
– effect   268ff., 290
– electrochemical property   141ff.
– electronic structure   218
– extraction of solutes   126
– functionalized   495
– gas solubility   103ff.
– halide impurity   30
– halide-free   31ff.
– hydrophobic   33
– innocent solvent   377f.
– Lewis acid-based   13, 294
– ligand/ligand precursor   380ff.
– ligand optimization   411
– material synthesis   609ff.
– melting point   60
– molecular structure   175ff.
– multiphasic reaction   464ff.
– neat   226f.
– neutral   292f.
– non-haloaluminate   87, 168
– non-haloaluminate imidazolium (IM)-based   158
– organic synthesis   265ff.
– phase diagram   57
– phase organic synthesis (IoLiOS)   507
– physical property   57ff.
– polymer synthesis   619ff.
– protic impurity   32
– purification   18
– purity   680
– pyridinium-based   274
– quality   27
– recycling   42, 684
– room-temperature electrochemical potential window   151
– scale-up of synthesis   36
– second generation   14, 368
– silica hybrid   615
– simulation   221ff.
– solid support   528
– solubility   89
– solvation   89
– stability   449, 682
– structure   193, 226ff.
– supercritical carbon dioxide   561
– supported ionic liquid phase (SILP)   373f.
– supported peptide synthesis (ILSPS), see ionic liquid supported peptide synthesis
– sustainability   20
– task-specific (TSIL), see TSIL
– thermodynamic property   104, 230f.
– toxicity   38, 683, 700
– transition metal catalysis   367ff.
– trialkylammonium-based   133
– 1,2,3-trialkylimidazolium-based   381, 696
– volatility   698
– viscosity   72ff., 268
– water   33ff.
IR spectroscopy   388
iridium
   – (0) nanocluster   385
   – $[Ir(COD)Cl]_2$ 446
   – XYLIPHOS   402
iron
   – (III)   356
   – chloride   214
   – (III) chloride   306f.
   – electrodeposition   581
   – $[HFe(CO)11]^-$ 393
   – nitrate   316
   – (III) porphyrin   406ff.
   – (III) triflimide   306
ISIS   177f.
isomerization reaction   312ff.
isoprene (IP)   329
2-(4-isobutylphenyl)propionic acid   554
4-isopropyloxazolidinone   338
isoquinoline   9
isothiocyanato   518

j
Jacobsen catalyst   409

k
Kamlet-Taft solvent parameter   281
ketone   399
Knoevenagel reaction   48, 346f.

l
lab on chip   522
lanthanum   213, 582
   – (III) triflate   337
Leuckart-Wallach reaction   493
Lewis acid   315, 344ff., 450, 533ff.
   – ionic liquid   13, 449
ligand
   – design   559
   – optimization   411

β-Lilial 447
Lindemann tube 184
linear alkyl benzene (LAB) 299, 483
lipase 651ff.
– B 657
lipophilicity 97
liquid clathrate 2
liquid crystal 185
– glass-forming 260
– liquid biphasic catalysis 375f.
– range 59
lithium
– bis(trifluoromethylsulfonyl)imide [Tf_2N] 491, 563
– 1-n-butyl-2,3-dimethylimidazolium [BMMIM] bis(trifluoromethylsulfonyl)imide/ lithium bis(trifluoromethylsulfonyl) imide 144
– electrodeposition 580

m

manganese
– (II) triflimide 306
– (III) porphyrin 353f.
– (III) (salen) complex 409, 697
Mannich reaction 349
mass spectrometry
– inductively coupled plasma (ICP-MS) 31
mass spectroscopy 389
MCM-41 544
MD, see molecular dynamics
mesylate 539
metal complex 91
metal salt solubility 90
metal triflate 315
metal triflimide salt 306
metathesis reaction 490
– incomplete 33
methacrolein 328
methacrylic acid 329
1-methoxy-ethyl-3-methylimidazolium [MOEMIM]$^+$ 652
1-methoxy-2-methyl-1-trimethylsiloxypropene 343
2-methoxy-naphthalene 319
(S)-(+)-2-(6-methoxy-2-naphtyl) propionic acid 653
methyl methacrylate (MMA) 625ff.
methyl tosylate 12
methyl triflate 12
methyl trifluoroacetate 12

methyl vinyl ketone (MVK) 330
methyl-6-amino-5-cyano-4-aryl-2-methyl-4H-pyran-3-carboxylate 504ff.
methyl-tert-butyl ether (MTBE) 361
1-methyl-3-butylimidazolylidine 348
2-methyl-2,3-dihydrobenzo[b]furan 333
3-methyl-4,7-dihydroxyindanone 303
(±)-methyl-3-hydroxy-2-methylenebutanoate 396
1-methyl-3-methylimidazolium [MMIM]$^+$ 649
– bis(trifluoromethylsulfonyl)imide 95
– chloride 223ff., 241ff.
– hexafluorophosphate 182, 223ff.
– triflate [TfO] 13
1-methyl-3-nonyl-imidazolium cation 262
1-methyl-3-octylimidazolium chloride [OMIM]Cl 36
1-methyl-3-pentylimidazolium 197
– bis(trifluoromethylsulfonyl)imide 197
2-(2-methyl-2-propenyl)phenol 333
2-methyl-2-propenyl phenyl ether 333
methylacrylate (MA) 440
4-methylbenzoyl chloride 303
methylene insertion reaction 340
N-methylimidazole 36
2-methylimidazolium 341
[1-(3-methylimidazolium)butanesulfonic acid]
– trifluoromethanesulfonate 334
2-methylindole 324
2-methylnaphthalene 299
N-methylpyrrole 434
1-methylpyrrolidine 9
N-methylpyrrolidone (NMP) 472
methylsulfate 645, 655
methyltrioxorhenium (MTO) 356, 408
methylviologen MV^{2+} 269
Michael reaction 336, 495
microwave dielectric spectroscopy 131
molecular dynamics (MD) 207ff., 220
– ab initio quantum chemical method 210f.
– classical 209
molecular reorientational dynamics 255ff.
molecular structure 175ff.
molecular solvent 515
molten salt 1, 185
molybdenum
– dioxomolybdenum(VI) complex 354
– metathesis 444
Monte Carlo (MC) method 220

morphine dehydrogenase (MDA)   351
1-(*N*-morpholino)-2-chloroethane   324
multiphasic reaction   463ff.
mutual diffusion   250
    – coefficient   252

n
Nafion-H   348
nanoparticle   609
    – functional   609
nanoscale process   587
nanostructure   609
    – inorganic   609ff.
naphthalene (N)   268
2-naphthol   356
2-naphthoxide   321
naproxen   653
natural atomic orbital (NAO)   218
neutron cell   178
neutron diffraction   175
neutron reflectivity   199
neutron source   177
    – pulsed   177
NHC, see carbene
nickel
    – acetylacetonate   404
    – bis-(tricyclohexylphosphine-nickel chloride)   550
    – complex   477ff.
    – (η-4-cycloocten-1-yl)(1,5-diphenyl-2,4-pentandionato-O,O′)nickel   385
    – (η-4-cycloocten-1-yl)(1,1,1,5,5,5-hexafluoro-2,4-pentandionato-O,O′)nickel   432, 483
    – dimerization   430
    – electrodeposition   584
    – [(HCOD)Ni(hfacac)]   433f., 483
    – [HNi(olefin)]   473
    – [(mAllyl)Ni(dppmo)][SbF$_6$]   437
    – [mallNi(dppmo)][SbF$_6$]   437
    – [(methally)Ni(Ph$_2$PCH$_2$PPh$_2$(O))][SbF6]   473
    – [Ni(MeCN)6][BF4]   434
    – oligomerization   430
    – (II) salt   478
    – (II) triflimide   306
nicotinamide diphosphate (NADP)   352
(S)-nicotine   654
nitrate   407, 468
nitrile-butadiene rubber (NBR)   395
nitroxo (NO$^+$)   288
    – hexafluorophosphate   288
    – tetrafluoroborate   288

NMR
    – spectroscopy   384
    – viscosity relationship   264
nuclear Overhauser enhancement (NOE)   256
    – equation   261
nucleophile   449
nucleophilic reaction   319
nucleophilic substitution   275, 319ff.
nucleophilicity   283f.

o
1-octadecyl-3-methylimidazolium [C$_{18}$MIM]$^+$ 200
    – hexafluorophosphate   201
1-octene   416f.
octyl tosylate   12
1-*n*-octyl-3-methylimidazolium [OMIM]$^+$ 93, 110, 300 [C8MIM]+
    – bis(trifluoromethylsulfonyl)imide   93, 110, 134
    – hexafluorophosphate   98, 122, 201f., 239, 561, 647
    – tetrafluoroborate   122
1-*n*-octyl-3-methylimidazolium [OMIM] bromide
    – aluminumchloride   301
1-*n*-octyl-3-methylpyrdidinium [OMPy]$^+$ 235
    – tetrafluoroborate   393
1-octyl-3-(3-triethoxysilylpropyl)-imidazolium hexafluorophosphate   537
octylpolyethyleneglycol-phenyl-phosphite   417
olefin   388ff., 673f.
    – metathesis   440
    – polymerization   627ff.
oleic acid   314
oligomerization   430, 619, 673
    – olefin   673
optimized potentials for liquid simulation (OPLS)   225
organic compound   96
organic reaction
    – stoichiometric   292ff.
organic synthesis   265ff.
    – selectivity   265ff.
organometallic compound   340, 570
organometallic synthesis   569
Osborn complex   391
osmium
    – [H$_3$Os$_4$(CO)$_{12}$]-   393
[OTf], *see* trifluoromethanesulfonate

oxidation 284, 404ff.
– alcohol 405
– bromide ion 284
– functional group 350ff.
oxide formation 573
oxim 406
oxime carbapalladacycle ionic liquid 550
oxycodone 351, 656
oxygene 117
oxyhalide 619

p
palladium 426, 472, 629f.
– (II) acetate 382
– acetylacetonate 392, 440ff., 550ff.
– (II) chloride 382, 406, 419, 438, 536
– (IV) compound 382
– electrodeposition 585
– hydroamination 375
– imidazolylidene 382
– nanoparticle 444ff.
– nanoparticle formation 420
– [(MIM)$_2$Pd(BMimy)Cl]$^+$ 427
– [Pd(bipy)$_2$][PF$_6$] 629
– [Pd(1,1′-bis-(diphenylphosphino)-ferrocene)][CF$_3$SO$_3$]$_2$ 550
– [PdBr$_2$(BMimy)$_2$] 420
– [Pd$_2$(μ-Br)$_2$Br$_2$(BMimy)$_2$] 420
– [PdCl$_2$(PhCN)$_2$]/Ph$_3$As/CuI 428
– [PdI$_2$(methylbenzthiazolylidene)$_2$] 447
– [PdX$_2$(BBimy)$_2$] 427
– Pd(OAc)$_2$ 427, 474
– [Pd(OAc)$_2$(PPh$_3$)$_2$] 390
– [Pd(OAc)$_2$(bipy)] 629
– [Pd(PPh$_3$)$_4$] 425f., 474
– [Pd(PPh$_3$)Cl$_2$(BMimy)] 425
PAO, see polyalphaolefin
PBD, see polybutadiene
PEG, see polyethylene glycol
pent-1-yne 341
3-pentenoic acid methyl ester 373
1-pentyl-3-methylimidazolium [PentMIM]$^+$ 652
pentylpyridium perbromide 288
perfluorinated solvent 368
perfluorobutylsulfonate 539
peroxotungstate 408, 537
phenol 300
phenoxyhalodiazirine 321
E-3-phenylacrylic acid butyl ester 542
phenylethyne 341

phenylguanidinium 473
phenylhalodiazirine 321
Z-phenylhydrazone 282
phenyliodoso diacetate (PhI(OAc)$_2$) 353, 409
phosgene 18
phosphine 9, 559, 680
phosphonium triflimides 300
phosphotungstic acid 406
photolysis 320
phthalimide 322
physisorption 539
piano stool complex 573
Pictet-Spengler reaction 302
pigment paste 678
platinum
– (II) 407
– (II) (2,6-bis(aminomethyl)pyridine)Cl]$^+$ 388
– (II) complex 330, 387
– [HPt(SnCl$_3$)$_4$]$^{3-}$ 474
– [(PPh$_3$)$_2$PtCl$_2$] 386
– (PR$_3$)$_2$PtCl$_2$ 379
– [Pt(NH$_3$)$_4$]Cl$_2$ 390
– Pt(PPh$_3$)$_4$ 381
– cis-[Pt(PPh$_3$)$_2$Cl(SnCl$_3$)] 410
– [Pt(PPh$_3$)$_2$(SnCl$_3$)$_2$] 410
PMIM, see 1-n-propyl-3-methylimidazolium [PMMIM]
– (trifluoromethylsulfonyl)imide 153
polarity 130, 265
– scale 138
polarizability 279ff.
polarizing optical microscopy (POM) 189
poly(1-butyl-4-vinylpyridinium bromide) 624
poly(1-ethyl-3-vinylimidazoliumbis (trifluoromethanesulfonyl)imide 624
poly(glycolic acid) (PGA) 635
poly(methyl methacrylate) (PMMA) 625f.
poly(1,3,4-oxadiazole) (POD) 635
poly(para-phenylene) (PPP) 634ff.
poly(vinylidenefluoride)-hexafluoropropylene 472, 553
poly(4-vinylpyridine) (PVP)
– methyl trioxorhenium (MTO) 355
polyalphaolefin (PAO) 674
polyamide 634
polybutadiene (PBD) 394f.
polyethylene glycol (PEG) 490
– ionic liquid phase (PEG$_n$-ILP) 498
polyimide 634
polymer 395f.

– conductive 633
– synthesis 619ff.
polymerization
    – atom-transfer radical (ATRP) 632
    – cationic 619
    – electrochemical 633
    – enzymatic 634f.
    – group transfer (GTP) 636
    – metathesis 630
    – radical 631
    – transition metal-catalyzed 627f.
    – Ziegler-Natta 627
polystyrene (PS) 355, 490
    – methyl trioxorhenium (MTO) 355
polyvinyl alcohol 490
potassium
    – $K_3Co(CN)_5$ 393
    – trithiocarbonate 326
processibility 452
product isolation 451
propane-1,3-dinitrile 346
1-n-propyl-3-methylimidazolium [PMIM]$^+$ 563
    – hexafluorophosphate 300
    – $[PhP(C_6H_4SO_3)_2]^{2-}$ 563
N-propyl-2-pyridylmethanimine 631
proton donor 283
Pseudomonas cepacia 654
purity 681
pyridine 9
pyridinium [HPY]$^+$, [PyH]$^+$ 278, 322ff., 382, 474
    – chloride 319
    – chloroaluminate(III) melt [PyH]Cl-$AlCl_3$ 299ff.
    – salt 28
pyrrole 434

q
quaternization reaction 9ff.
quinoline 347

r
radial distribution function (RDF) 227
reaction
    – chemical 138
    – gas 124
reactor source 178
recyclability 451
recycling 42, 684
red oil 2, 483
reduction 356

reflectivity 199
refractive index 137
regioselectivity 424, 448, 474, 651
Reichardt's dye 134
reorientational correlation time 257ff.
reversible addition-fragmentation chain transfer (RAFT) 626
rhenium
    – diperoxo complex 388
    – methyltrioxorhenium(VII) (MTO) 355, 407
    – monoperoxo complex 388
rhodacarbonate catalyst 399
rhodium 371
    – (I) complex 396, 629
    – hydroamination 375
    – hydroformylation 373f., 413, 472
    – hydrogenation 374
    – $[HRh(CO)_4]$ 563
    – $[HRh(CO)_2(sulfoxantphos)]$ 547
    – $[HRh(CO)(PPh_3)_3]$ 417
    – $[HRh(PPh3)2(diene)][PF_6]$ 473
    – NORBOS 545
    – $[Rh(acetylacetonate)(CO)_2]$ 388, 410ff., 543
    – [Rh(1,1′-bis-(diphenylphosphino)-ferrocene)(2,5-norbornadiene)] perchlorate 550
    – $[Rh(CO)_2I_2]^-$ 383
    – $[Rh(cod)(-)-(diop)][PF_6]$ 396
    – $RhCl_3$ 446
    – $[RhCl(PPh3)3]$ 548, 561
    – $[RhCl(TPP)_3]$ 416
    – $[Rh(norbornadiene)(PPh_3)_2]^+$ 391, 548
    – sulfonated phosphine ligand 472
    – TPPTS 465, 544
    – transition-metal complex 473
ring-closing metathesis 440, 630
ring-opening metathesis polymerization (ROMP) 440, 630
ring-opening polymerization (ROP) 635
[RMIM]$^+$, *see* 1-alkyl-3-methylimidazolium
Robinson annulation reaction 346
room temperature
    – ionic liquid 695
    – molten salt 2
[RPY] bis(trifluoromethylsulfonyl)imide 629
RTIL 513
ruthenium 630f.
    – (III) 355
    – allenylidene precatalyst 630

– catalyst   631
– chloride   393
– complex   343
– (II) complex   396
– first generation   441
– $[H_4Ru_4(C_6H_6)_4][BF_4]_2$ 473
– nanoparticle   446
– ring-closing metathesis   441
– ring-opening polymerization   441
– $[Ru_6C(CO)_{16}]^{2-}$ 393
– $[RuCl_2(S)\text{-BINAP}]_2\,NEt_3$ 396ff.
– [RuCl(S)-BINAP]Cl (chloro-[(S)-(-)-2,2′-bis(diphenylphosphino)-1,1′-biphenyl](isopropylbenzene)-ruthenium(II)-chloride)   549
– $[RuCl_2(dppe)]$ 562
– $(RuCl_2(PPh_3)_3)$ 393
– $[Ru(\eta^6\text{-p-cymene})(\eta^2\text{-}TRIPHOS)Cl][PF6]$ 395
– $[Ru(\eta^3\text{-2-methylallyl})_2(\eta^2\text{-COD})]$ 400
– [Ru(OAc)2(tolBINAP)]   560
– self-cross metathesis   443
– TPPTS   396
RTIL   501

s

saccharinate   22
Salen-Al   324
salt melting point   62
samarium (III) trifluoromethanesulfonate   344
SAPC, see supported aqueous-phase catalysis
scandium
    – (III) triflate   300, 330ff.
    – (III) trifluoromethanesulfonate   332ff.
    – trifluoromethanesulfonate   330
scanning tunnelling microscope (STM)   587
Schiff-base   359
Scholl reaction   310
Selectfluor™   288
self-cross metathesis   442
self-diffusion coefficient   166, 250ff.
separation   92
SFVS, see sum-frequency vibrational spectroscopy
Shell higher olefins process (SHOP)   429, 464
silica-supported vanadium(V)-oxide-alkali pyrosulfate sulphur dioxide oxidation catalyst   532
silica hybrid   615

silicon   588, 602
– chloride   602
SILP, see supported ionic liquid phase
silver
    – electrodeposition   584
    – 1-ethyl-3-methylimidazol-2-ylidene (EMIMY) silver(I) chloride   636
silyl ketene acetyl   636
simulation
    – ab initio   239
    – ionic liquid   221ff., 239
    – mixture   236
    – solution   220ff., 236
    – surface   239
small angle neutron diffractometer for amorphous and liquid sample (SANDALS)   177
sodium
    – $[BARF]^-$ 565
    – electrodeposition   579
sodium chloride
    – aluminumchloride   303ff.
    – potassiumchloride-aluminumchloride   310
solid support   529
solid-phase organic synthesis (SPOS)   500
solubility   89ff.
solvation   89
solvent   671ff.
    – green   558, 684
    – low vapor pressure (SLP)   555
    – supramolecular   610
    – transition metal catalysis   372ff.
solvent-separated ion pair   289
solute-anion interaction   289
Sonogashira reaction   509
sorbitol   635
spectroscopic gas solubility measurement   107
stereoselectivity   282f., 396
    – α 283
    – β 283
Stille coupling   429
stoichiometric gas solubility measurement   106
styrene-butadiene rubber (SBR)   395
substitution
    – electrophilic   287
    – nucleophilic   275ff., 319ff.
    – S_N1 278ff.
    – S_N2 278, 320ff.
sulfonylalkane   538
sulfoxantphos   388, 417

sum-frequency vibrational spectroscopy (SFVS) 530
supercritical fluid 126
– extraction 293
supported aqueous-phase catalysis (SAPC) 465
supported catalyst
– low temperature 533
supported ionic liquid catalyst (SILC) 527ff., 555
supported ionic liquid phase (SILP) 54, 373f., 403, 418, 452, 527, 555
– catalyst 446, 526ff., 540ff.
– Heck catalyst 422
– Ru-(S)-BINAP 549
supported liquid membrane (SLM) 658
supported molten salt (SMS) catalyst 527ff.
high temperature 530
supported organic synthesis (SPOS) 487ff.
– TSIL 502
surfactant
– biodegradable 23
Suzuki cross coupling 425
Suzuki-Miyaura coupling 509ff.
synthesis
– ionic liquid 7
– organic 265ff.

t
task-specific onium salt (TSOS) 498ff., 513ff.
– supported organic synthesis 515
TBHP, see tert-butyl hydroperoxide
TBP, see tri-n-butylphosphate
tellurium 583
– electrodeposition 583
tetraalkylammonium 8, 381, 418
– ionic liquid 696
tetraalkylphosphonium 8, 381, 418
– ionic liquid 696
tetrabutylammonium [Bu_4N], (TBA) 132, 193, 321
– bromide 349
– fluoride (TBAF) 320ff.
– tribromide 346
– tribromomanganate 193
tetrabutylphosphonium 16, 132, 321
– bromide 369
tetrachloronickel 194
tetrachlorooxovanadate 194
tetrachloropalladiate 200
tetradecane 356
tetraethylammonium
– chloride [Et_4N]Cl 134, 622
– trichlorostannate 369, 409, 474
tetraethylorthosilicate (TEOS) 535, 553
tetrafluoroborate 17ff., 31ff., 109f., 124, 153, 201, 275ff., 322, 341ff., 352ff., 368ff., 381f., 393ff., 408ff., 426ff., 445, 467, 513, 542, 559, 587, 612, 626ff.
tetrahydrofuran (THF) 320
1,1,3,3-tetramethyl-guanidinium [TGA] 446
– lactate TSIL 122
– lactate (TMGL) 544ff.
thermodynamic property 230f.
1-(2-thienyl)-4,4,4-trifluoro-1,3-butanedione (Htta) 94, 197
thioamidation 309
tin
– (II) 357
– (II) chloride 294
– catalyst 536
– electrodeposition 585
– (IV) salt 358
titanium 627
– cyclopentadiene 379, 627
– oxide 575
– Ti(O-^{i}Pr)$_4$/TADDOL 324
TMEDA
– copper(I) chloride 355
TMGL, see 1,1,3,3-tetramethyl-guanidinium lactate
toluene 315
p-toluenesulfonic acid 302
N,N-p-tolyl-2,3-dimethyl-1,4-diazabutadiene 353
[N-(p-tolylsulfonyl)imino]phenyliodinane 354
Tosi-Fumi force field 222
tosylate 539, 626
(R)-2-tosyloctane 320
toxicity 38, 700
TPPMS 416f.
TPPTS, see trisulfonated triphenylphosphine ligand
transesterification 35, 504f.
transition metal catalysis 368ff.
– application of ionic liquid 390ff.
– homogeneous 369
– nanoparticulate 444
transition state
– dipolar 274
– isopolar 268
– radical 268
translational diffusion 249ff.
transport

– number 169
– property 165
Traseolide® 305
tri-n-butylphosphate (TBP) 93, 197
trialkyl amine 9
trialkylammonium-based ionic liquid 133
1,2,3-trialkylimidazole 338
1,2,3-trialkylimidazolium-based ionic liquid 382, 696
trialkylsulfonium iodide 8
triarylphosphine 410
tributylamin Bu_3N 281
tributylammonium 133
tributylhexylammonium [$NBu_3(C_6H_{13})$] bistrifluoromethanesulfonylimide 302
tributylmethylammonium [Bu_3MeN]$^+$ 271
– bis(trifluoromethylsulfonyl)imide 271
trichloroacetimidate 325
1-(triethoxysilyl)propyl-3-methylimidazolium chloride 536
triethylammoniumclorid [$(C_2H_5)_3HN$]Cl
– aluminumchloride 299, 318
triethylmethylammonium methylsulfate 655
triethylsulfonium bromide [$(C_2H_5)_3S$]Br
– aluminumchloride 300
triflate [TfO], [Tf]$^−$ 12f., 226, 300, 318, 330ff., 539, 626
triflimide, see bis(trifluoromethylsulfonyl)imide
trifluoroacetate 328, 467
– [$HNet(Pr^i)_2$] 315
trifluoroacetic acid (TFAH) 315
trifluoromethanesulfonate [OTf] 281ff., 315ff., 332ff., 397, 438ff., 468, 513, 540, 650
1-(trimethoxysilylpropyl)-3-methylimidazolium 541
[3-(trimethoxysilyl)propyl]octadecyldimethylammonium chloride 536
trimethylsiloxy group 352
trimethylsulfonium 16
– iodide 339
triphenylphosphine 553
– guanidinium-modified 412
– sulfonated 92
[3-(triphenylphosphonium)propanesulfonic acid]
– tosylate 334
2,4,6-triphenylpyridinium-N-4-(2,6-diphenylphenoxide) betaine 134

tris(4,4′-bipyridyl)ruthenium [$Ru(bpy)_3$]$^{2+}$ 269
trisulfonated triphenylphosphine ligand
– (1-butyl-3-methyl-methylimidazolium) salt (TPPTI) 542
– trisodium salt (TPPTS) 543
trolox 269
TSIL (task-specific ionic liquid) 45ff., 487ff., 498ff.
– first generation 502
– second generation 509
TSOS, see task-specific onium salt
tungsten
– [$Cl_2W=NPh(PMe_3)_3$] 435
– [$HWOs_3(CO)_{14}$]$^−$ 392
– metathesis 443
– [$\{W(=O)(O_2)_2(H_2O)\}2(\mu\text{-}O)$]$^{2−}$ 537

u
uranium
– dioxouranium(VI) salt 196, 213, 237
urea-hydrogen peroxide (UHP) 407
UV-Vis spectroscopy 387

v
(S)-valinol 493
vanadium(V) 284
– silica-supported vanadium(V)-oxide-alkali pyrosulfate sulphur dioxide oxidation catalyst 532
viscosity 72ff.
– measurement 73
Vogel-Tammann-Fulcher (VTF) equation 153, 252
volatile organic compound (VOC) 37, 90
volatility 698

w
Wacker oxidation 406, 532
Walden product 155f.
Wang-type linker 499
water 33ff.
– poor condition 615
Wilke's complex 564
Wilkinson type complex 417
Wilkinson's catalyst 561
Williamson alkylation 499
Wohl-Ziegler α-bromination 318

x
X-ray diffraction 184
X-ray reflectivity 199
xantphos type ligand
– phenoxaphosphino-modified 416

XPS spectroscopy 390
xylene 297

y
ytterbium 213
    – trifluoromethanesulfonate 315, 343f.

z
zinc
    – (II) chloride 294, 471
    – chloride 1-ethyl-3-methylimidazolchloride 584
    – electrodeposition 584
    – hydroamination 375
    – (II) iodide 329
    – (II) trifluoromethanesulfonate [OTf] 341, 549
    – α-zinc bromide ester 340
zinc telluride 586
    – electrodeposition 586